GEOLOGICAL ASSOCIATION OF CANADA SPECIAL PAPER 25

PRECAMBRIAN SULPHIDE DEPOSITS

H.S. Robinson Memorial Volume

Edited by

R.W. Hutchinson, C.D. Spence and J.M. Franklin

1982

R.W. Hutchinson,
Department of Geology,
University of Western Ontario,
London, Ontario
N6A 5B7

C.D. Spence,
Rio-Canex Inc.,
800 West Pender, Suite 520,
Vancouver, B.C.
V6C 2V6

J.M. Franklin,
Geological Survey of Canada,
601 Booth Street,
Ottawa, Ontario
K1A 0E8

International Standard Book Number: 0-919216-21-8

Geological Association of Canada
Department of Earth Sciences
University of Waterloo
Waterloo, Ontario
Canada N2L 3G1

Printed by: Johanns Graphics, Waterloo, Ontario

CONTENTS

PREFACE

This Special Paper has been partially funded by a bequest to the Geological Association of Canada from H.S. Robinson, a prominent Canadian mining geologist whose interest in Precambrian sulphide deposits prompted his gift. D.R. Derry, a former president of the Association, who knew Robinson, obtained the frontispiece photograph, and prepared the biographical note which accompanies it. Special grants from Cominco Limited, Selco Mining Corporation Limited and Shell Canada Resources Limited provided for inclusion of special illustrations in the volume and this assistance is gratefully acknowledged.

The volume owes its conception mainly to Paul M. Kavanagh who, as General Chairman of the "Toronto 78" joint annual meetings of the Geological Association of Canada, Mineralogical Association of Canada and the Geological Society of America, thought that a symposium describing and discussing a number of Precambrian sulphide deposits would be a valuable contribution. He interested the G.A.C. into sponsoring such a symposium to be called appropriately, the Robinson Symposium. Twelve descriptive papers on Precambrian deposits, including discussions of their genesis were invited and presented at the Symposium, held on October 23 and 25, 1978 during the "Toronto 78" meetings. Paul Kavanagh further proposed that the twelve and other invited papers should be published by the G.A.C. as a Special Paper to be known as the H.S. Robinson Memorial Volume. To the original twelve papers, eleven others were added on suggestions by the editors and Kavanagh, six describing additional deposits, four dealing with Precambrian regional geology and a general review paper. The purpose was to cover a broad spectrum of deposits and their geological settings, both geographically throughout the Precambrian of North America, and by commodities. A. Sutherland Brown also provided helpful advice and suggestions during initial planning for the volume.

Many important Precambrian sulphide deposits are not considered in the volume. Absent are those that have been extensively studied and for which thorough descriptions and up-to-date genetic interpretations have recently been published. Specifically included are newly discovered deposits and older ones for which only limited or out-dated information has been available.

In addition to the support of the Robinson Foundation, funds for administrative and editorial expenses have been supplied by the Geological Association of Canada. Thanks for this aid are tendered to the Association, and particularly to W.G.E. Caldwell, president, and J.W. Kramers, chairman of the Association's editorial committee for their encouragement and support. Maureen E. Dickson Czerneda, G.A.C. Managing Editor has provided continuing and invaluable assistance during three years of the volume's gestation. Mrs. Joan Crockett has given continued patient administrative and secretarial assistance to the senior editor during the same lengthy time period.

All of the French language translations of the abstracts for the papers in the volume have been prepared by Andrée F. de Rosen-Spence who kindly offered this assistance, for which the editors are particularly grateful. The editors would like to thank all of the contributing authors for their submissions and for their willingness to revise early drafts of their manuscripts as suggested by the reviewers. Some of the manuscripts were submitted more than three years before the appearance of the volume, which was delayed mainly for submission of later manuscripts but also by what may have been unusually rigorous review procedures and editorial standards. The editors appreciate the patience and co-operation of the authors and regret the delay and any inconvenience it has caused. Finally, our thanks are extended to the many reviewers who have helped improve the papers presented here and who are too numerous to be named individually.

R.W. Hutchinson
C.D. Spence
J.M. Franklin
Editors
February 4, 1982

HOWARD S. ROBINSON

In the period from five years before World War II until 1950, from my recollection, almost any new mineral discovery of significance in Canada would result in a visit from the smiling and modest figure of Howard (known as "Bob" or "Robbie") Robinson, representing McIntyre Porcupine Mines.

He was born and educated in U.S.A. and eventually retired there. As far as I know, (from Mr. W.B. Dix, former Comptroller of McIntyre) he first arrived in Canada in 1919 when he went to work for Richard Innis, Manager of McIntyre Porcupine Mine in Schumacher, Ontario. There was less specialization in those days and his first duty was as Mill Engineer and relieving members of the engineering staff underground. Soon, however, he concentrated on the geological aspect, both underground and in the field. One of his early activities was in connection with the development of a silver property in Gowganda area, Ontario. He was active in the gold rush at Red Lake, Ontario in 1925 and 1926, and McIntyre was one of the backers of the Howey brothers in that area. In 1934, it was a team sent in by Robinson to the Mud Lake area of northwest Quebec who took up the ground that was put into production of gold in 1936 as the Belleterre Mine.

Howard Robinson moved back to U.S.A. about 1950, on account of his wife's illness, but he continued to commute to Canada until his retirement in 1957.

Robinson was one of the group of geologists and field engineers who formed the Toronto Geological Discussion Group in 1945 and two years later, the Geological Association of Canada, of which he was a founder member. His continued interest in the latter organization was shown in tangible form by his substantial bequest to it in his will.

The present volume is a most suitable subject for support from the fund instituted in his name and constitutes a memorial to a geologist who was equally at home in the mine and in field exploration.

Duncan R. Derry
July 18th, 1981

REGIONAL GEOLOGICAL PAPERS

H.S. Robinson Memorial Volume, Precambrian Sulphide Deposits, edited by R.W. Hutchinson, C.D. Spence and J.M. Franklin, Geological Association of Canada Special Paper 25, 1982

COMPARATIVE METALLOGENY OF THE SUPERIOR, SLAVE AND CHURCHILL PROVINCES

J.M. Franklin and R.I. Thorpe
Geological Survey of Canada, 601 Booth Street, Ottawa, Ontario K1A 0E8

ABSTRACT

The Slave, Superior and Churchill provinces of the Canadian Shield contain major copper-zinc, gold, nickel and iron deposits, primarily in volcanic-dominant 'greenstone' belts of Archean and Proterozoic age. In addition, sandstone-hosted and carbonate-hosted lead-zinc occurrences are present in shelf-sedimentary sequences such as the Wollaston belt and Belcher Island fold belt. Other significant deposit types, not treated here in any detail, include unconformity-related and clastic-hosted uranium deposits, silver and lead-zinc veins in cratonic-cover sequences, deposits associated with alkaline rocks and pegmatites, and the Chibougamau copper-vein deposits.

All massive sulphide deposits are of the Cu-Zn type, with Pb:Zn less than 0.1. Most occur in the second or higher cycles of multi-cyclic volcanic sequences, and many occur in the lower half of their host cycle. Most of the principal districts have prominent subvolcanic intrusions at the base of the host cycle. These are interpreted to represent a local thermal energy source that generated high temperature (~350°C) geothermal systems containing the metalliferous source fluids.

Gold deposits occur primarily in areas of prominent crustal instability, such as intravolcanic rift zones. Although most of the gold in intrusion- and volcanic-hosted deposits occurs in veins, and evidently formed at high crustal levels at least a short time after host-rock deposition, a few iron-formation hosted deposits may be syngenetic. Little evidence is available in the epigenetic deposits to indicate local re-concentration of gold from anomalously rich nearby strata; rather the gold was derived from lower crustal environments, by development of high temperature hydrothermal fluids in a very low water-rock regime. The fluids were focussed by intravolcanic, tensional faults.

Iron deposits are not directly related to massive sulphide-generating hydrothermal fluids. Their low base metal content and large areal extent require a large volume of relatively low temperature transporting fluid. Very large-volume sea-water circulation through volcanic rocks, driven by heat from the cooling pile, may have supplied the required iron-rich fluid. Superior-type iron-formation appears to be developed primarily in intracratonic rifts.

Nickel deposits are magmatic products, produced by the separation of a sulphide liquid from either tholeiitic or komatiitic melts. The high amount of sulphur required to produce the sulphide liquid may have been derived by contact metamorphism of sulphide-rich strata, primarily sulphide-facies iron-formation, which occurs in the underlying (or adjacent) strata.

Significant differences in the metallogeny of each of the geological provinces include: 1) the paucity of major gold deposits in both Proterozoic and Archean greenstone belts of the Churchill Property; 2) a lack of iron-formation in the Proterozoic volcanic and sedimentary rocks of the Churchill province; 3) an absence of both komatiite- and tholeiite-associated nickel deposits in Slave Province; and 4) a characteristically more radiogenic component of the lead, and a generally more lead-rich nature, in massive sulphide deposits in Slave Province.

Respectively the reasons for these differences may be: 1) a different tectonic style, possibly involving compressional, subduction-related deformation in the Proterozoic regime of southern Churchill Province, which did not permit either the development or the focussing of gold-rich, base-metal-poor, solutions; 2) a lack of tensional (rift-related) tectonism during major volcanism in the Flin Flon, LaRonge and Lynn Lake belts, which inhibited large-scale seawater circulation through the volcanic pile; 3) either a different tectonic style (lack of rifting), insufficient lower-mantle melting, or excessively thick crust inhibiting the formation of komatiitic flows and related intrusions; and 4) thick, relatively mature crust underlying the volcanic strata contributing to the formation of the volcanic rocks, which in turn were leached to form massive sulphide deposits containing galena relatively enriched in radiogenic Pb. Alternatively, metals for the massive sulphide deposits may have been derived from either sedimentary rocks or older basement underlying the volcanic strata.

RÉSUMÉ

Les provinces des Esclaves, du Supérieur et de Churchill dans le Bouclier Canadien renferment des gisements importants de cuivre-zinc, d'or, de nickel et de fer, ceci essentiellement dans les ceintures de roches vertes à dominante volcanique d'âges archéen et protérozoïque. Il existe aussi des dépôts de plomb-zinc dans des grès et carbonates appartenant à des séquences sédimentaires de platforme, telles les séquences des ceintures plissées de Wollaston et de Belcher Island. Les autres types de gisements importants incluent les dépôts d'uranium liés à des discordances et ceux encaissés dans des roches clastiques, les veines d'argent et de plomb-zinc des couvertures sédimentaires, les dépôts associés aux roches alcalines et pegmatites et les gisements filoniens de cuivre de Chibougamau.

Les amas sulfurés à métaux de base sont du type cuivre-zinc avec plomb:zinc <0.1. La pluspart se trouvent dans le second cycle volcanique ou, dans le cas de séquences volcaniques à cycles multiples, dans les cycles les plus élevés et plusieurs sont situés dans la moitié inférieure de leur cycle-hôte. Dans la pluspart des principaux districts miniers on trouve des intrusions subvolcaniques à la base du cycle-hôte. Celles-ci sont interprétées comme représentant les sources d'énergie thermale locales qui ont donné naissance aux systèmes géothermaux de haute température (~350°C) contenant les fluides métallifères.

Les gisements d'or se trouvent essentiellement dans des zones d'instabilité crustale marquée, telles que les zones de rift intravolcaniques. Bien que la pluspart des gisements d'or soient sous forme de veines recoupant des intrusions ou des roches volcaniques et se soient manifestement formés à des niveaux crustaux élevés quelque temps après les roches encaissantes, il existe aussi quelques dépôts d'or dans des formations ferrifères qui pourraient être syngénétiques. Il n'y a guère de preuves, dans les gisements épigénétiques, indiquant une reconcentration locale de l'or à partir de couches voisines anormalement riches, il semble plutôt que l'or dérive de zones crustales plus profondes où il aurait été lessivé par des fluides hydrothermaux de haute température dans un système où le rapport eau/roche est faible. Les fluides se seraient dirigés vers des failles de tension intravolcaniques.

Les gisements de fer ne sont pas directement apparentés aux fluides hydrothermaux qui sont à l'origine des amas sulfurés. Leur faible teneur en métaux de base et leur vaste étendue demande au contraire un grand volume de fluides à relativement basse température. Une large circulation d'eau de mer à travers les roches volcaniques, activée par la chaleur de l'empilement volcanique en voie de refroidissement, pourrait avoir donné le fluide riche en fer nécessaire. Les formations ferrifères du type Supérieur paraissent s'être développées essentiellement dans des rifts intracratoniques.

Les gisements de nickel sont magmatiques, et produits par la séparation d'un liquide sulfuré à partir d'un magma tholeitique ou komatiïtique. La grande quantité de soufre nécessaire pour produire ce liquide sulfuré pourrait dériver par métamorphisme de contact de strates sulfurées, essentiellement de formations ferrifères à faciès sulfuré, situées dans les strates sous-jacentes ou adjacentes.

Les principales différences dans la métallogénie de chacune de ces provinces géologiques comprennent: 1) rareté de gisements d'or importants dans les ceintures de roches vertes protérozoïques et archéennes de la province de Churchill; 2) absence de formations ferrifères dans les séquences sédimentaires et volcaniques protérozoïques de la province de Churchill; 3) absence de gisements de nickel associés aux komatiïtes et aux tholeiïtes dans la province des Esclaves; et 4) plomb caractéristiquement plus radiogénique et d'une façon générale en teneur plus élevée dans les amas sulfurés de la province des Esclaves.

Les causes de ces différences peuvent être respectivement: 1) un style tectonique différent, comprenant peut-être des déformations causées par une compression liée à une zone de subduction protérozoïque dans le sud de la province de Churchill et qui n'a pas permis le développement ou mettre au point de solutions riches en or mais pauvres en métaux de base; 2) absence de tectonique de tension (liée aux rifts) pendant le volcanisme majeur des ceintures de Flin Flon, LaRonge et Lynn Lake, qui n'a pas permis une large circulation d'eau de mer à travers les empilements volcaniques; 3) soit un style tectonique différent (absence de fossés), soit fusion de manteau inférieur insuffisante ou encore croûte excessivement épaisse empêchant la formation de coulées de komatiïte et d'intrusions associées; 4) croûte épaisse, relativement évoluée, sous les strates volcaniques et contribuant à leur formation. Les roches volcaniques auraient été lessivées ensuite pour former les dépôts d'amas sulfurés contenant de la galène relativement enrichie en plomb radiogénique. Une autre alternative serait que les métaux formant les amas sulfurés aient été dérivés soit de roches sédimentaires, soit d'un socle plus ancien supportant les strates volcaniques.

INTRODUCTION

This paper considers the metallogeny of the Slave, Churchill and Superior provinces (Fig. 1) of the Canadian Shield. The Bear and Southern provinces are not considered and a separate paper (Sangster and Bourne, this volume) deals with Grenville metallogeny. The Precambrian metallogeny of the Arctic islands is not included because these areas have very few deposits, and their tectonics and stratigraphy are not well known. The Slave, Churchill and Superior provinces contain a remarkably broad spectrum of ore deposit types, in comparison with most other Shield areas of the world, due, in part, to the broad age range and exceptional variation of litho-tectonic environments represented by the supracrustal sequences of all three provinces. A brief description of the principal tectonic elements associated with most of the major deposit types in each province is a useful introduction to the deposits described in this volume.

All mineral deposit types in each province cannot be reviewed in this brief paper, thus volcanic-hosted massive sulphide, gold, nickel-copper and iron deposits are emphasized; the first three because papers in this volume describe selected deposits of these types. Iron deposits were added to provide a broader picture of the metallogeny of the three Precambrian provinces. Other deposit types, such as sandstone-hosted lead-zinc deposits of the Wollaston belt (Fig. 2), are considered because they illustrate specific metallogenetic differences between provinces. Important deposit types not treated include those of uranium and silver.

Few reviews of the metallogeny of the Precambrian Shield are available. Hutchinson (1981a), Gross (1977), and Boyle (1976) summarized the genetic history and relationships of some major deposit types. Descriptions of many of the mineral

Figure 1. Index map of geological provinces of the Canadian Shield.

deposits are available in Douglas (1970). Hutchinson *et al*. (1971) interpreted the metallogenesis of the Abitibi belt, and provided a model for localization of gold, base-metal and iron deposits in a volcanic framework. In addition, numerous reviews of specific deposit types are available, as for massive sulphide (Sangster and Scott, 1976), iron (Gross, 1965), gold (Boyle, 1979), and nickel deposits (Naldrett, 1973).

Data on lead isotope compositions of some massive sulphide and gold deposits provide general indications of age and genesis and a basis for comparison between deposits, districts, and geological provinces.

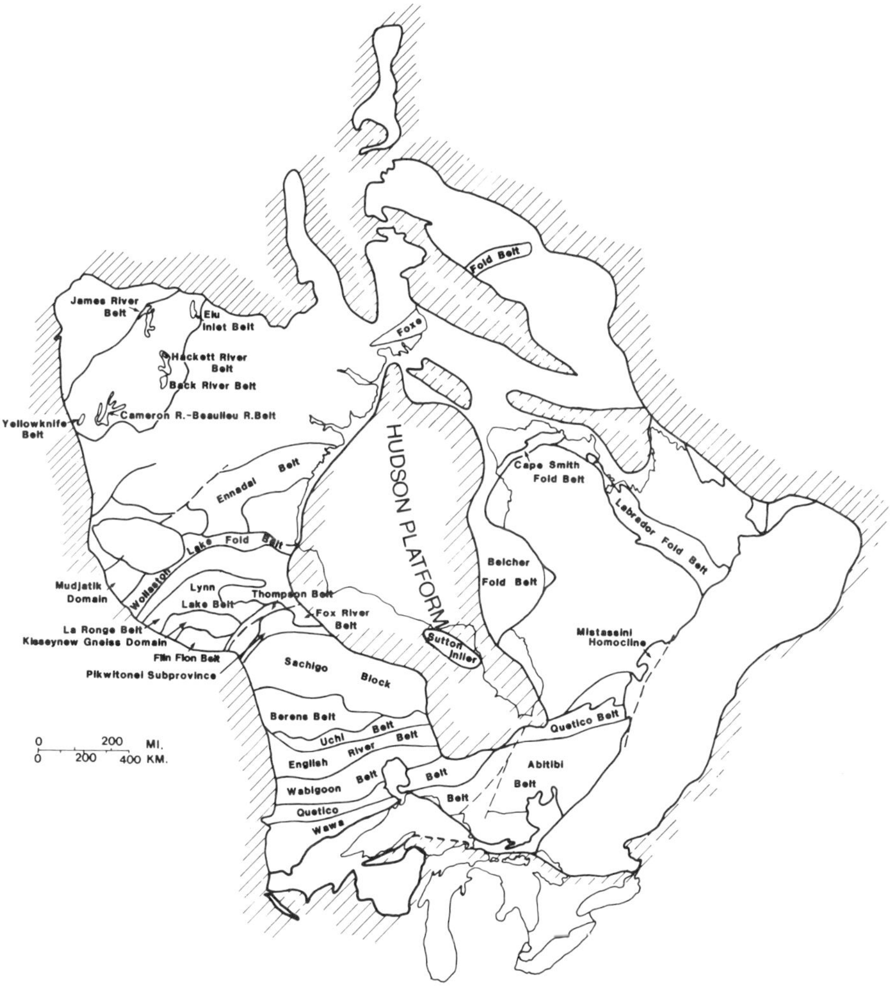

Figure 2. Index map showing many of the greenstone and gneiss belts of the Canadian Shield.

SUPERIOR PROVINCE

General Geology

This Archean portion of the Canadian Shield has been extensively studied for at least 50 years and its petrological and structural characteristics are reasonably well known. However, as in other Archean areas of the world, a uniformly applicable tectonic model is not yet available. Superior Province geology has been reviewed by Goodwin *et al.* (1972). Its geological characteristics provide a comparative standard for other shield areas discussed subsequently.

The Superior Province is composed of 13 sub-provinces (Stockwell, 1970) with numerous east-trending "greenstone belts" containing abundant volcanic, and significant but subordinate sedimentary strata. These belts, ranging up to 600 km in length, are separated by intervening "gneiss belts" composed of metasedimentary rocks, tonalitic gneiss and felsic plutonic complexes. The greenstones form approximately 12% of the total area of Superior province; metasedimentary rocks may form about 3%, although accurate measurements are very difficult as they are commonly mapped as "gneiss" and not clearly distinguished as paragneiss. Thus significantly more Archean sedimentary rock (15 to 20%) may be present.

The structural style of Superior Province is complex; recent studies (Poulsen *et al.*, 1980; Schwerdtner *et al.*, 1979) demonstrate that at least two periods of folding have affected most areas, although some areas appear to have less complex steeply dipping homoclinal sequences.

THE GREENSTONE BELTS

These dominantly volcanic belts, which consist of sigmoidal segments, are deformed into a series of steeply plunging isoclinal folds. The earliest folds in thick volcanic sequences are in general very large, with homoclinal limbs commonly containing several km of steeply dipping strata, as at Sturgeon Lake in Wabigoon Belt (Franklin *et al.*, 1977), Larder Lake - Noranda (Wilson, 1962) and Lake St. Joseph in the Birch-Uchi belt (Clifford, 1969). The large isoclines are themselves deformed into gentle warps and synclines. In thinner parts of the volcanic sequences, the first isoclinal folding is represented by many relatively smaller folds. Schwerdtner *et al.* (1979) showed that the first phase of deformation, which formed the steeply plunging folds, is related to the rise of boundary and inter-belt gneiss diapirs; the latter are usually mechanically remobilized tabular batholiths, but may also include basement gneiss components. The second phase of deformation is related to large scale shearing manifest in major "transcurrent" faults, now represented by major mylonite zones (e.g., the Quetico Fault in Northwestern Ontario). Poulsen *et al.* (1980) noted that the earliest phase of folding along the boundary between the Wabigoon volcanic belt and the Quetico belt involves recumbent folding, with subsequent refolding into upright folds which form the dominant "grain" of the area. Thurston and Breaks (1978) noted that both the Bee Lake and Red Lake volcanic belts of the Uchi subprovince are allocthonous fragments in gravity-driven nappe structures.

Goodwin *et al.* (1972) examined the lithological complement of Archean greenstone belts of Superior Province. The writers re-measured the Abitibi and Wabigoon belts; resulting data (Table I) are identical to those of Goodwin *et al.* (1972)

TABLE I
PROPORTIONS OF MAJOR LITHOLOGIES IN SELECTED GREENSTONE BELTS OF SUPERIOR PROVINCE

	Abitibi[1]	Uchi[1]	Wabigoon	Big Trout Lake[2]
Volcanic rocks	72.7		65.2	62.8
Sedimentary rocks	23.6		30.2	20.9
Mafic-ultramafic intrusions	3.7		4.6	16.3
Volcanic rocks				
Mafic	89.3	82.7		94.1
Felsic	10.7	17.3		5.9

[1]after Goodwin, 1977
[2]part of the Sachigo Belt

for the former, but differ slightly for the latter, as its entire lateral extent was included. The data were normalized omitting the "granitoids", which were at least in part emplaced during and after the first major deformation, hence not clearly related to the supracrustal rocks. Synvolcanic granitoid bodies, which otherwise might be considered part of the felsic volcanic group, were thereby also eliminated.

The volcanic rocks are commonly bimodal in SiO_2 content, very few containing 62% to 68% SiO_2. Mafic rocks, dominantly basalt and andesite, comprise approximately 89% of the volcanic rock (Table I). The less abundant felsic rocks range from approximately 6% in the Big Trout Lake area (Sachigo belt) to over 17% in the Uchi belt. Major element analyses (Goodwin, 1977) indicate that although most of the volcanic rocks belong to a tholeiitic trend, significant portions of most belts are calc-alkaline (43% in the Timmins-Noranda portion of the Abitibi belt and 29% in the Uchi belt). Recently, MacGeehan and MacLean (1980) and Gibson (1979) demonstrated that, at least locally, some calc-alkaline trends result from alteration of tholeiitic rocks, hence that some 'calc-alkaline' rocks are not necessarily of calc-alkaline parentage. However, in the Blake River Group of the Abitibi belt, the highly immobile rare earth elements indicate that both tholeiitic and calc-alkaline rocks are present (Smith, 1980); the latter are more common higher in the succession. Smith suggested a more evolved crustal rather than an upper mantle source for the calc-alkaline rocks.

Ultramafic lavas form a small ($< 1\%$) but important portion of some belts. These rocks, where best preserved, commonly have spinifex texture (Pyke *et al.*, 1973; Gélinas *et al.*, 1977), are intercalated with Mg-tholeiitic basalts, and associated with subvolcanic intrusions of similar, generally peridotitic, composition. Peridotitic komatiites are believed to be the product of high temperature melting of large amounts of mantle material, and a deep (~400 km.) source; high-Mg basaltic komatiites, however, originate from different mantle sources (Nesbitt *et al.*, 1979).

Most of the volcanic rocks are subaqueous, pillowed mafic flows predominating. A few sequences have local subaerial portions such as near Woman Lake in the Uchi belt (Thurston, 1979), in several portions of the Favourable Lake belt (Ayres, 1977), and near the Detour Mine, Quebec (Deptuck *et al.*, this volume).

The stratigraphic disposition of mafic, felsic and ultramafic volcanic rocks varies considerably from belt to belt. Most larger belts are multi-cyclic. Mafic, and less commonly ultramafic rocks predominate near the base of each cycle, with areally restricted felsic rocks near the top (Goodwin, 1977); however, tholeiitic rocks may occur anywhere in a belt, and only the Abitibi belt, the largest single greenstone belt in the Canadian Shield, contains all members of Goodwin's general stratigraphic order.

Most volcanic rocks of the Superior Province formed in a remarkably short time interval. U-Pb dating (Krogh and Davis, 1971; Nunes and Pyke, 1980) indicates that most volcanism occurred between 2700 and 2750 Ma; however the lowest cycle in the Uchi belt (Nunes and Thurston, 1980) is significantly older (2959 $\pm$ 2 Ma), and a tuff-breccia unit from the lower sequence at North Spirit Lake is 3013 Ma (Nunes and Wood, 1980).

Sedimentary lithofacies in greenstone belts range from volcanic-derived epiclastic rocks to distal turbidites. Thin but laterally persistent pyritic-graphitic shale, commonly intercalated with volcanic rocks, contains a tuffaceous component. Non-carbonate carbon is typically below one per cent (Cameron and Garrels, 1980) but ranges up to 12% (Franklin, 1975). Sulphur and carbon isotope data (Shegelski, 1978), the latter from the graphitic strata, suggest that organic activity was important in producing the sulphide and carbon.

In some volcanic domains and the gneiss belts which bound them, turbidite sequences are common. Although their stratigraphic relationships with the volcanic rocks are controversial, in most places, including the famous Coutchiching area (Wabigoon belt; Lawson, 1913), they overlie the volcanic rocks (Poulsen *et al.*, 1980); rarely, as at Savant Lake (Wabigoon belt), they are laterally equivalent to volcanic strata (Shegelski, 1978).

Carbonate sedimentary rocks are sparse in Superior province. However, at Atikokan (Wabigoon belt; Shklanka, 1972), North Spirit Lake (Sachigo belt; Wood, 1977), Favourable Lake (Sachigo belt; Ayres, 1968) and Woman Lake (Uchi belt; Thurston, 1979) marble forms locally distinctive marker horizons. The Atikokan marble is stromatolitic.

Age data on Archean metasedimentary terranes are sparse, but agree generally with data on adjacent volcanic rocks. Thus in the Rainy Lake region of Ontario and Minnesota, Peterman *et al.* (1972) obtained a Rb-Sr isochron age of 2560 Ma, comparable with a Rb-Sr isochron age of 2540 $\pm$ 45 Ma, and a zircon $^{207}Pb/^{206}Pb$ (minimum) age of 2590 Ma for nearby volcanic rocks (Hart and Davis, 1969).

THE GNEISS BELTS

The gneiss belts are sedimentary troughs containing highly variable amounts of psammitic to semi-pelitic sedimentary rocks, including those immediately adjacent to the volcanic belts, migmatites which are in part basement rocks to both sedimentary and volcanic rocks, and large granitoid batholiths. These troughs are laterally continuous. The Quetico belt is over 800 km long, 60 km wide (Pirie and Mackasey, 1978), and is, in its western part, a broad antiform with sedimentary rocks comprising approximately 50% of its area. Its core is composed of migmatite and tonalitic intrusions. The English River belt, 800 km long and 75 km wide, is mainly sedimentary rock in its northern half, but is dominantly plutonic to the south (Thurston and Breaks, 1978).

Metamorphic grade in the gneiss belts is typically amphibolite facies, generally higher than in adjacent volcanic belts. For example, Pirie and Mackasey (1978) noted a progressive increase in metamorphic grade from the margins to the centre of the Quetico belt. Similarly, the metamorphic grade of the English River belt increases from north to south within its paragneiss zone; this increase is accompanied by an increasing volume of metatexite (product of partial melting), and finally diatexite (product of nearly complete melting), which predominates in the southern plutonic zone (Thurston and Breaks, 1978).

Structurally the gneiss belts are very complex, but relatively tight isoclinal folds with upright axial planes and subhorizontal axes are most common.

Age data for the gneiss belts are sparse. However, tonalitic gneiss basement has been identified in the Berens and English River belts; U-Pb dates are in excess of 3000 Ma (Krogh *et al.*, 1976). Similarly old tonalitic gneiss has been identified in the Eastmain area, Quebec (3060 Ma, Verpaelst *et al.*, 1980).

Trondhjemitic to granodioritic orthogneiss forms basement to the volcanic sequence at Lake of the Woods (Wabigoon belt; Clark *et al.*, 1981). This gneiss is very similar to that which forms basement to sedimentary rocks in the gneiss belts, suggesting that much of the Superior Province was composed of 3,000 Ma-old gneiss prior to volcanism. Thus most, if not all, of the Superior Province greenstone belts may have formed in ensialic tectonic zones.

METALLOGENY

The Superior province probably has provided the most varied and richest mineral production of any Archean terrane in the world. The principal ore deposit types are volcanic-hosted massive sulphide deposits, stratabound and intrusion-hosted gold deposits, ultramafic- and gabbro-hosted nickel-copper deposits, and iron-formation. Other deposits include the unusual copper veins at Opemiska and in the Chibougamau district. Examples of the first three and last of these are described in this volume. In addition, minor production has resulted from prophyry copper and pegmatite deposits. Deposits genetically associated with Aphebian and Helikian strata occur in the area of Nipigon and Cobalt plates, and the Mistassini homocline. These strata are probably tectonically related to the Southern Province, but form cratonic cover on the Archean.

Massive Sulphide Deposits

Tonnage and grade data are available for 72 massive sulphide deposits in Superior Province. These occur mainly in clusters or districts (Table II; Fig. 3) and in most volcanic belts, although the thin discontinuous belts in the northwest contain only a few occurrences. The deposits are generally within volcanic and volcaniclastic rocks. Some deposits previously described as sedimentary-hosted, such as those at Manitouwadge (Pye, 1960) and the Marshall Lake occurrences (Langford, 1955), are now thought to be in dominantly volcanic terranes (Friesen *et al.*, this volume; Amukun, 1977). These deposits and their geological settings exhibit the following common characteristics; differences are principally in the proportions of volcanic and volcaniclastic rocks.

1) The deposits generally occur in relatively thick, multi-cyclic volcanic piles, typically in the second or higher cycle. However, some deposits, such as those in Manitouwadge district, occur near the base of a relatively thin single-cycle sequence. Deposits occur near the base of the host cycle; the Mattabi deposit, Sturgeon Lake, near the base of the second cycle (Franklin *et al.*, 1977); the Corbet and Millenbach deposits, Noranda (Knuckey *et al.*, this volume) near the base of the second cycle of the Blake River Group (Spence and de Rosen Spence, 1975).

2) At most deposits the immediate footwall rock is a relatively massive felsic flow or compact tuff. This unit is usually thin, approximately 25 m at Sturgeon Lake, 50 m at Millenbach, and 100 m at Garon Lake. Although the felsic footwall has been emphasized (Roscoe, 1965; Sangster, 1972; Hutchinson, 1973) recent studies have

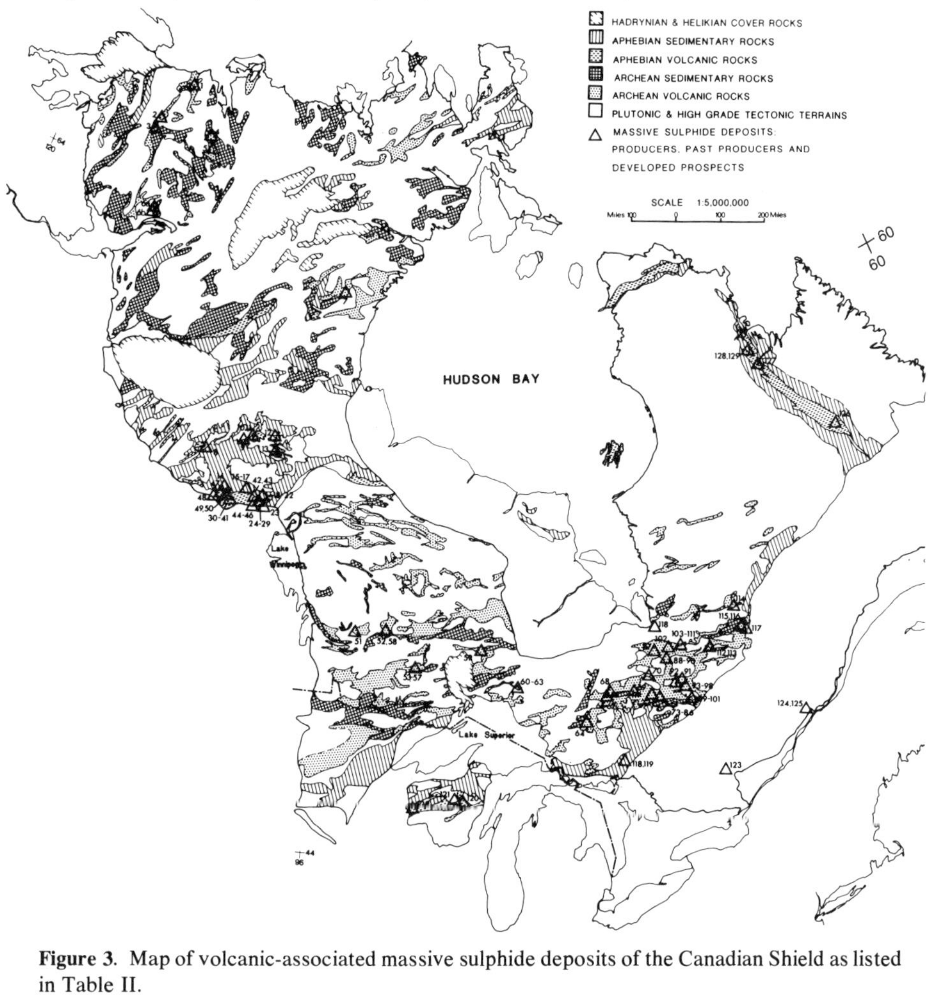

Figure 3. Map of volcanic-associated massive sulphide deposits of the Canadian Shield as listed in Table II.

examined the nature and composition of thicker footwall sequences. These commonly contain abundant fragmental rock (Franklin, 1976) ranging from volcanic-derived epiclastic breccia at Sturgeon Lake (Franklin *et al.*, 1975), to hydrothermally altered and brecciated thick flows at Noranda (Gibson, 1979) and both pyroclastic and epiclastic breccia at Kidd Creek (Walker *et al.*, 1975). In the north limb of the Mattagami Lake structure (MacGeehan and MacLean, 1980), however, the footwall sequence is composed dominantly of mafic flows, with a very thin felsic pyroclastic unit.

3) Most deposits lie close to their exhalative vents, and immediately over discordant alteration pipes characterized by alkali depletion, and abundant chlorite and lesser sericite in the upper portions (Roberts and Reardon, 1978; Riverin and Hodgson, 1980). The Mattabi deposit (Franklin *et al.*, 1975) has unusual alteration, with chlorite subordinate to sericite, and quartz enrichment; siderite and chloritoid dominate in the upper part of the pipe. A few small deposits, as the Lyon Lake and Creek zones at Sturgeon Lake, are interpreted as "distal" to their alteration pipes or exhalative vents (Harvey and Hinzer, 1981), but are generally within a few km of possible source vents.

4) Many, and possibly all of the deposits have a semi-conformable alteration zone up to several hundred m below the deposit in which the discordant alteration pipe stratigraphically beneath the deposit may terminate. At Sturgeon Lake (Franklin *et al.*, 1977) this lower zone is within the footwall epiclastic sequence and is extensively depleted in alkali elements. At Millenbach it lies within the Amulet "rhyolite", and consists of variably silicified, partly spilitized mafic rock (Knuckey *et al.*, this volume). At Matagami, Quebec the Garon Lake deposit is on the north limb of a major anticline (Sharpe, 1968), and overlies the Garon Lake rhyolite, which is up to a few hundred m thick. The Garon Lake basalts, beneath this rhyolite, have irregularly distributed, stratiform, silicified and spilitized zones (MacGeehan and Maclean, 1980).

5) Cu:Zn:Pb ratios in Superior Province deposits are relatively consistent compared with those in Churchill Province deposits, with Zn>Cu>Pb. A few deposits contain unusually high Au (Horne, Quemont, Lemoine, Taché Lake, each with approximately 6.2 g/tonne), but in general Au content is approximately 0.7 g/tonne.

6) Most districts have subvolcanic intrusions, commonly sills or irregular bodies near the base of the host cycle or near the top of the underlying cycle. These are commonly tonalitic or trondjhemitic, like the Beidelman Bay sill at Sturgeon Lake (Franklin *et al.*, 1977) and the Flavrian and Powell intrusions at Noranda (Goldie, 1979), but rarely are mafic, like the Bell River gabbro sills at Matagami Lake (MacGeehan and MacLean, 1980). These intrusions are possible heat sources for massive sulphide generation (Franklin, 1978) and may have important exploration significance. The major and minor element compositions, including rare-earth elements, of felsic subvolcanic intrusions are almost identical with those of their accompanying felsic volcanic rocks (Campbell *et al.*, 1981). The intrusions are commonly slightly fractionated, and more mafic (quartz dioritic) on their edges. They have narrow metamorphic haloes, commonly contain inclusions of country rock, and are affected by the earliest regional deformation (Poulsen and Franklin, 1981). Porphyry copper occurrences are present within some of them (Friske *et al.*, 1979, Kirkham and Thorpe, 1973).

TABLE II
VOLCANIC-HOSTED MASSIVE SULPHIDE DEPOSITS OF THE CANADIAN SHIELD

No.	N.T.S.	Latitude	Longitude	Deposit Name	Status*	Short Tons	Cu%	Zn%	Pb%	Ag oz/ton	Au oz/ton
	Slave Province										
1a	76M/07	67°22′50″	110°51′20″	High Lake AB zone	DP	2,400,000	5.4	1.1			0.033
1b	76M/07	67 22 35	110 51 18	High Lake D zone	DP	2,800,000	2.0	3.6	0.2	1.213	0.022
2a	86I/02	66 02 35	112 45 15	No. 41 Deposit (Texasgulf)	DP	320,000	1.57	4.12	0.2	0.52	
2b	86I/02	66 02 05	112 42 30	Hood River No. 10	DP	820,000	4.08	3.61		0.77	
3	86H/10	65 38 00	112 47 45	Izok Lake	DP	12,150,000	1.85	13.7	1.45	2.05	
4a	76F/16	65 55 05	108 21 59	Hackett River A	DP	5,000,000	0.25	8.5	1.5	7.72	0.055
4b	76F/16	65 55 00	108 27 39	East Cleaver Lake	DP	4,000,000	0.46	7.07	1.04	5.84	0.015
4c	76F/16	65 56 46	108 26 15	Boot Lake	DP	5,000,000	0.29	4.97	0.99	5.86	0.014
5	76G/12	65 36 40	107 56 11	Yava	DP	800,000	0.4	4.6	1.5	4.2	0.035
6a	75M/02	63 01 57	110 56 57	Indian Mtn. Lake (BB zone)	DP	1,000,000	0.2	10.3	0.85	2.70	
6b	75M/02	63 01 57	110 57 48	Kennedy Lake (West Cu)	DP	70,000		6.0	1.0	5.51	
	Churchill Province										
7	65H/16	61 46 25	96 12 10	Gemex	DP	est. <1,000,000					
8	64D/04	56 08 00	103 40 00	Brabant (McKenzie)	DP	4,330,000	0.64	4.43		0.154	0.002
9	64C/12	56 38 07	101 39 06	Fox Lake	P	14,792,000	1.81	1.63			
10	64C/14	56 50 00	101 00 30	DH-FL Groups	DP	500,000	0.90	2.20			
11	64C/14	56 49 45	101 01 30	Z Deposit (Sherritt Gordon)	DP	153,000	1.11	2.49			0.018
12	64B/13	56 53 06	99 55 00	Knobby Lake (MacBride Lake)	DP	533,540	0.35	8.77			
13	64B/5	56 28 30	99 38 14	Ruttan Lake	P	45,614,000	1.46	1.36		0.22	0.011
14	63M/01	55 05 45	102 13 35	Schotts Lake	DP	1,000,000	1.0				
15	63N/02	55 09 50	100 58 20	Jungle Lake	DP	3,700,000	1.42	1.1			
16	63N/03	55 08 22	101 06 25	Sherridon Mine	PP	8,531,500	2.37	2.28		0.606	0.022
17	63N/03	55 09 30	101 02 30	Bob Lake	DP	2,380,000	1.33	1.80		0.30	0.011
18	63N/01	55 01 31	100 02 38	Wim	DP	1,090,000	2.91				0.052
19	63J/13	54 51 23	99 55 07	Stall Lake (Rod-Stall Lake Mine)	DP	672,640	5.38	2.28			
20	63J/13	54 57 15	99 43 24	Osborne Lake	P	3,458,000	3.69	1.53			
21	63J/13	54 51 37	99 59 35	Anderson Lake	P	3,244,900	3.79			0.23	0.014
22	63J/13	54 51 20	99 56 28	Stall Lake	P	5,721,400	4.76	0.65		0.34	0.043

23	63J/12	54 38 57	99 52 24	Copper Man	DP	244,000	2.63	4.46			
24	63K/16	54 51 18	100 11 00	Bomber	DP	753,580	0.04	1.00		0.276	0.003
25	63K/16	54 46 48	100 10 54	Pot Lake	DP	112,000	1.43	4.50		0.606	0.012
26	63K/16	54 50 00	100 06 54	Chisel Lake	P	5,952,200	0.56	11.61	0.82	1.38	0.053
27	63K/16	54 49 54	100 06 12	Ghost Lake	P	614,743	1.40	8.68	0.75	1.31	0.048
28	63K/16	54 49 47	100 06 36	Lost Lake (One)	P	92,600	1.45	4.90	1.0(?)	2.612	0.094(?
29	63K/16	54 49 42	100 02 00	Joannie Lake	DP	480,000	1.28				
30	63K/12	54 35 05	101 59 52	Coronation Mine	PP	1,412,860	4.26	0.15		0.165	0.066
31	63K/13	54 46 02	101 52 56	Flin Flon	P	64,391,864	2.31	4.25		1.378	0.09(?)
32	63K/12	54 43 36	101 49 54	Mandy	PP	137,000	7.3	12.9		1.984	0.099(?
33	63K/13	54 45 43	101 34 23	Don Jon	PP	87,440	3.09			0.44	0.027
34	63K/12	54 43 11	101 49 34	Schist Lake	PP	2,033,000	4.33	7.17		1.16	0.044
35	63K/13	54 49 45	101 49 18	Embury (Trout Lake)	DP	3,600,000	2.6	4.3			
36	63K/12	54 38 33	101 50 14	Westarm	P	710,000	4.63	0.60			
37	63K/12	54 42 48	101 43 12	White Lake	P	601,200	2.27	4.83		1.05	0.02
38	63K/12	54 43 14	101 42 39	Cuprus	PP	509,050	3.24	6.40		0.93	0.044
39	63K/12	54 42 18	101 40 01	Centennial	P	1,400,000	2.06	2.60		0.77	0.044
40	63K/13	54 45 56	101 36 31	Pine Bay	DP	1,500,000	1.30				
41	63K/13	54 45 54	101 34 42	North Star	PP	264,420	6.10			0.25	0.011
42	63K/14	54 56 18	100 29 24	Hudvam (Vamp Lake)	DP	401,200	1.50	1.70		0.474	0.011
43	63K/16	54 51 18	100 29 24	Dickstone	PP	1,169,200	2.58	3.29		0.44	0.015
44	63K/09	54 34 24	100 24 00	Spruce Point	DP	500,000	2.80	4.50			
45	63K/10	54 44 54	100 35 30	Rail Lake	DP	325,000	3.00	0.70			
46	63K/10	54 38 13	100 32 56	Reed Lake (Hudson Bay Mining)	DP	875,000	2.30	0.70			
47	63L/16	54 56 45	102 23 15	Wildnest Lake	DP	2,186,000	0.61	1.20			
48	63L/10	54 44 17	102 45 05	Ramsay Showing (Quandt)	DP	710,000	2.04	1.60			
49	63L/09	54 40 37	102 01 43	Flexar Mine	PP	337,400	3.77	0.47		0.21	0.044
50	63L/09	54 39 41	102 01 58	Birch Lake	PP	307,265	6.17			0.132	0.003
	Superior Province										
51	52L/06	51 00 32	94 12 00	Trout Bay Cu	DP	139,000	1.42	7.56	0.24	1.79	0.008
52	52N/02	51 06 35	92 40 45	South Bay Mine	P	1,730,363	1.6	11.00		2.40	
53	52G/15	49 52 39	90 58 55	Mattabi	P	11,649,511	0.83	7.98	0.76	3.47	0.011
54	52G/15	49 52 15	90 53 04	Sturgeon Lake (Boundary)	P	2,254,600	2.74	10.01	1.41	6.25	0.022
55	52G/15	49 53 12	90 53 22	Lyon Lake (Mattagami)	P	3,945,000	1.24	6.53	0.63	3.42	0.01

No.	N.T.S.	Latitude	Longitude	Deposit Name	Status*	Short Tons	Cu%	Zn%	Pb%	Ag oz/ton	Au oz/ton
56	52G/15	49 52 57	90 53 02	Creek Zone (Mattagami)	DP	908,000	1.66	8.16	0.88	4.62	0.014
57	52G/15	49 52 30	90 58 30	F Group	DP	630,000	0.98	8.10	0.49	1.80	
58a	52K/15	50 58 00	92 52 40	Copper-Lode E	DP	276,000	0.66	4.8		0.43	
58b	52K/15	50 58 45	92 57 30	Copper-Lode Main	DP	425,600	1.56			1.08	
59	42L/05	50 25 20	87 35 13	Kendon Copper	DP	267,000	1.45	4.76			
60	42F/04	49 09 15	85 47 32	Geco	P	55,487,000	1.86	3.45		1.46	
61	42F/04	49 09 20	85 49 30	Willroy	PP }	8,714,600	0.94	4.62	0.18	1.92	
62	42F/04	49 10 45	85 53 00	Willecho	PP }						
63	42F/04	49 11 00	85 50 00	Big Nama Creek	PP	198,810	0.83	4.16		1.15	
64	41O/10	47 42 48	82 39 30	Grandora (Shunsby)	DP	2,500,000	1.20	1.28			
65	42A/12	48 36 10	81 35 30	Kam Kotia	PP	6,606,140	1.12	1.22		0.114	0.001
66	42A/12	48 34 45	81 35 15	Jameland Mine	PP	509,360	0.99	0.88		0.091	0.001
67	42A/12	48 32 10	81 34 12	Canadian Jamieson	PP	639,000	2.26	3.29		0.88	0.009
68	42A/11	48 41 30	81 21 55	Kidd Creek	P	157,866,800	2.46	6.00	0.2	2.25	
69	42A/09	48 36 00	80 12 40	Potter Mine (Munro Cu)	PP	600,000	2.05	1.67			
70	32E/03	49 00 07	79 21 58	Normetal	PP	11,143,800	2.15	5.12		1.455	0.018
71	32D/06	48 26 27	79 22 21	Magusi	DP	4,100,000	1.2	3.6		0.91	0.032
72	32D/06	48 26 30	79 20 40	New Insco	DP	1,148,000	2.11				
73	32D/06	48 18 00	79 04 54	Corbet Mine	P	3,117,000	2.99	2.13		0.68	0.027
74	32D/06	48 21 48	79 04 42	Vauze Mine	PP	385,000	2.90	0.94		0.772	0.021
75	32D/06	48 20 28	79 04 52	East Waite	PP	1,500,000	4.13	3.26		1.003	0.058
76	32D/06	48 18 45	79 04 31	Waite Amulet	PP	1,245,000	4.70	2.97	0.038	0.694	0.035
77a	32D/06	48 19 06	79 04 50	Amulet F	PP	280,000	3.54	3.40		0.149	0.017
77b	32D/06	48 18 34	79 04 36	Amulet C	PP	600,000	2.20	8.50		2.789	0.019
77c	32D/06	48 18 35	79 04 06	Amulet A and Lower A	PP	5,300,000	5.12	5.47		1.499	0.047
78	32D/06	48 21 00	79 03 24	Norbec, Lake Dufault	PP	4,019,400	2.78	4.76		1.140	0.018
79	32D/06	48 18 30	79 03 48	Lake Dufault No. 2 Deposit	PP	100,000	0.50	9.80		3.307	0.044
80	32D/06	48 18 04	79 03 07	Millenbach Mine	P	3,533,000	3.43	4.23		1.521	0.026
81	32D/06	48 15 16	79 00 41	Horne Mine	P	62,300,000	2.176				0.165
82	32D/06	48 15 30	79 00 25	Quemont	PP	15,349,000	1.31	2.43	0.018	1.003	0.173
83	32D/07	48 15 54	78 57 54	Delbridge	PP	483,000	0.49	9.40		3.086	0.088

84	32D/07	48 16 00	78 55 00	D'Eldona Gold	PP	90,000		7.70		2.205	0.187
85	32D/07	48 19 32	78 57 12	West MacDonald	DP	10,046,000	0.014	2.69		0.061	
86	32D/07	48 23 20	78 53 45	Mobrun	DP	3,041,000	0.62	2.32		0.683	0.057
87a	32E/15	49 47 00	78 55 00	Detour (Selco) Al Zone	DP	35,400,000	0.39	2.30		1.15	0.01
87b	32E/15	49 47 00	78 55 00	Detour (Selco) A2 Zone	DP	9,750,000	2.26	1.24		0.054	0.025
87c	32E/15	49 47 00	78 55 00	Detour (Selco) B Zone	DP	3,315,000	4.49	0.80		1.27	0.04
88	32E/08	49 27 12	78 21 02	Joutel Copper (Cu Zone)	PP	1,148,200	2.16				
				Joutel Copper (Zn Zone)	PP	410,000		8.90			
89	32E/08	49 26 43	78 23 03	Mines de Poirier	PP	3,639,780	1.514	0.96		0.054	
90	32E/08	49 26 00	78 17 00	Cons. Northern Expl.	DP	1,099,110	0.73	6.95		1.102	
91	32E/08	48 42 30	77 45 36	Trinity Property	DP	147,000	1.18	0.74			
92	32D/09	48 36 00	78 04 00	New Formaque (Conigo)	DP	2,452,000	1.10	0.31		0.86	
93	32C/12	48 31 00	77 43 00	Frebert Mine	DP	1,900,000		1.95		3.70	
94	32C/12	48 31 18	77 40 30	Barvue Mine	PP	9,720,230		3.5		1.433	
95	32C/05	48 25 47	77 40 54	Belfort (Roymont)	DP	250,000	0.21	7.0	0.12	0.672	
96	32C/05	48 25 24	77 39 42	Cons. Mogador (Vendome)	DP	1,230,000	0.47	7.30	0.34	1.786	0.037
97	32C/05	48 26 00	77 42 00	Barvallee Mine	DP	216,000	1.23	5.71		1.565	
98	32C/12	48 31 24	71 41 06	Cons. Pershcourt	DP	5,117,400		2.42		2.68	
99	32C/04	48 05 54	77 31 12	Louvem (Zn Ore)	P	2,550,000	0.21	6.29		1.07	0.04
				Louvem (Cu Ore)	PP	1,045,200	1.99				
100	32C/04	48 05 12	77 36 36	Manitou Barvue	PP	12,930,240	0.277	2.71	0.223	2.10	0.028
101	32C/04	48 04 15	77 42 34	East Sullivan	PP	16,482,000	0.96	0.47		0.231	0.008
102	32E/16	49 46 29	78 09 19	La Gauchetière (Phelps Dodge)	DP	1,700,000	1.1	4.9		0.55	
103	32F/12	49 42 30	77 42 45	Orchan Mine	P	5,159,000	1.05	9.93		0.66	0.007
104	32F/12	49 43 18	77 43 00	Mattagami Lake	P	21,556,000	0.65	8.40		1.05	0.017
105	32F/12	49 41 23	77 41 31	Bell Allard	PP	258,124	0.96	8.15		0.584	0.006
106	32F/13	49 47 18	77 50 05	New Hosco	PP	2,263,500	1.33	0.99		0.132	0.001
107	32F/12	49 38 48	77 43 18	Lynx (Obaska)	DP	225,000	1.60	0.35			
108	32F/13	49 46 35	77 39 45	Norita (Radiore A)	P	4,500,000	2.16	3.57		0.8	0.02
109	32F/13	49 46 12	77 37 30	Bell Channel No. 1	DP	90,000	1.95	0.57			
110	32F/13	49 45 24	77 34 06	Garon Lake	PP	539,415	1.59	2.17		0.132	0.004
111	32F/12	49 44 38	77 46 38	Bell Allard (Radiore 2)	DP	153,000	2.00	1.00			
112	32F/08	49 29 54	76 09 48	Coniagas Mine	PP	966,860		9.10	0.87	5.06	
113	32F/08	49 26 48	76 09 30	Empire Oil (Le Tac)	DP	365,000	0.225	2.45		0.16	

No.	N.T.S.	Latitude	Longitude	Deposit Name	Status*	Short Tons	Cu%	Zn%	Pb%	Ag oz/ton	Au oz/ton
114	32J/10	50 38 50	74 38 30	Frotet Lake (Lessard)	DP	1,464,000	1.73	2.96		1.21	0.021
115	32G/15	49 51 52	74 37 47	Selco-Scott	DP	747,000	0.55	6.91		0.38	0.006
116	32G/16	49 51 36	74 24 23	Taché Lake (Antoinette)	DP	708,310	0.33	5.26		0.33	0.06
117	32G/16	49 45 44	74 06 10	Lemoine (Patino)	P	741,900	4.81	10.64		2.56	0.141
	Southern Province										
118	41I/11	46 32 15	81 15 30	Errington	DP	13,662,000	1.20	3.87	0.99	1.59	0.024
119	41I/11	46 31 11	81 21 20	Vermilion Lake	DP	5,965,000	1.13	3.89	1.02	1.43	0.025
120	Wisconsin	45 40	88 45	Crandon	DP	70,000,000	1.0	6.5			
121	Wisconsin	45 45	89 10	Pelican R.	DP	2,300,000	1.0	4.5			
122	Wisconsin	45 25	91 00	Flambeau	DP	~5,000,000	~3.75				
	Grenville Province										
123	31F/10	45 42 00	76 41 00	Calumet Mine	PP	4,053,000	0.116	5.95	1.66	3.90	0.021
124	31I/16	46 50 30	72 21 00	Montauban	DP	862,240		3.39	1.07	1.224	0.017
125	31I/16	46 49 30	72 20 54	Tetreault	PP	2,655,000		4.53	1.56	2.76	0.022
	Churchill Province (Labrador Trough)										
126	23O/01	55 03 12	66 15 54	Frederickson L.	DP	308,000	0.77	4.38		1.23	0.02
127	24F/11	57 39 42	69 27 00	Boylen	DP	1,169,165	0.70	6.86	1.03	1.59	0.03
128	24K/04	53 19 19	69 52 06	Soucy	DP	6,000,000	1.34	1.87		0.54	0.047
129a	24K/05	53 15 36	69 54 18	Prud'homme No. 1 (Main zone)	DP	4,745,000	2.04	2.66		0.953	0.04
129b	24K/05	53 15 36	69 54 18	Prud'homme No. 1 (north zone)	DP	562,500	2.05	2.30		1.01	0.05

* The status of the deposit is recorded as producer (P), past producer (PP), or developed prospect (DP). The deposits are plotted in Figure 3.

Gold Deposits

Except for Yellowknife, NWT., all the major gold producing districts of the Canadian Shield are in the Superior Province. The cumulative gold production from the deposits in the Superior province exceeds 4,355,000 kg. The major deposits and main producing districts are listed in Tables III and IV, and their distribution is shown in Figure 4. The deposits exhibit a wide range of geological characterisitcs and can be variably classified depending on which of these are emphasized. They are here subdivided empirically, on form, major lithologic association, and current genetic concepts, into the following: 1) syngenetic deposits; 2) pluton-hosted deposits; 3) volcanic-hosted vein and shear zone deposits; and 4) "porphyry gold" deposits (discussed with porphyry Cu (-Mo) deposits).

The Chadbourne breccia pipe-hosted gold deposit at Noranda does not fit into this classification and is unique in the Canadian Shield. The pipe is elliptical, 300 m by 110 m, cuts andesite and rhyolite, and consists of fragments of these in a matrix of quartz, calcite and minor plagioclase (Walker and Cregheur, 1980). The orebody is funnel shaped with a surface diameter of 76 m.

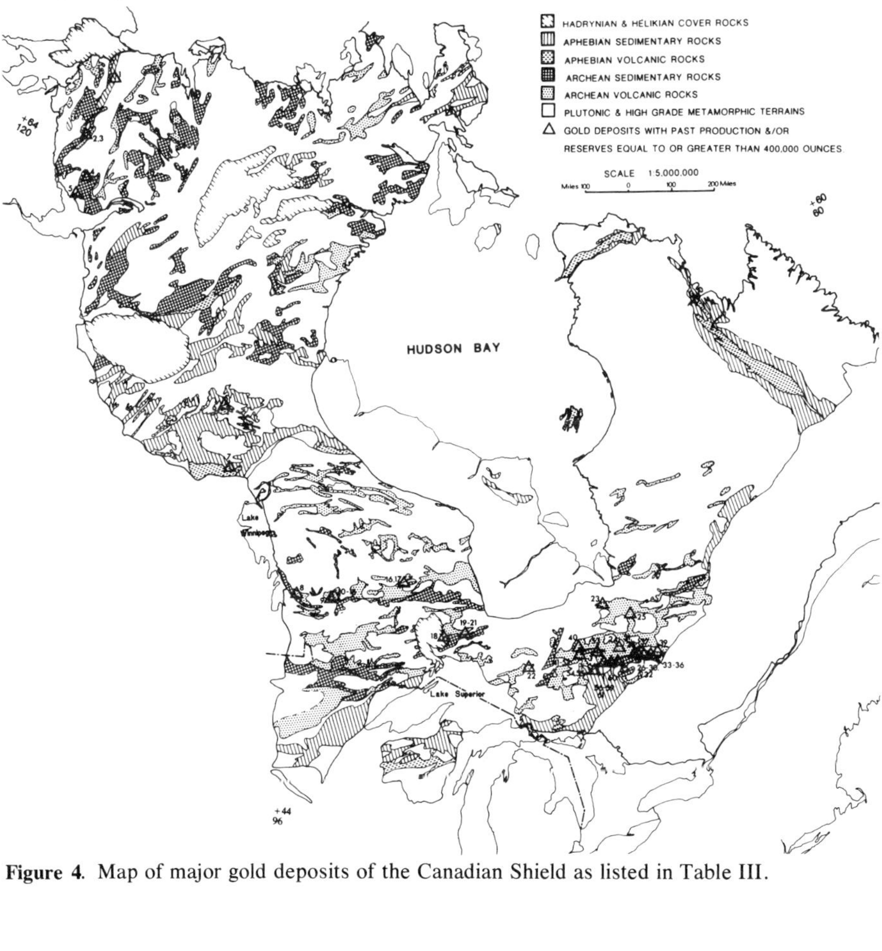

Figure 4. Map of major gold deposits of the Canadian Shield as listed in Table III.

TABLE III

PRODUCTION AND RESERVES OF GOLD DEPOSITS OF THE CANADIAN SHIELD FOR WHICH PRODUCTION, RESERVES OR A COMBINATION OF THE TWO TOTAL AT LEAST 400,000 OUNCES OF GOLD. THESE DEPOSITS ARE SHOWN ON FIGURE 4

No.	N.T.S.	Latitude	Longitude	Property	Production Tons	Production Ounces Au	Reserves Tons	Reserves Ounces Au	Comments
	Slave Province								
1	86J/06	66°26′22″	114°22′03″	Con-Rycon, N.W.T.	5,528,733	3,088,531	1,800,000	936,000	Prod. thru 1979. Res. end 1979.
2	86B/06	64 22 30	115 06 30	Indin Lake BOB, N.W.T.			13,000,000	1,118,000	
3	86B/06	64 22 45	115 06 30	Indin Lake PR, N.W.T.			20,000,000	1,060,000	
4	85P/04	63 11 07	113 53 30	Discovery, N.W.T.	1,033,000	1,028,970			Prod. thru May, 1969.
5	85I/12	62 30 10	114 21 25	Giant Yellowknife, N.W.T.	10,528,525	5,403,321	2,054,000	431,340	Prod. thru 1979. Res. end 1979.
	Churchill Province								
6	64C/15	56 53 56	100 57 31	Agassiz, Manitoba			2,330,000	745,600	Res. end 1979.
7	63K/16	54 53 12	100 01 24	Nor Acme, Manitoba	5,396,304	627,103	3,239,000	497,855	Prod. thru 1958. Res. end 1974.
	Superior Province								
8	52M/04	51 01 30	95 40 42	San Antonio-New Forty-four, Man.	4,390,567	1,178,538	724,393	166,610	Prod. thru 1968 but not given for 1967 & 68. Res. end 1974.
9	52K/13	50 58 00	93 54 50	Madsen Mine, Ontario	8,600,000	2,416,600	313,475	78,533	Prod. to June 15, 1976, closing. Res. at closing.
10	52N/04	51 04 42	93 49 30	McKenzie, Ontario	2,354,000	635,580			Prod. thru 1956.
11	52N/04	51 04 36	93 48 18	Cochenour Willans, Ontario	2,229,000	1,151,789			Prod. thru 1971.
12	52N/04	51 04 00	93 47 30	Wilmar, Ontario	193,435	60,771.5	3.15 to 4.5×10^6	(0.076 to 0.131 oz/ton)	Prod. thru 1971 + 1975. Res. end 1975.
13	52N/04	51 03 45	93 44 30	Campbell Red Lake, Ontario	7,750,981	4,716,516	1,976,500	1,298,560	Prod. thru 1979. Res. end 1979.

14	52N/04	51 03 55	93 43 50	Dickenson, Ont. (includes Prod. from Robin Red Lake Prop.)	4,195,154	1,973,994	389,740	206,172	Prod. thru 1979. Res. end 1979.
15	52N/04	51 01 00	93 49 30	Howey Mine, Ontario	4,631,000	416,790			Prod. thru 1957.
16	52C/09	51 30 30	90 03 20	Pickle Crow, Ontario	3,218,000	1,448,100			Prod. thru 1966.
17	52O/08	51 29 15	90 04 20	Central Patricia, Ontario	1,729,000	587,860			Prod. thru 1951.
18	52H/09	49 37 34	88 02 08	Leitch Mine, Ontario	964,000	886,880			Prod. to 1968.
19	42E/10	49 40 46	86 57 46	MacLeod Mosher, Ontario	13,769,915	1,879,828	869,873±	93,076±	Prod. to 1970.
20	42E/10	49 40 42	86 56 10	MacLeod Cockshutt, Ontario					
21	42E/10	49 41 52	86 51 22	Little Long Lac, Ontario	1,780,000	605,200			Prod. thru 1956.
22	42B/05	48 22 33	83 52 25	Renabie, Ontario	3,633,386	762,930	800,000	173,600	Prod. thru 1969. Res. end 1978.
23	32L/04	50 00 50	79 42 20	Amoco (Detour R.), Ontario			27,733,000	3,466,600	Reserves end 1980.
24	32D/11	48 30 30	79 14 30	Beattie-Duquesne, P.Q.	10,614,000	1,188,768			Prod. thru 1956.
25	32E/08	49 29 30	78 21 42	Agnico-Eagle, P.Q.	1,942,765	346,212	1,305,533	296,356	Prod. thru 1979. Res. end 1979.
26	32D/01	48 06 53	78 00 37	Malartic Goldfield, P.Q.	9,874,175	1,698,328			Prod. thru 1965.
27	32D/01	48 07 47	78 07 50	Canadian Malartic Mines, P.Q.	10,947,270	1,076,125			Prod. thru 1965.
28	32D/01	48 07 39	78 06 07	East Malartic, P.Q.	19,788,497	2,840,134	234,799	18,080	Prod. thru 1979. Res. end 1979.
29	32D/01	48 09 11	78 02 06	Camflo, P.Q.	5,768,309	1,263,056	2,297,420	330,836	Prod. thru 1979. Res. end 1979.
30	32D/01	48 09 49	78 01 50	Malartic Hygrade, P.Q.			2,200,000	550,000	Inferred, res. end 1973.
31	32D/01	48 14 00	78 23 48	O'Brien Mine, P.Q.	1,310,356	589,081			Prod. thru 1956.
32	32D/01	48 08	78 06 12	Barnat, P.Q.	9,318,882	1,206,847	105,000	9,450	Prod. thru 1970. Additional 200,000 tons prod. 1976-77. Res. end 1979.
33	32C/04	48 07 40	77 50 17	Sullivan Cons., P.Q.	5,085,518	1,134,350	344,000	74,992	Prod. to 1968. Res. at closing in 1968.
34	32C/04	48 05 58	77 45 27	Sigma, P.Q.	17,753,668	3,024,945	1,239,900	250,460	Prod. thru 1979. Res. end 1979.
35	32C/04	48 05 38	77 45 53	Lamaque, P.Q.	23,541,000	4,370,100	387,000	52,245	Prod. to end 1979. Res. end 1979.

No.	N.T.S.	Latitude	Longitude	Property	Production Tons	Production Ounces Au	Reserves Tons	Reserves Ounces Au	Comments
36	32C/04	48 07 13	773855 01	Kiena, P.Q.			2,767,000	643,430	Proven and probable res. end 1965.
37	32D/08	48 15 24	78 12 12	Doyon, P.Q.	4,901	3,864	4,356,600	764,800	Res. end 1975. 1979, test runs on 50,000 tons of ore, now in prod.
38	32D/08	48 15 06	78 28 12	Thompson-Bousquet, P.Q.	120,509	27,790	1,514,406	305,555	Prod. started July, 1979. Prod. to end of 79; res. end 1979.
39	32C/03	48 05 48	77 25 12	Bevcon, P.Q.	3,493,242	406,409	915,000	137,250	Prod. thru 1965. Res. end 1965.
40	42A/11	48 23 00	81 07 30	Owl Creek, Ontario			>2,500,000	375,000	Res. 1980.
41A	42A/06	48 28 26	81 18 06	Hollinger, Ontario	66,141,230	19,389,191	650,000	43,200	Prod. thru 1968 + 1978. Res. 1978 (after 1978, Prod. + Res. combined with Pamour).
41B	42A/06	48 28 07	81 18 54	Crown Mine, Ontario	1,791,000	551,628			Prod. thru 1941.
41C	42A/06	48 28 07	81 18 21	Vipond, Ontario	1,565,218	414,367			Prod. thru 1941.
42	42A/06	48 29 18	81 17 30	McIntyre (Schumacher), Ontario	34,372,465	10,456,810	670,045	41,599	Prod. thru 1978. Res. end 1978. (After '78, Prod. + Res. combined with Pamour.)
43	42A/06	48 27 45	81 14 33	Dome, Ontario	38,120,614	10,466,359	1,896,000	409,536	Prod. thru to 1979. Res. end 1978.
44	42A/11	48 31 32	81 06 54	Pamour, Ont. (Pamour #1)	25,500,000	2,837,000	1,398,000	141,188	Approx. Prod. thru 1979. Res. end 1978. (After 1978 Prod-Res. is combined with other prop.
45	42A/11	48 31 06	81 08 36	Hallnor, Ont. (Pamour #2)	4,000,000	1,580,600			Approx. Prod. thru to 1971.

46	42A/06	48 26 30	81 16 42	Aunor, Ont. (Pamour #3)	8,113,600	2,459,738	196,740	39,348	Prod. to end 1979. Res. end 1978.
47	42A/06	48 27 30	81 15 48	Paymaster, Ontario	5,607,000	1,005,291	519,000	129,750	Prod. thru 1966. Res. end 1966.
48	42A/06	48 27 17	81 14 21	Preston, Ontario	6,287,000	1,540,315	significant tonnage	≈ 0.15 oz/ton	Prod. thru 1968 – includes Prod. from Midcamp prop. Rehabilitated 1980.
49	42A/11	48 30 42	81 10 24	Reef Mine, Ontario	2,146,514	498,982		16,500	Prod. thru 1941.
50	42A/06	48 29 18	81 16 15	Coniaurum, Ontario	4,464,000	1,116,000			Prod. thru 1961.
51	42A/06	48 26 30	81 16 18	Ankerite, Ontario	5,269,631	1,014,924			Prod. thru 1956.
52	42A/06	48 26 10	81 17 34	Delnite, Ontario	3,847,000	920,424			Prod. thru 1964.
53	42A/08	48 28 45	80 17	Ross Mine, Ontario	5,192,000	845,707	246,000	37,884	Prod. thru 1979. Res. end 1978.
54	41P/15	47 56 50	80 41 35	Young-Davidson, Ontario	6,128,232	577,681			Prod. thru 1956.
55	41P/15	47 57 30	80 40 30	Matachewan Cons., Ont.	3,545,200	371,695			Prod. thru 1953-1979.
56	42A/01	48 08 50	80 02 50	Lakeshore, Ontario	16,773,458	8,504,642	3,636,000	371,950	Prod. thru 1967. Res. include 3,302,460 tons tailings at 0.07 oz/ton Au.
57	42A/01	48 07 45	80 03 25	Tech-Hughes, Ontario	9,683,095	3,686,828			Prod. thru 1968.
58	42A/01	48 08 25	80 04 20	Macassa, Ontario	5,229,607	2,332,788	411,025	210,028	Prod. thru 1979. Res. end 1978.
59	42A/01	48 08 45	80 03 55	Kirkland Lake Gold, Ont.	3,140,283	1,172,941			Prod. thru 1960.
60	32D/04	48 08 28	79 49 22	Upper Canada Mines, Ont.	4,734,617	1,398,291			Prod. thru 1971.
61A	42A/01	48 09 05	80 02 44	Wright-Hargreaves, Wasamac No. 1, Ontario	9,734,402	4,817,680	471,000	77,715	Prod. thru 1965.
61B	42A/01	48 09 00	80 01 01	Sylvanite, Ontario	5,049,536	1,667,520			Prod. thru 1961.
61C	42A/01	48 09 36	80 00 00	Toburn, Ontario	1,186,316	570,659			Prod. thru 1953.
62	32D/04	48 08 17	79 34 48	Kerr Addison, Ontario	35,726,118	9,889,679	444,926	111,232	Prod. thru 1979, Res. end 1997.

TABLE IV

PRODUCTION OF THE PRINCIPAL GOLD CAMPS IN SUPERIOR PROVINCE TO THE END OF 1979*

Camp	Tons	(Tonnes)	Ounces	(Kilograms)
Bissett, Manitoba	5,420,600	(4,917,500)	1,535,134	(47,748)
Red Lake, Ontario	33,033,791	(29,967,760)	12,229,727	(380,387)
Pickle Crow, Ontario	4,991,614	(4,528,320)	2,109,983	(65,628)
Geraldton-Beardmore, Ontario	23,192,645	(21,040,020)	4,709,785	(146,491)
Timmins, Ontario	182,938,299'	(165,958,890)	56,375,356	(1,753,470)
Matachewan, Ontario	10,143,950	(9,202,440)	1,031,293	(32,077)
Kirkland Lake-Larder Lake, Ontario	98,285,488	(89,163,120)	35,108,723	(1,092,003)
Rouyn-Noranda, Quebec	27,551,000	(24,993,850)	3,657,700	(113,765)
Val d'Or-Malartic, Quebec	121,063,184	(109,826,700)	20,335,129	(632,493)
Totals	506,620,571	(459,598,600)	137,092,830	(4,264,062)

*Deposits are excluded for which gold was not the principal product.

Syngenetic Deposits. These deposits are syngenetic precipitates, mainly chemical sedimentary rocks including various facies of iron-formation. Gold deposits are associated with carbonate-oxide iron-formation in the Geraldton and Pickle Crow districts, Ontario. The Amoco deposit, Detour Lake, Ontario, is an exhalative chert-sulphide variety. However, none of these are as convincingly of exhalative origin as the Lupin deposit at Contwoyto Lake in the Slave Province, which is an arsenide-sulphide-silicate iron-formation. In many cases (e.g. Geraldton) gold hosted by iron-formation is associated with remobilized or introduced arsenopyrite and pyrite that may have replaced oxide or carbonate-oxide iron-formation as described for a thin iron-formation in the Camflo mine, Quebec. This replacement may have been chemically induced (Latulippe, 1980), the gold deposited in association with iron oxides because oxidation of the transporting fluids led to instability of soluble gold-thiosulphate complexes (J.W. Lydon, pers. commun., 1980).

The Agnico-Eagle deposit, Quebec, however (Barnett, 1979; Barnett *et al.*, this volume), is perhaps the clearest example in the Superior Province of a syngenetic, iron-formation-hosted gold deposit. Its geological, mineralogical and geochemical characteristics provide convincing evidence of its exhalative, synsedimentary origin.

Massive to disseminated stratiform pyritic gold deposits are also probably of syngenetic exhalative origin. Massive deposits, like Dumagami, Quebec, resemble volcanic-hosted massive sulphide deposits; aside from their gold content they are essentially barren examples of the latter. Disseminated deposits, like Doyon, Quebec, in moderately coarse pyroclastic volcanic rocks may be synvolcanic and possibly also exhalative. Siliceous "Flow Ore" at Kerr Addison mine occurs in bedded cherty strata with graphite and pyrite nodules.

Pluton-Hosted Gold Deposits. Host rocks for these deposits range from gabbro, as at the San Antonio mine, Manitoba to syenite in the Kirkland Lake district, Ontario. The deposits are different from those termed "porphyry gold" because mineralization is controlled by prominent veins or shear zones in the former but is disseminated and/or controlled by minor fractures in the latter.

In the San Antonio mine, 132 quartz veins, essentially confined to a gabbro or disbase sill, were mined. This sill, 75 to 150 m thick, intruded a sequence of volcanic and sedimentary rocks.

Gold deposits in syenite in the Kirkland Lake district are in very steep major faults and offshoots that branch from these at acute angles. Many of the faults, despite significant displacement, and the ore shoots they contain, are very inconspicuous. Other veins, however, contain quartz and in some cases visible gold and associated tellurides.

Many deposits occur in small felsic plutons, like the pipe-like Sunbeam-Kirkland deposit, Manitoba, in the Falcon Lake quartz monzonite stock, a body about 3.6 by 2.4 km (Stewart, 1980). Near Val d'Or, Quebec the Belmoral deposit is in the central part of the Bourlamaque batholith, whereas the Sullivan Consolidated, Perron, Courvan, New Pascalis and New Formaque (Bras d'Or) deposits are within its margins, and the Siscoe and Bevcon deposits in satellitic plutons (Latulippe, 1980). Ore occurs in 5 chimney-like diorite-granodiorite intrusions at the Lamaque mine. More than 16.3 million tonnes have been produced from the main plug, and more than 5.4 million tonnes from three others nearby (Latulippe, 1980). The main Camflo orebody lies within a chimney-like 90 by 180 m monzonite porphyry and is probably partly of "porphyry gold" type.

Chemical and isotopic studies (Fyfe and Kerrich, 1979) revealed that gold-bearing felsic bodies exhibit anomalous enrichment of Na_2O, total volatiles, and ^{18}O, even where remote from Au-bearing veins, compared to "normal" values for these components in unmineralized felsic intrusions. These anomalous characteristics reflect pervasive albitization and the presence of oxide, carbonate and in some cases sulphate minerals. They were attributed to reaction of the felsic bodies with heated seawater in ore-forming geothermal systems, during thermally driven convective cooling. Unmineralized stocks were considered not to have interacted strongly with seawater.

Volcanic-Hosted Vein and Shear Zone Gold Deposits. These deposits are highly variable in geological characteristics. Some general, but not universal characteristics are: 1) location within Archean greenstone belts, 2) intense wallrock carbonatization and/or associated carbonate-bearing veins, 3) association with major crustal faults, 4) presence of associated felsic or alkalic, possibly sub-volcanic, intrusions, and 5) association with ultramafic rocks.

With the single, notable exception of the Renabie mine, Ontario, all these deposits occur in greenstone belts of the Superior Province but are essentially absent in surrounding gneissic-granitic terranes. Within the greenstone belts, most gold deposits are associated with basaltic volcanic rocks, both flows and pyroclastics, although association with a wide variety of other lithologies, including ultramafic to felsic igneous rocks, is not unusual.

Intense carbonatization of country rocks, as in parts of the Timmins, Larder Lake, Malartic, Cadillac and Val d'Or districts, complicates geological interpretation because it obscures the primary characteristics of the associated rocks. Some intensely carbonatized zones are stratigraphically controlled, whereas others are structurally controlled and discordant (Karvinen, 1978; Fig. 6 in Whitehead *et al.*, 1979). In the Timmins area both have been described by Fyon and Karvinen (1978).

Some gold-bearing carbonate zones are of syngenetic chemical sedimentary origin. However, relict spinifex texture in green and buff carbonate rock in the Timmins, Matachewan and Larder Lake areas, together with results of recent work at Red Lake (MacGeehan and Hodgson, 1980b; Rigg and Helmstaedt, 1980) and at Larder Lake (Tihor and Crocket, 1977) suggest that many gold-associated carbonate zones are due to carbonatization and are not carbonate exhalites.

Carbonatized rocks in the Timmins area exhibit increased CO_2 and As contents, and high CO_2/CaO molar ratios (Davies *et al.*, 1980). The most favorable areas are marked by CO_2/CaO ratios greater than 1.5 and As values greater than 10 ppm. Some favorable localities also have >0.75% K_2O and >20 ppm Rb. Weak carbonatization results in partial conversion of CaO in silicate minerals to calcite, more intense carbonatization converts CaO and also silicate MgO and FeO, at least in part, to Ca-Mg-Fe carbonate minerals. Carbonatization is also generally accompanied by gold enrichment above a background of about 2 ppb, depletion of Na_2O, reduction of Fe in the rock, and, at least near gold-bearing veins, a change in $\delta^{18}O$ from 8 per mil to 11 per mil (Kerrich *et al.*, 1979a). Y, Zr, Ti and Cr remained essentially immobile during carbonate alteration (Davies *et al.*, 1979; Davies and Whitehead, 1980).

Interbedded basaltic and carbonate-rich metavolcanic rocks in the Pamour mine have been converted by intense recrystallization and metasomatism into locally fuchsitic carbonate rock, talc-carbonate rock and altered basalt (Jak *et al.*, 1978). The first two have been identified (James and Whitehead, 1978) as peridotitic komatiite by their chemical compositions, relict primary textures and comparison with less altered stratigraphic equivalents elsewhere in the region. During carbonatization of these rocks SiO_2, MgO, K_2O and Na_2O were depleted, whereas CaO was added.

Many important deposits near Timmins are closely associated with the major Porcupine-Destor fault, many in the Larder Lake area with the Larder Lake break, and many in the Malartic and Cadillac districts with the Cadillac fault. The precise role of these faults is not known but they and subsidiary structures are the loci of intense carbonatization and gold deposition; they thus appear to have served as major channelways for egress of CO_2-rich mineralizing fluids.

In addition to the deposits in alkalic and felsic plutons, other gold deposits are near such bodies. At Timmins major gold deposits lie near the Pearl Lake porphyry and other small felsic bodies (Karvinen, 1976, 1980) and surround the McIntyre porphyry copper orebody (Kirkham and Thorpe, 1973). At Red Lake, seven past producers, the Madsen, Starratt-Olsen, Gold Eagle, Hasaga, McKenzie, Red Lake Gold Shore and Howey, were located around the Dome, McKenzie and Faulkenham Lake stocks. Deposits at Malartic, Quebec, are predominantly gold-bearing quartz veins associated with sill-like and irregular porphyry intrusions.

The association of ultramafic rocks with gold deposits, particularly in the Timmins (Pyke, 1976), Matachewan and Larder Lake districts, is striking. Spinifex-textured ultramafic rocks also occur at the Barnat Mine, Malartic area, Quebec (Issigonis, 1980). At (a) the Beaumont occurrence, 9 km northeast of Timmins, (b) a prospect in Yarrow Township, west of Matachewan, (c) a locality near the former Omega gold mine, and (d) the Kerr Addison mine near Larder Lake (Tihor and Crocket, 1977), spinifex texture is preserved in completely carbonatized ultramafic rocks. This, together with their petrochemistry indicates that these rocks were originally peridotitic komatiites or similar komatiitic lavas. Near most major deposits spinifex texture has been completely destroyed, perhaps by structural deformation. Carbonatization formed both "buff and green carbonate" rocks; the main difference is the presence of fuchsite in the latter, perhaps reflecting a lack of K_2O in the fluids that formed the former. The recognition of green carbonate rocks, like those at Kerr Addison, as carbonatized ultramafic rocks has been gradual. Prime evidence (Tihor and Crocket, 1977) is provided by their contents of Cr, Ni and major elements, which are appropriate for an ultramafic parent, as well as spinifex texture in talc-chlorite and fuchsitic quartz-carbonate rock. Tihor (1978) found that magnesite is the dominant carbonate in many carbonate rocks of the Kirkland Lake-Larder Lake district. Kerr Addison green carbonate ore is anomalous in W, As, B, F, and Hg (Downes, 1980), suggesting that these elements and gold were introduced by a volatile-rich fluid, possibly related to felsic magmatism. Presumably K_2O was also introduced, as reflected by the presence of fuchsite.

Fuchsite was commonly formed during the carbonatization of ultramafic rocks, especially near significant gold deposits. Although there are exceptions, the fuchsite is considered indicative that the unaltered rock was ultramafic. For example, although supporting chemical data are unavailable, fuchsite-rich "rods" in intensely deformed fragmental rock in the Matachewan Consolidated deposit probably represent ultramafic (volcanic?) fragments. Many fuchsite-bearing rocks associated with gold deposits contain gersdorffite (NiAsS) and/or other Ni-S minerals and nearly as many contain chromite grains. One specimen, a cobble from conglomerate in the Pamour Mine, Timmins, is exceptional and contains pyrrhotite with enclosed pentlandite flames, as well as native gold and tellurides. The timing of carbonate-fuchsite alteration and mineralization is not known but it seems likely that both occurred prior to deposition of the Temiskaming conglomerate.

Nickel (-Copper) Deposits

Most Archean nickel deposits both in Canada and Australia are magmatic, formed by segregation of an immiscible sulphide melt from a host mafic or ultramafic magma (Groves *et al.*, 1979), although recently (Lusk, 1976; Gale *et al.*, 1980; Robinson and Hutchinson, this volume) an "exhalative", presumably hydrothermal origin has been proposed for some. The deposits may be divided into two groups (Table V), one generally associated with tholeiitic, and the other with komatiitic rocks. These closely correspond to groups III A and III B of Naldrett and MacDonald's (1980) fourfold classification of the petro-tectonic settings of nickel deposits. The principal Canadian districts and deposits that have produced nickel, or contain significant reserves, are shown in Figure 5.

The tholeiitic host rocks for group III A deposits are extremely variable in composition, possibly due to variable degrees of differentiation. The Montcalm and Lorraine deposits are associated with gabbro, whereas the remainder are in or near ultramafic rocks. These ultramafic rocks have higher Fe/Fe + Mg ratios than those of the komatiitic suite, and may represent cumulate differentiates from tholeiitic parent melts. Overall, the compositions of the host-rocks to "tholeiitic" nickel deposits exhibit a broad range, possibly resulting from widely variable amounts of differentiation. Thus, if sufficient chemical data to determine the petrogenetic affinity of

TABLE V

NICKEL PRODUCERS AND SELECTED PROSPECTS OF THE CANADIAN SHIELD OTHER THAN THOSE AT SUDBURY, ONTARIO AND THE INTERNATIONAL NICKEL CO. MINES OF THE THOMPSON BELT, MANITOBA

	Structural Province	Production Period From	To	Production (ore tonnes)	Grade Ni(%)	Grade Cu(%)	Reserves (ore tonnes)	Grade Ni(%)	Grade Cu(%)
Type III A: hosted by tholeiitic, gabbroic (G) to ultramafic (U), intrusions									
Rankin, N.W.T. (U)	Churchill	1957	1962	368,093	3.1	0.9			
Lynn Lake (A,EL), Man. (G)	Churchill	1954	1976	19,261,675	1.2	0.5			
Dumbarton, Man. (G)	Superior	1969	1974	1,438,500	0.62	0.33			
Gordon Lake, Ont. (U)	Superior	1962	1972	1,438,827	0.85	0.60	1,541,900	0.50	minor
Thierry, Ont. (U)	Superior	1976	—	2,803,892 (thru 1979)	0.10	1.3			
Shebandowan, Ont. (U)	Superior	1972	—	est. 6,804,000 (thru 1980)	1.5(?)	0.8(?)			
Kanichee (Ajax), Ont. (G)	Superior	1974	1976	255,445	0.5	0.72			
Lorraine, Que. (G)	Superior	1965	1968	594,318	0.57	1.32			
Montcalm, Ont. (G)	Superior						4,081,500	1.41	0.66
Maskwa West, Man. (U)	Superior						816,300	1.34	0.23
Renzy, Que. (G)	Grenville	1969	1972	716,015	0.69	0.72	1,207,950	0.70	0.70
Somex (Lac Edouard), Que. (G)	Grenville	1973	1974	est. 30,800	1.5	0.5 Res.	10,880	1.5	0.5
Type III B: hosted by komatiitic lavas and related intrusions									
Manibridge, Man.	Churchill	1971	1977	est. 1,225,000	2.0	0.15			
Texmont, Ont.	Superior	1971	1972	178,550	0.96		2,857,050	0.93	
Langmuir, Ont.	Superior	1973	1978	1,010,636	1.43		489,780	1.3	
McWatters, Ont.	Superior						583,700	1.04	
Redstone R. (Draper Prop.), Ont.	Superior								
Hart, Ont.	Superior								
Alexo, Ont.	Superior	1912 1943	1919 1944	51,510	4.5	0.65			
Marbridge, Que.	Superior	1962	1973	702,363	2.82	0.1			

ultramafic-hosted deposits is lacking, assignment of them to this group may be hazardous. However, the 'tholeiite-associated' group are not in, or spatially associated with spinifex-bearing rocks. Their sulphide zones usually have Ni/(Ni + Cu) ratios <0.9.

Examples include the Gordon Lake and Kanichee deposits, which contain zones of disseminated, immiscible-sulphide blebs, as well as more massive sulphide zones, in ultramafic rocks. The Maskwa West deposit consists of disseminated sulphides in metaperidotite at the base of the differentiated ultramafic to gabbroic Bird River sill (Coats *et al.*, 1979). However, the eastern zone at the Dumbarton mine, 1.2 km southwest, consists of disseminated sulphide in andesite at or near the contact between andesite and granite (Karup-Møller and Brummer, 1971). The Lorraine deposit apparently occurs within mafic volcanic rocks adjacent to a gabbroic sill; a simple magmatic origin for the latter two deposits is not as easily demonstrable as for the others.

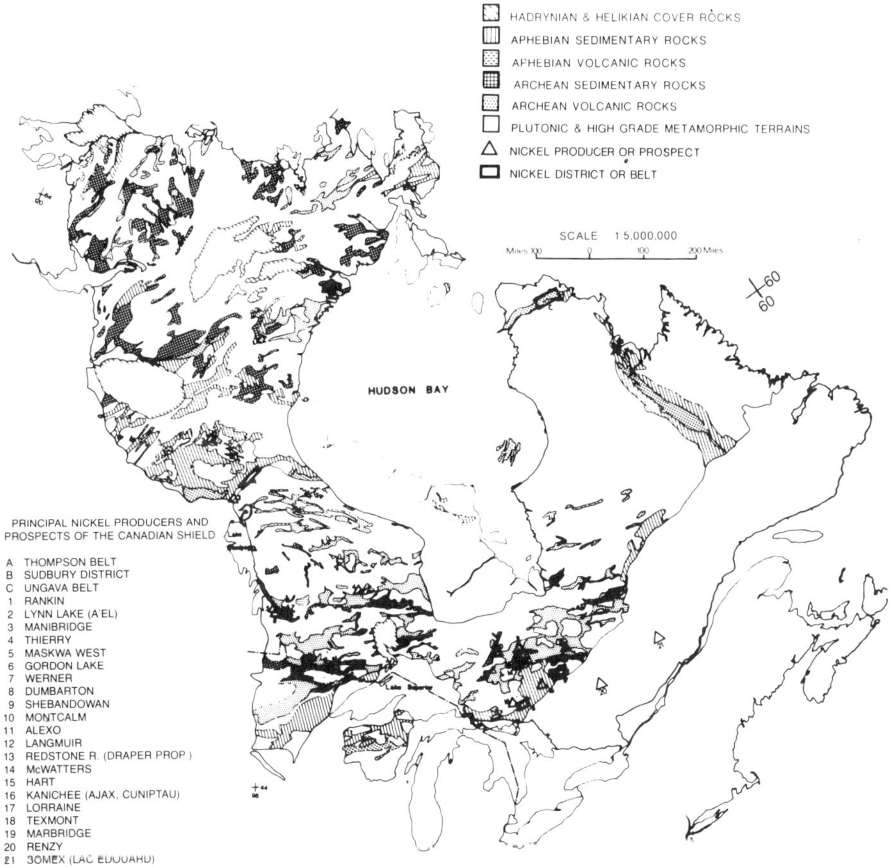

Figure 5. Map showing location of principal nickel producers and prospects of the Canadian Shield.

The Shebandowan deposit is a thin continuous zone of semi-massive sulphides containing breccia fragments of both the host ultramafic intrusion and adjacent mafic volcanic and granitic rocks. The sulphides occur at the top of the intrusion (Morton, 1979) and were apparently emplaced later than it. The ore composition is compatible with an immiscible sulphide liquid derived from a gabbroic melt. The Thierry deposit (Patterson, 1980) is a semi-continuous zone of sulphide blebs, streaks and veinlets in a mylonitic shear zone in intensely deformed ultramafic rock. During the intense shearing, sulphides underwent major redistribution, with addition of potash to the wallrocks resulting in formation of a chlorite-biotite-amphibole host for part of the ore. The remainder of the ore is irregularly distributed in brecciated ultramafic rock.

Deposits of the komatiite-associated group (III B) generally resemble those of the important Kambalda district in Western Australia (Gresham and Loftus-Hills, 1980). Nickel is much more abundant than copper (Ni/(Ni + Cu) >0.9), and the sulphides generally occur in massive to semi-massive, net-textured zones near the base of ultramafic sills and flows (Naldrett, 1973). As with the tholeiite-associated deposits, the sulphides apparently formed by segregation of an immiscible sulphide liquid. In these deposits however, segregation is typically more complete and the ratio of disseminated to massive sulphide is somewhat less than in group III A deposits. Spinifex-textured ultramafic rocks commonly either host the sulphides or occur nearby.

The deposits of the Timmins area (Eckstrand, 1973) typify Group III B deposits in Superior Province. For example, the Dundonald deposit (Muir and Comba, 1979) has net-textured and disseminated sulphides in an olivine cumulate zone in peridotitic flows, and semi-massive sulphides in basal chill zones and interflow sedimentary horizons. South of Timmins, Ontario, the Texmont, Langmuir, Hart and McWatters (Coad, 1977), and Redstone (Robinson and Hutchinson, this volume) deposits are also typical in terms of both host rocks and stratigraphic position, of this type. Rocks of the Shaw Dome south of Timmins form two major volcanic cycles, the lowermost Deloro Group, overlain by the Tisdale Group (Pyke, 1975). The former contains minor ultramafic and mafic volcanic rocks at its base, a central mafic volcanic unit, and an upper unit of intermediate and basaltic volcanic rocks, including several beds of sulphide - and oxide-facies iron-formation. The Tisdale Group has a major unit of ultramafic and komatiitic basalts at its base, with accompanying synvolcanic sills and minor iron-formation. Above this unit is the central formation of high-iron basalt, overlain by felsic volcanic rocks. Virtually all nickel deposits are in the lower part of the Tisdale group, whereas the gold deposits of the Timmins district are in the upper units of this group.

The Texmont deposit is in a spinifex-textured ultramafic flow at the base of the Tisdale Group, underlain by sulphide-bearing, felsic volcaniclastic rocks. The Langmuir deposit is similarly situated in ultramafic flows and sills above a thick sequence of volcanic-derived sedimentary rocks and abundant cherty, sulphide-facies iron-formation. A magmatic origin has been proposed for these deposits (Muir, 1979; Green, 1980; Eckstrand and Miller, 1973; Pyke 1975). Considering ore composition, sulphur isotopes and S-Se ratio, Green (1980) suggested that in the Langmuir deposit sulphur was assimilated from underlying iron-formations into the ultramafic melt generating a nickel-bearing immiscible sulphide liquid, which then rapidly separated from the ultramafic melt to form the ore. Pre-eruption assimilation of sulphur from subjacent sulphide-rich strata may be an important mechanism of formation for nickel deposits of this type (Groves *et al.*, 1979). Recently, Lusk (1976), Gale *et al.* (1980) and Robinson and Hutchinson (this volume) suggested that ultramafic-associated nickel ores are in part products of hydrothermal activity. Nickel-rich sedimentary rocks, with interbanded chert and nickeliferous sulphide are present at the Jan (Hutchinson, 1981c) and Juan deposits, Kambalda district in Australia, and in Canada (Robinson and Hutchinson, this volume), but their origin remains obscure. Elucidation of nickel solubility in hydrothermal fluids is required for clear understanding of the relative roles of hydrothermal versus magmatic processes in genesis of these deposits.

Figure 6. Map showing distribution of major iron formations of the Canadian Shield. Taken largely from Gross (1965, 1968 and 1970).

Iron Deposits

Iron-formations in Superior Province are of Algoma-type (Gross, 1965). Although iron-formation is present in or near all greenstone belts, only a few are of sufficient size or appropriate metallurgical characteristics to be mined or potential producers. In general, these are much smaller than the Superior-type iron-formations (Gross, 1965), examples of which occur in Churchill Province (Fig. 6).

Algoma-type deposits occur in a variety of geological settings and have variable internal characteristics. The facies classification of James (1954) is generally applicable, but some grouping, based primarily on setting, is possible.

1) *Turbidite-hosted deposits*: The Griffiths Mine, near Red Lake, Ontario, and the Savant Lake and Lake St. Joseph occurrences typify this setting. These deposits are in deep water, distal turbidites (Shegelski, 1978) which in most cases were derived from laterally equivalent volcanic terrane. Mafic flows are a very minor component of the turbidite sequence. Magnetite, the dominant ore mineral, forms massive to semi-massive beds; quartz is the dominant accessory. Cherty, banded iron-formation is relatively minor, as are sulphide and silicate facies. Accessory base metals are generally absent (Gross and McLeod, 1980). The turbidite basins may lie entirely within a volcanic domain as at Savant Lake, or may be marginal to major volcanic belts, as at Lake St. Joseph and the Griffiths mine.

2) *Volcanic-hosted deposits*: Characteristics of these deposits vary widely and there are two subtypes, carbonate-sulphide and banded cherty oxide. These are spatially associated in the Michipicoten district:

a) *Carbonate-sulphide subtype*: The principal deposits of this type are in the Michipicoten district (Goodwin, 1962) and the Atikokan area (Shklanka, 1972; Jolliffe, 1966). The Helen mine at Michipicoten contains moderately to poorly bedded siderite-pyrite, massive pyrite, and bedded ferruginous chert which overlie a felsic pyroclastic dome and are capped by pillowed mafic volcanic rocks (Goodwin, 1964). Jaspilitic oxide iron-formation is also present, possibly stratigraphically equivalent to the carbonate-sulphide facies, but is in sedimentary strata (Goodwin, 1962), mainly greywacke and conglomerate. The carbonate-sulphide zones are possibly a unique example of iron-formation underlain by a distinctive cross-cutting alteration pipe. The latter is composed dominantly of siderite with accessory chloritoid.

The stratigraphic sequence of the Steeprock deposits at Atikokan has a thin irregular conglomerate at its base, overlain by 120 m of locally stromatolitic limestone and 20 to 30 m of highly altered, sulphide-facies iron-formation. The latter is now primarily goethite and hematite, probably secondarily enriched from pyrite-siderite protore. Distinctive picritic ash-flow tuff caps the sulphide iron-formation; both mafic and felsic volcanic rocks overlie the picritic tuff. Sulphide-facies iron-formation is also common 10 km farther north at Finlayson Lake (Fenwick, 1976). Here, pyritic zones form discontinuous strata up to 2.5 km in length, contain minor chalcopyrite (<1%) and occur in a dominantly mafic volcanic terrane. Mixed oxide and sulphide iron-formations occur 18 to 20 km east of Atikokan (Grabowski, 1975). These are intimately associated with mafic intrusions which have assimilated some sulphur to form nickeliferous, net-textured sulphides.

In most greenstone belts, graphitic shale is commonly intercalated with volcanic rocks, generally comprising less than one per cent of the succession. It commonly contains up to 10% pyrite, locally massive, and magnetite. Pyrite is usually framboidal and associated with accessory chalcopyrite and hematite (Neilsen, 1974). Magnetite usually occurs as disseminated grains. These graphitic shale horizons are usually only a few tens of metres thick and contain small carbonate-rich zones.

Sulphide-facies iron-formations may contain sufficient accessory chalcopyrite to be of economic interest (Kirkham, 1979). Also, their cobalt content may be of interest; for example, sulphide-facies iron-formation east of Atikokan contains 0.1% Co and 0.5% Cu. Cobalt usually occurs in pyrite, which contains up to 0.2% Co (Grabowski, 1975).

b) *Cherty oxide sub-type*: These are of principal economic interest due to the relative ease in concentrating and pelletizing magnetite. The ore is typically thinly banded, cherty iron-formation and ranges up to a few hundred m thick and several km long. These iron-formations contain fine-grained magnetite comprising 20 to 40% of individual beds (Gross, 1970) with sulphide-, carbonate-, and silicate-facies as less dominant members. Host rocks are usually pillowed mafic volcanic rocks, as at Red Lake, but many of the larger deposits are in local accumulations of pelitic to semi-pelitic sedimentary rocks, as at the Sherman mine, Timagami, Ontario, or in felsic volcanic rocks as at the Adams mine near Kirkland Lake, and in the Kukatush area 70 km southwest of Timmins, Ontario (Milne, 1972).

Although bedded magnetite-chert predominates, mixed facies zones are common. For example, Milne (1972) described an upward change in the Radio Hill iron-formation near Kukatush from graphitic shale with disseminated sulphide (0 to 45 m) through 30 to 90 m of oxide-facies iron-formation, 10 to 75 m of minnesotaite-carbonate rock, to 0 to 25 m of sulphidic black shale. Association with sulphidic graphitic shale is common. Sulphur isotope data (Goodwin *et al.*, 1976; Shegelski, 1978) indicate that much of the sulphur may be biogenic, and much sulphide is diagenetic and commonly framboidal. Reduced sulphur may have been supplied during diagenesis by the breakdown or 'maturation' of organic compounds.

Although oxide-facies iron-formations occur at all stratigraphic levels within volcanic successions, they are commonest near the tops of major volcanic cycles. For example, at Sturgeon Lake (Franklin *et al.*, 1977) magnetite-bearing iron-formation occurs in a thin sedimentary unit which caps the volcanic pile. In the Timmins area, the large iron-formations occur at the top of the lower volcanic cycle, the Deloro Group (Pyke, 1975).

Deposits with Porphyry Characteristics

Numerous occurrences of copper and molybdenum in the Canadian Shield share some characteristics of Phanerozoic porphyry deposits (Riley *et al.*, 1971; Kirkham, 1972), although not all are in the Superior Province. Some contain economically significant gold, and some gold deposits of possible "porphyry gold type" are also known. Many abstracts and short notes, but very few comprehensive reports, have

TABLE VI

ARCHEAN AND PROTEROZOIC Cu AND Mo DEPOSITS AND OCCURRENCES WITH SOME PORPHYRY COPPER CHARACTERISTICS

	Deposit or occurrence	Host or associated body (bodies) and its character	Grade of mineralization	Size, tonnage of reserves	Remarks	References
Archean Deposits						
Cu(Mo,Ag)	Tuk, Contwoyto Lake area, N.W.T.	Foliated porphyritic quartz monzonite pluton.			Cu-Mo mineralization, with pyrite, as fine disseminations, less commonly in quartz veins. Along SW margin of the pluton near its contact with gneissic rocks, probably originally felsic to intermediate tuffs.	Bain, 1980
Cu(Mo)	Lynx Bay area, Oxford L., Man.	Trondhjemite stocks intruded into massive tonalites.	Assays generally less than 0.25% Cu.	One zone exposed in a 60m by 150m area. Second zone at least 182m by 61m.	Two small subeconomic elliptical bodies. Propylitic alteration, surrounded by zone of saussuritization of feldspar. Mineralization in closely spaced fractures.	Haskins and Stephenson, 1974
Cu	Red Cross L., Man.	Granodiorite batholith and numerous adjacent quartz-feldspar porphyry dikes and sills.	Max. 0.21% Cu across 3m.	Erratic low grade mineralization over a considerable area.	Dike-rich border zone to batholith 800m wide. Pyrite and chalcopyrite in small fractures and as disseminations in porphyry dikes and the nose of a granodiorite pluton.	Elbers, 1976

Mo	Setting Net L., Ont.	Oval, epizonal, porphyritic granodiorite-quartz monzonite stock (area 5.7 km²).	0.06% MoS_2 in east part of zone.	Mo mineralization in area 2500m by 450m.	Chloritization of biotite, albitization, sericitization. Mineralization in zone of greatest alteration. Greenschist facies regional metamorphism. Deposit tilted on side. Mo in quartz veins.	Ayres and Averill, 1974; Ayres and Findlay, 1976; Wolfe, 1976
Cu(Mo, Au,Ag)	Lang L., Ont.	Small granodiorite stocks, 10 phases of porphyritic dikes and sills.	0.25 to 0.35% Cu at west end of zone.	Zone 700m long and up to 300m wide. Widespread low-grade mineralization.	Mineralization in part in flows, agglomerate and volcanic conglomerate. Potassic alteration. Argillic? alteration. Upper greenschist facies regional metamorphism. Mineralization disseminated and as fracture fillings.	Findlay, 1975; Findlay and Ayres, 1977
Mo	Pickle L. area, Ont.	Bamaji Complex: trondjhemitic, porphyroblastic part of a batholith.			Mineralization perhaps genetically related to a second batholith.	Colvine and Sutherland, 1979
Mo	Pidgeon, Sioux Lookout area, Ont.	Lateral L. stock; granodiorite.	0.08% Mo	14.3 million tonnes.	Mineralization in discrete quartz veins and related to pegmatites and aplites (i.e. it is intrusion-related but not porphyry style).	Kustra, 1979; Colvine and Sutherland, 1979; Northern Miner, March 6, 1980
Cu,Mo,Au	High Lake, Ewart Twp., Ont.	High Lake pluton: porphyritic granodiorite to trondhjemite.	Core sections assayed 0.10-1.35% Cu, 0.34-1.7 g/tonne Au.		Cu,Mo controlled by shear fractures, tension fractures, and quartz veins. Not considered of porphyry type by Pedora. Associated magnetite.	Pedora, 1976; Davies, 1965

	Deposit or occurrence	Host or associated body (bodies) and its character	Grade of mineralization	Size, tonnage of reserves	Remarks	References
Cu(Au)	Beidelman Bay, Ont.	Beidelman Bay sill: trondhjemite, quartz-diorite border phase.	Average grades low but not defined. Max. values obtained 0.2% Cu, 0.03% MoS_2 and 0.4 g/tonne Au.	Two zones, each 3000 square metres area	Broad zone of diss. Cu mineralization in western part of the intrusion. Sericitization, silicification, biotitization, tourmalinization. Three subeconomic zones associated with areas with metavolcanic xenoliths. Also minor mineralized breccia zones.	Trowell, 1974; Friske *et al.*, 1979 Poulsen and Franklin (1981)
Cu(Au,Mo)	McIntyre (Schumacher) mine, Ont.	Pearl L. porphyry: highly altered roughly lenticular quartz porphyry body.	Prod. approx. 9,072,000 tonnes to end 1978 at 0.64% Cu. Approx. 0.05% Mo, 1.0 g/tonne Au.	Reserves end 1978, 5,584,455 tonnes at 0.62% Cu. Greater than 73 million tonnes of subeconomic Cu-Au-Mo mineralization. Zone 120m wide has been traced vertically for at least 1525m.	Albitization, sericitization, carbonatization. Anhydrite-bearing zone surrounds Cu orebodies. Ore associated with high K/Rb ratio. Porphyritic host rock. Attitude of higher grade shoots may be controlled by later folding and deformation. Some mineralization is in the volcanic country rock adjacent to the Pearl Lake porphyry. Chalcopyrite and pyrite disseminated and in veinlets.	Pyke and Middleton, 1970; Griffis, 1962; Kirkham and Thorpe, 1973; Davies and Luhta, 1978; Davies, 1980
Cu-Mo	Ryan L.; Stancop; Matachewan, Ont.	Small irregular porphyritic syenite plugs.	Ryan L. estimated 0.4% Cu, 0.09% MoS_2.	On Ryan L. property estimated 3.2 million tonnes.	Associated potassic alteration, silicification, chloritization. Mineralization multistage in quartz vein stockworks; also dissemi-	Sinclair, 1979

Mo,Cu, Au(Ag)	Renault, Dasserat L. area, Que.	Elongate syenite body. In silicified porphyritic syenite and alkaline volcanic rocks.		Not defined.	Mo and Cu in quartz veinlets and as disseminations. Intense, pyritic alteration zone. Silicification, sericitization, chloritization. A little magnetite and fluorite present.	Kirkham, pers. commun., 1980
Cu,Mo	Don Rouyn, mine, Que.	Powell pluton: trondhjemite.	>0.25% Cu in small central zones.	1,450,000 tonnes mined/year.	Cu and Mo (latter minor to trace) decrease outward from a pyrite-bornite-chalcopyrite zone. Mined as siliceous flux. Elliptical pipe-like body. Sulphides disseminated, in fractures and in quartz-carbonate veinlets.	Goldie *et al.*, 1979; Goldie and Kotila, 1979
Cu	Talbot, Queylus Twp., Chibougamau, Que.	Chibougamau pluton: tonalite to diorite; quartz latite and quartz-feldspar porphyry dike systems.			Potassic alteration, tourmalinization. Stockwork-type quartz-carbonate veinlets. Associated magnetite. Narrow diatreme (?) breccia zones. Apparently similar occurrences on Devlin Peninsula and in Scott Twp.	Kirkham, 1972; Cimon, 1973; Cimon, 1979; Bureau *et al.*, 1979
Cu,Mo	Clark Lake (Garth Lake), Chibougamau Que.	Small felsic porphyry dikes and breccias.			1-10% pyrite, both disseminated and in veins, and associated quartz and magnetite veinlets over an area of about one square mile.	Kirkham, 1972
Proterozoic Deposits						
Cu-Mo	Whitefish L., Man.	Whitefish Lake porphyry: quartz monzodiorite body approx. 2900m by 300 to 600m.	Max. about 0.08% Cu.		Potassic alteration, sericitization. Mineralization in most altered part of body. Chalcopyrite, molybdenite and pyrite in fractures.	Baldwin, 1980a

	Deposit or occurrence	Host or associated body (bodies) and its character	Grade of mineralization	Size, tonnage of reserves	Remarks	References
Cu,Mo,Au	Missi Island, Amisk L., Sask.	Highly altered fine-grained porphyritic gabbro to quartz monzonite dikes, stocks of altered trondhjemite.	Only subeconomic grades known.		Mineralization and intensity of alteration of metavolcanic rocks related to abundance of porphyritic dikes. Extensive mineralized pyritic alteration zones. Greenschist facies metamorphism. Cu and sporadic Mo as disseminations and in pyrite veinlets in some porphyry dikes and country rocks.	Kirkham, 1974b; Ayres and Findlay, 1976; Chite and Ayres, 1977
Cu	Dion L. area, Man.	Boot Lake Pluton: polydiapiric dioritic to granodioritic pluton.		Very small occurrence.	Pluton 1830 Ma. Fracture-controlled alteration and mineralization.	Wallster, 1979
Cu(Mo,Ag)	Tribag (East Breccia), Ont.	Oval breccia body 700m by 300m in Archean greenstone.	Subeconomic, Est. 0.13% Cu, 0.03 to 0.05% MoS_2, 1.35 g/tonne Ag.	Estimated 113 million tonnes.	Tightly packed breccia of mafic volcanic and diabase fragments, lesser felsitic fragments. Granite batholith 550m to the north. Uniformly mineralized. Chloritization, sericitization, argillic alteration. Mo in late fractures; other sulphides disseminated in rock fragments.	Blecha, 1974
Cu(Mo)	Jogran, Ont.	Quartz-feld. porphyry body 120m by 200m (quartz monzonite to granodiorite)	0.19% Cu, 0.05% Mo.		Potassic and propylitic to quartz-sericite alteration. Mineralization is disseminated in quartz and quartz-carbonate veinlets	Blecha, 1974; Armbrust, 1979a, 1979b, 1979c

appeared on these porphyry deposits. Both Archean and Proterozoic examples are listed in Table VI. Among other prospects are the McCombe, Minnitaki Lake, Ontario; Hopkins, east of Lake Nipigon, Ontario; one in the Painkiller Lake area, Ontario; Olier, west of Noranda, Quebec (Kirkham, 1973); Savard, in the Bourlamaque Batholith, Quebec (Kirkham,1972); Mink Lake, Birch Lake area, Ontario; Kewashegamuk Lake, southeast of Dinorwic, Ontario; one in Elmhirst Township, north of Beardmore, Ontario (Riley *et al.*, 1971); and at Gutcher Lake, Wawa area, Ontario (Kustra, 1979; Studemeister and Colvine, 1979). Metal zoning in these deposits consists of a Cu(Mo)-rich core and an outer Au-rich zone (Kirkham and Thorpe, 1973; Pedora, 1976; Davies and Luhta, 1978; Studemeister and Colvine, 1979; Sinclair, 1979). The peripheral gold commonly occurs in individual veins rather than in minor fractures or disseminations.

Are there porphyry-type gold deposits in the Canadian Shield? Considering present gold price and heap leaching technology, certain low-grade gold occurrences may be of this type and merit consideration. Lawrence (1978) described disseminated gold mineralization in the Vunda River area, Viti Levu, Fiji as porphyry type. Although the occurrence may not be characteristic, because no intrusions have been identified in its vicinity, the gold mineralization, which occurs sporadically in shoshonitic volcanic rocks, is associated with a 6 km^2 altered area and is controlled by closely spaced fractures. These are significant characteristics of porphyry-deposits but in the examples here considered, gold commonly accompanies pyrite on fracture surfaces in porphyry bodies that are apparently little altered. Examples are dikes in the Indin Lake area, N.W.T., and a syenite dike on the Knutson Discovery property, Larder Lake area, Ontario (Thomson, 1943).

Ore from the Young-Davidson (Matachewan, Ontario) and Howie-Hasaga (Red Lake, Ontario) properites, and some at the Sigma (Val d'Or, Quebec) and Camflo (Malartic, Quebec) mines has characteristics of porphyry-type deposits. Gold and associated pyrite are disseminated and/or controlled by minor factures in porphyritic igneous bodies. Host rocks are variable; syenite at Young-Davidson, felsic feldspar prophyry dikes at Sigma, and various latites, including mafic porphyritic latite, at Howie-Hasaga. Unlike many porphyry Cu and Mo deposits, extensive hydrothermal alteration is apparently absent in the auriferous examples. This may be largely due to insufficient study; the No. 6 porphyry deposit in the Barnat mine, Malartic, Quebec, is essentially coextensive with a pyritic-potassic alteration zone in a 45 m thick porphyritic quartz-diorite to diorite sill which intrudes ultramafic flows (Issigonis, 1980). The deposit is surrounded by a poorly mineralized propylitic alteration zone. McMillan (1972) described extensive potassic (biotitic) alteration associated with veins and stockwork Cu-Au deposits at Opemiska, Quebec. These deposits are in a layered sill associated with sodic felsic dikes. Molybdenite and scheelite are significant associated minor minerals.

Other gold deposits with porphyry-gold-type characteristics are listed in Table VII. The Wilmar and Goldlund deposits are included although they may not be typical. Gold mineralization at Wilmar occurs in a quartz-pyrite-tourmaline vein system in a small granodiorite body. Mineralization at the Meston Lake property, Quebec, is nearly identical. At Goldlund, gold occurs in a parallel set of quartz ladder (?) veins in a granodiorite dike 10 to 90 m wide. The deposit contains 9 to 13.6 million

TABLE VII

LOW GRADE "PORPHYRY GOLD" DEPOSITS IN THE CANADIAN SHIELD

Property	N.T.S.	Reserves Tonnes	Contained Gold Ounces	Contained Gold Kilograms	Grade oz/ton	Grade gm/tonne
Indin Lake (BOB), N.W.T.	86B/6	12,793,400	1,118,000	34,770	0.086	2.8
Indin Lake (PR), N.W.T.	86B/6	18,143,700	1,060,000	32,970	0.053	1.7
Athona, Sask. (West Zone)	74N/8	2,832,700	206,300	6,417	0.066	2.1
Athona, Sask. (East Zone)	74N/8	72,600	16,300	507	0.204	6.6
Box Mine, Sask.	74N/7	2,335,100	162,065	5,041	0.063	2.0
Box Mine, Sask. (additional low grade ore)	74N/7	11,339,800	437,500	13,600	0.035	1.1
Wilmar, Red Lake, Ontario (porphyry zone)	52N/4	3,628,700	340,000	10,600	0.085	2.7
Goldlund, Dryden District, Ontario	52F/16	1,072,300	104,016	3,235	0.088	2.8
Raven River, Larder Lake area, Ontario (estimated material to 610 m depth)	32D/4	10,000,000	~450,000	~14,000	0.04-0.05	1.4-1.7

tonnes of low grade ore to a depth of 305 m (National Mineral Inventory record). Another occurrence of this type may be the Komis in the Waddy Lake area, Saskatchewan. On the Raven River property, Larder Lake, Ontario gold is essentially confined to two pipe-like altered syenite bodies that have intruded green carbonate, talc-chlorite schist, greenstone and lamprophyre (Thomson, 1943).

SLAVE PROVINCE

General Geology

In the Slave Province some twenty greenstone belts, with northerly to northeasterly trends comprise 6% of its area. These contain Archean volcanic and sedimentary rocks of the Yellowknife Supergroup which are similar to other Archean supracrustal sequences in the Superior Province and elsewhere in the world. These belts, especially in southern Slave Province, flank large sedimentary basins. The volcanic sequences are thin relative to the thick overlying sedimentary successions, and the proportion of volcanic rocks (15 to 20%) to sedimentary rocks (80 to 85%) is markedly lower than in Superior Province greenstone belts (McGlynn and Henderson, 1970). The volcanic sequences are also thin compared to those in major Superior Province belts. The Beaulieu River belt, for example, has up to 4200 m of pillowed basalt (Lambert, 1978a), but in the Yellowknife belt approximately 12 200 m of volcanic strata are exposed (Henderson and Brown, 1966; Baragar, 1966).

Volcanic rocks in the southerly belts are mainly mafic pillow lavas, whereas felsic volcanic rocks are much more abundant in some more northerly and easterly greenstone belts of the Slave Province. For example, in the southerly Cameron River belt 90% of the volcanic succession consists of basaltic pillow lavas and pillow breccias (Lambert, 1977). Lenticular felsic volcanic units mark the top of the basaltic sequence, and are overlain by andesitic pillow lavas and pillow breccias that form the top of the volcanic succession. Ultramafic volcanic rocks appear to be absent in

Slave Province; none of the rare and minor ultramafic bodies have yet been shown to be extrusive. Layered mafic sills are very rare.

The volcanic-sedimentary greenstone segments are flanked and separated by gneissic and granitoid rocks. In part these are believed to form basement to the supracrustal sequences. Specifically, in the Ross Lake (Davidson, 1972a), Benjamin Lake (Heywood and Davidson, 1969), and Point Lake (Stockwell, 1933; Henderson, 1975a) areas, volcanic rocks appear to lie unconformably on older granitoid basement.

In many belts the rocks represent a single cycle of volcanism but rarely, as in the Back River and Tumpline Lake belts (Lambert, 1977) two or more cycles are present. Striking characteristics in at least the southeastern and eastern Slave Province are the local presence of subaerial felsic volcanic rocks as in the Cameron River, Beaulieu River, Back River and Hackett River greenstone belts (Lambert, 1976, 1977, 1978b; Frith and Roscoe, 1980), caldera structures as in the Back River area (Lambert, 1976, 1978b), and associated limestone, and carbonatized breccia and pyroclastic units.

Carbonatized breccia and agglomerate occur in the major caldera structure in the Back River area (Lambert, 1976, 1978b), and are present southwest of the Cleaver Lake deposit in the Hackett River belt. Limestone is associated with volcanic rocks in the Benjamin Lake area (Heywood and Davidson, 1969), the Turnback Lake area (Davidson, 1972a), the Back River area (Lambert, 1976, 1978b), and the Hackett River area (Frith and Roscoe, 1980; Casselman and Mioduszewska, this volume). Henderson (1975b) reported a thin stromatolitic carbonate unit in the James River volcanic belt in the northern part of the province. The comparatively widespread distribution of carbonate rocks in the Slave Province is in distinct contrast to Superior Province, and is a feature in common with the Rhodesian Shield (Clay, 1979; K.A. Viewing, pers. commun.).

The greenstone belts, at least in the southern Slave Province, appear to be tectonically controlled by major linear faults that generally mark the boundary between sedimentary basins and granitoid terrane (McGlynn and Henderson, 1970; Lambert, 1977). Volcanic rocks were extruded along these faults and concurrently, but also subsequently, more abundant sedimentary rocks were deposited in the adjacent basins. Later these basin margins became the sites of extensive deformation as in the Beaulieu River belt (Lambert, 1977), and of diapiric granitic intrusion (McGlynn and Henderson, 1970; Drury, 1977).

For the Hackett River belt, Fyson and Frith (1979) questioned the hypothesis that the volcanic rocks were deformed into keel-like subsiding segments and that concurrent, granitoid plutons rose diapirically (Anhaeusser, 1975; Drury, 1977). Rather, they interpreted two main phases of deformation and inferred that some plutons were emplaced during the first, some during the second, and other, smaller plutons, following the second. They were unable to define the relative time of emplacement of granitoid bodies and associated migmatized supracrustal rocks that border the belt.

Rare-earth and other trace element data indicate (Drury, 1979) that three granodiorite plutons in and near the Yellowknife greenstone belt could have formed by mantle-depth melting of garnet-bearing basaltic source rocks. Similar data for the potassic, post-kinematic Prosperous Lake granite, which intrudes Yellowknife

Supergroup metasedimentary rocks suggest generation by a high degree of partial melting of the latter (Drury, 1979).

Sedimentary rocks in and near the volcanic belts are dominantly turbiditic greywacke and mudstone. The sedimentary rocks both conformably and unconformably overlie the volcanic successions (McGlynn and Henderson, 1970; Lambert, 1977), but interfingering relationships are known in places (Lambert, 1978a). Unconformable contacts are commonly marked by conglomerate (McGlynn and Henderson, 1970). In the Yellowknife belt, and also at Gordon Lake, greywacke contains abundant silicic and intermediate volcanic rock fragments (McGlynn and Henderson, 1970). This is anomalous considering the small proportion of felsic volcanic rock in the belts, suggesting that felsic volcanic units were originally much more extensive. Greywacke also contains minor but widespread granitic rock fragments, probably indicating exposure of basement terrane. Extensive paragneiss terranes in Slave Province are suspected to represent metasedimentary rocks of the Yellowknife Supergroup, mainly because the latter were clearly deposited, in part, in extensive regional basins.

Metallogeny

Volcanic-hosted massive sulphide deposits with proven reserves are at High Lake, Hackett River, Takiyuak Lake, Izok Lake and Indian Mountain Lake (Table II). The Yellowknife and Lupin (Contwoyto Lake) gold deposits are also of major importance. Smaller gold mines, the most important of which were the Discovery, Thompson-Lundmark and Tundra, were prior producers in southern Slave Province, and other small deposits and numerous occurrences are known.

Other deposit types are, for the most part, of little economic significance. Lithium, niobium-tantalum and minor tin occur in pegmatites in southern Slave Province. A recent, potentially significant tantalum discovery lies near Great Slave Lake in the Blatchford Lake alkaline complex. Molybdenum occurs in a leucocratic dike near Likely Lake in the Yellowknife belt, and disseminated molybdenite also occurs in diorite along the margin of the Elu Inlet volcanic belt at the northeastern tip of Slave Province. A copper-molybdenum discovery of probable porphyry type is near Contwoyto Lake (Bain, 1980). Silver and galena occur in fractures in, and in the matrix to brecciated rhyolitic pyroclastic rocks associated with a major caldera structure in the Back River area. A differentiated gabbroic body in the Contwoyto Lake area contains vanadium-bearing titaniferous oxides (Laporte *et al.*, 1978).

Massive Sulphide Deposits

Numerous important Zn-Cu (Pb, Ag) massive sulphide deposits occur in Slave Province. Although some are only partially explored, aggregate tonnage is at least 38 million tonnes containing approximately: zinc 3.1, lead 0.41 and copper 0.58 million tonnes, silver 4,300 and gold 19 thousand kg. These deposits, in common with similar deposits in Superior Province, occur in Archean volcanic rocks, particularly with the felsic volcanic members. The High Lake deposit, near Coronation Gulf, is of Cu-Zn type, thus very similar compositionally to Superior Province deposits. Other deposits in the Slave Province exhibit important differences.

In general the Slave Province deposits are richer in lead and silver than typical Superior Province examples of this type. This may be because felsic pyroclastic and interbedded sedimentary rocks are more abundant in the footwall sequences of the Slave deposits and served as sources of these elements (Divi *et al.*, 1980). Six deposits in the Hackett River area contain more than 19 million tonnes of ore, with a silver grade of approximately 170 g/tonne, comparable to that of the Kidd Creek deposit in the Superior Province. The lead content, approximately one per cent, of some Slave Province deposits is higher than that of most Superior Province deposits, although the atypical Mattabi, Ontario and Coniagas, Quebec deposits, have comparable lead contents. The Izok Lake deposit is especially rich in base metals for a deposit of this type, but is similar in this respect to the Lemoine deposit, Quebec, in the Superior Province.

Frith and Roscoe (1980) suggested that a hangingwall sequence of rhyolitic tuff and more massive quartz-eye bearing rhyolitic rocks at the Yava deposit was deposited in a caldera. Abrupt thickness changes in limy tuff near some of the Hackett River deposits (Frith and Roscoe, 1980) suggest it may also have been deposited, and carbonatized, in a caldera. In the Back River area, other major calderas, without known associated massive sulphide deposits (Lambert, 1976, 1978b; Economic Geology Division, 1980, p. 156) are associated with both submarine and subaerial volcanic rocks. Thin carbonate strata, in part carbonate iron-formation, and zones of carbonatized fragmental rocks also occur in these caldera structures. Elsewhere, as at High Lake, Indian Mountain Lake and Victory Lake-Turnback Lake, carbonate strata are associated with, or near massive sulphide deposits, although calderas have not been identified. These features suggest environmental differences for massive sulphide genesis compared to the Superior province where sedimentary carbonate strata are extremely rare, subaerial volcanic rocks are comparatively rare, and caldera structures have not been as clearly documented and are probably rare (however, see Deptuck *et al.*, this volume).

Lead Isotope Data for Massive Sulphide Deposits

Lead isotope data for volcanic-hosted massive sulphide deposits in Slave Province define a steep linear pattern on a $^{206}Pb/^{204}Pb$ vs $^{207}Pb/^{204}Pb$ plot (Fig. 7; Table VIII). The slope of 0.792 ± 0.025 for the secondary isochron is unusually steep. Data for Superior Province massive sulphide deposits, in contrast, plot in a comparatively tight cluster at the lower end of the Slave Province trend. Slave Province lead thus appears to have been variably contaminated by radiogenic lead from some crustal source.

To obtain the best age for a "crustal contaminating" component requires establishment of a reliable age for the Slave Province volcanic rocks. A lead isochron plot for zircon from volcanic and sedimentary rocks in the Yellowknife belt (Green, 1968; Thorpe, 1971) yields an age of 2685 ± 15 Ma. The Western Granodiorite, which intrudes the western margin of this belt, is approximately the same age (Thorpe, 1971; Cumming and Tsong, 1975). The Southeastern Granodiorite, which also intrudes sedimentary rocks of the belt, has a discordia age of 2610 to 2650 Ma (Green and Baadsgaard, 1971). Rb-Sr ages for several other granitic plutons in Slave Province are 2655 to 2660 Ma (Wanless and Loveridge, 1972; Frith *et al.*, 1977). A zircon

U-Pb age of 2667 ± 7 Ma was obtained (Lambert and Henderson, 1980) for rhyolite from the Back River volcanic complex in eastern Slave Province. It thus appears that Yellowknife Supergroup volcanism occurred at approximately 2670 Ma. This age agrees with a 2677 Ma zircon age for an Archean inlier in Churchill Province in the Kaminak area, Keewatin District (Wanless, 1979).

An age of 2670 Ma for the Slave massive sulphide deposits results in a calculated two-stage age of 4017 ± 60 Ma for the "contaminating crustal" component, although there is no direct evidence for such old crust in the Slave Province. The oldest reported ages are U-Pb zircon ages of 3030 to 3040 Ma for a grey gneissic tonalite boulder in a tabular diatreme in the Con Mine (Nikic *et al.*, 1975) and a zircon U-Pb age of 3155 Ma for basement rocks in the Point Lake area (Krogh and Gibbons, 1978). It is interesting that the lead isotope data for a small massive sulphide body at Point Lake (Fig. 7) do not fit the linear pattern defined by all other deposits, but rather suggest an age approximately 100 Ma older.

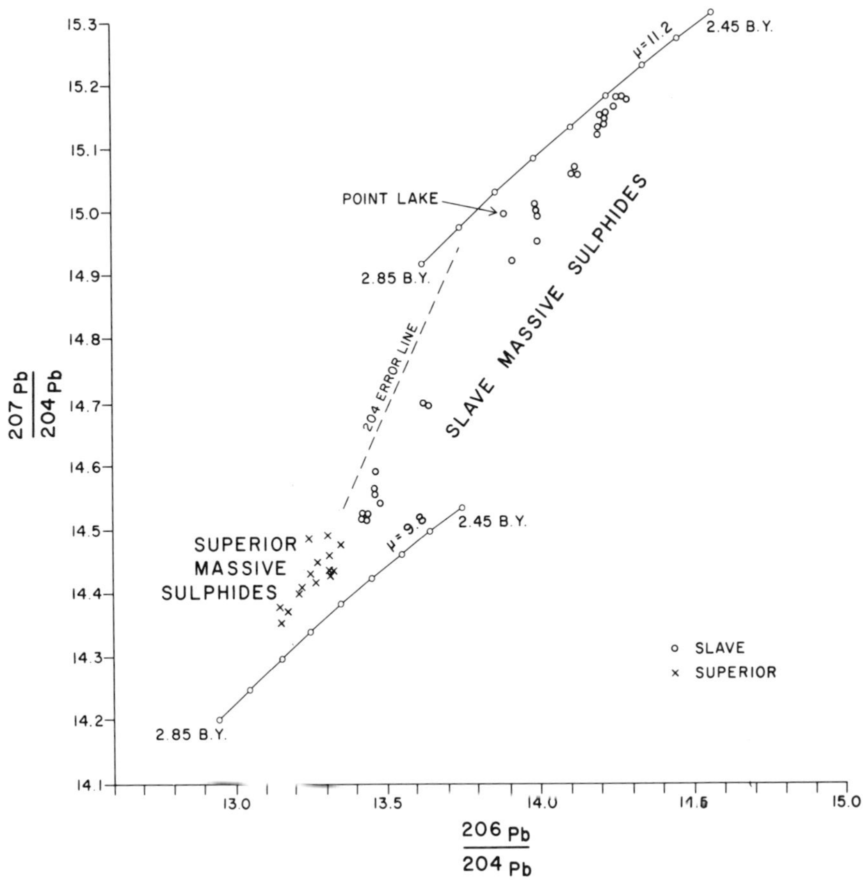

Figure 7. Lead isotope compositions of galena from Archean massive sulphide deposits in Slave and Superior provinces in relation to model curves based on the parameters of Cumming and Richards (1975).

The more radiogenic character of Slave Province galena, compared with galena from Superior Province, reflects a difference in source-rock history. If metal derivation for massive sulphide deposits is by leaching of the underlying strata (see "Deposit Characteristics and Provincial Comparisons: Massive Sulphide Deposits" later in paper), this difference suggests that Slave Province lead was derived either from sedimentary rocks (themselves derived from older, 4.0 Ga basement granitoid rocks), or from volcanic rocks that were generated by melting of recycled, continental-derived material, or by mantle melts that were contaminated by a significant amount of older crust. In either case, thicker, more mature or older continental crust may have underlain much of the Slave Province, compared with the known and presumed basement rocks beneath the Superior Province.

Gold Deposits

Gold production of the Northwest Territories, almost entirely from Slave Province, has totalled 311,000 kg. Most gold deposits are in Yellowknife Supergroup supracrustal rocks and can be subdivided into the Yellowknife type, associated with volcanic rocks, and the Contwoyto Lake type, associated with iron-rich sedimentary strata. The Coronation Gulf type differs in that its host rocks are gneissic, in part granitoid.

Yellowknife Type. Three of the four largest producers (Table III), Giant, Con and Negus, have accounted for more than 85% of the Northwest Territories gold production and represent the Yellowknife or "shear zone" type. The Yellowknife greenstone belt, composed predominantly of mafic flows and small, probably synvolcanic metagabbro bodies, extends northward from Great Slave Lake along the west sides of Yellowknife Bay and Walsh Lake for more than 40 km. Gold occurs in elongate quartz lenses in major shear zones which transect the strike and dip of the volcanic rocks for several miles. The volcanic rocks and ore-bearing shear zones are cut by late, north- to northwest-striking, steep to vertical, major faults. The most important, the West Bay fault, cuts the greenstone belt at an acute angle and has horizontal displacement of 4,919 m and vertical displacement of 479 m (Campbell, 1948). This fault marks the southern limit of the Giant ore system, which reappears on its other side as the Campbell zone, some 4.8 km to the south on the lower levels of the Con Mine.

The shear zones, up to 70 m wide, consist of carbonate-bearing chlorite and chlorite-sericite schists and are commonly considered early fault zones. Typical Giant ore contains 30 to 90% quartz and 5 to 10% metallic minerals (Brown and Dadson, 1953). It usually exhibits a ribboned appearance, caused by interlayered bands of quartz, carbonate, and sericite schist, except in the cores of some ore shoots where it is almost exclusively quartz. A complex assemblage of metallic minerals (Breakey, 1975) is sporadically distributed in the veins. Stibnite and antimony sulphosalts are generally indicative of high grade ore. Grey to black quartz, which formed earlier than, and is cut by, white quartz, contains fine carbon and/or graphite (Boyle, 1953).

The other main producers were the Discovery, Tundra (3251 kg from 170,282 tonnes) and Thompson-Lundmark (2188 kg from 121, 535 tonnes), all relatively simple quartz-vein deposits. The Discovery and Thompson-Lundmark veins occur in

TABLE VIII

LEAD ISOTOPE DATA FOR MASSIVE SULPHIDE DEPOSITS AND BASE METAL OCCURRENCES IN SLAVE PROVINCE AND MASSIVE SULPHIDE DEPOSITS IN SUPERIOR AND CHURCHILL PROVINCES (EXCEPT FOR DEPOSITS IN APHEBIAN VOLCANIC ROCKS OF SOUTHWESTERN CHURCHILL PROVINCE; SEE SANGSTER, 1978)

Sample No.	Property	$^{206}Pb/^{204}Pb$	$^{207}Pb/^{204}Pb$	$^{208}Pb/^{204}Pb$	Lab (Date)
Slave Province					
SP1570	Hackett River	13.43±0.081	14.52±0.131	33.21±0.432	G.S.C., 1971
SP1497	Indian Mtn. L.	13.62±0.082	14.70±0.132	33.38±0.434	G.S.C., 1971
SP1830	High Lake	13.424±0.062	14.52±0.102	33.15±0.292	G.S.C., 1972
TQ71-309	Homer Lake	14.101±0.014	15.058±0.023	33.93±0.068	Univ. of B.C., 1972
TQ71-267	Turnback Lake	14.209±0.014	15.154±0.023	33.954±0.068	Univ. of B.C., 1972
TQ71-308	Victory Lake	14.279±0.014	15.177±0.023	33.880±0.068	Univ. of B.C., 1972
TQ71-238	Homer Lake	14.115±0.141	15.058±0.152	33.929±0.346	Teledyne, 1974
TQ71-264	Turnback Lake	14.186±0.142	15.135±0.152	33.782±0.345	Teledyne, 1974
SP2323	No. 41 deposit	13.987±0.140	15.006±0.152	33.701±0.344	Teledyne, 1974
SP2322	Hood River No. 10	13.989±0.168	14.955±0.179	33.750±0.405	Teledyne, 1975
TQ71-97	Cleaver Lake	13.436±0.161	14.523±0.174	33.269±0.399	Teledyne, 1975
TQ71-238	Homer Lake	13.982±0.168	15.008±0.180	33.663±0.404	Teledyne, 1975
TQ72-13	Ross Lake	14.261±0.171	15.184±0.182	33.946±0.407	Teledyne, 1975
TQ72-25	Homer Lake (vein)	14.111±0.169	15.071±0.181	33.921±0.407	Teledyne, 1975
TQ72-38	Indian Mtn. Lake	13.635±0.164	14.697±0.176	33.403±0.401	Teledyne, 1975
TQ78-185	Yava	13.467±0.162	14.591±0.175	33.326±0.400	Teledyne, 1975
FR76-614	Izok Lake	13.988±0.106	14.996±0.114	33.710±0.256	Teledyne, 1977
FR76-615	Izok Lake	13.912±0.106	14.923±0.113	33.497±0.255	Teledyne, 1977
TQ75-22	Duck Lake	13.460±0.102	14.564±0.111	33.320±0.253	Teledyne, 1977
TQ74-684	Point Lake	13.881±0.014	14.995±0.021	33.766±0.061	G.S.C., 1976
RF78-15-66	O.K., Turnback L.	14.186±0.009	15.120±0.076	33.868±0.014	U. of Toronto, 1979
TQ78-2	Pail Gp. (Long L.)	13.478±0.081	14.541±0.087	33.287±0.200	Teledyne, 1978
TQ78-14	Pail Gp. (Long L.)	13.433±0.081	14.524±0.087	33.232±0.199	Teledyne, 1978
RF78-112	Turnback Lake	14.248±0.010	15.182±0.012	34.009±0.038	U. of Alberta, 1980
RF78-115	Turnback Lake	14.208±0.010	15.150±0.012	33.942±0.038	U. of Alberta, 1980
RF78-122	Turnback Lake	14.240±0.010	15.168±0.012	33.945±0.038	U. of Alberta, 1980
RF78-130	Turnback Lake	14.204±0.010	15.150±0.012	33.949±0.038	U. of Alberta, 1980
RF78-138	Turnback Lake	14.207±0.010	15.144±0.012	33.920±0.038	U. of Alberta, 1980

Sample No.	Property	$^{206}Pb/^{204}Pb$	$^{207}Pb/^{204}Pb$	$^{208}Pb/^{204}Pb$	Lab. (date) or Reference
Superior Province					
332	Willroy, Ont.	13.308±0.006	14.427±0.011	33.162±0.021	Ostic et al., 1967
369(T-658)	Lun Echo, Ont.	13.313±0.006	14.429±0.011	33.174±0.021	Ostic et al., 1967
370(T-659)	Geco, Ont.	13.316±0.010	14.431±0.002	33.203±0.005	Ostic et al., 1967
MG-38a	Willroy, Ont.	13.211±0.15	14.401±0.19	33.069±0.15	Stacey et al., 1969
SP1631	Mattabi, Ont.	13.31±0.08	14.46±0.13	33.12±0.43	G.S.C., 1971
SP1632	Trout Bay, Ont.	13.25±0.08	14.49±0.13	32.95±0.43	G.S.C., 1971
SP1633	South Bay, Ont.	13.27±0.08	14.45±0.13	33.11±0.43	G.S.C., 1971
TQ71-303	Mattagami L., Que.	13.146±0.060	14.381±0.101	33.173±0.292	G.S.C., 1972
SP1875	Sturgeon Lake, Ont.	13.343±0.013	14.478±0.022	33.116±0.066	Univ. of B.C., 1972
SP2507	Coniagas, Que.	13.174±0.109	14.372±0.122	32.993±0.313	Teledyne, 1976
TQ70-405	Normetal, Que.	13.150±0.100	14.353±0.109	32.960±0.250	Teledyne, 1977
KO77-98	Millenbach, Que.	13.307±0.080	14.490±0.087	33.161±0.199	Teledyne, 1978
TO74-554	Kidd Creek, Ont.	13.235±0.024	14.427±0.035	33.032±0.099	G.S.C., 1979
77JR-1	Kidd Creek, Ont.	13.246±0.013	14.434±0.015	33.096±0.033	B.R. Doe, 1978
FR80-1002	Millenbach, Que.	13.311±0.013	14.485±0.015	33.165±0.052	Geospec Cons., 1981

Sample No.	Property	$^{206}Pb/^{204}Pb$	$^{207}Pb/^{204}Pb$	$^{208}Pb/^{204}Pb$	Lab. (date) or Reference
Churchill Province					
SP1900	Spi Lake, N.W.T.	13.289±0.061	14.416±0.100	33.106±0.291	G.S.C., 1972
RF78-227	Gemex, N.W.T.	13.316±0.009	14.465±0.010	33.243±0.037	U. of Alberta, 1979
RF78-271	Rochon L., N.W.T.	13.425±0.024	14.537±0.035	33.313±0.100	G.S.C., 1979
RF78-271	Rochon L., N.W.T.	13.424±0.010	14.497±0.011	33.253±0.037	U. of Alberta, 1979
TQ78-27	Boylen, Labrador	15.222±0.027	15.245±0.037	35.143±0.105	G.S.C., 1979
RF66-836	Frederickson L., Lab.	15.298±0.015	15.222±0.015	35.222±0.035	B.R. Doe, 1980
RF66-837	Frederickson L., Lab.	15.004±0.015	15.197±0.015	35.256±0.035	B.R. Doe, 1980

sedimentary rocks of the Yellowknife Supergroup, although at Discovery these lie within mafic volcanic rocks. Some veins at the Tundra mine are along or near the contact between sedimentary and volcanic rocks, and others lie entirely in volcanic rocks.

Contwoyto Lake Type. Deposits of this type are arsenopyrite- (loellingite)-bearing pyrrhotitic and amphibole-rich metasedimentary rocks. The Lupin deposit at Contwoyto Lake, under development in March, 1981, is the principal one of this type, and is significant for three reasons. It has proven, probable and possible reserves of 2,450,000 tonnes grading 13 g/tonne Au (0.42 oz/ton) and will soon be an important producer. Secondly, it represents a type of mineralization that is widespread in Slave Province, occurring for example just west of Bathurst Inlet near "Turner" Lake and on the COT claims east of Bathurst Inlet. Finally, its characteristics match more closely the criteria for primary syngenetic iron-formation-hosted gold deposits than any other Canadian examples.

This ore is essentially silicate-sulphide iron-formation. Characteristics of syngenetic gold deposits formed as iron-formation remote from exhalative vents are: 1) strict stratigraphic control of ore, 2) lack of zoning, 3) pre-metamorphic deposition, indicated by similar metamorphic mineral assemblages in ore and in remote country rock, 4) absence of wallrock alteration, and 5) uniform distribution of gold values in the host strata. A significant exception to the fifth one is that gold, quartz, and some other components, may have been locally remobilized into veins, and the ore greatly thickened in fold hinges during structural deformation.

The Lupin ore, and similar rocks elsewhere in Slave Province, consists of thin beds of fine-grained pyrrhotite and amphiboles, including cummingtonite. Arsenides, restricted to some individual amphibole-rich beds, are not uniformly distributed but form coarse porphyroblasts up to 3 cm across, of intergrown arsenopyrite and loellingite (Bostock, 1968). Minor chert beds up to 15 cm thick and rare, thin iron oxide beds are evidence for chemical-sedimentary origin. The gold was probably a primary component of the chemical sediment.

The gold-bearing sedimentary strata do not now contain carbonate, although siderite may have been a primary component. However, a thin unit of carbonate iron-formation, with associated gold, is present along the Back River. It resembles carbonate-oxide iron-formation that hosts and is closely associated with gold deposits in the Geraldton-Beardmore and Pickle Crow districts, Superior Province. Both were probably formed by similar processes, probably as volcanic-associated chemical sedimentary rocks. Gold in these carbonate (±oxide) strata may, however, have been deposited by replacement because small- and large-scale replacement textures are common.

Coronation Gulf Type. Gold-quartz veins near Coronation Gulf are laterally persistent individual veins in gneissic rocks, partly orthogneiss. In general gneissic rocks are uncommon hosts for gold deposits although the Renabie deposit, Ontario in Superior Province is the most comparable example.

Lead Isotope Data for Gold Deposits

Lead isotope data for galena from gold deposits, mostly in the southern part of Slave Province, fit closely the steep linear distribution pattern defined from the massive sulphide deposits. Whether the data can serve to distinguish syn-volcanic deposits from those formed during post-volcanic metamorphism and deformation depends on the time lapse between volcanism and the later event.

Rb-Sr ages for metavolcanic rocks of the Yellowknife and Cameron River belts are 2570 ± 160 and 2574 ± 200 Ma, respectively (Green and Baadsgaard, 1971). The lead isochron age for the Southeastern granodiorite (Green and Baadsgaard, 1971) is similar. These data suggest a major orogenic-metamorphic event at approximately 2570 Ma. Lead isotope data for the southern Slave Province gold deposits fail to clearly record this event, but it may be reflected by veins in the Tree River area and at Hope Bay, east of Bathurst Inlet. However, only a few analyses are available for these areas and model ages of 2400 to 2550 Ma may be unreliable because the leads could belong to anomalous suites.

CHURCHILL PROVINCE

Significant mineral deposits in Churchill Province occur in Archean and Aphebian greenstone belts, Aphebian shelf-sequence sedimentary rocks, Aphebian rift-related (?) sedimentary and volcanic rocks, and Helikian sedimentary rocks. The principal Archean supracrustal domains, the Rankin-Ennadai belt, and the areas of the Prince Albert Group on the mainland and Mary River Group on Baffin Island (questionable Archean or Aphebian) lack producing mines. However, the Cullaton Lake gold deposit in the Rankin-Ennadai belt is under development and the Rankin Inlet nickel mine produced 400,000 tons of 3.1% Ni and 0.9% Cu ore. Large iron-formations in the Mary River and Prince Albert groups are important potential sources of iron ore.

Rankin-Ennadai Belt

The Rankin-Ennadai belt (Ridler, 1974; Ridler and Shilts, 1974a, 1974b) contains up to five mafic to felsic cycles. Age of the Archean sequence is approximately 2677 Ma (Wanless, 1979) according to zircon U-Pb studies. The Archean rocks are locally overlain by, and infolded with, Proterozoic sedimentary and minor volcanic rocks of the Hurwitz Group. The Archean volcanic rocks are compositionally similar to the Archean bimodal volcanic rocks in the Slave and Superior Provinces, with extensive tholeiitic basalt, and minor sodic rhyolite. Ultramafic rocks are rare. The belt contains several volcanic-hosted massive sulphide occurrences, stratabound gold deposits, and abundant Algoma-type iron-formation.

The massive sulphides include the Gemex (Heninga Lake), Spi Lake and Rochon Lake occurrences, the latter near Ennadai Lake. The Gemex deposit consists of three massive sulphide lenses in a zone 830 m long. The largest lens is about 180 m long and 10.5 m wide. The total size of the deposit is undetermined, but probably less than 1,000,000 tonnes in the three lenses.

The most important gold deposits are near Cullaton Lake. Here the B Zone deposit consists of mineralized quartz-magnetite iron-formation that is locally highly folded and fractured. The ore forms an irregular body consisting of quartz, chlorite, carbonate and sulphide. Chlorite and pyrrhotite are associated with rich gold, but coarse arsenopyrite, although relatively abundant, is not as closely associated. Four steeply plunging ore zones, transgressive to bedding of the iron-formation, have been recognized. The B zone contains 168,000 tonnes of proven ore at a grade of 25.4 g/tonne gold, and there is potential for an additional 91,000 tonnes. The nearby A zone is probably similar and may contain a similar tonnage, but is of lower grade. The iron-formation hosting the B zone strikes about N 10°W and is separated from the enclosing greywacke by narrow beds of carbonate-rich tuff. These sedimentary rocks, together with underlying mafic volcanic rocks, make up the Archean Henik Group (Eade, 1974).

The Shear Lake deposit, near Cullaton Lake, is in basal quartzite of the Proterozoic Hurwitz Group and was discovered in 1947. Gold, in part associated with pyrite, is in fractures in the quartzite. Fracturing and brecciation are related to a northwest-striking fault that cuts the quartzite. Drilling difficulties in fractured quartzite, and intense weathering of pyrite to earthy hematite have hampered evaluation of the deposit.

Iron-formation is widespread in the Rankin-Ennadai belt; all major occurrences are in the Archean Kaminak and Henik groups. Iron-formation is of Algoma type (Gross, 1965), and consists of banded chert-magnetite associated with greywacke, volcanic rocks, granitic gneiss and schist. The major deposits are at Mistake Bay, South Henik Lake and near the McConnell River. The latter deposit probably contains 360 million tonnes of potentially mineable rock grading 30 to 40% Fe (Moore, 1959). It is up to 200 m thick, the Mistake Bay deposit up to 60 m thick and the South Henik Lake deposit up to 90 m thick (Economic Geology Division, 1980, p. 37).

In addition to oxide-facies iron-formation, several carbonate- and sulphide-facies occurrences, in addition to the gold-bearing sulphide facies iron-formation at Cullaton Lake, are of metallogenic interest (Ridler and Shilts, 1974a, 1974b). The gold content of a carbonate-facies iron-formation near Turquetil Lake is generally low, mainly $<$ 20ppb (Shemeliuk, 1976).

Prince Albert Group

Rocks of this group form elongate belts, composed dominantly of basaltic and sedimentary rocks, with felsic types more abundant on Melville Peninsula. Quartzite of the sedimentary unit is fuchsitic. The basaltic rocks include komatiitic flows (Schau, 1977). Five major Algoma-type iron deposits occur in complexly folded sedimentary rocks of this group in western Melville Peninsula, and collectively contain more than 3 billion tonnes of 32 to 38% Fe. Similar iron-formations occur on the east side of Melville peninsula, where there are more than one billion tonnes grading 23 to 34% Fe in five areas (Economic Geology Division, 1980). Although minor chalcopyrite occurs in various rocks of this group (Schau, 1975) and nickel-copper sulphides occur in metapyroxenite (Eckstrand, 1975), no significant deposits of massive sulphide or gold, common in other Archean terranes, are known.

Proterozoic Domains

The Proterozoic rocks of Churchill Province comprise several highly deformed tectono-stratigraphic domains, and additional areas of undeformed cratonic cover. Each domain may represent a particular tectonic environment, but most have been inadequately studied to allow precise recognition of tectonic features such as rifts or volcanic arcs. The principal domains are:

1) LaRonge-Flin Flon-Lynn Lake greenstone belt composed of tholeiitic and felsic volcanic rocks, laterally equivalent turbidites, and overlying molasse-type red-beds.

2) the Wollaston domain, a shelf sequence of psammitic to pelitic sedimentary rocks, with some carbonate and very minor mafic volcanic rocks. The supracrustal remnants of the adjacent Mudjatik domain may be stratigraphically equivalent to some Wollaston Group rocks (Ray, 1978). The Hurwitz and Amer groups, Keewatin District, may be stratigraphically and tectonically related to the Wollaston rocks.

3) the Circum-Ungava geosyncline consisting of eugeoclinal and miogeoclinal domains, and extending more than 4000 km in strike length.

4) The highly metamorphosed and deformed Foxe fold belt of Melville Peninsula, and the Piling Group on Baffin Island, which consist of shelf sequence clastic and carbonate strata (Jackson and Taylor, 1972; Henderson *et al.*, 1979; Henderson and Tippett, 1980).

Although reliable age determinations are not available, three principal cratonic cover sequences are generally considered to be Paleohelikian: a) the Athabaska Formation of northern Saskatchewan and Alberta, a sequence of predominantly fluvial rocks, with a very minor carbonate component, b) the lithologically similar Thelon Formation in western Keewatin District, and c) a sequence on the Borden Peninsula of Baffin Island that includes the Society Cliffs Formation, composed of carbonate rocks, the Arctic Bay Formation, composed of shale and siltstone, and the Nauyat Formation, composed of basaltic flows.

Each of these domains and cratonic cover sequences has a distinctive suite of deposits and occurrences that is a product of its respective tectono-stratigraphic environment. Unconformity-related uranium deposits (not discussed) occur in the Athabaska and Thelon formations. The important Nanisivik lead-zinc deposit (Clayton and Thorpe, this volume) is in the Society Cliffs Formation.

LaRonge - Flin Flon - Lynn Lake Domain. Aphebian volcanic rocks in the LaRonge, Flin Flon and Lynn Lake belts have some similarities, but also some important contrasts, with typical Archean greenstone sequences. Volcanic rocks of the Amisk and Wasekwan groups are compositionally bimodal. The mafic suite is variable in composition with tholeiitic andesite dominant at Flin Flon (Stauffer *et al.*, 1975) and high alumina basalt dominant at Snow Lake (Walford and Franklin, this volume). The felsic suite is generally calc-alkaline, although many of the samples are slightly alkali depleted, rendering classification uncertain. Felsic rocks comprise over 50% of the volcanic rock at Snow Lake, but much less at Flin Flon, Lynn Lake and in the LaRonge belt. Ultramafic rocks are very minor, but differentiated gabbroic sub-volcanic (?) intrusions are prominent at Flin Flon and Lynn Lake and more numerous in all Aphebian greenstone belts of southern Churchill Province than in most Archean belts of Superior Province.

An Aphebian model lead age has been established for the Borys Lake occurrence in the LaRonge-Waddy Lake area (Sangster, 1978), but correlation of strata here with more easterly sequences is uncertain. In general this area contains thin, dominantly basaltic rocks, with felsic types confined to a few thin zones near Sulphide Lake.

Sedimentary sequences are of two types. Those flanking major volcanic accumulations, as at File Lake (Bailes, 1980) and in the Rusty Lake greenstone belt (Speakman *et al.*, this volume), are turbidites, dominantly greywacke. These may be correlative with the lower part of the Kisseynew metasedimentary gneiss complex (Bailes, 1971). Secondly, arkosic strata of the Missi and Sickle groups overlie the Amisk and Wasekwan volcanic rocks and their sedimentary equivalents. The arkoses are generally correlated with the upper, arkosic portion of the Burntwood Supergroup, and are fluviatile to shallow marine, clastic red-beds (Bailes, 1980). Rocks of the Sherridon Group (Goetz and Froese, this volume) are shelf-type sedimentary strata with minor volcanic components; their stratigraphic position relative to the Amisk and Missi groups is uncertain.

Mineral Deposits: The variety of mineral deposits in the LaRonge - Flin Flon - Lynn Lake domain is relatively limited compared to an Archean sequence of similar extent in Superior Province. Massive base metal sulphide and gabbro-hosted nickel sulphide deposits are the most important types. Gold production has been minor, and a few tungsten occurrences are known. Disseminated, sediment-hosted copper occurs in several areas.

Massive Sulphide Deposits: Massive sulphide deposits (Price, 1977) are in rocks that range from exclusively basaltic at the Coronation, Birch, Flexar, Westarm, and Centennial deposits, to those with minor felsic volcanic types at the Flin Flon mine (Koo and Mossman, 1975). At Snow Lake, felsic rocks predominate in the footwall of most deposits. In the Lynn Lake-Rusty Lake area, the Fox Lake deposit is in a dominantly mafic volcanic terrane, whereas the Ruttan deposit (Speakman *et al.*, this volume) is in the Wasekwan greywacke. Although the LaRonge-Waddy area has a few massive sulphide occurrences, such as the lead-zinc-silver mineralization near Borys Lake (Coombe, 1975), most sulphide occurrences are sulphide-facies iron-formations. One common feature of several Aphebian massive sulphide districts in southern Churchill Province is the presence of felsic subvolcanic intrusions near the base of the host volcanic sequences. These intrusions resemble their counterparts in Superior Province (see "Gold Deposits" mentioned earlier), although supporting age data are available only at Flin Flon (MacQuarrie, 1980) and Snow Lake (Bell *et al.*, 1975).

Most massive sulphide deposits are of 'proximal' type, formed immediately over exhalative vents, now seen as Cu-rich, chloritic stringer zones. A few, such as the White Lake and Westarm deposits, lack clearly associated alteration pipes, and occur within thin pelitic horizons; these may be 'distal' deposits (Sangster, 1972).

The complex structure at Flin Flon obscures the stratigraphic position of the massive sulphide bodies. The Chisel Lake deposit at Snow Lake (Price, 1977) lies stratigraphically higher than the Anderson and Stall deposits. At Dickstone in the File Lake segment of the Flin Flon - Snow Lake belt, two ore zones are separated by 175 m of rhyolite (Bailes, 1980). The Sherridon deposits (Goetz and Froese, this volume) have no associated alteration pipe and are unusual because they occur entirely in sedimentary rocks. They are compositionally similar to many of the Flin Flon deposits and may have been transported, either in solution or mechanically, to their present position.

Iron-Formation: Oxide-facies iron-formation is rare. A few bands are intercalated with mafic volcanic rocks, and thin iron-formations occur in the Burntwood Supergroup. Sulphide- and oxide-facies iron-formation are more common in the LaRonge area. The Dot Lake occurrence (Harper, 1975) contains both oxide and sulphide-facies interlayered with hornblende gneiss. Sulphide-facies iron-formation is relatively common. A pyritic zone in the Nistassini Lakes area (Harper, 1976) is semi-continuous for approximately 7 km and contains several gold occurrences and the Churchill Minerals copper occurrence. In the Sulphide Lake area (Coombe, 1976), gold occurs in quartz veins which cut a laterally extensive sulphide-facies iron-formation. Approximately 150 million tonnes of magnetite-rich rock grading 28% Fe occurs in the Choiceland deposit, 65 km east of Prince Albert, Sask., an Algoma-type iron-formation in Precambrian basement rocks beneath 1 km of Phanerozoic strata. Its age and geological nature have not been publicly documented.

Stratiform Copper Deposits: The Anglo-Rouyn copper deposit near Lac LaRonge (Forsythe, 1972) is a stratiform chalcopyrite-pyrite-pyrrhotite- (magnetite) zone in metamorphosed argillaceous and calcsilicate-bearing sandstone, and arkose. The sulphides can be traced for more than 16 km. The largest zone was 6 km by 60 m by 15 m. The sulphides are finely interlaminated with the clastic sedimentary rocks, but some zones consist of coarse, recrystallized laminated sulphides, and others contain numerous discordant, fracture-filling veins. The host rocks are not visibly altered. In total, 1.7 million tonnes of ore, grading 1.64% Cu, 5.19 g/tonne Ag, and 1.1 g/tonne Au were mined. The deposit is remarkably devoid of zinc, with a maximum of 300 ppm Zn in the most copper-rich portions (J. Randall, pers. commun., 1975). The lack of hydrothermal alteration, the disseminated but stratiform nature of the ore, and exceptionally low zinc content are consistent features of sandstone-type (sedimentary) deposits, although the gold content is much higher. Forsythe (1972) proposed an epigenetic origin, but Kirkham (1974a) considered it a recrystallized, syngenetic sandstone type.

Disseminated copper occurrences are numerous in the Russel Lake and Kadeniuk Lake areas of the Kisseynew sedimentary gneiss domain. These (Baldwin, 1976) are at the base of the Sickle Group meta-arkose, where it overlies greywacke of the Laurie River Group. Layers of amphibolite are present near most occurrences. Pyrite, pyrrhotite, chalcopyrite, bornite, native copper and molybdenite (Baldwin, 1980b) are disseminated in conformable, vertically zoned layers with pyrite above, chalcopyrite and bornite at the centre, and native copper plus minor bornite below. The copper content rarely exceeds 0.2%, and is typically 0.05%. The strata are highly metamorphosed, and the copper minerals are recrystallized, equant and uniformly distributed with quartz and feldspar grains. The copper-rich drill intersections range up to 7.5 m thick, and extend for strike lengths of 500 m or more. Although there are no detailed descriptions, these occurrences are superficially similar to sandstone-hosted deposits, like those in the Belt Series of Montana (Clark, 1971) and many shallow marine to deltaic sequences (Tourtelot and Vine, 1976). Such deposits originate by low temperature diagenetic processes. Although metamorphosed equivalents are rare, the Minto deposit (Pearson and Clark, 1979), Yukon Territory, and the Anglo Rouyn deposit may be examples.

Sandstone-type copper deposits are confined to post-Archean sequences and commonly occur in the basal beds of marine transgressive sequences overlying continental sedimentary strata (Kirkham, 1974a). However a continental to marine transition is not documented in the above examples, and further study is necessary to elucidate their sedimentation history.

Gold Deposits: Production from gold deposits in Churchill Province has been very small, less than 34,300 kg in comparison with more than 4.8 million kg from Superior Province. In Saskatchewan the Box Mine on Lake Athabasca produced 2197 kg between 1939 and 1942 from pyritic quartz veins in a fractured granite, and other mines produced small amounts. However, more than 103,000 kg have been recovered as by-product from volcanogenic massive base metal sulphide deposits in the Flin Flon region.

In Manitoba, 23,867 kg have been recovered from gold deposits in Churchill Province; in contrast, 58,271 kg have been won from Manitoba deposits within the Archean Superior Province. The Nor-Acme mine near Snow Lake was the largest producer in Churchill Province with 21,141 kg from 4,934,258 tonnes of ore. This deposit lies along and near the contact between felsic and mafic volcanic rocks, a setting very similar to that of massive sulphide deposits in the area. Small hornblendite intrusive (?) bodies occur throughout the mine area. The deposit consists of a quartz veinlet and stringer stockwork, with associated intense silicification. Gold accompanies acicular arsenopyrite, pyrite and pyrrhotite in an abundant gangue of sugary quartz that contains numerous inclusions of silicified wallrock. There is strong carbonate replacement in the mafic rocks, especially in the hornblendite. The deposit has been highly metamorphosed; staurolite schist is present at its west end. Part of the deposit may be a highly deformed, metamorphosed and remobilized chert-carbonate exhalite, but other parts consist of veins in hornblendite intrusions.

The Agassiz deposit in the Lynn Lake area, considered by Stewart (1980) as possibly of exhalative origin, also contains base metals. Reserves based on nearly 22,000 m of drilling are about 2 million tonnes grading 10 g/tonne Au. The east zone, which contains approximately 800,000 tonnes, averages 11.7 g/tonne Au, 1.3% Zn and 0.8%Pb. The deposit lies within a thin sequence of interbedded siliceous sedimentary rocks, pyroclastic beds (chloritic tuff) and mafic flows (amphibolite), flanked by more massive volcanic rocks. One of the pyroclastic beds is a cherty felsic tuff. Sulphides are disseminated in a sugary quartz-carbonate zone that conforms generally to the sedimentary and pyroclastic rocks. Transecting quartz stringers are common. The associated base metal content is unusual in Canadian Precambrian gold deposits, although the Berens River and Ross Mine deposits in Superior Province share this characteristic.

Minor gold production has come from the Henning-Maloney and Newcor Mines, 8 km south of Flin Flon. On the former property (now Flin Flon Mines Ltd.) gold occurs in diorite, both as disseminations and in quartz veins. The auriferous zone is a magnetite-rich mafic border phase of the Dion Lake tonalitic intrusion which also contains a small porphyry-copper occurrence (Wallster, 1979). The Newcor Mine, in Amisk basalt, has gold in a conformable, laterally extensive zone of arsenopyrite.

In the LaRonge area, gold occurs in sulphide iron-formation, and also in association with some intrusions. The Komis gold occurrence near Sulphide Lake is of the latter type; here, gold-bearing quartz veins in volcanic rocks are radially disposed about a quartz monzonite stock.

Nickel-Copper Deposits: There has been major production from nickel-copper deposits in the Lynn Lake portion of the Aphebian volcanic belts of Manitoba. From 1953 to 1976, Sherritt Gordon Mines Ltd. produced nearly 20 million tonnes of ore, grading 1.023% Ni and 0.535% Cu, from eleven separate orebodies in two mafic to ultramafic, pipe-like intrusions (Pinsent, in press). Ore occurs in noritic gabbro but other parts of the intrusions range from diorite to pyroxenite and peridotite (Pinsent, 1977).

The deposits consist in general of sulphide blebs which separated from a tholeiitic gabbro melt. The rocks are compositionally basaltic, and exhibit a tholeiitic Fe enrichment trend (Pinsent, in press). The orebodies include the following types: disseminated ore, plutonic breccia ore, sulphide breccia ore, massive ore, vein ore, all in gabbroic rocks, and rare siliceous felsite vein ore. Pinsent (1977) suggested that, in addition to magmatic immiscible liquid segregation, some of the breccia and vein ore types are products of hydrothermal redistribution of magmatically emplaced sulphides.

Wollaston Belt. The Wollaston Belt (Fig. 8) and adjacent Mudjatik domain contain highly metamorphosed remnants of shelf-sequence sedimentary rocks which lie unconformably on Archean granite (Ray and Wanless, 1980). Money *et al.* (1970) divided the Wollaston strata into the basal Sandfly Lake Group, composed of arkose, greywacke and minor, high Ti subaerial basalt, overlain by the Meyers Lake Group composed of quartzite, conglomerate and pelite. The uppermost Daly Lake Group is composed of arkose, greywacke and quartzite, with minor carbonate.

The most important deposits in the Wollaston and Mudjatik rocks are metamorphosed sandstone-hosted lead-zinc deposits, with fewer sandstone-hosted copper (Sangster and Kirkham, 1974) and uranium occurrences. The lead-zinc deposits resemble the sandstone-lead type (Bjørlykke and Sangster, 1981) which usually occur at or near a basement unconformity in orthoquartzite. The largest, the George Lake deposit (Karup-Møller and Brummer, 1970), contains at least 5 million tonnes of 2.6% Zn and 0.35% Pb (Coombe, 1977). Other occurrences contain different metals; at Duddridge Lake, a drab, copper-bearing uranium-rich, locally graphitic arkose is stratigraphically slightly separated from a nearby galena-bearing quartzite. Skutterudite is common in the graphitic rock. In the Johnson River area (Fig. 8) similar deposits are also relatively zinc-rich, but at Segment Lake, in the Mudjatik terrane, lead dominates over zinc (Pearson, 1977). In contrast to all the former, the Kaz, Janice, Juno Lake and Rafuse Lake sandstone-hosted occurrences in the Janice Lake area have disseminated chalcocite, chalcopyrite, native copper and bornite in variably hematitic arkose and conglomerate. A Pb-Zn zone is separated stratigraphically from these by only a few hundred m of sedimentary rocks.

In the Wollaston Lake region, the fold belt rocks contain discontinuous uraniferous zones. Although too low grade for mining, these may have provided a widespread, abundant uranium source for the important pitchblende deposits associated with the unconformity between these rocks and overlying cratonic sedimentary strata

Figure 8. Geology and mineral deposits (as listed in Table IX) of the Wollaston and LaRonge belts, Saskatchewan. Geology after Macdonald and Broughton (1980) and mineral deposit locations largely from Coombe (1975 and 1977).

TABLE IX

MINERAL DEPOSITS AND OCCURRENCES IN THE WOLLASTON AND LARONGE BELTS, SASKATCHEWAN, AS SHOWN ON FIGURE 7

No.	Mineral Occurrence	No.	Mineral Occurrence
1.	Borys Lake	22.	Rush Lake
2.	Esmay Lake	23.	Mallard Lake
3.	Bandur Lake	24.	Jojay Lake
4.	Brabant Lake	25.	WY Group
5.	Main Lake	26.	Hump Lake
6.	May Lake	27.	Bassett Lake
7.	Narrow Lake	28.	Waddy Lake
8.	Center Lake	29.	Sulphide Lake
9.	Tower Lake	30.	Foster River
10.	Nistoassini Lake	31.	Juno Lake
11.&12.	Nipew Lake	32.	Location of the ''Wathaman River Boulder''
13.	Gochager Lake	33.	Rafuse Lake
14.	Hayman Lake	34.	Janice-Kaz Lakes & the ''Kaz Boulder''
15.	Forbes Lake	35.	Mud Lake
16.	Neyrink Lake	36.	Johnson Lake
17.	Otter Lake	37.	George Lake
18.	Reef Lake	38.	E-18
19.	Fork Lake	39.	Simon Lake
20.	Star Lake	40.	Nekweaga Bay
21.	Tamar Lake		

of the Athabasca Formation (McMillan, 1977). The uraniferous source-rocks are apatite-rich (Chandler, 1978b; Hoeve and Sibbald, 1978), possibly metamorphosed phosphorites, and scapolitic albite-pyroxene (or amphibole) plagioclasites which may be metamorphosed evaporites.

Circum-Ungava Geosyncline. The Circum-Ungava Geosyncline (Davidson, 1972b) is a narrow, 4000 km long tectonic zone composed of the Labrador Trough, the Cape Smith-Wakeham Bay and Belcher fold belts, the latter comprised of the Ottawa and Belcher Islands and Richmond Gulf region. The Geosyncline may extend through the Sutton Inlier, westward to the Fox River belt (Scoates and Weber, 1977), eventually joining the Thompson belt (Baragar and Scoates, 1981).

The Labrador Trough and Belcher fold belt have 3 volcano-sedimentary cycles (Dimroth *et al.*, 1970). Rocks of the western side of the trough form a miogeoclinal shelf facies, but those to the east and north form a volcanic-dominanted 'eugeoclinal' assemblage. The latter also dominates in the Cape Smith-Wakeham Bay belt, but a miogeoclinal sequence is again dominant in the Ottawa and Belcher islands (Dimroth *et al.*, 1970), Richmond Gulf (Chandler and Schwarz, 1980) and the Sutton Inlier (Bostock, 1971).

Radiometric age determinations for rocks of the Circum-Ungava geosyncline and adjacent 'basement' indicate very ancient, pre 3.3 Ga, gneiss in the Nutak Province (Hurst *et al.*, 1975) east of the Trough, and pre 2.5 Ga rocks in the Superior Province to the west (Wanless, 1969). The depositional age of the Circum-Ungava geosynclinal strata is older than 1950 Ma (Fryer, 1972), and Schimann (1978) reported an age of 2.15 Ga for volcanic rocks in the Cape Smith Belt. Depositional ages in the Belcher fold belt and Richmond Gulf graben are probably similar, although most of the dated rocks may have been metamorphically reset (Chandler and Schwarz, 1980). No age data are available for the Sutton Inlier.

Tectonism during deposition of rocks of the Labrador Trough was restricted to vertical movement (Davidson, 1972b); compressional tectonics, with accompanying thrust faulting, affected all three sedimentary-volcanic cycles and post-dates deposition. Sedimentary facies suggest a depositional basin dominated by vertical movements, possibly indicating early rifting.

Tholeiitic basalt is the major volcanic rock throughout the Circum-Ungava belt Moore, 1977); calc-alkaline rocks are very minor. Komatiite occurs in the Cape Smith belt (Schwarz and Fujiwara, 1977). Compositional trends of most of the volcanic rocks are also typical of a rift-dominated environment (Moore, 1977). Within both the Labrador Trough and Cape Smith-Wakeham Bay belt, laterally persistent mafic to ultramafic sills, up to 300 m thick, are numerous.

Mineral Deposits of the Circum-Ungava Geosyncline. This extensive tectonic zone has relatively few deposit types; only iron is currently produced. Volcanic-hosted massive sulphide deposits, nickel deposits and a variety of veins make up most of the remaining deposits.

Iron Deposits: Large Superior-type iron-formations in this area are major sources of iron ore (Gross, 1965). The iron-formations are best developed in the western miogeoclinal shelf facies and are primarily oxide and silicate-carbonate facies (Gross, 1965). Much ore in the Schefferville, Quebec and Wabush, Nfld. areas was secondarily enriched, direct shipping type; however huge reserves of metataconite are presently being exploited. Similar, extensive Superior-type iron-formations occur in the Cape Smith-Wakeham Bay belt, the Nastapoka and Belcher Islands, and the Sutton Inlier.

Nickel-Copper Deposits: Deposits with Cu:Ni ratio greater than 1.0 in the Labrador Trough are disseminated, and (rarer) massive sulphides within "blotchy" glomeroporphyritic gabbro sills which intrude carbonaceous interflow metasedimentary strata. Examples include the small Chrysler No. 2 deposit, which contains 580,000 tonnes grading 1.79% Cu, 0.48% Ni and about 0.5% Zn, the Sogemines, Erikson No. 1 and No. 2, and Leslie No. 2 occurrences (Dugas, 1970).

In the Cape Smith-Wakeham Bay belt at least 15.4 million tonnes of mafic and ultramafic rock grading 2.23% Ni and 0.71% Cu have been outlined in 6 deposits, and eight additional major occurrences are known. The sulphides are immiscible magmatic segregations, mainly near and at the base of mafic and ultramafic sills and occur as disseminated droplets, net-textured, or massive accumulations. At the 2-3 deposit they form canoe-like troughs adjacent to steep-sided "deep zones" in the floor of the intrusions (Miller, 1977) The basal contact zone is clinopyroxenitic gabbro. Some disseminated or net-textured sulphides also occur in the wehrlitic-dunitic central core of the intrusions.

Massive Sulphide Deposits: Copper-zinc-lead massive sulphide deposits occur in the Labrador Trough in a sequence of pillowed mafic volcanic rocks, carbonaceous phyllite, and sills of massive and "blotchy" gabbro which contain the copper-nickel sulphides. Some of the massive sulphide deposits occur near gabbro sills, and the Chrysler No. 2 Ni-Cu-Zn deposit may in part be an assimilated massive sulphide body. Most deposits are in the northern part of the Trough, southwest of Ungava Bay (Sauvé and Bergeron, 1965; Gilders, 1975) where the Soucy, Prud'homme No. 1 and Boylen deposits, and several smaller bodies, are located. The small Frederickson Lake and Jimmick Lake deposits lie nearly 400 km farther southeast, approximately 45 km northeast of Schefferville. Zinc and lead grades are highest, 6.86% and 1.03%, respectively, in the Boylen deposit, which differs in that associated sedimentary rocks include cherty carbonate iron formation and carbonaceous phyllite. Lead isotope data for the Boylen and Frederickson Lake deposits (Table VIII) yield model ages of 1877 to 2057 Ma (Cumming-Richards model). No massive sulphide occurrences are known elsewhere in the Circum-Ungava geosyncline.

Vein and Carbonate-Hosted Deposits: Other metalliferous occurrences of the geosyncline are carbonate-hosted Pb-Zn bodies in shelf sequences at Richmond Gulf and the Nastapoka Islands, on the eastern side of Hudson's Bay. One of these, the Ruby Lake deposit at Richmond Gulf, has 525,000 tonnes of 1.26% Zn and 1.02% Pb. Other deposits include the Nancy Island deposit at Richmond Gulf with 267,000 tonnes of 2.15% Zn and 0.72% Pb and the Lake Monte deposit with 120,000 tonnes of 1.75% Pb with some Ag. These occur in brecciated algal-concretionary cherty limestone (Harwood, 1949). Silicification is associated with the sulphides. Galena, sphalerite and pyrite are the principal minerals, with minor chalcopyrite in places (Parks, 1949).

Copper occurs sporadically in stromatolitic dolomite at the base of the Nastapoka Group and also disseminated in a pyritic zone in sandstone at the top of the underlying Richmond Gulf Formation (Chandler, 1978a). Rare uranium occurrences are reported (Miller, 1978) near the unconformity between the Archean basement and arkose of both the Pachi and Richmond Gulf formations.

Thompson Belt. Thompson belt geology has been described by Peredery (this volume), Coats *et al.* (1972) and Baragar and Scoates (1981). The supracrustal rocks, the Ospwagan Group (Scoates *et al.*, 1977), are mainly metasedimentary, with lesser metavolcanic and ultramafic rocks. They occur as tightly folded remnants in Archean migmatite (Cranstone and Turek, 1976; Russell, 1981); the remnants increase in number and thickness, and decrease in metamorphic grade, toward the Thompson lineament.

The sedimentary rocks include pelite and psammite, with minor dolomite and iron-formation. The volcanic rocks are mainly pillowed basalt, with minor pyroxene spinifex and ultramafic varieties (Baragar and Scoates, 1981). Ultramafic rocks, dominantly serpentinite, are common near the western margin of the belt and were emplaced prior to regional metamorphism. Some are differentiated (Peredery, 1979).

Although Rb/Sr isochrons have been determined for various rocks in the Thompson Belt, none gives a reliable depositional age (Brooks, pers. commun., 1980). However, recent whole rock lead isochron age determinations of the Thompson nickel ores indicate emplacement at approximately 2350 Ma (G.L. Cumming and O.R. Eckstrand, pers. commun., 1980).

Mineral Deposits: The nickel deposits of the Thompson belt are described by Peredery (this volume) and are not considered here. The only other deposit type is Algoma-type iron-formation. Bands of iron-formation, including oxide, silicate and sulphide-facies are commonly spatially associated with nickel deposits. Carbonate beds are commonly intercalated with iron-formation near the Pipe Mine.

Fox River Belt. This belt forms part of the northeastern border of the Superior Province in Manitoba, and rocks here have been tentatively correlated with those of the Sutton Inlier to the east in northern Ontario, and the Thompson Belt to the west (Baragar and Scoates, 1981). The stratigraphic sequence includes approximately 7 km of metasedimentary rock in three separate units, dominantly pelitic with minor psammite, carbonate and iron-formation. In addition, it contains 5 km of pillowed basalt and komatiite (Scoates, 1981) divided into two units, and separated by a central metasedimentary unit. The latter contains the Fox River Sill which is in two parts, each approximately 70 km long (Baragar and Scoates, 1981), and 2 km thick. The sill is zoned, and each zone has distinctive cyclic members. Mineral deposits are generally lacking in the Fox River belt, although disseminated magmatic Ni-Cu sulphides occur locally in the gabbroic intrusions, and a few chalcopyrite-bearing veins are present in some sedimentary strata (Scoates, 1977).

DEPOSIT CHARACTERISTICS AND PROVINCIAL COMPARISONS

Massive sulphide, gold, nickel-copper and iron deposits differ between Slave, Churchill and Superior provinces in their relative importances, morphologies and/or compositions. Despite these differences, based on the characteristics common to each deposit type in the three provinces, a genetic model can be inferred for each.

Massive Sulphide Deposits

The compositions, morphologies and stratigraphic settings of massive sulphide deposits are quite similar in Slave, Churchill and Superior provinces. The deposits have variable Cu/Zn ratios, and are generally lead-poor (Fig. 9). Within districts, individual deposits commonly have similar metal ratios, as in the Matagami, Noranda and Sturgeon Lake districts, although there are exceptions. Deposits at Flin Flon comprise two compositional groups, and all but one at Snow Lake are very copper-rich.

Deposits may be separated into two groups, however, on the basis of lead content. Those in the Hackett River Belt of Slave Province (Casselman and Miodus-zewska, this volume), Sturgeon Lake (Franklin *et al.*, 1977), and a few individual deposits (Chisel Lake; Walford and Franklin, this volume) contain nearly 1.0% Pb with accompanying higher silver. These deposits have slightly different alteration characteristics, and are in strata typified by abundant carbonate, usually ferruginous, both in footwall alteration zones and regionally in footwall and hangingwall strata. Moreover, in the footwall alteration zone silicification is commonly more pervasive, and Mg enrichment less marked, in these deposits than beneath deposits with very low lead content like Millenbach (Riverin and Hodgson, 1980; Knuckey *et al.*, this volume). Also, well zoned alteration is more common in the alteration pipes beneath lead-poor deposits, which typically have large chloritic cores and sericitic outer margins. Deposits of the lead-rich group exhibit pervasive sericitic-siliceous alteration zones, surrounded by indistinct sericitic halos.

The two groups of deposits share similar regional settings as well as the following characteristics.

1) Although footwall strata range widely in composition, in many districts they are in part epiclastic, brecciated or highly vesicular (Sangster and Scott, 1976). The immediate footwall rock, extending for 10 to 150 m below the deposits, is usually relatively massive (Hodgson and Lydon, 1977).

2) Epizonal, synvolcanic intrusions are common near or at the base of the sequence. These are commonly felsic sills (e.g., Baldwin, 1980a) up to 1 km thick and 5 to 20 km long; near Matagami Lake, however, the synvolcanic sill is mafic (MacGeehan and MacLean, 1980). In some districts, felsic sills are prominent at the base of deposit-containing sequences, although felsic volcanic rocks are relatively minor. For example, in the Ruttan area (Baldwin, pers. commun.) a coeval sill lies at the base of a dominantly sedimentary succession.

3) A major, subconcordant alteration zone commonly occurs 50 to 500 m below many of the deposits. The alteration types here are widely variable, ranging from

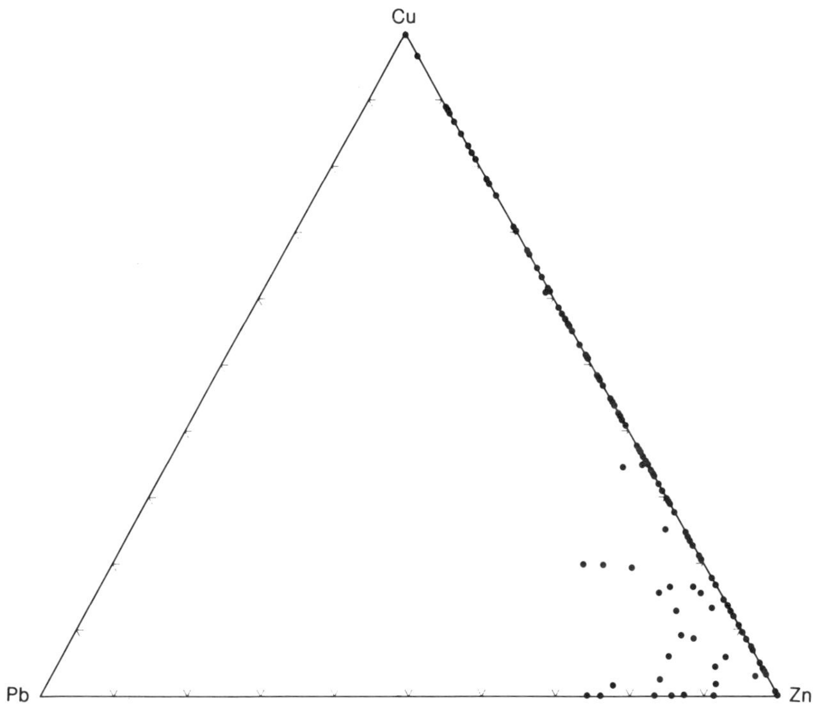

Figure 9. Weight proportions of base metals in massive sulphide deposits of the Canadian Shield.

silicification and Na addition in mafic rocks, as at Noranda (Gibson, 1979) and Matagami Lake (MacGeehan and MacLean, 1980), to very broad sodium depletion, with more local Mg and Fe addition in felsic volcanic, and felsic and mafic epiclastic rocks as at Sturgeon Lake (Franklin *et al.*, 1977) and Takiyuak Lake (Gill, 1976). These zones are commonly 5 to 10 km long, and 100 to 500 m thick. At Snow Lake (Walford and Franklin, this volume), Noranda, Matagami Lake and Sturgeon Lake, alteration pipes lead downward from the deposits to intersect these lower zones. Such zones may be present at other deposits, but have not yet been documented.

4) Many deposits occur in multi-cyclic volcanic piles, particularly in Superior Province. Age differences between these cycles have been investigated at Timmins (Nunes and Pyke, 1980), where the time between the first and second cycles may be only 8 to 10 Ma, and at Uchi Lake (Nunes and Thurston, 1980) where it is approximately 220 Ma. The deposits occur near the base of the second cycles in the South Sturgeon Lake (Franklin *et al.*, 1977) and Noranda (Spence and de Rosen Spence, 1975) districts. Although detailed stratigraphic relationships are not known in districts of the Churchill and Slave provinces, the deposits also apparently occur in higher volcanic cycles (Casselman and Mioduszewska, this volume; Frith and Roscoe, 1980). Thus the lower parts of the second or higher cycles appear most productive for massive sulphide deposits. This coincides with the position of the felsic subvolcanic Flavrian intrusion at Noranda (Goldie, 1979), the similar Beidelman Bay intrusion at Sturgeon Lake (Poulsen and Franklin, 1981), the unit 1-tonalite sill at Snow Lake (Walford and Franklin, this volume), the Cliff Lake intrusion at Flin Flon (Baldwin, 1980a) and of subvolcanic intrusions in several other districts. The synvolcanic mafic sills in the Matagami district are, however, at the base of the sequence, and the deposits occur in the lowest, as well as higher stratigraphic horizons.

Genetic Model. Extensive studies of important Precambrian and Phanerozoic massive sulphide districts and deposits (this volume; Ishihara, 1978; Tatsumi, 1970) suggest the following genetic model.

Regional Controls: The common positioning of deposits in the lower half of non-basal volcanic cycles and their spatial association with major, usually felsic, subvolcanic sills are important characteristics on a regional scale. The basic problems are to understand both the genesis of associated felsic rocks, their relationship to massive sulphide deposits, and the tectonic controls on formation of both associated volcanic rocks and the deposits themselves.

In all of the definitions of general volcanic stratigraphy in Superior Province (Goodwin, 1968; Wilson *et al.*, 1974; Jolly, 1975; Sopuck *et al.*, 1980) felsic rocks appear only in medial or upper cycles, and mafic and ultramafic rocks are more abundant in lower cycles. Although these schemes are not universally applicable, as is evident at Timmins where ultramafic rocks are abundant at the base of the upper cycle, and are underlain by felsic rocks of the lower cycle (Pyke, 1975), nevertheless felsic rocks are more common in the upper portion of any single cycle in Superior and Slave provinces. Recent chemical modelling of felsic melt petrogenesis in the Abitibi belt (Smith, 1980) indicates that these rocks formed from melting of lower crust. Thus the heat anomaly, possibly related to mantle convection (Lambert, 1980), necessary to provide sufficient mantle melting for mafic rock, which comprises over 90% of the volcanic piles, may have migrated upward with time into the lower crust, resulting in felsic melt generation. Alternatively, mafic rocks of earlier cycles may have subsided into older sialic crust (West and Mareschal, 1979) to initiate melting, and felsic melt generation. These alternatives of older tonalitic gneiss or greenstone sources are perplexing; Rb/Sr iostopic data suggest the latter, as the low initial ratio of most Archean felsic rocks (Moorbath, 1975) is best explained by either derivation of felsic melt from the mantle, a petrogenetically unlikely situation (Wyllie *et al.*, 1976; Smith, 1980), or from a mantle-derived, short-lived source such as basalt or greywacke.

Genetic Relationship to Volcanic Rocks: Massive sulphide deposits are products of rapid precipitation of metal sulphides from a hot aqueous solution, at and/or near the seafloor (Hutchinson, 1973, Sangster and Scott, 1976). The wide variability of their metal ratios indicates considerable variation in metal content and/or precipitation conditions of the hydrothermal solutions. Lydon (1977) and Hodgson and Lydon (1977) suggested that the metals are leached from the subjacent volcanic pile, concentrated in interpore water, collected in a reservoir and ultimately pumped to the surface by some mechanism such as seismic pumping (Lydon, 1981; Franklin *et al.*, 1981). According to this thesis, the metal ratios of the solution are controlled by source rock mineralogy. Oxygen isotope data (Heaton and Sheppard, 1976; Green *et al.*, 1980; Beaty and Taylor, 1982) confirms that meteoric water played a major role in forming these deposits. Generation of metal-rich brines by direct magmatic fractionation is incompatible with: 1) lack of in situ or upper crustal fractionation recorded for most felsic volcanic rocks, 2) the extreme range of metal ratios in comparison with the remarkably narrow range of felsic rock compositions, and 3) the poor correlation of deposit composition with composition of immediate host rocks. The leaching hypothesis, however, is compatible with these characteristics and also with the large, semi-concordant alteration zones beneath the deposits which may have been reservoirs where the fluids accumulated prior to expulsion onto the seafloor. Hodgson and Lydon (1977) emphasized the importance of an impermeable cap to prevent fluid loss; a cap is evident at many deposits, like the thin rhyolite immediately below the north limb deposits at Matagami (MacGeehan and MacLean, 1980), the Q.F.P. at Flin Flon (Koo and Mossman, 1975) and the footwall rhyolite at Sturgeon Lake (Franklin, 1976). Breaching of the cap by a synvolcanic fault would allow fluid discharge and massive sulphide deposition.

Relatively high fluid temperatures are necessary to obtain metal-rich solutions in equilibrium with their host rocks. Simple volcanism seems inadequate, as the melt-related heat is transferred rapidly to surface, and thus lost to the overlying water. Small subvolcanic domes, like those at Millenbach, (Knuckey *et al.*, this volume), Uchi (Pollock *et al.*, 1972), Flin Flon (Koo and Mossman, 1975) and the white rhyolite in the Kuroko district, Japan (Tatsumi and Clark, 1972) cannot provide sufficient heat to drive the hydrothermal system required for massive sulphide generation (Cathles, 1978). However, the large subvolcanic sills discussed above would provide sufficient heat to raise interpore water temperatures to the 200° to 400°C required. Their repeated presence beneath massive sulphide deposits suggests that they are of critical importance in deposit genesis and as exploration guides.

Some previously emphasized characteristics of Precambrian massive sulphide deposits merit comment, because, although prominent in specific deposits or districts, they may not be sufficiently common to warrant an important place in a generalized genetic model. These include:

a) Magma series of enclosing strata; Fox (1976) suggested a correlation between the presence of calc-alkaline (versus tholeiitic) rocks and lead content of deposits. However, this correlation is questionable, because use of major element data to determine magma series in Precambrian rocks is complicated by the combined effects of regional (synvolcanic) alkali depletion, and pervasive footwall alteration accompanying deposit formation.
b) Significance of rhyolite as a major component of the mine sequence. Although felsic volcanic rocks are immediately present at 82% of the massive sulphide deposits in the Canadian Shield, the remaining 18% are in either mafic volcanic rocks, sedimentary strata or epiclastic rocks (Divi *et al.*, 1980). Furthermore, at many deposits the thickness of the footwall rhyolite is limited to less than 100 metres. Rhyolite is an important, but not critical, component of the setting of volcanic-hosted massive sulphide deposits.
c) Relationship to iron-formation and "regional exhalite" (Ridler, 1973). Many early studies of massive sulphide deposits (Stanton, 1960) noted the close proximity of iron-formation. Although this relationship is obvious in the New Brunswick district (Stanton, 1959) it is less evident in the Shield. Oxide facies iron-formation is spatially associated only with the Manitouwadge deposits (Pye, 1960). Sulphide-rich tuffaceous horizons may be followed from some deposits such as near Noranda (Simmons *et al.*, 1973). However, in the Sturgeon Lake district (Franklin *et al.*, 1977) and the Uchi deposit, a sulphidic horizon extends beyond the ore for only a short distance. At Snow Lake and Flin Flon, iron-formation and sulphide-rich tuff are absent. Although a useful ore-guide, iron-formation is not an *essential* part of the depositional framework of massive sulphide deposits.
d) Hangingwall alteration: Boyle (1976) suggested an epigenetic origin for Precambrian massive sulphide deposits, based on inaccurate observation of "symmetrical alteration" about these deposits. Hangingwall alteration is generally only weakly developed and is not evident at many deposits. It consists of sericitized sedimentary rock or weak hydrothermal alteration for a few metres. In a few cases, as at Noranda (Sangster, 1972; Knuckey *et al.*, this volume), deposits are 'stacked' with the alteration pipe continuing above one deposit, and leading to another, stratigraphically higher. Hangingwall alteration is therefore of little consequence in a general genetic model for these deposits. Some caution should be exercised, however, in ignoring hangingwall alteration; the Kuroko deposits exhibit a hangingwall greenschist mineral assemblage (Izawa *et al.*, 1978; Utada *et al.*, 1974) over an area up to 5 km wide within rocks of zeolite facies metamorphic grade. Such a mineralogically defined zone may have a distinctive oxygen isotopic composition (Green and Ohmoto, 1980). For all of the Shield deposits this zone would be obliterated by subsequent metamorphism, but might be discernible by oxygen isotope studies.

Gold Deposits

Unquestionably gold deposits are significantly concentrated in Archean supracrustal rocks throughout the world and much less abundant in similar rocks of Proterozoic or younger age. A clear example is the paucity of deposits in the Proterozoic Flin Flon – Snow Lake greenstone belt, and in the Churchill Province generally, in comparison with several productive Archean belts in the Slave and Superior Provinces. Yet in many other characteristics, including its numerous massive sulphide deposits, the Flin Flon – Snow Lake belt is very similar to Archean belts.

The distinctive stratigraphic, structural and lithologic features of Archean gold deposits may result from a distinctive tectonic regime. The diachronous conglomerate that marks the facies change from volcanic to sedimentary rocks may be due to syn-sedimentary faulting. Sedimentary basins formed by progressive development of fault-bounded troughs may have subsided episodically but slowly, with continuous sediment infilling. Such basins are typical of intracratonic rifting, and many Archean gold districts may represent intravolcanic rift environments. Kirkland Lake, with its rapidly deposited sedimentary rocks (Hyde, 1978) and alkalic volcanic and plutonic rocks, is perhaps the most convincing example. Sedimentary basins at Timmins (Roberts, 1980) and Red Lake (Pirie, 1980; MacGeehan and Hodgson, 1980a) may also have been controlled by intravolcanic rifting. Major rifts in this environment may form the "main-breaks" so commonly associated with the Abitibi belt gold districts.

Other problems of gold metallogeny are the source of gold-rich, but base-metal-poor, hydrothermal fluids, the focussing mechanisms which directed these fluids into relatively restricted zones that form each of the gold districts, and the causes of gold precipitation.

Based on oxygen isotope and minor element data, Kerrich (1980) suggested that the concentration of gold deposits in the Archean may be related to high heat flow that enhanced metamorphic outgassing. He proposed this mechanism to account for the separation of rare and relatively immobile elements such as Au, As, Sb, W, B and Te, from the more abundant and soluble base metals (Cu, Pb, Zn). Metamorphic fluids represent low water/rock ratios and have very low chlorine contents, which favour transport of rare metals relative to base metals. High fluid temperatures and high pH, both favouring gold thiosulphate complexing in the presence of reduced sulphur, may be critical in gold transport without major transport of base metals (J.W. Lydon, pers. commun., 1980). Boyle (1961) also proposed a metamorphic genetic mechanism for the Yellowknife gold deposits.

Oxidation of the fluid may cause deposition. This could account for the association of gold with sulphide- and other facies of iron-formation. The iron reduction documented (Kerrich *et al.*, 1979a) in altered metabasalt adjacent to gold-bearing hydrothermal vein systems and stratiform carbonate units in the Dome mine, may also reflect oxidation of the mineralizing fluids.

This suggests that the world-wide association between gold and iron-formation may be epigenetic, the iron-formation having provided a suitable chemical environment for gold precipitation. However, the iron-formation-hosted Contwoyto Lake, N.W.T., and Agnico-Eagle, Quebec, deposits possess characteristics that are convincingly syngenetic. Other evidence also supports syngenetic concentration of gold in iron-formation. For example, in the Red Lake area sulphide, silicate and especially oxide facies iron-formations are enriched in gold. Oxide facies has a logarithmic mean content of 22 ppb for 19 samples, compared to a regional background value of

3.5 ppb in volcanic and detrital sedimentary rocks (Cowan, 1979; Cowan and Crocket, 1980). The Red Lake silicate facies iron-formation, 40% of which consists of mafic tuff, is, in addition to gold, rich in C, S, Ca, As, Sb, Cr, Ni and W, and variably enriched in Hg (Kerrich *et al.*, 1979b). Although the oxide facies is anomalous in gold relative to most rock types at Red Lake, the ore grades are at least 100 times greater. Derivation of gold from this iron-formation to produce the Red Lake deposits would require unrealistic quantities of iron-formation. The abundance of arsenopyrite in some iron-formation-hosted gold deposits, for example Cullaton, N.W.T. and Geraldton, Ontario suggests either epigenetic emplacement or local highly arsenical seafloor emanations, which appears necessary to account for the Contwoyto Lake type of deposit. Present scanty evidence favours both syngenetic and epigenetic formation of gold deposits in iron-formation. In most cases the possibility of an iron-formation source for epigenetic deposits has not been tested.

Fryer *et al.* (1979) suggested two stages of gold deposition at the Dome mine, Timmins district. The first involved deposition of gold-bearing exhalite consisting of bedded ferroan dolomite with subordinate chert, sulphides, and graphite at interflow horizons in the mafic volcanic rocks. The low Ni and Cr contents of these carbonate layers were not consistent with the hypothesis that they are carbonatized ultramafic rocks. However, Ti/Zr ratios of 120, together with high Mg, Cr and Ni abundances in magnesite-dolomite-quartz-chlorite schists which host banded quartz-albite-tourmaline veins, representing the second stage of gold deposition, are compatible with a primary komatiitic (ultramafic) composition. Roberts and Spiteri (1979) and Kerrich *et al.* (1979a) agree with Fryer *et al* . (1979) in interpreting the layered carbonate ("ankerite") veins as syngenetic ore beds.

In assessing the importance of syngenetic gold deposition at Dome mine, it is significant that fissure-filling veins at Red Lake, a number of which are parallel to foliation, consist of banded carbonate-'chert'-arsenopyrite (MacGeehan and Hodgson, 1980b). Deformation of similar epigenetic veins could produce syngenetic-appearing carbonate units. But conversely, epigenetic features can be produced in syngenetic deposits by deformation and metamorphism. Although the chemical evidence probably rules out formation of the layered carbonate zones by carbonatization of ultramafic rocks, some layered carbonate zones may have formed by carbonatization of basaltic or other rocks (Fyon and Karvinen, 1978). From our limited observations in the Timmins area we are not presently convinced that syngenetic gold deposits are of more than minor importance. Karvinen (1976) also concluded that carbonate rock of probable sedimentary origin occurs as thin lenses within two major carbonate-rich units and forms only a small proportion of all carbonate rock in the area. It may be concluded that some gold deposits are almost certainly syngenetic, while others, perhaps particularly those associated with pervasively carbonatized wallrocks, are likely epigenetic. However, epigenetic versus syngenetic origin for many gold deposits is a controversial problem whose resolution will require much study.

Oxygen isotope and fluid inclusion studies at the Dome and Aunor mines (Kerrich and Fryer, 1979) and fluid inclusion studies at the O'Brien Mine, near Cadillac, Quebec (Krupka *et al.*, 1977), indicate that the ore-forming fluids had temperatures of 300 to 450°C. Moreover, the calculated $\delta^{18}O$ of the mineralizing solution (~10 per mil) implies fluids of metamorphic origin (Kerrich and Fryer, 1979). Europium enrichment and predominance of Fe^{2+} near lode gold deposits indicates a reduced state of the solutions. Kerrich and Fryer therefore suggested that these fluids were generated by metamorphic outgassing at the greenschist-amphibolite transition.

In summary, large volumes of gold-rich fluid may have formed in the lower portions of Archean greenstone belts in rocks undergoing conversion from greenschist to amphibolite grade. The fluid may have been focussed along deep-penetrating faults that initially formed in intravolcanic rift-zones. Most of the gold was probably precipitated in subsurface fractures, and deposition may have occurred both during and after the principal deformation events.

Nickel Deposits

Nickel deposit genesis is related to ultramafic and mafic intrusions, and in some cases ultramafic flows. Immiscible sulphide liquid separation from the melts has been thoroughly studied (Usselman *et al.*, 1979). A better understanding of the fractionation history of the melt is important to explain the Cu/Ni ratio of the resulting sulphide (Duke and Naldrett, 1978) and the distribution of platinum group elements and gold in these deposits (Naldrett *et al.*, 1979).

Three aspects of regional geology are important to the understanding of nickel sulphide deposits; the mechanism of mafic (tholeiitic) and ultramafic (komatiitic) magma generation, the tectonic environment of magma emplacement, and possible sources of sulphur. The timing of sulphur saturation of the melts is also critical.

A large degree of partial melting of upper mantle material is required to form komatiitic liquids (Nesbitt *et al.*, 1979). The generally low stratigraphic position of these rocks within volcanic cycles indicates that deep mantle melting initiated the tectonism event which led to major effusive activity. Each major volcanic event (cycle) could have been initiated by rifting (Nesbitt *et al.*, 1979) of the older sial which surrounds most Archean greenstone belts. Alternatively, the melts must have been sufficiently buoyant to rise through tonalitic crust. Arndt (1977) indicated that at a smaller degree of partial melting, a picritic melt would separate from the mantle residuum and rise as a mantle diapir. The density of this melt at 20 kb, near the liquidus temperature, is 2.95. Extrapolating the density to the lower pressures of 2 to 5 kb, expectable pressures at the basal and central parts of a major volcanic edifice (Franklin *et al.*, 1977), and using a factor calculated from Kushiro *et al.* (1976), reveals a liquid density of 2.7 to 2.8, quite close to measured densities of tonalite (2.6 to 2.8). Thus diapiric emplacement of ultramafic melts is possible. Less dense tholeiitic melts could also rise diapirically.

A significant requirement for nickel-sulphide generation in sill or flow-dominated settings is the presence of sufficient sulphur in the melt to permit sulphur saturation prior to a major amount of silicate crystallization. The position of most productive ultramafic flow-related deposits near the base of non-basal volcanic cycles may be related to the sulphur source problem. Perhaps mantle-derived sulphur was insufficient in these melts, and sulphur assimilation from underlying sulphide-rich strata was essential to oversaturate the melt, forming immiscible sulphide. Most of the significant magmatic nickel-copper deposits occur near sulphide facies iron-formation, as at Thompson (Peredery, this volume), Timmins (Green, 1980; Robinson and Hutchinson, this volume), and in the Cape Smith-Wakeham Bay area (Miller, 1977). Sulphide assimilation is important in ore formation at Pipe mine (Naldrett *et al.*, 1979) and the Langmuir deposit (Green, 1980). Similarly, in the Duluth Complex, sulphur derived from the immediate country rocks (Virginia shale) dominates the immiscible sulphide phase (Mainwaring and Naldrett, 1977). Although local assimilation may not explain all magmatic nickel deposits, the spatial association of sulphur-rich strata such as iron-formation with the productive intrusions and flows may be genetically significant. Further evaluation of the role of sulphur derived from an upper-crustal source requires careful examination of such aspects as platinum group element distribution, S isotopes, S/Se ratios and perhaps other isotopic tracers.

Iron-Formation

The proximity of both Algoma and Superior type iron-formations to volcanic strata has prompted the suggestion of a volcanic-exhalative origin for them (James, 1954, Gross, 1965). The principal problems in applying exhalative concepts to iron-formation genesis are the lack of base metals and the enormous volume of iron. Even a relatively small iron deposit, the Griffiths mine, contains 1.8 x 10^8 tonnes of iron (Gross, 1970) compared with approximately 3 x 10^7 tonnes in very large massive sulphide deposits like Kidd Creek. The Michipicoten siderite-pyrite deposit at Wawa contains almost 1.6 x 10^8 tonnes of iron (Gross, 1970); the total iron content in the Wawa district exceeds 4.2 x 10^8 tonnes. The comparison with Superior-type iron-formations is more striking; the Carol East mine in the Labrador Trough has 5.6 x 10^8 tonnes of iron; the entire Wabush district contains approximately 4.6 x 10^{11} tonnes of iron (Gross, 1970). Iron-formations have much lower base metal contents than massive sulphide deposits. For example, all facies of the Algoma type iron-formation contain averages (447 samples) of 149 ppm Cu and 330 ppm Zn, with the oxide facies (352 samples) containing 96 ppm Cu and 330 ppm Zn (Gross and McLeod, 1980). The averages in all facies of Lake Superior type (201 samples) are 14 ppm Cu and 40 ppm Zn, with only 10 ppm Cu and 10 ppm Zn in the oxide facies.

In comparison with crustal iron content (e.g., 1 km^3 of basalt has over 2 x 10^8 tonnes Fe), the iron content of even the largest Superior type iron-formations is relatively small, and can be derived at a very low recovery factor from a small rock volume. Thus the main genetic problem is not a sufficient source of iron, but to identify a mechanism for transporting large volumes of iron in solution. Although at relatively high temperature (>200°C) iron, in the absence of reduced sulphur, is relatively soluble as a variety of complexes, so too are base metals (Edmond *et al.*, 1979). Hydrothermal activity along the East Pacific Rise consists in part of high temperature hot springs with abundant base metals (Francheteau *et al.*, 1979) but low temperature brines (10 to 25°C) are enriched dominantly in iron and manganese (Edmond *et al.*, 1979). The latter authors suggest that the loss of base metals from these solutions is due to subsurface mixing of high temperature brines with low temperature seawater.

Both types of iron-formation possibly formed from extensive low temperature emanations. Iron is most easily carried in the absence of abundant reduced sulphur (Edmond *et al.*, 1979), apparently only at low concentrations of 0.5 to 1 ppm at 8 to 12°C. Consequently a very large volume of water would be required for both types of iron-formation, from either a long-lived local source or a large number of emanation sites, relative to the time and number of sites required for formation of Cu-Zn massive sulphide deposits. Based on studies of Algoma-type iron-formation at Savant Lake, Shegelski (1978) suggested that iron minerals were precipitated over 100 sq.

km or more, in a distal turbidite basin adjacent to a major volcanic edifice. He envisaged rather steady precipitation of magnetite, and accumulation of iron-rich sediments interrupted by successive turbidity flows. The sedimentary rocks correlate with the upper volcanic cycle, and thus iron precipitation coincided with the waning stages of volcanism. No 'point source' of iron-rich fluid has been identified at Savant Lake, and possibly the excess iron came from either dewatering of the volcanic pile, or circulation of seawater through a large volume of supracrustal strata. An alternate, but not dissimilar mechanism to explain the Algoma-type iron-formation common in volcanic-associated turbidite basins might involve focussing of fluids near the volcanic margin by basin-boundary faults. At Savant Lake, for example, the greenstone-sedimentary rock interface is locally marked by high-energy sedimentary rocks accumulated as submarine fans (Shegelski, 1978) or debris flows, which formed adjacent to normal faults along the margin of a volcanic edifice.

Superior-type iron-formations are confined to Proterozoic sequences (Gross, 1965, 1980) and formed dominantly in shallow basins associated with major volcanically-active cratonic-margin or intracratonic rift systems. Gross (1980, p. 218) postulated that the abundance of Superior-type iron-formations in Proterozoic sequences is related to 'special worldwide tectonic features', and it is relevant that the oldest major, laterally extensive (>1000 km) rift systems are younger than 2500 Ma. Some of these may have been precursors to spreading centres. Modern spreading centres have abundant hydrothermal activity (Rise Project, 1980) resulting from prolific fluid flow through oceanic crust. Such fluids, in Precambrian time, could have leached iron from the strata near the rift axis, emerged along rift-related faults, and formed the major Superior type iron-formations.

Hutchinson (1981a) suggested that Archean oceans contained a very high ferrous iron content, derived from hydrothermal circulation in widespread Archean sequences. The onset of biogenically-induced oxygenation of the atmosphere in the early Proterozoic caused widespread iron precipitation in shallow basins. This mechanism requires no Proterozoic volcanic-related hydrothermal input, although this may have occurred locally.

Superior-type iron-formations formed over at least a 500 Ma interval, however; thus progressive oxygenation of the atmosphere may have been an important, but not dominant process in forming iron deposits. Furthermore, Proterozoic shelf sequences in the Lake Superior and Wollaston districts are similar in age, yet the latter contains no iron-formation, in distinct contrast to the former. A volcanic source thus must have some importance in forming the Superior-type iron-formation.

METALLOGENETIC SUMMARY

Superior Province, with its many deposit types and relatively easy access, has been explored much more thoroughly than either Churchill or Slave provinces. An economically based comparison is therefore misleading and the following discussion is based primarily on deposit types. This includes significant, but little understood, comparisons in terms of the deposit complement of each province and speculative, but, hopefully, challenging explanations for the major differences.

The metallogeny of both Slave and Superior provinces is dominated by greenstone belts of similar lithologic content and age, although the Slave belts may be

50 to 100 Ma younger. Superior Province greenstone belts like the Abitibi are larger, with more varied lithologic types than those in Slave Province, but are similar in deformational style. Both provinces have older basement rocks.

Both provinces contain massive sulphide deposits, stratabound and vein-type gold deposits, porphyry copper occurrences and iron-formations. Slave Province lacks significant nickel occurrences, and has few intrusion-hosted gold occurrences. The rarity of ultramafic rocks and differentiated mafic intrusions explains the former; the lack of trondhjemitic intrusions may explain the latter. Apparently tectonic conditions in Slave Province did not permit the high temperature, deep mantle melting necessary for ultramafic magma generation (Nesbitt *et al.*, 1979), or perhaps the melts were unable to rise through relatively thick crust. Coincidental sialic rifting (Nesbitt *et al.*, 1979) may have been important in Superior Province, and lacking in Slave. Layered gabbroic sills containing Cu-Ni occurrences in the Chibougamau, Matagami, Timagami and Fort Frances areas of Superior Province have no counterparts in Slave Province. Superior Province greenstone belts, with relatively thick, multi-cyclic volcanic complexes may have developed by sequential rifting. Rifting that accompanied the initiation of second and subsequent volcanic cycles may have been accompanied by emplacement of layered sills, as in Proterozoic and Phanerozoic rift zones (such as the Lake Superior area, and east Greenland). Slave Province seems to lack major multi-cyclic greenstone belts, and lack of sequential rifting may explain its paucity of both ultramafic flows and layered gabbroic sills, with accompanying nickel deposits.

Slave Province contains significant syngenetic, iron-formation-related gold deposits. These are also present, but not abundant, in Superior Province.

Some similarities and differences in both tectonic style and complements of mineral deposits are evident in comparing Churchill Province to both Slave and Superior provinces. The Archean portions of Churchill Province probably resemble Superior Province more than Slave in stratigraphy and lithology. The Rankin-Ennadai belt in Churchill is multi-cyclic, with abundant mafic intrusions and some ultramafic flows. The Prince Albert Group, with much thinner, elongate greenstone belts, is relatively simple in comparison with other Archean sequences, but does include numerous ultramafic rocks. The mineral deposit spectrum in the Archean supracrustal rocks of Churchill Province is broad, but mineable examples of many deposit types are rare. Massive sulphide deposits, vein- and stratabound-type gold deposits, nickel deposits and iron-formations are all represented only by occurrences that are at present, except for the Cullaton Lake gold deposit, subeconomic.

The Aphebian portions of southwestern Churchill Province are remarkably different from its Archean portions, as well as from other Archean provinces. The Flin Flon-Snow Lake-Lynn Lake-LaRonge greenstone belts and the Kisseynew metasedimentary gneiss terrane, although sufficiently similar to Archean belts to have been formerly grouped with them (Stockwell *et al.*, 1970), in fact have some significantly different lithologic characteristics, and mineral deposit types. These Aphebian volcanic belts in places have abnormally high ratios of felsic to mafic rock, are essentially lacking in ultramafic-komatiitic rocks, and contain significant clastic redbed sequences in both their volcanic and sedimentary domains. Basement rocks have not been identified within or adjacent to these greenstone belts, in spite of the presence of diapir-like structures as at Snow Lake (Froese and Moore, 1980). The Aphebian volcanic belts contain significant volcanic-related differentiated mafic intrusions, but these are pipe like or irregular, compared with extensive, relatively uniform sills in Superior Province.

The Aphebian greenstone belts contain massive sulphide deposits that are generally similar to those in Archean belts, but they are remarkably deficient in gold deposits and iron-formations. Gabbro-associated nickel-copper deposits in the Aphebian sequences have been important producers, compared with similar occurrences in Superior Province. Porphyry copper occurrences are very minor (Baldwin, 1980a) as they also are, with the prime exception of the McIntyre deposit, in Archean terranes, and ultramafic-related nickel deposits are absent. Sandstone-hosted copper occurrences are present in the Kisseynew metasedimentary terrane, and possibly at the Anglo-Rouyn mine; such deposits are unknown in Archean terranes.

The lack of ultramafic-associated nickel deposits is related to the absence of suitable host lithologies, but the reasons for lack of gold deposits and iron-formation are more perplexing. Rifting may control not only emplacement of ultramafic rocks, but also provide a focussing mechanism for sufficient volumes of gold-rich, base-metal-depleted fluids. Rifts may also engender voluminous circulation of seawater to produce iron-formation. Bailes (pers. commun., 1980) suggested that the lithologic distribution and tectonic style of the Aphebian greenstone belts and of the Wollaston belt were results of subduction related to plate tectonic movement. This conclusion may be supported by the chemical composition of the volcanic rocks (Walford and Franklin, this volume; Stauffer *et al.*, 1975; Bailes, 1980). Subduction, and a compressional regime were probably necessary for generation of the low-Ti, high-Al, calc-alkaline, and island-arc tholeiite magma types represented in the various Aphebian greenstone domains. Contrary to Hutchinson's (1981b) suggestion, rifting, generally associated with spreading, was probably absent from these Aphebian greenstone belts, thereby explaining the absence here of high-Ti melts and ultramafic rocks. The paucity of gold deposits and iron-formation may therefore be related to the lack of a rift-dominated tectonic regime.

Copper occurrences in the Kisseynew rocks resemble sandstone-type deposits in Phanerozoic terranes which formed in continental or stable marginal shelf sedimentary sequences. Oxidizing conditions, permitting copper remobilization, were provided by the subaerial red-bed environment which permitted passage of oxygenated groundwater. An intrastratal metal source (feldspar and ferromagnesian minerals) and biogenically-reduced seawater (?) sulphur led to formation of disseminated copper sulphide deposits. The Aphebian redbeds are molasse type, fluvial sequences indicative of essentially synchronous large scale uplift of the volcanic and sedimentary rocks. Such tectonic activity is typical of subduction-related compressional tectonics, and is very different from the vertical and tensional tectonics which dominated the Archean domains.

The Circum-Ungava geosyncline is very different from Archean terranes. It is in part an intracratonic rift (Dimroth *et al.*, 1970; Baragar and Scoates, 1981) of exceptional length, developed entirely within an Archean craton. Its contrasts in structural style and lithology compared to Archean domains are too numerous to list here, and its entire mineral deposit complement is typical of a rift regime. It has few base metal massive sulphide, and no significant gold deposits, but contains enormous iron-formations and major nickel deposits.

The lack of large base metal massive sulphide deposits is probably due to the absence of sufficiently localized synvolcanic heat sources for formation of hot, metal-rich brines. However, the geosyncline may not have been adequately explored for rift-related massive sulphide deposits of the shale-hosted type, like McArthur River, Australia. The vast amount of iron has been attributed (Gross, 1965) to volcanic exhalation processes. The exceptional volumes of fluid required for iron transport would necessarily be relatively cool, as discussed previously. Submarine rifting might have resulted in the draw-down of large volumes of seawater along the volcanic axis, forming a continuously recharged fluid circulation regime that leached large volumes of iron from the mafic volcanic rocks; expulsion of this fluid could account for major iron-formation development.

Nickel deposits of the Cape Smith-Wakeham Bay belt possibly formed when large volumes of mafic melt were injected through sulphide-facies iron-formation. This could have resulted in sulphurization of the intrusions (Mainwaring and Naldrett, 1977; Ripley and Rao, 1980), forming immiscible nickel-sulphide melts, alternatively sulphides might have been assimilated by the magma during intrusion.

ACKNOWLEDGEMENTS

Many mine geologists have given us underground and surface tours, thus providing first-hand knowledge of many of the deposits discussed here, and deserve particular thanks. Similarly, our colleagues in the Geological Survey of Saskatchewan, Manitoba Mines Branch, Ontario Geological Survey and Ministère des Richesses Naturelles du Québec have provided much appreciated detailed tours of various areas of the Canadian Shield.

We have drawn extensively on the collective expertise of our colleagues at the Geological Survey of Canada. Don Sangster not only critically read several drafts of the paper, but also provided much useful information. Similarly, Roger Eckstrand and Gordon Gross read sections of the paper and provided very useful data and expert comment. Murray Duke, Rod Kirkham and Dave Sinclair provided both descriptive and grade data. Richard Lancaster compiled much of the tabular information and reference material, and assisted in organizing the text, references and diagrams. W.J. Newlands undertook the planimetric measurements of volcanic belts and their major lithologies, and assisted in data compilation. Roy McLeod and Dave Garson provided computer assistance. Nguyen Khan Kim, Steven Green and Marilyn Atkinson assisted with data compilation and drafted the diagrams.

We thank R.H. McMillan, R.W. Hutchinson and C.J. Hodgson for reviewing the manuscript and offering numerous helpful suggestions for its improvement.

REFERENCES

Amukun, S.E., 1977, Geology of the Tashota Area, District of Thunder Bay: Ontario Geological Survey Report 147, 90p.

Anhaeusser, C.R., 1975, Precambrian tectonic environments: Annual Review of Earth and Planetary Sciences, v. 3, p. 31-53.

Armbrust, G.A., 1979a, Geology of the Jogran disseminated Cu-Mo deposit, Algoma District, Ontario; in Pye, E.G., ed., Ontario Geological Survey Geoscience Research Seminar, Abstracts, p. 1-2.

_______________, 1979b, Geology of the Jogran disseminated Cu-Mo deposit, Ryan Township, Ontario: in Pye, E.G., ed., Geoscience Research Grant Program, Summary of Research 1978-1979: Ontario Geological Survey Miscellaneous Paper 87, 152 p.

_______________, 1979c, Geology of the Jogran Disseminated Cu-Mo Deposit, Algoma District, Ontario: Geological Association of Canada/Mineralogical Association of Canada, Program with Abstracts, v. 4, p. 37.

Arndt, N.T., 1977, Thick, layered peridotite-gabbro lava flows in Munro Township: Canadian Journal of Earth Sciences, v. 14, p. 2620-2637.

Arndt, N.T., Naldrett, A.J. and Pyke, D.R., Komatiitic and iron-rich tholeiitic lavas of Munro Township, northeast Ontario: Journal of Petrology, v. 18, p. 319-369.

Ayres, L.D., 1968, Setting Net Lake area, District of Kenora: Ontario Department of Mines Summary of Field Work, 1968, p. 3-6.

_______________, 1977, Importance of stratigraphy in Early Precambrian volcanic terranes: cyclic volcanism at Setting Net Lake, northwestern Ontario: in Baragar, W.R.A., Coleman, L.C., and Hall, J.M., eds., Volcanic Regimes in Canada, Geological Association of Canada Special Paper 16, p. 243-264.

Ayres, L.D. and Averill, S.A., 1974, Setting Net Lake Stock: in University of Manitoba, Centre for Precambrian Studies, Annual Report, 1974, p. 202-205.

Ayres, L.D. and Findlay, D.J. 1976, Precambrian porphyry copper and molybdenum deposits in Ontario and Saskatchewan: Geological Survey of Canada Paper 76-1B, p. 39-41.

Bailes, A.H., 1971, Preliminary compilation of the geology of the Snow Lake-Flin Flon-Sherridon region: Manitoba Department of Mines and Natural Resources Geological Paper 1-71, 27p.

_______________, 1980, Geology of the File Lake area: Manitoba Department of Energy and Mines Geological Report 78-1, 134p.

Bain, D.J. 1980, An Archean volcanogenic Cu-Mo showing west of Contwoyto Lake, Northwest Territories: Canadian Institute of Mining and Metallurgy Bulletin, v. 73, no. 823, p. 34.

Baldwin, D.A., 1976, Evaluation of disseminated base metal environments; First Annual Report, Non-renewable Resources Evaluation Program: Manitoba Department of Mines, Resources and Environmental Management, Open File Report 77/1, p. 62-92.

_______________, 1980a, Porphyritic intrusions and related mineralization in the Flin Flon volcanic belt: Manitoba Department of Energy and Mines, Mineral Resources Division, Economic Geology Report ER79-4, 23p.

_______________, 1980b, Disseminated stratiform base metal mineralization along the contact zone of the Burntwood River metamorphic suite and the Sickle Group: Manitoba Department of Energy and Mines, Mineral Resources Division, Economic Geology Report ER 79-5, 20p.

Baragar, W.R.A., 1966, Geochemistry of the Yellowknife volcanic rocks: Canadian Journal of Earth Sciences, v 3, p. 9-30.

Baragar, W.R.A. and Scoates, R.J.F., 1981, Circum-Superior belt: a Proterozoic plate boundary?: in Kroner, A., ed., Precambrian Plate Tectonics, Developments in Geotectonics Series: New York, Elsevier, p. 297-330.

Barnett, E.S., 1979, Geology of the Agnico-Eagle gold mine, Quebec: M.Sc. Thesis, The University of Western Ontario, London.

Beaty, D.W. and Taylor, H.P., Jr., 1982, Some petrologic and oxygen isotopic relationships in the Amulet mine, Noranda, Quebec, and their bearing on the origin of Archean massive sulphide deposits: Economic Geology, v. 77, p. 95-108.

Bell, K., Blenkinsop, J. and Moore, J.M., 1975, Evidence for a Proterozoic greenstone belt from Snow Lake, Manitoba: Nature, v. 258, p. 689-701.

Bjφrlykke, A., and Sangster, D.F., 1981, An overview of sandstone-lead deposits and their relation to red-bed copper and carbonate-hosted lead-zinc deposits: 75th Anniversary Volume, Economic Geology, p. 179-213.

Blecha, M., 1974, Batchawana area - A possible Precambrian porphyry copper district: Canadian Institute of Mining and Metallurgy Bulletin, August, p. 71-76.

Bostock, H.H., 1968, Gold arsenopyrite-loellingite-pyrrhotite deposits in amphibolite, Itchen Lake-Contwoyto Lake area, District of Mackenzie: in Report of Activities, Part B, Geological Survey of Canada Paper 68-1B, p. 72-76.

______________, 1971, Geological Notes on Aquatuk River Map-Area, Ontario with Emphasis on the Precambrian Rocks: Geological Survey of Canada Paper 70-42, 57p.

Boyle, R.W., 1953, On the colour of black and grey quartz from Yellowknife, Northwest Territories, Canada: American Mineralogist, v. 38, p. 528-535.

______________, 1961, The geology, geochemistry and origin of the gold deposits of the Yellowknife district: Geological Survey of Canada Memoir 310, 193 p.

______________, 1976, Mineralization processes in Archean greenstone and sedimentary belts: Geological Survey of Canada Paper 75-15, 45p.

______________, 1979, The geochemistry of gold and its deposits: Geological Survey of Canada Bulletin 280, 584p.

Breakey, A.R., 1975, A mineralogical study of the gold-quartz lenses in the Campbell Shear, Con Mine, Yellowknife, N.W.T.: M.Sc. Thesis, McGill University, Montreal.

Brown, C.E.G. and Dadson, A.S., 1953, Geology of the Giant Yellowknife Mine: Canadian Institute of Mining and Metallurgy Bulletin, March 1953, p. 151-168.

Bureau, S., Guha, J., and Cimon, J., 1979, Breccia zone associated with Archean Porphyry Copper Deposits in the Chibougamau Region: Geological Association of Canada/ Mineralogical Association of Canada, Program with Abstracts, v. 4, p. 41.

Cameron, E.M. and Garrels, R.M., 1980, Geochemical compositions of some Precambrian shales from the Canadian Shield: Chemical Geology, v. 28, p. 181-197.

Campbell, I.H., Gorton, M.P., Scott, S.D., Franklin, J.M., and Thurston, P.C., 1981, Rare Earth Elements as a guide to massive sulfide exploration: Geological Association of Canada /Mineralogical Association of Canada/Canadian Geophysical Union, Program with Abstracts, v. 6, p. A-8.

Campbell, N., 1948, The West Bay fault, Yellowknife: Canadian Institute of Mining and Metallurgy, Jubilee Volume, Structural Geology of Canadian Ore Deposits, p. 244-259.

Cathles, L.M., 1978, Hydrodynamic constraints on the formation of Kuroko Deposits: Mining Geology, v. 28, p. 257-265.

Chandler, F.W., 1978a, Geological environment of Aphebian red beds of the north half of Richmond Gulf, New Quebec: in Current Research, Part A, Geological Survey of Canada Paper 78-1A, p. 107-110.

______________, 1978b, Geology of part of the Wollaston Lake fold belt, north Wollaston Lake, Saskatchewan: Geological Survey of Canada Bulletin 277, 25 p.

Chandler, F.W. and Schwarz, E.J., 1980, Tectonics of the Richmond Gulf Area, Northern Quebec - A Hypothesis; in Current Research, Part C, Geological Survey of Canada Paper 80 - IC, p. 59-68.

Chute, M.E. and Ayres, L.D., 1977, Missi Island volcanic centre, Saskatchewan: Report of Activities, Geological Survey of Canada, Paper 77-1B, p. 29-31.

Cimon, J., 1973, Possibility of an Archean porphyry copper in Quebec (Abst.): Canadian Mining Journal, v. 94, no. 4, p. 97.

________________, 1979, Relative age of emplacement of porphyry-copper mineralization in the local stratigraphy of Chibougamau Region, Quebec: Geological Association of Canada/Mineralogical Association of Canada, Program with Abstracts, v. 4, p. 43.

Clark, A.L., 1971, Stratabound copper sulfide in the Precambrian Belt Supergroup, northern Idaho and northwest Montana: International Association on Genesis of Ore Deposits, Tokyo-Kyota, 1970, Proceedings, Special Issue 3, p. 261-267.

Clark, G.S., Bald, R., and Ayres, L.D., 1981, Geochronology of orthogneiss adjacent to the Archean Lake of the Woods greenstone belt, northwestern Ontario: a possible basement complex: Canadian Journal of Earth Sciences, v. 18, p. 94-102.

Clay, A.N., 1979, An Archean limestone and volcanic association from a Rhodesian greenstone belt. Evidence of volcanogenic control in a sedimentary system: in Stocklmayer, V.R., ed., Annals of the Rhodesia Geological Survey, Vol. IV, 1978, Rhodesia Geological Survey, p. 55-63.

Clifford, P.M., 1969, Geology of the Western Lake St. Joseph Area: Ontario Department of Mines Geological Report 70, 61p.

Coad, P.R., 1977, Nickel sulphide deposits associated with ultramafic rocks of the Abitibi Belt and economic potential of mafic-ultramafic intrusions: Ontario Geological Survey Open File Report 5232, 105p.

Coats, C.J.A., Quirke, T.T., Bell, C.K., Cranstone, D.A. and Campbell, F.H.A., 1972, Geology and Mineral Deposits of the Flin Flon, Lynn Lake, and Thompson Areas, Manitoba and the Churchill-Superior Front of the Western Precambrian Shield: Guidebook to Field Excursion A31-C31, 24th International Geological Congress, Montreal, Canada.

Coats, C.J.A., Stockford, H.R. and Buchan, R., 1979, Geology of the Maskwa West nickel deposit, Manitoba: The Canadian Mineralogist, v. 17, p. 309-318.

Colvine, A.C., 1978, Early Precambrian porphyry deposits: in Milne, V.G., White, O.L., Barlow, R.B., and Robertson, J.A., eds., Summary of Field Work 1978 by the Ontario Geological Survey: Ontario Geological Survey Miscellaneous Paper 82, 235 p.

Colvine, A.C. and McCarter, P., 1977, Geology and mineralization of the Lateral Lake Stock, District of Kenora: in Milne, V.G., White, O.L., Barlow, R.B. and Roberton, J.A., eds., Summary of Field Work 1977 by the Geological Branch: Ontario Geological Survey Miscellaneous Paper 75, 208 p.

Colvine, A.C. and Sutherland, I.G., 1979, Archean porphyry deposits in Ontario: Geological Association of Canada/Mineralogical Association of Canada, Program with Abstracts, v. 4, p. 44.

Coombe, W., 1975, La Ronge-Wollaston base metals project: in Christopher, J.E. and Macdonald, R., eds., Summary of Investigations 1975, Department of Mineral Resources, Province of Saskatchewan, p. 101-109.

________________, 1976, La Ronge base metals project: in Christopher, J.E., and Macdonald, R., eds., Summary of Investigations 1976: Department of Mineral Resources, Province of Saskatchewan, p. 99-103.

________________, 1977, La Ronge-Wollaston Belts base metals project: George Hills, Johnson and Kaz Lakes and Geikie River areas: in Christopher, J.E. and Macdonald, R., eds., Summary of Investigations 1977: Department of Mineral Resources, Province of Saskatchewan, p. 85-104.

Cowan, P., 1979, The gold content of interflow metasedimentary rocks in the Red Lake area; M.Sc. Thesis, McMaster University, Hamilton.

Cowan, P. and Crocket, J.H., 1980, The gold content of interflow metasedimentary rocks from the Red Lake area, Ontario – A Preliminary Evaluation: in Current Research, Part B, Geological Survey of Canada Paper 80-1B, p. 129-133.

Cranstone, D.A. and Turek, A., 1976, Geological and geochronological relationships of the Thompson nickel belt, Manitoba: Canadian Journal of Earth Sciences, v. 13, p. 1058-1069.

Cumming, G.L. and Richards, J.R., 1975, Ore lead isotope ratios in a continuously changing earth; Earth and Planetary Science Letters, v. 28, p. 155-171.

Cumming, G.L. and Tsong, F., 1975, Variations in the isotopic composition of volatilized lead and the age of the Western Granodiorite, Yellowknife, Northwest Territories: Canadian Journal of Earth Sciences, v. 12, p. 558-573.

Davidson, A., 1972a, Granite studies in the Slave Province: in Report of Activities, Geological Survey of Canada Paper 72-1A, p. 109-115.

______________, 1972b, The Churchill Province: in Price, R.A. and Douglas, R.J.W., eds., Variations in Tectonic Styles in Canada: Geological Association of Canada Special Paper 11, p. 381-433.

Davies, J.C., 1965, High Lake-Rush Bay area: Ontario Department of Mines, Geology Report GR41, 57 p.

Davies, J.F., 1980, K-Rb patterns in an Archean "porphyry-type" copper deposit, Timmins, Ontario, Canada: Economic Geology, v. 75, p. 760-764.

Davies, J.F. and Luhta, L.E., 1978, An Archean "porphyry-type" disseminated copper deposit, Timmins, Ontario: Economic Geology, v. 73, p. 383-396.

Davies, J.F. and Whitehead, R.E.S., 1980, Further immobile element data from altered volcanic rocks, Timmins mining area, Ontario: Canadian Journal of Earth Sciences, v. 17, p. 419-423.

Davies, J.F., Grant, R.W.E., and Whitehead, R.E.S., 1979, Immobile trace elements and Archean volcanic stratigraphy in the Timmins mining area, Ontario: Canadian Journal of Earth Sciences, v. 16, p. 305-311.

Davies, J.F., Whitehead, R.E.S., Cameron, R.A., and Duff, D., 1980, Regional and local patterns of CO_2-K-Rb-As alteration: A guide to gold mineralization in the Timmins area (Abst.); Canadian Institute of Mining and Metallurgy Bulletin, v. 73, no. 820, p. 63.

Dimroth, E., Baragar, W.R.A., Bergeron, R., and Jackson, G.D., 1970, The filling of the Circum-Ungava geosyncline: in Baer, A.E., ed., Precambrian Basins and Geosynclines of the Canadian Shield: Geological Survey of Canada Paper 70-40, p. 45-142.

Divi, S.R., Thorpe, R.I. and Franklin, J.M., 1980, Use of discriminant analysis to evaluate compositional controls of stratiform massive sulphide deposits in Canada: Geological Survey of Canada Paper 79-20, 23 p.

Douglas, R.J.W., ed., 1970, Geology and Economic Minerals of Canada: Geological Survey of Canada, Economic Geology Report No. 1, 838 p.

Downes, M.J., 1980, The regional setting of the Kerr Addison mine: in The Canadian Institute of Mining and Metallurgy, Geology Division CIM Gold Symposium and Field Excursion, Val d'Or-Kirkland Lake-Timmins, 105 p.

Drury, S.A., 1977, Structures induced by granite diapirs in the Archaean greenstone belt at Yellowknife, Canada: implications for Archean geotectonics: Journal of Geology, v. 85, p. 345-358.

______________, 1979, Rare-earth and other trace element data bearing on the origin of Archaean granitic rocks from Yellowknife, Northwest Territories: Canadian Journal of Earth Sciences, v. 16, p. 809-815.

Dugas, J., 1970, Metallic mineralization in part of the Labrador Trough: Quebec Department of Natural Resources Special Paper 5, 23p.

Duke, J.M. and Naldrett, A.J., 1978, A numerical model of the fractionation of olivine and molten sulfide from komatiite magma: Earth and Planetary Science Letters, v. 39, p. 255-266.

Eade, K.E., 1974, Geology of Kognak River area, District of Keewatin, Northwest Territories: Geological Survey of Canada Memoir 377, 66p.

Eckstrand, O.R., 1973, Spinifex, ultramafic flows and nickel deposits in the Abitibi orogenic belt: in Volcanism and Volcanic Rocks: Geological Survey of Canada Open File 164, p. 111-128.

_______________, 1975, Nickel potential of the Prince Albert Group, N.W.T.; in Report of Activities, Part A, Geological Survey of Canada Paper 75-1A, p. 753-755.

Eckstrand, O.R. and Miller, A.R., 1973, A study of sulphide and spinifex distribution of the Texmont nickel deposit: in Report of Activities, Geological Survey of Canada Paper 73-1A, p. 125-126.

Economic Geology Division, 1980, Non-hydrocarbon mineral resource potential of parts of northern Canada; Geological Survey of Canada Open File 716, 377 p.

Edmond, J.M., Measures, C., Magnum, B., Grant, B., Szlater, F.R., Collier, R., Hudson, A., Gordon, L.I. and Corliss, J.B., 1979, On the formation of metal-rich deposits at ridge crests: Earth and Planetary Science Letters, v. 46, p. 19-30.

Elbers, F.J., 1976, Calc-alkaline plutonism, volcanism and related hydrothermal mineralization in the Superior Province of northeastern Manitoba; Canadian Institute of Mining and Metallurgy Bulletin, v. 69, no. 771, p. 1-18.

Fenwick, F.G., 1976, Geology of the Finlayson Lake area, District of Rainy River: Ontario Division of Mines, Geoscience Report 145, 86p.

Findlay, D.J., 1975, Lang Lake, an Archean porphyry copper-molybdenum deposit: in University of Manitoba, Centre for Precambrian Studies, 1975 Annual Report, p. 103-107.

Findlay, D.J. and Ayres, L.D., 1977, Lang Lake – an early Precambrian porphyry copper-molybdenum deposit in northwestern Ontario: Report of Activities, Part B, Paper 77-1B, p. 25-28.

Forsyth, D.A. 1981, Characteristics of the western Quebec seismic zone: Canadian Journal of Earth Sciences, v. 18, p. 103-119.

Forsythe, L.H., 1972, Anglo Rouyn copper mine, Ore Bay, Lac La Ronge, Saskatchewan: Geological Society of America Bulletin, v. 85, p. 3405-3414.

Fox, J.S., 1976, Some comments on the volcanic stratigraphy and economic potential of the West Amisk Lake area, Saskatchewan: Saskatchewan Research Council, Geology Division, Circular 9.

Francheteau, J., Needham, H., Choukroune, P., Siguret, M., Ballard, R., Fox, P., Normark, W., Carranzi, A., Cordoba, D., Guerrero, V., Rangin, C., Bougalt, P., and Hekinian, R., 1979, Massive deep sea sulfide ore deposits discovered by submersible on the East Pacific Rise, Project RITA, 21°N: Nature, v. 377, p. 523.

Franklin, J.M., 1975, Metallogenic relations at Sturgeon Lake: Geotraverse Workshop, University of Toronto, p. 39-1 to 39-25.

_______________, 1976, Role of laharic breccia in genesis of volcanogenic massive sulphide deposits: in Report of Activities, Part A, Geological Survey of Canada Paper 76-1A, p. 293-380.

_______________, 1977, Stratigraphic and Chemical Relations South Sturgeon Lake Area: in Proceedings of the 1977 Geotraverse Conference: The Precambrian Research Group, University of Toronto, p. 118-120.

_______________, 1978, Metallogeny of the South Sturgeon Lake area, Northwestern Ontario: Geological Association of Canada/Mineralogical Association of Canada/Geological Society of America, Program with Abstracts, v. 3, p. 403.

Franklin, J.M., Gibb, W., Poulsen, K.H., and Severin, P., 1977, Archean Metallogeny and Stratigraphy of the South Sturgeon Lake Area: Institute on Lake Superior Geology, 23rd Annual Meeting, Guidebook, 73p.

Franklin, J.M., Kasarda, J., and Poulsen, K.H., 1975, Petrology and chemistry of the alteration zone of the Mattabi massive sulphide deposit: Economic Geology, v. 70, p. 63-79.

Franklin, J.M., Lydon, J.W. and Sangster, D.F., 1981, Volcanic-associated massive sulphide deposits: 75th Anniversary Volume, Economic Geology, p. 485-627.

Friske, P., Poulsen, K.H., Franklin, J.M., 1979, The Beidelman Bay Porphyry Copper-Gold Occurrences, Northwestern Ontario: Geological Association of Canada/Mineralogical Association of Canada, Program with Abstracts, v. 4, p. 51.

Frith, Rosaline, Frith, R.A., and Doig, R., 1977, The geochronology of the granitic rocks along the Bear-Slave Structural Province boundary, northwest Canadian Shield: Canadian Journal of Earth Sciences, v. 14, p. 1356-1373.

Frith, R.A. and Roscoe, S.M., 1980, Tectonic setting and sulphide deposits of the Hackett River belt, Slave Province: Canadian Institute of Mining and Metallurgy Bull., v. 73, no. 815, p. 143-153.

Froese, E. and Moore, J.M., 1980, Metamorphism in the Snow Lake area, Manitoba: Geological Survey of Canada Paper 78-27, 16p.

Fryer, B.J., 1972, Age determinations in the Circum-Ungava Geosyncline and the evolution of Precambrian banded iron formations: Canadian Journal of Earth Sciences, v. 9, p. 652-663.

Fryer, B.J., Kerrich, R., Hutchinson, R.W., Pierce, M.G. and Rogers, D.S., 1979, Archean precious-metal hydrothermal systems, Dome mine, Abitibi greenstone belt. I. Patterns of alteration and metal distribution: Canadian Journal of Earth Sciences, v. 16, p. 421-439.

Fyfe, W.S. and Kerrich, R., 1979, Geochemistry and field relations of lode gold deposits in felsic igneous intrusions (Abst.): in Pye, E.G., ed., Geoscience Research Seminar, Abstracts, Ontario Geological Survey, 23 p.

Fyon, J.A. and Karvinen, W.O., 1978, Volcanic environment of gold mineralization in the Timmins area: in Milne, V.G., White, O.L., Barlow, R.B., and Robertson, J.A., eds., Summary of Field Work 1978 by the Ontario Geological Survey: Ontario Geological Survey Miscellaneous Paper 82, 235 p.

Fyson, W.K. and Frith, R.A. 1979, Regional deformation and emplacement of granitoid plutons in the Hackett River greenstone belt, Slave Province, Northwest Territories: Canadian Journal of Earth Sciences, v. 16, p. 1187-1195.

Gale, G.H., Baldwin, D.A., and Koo, J., 1980, A geological evaluation of Precambrian massive sulphide deposit potential in Manitoba: Manitoba Department of Energy and Mines, Mineral Resources Division, Economic Geology Report ER79-1.

Gélinas, L., Brooks, C., Perrault, G., Carignan, J., Trudel, P. and Grasso, F., 1977, Chemostratigraphic divisions within the Abitibi volcanic belt, Rouyn-Noranda district, Quebec: in Baragar, W.R.A., Coleman, L.C. and Hall, J.M., eds., Volcanic Regimes in Canada: Geological Association of Canada Special Paper 16, p. 265-295.

Gibson, H.L., 1979, Geology of the Amulet Rhyolite Formation, Turcotte Lake Section, Noranda Area, Quebec: M.Sc. Thesis, Carleton University, Ottawa.

Gilders, R., 1975, Geology and textural relationships of the Soucy Lake massive sulfide deposit: B.Sc. Thesis, University of Ottawa.

Gill, J.W., 1976, The Takiyuak Metavolcanic Belt: Geology, Geochemistry, and Mineralization: Ph.D. Thesis, Carleton University, Ottawa.

Goldie, R.J., 1979, Metamorphism of the Flavrian and Powell plutons, Noranda area, Quebec: Journal of Petrology, v. 20, p. 227-238.

Goldie, R.J. and Kotila, B., 1979, The Don Rouyn Porphyry Copper Deposit, Noranda, Quebec: Geological Association of Canada/Mineralogical Association of Canada, Program with Abstracts, v. 4, p. 54.

Goldie, R., Kotila, B. and Seward, D., 1979, The Don Rouyn Mine: An Archean porphyry copper deposit near Noranda, Quebec: Economic Geology, v. 74, p. 1680-1684.

Goodwin, A.M., 1962, Structure, stratigraphy and origin of iron formations, Michipicoten area, Algoma District, Ontario, Canada: Geological Society of America Bulletin, v. 73, p. 561-586.

—————, 1964, Geochemical studies at the Helen Iron Range: Economic Geology, v. 59, p. 684-718.

—————, 1968, Evolution of the Canadian Shield: Geological Association of Canada Proceedings, v. 19, p. 1-14.

—————, 1977, Archean volcanism in Superior Province, Canadian Shield: in Baragar, W.R.A., Coleman, L.C. and Hall, J.M., eds., Volcanic Regimes in Canada: Geological Association of Canada Special Paper 16, p. 205-241.

Goodwin, A.M., Ambrose, J.W., Ayres, L.D., Clifford, P.M., *et al.,* 1972, The Superior Province: in Price, R.A. and Douglas, R.J.W., eds., Variations in Tectonic Styles in Canada: Geological Association of Canada Special Paper 11, p. 527-624.

Goodwin, A.M., Monster, J. and Thode, H.G., 1976, Carbon and sulfur isotope abundances in Archean iron-formations and Early Precambrian life: Economic Geology, v. 71, p. 870-891.

Grabowski, G.P.B., 1975, The Geology and Geochemistry of the Atikokan Iron Mine: B.Sc. Thesis, Lakehead University, Thunder Bay.

Green, A.H., 1980, Komatiitic Nickel Deposits of the Abitibi Belt, Canada, with Special Reference to Langmuir: in Glover, J.E. and Groves, D.I., eds., Second International Archaean Symposium, Extended Abstracts, Perth, Western Australia, p. 60-61.

Green, D.C., 1968, Precambrian geology and geochronology of the Yellowknife area, N.W.T.: Ph.D. Thesis, University of Alberta, Edmonton, 166p.

Green, D.C. and Baadsgaard, H., 1971, Temporal evolution and petrogenesis of an Archean crustal segment at Yellowknife, N.W.T., Canada: Journal of Petrology, v. 12, p. 177-217.

Green, G.R. and Ohmoto, H., 1980, Oxygen isotope and alteration zonation in volcanic rocks from around the Fukazawa Kuroko deposit, Japan and its implication for mineral exploration (Abst.): Geological Society of America, Abstracts with Program v. 12, p. 436.

Green, G.R., Ohmoto, H., Date, J. and Takahashi, T., 1980, Oxygen isotope and alteration zonation in volcanic rocks from around the Fukazawa Kuroko deposit, Japan, and its implication for mineral exploration (Abst.): Geological Society of America, Abstracts with Program, v. 12, p. 436.

Gresham, J.J. and Loftus-Hills, G.D., 1980, Geology of the Kambalda Nickel Field: in Glover, J.E. and Groves, D.I., eds., Second International Archaean Symposium, Extended Abstracts, Perth, Western Australia, p. 60-61.

Griffis, A.T., 1962, A geological study of the McIntyre Mine: Canadian Institute of Mining and Metallurgy Bulletin, February, p. 76-83.

Gross, G.A., 1965, Geology of iron deposits in Canada - Volume 1, General Geology and Evaluation of Iron Deposits: Geological Survey of Canada, Economic Geology Report No. 22, 181p.

—————, 1968, Geology of iron deposits in Canada - Volume 3, Iron ranges of the Labrador Geosyncline: Economic Geology Report No. 22, 179p.

—————, 1970, Iron ore deposits of Canada and the West Indies: in Survey of World Iron Ore Resources, United Nations, New York, p. 237-269.

—————, 1977, Metallogenic evolution of the Canadian Shield; in Sidorenko, A.V., ed., Correlation of the Precambrian, Volume 2, p. 774-792.

—————, 1980, A classification of iron formations based on depositional environments: The Canadian Mineralogist, v. 18, p. 215-222.

Gross, G.A. and McLeod, C.R., 1980, A preliminary assessment of the chemical composition of iron formations in Canada: The Canadian Mineralogist, v. 18, p. 223-229.

Groves, D.I., Barrett, F.M. and McQueen, K.G., 1979, The relative roles of magmatic segregation, volcanic exhalation and regional metamorphism in the generation of volcanic-associated nickel ores of Western Australia: The Canadian Mineralogist, v. 17, pt 2, p. 319-336.

Guber, A.L. and Ohmoto, H., 1978, Deep sea environment of Kuroko formation as indicated by the benthic foraminifera from the Hokuroku District, Japan: Mining Geology, v. 28, p. 245-255.

Harper, C.T., 1975, Northern iron ore exploration: in Christopher, J.E., and Macdonald, R., eds., Summary of Investigations 1975: Department of Mineral Resources, Province of Saskatchewan, p. 94-100.

_______________, 1976, Northern iron ore exploration: in Christopher, J.E., and Macdonald, R., eds., Summary of Investigations 1976: Department of Mineral Resources, Province of Saskatchewan, p. 104-114.

Hart, S.R. and Davis, G.L., 1969, Zircon U-Pb and whole-rock Rb-Sr ages and early crustal development near Rainy Lake, Ontario: Geological Society of America Bulletin, v. 80, p. 595-616.

Harvey, J.D. and Hinzer, J.B., 1981, Geology of the Lyon Lake ore deposits, Noranda Mines Limited, Sturgeon Lake area, Ontario: Canadian Institute of Mining and Metallurgy Bulletin, v. 74, no. 833, p. 77-84.

Harwood, T.A., 1949, The lead-zinc deposits of Richmond Gulf: B.Sc. Thesis, University of Toronto, Toronto.

Haskins, R.A. and Stephenson, J.F., 1974, Geology and mineralization of western Oxford Lake and Carrot River (53L-13; 63I-16S, 9N): in Manitoba Department of Mines, Resources and Environmental Management, Mineral Resources Division, Summary of Geological Field Work, 1974, Geological Paper 2/74, p. 35-36.

Heaton, T.H.E. and Sheppard, S.M.F., 1976, Hydrogen and oxygen isotope evidence for sea-water-hydrothermal alteration and ore deposition, Troodos complex, Cyprus: in Volcanic Processes in Ore Genesis: Geological Society of London, Special Publication 7, p. 42-57.

Henderson, J.B., 1975a, Sedimentological studies of the Yellowknife Supergroup in the Slave Structural Province: in Report of Activities, Geological Survey of Canada Paper 75-1A, p. 325-330.

_______________, 1975b, Archean stromatolites in the northern Slave Province, Northwest Territories, Canada: Canadian Journal of Earth Sciences, v. 12, p. 1619-1630.

Henderson, J.F. and Brown, I.C., 1966, Geology and structure of the Yellowknife Greenstone Belt, District of Mackenzie: Geological Survey of Canada Bull. 141.

Henderson, J.R. and Tippett, C.R., 1980, Foxe Fold Belt in eastern Baffin Island, District of Franklin: in Current Research, Part A, Geological Survey of Canada Paper 80-1A, p. 147-152.

Henderson, J.R., Shaw, D., Mazuraki, M., Henderson, M., Green, R., and Brisbin, D., 1979, Geology of part of the Foxe Fold Belt, central Baffin Island, District of Franklin: in Current Research, Part A, Geological Survey of Canada Paper 79-1A, p. 95-99.

Heywood, W.W. and Davidson, A., 1969, Geology of Benjamin Lake map area, District of Mackenzie: Geological Survey of Canada Memoir 361.

Hodgson, C.J. and Lydon, J.W., 1977, Geological setting of volcanogenic massive sulphide deposits and active hydrothermal systems: Some implications for exploration: Canadian Institute of Mining and Metallurgy Bulletin, v. 70, no. 786, p. 95-106.

Hoeve, J. and Sibbald, T.I.I., 1978, On the genesis of Rabbit Lake and other unconformity-type uranium deposits in northern Saskatchewan, Canada: Economic Geology, v. 73, p. 1450-1473.

Hurst, R.W., Bridgwater, D., Collerson, K.D. and Wetherill, G.W., 1975, 3600 M.Y. Rb-Sr ages from very early Archean gneisses from Saglek Bay, Labrador: Earth and Planetary Science Letters, v. 27, p. 393-403.

Hutchinson, R.W., 1973, Volcanogenic sulfide deposits and their metallogenic significance: Economic Geology, v. 68, p. 1223-1246.

______________, 1981a, Mineral deposits as guides to supracrustal evolution: in O'Connell R.J. and Fyfe, W.S., eds., Evolution of the Earth: American Geophysical Union - Geological Society of America, Geodynamics Series, v. 5, p. 120-141.

______________, 1981b, Metallogenic evolution and Precambrian tectonics: in Kroner, A., ed., Precambrian Plate Tectonics: Amsterdam Elsevier Scientific Publishing Company, p. 733-759.

______________, 1981c, Contributed discussions: in Goldie, Ray and Bottrill, T.J., Seminar on Sea-Floor Hydrothermal Systems: Geoscience Canada, v. 8, p. 99.

Hutchinson, R.W., Ridler, R.H., and Suffel, G.G., 1971, Metallogenic relationships in the Abitibi Belt, Canada: a model for Archean metallogeny: Canadian Institute of Mining and Metallurgy Bulletin, v. 64, p. 106-115.

Hyde, R.S., 1978, Sedimentology, volcanology, stratigraphy and tectonic setting of the Archean Timiskaming Group, Abitibi greenstone belt, Northeastern Ontario, Canada: Ph.D. Thesis, McMaster University, Hamilton.

Ishihara, S., 1978, Metallogenesis in the Japanese Island-Arc System: Journal of the Geological Society, v. 135, pt. 4, p. 389-406.

Issigonis, M.J., 1980, Occurrence of disseminated gold deposits in porphyries in Archean Abitibi belt, northwest Quebec, Canada: Institution of Mining and Metallurgy, Transactions Section B (Applied Earth Science), v. 89, p. 157-158.

Izawa, E., Yoshida, T., and Saito, R., 1978, Geochemical Characteristics of Hydrothermal Alteration around the Fukazawa Kuroko Deposit, Akita, Japan: Mining Geology, v. 28, p. 325-335.

Jackson, G.D. and Taylor, F.C., 1972, Correlation of major Aphebian rock units in the northern Canadian Shield: Canadian Journal of Earth Sciences, v. 9, p. 1650-1669.

Jak, A., James, R., and Whitehead, R.E., 1978, Petrography and geochemistry of the volcanic carbonatized basaltic and komatiitic metavolcanic rocks, Pamour Mine, Timmins, Ontario (brief summary of research in progress): Canadian Geophysical Bulletin, v. 31, p. 125.

James, H.L., 1954, Sedimentary facies of iron-formation: Economic Geology, v. 49, p. 235-293.

James, R.S. and Whitehead, R.E., 1978, Petrology and geochemistry of peridotitic komatiites from the Goose Lake Formation, Timmins, Ontario (brief summary of research in progress): Canadian Geophysical Bulletin, v. 31, p. 125-126.

Jolliffe, A.W., 1966, Stratigraphy of the Steeprock Group, Steep Rock Lake, Ontario: in Goodwin, A.M., ed., Precambrian Symposium The Relationship of Mineralization to Precambrian Stratigraphy in Certain Mining Areas of Ontario and Quebec: Geological Association of Canada Special Paper 3, p. 68-75.

Jolly, W.T., 1975, Subdivision of the Archean lavas of the Abitibi area, Canada from Fe-Mg-Ni-Cr relations: Earth and Planetary Science Letters, v. 27, p. 200-210.

Karup-Møller, S. and Brummer, J.J., 1970, The George Lake Zinc Deposit, Wollaston Lake Area, Northeastern Saskatchewan: Economic Geology, v. 65, p. 862-874.

______________, 1971, Geology and sulphide deposits of the Bird River claim group, southeastern Manitoba: in Turnock, A.C., 1972, Geoscience Studies in Manitoba: Geological Association of Canada Special Paper 9, p. 143-153.

Karvinen, W.O., 1976, Distribution of carbonate-rich rocks, porphyries and gold deposits, Timmins area: in Summary of Field Work 1976 by the Geological Branch: Ontario Ministry of Natural Resources, Division of Mines Miscellaneous Paper 67.

_______________, 1978, The Porcupine camp – A model for gold exploration in the Archean: Canadian Mining Journal, September, p. 48-53.

_______________, 1980, Geology and evolution of gold deposits, Timmins area, Ontario: in R.G., Roberts, ed., Genesis of Archean, Volcanic-Hosted Gold Deposits: Ontario Geological Survey Open File Report 5293, 387p.

Kerrich, R., 1980, Separation of metals in Archean lode gold deposits: implications of low water/rock source regions, and high terrestrial heat flow (Abst.): Canadian Institute of Mining and Metallurgy Bulletin, August, p. 64.

Kerrich, R. and Fryer, B.J., 1979, Archean precious-metal hydrothermal systems, Dome mine, Abitibi greenstone belt II. REE and oxygen isotope relations: Canadian Journal of Earth Sciences, v. 16, p. 440-458.

Kerrich, R., Hutchinson, R.W. and Hodder, R.W., 1979a, Field relations and geochemistry of Au, Ni and Cr deposits in ultramafic-mafic volcanic rocks: in Pye, E.G., ed., Geoscience Research Grant Program, Summary of Research 1978-1979: Ontario Geological Survey Miscellaneous Paper 87, 152p.

Kerrich, R., Robinson, D., Hodder, R.W. and Hutchinson, R.W., 1979b, Field relations and geochemistry of Au, Ni, and Cr deposits in ultramafic-mafic volcanic rocks: in Pye, E.G., ed., Geoscience Research Seminar Abstracts: Ontario Geological Survey, 23p.

Kirkham, R.V., 1972, Geology of copper and molybdenum deposits: Geological Survey of Canada Paper 71-1A, p. 82-87.

_______________, 1973, Geology of copper and molybdenum deposits in Canada: in Report of Activities, Geological Survey of Canada Paper 73-1A, p. 126-127.

_______________, 1974a, A Synopsis of Canadian Stratiform Copper Deposits in Sedimentary Sequences: in Bartholome, P., ed., Gisements Stratiformes et Provinces Cupriféres: Societé Géologique de Belgique, Liege, p. 367-382.

_______________, 1974b, Geology of copper and molybdenum deposits in Canada: in Report of Activities, Geological Survey of Canada Paper 74-1A, p. 377-379.

_______________, 1979, Copper in Iron Formation: in Current Research, Part B, Geological Survey of Canada Paper 79-1B, p. 17-22.

Kirkham, R.V. and Thorpe, R.I., 1973, Studies of gold-copper deposits suggest red metal may be used as guide to gold: The Northern Miner, November 29, p. 55,57.

Koo, J. and Mossman, D.J., 1975, Origin and Metamorphism of the Flin Flon Stratabound Cu-Zn Sulfide Deposit, Saskatchewan and Manitoba: Economic Geology, v. 70, p. 48-62.

Krogh, T.E. and Davis, G.L., 1971, Zircon U-Pb ages of Archean metavolcanic rocks in the Canadian Shield: Annual Report Geophysical Laboratory, Carnegie Institution, Washington Yearbook, 1970-71, p. 241-242.

Krogh, T.E., Davis, G.L., Ermanovics, I., and Harris, N.B.W., 1976, U-Pb isotopic ages of zircons from the Berens Block and English River gneiss belt: Proceedings of the 1976 Geotraverse Conference, Precambrian Research Group, University of Toronto, March 15-16, p. 46.

Krogh, T.E. and Gibbons, W., 1978, U-Pb isotopic ages of basement and supracrustal rocks in the Point Lake area of the Slave Structural Province, Canada: Geological Association of Canada/Mineralogical Association of Canada, Program with Abstracts, v. 3, p. 438.

Krupka, K.M., Ohmoto, H., and Wickham, F.E., 1977, A new technique in neutron activation analysis of Na/K ratios of fluid inclusions and its application to the gold-quartz veins at the O'Brien mine, Quebec, Canada: Canadian Journal of Earth Sciences, v. 14, p. 2760-2770.

Kushiro, I., Yoder, H.S., Jr., and Mysen, B.O., 1976, Viscosities of basalt and andesite melts at high pressures: Journal of Geophysical Research, v. 81, p. 6351-6356.

Kustra, C.R., 1979, Exploration possibilities in Ontario: Canadian Mining Journal, v. 100, no. 5, p. 73.

Lambert, M.B., 1976, The Back River volcanic complex, District of Mackenzie: in Report of Activities, Part A, Geological Survey of Canada Paper 76-1A, p. 363-365.

————————, 1977, Anatomy of a greenstone belt-Slave Province, N.W.T.: in Baragar, W.R.A., Coleman, L.C., and Hall, J.M., eds., Volcanic Regimes in Canada: Geological Association of Canada Special Paper 16, p. 331-340.

————————, 1978a, The Beaulieu River volcanic belts, Slave Province, N.W.T. (brief summary of research in progress): Canadian Geophysical Bulletin, v. 31, p. 128.

————————, 1978b, The Back River volcanic complex - a cauldron subsidence structure of Archean age: Geological Survey of Canada Paper 78-1A, p. 153-157.

Lambert, M.B. and Henderson, J.B., 1980, A uranium-lead age of zircons from volcanics and sediments of the Back River volcanic complex, eastern Slave Province, District of Mackenzie: Current Research, Geological Survey of Canada, Paper 80-1C, p. 239-242.

Lambert, R. St. J., 1980, Thermal History of the Earth: The Layered Heterogeneous Model: in Glover, J.E. and Groves, D.I., eds., Second International Archaean Symposium, Extended Abstracts, Perth, Western Australia, p. 92-93.

Langford, F.F., 1955, Geology of the Gripp Lake Area, Thunder Bay District, Ontario: in Sixty-Seventh Annual Report, Ontario Department of Mines, v. 67, pt. 3, 22p.

Laporte, P.J., Gibbins, W.A., Hurdle, E.J., Lord, C., Padgham, W.A. and Seaton, J.B., 1978, Mineral Industry Report 1975, N.W.T.: Department of Indian and Northern Affairs, EGS 1978-5, 187p.

Latulippe, M., 1980, An overview of the geology of gold prospects and developments in N.W. Quebec: in The Canadian Institute of Mining and Metallurgy, Geology Division CIM Gold Symposium and Field Excursion, Val d'Or-Kirkland Lake-Timmins, 105p.

Lawrence, L.J., 1978, Porphyry type gold mineralization in shoshonite at Vunda, Fiji: Proceedings of the Australasian Institute of Mining and Metallurgy, No. 268, p. 21-31.

Lawson, A.C., 1913, The Archean geology of Rainy Lake re-studied: Geological Survey of Canada Memoir 40, 111p.

Lusk, J., 1976, A possible volcanic-exhalative origin for lenticular nickel sulphide deposits of volcanic association, with special reference to those in Western Australia: Canadian Journal of Earth Sciences, v. 15, p. 451-458.

Lydon, J.W., 1977, The significance of metal ratios of hydrothermal ore deposits: Ph.D. Thesis, Queen's University, Kingston, Ontario.

Lydon, J.W., 1981, Contributed discussions: in Goldie, Ray and Bottrill, T.J., Seminar on Sea-Floor Hydrothermal Systems: Geoscience Canada, v. 8, p. 101-103.

Lydon, J.W. and Hodgson, C.J., in prep., Hydrothermal systems and genetic models for volcanogenic massive sulphide deposits.

Macdonald, R. and Broughton, P., 1980, Geological Map of Saskatchewan, Provisional Edition: Saskatchewan Geological Survey, Sheet 1 (North).

MacGeehan, P.J. and Hodgson, C.J., 1980a, Relationship of gold mineralization to volcanic and metamorphic features in the area of Campbell and Dickenson mines, Red Lake District: in Genesis of Archean, Volcanic-Hosted Gold Deposits, Ontario Geological Survey Open File Report 5293, p. 212-242.

————————, 1980b, The morphology, mineralogy, chemistry and zonation of gold-bearing zones in the Campbell and Dickenson mines, Red Lake district, Ontario (Abst.): Canadian Institute of Mining and Metallurgy Bulletin, August, p. 64.

MacGeehan, P.J. and MacLean, W.H., 1980, An Archaean Sub-Seafloor Geothermal System, 'Calc-alkali' Trends, and Massive Sulphide Genesis: Nature, v. 286, no. 5775, p. 767-771.

McMillan, R.H., 1972, Petrology, geochemistry and wallrock alteration at Opemiska - a vein copper deposit crosscutting a layered Archean ultramafic-mafic sill: Ph.D. Thesis, University of Western Ontario, London.

————————, 1977, Uranium in Canada: Canadian Petroleum Geology Bulletin, v. 25, no. 6, p. 1222-1249.

MacQuarrie, R., 1980, Absolute age of the Flin Flon deposit: Canadian Institute of Mining and Metallurgy Bulletin, v. 73, no. 820, p. 53.

Mainwaring, P.R. and Naldrett, A.J., 1977, Country rock assimilation and the genesis of Cu-Ni sulfides in the Water Hen Intrusion, Duluth Complex, Minnesota: Economic Geology, v. 72, p. 1269-1284.

McGlynn, J.C. and Henderson, J.B., 1970, Archean volcanism and sedimentation in the Slave Structural Province: in Baer, A.J., Sumposium on basins and geosynclines of the Canadian Shield, Geological Survey of Canada Paper 70-40, p. 31-44.

Miller, A.R., 1977, Petrology and geochemistry of the 2-3 ultramafic sill and related rocks, Cape Smith-Wakeham Bay Fold Belt, Quebec: Ph.D; Thesis, University of Western Ontario, London.

______________, 1978, Uranium reconnaissance of the northern Richmond Gulf area, New Quebec: in Current Research, Part A, Geological Survey of Canada Paper 78-1A, p. 111-117.

Milne, V.G., 1972, Geology of the Kukatush-Sewell Lake area, District of Sudbury: Ontario Division of Mines, Geological Report 97, 116p.

Money, P.L., Baer, A.J., Scott, B.P. and Wallis, R.H., 1970, The Wollaston Lake belt, Saskatchewan, Manitoba, Northwest Territories: in Baer, A.J., ed., Symposium on Basins and Geosynclines of the Canadian Shield: Geological Survey of Canada Paper 70-40, p. 170-200.

Moorbath, S., 1975, Evolution of Precambrian Crust from Strontium Isotopic Evidence: Nature, v. 254, p. 395-397.

Moore, G.W., 1959, Iron in Keewatin: Western Miner and Oil Review, October, p. 132.

Moore, J.M., 1977, Orogenic volcanism in the Proterozoic of Canada: in Baragar, W.R.A., Coleman, L.C., and Hall, J.M., eds., Volcanic Regimes in Canada: Geological Association of Canada Special Paper 16, p. 127-148.

Morton, P., 1979, Volcanic Stratigraphy in the Shebandowan Ni-Cu Mine Area, Ontario: in Current Research, Part B, Geological Survey of Canada Paper 79-1B, p. 39-43.

Muir, J.E. and Comba, C.D.A., 1979, The Dundonald deposit: an example of volcanic-type nickel-sulfide mineralization: The Canadian Mineralogist, v. 17, pt. 2, p. 351-359.

Muir, T.L., 1979, Discrimination between extrusive and intrusive Archean ultramafic rocks in the Shaw Dome area using selected major and trace elements: Canadian Journal of Earth Sciences, v. 16, p. 80-90.

Naldrett, A.J., 1973, Nickel sulphide deposits - their classification and genesis, with special emphasis on deposits of volcanic association: Canadian Institute of Mining and Metallurgy Transactions, v. 76, p. 183-201.

Naldrett, A.J. and Duke, J.M.,1980, Platinum Metals in Magmatic Sulfide Ores: Science, v. 208, p. 1417-1424.

Naldrett, A.J. and Macdonald, A.J., 1980, Tectonic Setting of Ni-Cu Sulphide Ores: Their Importance in Genesis and Exploration: in Strangway, D.W., ed., The Continental Crust and its Mineral Deposits: Geological Association of Canada Special Paper 20, p. 633-658.

Naldrett, A.J., Hoffman, E.C., Green, A.H., Chou, C.L., Naldrett, S.R. and Alcock, R.A., 1979, The composition of Ni-sulfide ores with particular reference to their content of PGE and Au: The Canadian Mineralogist, v. 17, p. 403-415.

Nesbitt, R.W., Sun, S.-S., and Purvis, A.C., 1979, Komatiites: geochemistry and genesis: The Canadian Mineralogist, v. 17, pt. 2, p. 165-186.

Nielsen, P.E., 1974, The Geochemistry of the Lyon Lake-Claw Lake Sulphide Bearing Graphitic Shale, Sturgeon Lake Area, Ontario: B.Sc. Thesis, Lakehead University, Thunder Bay, Ontario.

Nikic, Z., Baadsgaard, H., Folinsbee, R.E. and Leech, A.P., 1975, Diatreme containing boulders of 3030 m.y. old tonalite gneiss, Con Mine, Yellowknife, Slave Craton (Abst.): Geological Society of America, Abstracts with Program, vol. 7, no. 7, p. 1213-1214.

Nunes, P.D. and Pyke, D.R., 1980, Geochronology of the Abitibi metavolcanic belt, Timmins-Matachewan area - Progress Report: in Pye, E.G., ed., Summary of Geochronology Studies 1977-1979: Ontario Geological Survey Miscellaneous Paper 92, 45p.

Nunes, P.D. and Thurston, P.C., 1980, Two hundred and twenty million years of Archean evolution: a zircon U-Pb age stratigraphic study of the Uchi-Confederation Lakes greenstone belt, northwestern Ontario: Canadian Journal of Earth Sciences, v. 17, p. 710-721.

Nunes, P.D. and Wood, J., 1980, Geochronology of the North Spirit Lake area, District of Kenora - Progress Report: in Pye, E.G., ed., Summary of Geochronology Studies 1977-1979: Ontario Geological Survey Miscellaneous Paper 92, 45p.

Ostic, R.G., Russell, R.D. and Stanton, R.L., 1967, Additional measurements of the isotopic composition of lead from stratiform deposits: Canadian Journal of Earth Sciences, v. 4, p. 245-269.

Parks, T., 1949, A report on the geology of the Nastapoka Group of sediments (Hudson's Bay) with its contained lead and zinc-bearing strata: B.Sc. Thesis, University of Toronto, Toronto.

Patterson, G.C., 1980, Geology of the Kapkichi Lake ultramafic-mafic bodies and related Cu-Ni mineralization of Pickle Lake, Ontario: Ph.D. Thesis, Carleton University, Ottawa, Ontario.

Pearson, D.E., 1977, The geology of the Mudjatik area: Saskatchewan Dept. of Mineral Resources, Report No. 166, 31 p.

Pearson, W.N. and Clark, A.H., 1979, The Minto Copper Deposit, Yukon Territory; A Metamorphosed Orebody in the Yukon Crystalline Terrane: Economic Geology, v. 74, p. 1577-1599.

Pedora, J.M., 1976, Mineralization of the High Lake pluton and adjacent country rocks; M.Sc. Thesis, University of Manitoba, Winnipeg.

Peredery, W.V., 1979, Relationship of ultramafic amphibolites to metavolcanic rocks and serpentinites in the Thompson Belt, Manitoba: The Canadian Mineralogist, v. 17, p. 187-200.

Peterman, Z.E., Goldich, S.S., Hedge, C.E. and Yardley, D.H., 1972, Geochronology of the Rainy Lake Region, Minnesota-Ontario: in Studies in Mineralogy and Precambrian Geology: Geological Society of America Memoir 135, p. 193-215.

Pinsent, R.H., 1977, NM7504-B, The Lynn Lake Ni-Cu Deposits: in Canada-Manitoba Non-Renewable Resources Evaluation Program, Manitoba Department of Mines, and Department of Energy, Mines and Resources Canada, NREP 2nd Annual Report 1976-1977, Part A - Technical Report, p. 33-42.

_______________, in press, Nickel-Copper Mineralization in the Lynn Lake Gabbro: Manitoba Resources Division, Economic Geology Report ER79-3.

Pirie, J., 1980, Regional geological setting of gold mineralization in the Red Lake area, NW Ontario: in Genesis of Archean, Volcanic-Hosted Gold Deposits, Ontario Geological Survey Open File Report 5293, p. 303-358.

Pirie, J. and Mackasey, W.O., 1978, Preliminary examination of regional metamorphism in parts of Quetico metasedimentary belt, Superior Province, Ontario: in Metamorphism in the Canadian Shield: Geological Survey of Canada Paper 78-10, p. 37-47.

Pollock, G.D., Sinclair, I.G.L., Warburton, A.F. and Wierzbicki, V., 1972, The Uchi orebody - a massive sulphide deposit in an Archean siliceous volcanic environment: in Mineral Deposits, 24th International Geological Congress, Section 4, 553p.

Poulsen, K.H., Borradaile, G.J. and Kehlenbeck, M.M., 1980, An inverted Archean succession at Rainy Lake, Ontario: Canadian Journal of Earth Sciences, v. 17, p. 1358-1369.

Poulsen, K.H. and Franklin J.M., 1981, Copper and gold mineralization in an Archean trondhjemitic intrusion Sturgeon Lake, Ontario: Current Research, Geological Survey of Canada Paper 81-1A, p. 9-14.

Price, D., 1977, Tour Guide, Flin Flon-Snow Lake Geology: Canadian Institute of Mining and Metallurgy, 55p.

Pye, E.G., 1960, Geology of the Manitouwadge area: Ontario Department of Mines 66th Annual Report, v. LXVI, pt. 8, 1957, p. 113.

Pyke, D.R., 1975, On the relationship of gold mineralization and ultramafic volcanic rocks in the Timmins area: Ontario Division Mines Miscellaneous Paper 62, 23p.

______________, 1976, On the relationship between gold mineralization and ultramafic volcanic rocks in the Timmins area, northeastern Ontario: Canadian Institute of Mining and Metallurgy Bulletin, v. 69, no. 773, p. 79-87.

Pyke, D.R. and Middleton, R.S., 1970, Distribution and characteristics of the sulphide ores of the Timmins area: Ontario Department of Mines Miscellaneous Paper 41, p. 13.

Pyke, D.R., Naldrett, A.J. and Eckstrand, O.R., 1973, Archean ultramafic flows in Munro Township, Ontario: Geological Society of America Bulletin, v. 84, p. 955-978.

Ray, G.E., 1978, Reconnaissance geology: Wollaston Lake (West) area: in Christopher, J.E. and Macdonald, R., eds., Summary of Investigations 1978: Saskatchewan Geological Survey Paper 78-10.

Ray, G.E. and Wanless, R.K., 1980, The age and geological history of the Wollaston, Peter Lake, and Rottenstone domains in northern Saskatchewan: Canadian Journal of Earth Sciences, v. 17, p. 333-347.

Ridler, R.H., 1966, Petrographic study of a Crow Lake ultrabasic sill, Keewatin volcanic belt, Northwestern Ontario: M.Sc. Thesis, University of Toronto, Toronto.

______________, 1973, Volcanic stratigraphy and metallogeny; Rankin Inlet-Ennadai Belt, District of Keewatin: in Report of Activities, Part A, Geological Survey of Canada Paper 73-1A, p. 165-174.

______________, 1974, Volcanic stratigraphy and metallogeny of the Kaminak Group, Spi Lake area, District of Keewatin: in Report of Activities, Part A, Geological Survey of Canada Paper 74-1A, p. 181-185.

Ridler, R.H. and Shilts, W.W., 1974a, Mineral potential of the Rankin Inlet, Ennadai Belt: Canadian Mining Journal, v. 95, no. 7, p. 32-42.

______________, 1974b, Exploration for Archean polymetallic sulphide deposits in permafrost terrains: An integrated geological/geochemical technique; Kaminak Lake area, District of Keewatin: Geol. Survey of Canada Paper 73-34, 33p.

Rigg, D.M. and Helmstaedt, H., 1980, Relations between structures and gold mineralization in the Campbell Red Lake and Dickenson Mines, Red Lake District, Ontario: in Genesis of Archean, Volcanic-Hosted Gold Deposits, Ontario Geological Survey Open File Report 5293, p. 243-279.

Riley, R.A., King, H.L. and Kustra, C.R., 1971, Mineral exploration targets of northwestern Ontario; Ontario Department of Mines and Northern Affairs Miscellaneous Paper 47, 72p.

Ripley, E.M. and Rao, B.V., 1980, Oxygen isotopic studies of the Dunka Road Cu-Ni Deposit, Duluth Complex, Minnesota: in Geological Society of America, Abstracts with Program, v. 12, no. 7, p. 510.

RISE Project Group, 1980, East Pacific rise: Hot springs and geophysical experiments: Science, v. 207, no. 4438, p. 1421-1432.

Riverin, G. and Hodgson, C.J., 1980, Wall-rock alteration at the Millenbach Cu-Zn mine, Noranda, Quebec: Economic Geology, v. 75, p. 424-444.

Roberts, R.G., 1980, The volcanic-tectonic setting of gold deposits in the Timmins area: in Genesis of Archean, Volcanic-Hosted Gold Deposits, Ontario Geological Survey Open File Report 5293, p. 359-387.

Roberts, R.G. and Reardon, E.J., 1978, Alteration and ore-forming processes at Mattagami Lake mine, Quebec: Canadian Journal of Earth Sciences, v. 15, p. 1-21.

Roberts, R.G. and Spiteri, J.G., 1979, Volcanic-tectonic setting of gold deposits in the Timmins District: in Pye, E.G., ed., Geoscience Research Grant Program, Summary of Research 1978-1979: Ontario Geological Survey Miscellaneous Paper 87, 152p.

Roscoe, S.M., 1965, Geochemical and isotope studies, Noranda and Matagami areas: Transactions of the Canadian Institute of Mining and Metallurgy, v. 58, p. 965-971.

Russell, J.K., 1981, Metamorphism of the Thompson nickel belt gneisses: Paint Lake, Manitoba: Canadian Journal of Earth Sciences, v. 18, p. 191-209.

Sangster, D.F., 1972, Precambrian volcanogenic massive sulphide deposits in Canada: A Review: Geological Survey of Canada Paper 72-22, 44p.

__________, 1978, Isotopic studies of ore leads of the circum-Kisseynew volcanic belt of Manitoba and Saskatchewan: Canadian Journal of Earth Sciences, v. 15, p. 1112-1121.

Sangster, D.F. and Kirkham, R.V., 1974, Disseminated base metal mineralization along the Wollaston Lake fold belt, Saskatchewan: Report of Activities, Geological Survey of Canada Paper 74-1A, p. 143-144.

Sangster, D.F. and Scott, S.D., 1976, Precambrian stratabound massive Cu-Zn-Pb sulphide ores of North America: in Wolf, K.H., ed., Handbook of Strata-Bound and Stratiform Ore Deposits, v. 5: New York, Elsevier, p. 129-222.

Sauvé, P. and Bergeron, R., 1965, Gerido Lake - Thevenet Lake area, New Quebec: Quebec Department of Natural Resources, Report No. 104.

Schau, M., 1975, Komatiitic and other ultramafic rocks in the Prince Albert Group, Hayes River region, N.W.T.: in Report of Activities, Part A, Geological Survey of Canada Paper 75-1A, p. 363-367.

__________, 1977, "Komatiites" and quartzites in the Archean Prince Albert Group: in Baragar, W.R.A., Coleman, L.C., and Hall, J.M., eds., Volcanic Regimes in Canada: Geological Association of Canada Special Paper 16, p. 341-354.

Schimann, K., 1978, Geology of the Wakeham Bay area, eastern end of the Camp Smith Belt, New Quebec: Ph.D. Thesis, University of Alberta, 476p.

Schwarz, E.J. and Fujiwara, Y., 1977, Komatiitic basalts from the Proterozoic Cape Smith Range in northern Quebec, Canada: in Baragar, W.R.A., Coleman, L.C., and Hall, J.M., eds., Volcanic Regimes in Canada: Geological Association of Canada, Special Paper 16, p. 193-201.

Schwerdtner, W.M., Stone, D., Osadetz, K., Morgan, J., and Stott, G.M., 1979, Granitoid complexes and the Archean tectonic record in the southern part of northwestern Ontario: Canadian Journal of Earth Sciences, v. 16, p. 1965-1977.

Scoates, R.J.F., 1977, Structure and textures of Paleohelikian (?) volcanic rocks of the Fox River area, northeastern Manitoba: Geological Association of Canada/Mineralogical Association of Canada, Program with Abstracts, v. 2, p. 47.

__________, 1981, Volcanic rocks of the Fox River belt, N.W. Manitoba: Mineral Resources Division, Manitoba Department of Mines, Geological Paper GP-81.

Scoates, R.J.F., Macek, J.J., and Russell, J.K., 1977, Thompson Nickel Belt Project: in Report of Field Activities, 1977, Manitoba Department of Mines, p. 47-53.

Scoates, R.J.F. and Weber, W., 1977, Fox River Regional Correlation – Part A: Dafoe-Fox River: in Report of Field Activities 1977, Manitoba Department of Mines, p. 62-63.

Sharpe, J.I., 1965, Field Relations of Matagami Sulphide Masses Bearing on their distribution in Time and Space; Canadian Institute of Mining and Metallurgy, v. 58, no. 64, p. 951-964.

__________, 1968, Geology and sulphide deposits of the Matagami area: Quebec Department of Natural Resources, Geological Report 137, 122p.

Shegelski, R.J., 1978, Stratigraphy and geochemistry of Archean Iron Formations in the Sturgeon Lake-Savant Lake Greenstone Terrain, N.W. Ontario: Ph.D. Thesis, University of Toronto, Toronto.

_______________, 1979, Geologic evolution of the Shebandowan-Thunder Bay Greenstone Terrain: Geological Association of Canada/Mineralogical Association of Canada, Program with Abstracts, v. 4, p. 78.

Shemeliuk, V.A.B., 1976, Compositional variations in an iron-rich sequence of carbonate and siliceous sediments, Turquetil Lake area, N.W.T.: M.Sc. Thesis, University of Manitoba, 97p.

Shklanka, R., 1972, Geology of the Steep Rock Lake area, District of Rainy River: Ontario Geological Survey, Geological Report 97, 114p.

Simmons, B.P. and the Geological Staff, Lake Dufault Div., Falconbridge Copper Limited, 1973, Geology of the Millenbach Massive Sulphide Deposit, Noranda, Quebec: Canadian Institute of Mining and Metallurgy, v. 66, no. 739, p. 67-78.

Sinclair, W.D., 1979, Copper-molybdenum occurrences of the Matachewan area, Ontario: in Current Research, Part A, Geological Survey of Canada, Paper 79-1A.

Smith, I.E.M., 1980, Geochemical Evolution in the Blake River Group, Abitibi Greenstone Belt, Superior Province: Canadian Journal of Earth Sciences, v. 12, p. 1292.

Sopuck, V.J., Lavin, O.P. and Nichol, I., 1980, Lithogeochemistry as a guide to identifying favourable areas for the discovery of volcanogenic massive sulphide deposits: Canadian Institute of Mining and Metallurgy Bulletin, v. 73, no. 823, p. 152-166.

Spence, C.D. and de Rosen-Spence, A.F., 1975, The place of sulfide mineralization in the volcanic sequence at Noranda, Quebec: Economic Geology, v. 70, p. 90-101.

Stacey, J.S., Delevaux, M.E. and Ulrych, T.J., 1969, Some triple-filament lead isotope ratio measurements and an absolute growth curve for single-stage leads: Earth and Planetary Science Letters, v. 6, p. 15-25.

Stanton, R.L., 1959, Mineralogical features and possible mode of emplacement of the Brunswick Mining and Smelting orebodies, Gloucester County, New Brunswick: in Symposium on the Genesis of Massive Sulphide Deposits: Canadian Institute of Mining and Metallurgy Bulletin, v. 52, p. 631-643.

_______________, 1960: General features of the conformable pyritic orebodies, Part I - Field association; Part II - Mineralogy: Canadian Institute of Mining and Metallurgy Bulletin, v. 53, p. 24-29, 66-74.

Stauffer, M.R., Mukherjee, A.C. and Koo, J., 1975, The Amisk Group: An Aphebian island arc: Canadian Journal of Earth Sciences, v. 12, p. 2021-2035.

Stewart, J.W., 1980, Gold mines of Manitoba: Province of Manitoba, Department of Energy and Mines, Education Series ES80-1, 16p.

Stockwell, C.H., 1933, Great Slave Lake-Coppermine River Area, N.W.T.: Geological Survey of Canada, Annual Report 1932, Part C, p. 37-63.

Stockwell, C.H. *et al.,* 1970, Geology of the Canadian Shield: in Douglas, R.J.W., ed., Geology and Economic Minerals of Canada: Geological Survey of Canada, Economic Geology Report No. 1, p. 43-150.

Studemeister, P. and Colvine, A.C., 1978, Geology and mineralization of the Gutcher Lake Stock: in Milne, V.G., White, O.L., Barlow, R.B., and Robertson, J.A., eds., Summary of Field Work 1978 by the Ontario Geological Survey: Ontario Geological Survey Miscellaneous Paper 82, 235 p.

Studemeister, P.A., and Colvine, A.C., 1979, Copper, gold and molybdenum distribution in a volcanic-plutonic complex, Abotossaway Township, District of Algoma, Ontario: Geological Association of Canada/Mineralogical Association of Canada, Program with Abstracts, v. 4, p. 81.

Sutherland, I.G., 1978, Molybdenum mineralization in the Slate Falls area: in Milne, V.G., White, O.L., Barlow, R.B., and Robertson, J.A., eds., Summary of Field Work 1978 by the Ontario Geological Survey: Ontario Geological Survey Miscellaneous Paper 82, 235 p.

Tatsumi, T., ed., 1970, Volcanism and ore genesis: Tokyo, University of Tokyo Press, 448p.

Tatsumi, T. and Clark, L.A., 1972, Chemical composition of acid volcanic rocks genetically related to formation of the Kuroko deposits: Journal of Geological Society of Japan, v. 78, p. 191-201.

Thomson, J.E., 1943, Geology of McGarry and McVittie Townships, Larder Lake area: Fiftieth Annual Report of the Ontario Department of Mines, v. 50, no. 7, 1941.

Thorpe, R.I., 1971, Comments on rock ages in the Yellowknife area, District of Mackenzie: Geological Survey of Canada Paper 71-1, Part B, p. 76-79.

Thurston, P.C., 1976, Geology of the North Onaman area, NTS 42L/SW, Thunder Bay District, Ontario: Ontario Division of Mines Open File Report 5197, p. 118.

________________, 1979, Subaerial volcanism in the Archean Uchi-Confederation volcanic belt: Pye, E.G., ed., in Geoscience Research Seminar Abstracts: Ontario Geological Survey, 23p.

Thurston, P.C. and Breaks, F.W., 1978, Metamorphic and tectonic evolution of the Uchi-English River Subprovince: in Metamorphism in the Canadian Shield: Geological Survey of Canada Paper 78-10, p. 49-62.

Tihor, L.A. and Crocket, J.H., 1977, Gold distribution in the Kirkland Lake-Larder Lake area, with emphasis on Kerr Addison-type ore deposits - a progress report: in Report of Activities, Geological Survey of Canada Paper 77-1A.

Tihor, S.L., 1978, The mineralogical composition of the carbonate rocks of the Kirkland Lake-Larder Lake gold camp: M.Sc. Thesis, McMaster University, Hamilton.

Tourtelot, Elizabeth B. and Vine, James D., 1976, Copper deposits in Sedimentary and Volcanogenic Rocks: United States Geological Survey Professional Paper 907C, 34p.

Trowell, N.F., 1974, Geology of the Bell Lake-Sturgeon Lake area, Districts of Kenora and Thunder Bay: Ontario Division of Mines Geological Report GR 114, 67p.

Usselman, T.M., Hodge, D.S., Naldrett, A.J. and Campbell, I.H., 1979, Physical constraints on the characteristics of nickel-sulfide ore in ultramafic lavas: The Canadian Mineralogist, v. 17, p. 361-372.

Utada, M., Minato, H., Ishikawa, T. and Yoshizaki, Y., 1974, The Alteration Zones Surrounding Kuroko-Type Deposits in the Nishi-Aizu District, Fukushima Prefecture, with Emphasis on the Analcime Zone as an Indicator in Exploration for Ore Deposits: in Ishihara, S., ed., Geology of Kuroko Deposits: Mining Geology Special Issue 6, p. 291-302.

Van Hees, E.H.P., 1979, Auriferous ankerite vein genesis in the Aunor mine, Timmins, Ontario: M.Sc. Thesis, University of Western Ontario, London.

Verpaelst, P., Brooks, C. and Franconi, A., 1980, The 2.5 Ga Duxbury massif, Quebec: a remobilized piece of pre-3.0 Ga sialic basement (?): Canadian Journal of Earth Sciences, v. 17, p. 1-18.

Walker, R.R., Matulich, A., Amos, A.C., Watkins, J.J. and Mannard, G.W., 1975, The geology of the Kidd Creek Mine: Economic Geology, v. 70, p. 80-89.

Walker, S. and Cregheur, P., 1980, The Chadbourne breccia pipe (Abst.): Canadian Institute of Mining and Metallurgy Bulletin, August, p. 60.

Wallster, D.E., 1979, Geology reviewed for the Dion Lake Region with specific reference to a possible Aphebian porphyry-type copper deposit: B.Sc. Thesis, University of Western Ontario, London.

Wanless, R.K., 1969, Isotopic age map of Canada: Geological Survey of Canada Map No. 1256A.

________________, 1979, Geochronology of Archean rocks of the Churchill Province: Geological Association of Canada/Mineralogical Association of Canada, Program with Abstracts, v. 4, p. 85.

Wanless, R.K. and Loveridge, W.D., 1972, Rubidium-strontium isochron age studies, report 1: Geological Survey of Canada Paper 72-23.

West, G.F. and Mareschal, J-C., 1979, A model for Archean tectonism. Part 1. The thermal conditions: Canadian Journal of Earth Sciences, v. 16, p. 1942-1950.

Whitehead, R.E.S., Cameron, R.A. and Davies, J.F., 1979, The relation of gold deposits to concentrations of CO_2 in volcanic rocks: in Pye, E.G., ed., Summary of Research 1978-1979, Geoscience Research Grant Program: Ontario Geological Survey Miscellaneous Paper 87, p. 108-110.

Wilson, M.E., 1962, Rouyn-Beauchastel Map-areas, Quebec: Geological Survey of Canada Memoir 315, 140p.

Wilson, H.D.B., Morrice, M.G., and Ziehlke, D.V., 1974, Archean Continents: Geoscience Canada, v. 1, no. 3, p. 12-20.

Wolfe, W.J., 1976, Geochemical and biogeochemical exploration of molybdenum-copper mineralization at Setting Net Lake, District of Kenora (Patricia Portion): Ontario Division of Mines, Geoscience Report 156, 28p.

Wood, J., 1977, Geology of the North Spirit Lake area, District of Kenora (Patricia Portion): Ontario Division of Mines, Ministry of Natural Resources, Geoscience Report 150, 60p.

Wyllie, P.J., Huang, W.L., Stern, C.R. and Maalφe, S., 1976, Granitic magmas: possible and impossible sources, water contents and crystallization sequences: Canadian Journal of Earth Science, v. 13, p. 1007-1019.

MS received August 1981;
Accepted for publication January 1982

Precambrian Sulphide Deposits, H.S. Robinson Memorial Volume, edited by R.W. Hutchinson, C.D. Spence and J.M. Franklin, Geological Association of Canada Special Paper 25, 1982

GEOLOGY OF THE GRENVILLE PROVINCE, AND REGIONAL METALLOGENESIS OF THE GRENVILLE SUPERGROUP

A.L. Sangster
Sulpetro Minerals Ltd., P.O. Box 1123, Truro, Nova Scotia B2N 5H1

James Bourne
Département des Sciences de la Terre, Université du Québec à Montréal, Montréal, Québec H3C 3P8

ABSTRACT

The Grenville Province is a high grade metamorphic terrain which constitutes the southeastern margin of the exposed Canadian Shield. Here, rocks of Archean to Helikian age have been metamorphosed several times. They now have an average K-Ar age of 950 Ma. In southeastern Ontario and southwestern Quebec, rocks of Archean to Aphebian age form an older basement to rocks of the Grenville Supergroup of Helikian age. Rocks of the Grenville Supergroup are the most studied and contain the greatest concentration of mineral occurrences in the Grenville Province.

The Grenville Supergroup is composed of a thick basal series of meta-volcanic rocks overlain by a thick series of intercalated marbles and arkosic metasedimentary rocks. Metamorphism is only locally below amphibolite facies. The Grenville Supergroup is cut by rocks of at least two intrusive suites and is overlain by younger sedimentary rocks, the Flinton Group, which are in turn metamorphosed.

Oxide and sulphide facies iron-formation, pyritic fahlbands and exhalative volcanogenic base metal sulphide deposits are related to the early Grenvillian volcanism. Stratabound zinc deposits formed within more distal shelf carbonate rocks laterally equivalent to the volcanic lithofacies. The first major period of intrusion and metamorphism generated magmatic Cu-Ni occurrences in gabbroic rocks and magnetite skarn deposits in limestones near the granitic intrusions. A second period of intrusion and metamorphism formed large, low grade magmatic magnetite-ilmenite deposits in the mafic intrusions and minor intrusive contact skarns. Iron skarns were also produced by metamorphic reaction along marble-amphibolite contacts. Post-Flinton metamorphism locally produced gold-quartz and quartz-carbonate-base metal sulphosalt veins. Late calcite-barite-fluorite and calcite-barite-galena veins cut both Grenville and Paleozoic cover rocks, and are related to Cretaceous graben development in the Ottawa valley.

RÉSUMÉ

La province du Grenville est un terrain à métamorphisme élevé qui constitue la bordure affleurant au sud-est du Bouclier Canadien. Les roches y sont d'âge archéen à hélikien et ont été métamorphisées à plusieures reprises. Elles ont maintenant un âge moyen K-Ar de 950 millions d'années. Dans le sud-est de l'Ontario et le sud-ouest du Québec, les roches d'âge archéen à aphébien servent de socle aux roches du supergroupe du Grenville d'âge hélikien. Les roches du supergroupe du Grenville ont été les plus étudiées et contiennent le plus grand nombre d'indices minéralisés de la province du Grenville.

Le supergroupe du Grenville est composé d'une série inférieure de roches métavolcaniques recouvertes par une épaisse série du marbres et de méta-arkoses. Le degré de métamorphisme est élevé et ne descend que localement au-dessous du faciès amphibolite. Le supergroupe du Grenville est recoupé par au moins deux séries intrusives et recouvert par des roches sédimentaires du groupe de Flinton qui ont été à leur tour métamorphisées.

Des formations ferrugineuses à faciés sulfuré et oxydé, des fahlbandes pyriteuses et des amas sulfurés de métaux de base sont apparentés au volcanisme précoce grenvillien. Des dépôts de zinc liés aux strates se sont formés dans des carbonates de plateforme éloignée qui sont les équivalents stratigraphiques latéraux du facies volcanique. La première période majeure d'intrusion et de métamorphisme est à l'origine d'indices de Cu-Ni magmatique dans des roches gabbroïques et de dépôts de skarns à magnétite dans des calcaires proches d'intrusions granitiques. Une seconde période d'intrusion et de métamorphisme a formé les grands dépôts magmatiques à faibles teneurs de magnétite-ilménite dans les intrusions mafiques et quelques skarns au contact d'intrusions. Des skarns ferrifères ont aussi été produits par réaction métamorphique le long de contacts marbre-amphibolite. Le métamorphisme post-Flinton est à l'origine de veines locales de quartz aurifère et de sulfosels de métaux de base. Les veines tardives à calcite-barytine-fluorine et à calcite-barytine-galène recoupent aussi bien les roches de converture du Grenville que celles du Paléozoïque et sont associées au développement du graben de la vallée d'Ottawa, pendant le Crétacé.

GEOLOGY OF THE GRENVILLE PROVINCE

Introduction

The Grenville Province is located at the southeastern extremity of the Canadian Shield. It is a crudely rectangular area 1000 miles long by 300 miles wide, extending from Georgian Bay on Lake Huron to the Atlantic coast of Labrador. It should be considered as the largest exposed area of a more extensive orogenic zone, extending from Mexico to Norway (Fig. 1). Regional mapping is generally at the reconnaissance stage. Some portions of the eastern Grenville Province have yet to be mapped. Not more than 100 000 square miles have been subject to detailed (1:50 000 or better) investigation. The Grenville Province is therefore considered as a large, poorly exposed and under-examined portion of the Canadian Shield.

Age of the Grenville Province

It has been long realized that the Grenville Province consists in part of re-worked older crust that is significantly older than the 950 Ma old K-Ar age commonly associated with the Grenvillian event. Hence a preliminary step in the analysis of the tectonic evolution of this huge region is to subdivide the Grenville Province into areas of differing primary age. Unfortunately this phase of study of the Grenville Province is still in its infancy.

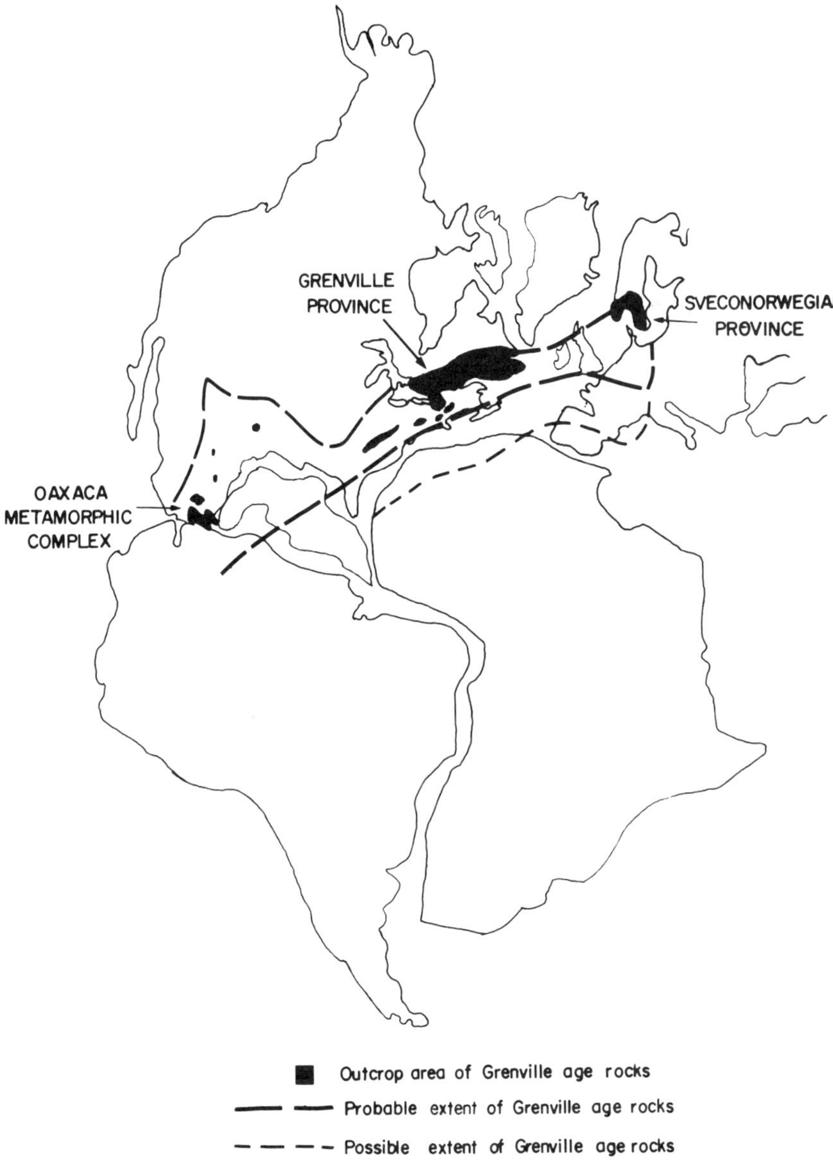

Figure 1. Pre-drift reconstruction of North America, Africa and South America after Martignole (1975). Outcrop areas of Grenville age (K-Ar on biotite of 950+150 Ma) gneisses, mainly after King (1969). Only the largest outcrop areas are labelled.

This problem is not merely of academic interest. As an example, the east-west trending Archean greenstone belts of the Abitibi region of Quebec are truncated by the Grenville Front. There is no reason to assume that the terrain southeast of the Grenville Front, within the Grenville Province, might not be equally favourable ground for gold and/or base metal exploration. That there is a possibility of success in this region was recently emphasized by Sethuraman (1979), and the knowledge of the areal extent of Archean rocks within the Grenville Province thus assumes considerable economic importance. Knowledge of the existence of "older rocks" within the Grenville Province comes from two sources: 1) geological, and 2) radiometric.

1) Geological. Older rocks can be traced into the Grenville Province across the Grenville Front. Examples of such areas include: a) The Surprise Lake area southwest of Chibougamau (SL, Fig. 2), where Archean greenstone belts have been observed to grade into higher grade metamorphic equivalents within the Grenville Province (Déland, 1956); b) The south end of the Labrador Trough (LT, Fig. 2), where higher grade equivalents of the trough rocks are found within the Grenville Province (Duffell and Roach, 1959; Jackson, 1976); and c) The Sudbury area (S, Fig. 2), where higher grade equivalents of the Southern Province rocks were considered to be exposed within the Grenville Province (Quirke and Collins, 1930).

In all cases, the "old rocks" cannot be traced far into the Grenville Province because the generally higher grade of metamorphism makes their recognition difficult.

2) Radiometric. Rb-Sr and U-Pb methods have succeeded in partially delimiting the areal extent of older rocks within the Grenville. Frith and Doig (1975), Doig (1977), and Krogh and Davis (1969), have all obtained Archean ages from Grenville Province rocks. These have all been within 60 km of the Grenville Front. Within this zone, ages as old as 3100 Ma have been reported. Dates on rocks more distant from the front have generally yielded Aphebian or Helikian ages.

In the Lake Nipissing area, Krogh obtained an 1800 Ma old age on metagreywackes known as the French River paragneiss. This is interpreted as the age of the maximum intensity of metamorphism, hence the age of the metasedimentary suite is older and may be time equivalent to the Huronian rocks to the northwest, in agreement with Quirke and Collins. These rocks, together with plutonic rocks of a variety of compositions intrusive into them, comprise the basement to the Grenville Supergroup discussed below, the oldest known members of which yield a date of 1310 Ma.

The geologic and radiometric evidence thus indicate a complicated evolution for the southwestern portion of the Grenville Province. The northeastern reaches of the mobile belt are more poorly known. Within this area the major time markers are the metamorphosed anorthosites. The commonly accepted time of emplacement for the majority of these bodies is between 1400 and 1100 Ma. In the eastern Grenville Province, the anorthosites are little deformed and metamorphosed, whereas the rocks into which they are emplaced are upper amphibolite facies gneiss and migmatite (Bourne, 1978). This notable difference in metamorphic grade suggests that the gneisses may be considerably older than the anorthosites. It is likely that these gneisses were deformed during the Hudsonian Orogeny and are older than 1650 to 1700 Ma.

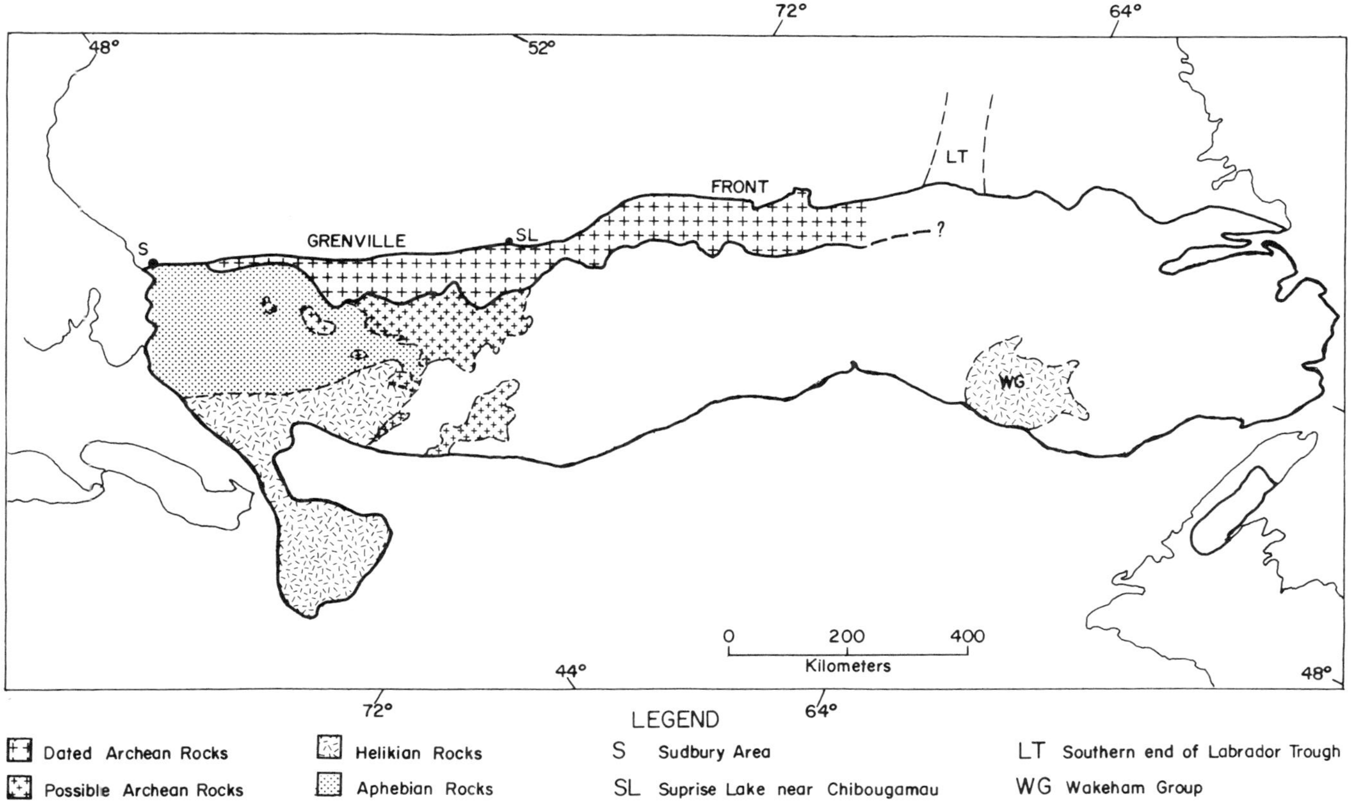

Figure 2. Known distribution of Archean, Aphebian and Helikian rocks in the Grenville Province (after Baer *et al.*, 1978). The heavy line is the southeastern limit of an aeromagnetic low, between which and the Grenville Front all the Archean ages within the Grenville Province have so far been reported.

Metamorphism in the Grenville Province

The metamorphic grade in the Grenville Province has recently been compiled as part of the Metamorphic Map of the Canadian Shield (Bourne, 1978; Sharma *et al.*, 1978; Rondot, 1978). The resulting picture (Fig. 3) is consistent with the widely held view that the Grenville Province is a zone of high grade metamorphic rocks.

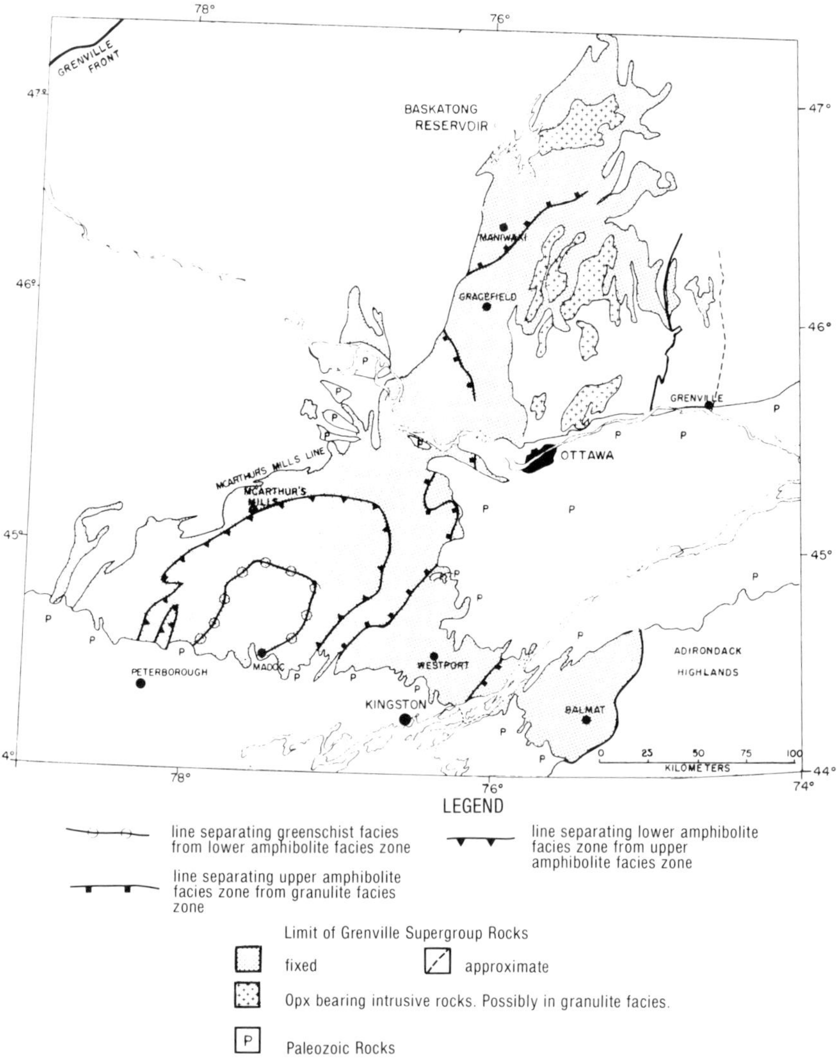

Figure 3. Distribution of metamorphic grade in rocks of the Grenville Supergroup.

A small chloritoid zone area, (M, Fig. 3) which at Madoc Ontario also contains chloritoid – staurolite assemblages, (Thompson, pers. commun., 1980) is the only positively recognized area of greenschist facies metamorphism in the Grenville Province. The boundaries of this zone as shown on Figure 3 are taken from Lumbers (1964). The Wakeham Group rocks of eastern Quebec (WG, Fig. 2) have also been metamorphosed to a relatively low metamorphic grade, however, mineral assemblages characteristic of the greenschist facies have not yet been found.

The remainder of the Grenville Province is mainly in the upper amphibolite to granulite facies of metamorphism. This is certainly true for that portion of southwestern Quebec underlain by the rocks of the Grenville Supergroup. Most muscovite from these rocks is believed to be of secondary origin. Sillimanite plus potash feldspar assemblages are common throughout, but not ubiquitous. Figure 3 shows the distribution of metamorphic grade in the area underlain by the Grenville Supergroup rocks. The grade increases from the low near Madoc to granulite facies near Westport (W, Fig. 3). From Westport, the grade diminishes toward the southeast, possibly reaching middle amphibolite facies conditions near Balmat, New York (B, Fig. 3) before increasing again to granulite facies in the Adirondack Highlands.

Tectonic Framework of Southeastern Ontario and Southwestern Quebec

The greatest concentration of geological and radiometric data in the Grenville Province in Canada is in southeastern Ontario. This is the only portion of the Grenville Province where individual formations in the supracrustal rocks have been recognized throughout any significant area. It is similarly the only area where the geological evolution has been established with any degree of certainty. In this region the Grenville Supergroup has been subdivided into three groups and 15 formations (Table I.)

Hermon and Mayo Groups. Silver and Lumbers (1965) reported an age of 1310±15 Ma on a felsic metavolcanic member of the lowermost Tudor Formation. This date has been widely considered as the age of the base of the Grenville Supergroup in southeastern Ontario, although the base itself is not exposed. It is significant in this regard that no ages greater than 1310 Ma have as yet been reported from Grenville Supergroup rocks.

Figure 4 shows the hypothetical relationship between the formations comprising the Hermon and Mayo Groups as envisaged by Lumbers. It is noteworthy that the upper portions of the Hermon are intercalated with the lower portions of the Mayo. Hence, although the former is in general older than the latter, the two groups are in part contemporaneous.

The predominant metasedimentary rocks are feldspathic metasandstone, possibly derived from turbidites, and marble, derived from both calcitic and dolomitic limestones.

Intrusive Rocks. The Hermon and Mayo Groups have been intruded by four distinct groups of intrusive rocks: 1) biotite diorite series, 2) mafic sills and dykes, 3) late gabbro and diorite, as well as nepheline syenite, and 4) quartz monzonite and potassic granite.

The following summary, (paraphrased from Lumbers, 1967a, p. 21) remains the most succinct statement about these igneous bodies.

TABLE I

TABLE OF FORMATIONS FOR THE PORTION OF THE GRENVILLE SUPERGROUP EXPOSED IN SOUTHEASTERN ONTARIO. (HERMON AND MAYO GROUPS AFTER LUMBERS, 1967b, FLINTON GROUP AFTER MOORE AND THOMPSON, 1972; MOORE AND THOMPSON, 1980)

GROUP	FORMATION	DESCRIPTION
	Stewart	marble, graphitic marble
	Madoc	pelite
	Fernleigh	thin-bedded black shale and limestone, pyritic in part
	Myers Cave	dolostone, limestone, dolostone conglomerate, black pyritic shale
Flinton		
	Lessard	calcareous and non-calcareous feldspathic sandstone
	Bishops Corners	Basal, hematitic quartzite (locally cross-bedded) quartzite-pebble conglomerate and shale
	UNCONFORMITY	
	Lasswade	marble, minor calcareous metasandstone and metasiltstone, rare recrystallized chert
	Apsley	poorly sorted feldspathic metasandstone with upper and lower members of calcareous metasandstone and metasiltstone
Mayo		
	Dungannon	80% marble; remainder mainly calcareous metasandstone and metasiltstone, oorly sorted feldspathic metasandstone, well sorted quartz-rich metasandstone, and rare recrystallized chert; mafic metavolcanic flows and iron-formation rare near base of formation overlying Tudor metavolcanic rocks
	Burnt Lake	mainly rhyolitic, trachytic, dacitic, and andesitic metavolcanic rocks with minor metabasaltic flows; lensoid marble units and sandy siliceous metasedimentary rocks found locally
	Turriff	75% pillowed basaltic and andesitic metavolcanic flows; remainder dacitic and felsic flows and pyroclastic rocks, and minor iron-formation
Hermon		
	Vansickle	mainly poorly sorted feldspathic metasandstone locally with abundant marble, felsic and mafic Hermon metavolcanic rocks, metaconglomerate, and well sorted quartz-rich metasandstone; arkosic metasedimentary rocks commonly associated with felsic metavolcanic rocks
	Oak Lake	mainly metamorphosed felsic pyroclastic rocks and arkose with some mafic and felsic metavolcanic flows; rare marble and metaconglomerate in upper part of formation
	Tudor	mainly metamorphosed, pillowed, basaltic and andesitic flows

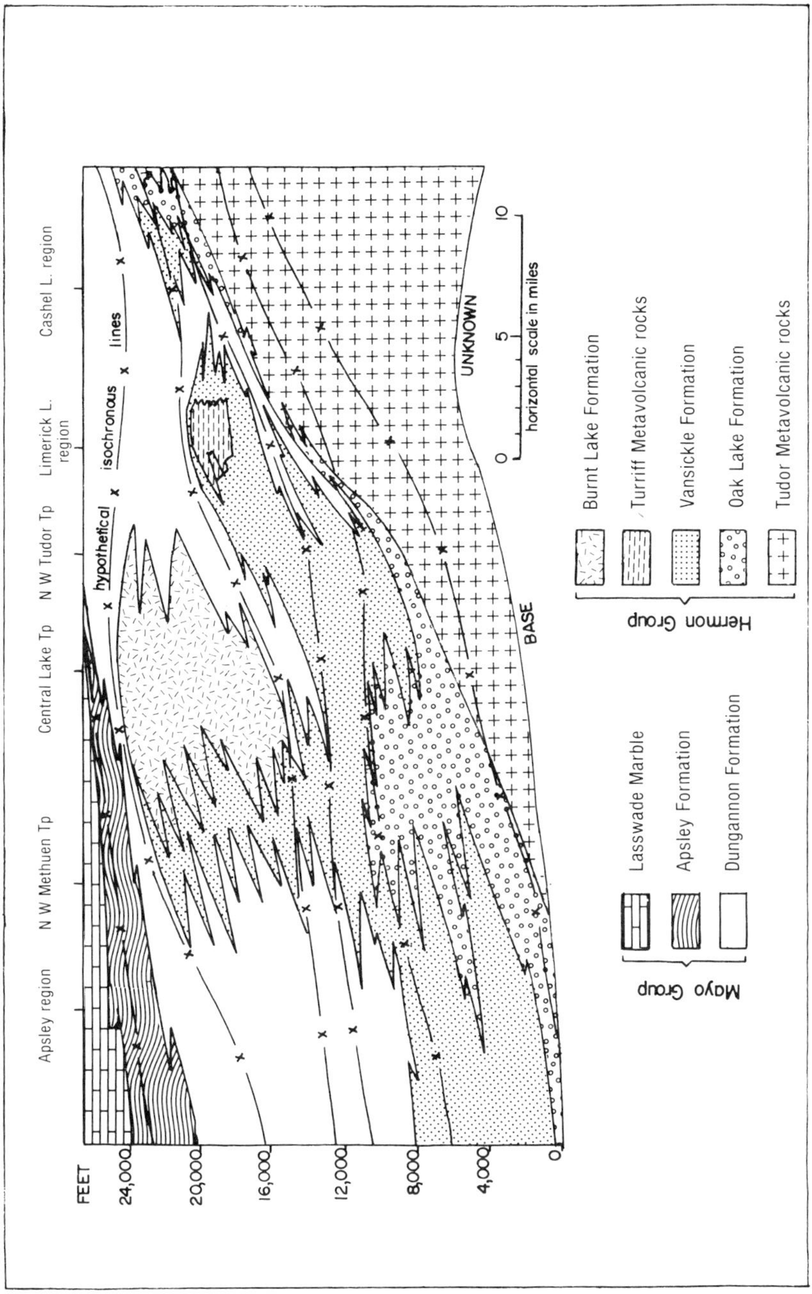

Figure 4. Section showing hypothetical relationship between the Hermon and Mayo Groups (Lumbers, 1967b).

(The oldest plutonic rocks form a biotite diorite series (group 1) characterized by trondhjemite-sodic granite differentiation trends and restricted spatially to those parts of the area rich in metavolcanic rocks. This series comprises a) an albite granite and syenite group, b) a diorite group, and c) a trondhjemite and sodic granite group. Plutons of the albite granite and syenite group were emplaced under epizonal conditions before the culmination of the low and intermediate rank regional metamorphism, but after the development of the principal fold pattern. Plutons of the diorite group and the trondhjemite and sodic granite group were emplaced mainly under the conditions of the mesozone. The biotite diorite series forms a distinct age group, about 1250 ± 25 Ma (Silver and Lumbers, 1965), and some epizonal stocks of both the albite granite and syenite group and the trondhjemite and sodic granite group are similar in isotopic age to meta-rhyolite from the upper part of the Hermon Group; volcanism of the Hermon Group and plutonic activity of the biotite diorite series may be related genetically.

It is interesting that in localities where the diorite group is undeformed, the inclusions found in them are highly deformed amphibolites. This suggests the existence of a metamorphic episode (herein called M^1) which predates the diorite.

Following the emplacement of the biotite diorite series, mafic sills and dykes (group 2) were emplaced. Shortly after this period of mafic intrusion and uplift, a few late gabbro and diorite stocks (group 3) were emplaced and are characterized by early, slightly undersaturated, mafic phases and late potassic syenite phases. Nepheline syenite masses found in the area may have been emplaced contemporaneously with the late gabbro and diorite stocks and both these plutonic groups could be related genetically.

Lastly, a series of granite to quartz monzonite plutons were emplaced (group 4). These rocks have been described as 'catazonal' (Lumbers, 1967a; Wynne-Edwards, 1972, p. 315) because they are in large part foliated, although some, such as the Westport pluton, are predominantly massive. Silver and Lumbers (1965) considered that both the foliated and massive plutons were emplaced at approximately the same time, at approximately 1125 Ma. This age estimate has been supported by more recent U-Pb age determinations (Bell and Blenkinsop, 1979) which indicate an age of 1104 ± 25 Ma for the Addington gneissic quartz monzonite – a member of the quartz monzonite group of Lumbers. The determined age probably reflects the time of attainment of the maximum metamorphic temperature and pressure (the M_1 event). Lumbers opined that some members of this group may be synchronous with the geologically older nepheline syenite, and late gabbro and diorite groups.

The predominantly massive Westport pluton appears to be anomalously younger than the other members of the foliated quartz monzonite group. Krogh and Hurley (1968) reported an age of 1016 Ma for this body. Using the = 1.42 x 10^{-11} yr^{-1} decay constant (Steiger and Jäger, 1977) the age becomes 995 Ma. This age was interpreted by Wynne-Edwards (1972) as the age of late syntectonic igneous activity, although Krogh and Hurley considered this intrusion to be a post-tectonic body. The picture pertaining to the group 4 granites and related plutons is therefore unclear. The available evidence suggests the possibility of more or less continuous igneous activity from perhaps 1225 Ma, the age of the group 3 intrusions, to 995 Ma and possibly younger.

Flinton Group. A major problem in the Precambrian geology of southeastern Ontario has been the interpretation of the significance of the quartz-pebble conglomerate and sandstone in the Madoc area. Miller and Knight (1914) considered them to be among the oldest members of the "Hastings Series", and that the Hastings Series was collectively younger than the biotite diorite series of Lumbers. Alternatively, Ambrose and Burns (1956) considered that the conglomerates are intraformational within the Grenville Supergroup and of local significance only. The latter position was apparently adopted by Lumbers (1967b). Moore and Thompson (1972) recently reviewed the evidence and favoured the former hypothesis. They urged the adoption of the term 'Flinton Group' for the rocks above and including the conglomerate, because the Hastings Series, as originally defined, included some rocks below the conglomerate as well. They noted that, although the quartz monzonites and potassic granites, as well as the members of the biotite-diorite series, cut the rocks below the conglomerate, only late pegmatites cut the Flinton Group. They concluded that the Flinton Group is significantly younger than the Hermon and Mayo Groups, and that the period of deposition of the Flinton Group is bracketed between the age of the quartz monzonites and potassic granites (1125 Ma), and that of the late pegmatites (1050 Ma), although as noted above, some of the granite plutons appear to be considerably younger in age. The Flinton Group has been subdivided into seven formations (Table I). The group consists mainly of clastic sedimentary rocks, although carbonate rocks are well represented.

These aspects of the Precambrian geology of southeastern Ontario are summarized in Table II, which shows three distinct metamorphic events separated by approximately 175(?) Ma. Lumbers considered M'_1 and M''_1 to represent two separate peaks of intensity during one protracted episode of regional deformation and metamorphism (M_1). However he considered the Flinton Group to be a part of the Grenville Supergroup, whereas it now appears that the evidence in favour of a younger age for the Flinton Group is compelling. Both Lumbers, and Moore and Thompson found that northeasterly trending structures are the youngest in the area (Table II). Because the Flinton Group rocks have themselves been metamorphosed and deformed, there is strong evidence of at least two main periods of orogenic events in southeastern Ontario. It is possible that the northeasterly trending structures mapped by Moore and Thompson, and which must have formed during the second of these events, completely obliterated antecedent structures in Lumbers' area. Clearly more detailed mapping as well as a major radiometric dating program are required to shed further light on this problem.

It is probable that a large percentage of the M_1 metamorphic mineral assemblages were destroyed during the later post-Flinton (M_2) event, however the exact amount is unknown. This is most likely to have occurred in those areas where the M_1 metamorphic grade was low and the M_2 metamorphic grade was high. Our current state of knowledge is insufficient to assign any given metamorphic minerals assemblage to either the M_1 or the M_2 metamorphic episodes.

TABLE II
PROPOSED SEQUENCE OF GEOLOGICAL EVOLUTION OF THE GRENVILLE SUPERGROUP IN SOUTHEASTERN ONTARIO. (COMPILED FROM LUMBERS, 1967b; MOORE AND THOMPSON, 1972)

Time (Ma B.P.)	Event	Comment
1000	Westport and related plutons	catazonal?!
	D_4-gentle warping	
1050	Emplacement of pegmatite	
	M_2-intermediate grade metamorphism	
	D_3-northeast trend	
	D_2-isoclinal folds	
	Deposition of mature to immature clastic sedimentary rocks	Flinton Group
	– M''_1 high grade metamorphism	
1125	– Emplacement of quartz monzonite and potassic granites (catazonal)	
	– Emplacement of late gabbro, diorite and nepheline syenite	
	– Emplacement of mafic sills and dykes	
	– Emplacement of diorite, trondhjemite and soda granite (mesozonal)	BDS, units 2 & 3
	– M'_1, low grade to intermediate metamorphism	
1250		
	– D_1, northeast trend.	
	– Emplacement of albite granites and syenites (epizonal)	BDS, unit 1
1310	– Deposition of basic to intermediate volcanic rocks, carbonates, feldspathic metasandstones (base not exposed)	Hermon and Mayo Groups = Grenville Supergroup

BDS = biorite diorite series of Lumbers (1967).
D = period of deformation.
M = metamorphic event.
M′ = subdivision of metamorphic event

GRENVILLE METALLOGENY

Introduction

Major mineral production in the Canadian portion of the Grenville Province has been restricted to iron at Lac Gagnon and Marmora, titanium at Lac Allard and uranium at Bancroft. Minor base metal, gold and talc production have been recorded. In the United States, zinc in the Balmat-Edwards district, talc at Gouverneur and iron in the Adirondack region, all in New York state are important commodities currently being produced.

Occurrences of Cu, Pb, Zn, Au, Ni, are common only in the southwestern part of the Grenville Province. They are similar in morphology to important ore deposit types in other areas of the Shield and may be classified as follows:

1) Stratabound deposits	a) Oxide facies iron-formation
	b) Sulphide facies iron-formation
	c) Zinc deposits in carbonate rocks
2) Magmatic deposits	a) nickel-copper
	b) magnetite-ilmenite
3) Metasomatic deposits	a) contact magnetite skarns
	b) metamorphic magnetite skarns
	c) magnetite in gneissic rocks
4) Quartz-vein deposits	gold, arsenopyrite, base metal sulphosalt veins
5) Post Ordovician veins	a) Calcite-fluorite-barite veins
	b) Calcite-barite-galena veins
	c) Calcite-celestite veins
6) Miscellaneous Mineral Occurences	

1) Stratabound Deposits

Quartz-magnetite and pyritic iron-formations, and zinc concentrations in carbonate rocks form the majority of the stratbound mineral occurrences (Fig. 5). The quartz-magnetite iron-formations are typically small and of mineralogical interest only. The pyritic deposits are somewhat larger, have little base metal content, and were mined for sulphur early in this century. All of the significant base metal production from Grenville Supergroup rocks has come from concordant zinc or zinc-lead deposits within Grenville marble. The most important of these are New Calumet and Tetrault in Quebec, Long Lake in Ontario and Balmat-Edwards in New York State.

1a) Oxide Facies Iron-Formation. Quartz-magnetite iron-formations are not abundant (Fig. 5) and form small bodies of little economic importance. Deposits are restricted to the Hastings basin although some of the metamorphic deposits discussed below may have similar affinities. The largest and best known is the Belmont iron-formation where the mineralization forms a sharply defined bed of grey cherty quartz containing bands of magnetite and specular hematite. The iron-formation is 15 m thick, is exposed over a strike length of 0.4 km and contains up to 24% iron (Miller and Knight, 1914). Gross (1967) identified metamorphosed oolites and granules in the Belmont iron-formation, indicating a similarity to Proterozoic Superior-type sedimentary iron-formations.

1b) Sulphide Facies Iron-Formation. The massive pyrite and pyritic schist deposits of the Grenville Supergroup rocks are sulphide iron-formations with only minimal associated base metals. The largest and highest grade occurrences (Canadian Sulphur Ore, Clare River, Fig. 5) are associated with metasedimentary rocks at the top of the Tudor volcanic pile. Smaller and lower grade concentrations occur within both the clastic metasedimentary rocks and marbles overlying the volcanic rocks. The latter may be in rocks which are facies equivalents of the upper part of the metavolcanic sequence (Fig. 5).

Typically pyrite is interbanded with quartz and is massive to heavily dissemi-

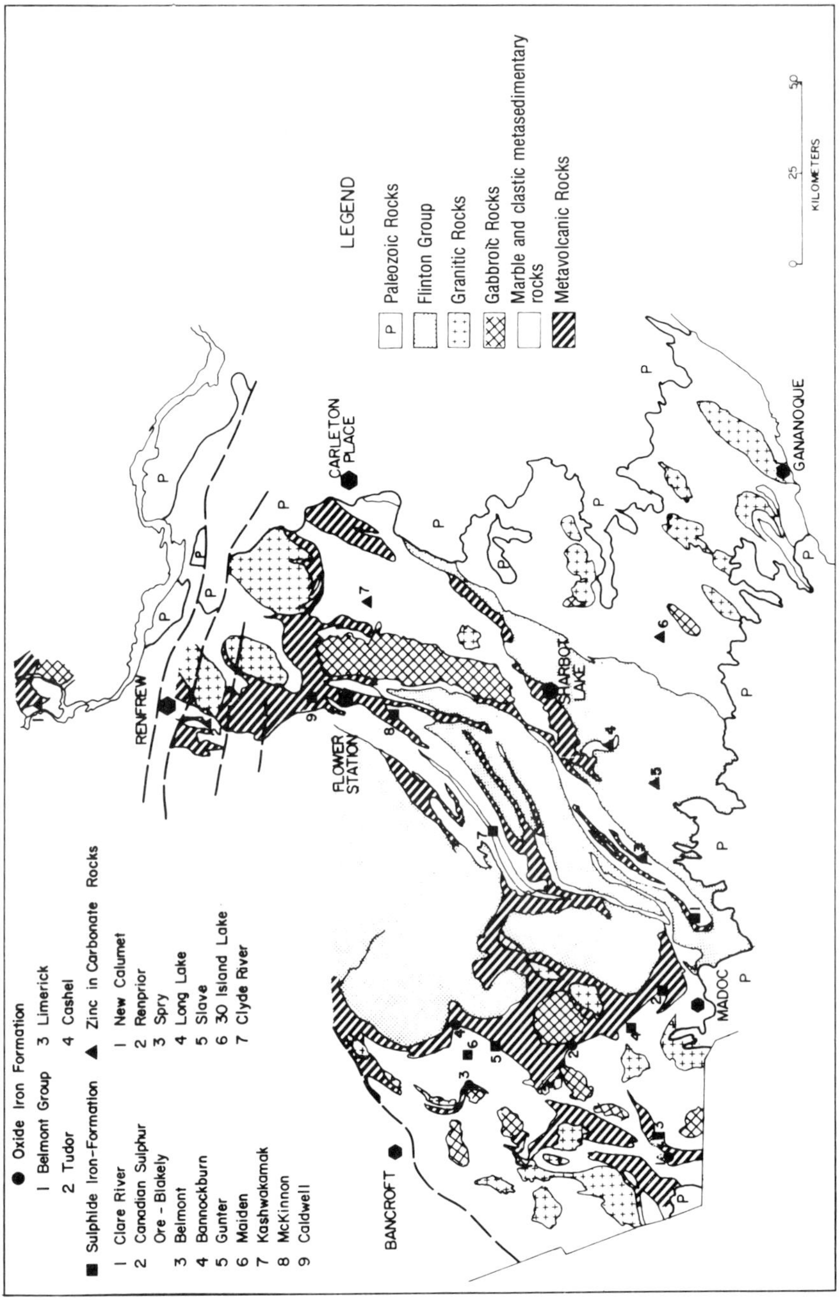

Figure 5. Selected stratabound mineral occurrences.

nated in a quartz matrix. Carbonaceous material may be present and is associated with biological forms at Canadian Sulphur Ore (Sangster, 1970). The quartz-hosted zones may change along strike into pyritic quartz-muscovite schists and quartz-feldspar-muscovite schists. A decrease in muscovite and increase in biotite along strike from the Clare River occurrence is accompanied by a decrease in pyrite content, suggesting a pre-metamorphic sulphide concentration.

Base metal sulphides are present only locally, and usually in minor amounts in these pyritic strata (Sangster, 1970; Hewitt, 1967; Lumbers, 1964). Arsenic is generally enriched in pyrite (1000 to 5000 ppm). Sphalerite is abundant within a small pyritic bed associated with rhyolite at the Blakely showing and minor chalcopyrite occurs with pyrrhotite at the McKinnon occurrence. Low concentrations of base metals occur in carbonaceous pyritic schists overlying the Tudor volcanic rocks near the Canadian Sulphur Ore occurrences, and in a rusty schist overlying the Belmont volcanic rocks near the boundary between Marmora and Belmont townships. In the latter, two zones, one 600 m long and 240 m wide, the other 900 m long and 90 m wide, contain 15 to 25% pyrite and up to 0.5% zinc and copper, in rusty graphitic schist.

The origin of the sulphide iron-formations is somewhat problematic. The pyritic iron-formations and related pyritic schists at the top of the Tudor volcanic rocks probably represent a sulphide facies of sedimentary iron formation (Goodwin, 1962). However, the absence of associated base metal sulphides and alteration indicates that these iron-formations are not genetically identical to the large pyritic massive base metal sulphide deposits of the Shield.

1c) Zinc Deposits in Carbonate Rocks. Zinc is widely distributed within Grenville supergroup marbles (Fig. 5). Historically, these carbonate-hosted zinc deposits have been considered as a single group. However, significant variations between occurrences enable division into four groups.

Group I deposits are characterized by zinc-lead-silver mineralization in thin marble or calc-silicate horizons within siliceous gneisses. New Calumet, the type example for this group (Sangster, 1967), consists of heavily disseminated to locally massive sphalerite with argentiferous galena, pyrrhotite and minor pyrite and chalcopyrite within a concordant calc-silicate marble horizon between hanging wall amphibolite and underlying quartzofeldspathic gneiss and migmatite. The host carbonate is only as extensive as the known mineralization.

Group II deposits are commonly small pods or lenses of sphalerite with only traces of galena and occur within a very restricted horizon of magnesian marble or dolomite within a calcite marble terrain. Calc-silicate minerals may be present but are not extensively developed. The Long Lake, Slave Lake and Thirty Island Lake occurrences are examples. The regional environment of these deposits is poorly known because both Long Lake and Slave Lake occur in small roof pendants of marble within intrusive bodies. They do however appear to flank the main Grenville volcanic axis.

Group III deposits are compositionally similar to group II deposits except that dolomite and calc-silicate dolomite are much more abundant both as host to the mineralization and regionally. The mineralization is commonly more widespread than group II occurrences, and exhibits strong stratigraphic control. The Renprior and

Spry showings are examples, as are those in the Maniwaki area (Gauthier and Brown, 1979). Deposits of group III occur within marbles intercalated with or overlying rocks of probable volcanic origin.

Group IV deposits are those currently producing in the Balmat-Edwards district of New York State. (Lea and Dill, 1968; de Lorraine and Dill, this volume). The deposits are primarily zinc with only minor lead and are similar to group III deposits as they occur in an extensive dolomitic, calc-silicate terrain. However they are unlike any Canadian occurrences in their association with stratiform anhydrite which must indicate that the host rocks for the ores were formed in a stable shelf environment in contrast to the less stable central basin volcanic environment of groups I, II and III.

In spite of differences between individual types, Figure 6 shows that all four types are related to siliceous dolomite. The origin of these deposits is obscured by the superimposed high metamorphic grade, concordant form, and lack of apparent genetic association with intrusive rocks which are common to all occurrences.

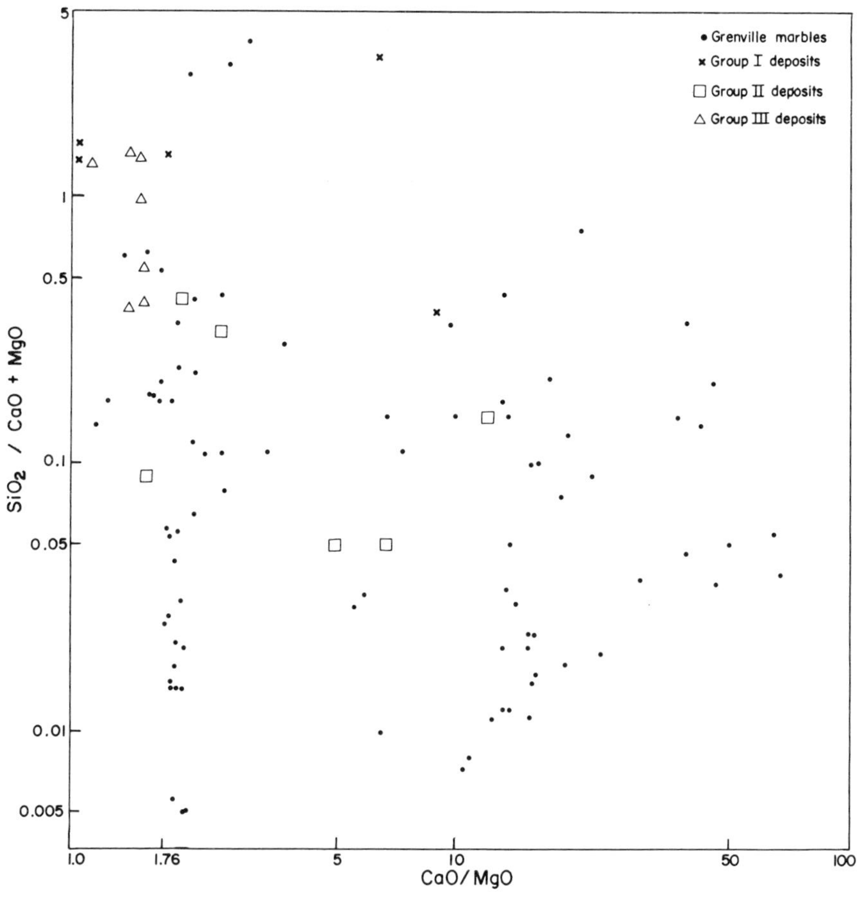

Figure 6. Comparison of compositional characteristics of siliceous dolomitic marbles associated with zinc deposits compared to unmineralized marble.

The New Calumet occurrence is unique in its restricted carbonate environment and in its content of antimony and silver which are more common in classic hydrothermal deposits than in deposits lacking igneous affiliations in carbonate rocks (Ohle, 1967). The host rocks are volcanic (Sangster, 1967). The New Calumet deposit may be similar to the Errington deposit in the Sudbury basin where base metal sulphides are present within a volcanogenic chert-carbonate horizon believed to have been deposited by hot springs at the contact between the Onaping andesitic tuff and the overlying Onwatin Slate (Thompson, 1956). Subsequent metamorphism of such a sulphide zone to upper amphibolite facies would produce a calc-silicate assemblage similar to that at New Calumet.

At Balmat, concepts explaining the origin of the deposit have been summarized and elaborated by Lea and Dill (1968), and de Lorraine and Dill, (this volume). Early investigations (Newland, 1916; Smythe, 1918; Brown, 1936, 1947) proposed hydrothermal replacement of marbles by zinc sulphides carried in hot gases or solutions evolved from intruding magmas. More recently, Lea and Dill concurred with later revisions of these views by Brown (1965) and Solomon (1963) that the Balmat deposits were formed in a shallow basin where carbonate rocks were being precipitated along with accumulations of fine-grained sulphides. The sulphides were then lithified with the sedimentary sequence and later deformed by folding and metamorphism to give their current distribution and character.

The Balmat deposit is in some morphological aspects similar to the Group III deposits except that the latter are associated spatially with amphibolitic volcanic rocks, and no anhydrite has been encountered in 3000 m of drilling at the Renprior occurrence. Therefore Balmat is not a group III deposit, and was formed in a different environment.

A syngenetic origin is suggested at Spry and Slave Lakes (Fig. 5) by the stratigraphic persistance of mineralization for hundreds of metres but with thicknesses of only a few centimetres. Highly contorted compositional banding between sulphides and gangue at Thirty Island Lake (Harding, 1947) and the concordant, lenticular nature of the sphalerite at Long Lake (Uglow, 1916) also indicate that these deposits are premetamorphic.

Favourable sedimentary or diagenetic ore-forming conditions similar to those at Balmat may have existed locally or briefly within the carbonates associated with or marginal to the central volcanic terrain. Although this suggests some stability in the central axis, it is felt that stability was insufficient in these areas to allow the formation of a larger deposit as was possible under more stable shelf conditions at Balmat.

Recently published lead isotope data for the Grenville Supergroup occurrences (Fletcher and Farquhar, 1977; Fletcher, 1979) is significant because it broadly supports the classification of zinc deposits in carbonate rocks presented above. Two and possibly three distinct groups of Pb-Zn occurrences were distinguished within the Central Metasedimentary Belt. The first was Balmat, clearly different from all the others. The remainder can possibly be subdivided into: deposits in the Mont Laurier Basin and deposits in the Hastings Basin. Fletcher (1979, p. 139) stated that the chemical characteristics of the second and third groups are those which, "in other regions correlate with basic magmatism and crustal rifting. This effectively precludes the possibility that the mineralized lead in these portions of the Grenville Supergroup is of continental origin".

2) Magmatic Deposits

Magmatic deposits of nickel-copper and iron-titanium (Fig. 7) are associated respectively with early (biotite-diorite) and late gabbroic rocks that cut the Grenville Supergroup.

2a) Nickel-Copper. Magmatic segregations of iron-nickel-copper sulphides occur within basic intrusions associated with the earliest (biotite-diorite) period of intrusion into the Grenville Supergroup. The deposits are poorly exposed in general. However, samples of mineralization from the lower metamorphic grade occurrences have magmatic textures.

The Bonter, Lingham Lake and Macassa occurrences (Fig. 7) are basic to ultrabasic intrusions which cut greenschist and lower amphibolite facies metamorphic rocks within the Hastings basin. The Macassa is the largest and contains 2,000,000 tons of 1.0% Ni and 0.25% Cu to 300 m depths (Shklanka, 1969). Lumbers (pers. comm.) described the deposit as a broad, irregular zone of disseminated pyrite, pyrrhotite and chalcopyrite within peridotite of the Thanet complex. The Lingham Lake occurrence is smaller but similar. On the Bonter property, copper-nickel sulphides occur within a pyroxenite dike over a strike length of 105 m and width of 15 to 22 m. On the north rim of the Hastings basin, nickel-copper sulphides (1% combined) occur within Raglan Hills gneissic metagabbro. Again the occurrence is not large. The deposit is probably similar to those in the lower grade terrain. Mosaic textures and subgrain development in pyrrhotite, and poikiloblastic development of pyrite indicate that the sulphides underwent high grade metamorphism with their host rocks.

More recently nickel has been mined from a small deposit at Lac Renzy, 190 km north of Ottawa. The deposit is beyond the area considered here but is mentioned because nickel-copper sulphides occur in a relatively fresh pyroxenite sill cutting high grade gneisses. This may indicate a potential for nickel in the younger basic instrusions elsewhere.

2b) Magnetite-Ilmenite. Magmatic magnetite-ilmenite deposits of the region have been extensively described by Rose (1958), Hewitt (1962), Krogh (1961), and Gross (1967). This type of mineralization is widespread in southeastern Ontario. Most of the occurrences are small. Large tonnage deposits do exist, but low grades and metallurgical problems have precluded production.

The two best known deposits are the Uffframville and Matthews-Chaffey (Fig. 7), and the latter is characteristic of the type. The Matthews-Chaffey deposit lies within the Newboro gabbro approximately 40 km north-northwest of Gananoque. The deposit lies within a zone 1000 m long by 100 m wide and is estimated to contain 45 million tonnes of material grading 25% iron and 7% titanium. The deposit occurs within a banded gabbroic anorthosite which was emplaced into a high grade metamorphic gneiss-marble terrain. Monzonite and quartz monzonite either cut or show gradational contact with gabbro (Gross, 1967). The gabbro is dark grey to green and is composed of andesine-labradorite feldspar and augite, with accessory hornblende, biotite and apatite. Magnetite, ilmenite and laths of ilmenite exsolution in magnetite, together with mafic silicates, occur as layers alternating with plagioclase to form a banded arcuate body which parallels the south and east side of the intrusion. Relationships between the compositional bands appear gradational (Gross,

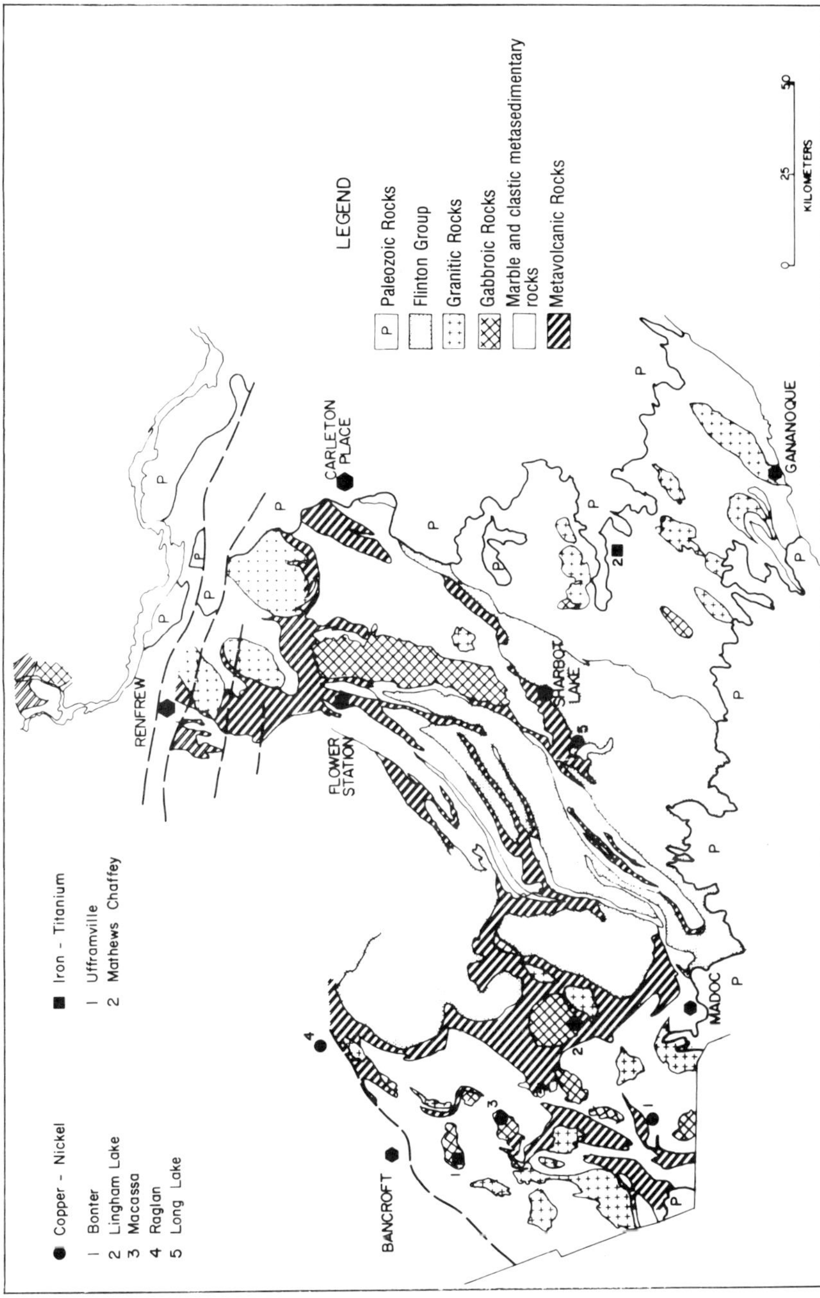

Figure 7. Selected magmatic deposits and occurrences.

1967) indicating accumulation by gravitational settling. Krogh (1961) concluded that the major portion of the oxide body formed from the injection of a late, mafic-rich magmatic fluid along closely spaced arcuate fractures. The intrusion is time-correlative with the "late gabbro series" (Lumbers, 1967a). The Ufframville deposit is of similar age, dimensions and grade. Many smaller occurrences are in part related to rocks of the early biotite-diorite plutonic series.

3) Metasomatic Deposits

Iron deposits of metasomatic origin are extensively developed within rocks of the Grenville Supergroup (Shklanka, 1968; Gross, 1967; Rose, 1958; Fig. 8). The deposits occur in all metamorphic grade terrains and differ with the metamorphic grade of their host rocks. The deposits, previously considered to be entirely of contact metasomatic origin are divided into three groups: a) contact metasomatic skarns, b) metamorphic metasomatic skarns at marble-silicate gneiss contacts, and c) metamorphic metasomatic iron deposits within silicate gneiss.

3a) Contact Metasomatic Skarns. These deposits form tabular to podiform bodies of massive to disseminated magnetite within marbles at or near their contact with an intrusion. The host is usually a diopside-amphibole-epidote-bearing skarn. Pyrite is common with rare chalcopyrite. Although deposits occur at the contacts of intrusions of all compositions from syenite through granite to gabbro, the majority of the occurrences lie at the contact of early biotite-diorite series intrusions. The most significant of the deposits is the Marmoraton Mine (Park, 1967) which produced iron ore pellets from 1955 to 1979.

The Bristol Mine at Hilton, Quebec (Gross, 1967) and the Coe Hill and Ridge Properties south of Bancroft (Shklanka, 1968) are spatially related to younger intrusive complexes, synchronous with or younger than M''_1 and form a second age of skarn development. Sulphides, including pyrrhotite and chalcopyrite are associated with the Ridge and Coe Hill magnetite skarns.

3b and 3c) Metamorphic Metasomatic Deposits. Metamorphic metasomatic skarns of both types (3b and 3c) are restricted to higher metamorphic grade terrains and are most abundant and largest in the Sharbot Lake – Ottawa Valley area. Production recorded from these deposits is of historic interest only. Grades are comparable with those in contact skarn deposits but the tonnages are smaller.

Deposits lying at the contact between basic gneiss (or amphibolite) and marble consist of massive to heavily disseminated magnetite within pyroxene-hornblende skarn. Accessory hornblende is more abundant than in contact skarns and epidote is absent. Deposit sizes range from a few metres in strike length up to 1000 m long by 130 m wide for the Calabogie deposit (Rose, 1958). These deposits are the result of reaction between marble and basic gneiss during metamorphism.

Metamorphic metasomatic magnetite is a common accessory in basic gneiss and locally reaches concentrations of up to 50 or 60% iron (Radenhurst, Radnor, Parks occurrences, Fig. 8). The zones are tabular and are generally of lower grade than those described above. The host rocks are amphibolite and hornblende gneiss found throughout the Grenville Province.

The restriction of these deposits to high grade metamorphic rocks with the absence of an observable intrusive association appears to rule out genesis by direct

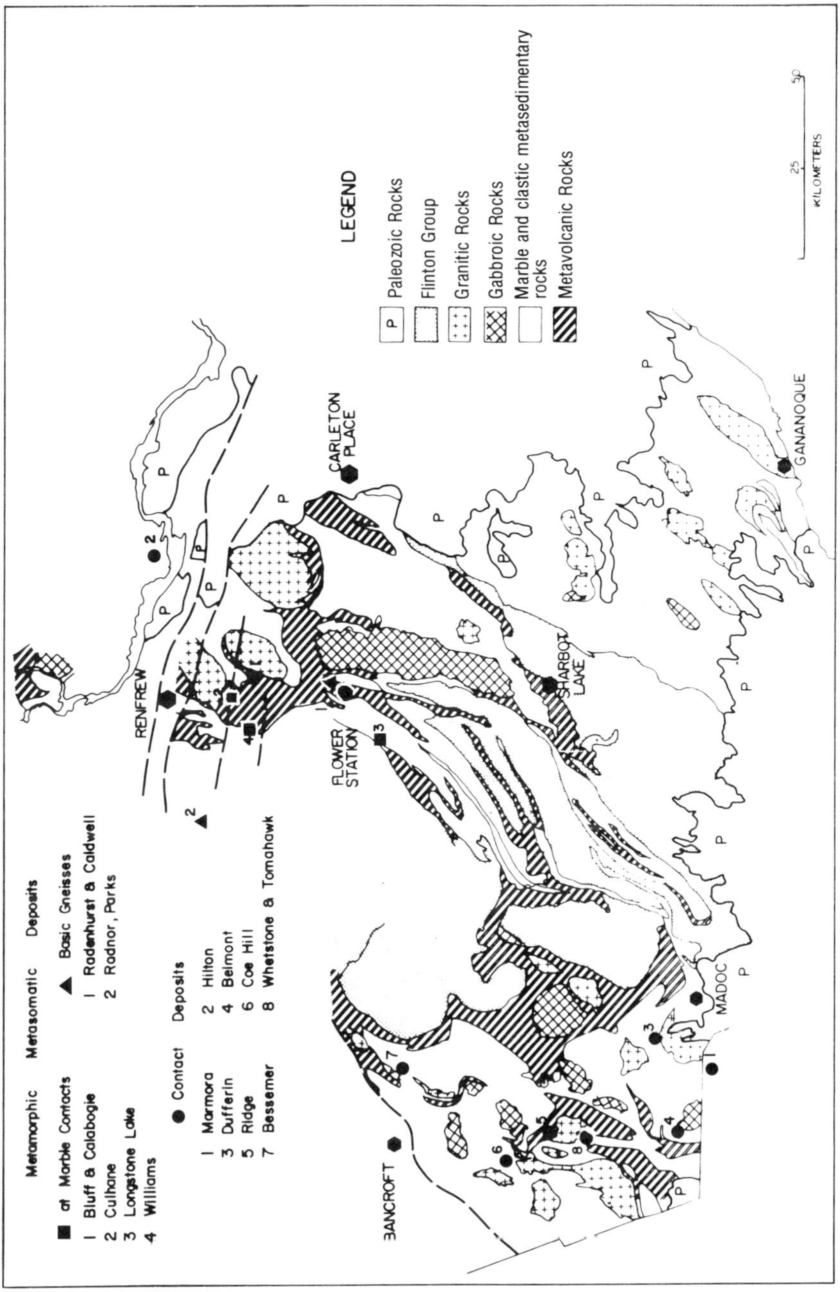

Figure 8. Selected metasomatic deposits.

magmatic processes. The Scott magnetite deposit, New York, a massive magnetite deposit in amphibolitic rocks (Hagner *et al.*, 1963) exhibits a sharp decrease of iron in the amphibolite adjacent to the magnetite body. Iron in the ore zone was therefore apparently derived from the host amphibolite. Collins (1969) indicated that during metamorphic recrystallization, mafic silicates rich in iron recrystallized to form calcium-magnesium pyroxene and amphibole, with iron released as magnetite to form ore zones.

4) Quartz Vein Deposits

Quartz veins containing gold, arsenic, and base metals (Fig. 9) are common within the greenschist and lower amphibolite facies metamorphic rocks around the Elsevier-Ashby dome, north and northeast of Madoc and adjacent to the west contact of the Deloro pluton a few kilometres west of Madoc. Past mineral production from these veins is not significant. Deloro Gold Mines Ltd. produced 2,000 tonnes of arsenic and an unknown quantity of gold between 1899 and 1903 (Hurst, 1927). Harding (1942) reported production of $10,000 in gold from the Addington Mine up to 1921. Two classes of veins are recognized: 1) quartz veins with arsenopyrite, chalcopyrite, gold and traces of base metals within volcanic and clastic sedimentary rocks, and 2) quartz-carbonate veins with sphalerite, galena and antimony sulphosalts in carbonate rocks.

Fourteen occurrences of gold-arsenopyrite-bearing quartz veins are known west of the Deloro pluton (Fig. 9). The veins are small and discontinuous and lie both within the margin of the granite and within greywacke and limestone west of the contact. They are parallel to the granite contact and to the foliation within the metasedimentary rocks. The veins cut granite dikes at the Deloro mine (Hurst, 1927), and are therefore younger than the pluton. The sulphide content of the veins is low. Disseminated, euhedral arsenopyrite is most abundant and is associated with chalcopyrite, pyrrhotite and pyrite. Nonsulphide minerals include native gold, ankerite, biotite, muscovite, fluorite, zircon, hematite, magnetite, and tourmaline (Hurst, 1927).

The main area of vein development lies within the Tudor volcanic rocks, overlying marbles and sedimentary rocks of the Flinton Group (Moore, 1967; Moore and Thompson, 1972) on the east side of the Elsevier Ashby dome. Deposits in this area have been extensively described by Meen (1942) and Harding (1942). Throughout this region and irrespective of host rock, the veins are parallel to the foliation within the enclosing rocks.

Veins in the volcanic rocks show two associations. One group, including the Addington (Golden Fleece) and Ore Chimney properties, lies in or near the contact between the Tudor volcanic rocks and conglomerates of the Flinton Group. To the north, the O'Donnell vein group occurs within the volcanic rocks. The second group of veins is within a carbonate terrain and commonly lies adjacent to contacts between dolomitic marble and structurally more competent clastic units. Some occurrences (Mazinaw, Barrie, Gough, Ardoch; Fig. 9) are probably within rocks of the Flinton Group. The veins are generally narrow, discontinuous, range from 0.1 to 2 m in width and occur over strike lengths of a few to a few hundred metres. The largest vein in the Addington mine was traced for 1000 m with a maximum width of 10 m in volcanic rocks and conglomerate.

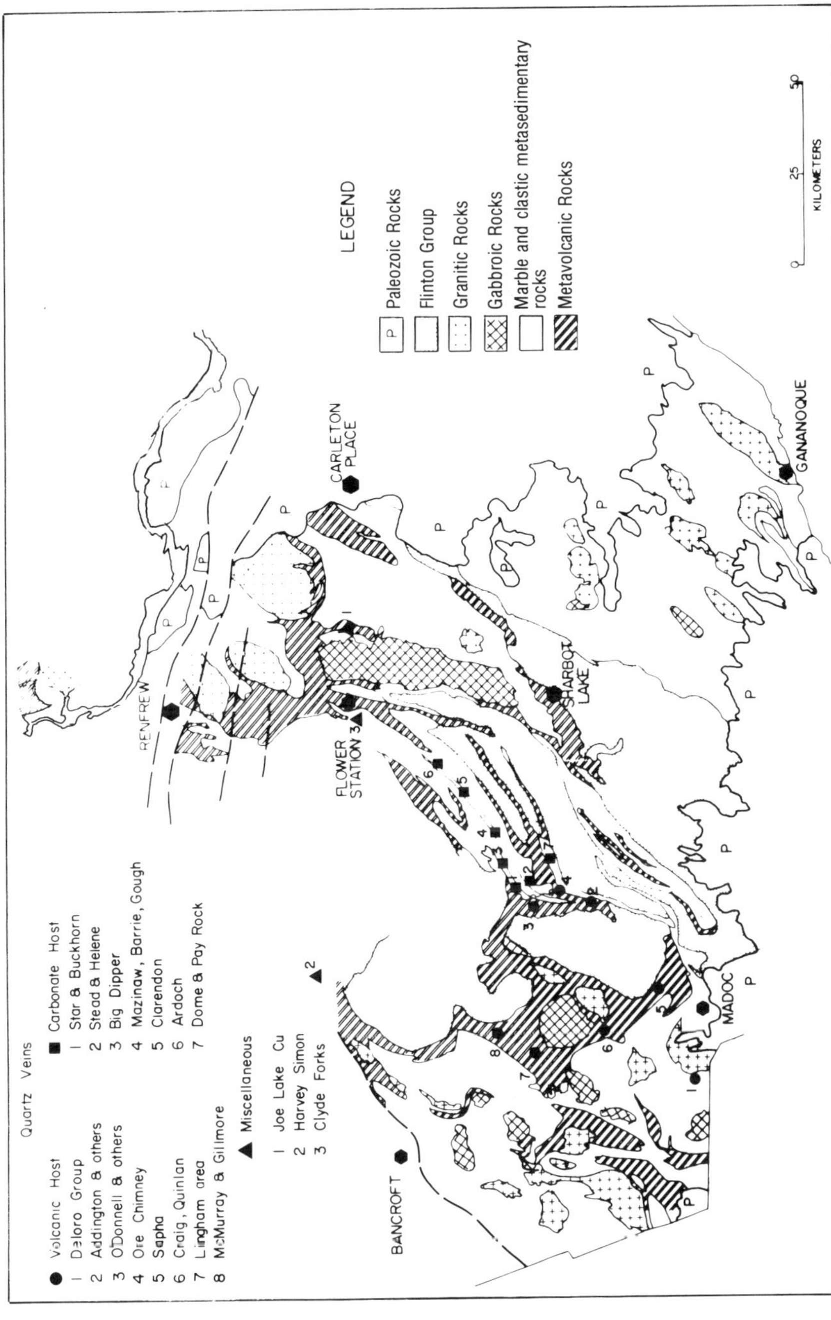

Figure 9. Quartz vein deposits and occurrences.

Examination of old dumps and exposures shows that mineralization is sparse in all veins although occasional impressive samples can be found. In volcanic-hosted veins, arsenopyrite, pyrite, pyrrhotite and chalcopyrite are common with sparse accessory sphalerite, galena, gold, menhegenite, and valleriite. Veins in carbonate terrain contain sphalerite, galena, pyrite and chalcopyrite with minor tetrahedrite, jamesonite and bournonite. Gold, arsenopyrite, chalcocite, bismuth, bismuthinite, and scheelite are recorded. The overlap in mineralogical composition between veins in volcanic and carbonate host rock suggests a common genesis.

Host rock compositions have affected vein compositions. Boyle (1961) demonstrated that lateral diffusion of vein components from their host rock into structurally prepared sites explained the origin of the arsenical gold deposits at Yellowknife. In the Grenville Supergroup, derivation of arsenic and related elements from the volcanic rocks and base metals from the carbonate rocks is compatible with the chemistry of these rocks. The period of vein formation was probably during the second and major metamorphic event (M_2) when shearing related to ductility contrasts along volcanic-metasedimentary rock and marble-metasedimentary rock contacts provided suitable site of deposition for elements mobilized by steep thermal gradients related to the metamorphism. The veins lie within the lowest metamorphic grade rocks (Lumbers, 1964), and grade of metamorphism increases sharply to the southeast.

The arsenopyrite-gold-quartz veins west of the Deloro granite are unusual because they contain the mineral assemblage of a volcanic host but lie within sedimentary rocks and the biotite-diorite series of intrusions. According to the stratigraphic sequence (Lumbers, 1967a), these rocks overlie volcanic rocks and it seems probable that the vein components were mobilized to structural sites stratigraphically above their source area.

5) Post Ordovician Veins

A large number of low-temperature fissure veins cut both Grenville marbles and Ordovician carbonate rocks around the edge of the Grenville Province in Southeastern Ontario and Southern Quebec (Fig. 10). The veins trend parallel to faults of the Ottawa valley graben system to which they are believed to be genetically related. The veins are classified into three types: a) calcite-fluorite-barite veins, b) calcite-barite-galena veins, and c) calcite-celestite veins. Economically, the veins are now of little significance. Fluorite was produced from several veins at Madoc, primarily during wartime, and the Kingdon mine on Chats Island in the Ottawa River produced 900,000 tons of 8.5% lead ore from a sinuous, narrow calcite-galena vein.

5a) Calcite-Fluorite-Barite Veins. The fluorite district of southeastern Ontario is a 16 kilometre-long, northwesterly trending swarm of small veins whose center passes about one kilometre southwest of Madoc, Ontario (Wilson, 1929; Guillet, 1964). About one-third of the veins lie within Grenville crystalline metamorphic rocks, mainly low grade crystalline limestone, and two thirds are in flat-lying, undeformed Ordovician limestone cover. Noteworthy is the fact that 83% of past production is from those veins within a Grenville host.

The veins contain fluorite (60%), barite (25%) and calcite (15%) with accessory celestite, quartz, marcasite, pyrite, and sphalerite. The fluorite is coarse-grained and green, clear or amber. It is commonly interbanded with pink or buff barite. The

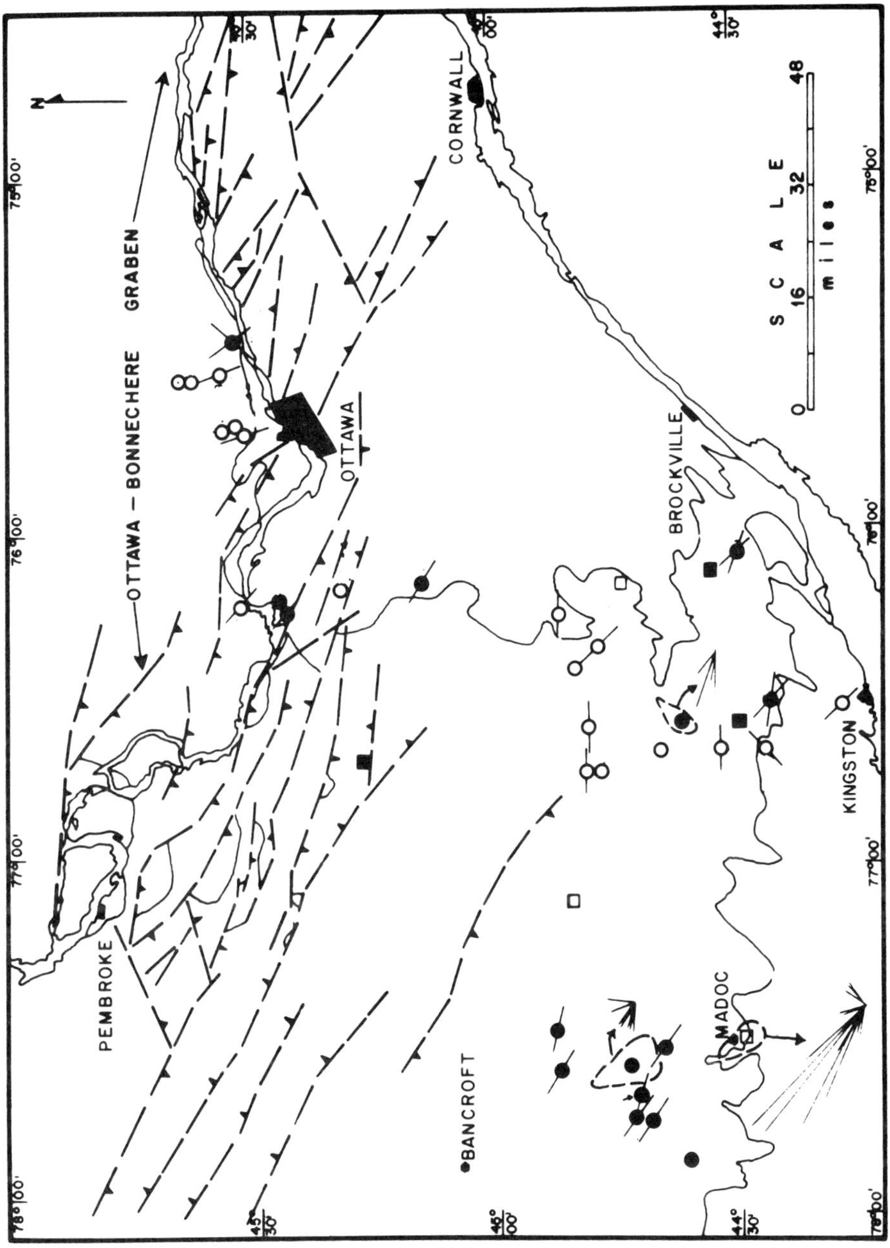

Figure 10. Distribution and strike direction of Ottawa-Bonnechere graben faults and post-Precambrian vein deposits. The various types of deposits are indicated as follows: calcite-barite-fluorite veins, open squares: celestite veins, solid squares; calcite-barite veins, open circles: and calcite-barite-galena veins, solid circles. The strike direction of the veins is indicated by bars on the deposit symbol or by rose diagrams adjacent to vein districts.

banding is developed through crustification in open cavities and is accompanied by coxcomb textures, vugs, and banded, botryoidal barite and calcite. There is no consistent order to the sequence of deposition. Fluid inclusion work by Rupert (1963) and McCartney (1964) showed that temperatures of deposition varied from 75°C to 150°C.

The veins average 100 m long and less than 3 m thick although dimensions of 330 m by 4 m are known. Dips are steep. The ore is discontinuous, being concentrated in shoots with a southerly plunge (Guillet, 1964). Wallrock alteration is negligible.

5b) Calcite-Barite-Galena Veins. Calcite-barite-galena veins are more widespread and more common than the fluorite veins. They occur primarily in Grenville marbles with only a few in the Ordovician cover rocks and cut marbles of all metamorphic grades from greenschist facies to granulite facies.

Calcite, barite and galena are the most common minerals. White to pinkish, coarse-grained, calcite is the predominant mineral accompanied usually by white to salmon-pink barite. Galena occurs in places as sparse disseminations or coarse aggregations. Sphalerite is common locally and traces of marcasite, pyrite and chalcopyrite are found. Celestite occurs with some of the Ottawa Valley occurrences. These veins generally have a northwesterly strike and vertical dips, although there is a more pronounced scattering than shown by the fluorite veins.

5c) Calcite-Celestite Veins. A few veins, similar to those described above except for the presence of abundant celestite, occur in both the Ottawa Valley and Kingston-Sharbot Lake areas. These veins were probably generated by circulating meteoric waters during the early stages of development of the Ottawa-Bonnechere Graben. They show good correlation in attitude with the strike direction of faults associated with the graben (Fig. 11) and the host structure probably developed by extension in regional doming during the early phases of development of the graben (Sangster, 1970). This would place their age as latest Jurassic or earliest Cretaceous. Wilson (1929) proposed that the fluorite-bearing deposits were formed by circulating meteoric waters, based on the absence of known intrusive activity and the suspected low temperatures of deposition, later confirmed by Rupert (1963) and McCartney (1964). Sangster (1970) later suggested that this mechanism applied to all veins of the three types and speculated that veins containing fluorite may have derived their materials from Paleozoic rocks whereas the components of the lead-bearing veins may have had a source in Grenville rocks.

6) Miscellaneous Occurrences

Numerous small occurrences have not been included in the foregoing discussions due to difficulty in assigning them to a particular class. These include copper occurrences such as veins at Joe Lake (Smith, 1956). Two occurrences merit description. Cochrane (1964) described a concordant body of pyrrhotite up to 9 m thick and 150 m long which contains 225,000 tonnes assaying 1.1% copper (Shklanka, 1969). The occurrence is within hornblende gneiss near the northern margin of Grenville Supergroup exposure in Lyndoch township. The deposit seems to have volcanogenic affinities but is not classified here because of a lack of data. Another showing of interest occurs in marble 3 km west of Clyde Forks. It consists of disseminated tetrahedrite and chalcopyrite in a conformable barite-bearing horizon in

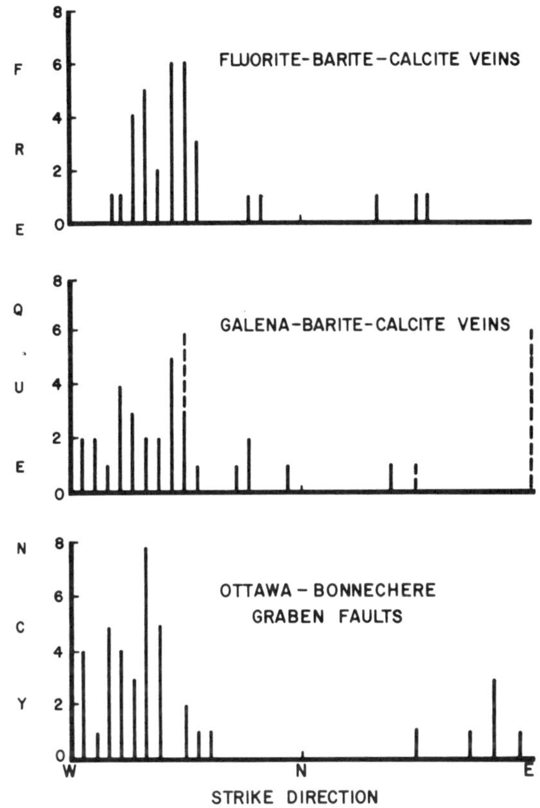

Figure 11. Strike direction of fluorite-barite-calcite veins, galena-barite-calcite veins, and Ottawa-Bonnechere graben faults. Dashed bars indicate approximate vein attitudes.

dolomite. Mercury is present in anomalous amounts in the tetrahedrite. The mineralization is known over a width of 1.5 to 3.5 m in a zone 180 m long. These deposits require further study and are not classified here.

DISCUSSION

Two hypotheses have recently been proposed to explain the origin of the Grenville Supergroup rocks. Firstly, Brown *et al.* (1975) suggested that ultramafic rocks located beneath the base of the metavolcanic sequence in southeastern Ontario represent material derived from the ocean floor. They suggested that an ensimatic subduction model may have been involved in the emplacement of these rocks. According to this hypothesis, the granodiorites are considered to have been derived from the partial melting of the downgoing slab and the marbles of the Dungannon and Lasswade formations were presumably deposited in the shallow continental sea behind the island arc. It was further suggested that the contentious conglomerate at the base of the Flinton Group represents the first good evidence of sediments derived from a continental source.

Secondly, Baer (1976) pointed out that the age of the Grenville Supergroup rocks is approximately the same as that of rocks in a number of areas located adjacent to the Grenville Front and believed to be rift-related (Barager, cited in Baer, 1974). He suggested that the Grenville Supergroup rocks were deposited within an aulacogene whose margins corresponded roughly to the current eastern and western margins of the supergroup, and which ran in a north-northeasterly direction to intersect the Grenville Front. He considered that the upper surface of the fault blocks would be the locus of relatively quiescent carbonate sedimentation whereas the fault zone itself would serve as the locus of emplacement of volcanic material and of deposition of course clastics. According to this hypothesis, the deposition of the Hermon and Mayo Groups in the latter environment and the Flinton Group in the former would have been roughly contemporaneous.

The classical metallogenic work by Bilibin (1955) provides a basis for modern metallogenic investigations in orogenic belts. His study of the endogenous and exogenous mineral deposits of post-Precambrian geosynclinal belts in the Soviet Union noted several recurrent features: 1) Similar mineral complexes occur in an orderly sequence in widely separated orogenic belts of different ages; 2) Similar intrusive complexes occur in an orderly sequence in widely separated orogenic belts of different ages; 3) Some types of recurring mineral districts in (1) show a close association to recurring intrusive complexes in (2); and 4) the intrusive complexes and the corresponding mineral complexes of different character succeed each other in time in a similar sequence within orogenic belts, despite different absolute ages. By relating these recurrent features to the stages of tectonic development in individual orogenic areas, Bilibin devised a five stage metallogenic model. Each stage is characterized by specific suites of intrusive and extrusive rocks, and specific types of sedimentation, with each having a relationship to specific metallic mineral assemblages.

In attempting to apply the Bilibin metallogenic model to the area underlain by the Grenville Supergroup rocks, one is immediately faced with the choice of which of the two possible tectonic hypotheses is most applicable. The writers tentatively prefer the hypothesis of Brown *et al.* (1975) for the Hermon and Mayo Groups and for the associated intrusive suites from the biotite-diorite series to the relatively younger potassic granite group. The problems with Baer's hypothesis are:

1) if the Hermon and Mayo Groups were deposited in an aulacogene, then their age and that of the rifting should be similar. The available evidence indicates that the oldest rock attributable to rifting is the Blue Mountain nepheline syenite, whose age was reported by Krogh and Hurley (1968) to be 1285 Ma Rb-Sr (or 1258 Ma using $\lambda = 1.42 \times 10^{-11}yr^{-1}$). Thus the earliest evidence for rifting appears to post-date formation of the Hermon and Mayo Groups which are theorized to have formed within and adjacent to the rift. Admittedly the ages are close and the argument circuitous: further work may overcome this objection.

2) the lead isotope data discussed above indicate that, except for the Balmat deposit, there has been little significant contribution to the Grenville Supergroup rocks by continentally derived material. This data agrees with the hypothesis of Brown *et al.*

3) the Flinton Group is not cut by intrusive rocks other than pegmatite, whereas the Hermon and Mayo Groups are cut by a variety of intrusive rocks of different composition. This suggests that there is a considerable difference in age between the two groups, which conflicts with Baer's hypothesis.

The intrusive suites have not been completely defined with respect to age of emplacement, however what little is known appears to accord roughly with a simplied Bilibin model. The biotite-diorite series of Lumbers (1967a) is equivalent to the tonalites and plagiogranites of the Bilibin metallogenic cycle which are commonly related to the early stages of tectonic evolution in a mobile belt. The potassic granite group of Lumbers appears to correspond to the main orogenic granites of Bilibin which are related to later major metamorphic and tectonic events.

The Flinton Group remains difficult to interpret. If, as emphasized by Moore and Thompson (1972), the Flinton is later than the potassic granite group of intrusions, one possibility might be to suggest that the Flinton was deposited within an aulacogene, as proposed by Baer for the Grenville Supergroup itself. The presence of nepheline syenite could be interpreted as evidence for the former existence of such a structure. Alternatively, the presence of granites *described as catazonal* which appear to be younger than 1050 Ma (the minimum age for the Flinton) might explain why the Flinton is not cut by these plutons: this may be due to its geographic location rather than to some fundamental geological relationship. More systematic age data are required before this complex problem can be satisfactorily resolved.

The mineral deposits bear a distinct temporal relationship to the developmental sequence outlined above (Fig. 12). The oxide and sulphide facies iron-formations are integral units of their respective stratigraphic sequences on top of, and lateral to, the Tudor volcanics. The New Calumet deposit is probably the best representative of clearly exhalative, volcanogenic-fumarolic processes. Small concentrations of zinc-bearing sulphides formed locally with siliceous dolomitic carbonates in areas of major carbonate sedimentation within and marginal to the central volcanic belt. Larger deposits like Balmat formed in a restricted sub-basin to the southeast in conjunction with anhydrite. This tectono-stratigraphic relationship is unusual during early tectonic stages.

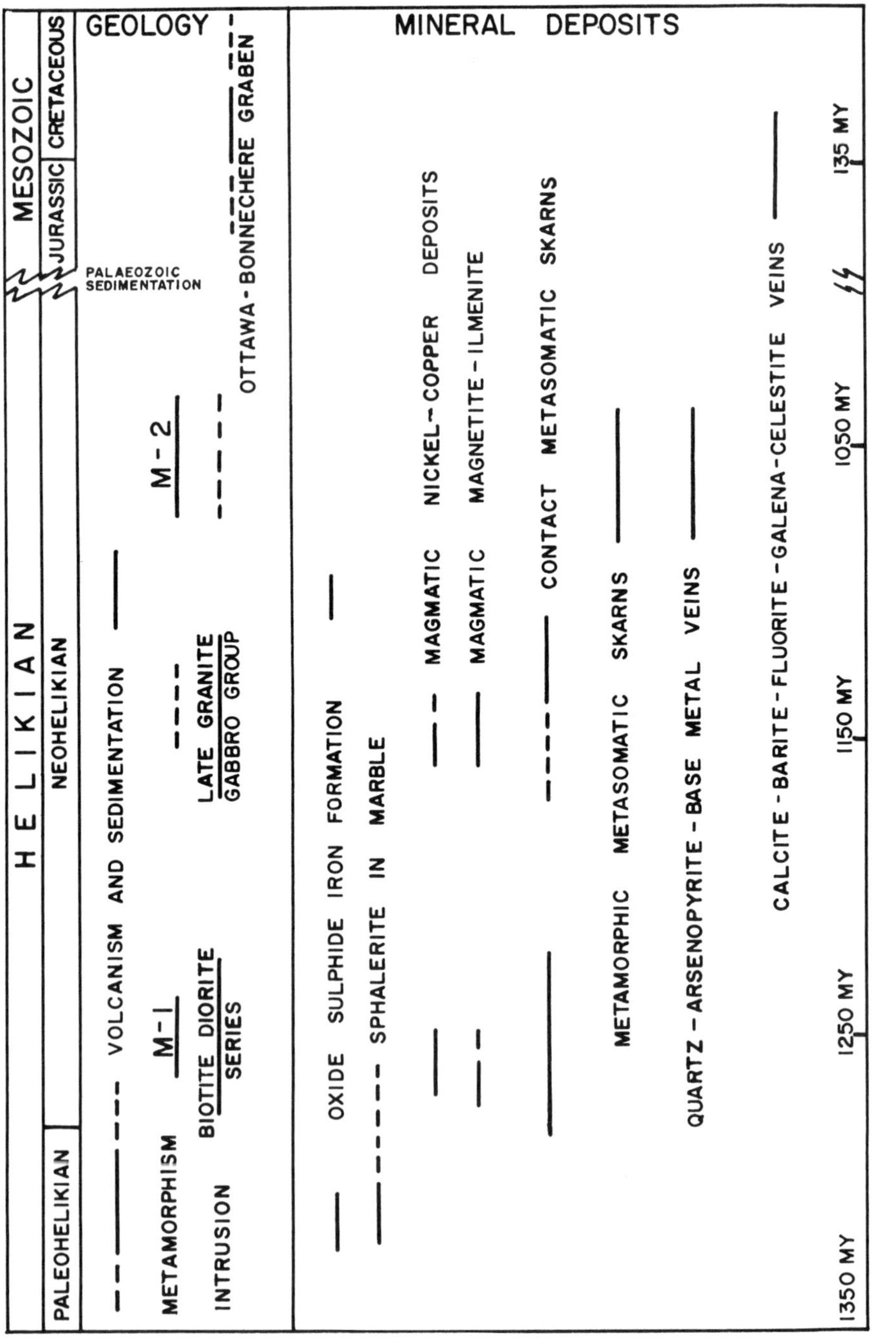

Figure 12. Generalized relationship of geology and mineral deposits in the Grenville Province of southeastern Ontario.

With the onset of initial magmatism in the form of gabbroic to granitic intrusions of the biotite-diorite series (Lumbers, 1967a), deposits of magmatic and contact metasomatic origin formed. Magmatic copper-nickel segregations formed within peridotitic phases of early gabbros along with very minor iron-titanium segregations. Contact metasomatic magnetite deposits formed within carbonate rocks adjacent to biotite-diorite series intrusions of all compositions. There is no geological record to indicate the events between the early intrusive suite at 1250 Ma and the main gabbro and potassic granite intrusive episode at 1125 Ma. During emplacement of the latter intrusive suite, ore forming processes similar to those related to the early group were operative. Large, low grade magmatic iron-titanium deposits formed by segregation in the late gabbros. The nickel-copper deposit mined at Lac Renzy may also be related to this stage of intrusion. Contact skarns also are related to this period of intrusion but are less abundant than in the early stage.

The deposition of the Flinton Group was accompanied by the formation of low-grade pyritic shale deposits. Major metamorphism after deposition of the Flinton Group (M_2) induced metamorphic reactions involving iron silicates and carbonates and produced bodies of magnetite within mafic gneisses and at mafic gneiss – marble contacts. Simultaneously, in areas not greatly affected by metamorphism, quartz and quartz – carbonate – gold – arsenopyrite-base metal sulphide veins were formed by processes of lateral secretion. Finally, in Jurassic-Cretaceous time and following Palaeozoic carbonate sedimentation upon the Grenville craton, the region was domed prior to development of the Ottawa-Bonnechere graben. Vein deposits containing calcite-barite, fluorite, celestite and base metals formed in tension fractures related to this sytem.

ACKNOWLEDGEMENTS

The writers are indebted to many people in the preparation of this paper. The summary of mineral deposits is an updated condensation of a thesis submitted by A.S. to Queen's University. In this regard A.S. acknowledges the financial support of the St. Joe Minerals Corp. in collecting field data and E.R. Lea for his interest in the Project. At Queen's the project was realized through the encouragement of the late W.D. McCartney. The data is published with the permission of St. Joseph Explorations Limited.

The summary of Grenville geology is in part from a thesis by J.B. submitted to Queen's University. He benefitted greatly from the active interest and support of H.R. Wynne-Edwards (the godfather of Grenville geology) who directed the project so ably. Numerous additions and modifications have resulted from lengthy discussions with former colleagues at the Geological Survey of Canada, in particular Peter Thompson.

NOTE ADDED IN PROOF

An excellent and detailed account of the stratigraphy, petrology, structure and age of the Flinton Group has recently been published (Moore and Thompson, 1980). Unfortunately it was not possible to incorporate their data into the main body of this report. The highlights of their article are:

1) a detailed description of each formation of the Flinton Group.

2) they suggest the removal of the Flinton Group from the Grenville Supergroup, which henceforth would consist only of the Hermon and Mayo Groups.
3) they recommend the use of the terms "Elzevirian" and "Ottawan" orogenies for the M_1 and M_2 events discussed herein. They also recommend that the term "Grenvillian orogeny" be retained to refer in a general way to the Elzevirian, Ottawan, and possibly other, as yet unrecognized deformational events of Neo-helikian age in this area.
4) they also support the Brown et. al. (1975) model for the formation of the Hermon and Mayo Groups, and criticize the Baer (1976) aulacogen model, albeit for different reasons.
5) they give substance to the offhand suggestion made here that the Flinton Group formed in an extensional tectonic environment, although they avoid the use of the term "aulacogen".

Their summary of the regional tectonic history of the Precambrian geology of southeastern Ontario is substantively similar to that presented here.

REFERENCES

Ambrose, J.W. and Burns, C.A., 1956, Structures in the Clare River Syncline: A demonstration of granitization: in Thompson, J.E., ed., The Grenville Problem: Proceedings of the Royal Society of Canada, Special Publication 1, p. 42-52.

Baer, A.J., 1974, Grenville Problems Reviewed: Geotimes, July 1974, p. 17.

________________, 1976, The Grenville Province in Helikian times: A possible model of evolution: Philosophical Transactions, Royal Society of London, Series A, v. 280, p. 499-515.

Baer, A.J., Poole, W.H., and Sandford, B.V., 1978, Riviere Gatineau: Geological Survey of Canada Map 1334A.

Bell, K.C. and Blenkinsop, J., 1979, Rubidium-strontium studies in the Grenville Province of southeastern Ontario: in Davidson, A., Britton, J.M., Bell, K., and Blenkinsop, J., Regional Synthesis of the Grenville Province of Ontario and Western Quebec: Geological Survey of Canada Paper 79-1B, p. 153-172.

Bilibin, U.A., 1955, Metallogenic provinces and metallogenic epochs: Queen's College Geological Bulletin, Queen's College Press, Flushing, New York, 35 p.

Bourne, J.H., 1978, Metamorphism in the eastern and southwestern portions of the Grenville Province: in Fraser, A. and Heywood, W.W., eds., Metamorphism in the Canadian Shield: Geological Survey of Canada Paper 78-10, p. 315-328.

Boyle, R.W., 1961, Geology, geochemistry, and origin of the gold deposits of the Yellowknife District, Northwest Territories: Geological Survey of Canada Memoir 310, 193 p.

Brown, J.S., 1936, Structure and primary mineralization of the zinc mine at Balmat, New York: Economic Geology, v. 31, p. 233-258.

________________, 1947, Porosity and ore deposition at Edwards and Balmat, New York: Geological Society of America Bulletin, v. 58, p. 505-546.

________________, 1965, Oceanic lead isotopes and ore genesis: Economic Geology, v. 60, p. 47-68.

Brown, R.L., Chappel, J.F., Moore, J.M., Jr., and Thompson, P.H., 1975, An ensimatic island arc and ocean closure in the Grenville Province of Southeastern Ontario, Canada: Geoscience Canada, v. 2, p. 141-144.

Cochrane, D.R., 1964, A copper deposit near Denbeigh, Ontario: M.Sc. Thesis, Queen's University.

Collins, L.G, 1969, Regional recrystallization and the formation of magnetite concentrations, Dover magnetite district, New Jersey: Economic Geology, v. 64, p. 17-33.

Déland, A., 1956, The boundary between the Temiskaming and Grenville subprovinces in the Surprise Lake area, Quebec: Geological Association of Canada Proceedings, v. 8, p. 127-141.

Doig, R., 1977, Rb-Sr geochronology and evolution of the Grenville province in northwestern Quebec, Canada: Geological Society of America Bulletin, v. 88, p. 1843-1856.

Duffell, S. and Roach, R.A., 1959, Mount Wright, Quebec – Newfoundland: Geological Survey of Canada Map 6-1959.

Fletcher, I.R., 1979, A lead isotopic study of lead-zinc mineralization associated with the central metasedimentary belt of the Grenville Province: Ph.D. Thesis, University of Toronto.

Fletcher, I.R., and Farquhar, R.M., 1977, Lead isotopes in the Grenville and adjacent Paleozoic formations: Canadian Journal of Earth Science, v. 14. p. 56-66.

Frith, R.A. and Doig, R., 1975, Pre-Kenoran tonalitic gneisses in the Grenville Province: Canadian Journal of Earth Science, v. 12, p. 844-849.

Goodwin, A.M., 1962, Structure, stratigraphy and origin of iron formation, Michipicoten area, Algoma District, Ontario, Canada: Geological Society of America Bulletin, v. 73, p. 561-586.

Gross, G.A., 1967, Geology of iron deposits in Canada, Volume III: Geological Survey of Canada, Economic Geology Report 22, 181 p.

Guillet, G.R., 1964, Fluorspar in Ontario: Ontario Department of Mines, Industrial Minerals Report No. 12, 68 p.

Hagner, A.F., Collins, L.G., and Clemency, C.V., 1963, Host rocks as a source of magnetite ore, Scott Mine, Sterling Lake, New York: Economic Geology, v. 58, p. 730-768.

Harding, W.D., 1942, Geology of Kaladar and Kennebec townships: Ontario Department of Mines, v. 51, pt. 4, p. 51-74.

____________, 1947, Geology of the Olden-Bedford area: Ontario Department of Mines, v. 56, pt. 6, 100 p.

Hewitt, D.F., 1962, Geology of Wollaston Township: Ontario Department of Mines, Geological Report 11, p. 29-55.

____________, 1967, Pyrite deposits of Ontario: Ontario Department of Mines and Mineral Resources Circular No. 5, 64 p.

Hurst, M.E., 1927, Arsenic bearing deposits of Canada: Geological Survey of Canada, Economic Geology Report 4, 181 p.

Jackson, G.D., 1976, Opocopa Lake (east half), Quebec-Labrador: Geological Survey of Canada Map 1417A.

King, P.B., 1969, The tectonics of North America – a discussion to accompany the Tectonic Map of North America – scale 1:5,000,000: United States Geological Survey Professional Paper 628, 95 p.

Krogh, T.E., 1961, The titaniferous magnetite deposit of the Newboro District: M.Sc. Thesis, Queen's University.

Krogh, T.E. and Davis, G.L., 1969, Geochronology of the Grenville Province: Carnegie Institute of Washington Yearbook, v. 67, p. 224-233.

Krogh, T.E. and Hurley, P.M., 1968, Strontium isotope variation and whole rock isochron studies, Grenville Province of Ontario: Journal of Geophysical Research, v. 73, p. 7107-7125.

Lea, E.R. and Dill, D.B., 1968, Zinc deposits of the Balmat-Edwards District: in Ridge, John D., ed., Ore Deposits in the United States, 1933-1967: Graton-Sales Volume I, American Institute of Mining, Metallurgical and Petroleum Engineers, p. 20-48.

Lumbers, S.B., 1964, Preliminary report on the relationship of mineral deposits to intrusive rocks and metamorphism in part of the Grenville Province of southeastern Ontario: Ontario Department of Mines, Preliminary Report 1964-4, 37 p.

———————, 1967a, Stratigraphy, plutonism, and metamorphism in the Ottawa River remnant in the Bancroft-Madoc area of the Grenville Province of southeastern Ontario: Ph.D. Thesis, Princeton University, 331 p.

———————, 1967b, Geology and mineral deposits of the Madoc-Bancroft area: in Jenness, S.E., ed., Guidebook to Parts of Eastern Ontario and Western Quebec: Geological Association of Canada, Annual Meeting, Kingston, Ontario, p. 13-30.

Martignole, J., 1975, Le Precambrien dans le sud de la province Tectonique de Grenville (Bouclier Canadien): Publication de la Département de Géologie, Université de Montréal, 405 p.

McCartney, W.D., 1964, Barite-fluorite study: Geological Survey of Canada, Paper 64-2, p. 75-78.

Meen, V.B., 1942, Geology of the Grimsthorpe-Barrie area: Ontario Department of Mines, v. 51, pt. 4, p. 1-50.

Miller, W.G. and Knight, C.W., 1914, The Precambrian geology of southeastern Ontario: Ontario Bureau of Mines, vol. 22, pt. II, 152 p.

Moore, J.M. Jr., 1967, Grenville structure, stratigraphy and metamorphism, southeastern Ontario: Northbrooke-Ompah area; in Guidebook to geology of parts of eastern Ontario and western Quebec: Geological Association of Canada Guidebook, Kingston, p. 59-71.

Moore, J.M., Jr. and Thompson, P.A., 1972, The Flinton Group, Grenville Province, eastern Ontario, Canada: Proceedings 24th International Geological Congress, section 1, p. 221-229.

———————, 1980, The Flinton Group: A late Precambrian metasedimentary succession in the Grenville Province of eastern Ontario: Canadian Journal of Earth Sciences, v. 17, p. 1685-1707.

Newland, D.H., 1916, The new zinc mining district near Edwards, New York: Economic Geology, v. 11, p. 623-644.

Ohle, E.L., 1967, Origin of ore deposits of the Mississippi Valley type: in Genesis of Stratiform Lead-Zinc-Barite-Fluorite Deposits: Economic Geology Monograph 3, p. 33-39.

Park, F.B., 1967, Genesis of the Marmoraton pyrometasomatic iron deposit, Marmora, Ontario: Ph. D. Thesis, Queen's University.

Quirke, T.T. and Collins, W.H., 1930, The disappearance of the Huronian: Geological Survey of Canada Memoir 160, 129 p.

Rondot, Jehan, 1978, Stratigraphie et métamorphismes de la région du St. Maurice: Geological Survey of Canada Paper 78-10, p. 329-352.

Rose, E.R., 1958, Iron deposits of eastern Ontario and adjoining Quebec: Geological Survey of Canada Bulletin 45, 120 p.

Rupert, R.J., 1963, A study of the fluorospar mines near Madoc, Ontario: B.Sc. Thesis, Queen's University.

Sangster, A.L., 1967, Metamorphism of the New Calumet sulphide deposit, Quebec: M.Sc. Thesis, Carleton University.

———————, 1970, Metallogeny of base metal, gold and iron deposits of the Grenville Province of southeastern Ontario. Ph.D. Thesis, Queen's University.

Sethuraman, K., 1979, Discovery of an Archean volcanogenic environment in the Grenville structural province, Echouani area, Quebec: Geological Association of Canada, Program with Abstracts, v. 4, p. 78.

Sharma, K.N.M., Clark, T., Franconi, A., and Laurin, A., 1978, Metamorphism in the central Grenville Province: Geological Survey of Canada Paper 78-10, p. 353-356.

Shklanka, R., 1968, Iron deposits of Ontario: Ontario Department of Mines and Mineral Resources, Circular 11, 489 p.

________________, 1969, Copper, nickel, lead and zinc deposits of Ontario: Ontario Department of Mines and Mineral Resources Circular 12, 396 p.

Silver, L.T. and Lumbers, S.B., 1965, Geochronological studies in the Madoc-Bancroft area of the Grenville Province (Abst.): Geological Society of America, Special Publication 87, p. 156.

Smythe, C.H., 1918, Genesis of the zinc ores of the Edwards district, St. Lawrence County, New York: New York State Museum Bulletin No. 201, p. 7-39.

Smith, B.L., 1956, Geology of the Clorendon-Dalhousie area: Ontario Department of Mines, v. 65, pt. 7, p. 1-46.

Solomon, P.J., 1963, Sulphur isotopic and textural studies of the ores at Balmat, New York and Mt. Isa, Queensland: Ph.D. Thesis, Harvard University.

Steiger, R.H. and Jäger, E., 1977, Subcommission on geochronology: Convention on the use of decay constants in geo- and cosmochronology: Earth and Planetary Science Letters, v. 36, p. 359-362.

Thompson, J.E., 1956, Geology of the Sudbury Basin, District of Sudbury: Ontario Department of Mines, v. 65, pt. III, p. 1-56.

Uglow, W.L., 1916, Lead and zinc in Ontario and eastern Canada: Ontario Bureau of Mines, v. 25, pt. 2, 56 p.

Wilson, M.E., 1929, Fluorspar deposits of Canada: Geological Survey of Canada Economic Geology Report #6, 97 p.

Wynne-Edwards, H.R., 1972, The Grenville province: in Price, R.A., and Douglas R.J.W., eds., Variations in Tectonic Styles in Canada: Geological Association of Canada Special Paper 11, p. 263-334.

MS received May 1980;
Accepted for publication March 1981.

Precambrian Sulphide Deposits, H.S. Robinson Memorial Volume, edited by
R.W. Hutchinson, C.D. Spence and J.M. Franklin, Geological Association of Canada
Special Paper 25, 1982

THE PURCELL SUPERGROUP IN SOUTHEASTERN BRITISH COLUMBIA: SEDIMENTATION, TECTONICS AND STRATIFORM LEAD-ZINC DEPOSITS

Trygve Höy
British Columbia Ministry of Energy, Mines, and Petroleum Resources,
Victoria, British Columbia V8V 1X4

ABSTRACT

Rocks of the Purcell Supergroup are exposed within the Purcell Mountains and the Clark, Hughes, Lizard and Galton Ranges of the Rocky Mountains in southeastern British Columbia and southwestern Alberta. They constitute a series of sedimentary wedges that, in general, thicken and become more argillaceous westward. In the eastern exposures in the Rocky Mountains thickness and facies changes are dramatic and the bulk of Purcell rocks are shallow-water shelf, tidal flat, flood plain or deltaic deposits. In the Purcell Mountains, changes are much more subtle and shallow-water facies are restricted to younger Purcell rocks which have prograded westward over older, deep-water turbidite facies.

The Purcell Mountains are cut by a number of northeast-trending transverse faults. Mesozoic granites and Precambrian basic volcanic rocks, gabbroic sills and a granite body are aligned with these faults. The faults follow the locus of Precambrian structures that controlled the deposition of the late Proterozoic Windermere Supergroup, and may locally have modified the depositional pattern of older Purcell rocks.

Stratiform lead-zinc deposits in the Aldridge Formation of the Purcell Supergroup are preferentially located in the vicinity of the transverse fault zones. Two of the deposits occur in somewhat thickened sections (Sullivan and Kootenay King) and a number are associated with intraformational conglomerates suggesting that local basin development associated with syn-depositional faulting may have been a localized control of mineralization.

RÉSUMÉ

Les roches du supergroupe de Purcell, d'âge protérozoïque, affleurent dans les montagnes de Purcell et les chaînes de Clark, Hughes, Lizard et Galton des Montagnes Rocheuses, dans le sud-est de la Colombie Britannique et le sud-ouest de l'Alberta. Elles constituent une série de coins sédimentaires qui, en général, s'épaississent et deviennent plus argileux vers l'ouest. Dans les affleurments orientaux des Montagnes Rocheuses les variations d'épaisseur et de faciès sont dramatiques et la pluspart des roches du supergroupe de Purcell représentent des dépôts peu profonds de plate-forme, de bas-fonds de la zone des marées, de plaine d'inondation ou de delta. Dans les montagnes de Purcell, les changements sont beaucoup plus subtiles et les faciès d'eau peu profonde sont limités aux roches les plus récentes du supergroupe de Purcell qui ont transgressé vers l'ouest sur des faciès plus anciens de turbidite d'eau profonde.

Les montagnes de Purcell sont coupées par un certain nombre de failles transverses nord-est. Les granites mésozoïques ainsi que des roches volcaniques, des sills de gabbro et un corps granitique, tous ceux-ci d'âge précambrien, sont alignés avec ces failles. Les failles suivent le lieu de structures précambriennes qui avaient contrôlé le dépôt du supergroupe de Windermere d'âge Protérozoïque tardif et qui peuvent avoir modifié localement le mode de dépôt des roches plus anciennes de Purcell.

Les dépôts stratiformes de Pb-Zn dans la formation d'Aldridge du supergroupe de Purcell, sont préférentiellement localisés au voisinage des zones de failles transverses. Deux de ces dépôts se trouvent dans des sections un peu épaissies (Sullivan et Kootenay King) et un certain nombre sont associés à des conglomérats intraformationnels suggérant que le développement local d'un bassin associé à des failles contemporaines des dépôts, ait pu être un contrôle localisé de la minéralisation.

INTRODUCTION

The Precambrian Purcell Supergroup comprises dominantly fine-grained clastic rocks exposed in the core of the Purcell anticlinorium in southeastern British Columbia. The Purcell anticlinorium (Fig. 1) is within the Columbian orogen, the eastern of two tectono-stratigraphic divisions of the Cordilleran orogen in western Canada (Wheeler *et al.*, 1972). To the west, the Purcell anticlinorium merges with the Kootenay Arc, a north-trending arcuate structural zone developed in a succession of rocks ranging in age from late Proterozoic to early Mesozoic. West of the Kootenay Arc is the Shuswap Metamorphic Complex, a terrain of high grade metamorphic rocks that constitutes the core of the Columbian orogen. East of the Purcell anticlinorium is the Rocky Mountain Foreland Thrust and Fold belt consisting of well-layered miogeosynclinal sedimentary rocks cut by generally "south-west dipping, concave upward, locally folded thrust faults" (Wheeler, 1970, p. 155).

The Purcell anticlinorium is transected by a number of steep transverse and longitudinal faults. The transverse faults appear to have been active periodically since at least late Proterozoic time and played an important role in controlling the type, distribution and thickness of late Proterozoic to early Paleozoic sediments (Lis and Price, 1976). Based on recent detailed studies on the east side of the Rocky Mountain Trench in the vicinity of the Kootenay King deposit (Höy, 1978), it is suggested that movement on one of these faults, the eastern extension of the St. Mary fault, began earlier and locally affected deposition of Purcell rocks. The transverse faults in the Purcell Mountains may also have been periodically active in pre-Windermere time and may be the cause of local occurrences of intraformational conglomerates in Purcell sedimentary rocks. Stratiform base metal deposits are commonly associated with these conglomerates, suggesting a genetic link between mineralization and synsedimentary faulting.

The base of the Purcell Supergroup is not exposed in southeastern British Columbia. It is inferred to rest unconformably on basement gneiss that in the Alberta foothills and in Montana has been dated at 1,700 to 1,900 Ma (Burwash *et al.*, 1962; Obradovich and Peterman, 1968); a minimum age of Purcell rocks is defined by a 750 Ma date for thermal metamorphism that accompanied the East Kootenay Orogeny, a period of regional uplift and associated block faulting that affected Purcell rocks (Leech and Wanless, 1962). Numerous gabbroic sills and dykes, with dates ranging from 1,665 Ma. (Hunt, 1962) to 760 Ma (Obradovich and Peterman, 1968) intrude the lower part of the Purcell succession. On the western flank and northern crest of the Purcell anticlinorium, Purcell rocks are overlain unconformably by coarser clastic rocks of the Windermere Supergroup of late Proterozoic age which in

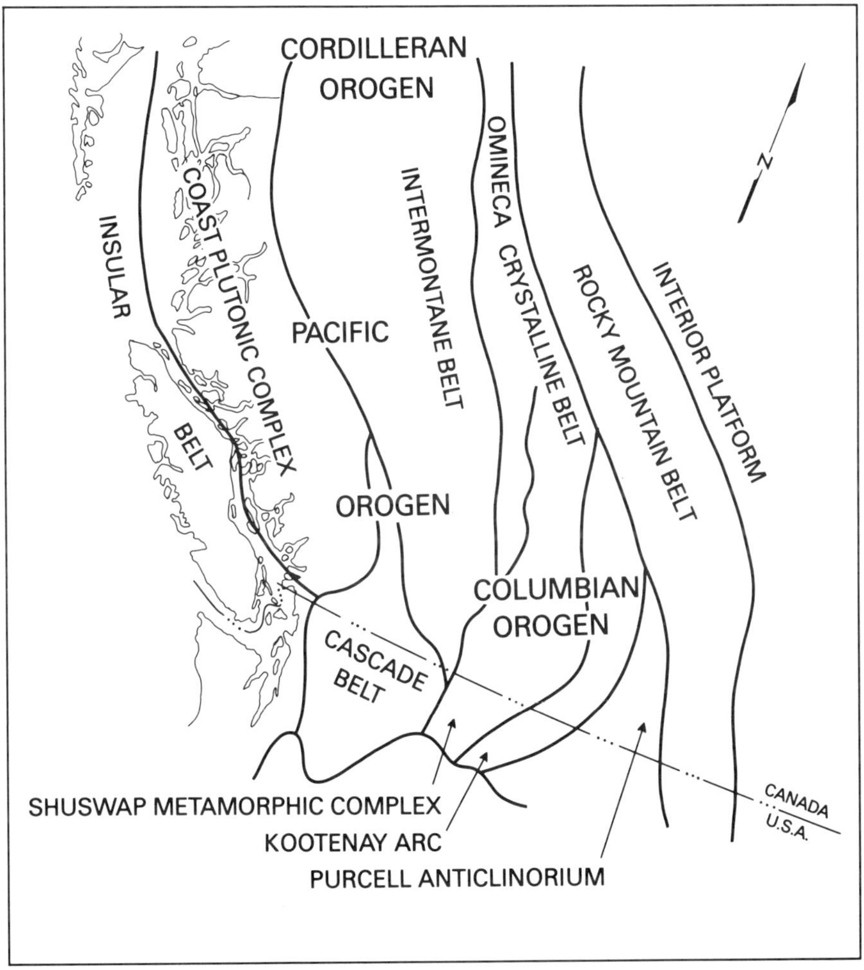

Figure 1. Regional tectonic map.

turn are overlain by well sorted lower Cambrian quartzite. Along most of the eastern flank of the Purcell anticlinorium, Purcell rocks are overlain unconformably by lower Paleozoic quartzite and carbonates.

Purcell rocks, and correlative Belt Supergroup rocks in the United States, were deposited in an epicratonic reentrant of a sea that extended along the western edge of the North American Precambrian craton (Harrison, 1972). A large proportion of Belt-Purcell rocks are deltaic or shallow-water tidal flat deposits with generally a southern or eastern source area (Price, 1964; Winston, 1977). Deeper-water turbidite deposits (Bishop *et al.*, 1970) underlie these shallow-water deposits in the central and western part of the "Belt basin".

Belt-Purcell rocks contain a number of stratabound metal occurrences and numerous vein deposits. The two major types of stratabound deposits in the Purcell Supergroup in Canada, lead-zinc and copper, have significant differences. Stratabound lead-zinc deposits, of which the Sullivan deposit is by far the largest, generally occur in more concentrated form, have a more limited areal distribution and are restricted to rocks of deeper water, more argillaceous facies. Stratabound copper occurrences are lower grade, generally more widespread, and were formed in fluvial or shallow-water environments. Stratabound copper occurrences are described by Morton *et al.* (1973) and Collins and Smith (1977) and will not be discussed further in this report.

This paper summarizes the stratigraphy and depositional environment of Purcell rocks and presents a model for the control of clastic-hosted stratiform lead-zinc deposits. Recent studies on the east side of the Rocky Mountain Trench (Höy, 1978, 1979; McMechan, 1979) stress the use of sedimentary structures, local facies changes and thickness variations to determine the depositional environment of Purcell rocks. Höy (1978) suggested that rapid lateral facies changes in Purcell rocks in the Kootenay King area east of the Rocky Mountain Trench are related to movements on easterly and northeasterly trending growth faults. This paper attempts to show that the distribution of stratiform Pb-Zn deposits may also in part be related to early transverse faults. The recognition of these early structures and their growth history, now recorded in the sedimentary succession may, therefore, be important in guiding future exploration for these deposits and in better understanding their origin.

STRATIGRAPHY AND DEPOSITIONAL ENVIRONMENT OF PURCELL ROCKS

Purcell rocks are exposed in three tectonically distinct areas in southeastern British Columbia and southwestern Alberta. From east to west, these are in the Clark Range within the Lewis thrust sheet, in the Hughes, Lizard and Galton Ranges on the east side of the Rocky Mountain Trench, and in the Purcell Mountains within the Purcell anticlinorium (Fig. 2). Correlations among these areas are illustrated in Figure 3. In the Clark Range, arenaceous and carbonate facies are relatively more dominant and formations are thinner and more distinctive lithologically. In the Purcell Mountains, formations are thicker and rocks somewhat more argillaceous, whereas rocks just east of the trench are transitional between the areas to the southeast and west.

Errata — Special Paper No. 25

Figure missing on page 292

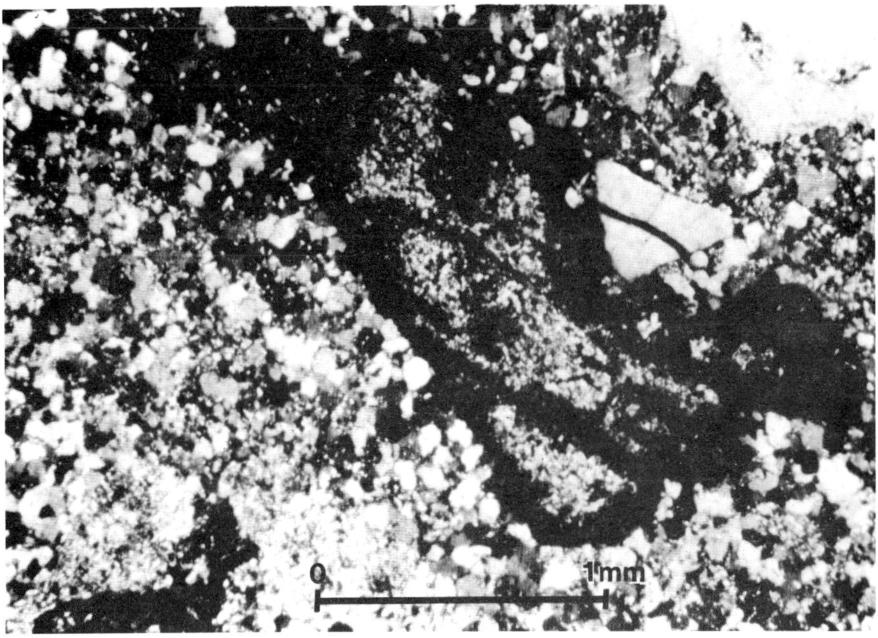

Figure 23. Alteration of a feldspar phenocryst in Q.F.P. The phenocryst is set in a matrix of microcrystalline quartz. The centre of the phenocryst is occupied by an aggregate of sericite (pale grey) and quartz. The dark rim and zones within the pseudomorphed phenocryst are chlorite replacing sericite. Crossed nicols.

Figure 9 on page 310 should read
Metal zoning, section 800N.

Figure 10 on page 311 should read
Plan illustrating iron mineral zoning.

Figure 17 on page 720 should read
Photomicrograph (reflected light) showing oriented exsolution lamellae of chalcopyrite in bornite. Darker grey digenite rims chalcopyrite lamellae and occurs as irregular intergrowths throughout bornite.

Figure 18 on page 722 should read
Core sample of copper-bearing cryptalgal-laminite interbedded with finely dolomitized, silty micrite from June Creek deposit. Sample location is 2 m above main cryptalgal-laminated unit at base of Coppercap Formation.

Clark Range

Stratigraphy. The Lewis thrust fault forms the base of the Purcell sequence in the Clark Range. The following descriptions are summarized from Douglas (1952) and Price (1962, 1964 and 1965). The total thickness of exposed Purcell rocks ranges from approximately 1500 m in southeastern Clark Range (Douglas, 1952) to greater than 5000 m in southwestern Clark Range (Price, 1964). The Waterton Formation at the bottom of the exposed section consists of laminated limestone and dolomite, stromatolitic dolomite and cherty dolomite. It is overlain by more argillaceous dolomite of the Altyn Formation. Striking facies and thickness variations in the Altyn Formation cause difficulties in correlation throughout the Clark Range but show trends that can be used in correlating with formations to the northwest and west. The Altyn Formation is overlain by green argillite, quartzite and sandy or argillaceous dolomite of the Appekunny Formation. Westward, the thickness, proportion of red

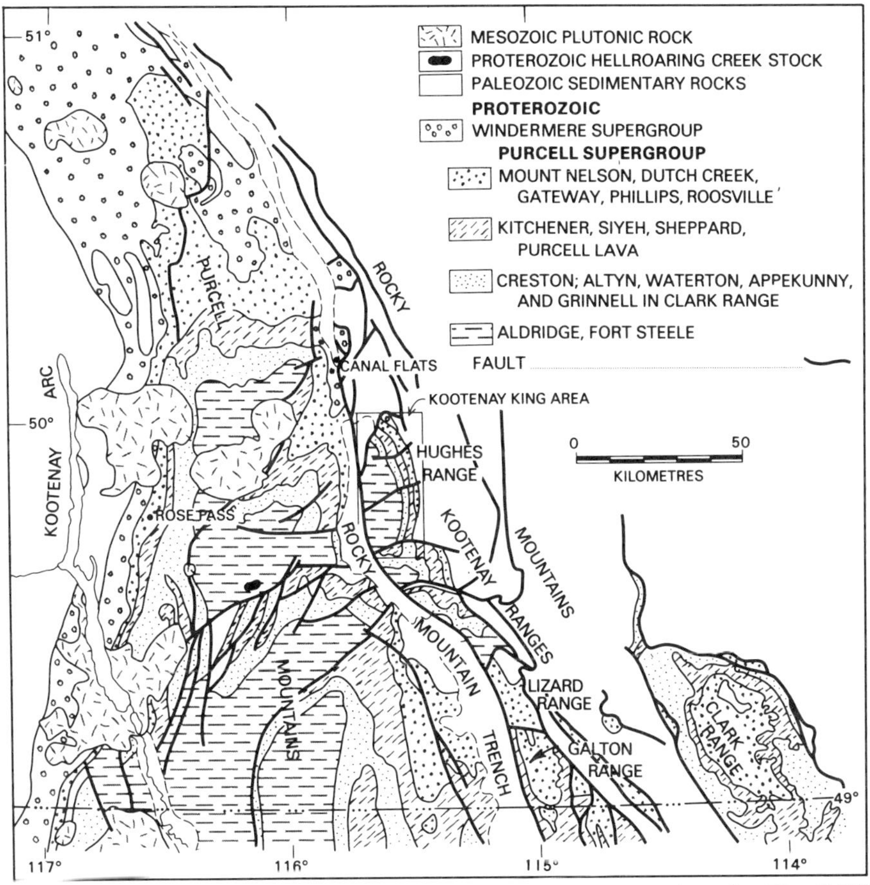

Figure 2. Geological map showing distribution of Purcell Supergroup; data after Rice (1941), Douglas (1952), Leech (1952, 1958a, 1960), Reesor (1958, 1973), Price (1962, 1965), Höy (1978, 1980), and Okulitch and Woodsworth (1977).

argillite and the number of ripple marks and mudcracks decrease. The Grinnell Formation, a red argillite and argillaceous siltstone that contains numerous mudcracks and ripple marks overlies the Appekunny Formation. It is transitional into argillaceous and arenaceous dolomite and dolomitic sandstone at the base of the overlying Siyeh Formation. Algal stromatolites occur in the central part of the Siyeh and dolomitic argillite and green argillite dominate in the top part. The overlying Purcell lava, which is the most important time-stratigraphic marker in the eastern exposures of the Purcell Supergroup, consists of a series of chloritized andesite flows. The Sheppard Formation above the Purcell lava consists of dolomite, stromatolitic dolomite, sandstone, siltstone and argillite. It grades upward into dark red argillaceous siltstone and argillite of the lower member of the Gateway Formation. The upper member of the Gateway consists of greenish argillite, dolomitic argillite, and sandy dolomite. It is overlain in turn, by red and purplish-red sandstone and siltstone of the Phillips Formation and green argillite, dolomitic argillite, siltstone and sandstone of the Roosville Formation. Clean quartz sandstone of the Middle Cambrian Flathead Formation rests unconformably on the Roosville Formation. The unconformity cuts down section into the Phillips Formation to the east and northwest.

Depositional Environment. Purcell rocks in the Clark Range are dominantly shallow-water facies deposited in a shelf or tidal flat environment (Price, 1964). The Waterton Formation is a shallow-water carbonate shelf deposit. The Altyn Forma-

PURCELL MOUNTAINS	HUGHES–LIZARD–GALTON RANGES	CLARK RANGE
CAMBRIAN		
Hamill-Cranbrook	Jubilee Eager Cranbrook	Flathead
		(unconformity)
WINDERMERE SUPERGROUP		
Horsethief Creek Toby	Horsethief Creek Toby	
		(unconformity)
PURCELL SUPERGROUP		
Mount Nelson		
Dutch Creek	Roosville Phillips Gateway	Roosville Phillips Gateway Sheppard
Kitchener-Siyeh	Purcell lava	Purcell lava
	Kitchener-Siyeh	Siyeh
Creston	Creston	Grinnell Appekunny
Aldridge	Aldridge Fort Steele	Altyn Waterton
		(Lewis thrust)
(base not exposed)	(base not exposed)	

Figure 3. Correlation of Purcell Supergroup in southeastern British Columbia; after Leech (1958a, 1958b, 1960), Price (1964), and Edmunds (personal communication, 1976).

tion thickens and becomes more argillaceous westward suggesting a transition from a shallow-water carbonate platform environment in the east to a deeper-water environment characterized by deposition of dark terrigenous muds to the west. Similarly, the Appekunny and overlying Grinnell Formations thicken southwestward changing from non-marine flood-plain or brackish tidal flat deposits to deeper-water marine deposits. Carbonates of the Siyeh and Sheppard Formations represent deposition of calcareous muds on an extensive shallow-water platform or shelf that was intermittently emergent. An influx of terrigenous clastics on to extensive tidal flats marks the start of Gateway deposition. The oxidized siltstone and sandstone of the Phillips Formation and the abundance of mudcracks and ripple marks in the Phillips and overlying Roosville also suggest a tidal flat environment.

Hughes, Lizard and Galton Ranges

Stratigraphy. Purcell rocks are exposed in the Hughes, Lizard and Galton Ranges along the east side of the Rocky Mountain Trench. Descriptions of all or some of these rocks have been published by Rice (1937), Leech (1958b and 1960), Price (1962) and more recently by Höy (1978 and 1979) and McMechan (1978 and 1979). The following discussion is confined largely to the Kootenay King area within the Hughes Range (Fig. 2). The Fort Steele Formation is at the base of the exposed succession in the Kootenay King area. It comprises at least three sequences, several hundred metres thick, that grade upwards from coarse-grained, cross-bedded quartzites at the base to finely laminated siltstones containing numerous mudcracks at the top.

Conformably overlying the Fort Steele Formation and transitional with it is approximately 3800 m of rusty weathering argillite, siltstone and shale of the Aldridge Formation. The Aldridge Formation comprises three main divisions in the Kootenay King area. The lowest division, Al, is dominantly laminated and ripple cross-laminated siltstone, finely laminated dark argillite, quartzite and, near the base, a dolomite member containing intercalated chert. Quartzite beds that contain coarse (to several cm diameter) siltstone and argillite clasts near their base and occasional northerly trending flute casts are interlayered with a buff-weathering, commonly graded and lenticular-bedded siltstone near the middle of Al. Individual quartzite beds tend to become thinner and finer-grained northward.

The middle division of the Aldridge Formation, A2, is approximately 500 m thick in the Kootenay King area (Höy, 1978) whereas similar rocks to the south are more than 2800 m thick (McMechan, 1979). It consists of interlayered siltstone-quartzite beds separated by dark, rusty weathering argillite. The quartzite beds are generally massive, although the top few centimetres may be crudely layered and finer-grained than the base. Flute casts on the base of some quartzite beds indicate northward transport.

The upper division, A3, comprises 500 m to 800 m of finely laminated, dark argillite and siltstone. It is transitional into green quartzite and siltstone at the base of the overlying Creston Formation.

The Creston Formation, approximately 1500 m thick, consists dominantly of green and white siltstone and quartzite with numerous mudcracks, cross-beds and ripple marks. Thick units of purple siltstone with thin white cross-bedded and graded quartzite layers are abundant in the middle part of the Creston Formation within the Kootenay King area. The top of the formation is green-tinged, finer-grained, and more evenly bedded. Argillaceous dolomite beds appear in the transition zone with the overlying Kitchener-Siyeh Formation. Because of confusion concerning the correct usage of the terms Kitchener and Siyeh, rocks between the Creston Formation and the Purcell lava in the Hughes, Lizard and Galton Ranges, and in the Purcell Mountains, are informally referred to as the Kitchener-Siyeh formation. They are correlative with the middle and upper members, respectively, of Siyeh Formation in the Clark Range (Price, 1964). McMechan *et al.* (1980) propose to clarify this confusion by introducing two new formation names. The "Van Creek Formation" includes dominantly siltite and argillite between the Kitchener Formation and the base of the overlying "Nicol Creek Formation". The "Nicol Creek Formation" includes volcaniclastic rocks, flows and tuffs between the Van Creek Formation and the overlying Sheppard or Gateway Formations. In the lower part of the Kitchener-Siyeh, carbonate dominates and consists of interlayered silty dolomite, dolomitic siltstone, and massive dolomite; the upper part consists of interlayered siltstone and argillite. The Kitchener-Siyeh formation contains numerous graded beds, siltstone-filled scour channels, mud-chip breccias, mudcracks and cross-beds. The overlying Purcell lava is made up of a number of sequences of chloritized porphyritic and amygdaloidal andesitic flows, separated by siltstone. The lavas are overlain by a unit of siltstone and argillite that grades upwards into interlayered dolomite and siltstone, and then into a succession of interlayered stromatolitic dolomite, oolitic limestone, sandy dolomite and green or locally purple siltstone and argillite, and finally into thick, massive dolomite and siltstone beds that mark the top of this carbonate-rich succession. This succession is correlative with rocks of the combined Sheppard Formation and Gateway Formation (member A of the Kintla Formation) in the Clark Range (Price, 1964) and has been termed the Gateway Formation by Rice (1937) and Leech (1958b). Purple argillite and siltstone of the Phillips Formation and dolomite, argillite and green siltstone of the Roosville Formation overlie the Gateway Formation.

A sub-Cambrian unconformity truncates rocks of the Gateway Formation northward throughout most of the Hughes Range north of the Dibble Creek fault (Fig. 2). However, in the northern part of the Kootenay King area, the unconformity cuts up-section exposing the Phillips Formation and the overlying Roosville Formation. The Phillips Formation also is exposed just north of the Dibble Creek fault (McMechan, 1979). South of the fault, a sub-Devonian unconformity, although irregular, generally cuts up-section towards the south. Immediately south of the fault, Kitchener-Siyeh rocks and the Purcell lava unit occur at the unconformity whereas farther south, in the Lizard Range, the Gateway Formation occurs at the unconformity (McMechan, 1979). Southward in the Galton Range, more than 1000 m of Roosville strata occur beneath the sub-Cambrian unconformity (Leech, 1960; Price, 1962; Norris and Price, 1966).

Depositional Environment. As in the Clark Range, the bulk of Purcell rocks in the Hughes Range record deposition in a shallow-water tidal flat or flood-plain environment. However, lithology of the Aldridge Formation indicates that it was deposited in deeper water.

The Fort Steele Formation at the base of the Purcell Supergroup is unique to the Kootenay King area. It consists predominantly of braided fluvial deposits derived from a source area to the south (Höy, 1978). A marine transgression is apparent in lower Aldridge rocks, and the alluvial fan deposits of the Fort Steele are overlapped by intertidal and subtidal mud flat deposits, which give way upward to slightly crenulated and laminated carbonates that are similar to subtidal algal mat deposits. Overlying laminated siltstone and argillite, correlative with the middle Aldridge of the Purcell Mountains (Cominco geologists, pers. commun.) represent continuing transgression. In the Kootenay King area (Fig. 2), the lower division of the Aldridge Formation, Al, thins from approximately 2500 m in the south to less than 1500 m only 15 km to the north. In addition, the character of lower division rocks also changes northward. Thick sections of quartz-rich siltstone in the south give way to finer-grained and more dolomitic siltstone to the north and finally grade into laminated black argillite. As well, thick, coarse-grained quartzites that contain fine to coarse, angular clasts near their base become thinner and finer-grained northward. The thinning and fining of sediments to the north, reinforced by a northerly current transport direction, suggested (Höy, 1978) that the Kootenay King area formed in a fault-bounded basin with south-side-up in middle Aldridge time. The Boulder Creek fault, an eastern extension of the St. Mary Fault which was active in Windermere time (Lis and Price, 1976), is located here and was, therefore, postulated to be the locus of an earlier fault that defined the southern limit or hinge line of the basin. However, there was a thick succession of correlative turbidites south of the Kootenay King area and Boulder Creek fault zone (McMechan, 1979) in middle Aldridge time, indicating that there may have been a major structural basin here. The Kootenay King area, therefore, appears to be a local fault-bounded, sub-basin on the northern rim of a large structural basin. Coarse sandstone and dolomitic siltstone near the southern end of the Kootenay King area may be shoal deposits on a topographic high near the basin margin, whereas the bulk of sedimentary rocks in the Kootenay King area may be over-spill deposits into a small clastic sub-basin. Coarse quartzite layers, interpreted to be channel deposits in the inner part of the submarine fan (Höy, 1978), may have been transported along channels that cut through a barrier shoal complex, and deposited in the Kootenay King sub-basin. Channel erosion may have been initiated by fault movements in the Boulder Creek hinge line. The Kootenay King, a stratiform lead-zinc deposit, is in the upper part of the thickest and coarsest of these quartzite sequences. Quartzite interlayered with conglomerate just north of the transverse Lewis Creek-Nicol Creek fault zone (See Fig. 6 for location) also thins, becomes finer-grained and dies out to the north suggesting that this zone also was tectonically active in Aldridge time and locally controlled sedimentation.

The middle division of the Aldridge Formation consists of proximal turbidite deposits, transported in a northerly direction. The first evidence of a marine regression appears near the top of the Aldridge Formation (Höy, 1978) and green siltstone and shale near the base of the overlying Creston Formation were deposited in a peritidal environment. Overlying purple siltstones that contain numerous mudcracks, rip-up clasts and ripple marks are tidal flat or flood-plain deposits similar to those described in the Clark Range. The carbonate rocks of the Kitchener-Siyeh, and the mudcracked siltstone and shale of the Gateway, Phillips and Roosville Formations, were also formed on extensive tidal flats or flood plains (Price, 1964).

Purcell Mountains

Stratigraphy. The oldest Purcell rocks exposed in the Purcell anticlinorium are thinly to irregularly bedded, rusty-weathering argillite, siltstone and quartzite of the lower Aldridge Formation (Fig. 3). They are overlain by about 3000 m of interbedded grey arenaceous quartzite and thinly bedded siltstone and argillite of the middle Aldridge Formation.

The lower Aldridge – middle Aldridge contact, normally located where grey quartzite becomes predominant, is not at the same stratigraphic level throughout the Purcell Mountains. In the Dewar Creek area just south of the White Creek batholith, Reesor (1958) noted that the contact is somewhat arbitrary as several thin successions of grey quartzite occur in the upper part of the lower Aldridge. Farther south in the Purcell Mountains, the contact is even less distinct and grey quartzite layers, similar to those that characterized the middle Aldridge, occur at much lower levels. The contact at some locations is marked by a discontinuous sheet of conglomerate (Edmunds, 1977). This conglomerate is most noticeable in the footwall of the Sullivan deposit and the stratiform Hilo lead-zinc showing.

The quartzite beds in the middle Aldridge are generally massive, but contain sedimentary structures that suggest that they are turbidite deposits (Bishop *et al.*, 1970; Edmunds, 1973). Within the interlayered siltstone-argillite successions in the middle Aldridge are dark and light coloured laminated siltstone sequences several metres thick that can be matched, lamination for lamination, across distances of several hundred kilometres (Edmunds, 1973; Huebschman, 1973). They are reliable stratigraphic markers that allow correlation of otherwise nondescript Aldridge metasedimentary rocks. Their recognition in the Hughes Range (F.R. Edmunds, pers. commun., 1976) forms the basis of the correlation of the bulk of unit Al and all of A2 with the middle Aldridge of the Purcell Mountains and consequently allows, but does not dictate, the correlation of the lower part of Al and the Fort Steele Formation with the lower Aldridge.

The upper Aldridge Formation comprises a 300 m to 400 m succession of thinly bedded, rusty-weathering argillite and siltstone. The overlying Creston Formation consists of varicoloured argillaceous quartzite, siltstone and argillite that contain numerous shallow-water sedimentary structures. The Creston is overlain by a carbonate-rich unit that grades upward into a more argillaceous unit. These two units are grouped together and referred to informally as the Kitchener-Siyeh formation (e.g. see Reesor, 1973; Rice, 1941). The Purcell lavas overlie the Kitchener-Siyeh formation on the southeastern flank of the Purcell anticlinorium but are not present throughout most of the Purcell Mountains. However, a tuffaceous horizon occurs at the same stratigraphic level south of the White Creek batholith 40 km to the west of Purcell lava exposures (Rice, 1941, p. 10; Reesor, 1958, p. 21). Rocks of the overlying Dutch Creek and Mount Nelson Formations become somewhat darker and more argillaceous westward. On the northeast side of the Purcell anticlinorium, near 50°N latitude, the Dutch Creek Formation consists mainly of grey and green argillite, dolomite and quartzite (Leech, 1958a). Farther northwest it is made up of varicoloured slates with abundant quartzite and some limestone and dolomite at the base (Reesor, 1973) whereas at Rose Pass, on the west side of the Purcell anticlinorium, finely laminated grey to black argillite dominates even though quartzite and limestone beds are still abundant in the basal part. The Mount Nelson Formation also changes to the west from dominantly thick successions of dolomite and dolomitic limestone (and associated oolitic and stromatolitic dolomite) in the Lardeau map area (Reesor, 1973; Atkinson, 1976) to mainly laminated argillite and argillaceous limestone in the vicinity of Rose Pass (Rice, 1941). A prominent white basal quartzite thickens and becomes more argillaceous westward, and mudcracks and ripple marks, which are common in the northeastern exposures, are rare in the west. The Toby conglomerate at the base of the Windermere Supergroup unconformably overlies the Dutch Creek or Mount Nelson Formations on the western and northeastern flanks of anticlinorium.

Depositional Environment. Thick accumulations of finely laminated, rusty-weathering dark argillite and siltstone of the Aldridge Formation, that lack sedimentary structures indicative of subareal exposure or wave action, were deposited in deeper water parts of the "Belt basin". Middle Aldridge time is marked by the introduction of extensive and thick accumulations of turbidite deposits with local intraformational conglomerates. The conglomerates appear to be localized near late fault zones, such as the Hall Lake fault and between the St. Mary and Kimberley faults (Fig. 6), suggesting that these faults, or perhaps more northerly trending splay faults may coincide with much older faults that produced local scarps in middle Proterozoic time (see section entitled "Transverse structures"). Movement on faults in middle Aldridge time might have triggered turbidite deposition. Edmunds (1973, 1977) noted that sole markings on the turbidite beds trend north-south, defining a north-south basinal axis through the Purcell Mountains.

Purcell rocks overlying the Aldridge Formation are largely shallow-water deposits. There is a subtle westward change to thicker, more argillaceous successions with fewer sedimentary structures indicative of wave action or subareal exposure, suggesting deeper-water conditions to the west.

Summary – The Purcell Supergroup in Southeastern British Columbia

Purcell rocks in Canada are on the northern edge of a large northwest-trending basin that developed in an epicratonic re-entrant along the eastern edge of the Cordilleran miogeocline (Harrison, 1972). They constitute a series of sedimentary wedges that, in general, thicken and become more argillaceous westward. In the eastern exposures in the Clark Ranges and in the ranges just east of the Rocky Mountain Trench, thickness and facies changes are pronounced and the bulk of Purcell rocks are shallow-water shelf, tidal flat, flood-plain or deltaic deposits. In the Purcell Mountains, changes are much more subtle, and shallow-water facies are restricted to younger Purcell rocks which have prograded westward over older deep-water facies. These relations are documented clearly by Price (1964), who compared the depositional environment of Purcell rocks to the present Gulf Coast miogeocline, with sediments derived from a large, low relief drainage basin in the craton, and deposited on the western margin of the craton on and adjacent to the flood-plain of a large subsiding delta.

Correlation of Purcell rocks between the Purcell Mountains and the Hughes, Lizard and Galton Ranges, is taken from Leech (1958a, 1958b and 1960). Based on

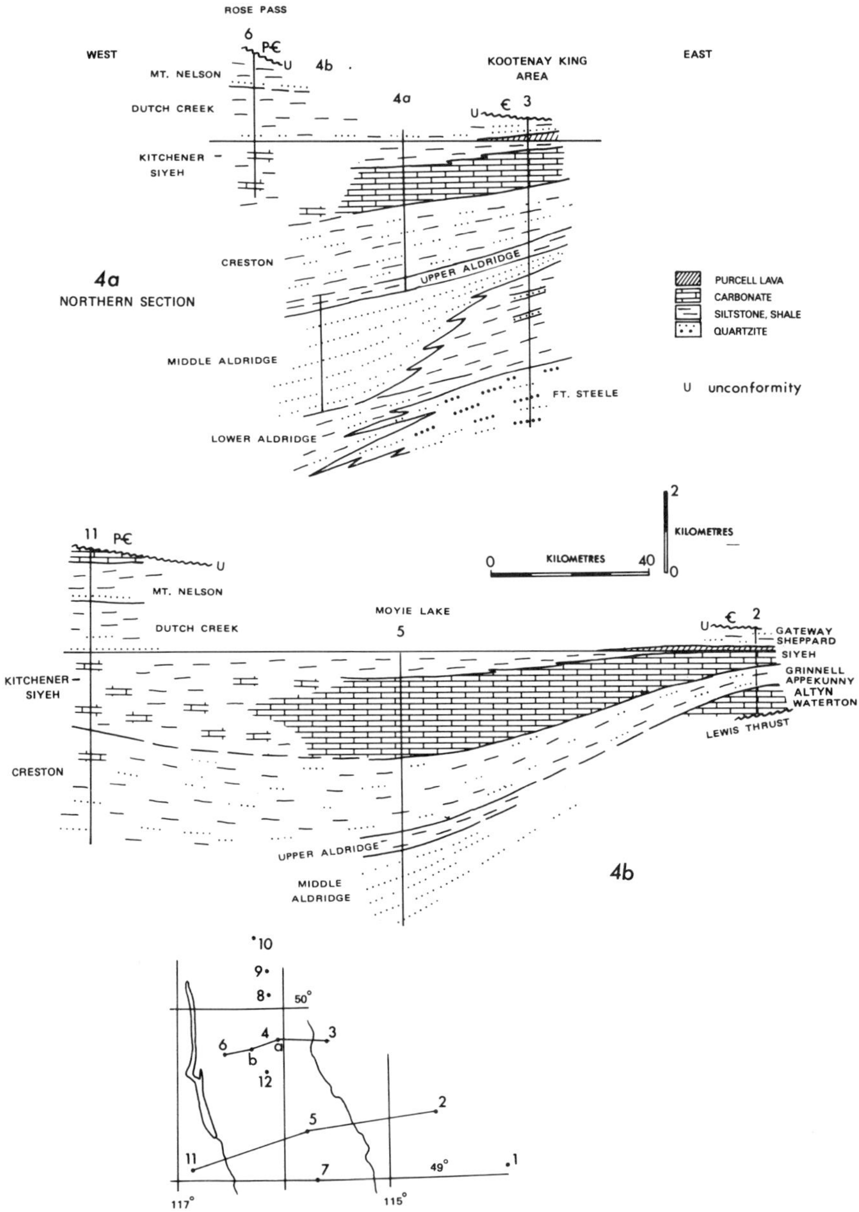

Figure 4. Schematic, east-west stratigraphic sections, Purcell Supergroup, based on correlations by Price (1964). Numbered sections are: Section 1 – Waterton area, averages of thickness estimates by Douglas (1952); Section 2 – Flathead area, measured sections, Price (1965); Section 3 – Hughes Range, measured sections (Höy, 1978, and unpublished data); Section 4 – Dewar

marker data by Cominco geologists (previously referred to under Hughes, Lizard and Galton Ranges: Stratigraphy heading), the basal part of the Aldridge Formation and the Fort Steele Formation in the Hughes Range are tentatively correlated with the lower Aldridge of the Purcell Mountains. The correlation of these successions with the Clark Range succession is after Price (1964).

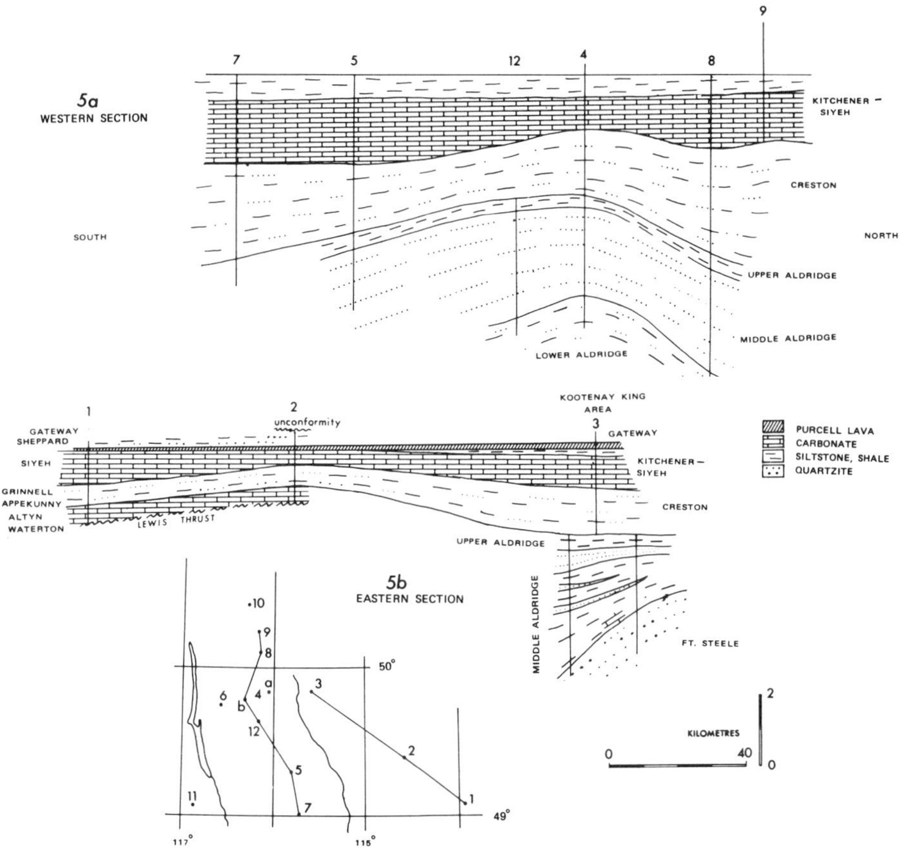

Figure 5. Schematic, north-south stratigraphic sections, Purcell Supergroup, based on correlations by Price (1964); see caption to Figure 4 for details of individual numbered sections.

Figure 4 (cont'd)

Creek area, Purcell Mountains, measured sections (Reesor, 1958); Section 5 – Moyie Lake area, Purcell Mountains, Upper Aldridge and Creston thicknesses are estimated from unpublished maps by Höy; Kitchener-Siyeh thicknesses are estimates by Rice (1937); Section 6 – Rose Pass area, measured sections, Rice (1941); Section 7 – U.S.A. border area, Purcell Mountains, estimated from map of Leech (1960); Sections 8 and 9 – Purcell Mountains, estimates by Reesor (1973); Section 10 – Toby Creek area, Purcell Mountains, measured sections of upper 300 m of Dutch Creek Formation (Atkinson, 1976) and upper 400 m of Mt. Nelson Formation (Atkinson, 1977) (not plotted in Figures 4 or 5); Section 11 – Selkirk Mountains, estimated from map of Rice (1937); Section 12 – St. Mary Lake area, Purcell Mountains, estimates by Leech (1952).

The schematic stratigraphic sections through the Purcell basin (Figs. 4 and 5) illustrate regional facies and thickness trends. In the southern section (Fig. 4b) a relatively thin, shallow-water carbonate platform in the east thickens and becomes more argillaceous westward. It is homotaxial with a thick succession of dark, laminated argillite and siltstone, and turbidite deposits of the Aldridge Formation in the Purcell Mountains. Farther north (Fig. 4a), coarse fluvial sandstone of the Fort Steele Formation appears to be supplanted to the west by deep-water, finely laminated argillite and siltstone, and quartz sandstone of the lower Aldridge. Coarse-grained quartzite and locally conglomerate in calcareous siltstone and argillite overlying the Fort Steele Formation, that are interpreted to be channel deposits in the inner part of a submarine fan, may have been the local source of turbidite deposits in correlative middle Aldridge rocks to the north and west. With continued subsidence, the basin of turbidite deposition expanded and overlapped the inner fan deposits in the Kootenay King area. Shallow-water clastic facies of the Appekunny and Grinnell Formations in the Clark Range and the Creston Formation just east of the Rocky Mountain Trench thicken westward, and the relatively thin succession of dominantly carbonate rocks of the overlying Kitchener-Siyeh formation similarly thickens and becomes more argillaceous to the west (Figs. 4a and 4b).

Purcell rocks overlying the Purcell lava in eastern exposures are divisible into a number of lithologically distinct formations that are primarily tidal flat or flood-plain deposits. Farther west in the Purcell Mountains, correlative rocks are thicker, contacts between formations are more gradational, and dark argillaceous facies more prominent. These are generally deeper water shelf-type sedimentary rocks.

TRANSVERSE STRUCTURES

The Purcell Mountains exhibit a pronounced northeast-trending structural grain, delineated by late transverse faults with attendant localization of granitic intrusions. This grain is inherited from older, fundamental faults that were active during sedimentation in late Proterozoic Windermere time and lower Paleozoic time (Leech, 1958b; Price and Lis, 1975; Lis and Price, 1976). It is suggested that these faults were also active in pre-Windermere time and locally modified the regional facies trends of Purcell rocks and partially controlled the location of stratiform Pb-Zn deposits.

Kanasewich *et al.* (1969) hypothesized that a southwest-trending Precambrian rift, identified beneath 2.5 km of flat-lying Paleozoic and Mesozoic sedimentary rocks in Alberta, continued beneath the Rocky Mountains into southeastern British Columbia. The evidence for the western extension of this rift is based primarily on a Bouguer gravity low and a magnetic lineation, both of which trend southwesterly, south of Kimberley. As pointed out by Price and Lis (1975), the rift coincides with two late transverse fault zones, the St. Mary and Boulder Creek faults on the north and the Moyie and Dibble Creek faults to the south. A granitic intrusion, the Bayonne batholith, occurs at the western extension of the St. Mary fault (Fig. 6). To the north, a third structural zone is less well defined, but is also inferred to transect the general north to northwest structural grain. It coincides with a series of late northeast-trending faults and a synclinal structure south of Canal Flats, with part of the Hall Lake fault, and with the White Creek batholith.

The Moyie-Dibble Creek fault system coincides with the southern transverse zone. The Dibble Creek fault, an eastern extension of the Moyie fault, is a north-dipping right-hand reverse fault that appears to follow the locus of an earlier fault along which the north side was relatively down-dropped (Leech, 1958b) in pre-late Devonian time. North of the fault, upper Devonian rocks rest unconformably on a thick sequence of lower Paleozoic strata, whereas south of it upper Devonian rocks rest unconformably on the Kitchener-Siyeh formation of the Purcell Supergroup (Leech, 1958b). A similar relationship exists across the Moyie fault. A thick sequence of Cambrian rocks lies unconformably on Upper Purcell rocks north of the fault, whereas Devonian rocks rest unconformably on the Purcell succession south of it (Leech, 1972).

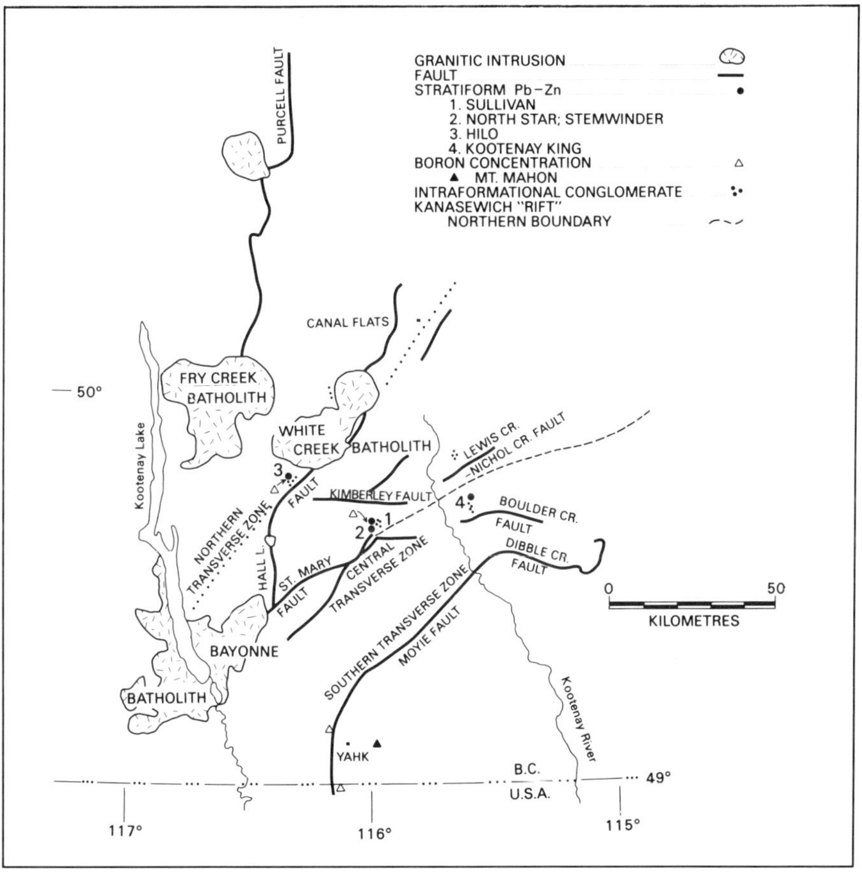

Figure 6. Location of northeast trending structures (slightly modified from regional maps by Rice, 1941; Leech, 1958a and 1960; and Reesor, 1973); and distribution of stratiform Pb-Zn occurrences (from British Columbia Ministry of Mines Mineral Inventory maps), boron concentrations (Ethier and Campbell, 1977 and Gifford, 1971) and intraformational conglomerate (Gifford, 1971; Ransom, 1977; Höy, 1978).

The St. Mary-Boulder Creek fault system coincides with the central transverse structural zone. The St. Mary fault, a north-dipping right-hand reverse fault, follows the locus of a late Proterozoic fault along which the north side was relatively down-dropped (Lis and Price, 1976). North of the fault, approximately 9 km of pelite, sandstone and conglomerate of the Windermere Supergroup lie unconformably on the Upper Purcell Mount Nelson Formation, and are in turn overlain unconformably by Lower Cambrian quartzites of the Hamill and Cranbrook Formations. South of the fault, the Windermere is completely missing and the lower Cambrian strata rest unconformably on middle Purcell rocks.

The western extension of the central structural zone extends into the Kootenay Arc southwest of Kootenay Lake (Fig. 6) where it is marked by a thick sequence of andesitic volcanic rocks, the Irene Formation, in the Windermere succession. Farther west, a thick middle Cambrian limestone, the Nelway Formation (Fyles and Hewlett, 1959), may have developed on a relatively uplifted area south of the structural zone. The limestone dies out to the north, supplanted by more argillaceous rocks of the Lardeau Group in the Riondel area (Höy, 1980).

Early movements on associated faults in the northern transverse structural zone have not been documented. The zone is inferred to coincide with northeast-trending structures in the Canal Flats area and a northeast-trending segment of the Hall Lake fault (Fig. 6). It may extend southwestward to just north of the Bayonne batholith where a tectonic high may have developed south of the zone in post-Purcell/pre-Windermere time; immediately north of the batholith, the Toby Formation at the base of the Windermere rests unconformably on the Dutch Creek Formation whereas 10 km farther north it rests on the younger Mount Nelson Formation (Rice, 1941). This northern zone may merge with the St. Mary-Boulder Creek transverse zone in the region of the Bayonne batholith.

The suggestion that these transverse zones were active structural features in pre-Windermere time and that they influenced the deposition of Purcell sediments and stratiform mineral deposits is based primarily on a comparison with the Kootenay King area east of the Rocky Mountain Trench. Here, dramatic thickness and facies changes northward from the southern margin of the map area, from the vicinity of the Boulder Creek fault, can be explained by fault movements on the margins of the depositional basin in Aldridge time. The coarse sands in the channel deposits in the middle Aldridge may be derived locally, perhaps from a barrier sand complex to the south or from overspill of turbidites from south of the Boulder Creek fault, and deposited in a local fault-bounded structural basin to the north. Conglomerate, consisting of angular clasts of siltstone and argillite in the basal part of the channel sand deposits, probably represents both basal rip-up debris and local slump debris.

Coarse conglomerate-quartzite layers 14 km to the north that thin northward, accompanied by a rapid decrease in clast and grain size, also suggest that local Purcell-age faulting influenced sedimentation. This inferred Purcell fault zone coincides with the late Nicol Creek and Lewis Creek faults (Fig. 6; Höy, 1978).

McMechan (1979) ascribed stratigraphic changes in Purcell sedimentary rocks between the Dibble Creek fault and Boulder Creek fault to "early structures" and indicated that differential basin subsidence and sedimentation occurred concurrently in this area. McMechan (1979, p. 6) noted, however, that stratigraphic contrasts in Purcell rocks across the Dibble Creek fault may be due to "both stratigraphic changes related to the evolution of this ancient structure and to juxtaposition of originally separate terrains by fault movement".

In summary, the Boulder Creek fault and perhaps the Dibble Creek fault may have been intermittently active in Purcell time, and may have locally controlled or modified Purcell sedimentation. The Boulder Creek fault may have been the structural hinge line of a large basin that coincided with the Kanasewich rift zone. Local, short-lived tectonic sub-basins developed along its northern margin in the Kootenay King area. Intraformational conglomerate and coarse sandstone, shed northward, are the most obvious records of the development of this growth fault.

By analogy with the Kootenay King area, intraformational conglomerate within the Aldridge Formation in the Purcell Mountains may be indicative of local faulting during Purcell sedimentation. Edmunds (1977, p. 24) attributed "slump conglomerates" near the lower Aldridge-middle Aldridge contact to the "collapse of local submarine fault scarps". Many of these conglomerate occurrences (Fig. 6), described in the following section, are preferentially located adjacent to major transverse structures. Lithologic or thickness changes in Purcell rocks that can be attributed to syndepositional faulting in the Purcell Mountains are not as apparent. Reesor (1958) noted minor changes in the basal member of the Creston Formation across the Hall Lake fault (which might merely be related to juxtaposition of more separate terrains by late movements on the fault), and an increase in the thickness of the middle turbidite unit of the Aldridge Formation westward from the fault.

RELATIONSHIP BETWEEN FAULTING, SEDIMENTATION AND STRATIFORM MINERAL DEPOSITS

A number of stratiform lead-zinc deposits in Aldridge sedimentary rocks are preferentially located in the vicinity of transverse fault zones (Fig. 6). The Sullivan (this volume) and the Kootenay King (this paper), occur in somewhat thickened sections and a number are associated with intraformational conglomerate suggesting that local basin development associated with syndepositional faulting was a localized control of mineralization.

The thickest and coarsest conglomerates occur close to the lower Aldridge-middle Aldridge contact. Conglomerate at this horizon is found beneath the stratiform Sullivan deposit and below the Hilo mineral showing. Sullivan, which is located between the St. Mary and Kimberley faults, has an intraformational conglomerate up to 60 m in thickness in the footwall of the orebody (Ransom, 1977). It consists of rounded to angular clasts of argillite and siltstone in an argillaceous matrix. Locally it is cut by a chaotic breccia that occurs in north-trending zones and is crossed by alteration zones associated with a fracture system that underlies the central, more massive part of the orebody (Ransom, 1977). Hilo, which is located just north of the Hall Lake fault, has a conglomerate-diamictite layer in excess of 100 m thick in the footwall of the showing (Gifford, 1971). The conglomerate underlies a thin stratiform pyrrhotite-laminated zone containing minor lead and zinc (Gifford, 1971). The conglomerate layer consists of locally derived siltstone and argillite clasts in an argillaceous to silty matrix. As noted previously, conglomerate occurs in the basal part of the quartzite that underlies the Kootenay King deposit in the Hughes Range. The quartzite is considered to be a submarine channel deposit that developed in response to fault movements along the basin margin.

A second weaker line of evidence relating stratiform mineralization to syndepositional faulting is the association between mineralization and high boron (tourmaline) concentrations. Ethier and Campbell (1977) suggested that boron in Aldridge sedimentary rocks is concentrated near basin fractures. Anomalous boron concentrations (Fig. 6) occur in the vicinity of the Hilo occurrence (Gifford, 1971), the Sullivan, Stemwinder and North Star deposits, and in the Yahk area (Ethier and Campbell, 1977). All but one of the occurrences are, on a regional scale, in the vicinity of known transverse fault zones. The exception, the Mt. Mahon showing, is associated with an intraformational conglomerate. Local controls, as appears evident for example, at the Sullivan deposit by the distribution of north-trending zones of chaotic breccia (Ransom, 1977) may be north-trending fractures that produced local sub-basins in which sulphides accumulated.

In summary, the preferential distribution of a number of stratiform lead-zinc deposits adjacent to faults that appear to have been active in Belt-Purcell time, the common occurrence of intraformational conglomerates in the footwalls of the deposits, and the association with anomalous boron concentrations suggest that these deposits may be genetically linked to transverse faults and perhaps associated splay faults that were intermittently active during Purcell time and locally modified the depositional environment of Purcell rocks. A somewhat higher concentration of Precambrian gabbro sills in the vicinity of the transverse faults, the occurrence of a thick succession of basic volcanic rocks of upper Proterozoic age west of the St. Mary fault, and the preferential alignment of Mesozoic batholiths in the fault zones indicate that they coincide with deep crustal structures. These structures might have controlled the outflow of metal-charged formation fluids or deep crustal or mantle fluids, leading to the formation of stratiform sulphide mineralization in their vicinity.

ACKNOWLEDGEMENTS

This paper is, in part, a review of the stratigraphy and depositional environment of the Purcell Supergroup in southeastern B.C., and therefore relies heavily on published material by other workers, particularly an excellent regional synthesis paper by Price (1964). Ideas presented in sections titled "Transverse structures" and "Relationship between faulting, sedimentation and mineral deposits" developed through discussions with a number of Cominco geologists involved in both Sullivan mine geology and regional exploration in southeastern B.C. and with R.A. Price of Queens University. The support and advice of these geologists during mapping of the Kootenay King area is greatly appreciated. Although these persons influenced this paper, they do not necessarily support or agree with all the interpretations presented in it.

The paper benefited considerably from reviews by A. Sutherland Brown and A. Panteleyev of the B.C. Ministry of Energy, Mines and Petroleum Resources, and by two referees appointed by the editors of this Special Paper.

REFERENCES

Atkinson, S.J., 1976, Geology of the Toby Creek area, British Columbia: British Columbia Ministry of Energy, Mines and Petroleum Resources Preliminary Map 23.

______________, 1977, Surface geology of the Paradise basin: in Geology in British Columbia, 1975: British Columbia Ministry of Mines and Petroleum Resources, p. 7-12.

Bishop, D.T., Morris, H.C., and Edmunds, F.R., 1970, Turbidites and depositional features in the lower Belt-Purcell Supergroup (Abst.): in Geological Society of America Program, Boulder, Colorado, p. 797.

British Columbia Ministry of Energy, Mines and Petroleum Resources: Mineral Inventory Maps 82F and 82G.

Burwash, R.A., Baadsgaard, H., and Peterman, Z.E., 1962, Precambrian K-Ar dates from the western Canada sedimentary basin: Journal of Geophysical Research, v. 67, p. 87-124.

Collins, J.A. and Smith, L., 1977, Genesis of cupriferous quartz arenite cycles in the Grinnell Formation (Spokane equivalent), Middle Proterozoic (Helikian) Belt-Purcell Supergroup, eastern Rocky Mountains, Canada: Bulletin of Canadian Petroleum Geology, v. 25, p. 713-735.

Douglas, R.J.W., 1952, Preliminary map, Waterton, Alberta: Geological Survey of Canada Paper 52-10.

Edmunds, F.R., 1973, Stratigraphy and lithology of the lower Belt Series in southern Purcell Mountains, British Columbia: in Belt Symposium Volume 1: Idaho Bureau of Mines and Geology, p. 230-234.

______________, 1977, Kimberley to Creston, Stratigraphy and lithology of the lower Belt Series in the Purcell Mountains, British Columbia: in Höy, T., ed., Lead-zinc deposits of southeastern British Columbia: Geological Association of Canada Fieldtrip Guidebook 1, p. 22-32.

Ethier, V.G. and Campbell, F.A., 1977, Tourmaline concentrations in Proterozoic sediments of the southern Cordillera of Canada and their economic significance: Canadian Journal of Earth Sciences, v. 14, p. 2348-2363.

Fyles, J.T. and Hewlett, C.G., 1959, Stratigraphy and Structure of the Salmo Lead-Zinc Area: British Columbia Ministry of Mines and Petroleum Resources Bulletin 41, 162 p.

Gifford, R.G., 1971, Geological Survey, Hilo claim group, Fort Steele Mining Division: British Columbia Ministry of Mines and Petroleum Resources Assessment Report 3300, 12 p.

Harrison, J.E., 1972, Precambrian Belt basin of northwestern United States: Its geometry, sedimentation and copper occurrences: Geological Society of America Bulletin, v. 83, p. 1215-1240

Höy, T., 1978, Geology of the Estella-Kootenay King Area, Southeastern British Columbia: British Columbia Ministry of Mines and Petroleum Resources Preliminary Map 28.

______________, 1979, Geology of the Estella-Kootenay King Area, Hughes Range, Southeastern British Columbia: British Columbia Ministry of Energy, Mines and Petroleum Resources Preliminary Map 36.

______________, 1980, Geology of the Riondel Area, Central Kootenay Arc, Southeastern British Columbia: British Columbia Ministry of Mines and Petroleum Resources Bulletin 73, 89 p.

Huebschman, R.P., 1973, Correlation of fine carbonaceous bands across a Precambrian stagnant basin: Journal of Sedimentary Petrology, v. 43, p. 688-699.

Hunt, G., 1962, Time of Purcell eruption in southeastern British Columbia and southwestern Alberta: Journal of Alberta Society of Petroleum Geologists, v. 10, p. 438-442.

Kanasewich, E.R., Clowes, R.M., and McCloughan, C.H., 1969, A buried precambrian rift in western Canada: Tectonophysics, v. 8, p. 513-527.

Leech, G.B., 1952, Preliminary Map, St. Mary Lake, British Columbia: Geological Survey of Canada Paper 52-15.

_______________, 1958a, Canal Flats, Kootenay District, British Columbia: Geological Survey of Canada Map 24-1958.

_______________, 1958b, Fernie Map-Area, West Half, British Columbia: Geological Survey of Canada Paper 58-10.

_______________, 1960, Geology Fernie (west half), Kootenay District, British Columbia: Geological Survey of Canada Map 11-1960.

_______________, 1972, Western Rocky Mountains and Southern Rocky Mountain trench: in Glass, D.J., ed., Geology of the Southern Canadian Cordillera: International Geological Congress Field Excursion A03-C03, p. 21-26.

Leech, G.B. and Wanless, R.K., 1962, Lead isotope and potassium argon studies in the East Kootenay district of British Columbia: in Petrologic studies – a volume in honour of A.F. Buddington: Geological Society of America, p. 241-279.

Lis, M.G. and Price, R.A., 1976, Large-scale block faulting during deposition of the Windermere Supergroup (Hadrynian) in southeastern British Columbia: Geological Survey of Canada Paper 76-1A, p. 135-136.

McMechan, M.E., 1978, Geology of the Precambrian Purcell Supergroup, Cranbrook-Fort Steele area (82G/11,12): in Geological Fieldwork 1977: British Columbia Ministry of Mines and Petroleum Resources, p. 73-79.

_______________, 1979, Geology of the Mount Fisher-Sand Creek Area: British Columbia Ministry of Energy, Mines and Petroleum Resources Preliminary Map 34.

McMechan, M.E. and Höy, T., and Price R.A., 1980, Van Creek and Nicol Creek Formations (New); A revision of the stratigraphic nomenclature of the Middle Proterozoic Belt-Purcell Supergroup, southeastern British Columbia: Bulletin of the Canadian Society of Petroleum Geologists, v. 28, p. 542-558.

Morton, R., Goble, E., and Goble, R.J., 1973, Sulphide deposits associated with Precambrian Belt-Purcell strata in Alberta and British Columbia, Canada: in Belt Symposium, v. 1: Idaho Bureau of Mines and Geology, p. 159-179.

Norris, D.K. and Price, R.A., 1966, Middle Cambrian litho-stratigraphy of southeastern Canadian Cordillera: Bulletin of the Canadian Society of Petroleum Geologists, v. 14, p. 385-404.

Obradovich, J.D. and Peterman, Z.E., 1968, Geochronology of the Belt Series, Montana: Canadian Journal of Earth Sciences, v. 5, p. 737-747.

Okulitch, A.V. and Woodsworth, G.J., 1977, Kootenay River Map-Area, B.C., Alberta and U.S.A.: Geological Survey of Canada Open File 481.

Price, R.A., 1962, Fernie map-area, east half, Alberta and British Columbia: Geological Survey of Canada Paper 61-24.

_______________, 1964, The Precambrian Purcell System in the Rocky Mountains of southern Alberta and British Columbia: Bulletin of Canadian Petroleum Geology, v. 12, p. 399-426.

_______________, 1965, Flathead Map-Area, British Columbia and Alberta: Geological Survey of Canada Memoir 336, 221 p.

Price, R.A. and Lis, M.G., 1975, Recurrent displacements on basement-controlled faults across the Cordilleran miogeocline in southern Canada: Geological Society of America, Abstracts with Programmes, p. 1234.

Ransom, P.W., 1977, Geology of the Sullivan orebody: *in* Höy, T., ed., Lead-zinc Deposits of Southeastern British Columbia: Geological Association of Canada Fieldtrip Guidebook 1, p. 7-21.

Reesor, J.E., 1958, Dewar Creek Map-Area with Special Emphasis on the White Creek Batholith, British Columbia: Geological Survey of Canada Memoir 292, 78 p.

_______________, 1973, Geology of the Lardeau Map-Area, East Half, British Columbia: Geological Survey of Canada Memoir 369, 129 p.

Rice, H.M.A., 1937, Cranbrook Map-Area, British Columbia: Geological Survey of Canada Memoir 207, 67 p.

______________, 1941, Nelson Map-Area, East Half: Geological Survey of Canada Memoir 228, 86 p.

Wheeler, J.O., 1970, Summary and discussion: in Wheeler, J.O., ed., Structure of the Southern Canadian Cordillera: Geological Association of Canada Special Paper 6, p. 155-166.

Wheeler, J.O., Aitken, J.D., Berry, M.J., Gabrielse, H., Hutchison, W.W., Jacoby, W.R., Monger, J.W.H., Niblett, E.R., Norris, D.K., Price, R.A., and Stacey, R.A., 1972, The Cordilleran structural province: in Price, R.A., ed., Variations in Tectonic Styles in Canada: Geological Association of Canada Special Paper 11, p. 1-82.

Winston, D., 1977, Alluvial fan, shallow water, and subwave base deposits of the Belt Supergroup near Missoula, Montana: Rocky Mountain Section, Geological Society of America Fieldguide 5, 41 p.

MS received December 1978;
Accepted for publication February 1980.

Precambrian Sulphide Deposits, H.S. Robinson Memorial Volume, edited by R.W. Hutchinson, C.D. Spence and J.M. Franklin Geological Association of Canada Special Paper 25, 1982

PRECAMBRIAN OF THE MACKENZIE FOLD BELT – A STRATIGRAPHIC AND TECTONIC OVERVIEW

J.D. Aitken
Geological Survey of Canada, 3303-33rd Street N.W.,
Calgary, Alberta T2L 2A7

ABSTRACT

The Proterozoic rocks of the Mackenzie Mountains are host to stratabound deposits of zinc, copper and iron. A major unconformity[1] separates the exposed Proterozoic succession into two dissimilar successions which probably correspond to the Purcell and Windermere Supergroups of the southern Canadian Cordillera[1]. The older succession ('Mackenzie Mountains Supergroup') consists of chemical and mature clastic sedimentary formations of platformal origin that thicken gradually southwestward. The younger succession ('Ekwi Supergroup') is dominated by immature clastic rocks laid down in southwestward-thickening wedges in a continental-margin environment. Volcanic rocks are rare in both successions.

RÉSUMÉ

Les roches protérozoïques des Montagnes de Mackenzie sont les hôtes de dépôts de zinc, cuivre et fer liés aux strates. Une discordance majeure[1] devise la succession protérozoïque telle qu'exposée, en deux successions dissemblables qui correspondent probablement aux supergroupes de Purcell et de Windermere de la Cordillère Canadienne méridionale[1]. La succession la plus ancienne ('supergroupe des Montagnes de Mackenzie') consiste en formations sédimentaires chimiques et clastiques évoluées déposées sur une plateforme, et qui s'épaississent graduellement vers le sud-ouest. La succession plus récente ('supergroupe d'Ekwi') est dominée au contraire par des roches clastiques immatures déposées en biseaux s'épaississant eux-aussi vers le sud-ouest, et témoignant d'un environnement de bordure continentale. Les roches volcaniques sont rares dans les deux successions.

INTRODUCTION

Two of the papers in this volume deal with stratabound deposits of base metal sulphide minerals in Precambrian strata of the Mackenzie Mountains. This paper is intended to provide a background to the papers by Ruelle and Hewton (this volume), by presenting a brief overview of the history of Precambrian depositional events, and of the deformational and erosional events that govern the present distribution and exposure of Precambrian formations in these mountains.

STRATIGRAPHY

Introduction

Proterozoic rocks exposed in the Mackenzie Mountains form two major sucessions that are distinct in lithology and distribution and generally separated by an unconformity (Fig. 1).

The older succession, informally named the 'Mackenzie Mountains Supergroup' (Young *et al.,* in press) consists of carbonate, evaporite and mature detrital strata of platformal origin. It is about 5 km thick. The younger succession, informally named the 'Ekwi supergroup' (Young *et al.,* in press) is dominated by immature detrital and carbonate strata in its upper part. It, too, is about 5 km thick. Rocks of both supergroups are as fresh as those of the overlying Paleozoic succession. Exposures are excellent, and the tectonic style permits the measurement of long, unbroken stratigraphic sections.

The Older Proterozoic Succession ('Mackenzie Mountains Supergroup'[1]). The oldest exposed unit in the Mackenzie Fold Belt is a thick, cherty dolomite unit with abundant stromatolites, informally designated 'Map-unit H1' (Fig. 1). Fluorite occurs widely in the unit and rare pyrobitumen is found in vugs. From the profile of folds in which 'Map-unit H1' is involved, it is clear that bedded rocks thousands of metres thick intervene between 'H1' and any rigid basement. Correlation of algal stromatolites of 'Map-unit H1' with the stromatolite successions of the USSR suggests a basal Middle Riphaean age, about 1350 ± 50 Ma, but regional correlations (Young *et al.,* in press) show that it must post-date the 1200 Ma-old Coppermine lavas. Carbonate deposition ended abruptly with an influx of fine black pyritic clastic strata of the basal Tsezotene Formation (Htz), coincident with accelerated subsidence. The Tsezotene is a coarsening-upward, shallowing-upward sequence, partly red in its upper part; it passes into the sandstone-dominated Katherine Group (Hk; Fig. 2), a complex of interleaved marine-littoral and fluvial-deltaic sequences (Aitken *et al.,* 1978a).

The Katherine Group is succeeded by the thick and varied Little Dal Group, whose lower formations record a sharp decrease in sand supply and reduction in fine terrigenous supply, and subsequent deepening. Deepening was most pronounced in the northwest, where the deepwater 'basinal sequence' (Hb) is characterized by limestone-fine detrital rhythmite which encloses a belt of giant reefs of stromatolitic limestone and dolomite up to 300 m high and several km in greatest dimension (Fig. 3). Because overlying units maintain their platformal character above the 'basinal sequence', it is clear that the latter was deposited in an intraplatform basin. The reef belt is of interest because it coincides geographically with the Gayna River zinc deposits, however, the bulk of the sphalerite is in the overlying 'grainstone subunit' (Hgr), and relatively little in the reefs proper.

In the southeast, at the type section of the Little Dal Group in Glacier Lake map-area, for example, deepwater rhythmites are missing, and the corresponding interval is dominated by ooid and intraclast grainstones with numerous algal stromatolite biostromes.

The 'grainstone subunit' was deposited when the intra-platform basin filled up to shallow subtidal and intertidal depths, as recorded by abundant, dolomitized ooid grainstone, cryptalgal laminite, and stromatolites. A second lithofacies within the 'grainstone subunit', known as the "platy dolomite facies" contains abundant mud-

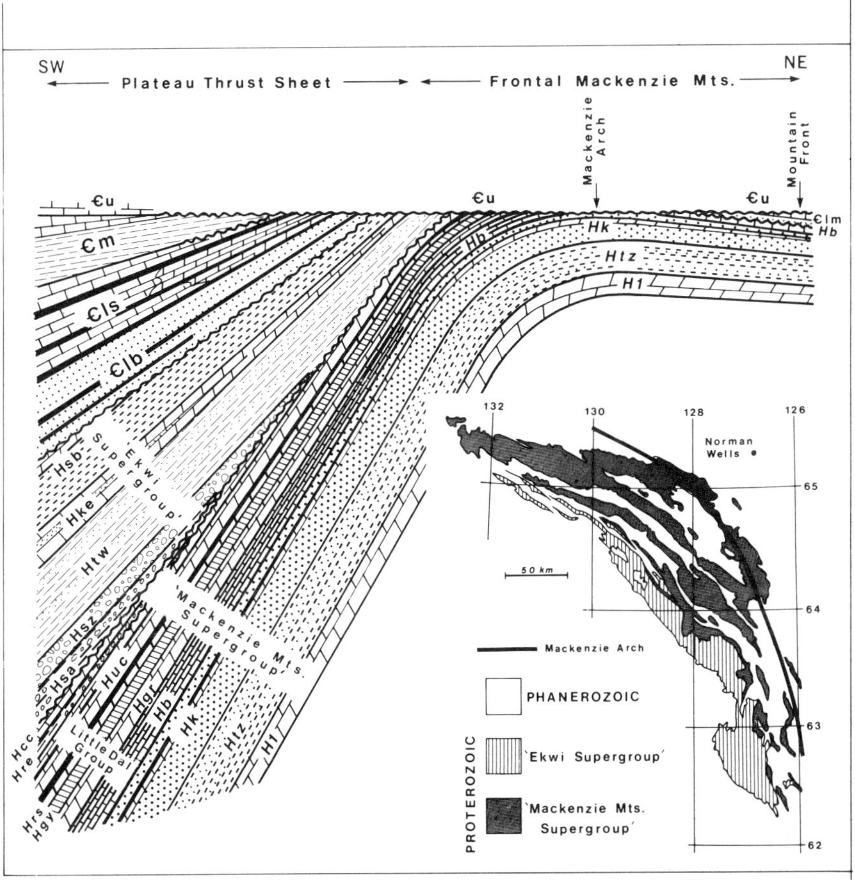

Figure 1. Stratigraphic relationships of Precambrian rocks, Mackenzie Mountains. H1 – 'Map-unit H1'; Htz – Tsezotene Fm.; Hk – Katherine Gp.; divisions of Little Dal Group: Hb – 'basinal'; Hgr – 'grainstone'; Hgy – 'gypsum'; Hrs – 'rusty shale'; Huc – 'upper carbonate'; Hre – Redstone River Fm.; Hcc – Coppercap Fm.; Hsa – Sayunei Fm.; Hsz – Shezal Fm.; Htw – Twitya Fm.; Hke – Keele Fm.; Hsb – Sheepbed Fm.; Єlb – Backbone Ranges Fm.; Єls – Sekwi Fm.; Єm – Middle Cambrian; Єu – Upper Cambrian; Єlm – Lower and Middle Cambrian.

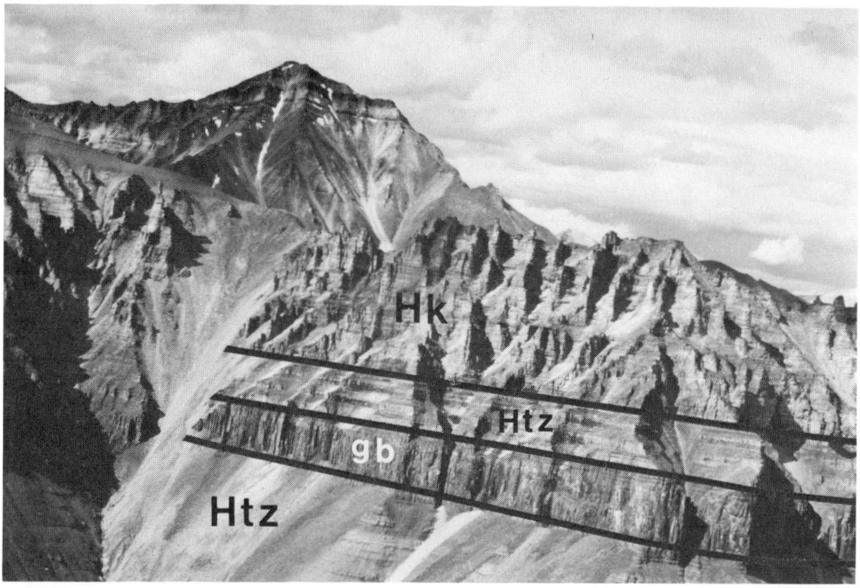

Figure 2. Tsezotene Formation (Htz) and Katherine Group (Hk) at Lat. 64°13′, Long. 128°02′. A gabbro sill (gb) is intruded near the top of the Tsezotene.

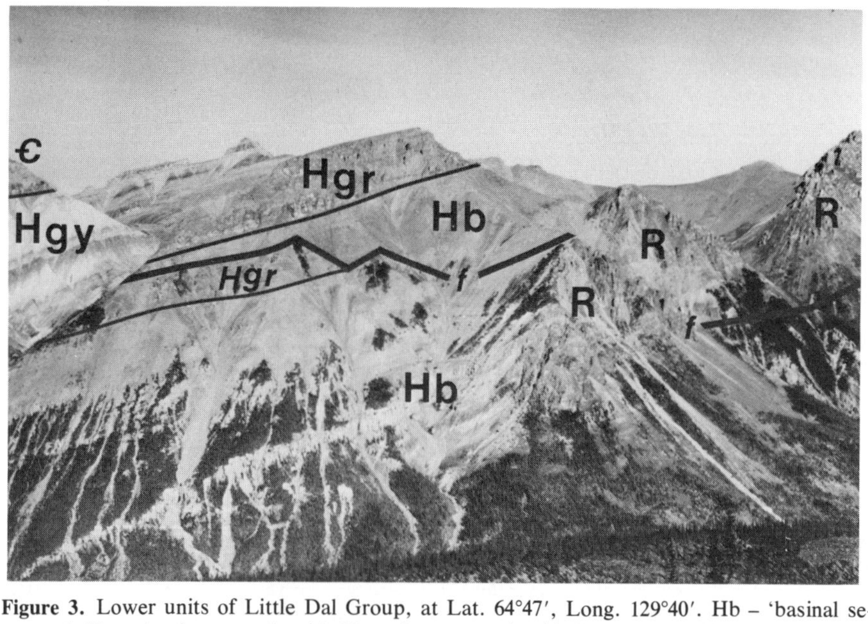

Figure 3. Lower units of Little Dal Group, at Lat. 64°47′, Long. 129°40′. Hb – 'basinal sequence'; Hgr – 'grainstone subunit'; Hgy – 'gypsum subunit'; R – stromatolitic reef; є – Upper Cambrian carbonate strata (unconformable on Little Dal); f – fault, with gabbro dyke emplaced along it.

cracks and ripple-marks, with salt-crystal casts near the top, and records high intertidal to supratidal deposition. This lithofacies forms an upper member of the 'grainstone subunit' in the northwest, but occurs throughout the subunit in the southeast.

The thick-bedded, massive, dolomitized ooid grainstones are the principal host

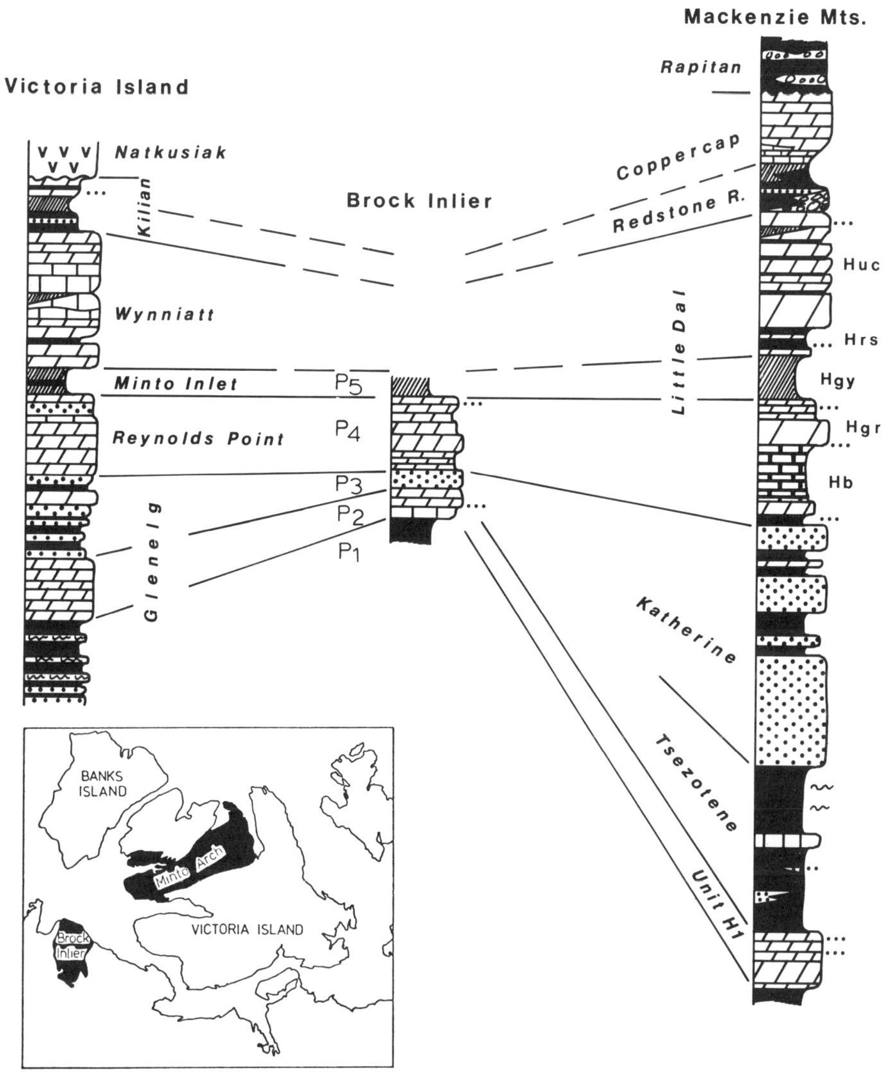

Figure 4. Lithostratigraphic correlations, Victoria Island – Brock Inlier – Mackenzie Mountains. Little Dal subunits: Hb – 'basinal'; Hgr – grainstone'; Hgy – 'gypsum'; Hrs – 'rusty shale'; Huc – 'upper carbonate'.

to sphalerite at Gayna River. The Mississippi Valley-type mineralization occurs in beds intensely veined by coarse white dolomite (although similar veins are common elsewhere without zinc mineralization). Interestingly, rare pyrobitumen occurs in the 'grainstone subunit' and the underlying 'basinal sequence'. The 'grainstone subunit' also is host to pockets of chalcocite-barite mineralization near the junction of Ekwi and Keele Rivers.

Following deposition of the 'grainstone subunit', a blanket of gypsum and anhydrite (Hgy), locally more than 500 m thick, was laid down across the exposed part of the basin. The extent of this evaporite deposit is vast; it is map-unit "P5" of Brock Inlier and the Minto Inlet Formation of Victoria Island. (Fig. 4; Cook and Aitken, 1969; Aitken *et al.*, 1978a, 1978b; Young, 1977).

Because the Proterozoic record in Brock Inlier (the most craton-ward area with which correlations can be made) ends with deposition of the 'gypsum subunit', it is worth reviewing the record to this point. The record is one of supracratonic platform deposition that was essentially without interruption and that extended from the formations of Brock Inlier to their much thicker equivalents in the Mackenzies, generally with little change in lithofacies (Fig. 4). In the Mackenzie Mountains, isopachs are northwest-trending, arcuate, and congruent with the tectonic arc (Aitken and Long, 1978). Southwestward thickening in the lower part of the supergroup reaches as much as ten times (relative to Brock Inlier) for some correlatable lithologic units, but in the upper part, is less than two times. Thus, given the near-constancy of lithofacies, the aspect of the "Mackenzie Mountains supergroup" is more platformal than miogeoclinal. Platform conditions persisted throughout the region in which exposures occur to the end of Little Dal time.

Following deposition of the 'gypsum subunit', Little Dal sedimentation continued with a minor sharp pulse of detrital material, recorded as the shale, sandstone and stromatolitic carbonate of the 'rusty shale subunit' (Hrs; Fig. 1), all deposited under peritidal conditions. The rusty shales are succeeded by a thick, carbonate-dominated formation, the 'upper carbonate subunit' (Huc; Figs. 5, 7), characterized by cyclical, stromatolitic, pure and impure dolomites. The last events recorded by the Little Dal Group are a brief episode of basaltic lava eruption, followed by renewed terrigenous supply seen in the shale, siltstone and sandstone beds that are interbedded with carbonates at the top of the 'upper carbonate'. The lava and clastic rocks possibly record an early pulse of the tectonism that characterized the environment of the succeeding Redstone River Formation.

The 'upper carbonate subunit' is host to sub-economic copper mineralization at Ekwi River. The mineralization lies within the Twitya Uplift (Cook and Aitken, 1978), where Siluro-Devonian strata rest unconformably on the 'upper carbonate subunit'.

A change in the stromatolite assemblages low in the 'upper carbonate subunit' corresponds to that at the Middle Riphaean – Upper Riphaean boundary, dated at 950± 50 Ma in the Soviet Union. The age significance of this change remains debatable.

In contrast to the almost regional persistence of individual lithologic units in the pre-Redstone River succession, rock types and thicknesses vary markedly within the Redstone River; stability was replaced by tectonic unrest. Although the base of the

Redstone River appears conformable at many localities, the Redstone conglomerates, where present, consist largely of debris from the top of the Little Dal Group and record nearby paleotopographic "highs" interpreted by Eisbacher (1977) as fault scarps. On at least one of these structural "highs", the base of the Redstone-Coppercap cycle rests on the *lower* Little Dal.[1]

The Redstone River Formation (Hre), in common with the overlying Coppercap Formation (Hcc; Fig. 5) occupies a series of embayments that open towards the southwest. In the axes of the embayments, gypsum was followed by fine-grained clastic red beds; toward the margins, these interfinger with fanglomerates (Eisbacher, 1977; Jefferson, 1978; Ruelle, this volume).

Redstone River clastic deposition gave way to carbonate deposition recorded by the Coppercap Formation (Eisbacher, 1977; Jefferson, 1978; Ruelle, this volume) as the high structural blocks supplying the Redstone detritus were either worn down or submerged by rising sea level. The Coppercap is dominated by thin-bedded and laminated lime muds (now largely preserved as dolomite) with subordinate intraclast grainstones and black, calcareous, pyritic shales, and near the base, anhydrite and gypsum. Peritidal deposition is recorded only in the lower Coppercap, and is succeeded by carbonates deposited in a subtidal, quiet, reducing environment (Ruelle, this volume). Continuing tectonic activity is recorded, first, by the superposition of axes of thick Coppercap deposition on those of thick Redstone River deposition, and second, by the spectacular development of gravity slides, olistoliths, and debris-flow breccias, all of which contain displaced planar, domal and rare columnar algal stromatolites at some localities.

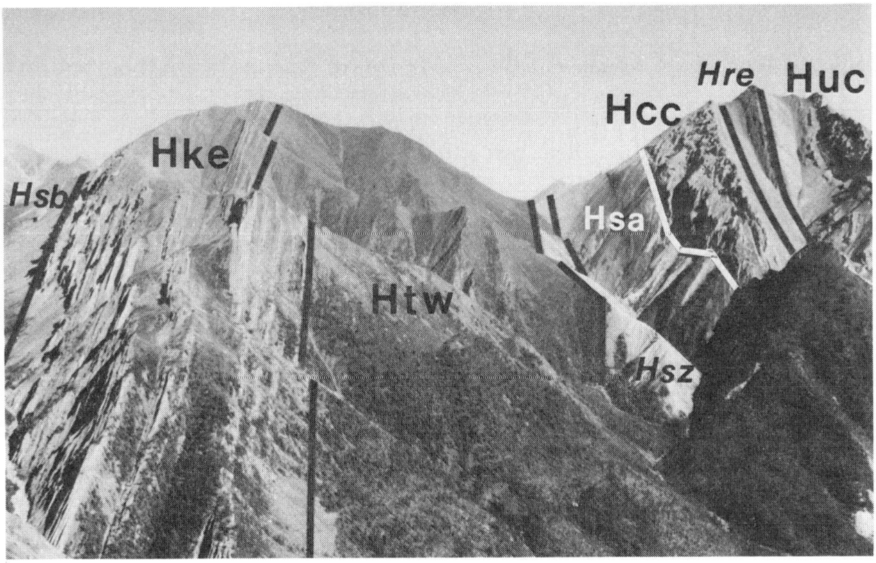

Figure 5. 'Mackenzie Mountains' and 'Ekwi' supergroups at Lat. 64°47′, Long. 129°40′, Huc – 'upper carbonate subunit' (Little Dal Group); Hre – Redstone River Formation (a depositional lens at this locality); Hcc – Coppercap Formation; Hsa – Sayunei Formation; Hsz – Shezal Formation; Htw – Twitya Formation; Hke – Keele Formation; Hsb – Sheepbed Formation.

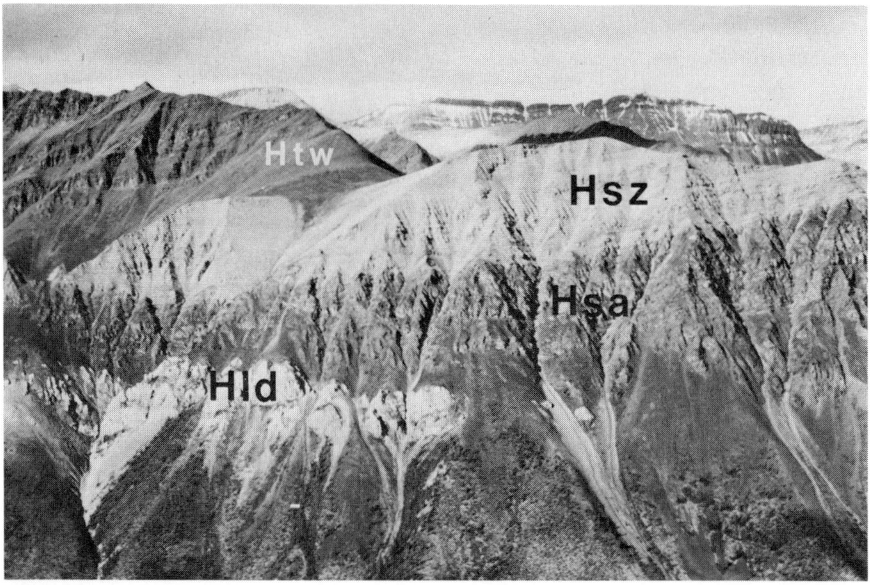

Figure 6. 'Ekwi supergroup' unconformable on 'Mackenzie Mountains supergroup', at Lat. 64°30′, Long. 129°16′. Note erosional thinning of Little Dal dolomites (Hld) from left to right, and depositional thinning of Sayunei clastics (Hsa) from right to left, onlapping paleotopography. Hsz – Shezal Formation; Htw – Twitya Formation. Foot of outcrop approximates Plateau Fault.

Figure 7. 'Mackenzie Mountains' and 'Ekwi' supergroups at Lat. 64°51′, Long. 129°40′. Hld – Little Dal Group ('upper carbonate subunit'); Htw – Twitya Formation; Hke – Keele Formation; Hsb – Sheepbed Formation; Єb – Backbone Ranges Formation.

The disseminated iron and copper sulphides of the Redstone Copper Belt (Ruelle, this volume) occur in cryptalgal, laminated carbonates of the upper Redstone, the basal Coppercap, and the transition zone between them. This mineralization is stratabound and regional in extent, although by no means everywhere of "ore grade".

The Younger Proterozoic Succession ('Ekwi Supergroup'). The sediments deposited upon pronounced sub-Rapitan paleotopography (Figs. 4, 6) are maroon and green, laminated, turbiditic argillite and siltstone of the Sayunei Formation (Hsa; Eisbacher, 1978), with units of pebbly mudstone recording debris-flows. The presence of glacial dropstones in this quiet and probably deepwater environment is widely accepted. Although the fine-grained turbidites are iron-rich regionally, jaspilitic iron-formation occurs only discontinuously near the top of the Sayunei. Of the iron deposits, the one at Snake River (20 billion tons) probably outweighs the sum of all the others. The iron-formation may have accumulated in local depressions or sub-basins (Young, 1976).

Above the Sayunei Formation, probably conformable with it and overlapping it locally to lie upon pre-Rapitan rocks is the glaciogene Shezal Formation (Hsz; Figs. 5, 6). At some localities the Shezal consists almost entirely of calcareous and dolomitic tillite that is buff and locally red and green in colour, whereas at others up to 12 distinct sheets of diamictite alternate with cross-laminated sandstone and siltstone, and minor shale (Eisbacher, 1978).

The abrupt contact between black, pyritic shale or dark bituminous limestone at the base of the Twitya Formation (Htw) and the underlying, glaciogene beds of the Shezal, is evidence for either a geologically instantaneous climatic change (Eisbacher, 1978) or a disconformable contact. The Twitya consists of dark-coloured mudrocks, siltstone and fine-grained sandstone, in which evidence of sliding and slumping on depositional slopes is abundant. Diminution of grain-size southwestward (basin-ward) is pronounced. The Twitya represents deposition either in a prodeltaic environment or on a continental slope. It overlaps the underlying formations to lie upon pre-Rapitan strata at the northeastern limit of the 'Ekwi supergroup' (Fig. 7).

Diminution of clastic supply toward the end of Twitya time permitted the deposition of tongues of shallow-water carbonate strata, marking the base of the Keele Formation (Hke), recently included in the Rapitan Group (Eisbacher, 1978). Geographic variation in the makeup of the Keele is extreme, particularly in the ratio of sandstone and shale to carbonate rocks. The carbonates are almost entirely grainstones (notably ooid grainstones) and cryptalgal laminites. Lenses of sandstone and quartz-pebble conglomerate within extensive limestone and dolomite units are interpreted as gravel-filled channels transecting a narrow carbonate platform, and probably record regressions. Where the outer edge of the Keele platform can be observed, enormous carbonate olistoliths have broken off and slid downslope southwestward to be buried in the fine-grained detrital rocks of the Twitya.

An abrupt contact with black shales of the overlying Sheepbed Formation (Hsb; Fig. 7), marks sudden drowning of the Keele platform. Although finer-grained than the Twitya (siltstone and sandstone are insignificant), the Sheepbed and an unnamed and undescribed carbonate unit at its top repeat the platform-edge relationship seen between the Twitya and Keele Formations. The 'Sheepbed carbonate' occurs in a platformal facies with prominent cryptalgal laminites, stromatolites and grainstones, in an off-platform facies of uniformly bedded lime mudstones and derived dolomites, and as large, gravity-displaced olistoliths of both facies. The Sheepbed platform-edge coincides geographically with the Keele platform-edge.

RELATIONSHIP BETWEEN OLDER AND YOUNGER SUCCESSIONS

The contact between 'Mackenzie Mountains' and 'Ekwi' Supergroups records major changes in the configuration and behaviour of both the area of deposition and the land areas adjoining it:

a) The basin margin retreated hundreds of km to the southwest [with minor exceptions the northeastward limit of the younger succession is the trace of Plateau Fault (Aitken and Long, 1978; Aitken *et al.*, 1982), where the older succession disappears southwestward beneath the younger].

b) Differential subsidence became much more pronounced; the younger succession and the succeeding Lower Cambrian succession were deposited as southwestward-thickening wedges (Fig. 1).

c) Sedimentation became predominantly detrital, and intially, coarse-grained.

All authors agree that the contact is at least locally unconformable. At Mountain River (Fig. 6), the sub-Sayunei contact, with visible paleotopography, cuts from the Coppercap Formation down to the 'rusty shale subunit' of the Little Dal Group within a few kilometres. Near Dal Lake, the Sayunei lies upon strata well down in the *lower* Little Dal. On the other hand, as noted by Jefferson (1978), the contact is locally conformable. These relationships apparently record continuous deposition in paleostructural "lows", while erosion took place on the "highs", a condition previously noted to have prevailed during the deposition of the Redstone River Formation. The tectonic movements undoubtedly took place on the many northerly to northeasterly trending faults that are truncated beneath one or other of the Rapitan formations. Young *et al.*, (in press) identify the sub-Rapitan contact as "the major stratigraphic discontinuity in the Proterozoic sequence in the Mackenzies"; this conclusion is probably basically correct, but it should be understood that the discontinuity records taphrogeny that commenced at about the time of eruption of the Little Dal lavas, and was active during the Redstone-Coppercap cycle at the end of the older Mackenzie Mountains succession, as well as during subsequent deposition of the Sayunei and possibly, younger formations.

IGNEOUS EVENTS

Volcanism plays a very minor role in the history of the exposed Precambrian succession of Mackenzie Mountains.

Up to five flows of basaltic lava occur at the top of the Little Dal Group in the region south of Keele River. Andesine, talc and glauberite in the mudrocks of the Redstone Formation may record distant volcanism (Ruelle, this volume).

Dykes and sills of gabbro intruding the 'Mackenzie Mountains Supergroup' are widespread. The sills occur no higher than the Tsezotene-Katherine contact (Fig. 2), but large dykes cut as high as the 'grainstone subunit' of the Little Dal (Fig. 3). Two of the sills have been dated at about 770 Ma by the rubidium-strontium isochron method (Armstrong *et al.*, 1982). It has been assumed widely that the gabbroic intrusions belong to the same igneous episode as the Little Dal lavas. This view is supported by the positions of the dykes, many of which lie along pre-Rapitan faults. Nevertheless, more detailed petrologic work is required before the kinship of intrusions and lavas can be accepted as proven.

One basic dyke cuts the Redstone River, Coppercap and basal Sayunei Formations near Coates Lake (Jefferson, 1978). Chloritic diatremes cut the Redstone River and Coppercap near Coates Lake and at Keele River. It is not established whether or not these bodies are related to the gabbros that elsewhere are restricted to the Little Dal and older formations. Because the former are local and the latter regional, a relationship is doubtful; some chloritic diatremes in the region are as young as Ordovician (C.I. Godwin, pers. commun., 1979).

RELATIONSHIP WITH THE PALEOZOIC SUCCESSION

Published descriptions of the relationship between the earliest Cambrian strata of the region and the Sheepbed Formation tend to suggest that the contact is nearly conformable. Four lines of evidence suggest otherwise:

a) Faint but distinct angularity, visible in favourable views.

b) The discontinuous distribution of the 'Sheepbed carbonate' – as erosional outliers?

c) Grit-filled solution-openings at the top of the 'Sheepbed carbonate'. These extend downward several metres from the contact with the overlying Lower Cambrian rocks.

d) Given the obvious inheritance from the Keele platform-edge to the Sheepbed platform-edge, the fact that the overlying, Lower Cambrian Backbone Ranges and Sekwi Formations (shallow-water detrital and carbonate units, respectively) pass across the underlying platform-edges without pronounced change in facies or thickness implies that the Sheepbed shales were largely compacted *before* they were loaded by the basal Cambrian sediments.

It is therefore inferred that a considerable thickness of sediments overlying the youngest preserved Proterozoic formations was removed by erosion prior to the onset of Cambrian deposition.

POST-DEPOSITIONAL EVENTS

The present configurations of the Proterozoic successions of the Mackenzie Mountains, the older succession as a southwest-thickening slab, the younger as an abruptly southwest-thickening wedge, result from both depositional and subsequent erosional events. Northeastward changes to proximal facies in both the Redstone River (Eisbacher, 1977; Jefferson, 1978) and Twitya Formations (Aitken *et al.*, 1973) strongly suggest that the Proterozoic deposits younger than Little Dal Group never extended a great distance northeast of their present limit. It follows that the pre-Redstone succession probably underwent erosional bevelling not only during the sub-Rapitan break, but also during deposition of the younger succession.

Mackenzie Arch (Fig. 1), which profoundly affects the present-day distribution of Proterozoic rocks in the Mackenzies, is not reflected in pre-Rapitan depositional patterns. It possibly first became active during the break between older and younger Proterozoic successions, although its trend diverges from that of the interpreted zero-edge of the younger succession. By the Early Cambrian, the older succession was deeply bevelled over the Arch, and the Arch remained high. A final pulse of uplift at about the end of Middle Cambrian brought total erosional bevelling well down into the Katherine Group, along the crest of Mackenzie Arch.

From the Late Cambrian until the great mountain-building episode of the mid-Cretaceous to early Tertiary, the Proterozoic strata of Mackenzie Mountains remained essentially undisturbed and buried, except for localized exposure during the Middle to Late Ordovician in Redstone Arch (Gabrielse and co-workers, 1973) and during the Middle to Late Silurian in Twitya Uplift (Cook and Aitken, 1978).

The terminal orogeny produced structures of fairly simple form and grand scale: in the frontal Mackenzies, mainly broad and extremely long box-folds with kink-band-like hinges and straight limbs; deeper in the mountains, mainly widely spaced thrust faults. These structures and the post-orogenic sculpturing they have undergone have created ideal conditions in which to explore for stratigraphically controlled mineral deposits.

Notes

1. The correlation between Purcell and 'Mackenzie Mountains' Supergroups can no longer be defended. G.M. Young has demonstrated (Transactions Royal Society Edinburgh, v. 70, p. 323-326) that the 'Mackenzie Mountains Supergroup' must post-date the 1200 Ma Coppermine Lavas. M.E. McMechan and R.A. Price (Canadian Journal of Earth Sciences, in press) have demonstrated that Purcell sedimentation was ended by 1300 Ma.
2. The position and character of the boundary between the older and younger successions of the Mackenzie Mountains are now much more fully understood (see J.D. Aitken, Geological Survey Canada Paper 81-10, p. 47 - 72).

REFERENCES

Aitken, J.D., 1977, New data on correlation of the Little Dal Formation and a revision of Proterozoic map-unit "H5": Report of Activities, Part A, Geological Survey of Canada Paper 77-1A, p. 131-135.

Aitken, J.D. and Cook, D.G., 1974a, Geology of Parts of Mount Eduni (106A) and Bonnet Plume map-areas, District of Mackenzie: Geological Survey of Canada Open File Report 221.

——————, 1974b. Caracajou Canyon Map-Area (96D), District of Mackenzie, Northwest Territories: Geological Survey of Canada Paper 74-13.

Aitken, J.D. and Long, D.G.F., 1978, Mackenzie tectonic arc – Reflection of early basin configuration?: Geology, v. 6, p. 626-629.

Aitken, J.D., Cook, D.G., and Yorath, C.J., 1982, Upper Ramparts River and Sans Sault Rapids Map-Areas (106G, H), District of Mackenzie: Geological Survey of Canada Memoir 388, with maps 1452A and 1453A.

Aitken, J.D., Long, D.G.F., and Semikhatov, M.A., 1978a. Progress in Helikian stratigraphy, Mackenzie Mountains: in Current Research, Part A, Geological Survey of Canada Paper 78-1A, p. 481-484.

——————, 1978b, Correlation of Helikian strata, Mackenzie Mountains – Brock Inlier – Victoria Island: in Current Research, Part A, Geological Survey of Canada Paper 78-1A, p. 485-486.

Aitken, J.D., Macqueen, R.W., and Usher, J.L., 1973, Reconnaissance Studies of Proterozoic and Cambrian Stratigraphy, Lower Mackenzie River Area (Operation Norman), District of Mackenzie: Geological Survey of Canada Paper 73-9.

Armstrong, R.L., Eisbacher, G.H., and Evans, P.D., 1982, Age and stratigraphic-tectonic significance of Proterozoic diabase sheets, Mackenzie Mountains, northwestern Canada: Canadian Journal of Earth Sciences, v. 19, p. 316-323.

Cook, D.G. and Aitken, J.D., 1969, Erly Lake, District of Mackenzie (97A); Geological Survey of Canada, Map 5-1969.

_______________, 1978, Twitya Uplift: a pre-Delorme phase of the Mackenzie Arch: in Current Research, Part A, Geological Survey of Canada Paper 78-1A, p. 383-388.

Delaney, G.D., Jefferson, C.W., Yeo, G.M., McLennan, S.M., Bell, R.T., and Aitken, J.D., in press. Some Proterozoic sediment-hosted metal occurrences of the northeastern Canadian Cordillera: Society of Economic Geologists, Couer d'Alene Field Conference, Wallace, Idaho, Nov. 3-5, 1977, Idaho Bureau of Mines.

Eisbacher, G.H., 1976, Proterozoic Rapitan Group and related rocks, Redstone River area, District of Mackenzie: in Report of Activities, Part A, Geological Survey of Canada Paper 76-1A.

_______________, 1977, Tectono-stratigraphic framework of the Redstone Copper Belt, District of Mackenzie: in Report of Activities, Part A, Geological Survey of Canada Paper 77-1A.

_______________, 1978, Re-definition and Subdivision of the Rapitan Group, Mackenzie Mountains: Geological Survey of Canada Paper 77-35.

Gabrielse, H.R., 1972, Younger Precambrian of the Canadian Cordillera: American Journal of Science, v. 272, p. 521-536.

Gabrielse, H.R., Blusson, S.L., and Roddick, J.A., 1973, Geology of Flat River, Glacier Lake and Wrigley Lake Map-Areas, District of Mackenzie and Yukon Territory: Geological Survey of Canada Memoir 366, Pts. I and II.

Gross, G.A., 1965, Iron-formation of Snake River area, Yukon and Northwest Territories: in Report of Activities, Part A, Geological Survey of Canada Paper 65-1A, p. 143.

Jefferson, C.W., 1978, Stratigraphy and Sedimentology, Upper Proterozoic Redstone Copper Belt, Mackenzie Mountains, Northwest Territories – a preliminary report: Canada Department of Indian and Northern Affairs, Mineral Industry Report 1975, p. 157-169.

Kirkham, R.V., 1974, A synopsis of Canadian stratiform copper deposits in sedimentary sequences: in Bartholomé, P., ed., Gisements Stratiformes et Provinces Cuprifères: Societe Géologique de Belgique, Liège, p. 367-382.

Stewart, J.H., 1972, Initial deposits in the Cordilleran geosyncline; evidence of a later Precambrian (< 850 m.y.) continental separation: Geological Society of America Bulletin, v. 83, p. 1345-1360.

Stuart, R.A., 1963 Geology of the Snake River Iron Deposit: Canada Department of Indian Affairs and Northern Development, Yellowknife, Open File Report.

Young, G.M., 1976, Iron-formation and glaciogenic rocks of the Rapitan Group, Northwest Territories, Canada: Precambrian Research, v. 3, p. 137-158.

_______________, 1977, Stratigraphic correlation of Upper Proterozoic rocks of northwestern Canada: Canadian Journal of Earth Sciences, v. 14, p. 1771-1787.

Young, G.M., Jefferson, C.W., Long, D.G.F., Delaney, G.D. and Yeo, G.M., in press, Upper Proterozoic stratigraphy of northeastern Canada and Precambrian history of North American Cordillera: Society of Economic Geologists, Couer d'Alene Field Conference, Wallace, Idaho, Nov. 3-5, 1977, Idaho Bureau of Mines.

MS received April 1979;
Accepted for publication December 1979.

CASE HISTORY PAPERS

Precambrian Sulphide Deposits, H.S. Robinson Memorial Volume, edited by R.W. Hutchinson, C.D. Spence and J.M. Franklin, Geological Association of Canada Special Paper 25, 1982

GEOLOGY AND NICKEL SULPHIDE DEPOSITS OF THE THOMPSON BELT, MANITOBA

Walter V. Peredery and Geological Staff
Inco Metals Company, Geological Research Group, Copper Cliff, Ontario P0M 1N0

ABSTRACT

The Thompson Belt is a linear NE-trending tectonic feature located on the boundary between the Superior and Churchill Provinces in north-central Manitoba. The Belt is comprised largely of gneisses, metasedimentary, metavolcanic, and ultramafic (UM) rocks, and felsic plutons. Most of the metasedimentary, metavolcanic and UM rocks are confined to the western side of the Belt which also contains the known nickel deposits.

The gneisses are felsic to intermediate in composition and could, at least in part, be paragneisses. They are generally stratiform in character, variously migmatized, contain basic components and have complex tectonic and metamorphic histories. They show evidence of an earlier Archean granulite facies of metamorphism and a pervasive retrograde metamorphism to amphibolite facies, interpreted as a product of the Hudsonian tectonic event.

The metasedimentary-metavolcanic assemblages form long, thin, complexly deformed bands. Some, such as the Pipe group, appear to be younger than the gneisses, and others, such as the Thompson group, may be coeval with the gneisses. The metasedimentary sequences in both groups include siliceous, calcareous, pelitic and ferruginous rocks which have mineralogical assemblages indicative of high grade and complex metamorphic histories. Structurally, at least two folding events are recognized. The earlier folding produced tight, subhorizontally plunging synclinal structures. Later cross folding gave rise to subvertically plunging folds attributed to the Hudsonian event.

The metavolcanic pile consists largely of pillowed and massive metabasaltic flows. They are completely recrystallized to hornblende-feldspar amphibolites so that no primary textures are evident. Closely associated with these rocks are magnesian metabasalts and minor ultramafic flows. The magnesian metabasalts are recrystallized to amphibole-chlorite-feldspar schists but locally, relict quench textures are preserved. The ultramafic flows are pillowed and recrystallized to ultramafic amphibolites. Field relationships in general suggest that the metavolcanic rocks are coeval with the metasedimentary strata.

The UM are divided into two groups – serpentinites and ultramafic amphibolites. The serpentinites occur as sheet-like or lenticular concordant bodies in the gneisses, metasedimentary and metavolcanic rocks. They range in composition from dunite through peridotite to pyroxenite and show chemical differentiation trends. The ultramafic amphibolites also occur as lenticular concordant bodies in the gneisses and metavolcanic rocks but are scarce or lacking in the metasedimentary members. They range in composition from picrite to pyroxenite and may have minor peridotitic phases. They do not carry significant sulphide mineralization. The general character of the UM indicates that many of them were probably originally intruded as sills. Since they were involved in folding together with the metasedimentary and metavolcanic rocks they are either early Hudsonian or pre-Hudsonian in age.

The nickel sulphide deposits are genetically and spatially related to the serpentinites. However, their present distribution is largely the result of remobilization during the long and complex tectonic history of the Thompson Belt. The sulphides occur as interstitial grains in the serpentinites, as massive and inclusion-bearing sulphides on the contact between the serpentinites and the country rocks, and as stringers or veins in the serpentinites and country rocks. Some sulphides, particularly those at Thompson mine, were extensively mobilized along the pelitic member of the metasedimentary sequence.

RÉSUMÉ

La ceinture de Thompson est une structure tectonique linéaire de direction nord-est et située à la limite des Provinces Supérieure et de Churchill dans le centre nord du Manitoba. La ceinture est largement composée de gneiss, métasédiments, métavolcanites, ultramafites et plutons felsiques. La pluspart des métasédiments, métavolcanites et ultramafites sont confinés à la partie occidentale de la ceinture qui contient aussi les gisements de nickel connus.

Les gneiss sont de composition intermédiaire à felsique et pourraient, au moins en partie, être des paragneiss. Ils sont généralement de caractère stratiforme et migmatisés à des degrés divers, contiennent des composants basiques et ont des histoires tectonique et metamorphique complexes. Ils montrent des signes d'un faciès granulite précoce d'âge archéen et d'un métamorphisme rétrograde général, interprété comme un produit de la phase tectonique hudsonienne.

Les assemblages métasédimentaires et métavolcaniques forment de longues et minces bandes déformées de façon complexe. Certains, tel le groupe de Pipe, semblent plus jeunes que les gneiss, alors que d'autres, tel le groupe de Thompson, sont peut-être contemporains des gneiss. Les séquences métasédimentaires dans les deux groupes comprennent des roches siliceuses, calcaires, pélitiques et ferrugineuses avec des assemblages minéralogiques indiquant une histoire métamorphique complexe. Structuralement, on y reconnait au moins deux phases de plissement. La première phase produisit des structures synclinales plongeant subhorizontalement. Des plissements transverses ultérieurs donnèrent des plis plongeant subverticalement attribués à la phase hudsonienne.

L'empilement volcanique consiste surtout en coulées métabasaltiques massives et en coussins. Elles sont complètement recristallisées en amphibolites à hornblende et feldspath si bien qu'aucune structure primaire n'y est évidente. Des métabasaltes magnésiens et des coulées ultramafiques mineures sont intimement associés à ces roches. Les métabasaltes magnésiens ont recristallisé en schistes à amphibole, chlorite et feldspath, cependant des structures de trempe ont été préservées localement. Les coulées ultramafiques sont en coussins et recristallisées en amphibolites ultramafiques. Les relations de terrain suggèrent généralement que les roches métavolcaniques et métasédimentaires sont contemporaines.

Les ultramafites sont divisées en deux groupes – serpentinites et amphibolites ultramafiques. Les serpentinites forment des corps en feuillet ou lenticulaires, concordants dans les gneiss, métavolcanites, et métasédiments. Elles varient en composition de dunite à pyroxénite en passant par péridotite et montrent des courbes de différenciation chimique. Les amphibolites ultramafiques se trouvent aussi en corps lenticulaires mais sont rares ou absentes dans les membres métasédimentaires. Elles varient en composition de picrite à pyroxénite et peuvent avoir des phases péridotiques mineures. Elles ne contiennent aucune minéralisation sulfurée d'importance. Le caractère général des ultramafites indiquent qu'à l'origine beaucoup d'entre elles furent probablement injectées en sills. Elles ont subies les mêmes plissements que les métasédiments et les métavolcanites, et sont par conséquent d'âge pré-hudsonien on hudsonien précoce.

Les gisements de sulfures de nickel sont associés spatialement et génétiquement aux serpentinites. Cependant, leur répartition actuelle est largement le résultat d'une remobilisation lors de la longue et complexe histoire tectonique de la ceinture de Thompson. Les sulfures apparaissent en grains intersticiels dans les serpentinites, en sulfures massifs et contenant des inclusions au contact des serpentinites et des roches encaissantes, ainsi qu'en veinules ou veines dans les serpentinites et les roches encaissantes. Certains sulfures, particulièrement ceux de la mine de Thompson, ont été considérablement mobilisés le long du membre pélitique de la séquence métasédimentaire.

INTRODUCTION

The historical development of the geology along the boundary between the Superior and Churchill provinces in northern Manitoba has been summarized by Bell (1971a). Bell (1971b) correlated the geology with geophysical and geochronological data producing a comprehensive picture of the region and outlined the Thompson Belt structure as it is known now.

Inco Metals Company (formerly The International Nickel Company of Canada, Limited) was attracted to the area in 1946 because of known nickel occurrences at Lynn Lake, Ospwagan Lake and elsewhere. By 1948, geological data indicated that this was an area of major interest and, with the help of an aeromagnetic survey carried out by the Company, the Thompson Belt came into focus. This survey outlined not only the Belt but also numerous anomalous areas within it which became prime targets for exploration. In 1951 the Company carried out its first airborne electromagnetic survey. This added impetus to exploration and culminated in nickel sulphide discoveries in the Belt.

In this paper the regional and tectonic setting of nickel deposits in the Belt are described with emphasis on the importance of the metasedimentary sequences as hosts of ultramafic rocks and nickel sulphides, and the effects of subsequent tectonism.

REGIONAL SETTING OF THOMPSON BELT

The Thompson Belt is a complexly deformed Precambrian mobile belt, between the Churchill and Superior provinces, in northern Manitoba (Fig. 1). It has a distinctive geophysical expression (Bell, 1971b) which has been used to trace it under the Paleozoic cover to the south (Roth, 1975, Kornik, 1969; Gibb, 1968; Kornik and McLaren, 1966; Innes, 1960). Its northern extension is less well defined and is subject to debate.

K-Ar dating indicates a Hudsonian (1.8 Ga) age for the last major tectonism, but Rb-Sr dates suggest that both Proterozoic and Archean rocks may be present in the

Thompson Belt (Cranstone and Turek, 1976). The Belt trends northeasterly and is in part fault-bounded on both sides (Fig. 1). To the northwest, it is in contact with the Churchillian Kisseynew paragneisses of greywacke and arkosic compositions, locally with some siliceous, calcareous and ferruginous metasedimentary rocks and minor

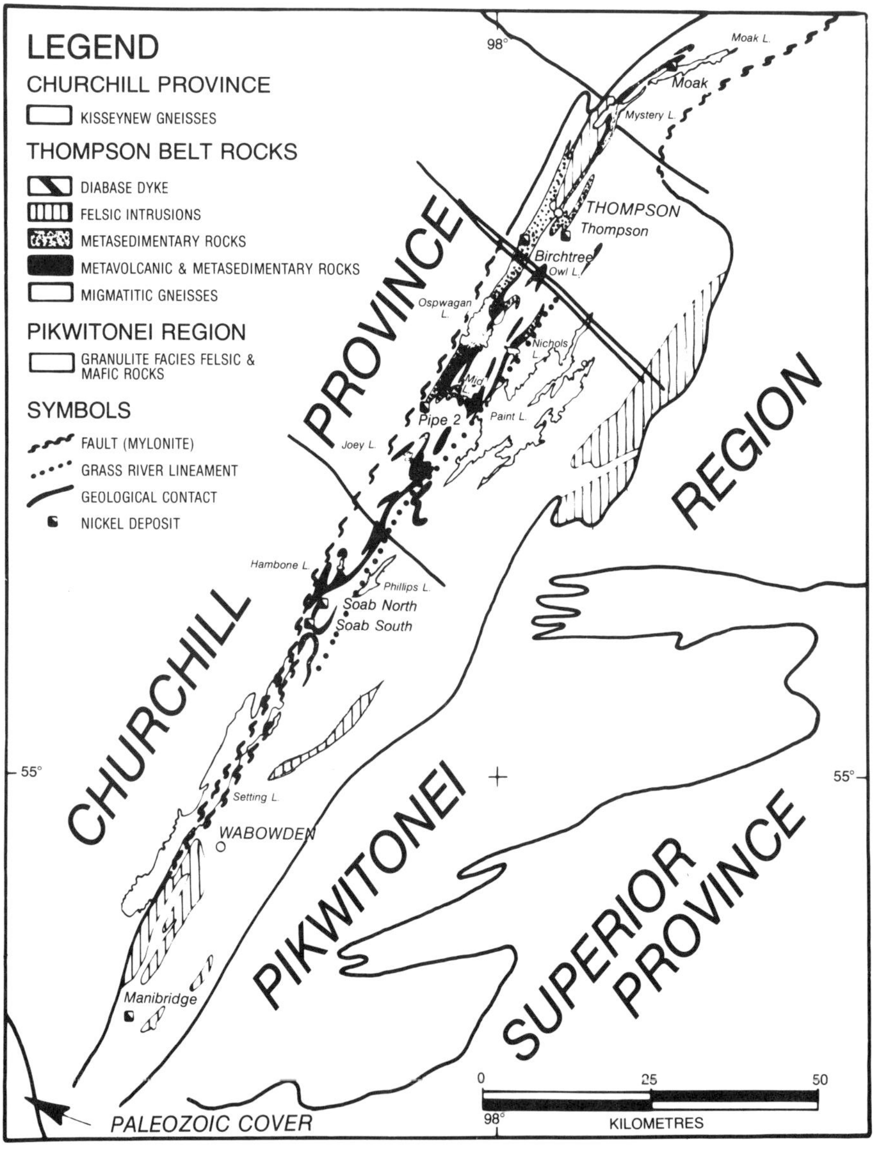

Figure 1. Regional geology of the Thompson Belt.

mafic metavolcanic rocks. The Kisseynew gneisses form a large domal structure about 400 km in diameter and are extensively migmatized, particularly in the central portion of the dome. Partially flanking these gneisses are the older metavolcanic-metasedimentary Wasekwan Group in the Lynn Lake area and the metavolcanic Amisk and metasedimentary Missi Groups near Flin Flon (McRitchie, 1974). Rb-Sr dating on Amisk metavolcanic rocks yielded Hudsonian (1.7 Ga) ages (Mukherjee *et al.*, 1971; Bell *et al.*, 1975), but at Hanson Lake, metavolcanic rocks yielded an Archean age of 2.5 Ga (Coleman, 1970).

The Kisseynew gneisses are generally metamorphosed to upper amphibolite facies but granulite facies have been reported also (Elphick, 1972; McRitchie, 1977). They have a northeasterly trending Hudsonian fabric superimposed on an earlier cryptic east-west fabric (McRitchie, 1974). The Hudsonian fabric is characterized by open folds which become tighter close to the Thompson Belt. Isotopic data from the Kisseynew gneisses indicate mainly Hudsonian ages. A K-Ar age of 2.67 Ga from Granville Lake, south of Lynn Lake, is reported by Stockwell (in Leech *et al.*, 1963, p. 65) but there is some question whether the particular samples came from that area (McRitchie, pers. commun.).

To the southeast, the Thompson Belt has a faulted contact with the ''Pikwitonei Province'' which has been interpreted by Bell (1971b) as the basement of the Superior Province. Recent field studies by Weber and Scoates (1978), however, seem to indicate that the Pikwitonei rocks are continuous with the Superior, casting doubt on Bell's interpretation. The Pikwitonei comprises felsic and mafic granulite facies gneisses intruded by numerous mafic to ultramafic dykes. The gneisses have a complex metamorphic history with a weak but distinct Hudsonian overprint (Hubregtse, 1977, 1978). Rb-Sr studies are currently being carried out at the University of Montreal to determine the age of the Pikwitonei rocks.

The Thompson Belt is characterized by a strong northeasterly fabric and subvertical dips and plunges. The Belt is underlain by migmatitic gneisses containing linear bands of metavolcanic and metasedimentary rocks (Fig. 1) which show evidence of a complex tectonic history and are intruded by mafic, ultramafic and felsic bodies of various ages (Table I). The metasedimentary and metavolcanic rocks are closely associated and occur mainly on the western side of the Thompson Belt. This area is an important locus of ultramafic magmatism and nickel sulphide mineralization. The metasedimentary rocks in particular, controlled the emplacement of ultramafic magma and sulphides, but subsequent tectonism played an equally important role in the final distribution of ultramafic rocks and nickel sulphides. The emphasis in this paper is on the two principal metasedimentary assemblages, the Pipe and the Thompson groups (named after the two mines) which host the major nickel deposits in the Thompson Belt.

GEOLOGY OF THE THOMPSON BELT

The Thompson Belt has distinctive geophysical, structural and lithological features, which set it apart from adjacent geological terranes (Bell, 1971b). In this paper, the rock groups of the Thompson Belt outlined by Coats *et al.*, (1972) have been subdivided and a number of rock types, not recognized earlier, have been distinguished (Table I). The gneissic assemblage, which is the most abundant in the Belt,

TABLE I
LITHOLOGIC GROUPS OF THE THOMPSON BELT

			Lithologic Groups of Coats *et al.*, 1972
Diabase Dykes			12
Hudsonian Felsic Intrusions and Pegmatites			9 and 10
Ultramafic Rocks:	Serpentinites Metapicrites		11
Metasedimentary Rocks:	Pipe Metasedimentary Group Thompson Metasedimentary Group		5
Metavolcanic Rocks and Related Metagabbros:	Metabasalts Magnesian Metabasalts and Ultramafic Flows Metagabbros		6 and 8
Gneisses:	Migmatitic Gneisses Paragneisses		7
	Amphibolite Gneisses	– dykes sills, metavolcanic rocks (?) and gabbroic layered complex (GLC)	
	Metasedimentary – Rocks	siliceous, pelitic, calcareous and ferruginous	
	Orthogneisses		

underlies most of the eastern magnetically flat portion, and in the western magnetically anomalous part it appears to be the basement on which the sedimentary and volcanic rocks were deposited. In the gneisses there are thin units of mafic and metasedimentary rocks which are apparently part of the gneissic assemblage. The gneissic assemblage has been variably migmatized and intruded by concordant bodies of felsic orthogneisses. Both the gneisses and the Pipe metasedimentary band are intruded by Hudsonian felsic plutons. Conformable, lenticular ultramafic bodies occur in the gneissic and in the metavolcanic and metasedimentary assemblages. Northwest-trending, near-vertical diabase dykes cut most of the rocks at a number of localities in the Belt.

Gneissic Assemblage

The Thompson Belt gneisses have been variously described as granoblastic gneisses, layered migmatitic gneisses and lit-par-lit gneisses with an emphasis on the migmatitic component. Though anatexis has obviously been important in their formation, the original host rocks are locally still recognizable. Emphasis is placed on the recognition of host rocks rather than the migmatitic component, but it should be understood that this component is nearly always present. On this basis the gneisses can be subdivided into five major lithological units: 1) migmatitic gneisses, 2) paragneisses, 3) amphibolitic gneisses – dykes, sills and gabbroic layered complex (GLC), 4) siliceous, pelitic, calcareous, and ferruginous metasedimentary rocks, and 5) orthogneisses.

The first unit comprises all the gneisses of uncertain origin. They are generally highly migmatized and are the most abundant of all the gneisses. The restite fraction is variable and composed of biotite, hornblende, feldspar, quartz and commonly garnet. Locally, relict clinopyroxene and/or orthopyroxene mantled by hornblende are present suggesting that perhaps much of the restite was originally in the granulite facies of metamorphism. In the mobilizate at least three different fractions are recognized: an older, deformed, medium-grained, granular rock of granodioritic composition, and two younger fractions, one a pink or grey aplite, the other a coarse-grained pegmatite. Both the aplite and the pegmatite are generally undeformed, but locally are sheared and mylonitized.

Two main varieties of paragneiss are recognizable: a relatively pure quartzo-feldspathic gneiss, and an impure quartzo-feldspathic gneiss with variable proportions of mafic minerals. In the first type there is a distinct banding due to grain size variation and variable proportions of quartz and feldspar. Some gneisses, as at Soab Lake, display grain size variations suggestive of graded bedding (Fig. 2). Plagioclase and quartz are the essential minerals; biotite and/or garnet are generally present as minor minerals, and locally secondary potassic feldspar or muscovite is present. The mineral assemblage indicates amphibolite facies of metamorphism. Elsewhere, as at

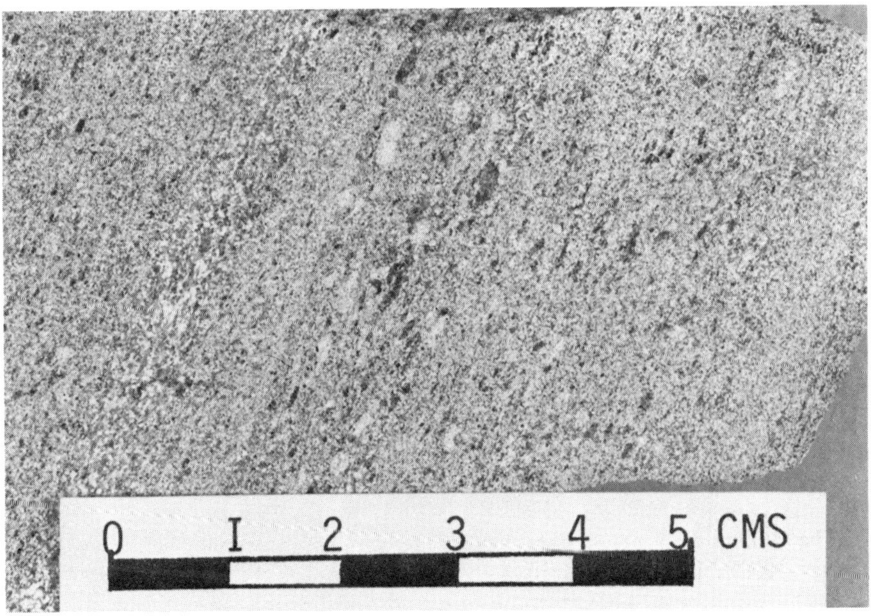

Figure 2. Grain size variation suggestive of graded bedding in a quartzofeldspathic paragneiss from Soab Lake area.

Phillips Lake, the quartzo-feldspathic gneisses show granulite to amphibolite facies metamorphism. The mafic constituents include two pyroxenes plus garnet, all showing partial retrogression to amphibole, and the felsic constituents are recrystallized.

In the impure quartzo-feldspathic paragneisses, the mafic content ranges from 10 to 30% and includes biotite, garnet, amphibole and locally sillimanite or pyroxenes. The rocks show mineralogical layering, or probably primary banding, and in places are closely associated with distinctly metasedimentary siliceous members. Mineralogically the impure paragneisses are quite similar to the restite in the highly migmatized gneisses.

The minor but ubiquitous amphibolite component of the gneisses represents several rock types. The amphibolites with fine-grained margins and coarse-grained interiors are probably basic dykes or sills. They are up to 30 m wide and appear to be highly boudinaged bodies with a strong foliation developed parallel to the regional fabric. They consist of hornblende + plagioclase ± garnet ± biotite, an assemblage which is indicative of amphibolite facies of metamorphism. They probably post-date granulite facies of metamorphism.

Many of the amphibolitic components of the gneisses display mineralogical banding and are referred to here as the "gabbroic layered complex" (GLC). They are ubiquitous in the Thompson Belt gneisses, and range in size from less than a metre to mappable units several hundred metres in width and up to 3 km in length as at Paint Lake. The GLC is composed of hornblende and plagioclase but garnet, orthopyroxene and clinopyroxene relicts are common. Judging by the mineralogy the GLC had undergone granulite facies metamorphism and has been partially or completely retrogressed to amphibolite facies. The GLC ranges in composition from gabbro to pyroxenite and locally anorthosite. Layering in the GLC is at an angle to the regional fabric in the gneisses suggesting that the GLC may represent tectonic slices from a large layered pluton at depth, or highly deformed smaller intrusions.

Recognizable metasedimentary components in the gneisses include siliceous, pelitic, calcareous and ferruginous rocks which occur as boudinaged lenses and narrow discontinuous bands up to 10 to 20 m wide. In any band, one or more of these rock types may be present. The siliceous metasedimentary rocks are commonly quartzites (more than 90% quartz) and feldspathic quartzites (10 to 30% feldspar), some of which are garnetiferous. The pelitic members are generally schistose and contain sillimanite and garnet, or cordierite and anthophyllite. The calcareous rocks have two distinct assemblages. One consists essentially of diopside, microcline and mica. The other consists of diopside, olivine and colourless mica in a carbonate matrix. Both assemblages are similar to the skarns in the Thompson group. An olivine-bearing skarn at Paint Lake has, in addition, garnet which has not been recognized in skarns elsewhere. The ferruginous rocks are coarse-grained, mineralogically banded iron-formations. Both oxide and silicate facies are recognized, the latter commonly with clinopyroxene, orthopyroxene, garnet and cummingtonite-grunerite amphibole. Disseminated sulphides, pyrrhotite and pyrite, occur commonly in these rocks and locally massive sulphide facies have been observed.

The orthogneisses are granodioritic and of two ages. The older gneisses are relatively homogeneous, coarse-grained rocks, composed of plagioclase, potassic feldspar, quartz, orthopyroxene, clinopyroxene, biotite and amphibole. They are particularly well preserved in the Paint Lake area. Strongly tectonized and migmatized equivalents of such gneisses are mineralogically similar to, and could constitute a significant component of the migmatitic gneisses.

The younger granodioritic orthogneisses cut the older orthogneisses and all other gneisses and occur as small, concordant, lenticular bodies which appear to have been folded. They are medium-grained, foliated, homogeneous and are composed of plagioclase and quartz, and minor potassic feldspar, hornblende, garnet and biotite. Mineralogically they are similar to the granodioritic fraction in the migmatitic gneisses and may be related to this migmatization event.

Metasedimentary Rocks

Pipe and Thompson Metasedimentary Assemblages. In establishing the lithological groupings in the Thompson Belt, Coats et al. (1972) included both Pipe and Thompson metasedimentary assemblages within Group 5, (Table I). In this paper geographic names have been assigned to these two metasedimentary groups for easier reference. Weber and Scoates (1978) used the expression "Ospwagan group" in describing the metasedimentary assemblage at Ospwagan Lake which is equivalent to the Pipe assemblage. The latter expression is preferred here since at Pipe mine an easily accessible complete section is presented.

The two metasedimentary groups, which are separated by gneisses on surface and at depth, are similar but there are a number of differences. In this paper the differences are emphasized because these are significant to the interpretation of the Thompson Belt.

The stratigraphic succession of the Pipe and Thompson metasedimentary groups is compared in Figure 3. On the basis of primary sedimentary features, the stratigraphic top is in the direction from pelitic rocks to iron-formation in the Pipe group. In the Thompson group the sequence of rocks is similar to that at Pipe, but there are no reliable sedimentary stratigraphic top indicators. The stratigraphic succession in both the Pipe and Thompson groups is from siliceous to calcareous to pelitic, and finally to ferruginous rocks. The calcareous and pelitic rocks contain minor siliceous layers. Dolomite and skarn are associated with the lower quartzites and iron-formation in the Pipe group, but in the Thompson group skarns are confined to a zone between quartzite and pelitic schist. In the Thompson group the width of the calcareous unit varies considerably; it may be up to a couple of hundred metres wide or altogether absent. Pipe group calcareous members are thinner, but are continuous over tens of kilometres. The siliceous metasedimentary rocks of the Thompson group are commonly feldspathic, whereas those of the Pipe group are more micaceous. Meta-arkoses occupy the core of the Thompson group structure, whereas in the core of the Pipe group micaceous quartzites are dominant. Very siliceous quartzites are associated with the iron-formation in both groups.

The contact between the metasedimentary rocks and gneisses is sharply defined where it is relatively undisturbed. Elsewhere, the contact is difficult to place. It could be described as a complex zone marked by intense mylonitization and variable development of feldspar porphyroblasts. The zone is up to 100 m wide, with siliceous metasedimentary rocks apparently interbanded with gneisses. The gneisses have

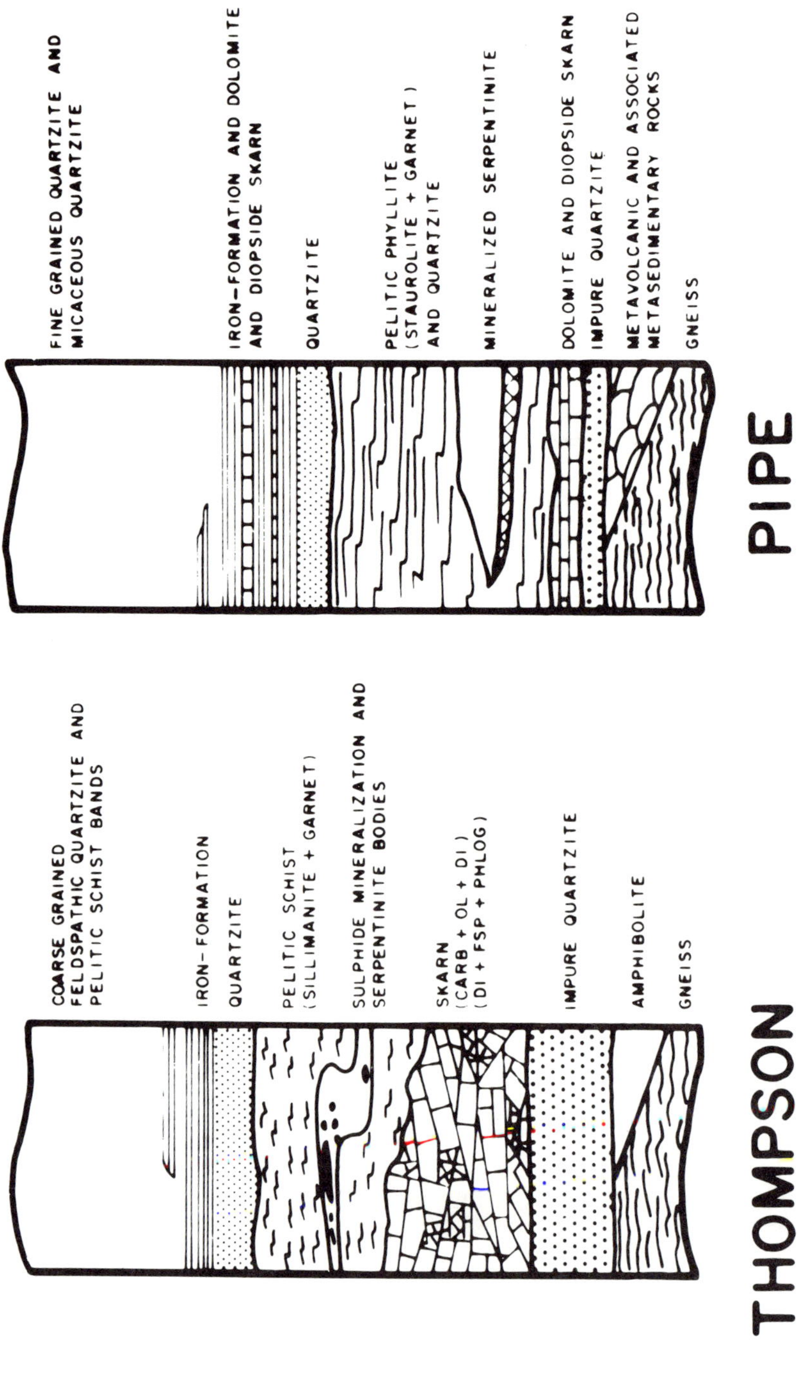

Figure 3. Stratigraphic successions in Pipe and Thompson metasedimentary groups.

blastomylonitic textures and in extreme cases are superficially similar to the siliceous metasedimentary units. As the intensity of mylonitization decreases, the original character of the rock becomes more evident.

Siliceous Metasedimentary Rocks. The siliceous metasedimentary rocks are banded, fine- to medium-grained meta-arkoses, micaceous quartzites and very siliceous metacherts. The meta-arkoses of the Thompson group carry up to 30% modal plagioclase and microcline, and up to 10 per cent biotite and/or muscovite. In the core of the Thompson group structure, the meta-arkoses are coarse-grained and carry up to 50% sodic plagioclase, and usually less than 10% microcline. The microcline appears to be secondary. Sillimanite is a minor constituent associated with micaceous slivers in the meta-arkoses. Minor garnet may also be present. Accessory minerals include apatite, zircon and sphene. The micaceous quartzites of the Pipe group carry up to 15% biotite and/or muscovite and minor feldspar. The metacherts are composed mainly of quartz but may contain minor feldspar, mica, amphibole, or zoisite.

Quartz in many of the siliceous metasedimentary rocks shows either very fine granular recrystallization or undulatory extinction and elongation; the feldspars are generally intact but commonly show partial alteration to sericite and myrmekite. This type of deformation and alteration is common to many of the quartzo-feldspathic rocks of the Thompson Belt.

Calcareous Metasedimentary Rocks. The calcareous rocks have been subdivided into three main types: 1) dolomites, 2) diopside-microcline skarns, and 3) olivine-diopside-carbonate skarns.

The first and second types occur in the Pipe group and the second and third types

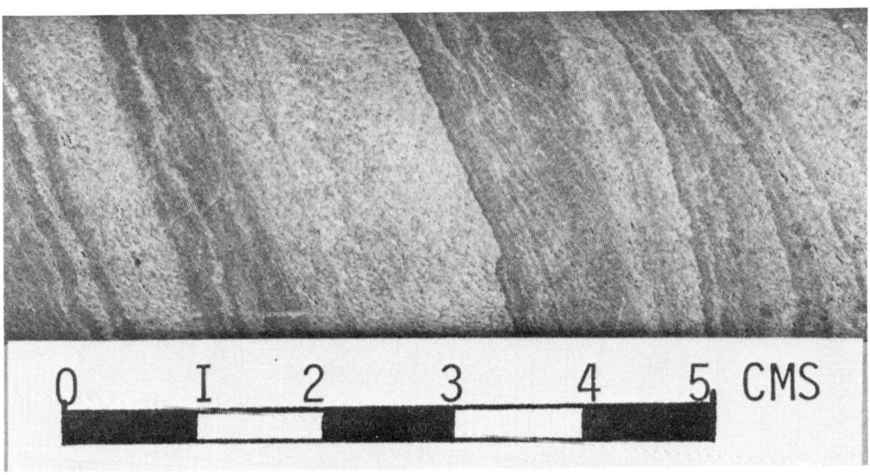

Figure 4. Compositional gradational banding in impure calcareous metasedimentary rock from Pipe group. The dark gray mica-rich layers consistently have alternate sharp and gradational contacts with light gray, relatively pure dolomite.

are common in the Thompson group. In the Pipe group both the dolomites and the diopside-microcline skarns are commonly finely banded (1 to 5 cm), and some display compositional gradations from dolomite to micaceous calcareous rock (Fig. 4). This banding is thought to represent original bedding and could possibly serve as a stratigraphic top indicator. The calcareous rocks of the Pipe group show dragfolding, boudinaging and plastic deformation but few, if any, microscopic deformational features. This is a feature typical of carbonate rocks in general and also applies to the Thompson group skarns.

In the Thompson group skarns foliation is accentuated by the development of secondary phlogopite, particularly in the diopside-microcline skarns, but no primary bedding is evident. There is much flowage in the skarns as indicated by contortion and brecciation, and pinching and swelling (up to 100 m in width) along the length of the calcareous unit. Both diopside-microcline and olivine-bearing skarns commonly appear to have been mixed because of the flowage. Where the two skarn types are relatively undisturbed, the diopside-microcline skarn is generally in contact with the pelitic rocks. Locally, relatively pure carbonate rocks composed of dolomite plus calcite are present.

The diopside-microcline skarn is essentially bimineralic in composition in the Pipe group, but in the Thompson group is more variable. In addition to diopside and microcline, carbonate, plagioclase, scapolite and quartz may be important constituents and minor sphene, apatite and rarely ziron may be present. Generally, the alteration products are tremolite and chlorite after diopside, and phlogopite after diopside and feldspar in both groups.

The olivine-diopside-carbonate skarn is common in the Thompson group calcareous unit, but has not been observed in the Pipe group. The olivine measures up to 2 mm in diameter and is subhedral, equant and partially serpentinized, and the diopside is subhedral and fresh or partially altered to tremolite. In addition to olivine and diopside, colourless mica may be present, and locally minor chondrodite. The total mafic content varies considerably, ranging from a few per cent to 40%.

A variety of skarn composed essentially of diopside occurs commonly on the contacts between diopside-microcline or olivine-bearing skarns and country rocks and probably represents a reaction zone between calcareous and siliceous rocks.

Pelitic Rocks. The pelitic unit, or "biotite schist" as it is referred to in the mines, is of particular interest because it is the locus of nickel sulphide mineralization (Thompson and Birchtree mines), or of ultramafic lenses with which sulphide mineralization is associated (Pipe 2 mine). Although previously no distinction has been made between Pipe and Thompson types, there are significant differences in their physical characteristics and in the types of aluminous silicates present.

The pelitic rocks in the Pipe group are mineralogically banded phyllites, whereas in the Thompson group they are generally coarse-grained foliated, relatively homogeneous schists. Pegmatite stringers are abundant in the schists but are absent or rare in the phyllites. In the Thompson schists the key minerals are commonly garnet, sillimanite and graphite whereas in the Pipe phyllites they are garnet, andalusite and staurolite. These latter assemblages indicate a lower grade of metamorphism in the Pipe group than the Thompson group pelites. Only locally is sillimanite found in the Pipe group.

Minor cordierite occurs at Thompson and Birchtree mines. Also, at Birchtree corundum and hercynitic spinel are recognized. The bluish-grey corundum crystals are up to a centimetre in diameter and occur in a biotite-rich rock in contact with an ultramafic body. The spinel occurs as single crystals or in clusters in the pelite and in the sulphides.

Within the pelitic schists at Thompson mine there are remnants of medium-grained, massive, homogeneous pelitic rock which consists of biotite, feldspar and quartz. On recrystallization it becomes coarse-grained and acquires the mineralogy, and pegmatitic fraction typical of the enclosing pelitic schist. It is surmised that much of the Thompson group pelitic schist probably originated from this massive parent rock.

Iron-Formations. The iron-formations are commonly mineralogically banded and continuous over long distances. Three facies are recognized: oxide, silicate and sulphide.

The oxide facies is composed mainly of alternating bands of magnetite and quartz. The silicate facies is mineralogically banded, with layers of various proportions of cummingtonite-grunerite, hornblende, magnetite, quartz, garnet and clinopyroxene alternating with siliceous bands. Apatite and tourmaline are common accessory minerals. At Thompson mine, orthopyroxene is also present in the silicate facies. In both the oxide and silicate facies some regular variation in band thickness suggests cyclicity. The siliceous bands become progressively thicker towards the margins of the iron-formation and mark the transition into the adjacent quartzites.

The silicate facies is the dominant type of iron-formation in the Thompson Belt metasedimentary assemblages. The oxide facies units are up to several metres wide and generally occur between layers of silicate facies. Both the oxide and silicate types contain up to 20% finely disseminated sulphides which appear to be primary, and where more than 10% sulphides are present, these are designated sulphide facies in this paper. Pyrrhotite is the dominant sulphide, and minor chalcopyrite and pyrite may be present.

The Pipe iron-formation, though similar, is generally finer grained than that of the Thompson group, and orthopyroxene has not been observed in the former.

Metavolcanic Rocks, Related Intrusions and Associated Metasedimentary Rocks

The metavolcanic rocks and related metagabbros have been recently described (Peredery, 1979). Chemically the rocks are basalts with komatiitic affinities (up to 11% MgO and less than one per cent TiO_2). The volcanic pile consists of massive and pillowed flows, minor volcanic breccias and finely banded tuffs, and is intruded by co-magmatic metagabbroic and metapicritic sills showing partial differentiation. The metabasalts and metagabbros are recrystallized to amphibolites and belong to the middle to upper amphibolite facies of metamorphism. Garnet and/or clinopyroxene are present locally, particularly in the highly deformed members.

Associated with the metabasalts is a more magnesian metabasalt characterized by the presence of quench textures. These rocks are composed of actinolitic amphibole (pseudomorphic after fibrous skeletal pyroxenes) in a felsic devitrified matrix. The magnesian metabasalt pile consists of alternating pillowed and massive flows and is intruded by sill-like bodies of metagabbro and metapicrite. Chemically, the magnesian metabasalts are komatiitic, as characterized by high magnesia (12 to 15% MgO) and less than one per cent TiO_2.

At Ospwagan Lake pillowed ultramafic flows are closely associated with the magnesian metabasalts. The flows are fine-grained, have a felty texture and are composed essentially of actinolitic amphibole. They have about 19 per cent MgO, generally less than one per cent TiO_2, and are interpreted as pyroxenitic komatiites. Chemically and mineralogically, they are similar to the metapicrite sill-like bodies and are considered to be co-magmatic with them. Some of the sill-like bodies could be massive flows.

The relationship between the metavolcanic rocks and the adjacent metasedimentary rocks and gneisses is structurally complex. However, from pillow lava top determinations, and other stratigraphic top data, it appears that the metavolcanic and Pipe metasedimentary rocks are coeval and stratigraphically above the gneisses.

In the Ospwagan Lake area, turbidite and conglomeratic metagraywackes which have not been found elsewhere, are closely associated with magnesian metabasalts. These metasedimentary strata occur immediately below or to the east of the metavolcanic band. The turbidite unit displays rhythmic layering and well preserved graded bedding (Fig. 5), and is stratigraphically above the conglomeratic metagraywacke.

Figure 5. Graded bedding in metagraywacke from Lower Ospwagan Lake. There is a distinct and consistent decrease in grain size and a corresponding increase in the micaceous component toward the left side of the figure, or upper part of the beds.

The turbidite stratigraphic tops face to the west in agreement with the pillow lava top determinations.

The turbidite crops out only very sparsely. However, diamond drill hole data indicate it is a few metres thick and continuous over a distance of 3 km. A typical graded bed shows a progressive decrease in the grain size of quartz and feldspar from a few millimetres at the base to silt size at the top, and a corresponding increase in the mica content from base upward. The graded beds vary in thickness from a few centimetres to about a metre. A steeply dipping penetrative cleavage and chevron style folding developed in these turbidites are attributed to Hudsonian deformation.

The conglomeratic metagraywacke associated with the turbidite has been traced for about 4 km, and is up to 100 m wide. The proportion of rock fragments to matrix is about 1:20. Subangular, pebble- to boulder-sized rock fragments are generally elongated in a subvertical direction, and are of several rock types including grey quartzites, biotite-quartz-feldspar gneisses, amphibolites and relatively pure calcareous rocks. Locally there are abundant cobble-sized calcareous concretions, some of which are cored by fragments of amphibolite. Generally, there is little or no sorting of the pebbles so that bedding is not obvious. Associated with the conglomeratic metagraywacke are minor quartzite beds up to 3 m wide.

The matrix of the conglomeratic metagraywacke consists of a variable proportion of oval, recrystallized, quartz grains up to several millimetres in diameter, and a finer-grained, schistose groundmass of biotite, chlorite, feldspar, quartz and carbonate. In places, the matrix has euhedral laths of plagioclase, and minute crystals of plagioclase invade the margins of the larger quartz grains producing a micrographic texture. These textures are metamorphic in origin but whether they are related to the regional metamorphism or the overlying magnesian metabasalt flows is not certain.

The contact between the conglomeratic metagraywacke and the metasedimentary rocks to the east is not exposed. A long lenticular body of serpentinite and irregular masses of metagabbro occur along this contact on the south shore of Ospwagan Lake.

Ultramafic Rocks

Most of the ultramafic rocks in the Thompson Belt are located in a zone about 6 km wide along the western side of the Belt. They occur in the metasedimentary and metavolcanic rocks and gneisses, but the largest bodies are found within the metasedimentary rocks. The ultramafic bodies are generally lensoid or tabular in shape and measure from less than one metre to about a kilometre in width (Table II). The contact between the ultramafic bodies and the host rocks is usually conformable, but the actual contact rocks are so deformed and altered that their true character is uncertain. The interiors of the lenses are generally altered but less deformed.

Some of the ultramafic bodies appear to have been folded together with their enclosing host rocks. Faulting of the ultramafic rocks in some cases offsets or even separates them into blocks; some are extensively fragmented. Good examples of fragmentation are observed at the Thompson mine where clusters of angular blocks of ultramafic rocks occur either in sulphide matrix or in pelitic schist.

The ultramafic rocks of the Thompson Belt fall into two main groups: serpenti-

TABLE II

STATISTICAL DATA ON THE SERPENTINIZED ULTRAMAFIC (UM) ROCKS AND NICKEL SULPHIDE MINERALIZATION OF THE THOMPSON BELT.

	Thompson	Birchtree	Pipe 1	Pipe 2	Soab North	Soab South	Moak	Mystery North	Mystery South	Nichols	Kipper	South Manasan	Burntwood River	Ospwagan	Hambone	Grass River
Size of UM bodies																
(width in meters)																
1–10	X	X	X		X	X										
10–250	X	X	X	X									X		X	X
250–1000				X			X	X	X	X	X	X				
1000+														X		
UM rock types																
dunite	X			X	X		X		X	X		X		X		
peridotite	X	X	(X)*	X	X	X	X	(X)	X	X	(X)	X	(X)	X	(X)	(X)
pyroxenite	X	X		X	X	X	X		X					X		
*MgO/(MgO+FeO)***	.75-.83	.55-.82		.40-.86	.72-.82		.54-.81		.61-.81	.74-.86	.71-.84	.84-.88	.67-.89	.70-.84		
Sulphides																
massive	X	X		X												
inclusion-bearing	X	X	X	X	X	X										
mineralized country rock	X	X													X	X
mineralized UM rock	X	X	X	X	X		X	X	X	X	X	X	X	X		
UM rock/sulphides	1.4	7	6	135	14	1.1	90	–	400	56	17	125	14	–	1.1	0.4
Po/Ni	18	22	19	35	16	23	26	–	–	–	–	–	–	–	–	–
Ni/Cu	14	16	12	12	14	13	11	11	13	12	12	20	18	13	10	11
Ni/(Ni+Cu)	0.93	0.94	0.92	0.92	0.93	0.93	0.92	0.92	0.93	0.92	0.92	0.95	0.95	0.93	0.91	0.91
Ni/Co	51	38	25	22	48	45	38	–	–	–	–	–	–	39	–	–
Pt/Pd	0.49	0.81	0.48	1.46	0.86	0.87	0.89	0.28	1.27	–	–	0.45	–	–	–	–

NOTE: All ratios are calculated on a weight per cent basis

* (X) partially investigated serpentinites

** Total iron calculated as FeO

nite and metapicrite. They occur in metasedimentary, metavolcanic and gneissic terranes. Both barren and mineralized serpentinites occur in all these terranes, but the metasedimentary terrane contains the important nickel sulphide deposits. The metapicrites do not carry significant sulphide mineralization.

Serpentinites. As a result of extreme alteration and deformation many ultramafic rocks of the Thompson Belt have been converted to serpentinite. The extent of alteration or deformation is variable. Several rock types are recognizable in the serpentinites or inferred from secondary mineral assemblages. These include dunite, peridotite, pyroxenite and picrite (Fig. 6) of which peridotite and dunite are by far the most abundant.

The serpentinites are layered. The layers are wide, indistinct and commonly have gradational contacts, particularly between peridotite and dunite. However, sharp or sharply gradational contacts between peridotite and pyroxenite also exist. The pyroxenites or picrites generally occur at one of the serpentinite margins.

Dunites are medium- to coarse-grained, and are composed primarily of subhedral to euhedral, equant to elongate serpentine pseudomorphs after olivine. The elongate grains are commonly aligned, producing a primary-looking igneous foliation which is quite distinct from the secondary foliation imposed by tectonism (Fig. 7). In some cases up to several per cent of very coarse-grained, equant "phenocrysts" of olivine, or clusters of "phenocrysts", occur in the dunite or in the peridotite (Fig. 8).

Figure 6. Thompson Belt serpentinite consists of dunite and peridotite with minor pyroxenite and picrite. From left to right, totally serpentinized dunite with finely disseminated sulphides, partially serpentinized poikilitic harzburgite, relatively fresh orthopyroxenite and recrystallized picrite (olivine (black) + amphibole + orthopyroxene).

Figure 7. Primary and tectonic foliations in serpentinite. In the primary type (left) foliation results from the alignment of elongate coarse-grained serpentinized olivine crystals. Tectonic foliation (right) is produced by shearing and fine-grained recrystallization in a pyroxene-rich peridotite. The dark gray lenticular structures are richer in serpentinized olivine than the matrix.

Poikilitic harzburgites are the most common among a number of mineralogical and textural varieties of peridotite. They are characterized by large orthopyroxenes enclosing fine equant grains of olivine and are in various stages of serpentinization. Relicts of orthopyroxene and olivine are found in some of the less altered harzburgites. The pyroxene to olivine ratio, the degree of poikilitic intergrowth, and the grain size are variable so that harzburgites are amenable to further subdivision.

Poikilitic textures persist in the lherzolites and wehrlites associated with harzburgite, but these peridotites are not as well documented. Clinopyroxene relicts are rare, but at Pipe 2, wehrlite east of the serpentinite has well preserved clinopyroxene interstitial to the olivine, and orthopyroxene occurs as very coarse cumulus grains with minor poikilitically included olivines.

Pyroxenites which comprise olivine orthopyroxenite and orthopyroxenite are the least serpentinized of all the ultramafic rocks. The poikilitic textures found in harzburgite persist in pyroxenite. The change from harzburgite to olivine orthopyroxenite is marked by a rapid increase in the orthopyroxene and a corresponding decrease in the olivine contents. The picrites, locally referred to as ultramafic amphibolites, are extensively altered and recrystallized to amphibole-rich rocks. They consist of various proportions of amphibole, olivine, orthopyroxene and olive-green spinel. Mineralogically and texturally they are similar to the metapicrite sill-like bodies. It should be noted that serpentinites with associated metapicrites are typical of the Thompson band metasedimentary terrane whereas serpentinites with associated pyroxenite are typical of Pipe band metasedimentary terrane.

Recently, primary plagioclase-olivine-bearing pyroxenites have been identified associated with Pipe 2 mine and Ospwagan Lake serpentinites. Serpentinized olivine

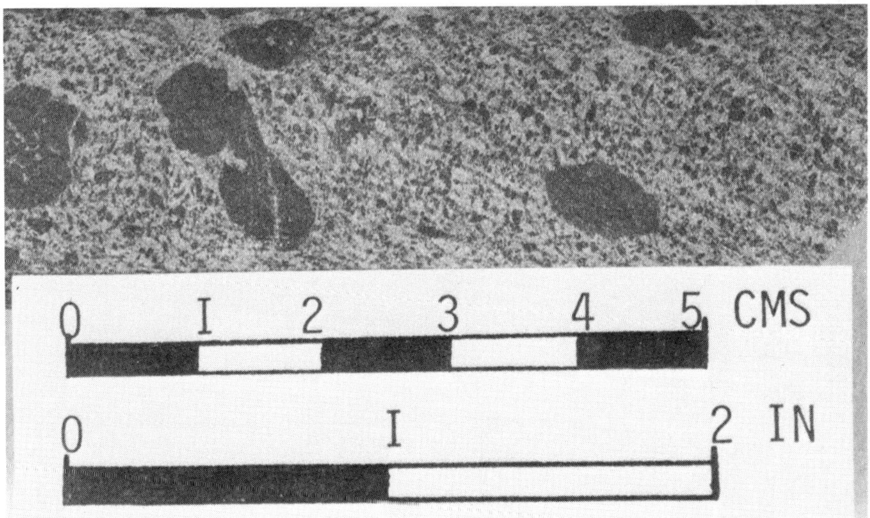

Figure 8. Very coarse euhedral crystals or "phenocrysts" of serpentinized olivine in serpentinite from Ospwagan Lake area. The matrix is composed of serpentine pseudomorphs after olivine (black) and pyroxene (grey) extensively altered to amphibole and chlorite. Note alignment of small olivine crystals (black) in the matrix and a diagonal secondary foliation.

is poikilitically enclosed in coarse grains of orthopyroxene, clinopyroxene and also within plagioclase. The plagioclase is bytownite, and occurs interstitially to the pyroxenes, constituting up to several modal per cent of the rock. The clinopyroxene has myrmekitic patches which may represent an exsolution texture. The orthopyroxene lacks this texture. Wherever plagioclase is in contact with clinopyroxene, a reaction zone of amphibole is present. In the early stages of development the amphibole has a radial pattern; in the more advanced stages of alteration the amphibole becomes feathery and felty and completely replaces the plagioclase.

Metapicrites. Included in this group are a number of ultramafic rocks characterized by abundant secondary amphibole, porphyroblastic orthopyroxene and anhedral olivine, and locally referred to as ultramafic amphibolites. Texturally these rocks are metamorphic. Their normative mineralogy suggests the original rocks were mainly feldspar-bearing, olivine-rich picritic rocks and some were pyroxenites. Much of the amphibole has probably been derived by a reaction between plagioclase and clinopyroxene under hydrous conditions as described in the preceding paragraph. The metapicrites are described in detail in an earlier publication (Peredery, 1979) and are considered to be in part subvolcanic intrusions, and in part ultramafic flows, some of them pillowed. They are closely associated with the metabasalts and magne-

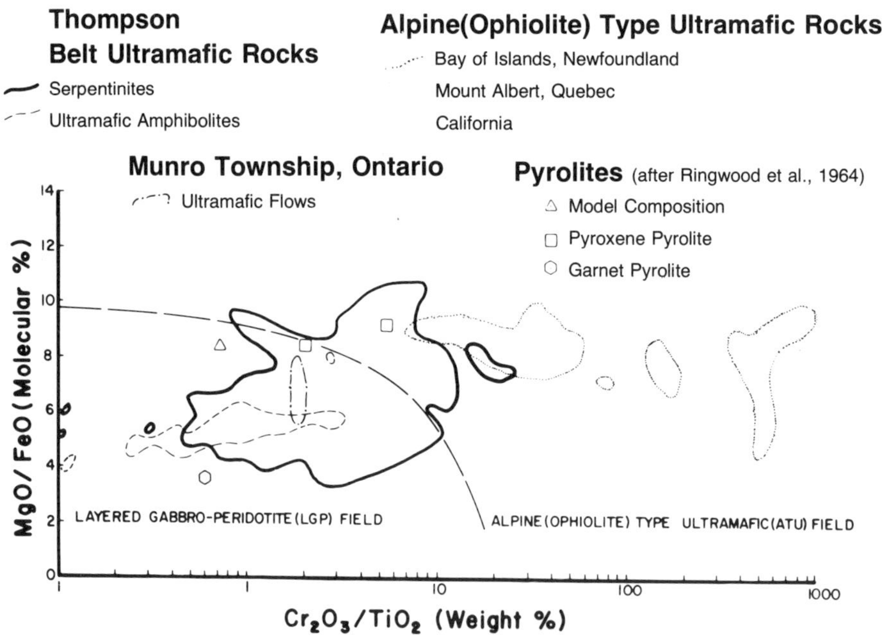

Figure 9. MgO/FeO (Total Fe calculated as FeO) versus Cr_2O_3/TiO_2 plot (modified after Suturin *et al.*, 1972) of Thompson Belt and other ultramafic rocks. Most of the Thompson Belt serpentinites fall in the layered gabbro peridotite (LGP) field and some close to the statistical borderline between the LGP and alpine type ultramafic (ATU) fields. The ultramafic amphibolites of the Thompson Belt and Munro Township spinifex ultramafic flows fall in the LGP field. Data on the alpine ultramafic rocks from the Bay of Islands in Newfoundland, California and Mount Albert, Quebec plot in the ATU field as expected.

sian metabasalts, are differentiated with mineralogical layering, and chemically are komatiitic.

Metapicrites associated with serpentinites are interpreted as fractional differentiates of an ultrabasic magma. They are chemically similar to the metapicritic intrusion, flows and metabasalts, and appear to be genetically related to them.

Chemistry of Serpentinites. In Figure 9 alpine type serpentinites and mantle portions of ophiolite complexes are compared to the Thompson Belt ultramafic rocks using an MgO/FeO (molecular) versus Cr_2O_3/TiO_2 (weight) plot (modified after Suturin *et al.*, 1972). It is commonly possible with this plot to distinguish alpine type ultramafic rocks (ATU, and mantle portions of ophiolite) from those of gabbro-peridotite association, as in layered gabbro-peridotite complexes (LGP). Many of the Thompson Belt ultramafic rocks fall on the LGP side of the diagram, but many straddle the boundary between the two fields and some overlap into the ATU field. Published data on ultramafic rocks from ophiolite complexes (Irvine and Findlay, 1972; Bailey and Blake, 1974) fall in the ATU field, as expected. Pyke's *et al.* (1973) chemical data on the ultramafic flows from Munro Township, Ontario, plot in the LGP field.

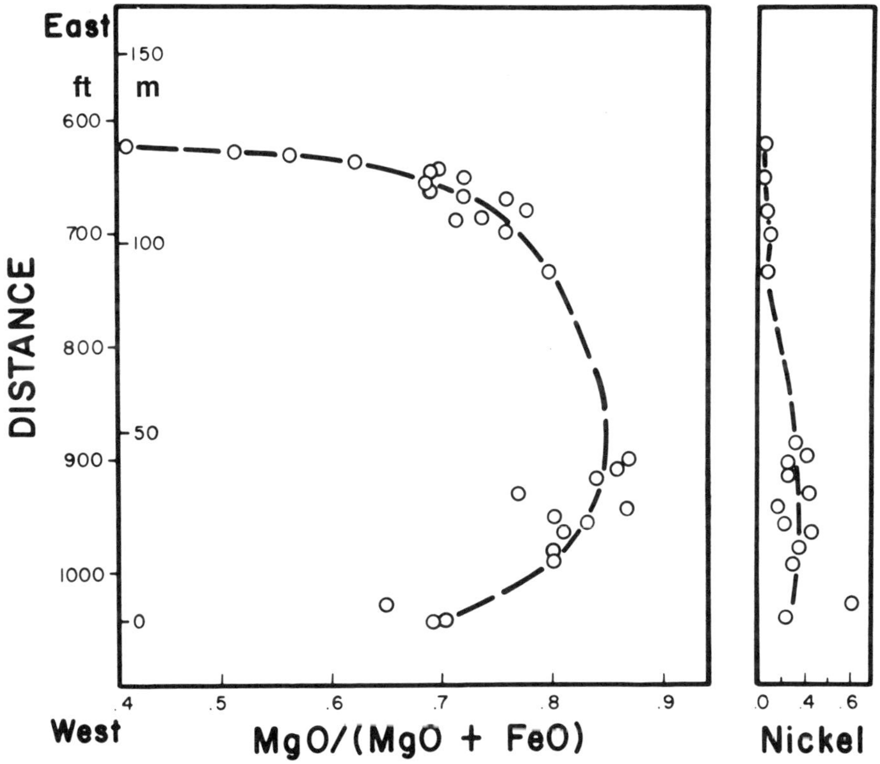

Figure 10. MgO/(MgO+FeO) versus distance plot across the Pipe 2 serpentinite. The differentiation curve indicates a progressive depletion in magnesia from west to east across the serpentinite, particularly on its eastern side. Corresponding nickel variation (in weight per cent in all figures) shows a similar trend.

Further similarities between the Thompson Belt serpentinites and the LGP association are shown by other chemical ratios. Variations of the MgO/(MgO+FeO) ratio across the serpentinite body at Pipe 2 mine, is shown in Figure 10. There is a distinct enrichment in iron from west to east consistent with a differentiated, originally sill-like intrusion. Peridotite and dunite with MgO/(MgO+FeO) ratios up to 0.87 occur on its west side, and on its east side poikilitic olivine orthopyroxenites with a ratio of about 0.75 grade into an altered marginal pyroxenite with a ratio of about 0.40. The sequence would place the magmatic top in the Pipe 2 serpentinite to the east. The mineralized ultramafic body at Soab North mine (Fig. 11) shows a similar trend with MgO/(MgO+FeO) ratio varying from 0.70 to 0.82. A relatively thick altered pyroxenitic margin exists on the east side of the serpentinite suggesting the magmatic top is in that direction.

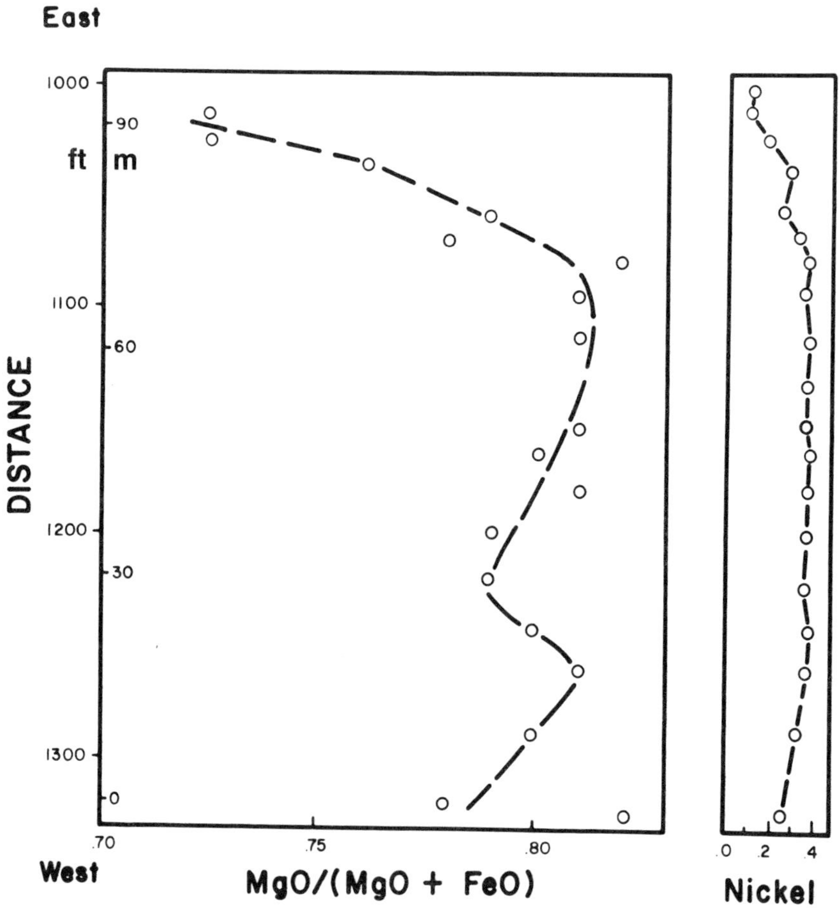

Figure 11. MgO/(MgO+FeO) versus distance plot across the Soab North serpentinite. The differentiation trend is similar to that in Pipe 2 serpentinite.

Serpentinites at Birchtree mine occur as relatively small lenses up to 100 m wide strung out along the pelitic member. Chemical data on two of these lenses are plotted in Figure 12. Data from one lens indicate a distinct decrease in the MgO/(MgO+FeO) ratio from west to east as the rocks change in composition from dunite to peridotite (lower part of Fig. 12). Data from another lens show a similar decreasing ratio from west to east, but the rocks range in composition from peridotite to pyroxenite (upper part of Fig. 12) and in terms of MgO/(MgO+FeO) ratios form a continuous trend with the other serpentinite suggesting that perhaps the two originally constituted one larger body. The trend at Birchtree is similar to that in Pipe 2 serpentinite and in both cases the serpentinites are in the same stratigraphic position and have east-facing magmatic tops. This is consistent with the interpretation that ultrabasic magma was originally intruded as sills into the Pipe metasedimentary pile.

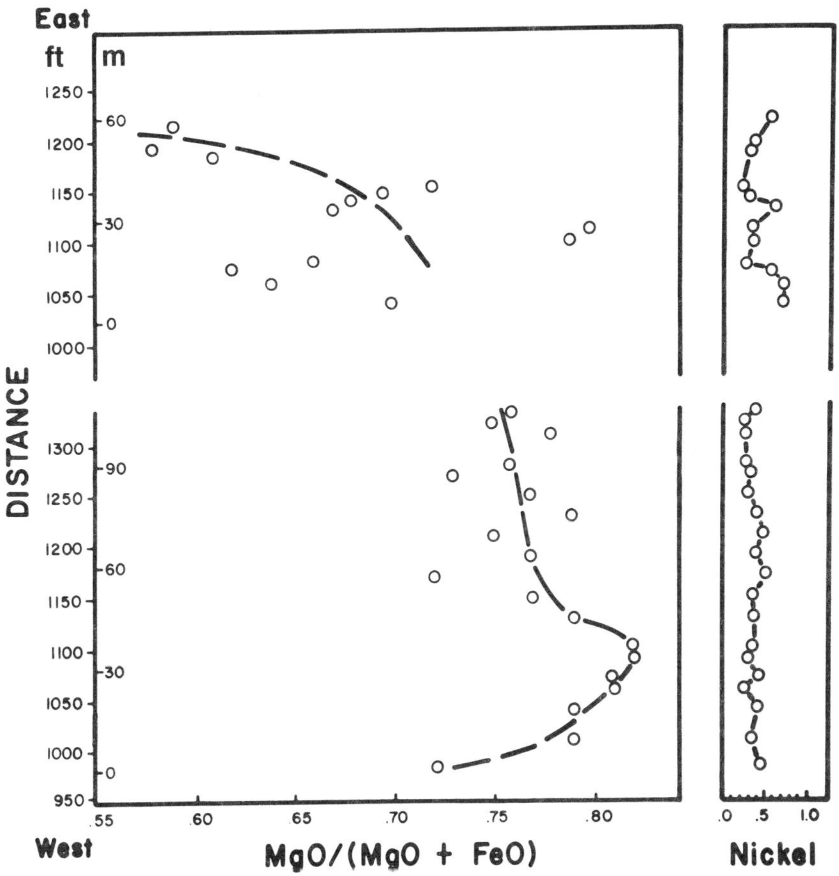

Figure 12. MgO/(MgO+FeO) versus distance plots across two serpentinite lenses at Birchtree mine. Continuity of the trends suggests a pattern similar to that in serpentinites at Pipe 2 and Soab North mines.

Serpentinites in the Thompson metasedimentary sequence occur in the pelitic member as small lenses, some of them highly disrupted. The rocks range in composition from peridotite to pyroxenite or picrite. The MgO/(MgO+FeO) variation across one such lens is illustrated in Figure 13. There is a distinct differentiation trend, or depletion in magnesia from west to east, with a less basic rock at the western margin

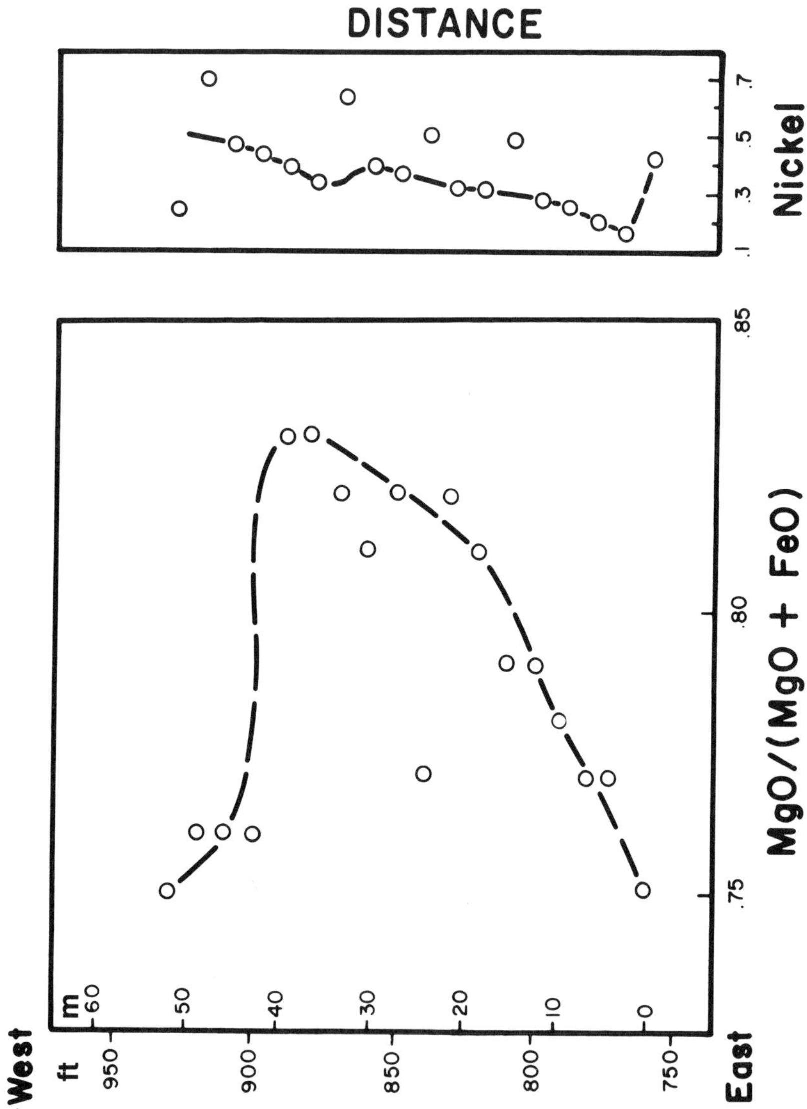

Figure 13. MgO/(MgO+FeO) versus distance plot across a serpentinite lens at Thompson mine. The steep slope of the curve indicates a relatively rapid differentiation trend. Rocks at the western margin could be interpreted as part of another magmatic cycle, or a marginal phase.

of the serpentinite. The trend is similar to that in the differentiated sills described above and suggests that originally this ultramafic body was also intruded as a sill.

The MgO/(MgO+FeO) variation across a lenticular seven km long serpentinite body in the gneissic terrane from the Ospwagan Lake area is illustrated in Figure 14. Throughout the main part of the body the ratio is relatively uniform (0.79), but towards its western side, where there is pyroxenite, the ratio decreases to about 0.74. In terms of basicity and differentiation, this serpentinite is essentially similar to those at Pipe 2 and Soab mines but contains no significant nickel sulphide mineralization.

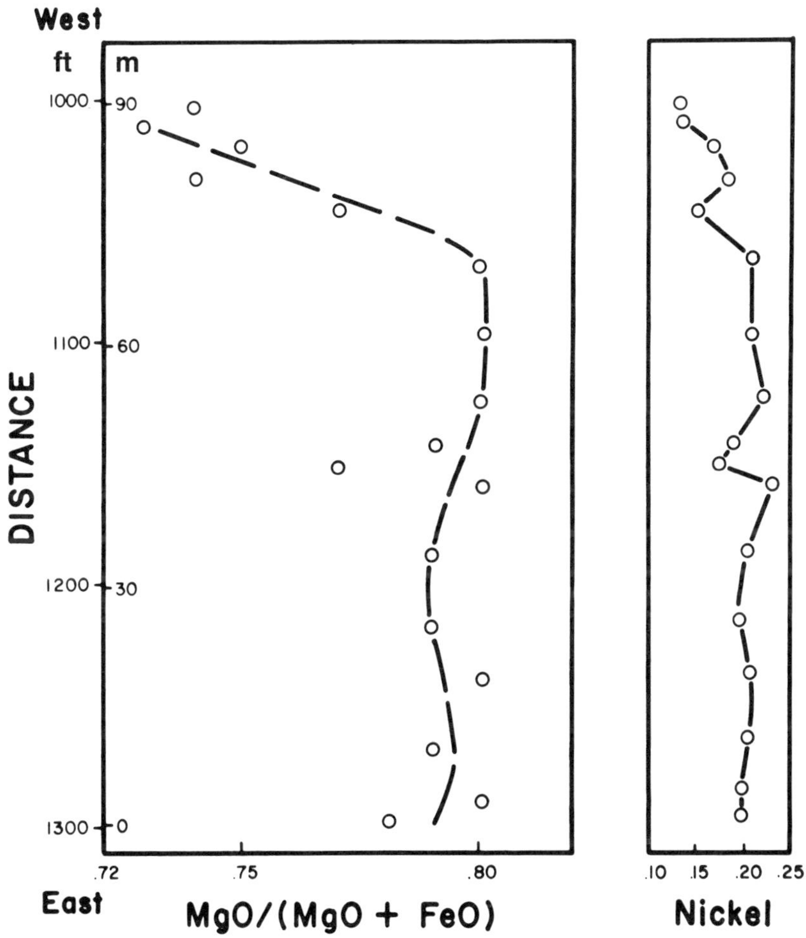

Figure 14. MgO/(MgO+FeO) versus distance plot across a serpentinite from the gneissic terrane in the Ospwagan Lake area. The curve is similar to that in Pipe 2 serpentinite.

Structural Geology

The dominant structural fabric in the Thompson Belt trends northeasterly, subparallel to that in the adjacent Churchill province. The Belt rocks are steeply dipping and tightly folded. The dominant folds are characteristically subvertically plunging. K-Ar isotopic data (Bell, 1971b) indicate a Hudsonian age (1.8 Ga) for the last major tectonism in the Belt and together with the dominant structural fabric reflect the Hudsonian orogeny typical of the Churchill province. However, structural complexity in the Thompson Belt rocks indicates that they are much older than the Hudsonian event and this is substantiated by the recent Rb-Sr Archean date of 2.8 Ga (Cranstone and Turek, 1976).

Interpretation of large scale structures in the gneisses is difficult because of lack of good marker horizons. However, in the south half of the Belt, on its western side, domal structures 5 km in diameter and larger are present (Coats *et al.*, 1972). They are outlined by the metavolcanic-metasedimentary bands and also in part by geophysical trends in the gneisses. The metavolcanic-metasedimentary sequences are thin, very tightly folded, appear to be stratigraphically above the gneisses and could be interpreted as deeply infolded keel-like structures.

Large fold structures, up to several kilometres in amplitude, are suggested by geophysical trends in the gneisses and some of them have been mapped in detail. They are isoclinal, tight, steeply dipping and have the same subvertical plunges as

Figure 15. Two sets of cross cutting lineations (mineral crenulations) in pelitic phyllite from Ospwagan Lake area. One set (L2) plunges steeply to the northeast and is attributed to the Hudsonian cross folding. The other set (L1) is subhorizontal and is attributed to the earlier synclinal folding of Pipe metasedimentary band. The bar is 5 cm.

most of the smaller linear structures. Their dominant character suggests they are related to the last major tectonism in the area, the Hudsonian event.

The attitude of the gneisses prior to superimposition of the Hudsonian fabric is not certain but there are, in places, earlier trends on which the later north-northeast fabric has been superimposed. At Paint Lake the earlier fabric strikes northeasterly and has subhorizontal linear structures; at Mid Lake and south of Joey Lake it strikes in a northwesterly direction. This variation suggests that the earlier fabric was a complex one similar to that in the adjacent Pikwitonei region.

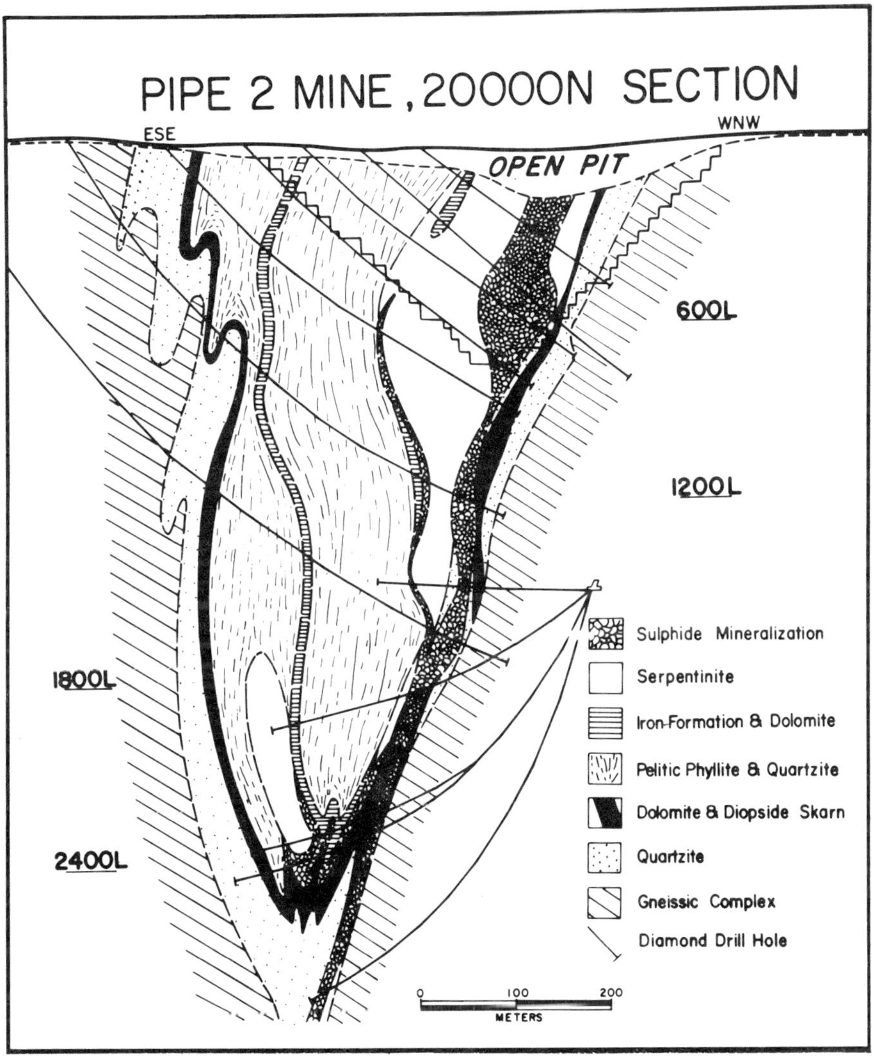

Figure 16. Vertical section, 20000N, of the Pipe 2 mine.

Cross cutting fabric relationships, including refolding of folds, cross cutting lineations and other features, indicate in general that the older fold structures were subhorizontal and have been overprinted by the vertical ones. These relationships are particularly evident in the metasedimentary rocks. (Fig. 15). At Pipe 2 mine, dragfolds, crenulations and rodding structures plunge at about 80° in a north-northeasterly direction reflecting the dominant Hudsonian fabric. This fabric is superimposed on earlier linear elements which plunge at shallow angles to the southwest. A few kilometres northeast of Pipe mine, in the Ospwagan Lake area, the earlier linear elements are nearly flat or plunge gently to the northeast.

On the basis of this structural data and stratigraphy the Pipe metasedimentary group is interpreted as a very tight sub-horizontal, synclinal fold which has been cross folded in the Hudsonian event. The axial planes of both the earlier fold and Hudsonian folds strike northeasterly so that the angle between the axial planes is very small (10 to 20°). Both axial planes are subvertical. The synclinal interpretation is based on the presence of similar rock sequences on both limbs of the fold and on the stratigraphic tops determined from graded bedding and pillow lavas. At Pipe 2 mine, a vertical section indicates a syncline (Fig. 16), and at Moak Lake a synclinal structure is also indicated (Fig. 17). The plunge of the syncline at Pipe 2 mine is indicated by the earlier lineations, which plunge at a shallow angle to the southwest. However, in plan view the same fold at Pipe 2 mine has a closure to the south. This fold is subvertically plunging and Hudsonian in age. The dominant linear fabric is related to it and is also steeply plunging.

Cross folding also accounts for the configuration of the Thompson metasedimentary group. At Thompson mine, the axial plane of the earlier folding strikes north-northeast, but the axial plane of the cross folding strikes nearly north-south. The dominant linear fabric related to cross folding plunges south at about 60° as exemplified by the large complex fold in the south end of the mine. In the north part of the mine, the linear fabric is subhorizontal and by analogy to that in the Pipe metasedimentary band, is interpreted as earlier.

In vertical section, the south-plunging complex fold at the south end of the Thompson mine, previously interpreted as an anticline (Zurbrigg, 1963), is an antiform overturned to the east. Although there is no question about the antiformal structure, it is not clear whether the Thompson metasedimentary band itself represents a syncline or an anticline. On the basis of general similarities in the stratigraphic succession between the Thompson and Pipe sedimentary sequences, the Thompson assemblage could be a synclinal structure.

At Thompson mine, the sulphides within the pelitic schist commonly define very tight drag folds with amplitudes up to 500 m and up to several kilometres long. In the nose of the antiformal structure they plunge south at about 60° and are apparently related to the antiform in the south end of the mine. Farther north, similar drag folds are near-horizontal to moderately northeast plunging and their configuration suggests that the Thompson metasedimentary group is anticlinal. It should be pointed out that, despite the magnitude of the sulphide drag fold in the pelitic schist, this structure is not always reflected in the adjacent metasedimentary units. It appears that drag folding was confined mainly to the pelitic schist member, probably due to its incompetent character.

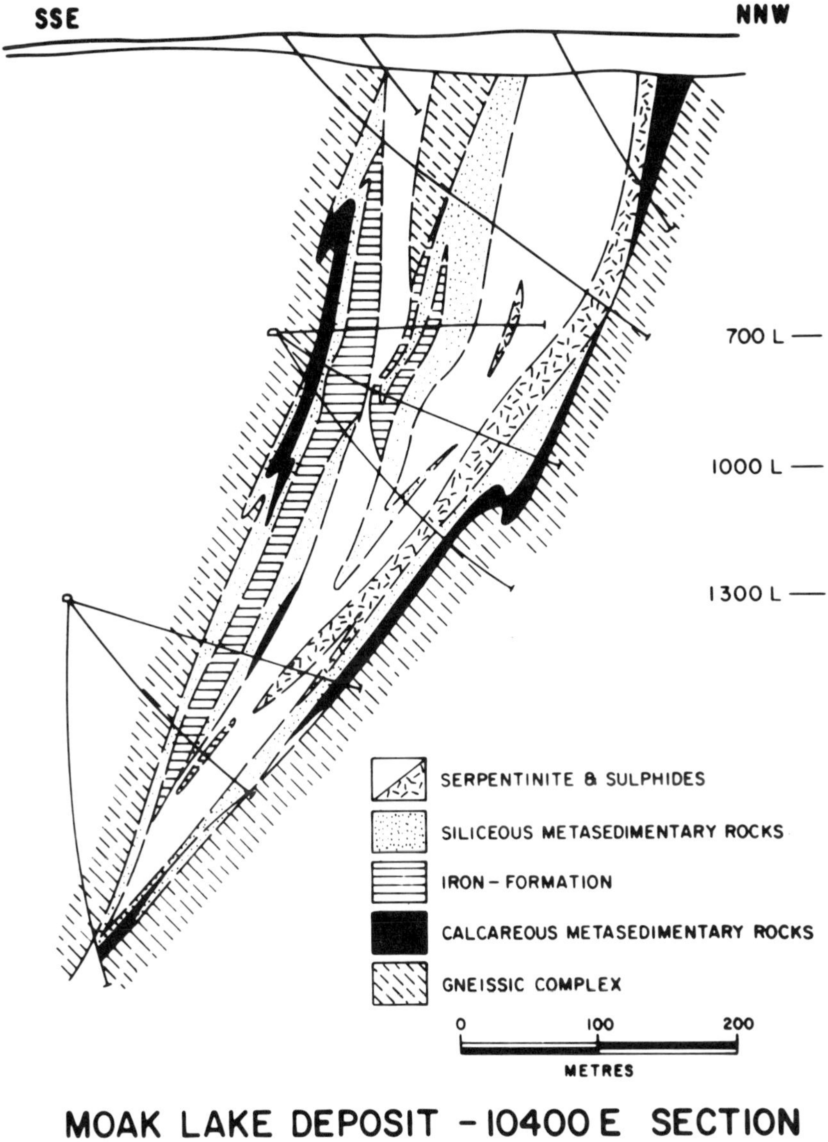

Figure 17. Vertical section, 10400E, of the Moak Lake deposit.

SULPHIDE MINERALIZATION

There are two operating Inco mines, Thompson and Pipe 2, in the Thompson Belt. Operations at four other Inco mines, Birchtree, Soab North, Soab South and Pipe I have been temporarily suspended. Falconbridge's Manibridge mine, operational since 1971, was closed in April 1977. In addition to these, there are a number of subeconomic sulphide deposits and numerous sulphide occurrences within the Belt. The nickel sulphides occur in close association with the serpentinites and in the adjacent rocks. The sulphide mineralogy is simple and consists of pyrrhotite (both monoclinic and hexagonal), pentlandite, minor·chalcopyrite and cubanite and traces of sphalerite. Niccolite and other arsenides occur in small quantities in most of the sulphides, and locally are concentrated in arsenide-tourmaline-rich veins. Minor secondary alteration minerals include pyrite, violarite, mackinawite, tochilinite and iron oxides. Sphalerite and galena are associated with late, undeformed carbonate veins.

Most of the nickel sulphide mineralization is metamorphosed and recrystallized to a variable degree and displays metamorphic banding. Alteration is related to late fracturing and mylonitization.

Distribution of nickel (mainly as nickel sulphides) indicates that about 75% of all known metal in the Thompson Belt is directly associated with serpentinites; the remaining 25% occurs with metasedimentary rocks and gneisses.

Mineralization Associated with Serpentinite

There are six main modes of occurrence of sulphides directly associated with the serpentinite bodies: 1) primary interstitial sulphides, 2) sulphide veins, 3) sulphide breccia, 4) massive sulphides. 5) dust-like disseminated secondary sulphides, and 6) interstitial spinifexoid sulphides.

The *interstitial sulphide* type of mineralization is considered to be primary and is common in the serpentinites of the Thompson Belt. The interstitial sulphides constitute up to several modal per cent of the rock (Fig. 18). In some deposits, as at Mystery Lake South, euhedral chromian spinel is associated with the interstitial sulphides. In highly deformed serpentinites the interstitial character of sulphide is more difficult to identify.

The *sulphide vein* type of mineralization is found in nearly all mineralized serpentinites, but is more abundant in some than in others. In some serpentinites, as on the east side of the open pit at Pipe 2 mine, it constitutes the ore. The veins range from a few millimetres to a metre in thickness and usually have reaction borders. In the serpentinites these reaction borders consist of coarse-grained pyroxene-rich rock which may be partially or completely altered to amphibole or chlorite. In the pyroxenites, the reaction borders are commonly of amphibole and/or chlorite. These are generally about the same width as the sulphide vein and commonly contain disseminated sulphides.

The *sulphide breccia* type consists of variously sized inclusions of serpentinite and/or other host rocks in a sulphide matrix. At the Pipe 2 open pit this type of mineralization constitutes the bulk of the ore (Fig. 16). It occurs on the western side of the serpentinite and has inclusions mainly of serpentinite. The reaction borders around the serpentinite inclusions are subtle, but in some cases distinct amphibolitic or talcose rims have been observed.

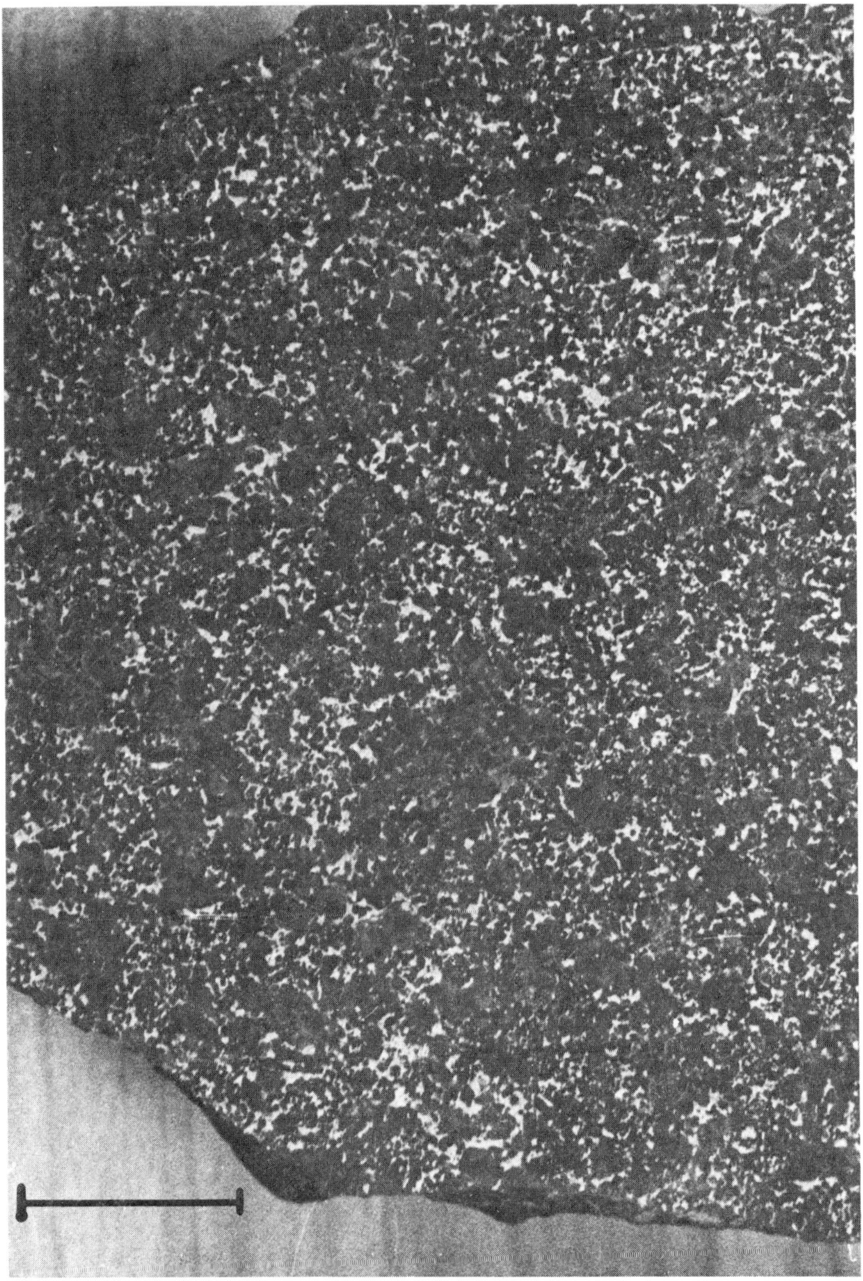

Figure 18. Primary interstitial pyrrhotite and pentlandite in partially serpentinized peridotite from Mystery Lake South deposit. The bar is 2 cm.

Inclusions in the sulphide breccia at Thompson mine are mainly of pelitic schist but in places are mainly of serpentinite. The sulphide matrix has been metamorphosed and is generally foliated and characterized by the development of pentlandite-rich bands up to one centimetre in width, and pentlandite-rich envelopes around the inclusions. The foliation is more or less parallel to the regional fabric, but on a local scale may be quite variable. The pentlandite-rich bands are usually coarser-grained than the pyrrhotite matrix and are parallel to the foliation in the pyrrhotite matrix.

Some sulphides do not possess any apparent foliation fabric. These tend to be coarse-grained, and are somewhat richer in pentlandite than other nearby sulphides. Whether they represent original sulphides not affected by metamorphism, or sulphides reconstituted by metamorphism, is not certain. Considering the intense deformation in the country rocks, and the relatively low competency of massive sulphides, the latter alternative is more likely.

The position of *massive sulphides* with respect to serpentinite bodies is variable but seems to depend on the degree of deformation. In some cases, massive sulphides are confined entirely to one margin of the serpentinite (Moak Lake deposit) and conform to the magmatic differentiation model. In others, they occur on both sides of serpentinite lenses (Birchtree mine) and in still others they are concentrated at the ends of the serpentinites (Thompson and Birchtree mines). In the extreme, they are not connected with serpentinites but are in the same stratigraphic or tectonic zone (Thompson mine and Hambone deposit). The massive sulphides are commonly foliated with pentlandite-rich bands similar to those described above, and carry minor inclusions of serpentinite and country rocks (Fig. 19). Generally pentlandite is coarse-grained and equigranular, but the pyrrhotite matrix varies in grain size from fine to coarse and is equigranular or foliated.

The *dust-like disseminated sulphide* type of mineralization occurs in practically all of the serpentinites in the Thompson Belt. The sulphide grains vary from a few microns to half a millimetre in size and are commonly monomineralic, consisting of either pyrrhotite or pentlandite. The very fine grain size suggests these sulphides are largely of secondary origin, probably derived from primary silicates during the serpentinization process. The nickel grades in this type of mineralization vary from 0.2 to 0.5%. The nickel content in the secondary silicates is generally less than 0.05%.

The term *interstitial spinifexoid* is applied to a particular variety of mineralization that consists of very coarse-grained, lath-like to lozenge-shaped crystals of olivine and interstitial sulphides (Fig. 20). The olivine crystals average several centimetres but some are up to one metre in length. They are generally altered to serpentine and magnetite. At Birchtree mine, this type of mineralization occurs on the boundary zone between massive sulphides and serpentinite lenses. It is up to four metres wide and is consistently on the western side of the serpentinite lenses. The massive sulphides occur also on the eastern side of the lenses but no interstitial spinifexoid mineralization has been found there. Because magmatic tops, as interpreted, face east at Birchtree mine, the interstitial spinifexoid mineralization is localized, curiously enough, along the magmatic bottom of the serpentinite bodies.

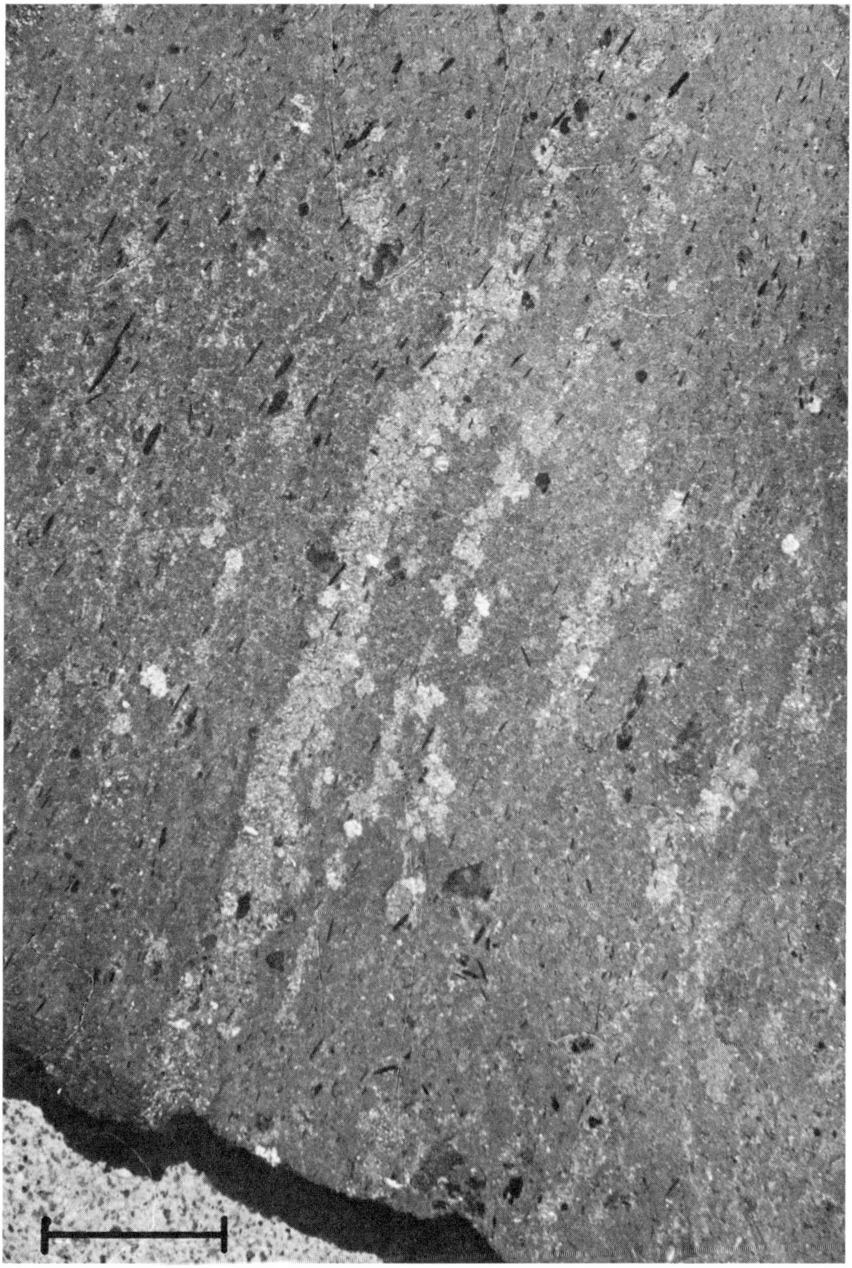

Figure 19. Typical foliated massive nickel sulphide ore. Penlandite is segregated into layers (diagonally trending bands) in a pyrrhotite matrix with inclusions of mica (black rods) and siliceous metasedimentary rocks (dark gray granules). The bar is 2 cm.

Mineralization Associated with Metasedimentary Rocks

Most of the sulphide deposits of this type have inclusions of the country rocks. The proportion of sulphides to inclusions is variable, but in every case ultramafic rocks are either closely associated with the mineralization or are found in the vicinity. Of the total nickel sulphides in metasedimentary rocks, a large proportion occurs in the Thompson mine where all of the sulphides are confined to the pelitic member. Closely associated with the sulphides are small lenses of ultramafic rocks measuring up to 60 m in thickness and locally numerous blocks of serpentinite up to several tens of metres in diameter are enclosed in the ore. These blocks vary in composition from serpentinized dunite to peridotite to partially altered pyroxenite. Some show lithological banding with sharp contacts, and some carry interstitial-looking sulphides. Most blocks have alteration envelopes of very coarse sheaf-like anthophyllite and/or tremolite and/or biotite. Lithologically these disrupted blocks have characteristics that are akin to the serpentinites elsewhere in the Belt. The paucity of serpentinites in the Thompson mine is notable and it is estimated that, in terms of weight per cent, there is as much sulphide as ultramafic rock (Table II).

The sulphide mineralization in the pelitic rocks can be subdivided into three main types: massive sulphide, inclusion-bearing sulphide and mineralized schist.

The *massive sulphide* type is generally foliated and is essentially similar to the massive sulphide associated with serpentinites.

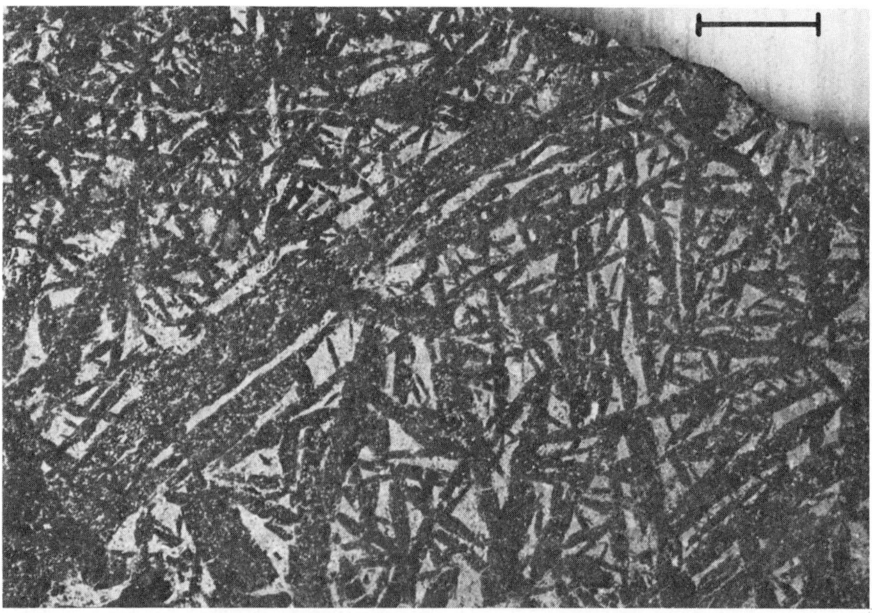

Figure 20. Interstitial spinifexoid type of mineralization from Birchtree mine. The large black crystals are of partially serpentinized olivine. Pyrrhotite and pentlandite occur interstitially and also as fine inclusions in olivine. The bar is 2 cm.

The *inclusion-bearing sulphides* are equivalent to the sulphide breccia type described above. The inclusions are commonly of the enclosing host rock and include in order of abundance: pelitic schist, coarse grained pegmatite, siliceous metasedimentary rock, quartz vein fragments, garnet-rich pelitic schist, garnet-amphibole (tremolite-actinolite) schist and minor ultramafic rocks. The ore at Birchtree mine, in addition, carries exotic inclusions composed of hercynitic spinel, calcic plagioclase and garnet. Calcareous metasedimentary rocks and iron-formation rarely occur as inclusions in the sulphides.

In *mineralized schist* stringers of sulphides occur parallel to the schistosity. The amount of sulphide present varies considerably, and in the extreme, this type of mineralization overlaps with the inclusion-bearing sulphide type. Mineralized schist is the least abundant of the three types. Although the sulphide stringers are normally conformable with the schist, locally there are cross cutting veins with chloritized country rock margins. The veins commonly have pentlandite concentrated at the borders and as internal stringers. Some are enriched in chalcopyrite.

At Thompson mine, minor but peculiar *brecciated sulphides* occur locally (Fig. 21). The breccia fragments are composed of very fine-grained massive pyrrhotite, with minor graphite and silicates, and occur in a coarse-grained pyrrhotite-pentlandite matrix. The fragments are angular to subangular, up to several centimetres in diameter, and form zones up to 2 m in width in massive sulphides. Fine-grained pentlandite-rich material occurs as selvages and also as veinlets in the sulphide fragments.

Figure 21. Brecciated, fine-grained massive pyrrhotite in a coarse-grained matrix of pyrrhotite and pentlandite, from Thompson Mine. The fine-grained pyrrhotite contains minor graphite and silicates and apparently predates the nickeliferous sulphides. The bar is 2 cm.

This occurrence, and many other non-nickeliferous sulphides in various country rocks described earlier point to the existence of barren sulphides predating nickeliferous sulphides. The barren sulphides may have played an important role in the genesis of later nickeliferous sulphide mineralization related to the ultramafic magmatism in the Thompson Belt.

Structural Control of Mineralization in the Pelitic Unit. At the Thompson mine, although the general character of the sulphides is conformable, the ore is controlled by drag folds (Fig. 22) being thicker at the nose and attenuated on the limbs. The sulphides are generally coarser-grained and richer in nickel, copper and cobalt in the nose of the structure than on the limbs (Fig. 23). Weighting all the samples equally, the mean nickel content is 7.47 weight per cent with s.d. (standard deviation) = 1.58 weight per cent in the nose area, whereas on the limb the mean nickel content is 5.44 weight per cent (s.d. = 0.87 weight per cent). The difference between the two means is not simply due to dilution by country rock inclusions, and can be demonstrated by calculating the ratio of pyrrhotite to Ni (Po/Ni) for both the nose and limb ore. The nose ore has a mean Po/Ni value of 8.55 (s.d. = 0.74) as opposed to a mean Po/Ni value of 10.63 (s.d. = 0.98) for the limb ore. The two values are significantly different which also indicates enrichment in nickel sulphide relative to iron sulphide in the nose of the fold.

Metal Ratios. Although there is considerable variation on a local scale, the Ni/(Ni+Cu) or Ni/Cu ratios vary little in the sulphide deposits of the Thompson Belt

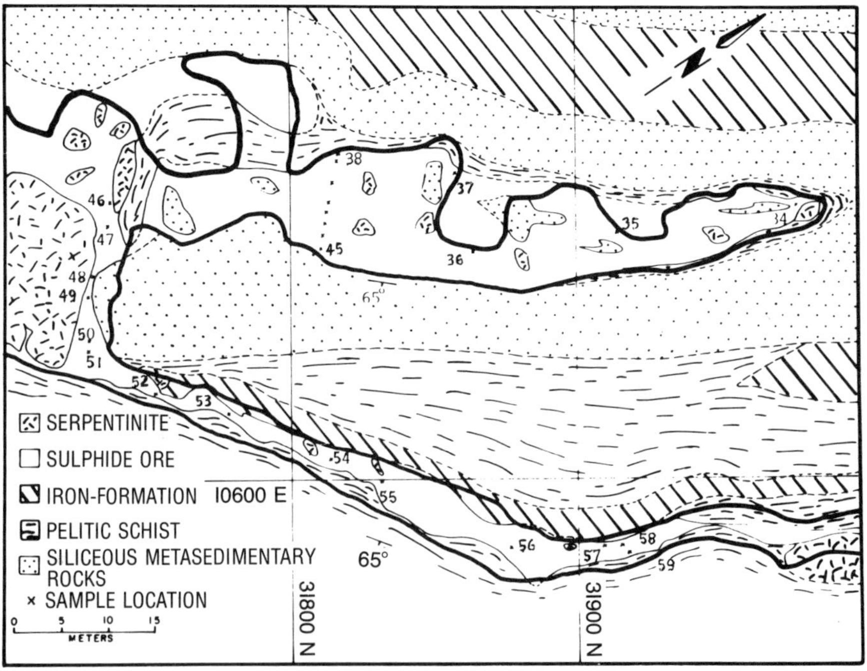

Figure 22. Sulphide dragfold structure in the pelitic schist at Thompson mine, 2000 level, 319 stope. The drag fold plunges to the south at about 50°. Sample locations (34 to 59) as indicated.

(Table II), whether the sulphides are in the metasedimentary rocks or associated with the serpentinites. This would seem to point to a common parent source for most, if not all, the nickeliferous sulphides in the Belt.

At Pipe 2 mine, where most of the sulphide mineralization is directly associated with the serpentinite, the average Ni/Cu ratio is 12, whereas in the Thompson mine

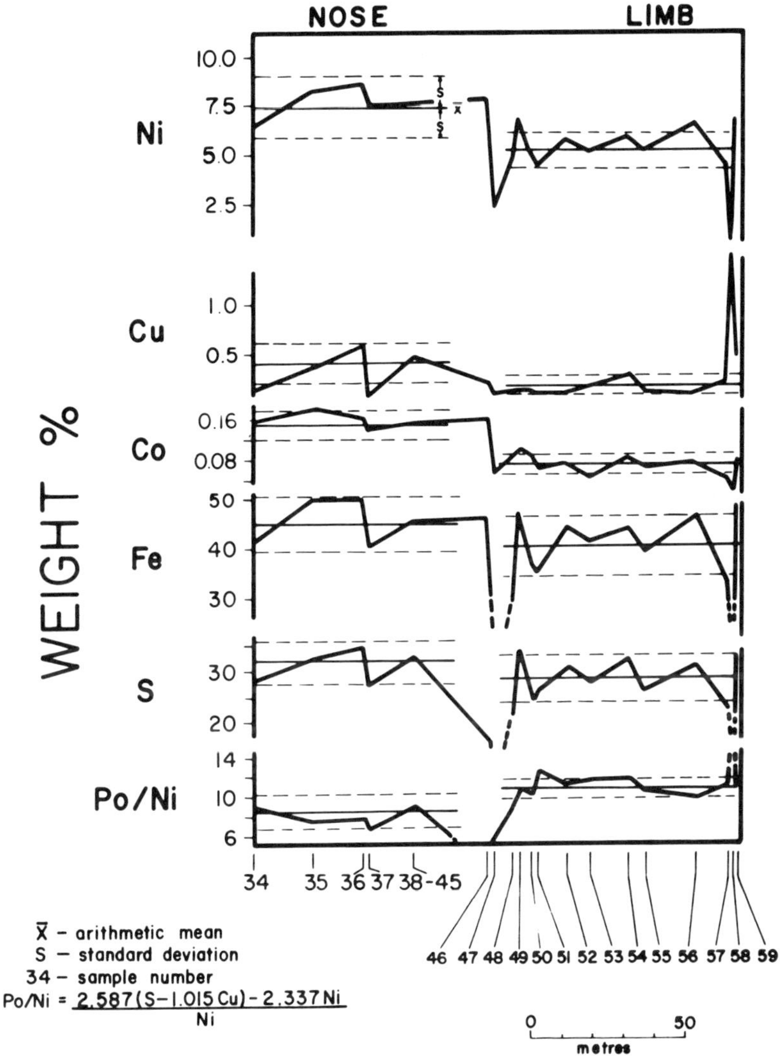

Figure 23. Variation of metals, sulphur and pyrrhotite/nickel (Po/Ni) ratio in the nose and limb of drag fold in the pelitic schist at Thompson mine, 2000 level, 319 stope. Concentration of metals is significantly higher in the nose than in the limb of the drag fold. See Figure 22 for sample locations.

metasedimentary environment it is close to 14, suggesting a slight depletion in copper. In view of the intensity of deformation in the Thompson mine, and considering that copper is more mobile than nickel, it could well be that some copper has migrated into the country rocks. This is supported by the presence of chalcopyrite tension fracture veins in the massive sulphides and adjacent country rocks at the mine.

DISCUSSION

Structural and mineralogical studies indicate that the Thompson Belt has had a complicated tectonic history. Although most of the rocks have been involved in Hudsonian tectonism and have a strong Hudsonian fabric, there is structural evidence of an earlier, Archean tectonic event. This is supported by the recent Rb-Sr studies by Cranstone and Turek (1976) which indicate the presence of Archean age (2820 ± 135 Ma) rocks in the gneissic assemblage. Rance (1966) had previously arrived at a similar conclusion from his structural study of the southern part of the belt. Thus, both isotopic and the structural data indicate that the gneissic terrane is of Archean age. The cross folding event is taken as Hudsonian. Evidence of cross folding in both Pipe and Thompson groups implies that these rocks are either early or pre-Hudsonian in age. Whether they are Archean or Aphebian is not certain.

The Thompson Belt rocks have a complex metamorphic pattern. The dominant metamorphic grade is amphibolite facies and is taken as the imprint related to the Hudsonian event. The island-like domains of granulite facies gneisses in an amphibolite facies terrane were previously interpreted by Cranstone (1970) as due to Hudsonian progressive metamorphism. Although it is true that some of the granulite facies rocks are rather fresh, others have been obviously retrogressed during Hudsonian tectonism. The high grade domains are interpreted as relicts of an older event, dated at 2820 ± 135 Ma (Cranstone and Turek, 1976), and are correlatable with the granulite facies metamorphism in the Pikwitonei region.

The gneissic assemblage in the Thompson Belt has recognizable supracrustal and intrusive components of more than one age. The supracrustal rocks include siliceous, calcareous, pelitic, and ferruginous metasedimentary rocks and possibly some metavolcanic rocks. The felsic gneisses underlying much of the Thompson Belt, are also interpreted to be largely supracrustal rocks which could have been derived from graywackes, volcanic rocks, or both.

Within the gneisses there are some relatively homogeneous felsic plutonic rocks, but whether they are intrusions or basement rocks is not certain. Some of these show evidence of granulite facies metamorphism and are therefore older than the amphibolite facies metamorphic event. However, certain orthogneisses do not show any direct evidence of granulite facies metamorphism. They have garnet and hornblende but no orthopyroxene. They are probably younger than granulite facies metamorphism but are older than the Hudsonian event because they are foliated and folded by the Hudsonian deformation. The little deformed Burntwood felsic pluton emplaced into the Pipe metasedimentary band has Hudsonian K-Ar ages and is considered to be related to the Hudsonian event (Bell, 1971b).

The amphibolite component of the gneisses could represent volcanic rocks, mafic sills or dykes, or retrogressed gabbroic layered complex (GLC). The layered character of the amphibolites and evidence of granulite facies in these rocks suggest that a significant proportion of amphibolites could have originated by retrogressive metamorphism of the GLC. The GLC rocks show evidence of granulite facies metamorphism and are therefore considered to be as old as the gneisses. Their distribution, layered character and mineralogy indicate they could be tectonic slices of a very large layered pluton or a number of smaller intrusions. GLC rocks occur throughout the Thompson Belt and similar rocks are also reported from Pikwitonei region (Hubregtse, 1977, 1978).

The metasedimentary rocks in the Pipe and Thompson groups are similar in some respects, but there are significant compositional and metamorphic differences between them as well as in their order of succession. Stratigraphically, the Pipe metasedimentary and associated metavolcanic rocks lie above the gneisses. The key mineral assemblages in the Pipe metasedimentary strata generally denote middle to upper amphibolite facies metamorphism. The pelitic phyllites have staurolite, andalusite, garnet and locally sillimanite; the iron-formations carry cummingtonite-grunerite and garnet; and the calcareous metasedimentary members contain mica, tremolite and diopside. The stratigraphic position of the Thompson group metasedimentary rocks with respect to the gneisses is not certain. The mineral assemblages in these metasedimentary rocks indicate an earlier granulite facies and later pervasive retrogressive amphibolite facies metamorphism. The pelitic schists have biotite, sillimanite and garnet; the iron-formations have cummingtonite-grunerite, garnet and orthopyroxene (and fayalite, as at Pipe 1 mine), and the calcareous rocks carry mica, tremolite, diopside, and olivine. These metasedimentary rocks are lithologically and mineralogically similar to the high grade metasedimentary units forming part of the gneissic assemblage. Thus, in terms of relict mineralogies, the Thompson metasedimentary group is similar to the gneissic assemblage and both had similar metamorphic histories. The metamorphic assemblages in the Pipe metasedimentary rocks do not appear to have reached the high grade metamorphism of the gneisses or the Thompson group metasedimentary strata. The Hudsonian retrogressive metamorphism has a tendency to camouflage any differences between the Pipe and Thompson groups.

In the Pipe group, the lithological succession of quartzite-dolomite-phyllite-iron-formation-dolomite-quartzite has great lateral continuity, being present at Pipe 2, Birchtree and as far north as Moak Lake, a total distance of some 65 km. The metasedimentary sequence of the Thompson group, although in general comparable to that of the Pipe group, has some differences in lithology and succession. In the Thompson group, the sequence is quartzite-skarn-schist-iron-formation-quartzite and although the lower part of the sequence is apparently similar to that of the Pipe group, the character of the calcareous unit is different. In the Thompson group, it is a skarn which has no primary banding. Judging by the silicate content, it could have been originally a siliceous calcareous mud, as opposed to the Pipe group's relatively pure dolomitic rocks with primary banding still well preserved in places. In the Pipe sequence, dolomite interbanded with the iron-formation is a typical feature not present in the Thompson assemblage. Furthermore, the metasedimentary strata overlying the iron-formation in the Thompson group are gneissic-looking, impure feldspathic rocks interbanded with minor pelitic bands, whereas the metasedimentary rocks overlying the iron-formation of the Pipe group are siliceous and fine grained.

The possibility that sedimentary facies variation is responsible for these differences is considered to be unlikely, because of the abrupt change in lithological succession between the two groups over a short distance. It is also noteworthy that the stratigraphic succession in each band is typical of that band and is relatively uniform over great lateral extent. In conclusion, the above discussed differences, combined with an apparent difference in the degree of metamorphism, imply that the Thompson and Pipe metasedimentary groups are not necessarily correlatable. The Thompson metasedimentary strata are very similar to other complexly deformed metasedimentary bands in the gneisses. In some of these, such as the metasedimentary bands at Owl, Mid, and Nichols Lakes, the stratigraphic succession is similar to and possibly correlatable with the Thompson group.

Volcanism and sedimentation appear to have been coeval in the Thompson Belt. The metavolcanic rocks range in composition from basalts to ultramafic flows and some have original quench texture preserved. Chemically, they are komatiitic and could be interpreted as precursors of the ultrabasic magmatism that gave rise to the serpentinites (Peredery, 1979). Field relationships, petrography and chemistry suggest that the metapicrites are coeval with the metavolcanic rocks and may be older than the serpentinites.

Emplacement of serpentinites in the Thompson Belt has been regarded by some investigators to be related to the major faults in the region (Wilson and Brisbin, 1961). Although this may have been the original mechanism of intrusion from depth, the regional setting demonstrates that most of the serpentinites, especially those within the Pipe metasedimentary group, were emplaced as conformable sills. Here, stratigraphic control is evident and differentiation trends in the serpentinites are in agreement with stratigraphic tops in the host rocks. Within the gneisses, the conformable nature of many of the serpentinites is evident but the sill-like character of the intrusions is difficult to demonstrate.

Some serpentinites are highly deformed and segmented and are indeed associated with faults. This could mean that some faults postdate emplacement of serpentinites or were reactivated after their emplacement. In the latter case, the fault could be the feeder structure. The Grass River fault (Fig. 1) may represent such a rejuvenated feeder structure.

Because the emplacement of serpentinites in the metasedimentary pile(s) is stratigraphically controlled it follows that the age of emplacement must be either early or pre-Hudsonian because the serpentinites have been affected by this folding event. They certainly have been involved in, and therefore predate the Hudsonian cross folding. More often, however, the serpentinite lenses do not show any folding, but have been affected by shearing, faulting and in some cases extreme fragmentation as in the mine environments.

The serpentinites of the Thompson Belt are much more like layered gabbro-peridotite complexes (LGP) than alpine type ultramafic rocks (ATU). Their occurrence as long, conformable, sheet-like bodies varying from dunite to peridotite to pyroxenite is similar to LGP and argues against their genetic relationship to the ATU. The presence of sulphide mineralization closely associated with serpentinites in the Thompson Belt is a feature commonly found in LGP but is most uncharacteristic of ATU. Furthermore, the presence of primary plagioclase in some pyroxenitic differentiates of Thompson Belt serpentinites is characteristic of LGP plutons. For these reasons the Thompson Belt serpentinites are considered to have more in common with LGP complexes than alpine type serpentinite. It is suspected that plagioclase was much more abundant in the original Thompson Belt pyroxenites, but most of it reacted with clinopyroxene under hydrous metamorphic conditions giving rise to tremolite-actinolite, a common alteration product.

Chemically, the Thompson Belt serpentinites show differentiation, typical of LGP but not of ATU. In terms of MgO/FeO versus TiO_2/Cr_2O_3, the Thompson Belt serpentinites fall mainly into the LGP field. Many data points, however, fall close to the borderline and some overlap into the ATU field. The rocks close to the borderline are similar in terms of these four elements to Ringwoods *et al.* (1964) theoretical pyrolite compositions, some of which also plot close to the borderline between the two fields. This is significant, because the line separating the LGP and ATU rocks corresponds to the theoretical proportion of magnesium, iron, chromium and titanium in the mantle before melting. The implication is, therefore, that the presence of ultramafic rocks of near pyrolite composition in the Thompson Belt may indicate a near-complete melting of the mantle material and its emplacement into the upper crust of the earth.

The variable basicity of serpentinites in the Thompson Belt suggests either a difference in the degree of melting of the original mantle rock(s) or different magmatic histories in the ascent of magma through the earth's crust. Melting could have been initiated as a result of some disturbance in pre-Hudsonian time. Whatever the cause, the disturbance was of catastrophic magnitude considering the total volume of serpentinites and their distribution along the Belt. Ultrabasic magma rose along the deep-seated structural weaknesses in the crust and ascended until it reached the upper levels. Here, new weaknesses were found and the magma was injected laterally as sill-like bodies into the flat-lying sedimentary pile and the basement rocks. The magma may have been charged with early-formed crystals of olivine, judging from the relatively unvarying chemical profiles of serpentinite bodies. However, normal differentiation followed intrusion, giving rise to pyroxenitic differentiates. The immiscible sulphide liquid percolated to the lower portions of the intrusions and solidified as interstitial grains. The interstitial sulphides in many serpentinite bodies, and massive sulphides closely associated with the serpentinites in Thompson Belt support the above mode of emplacement.

The disturbance that initiated mantle melting probably continued long after emplacement of magma into the upper levels of the crust and culminated in synclinal folding of the sedimentary pile. The ultramafic sills were involved in this folding, and the associated sulphides, particularly the massive sulphides, were partially redistributed. Injection of mobilized sulphides appears to have occurred mainly in the parent rocks, as most of the cross cutting sulphide veins are in the serpentinites. Sills appear to have been not only fractured but also brecciated during folding and the brecciated rocks served as excellent traps for the mobilized sulphides, as at Pipe 2 mine.

Later, during Hudsonian orogeny the earlier structures were cross folded, and a new tectonic fabric superimposed. Much of the deformation and shearing, faulting, and mylonitization in the Thompson Belt is ascribed to this event.

The effect of Hudsonian overprint on sills and associated sulphides is variable. At Pipe 2 mine, both sill and associated sulphides were relatively little affected except for some sulphide remobilization and the development of pentlandite-rich stringers in the massive sulphides. The sill and the contained sulphides behaved essentially as a unit. This unit could have been mobile during Hudsonian tectonism but apparently is less deformed than the enclosing host rocks. At Birchtree mine, the Hudsonian deformation was more disruptive, breaking the sill into several smaller bodies and causing considerable migration of sulphides. As the sulphides were mobilized they enveloped the serpentinite lenses and migrated laterally along the phyllitic member which hosts the serpentinite bodies. Rich mineable sulphide concentrations were formed in drag folds in the phyllitic member.

Thompson mine provides the most spectacular example of the effect of Hudsonian deformation on an ultramafic sill and associated sulphides. Originally, the sill was evidently highly brecciated. Its fragments occur as inclusions in the sulphides, and a few small, lenticular bodies survive in the pelitic schist. Evidence of layering in the ultramafic inclusions and lenses indicates that they were originally part of a much larger igneous layered body.

The Hudsonian event appears to have affected the sulphides in several ways. In the initial stages of deformation, mobilization occurred laterally along the path of least resistance, the incompetent pelitic schist member. In this process rock component may have been added to the sulphides which were distributed over a large area. In the final stages, concentration occurred in the noses of fold structures and by repetition of the sulphide member in the drag folds. Evidence of deformation in the sulphides are the concentration of pentlandite-rich layers in a pyrrhotite-rich matrix, and their parallelism to regional fabric. Enrichment of nickel in the noses of drag folds could also have resulted from deformation. The present distribution of sulphides in Thompson mine suggests that sulphides may have migrated as far as several kilometres. This is not unreasonable considering that at Birchtree mine the sulphide body in the pelitic member stretches for about 2 km.

The most anomalous feature of the Thompson orebody is the paucity of ultramafic rocks in relation to the total amount of sulphides. This fact has led to a proliferation of different models to explain the mineralization. Ideas, such as volcanogenic (Kambalda type), volcanogenic exhalative, hydrothermal replacement and massive sulphide flow have been suggested as possible modes of origin. In this study, much of the evidence indicates that the serpentinites and nickel sulphides in the Thompson mine are related in spite of the low ultramafic rock/sulphide ratio. Perhaps the main mass(es) of the serpentinite(s) has (have) been eroded, or perhaps there are some ultramafic rocks at depth not yet discovered.

In an arbitrary scheme of deformation for the Thompson Belt orebodies, the Pipe orebody is relatively undeformed, the Birchtree orebody is moderately to strongly deformed, and the Thompson orebody represents the extreme end member of highly deformed serpentinite-related nickel sulphide deposits of the Thompson Belt.

The source of sulphur in the sulphides is a question practically untouched in the literature on the Thompson Belt. Sulphide, as a minor rock-forming constitutent, is abundant in many of the major rock units in the Thompson Belt. Pyrrhotite and pyrite occur in the granulite facies gneisses, in the associated siliceous metasedimentary rocks, iron-formation and graywacke members, and in the amphibolites. Minor sulphides are ubiquitous in both the Pipe and Thompson metasedimentary assemblages. In general, the Thompson Belt rocks have an anomalously high sulphur content which could have been, at least in part, a source of sulphur for the nickeliferous sulphides.

In summary, the nickel sulphides of the Thompson Belt are clearly genetically linked with the ultrabasic magmatism. The magma was originally intruded as sills in the sedimentary-volcanic pile and possibly as sills and dykes in the basement gneisses. As differentiation proceeded the immiscible sulphide fraction settled under normal gravity conditions to form pools of sulphide at the base and was also retained as interstitial mineralization in the lower portion of sills. Subsequently this entire sequence was folded and probably some mobilization of the sulphides occurred. Later, in Hudsonian time, nearly all rocks of the Belt were deformed again. The ultramafic sills underwent extensive serpentinization and the sulphides were mobilized, recrystallized and concentrated in low pressure areas, such as drag folds. Variations in the degree of mobilization of sulphides gave rise to a wide array of structural characteristics of nickel sulphide deposits in the belt.

ACKNOWLEDGEMENTS

Many Company geologists worked in the area over the years and contributed to the understanding of the Belt. The cooperation and effort of the geological and technical staff of Inco Metals Company, Manitoba Division, is appreciated and acknowledged. In particular, the contributions of J. Cuthill, M. Parr and P. Rush who were transferred out of the Manitoba Division before the manuscript was completed, are recognized. The manuscript was reviewed by company geologists outside of Manitoba Division, in particular by R. A. Alcock, T. Podolsky, T. T. Quirke, Jr., and R. J. Worsfold. Their critical comments are very much appreciated. Many thanks go to Anne Iveta, Dorothy Jarvis and Diana Gletnak for their efforts in typing various versions of the manuscript, to Lyle Fennell for drafting the illustrations and to Jim Stadnick for photography.

The paper is published with the permission and support of Inco Metals Company.

REFERENCES

Bailey, E.H., and Blake, M.C., Jr., 1974, Major chemical characteristics of Mesozoic Coast Range ophiolite in California: Journal of Research of the United Stated Geological Survey, v. 2, p. 637-656.

Bell, C.K., 1971a, History of the Superior-Churchill boundary in Manitoba, in Turnock, A.C., ed., Geoscience Studies in Manitiba: Geological Association of Canada Special Paper 9, p. 5-10.

________________, 1971b, Boundary geology, Upper Nelson River area, Manitoba and northwestern Ontario: in Turnock, A.C., ed., Geoscience Studies in Manitoba: Geological Association of Canada Special Paper 9, p. 11-39.

Bell, K., Blenkinsop, J., and Moore, J.M., 1975, Evidence for a Proterozoic greenstone belt from Snow Lake, Manitoba: Nature, v. 258, p. 698-701.

Coats, C.J.A., Quirke, T.T., Jr., Bell, C.K., Cranstone, D.A., and Campbell, F.H.A., 1972, Geology and mineral deposits of the Flin Flon, Lynn Lake and Thompson areas, Manitoba, and the Churchill-Superior front of the western Precambrian Shield: Guidebook, Field Excursion A31-C31, 24th International Geological Congress, Montreal, p. 1-96.

Coleman, L.C., 1970, Rb/Sr isochrons for some Precambrian rocks in the Hanson Lake area, Saskatchewan: Canadian Journal of Earth Science, v. 7, p. 338-345.

Cranstone, D.A., 1970, Manitoba nickel belt: Manitoba Mines Branch Geological Paper No. 4/70, p. 30-32.

Cranstone, D.A., and Turek, A., 1976, Geological and geochronological relationships of the Thompson nickel belt, Manitoba: Canadian Journal of Earth Science, v. 13, p. 1058-1069.

Elphick, S.C., 1972, Geology of the Mynarski-Notigi Lakes area: Manitoba Mines Branch Publication 71-2C.

Gibb, R.A., 1968, A geological interpretation of the Bouger anomalies adjacent to the Churchill-Superior boundary in northern Manitoba: Canadian Journal of Earth Science, v. 5, p. 439-453.

Hubregtse, J.J.M.W., 1977, Sipiwesk Lake-Wintering Lake area: Manitoba Mines Branch, Report on Field Activities, 1977, p. 73-79.

______________, 1978, Sipiwesk Lake-Landing Lake-Wintering Lake area: Manitoba Mines Branch, Report on Field Activities, 1978, p. 54-62.

Innes, M.J.S., 1960, Gravity and isostasy in northern Ontario and Manitoba: Ottawa, Dominion Observatory Publication, v. 21, No. 6, p. 265-338.

Irvine, T.N., and Findlay, T.C., 1972. Alpine-type peridotite with particular reference to the Bay of Islands Igneous Complex: in Irving, E., ed., The Ancient Oceanic Lithosphere: Publication of the Earth Physics Branch, Department of Energy Mines and Resources, Ottawa, v. 42, no. 3, p. 97-128.

Kornik, L.J., 1969, An aeromagnetic study of the Moak Lake-Setting Lake structure in northern Manitoba: Canadian Journal of Earth Science, v. 6, p. 373-381.

Kornik, L.J. and MacLaren, A.S., 1966, Aeromagnetic study of the Churchill-Superior boundary in northern Manitoba: Canadian Journal of Earth Sciences, v. 3, p. 547-557.

Leech, G.B., Lowdon, J.A., Stockwell, C.H., and Wanless, R.K., 1963, Age determinations and geological studies: Geological Survey of Canada Paper 63-16.

McRitchie, W.D., 1974, The Sickle-Wasekwan debate: a review: Manitoba Mines Branch, Geological Paper 1/74, 23p.

______________, 1977, Reindeer Lake-Southern Indian Lake: Manitoba Mines Branch, Report on Field Activities 1977, p. 13-18.

Mukherjee, A.C., Stauffer, M.R., and Baadsgaard, H., 1971, The Hudsonian Orogeny near Flin Flon, Manitoba. A tentative interpretation of Rb/Sr and K/Ar ages: Canadian Journal of Earth Sciences, v. 8, p. 939-946.

Peredery, W.V., 1979, Relationship of the ultramafic amphibolites to the metavolcanics and serpentinites in the Thompson Belt, Manitoba: The Canadian Mineralogist, v. 17, p. 187-200.

Pyke, D.R., Naldrett, A.J., and Eckstrand, O.R., 1973, Archean ultramafic flows in Munro Township, Ontario: Geological Society of America Bulletin, v. 84, p. 955-978.

Rance, H., 1966, Superior-Churchill structural boundary, Wabowden, Manitoba: Ph.D. Thesis, University of Western Ontario.

Ringwood, A.E., MacGregor, I.D., AND Boyd, F.R., 1964, Petrological composition of the upper mantle: Carnegie Institute, Annual Report of the Director of the Geophysical Laboratory 1963-1964, p. 147-152.

Roth, J., 1975. Exploration of the southern extension of the Manitoba Nickel Belt: The Canadian Institute of Mining and Metallurgy Bulletin, v. 68, pl 73-80.

Suturin, A.N., Kornakov, Yu. N., and Parshin, A.P., 1972, Identification of the formational type of ultramafic massifs by a complex of geochemical characteristics: Geochemistry International, p. 229-233.

Weber, W., and Scoates, R.F.J., 1978, Archean and Proterozoic metamorphism in the northwestern Superior Province and along the Churchill-Superior boundary, Manitoba: in Fraser, J.A. and Haywood, W.W., eds., Metamorphism in the Canadian Shield: Geological Survey of Canada Paper 78-10, p. 5-16.

Wilson, H.D.B. and Brisbin, W.C., 1961, Regional structure of the Thompson-Moak Lake Nickel Belt: The Canadian Institute of Mining and Metallurgy Bulletin, v. 54, p. 815-822.

Zurbrigg, H.F., 1963, Thompson Mine geology: The Canadian Institute of Mining and Metallurgy Bulletin, v. 56, p. 451-460.

MS received April 1979;
Accepted for publication August 1979.

Precambrian Sulphide Deposits, H.S. Robinson Memorial Volume, edited by
R.W. Hutchinson, C.D. Spence and J.M. Franklin, Geological Association of Canada
Special Paper 25, 1982

EVIDENCE FOR A VOLCANOGENIC-EXHALATIVE ORIGIN OF A MASSIVE NICKEL SULPHIDE DEPOSIT AT REDSTONE, TIMMINS, ONTARIO

D.J. Robinson
Westmin Resources, 390 Bay Street, Toronto, Ontario M5H 2Y2

R.W. Hutchinson
Department of Geology, University of Western Ontario,
London, Ontario N6A 5B7

ABSTRACT

The Archean Redstone stratabound volcanogenic nickel sulphide deposit is located on the southern flank of the Shaw Dome, and lies within steeply dipping, southwest-facing dacitic tuff of the Deloro Group, which is overlain by komatiitic flows, with spinifex tops, of the Tisdale Group. A main layer of massive Fe-Ni sulphide minerals is enclosed within a zone of highly altered rocks along the komatiite-dacitic tuff contact, and exhibits along strike lateral zoning in the nickel content of pentlandite. Smaller lenses of massive Fe-Ni sulphide occur above the komatiitic flows. The composition of pentlandite varies systematically with the sulphide assemblage, from a low nickel tenor in association with pyrrhotite to a high nickel tenor with millerite. The deposit is characterized by a heterogeneous distribution of platinum group elements and Au, Ni and Cu. A sulphide-facies iron-formation is located at or near the top of the dacitic volcanic sequence, flanks the main Fe-Ni sulphide layer and grades laterally into it.

Metasomatic chloritic rock caps the massive Fe-Ni sulphide layer along the contact between the komatiitic flows above the altered dacitic tuff. This chlorite rock was derived from dacitic volcanic rock by hydrothermal alteration involving leaching of SiO_2 and introduction of MgO and FeO, assuming relative immobility of A1, Ti, Zr and Y, during serpentinization and talc-carbonate alteration of the adjacent ultramafic rock. The mineral and metal zoning within the sulphide layer, the presence of disseminated and massive Fe-Ni sulphides within the dacitic tuff and of massive Fe-Ni sulphides above komatiitic flows are all incompatible with a process of in situ gravitational settling of homogeneous immiscible sulphide blebs from a komatiitic magma. Deposition of nickel-bearing sulphides into a paleotopographic depression, by volcanic-exhalative processes accompanied by minor accumulation of volcanic sediment prior to eruption of komatiitic lava, provides the most satisfactory explanation for the presence of intercalated sulphides and dacitic tuff, for the abundant quartz-carbonate assemblage within the Fe-Ni sulphide layer and for the close spatial association with sulphide-facies iron-formation.

RÉSUMÉ

Le gisement de nickel de Redstone, d'âge archéen, lié aux strates et volcanogénique, est situé sur le flanc sud du dôme de Shaw dans le tuf dacitique du groupe de Deloro qui est recouvert par les coulées de komatiite à sommets de spinifex du groupe de Tisdale; les pendages sont forts et orientés vers le sud-ouest. Une couche principale de sulfures de Fe-Ni massifs est enclose dans une zone fortement altérée le long du contact komatiite-tuf dacitique et montre une zonalité latérale dans la teneur en nickel de la pentlandite. Des lentilles de sulfures de Fe-Ni massifs existent au-dessus des coulées de komatiite. La composition de la pentlandite varie avec l'assemblage des sulfures d'une teneur en nickel basse lorsqu'associée à de la pyrrhotite à une teneur élevée lorsqu'associée à de la millérite. Le gisement est aussi caractérisé par la distribution hétérogène des platinoïdes, de l'or, du nickel et du cuivre. Une formation ferrugineuse à faciès sulfuré est située au, ou près du, sommet de la séquence dacitique, sur les flancs de la couche principale sulfurée à laquelle elle passe latéralement graduellement.

Une roche chloriteuse, d'origine métasomatique, recouvre la couche sulfurée principale et sépare les coulées de komatiite du tuf dacitique sous-jacent. Cette roche chloriteuse résulte de l'altération hydrothermale du tuf dacitique par lessivage de SiO_2 et introduction de FeO et MgO si l'on suppose que Al, Ti, Zr et Y soient restés sur place pendant la serpentinisation et l'altération en talc-carbonate des roches ultramafiques sus-jacentes.

La zonalité minérale et métallique dans la couche sulfurée principale ainsi que la présence de sulfures de Fe-Ni disséminés et massifs dans le tuf dacitique et de sulfures de Fe-Ni massifs au-dessus de coulées de komatiite sont incompatibles avec le processus de ségrégation gravitationelle à partir d'un magma komatiitique de gouttelettes homogènes de sulfures immiscibles. Le dépôt de sulfures de Fe-Ni dans une dépression paléovolcanique, par un processus volcano-exhalatif accompagné du dépôt d'un peu de sédiment volcanique, ceci avant l'éruption des coulées de komatiite, fournit l'explication la plus satisfaisante de la présence d'intercalations de tufs dacitiques et de sulfures, de l'assemblage quartz-carbonate abondant dans la couche sulfurée principale et de l'association spatiale étroite avec une formation ferrugineuse à faciès sulfuré.

INTRODUCTION

Archean volcanic-hosted Fe-Ni sulphide deposits have been identified within Precambrian shields throughout the world and many of these deposits have common geological relationships in terms of age, ore mineralogy, ore form, associated volcanic, chemical and clastic sedimentary rocks, rock alteration, etc. (Marston and Kay, 1980; Groves *et al.*, 1979; Naldrett, 1979, 1973; Kerrich *et al.*, 1979; Muir and Comba, 1979; Williams, 1979; Green, 1978; Coad, 1977; Barrett *et al.*, 1977; Lusk, 1976a,b; Ross and Hopkins, 1975; Miller and Smith, 1975; and others). Many of these deposits are thought to be magmatic accumulations formed by gravitational settling of immiscible sulphide blebs from an ultramafic magma. This paper provides a detailed account of field relations and geochemistry of the Redstone Fe-Ni deposit, which is similar to other Canadian and Australian volcanic-hosted Fe-Ni sulphide deposits, with emphasis on the derivation of a sequence of contact metasomatic rocks which surround the zoned massive Fe-Ni sulphide layer. In light of this study a volcanic-exhalative process is proposed as an alternative method of ore formation for the Fe-Ni sulphides at the Redstone deposit and, through comparison and inference, perhaps for other similar deposits elsewhere.

ANALYTICAL METHODS

Minerals were analyzed by a MAC 400 electron microprobe fitted with the KRISEL automation system using appropriate natural minerals for calibration standards. The data were reduced on line with the MAGIC IV data reduction program. Operating conditions were 15 kv accelerating voltage and 0.03 u amp sample current. Major element oxide analyses of whole rocks were conducted by X-ray fluorescence techniques at X-ray Assay Labs, Toronto, and at the University of Western Ontario. Trace element analyses were also obtained by XRF at U.W.O. using USGS rock standards and mass absorption corrections. Platinum group elements and Au were analyzed at X-ray Assay Labs by neutron activation analysis.

REGIONAL GEOLOGICAL SETTING

Redstone is a small Archean nickel sulphide deposit located approximately 25 km southeast of Timmins, Ontario within the Abitibi greenstone belt. The belt is an east-west-trending tectonic zone composed predominantly of metavolcanic, subvolcanic and metasedimentary rocks, with coeval ultramafic, mafic and felsic intrusions surrounded by large granitic batholiths (Goodwin and Ridler, 1970; Fig. 1). Archean gneissic and metasedimentary rocks form its boundaries to the north and south, whereas Grenville gneisses and Kapuskasing crystalline rocks truncate the belt to the east and west respectively.

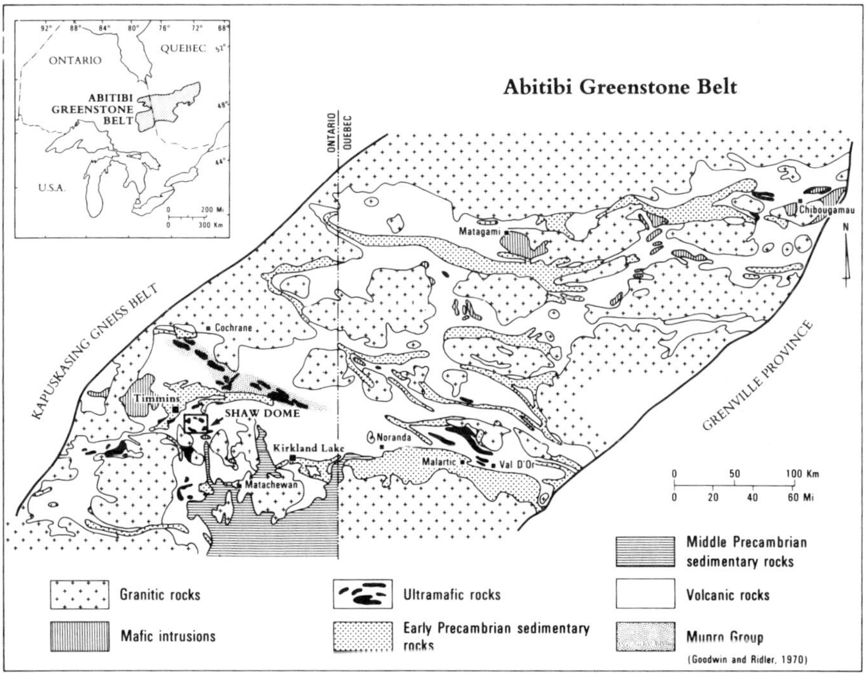

Figure 1. Regional geology of the Abitibi Greenstone Belt (after Goodwin and Ridler, 1970). Outlined area of Shaw dome is shown in Figure 2.

Numerous centres of eruptive and subvolcanic rocks are recognized on a regional scale which display a progressive upward change in volcanism from a lowermost ultramafic to mafic base to a felsic top, with associated intercalated volcaniclastic and chemical sedimentary rocks (Goodwin, 1977). A petrochemical progression has been recognized within these centres from lower ultramafic to mafic komatiitic flows and sills intermixed with and overlain by Fe-rich tholeiitic basaltic pillow lavas to upper Al-rich calc-alkaline lavas. Intermediate to felsic flows and pyroclastic rocks of calc-alkaline affinity are concentrated at the top of the succession (Arndt, 1976).

The volcanic stratigraphic succession in the Timmins district is subdivided into two ultramafic to felsic cycles referred to as the Deloro and Tisdale Groups. The Deloro Group is interpreted as the older (Pyke *et al.*, 1978), and is confined mainly to the Shaw dome (Fig. 2). The dome is a large, east-west-trending elliptical structure

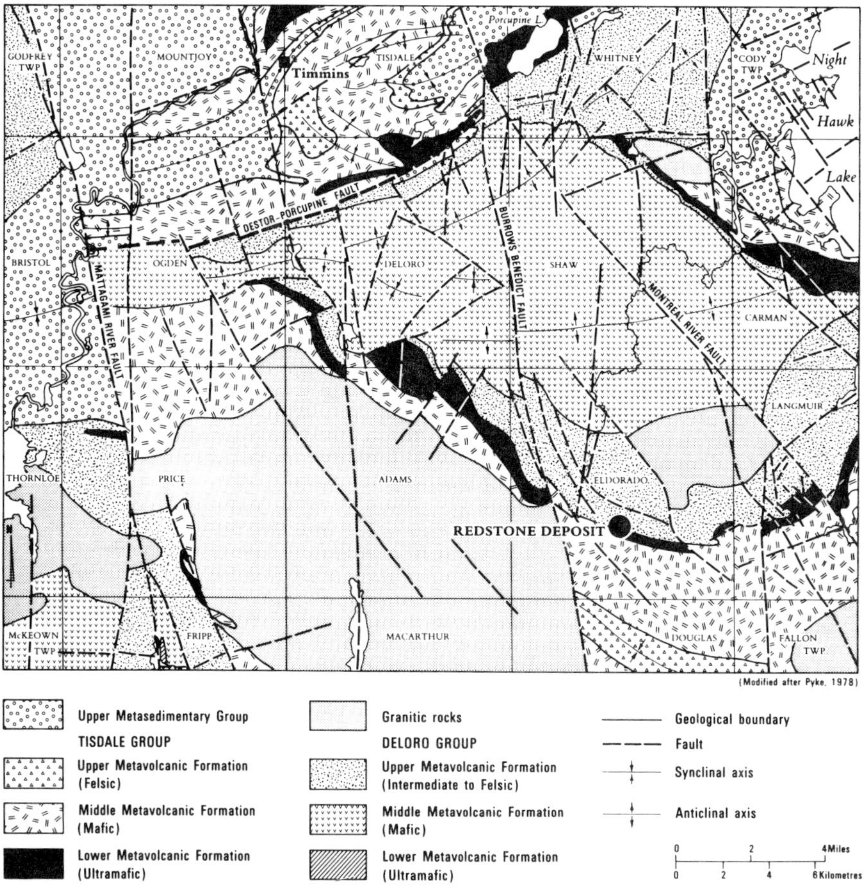

Figure 2. Geology of the Shaw dome (after Pyke *et al.*, 1978). Komatiitic rocks are shown in solid pattern and outline the perimeter of a large volcanic complex. Redstone location is indicated.

with a core of calc-alkaline basaltic pillow lavas overlain by intermediate to felsic flows and pyroclastic rocks. Sulphide-facies iron-formation occurs regionally at or near the contact between the felsic rocks of the Deloro Group, and marks a period of chemical sedimentation that preceded the major change in volcanism. Thus felsic pyroclastic rocks and the iron-formation are interdigitated with and overlain by a discontinuous but conformable sequence of ultramafic to mafic komatiitic flows of the Tisdale Group (Muir, 1979; Pyke *et al.*, 1978, Pyke, 1975). Komatiitic flows are in turn overlain by a thick sequence of tholeiitic basalt and an uppermost volcaniclastic unit of calc-alkaline composition. Metasedimentary rocks referred to as the Porcupine Group, consisting largely of a turbidite sequence of interlayered graywacke, siltstone and lesser conglomerate, are time-equivalent with the upper portion of the Deloro Group and all of the Tisdale Group (Pyke *et al.*, 1978).

Concordant sills, dunitic to peridotitic in composition, occur within the intermediate to felsic metavolcanic rocks of the Deloro Group and are interpreted as possible magma reservoirs for the komatiitic flows by Pyke *et al.* (1978). Late Archean granodioritic and trondhjemitic stocks have intruded the region and finally, all units are cut by a northeast-trending swarm of diabase dykes of middle to late Precambrian age. Regional metamorphic grade is prehnite-pumpellyite to greenschist facies except in aureoles of granitic intrusions where amphibolite facies is observed (Green, 1978; Arndt, 1976; Pyke, 1975).

The Redstone, Langmuir, McWatters and Hart nickel sulphide deposits are situated around the southern flank of the Shaw dome along or near the contact between felsic metavolcanic rocks of the Deloro Group and overlying komatiitic flows of the Tisdale Group and are spatially associated with sulphide-facies iron-formation (Kerrich *et al.*, 1979; Green, 1978; Coad, 1977; Pyke and Middleton, 1971).

DETAILED PROPERTY GEOLOGY

Dacitic tuff and minor flows of the Deloro Group are the oldest rocks in the study area and form the footwall and in places hanging wall to a main Fe-Ni sulphide layer termed the R zone. Sulphide-facies iron-formation occurs at the top of the dacitic tuff above and overlapping the R zone and grades laterally into a small lens of massive nickeliferous sulphide designated as the S zone. The iron-formation is underlain by a hornfelsic rock containing manganese garnet and stilpnomelane (Robinson *et al.*, 1980). The mineralogically zoned, stratabound, massive Fe-Ni sulphides of both R and S zones occur at and near the top of dacitic tuff (unit 1; Figs. 3,4,5), overlain by komatiitic flows (unit 2).

Komatiitic flows of the Tisdale Group form the hanging wall of the Fe-Ni sulphide layer and are subdivided into two units based on stratigraphic position, physical and chemical characteristics. The lowermost komatiitic rocks (unit 2) immediately overlying the R zone are massive, thick (25 m) flows, very rich in MgO. These rocks are komatiitic peridotite and peridotitic pyroxenite, with rare dunitic rocks according to the classification of Arndt *et al.* (1977). These unit 2 flows are situated within a paleotopographic depression as illustrated in Figure 3, and are interdigitated with dacitic tuff and quartz-feldspar porphyry to the west (Figs. 4,5), here indicating contemporaneous volcanism of the Deloro and Tisdale Groups. Contact metasomatic rocks composed of nearly monomineralic chlorite, biotite, actino-

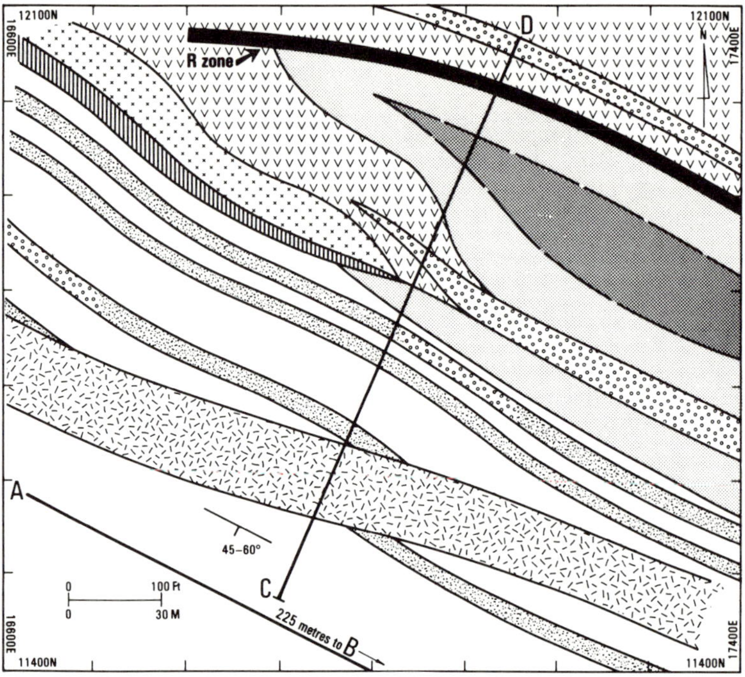
12100N
16600E
R zone
D
12100N
N
17400E
A
45–60°
0
100 Ft
0
30 M
C
225 metres to B
16600E
11400N
17400E
11400N

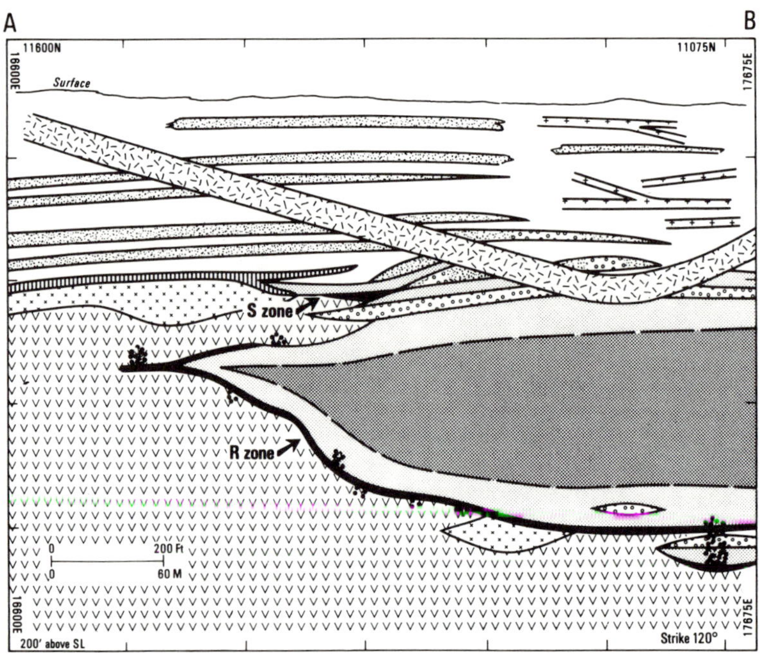
A
B
11600N
16600E
Surface
11075N
17675E
S zone
R zone
0
200 Ft
0
60 M
16600E
200′ above SL
Strike 120°
17675E

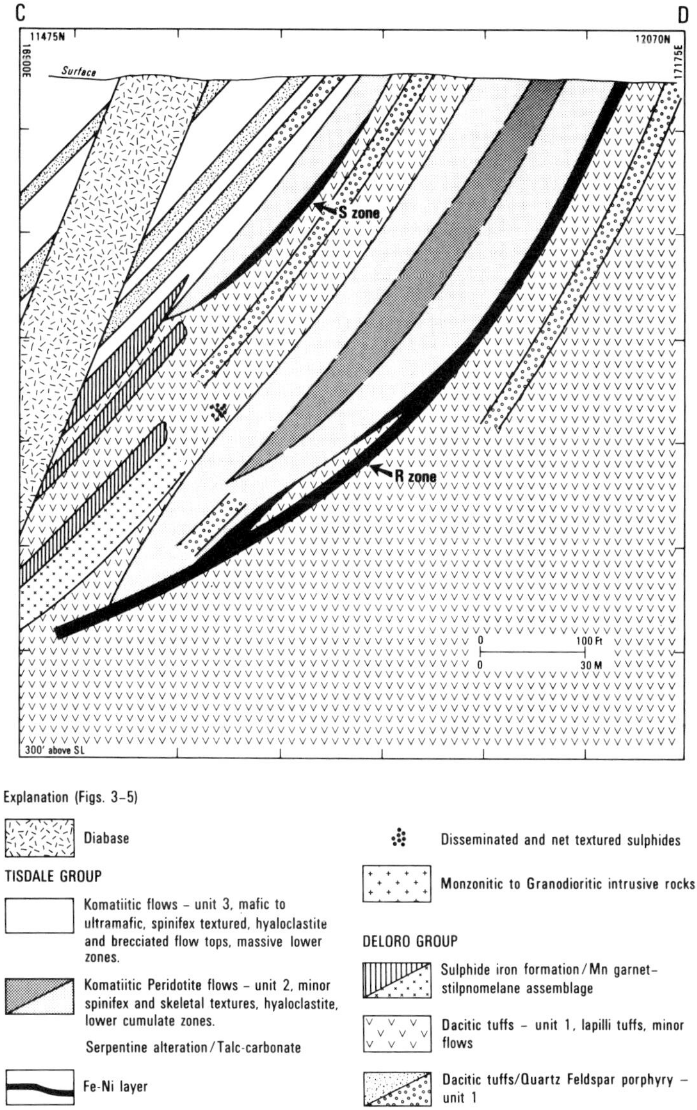

Figure 3. Geological sketch of bedrock surface of the Redstone deposit. The units strike at 120° and dips are southwest from 45 to 60°. Geology in section A-B is shown in Figure 4, section C-D in Figure 5.

Figure 4. Vertical longitudinal geological sketch of the Redstone deposit along A-B. Massive Fe-Ni sulphide is labelled R and S zones. Illustrated is the close spatial association of sulphide-facies iron-formation, garnet-stilpnomelane rock and Fe-Ni sulphide.

Figure 5. Cross section of the Redstone deposit along C-D.

lite or talc occur along the footwall, internal and hanging wall contacts of these komatiitic flows with the intercalated dacitic tuff or quartz-feldspar porphyry.

These komatiitic extrusive rocks of unit 2 and sulphide-facies iron-formation are stratigraphically overlain by a sequence of thin (10 m), massive to spinifex-textured flows (unit 3). These flows are komatiitic peridotitic pyroxenite to basalt in composition (Arndt *et al.*, 1977). Dacitic crystal tuff and concordant quartz-feldspar porphyry are also interbedded with the komatiitic flows of unit 3.

Locally the area has been intruded by lenses and dykes of coarse-grained monzonite and by rare lamprophyre dykes, and all rocks are cut by a northwest-trending diabase dyke.

STRUCTURE

The local trend of the southwest-facing host rocks at Redstone is 120° and dips are southwesterly at 45° to 60° (Figs. 3,4). Regionally the metavolcanic rocks are draped about an east-west-trending axis which traverses the centre of the Shaw dome. No other large scale folds have been delineated in the central dome area but isoclinal folds have been recognized near the outer margin of the dome (Green, 1978; Coad, 1977; Pyke, 1975; Pyke and Middleton, 1971). Local isoclinal folding has been suggested as an explanation for the intercalation of dacitic tuff and komatiitic flows at Redstone (Coad, 1979; pers. commun.). However, this explanation is not favoured by the writers because dacitic tuff and quartz-feldspar porphyry are interfingered and interbedded with komatiitic rocks throughout the stratigraphic section on both regional and local scales. Moreover, the contacts of these rocks are concordant and parallel to the regional trend. Bedding attitudes of dacitic tuff surrounding the proposed 'fold nose' of the komatiitic rocks are also concordant and parallel to stratigraphic contacts, and strike-slip shearing or folding was not observed. Finally, the sulphide-facies iron-formation which overlies, and is interfingered with dacitic tuff and komatiitic peridotite, is laterally continuous and conformable with these units and shows no evidence of isoclinal folding.

PETROLOGY AND GEOCHEMISTRY OF THE HOST ROCKS AND Fe-Ni SULPHIDES

Dacite (Unit 1)

The dacitic rocks of the Deloro Group are predominantly well bedded to massive, very fine-grained waterlain tuff with rare flows. Minor lapilli tuff, ash-flow or ash-fall rocks and crystal tuff are also present. Numerous concordant and discordant quartz-feldspar porphyry bodies occur throughout the volcanic pile, some of which grade laterally and vertically into crystal tuff. Phenocryst size and abundance increase toward the centre of the porphyry. Spotted texture within well bedded tuff may represent alteration, primary pumice fragments or varioles (Thurston and Ridler, pers. commun.; Plate I, Fig. 18). The spots are 2 mm to 1 cm in diameter, rounded to subelliptical and generally dark green within a buff ash matrix. The spotted texture is confined to a few metres of vertical extent but appears to be continuous for tens of metres along strike. The spots have well defined boundaries and in places appear to coalesce.

The dacitic tuff is composed of very fine-grained quartz and feldspar which has been extensively altered to sericite, epidote, chlorite and biotite. Fe-rich actinolite (Table II, nos. 15,16), and pyrite are accessory minerals. Quartz and carbonate occur in the tuff as banded lenses, anastomosing fracture-filling networks and disseminated grains, spatially associated with the R zone (Plate I, Figs. 19,20). Pentlandite and gersdorffite blebs and veinlets are common above and below the R zone, within dacitic tuff.

Chemically, the dacitic tuff at Redstone is of calc-alkaline affinity (Figs. 6A,6D) and comparable to average dacite composition in the Abitibi greenstone belt (Goodwin, 1977, Table V). Two samples which plot just outside the calc-alkaline field (Table I, nos. 31,32; Figs. 6A,6D), are spatially associated with, or contain anomalous concentrations of Fe and Ni and are hydrothermally altered from an original calc-alkaline parent. Contents of Na_2O, K_2O, CaO, FeO, MgO and SiO_2 along with Sr, Ba, Rb, Ni and Cr are variable and this is also attributed to hydrothermal alteration (Table I; nos. 28 to 32). In a geochemical study of the Abitibi greenstone belt, Goodwin (1977) recognized that CaO and alkali elements are particularly mobile during alteration and some mobility of other elements may also be assumed. Despite extensive alteration undergone by the dacitic tuff at Redstone, elemental ratios of Al, Ti, Zr and Y are uniform and consistent with this rock classification.

Iron-Formation

Sulphide-facies iron-formation occurs at the stratigraphic top of the dacitic rocks, locally intercalated with dacitic tuff, and the overlying komatiitic flows, and has been mapped as a semi-continuous stratigraphic marker horizon encircling the Shaw dome. Sulphide-facies iron-formation is spatially associated with nickel sulphide deposits at Langmuir, McWatters, Hart and Redstone (Kerrich *et al.*, 1979; Green, 1978; Coad, 1977; Pyke, 1975). Iron-formation at Redstone overlies the R zone to the west (Figs. 3,4,5) and occurs regionally east of the deposit at the same stratigraphic horizon as the Fe-Ni sulphides but where the latter are absent. Sulphide-facies iron-formation may be traced laterally into the S zone, a small lens of nickeliferous massive sulphide above the R zone (Figs. 4,5).

The iron-formation is composed of massive pyrrhotite and pyrite layers with alternating grey to white chert bands (Plate II, Figs. 21,22). A vertical zoning is present within the iron-formation which has a pyrrhotite-rich base, a midsection of white chert containing minor bands and veinlets of both pyrrhotite and pyrite, and a pyrite-rich top. Average grain size of the sulphides is less than 0.1 mm, however minor accumulations of 1 mm diameter pyrite euhedra occur within finer-grained pyrite- and pyrrhotite-rich layers.

Sulphide-facies iron-formation is invariably underlain by a hornfelsic rock composed of medium-grained, pink, manganese garnet megacrysts and pale green to brown stilpnomelane sheaves, 1 mm to 0.5 cm, within a blue-green ferro-tschermakitic hornblende matrix (Leake, 1978); chlorite, biotite, calcite, pyrite and accessory epidote, quartz and feldspar are also present. The garnets have complex normal, oscillatory and reverse concentric zoning of Fe and Mn abundances and to a lesser extent Ca and Mg (Robinson *et al.*, 1980). Stilpnomelane-rich bands are interbedded with chert near the base of the sulphide-facies iron-formation. Garnet-

TABLE I
ROCK ANALYSES

Sample No.	1	2	3	4	5	6	7	8	9	10	11	12	13	14	15	16	17
SiO_2 wt.%	34.2	38.4	38.7	38.6	39.1	37.8	36.0	36.7	38.5	53.2	49.3	53.2	48.1	53.3	45.6	56.4	44.3
Al_2O_3	1.63	2.07	2.52	2.16	1.89	1.39	0.81	0.52	3.38	2.95	1.86	2.45	5.12	3.43	7.19	1.49	8.51
CaO	6.03	2.15	0.64	1.33	0.44	0.05	0.06	0.00	8.42	0.99	4.39	0.74	6.06	4.71	6.85	4.35	6.88
MgO	33.5	36.7	36.6	38.3	39.9	39.8	40.5	41.3	26.5	28.8	28.7	29.6	24.7	27.0	23.8	28.5	26.0
Na_2O	0.00	0.21	0.00	0.00	0.00	0.00	0.00	0.11	0.05	0.02	0.01	0.01	0.03	0.00	0.07	0.00	0.00
K_2O	0.00	0.01	0.00	0.01	0.01	0.01	0.00	0.00	0.01	0.01	0.01	0.01	0.01	0.01	0.01	0.01	0.01
FeO*	6.14	7.46	7.08	7.12	6.75	6.71	6.10	4.41	8.14	6.71	4.92	6.00	8.97	5.54	8.31	4.28	7.27
MnO	0.21	0.08	0.11	0.09	0.08	0.05	0.05	0.02	0.17	0.08	0.08	0.05	0.21	0.11	0.17	0.08	0.18
TiO_2	0.08	0.14	0.13	0.14	0.13	0.11	0.04	0.04	0.20	0.18	0.11	0.15	0.22	0.15	0.36	0.07	0.19
P_2O_5	0.00	0.01	0.00	0.00	0.01	0.01	0.00	0.00	0.01	0.01	0.01	0.01	0.02	0.02	0.03	0.00	0.01
LOI	17.6	13.5	14.2	13.3	12.8	14.8	16.2	17.1	14.8	5.88	10.6	6.7	5.38	4.88	6.75	4.76	7.51
Total	100.1	101.6	100.8	101.9	101.8	101.6	100.5	100.6	101.1	99.5	100.5	100.3	99.8	99.7	100.0	100.4	101.7
Cr ppm	1619	2133	2103	1962	1726	1737	2177	1580	2231	1732	1161	1263	2486	1696	2087	1492	2253
Ba	7	1	5	4	3	5	nd	nd	7	nd	5	nd	11	nd	2	nd	7
Ni	1876	2075	1929	2537	2120	1916	2110	2301	1409	1903	2079	2028	4661	2315	1931	2126	2901
Nb	9	6	9	9	12	10	18	8	4	6	4	9	4	8	nd	6	4
Zr	8	11	11	13	11	12	8	6	15	11	11	11	17	29	34	10	14
Y	6	5	6	5	7	6	6	4	7	7	3	7	12	9	14	5	7
Sr	34	5	4	36	5	1	2	nd	55	nd	21	7	4	3	1	2	11
Rb	2	nd	1	1	nd	1	nd	nd	3	3	2	2	1	2	2	nd	2
Pb	5	5	3	4	6	7	4	4	3	8	9	1	1	6	5	7	6
Co	89	90	154	107	70	62	75	43	78	78	75	83	115	72	80	68	78
Zn	4	1	17	16	9	3	20	10	16	19	14	13	30	39	45	4	57
V	28	41	46	46	36	28	22	14	67	56	32	55	na	47	104	25	85
SG	2.71	2.69	2.70	2.67	2.65	2.60	2.52	2.47	2.86	2.83	2.82	2.82	2.89	2.85	2.89	2.84	2.85

Sample No.	18	19	20	21	22	23	24	25	26	27	28	29	30	31	32	33	34
SiO_2 wt.%	28.2	26.6	30.1	35.7	42.6	41.8	40.0	39.0	43.3	48.8	66.2	67.3	62.8	67.2	63.6	40.1	44.2
Al_2O_3	19.1	20.5	18.3	21.6	6.17	6.59	9.54	6.07	8.68	9.79	13.9	17.8	16.7	16.7	15.3	9.24	8.60
CaO	1.50	0.56	0.96	2.94	5.38	5.48	5.07	8.63	11.5	9.17	6.20	2.56	6.31	5.55	6.63	5.43	5.09
MgO	19.3	20.1	20.2	18.5	22.8	24.3	21.1	17.4	16.5	12.5	1.26	1.11	1.34	0.76	1.84	4.29	3.70
Na_2O	0.07	0.03	0.02	0.00	0.00	0.00	0.03	0.09	0.79	1.96	5.30	1.01	3.86	1.24	2.50	0.33	0.44
K_2O	0.34	0.00	2.57	3.97	0.04	0.76	0.73	0.00	0.05	0.10	0.36	3.76	1.60	2.49	1.20	2.80	1.14
FeO*	20.2	20.1	15.7	8.09	10.6	11.0	12.2	8.12	10.5	10.3	2.43	2.67	2.92	3.00	6.38	28.7	26.9
MnO	0.40	0.34	0.23	0.19	0.15	0.14	0.15	0.20	0.22	0.20	0.09	0.05	0.08	0.12	0.13	1.20	1.66
TiO_2	0.69	0.46	0.73	0.79	0.34	0.38	0.53	0.31	0.51	0.55	0.43	0.54	0.50	0.48	0.46	0.32	0.32
P_2O_5	0.10	0.18	0.14	0.13	0.03	0.02	0.03	0.02	0.07	0.05	0.09	0.12	0.11	0.11	0.06	0.06	0.06
LOI	9.28	9.95	8.22	7.03	12.7	9.65	10.2	16.5	7.10	6.12	4.25	2.42	3.71	2.31	1.94	2.87	2.87
Total	101.5	101.2	98.9	99.7	102.0	101.3	101.0	97.2	100.4	100.7	100.8	99.7	100.2	100.2	98.4	98.5	97.9
Cr ppm	143	34	749	155	2806	3076	3401	2566	2199	1732	85	124	102	77	189	32	52
Ba	149	8	2025	1684	190	146	28	n.d.	n.d.	9.0	537	534	295	533	284	285	178
Ni	244	1392	6201	792	1320	1326	790	859	606	376	94	102	132	99	652	43	26
Nb	7	4	5	7	8	2	5	8	9	7	9	11	10	13	16	12	10
Zr	195	143	174	208	20	22	27	23	46	42	121	152	141	138	136	81	76
Y	36	13	21	23	12	12	9	8	15	13	17	21	21	19	17	28	25
Sr	49	5	25	100	116	39	40	1	6	6	250	74	140	91	152	16	16
Rb	15	3	81	114	1	42	52	50	64	70	12	101	50	55	46	97	70
Pb	10	9	1	7	6	8	4	3	n.d.	4	3	10	15	4	12	30	11
Co	80	113	122	56	104	112	111	85	87	81	34	44	44	55	31	96	87
Zn	103	127	92	55	27	33	52	50	64	70	18	5	39	22	57	24	13
V	132	64	127	118	116	121	n.a.	106	159	197	68	119	79	73	na	52	45
SG	2.89	2.90	2.88	2.87	2.85	2.84	2.84	2.83	2.95	2.89	2.70	2.79	2.75	2.83		3.10	2.97

Rock Analyses; nos. 1 to 8, serpentinized komatiitic flows of unit 2, massive; nos. 9 to 12 talc-carbonate-rich komatiitic flows of unit 2, massive; nos. 13 to 17, CMR-actinolite-assemblage, massive; nos. 18 to 21, CMR-chlorite-assemblage, minor biotite, massive; nos. 22 to 24, sheaf spinifex-textured komatiite flows of unit 3; nos. 25 to 27, komatiitic basalt of unit 3, massive; nos. 28 to 32, calc-alkaline dacitic tuff, massive to tuffaceous; nos. 33, 34, garnet-stilpnomelane rock, hornfelsic texture, massive; FeO* – total iron reported as FeO, CMR – contact metasomatic rock, n.a. – not analyzed; n.d. – not detected; SG – specific gravity.

TABLE II
MICROPROBE ANALYSES OF CHLORITE, CALCIC-AMPHIBOLE, SERPENTINE AND TALC

	1	2	3	4	5	6	7	8	9	10	11	12
SiO_2	26.52	25.40	26.92	26.79	28.60	29.59	29.49	31.57	58.32	57.63	55.90	57.14
TiO_2	.10	.01	.00	.01	.01	.00	.00	.01	.00	.02	.08	.00
Al_2O_3	19.12	20.11	19.69	20.05	18.88	20.77	20.37	13.46	.07	.14	2.85	.17
FeO*	29.86	26.92	27.28	22.38	17.55	12.85	10.12	9.39	4.06	2.84	6.38	6.87
MnO	1.22	.29	.42	.52	.27	.16	.12	.26	.16	.21	.16	.25
MgO	11.05	13.63	14.01	17.44	21.93	23.94	26.92	29.33	22.08	21.70	19.87	19.85
CaO	.10	.00	.00	.00	.04	.00	.00	.02	12.89	13.51	12.70	13.20
Na_2O	.01	.00	.00	.00	.00	.00	.00	.00	.04	.00	.31	.08
K_2O	.20	.00	.00	.01	.00	.00	.03	.00	.01	.00	.01	.00
Cr_2O_3	.06	.18	.00	.10	.34	.08	.50	2.77	.01	.04	.09	.08
NiO	na	.55	.19	.44	na	.79	na	.53	na	.04	na	na
Total	88.25	87.10	88.51	87.73	87.62	88.20	87.53	87.33	97.64	96.12	98.34	97.64

Structural formulae calculations –

Cation proportions on the basis of 28, 23, 14 and 12 oxygens respectively.

	1	2	3	4	5	6	7	8	9	10	11	12
Si	5.721	5.467	5.673	5.570	5.776	5.785	5.723	6.204	8.029	8.034	7.742	7.992
Al^{IV}	2.279	2.533	2.327	2.430	2.224	2.215	2.277	1.796	.000	.000	.258	.008
Al^{VI}	2.581	2.566	2.563	2.482	2.270	2.570	2.381	1.321	.011	.024	.208	.021
Ti	.016	.002	.000	.001	.002	.000	.000	.001	.000	.003	.008	.000
Fe	5.387	4.845	4.808	3.892	2.964	2.101	1.643	1.544	.467	.331	.739	.804
Mn	.223	.052	.075	.092	.046	.027	.019	.043	.019	.025	.019	.030
Mg	3.554	4.371	4.401	5.404	6.601	6.977	7.787	8.597	4.531	4.509	4.102	4.137
Cr	.009	.031	.000	.016	.054	.013	.076	.430	.001	.004	.009	.009
Ni	.000	.095	.032	.073	.000	.125	.000	.083	.000	.004	.000	.000
Sum Al^{VI} to Ni	11.772	11.962	11.879	11.961	11.938	11.813	11.906	12.014	5.029	4.899	5.085	5.001
Ca									1.902	2.018	1.885	1.978
Na									.010	.000	.082	.021
K									.001	.000	.002	.000
F/M	1.579	1.121	1.110	.737	.456	.305	.213	.185	.107	.079	.185	.201
	← chlorite (1–8) →								← calcic (9–12)			

Microprobe analyses of Chorite, Calcic-Amphibole, Serpentine and Talc, analyses represent average determinations for 6 to 10 spot determinations; nos. 1, 3 chlorite in dacitic tuff, nos. 2, 4, 5, 6, 7, chlorite from chlorite-assemblage, contact metasomatic rocks, no. 8, chlorite in talc-carbonate

	13	14	15	16	17	18	19	20	21	22	23	24
SiO_2	54.82	53.69	51.21	48.67	43.85	42.54	42.42	44.26	62.06	60.61	60.53	58.16
TiO_2	.10	.04	.09	.13	.00	.00	.01	.01	.00	.02	.01	.00
Al_2O_3	3.00	1.96	4.83	6.16	1.70	2.25	1.53	1.39	.20	.29	.00	1.89
FeO*	7.15	11.37	15.51	16.57	7.36	6.63	4.83	1.67	1.47	2.52	3.53	4.25
MnO	.44	.41	.58	.69	.04	.01	.00	.02	.01	.01	.05	.02
MgO	19.31	15.75	12.73	12.52	36.19	37.21	38.96	40.98	29.98	29.34	27.55	28.56
CaO	12.70	12.54	12.48	10.89	.04	.00	.10	.03	.00	.02	.00	.11
Na_2O	.39	.27	.51	.40	.08	.00	.01	.00	.00	.02	.00	.00
K_2O	.04	.03	.08	.08	.00	.00	.01	.00	.00	.01	.01	.00
Cr_2O_3	.25	.02	.08	.07	.12	.05	.12	.08	.01	.03	.02	.01
NiO	na	na	na	na	.21	na	.12	.18	na	na	na	na
Total	98.20	96.08	98.10	96.17	89.59	88.69	88.11	88.62	93.74	92.87	91.72	93.00

	13	14	15	16	17	18	19	20	21	22	23	24
Si	7.663	7.816	7.476	7.283	4.058	3.967	3.958	4.030	4.370	4.337	4.398	4.206
Al^{IV}	.337	.184	.524	.717	.000	.033	.042	.000	.000	.000	.000	.000
Al^{VI}	.157	.152	.307	.368	.185	.214	.126	.149	.016	.025	.000	.161
Ti	.011	.004	.010	.015	.000	.000	.001	.001	.000	.001	.001	.000
Fe	.836	1.385	1.894	2.074	.570	.517	.377	.127	.087	.151	.214	.257
Mn	.052	.050	.072	.087	.003	.001	.000	.002	.000	.001	.003	.001
Mg	4.023	3.419	2.770	2.792	4.992	5.172	5.419	5.562	3.147	3.129	2.983	3.079
Cr	.028	.003	.009	.008	.009	.003	.009	.006	.000	.002	.001	.001
Ni	.000	.000	.000	.000	.016	.000	.009	.013	.000	.000	.000	.000
Sum Al^{VI} to Ni	5.107	5.013	5.062	5.344	5.775	5.907	5.941	5.860	3.251	3.308	3.203	3.499
Ca	1.902	1.955	1.952	1.745								
Na	.106	.077	.143	.117								
K	.007	.005	.015	.015								
F/M	.221	.420	.710	.774	.115	.100	.070	.023	.028	.048	.073	.084
	amphibole →				← serpentine →				← talc →			

komatiite; nos. 9 to 14, calcic-amphibole in komatiite, nos. 15, 16, calcic-amphibole in dacitic tuff; nos. 17 to 20, serpentine in komatiite of unit 2A, nos. 21 to 24, talc in komatiite of unit 2B; FeO* – total iron reported as FeO; n.a. – not analyzed.

stilpnomelane rock occurs as a laterally continuous stratigraphic unit below sulphide-facies iron-formation, flanking and in places footwall to the R zone, and is intercalated with dacitic tuff (Figs. 4,5). This rock may represent a primary silicate-facies iron-formation or, alternatively, a secondary footwall alteration rock generated during exhalative deposition of the overlying sulphide-facies iron-formation. Comparable manganiferous garnet-, and stilpnomelane-bearing rocks are regarded as silicate-facies iron-formation or low grade metamorphic equivalents of primary chemical sedimentary strata (Stanton and Vaughan, 1979; Floran and Papike, 1977; Button, 1976; Ayres, 1972).

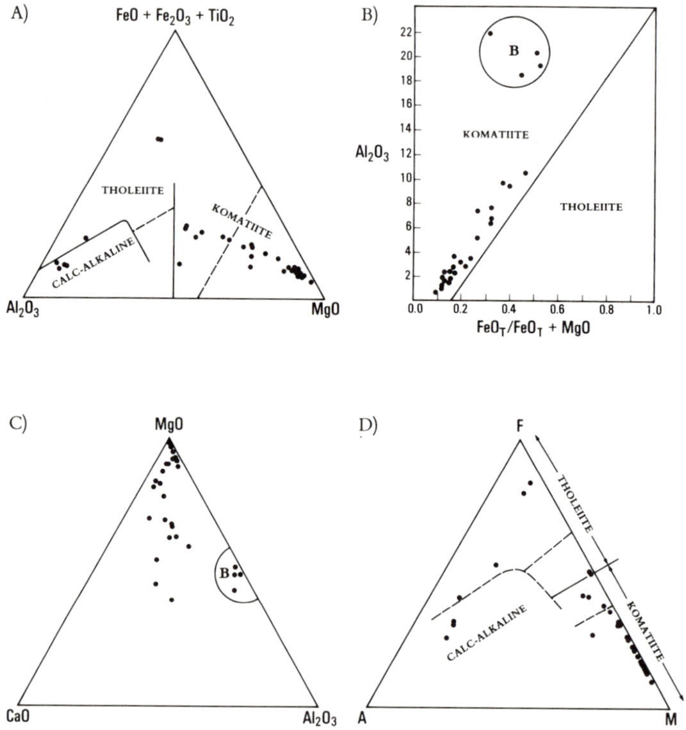

Figure 6. A) Jensen Cation Plot, showing a continuum from ultramafic to mafic compositions in the komatiite suite. Garnet-stilpnomelane rock plots within the tholeiite field. Footwall volcanic rocks plot within or near the calc-alkaline field. B) Plot of Al_2O_3 vs FeO_T/FeO_T + MgO) (after Arndt *et al.*, 1977). FeO_T represents total iron. Redstone komatiitic rocks form a continuum from ultramafic to mafic compositions. The chlorite-assemblage from the contact metasomatic sequence is distinguished from the komatiitic suite at Redstone by its large $A1_2O_3$ concentration, and plots in field B. C) MCA plot of MgO-CaO-$A1_2O_3$ for Redstone komatiitic flows of ultramafic to mafic composition. Chlorite-assemblage from the contact metasomatic sequence plots in field B. D) AFM plot of $Na_2O + K_2O$ – FeO – MgO (field boundaries from Pyke *et al.*, 1978). Redstone komatiitic rocks plot along the MgO-FeO axis, garnet-stilpnomelane rocks plot within the tholeiite field and footwall volcanic rocks plot in or near the calc-alkaline field.

TABLE III
CHEMICAL ANALYSES OF SULPHIDE-FACIES IRON FORMATION

		1	2	3	4	5	6
wt%	SiO_2	76.6	54.0	53.1	24.7	61.2	80.8
	Al_2O_3	0.40	3.36	2.30	1.04	0.43	0.47
	CaO	0.00	0.74	0.00	0.01	0.00	0.00
	MgO	0.08	1.11	3.10	0.64	0.59	0.45
	Na_2O	0.00	0.70	0.00	0.00	0.00	0.00
	K_2O	0.00	0.18	0.30	0.00	0.00	0.00
	FeO*	13.06	23.1	19.0	33.9	20.1	10.9
	MnO	0.04	0.37	0.05	0.39	0.19	0.06
	TiO_2	0.00	0.09	0.11	0.03	0.00	0.01
	LOI	4.21	6.84	11.7	7.24	8.95	4.85
	S	6.30	10.8	12.1	25.0	10.2	5.95
ppm	Cr	990	1187	527	1242	1693	1304
	Ba	24	96	49	39	9	29
	Ni	131	95	270	133	144	158
	Nb	nd	8.2	11.8	1.4	4.8	10.9
	Zr	53	54	51	45	44	56
	Y	nd	10	7	5	4	3
	Sr	nd	11	5	2	4	2
	Rb	nd	10	25	6	8	7
	Pb	nd	9	3	nd	7	33
	Co	54	57	56	46	47	28
	Zn	nd	14	19	nd	7	16
	Cu	77	nd	327	98	122	97

Chemical analyses of sulphide-facies iron-formation. FeO* – reported as total iron, n.d. – not detected.

Sulphide-facies iron-formation is depleted in precious metals, Cu, Pb, Zn and Ni (Table III). However significant, Cr concentrations (527 to 1693 ppm) were detected and are comparable to those in nickeliferous sulphide. The tuffaceous component within the iron-formation is generally minor as evidenced by small Al_2O_3, TiO_2, MgO, CaO, MnO and alkali concentrations. Garnet-stilpnomelane rock is extremely enriched in FeO, MnO, MgO, Co and Y, and depleted in SiO_2, Al_2O_3, TiO_2, Na_2O, Cr, Ba, Ni, Zr and Sr in comparison to dacitic tuff (Table I, nos. 33,34). These samples plot in the field of Fe-rich tholeiitic basalt (Figs. 6A,6D).

Komatiitic Rocks (Units 2 and 3)

Komatiitic peridotite flows of unit 2 are generally massive, fine-grained, and contain minor sheaf and random-type spinifex textures, similar to those described by Arndt *et al.* (1977) and Pyke *et al.* (1973); hyaloclastite and cumulate-textured members are also present. Random skeletal textures and large 10 to 15 cm vesicle-like structures filled with talc-carbonate minerals are best observed on weathered outcrop surfaces (Plate I, Fig. 16). The flows pinch out laterally from east to west and downdip (Figs. 3,4,5). In general, rocks of unit 2 have a serpentinized core (unit 2A), symmetrically enveloped by a talc-carbonate margin (unit 2B).

The serpentinized core is massive, green-black, generally fine-grained, and composed predominantly of serpentine, with minor magnetite, magnesite, stichtite and hematite. Cumulate olivine grains are completely pseudomorphed by serpentine, with a fine magnetite dusting around their rims or along fractures. Chrysotile, magnetite, magnesite and rarely hematite occur as crosscutting veinlets and disseminated grains. Stichtite occurs as mauve 2 to 3 mm porphyroblasts in a serpentine matrix. Serpentinized orthopyroxene and clinopyroxene, which are distinguished by larger chromium and aluminum contents (Dungan, 1979), have not been detected in any of the komatiitic rocks.

The serpentinized core grades laterally and vertically into the talc-carbonate margin. Relict cumulate textures composed of pseudomorphic serpentine and magnetite exhibit sequential replacement by talc and carbonate along original olivine grain boundaries and internal fractures. The modal abundances of serpentine and magnetite decrease progressively toward the margins of unit 2, with a corresponding increase in talc and carbonate minerals. The FeO/MgO ratio of serpentine varies proportionatly with the bulk composition of the host rock and in turn with the modal percentage of talc, thereby reflecting the lateral and vertical changes in magnesium content of the flows (Table II, nos. 9 to 12). The marginal zone is composed predominantly of talc, magnesite and dolomite, with minor magnesium-rich actinolite, chlorite, rare serpentine and magnetite.

The komatiitic flows of unit 3 are very fine-grained, grey-green and have ubiquitous chlorite and talc-carbonate alteration. These flows contain prominent sheaf and random-type spinifex textures, hyaloclastite zones and massive members (Plate I, Fig. 15). They are composed of talc, chlorite, dolomite, with accessory quartz, actinolite, biotite, epidote, rutile, sphene, pyrite and rare serpentine. A number of narrow (5 m) dacitic crystal tuff and quartz-feldspar porphyry bodies are intercalated with the ultramafic rocks of unit 3 (Fig. 4).

The komatiitic rocks at Redstone range in composition from dunite to basalt, with increasing iron, titanium, aluminum and silica and corresponding decrease in magnesium (Fig. 6A; Table I, nos. 1 to 12, 22 to 27). These rocks exhibit low FeO/(FeO + MgO) ratios for a given Al_2O_3 content (Fig. 6B), and contain very little TiO_2. An MCA diagram (Fig. 6C), illustrates the highly magnesian nature of the komatiitic lavas and their variable CaO/Al_2O_3 ratios. The latter characteristic may be a function of either hydrothermal alteration or primary magmatic fractionation. An AFM diagram (Fig. 6D) demonstrates the alkali-deficient nature of the komatiitic flows and the linear progression from ultramafic to mafic composition as a function of FeO/MgO ratio. It has been suggested that the chemical differences amongst altered komatiitic rocks at Redstone may represent a primary magmatic compositional variation as opposed to a progressive alteration phenomenon (Eckstrand, written commun.). It is apparent from this study that these chemical and mineralogical differences are due in part to *both* of these; they represent primary magmatic characteristics which have subsequently been hydrothermally altered.

The komatiitic rocks have low abundances of trace elements but high Ni and Cr, and moderate Co contents (Table I). Spinifex-textured flows of unit 3 (Table I, nos. 22, 23), contain more Al_2O_3, CaO, TiO_2, FeO and MnO, but less MgO than the komatiitic rocks of unit 2. Cr contents are greatest in the spinifex-textured flows of unit 3, whereas Ni concentrations are lower here than in the pseudomorphosed olivine-cumulate rocks of unit 2. Despite widespread and pervasive hydrothermal alteration, Redstone komatiitic rocks are chemically comparable to similar rocks studied by Green (1978), Coad (1977), and Pyke (1975) in the Shaw dome area and by Arndt *et al.* (1977) and Pyke *et al.* (1973) in Munro township.

Contact Metasomatic Rocks

A sequence of contact metasomatic rocks occurs at the interface between dacitic tuff or quartz-feldspar porphyry and komatiitic flows of unit 2, along footwall, internal and hangingwall contacts of the latter. These altered rocks are narrow, less than 2 m thick collectively but laterally continuous, and occur in an orderly sequence of rather abrupt mineralogical changes from dacitic tuff on one side of the contact through biotite-, chlorite-, actinolite-, and talc-rich assemblages into talc-carbonate rocks on the other side (Fig. 7). The altered rocks are nearly monomineralic, with only minor accessory minerals and contacts between the assemblages are sharp. However mineralogical overlap is common and one or more assemblages may be absent. The R zone is directly overlain by the chlorite assemblage which thus separates it from the komatiitic flows and is underlain by the biotite assemblage and/or dacitic tuff along the footwall contact of unit 2. The chlorite assemblage also separates massive Fe-Ni sulphides from the komatiitic flows along hangingwall contacts of unit 2.

Similar metasomatic assemblages are associated with ultramafic rocks in both nickel sulphide-bearing and non-sulphide-bearing environments, and have been described by Marston and Kay (1980), Barrett *et al.* (1977), Evans (1977), Ross and Hopkins (1975), Frost (1975), Springer (1974), Curtis and Brown (1971, 1969), Wilson *et al.* (1969), and Phillips and Hess (1936).

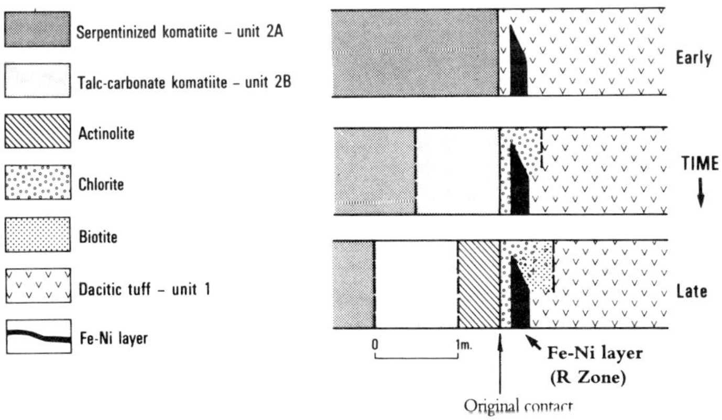

Figure 7. Sketch illustrating the progressive development of the contact metasomatic sequence with time. The original komatiite/dacitic tuff contact is located at the actinolite/chlorite interface. The Fe-Ni sulphide layer is separated from komatiitic rocks by the chlorite-assemblage.

The talc and actinolite assemblages at Redstone are restricted to komatiites whereas chlorite and biotite penetrate the adjacent dacitic tuff. The talc unit is very fine-grained, massive, weakly schistose and contains minor prismatic actinolite crystals or very fine-grained chlorite or serpentine. The talc composition is relatively constant, with minor variation in the Fe/(Fe + Mg) ratio from 2 to 12 per cent (Table II, nos. 21 to 24), which is dependant upon the bulk composition of the host rock and in turn upon the modal proportion of serpentine, actinolite or chlorite. The actinolite assemblage is coarse-grained with radiating sheaves of prismatic actinolite up to 1 cm long in random orientation (Plate I, Fig. 17), and interstitial chlorite. It also contains matrix-textured Fe-Ni sulphides (Plate III, Fig. 33). The composition of calcic amphibole at Redstone varies directly with the bulk composition of the host rock and forms a continuum from very Mg-rich calcic amphibole within komatiitic rocks through intermediate compositions in dacitic tuff to very Fe-rich varieties hosted within garnet-stilpnomelane iron-formation (Fig. 8A). The composition of calcic amphibole within komatiitic rocks including the contact metasomatic assemblage ranges from tremolite to actinolite (Leake, 1978; Table II, nos. 9 to 14), but is mainly low-aluminum actinolite (X_{Mg} = 0.85). Dacitic tuff contains a fine-grained, prismatic actinolite as a minor accessory, which is distinguished from the former amphibole by

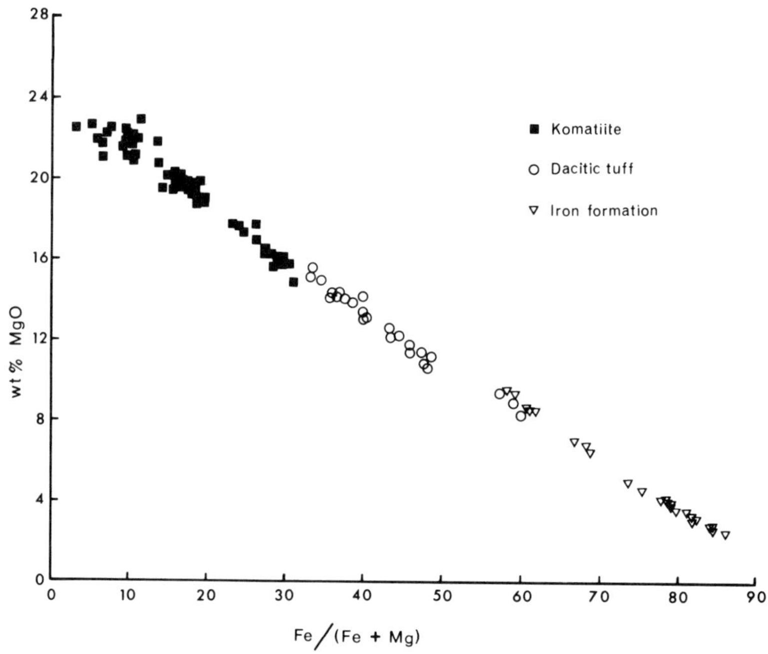

Figure 8A. Plot of MgO vs. Fe/ (Fe + Mg) ratio based on structural formulae for calcic amphiboles at Redstone. The composition of calcic amphibole varies directly with bulk composition from Mg-rich varieties in komatiite through intermediate compositions in dacitic tuff to Fe-rich amphiboles in iron-formation.

its higher content of Al and Fe and lower Si and Mg (Table II, nos. 15 and 16). Calcic amphibole hosted within garnet-stilpnomelane iron-formation is extremely enriched in Fe and depleted in tetrahedrally coordinated Si atoms per formula unit, in contrast to the tremolite-actinolite series.

The chlorite assemblage is massive, weakly schistose, overlies the R zone and contains small lenses of Fe-Ni-Cu sulphides. The Fe/(Fe + Mg) ratio of chlorite varies systematically with the host rock and forms a continuum from a Mg-rich variety in komatiites through intermediate compositions in the chlorite assemblage, to very Fe-rich chlorite hosted within dacitic tuff and iron-formation (Fig. 8B). Chlorite in talc-carbonate komatiites is however distinguished from those in the other assemblages by its lower Al and higher Cr contents. Chromian chlorite (Table II, no. 8), is associated with disseminated chromian spinel and magnetite in a talc-carbonate matrix.

The biotite assemblage lies farthest from the komatiites in the contact metasomatic sequence, overlaps the chlorite assemblage and infiltrates the adjacent

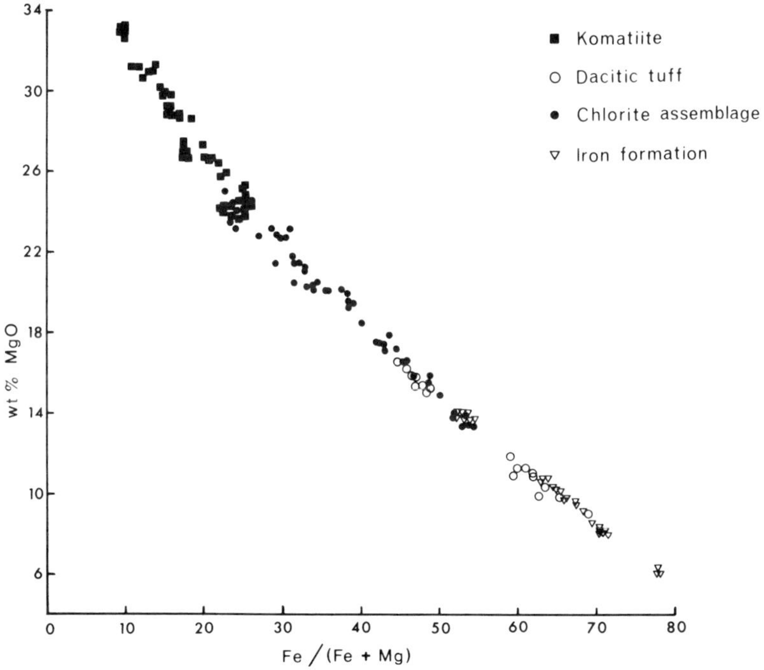

Figure 8B. Plot of MgO vs. Fe/ (Fe + Mg) ratio based on structural formulae for chlorite at Redstone. The composition of chlorite varies from very Mg-rich examples hosted in komatiite to Fe-rich chlorite in iron-formation. Note that the chlorite-assemblage from the contact metasomatic assemblage overlaps with chlorite hosted in komatiite and dacitic tuff. Komatiitic-hosted chlorite compositions generally contain less Al_2O_3 than the other varieties.

dacitic tuff or porphyry. The mica compositions become more magnesian toward the talc-carbonate rock, and phlogopite occurs with epidote adjacent to ultramafic rock whereas barian muscovite is present in dacitic tuff.

Chemically the chlorite assemblage is enriched in FeO, Al_2O_3, TiO_2, MnO, P_2O_5, Zr and Y, and non-mineralised samples are depleted in Ni and Cr, relative to the komatiitic flows (Table I, nos. 18 to 21). The chlorite assemblage plots along the komatiitic trend on the Jensen Cation and AFM diagrams (Figs. 6A, 6D), but it may be distinguished from the komatiites on the MCA and Al_2O_3 vs. FeO/(FeO + MgO) diagrams, where it is represented by field B (Figs. 6B, 6C).

Iron-Nickel Sulphide Distribution and Form

Nickel sulphides at Redstone occur in four forms: 1) massive; 2) matrix-textured (Ross and Hopkins, 1975); 3) disseminated; and 4) xenolithic. The principal nickel sulphide mineral is pentlandite, with minor millerite, violarite, gersdorffite and trace amounts of niccolite. Pyrite and pyrrhotite constitute 5 to 50 per cent of the sulphides and chalcopyrite is a ubiquitous accessory. Chlorite, biotite, actinolite, talc, chert, calcite, siderite, manganoan calcite and magnetite comprise up to 50 per cent of the massive sulphides within which they form bands, layers and inclusions.

The most important form is a massive, stratabound Fe-Ni sulphide body referred to as the R zone which is conformable with its enclosing host rocks, strikes 120° to 130° and plunges southeasterly from a subcropping expression to several hundred metres depth. This stratabound sulphide layer extends to the west (Fig. 4), and down-dip (Fig. 5), tens of metres or more beyond the limits of the overlying komatiitic flows and is here enclosed within dacitic tuff. Massive Fe-Ni sulphides are also present *on top* of komatiitic peridotite, but are separated from the latter by narrow lenses of dacitic tuff or by the chlorite assemblage, at the western extremity of the deposit. A small lens of massive Fe-Ni sulphide, labelled S zone, occurs stratigraphically above the R zone and laterally along strike from sulphide-facies iron-formation (Figs. 4,5). Finally a small massive Fe-Ni sulphide lens occurs below the R zone in the eastern portion of the deposit.

The R zone displays mineralogical and metal zoning laterally from west to east (Fig. 9). Four major zones are composed of: A) pendlandite + pyrite + pyrrhotite; B) pentlandite + pyrite; C) pentlandite + pyrite + millerite; and D) pentlandite + pyrite + violarite. Minor mineralogical overlap occurs between millerite- and violarite-bearing zones C and D. The S zone contains the same sulphide minerals, excluding violarite, and similar mineral zones A and C. The composition of pentlandite varies sympathetically with the other sulphide minerals present in each zone, and ranges from a low nickel tenor where associated with pyrrhotite to a high nickel tenor where associated with millerite (Fig. 9). Marston and Kay (1980), Williams (1979), Barrett *et al.* (1977), Keele and Nickel (1974), Misra and Fleet (1973) and Graterol and Naldrett (1971) have all described zoning of nickel sulphide minerals and contrasting nickel tenors of pentlandite or nickel sulphides within a single deposit. At Redstone chalcopyrite, gersdorffite and magnetite are accessory minerals in all massive Fe-Ni sulphide zones.

Pentlandite occurs as fine to medium, euhedral to subhedral, octahedra within a finer-grained matrix of subhedral pyrrhotite grains (Plate III, Fig. 29), and rare pentlandite exsolution lamellae occur within pyrrhotite (Plate III, Fig. 30). Millerite forms anhedral patches within pentlandite. Violarite rims and progressively replaces pentlandite octahedra and is of supergene origin (Plate III, Fig. 31). Gersdorffite occurs as subrounded, anhedral to subhedral blebs, up to 100 μ in diameter, in pentlandite and pyrrhotite, surrounds rare niccolite grains, and is most abundant and commonly associated with carbonate-rich sulphides. Chalcopyrite forms anhedral grains and is generally interstitial to more euhedral minerals such as pyrrhotite or pentlandite. Chalcopyrite rims euhedral pyrite cubes and occurs as rare blebs in pyrite (Plate III, Fig. 32).

Mineralogical layering consisting of bands of subhedral to euhedral pyrite alternating with bands of pentlandite-pyrrhotite is present within the R zone. Both primary and metamorphic sulphide banding are believed to be present (Plate II, Figs. 27, 28 respectively). Primary banding is recognized by marked contrasts in mineral composition between pyritic and pyrrhotite-pentlandite-rich layers, relatively continuous layers, and by sharp upper and lower contacts between the layers of fine-grained pyrite, from 1 mm to 1 cm thick, within massive Fe-Ni sulphides. In contrast, the metamorphic banding consists of less regular, discontinuous layers of medium- to

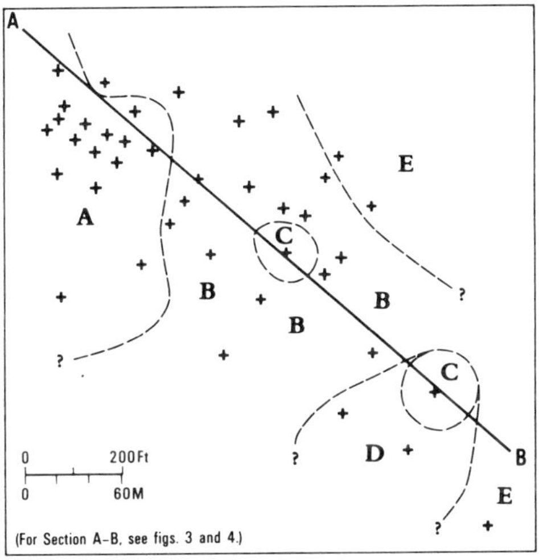

	Mineral Assemblage	Pn Composition (wt %)				
		Ni	Fe	Co	S	Total
A	pn + py + po	35.58	30.85	0.42	33.56	100.41
B	pn + py	35.98	29.91	0.33	33.66	99.88
C	pn + py + ml	42.04	24.57	0.34	32.95	99.90
D	pn + py + viol	38.56	27.45	0.28	32.72	99.01
E	no sulphides					
	cpy + gf are in A to D	+ Sample location				

Figure 9. Mineral zoning in the Fe-Ni sulphide layer: plan view showing the R zone which plunges southeasterly from a subcropping expression at A to several hundred metres depth at B. Mineralogical zones A and B are characterized by low nickel tenor and zones C and D, by high nickel tenor in pentlandite. Zone D also contains secondary violarite.

coarse-grained recrystallized pentlandite and pyrrhotite, with irregular contacts between bands. Quartz and carbonate are very abundant within the massive and disseminated Fe-Ni sulphides, where they occur interbanded with sulphides, and as disseminated grains and crosscutting veinlets (Plate II, Figs. 23, 24).

Matrix-textured sulphides are rare, at the base of komatiitic periodotite, near the eastern limit of the R zone. Here they form a continuous interlocking network of pentlandite, pyrite, chalcopyrite and millerite within a matrix of actinolite or talc (Plate II, Fig. 25). Actinolite megacrysts are riddled with Fe-Ni sulphide inclusions and contain sulphides preferentially along cleavage planes. Large, optically continuous prismatic actinolite grains are transected by sulphides and actinolite euhedra occur within and crosscut the sulphides (Plate III, Figs. 32, 33).

Disseminated Fe-Ni sulphide blebs occur in the basal portion of the komatiitic peridotite but are laterally and vertically discontinuous. The sulphide bleb size and modal abundance does not increase progressively toward the base of the komatiitic peridotite, as at Kambalda (Ross and Hopkins, 1975), rather the size and distribution of sulphide blebs is erratic. Pentlandite is predominant with lesser amounts of pyrite and magnetite within a talc-carbonate matrix, and these sulphide blebs are commonly mantled by carbonate sheaths (Plate III, Fig. 34). Disseminated blebs of pentlandite, gersdorffite, pyrite and chalcopyrite, enveloped by carbonate minerals are abundant within *dacitic* tuff above and below the R zone (Plate III, Fig. 35). The footwall tuff is extensively mineralized with Fe-Ni sulphides, both in veinlets and disseminated blebs, generally within a few metres of the R zone. A 15 m section of disseminated nickeliferous sulphides is vertically continuous through komatiitic rocks, quartz-feldspar porphyry and dacitic tuff, at the eastern limit of the deposit (Fig. 4).

The xenolithic form of mineralization consists of massive Fe-Ni sulphides which contain inclusions of chlorite, biotite, actinolite and talc. Xenoliths are schistose, sigmoid-shaped and range from a few mm to 30 cm in maximum dimension (Plate II, Fig. 26). There is a general increase in xenolith abundance from west to east corresponding to increasing thickness of the komatiitic rock. Xenoliths from the contact metasomatic sequence are rare to absent at the western limit of the R zone. The xenoliths appear to have been rotated and slumped, indicating local remobilization and plastic deformation of the massive sulphides (Plate III, Fig. 36).

Platinum Group Elements, Au, Ni and Cu

A representative suite of Fe-Ni sulphides was analyzed for PGE, Au, Ni and Cu, to assess their abundance and distribution in the different sulphide zones and minerals (Table IV, Figs. 10, 11). Ni and Cu vary widely both in concentration and relative proportions as illustrated by the scatter of data points in Figure 10. This broad variability is partially a function of the proportion of sulphides relative to silicate and carbonate minerals, but also reflects the diverse sulphide mineralogy and composition. The calculated content of Ni metal in pure (100%) sulphide composition varies from a maximum tenor in the millerite-bearing zone to a minimum in the pyrrhotite-rich zone. This relationship is also reflected in the varying composition of pentlandite in the same two zones.

Analyzed values for the platinum group elements and Au (Table IV), have been recalculated to a Ni content of 14 per cent and then normalized relative to chondritic

TABLE IV
ANALYSES OF PLATINUM GROUP ELEMENTS, Au, Ni, Cu As, AND CaO IN SULPHIDE ASSEMBLAGES AT REDSTONE

Zone	Sample No.	Ni wt%	Cu wt%	Rh ppb	Pd ppb	Pt ppb	Ir ppb	Os ppb	Ru ppb	Au ppb	% sulphide (est.)	As ppm	CaO wt %
A	1	12.2	0.44	360	310	250	310	310	1500	70	80	1020	3.98
A	2	2.68	0.10	60	310	50	35	33	180	23	<10*	na	nd
A	3	19.0	0.22	400	520	330	305	305	1320	85	90	na	2.94
A	4	4.92	0.21	70	800	420	54	55	225	83	<20	na	2.23
A	5	9.09	0.51	265	1625	410	175	180	835	125	50	3280	12.55
A	6	13.3	0.49	420	1375	860	325	335	1475	310	80	na	9.29
B	7	21.13	0.23	420	1970	1435	375	365	965	565	75	330	2.80
B	8D	21.13	0.23	420	1830	1430	375	370	1025	380	75	na	2.80
B	9	3.80	0.06	73	760	520	84	90	120	135	15	na	1.63
B	10	18.4	0.90	325	590	750	400	425	550	210	60	620	2.62
B	11	10.35	0.97	210	6020	955	220	260	400	1150	50	na	18.5
B	12	0.76	0.08	17	135	85	31	34	30	43	< 5	na	na
B	13	2.11	0.18	31	320	275	32	32	40	81	<10*	35	na
B	14D	18.4	0.90	375	560	750	440	445	620	300	60	700	2.62
B	15	19.92	0.60	465	2350	945	560	550	830	2120	90*	na	na
C	16	27.2	0.39	400	1010	1650	460	455	745	1300	85	720	3.74
C	17D	27.2	0.39	385	1230	2340	445	440	645	960	85	670	3.74
C	18	12.96	0.14	70	1015	3675	26	25	40	550	35	na	0.67
C	19	7.34	0.67	76	510	550	43	32	60	160	25	na	6.50
C	20	1.17	0.03	16	200	105	12	8	40	40	<10*	10	5.28
C	21	4.05	0.75	69	260	370	105	125	195	24	15	na	5.41
D	22	8.78	0.12	62	915	225	38	22	20	730	30	44	nd
D	23	4.73	0.16	110	880	425	155	180	310	190	20	24	4.46
1F	24	95 #	–	3	15	15	1.6	< 5	< 5	20		na	0.90
1F	25	133 #	98 #	5	16	15	4.0	5	10	32		na	0.00
1F	26	270 #	327 #	1	10	10	1.1	< 5	< 5	17		3	0.00
1F	27	90 #	–	4	55	35	4.0	< 5	5	30		140	na
1F	28	238 #	–	6	110	80	1.9	< 5	< 5	46		44	na

Analyses of Platinum Group Elements, Au, Ni, Cu, As and CaO in sulphide assemblages at Redstone; nos. 1 to 6, pentlandite + pyrite + pyrrhotite, zone A; nos. 7 to 15, pentlandite + pyrite, zone B; nos. 16 to 21, pentlandite + pyrite + millerite, zone C; nos. 22, 23, pentlandite + pyrite + violarite, zone D; nos. 24 to 28, sulphide-facies iron-formation, nos. 8D, 14D, 17D, duplicate analyses of samples 7, 10, 16 respectively; 95# – analyses in ppm; <10* – visual estimate of sulphides, remaining sulphide % estimates are based on total major and trace element analyses; n.a. – not analyzed; n.d. – not detected; PGE analyses at X-Ray Assay Lab., Toronto.

abundances according to the procedure outlined by Naldrett *et al.* (1979) and Ross and Keays (1979). The data were recalculated to a constant Ni content to facilitate comparison of samples of differing sulphide proportions, and to assess the relative variation of PGE with Ni. The resulting PGE and Au data from Redstone have been grouped into zones A through D according to mineralogical zoning and are plotted in Figure 11, along with typical abundances for komatiite-hosted deposits (Naldrett *et al.*, 1980) and average abundances from Redstone. The relative abundance of precious metals at Redstone is Pd>Pt>Ru>Au>Os>Ir>Rh. Redstone PGE and Au abundances are characterized by a positive slope from Os to Au, relative to chondrites, and plot generally in an overlapping field with typical komatiite-hosted deposits (Fig. 11). Precious metals at Redstone are depleted in Ru and to a lesser extent Pd, and enriched in Ir, Rh and Au, relative to typical komatiites.

A heterogeneous distribution of PGE and Au is apparent throughout the deposit and these precious metals do not correlate well with Ni or Cu. In general the sulphide zones are depleted in Os, Ir, Ru and Pt, and enriched in Rh, Pd and Au relative to chondritic abundances (Fig. 11). Pyrrhotite-bearing sulphides from zone A, are enriched in Ru and Rh, and depleted in Pt relative to the other zones. Millerite-rich samples from zone C are depleted in Os, Ir, Ru, Rh, Pd and Au, and enriched in Pt. Violarite-bearing samples from zone D contain similar abundances of PGE, but are

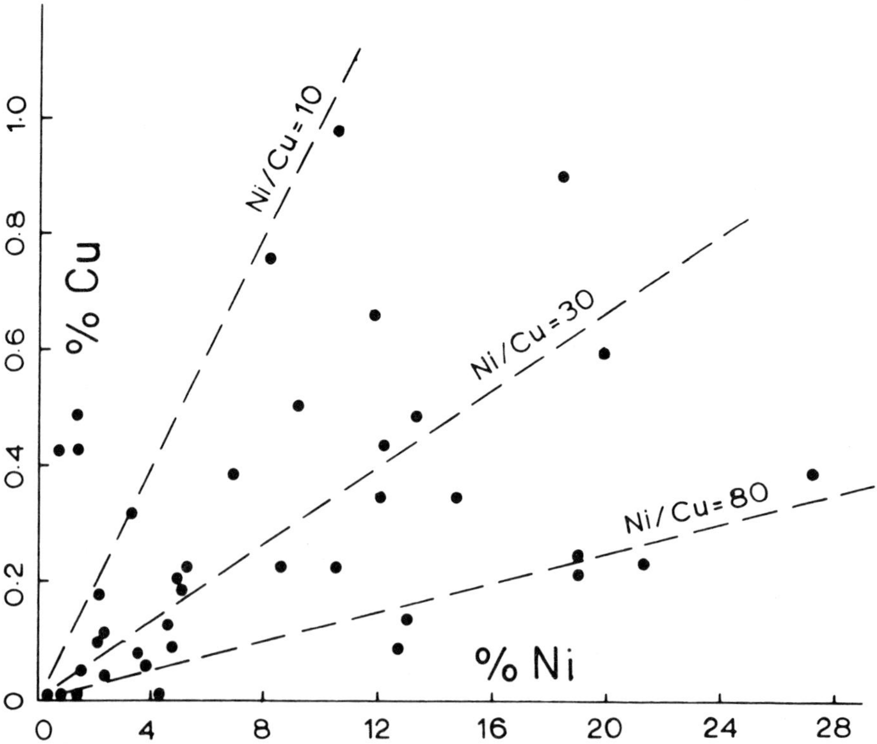

Figure 10. Plot of Ni-Cu in wt per cent from Fe-Ni sulphides at Redstone.

enriched in Au relative to the average concentration at Redstone. The sulphide-facies iron-formation is depleted in PGE and Au, however samples 27 and 28 (Table IV) contain anomalous amounts of Pd and Pt.

A few sulphide-rich samples at Redstone were analyzed for As and CaO, which are concentrated in gersdorffite and carbonate minerals respectively (Table IV). Arsenic was detected in each of the Fe-Ni sulphide zones A to D and in iron-formation. Twelve of thirteen As-bearing samples are also enriched in Pd and/or Pt. Similarly, carbonate-rich Fe-Ni sulphide samples (Table IV, nos. 5,6,11) are anomalously high in Pd, and Au is also enriched in sample 11.

ROCK ALTERATION

Simple inspection of major and trace element data is unsatisfactory for evaluating chemical changes and element mobility during alteration. Therefore rock alter-

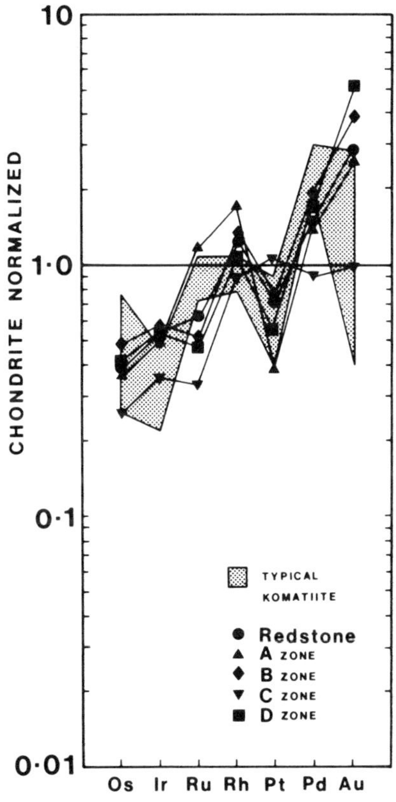

Figure 11. Log plot of PGE and Au for the Redstone deposit compared wtih typical komatiitic-hosted deposits. The field for typical komatiite is from Naldrett *et al.* (1980). Redstone samples are grouped according to the sulphide mineral assemblage, zones A to D as outlined in Figure 9. The average concentration for Redstone is shown as a solid circle. Redstone analyses (Table IV), have been recalculated to a constant nickel content (14%) and normalized by dividing chondritic abundances.

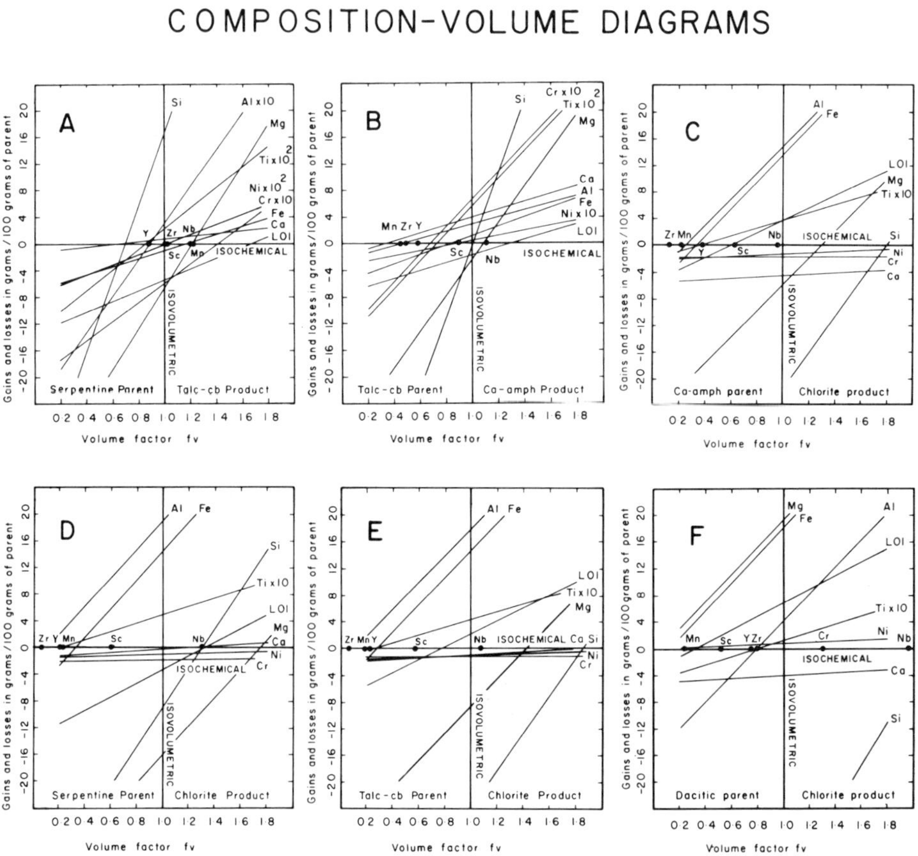

Figure 12. Composition-volume diagrams (after Gresens, 1967). The geochemical-volume relationships are illustrated for the transformation of an 'unaltered' parental rock to an altered product; for the komatiites, contact metasomatic rocks and dacitic tuffs at Redstone. Al, Ti, Y and Zr are regarded as the least mobile elements and therefore represent the best estimate for $\triangle$V for given alteration reactions. A1, Ti, Ni and Cr are multiplied by 10 or 10^2 in some diagrams to accentuate their intersections on the isochemical axis. Volume factors of Y, Zr, Sc, Nb and Mn are plotted as points on the isochemical axis. Sample compositions used in mass balance calculations include: serpentinized peridotite – Table I, nos. 1, 2, 3, 4; talc-carbonate rock – nos. 10, 11, 12; actinolite-assemblage – nos. 13 to 17; chlorite-assemblage – nos. 18 to 21; dacitic tuff – nos. 28 to 32.

ation at Redstone was evaluated using mass balance calculations (Gresens, 1967), which incorporate specific gravity, volume change and measured component concentrations as variables. Major and selected trace elements are graphically illustrated on composition-volume diagrams (Fig. 12), which display component gains and losses in grams/100 grams of parent rock as a function of volume change. Any comparison of an altered rock with its unaltered precursor involves a basic assumption that either the volume change or geochemical behaviour of one element can be established. Al, Ti, Zr and Y are widely recognized as relatively immobile during low grade metamorphism (Davies and Whitehead, 1980; Davies *et al.*, 1979; Nesbitt *et al.*, 1979; Floyd and Winchester, 1978; Humphris and Thompson, 1978; Condie *et al.*, 1977; Nesbitt and Sun, 1976; Pearce and Cann, 1973; and others). Therefore calculated volume factors for these elements are assumed to provide the best estimate of $\triangle V$ for given alteration reactions at Redstone.

The profiles of Al, Ti, Ni and Cr are graphically accentuated by a factor of 10 or 10^2 in some diagrams (Fig. 12), to emphasize their magnitudes of change and to accentuate their intersection points on the isochemical axis. Volume factors of Mn, Zr, Y, Nb and Sc are plotted as points along the isochemical axis. The net gains or losses in grams/100 grams of these elements, along with those of Ti, Ni, Cr and others may be quite small and calculated profiles for them therefore are subparallel or overlap the isochemical axis. However, the percentage change in these elements for given alteration reactions may be quite large, therefore placing geochemical constraints on possible precursor rocks.

Element mobility and volume relationships are shown for talc-carbonate alteration of a serpentinized komatiitic peridotite and also for the development of the actinolite- and chlorite-assemblages from the contact metasomatic sequence. Several samples of komatiite, dacitic tuff, actinolite- and chlorite-assemblages were analyzed and compared to establish the average chemical changes during each stage of the progressive alteration.

Talc-Carbonate Alteration

Talc-carbonate alteration of a serpentinized komatiitic peridotite produces extreme variations in SiO_2, MgO and volatile contents for any specified volume, as indicated by the steep slope of the profiles for these components (Fig. 12A). The profiles of Al, Ti, Zr and Y intersect the isochemical axis at or slightly below $fv = 1$. This suggests that the talc-carbonate alteration occurred at approximately constant volume, perhaps under conditions of extensive hydrothermal leaching. The data indicate that large enrichments in SiO_2 and large depletions in MgO and volatile components accompanied this process. Fe is slightly depleted whereas Ca is enriched. Ni content appears to be constant and Cr undergoes depletion during talc-carbonate alteration. Changes in Ba, Sr, Rb, Pb, Zn, P and alkali element content are small and variable during the change from serpentinized rock to talc-carbonate derivatives.

Contact Metasomatism

1) Actinolite-Assemblage. Elemental changes for the transformation of talc-carbonate ultramafic rock into the actinolite rock are represented in Figure 12B. The profiles of Si, Mg and volatiles intersect the isochemical axis at fv ⩾ 1. Profiles and volume factors of Al, Ti, Zr, Y along with Fe, Ca, Mn, Ni and Cr indicate that large volume reductions are required for their isochemical behaviour accompanied with en masse hydrothermal leaching of Si and Mg. Fe, Ni and Cr appear to undergo slight depletion during this reaction whereas Ca is enriched.

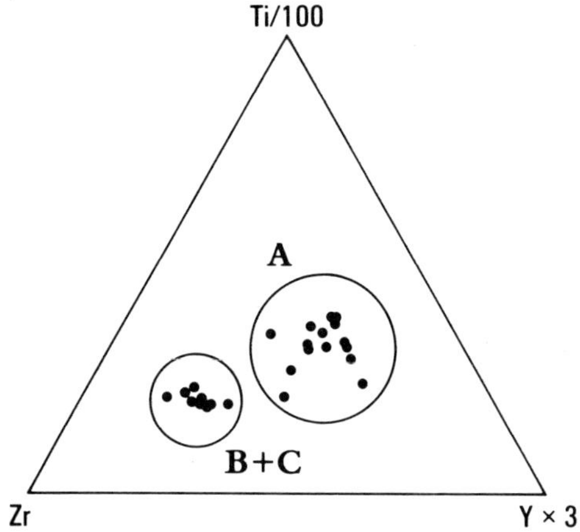

Figure 13. Plot of Zr-Ti/100-Yx3 (after Pearce and Cann, 1973). Redstone komatiitic rocks plot in field A. The chlorite-assemblage and dacitic tuff plot in field B + C. Komatiitic rocks have low concentrations of Zr, Ti and Y, with a large scatter, whereas the chlorite-assemblage and dacitic tuff have high and comparable concentrations of these elements, with a smaller dispersion of samples.

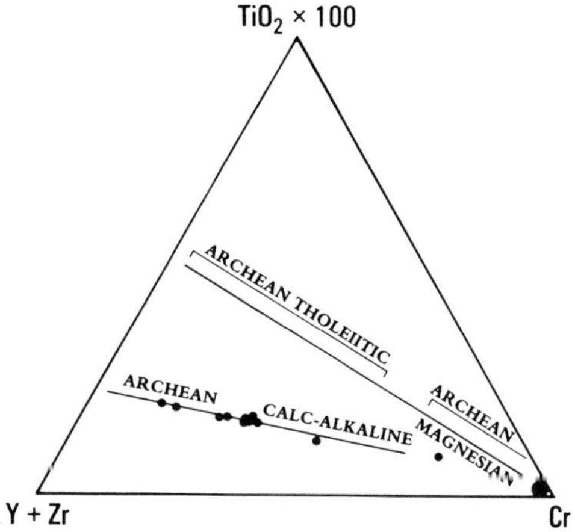

Figure 14. Plot of $Y+Zr–TiO_2\times100–Cr$ (after Davies *et al.*, 1979). Redstone komatiitic rocks plot at the Cr apex along the trend for Archean tholeiitic and magnesian rocks; footwall dacitic tuff, garnet-stilpnomelane rocks and the chlorite-assemblage plot along the trend for calc-alkaline rocks.

2) Chlorite-Assemblage. The chlorite-assemblage is compared to the actinolite-assemblage, serpentinized- and talc-carbonate komatiitic rock, and to dacitic tuff in figures 12C to 12F respectively. The transformation to chlorite rock from a komatiitic parent (Figs. 12C to 12E), would require an extreme volume reduction of 80 to 90 per cent, assuming Al, Ti, Zr and Y immobility, accompanied by extensive hydrothermal leaching of Si and Mg with lesser depletions of Ni and Cr. Ca and volatile components are also depleted.

The chemical changes required during generation of chlorite rock from dacitic tuff are illustrated in Figure 12F. The profiles of Al, Ti, Zr and Y cluster on the isochemical axis at $fv = 0.8$. Fe and Mg are strongly enriched and Si depleted, for a constant content of Al, Ti, Zr and Y. Ca and Cr are depleted and Ni slightly enriched for this transformation. Absolute contents of Al, Ti, Mn, P, Y, Zr, Ni, Cr, Ba, Pb, Co and Zn are similar in both dacitic tuff and the chlorite-assemblage.

Redstone host rocks fall into two fields on a ternary trace element diagram Zr – Ti/100 – Y.3 (Fig. 13). Field A represents komatiitic rocks including serpentine-, talc-carbonate- and actinolite-bearing varieties. Fields B + C encompasses calc-alkaline rocks, including dacitic tuff and the chlorite-assemblage from the contact metasomatic sequence. Dacitic tuff and the chlorite-assemblage have comparable and high contents of Zr, TiO_2 and Y.

A discriminatory diagram Y + Zr – TiO_2.100 – Cr distinguishes between Redstone komatiitic rocks and dacitic tuffs, chlorite-assemblage of the metasomatic sequence and garnet-stilpnomelane rock (Fig. 14). Redstone komatiitic rocks are restricted to the Cr apex of the diagram, along the trend established for Archean magnesian and tholeiitic rocks (cf. Davies *et al.*, 1979). Dacitic tuff, garnet-stilpnomelane rock and the chlorite-assemblage plot along the calc-alkaline trend.

DISCUSSION

Contact Metasomatism

The metasomatic alteration zones along the contacts between komatiite and dacitic tuff or metabasalt are spatially associated with Fe-Ni sulphide bodies at Redstone and Kambalda, Western Australia (Marston and Kay, 1980; Barrett *et al.*, 1977). Similar metasomatic zones not associated with Fe-Ni sulphide concentrations are known in Washington (Frost, 1975), Scotland (Curtis and Brown, 1971, 1969) and in the Appalachians (Phillips and Hess, 1936). Contact metasomatism and Fe-Ni sulphide deposition are therefore not necessarily genetically related, however the alteration processes that formed the chlorite-assemblage as indicated from the chemical data at Redstone are significant to, and place some constraints on possible mechanisms of sulphide emplacement.

Frost (1975) and Curtis and Brown (1971, 1969) suggested that a similar contact metasomatic assemblage represents progressive alteration of ultramafic rock, and that the original ultramafic/adjacent rock contact was located at the chlorite/adjacent rock interface. Phillips and Hess (1936) proposed that both chlorite and biotite assemblages represent altered felsic rock. They suggested that the original ultramafic/adjacent rock interface was situated at the chlorite/actinolite assemblage boundary (Fig. 7). This interpretation is in better agreement with the mass balance calculations and trace element data at Redstone. Partial chloritic and biotitic alteration of dacitic tuff is also indicative of a felsic precursor for the chlorite and biotite assemblages, whereas the progressive increase of talc followed by actinolite at the expense of serpentine in the komatiites supports an ultramafic parent for these assemblages. Dacitic tuff has been subjected to extreme Mg metasomatism during serpentinization of olivine-rich rocks, similar to that described by Turner and Verhoogen (1960), resulting in transformation into massive chloritic rock. Phillips and Hess (1936) also observed that the chemical changes during alteration of adjacent felsic volcanic and ultramafic rocks counterbalanced one another in respective gains and losses from each.

An alternative possibility is that the contact metasomatic sequence represents a primary magmatic chemical variation. However, the progressive sequence of the contact assemblages, their spatial position with respect to the komatiites on footwall, internal and hangingwall contacts, and the progressive mineralogical replacement in komatiites and dacitic tuff toward the contact all strongly support a metasomatic origin. The chemical and mineralogical variations within the komatiitic suite from serpentine, through talc-carbonate to actinolite rocks reflect in part magmatic differentiation which has been subsequently metasomatized.

At Redstone massive Fe-Ni sulphide is capped by the chlorite-assemblage along the footwall of the komatiites and underlain by the chlorite-assemblage on hanging wall contacts. In other words, the chlorite-assemblage separates the massive Fe-Ni sulphides from altered komatiite (Fig. 7). Sulphide ores from Kambalda, W. Australia occur in metasomatic assemblages that developed during metamorphism between mineralized ultramafic rocks and chemically contrasting rocks such as footwall metabasalt and/or metasedimentary rocks (Barrett *et al.*, 1977). Marston and Kay (1980) suggested that a thick sulphide layer at an ultramafic/adjacent rock contact would form a chemical barrier to reaction and diffusion of elements between the contrasting silicate rocks, thereby explaining local absence of the metasomatic assemblages in the underlying footwall rocks. This suggestion appears to apply well at Redstone where it explains the presence of non-metasomatized dacitic tuff intercalated with the sulphide layer, and unaltered dacitic tuff beneath the sulphide layer (Fig. 7). On this assumption however, the R zone at Redstone must have been overlain by a thin sequence of dacitic tuff prior to emplacement of the overlying komatiitic lavas, thereby ruling out the possibility of gravitational settling of immiscible sulphide droplets in situ from an ultramafic melt as the genetic process for the R zone.

Genetic Hypotheses

Four major hypotheses have been proposed to explain the generation of volcanic-hosted nickel sulphide deposits: 1) magmatic-gravitational settling (Marston and Kay, 1980; Groves *et al.*, 1979; Ross and Hopkins, 1975; Naldrett, 1973; Ewers and Hudson, 1972; and others); 2) contact metasomatic replacement (Fleet, 1979, 1977); 3) metamorphically/tectonically modified magmatic process (Barrett *et al.*, 1977; Groves *et al.*, 1976); and 4) volcanic-exhalative (Lusk, 1976a, 1976b).

1) Magmatic-Gravitational Settling. The universal occurrence of nickel sulphide accumulations at the basal contacts of ultramafic rocks is the broad basis for the

PLATE I

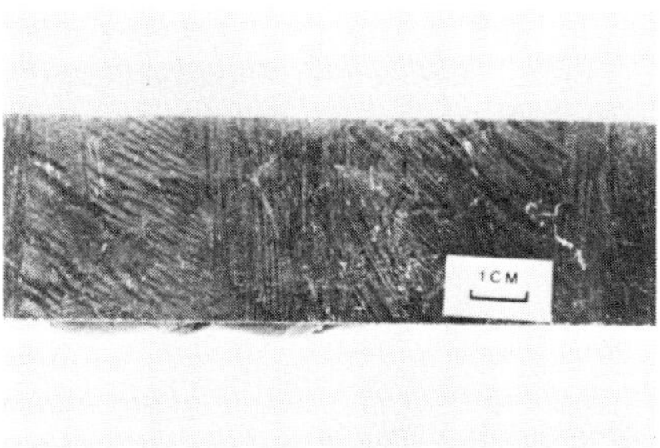

Figure 15

Figure 16

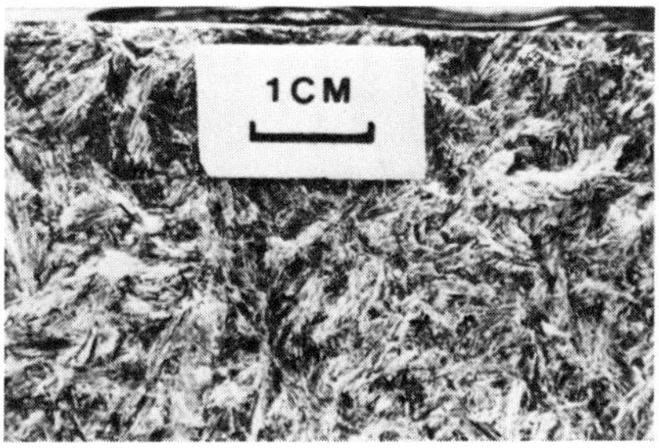

Figure 17

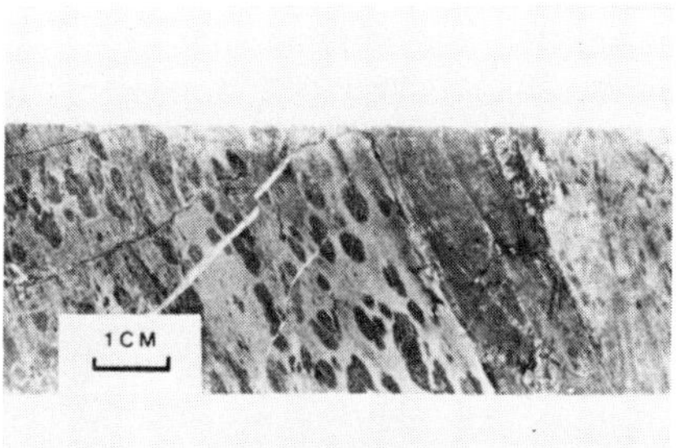

Figure 18

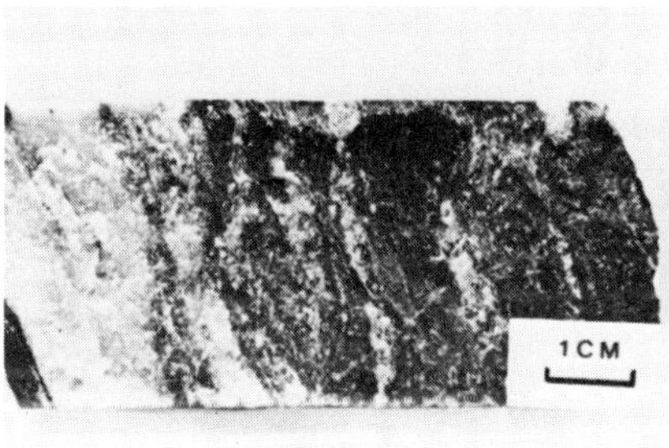

Figure 19

Figure 20

EXPLANATION OF PLATE I

Figure 15. Sheaf spinifex textured komatiitic flow of unit 3.

Figure 16. Talc-carbonate altered komatiitic flow of unit 2B, with elliptical vesicles.

Figure 17. Coarse-grained actinolite from the contact metasomatic sequence.

Figure 18. Spotted dacitic tuff of unit 1. Spots are 2 mm to 1 cm diam., appear to coalesce in some examples; spotted band is laterally continuous for tens of metres.

Figure 19. Footwall dacitic tuff of unit 1, with banded carbonate layers; overlain by massive Fe-Ni sulphide.

Figure 20. Footwall dacitic tuff of unit 1, with a network of quartz-carbonate veinlets.

magmatic hypothesis. Gravitational settling of a homogeneous immiscible sulphide liquid from an ultramafic melt in situ has been invoked to explain massive, net-textured and disseminated sulphide zones concentrated toward and at the base of the ultramafic host rock (Naldrett, 1973). Other detailed evidence considered conclusive of magmatic origin for most deposits (Naldrett, 1979; Groves *et al.*, 1979), includes lack of significant footwall rock alteration, consistent Ni/Cu ratios, abundant platinum group elements, relatively low Pd/Ir ratios, dissimilar geochemistry of massive ores and along-strike exhalative sulphidic metasedimentary rocks, absence of recognized hydrothermally concentrated elements such as Zn, and the presence of associated ferrochromite minerals.

Other magmatic hypotheses involve minor variations in the *timing* of separation of immiscible sulphide liquid from ultramafic magma. Ross and Hopkins (1975) attributed massive and disseminated nickel sulphide deposition to alternating pulses of sulphide melt and ultramafic magma. They suggested that segregation of molten sulphide, olivine crystals and a picritic liquid during vertical upwelling of magma would result in early eruption of sulphide melt, late eruption of olivine-rich liquid and intervening extrusion of silicate liquid, olivine crystals and sulphide droplets. Initial eruption of the sulphide melt followed later by the ultramafic flows would allow time for the massive sulphide to develop an upper chilled surface before superposition of the silicate liquid.

In slight modifications of the Ross and Hopkins model, Marston and Kay (1980) and Groves *et al.* (1979) suggested that olivine-charged komatiitic magma and sulphide liquid erupted *simultaneously* as a single flow. During eruption, gravitational

---

EXPLANATION OF PLATE II

Figure 21. Sulphide-facies iron-formation, massive pyrrhotite, minor porphyroblastic pyrite and sigmoid-shaped chert slump structure.

Figure 22. Cherty iron-formation with pyrite bands and veinlets.

Figure 23. Carbonate-hosted massive Fe-Ni sulphide containing an S-shaped quartz-carbonate-sulphide band; disseminated quartz and carbonate within a pentlandite + pyrite matrix, sample no. 11, Table IV.

Figure 24. Dacitic tuff of unit 1 with quartz-carbonate bands and pentlandite.

Figure 25. Matrix-textured sulphide, pentlandite + pyrite + millerite within an actinolite matrix, sample no. 18, Table IV.

Figure 26. Xenolithic massive sulphide, with chlorite, actinolite and talc inclusions within a pentlandite + pyrite matrix, sample no. 10, Table IV.

Figure 27. Primary sulphide banding; 1 cm pyrite (py) band within massive pentlandite (pn) + pyrite matrix, xenoliths of chlorite, actinolite and talc.

Figure 28. Metamorphic mineral banding of massive pentlandite + pyrite + pyrrhotite. Primary alternating layers of pentlandite and pyrrhotite have recrystallized during regional low grade metamorphism, sample no. 1, Table IV.

PLATE II

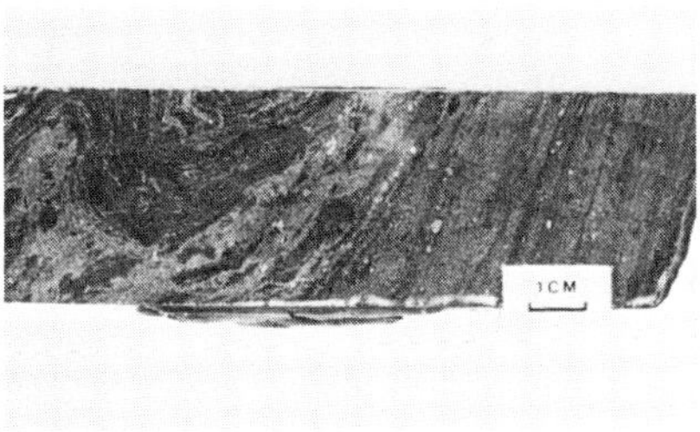

Figure 21

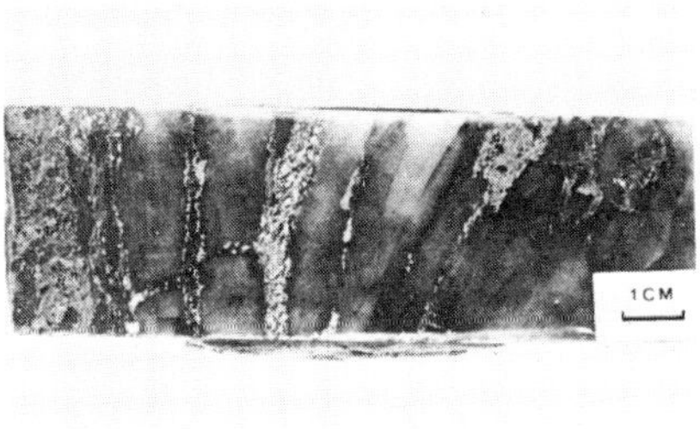

Figure 22

Figure 23

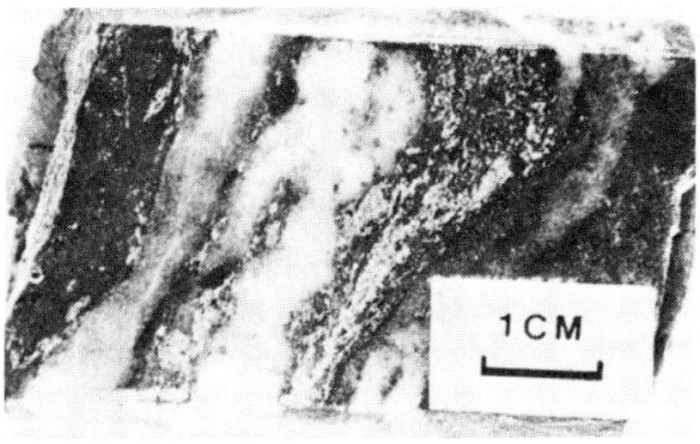

Figure 24

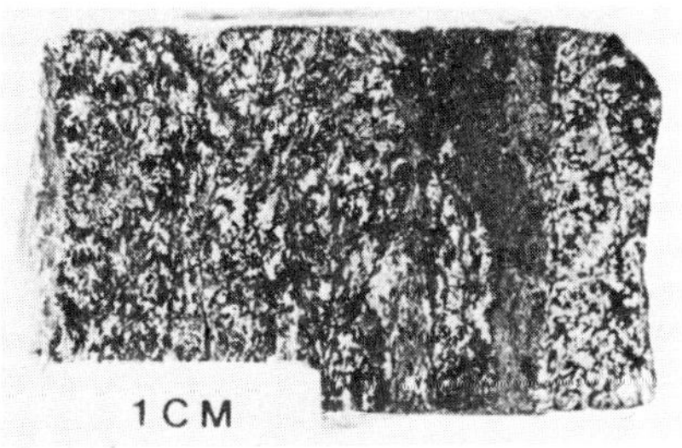

Figure 25

Figure 26

Figure 27

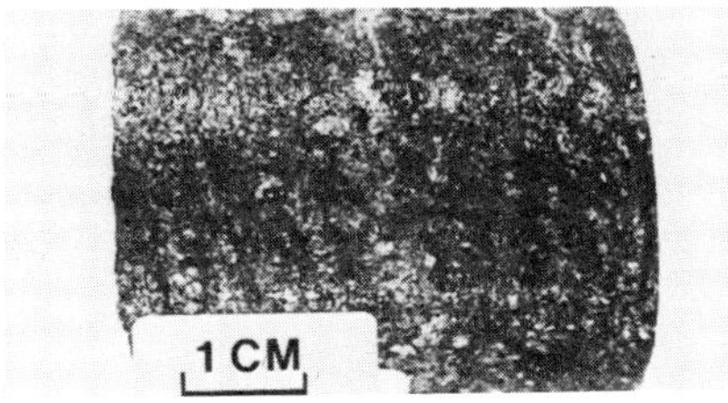

Figure 28

separation of a low viscosity sulphide liquid occurred, and this liquid was extruded before more viscous, olivine-crystal-rich komatiitic magma because of viscosity and density contrasts and differential flow rates. Partial recrystallization of the sulphide horizon was again invoked prior to extrusion of the ultramafic lavas to prevent any displacement or reincorporation of the sulphides into the overlying olivine-crystal-rich lava. Marston and Kay also reported the presence of interpillow, primary massive Fe-Ni sulphide in footwall metabasalt at the Juan shoot, which implies that the eruption of these two was simultaneous and preceded komatiitic volcanism.

Applying the magmatic hypothesis to the R zone at Redstone, the absence of a laterally continuous disseminated sulphide zone at the base of the komatiite requires rapid settling and nearly complete separation of immiscible sulphide blebs within the cooling komatiitic magma, thus avoiding entrapment of intercumulate nickel sulphides to form disseminated ore. But deposition of nickel sulphide liquid into a paleogeographic depression, accompanied by minor accumulation of volcaniclastic sediment must have occurred *prior* to eruption of komatiitic lava, in order to explain the observed intercalation of dacitic tuff and R zone sulphides, and the subsequent development of the metasomatic cap rock.

Mineralogical and chemical variations in Fe-Ni sulphide assemblages may be attributed to separate or multiple sources of immiscible magmatic sulphide liquids of contrasting composition, to later hydrothermal alteration of primary magmatic sulphides, or alternatively a non-magmatic origin. Marston and Kay (1980) described two separate arcuate belts of contrasting nickel tenor at Juan Shoot, Kambalda. They suggested that these belts represent *separate* sulphidic komatiitic magmas, extruded into original seafloor depressions. They also reported an unexpectedly broad range in Ni content of pentlandite from 26.4 to 28.2 atomic per cent, within an equilibrium assemblage of pentlandite + pyrite + monoclinic pyrrhotite.

EXPLANATION OF PLATE III

Figure 29. Massive Fe-Ni sulphide, subhedral to euhedral pentlandite (pn) and subhedral to anhedral pyrrhotite (po).

Figure 30. Exsolution lamellae of pentlandite (pn, white) in pyrrhotite (po, grey).

Figure 31. Subhedral pentlandite octahedra (pn) replaced by violarite rims (viol).

Figure 32. Euhedral pyrite cubes (py) mantled by anhedral chalcopyrite rims (cpy) in a pentlandite matrix (pn). Euhedral actinolite megacrysts form the silicate gangue.

Figure 33. Matrix-textured sulphides. Euhedral actinolite megacrysts within and crosscut by pentlandite (pn).

Figure 34. Disseminated pentlandite in talc-carbonate matrix. Note carbonate rims around sulphide blebs.

Figure 35. Disseminated pentlandite (pn), gersdorffite (gf) and carbonate within dacitic tuff.

Figure 36. Xenolithic massive sulphide. Chlorite, talc and actinolite fragments are S-shaped due to plastic deformation.

PLATE III

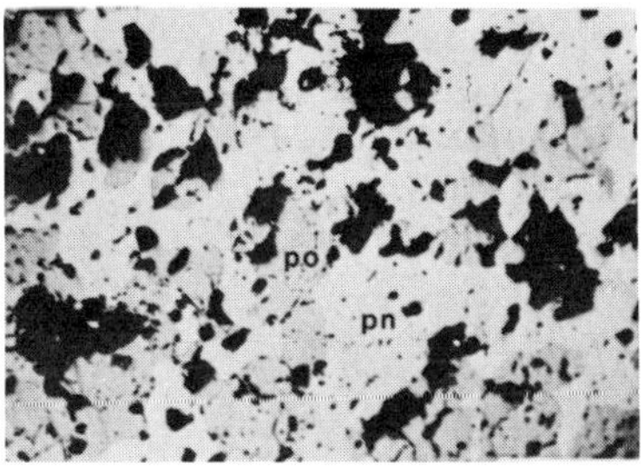

Figure 29

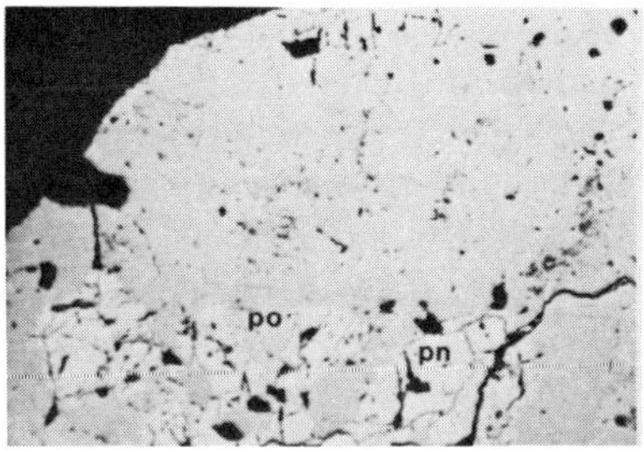

Figure 30

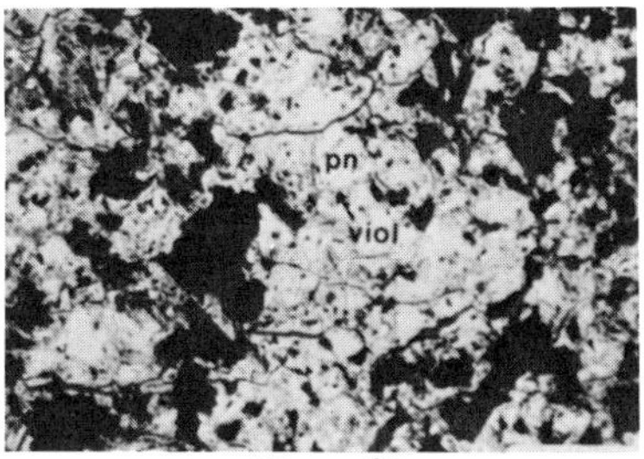

Figure 31

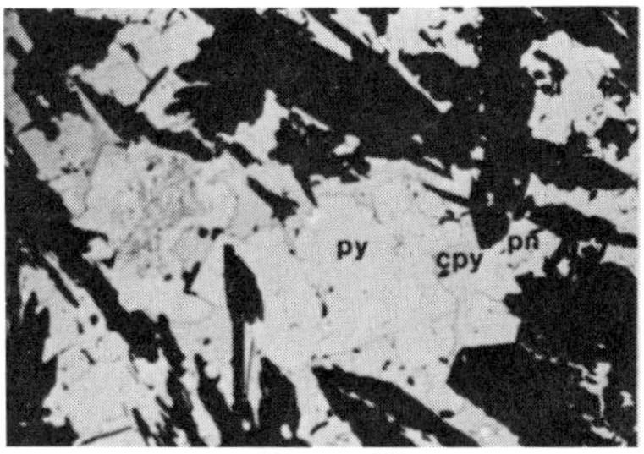

Figure 32

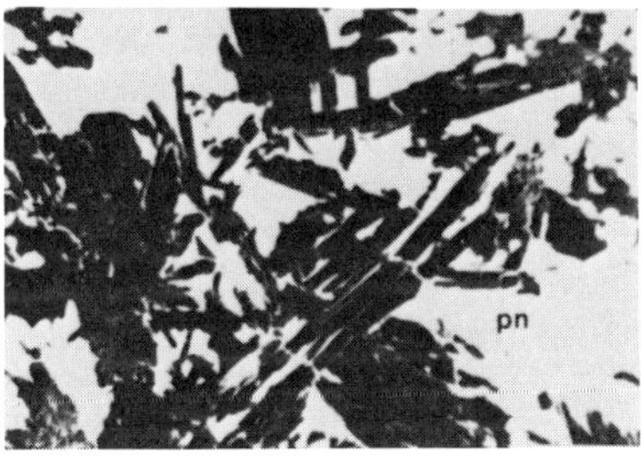

Figure 33

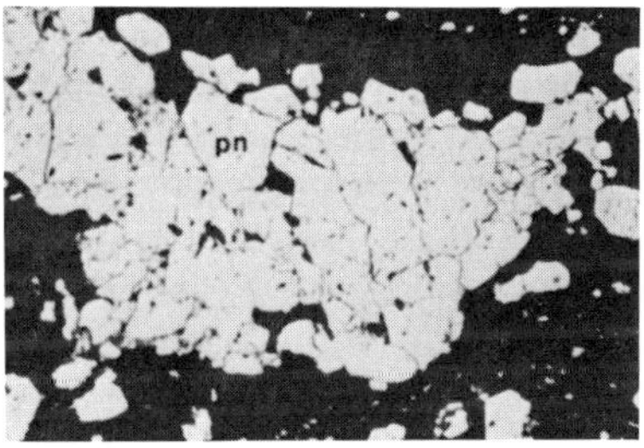

Figure 34

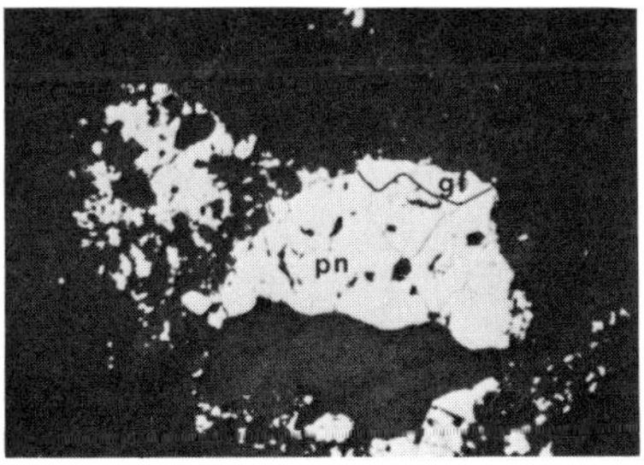

Figure 35

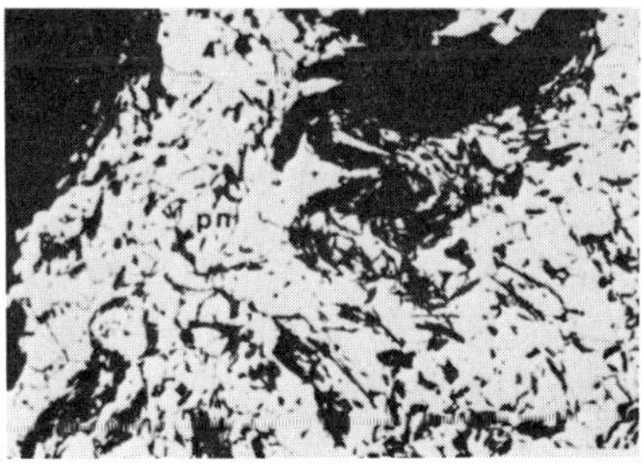

Figure 36

0·5 mm

Hydrothermal processes have been proposed as an explanation for the variation in nickel sulphide composition within a single deposit by Naldrett (1979) and Fleet (1979). However the relative timing of these hydrothermal processes with respect to nickel sulphide deposition is disputed. Four nickel sulphide deposits at Marbridge, Quebec exhibited a wide range in Ni/(Ni+Fe) ratio (Naldrett, 1979; Graterol and Naldrett, 1971; Buchan and Blowes, 1968). Naldrett (1979) suggested that the Marbridge ores were not simple magmatic accumulations but, rather, original magmatic deposits later modified and upgraded by hydrothermal activity during serpentinization according to the model proposed by Eckstrand (1975). However, Eckstrand also pointed out that massive sulphide assemblages, like those at Redstone, are self-buffered and therefore not greatly affected by alteration reactions. Furthermore, in an evaluation of platinum group elements, Ross and Keays (1979) reported that the sulphides within the Kambalda deposits have experienced very limited modifications across a wide range of metamorphic settings and associated ultramafic rock alteration assemblages. Therefore, the heterogeneous distribution of PGE and Au combined with their poor correlation with Ni or Cu, broad range in pentlandite compositions and diverse nickel sulphide mineralogy at Redstone may reflect an original geochemical signature which has not been subsequently modified during alteration of the enclosing host rocks.

2) Contact Metasomatic Replacement. Fleet (1979, 1977), suggested that the heterogeneous proportions of pentlandite to chalcopyrite at Frood, and inclusions of silicate minerals within sulphide blebs are inconsistent with direct formation from immiscible sulphide droplets but indicative of an epigenetic hydrothermal replacement process. At Redstone contact metasomatic replacement involving assimilation of pre-existing iron-formation must be considered as a possible ore-forming mechanism. The sulphide-facies iron-formation, although semi-continuous for considerable distances along or near the same stratigraphic horizon as the Fe-Ni sulphide layer, is conspicuously absent across a large interval of the R zone, whereas garnet-stilpnomelane rock underlies both sulphide assemblages. Sulphide-facies iron-formations are also associated with many Fe-Ni sulphide deposits in Canada, at Langmuir, Hart, McWatters and Redstone, and in W. Australia at Windarra. Furthermore, the juxtaposition of the contact metasomatic sequence at Redstone, on footwall and hanging wall contacts of the komatiites and coincident with Fe-Ni sulphides may indicate a replacement process.

However, en masse metasomatic replacement does not adequately explain: 1) the presence of massive Fe-Ni sulphides at Redstone, hosted within dacitic tuff and not spatially associated with the contact metasomatic sequence; 2) spatial overlap of sulphide-facies iron-formation with the R zone and the gradational relationship with the S zone; 3) the presence of carbonate-hosted disseminated Fe-Ni sulphide blebs within dacitic tuff above and below the R zone; 4) Fe-Ni sulphide deposits not associated with sulphide-facies iron-formation and 5) contact metasomatic sequences associated with ultramafic rocks which do not contain Fe-Ni sulphides.

3) Metamorphically Modified Magmatic Process. Groves *et al.* (1979) and Barrett *et al.* (1977) stressed the roles of structural deformation and metamorphic modification of pre-existing magmatic accumulations in the deposits of Western Australia. They proposed a possible mechanism of upgrading disseminated Fe-Ni sulphides into matrix and massive ore by amphibolite facies metamorphic processes. Further, the formation of metamorphic layering, tectonic fabrics, sheared contacts, breccia ore, variable Fe/Ni ratios, offset massive Fe-Ni sulphides, and remobilization of massive sulphides were all attributed to post-magmatic structural and metamorphic phenomena.

Fe-Ni sulphide assemblages and enclosing rocks at Redstone and elsewhere in the same region have been structurally and metamorphically modified, but only at greenschist rank (Green, 1978; Coad, 1977; Pyke, 1975; this study), a much less intense metamorphism than that proposed for the Australian deposits. The proportion and distribution of komatiite-hosted disseminated Fe-Ni sulphides at Redstone are small and erratic, and not compatible with metamorphic upgrading. Xenoliths of silicate rocks have clearly been incorporated into massive sulphide from adjacent metasomatic assemblages and dacitic tuff, presumably as a consequence of local sulphide remobilization and plastic deformation. Yet metasomatic xenoliths are numerous where the wall rocks are komatiitic, but rare or absent along strike or down-dip where the sulphides are hosted in dacitic tuff. This suggests that remobilization of sulphides was very local and that the western section of the R zone (Fig. 4) was not in contact with the komatiitic rocks. Local folding and faulting has been proposed to explain the interfingering of dacitic tuff and komatiitic peridotite (Figs. 4,5; Coad, 1980, pers. commun.) as discussed above. However, the dacitic and komatiitic rocks are interfingered and intercalated on both regional and local scales and the overlying rocks, including sulphide-facies iron-formation are stratiform and not isoclinally folded.

A metamorphically modified magmatic hypothesis does not account for: 1) the juxtaposition of the contact metasomatic sequence which caps the R zone and implies that the Fe-Ni sulphides were emplaced prior to the komatiitic lavas; 2) the presence of carbonate-hosted Fe-Ni sulphides within dacitic tuff; 3) the heterogeneous distribution of PGE, Au, Ni and Cu; 4) interpillow Fe-Ni sulphide rims at Juan and Ken Shoots, Kambalda and 5) the stratabound nature of the R zone and sulphide-facies iron-formation which are concordant with their enclosing host rocks.

4) Volcanogenic-Exhalative Hypothesis. Evidence of exhalative activity at many massive nickel sulphide deposits are the associated iron-formations and intercalated sulphidic metasedimentary rocks, and exhalative activity was suggested as an alternative ore-forming process (Lusk, 1976a). The following relationships at Redstone and elsewhere, some of which have been shown to be inconsistent with magmatic processes involving gravitational settling, may be more adequately explained by exhalation of volcanogenic hydrothermal nickel-bearing solutions onto the sea floor accompanied by deposition of chemical and volcaniclastic sediment: 1) nickel mineral and metal zoning within a single massive sulphide layer; 2) juxtaposition of the chlorite assemblage which caps the R zone; 3) non-metasomatized dacitic tuff intercalated with, and enclosing the R zone; 4) the concordant and stratabound distribution of Fe-Ni sulphides which are stratigraphically controlled both locally at Redstone and regionally as in other deposits around the Shaw dome, and coincident with sulphide-facies iron-formation; 5) the spatial association of Fe-Ni sulphides with sulphide-facies iron-formation, intercalated carbonate and chert; 6) massive Fe-Ni sulphides hosted within dacitic tuff or altered equivalents; 7) carbonate-hosted disseminated Fe-Ni sulphide blebs in dacitic tuff above and below the R zone; 8) rare and discontinuous disseminated Fe-Ni sulphide blebs in the base of komatiitic flows; 9) pyrite layering within Fe-Ni sulphide; 10) arsenic-rich nickel minerals such as gersdorffite and niccolite; 11) interpillow Fe-Ni sulphide rims within footwall metabasalt at Juan and Ken Shoots, Kambalda; and 12) the presence of nickel-bearing sulphides not associated with komatiitic rocks such as those at Key Lake, Saskatchewan, King Island, Tasmania and the Kuroko deposits of Japan.

Nickel sulphide deposits of this type are commonly spatially associated with chemical sedimentary rocks, particularly banded iron-formation. Nickel sulphides occur atop, within and below chemical sedimentary strata in Western Australia at Windarra and Sherlock Bay (Groves *et al.*, 1979; Barrett *et al.*, 1977; Miller and Smith, 1975); in Rhodesia (Williams, 1979); and at the Langmuir, McWatters, Hart, Dundonald and Redstone deposits in Canada (Muir and Comba, 1979; Green, 1978; Coad, 1977; Pyke, 1975; Pyke and Middleton, 1971; this study).

Chemical sedimentary strata in Western Australia are markedly enriched in Cu and Zn relative to Ni in comparison to Ni ores (Groves *et al.*, 1979), and were therefore considered unrelated genetically. Naldrett (1979) concluded that the paucity of Zn in Ni-Cu ores, coupled with its presence in surrounding host rocks, is indicative of a magmatic origin for these ores. Considering the well known and pronounced lateral and vertical metal zoning in exhalative deposits (Sangster and Scott, 1976; Hutchinson, 1973), this argument may not be well founded. Moreover, chemical sedimentary rocks and Fe-Ni sulphides from Redstone and Windarra (Groves *et al.*, 1979) are both depleted in Zn and therefore such variations in Zn are not a reliable or distinctive genetic criterion between exhalative and magmatic deposits. Anomalous zoning of Mn, in garnet from the garnet-stilpnomelane rock and in carbonate from the nickeliferous sulphide body itself, are strongly indicative of exhalative environments where they are common and distinctive (Barnett *et al.*, this volume; Hamilton *et al.*, this volume; Hutchinson, 1979; Stanton and Vaughan, 1979; Bischoff, 1969).

At Redstone, as in many metamorphosed massive base metal sulphide bodies, distinction of primary compositional layering due to bedding from metamorphic banding due to later plastic deformation and flowage of the sulphides (Buerger, 1928), is difficult. In general however, the former tends to be more regular, planar and continuous, whereas the latter is thin, irregular and discontinuous. Both are commonly present in the same body due to metamorphic overprint of the latter on the former and in view of this it is perhaps simplistic and incorrect to ascribe all observed layering to one or the other of these processes. In the case of massive Fe-Ni sulphide bodies, Lusk (1976a) considered the relative abundance of pyrite and millerite together with prominent mineralogical layering to be indicative of volcanic exhalative processes. Conversely Groves *et al.* (1976) suggested that the presence of millerite and other relatively sulphur-rich nickel minerals in disseminated ores was due to oxidation of pre-existing Fe-Ni sulphides, and that sulphide banding was a metamorphic phenomenon. At Redstone, regular, pyritic layering due to primary bedding *and* metamorphic banding due to plastic deformation, flowage and/or recrystallization of the sulphides, with alternating pentlandite- and pyrrhotite-rich bands, are *both* present and are distinguishable from one another (Plate II, Figs. 27,28).

Moreover, concordant accumulations of massive and disseminated iron-nickel sulphides and discordant fracture-filling sulphides occur within altered dacitic tuff, and are underlain by the chlorite-assemblage along the *hanging wall* contact of the komatiitic flows. These sulphide concentrations are incompatible with gravitational sulphide settling in situ, or during komatiitic emplacement.

Mineral and metal zoning in nickel sulphide deposits place constraints on their genesis. Pentlandite compositions display a sympathetic variation in Fe-Ni ratio with, and dependant upon, associated sulphide minerals. Graterol and Naldrett (1971) and Misra and Fleet (1973) showed that pentlandite equilibrates with its associated minerals following breakdown of the monosulphide solid solution below 250°C. Pentlandite at Redstone has equilibrated with its corresponding associated sulphides to form pentlandite of low Ni tenor with pyrrhotite in some places, but pentlandite of high Ni tenor with millerite elsewhere within the same horizon. Variation in mineralogical and chemical composition such as these at Redstone are inconsistent with separation of homogeneous immiscible sulphide liquid from a komatiitic magma. As Fleet (1977) pointed out, an important corollary of the magmatic hypothesis is that at the time of initial sulphide crystallization as a high temperature monosulphide solid solution, sulphide blebs from the same portion of an orebody should all have the same composition.

Finally, mineralogical comparison between Redstone and certain other nickeliferous deposits may be genetically significant, and in any case clearly indicates that not all nickel sulphides are of magmatic or magmatic/metamorphic origin. The Key Lake uranium-nickel deposits contain up to 20 per cent Ni, in the form of nickel arsenides and sulphides and are considered to have formed by non-magmatic processes (Gatzweiler *et al.*, 1979). Fe-Ni sulphides at Redstone contain a similar assemblage of millerite, gersdorffite and rare niccolite. At Black Swan, Western Australia, Groves *et al.* (1974) and Groves and Keays (1979) suggested that As was added during carbonate alteration of the ultramafic rocks, thereby explaining the presence of gersdorffite. Alternatively however, the As-rich mineral assemblage at Redstone might have resulted from primary carbonate-rich exhalations that preceded the eruptions of the komatiitic flows. Millerite and vaesite are reported as accessory minerals within the Kuroko deposits of Japan (Lambert and Sato, 1974), which are widely accepted as of volcanogenic exhalative origin, and pentlandite is abundant in the contact metasomatic King Island scheelite deposit in Tasmania (Burchard, 1977). Gaal *et al.* (1975) reported a polymetallic sulphide deposit at Outokumpu, Finland, which contained 28 million metric tons of 3.5 to 4.0 per cent Cu, 1 per cent Zn, 0.02 to 0.5 per cent Ni and small amounts of Co, Ag, Au and Se, hosted within quartzite, skarn, serpentinite and black schist. This metal assemblage and host rocks, along with previously cited examples are all inconsistent with the magmatic hypothesis, and indicative of aqueous transport of nickel. Some of them are best explained by a volcanic-exhalative hypothesis.

CONCLUSIONS

Contact metasomatic rocks composed of chlorite, biotite, actinolite or talc are developed along komatiite/dacitic tuff contacts. These rocks represent progressive stages of alteration of both the dacitic and komatiitic volcanic rocks by extensive, probably deuteric seafloor, hydrothermal leaching. This resulted in extensive exchange of components across the contact, including leaching of SiO_2 and precipitation of MgO and FeO in the dacitic rock. Nickel sulphide mineral and metal zoning within one sulphide layer together with the juxtaposition of the contact metasomatic assemblages, massive Fe-Ni sulphides atop komatiitic flows, and rare and discontinuous disseminated Fe-Ni sulphides in the base of komatiitic flows at Redstone are inconsistent with gravitational settling of a homogeneous immiscible sulphide liquid from a komatiitic magma. The concordant stratabound distribution of the sulphide layer, spatial association with sulphide-facies iron-formation and intercalated carbonates, chert and dacitic tuff in the Fe-Ni sulphide layer, in addition to carbonate-hosted disseminated Fe-Ni sulphide blebs in dacitic tuff are more compatible with a volcanogenic-exhalative mode of origin.

ACKNOWLEDGEMENTS

The assistance of Utah Mines Ltd. is gratefully acknowledged, in particular that of Dr. D. G. Cargill, Mr. L. Godbout and Mr. M. Young for providing the opportunity to examine the Redstone deposit, for financial assistance for rock geochemistry and for permission to publish the results of this investigation. Thanks are also extended to Mr. R. L. Barnett for his assistance with the electron microprobe. Drs. R. W. Hodder, R. Kerrich, M. E. Fleet and Mr. Gregg Morrison provided constructive criticisms of the manuscript and helpful suggestions during the research. The study will be the subject of a Ph.D. dissertation at the University of Western Ontario by D. J. R. who was supported by an Ontario Geological Survey research grant. Financial assistance has also been provided to R. W. H. from a Natural Sciences and Engineering Research Council of Canada research grant.

REFERENCES

Arndt, N.T., 1976, Ultramafic lavas in Munro township: economic and tectonic implications: in Strong, D.F., ed., Metallogeny and Plate Tectonics: Geological Association of Canada, Special Paper 14, p. 617-658.

Arndt, N.T., Naldrett, A.J. and Pyke, D.R., 1977, Komatiitic and iron-rich tholeiitic lavas of Munro township, Northeast Ontario: Journal of Petrology, v. 18, p. 319-369.

Ayres, D.E., 1972, Genesis of iron-bearing minerals in banded iron formation mesobands in the Dales Gorge Member, Hammersly Group, Western Australia: Economic Geology, v. 67, p. 1214-1233.

Barrett, F.M., Binns, R.A., Groves, D.I., Marston, R.J. and McQueen, K.G., 1977, Structural history and metamorphic modification of Archean volcanic-type nickel deposits, Yilgarn Block, Western Australia: Economic Geology, v. 72, p. 1195-1224.

Bischoff, J.L., 1969, Red Sea geothermal brine deposits: their mineralogy, geochemistry and genesis: in Degens, E.T., and Ross, D.A., eds., Hot Brines and Recent Heavy Metal Deposits in the Red Sea: New York, Springer Verlag, p. 368-401.

Buchan, R. and Blowes, J.H., 1968, Geology and mineralogy of a millerite nickel ore deposit, Marbridge No. 2 Mine, Malartic, Quebec: Canadian Institute of Mining and Metallurgy Bulletin, v. 61, p. 529-534.

Buerger, M.J., 1928, The plastic deformation of ore minerals: American Mineralogist, v. 13, p. 1-17, 35-51.

Burchard, U., 1977, Genesis of the King Island (Tasmania) scheelite mine: in Klemm, D.D. and Schneider, H.-J., eds., Time- and Stratabound Deposits: New York, Springer Verlag, p. 199-204.

Button, A., 1976, Transvaal and Hamersley basins – Review of basin development and mineral deposits: Minerals Science Engineering, v. 8, no. 4, p. 262-293.

Coad, P.R., 1977, Nickel Sulphide Deposits Associated with Ultramafic Rocks of the Abitibi Belt and Economic Potential of Mafic-Ultramafic Intrusions: Ontario Geological Survey Open File Report 5232, 105 p.

Condie, K.C., Viljoen, M.J. and Kable, E.J.D., 1977, Effects of alteration on element distributions in Archean tholeiites from the Barberton Greenstone Belt, South Africa: Contributions to Mineraology and Petrology, v. 64, p. 75-89.

Curtis, C.D. and Brown, P.E., 1969, The metasomatic development of zoned ultrabasic bodies in Unst, Shetland: Contributions to Mineralogy and Petrology, v. 24, p. 275-292.

______________, 1971, Trace element behaviour in the zoned metasomatic bodies of Unst, Shetland, Contributions to Mineralogy and Petrology, v. 31, p. 87-93.

Davies, J.F. and Whitehead, R.E.S., 1980, Further immobile element data from altered volcanic rocks, Timmins mining area, Ontario – Note: Canadian Journal of Earth Sciences, v. 17, p. 419-423.

Davies, J.F., Grant, R.W.E. and Whitehead, R.E.S., 1979, Immobile trace elements and Archean volcanic stratigraphy in the Timmins mining area, Ontario: Canadian Journal of Earth Sciences, v. 16, p. 305-311.

Dungan, M.A., 1979, A microprobe study of antigorite and some serpentine pseudomorphs: Canadian Mineralogist, v. 17, pt. 4, p. 771-784.

Eckstrand, O.R., 1975, The Dumont serpentinite: A model for control of nickeliferous opaque mineral assemblages by alteration reactions in ultramafic rocks: Economic Geology, v. 70, p. 183-201.

Evans, B.W., 1977, Metamorphism of Alpine peridotite and serpentinite: Annual Review Earth and Planetary Science, v. 5, p. 397-447.

Ewers, W.E. and Hudson, D.R., 1972, An interpretive study of a nickel-iron sulfide ore intersection, Lunnon Shoot, Kambalda, Western Australia: Economic Geology, v. 67, p. 1075-1092.

Fleet, M.E., 1977, Origin of disseminated copper-nickel sulfide ore at Frood, Sudbury, Ontario: Economic Geology, v. 72, p. 1449-1456.

______________, 1979, Partitioning of Fe, Co, Ni and Cu between sulfide liquid and basaltic melts and the composition of Ni-Cu sulfide deposits – a discussion: Economic Geology, v. 74, p. 1517-1519.

Floran, R.J. and Papike, J.J., 1978, Mineralogy and petrology of the Gunflint Iron Formation, Minnesota-Ontario: Correlation of compositional and assemblage variations at low to moderate grade: Journal of Petrology, v. 19, p. 215-288.

Floyd, P.A. and Winchester, J.A., 1978, Identification and discrimination of altered and metamorphosed volcanic rocks using immobile elements: Chemical Geology, v. 21, p. 291-306.

Frost, R.B., 1975, Contact metamorphism of serpentinite, chloritic blackwall and rodingite at Paddy-Go-Easy Pass, Central Cascades, Washington: Journal of Petrology, v. 16, pt. 2, p. 272-313.

Gaal, G., Koistinen, T. and Mattila, E., 1975, Tectonics and Stratigraphy of the Vicinity of Outokumpu, North Karelia, Finland, Including a Structural Analysis of the Outokumpu Ore Deposit: Geological Survey of Finland Bulletin 271, 67 p.

Gatzweiler, R., Lehnert-Thiel, K., Clasen, D., Tan, B., Voultsidis, V., Strand, J.G. and Rich, J., 1979, The Key Lake uranium-nickel deposits: Canadian Institute of Mining and Metallurgy Bulletin, v. 72, no. 807, p. 73-79.

Goodwin, A.M., 1977, Archean volcanism in Superior Province, Canadian Shield: in Baragar, W.R.A., Coleman, L.C. and Hall, J.M., eds., Volcanic Regimes in Canada: Geological Association of Canada Special Paper 16, p. 205-242.

Goodwin, A.M. and Ridler, R.H., 1970, The Abitibi orogenic belt: in Baer, A.J., ed., Symposium on Basins and Geosynclines of the Canadian Shield: Geological Survey of Canada Paper 70-40, p. 1-28.

Graterol, M. and Naldrett, A.J., 1971, Mineralogy of the Marbridge No. 3 and No. 4 nickel-iron sulfide deposits, with some comments on the low temperature equilibrium in the Fe-Ni-S system: Economic Geology, v. 66, p. 886-900.

Green, A.H., 1978, Evolution of Fe-Ni sulphide ores associated with Archean ultramafic komatiites, Langmuir Township, Ontario: Unpublished Ph.D. thesis, University of Toronto, Ontario, 355 p.

Gresens, R.L., 1967, Composition-volume relationships of metasomatism: Chemical Geology, v. 2, p. 47-65.

Groves, D.I. and Keays, R.R., 1979, Mobilization of ore-forming elements during alteration of dunites, Mt. Keith-Betheno, Western Australia: Canadian Mineralogist, v. 17, pt. 2, p. 373-389.

Groves, D.I., Barrett, F.M. and McQueen, K.G., 1979, The relative roles of magmatic segregation, volcanic exhalation and regional metamorphism in the generation of volcanic-associated nickel ores of Western Australia: Canadian Mineralogist, v. 17, pt. 2, p. 319-336.

Groves, D.I., Hudson, D.R. and Hack, T.B.C., 1974, Modification of iron-nickel sulfides during serpentinization and talc-carbonate alteration at Black Swan, Western Australia: Economic Geology, v. 69, p. 1265-1281.

Groves, D.I., Barrett, F.M., Binns, R.A., Marston, R.J. and McQueen, K.G., 1976, A possible volcanic-exhalative origin for lenticular nickel sulphide deposits of volcanic association with special reference to those in Western Australia – discussion: Canadian Journal of Earth Sciences, v. 13, p. 1646-1650.

Humphris, S.E. and Thompson, G., 1978, Trace element mobility during hydrothermal alteration of oceanic basalts: Geochimica et Cosmochimica Acta., v. 42, p. 127-136.

Hutchinson, R.W., 1979, Evidence for exhalative origin of Tasmanian tin deposits: Canadian Institute of Mining and Metallurgy Bulletin, v. 72, no. 808, p. 91-104.

—————, 1973, Volcanogenic sulfide deposits and their metallogenic significance: Economic Geology, v. 68, no. 8, p. 1223-1246.

Irving, T.N. and Baragar, W.R.A., 1971, A guide to the classification of the common volcanic rocks: Canadian Journal of Earth Sciences, v. 8, p. 523-548.

Jensen, L.S., 1976, A new cation plot for classifying subalkalic volcanic rocks: Ontario Division of Mines, MP 66, 22 p.

Keele, R.A. and Nickel, E.H., 1974, The geology of a primary millerite-bearing sulfide assemblage and supergene alteration at the Otter Shoot, Kambalda, Western Australia: Economic Geology, v. 69, p. 1102-1117.

Kerrich, R., Robinson, D.J., Hodder, R.W. and Hutchinson, R.W., 1979, Field relations of Au, Ni and Cr deposits in ultramafic-mafic volcanic rocks: in Pye, E.G., ed., Geoscience Research Seminar, Abstracts, Ontario Geological Survey, 23 p.

Lambert, I.B. and Sato, T., 1974, The kuroko and associated ore deposits of Japan: A review of their features and metallogenesis: Economic Geology, v. 69, p.1215-1236.

Leake, B.E., 1978, Nomenclature of amphiboles: Canadian Mineralogist, v. 16, pt. 4, p. 501-520.

Lusk, J., 1976a, A possible volcanic-exhalative origin for lenticular nickel sulfide deposits of volcanic association with special reference to those in Western Australia: Canadian Journal of Earth Sciences, v. 13, p. 451-458.

_______________, 1976b, A possible volcanic-exhalative origin for lenticular nickel sulfide deposits of volcanic association, with special reference to those in Western Australia – reply: Canadian Journal of Earth Sciences, v. 13, p. 1651-1653.

Marston, R.J. and Kay, B.D., 1980, The distribution, petrology and genesis of nickel ores at Juan Complex, Kambalda, Western Australia: Economic Geology, v. 75, p. 546-565.

Miller, L.J. and Smith, M.E., 1975, Sherlock Bay nickel-copper: in Knight, C.L., ed., Economic Geology of Australia and Papua, New Guinea. I. Metals: Australasian Institute of Mining and Metallurgy, p. 168-174.

Misra, K.C. and Fleet, M.E., 1973, The chemical composition of synthetic and natural pentlandite assemblages: Economic Geology, v. 68, p. 518-539.

Muir, J.E. and Comba, C.D.A., 1979, The Dundonald deposit: an example of volcanic-type nickel-sulphide mineralization: Canadian Mineralogist, v. 17, pt. 2, p. 351-359.

Muir, T.L., 1979, Discrimination between extrusive and intrusive Archean ultramafic rocks in the Shaw Dome area using selected major and trace elements: Canadian Journal of Earth Sciences, v. 16, p. 80-90.

Naldrett, A.J., 1973, Nickel sulphide deposits – their classification and genesis, with special emphasis on deposits of volcanic association: Canadian Institute of Mining and Metallurgy Bulletin, v. 66, no. 739, p. 45-63.

Naldrett, A.J., 1979, Partitioning of Fe, Co, Ni and Cu between sulfide liquid and basaltic melts and the composition of Ni-Cu sulfide deposits – a reply and further discussion: Economic Geology, v. 74, p. 1520-1528.

Naldrett, A.J., Innis, D.G. and Sowa, J.M., 1980, Platinum group element concentrations in some magmatic ores in Ontario: in Pye, E.G., ed., Geoscience Research Grant Program, Summary of Research, 1979-80: Ontario Geological Survey Miscellaneous Paper 93, 262 p.

Naldrett, A.J., Hoffman, E.L., Green, A.H., Chen-Lin Chou, Naldrett, S.R. and Alcock, R.A., 1979, The composition of Ni-sulfide ores, with particular reference to their content of PGE and Au: Canadian Mineralogist, v. 17, pt. 2, p. 403-416.

Nesbitt, R.W. and Sun, Shen-Su, 1976, Geochemistry of Archean spinifex-textured peridotites and magnesian and low-magnesian tholeiites: Earth and Planetary Science Letters, v. 31, p. 433-453.

Nesbitt, R.W., Sun, Shen-Su and Purvis, A.C., 1979, Komatiites: geochemistry and genesis: Canadian Mineralogist, v. 17, pt. 2, p. 165-186.

Pearce, J.A. and Cann, J.R., 1973, Tectonic setting of basic volcanic rocks determined using trace element analysis: Earth and Planetary Science Letters, v. 19, p. 290-300.

Phillips, A.H. and Hess, H.H., 1936, Metamorphic differentiation at contacts between serpentinite and siliceous country rocks: American Mineralogist, v. 21, p. 333-362.

Pyke, D.R., 1975, Geology of Adams and Eldorado Townships, District of Cochrane: Ontario Division of Mines, GR 121, 51 p.

Pyke, D.R. and Middleton, R.S., 1971, Distribution and characteristic of the sulphide ores of the Timmins area: Canadian Institute of Mining and Metallurgy Bulletin, v. 64, no. 710, p. 55-66.

Pyke, D.R., Naldrett, A.J. and Eckstrand, O.R., 1973, Archean ultramafic flows in Munro township, Ontario: Geological Society of America Bulletin, v. 84, p. 955-978.

Pyke, D.R., MacVeigh, J.G. and Middleton, R.S., 1978, Volcanic stratigraphy and geochemistry in the Timmins mining area: Geological Association of Canada Guidebook, p. 160-183.

Robinson, D.J., Kerrich, R., Barnett, R.L., Hodder, R.W. and Hutchinson, R.W., 1980, Mineralogical and chemical dispersion haloes in Archean volcanic-hosted Au and Ni deposits: in Pye, E.G., ed., Geoscience Research Seminar, Abstracts: Ontario Geological Survey, 21 p.

Ross, J.R. and Keays, R.R., 1979, Precious metals in volcanic-type nickel sulfide deposits in Western Australia. I. Relationship with the comparison of the ores and their host rocks: Canadian Mineralogist, v. 17, pt. 2, p. 417-436.

Ross, J.R. and Hopkins, G.M.F., 1975, The nickel sulfide deposits of Kambalda, Western Australia: in Knight, C.L., ed., Economic Geology of Australia and Papua, New Guinea. I. Metals: Australasian Institute of Mining and Metallurgy, p. 100-121.

Sangster, D.F. and Scott, S.D., 1976, Precambrian, strata-bound, massive Cu-Zn-Pb sulfide ores of North America: in Wolf, K.H., ed., Handbook of Strata-bound and Stratiform Ore Deposits, v. 6, p. 129-222.

Springer, R.K., 1974, Contact metamorphosed ultramafic rocks in the Western Sierra Nevada foothills, California: Journal of Petrology, v. 15, p. 160-195.

Stanton, R.L. and Vaughan, J.P., 1979, Facies of ore formation: A preliminary account of the Pegmont deposit as an example of potential relations between small ''iron formations'' and stratiform sulphide ores: Australasian Institute of Mining and Metallurgy, no. 270, p. 25-38.

Turner, F.J. and Verhoogen, J., 1960, Igneous and Metamorphic Petrology, Second Ed.: McGraw-Hill, 694 p.

Williams, D.A.C., 1979, The association of some nickel sulfide deposits with komatiitic volcanism in Rhodesia: Canadian Mineralogist, v. 17, pt. 2, p. 337-349.

Wilson, H.D.B., Kilburn, L.C., Graham, A.R. and Ramlal, K., 1969, Geochemistry of some Canadian nickeliferous ultrabasic intrusions: Economic Geology, Mono. 4, p. 294-309.

MS received July 1980;
Accepted for publication December 1980.

Precambrian Sulphide Deposits, H.S. Robinson Memorial Volume, edited by R.W. Hutchinson, C.D. Spence and J.M. Franklin, Geological Association of Canada Special Paper 25, 1982

STRUCTURE, METAL ZONING AND ALTERATION AT THE MILLENBACH DEPOSIT, NORANDA, QUEBEC

M.J. Knuckey, C.D.A. Comba, and G. Riverin
Corporation Falconbridge Copper, Post Office Box 2000,
Noranda, Quebec J9X 5B4

ABSTRACT

The Millenbach deposit occurs in a relatively undeformed sequence of submarine andesitic and rhyolitic flows of Archean age. Three rhyolite domes and fifteen massive sulphide lenses are localized by northeasterly and northwesterly trending synvolcanic faults. The largest lens is located at the intersection of the two fracture systems and developed during a period of domal formation.

Individual lenses have copper-rich cores grading outward to zinc-rich margins. In a cluster of lenses formed at the same stratigraphic horizon the average ratio of copper to copper plus zinc decreases steadily with increase in distance of individual lenses from the main hydrothermal vent. Where lenses have formed at successive horizons vertically above each other, the higher copper content of the lower lenses may be due, in part, to replacement and veining.

The alteration pipes underlying the various lenses, show a core of cordierite-anthophyllite or massive chlorite grading outward into a spotted biotite-rich zone. Normative calculations show that the core was initially chlorite which graded laterally into a sericitic outer zone. Regardless of initial rock type, volume was conserved and a distinctive metasomatic trend developed across the alteration pipe resulting in depletion of Ca and Na and enrichment of Fe, Mg, S, Cu and Zn; K and Al have generally been leached at the core and added at the margins. Similar chemical and mineral zoning is present on the selvedges of individual sulphide veinlets.

RÉSUMÉ

Le gisement de Millenbach se trouve dans une séquence relativement peu déformée de coulées d'andésite et rhyolite sous-marines, d'âge archéen. Trois dômes de rhyolite et 15 lentilles de sulfures massifs sont contrôlés par des failles synvolcaniques d'orientation nord-est et nord-ouest. La plus grande lentille est située à l'intersection de deux systèmes de fractures et s'est développée pendant une période de formation de dômes.

Chaque lentille a un coeur riche en cuivre passant graduellement à une bordure riche en zinc. Dans un groupe de lentilles formées au même niveau stratigraphique, le rapport moyen de cuivre à cuivre plus zinc décroît de façon continue à mesure que s'accroît la distance séparant des lentilles individuelles de l'orifice hydrothermal principal. Là où des lentilles se sont formées successivement les unes au dessus des autres, la teneur plus élevée en cuivre des lentilles inférieures peut en partie être due à des remplacements et des recoupements par veinules.

Les cheminées d'altération sous-jacentes aux lentilles montrent un coeur à cordiérite-anthophyllite ou à chlorite massive, passant graduellement à une zone tachetée riche en biotite. Les calculs normatifs montrent que ce coeur était à l'origine composé de chlorite et passait latéralement à une zone externe sériciteuse. Pendant l'altération, le volume des roches a été conservé quelque soit le type de roches et un gradient métasomatique distinctif s'est développé à travers la cheminée d'altération résultant en une perte de Ca et Na et un gain de Fe, Mg, S, Cu et Zn; K et Al ont généralement été lessivés dans le coeur et ajoutés aux bordures. Une zonalité chimique et minérale similaire se retrouve aux épontes des veinules sulfurées.

INTRODUCTION

The Millenbach and Corbet stratabound massive sulphide orebodies of Falconbridge Copper Limited (an amalgamation of Lake Dufault Mines Ltd., Opemiska Mines Ltd., and Falconbridge Quebec Ltd.) are located 8 km north of Rouyn-Noranda in the Abitibi region of northwestern Quebec about 560 km northwest of Montreal.

The Millenbach deposit containing more than 3 million tonnes grading 3.5% Cu, 4.5% Zn, 48 g Ag/t, 0.86 g Au/t was discovered in 1966 and is currently in production. The Corbet deposit was discovered in 1974 and is located 2200 m west of Millenbach. Pre-production ore reserves are estimated at 2.7 million tonnes grading 2.9% Cu, 2.0% Zn, 20.5 g Ag/t and 0.93 g Au/t. Both deposits lie between 750 and 950 m below surface.

This paper updates and amplifies earlier geological descriptions of the Millenbach Mine (Simmons *et al.*, 1973; Knuckey, 1975) particularly with respect to metal zoning and wall-rock alteration, and shows that similar mine-scale metal zoning also exists at the nearby Amulet Mine of Noranda Mines Limited. It emphasizes the role played by synvolcanic structures and, throughout, attempts to show the intimate relationship between volcanic and ore-forming events. The paper also provides background information on the Corbet deposit, described in the following paper (Knuckey and Watkins, this volume). For more formal descriptions of the stratigraphic succession the reader is referred to papers by Wilson (1941), Simmons *et al.* (1973), and de Rosen-Spence (1976).

GENERAL GEOLOGY

The Millenbach, Corbet and other deposits of the Noranda District occur within volcanic rocks of the Archean Abitibi Greenstone Belt of the Superior Province. This is a belt of volcanic, sedimentary and intrusive rocks 160 km in width extending for 760 km from the Grenville Province in the east to the Kapuskasing sub-Province in the west. A number of geologists have recognized several felsic volcanic centres within the belt including a major one in the Noranda District. These centres are characterized by thickening of the volcanic sequence, increased proportions of felsic and pyroclastic rocks, increased intrusive activity and a number of important mineral deposits.

The rocks of the Noranda District, described by Wilson (1941), Gilmour (1965), Dugas (1966), Spence (1967) and others, form part of the Blake River Group which comprises an estimated 12,200 m of mainly submarine volcanic rocks (Baragar, 1968). These range from tholeiitic and calc-alkalic basalt to rhyolite and appear to have accumulated in a number of interfingering mafic to felsic cycles (Spence, 1967; Spence and de Rosen-Spence, 1975). The central part of what may have been a caldera is now occupied by two "granitic" bodies and several smaller intrusions. The Flavrian granitic complex to the west is believed to be an older, semi-conformable, subvolcanic intrusion with a weakly developed thermal metamorphic aureole. In contrast, the Lake Dufault granodiorite farther east is a slightly younger, cross-cutting intrusion with a strong thermal aureole (Fig. 1).

The rocks are tilted moderately to the east and have been folded about east-west axes. Folding is broad and gentle in the central portion but is much tighter toward major shear zones marking the southern, northern and eastern margins of the complex. The area is extensively and complexly faulted, making stratigraphic correlation difficult. Faults with large apparent horizontal displacements trend northeasterly and northwesterly. Faults with smaller displacements strike east-west and north-south. A number of important thrust faults, dipping to the east, are known to the west and north of the Lake Dufault granodiorite. Many fault zones are intruded by a wide variety of dykes, in some cases probably feeders to overlying flows. Proterozoic diabase dykes, syenite stocks and lamprophyre dykes cut rocks of all ages.

Regional metamorphism is of lower greenschist grade and in general, original volcanic textures and structures are well preserved. A contact metamorphic aureole 1500 m in width surrounds the Lake Dufault granodiorite.

Economic Geology

Since the discovery of the Horne Mine in 1923 the Noranda Camp has yielded 17 base metal mines and 15 gold mines which have produced close to two million tonnes of copper, one million tonnes of zinc, 373,000 kg of gold and 1.8 million kg of silver. Three new base-metal deposits, the Magusi, New Insco and Corbet, and three new gold deposits, the Silverstack, Thompson-Bousquet and Dumagami, have been discovered in the past 8 years. Although Millenbach is the only mine currently in production three others are under active development.

Base metal deposits in the area are almost all stratabound bodies of massive sulphides occurring generally, but not exclusively, within or at the stratigraphic top of felsic units (Dugas, 1966). Most deposits are underlain by strong zones of chloritic alteration and in the Lake Dufault area by economic stringer sulphide mineralization within the alteration "pipes". Gilmour (1965), Beecham (1971), Kutina and Fabbri (1972) and others have noted the association of sulphide occurrences with regionally extensive fracture patterns.

Gold was a major by-product of the Horne and Quemont Mines but is of lesser importance in the other base metal deposits of the district. In addition, large gold-bearing quartz veins of the Eldrich and Elder mines in the Flavrian and adjacent Powell granites were mined as flux for the Noranda smelter. Other significant gold deposits occur in shear zones along the southern margin of the district. More recently some geologists have attempted to relate gold occurrences genetically, spatially and

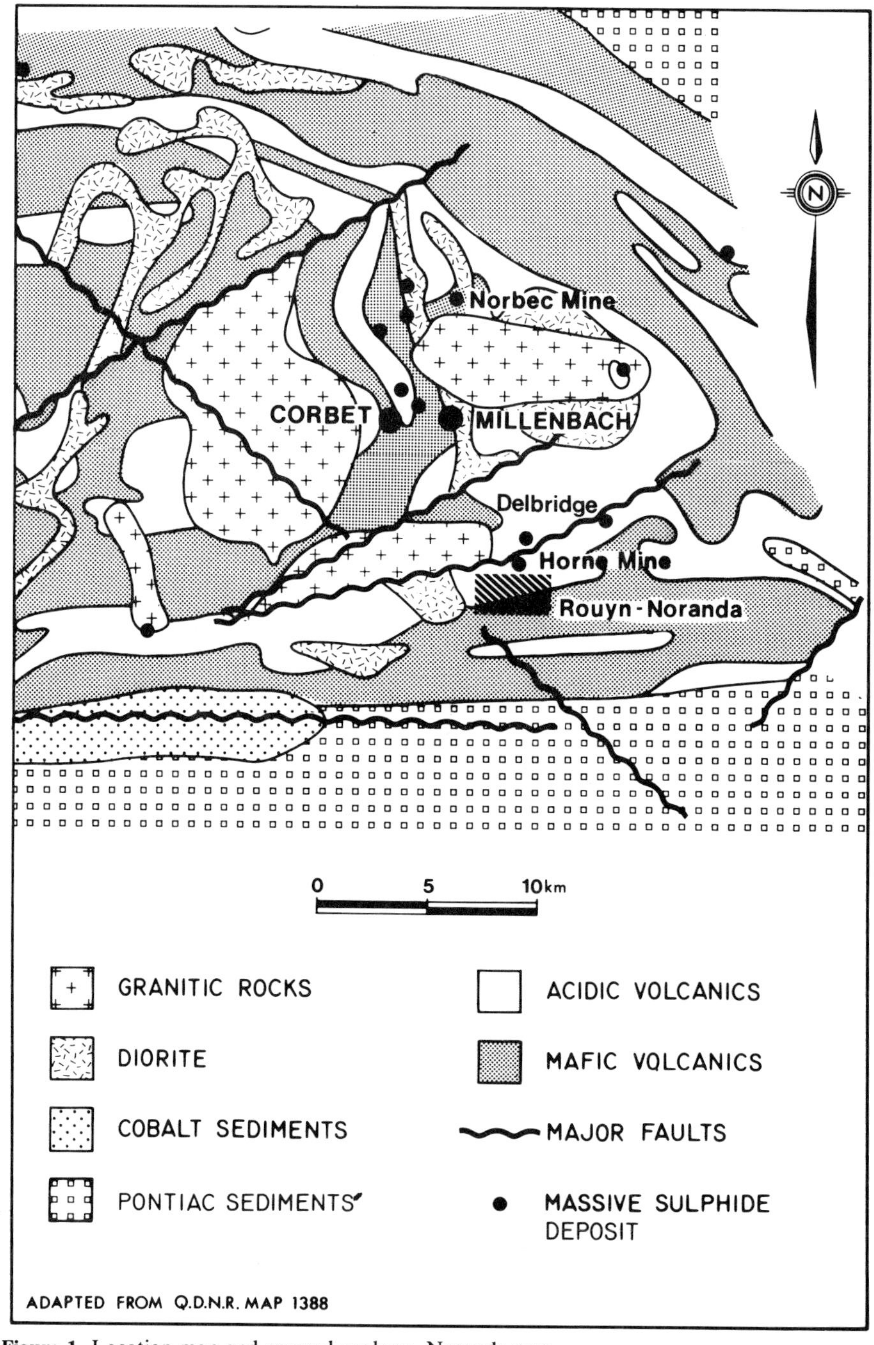

Figure 1. Location map and general geology, Noranda area.

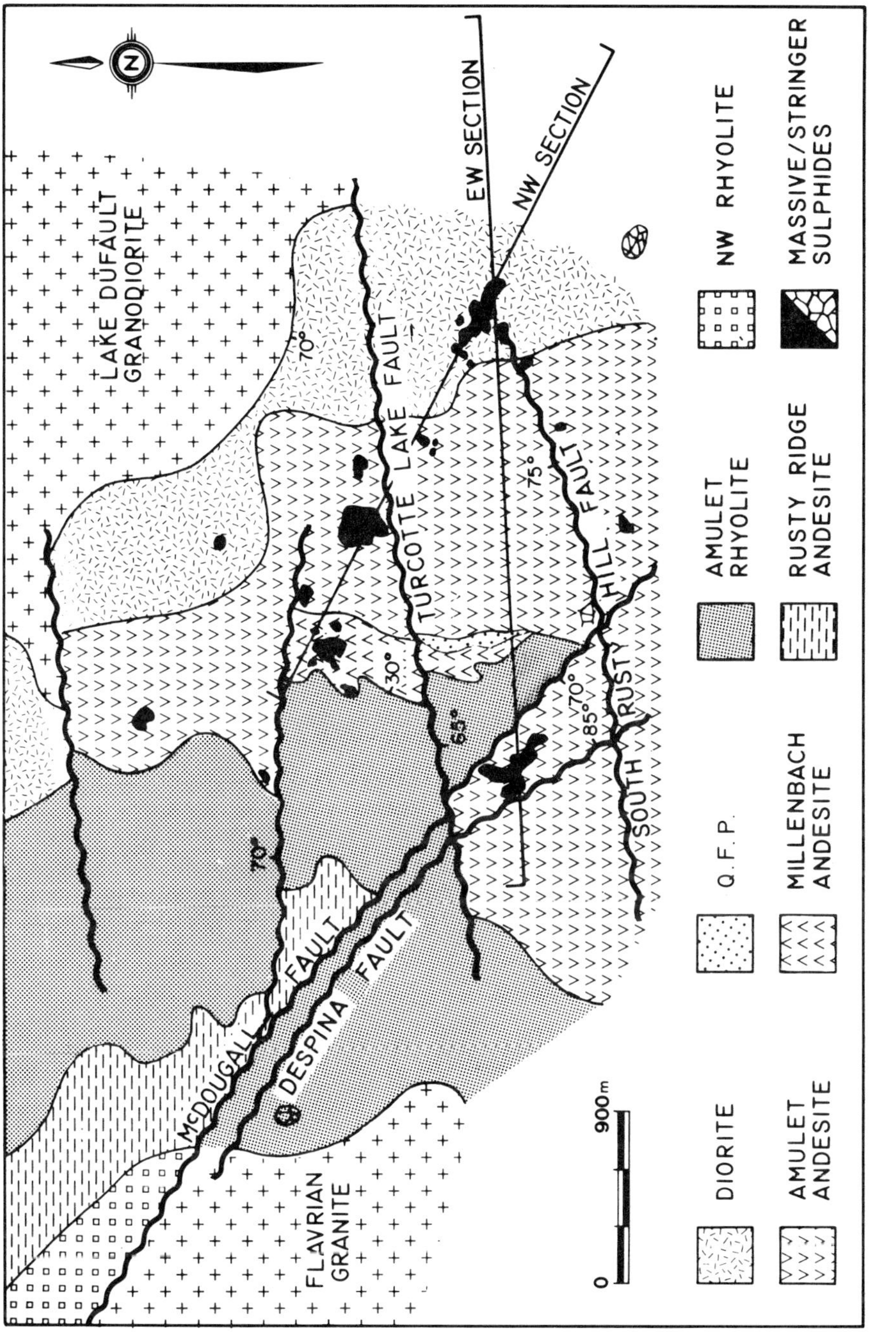

Figure 2. Surface geology in the area of the Amulet, Corbet and Millenbach Mines.

stratigraphically to the volcanic processes active throughout the area (Hutchinson, *et al.*, 1971; Sangster, 1972). The newly discovered stratabound, auriferous pyrite deposits of Silverstack and Thompson-Bousquet in felsic tuffs located at the southeastern margin of the Noranda District add credence to these ideas.

MINE AREA GEOLOGY

Introduction

The mine area (Fig. 1) lies 8 km north of the cities of Rouyn-Noranda and includes three main orebodies. These are the Amulet Lower A of 4.5 million tonnes formerly mined by Waite-Amulet and Lake Dufault Mines Ltd., the Millenbach of 3 million tonnes and the Corbet of at least 2.7 million tonnes both owned by Corporation Falconbridge Copper. Other small satellite sulphide lenses, the Amulet 'C' and 'F', D-68, D-266 etc., have been discovered clustered around the main deposits. The Bedford and South Rusty Hill (S.R.H.) deposits are sub-economic stringer sulphide zones.

The geology of the mine area is shown in Figure 2. The volcanic rocks lying between the Flavrian and Lake Dufault intrusions strike roughly north-south and dip to the east at 30 to 50°. Important faults are the synvolcanic, northwesterly trending, McDougall and Despina faults and the northwesterly trending South Rusty Hill fault. The Turcotte Lake fault is a younger reverse fault striking east-west and dipping south at 60 to 70°.

Stratigraphy

The stratigraphic succession is illustrated on the vertical sections between the Corbet and Millenbach Mines (Fig. 3) and between the Amulet and Millenbach Mines (Fig. 4). Except for the Millenbach Andesite, the names of the rock units are the same as those proposed by Spence and de Rosen-Spence (1975).

Flavrian Andesite. This is the oldest unit in the mine area and consists of a sequence of unknown thickness, but greater than 500 m, of massive or pillowed, amygdaloidal flows, flow breccias and volcaniclastic rocks. The Corbet orebodies occur as a system of stacked lenses in the upper 250 m of this unit (Knuckey and Watkins, this volume).

Northwest Rhyolite. This unit is a succession of massive to flow banded, spherulitic rhyolitic and rhyodacitic flows ranging in thickness from 15 to 335 m. Limited information suggests that they may form a domal structure trending northeasterly.

Rusty Ridge Andesite. This unit comprises a series of massive to well pillowed flows, locally strongly amygdaloidal, about 245 m thick at Corbet but much thicker to the northwest. It is overlain locally, particularly near the McDougall fault, by a discontinuous felsic pyroclastic horizon, named the Beecham Breccia, up to 10 m thick. Five kilometres north of the Corbet Mine this unit hosts significant stringer sulphide mineralization.

Amulet Rhyolite. This unit is mentioned by many earlier authors as it underlies the Amulet and Millenbach deposits. Detailed surface mapping and extensive diamond drilling have shown it to be an extremely complex, 600 m thick sequence of

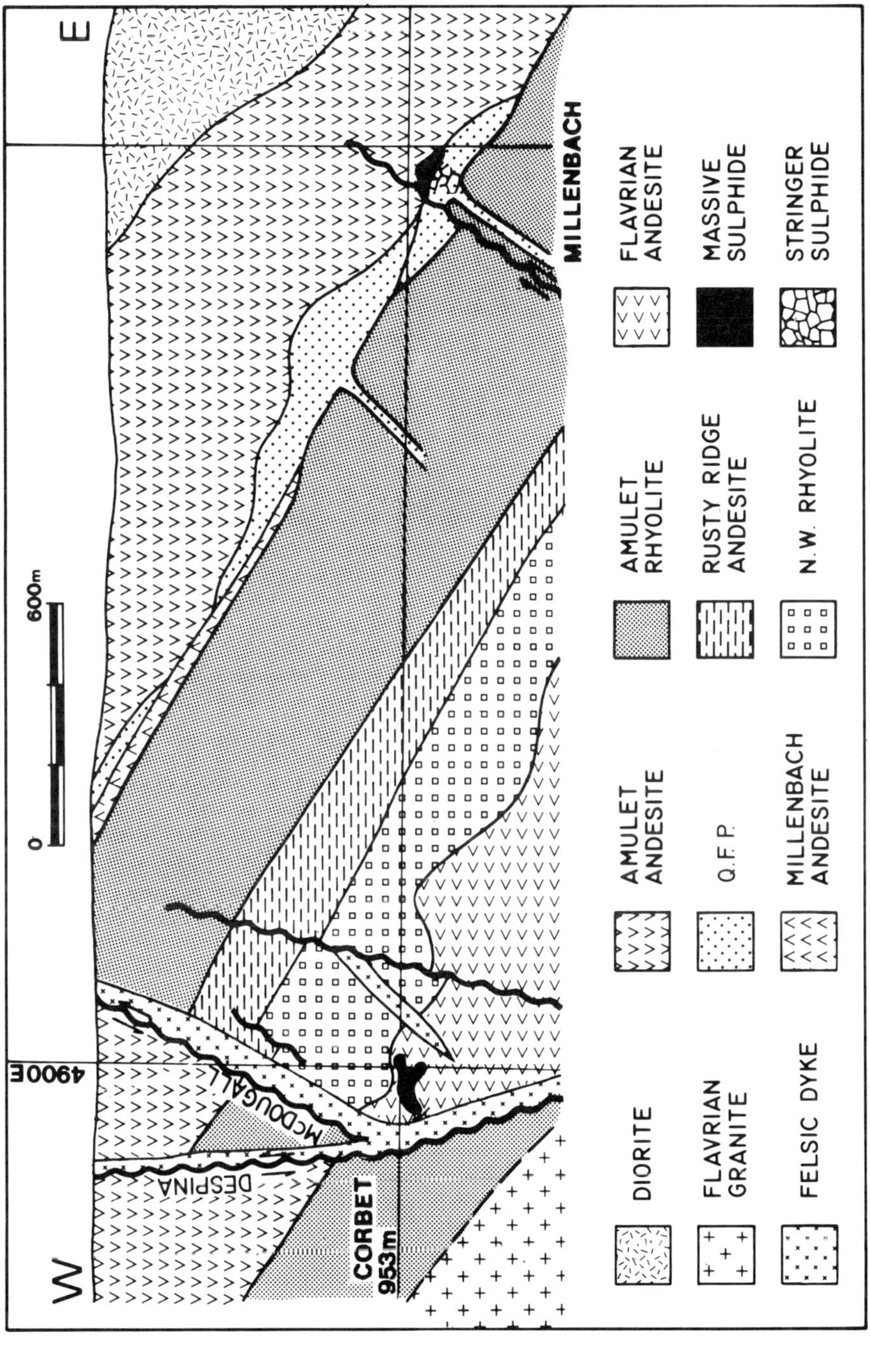

Figure 3. East-west section through the Millenbach and Corbet deposits, looking north.

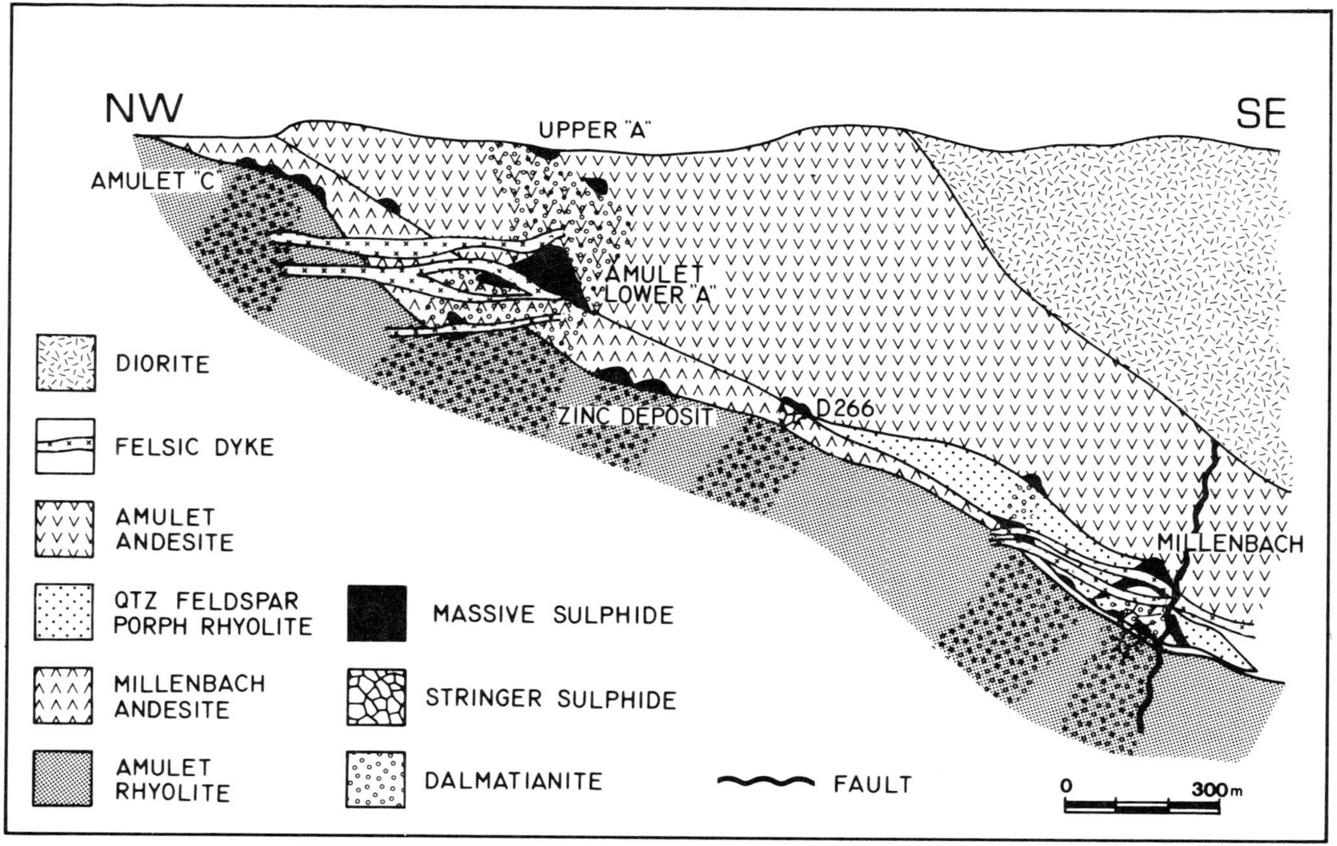

Figure 4. Northwest-southeast section through the Millenbach and Amulet deposits, looking northeast.

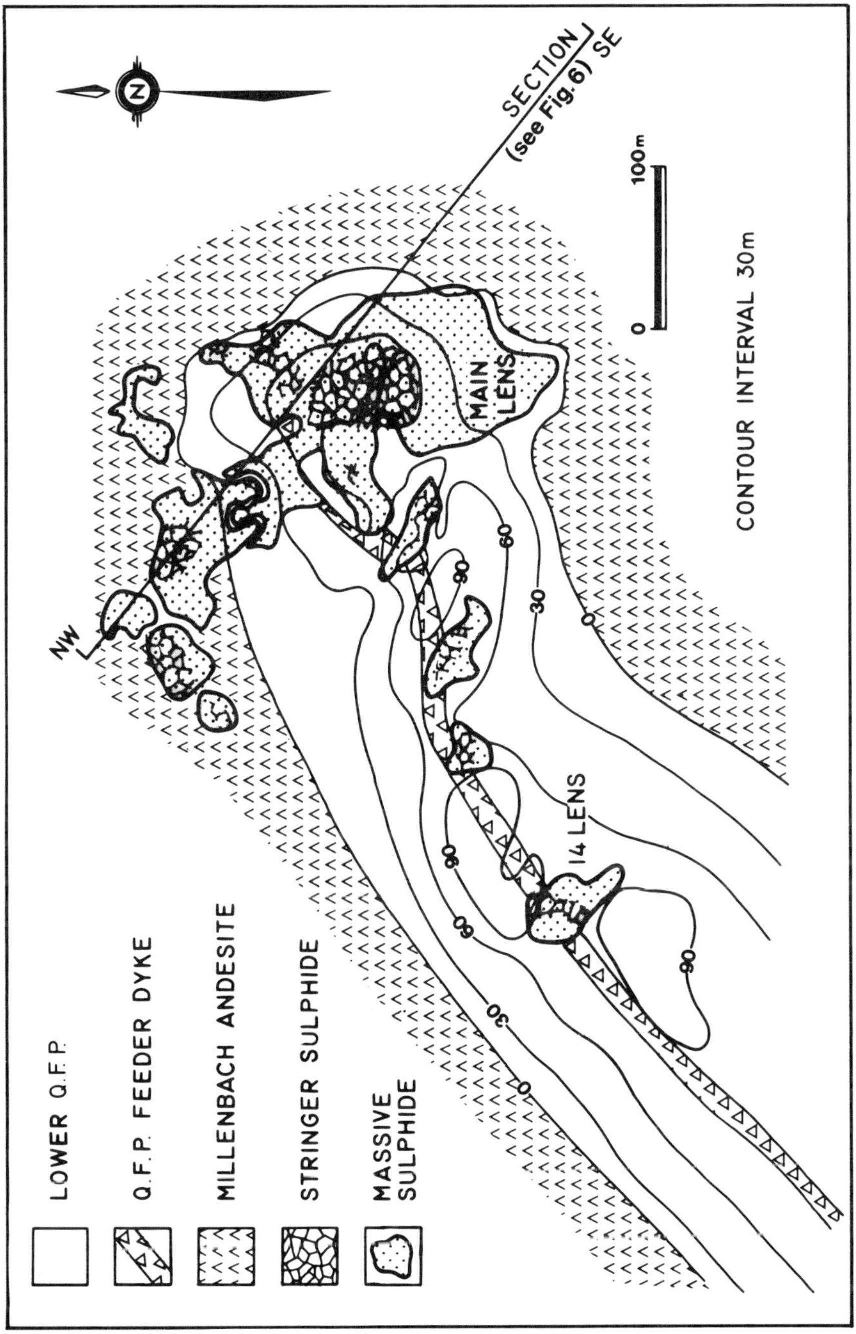

Figure 5. Isopach map of the Lower Q.F.P. Dome at the Millenbach Mine.

interfingering rhyolitic, andesitic and basaltic submarine flows and flow breccias. The rhyolitic rocks form relatively small domal features toward the base of the unit. This period of intermediate and felsic volcanism was accompanied by, or ended with very active, and widespread hydrothermal hot-spring activity resulting in: 1) pervasive silicification of the upper third of the Amulet Rhyolite (Gibson, 1979); 2) deposition of extensive, sulphidic cherty tuff and exhalite, known as the "C" Horizon, with an average thickness of about 30 cm containing variable quantities of base metals at the top of the Amulet Rhyolite; 3) formation of stratiform massive sulphide ore lenses at volcanic contacts at the Amulet Mine and the Lake Dufault Zinc Deposit (Fig. 4).

Millenbach Andesite. Ranging from 0 to more than 60 m thick at the Millenbach Mine, this unit thickens rapidly to the north and northwest and may be a southern extension of the more northerly Waite Andesite (Dugas, 1966). In composition it ranges from basalt to andesite. Pillows, columnar jointing and flow breccias have been recognized locally in a series of generally massive flows. Simmons *et al*. (1973) provided petrographic details of the volcanic sequence at the Millenbach Mine (although the Millenbach Andesite was then called dacite and elsewhere it has occasionally been called the Upper McDougall Dacite and also "Black Rhyolite"). Evidence suggests that the Millenbach Andesite was extruded from a vent located farther north, into an active hot spring field. Massive sulphide lenses of up to 230,000 tonnes at Millenbach and up to 4.5 million tonnes at the Amulet Mine were deposited over hydrothermal vents on the upper surface of the flows. Below the massive sulphides, many pillow interstices, borders, and concentric cooling fractures were infilled and/or replaced by chlorite and sulphides. Exhalites formed at this time horizon, known as the Lower A horizon, were restricted to the general area of the active hot springs; generally less or much less than 1000 m from the main hydrothermal vents.

Lower Q.F.P. Active volcanism in the mine area was renewed with the opening of an east-northeasterly striking fissure at Millenbach followed by extrusion of the Lower Q.F.P. This quartz-feldspar porphyritic rhyolite forms a hummocky ridge about 760 m long by up to 300 m wide and is a maximum of 110 m thick (Figs. 5 and 6). The main Millenbach massive sulphide orebody was deposited on the subaqueous surface of the Lower Q.F.P. together with a very local cherty exhalite, known as the Inter Q.F.P. Tuff, which is restricted to within a few tens of metres of the orebody.

Upper Q.F.P. Concomitantly or perhaps slightly later, a larger dome was extruded to the northwest of the Lower Q.F.P., overlapping it locally. A small lens of massive sulphides was deposited on the surface of this Upper Q.F.P. in an area of persistent hot spring activity just northwest of the main centre. A very local exhalite, the Top Contact Tuff, is associated with this lens.

D-68 Dome. At the same time, 1.5 km southwest of the Millenbach Mine (Fig. 7), a third, smaller, porphyritic rhyolite domal complex of two or more overlapping flows was extruded from fissures striking east-west to east-northeast. Mineralization is similar to, but weaker than that on the Q.F.P. domes. The D-68 Zone, not yet explored in detail from underground, appears to consist of two small lenses of about 100,000 tonnes stacked above each other and above one of the Q.F.P. feeder dykes. The Millenbach and D-68 porphyritic rhyolite domes are possibly time equivalent to

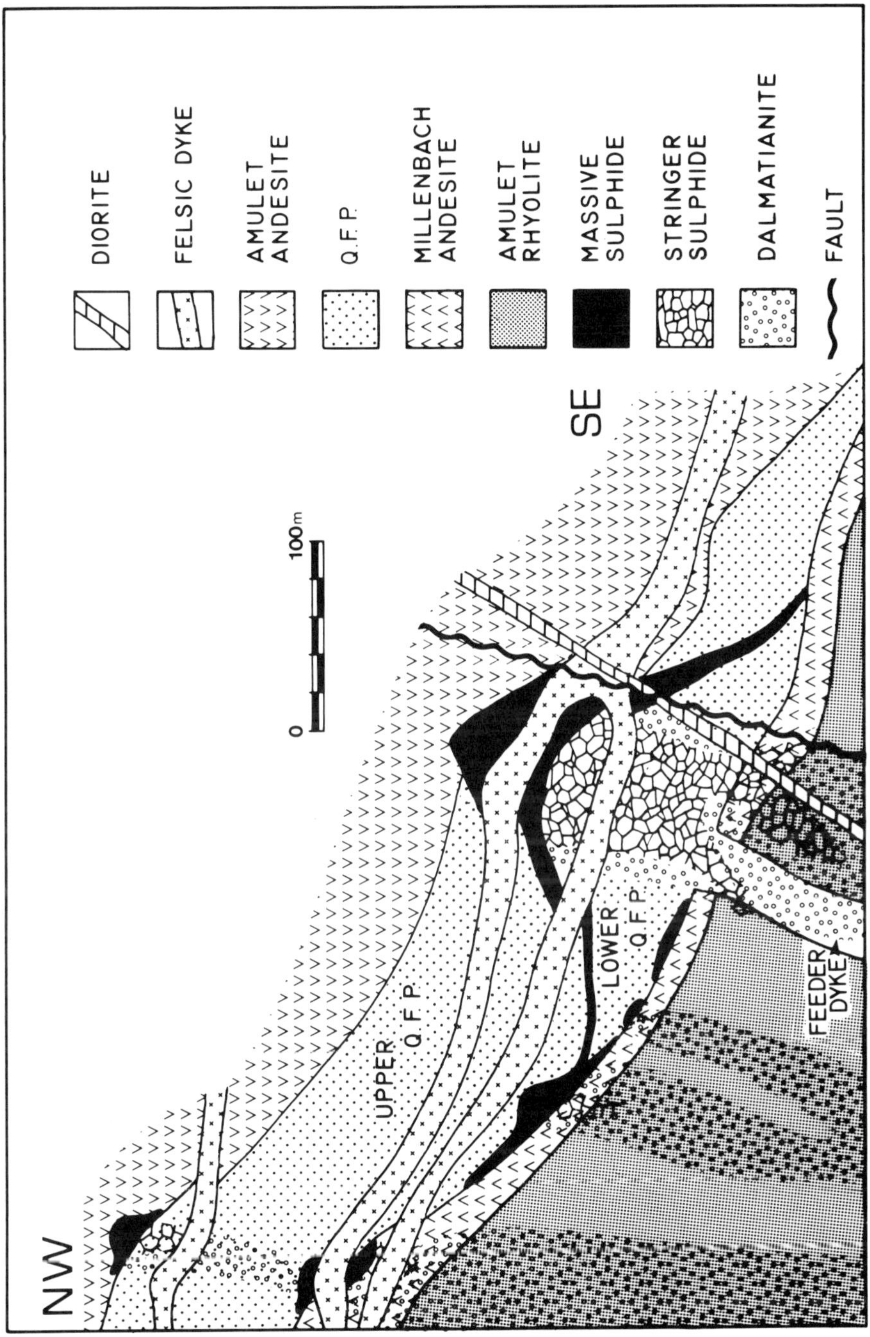

Figure 6. Northwest-southeast section through the Millenbach deposit, looking northeast.

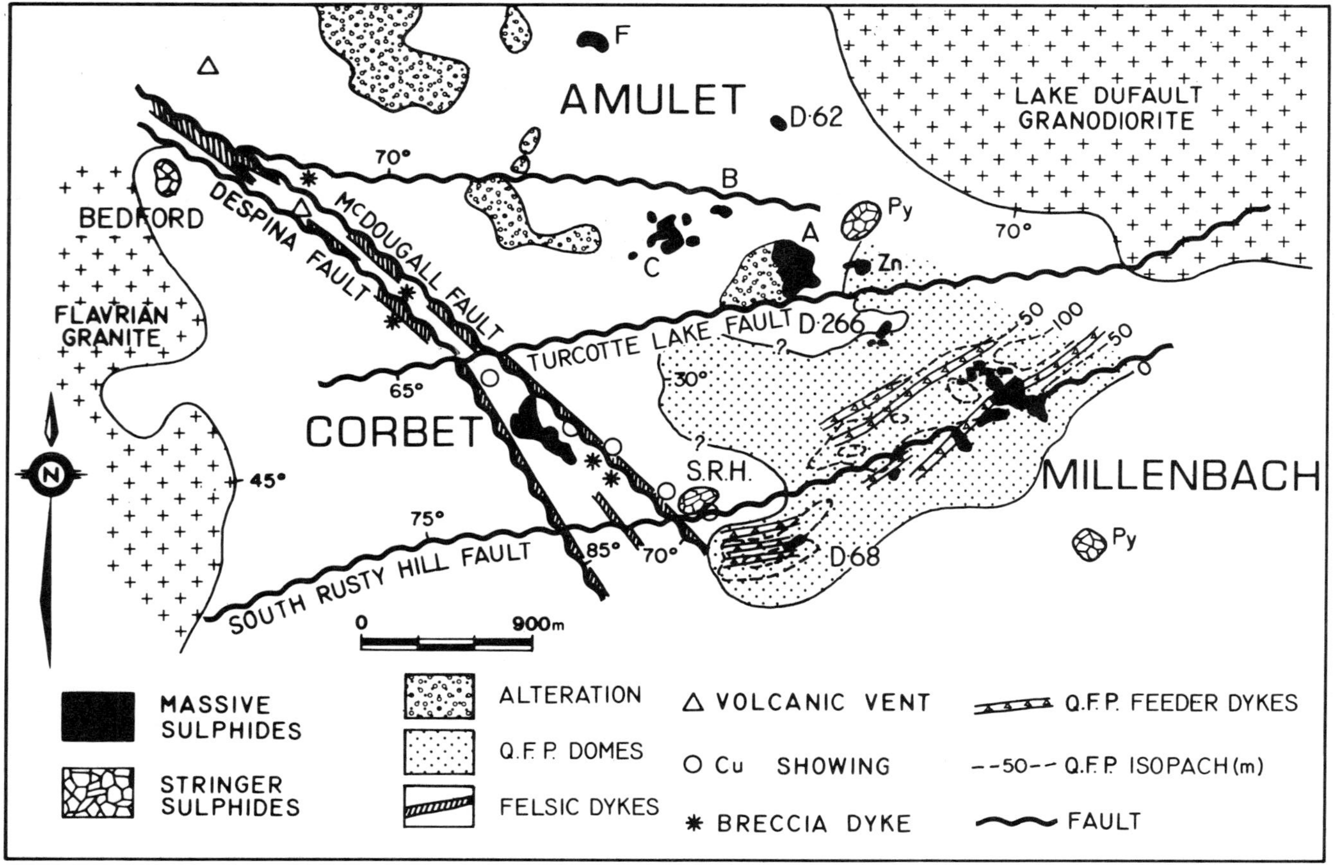

Figure 7. Mine Area Plan showing the relationships between ore deposits, alteration pipes, synvolcanic structures and the Q.F.P. domes.

the much thicker Waite Rhyolite that underlies the Norbec and adjacent deposits in the northern part of the district (Dugas, 1966; Purdie, 1967; Spence, 1975).

Amulet Andesite. This is a series of pillowed or massive, locally feldspar porphyritic flows totalling more than 610 m in thickness. At Millenbach these andesites are generally unaltered and unmineralized, indicating that hydrothermal activity in the Millenbach area had ceased prior to extrusion of the Amulet Andesite. However, a kilometre to the northwest, vigorous hydrothermal activity persisted to a later time because, at the Amulet Mine, high-grade orebodies, local exhalite and intense alteration are found for at least 300 m up into the Amulet Andesite.

Synvolcanic Structures

At Millenbach, massive sulphides were deposited at rock-seawater interfaces before, during and after the formation of the Q.F.P. domes. The bisection of an older massive sulphide lens, at the Lower A horizon, by the feeder dyke to the lower Q.F.P. dome and the alignment of later ore lenses vertically above this feeder dyke (Fig. 5) provide evidence that a deep-seated, northeasterly trending fracture system controlled not only the site of the Q.F.P. dome but also channelled the coeval, ore-forming hydrothermal solutions. Later movement along this same synvolcanic fracture system displaced portions of many of the ore lenses and was accompanied by the intrusion of a variety of felsic and mafic dykes, some of which may have fed overlying flows.

The sub-surface distribution and approximate combined thickness of the several Q.F.P. domes in the mine area are shown in Figure 7. Massive and stringer sulphide deposits are shown projected vertically: outcropping alteration pipes, probably sites of eroded massive sulphide deposits, are also shown. The feeder dykes, the alignment of the ore lenses and the long axes of the domes all trend northeasterly, subparallel to the South Rusty Hill fault.

Striking northwesterly, the McDougall-Despina fault system, with an associated swarm of flow banded rhyolite dykes, a number of porphyritic andesite dykes and somewhat larger quartz diorite intrusions, is also a synvolcanic structure. Several small polymictic breccia pipes and dykes are also located along this structure as well as a number of local volcanic vents of various ages. A number of copper showings and strongly anomalous base metal values along and adjacent to the McDougall-Despina structure provide evidence that it also served as a channelway for mineralizing hydrothermal fluids.

The McDougall-Despina fault is parallel to the trend of the Millenbach-Amulet deposits. A possible branch of the McDougall fault can be traced eastward on surface, as far as the Amulet (Fig. 7) but its continuation to the southeast in older rocks beneath Millenbach is conjectural. Structure is believed to be one of the key factors controlling the location of volcanic vents and consequently the local, detailed stratigraphic succession. Faults are fundamental in providing channelways for migrating ore-forming hydrothermal fluids in massive, relatively impermeable rock sequences. They are the common link between orebodies found in a variety of rock types at favourable stratigraphic horizons scattered through more than one thousand metres of the volcanic pile.

METAL ZONING

Introduction

The Millenbach deposit includes at least 15 individual, mushroom-shaped orebodies, each a characteristic steep-sided, conical cap of massive sulphides overlying an irregular pipe-like zone of stringer and disseminated sulphides. Orebodies range in size from 4,000 to 2 million tonnes and occur at three stratigraphic horizons (Fig. 6). More than 80% of the ore is found in several orebodies on top of the Lower Q.F.P. dome. Massive ore comprises from 24 to 100%, and averages about 46% of the total ore in any one orebody. Sulphides are defined as massive where they constitute more than 60% of the rock. The major constituents are pyrite, pyrrhotite, sphalerite, and chalcopyrite with interstitial quartz and minor amounts of magnetite, biotite and chlorite. At the base of the massive sulphides there is often a transition zone of semi-massive mineralization, containing 30 to 60% sulphides, which occurs as thick veins of massive sulphide with many inclusions of the underlying rock.

Orebody Scale Zoning

Within individual orebodies, metal zoning patterns are typically concentric (Fig. 8). A copper-rich core grades outward to a zinc-rich margin. Figures 8 and 9 show the percentage of copper to copper plus zinc (i.e., % Cu × 100/% Cu + % Zn); the high numbers indicate sulphides relatively rich in copper and the low numbers, sulphides relatively rich in zinc. The patterns shown are representative of all 15 orebodies even though there is a wide variation in the weighted mean grades of each (Table I). Most orebodies show four component parts:

Copper Spine. Relatively copper-rich ore occurs in the central part of the massive sulphides of all orebodies. The persistence of a high-grade copper spine containing up to 20% Cu and 5% Zn and apexing toward the peak of the lens, is characteristic of the major orebodies at Millenbach and also of the Norbec deposit located 5 km to the north. This copper spine is of massive chalcopyrite with interstitial sphalerite and pyrrhotite. A section located off-centre may not show this feature and gives the impression of vertical rather than lateral zoning as was previously indicated at the Norbec deposit (Purdie, 1967).

Breccia Ore. Surrounding the copper spine is a zone of mixed sulphides, often with the appearance of a breccia comprising blocks of sphalerite-rich and chalcopyrite-rich ore in a matrix of pyrite and pyrrhotite ± chalcopyrite. The pyrite and pyrrhotite content ranges from 35 to 60%.

Banded Ore. Crudely to finely banded pyrite-sphalerite ± chalcopyrite ore flanks and partly overlies the breccia ore. Pyrite content ranges from 40 to over 80%. This banded ore may end abruptly within a few metres or may grade into a banded, cherty, sulphide-bearing tuff or exhalite. Banded ore or exhalite sometimes also underlies the massive sulphide.

Stringer Pipe. Similar concentric metal zoning patterns exist in the underlying, cross cutting zone or pipe of vein and disseminated mineralization. A typical pipe at Millenbach has a diameter of 15 to 100 m and may be of mineable grade over a vertical range of 50 to 150 m. Copper contents of 3 to 6% at the top of the pipe decrease with depth. Below one per cent Cu, grades drop off rapidly within a few

TABLE I
MILLENBACH MINE: TOTAL TONNES AND GRADES (MASSIVE AND STRINGERS) UNDILUTED

	Massive Sulphides						Stringers						Combined Massive + Stringers					
Lens	Tonnes	Cu (%)	Zn (%)	Ag (g/t)	Au (g/t)	Cu x 100 / Cu + Zn	Tonnes	Cu (%)	Zn (%)	Ag (g/t)	Au (g/t)	Cu x 100 / (Cu + Zn)	Tonnes	Cu (%)	Zn (%)	Ag (g/t)	Au (g/t)	Cu x 100 / Cu + Zn
1 & 2	21,145	5.81	5.58	60.10	1.016	51	30,755	1.87	0.95	16.65	0.452	66	51,900	3.42	2.84	34.42	0.677	55
3 & 4	33,564	3.52	9.07	68.56	1.383	28	17,291	2.66	1.18	16.65	0.339	69	50,855	3.23	6.39	50.79	1.016	34
5	4,430	2.59	10.36	46.84	0.621	20	9,146	3.55	1.33	29.06	0.705	73	13,576	3.24	4.28	34.99	0.677	43
6	181,352	4.07	8.78	87.47	0.508	32	49,380	3.37	1.12	23.98	0.367	75	230,731	3.92	7.14	73.93	0.480	35
7	21,214	4.26	8.58	54.74	1.326	33							21,214	4.26	8.58	54.74	1.326	33
8, 9, 10, 12, 13	773,112	6.39	11.08	92.55	1.552	37	1,074,792	3.17	1.03	15.24	0.310	75	1,847,904	4.52	5.23	47.12	0.818	46
11 & 18	33,764	6.90	15.09	96.22	0.536	31	21,008	3.19	1.20	37.53	0.282	73	54,773	5.48	9.76	73.64	0.451	36
14	71,394	3.59	12.10	99.32	1.157	23	79,557	3.89	1.67	31.60	0.508	70	150,951	3.75	6.60	63.49	0.818	36
15 & 20	25,897	5.87	11.38	111.45	1.157	34	23,233	3.29	1.70	46.27	0.705	66	49,130	4.65	6.80	80.70	0.931	41
16 & 21	16,441	3.18	10.86	99.88	2.088	23	52,142	2.47	1.13	23.14	0.310	69	68,583	2.64	3.46	41.48	0.734	43
17	4,239	6.41	5.10	40.91	0.705	56	5,849	1.34	0.79	11.85	0.367	63	10,038	3.47	2.60	23.98	0.508	57
19	4,191	8.35	4.56	87.47	1.947	65							4,191	8.35	4.56	87.47	1.947	65
22 (D-266)	39,279	4.69	8.22	64.05	1.890	36	17,509	2.45	0.70	18.34	0.762	78	56,788	4.00	5.90	49.94	1.524	40
23	10,496	2.26	15.20	76.75	0.818	13	21,648	2.10	2.27	42.04	0.367	48	32,144	2.15	6.49	53.33	0.508	25
D-68													40,915	3.82	5.02	61.79	1.975	43
TOTAL	1,240,517	5.63	10.62	89.44	1.411	32	1,402,310	3.13	1.10	18.06	0.339	69	2,683,741	4.29	5.56	51.92	0.875	38
Zinc Deposit	73,000	0.17	8.65	45.26	0.69	2							73,000	0.17	8.65	45.26	0.69	2

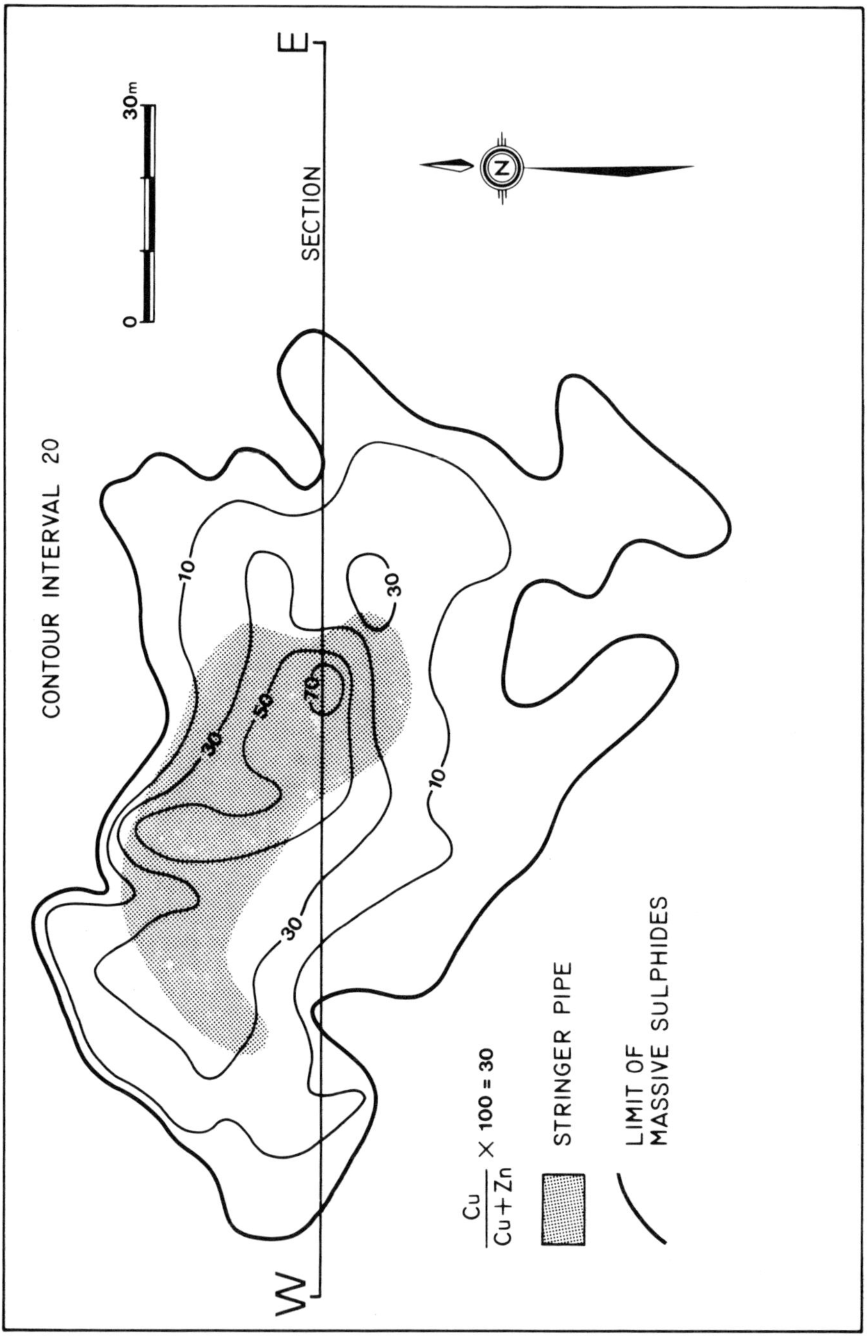

Figure 8. Plan #6 Lens, contoured Cux100/Cu+Zn metal ratios.

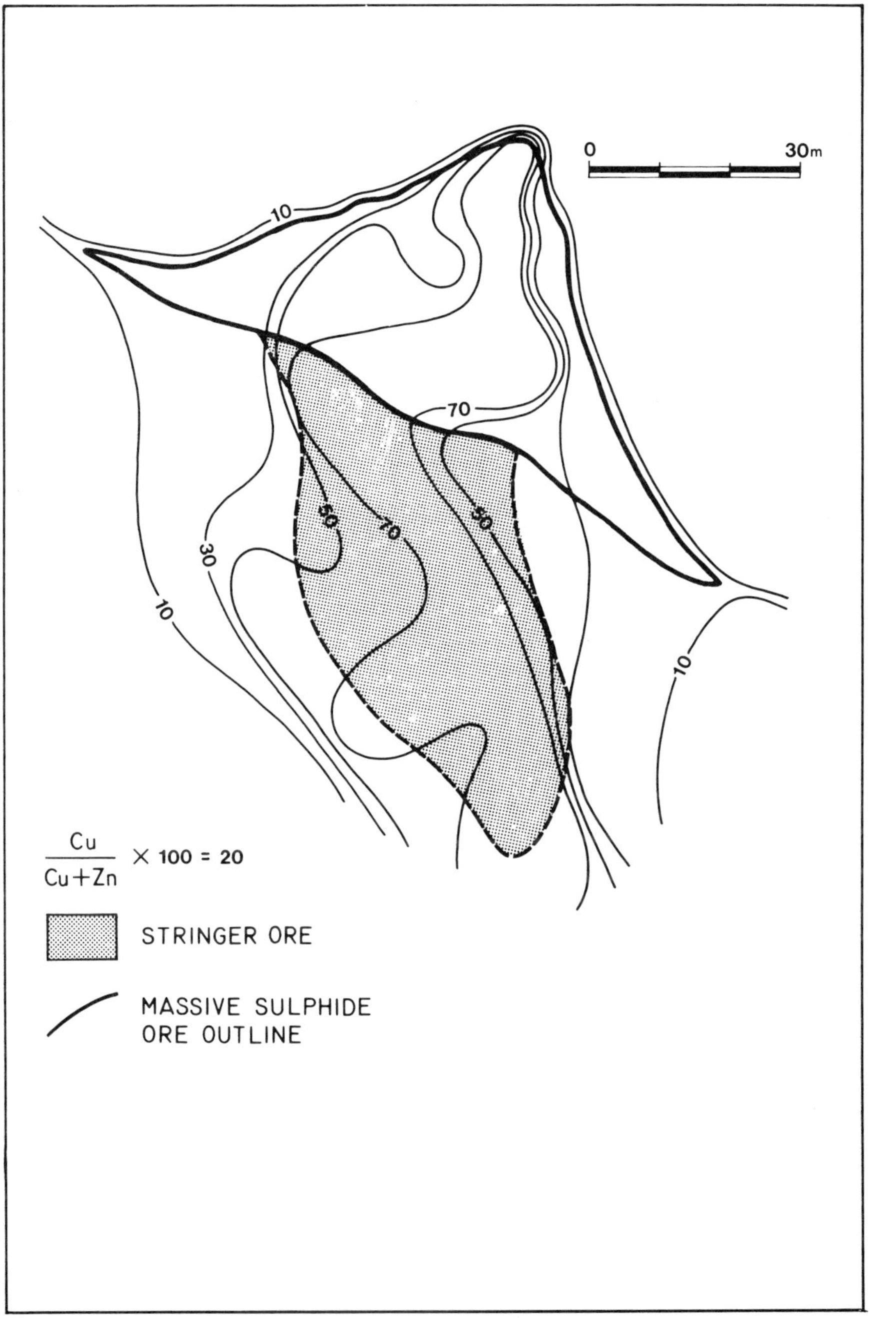

Figure 9. East-west section #6 Lens, contoured Cux100/Cu+Zn metal ratios.

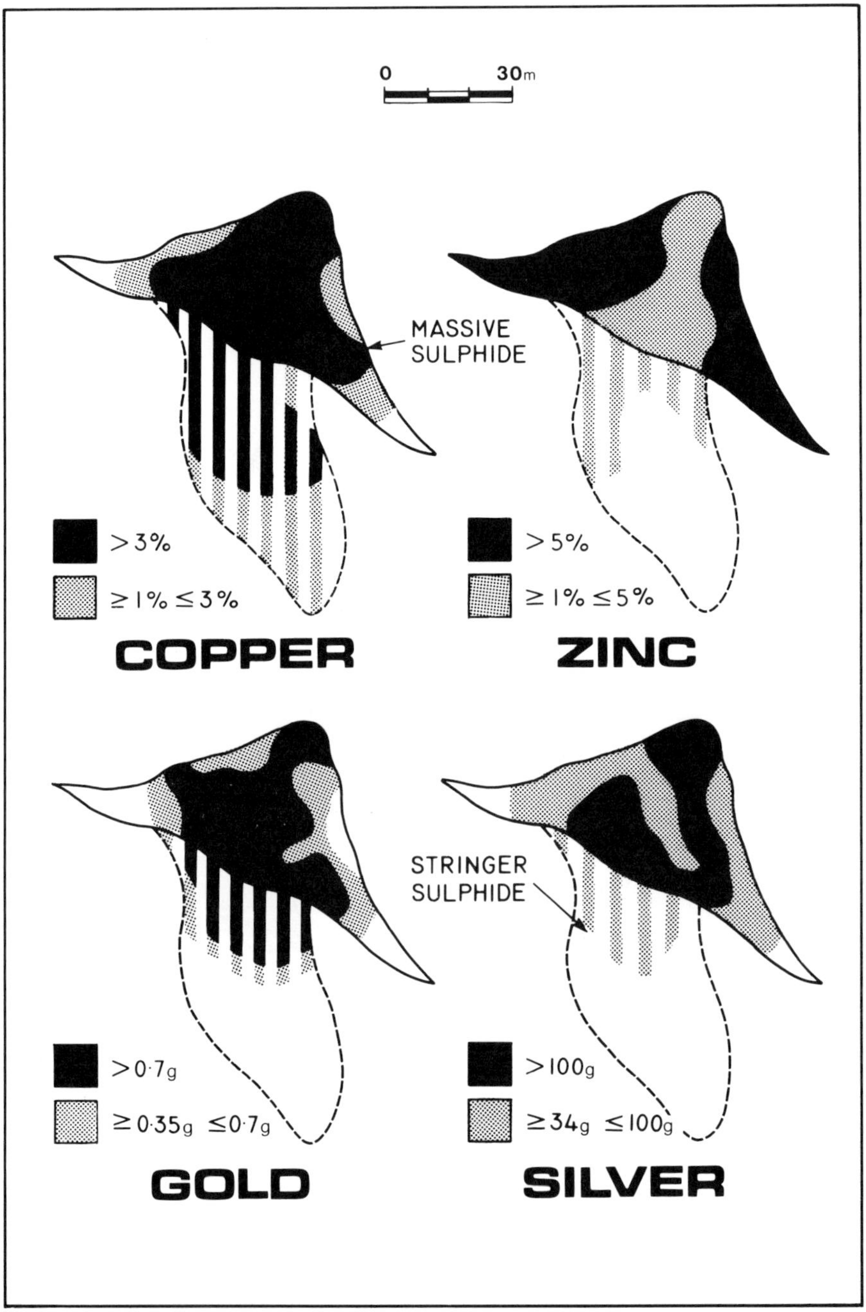

Figure 10. East-west isograde sections #6 Lens: showing the distribution of copper, zinc, gold and silver.

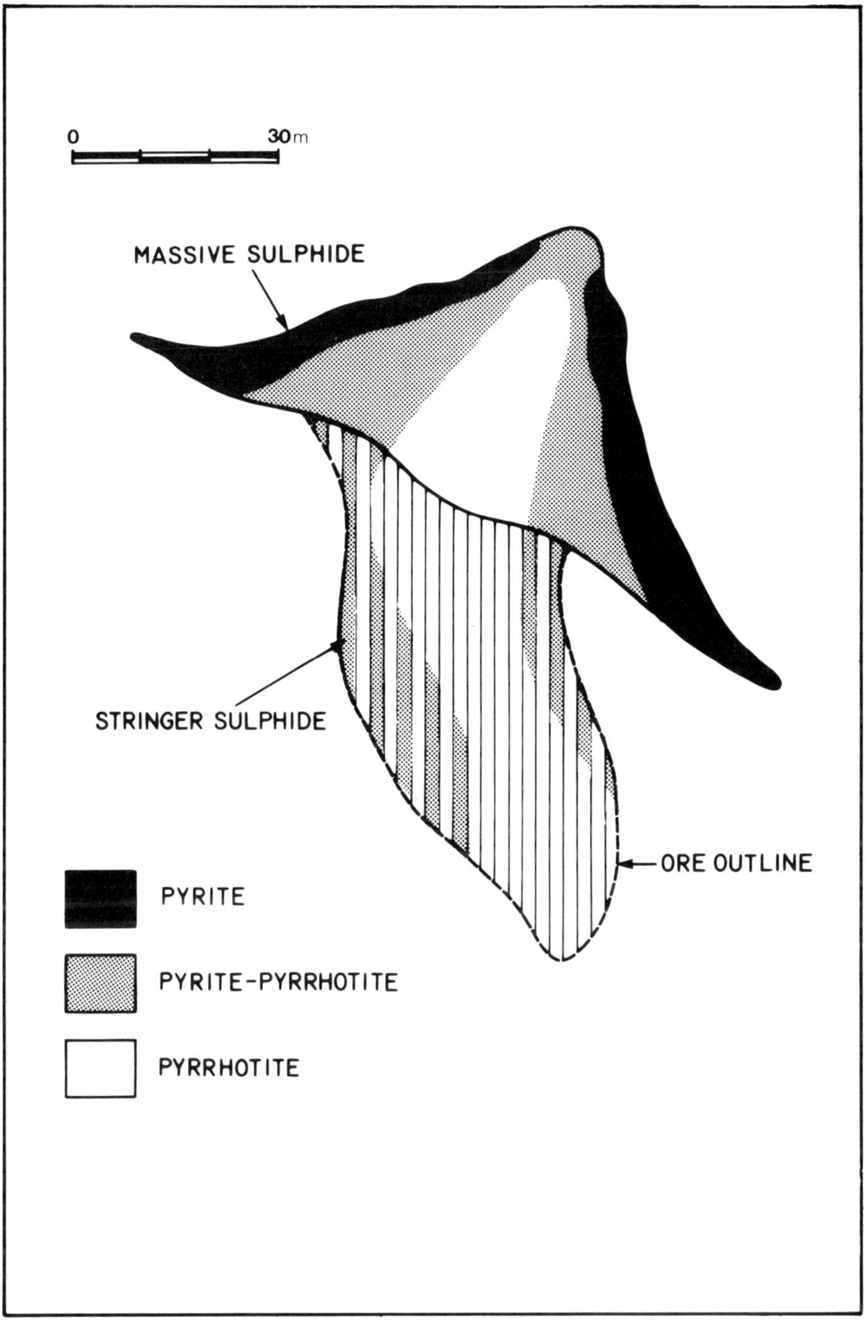

Figure 11. East-west section #6 Lens, iron mineral distribution.

metres both laterally and vertically. Surrounding the copper-rich core, an irregular fringe of zinc mineralization grading one to 3% Zn may persist for one to 10 m.

For comparative purposes isograde contours for copper, zinc, silver and gold are shown in Figure 10 and on the same section of #6 orebody as shown in Figure 9. Significant silver and gold values are usually restricted to massive sulphides and, in contrast to those at Corbet are commonly associated with higher copper grades. Figure 11 illustrates the distribution of iron minerals in #6 orebody. A vertical and upward trend from pyrrhotite to pyrite is present but the major shift is lateral, both in the stringer and the massive sulphides. The #6 orebody is unusual in that some slumping has occurred in the banded ore on the steeper southeast flank. Preconsolidation slump folds are well preserved locally in banded ore with an unusually high cherty-tuff component.

Two cross-sections through #16 orebody (Fig. 12) illustrate several interesting features. The lenticular rather than conical shape of the massive sulphide cap on the east-west section is due to the location of the section across the flank rather than through the centre of the cone. For its size, this orebody has a particularly elongated stringer zone. About 75 to 100 m below the massive sulphides the stringer zone intersects two tuffaceous exhalite layers known as the "Lower A" and "C" after the names of the orebodies occurring at these stratigraphic levels at the Amulet Mine. Where intersected by the stringer pipe these horizons have much higher per cent copper/copper + zinc ratios than normal (95 vs. 7). In this area the "C" exhalite has a much higher sulphide content than the "Lower A". Although no orebodies have been found in the "C" layer at Millenbach within the #16 orebody stringer pipe, this 30 cm thick exhalite of massive pyrite-pyrrhotite has been replaced by up to 55% of chalcopyrite. Normal background values rarely exceed 200 ppm copper.

Mine Scale Zoning

The 15 orebodies, each with its own massive cap and underlying stringer zone, are shown reconstructed and projected vertically onto a horizontal plan in Figure 13. Reconstruction has eliminated the effect of faults and dykes that dislocate almost every orebody. For mining purposes each dislocated block of ore was assigned a lens number. For example, lenses 8, 9, 10, 12 and 13 are all dilated or faulted portions of what was once Millenbach's single largest orebody of 1.85 million tonnes (Table I) and are known collectively as Main Lens. Some small lenses that developed at lower rock/sea-water interfaces are obscured in Figure 13 by larger overlying lenses. Northeast and northwest trends are apparent with the largest orebody located at their intersection. Eighty percent of mineable Millenbach ore was deposited within a 275 m diameter circle centred at the main hydrothermal vent (Fig. 13).

In the cluster of orebodies at Millenbach formed at the same stratigraphic horizon, the ratio of percent copper to copper plus zinc decreases steadily with increasing distance from the main hydrothermal vent, and this is true particularly for the massive sulphide component. The percent metal ratios are in the order of 45 to 55 in orebodies developed over major vents. Orebodies developed 300 m from the major vent have ratios in the 35 to 45 range whilst those 800 m away are relatively zinc-rich with ratios in the mid twenties. These more distant orebodies may contain a little galena, and higher zinc values occur in their stringer zones, however, they still possess relatively copper-rich cores. Southeast of Millenbach and east of Amulet,

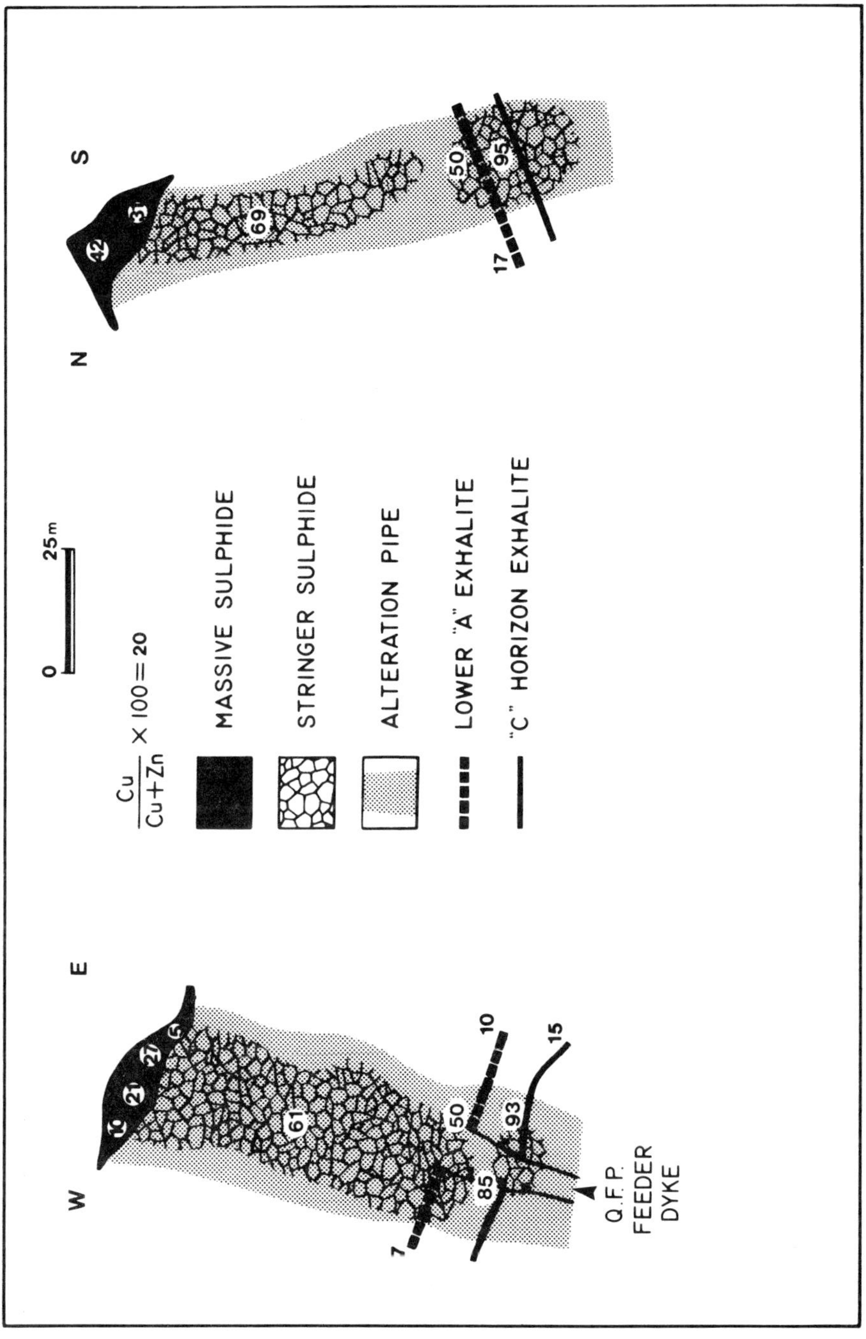

Figure 12. North-south and east-west sections #16 Lens showing metal ratios.

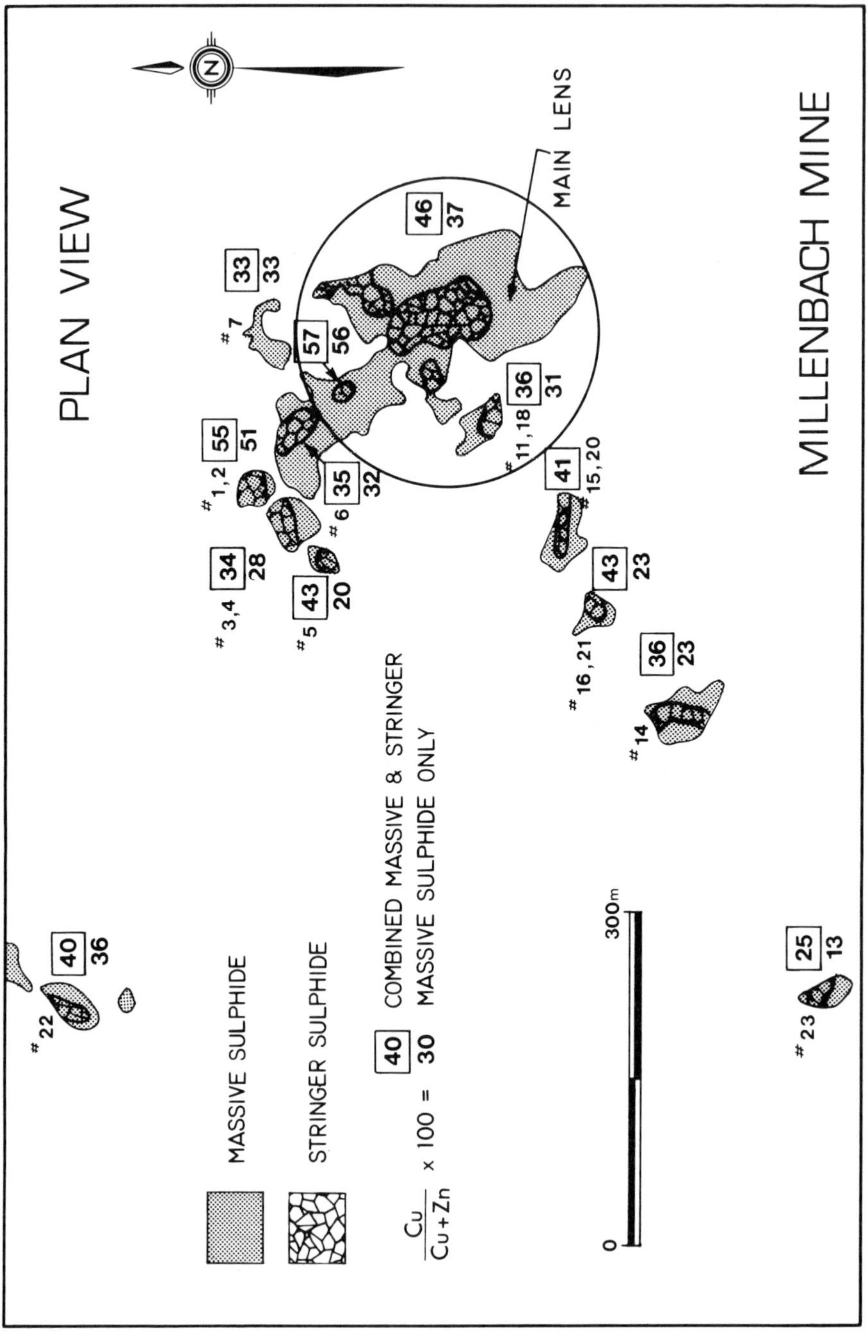

Figure 13. Plan of Millenbach deposit showing metal ratios.

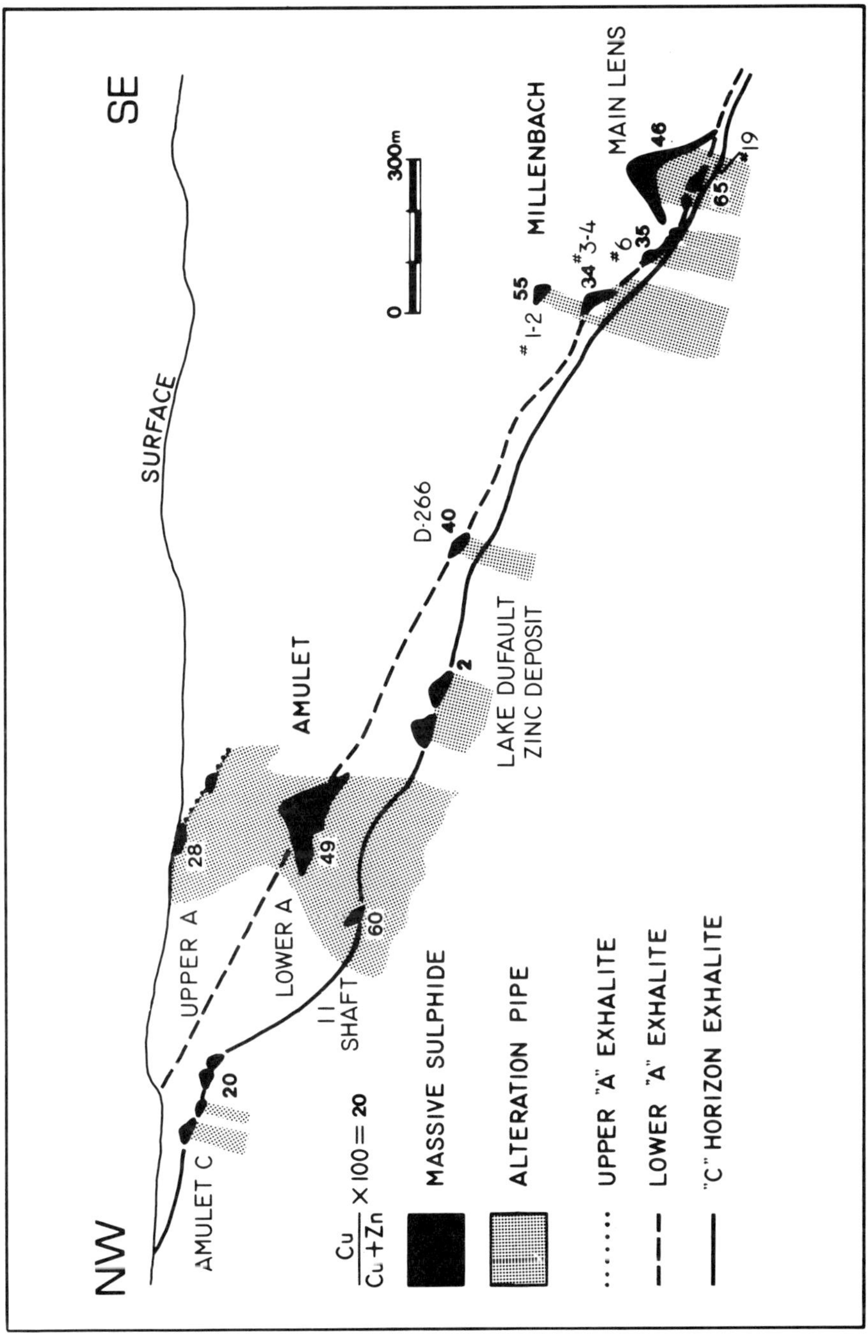

Figure 14. Northwest-southeast section through Millenbach and Amulet showing metal ratios for combined massive and stringer ore.

zones of heavily disseminated to semi-massive "barren" pyrite occur in the "C" exhalite. These zones rarely carry zinc values in excess of one per cent and have per cent metal ratios of less than 10. Underlying alteration is weak with minor soda depletion. Such barren zones probably represent deposition from much cooler solutions on the extreme fringe of a hydrothermal system.

Figure 14 is a northwest-southeast vertical section through various orebodies of the Amulet and Millenbach deposits. Percent metal ratios are plotted for combined massive and stringer ore. At Amulet, copper decreases with respect to zinc both stratigraphically upward and laterally. The lateral zoning of the lenses, though not of the exhalite itself, is more pronounced along the lowermost "C" exhalite where relatively distant orebodies with ratios of 2 and 20 flank the main hydrothermal pipe containing an orebody with a ratio of 60. The "Upper A" orebody, developed 300 m higher in the statigraphic sequence has a higher absolute copper content as well as a higher ratio, 28 vs 20 and 2, than either of the older, lower flanking orebodies in the "C" exhalite. It is suggested that vertical zonation in metal ratios in the main Amulet hydrothermal conduit from 60 to 49 to 28 is due to copper enrichment of earlier-formed massive sulphides by resurgent hydrothermal fluids that deposited the younger and stratigraphically higher orebodies. Field evidence in the #s 13 and 16 orebodies at Millenbach shows that this took place by fracturing and veining of the older sulphides and exhalite horizons, and to some extent by replacement of pyrite and pyrrhotite by chalcopyrite. Midway between the main Amulet and Millenbach hydrothermal vent areas a small ore lens known as D-266 (or #22 Lens) has a metal ratio of 40, perhaps characteristic of a small local centre. At Millenbach, lateral zonation of 65 to 34 is again more consistent and pronounced than any vertical change. The youngest lens in fact, is copper-rich with a metal ratio of 55.

Understated Zinc Content. The zinc content of these deposits tends to be understated for three reasons. First, because zinc is a less valuable metal, higher grades are required before zinc ores can be included in official reserves. Second, whereas copper ores occur in the thickest parts of the lens and are high grade and therefore mined, zinc ores tend to occur on the fringes of the lens, are usually much thinner and therefore may not be profitably mineable. Third, the source data for older deposits is often obtained from production statistics in which the zinc content of the original ore is frequently understated by as much as 20% due to lower zinc recoveries in milling which are typically only 70 to 80% compared with 90% for copper.

The metal ratios used in this paper for the Millenbach Mine are all derived from ore reserve data (Table I); those for the Amulet Mine are from production statistics, adjusted for differences in copper and zinc recoveries (B. Bancroft, pers. commun.). Metal ratios for combined massive and stringer mineralization (Figs. 13, 14) do not take into account zinc contained in the submarginal grade halos of the stringer zones. Inclusion of this zinc mineralization would not alter the ratios significantly except in the case of #16 orebody. Here, because of the abnormally large proportion of stringer ore, it is estimated that the combined ratio would decrease from 43 to 40.

Nature of the Massive Sulphide Precipitate

Evidence at Millenbach clearly supports Spence's conclusion (Spence, 1975) that massive sulphides were deposited as hard, encrusting sinters rather than as soft gels or muds. This evidence includes; 1) the steep-sided conical shapes of the orebodies, 2) their location on the tops of earthquake-prone rhyolite domes, 3) angular fragments of massive sulphides cemented by slightly later sulphides in the breccia ore, and 4) the general absence of pre-consolidation slump structures, except in #6 Lens as already discussed. In two locations, portions of the southern lobe of the Main lens and possibly #7 Lens, where slumping may have occurred, the massive sulphides appear to have moved as a single solid mass.

Other evidence at Millenbach supporting the above conclusion is the local presence of lithic andesitic tuff consisting of accumulations up to 15 m thick of angular fragments of Q.F.P. and massive sulphides in a fine-grained andesitic matrix. These massive sulphide fragments or blocks range in size from a centimetre to over 2 m in length and locally are sufficiently numerous to make ore. Lithic tuff occurs only on the south-east side of the orebodies with which it is associated and is a useful guide to ore on a mine scale. Elsewhere, small blocks of massive sulphide have been caught up in the overlying Amulet Andesite. The consistent position of the lithic tuff at Millenbach and also Norbec, suggests that one or more of the following may have happened: 1) regional tilting to the south-east may have de-stabilized the steep-sided orebodies giving rise to avalanches on the down-slope slide, 2) strong underwater currents, perhaps caused by renewed volcanic activity, carrying andesitic ash scoured the sea-bed and deposited part of their load in the lee of the Q.F.P. domes and prominent massive sulphide caps, and 3) andesitic flows encroaching from the northwest engulfed any remaining loose debris. The angularity and sharp margins of the blocks show that the massive sulphides must have been well indurated at this time.

Metal Content of the Hydrothermal Fluids

Before interpreting the metal zoning at Millenbach, it is useful to reiterate briefly the major features of metal distribution. These are that very little zinc occurs in the stringer pipe, that copper predominates over zinc in the copper spine, that the two metals occur in roughly equal proportions, or with zinc somewhat more abundant in the breccia ore, and finally, that zinc usually occurs in much greater quantities than copper in the banded massive ore on the fringes.

In contrast with the Norbec deposit, extensive metal-rich exhalites do not occur at Millenbach associated with the main ore horizon. At Norbec the ore horizon grades laterally into a sulphidic, base metal-rich, andesitic tuff averaging less than one metre in thickness. The tuff may or may not include a cherty exhalative component. Based on drill hole sampling on a 100 m grid over an area of 2.5 km^2, copper and zinc were precipitated in the same overall proportions as in the orebody (ratio 39) but with no discernible zoning pattern. The quantity of the two metals in the tuff within this area is equal to about 10% of that in the Norbec orebody. It may reasonably be concluded that these exhalites precipitated from metal-rich hydrothermal brines dispersing in sea water over a fairly wide area (Sato, 1972), but that such precipitation, per se, did not contribute significantly to the orebody.

Any interpretation of this metal zoning must take into account not only the juxtaposition of the copper and zinc-rich zones but also the sinter-like nature of the precipitates and the absence of copper-zinc zoning laterally along the more extensive "C" and Norbec exhalites deposited under relatively quiescent conditions.

The metal zoning pattern in a simple lens at Millenbach and similarly at the Amulet and Norbec Mines is thought, therefore, to reflect very local conditions and to result from the net effects of sudden changes in Eh, pH, pressure and temperature over relatively short distances (Large, 1977) rather than any major, time-related changes from copper to zinc-rich, in the composition of the primary hydrothermal fluids. The occurrence, in the same hydrothermal centre, of copper-rich orebodies higher in the stratigraphic column than zinc-rich orebodies also supports this interpretation. This condition is evident in comparing the Amulet Lower and Upper A orebodies with the Amulet C and Lake Dufault Zinc Deposits, and also in comparing the Millenbach #1-2 Lens with the #3-4, 5 and 6 Lenses (Figs. 4, 6, 13, 14). The evidence suggests that in a simple lens, massive sulphide formation terminated with the gradual plugging of the conduit by precipitation of chalcopyrite to form the stringer zone and copper spine. Meanwhile copper-depleted and therefore relatively zinc-enriched hydrothermal fluids, escaped to the seafloor through a decreasing number of fractures, and simultaneously deposited massive banded, sphalerite-rich ores radially from the spine on top of earlier-former massive sulphide breccia ore. The breccia ore is believed to have resulted from repeated refracturing and dislocation of sulphide sinters by hydrodynamic pressure including possibly, small steam-blast explosions. Shape and complexity of the orebodies were influenced, no doubt, by additional factors such as topography, frequency and strength of earthquakes, strength and persistence of the hydrothermal system and strength and direction of submarine currents.

Orebodies farther from the main vents show two distinct characteristics. First, they are relatively base metal poor, with lower or much lower grades than those formed close to the main hydrothermal vent. Second, their copper content relative to zinc and iron is low. It is therefore suggested that the metal zoning pattern, at mine scale, may be attributed to the precipitation of significant, but uneconomic quantities of base metals, particularly copper, in the conduits feeding the more distant lenses. This may be due to decreasing temperatures away from the main hydrothermal centre resulting from longer solution paths and possibly greater mixing with sea water. Some evidence of this precipitation is provided by deep exploration holes at all three mines (Amulet, Corbet and Millenbach). These holes penetrate to more than 1000 m below the main ore horizon at Millenbach and show base metal enrichment (deposition) rather than base metal leaching of the various rock units where these are intersected by the alteration pipes.

The evidence presented above does not support the view that the hydrothermal fluids have evolved from copper-rich to zinc-rich with time in the case of the Millenbach-Amulet hydrothermal field. Distance from a major hydrothermal vent appears to be more important in determining the metal ratios in orebodies at the same stratigraphic horizon. In particularly persistent vents where orebodies have formed vertically above each other at successively higher stratigraphic positions, the older lenses have become enriched in copper relative to zinc as the system of stacked lenses developed.

On a district scale the Corbet, Millenbach, Amulet, Old Waite, East Waite, Vauze, Norbec, Quemont and Horne orebodies (Spence and de Rosen-Spence, 1975) all showed essentially similar base metal zoning characteristics of relatively copper-rich ore in and over the main vent and zinc-rich ore farther away. Although somewhat higher stratigraphically, the small, zinc-rich Delbridge deposit (Boldy, 1968) showed the same general characteristics as some of the satellite orebodies at Millenbach and Amulet. The #1 lens at Delbridge contained 90,000 tonnes grading an unknown per cent Cu, 7.7% Zn, 68 g Ag/t, 5.8 g Au/t, and the #2 lens contained 372,000 tonnes grading 0.61% Cu, 9.66% Zn, 110 g Ag/t, 2.8 g Au/t). The #2 lens, composed of massive sulphides with a metal ratio of 7, was located above an alteration pipe with very weakly developed, uneconomic stringer sulphides. Nevertheless Delbridge showed concentric zoning with ratios of up to 24 in the centre of the lens.

The more isolated West MacDonald, 8 km east of Millenbach, and Mobrun, 15 km northeast of Millenbach, have metal ratios of 2 and 21 respectively. These deposits occur significantly higher in the stratigraphic sequence and may indicate some evolution toward zinc-rich fluid compositions with time. Alternatively they may fall into the above pattern and be relatively zinc-rich satellitic or distal orebodies of other major copper deposits yet to be discovered.

WALL-ROCK ALTERATION

Introduction

The alteration zones of two distinct massive sulphide bodies, the Main Lens and #14 Lens, were chosen for a study of the wall-rock alteration associated with economic mineralization at Millenbach (Fig. 5 and Table I). The results presented here are based on detailed mapping of alteration types in underground exposures and drill core, supported by petrographic work and chemical analyses of selected specimens of altered rocks. Only the main conclusions of the study are presented, but additional data are available elsewhere (Riverin, 1977; Riverin and Hodgson, 1980).

Metamorphism

Petrographic studies indicate that most primary alteration minerals have been recrystallized to hornblende-hornfels assemblages. This metamorphism was probably caused by the intrusion and cooling of the Lake Dufault granodiorite, as suggested by de Rosen-Spence (1969). The most notable effect of this metamorphism is perhaps the development of a spotted texture in rocks of particular bulk composition. Spotted rocks of this type were called "dalmatianite" by early workers in the area (Cooke, 1930; Wilson, 1935). At Millenbach, the minerals formed during metamorphism include cordierite, anthophyllite, biotite, andalusite, gahnite, hornblende and plagioclase, the specific assemblage depending on the bulk composition of the rocks, which in turn depends in part on pre-metamorphic metasomatism. The distribution of anthophyllite at Millenbach indicates that the metamorphic isograd defined by the reaction:

chlorite + quartz = cordierite + anthophyllite

lies between the Main and #14 Lenses (Fig. 13). Indeed, anthophyllite is present in the alteration zones of the Main Lens, but is absent around #14 Lens in rocks of the same bulk composition. The location of the isograd is consistent with the fact that #14 Lens is farther from the granodiorite intrusion than is Main Lens (Fig. 7), and

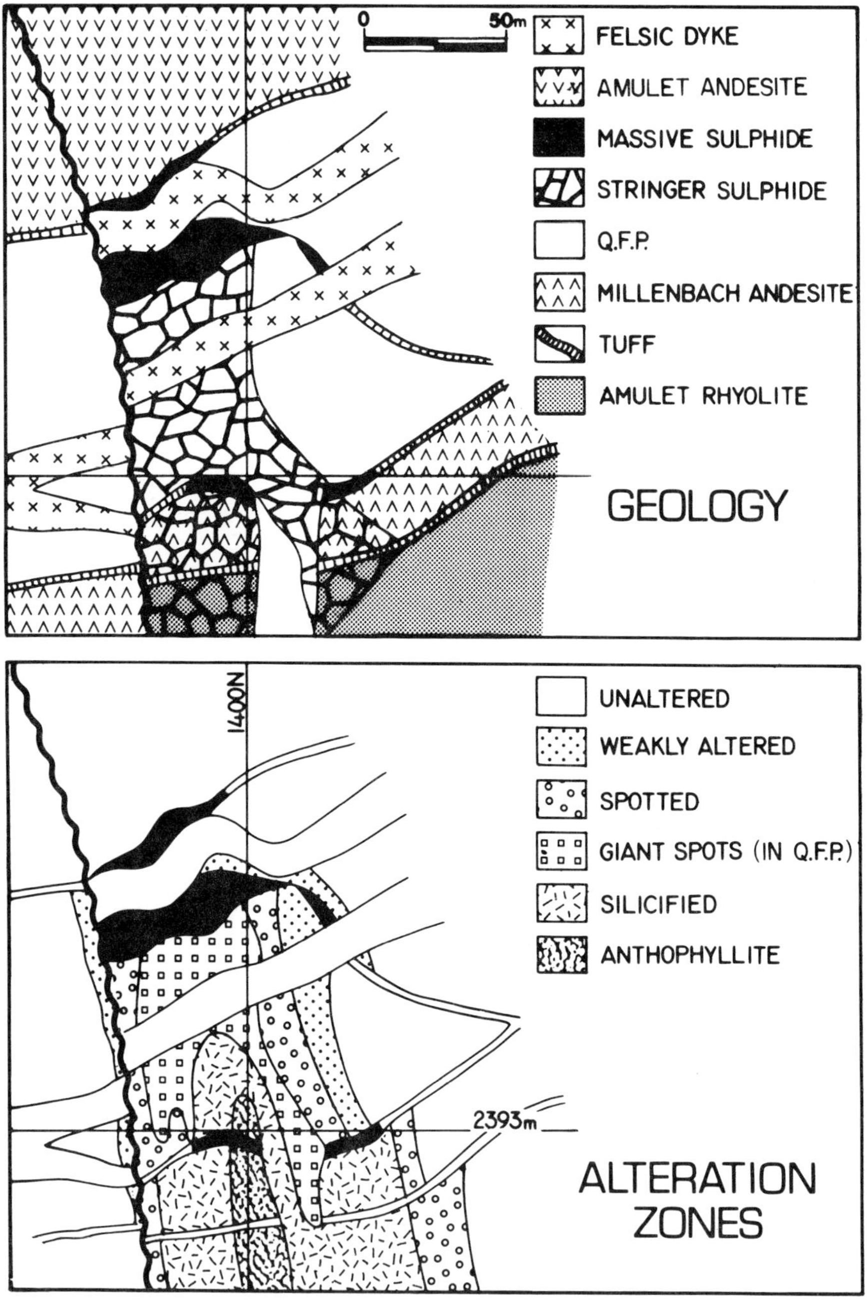

Figure 15. Geology and alteration zones around the Main Lens, Millenbach deposit: Section 12450E, looking west.

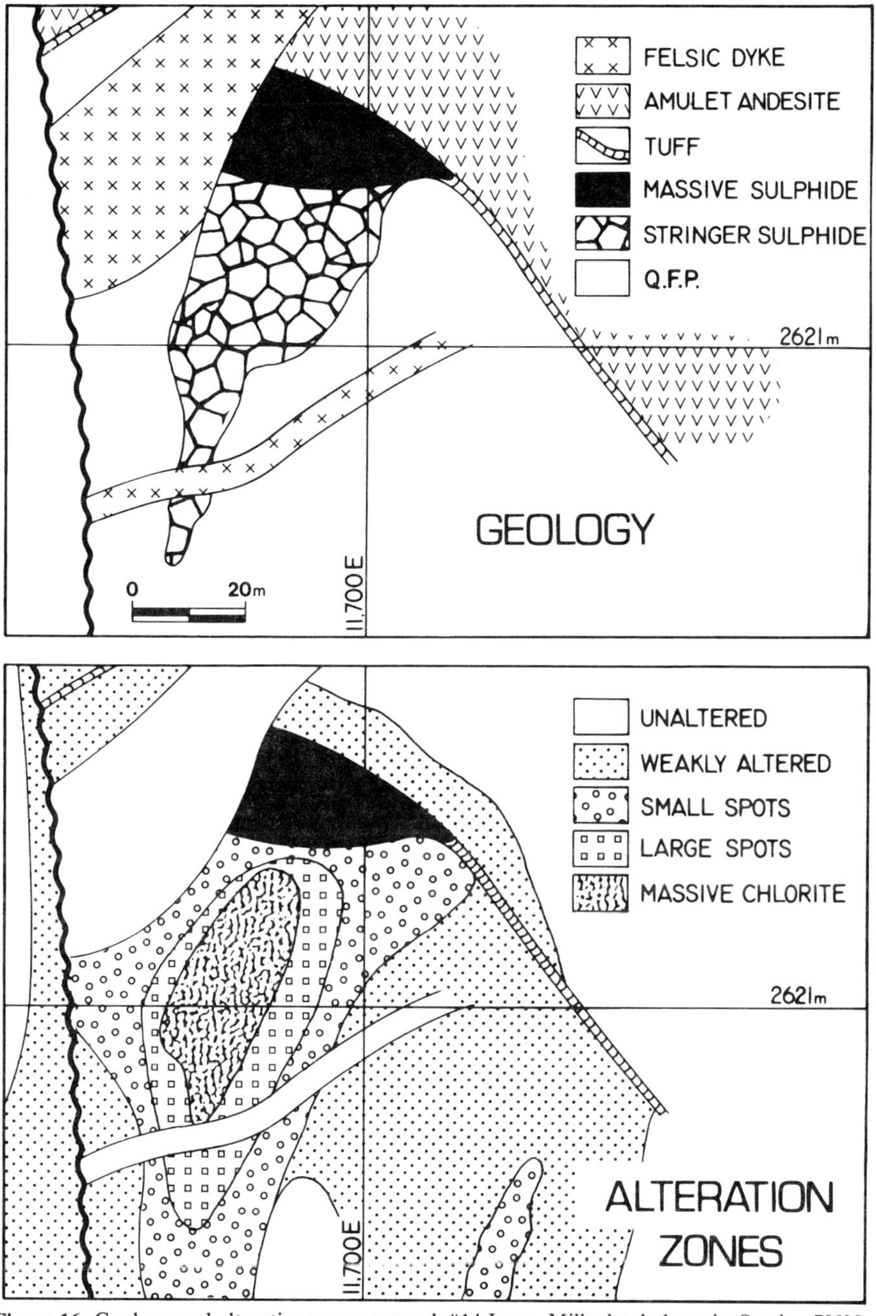

Figure 16. Geology and alteration zones around #14 Lens, Millenbach deposit: Section 700N, looking north.

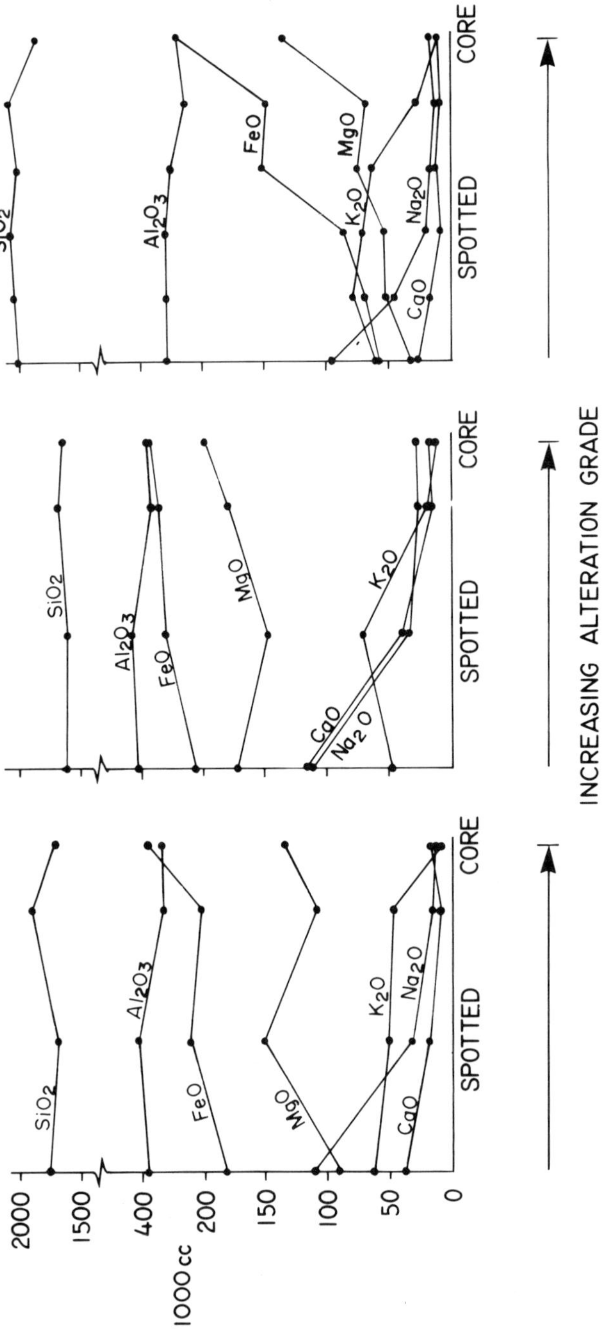

Figure 17. Chemical variations across alteration pipes assuming constant volume during metasomatism.

indicates that metamorphic temperatures of 555°C to 600°C must have prevailed in the Main Lens area (Akella and Winkler, 1966).

Distribution of Alteration Facies

Based strictly on field observations, a series of alteration facies have been defined and mapped in detail. Characteristically, altered rocks formed pipe-shaped zones underlying massive sulphide bodies, although weakly altered rocks also occur locally above the bodies. At the Main Lens, the alteration facies have a definite zonal distribution (Fig. 15). The core of the alteration pipe is characterized by cordierite and anthophyllite. This assemblage grades outward through a silicified zone into a biotite-rich zone characterized by a spotted texture (dalmatianite). The spotted zone grades finally into fresh rocks. This zonal distribution can be observed in all three rock types, Amulet Rhyolite, Millenbach Andesite and Q.F.P. At #14 Lens, the distribution of alteration facies is similar, except that massive chlorite forms the core of the pipe instead of cordierite-anthophyllite (Fig. 16)

Chemical Zoning

The nature and importance of metasomatism was studied by analyzing several representative samples of altered and unaltered rocks. Because of the difficulties involved in getting sufficient samples on a single vertical section, the average chemi-

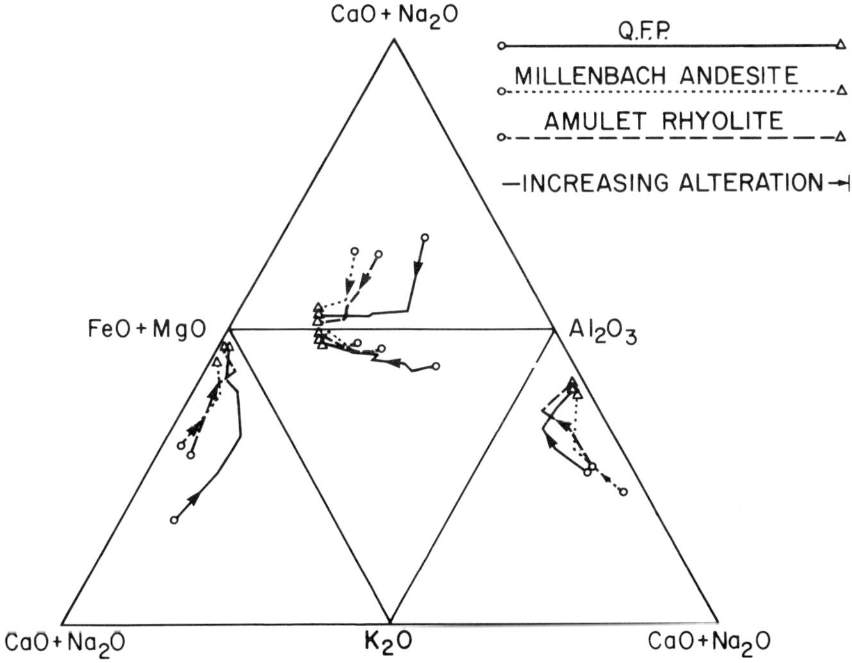

Figure 18. Chemical variations across alteration pipes on unfolded F-C-A-K tetrahedron. In all triangles rocks of the spotted zone are at the main inflexion point of the metasomatic trends.

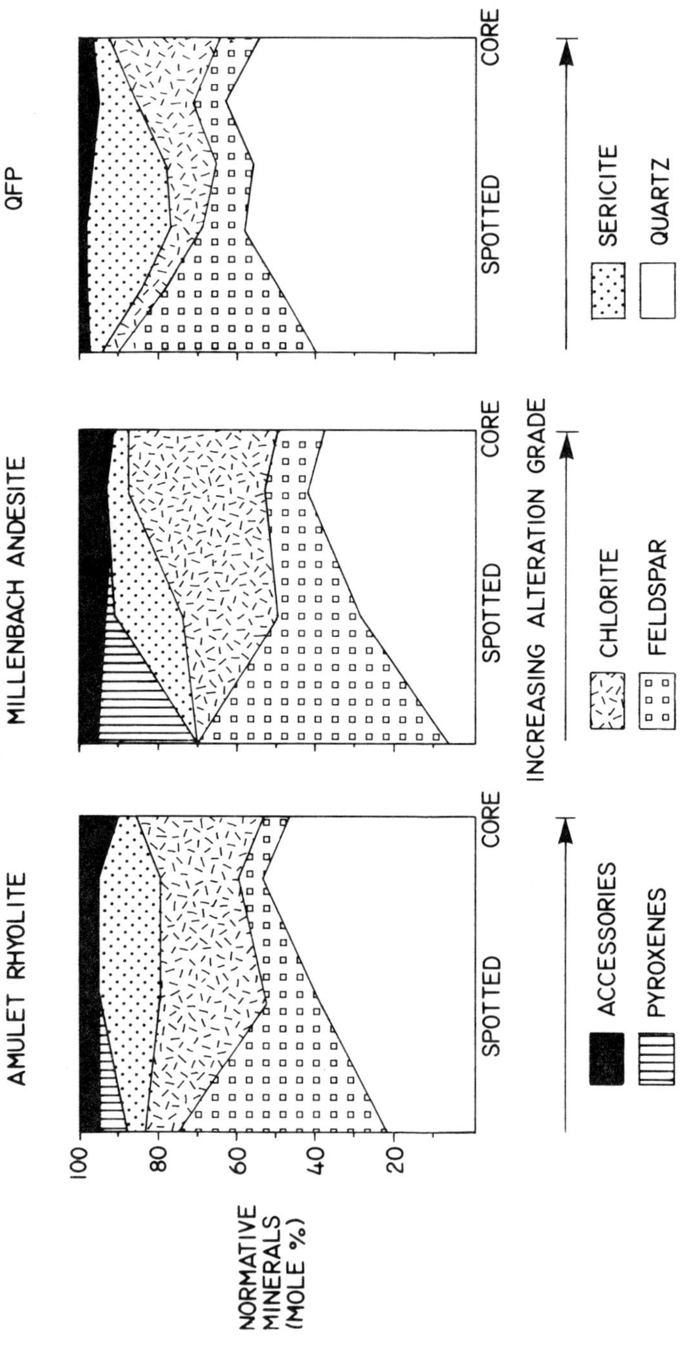

Figure 19. Normative mineral zoning across alteration pipes.

cal compositions of the various alteration zones were measured. Since the distribution of the alteration zones is known, the average composition of the zones can be used to determine chemical patterns across an idealized alteration pipe. Analytical data calculated on a sulphide-free basis are shown on Figure 17. Assuming that volume was unchanged during metasomatism, the data shown on Figure 17 represent the net exchange of constituents.

Leaching of CaO and Na_2O, and enrichment in FeO (total Fe) and MgO are the most prominent features of metasomatism at Millenbach. CaO and Na_2O are strongly depleted at the margin of the pipe and remain at a constant low value across the inner zones of the pipe. K_2O shows a slight enrichment at the margins of the pipe, followed by a strong depletion at the core of the pipe. SiO_2 and Al_2O_3 were the least mobile, although the latter tends to be slightly enriched at the margin of the pipe and slightly depleted at the core.

Figure 18 shows the same chemical data as Figure 17, but plotted in the unfolded tetrahedron (FeO + MgO) – (CaO + Na_2O) – Al_2O_3 – K_2O, called hereafter the F-C-A-K tetahedron. This chemical system was found to be the most applicable to the metasomatism at Millenbach. By plotting the average compositions of the alteration zones in this tetahedron, metasomatic trends have been defined. The main features shown by these trends are first, that (CaO + Na_2O) decreases abruptly, relative to other components, in the outer part of the alteration pipe, where the rock is altered from the fresh to the spotted facies. Second, in the inner part of the alteration pipe, (CaO + Na_2O) does not change appreciably relative to other components, but (FeO + MgO) increases markedly relative to K_2O and Al_2O_3, whereas the latter increases relative to K_2O. For all three rock types, the trends are similar and consist of two linear parts, each with a different orientation. This suggests that the chemical processes operating at the margins of the alteration pipe were different from those operating at the core. It is also notable that, in the F-C-A-K system, a uniform bulk composition has been achieved at the core of the pipe, regardless of the original rock type.

Normative Mineral Zoning

In order to estimate the proportion of original alteration minerals across an "idealized" alteration pipe, the chemical data was recalculated into normative alteration minerals. This, for example, allows the metamorphic cordierite-anthophyllite-biotite to be reconverted into the original chlorite-sericite-quartz. Figure 19 shows the proportion of various normative minerals across the idealized alteration pipe, using the chemical data of Figure 17. Figure 19 indicates that original (magmatic) feldspars and pyroxenes were completely destroyed at the margin of the alteration pipe. Chlorite shows an almost steady increase toward the centre of the pipe, whereas sericite reaches a maximum in the spotted facies and becomes steadily less abundant toward the core. The diagram also shows that in spite of relatively constant SiO_2 (Fig. 17), large quantities of quartz were produced simply by the breakdown of primary igneous minerals.

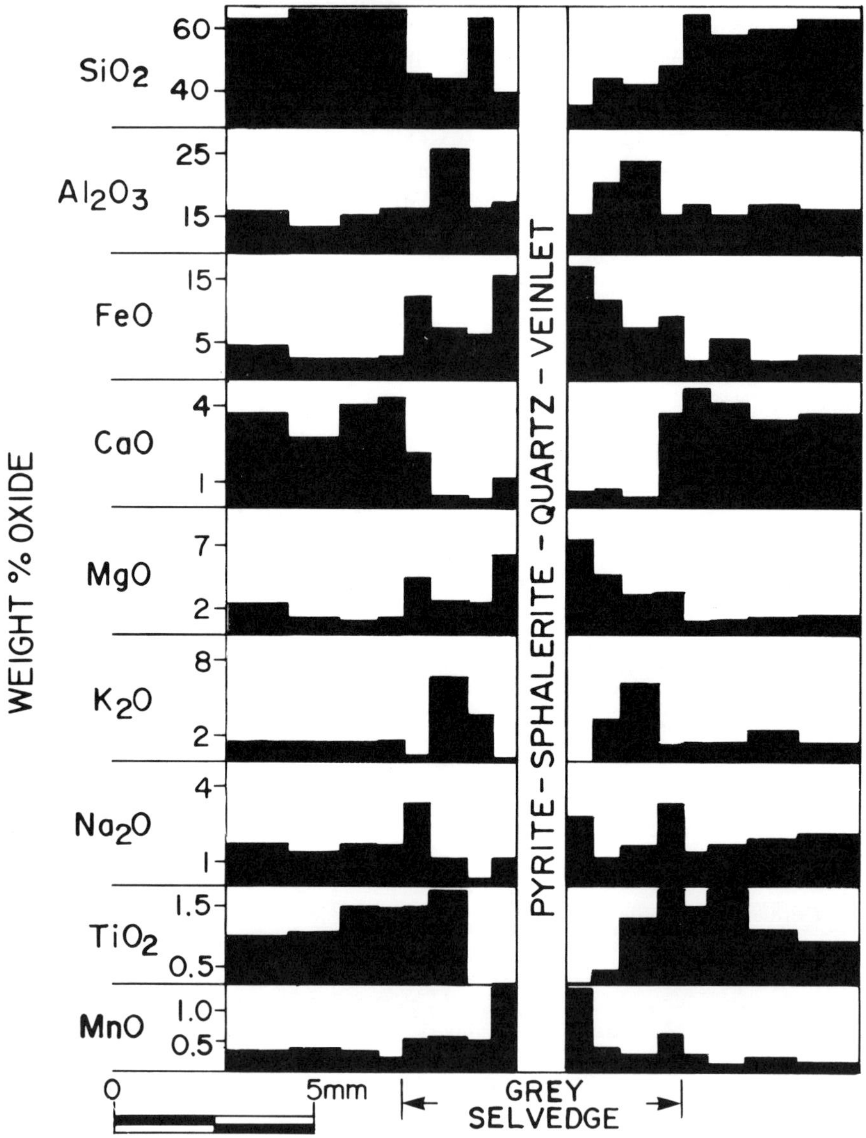

Figure 20. Electron microprobe analyses (weight %) across an alteration selvedge bordering a pyrite-sphalerite-quartz veinlet. "Grey selvedge" refers to a well defined alteration envelope, including both chloritic and sericitic layers, bordering sulphide veinlets in the spotted facies of Millenbach Andesite.

Alteration Selvedges Along Veinlets

Alteration mineral zoning along veinlet selvedges is analogous to that observed on the scale of the alteration pipes. A chloritic envelope is directly in contact with the sulphide veinlet, and grades abruptly outward through a narrow zone of intermediate composition into a sericitic zone (Riverin, 1977). Together these chloritic and sericitic layers form a grey selvedge along sulphide veinlets.

Figure 20 shows that the chemical composition across the selvedges is remarkably similar to that across the pipes. For example, CaO is depleted at a low value and remains so across the selvedge, whereas FeO andMgO are enriched across the selvedge toward the sulphide veinlet. Furthermore, both K_2O and Al_2O_3 reach peaks at the outer sericitic margin but are depleted at the inner chloritic envelope.

Figure 21 shows the same data as Figure 20, but plotted in the F-C-A-K tetrahedron and indicates that the metasomatic trends at the veinlet scale are similar to those developed at the scale of the alteration pipes. Most notable is the persistence of a sharp break in the metasomatic trends at both scales.

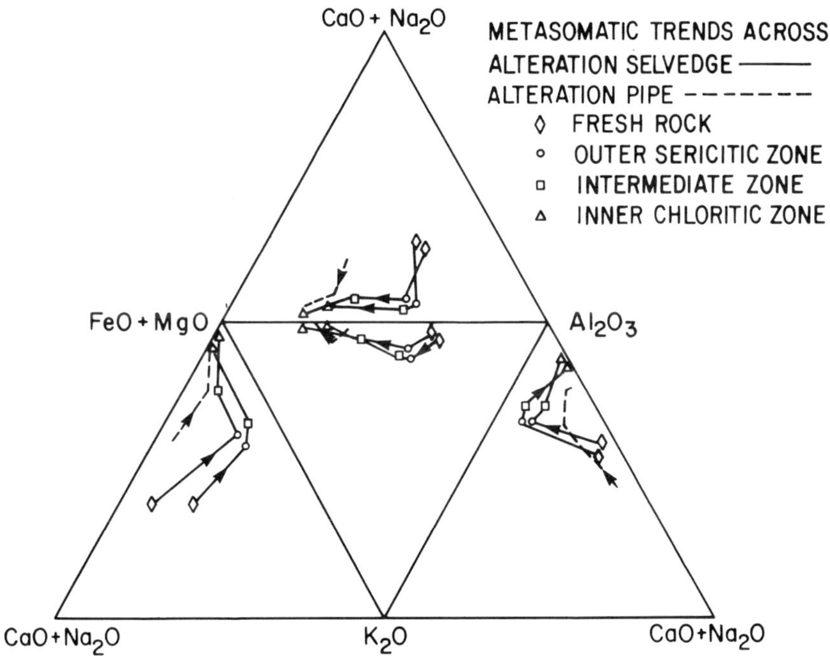

Figure 21. Chemical variations across an alteration selvedge bordering a pyrite-sphalerite-quartz veinlet, on the unfolded F-C-A-K tetrahedron.

Alteration Model

The best way to interpret these chemical and mineralogical data is to consider alteration by two different processes, or two types of alteration reaction as follows: 1) the breakdown of feldspar into sericite and quartz, and 2) the replacement of this sericite by chlorite. These two processes, taking place simultaneously but in different places, are respectively responsible for the formation of the sericitic margin and chloritic core in a developing alteration pipe (Fig. 22).

Model reactions leading to the development of the sericitic zone at the expense of the fresh rock, and of the chloritic zone at the expense of the sericitic zone, are shown below. Mineral formulae used in these reactions are listed in Table II.

The sericitization of plagioclase can be described by the constant volume reaction:

$$\text{2 andesine} + 0.48\ Al\,(OH)^-_4 + 1.16\ K^+ + 2.32\ H^+ = \\ \text{1.16 sericite} + \text{1.52 quartz} + Na^+ + Ca^{++} + 0.96\ H_2O \qquad (1)$$

Figure 19 indicates that chloritization of pyroxene also takes place in the outer zone of the pipe. This can be accounted for by the constant volume reaction:

$$\text{2 diopside} + 0.58\ Al\,(OH)^-_4 + 5.68\ H^+ = \\ \text{0.29 chlorite} + \text{3.3 quartz} + 2\ Ca^{++} + 0.55\ (Fe, Mg)^{++} + 2.84\ H_20 \qquad (2)$$

The main reaction occurring at the core of the pipe is the chloritization of the sericite already formed at the margins of the pipe. At constant volume, this could be represented by the reaction:

$$\text{3 sericite} + 10\ (Fe, Mg)^{++} + 32\ H_2O = \\ \text{2 chlorite} + 3\ Si(OH)_4 + 5\ Al\,(OH)^-_4 + 3\ K^+ + 22\ H^+ \qquad (3)$$

The chemical changes caused by reactions (1), (2) and (3) are summarized in Table III. The net results on rock composition will depend, of course, on the extent to which each reaction could proceed. The above reactions are consistent with the observed mineralogical and chemical data both at veinlet scale and at the scale of an alteration pipe.

One interesting fact is that the products ($Al(OH)^-_4$, K^+ and H^+) of model reaction (3) occurring at the core are exactly the reactants needed for model reaction (1) which proceeds at the margin of the pipe. This suggests that a single solution, by evolving chemically as it flows through and reacts with the rocks, produced the two alteration zones. The process can be visualized by considering an increment of $(Mg, Fe)^{++}$ -rich fresh solution that is fed through a main fracture. The fluid would cause the destruction of some sericite at the core-margin boundary according to reaction (3) and thus produce an incremental growth of the chloritic core, while being enriched in $Al(OH)^-_4$, K^+ and H^+. This chemically evolved solution, on reacting at the boundary of fresh and altered rocks would then produce an incremental growth of the sericite margin, according to reaction (1).

The above model implies that any point in space would first be subjected to sericitization before being affected by chloritization. Textural evidence strongly supports such a sequence of alteration reactions. For example, Figure 23 shows a feldspar phenocryst that was first completely sericitized and later partly replaced by chlorite. Assuming that the features observed at the scale of a phenocryst can also be

TABLE II
MINERAL FORMULAE AND MOLAR VOLUMES USED FOR BALANCING ALTERATION REACTIONS

Mineral	Formula	Density	Molar Volume (cm³)
andesine	$Ca_{.5}Na_{.5}Al_{1.5}Si_{2.5}O_8$	2.69	100
diopside	$Ca(Mg,Fe)Si_2O_6$*	3.40	67
chlorite	$(Mg, Fe)_5Al_2Si_3O_{10}(OH)_8$*	2.95	215
quartz	SiO_2	2.65	23
sericite	$KAl_3Si_3O_{10}(OH)_2$	2.80	142

*The Mg:Fe ratio has been assumed to be 1:1 in the calculation of molar volume.

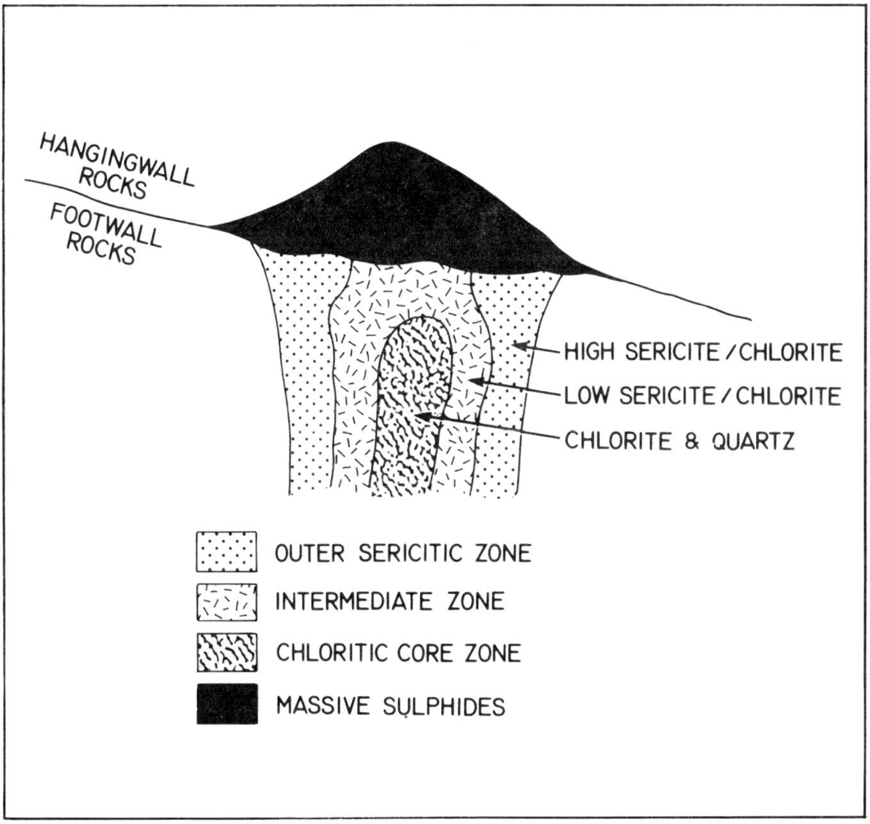

Figure 22. Alteration model associated with a volcanogenic sulphide lens.

TABLE III
SUMMARY OF CHEMICAL AND MINERALOGICAL CHANGES TAKING PLACE DURING THE DEVELOPMENT OF ALTERATION PIPES

Changes	Outer Zone	Inner Zone
Mineralogical	feldspars → sericite + quartz pyroxenes → chlorite + quartz	sericite → chlorite ± quartz
Rock Composition	Ca↓, Na↓, K↑, Al↑ Mg↓ (in mafic rocks)	Fe↑, Mg↑, K↓, Al↓ Si↓ (partly re-precipitated as quartz)
pH of Fluid	pH↑	pH↓

Figure 23. Alteration of a feldspar phenocryst in Q.F.P. The phenocryst is set in a matrix of microcrystalline quartz. The centre of the phenocryst is occupied by an aggregate of sericite (pale grey) and quartz. The dark rim and zones within the pseudomorphed phenocryst is chlorite replacing sericite. Crossed nicols.

projected at the scale of an alteration pipe, the sericitic outer zone most probably grew at the expense of fresh rock through reactions (1) and (2), while the chloritic core developed at the expense of the sericitic zone through reaction (3).

CONCLUSIONS

Significant new conclusions reached since the publication of the paper on the Millenbach deposit by Simmons *et al.* (1973) are as follows.

1) In the Mine Area synvolcanic faulting, trending E.N.E. and N.W., played an essential role in localizing both volcanic and hydrothermal activity.

2) At the Millenbach Mine the distribution of orebodies and rhyolite domes was controlled by the synvolcanic, east-northeasterly trending South Rusty Hill fault; the main hydrothermal vent and largest orebody were localized at its intersection with a northwesterly trending fracture zone.

3) Massive sulphide orebodies formed in an active volcanic environment on top of submarine lava flows before, during, and after the extrusion of quartz-feldspar porphyritic rhyolite domes.

4) Massive sulphides were deposited as hard, encrusting sinters and formed concentrically zoned, conical bodies up to 50 m thick directly above individual hydrothermal vents.

5) Metal zoning within orebodies, from copper at the centre to zinc at the margins, resulted from very local and abrupt changes in physical and chemical conditions taking place close to the rock/sea-water interface.

6) Metal zoning patterns on a broader scale show that in the cluster of orebodies formed at the same stratigraphic horizon, the average copper content, relative to zinc, of satellite lenses decreases steadily with increasing distance from the main hydrothermal vent.

7) In the Millenbach-Amulet hydrothermal field the most important factor governing the base metal content and ratio of an individual orebody appears to be distance from a major vent, rather than stratigraphic position or age. However, where stacked lenses formed directly above each other in the same vent, the older lenses were enriched in copper due to fracture-filling and replacement by chalcopyrite.

8) "Metamorphism has not significantly altered the composition of the rocks and therefore their present mineralogy and chemistry reflect their original mineralogy and chemistry" (Riverin and Hodgson, 1980).

9) "Chemical and mineral zoning are similar within vein selvedges and the alteration pipe as a whole. Also, the compositions of the most intensely altered rocks are always similar, irrespective of the original rock type. These features indicate that the zonal pattern of alteration at Millenbach is the result of the progressive reaction of fluids and rocks as the fluids moved upward within and outward from permeable channelways in the conduit system" (Riverin and Hodgson, 1980).

10) These conclusions are useful guides in exploration for, and interpretation of new volcanogenic deposits, particularly in flow (as contrasted with pyroclastic or volcano-sedimentary) environments.

ACKNOWLEDGEMENTS

Ore discoveries in Noranda by Falconbridge-associated companies are linked by the endeavours of many geologists, particularly: J.G. MacGregor (Amulet Lower "A", 1938), Dr. A.S. Dadson and D.H. Brown (D-68, 1956 and Norbec deposit, 1961), J. Boldy (Delbridge deposit, 1965), Dr. H. Squair (Millenbach deposit, 1966) and M.J. Knuckey (Corbet deposit, 1974). The authors wish to express their appreciation to Falconbridge management, past and present, for its forward-looking policies, and specifically permission to publish this paper. We acknowledge with thanks the contribution of mine staff. Also, grateful thanks are due to Dr. C.J. Hodgson of Queen's University for his supervision of theses by C.D.A. Comba and G. Riverin.

REFERENCES

Akella, J. and Winkler, H.G.F., 1966, Orthorhombic Amphiboles in Some Metamorphic Reactions: Contributions to Mineralogy and Petrology, v. 12, p. 1-12.

Baragar, W.R.A., 1968, Major element geochemistry of the Noranda volcanic belt, Quebec-Ontario: Canadian Journal of Earth Science, v. 5, p. 773-790.

Beecham, A., 1971, Internal Company Report: Falconbridge Copper Limited.

Boldy, J., 1968, Geological observations on the Delbridge massive sulphide deposit: Canadian Institute of Mining and Metallurgy Transactions, v. 71, p. 247-256.

Cooke, H.C., 1930, The Amulet Mine, Quebec: Canadian Institute of Mining and Metallurgy, Transactions, v. 33, p. 398-408.

Dugas, J., 1966, The relationship of mineralization to Precambrian stratigraphy in the Rouyn-Noranda area, Quebec: in Goodwin, A.M., ed., Precambrian Symposium: Geological Association of Canada Special Paper 3, p. 43-56.

Gibson, H., 1979, Geology of the Amulet Rhyolite Formation, Turcotte Lake Section, Noranda, Quebec: Unpublished M.Sc. Thesis, Carleton University, Ottawa, Ontario.

Gilmour, P., 1965, The origin of the massive sulphide mineralization in the Noranda District, northwestern Quebec: Geological Association of Canada Proceedings, v. 16, p. 63-81.

Hutchinson, R.W., Ridler, R.H., and Suffel, G.G., 1971, Metallogenic relationships in the Abitibi Belt: A model for Archean metallogeny: Canadian Institute of Mining and Metallurgy Bulletin, v. 64, p. 48-57.

Kutina, J., and Fabbri, A.G., 1972, Relationships of Structural Lineaments and Mineral Occurrences in the Abitibi Area of the Canadian Shield: Geological Survey of Canada Paper 71-9.

Knuckey, M.J., 1975, Geology of the Millenbach copper-zinc deposit, Noranda, Quebec. Paper presented at the SEG-AIME Annual General Meeting, New York (Abst.): Economic Geology, v. 70, p. 247.

Large, R. R., 1977, Chemical evolution and zonation of massive sulphide deposits in volcanic terrains: Economic Geology, v. 72, p. 549-572.

Purdie, J.J., 1967, Lake Dufault Mines Limited: in Canadian Institute Mining and Metallurgy Centennial Field Excursion, Northwestern Quebec – Norther Ontario. p. 53-57.

Riverin, G., 1977, Wall-Rock Alteration at the Millenbach Mine, Noranda Area, Quebec: Unpublished Ph.D. Thesis, Queen's University, Kingston, Ontario.

Riverin, G. and Hodgson, C.J., 1980, Wall-rock alteration at the Millenbach Cu-Zn Mine, Noranda, Quebec: Economic Geology, v. 75, p. 424-444.

de Rosen-Spence, A., 1969, Genèse des roches à cordiérite-anthophyllite des gisements cuprozincifères de la région de Rouyn-Noranda, Québec: Canadian Journal of Earth Science, v. 9, p. 1339-1345.

————————, 1976, Stratigraphy, Volcanic History and Chemistry of Archean Volcanic Rocks in the Noranda Area, Quebec: Ph.D. Thesis, University of Toronto.

Sangster, D.F., 1972, Precambrian Volcanogenic Massive Sulphide Deposits in Canada: A Review: Geological Survey of Canada Paper 72-22, 44 p.

Sato, T., 1972, Behaviour of ore-forming solutions in seawater: Mining Geology, v. 22, p. 31-42.

Simmons, B.D., and Geological Staff, 1973, Geology of the Millenbach massive sulphide deposit, Noranda, Quebec: Canadian Institute of Mining and Metallurgy Bulletin, v. 66, p. 67-78.

Spence, C.D., 1967, The Noranda area: in Centennial Field Excursion Volume, NW Quebec and NE Ontario: Canadian Institute of Mining and Metallurgy, p. 36-39.

————————, 1975, Volcanogenic features of the Vauze sulfide deposit, Noranda, Quebec: Economic Geology, v. 70, p. 102-114.

Spence, C.D. and de Rosen-Spence, A.F., 1975, The place of sulfide mineralization in the volcanic sequence at Noranda, Quebec: Economic Geology, v. 70, p. 90-101.

Wilson, M.E., 1935, Rock alteration at the Amulet Mine, Noranda District, Quebec: Economic Geology, v. 30, p. 478-492.

Wilson, M.E., 1941, Noranda District, Quebec: Geological Survey of Canada Memoir 229, 169 p.

MS received February 1980;
Accepted for publication January 1981.

Precambrian Sulphide Deposits, H.S. Robinson Memorial Volume, edited by R.W. Hutchinson, C.D. Spence and J.M. Franklin, Geological Association of Canada Special Paper 25, 1982

THE GEOLOGY OF THE CORBET MASSIVE SULPHIDE DEPOSIT NORANDA DISTRICT, QUEBEC, CANADA

M.J. Knuckey and J.J. Watkins
Corporation Falconbridge Copper, Post Office Box 2000,
Noranda, Quebec J9X 5B4

ABSTRACT

The Corbet massive sulphide deposit of about 2.7 million tonnes grading 2.9% Cu, 2.0% Zn, 20.5 g Ag/t and 0.93 g Au/t is situated near the top of a thick sequence of Archean submarine mafic lava flows, known as the Flavrian Andesite. It is particularly significant because it lies at a stratigraphic level 1000 m lower than other known volcanogenic sulphide deposits in the area, and because it occurs toward the top of a thick sequence of andesitic rocks rather than in the characteristic rhyolitic domal environment of the Noranda district. Sulphides were deposited rapidly in an active volcanic environment around coincident magmatic and hydrothermal vents.

Lateral mineralogical zoning within the massive sulphides, which overlie a chalcopyrite + pyrrhotite stringer zone, proceeds from magnetite + chalcopyrite to pyrrhotite + chalcopyrite, to pyrite + sphalerite ± chalcopyrite. Higher concentrations of Au, Ag and Sn occur in association with the pyrite + sphalerite ore. Bedded, cherty exhalite locally flanks and underlies a portion of the orebody. The pipe-like zone of stringer sulphides occurs within an irregular zone of strong chlorite alteration and plunges at 45 to 60 degrees to the northwest.

Rocks composed of unsorted volcaniclastic debris, the predominant ore host, are overlain in part by massive and pillowed andesite flows. Encroachment of these overlying flows onto a still hydrothermally-active vent, resulted in the formation of satellite sulphide lenses above the main Corbet orebody. A zone of sericitic alteration and weak mineralization in the overlying rhyolite is evidence that hydrothermal activity continued after the extrusion of the mafic lava sequence.

The McDougall and Despina faults are northwesterly trending synvolcanic structures that localized persistent magmatic and hydrothermal activity. The Corbet orebody occurs at the intersection of the McDougall with the east-northeasterly trending #1 Lens fault.

RÉSUMÉ

Le gisement de Corbest est un amas sulfuré d'environ 2,7 millions de tonnes à 2,9% de Cu, 2% de Zn, 20,5g d'Ag/t et 0,93g d'Au/t. Il est situé près du sommet d'une épaisse séquence de laves mafiques sous-marines archéennes, connues sous le nom d'Andésites de Flavrien, à un niveau stratigraphique de 1.000m inférieur à celui des autres dépôts sulfurés volcanogéniques de la région. Les sulfures paraissent avoir été déposés rapidement dans un milieu volcanique actif autour des orifices magmatique et hydrothermal superposés.

L'amas sulfuré qui recouvre un réseau en forme de cheminée de veinules à chalcopyrite et pyrrhotite, montre une zonalité minéralogique latérale qui passe de magnétite + chalcopyrite au centre à pyrrhotite + chalcopyrite, puis à pyrite + sphalérite + chalcopyrite. Les teneurs les plus élevées en Au, Ag et Sn sont associées au minerai à pyrite et sphalérite. Un chert exhalatif lité se rencontre localement sur les flancs et sous une partie du gisement. La zone en forme de cheminée à veinules sulfurées est elle-même située dans une zone de forte altération chloritique, de forme irrégulière, et qui plonge à 45-60° vers le nord-ouest.

La couche hôte prédominante est formée de débris volcaniclastiques d'andésite, non triés, et est recouverte en partie par des coulées andésitiques massives et en coussinets. La mise en place de ces coulées dans un milieu hydrothermal encore actif a résulté dans le dépôt de lentilles sulfurées satellites au-dessus du gisement principal de Corbet. Une zone d'altération séricitcuse faiblement minéralisée dans la rhyolite recouvrant les Andésites de Flavrien indique que l'activité hydrothermale continua après l'extrusion de la séquence de laves mafiques.

Les failles de McDougall et de Despina qui encadrent le gisement sont des structures synvolcaniques orientées nord-ouest qui ont canalisé une activité magmatique et hydrothermale persistante. Le gisement de Corbet est situé à l'intersection de la faille de McDougall et d'une autre faille, dite de la lentille #1, orientée est-nord-est.

INTRODUCTION

The Corbet deposit is located about 8 km north of the cities of Rouyn-Noranda in northwestern Quebec. It is 2200 m west of the Millenbach Mine described by Knuckey *et al.* (this volume). To date about 2.7 million tonnes of ore grading 2.9% copper and 2.0% zinc have been delineated. The ore contains low, but recoverable gold and silver values. Limited production commenced early in 1980.

Most of the previously known deposits in the central Noranda area are restricted to a relatively narrow stratigraphic interval, the "Mine Series" of Spence (1967), and are closely associated with rhyolite domes and sea floor, hot spring activity in the waning stages of a felsic volcanic cycle (Spence and de Rosen-Spence, 1975). The discovery of the Corbet deposit 1000 m deeper in the volcanic succession and located within the top 250 m of a thick andesitic lava sequence opens new targets for exploration and invites modifications to accepted models of ore genesis.

This paper is a preliminary description of the geological setting of the Corbet deposit with emphasis on the volcanic environment in which it formed and the metal zoning patterns that developed. Wall rock alteration is compared with that at the Millenbach Mine. The deposit is believed to be of volcanogenic origin and evidence is presented to support this interpretation. This paper therefore considers magmatic, hydrothermal and ore-forming events, as a continuum of geological activity in a volcanogenic system, rather than as isolated, independent phenomena.

The reader is particularly referred to the previous paper in this volume for information about the regional geology, stratigraphic succession in the mine area and synvolcanic structures linking the Corbet with the Millenbach and Amulet deposits (Figs. 1, 2, 3 and references in paper by Knuckey *et al.*, this volume).

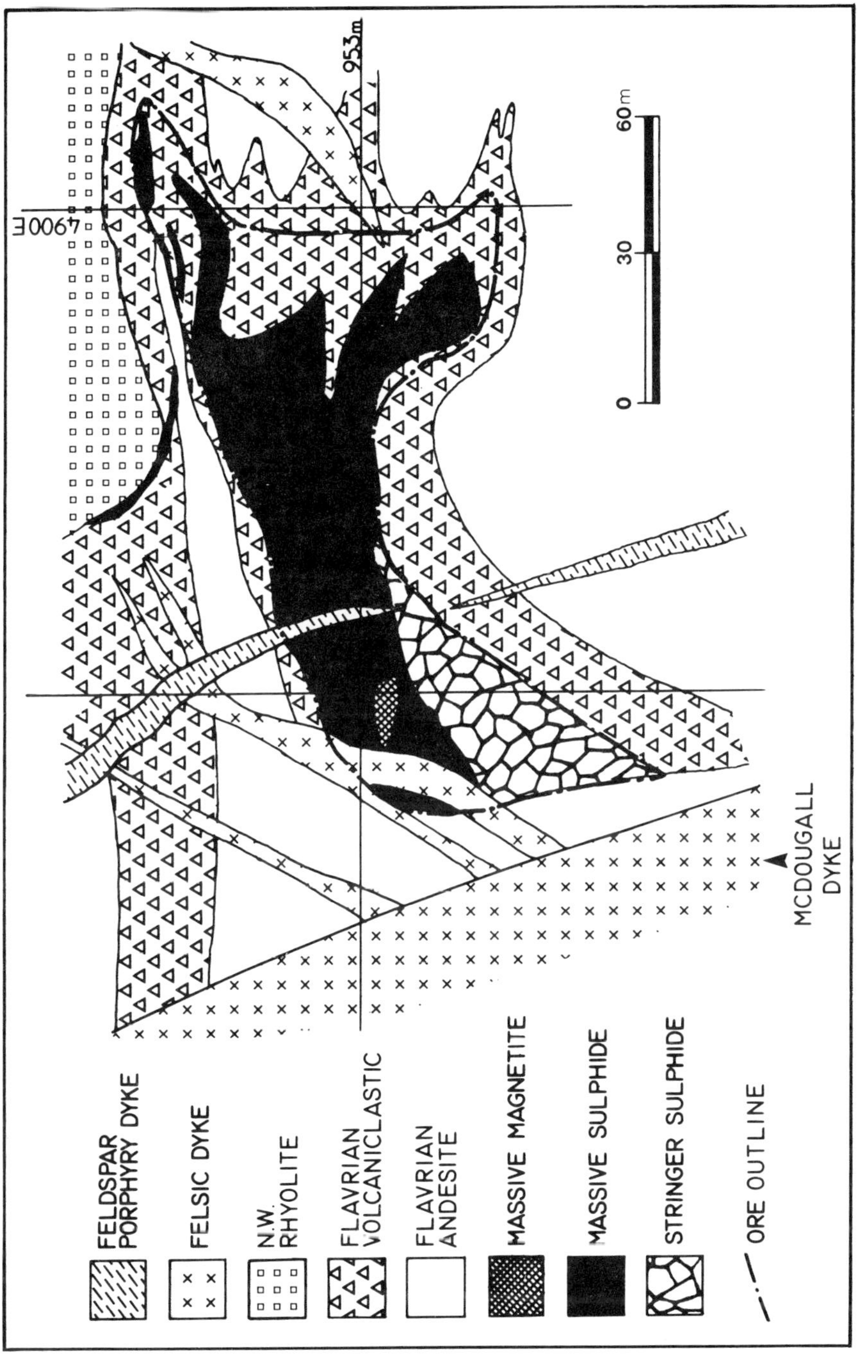

Figure 1. Geology, section 800N, looking north.

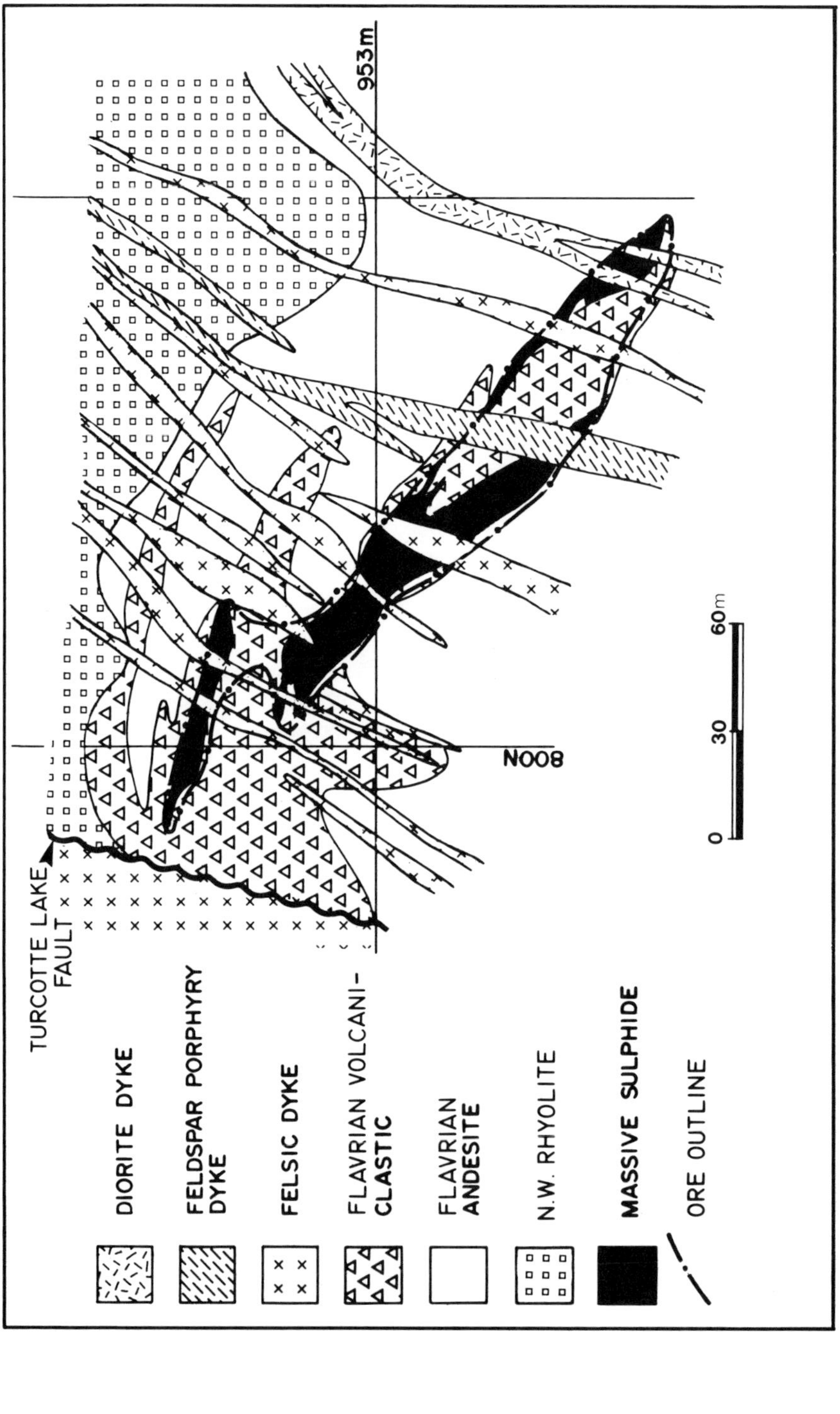

Figure 2. Geology, section 4900E, looking west.

THE CORBET OREBODY

The Corbet deposit is similar in many respects to the typical copper-zinc, volcanogenic, massive sulphide deposit of Archean age described by Sangster (1972). It is a stratabound accumulation of mineralogically-zoned, massive and semi-massive sulphides with an underlying zone of stringer sulphides occurring totally within submarine volcanic flows and volcaniclastic rocks. In common with other deposits of this type, evidence strongly indicates that the orebody formed from hot, metal-bearing brines discharging onto the sea floor and precipitation of metals as sulphides around a main hydrothermal vent. The Corbet massive sulphides were deposited as a series of lenses in the upper 250 m of the Flavrian Andesite (Figs. 1, 2).

Environment of Sulphide Deposition

Flavrian Andesite. This is a sequence of massive or pillowed, mafic to intermediate flows with minor flow and pillow breccias. The total thickness of the unit has not been established but it is known to be greater than 500 m. Two other types of breccia of explosive origin are important, because in this area they have been found only at Corbet and appear to be intimately associated with the magmatic and hydrothermal events leading to the formation of the orebody.

Massive and pillowed andesite apparently grade into flows brecciated in-situ which, in turn, give way to local accumulations of rock composed of highly vesiculated andesitic debris consisting of generally unsorted, angular to subangular fragments set in a micro-breccia matrix. Locally these breccias are weakly layered and occasionally graded (Gibson, 1980). This debris (Fig. 4) is mixed locally with abundant hyaloclastite consisting of vitric fragments that formed by the direct granulation of magma when quenched by sea-water. The volcaniclastic rocks shown in the figures may also include flow and pyroclastic breccias, containing both lithic and vitric fragments, because mapping has been insufficient to reveal the distribution of the breccia types in detail. Individual fragments may be amoeba-shaped, subangular or

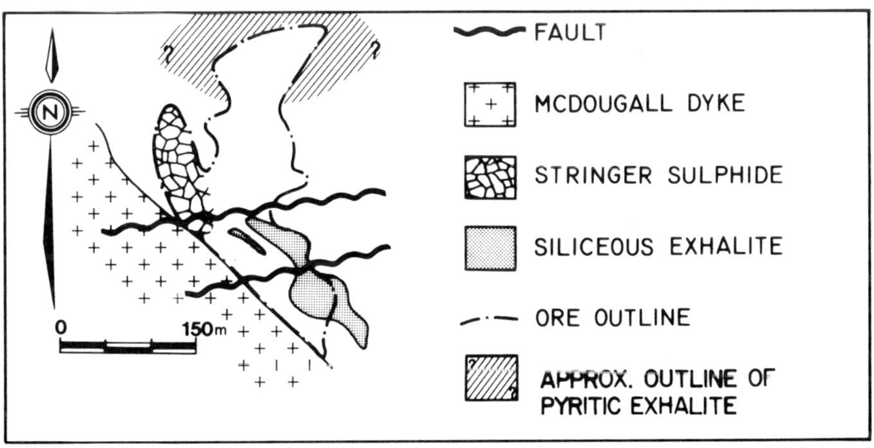

Figure 3. Plan showing the distribution of siliceous and pyritic exhalite.

shard-like and range from less than one to more than 30 cm in maximum dimension. This pile of chaotic, volcaniclastic debris locally reaches a thickness of up to 100 m and may have had an initial porosity as high as 40 per cent.

Hydrothermal Activity. The period of explosive volcanic activity, during which the volcaniclastic rocks were formed, was accompanied and/or followed by strong hydrothermal activity resulting locally in intensive alteration of the host rocks and the deposition of massive sulphides at the volcaniclastic/sea water interface (Figs. 1, 2, 5). Subsurface openings such as amygdules, interclastic voids and fractures were partly, or totally filled with silica and metallic sulphides (Fig. 6). From the limited underground exposures available to date, Gibson (1980) described the in situ breccias as "consisting of variably altered, angular fragments of andesite in a vein network matrix of quartz-pyrite-sphalerite and occasionally galena. Veins and veinlets separating fragments are zoned, sometimes with an outer margin of massive or colloform pyrite but more often with (inner) zones of quartz and sphalerite. Colloform textures and zoning within veins suggest open-space deposition".

Exhalite. Silica was also precipitated as a gangue in the massive sulphides and locally as a thin, cherty exhalite underlying and flanking the southeastern margin of the orebody (Figs. 3, 5). The exhalite is well bedded and consists of a microcrystalline mosaic of quartz displaying grain size gradation and soft sediment deformation. Pyrite, with or without sphalerite, occurs as thin bands parallel to bedding planes. Similar thin, non-cherty pyritic exhalite locally flanks the orebody to the

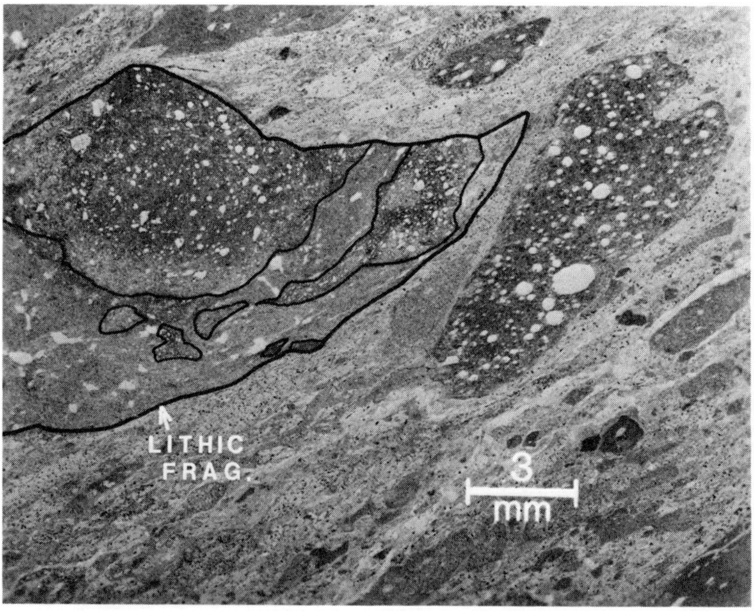

Figure 4. Andesitic volcaniclastic rock, Corbet Area, is characteristically heterogeneous consisting of lithic and hyaloclastic fragments. Large clast, left centre, contains fragments of previously consolidated volcaniclastic rock. Plane light.

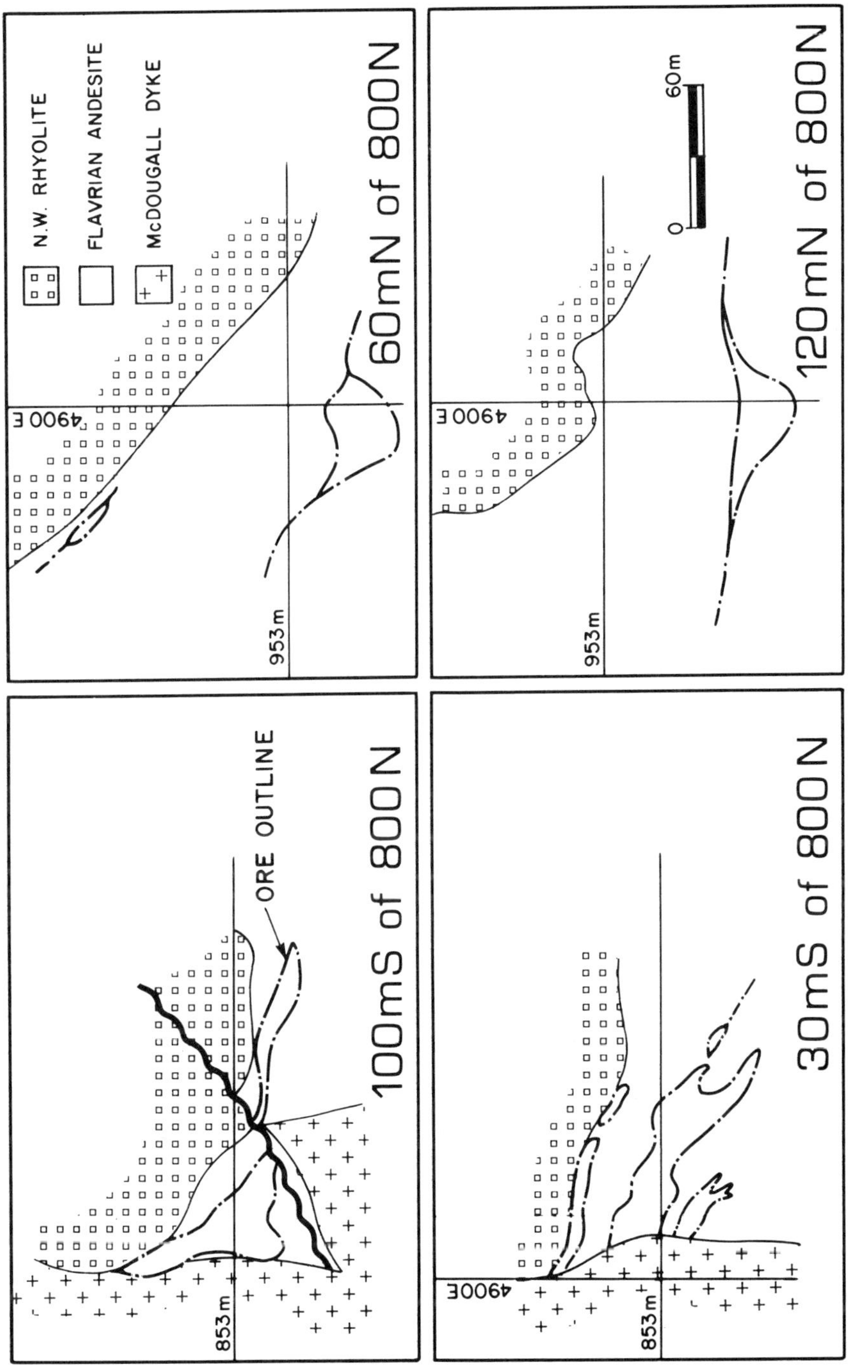

Figure 5. Geology sections through the Corbet orebody.

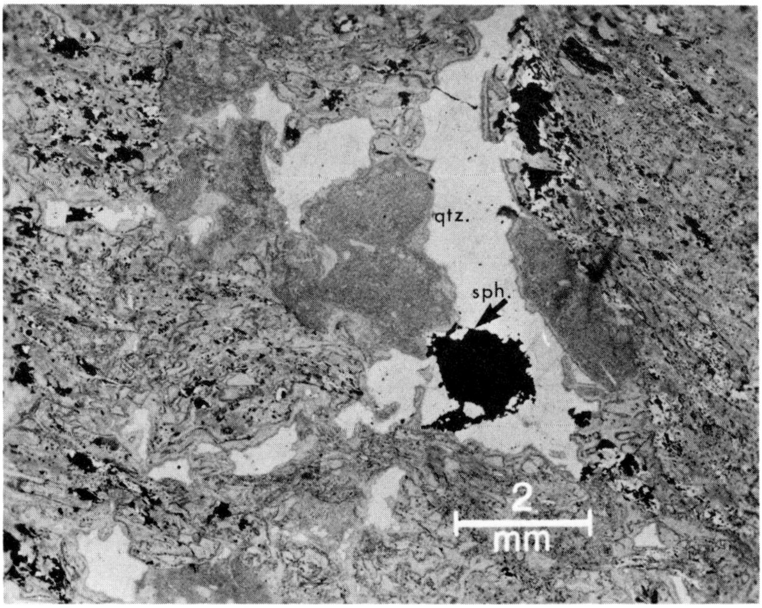

Figure 6. Quartz (qtz) followed by sphalerite (sph) was precipitated in the primary porosity shortly after the formation of the volcaniclastic rocks. Plane light.

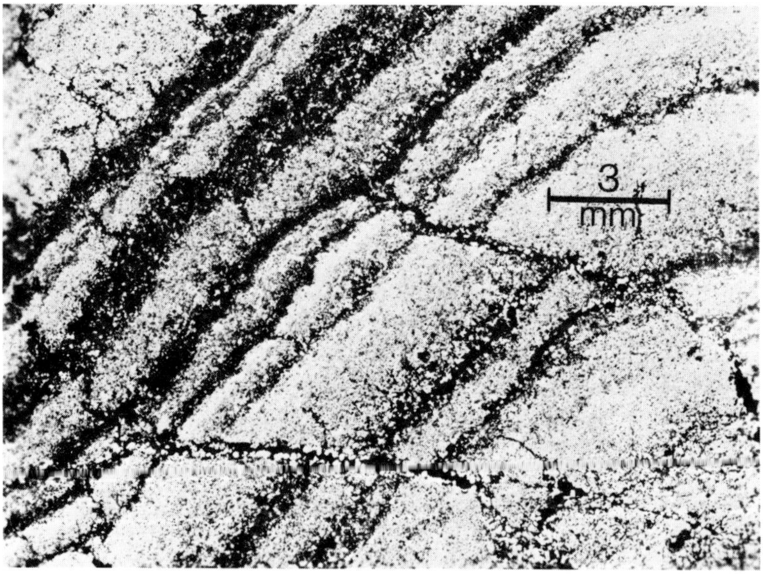

Figure 7. Siliceous exhalite displaying well developed bedding and grain-size gradation. Opaque minerals, pyrite and sphalerite, occupy bedding planes, chalcopyrite occurs as post-consolidation cross cutting veinlets. Plane light.

north (Fig. 3). Veinlets of chalcopyrite and quartz transect the sedimentary features (Fig. 7). In contrast to many other volcanogenic massive sulphide deposits no regionally extensive, siliceous, sulphidic or tuffaceous exhalites are known to be associated with the Corbet hydothermal vent. Possibly the volcanic environment was too active to permit their accumulation. Nevertheless, these local bedded cherts provide clear evidence of subaqueous deposition around hot spring centres as is the case elsewhere in the Noranda District.

Upper Flavrian Andesite. Following the main period of massive sulphide formation a renewal of the earlier andesitic volcanism formed thick flows of massive and pillowed andesite which encroached upon the Corbet vent from the northwest, covering the northern half of the orebody (Figs. 2, 5). The southeastern fronts of these onlapping, slightly younger flows are relatively thin and well pillowed. Where they covered active hot springs near the orebody, pillow interstices and borders were infilled and/or partly replaced by finely banded chert or by massive pyrite-sphalerite (cf. Millenbach Andesite, Knuckey *et al.*, this volume). In drill core such occurrences may be mistaken for widespread exhalites with greatly different implications of continuity.

Several smaller, satellitic massive sulphide lenses occur at two or more ill-defined flow contacts within the onlapping andesite flows. The top of the Flavrian Andesite sequence is, in some places, marked by a silicified, volcaniclastic breccia. In the area shown on Figures 1 and 2 the breccia in contact with the overlying N. W. Rhyolite consists of variably silicified and sericitized andesite scoria, and occasional sulphide clasts, in a fine-grained breccia and ash matrix. The ash component of the breccia is typically laminated and graded, filling voids between larger fragments, and is presumed to mark a hiatus in volcanic extrusive activity. A moderate-sized pod of massive sulphides occurs at this horizon (Figs. 1, 5). These smaller satellitic lenses, each underlain by an alteration pipe and stringer zone, clearly indicate that weaker hydrothermal activity persisted throughout the period of renewed andesitic eruption.

Quartz Diorite Intrusion. The Flavrian Andesite flows to the northwest of the Corbet mine are intruded by a number of andesitic and dioritic dykes and by a large irregular mass of quartz diorite. Based on drillhole information, the main quartz diorite intrusion has a northwest strike and a northeast dip. It is parallel to the Despina-McDougall structure and plunges to the southeast. Its footwall contact appears to be relatively straight and may be fault-bounded whereas the upper contact is hummocky and irregular. The quartz-diorite intrusion appears to be roughly concordant with the upper contact of the Flavrian Andesite but discordant with internal flow contacts (Gibson, 1980), and is similar to the Meritens phase of the Flavrian Granite (Goldie, 1978) which crops out between the Despina and McDougall faults 2300 m northwest of the Corbet deposit (Knuckey *et al.*, this volume, fig. 2). The quartz-diorite is clearly intrusive into the Flavrian Andesite stratigraphically below the main orebodies and, locally, a quartz-porphyritic phase was altered and mineralized by the Corbet hydrothermal activity.

Volcanic Centre. At least two features suggest that the Corbet orebody and interpreted hydrothermal vent developed at a topographically high, submarine location. First, the ore horizon represented by cherty exhalite to the southeast, massive sulphides in the centre and pyritic exhalite to the north is covered by up to 30 m of

Upper Flavrian Andesite to the southeast compared with 0 m in the centre and 150 m to the north (Fig. 3, 5). Second, a sharp steepening of the dip of the main massive sulphide lens toward its southeastern extremity indicates that the andesite cover is probably thickening in that direction also. The apparently crosscutting intrusive form of some of the massive Flavrian andesite close to the McDougall dyke (Fig. 1) and the occurrence of pyroclastic breccias suggest that a magmatic as well as a hydrothermal vent may have been located at, or close to the site of ore deposition.

To the northwest of the Corbet deposit, contours on the top contact of the Upper Flavrian Andesite indicate a doming of the andesite above the quartz diorite, the dome having a northwesterly trend. The coincidence of the dome with the underlying quartz diorite suggests that the doming occurred in response to the intrusion of the diorite (Gibson, 1980). On the east flank of the dome the present steep attitude of the stratigraphically highest massive sulphide lens (Fig. 5) at the Flavrian Andesite/N.W. Rhyolite contact, could be explained by subvolcanic intrusions causing endogenic growth of the andesite pile during or slightly later than the formation of this lens. Those intrusions may have steepened an original regional dip of 30° to the east to almost vertical. The overall impression is of continuing volcanism with alternating or coeval extrusive, subvolcanic and hydrothermal activity.

N.W. Rhyolite. This is a series of massive to locally flow banded and spherulitic, rhyolitic and rhyodacitic flows overlying the very irregular surface of the Flavrian Andesite. In the immediate area of Corbet the unit ranges in thickness from 250 to 325 m (Knuckey *et al.*, this volume, fig. 3). The N.W. Rhyolite directly overlies only the central-southern part of the orebody. In this area waning hydrothermal activity resulted in local sericitization and pyritization of the rhyolite flows for up to 100 m above the contact. In general however, the rhyolite is unaltered and unmineralized.

Structural Geology

The structural geology of the Corbet area, apart from primary volcanic features discussed above, is dominated by major faulting and dyking. Volcanic rocks on both sides of the McDougall fault strike northeasterly to northerly and dip east at an average of 30 degrees (Knuckey *et al.*, this volume, figs. 2, 3, 7). Local dips range from 10° to almost vertical due to tilting of primary volcanic slopes. Folding does not appear to be significant. The main Corbet orebody is dislocated into three lenses by three important faults respectively known as the McDougall, the Turcotte Lake and the #1 Lens faults (Fig. 8; Knuckey *et al.*, this volume, fig. 2).

The McDougall fault is a major dyke-filled fault zone striking northwesterly and dipping 60 to 80° southwest. The sub-parallel Despina fault dipping 80° east lies 320 m to the west on surface but converges with the McDougall at depth and to the northwest. Apparent vertical displacement is 675 m on the McDougall fault and up to 430 m on the Despina for a combined effective down-throw on the west side of about 1000 m using the top of the Amulet Rhyolite for correlation (Knuckey *et al.*, this volume, Fig. 3). South of the Turcotte Lake fault, the Corbet massive and stringer sulphide zones are faulted off on the west by the McDougall fault now occupied by the McDougall dyke (Fig. 8). West of the McDougall and Despina faults the Flavrian Granite occupies a position as high in the stratigraphy as the Rusty Ridge Andesite or

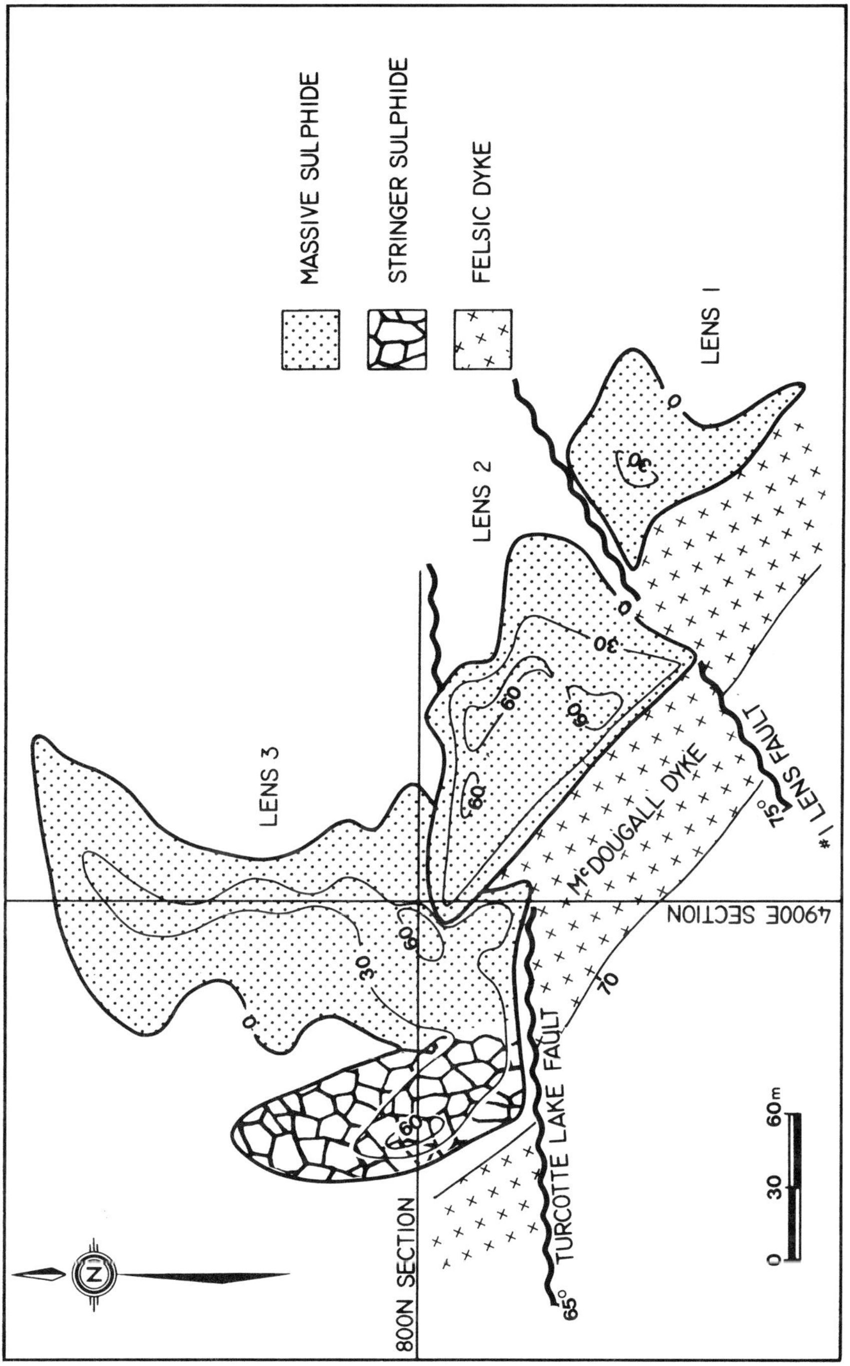

Figure 8. Ore isopach map of the Corbet deposit.

600 m above the stratigraphic position of the Corbet orebody (Knuckey *et al.*, this volume, Fig. 3). Discovery of the faulted offset of the Corbet orebody remains a challenge.

Both the McDougall and Despina faults are intruded by fine-grained rhyolite dykes whose margins in many places are strongly flow banded. The dykes vary from 2 to 100 m wide. Swarms of similar, parallel or sub-parallel felsic dykes are commonly indistinguishable from the main dykes. The actual faults themselves are rarely obvious. Polymictic breccia dykes and pipes, 1 to 2 m across, also occur along these structures. In the Corbet Mine area, a large quartz diorite intrusion and dykes of feldspar porphyry, andesite, diorite and lamprophyre, as well as local vents for the later Rusty Ridge Andesite, Beecham Breccia and Amulet Rhyolite described by Knuckey *et al.* (this volume) all indicate continuing igneous activity along these faults from the time of Flavrian Andesite extrusion at least until post Amulet Andesite eruption. The presence of several Cu-rich stringer zones, Cu-Au veins, and Cu-Zn occurrences along these same faults for more than 3000 m (Knuckey *et al.*, this volume) clearly indicates that they also channeled hydrothermal activity for a similarly prolonged time period.

The Turcotte Lake Fault is a high angle reverse fault striking east-west and dipping south at 60 to 70 degrees. It has left-handed movement with the north side displaced about 75 m relative to the south side. The fault varies from a gouge seam several centimetres wide to a shear zone, including associated carbonate up to 50 m wide. Massive, fine-grained felsic dykes, chemically similar to the McDougall dykes, occupy the fault zone locally in the mine area (Figs. 2, 8).

The #1 Lens Fault strikes east-northeast (070°), dips at 70 to 80° to the north and dislocates the south end of the orebody. Movement is strike slip with left-lateral displacement of about 25 m. The fault is a narrow shear zone up to 30 cm wide and is parallel to the South Rusty Hill fault, important at the Millenbach Mine. The Turcotte Lake and #1 Lens faults converge at depth but insufficient work has been done to determine their age relationships. Both faults are younger than, and displace the McDougall and Despina faults.

Sulphide Zones

In plan view, the main Corbet deposit forms an elongated orebody about 400 m long by up to 150 m wide (Fig. 8). Maximum thickness is about 70 m. The orebody consists of a number of stratiform lenses of massive sulphides, separated by semi-massive and disseminated mineralization in volcaniclastic rocks and underlain, centrally, by a pipe-like zone of crosscutting stringer and disseminated sulphides. The deposit is strongly zoned with a massive pyrrhotite-chalcopyrite core, centred over the zone of disseminated and stringer mineralization, and flanked by massive pyrite-sphalerite. In the core, massive sulphides are solid and continuous. Toward the east and southeast however, the presence of irregular, internal lenses of mineralized volcaniclastic rocks results in separate pods of massive sulphides along the foot- and hangingwalls of the orebody (Figs. 1, 2, 5).

Figure 1 is a typical vertical east-west cross section through the thickest part of the orebody. The contacts of the main lens are clearly not related to the Flavrian

Andesite/N.W. Rhyolite contact. After adjustment for the regional dip of 30 degrees to the east, it appears that the deposition of massive sulphides, mostly pyrrhotite-chalcopyrite but including a lens of magnetite-chalcopyrite, commenced in a topographic depression about 60 m in depth centred directly above the stringer zone. This depression or small crater may have been created by steam-blast explosions as suggested by the evidence of in-situ brecciation grading into chaotic volcaniclastic debris.

South of the Turcotte Lake fault the western part of the orebody, including the probable continuation of the stringer zone and massive sulphides that would have directly overlain it, have been faulted and "dyked" out along the McDougall fault (Fig. 8). The remaining eastern part, consisting of two pods of massive sulphide separated by mineralized volcaniclastic rock, dips to the northeast or east at 30 to 50 degrees. Dips steepen markedly toward the southeastern end. The upper contact of the ore zone is located from 0 to 60 m below the Flavrian Andesite/N.W. Rhyolite contact (Fig. 5).

Figure 2 is a vertical north-south section through an elongated tongue of massive and disseminated sulphides plunging to the north at 35 degrees. The tongue is centred on section 4900E, extends from 40 to 150 m north of section 800N and roughly corresponds with the 30 m isopach of lens 3 (Fig. 8). The tongue comprises at least two pods of massive sulphides, separated and partly underlain by mineralized andesitic volcaniclastic rocks, probably flow breccias. This tongue may be equivalent to the upper of two pods south of the Turcotte Lake fault. In cross-section the shape of the tongue is irregular but roughly approximates an inverted bell (Fig. 5). There is no evidence indicating that the attenuated (5:1) form of the tongue results from structural deformation and it is inferred therefore, that topography was a major influence in determining its shape and location. Toward the northern end of the tongue a thin bed of fine-grained 'muddy' pyritic exhalite, 15 to 60 cm thick and carrying from 200 to 3000 ppm copper and/or zinc extends laterally, east and west, from the upper contact of the tongue of massive sulphide for at least 100 m (Fig. 3).

To the northwest and southeast of the tongue several smaller lenses of relatively more zinc-rich massive and stringer sulphides occur within and at the top of the Upper Flavrian andesite. These, and the stringer zone beneath the main tongue, require further underground exploration and will not be described in this paper.

METAL AND MINERAL ZONING

The principal metallic minerals in decreasing order of abundance are pyrrhotite, pyrite, chalcopyrite, sphalerite, magnetite, galena and native gold. Mackinawite, tetradymite (Bi_2Te_2S) and krennerite ($AgTe_2$) have been identified in trace amounts (Muir, 1978). Non-sulphide gangue minerals include mainly quartz and chlorite.

Iron. The deposit shows a pronounced change from magnetite to pyrrhotite to pyrite with increasing distance from the hydrothermal vent represented by the stringer sulphide zone. Figure 9 is a reconstructed plan with offset due to east-west faulting removed, and illustrates the well developed iron mineral zoning. Figure 10 shows the iron mineral zoning on mine section 800N. A small lens of massive magnetite, carrying low copper values, lies directly over a portion of the stringer pipe. It is completely engulfed in massive, non-magnetic pyrrhotite-chalcopyrite ore. Toward

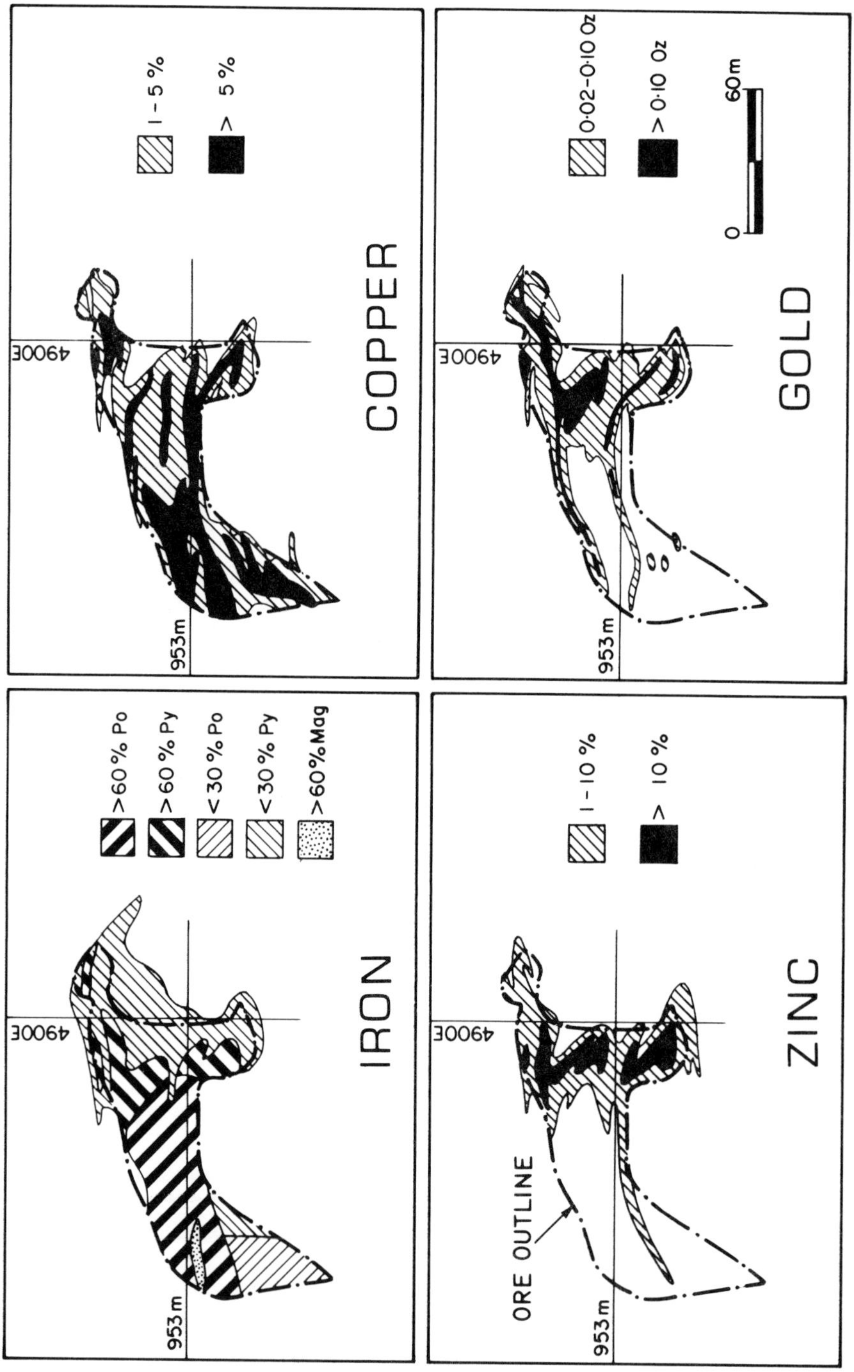

Figure 9. Plan illustrating iron mineral zoning.

the margins of the deposit the proportion of pyrite increases at the expense of pyrrhotite, resulting in a pyritic fringe. There is a strong correlation between copper and pyrrhotite, and between zinc and both pyrite and gold.

Copper. The distribution of copper on mine section 800N is shown in Figure 10. Copper values are highest in the pyrrhotite core and in the stringer zone. In the former, chalcopyrite occurs as equigranular aggregates in a mosaic of pyrrhotite and quartz. In the stringer zone it occurs with pyrrhotite in an anastomozing network of veinlets between and cutting strongly chloritized volcanic rock fragments. In the pyritic fringe to the massive sulphides it occurs as fine veinlets and disseminations. Copper is virtually absent in the mineralized volcaniclastic rocks between and below the massive sulphide pods that form the north-plunging tongue (Fig. 2).

Zinc. In sharp contrast with copper, zinc is mostly concentrated in the pyritic fringe of the massive sulphides. An exception is a band of relatively high grade zinc ore that can be traced along the base of the massive pyrrhotite-chalcopyrite zone even where it is underlain by stringers (Fig. 10). Sphalerite is almost always associated with pyrite and varies in colour from amber through reddish-brown to dark brown. Pyrite nodules up to 5 mm in diameter are common in a matrix of sphalerite and quartz. Crude banding of pyrite-sphalerite may occur locally but is rarely well developed. In the north-plunging tongue of ore, sphalerite occurs not only with chalcopyrite and pyrite in the two massive sulphide pods but unlike chalcopyrite, is also disseminated in ore-grade quantities in the volcaniclastic rocks between them. The underlying flow breccias are also mineralized in places with 0.5 to 1.0% Zn over thicknesses of as much as 20 m. Sphalerite occurs in the interstices of the breccia, in amygdules within the fragments and sometimes has diffused into the margins of individual clasts forming a dusty brown rim. The occurrence of such widespread disseminated sphalerite in rocks below an ore horizon is unusual in the central Noranda area. An adequate explanation must await more detailed underground exploration.

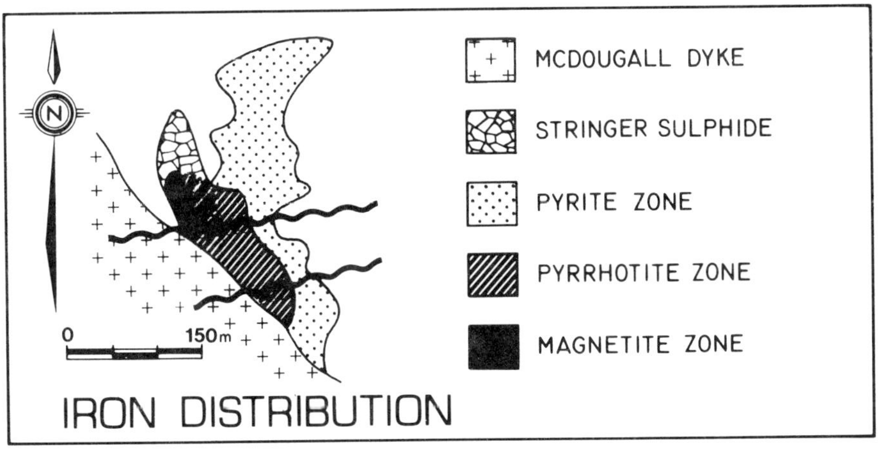

Figure 10. Metal zoning, section 800N.

Gold. The average gold content of the deposit is 0.93 g/t. Gold distribution (Fig. 10), whilst very similar to that of zinc is more closely related to that of pyrite. This is particularly evident in the north-plunging tongue of ore, where relatively high gold values occur in the massive sulphide pods but practically none in the intercalated zinc-bearing but essentially non-pyritic volcaniclastic rocks. Native gold has been identified as fine inclusions in pyrite nodules.

Lead. Galena occurs in minor amounts toward the stratigraphic top of the pyritic fringe.

Tin. Tin content averages 100 ppm in the pyrrhotite zone compared with 500 ppm in the pyritic fringe. No tin minerals have been identified.

WALLROCK ALTERATION

The host volcanic rocks have been altered by processes similar to those described at the Millenbach Deposit (Riverin, 1977; Knuckey *et al*., this volume) and at other volcanogenic massive sulphide occurrences (Sangster, 1972), however, at Corbet the massive sulphides are enveloped in a zone of strong chloritic alteration ranging from 10 to 30 m in thickness. Below the pyrrhotite-chalcopyrite core the chloritic alteration extends downward for at least 400 m as an irregular pipe. The exact projection of this chloritic pipe to depth is not well established but it appears to have a northwesterly plunge of 45° to 60°.

Least altered Flavrian Andesite is distinctly feldspar porphyritic with a fine-grained groundmass consisting of microlitic albite and chlorite (Fig. 11). In the core of the alteration pipe the host andesite has been intensely chloritized. Ghost volcaniclastic textures together with relict feldspar microlites pseudomorphed by quartz, and also quartz-sulphide filled amygdules, attest to its original andesitic nature. Vitric fragments are invariably chloritized whereas lithic ones are more variably altered.

As at the Millenbach deposit (Riverin, 1977), the chloritic alteration at the core of the pipe gives way outward to a zone of sericitic alteration and silicification. At Corbet the sericitic zone is about 10 m wide and has a bleached appearance, the rims of individual fragments being variably sericitized or silicified. In contrast to Millenbach, wallrock alteration at Corbet persists into the overlying rocks. Above the southern part of the orebody, and particularly where it is in contact with the main massive sulphides, the N.W. Rhyolite has been sericitized and pyritized for up to 100 m above the contact. Chloritic alteration in the N.W. Rhyolite is restricted to veins and small stringers. Above the northern part of the orebody small, chloritic alteration pipes penetrate the overlying Upper Flavrian Andesite flows and develop into small massive sulphide lenses at flow contacts as high stratigraphically as the base of the N.W. Rhyolite.

At variable depth, a zone of strongly spotted alteration occurs. It appears to be related, spatially, to the large quartz diorite intrusion, but its exact significance has not yet been established. The spotted zone is characterized by round or ovoid patches up to 5 mm in diameter of fine-grained sericite in a chloritic host (Fig. 12). The spots appear to have formed around feldspar phenocrysts. Away from the chloritic pipe the andesite has been epidotized with the development of epidote as large haloes to amygdules and as irregular patches. The exact distribution of the epidote zone and its relationship, if any, to the ore forming process is still unknown.

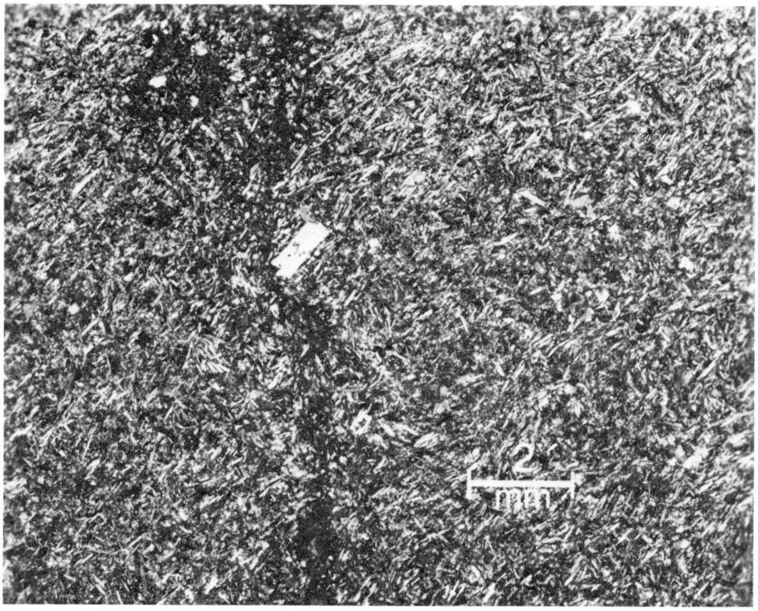

Figure 11. Flavrian Andesite occurring away from the deposit is feldspar porphyritic and microlitic. The ground mass is fine-grained chlorite. Plane light.

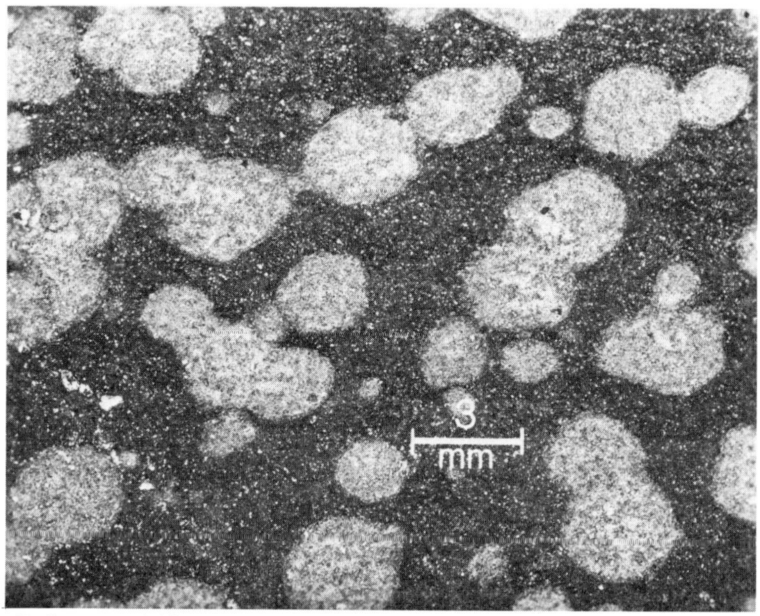

Figure 12. Strong sericitic, spotted alteration. Plane light.

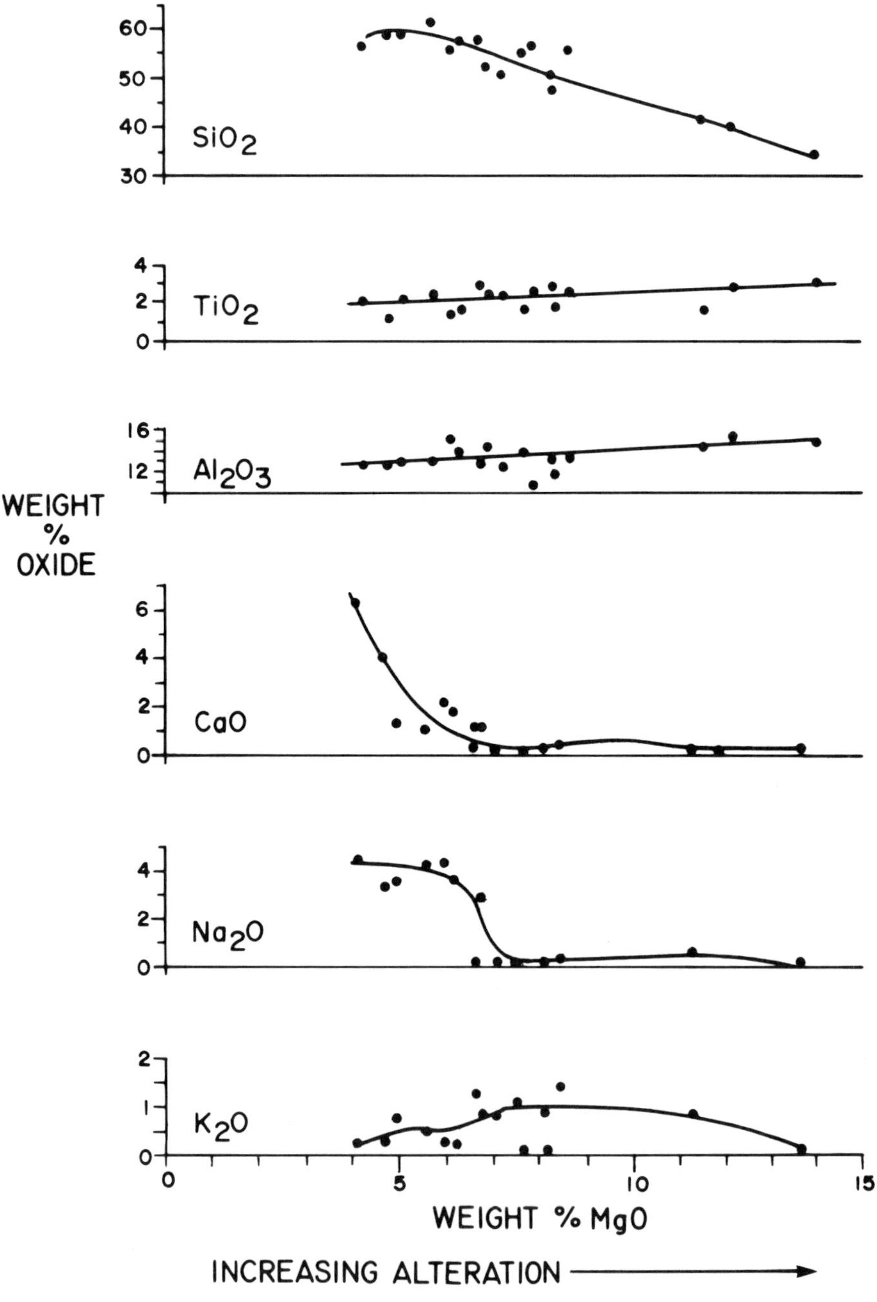

Figure 13. Trends in major oxides for the Flavrian Andesite using MgO as an index of alteration.

Progressive alteration of the Flavrian Andesite toward the Corbet orebody involves a significant increase in magnesium content. Roberts and Reardon (1978) used an approach in which the MgO content of the rocks in the vicinity of the Mattagami Lake Deposit, used as an index of alteration, was compared with the variations of other oxides. A similar approach is taken here (Fig. 13) to depict the chemical changes that occurred in the host rock due to the ore-forming process. A number of samples, displaying various degrees of alteration of the Flavrian Andesite in the area of the deposit were analysed for major oxides. Results (Fig. 13) show that SiO_2, Na_2O and CaO decrease markedly with increasing alteration intensity. K_2O shows a slight increase followed by a decrease in the direction of increasing alteration. Al_2O_3 and TiO_2 show no significant change.

At the Millenbach deposit Riverin (1977) displayed the compositions of altered rocks associated with the ore-forming process in the four-component system $(FeO+MgO) - (CaO+Na_2O) - (Al_2O_3) - (K_2O)$, the F-C-A-K system. By plotting the average chemical compositions of variably altered rocks on the unfolded F-C-A-K tetrahedron Riverin explained the metasomatic trends for the alteration patterns associated with individual sulphide lenses at the Millenbach deposit. The various analyses of the Flavrian Andesite (Fig. 13) were plotted on an unfolded F-C-A-K

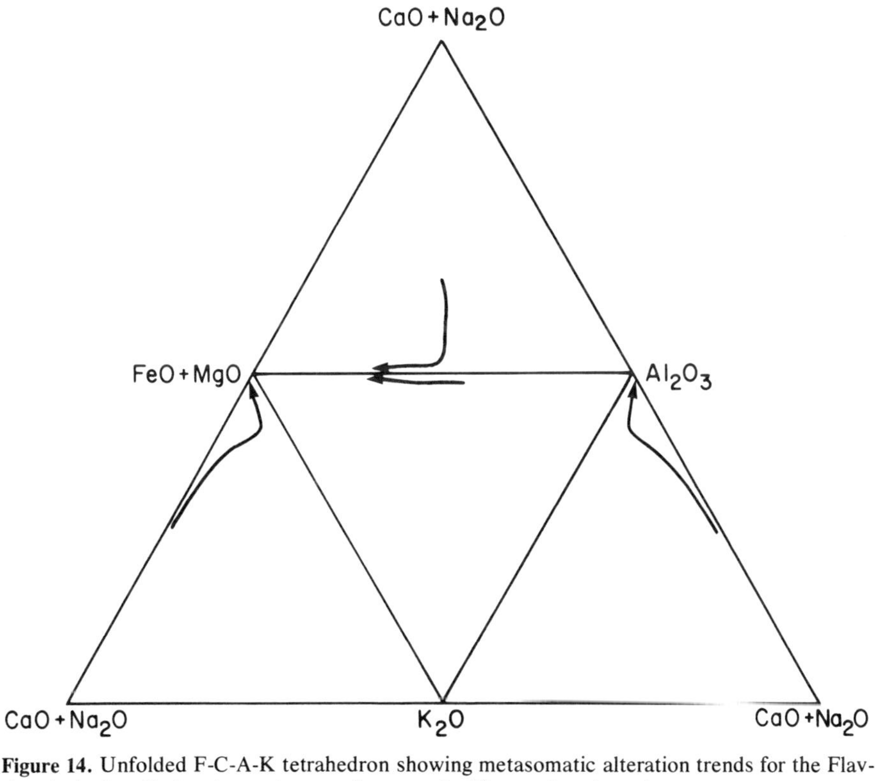

Figure 14. Unfolded F-C-A-K tetrahedron showing metasomatic alteration trends for the Flavrian Andesite, Corbet Area. Arrows show trend of increasing alteration.

tetrahedron (Fig. 14). The metasomatic trend for the Flavrian Andesite at the Corbet deposit is identical to that displayed in altered rocks at the Millenbach deposit (cf. Fig. 14 with Fig. 18, of Knuckey *et al.*, this volume), strongly suggesting that the hydrothermal processes were the same.

CONCLUSIONS

The presence of pillowed flows, hyaloclastites and well bedded, siliceous exhalite in close proximity to ore leaves little doubt that the Corbet deposit was formed in a submarine environment. The formation of volcaniclastic rocks, mixed with abundant hyaloclastic debris from highly vesiculated lava (Fig. 15) indicates relatively shallow water. Localized magmatic and steamblast explosions mixed lithic and vitric fragments around an active vent.

Hydrothermal fluids, perhaps generated by a subvolcanic quartz diorite intrusion, passed up through this highly permeable volcaniclastic pile precipitating quartz and stringer sulphides. At the sea water interface a well zoned massive sulphide deposit formed, initially within the topographic confines of a depression or small crater. Hydrothermal fluids penetrated through, or decanted over the rim of the crater to form a tongue of more distal-type ore down an irregular trough in the outer slope of a submarine andesitic volcano.

The absence of banding in the massive sulphides and the restricted distribution of the siliceous exhalite together with evidence that andesitic flows from the northwest encroached upon an active hydrothermal vent, strongly suggest that the orebody was deposited quickly in an active volcanic environment. This is similar to the

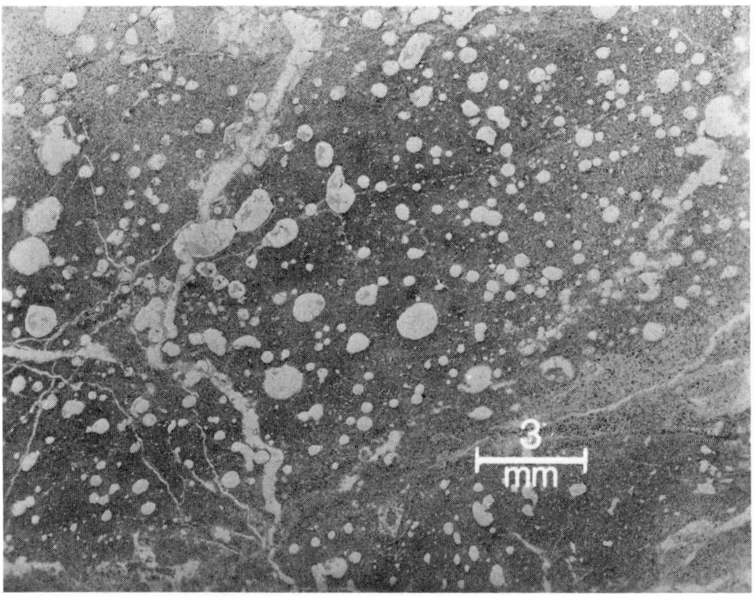

Figure 15. Strongly amygdaloidal Flavrian Andesite.

environment in which the Millenbach ore lenses are believed to have formed concomitantly with rhyolite domes (Knuckey *et al.*, this volume). It contrasts however with the depositional environment of the Norbec deposit 5 km to the north, which formed under quiescent conditions, with metal-bearing exhalites precipitating over an area of at least 10 km_2, no stacked ore-lenses, no alteration in overlying rocks and no evidence of coeval magmatism.

The hydrothermal vent associated with the orebody and the magmatic activity to the northwest of the deposit, lay adjacent to the northwesterly trending, dyke-filled McDougall fault. The McDougall structure was synvolcanic and for a considerable time influenced not only the location of extrusive magmatism in the area of the mine but also channelled the migrating, metal-charged hydrothermal fluids. The Corbet deposit occurs where hot, ore-forming brines discharged onto the sea floor.

The discovery of this well preserved, volcanogenic, massive sulphide deposit in andesitic rocks much deeper in the stratigraphic succession than previously known orebodies has stimulated new exploration and concepts. It has shown that despite intense exploration and study the Noranda District can produce new deposits in different environments. Much remains to be learned of the conditions that generate metal-bearing hydrothermal systems and the faults and fractures that channel them to the sea floor to form economic mineral deposits.

ACKNOWLEDGEMENTS

Thanks are due to Corporation Falconbridge Copper for permission to publish, and for its strong support throughout the preparation of this paper.

REFERENCES

Gibson, H., 1980, Internal company report: Corporation Falconbridge Copper.

Goldie, Raymond, 1978, Magma mixing in the Flavrian pluton, Noranda area, Quebec: Canadian Journal of Earth Science, v. 15, p. 132-144.

Muir, J.E., 1978, Falconbridge Metallurgical Laboratories Mineralogical Report No. 1081.

Riverin, G., 1977, Wall-rock Alteration at the Millenbach Mine, Noranda, Quebec: Unpublished Ph.D. Thesis, Queen's University, Kingston, Ontario.

Roberts, R.G., and Reardon, E.J., 1978, Alteration and ore-forming processes at Mattagami Lake Mine, Quebec: Canadian Journal of Earth Science, v. 15, p. 1-21.

Sangster, D.F., 1972, Precambrian Volcanogenic Massive Sulphide Deposits in Canada: A Review: Geological Survey of Canada Paper 72-22, 44 p.

Spence, C.D., 1967, The Noranda area: in Centennial Field Excursion Volume, NW Quebec and NE Ontario: Canadian Institute of Mining and Metallurgy, p. 36-39.

Spence, C.D. and de Rosen-Spence, A.F., 1975, The place of sulfide mineralization in the Volcanic Sequence at Noranda, Quebec: Economic Geology, v. 70, p. 90-101.

MS received February 1980;
Accepted for publication January 1981.

Precambrian Sulphide Deposits, H.S. Robinson Memorial Volume, edited by R.W. Hutchinson, C.D. Spence and J.M. Franklin, Geological Association of Canada Special Paper 25, 1982

GEOLOGY OF THE DETOUR ZINC-COPPER DEPOSITS, BROUILLAN TOWNSHIP, QUEBEC

R. Deptuck
Selco Inc., 55 University Ave., Suite 1700,
Toronto, Ontario M5J 2H7

H. Squair
Selco Inc., Western Canada Exploration Office,
535 Thurlow St., Suite 402, Vancouver, B.C. Y6E 3L2

V. Wierzbicki
Selco Inc., 55 University Ave., Suite 1700,
Toronto, Ontario M5J 2H7

ABSTRACT

The Detour deposit, discovered in 1974 occurs within acid pyroclastic and volcaniclastic rocks which form part of the Matagami section of the Archean Abitibi orogenic belt. Three zones, A-1, A-2 and B, have been outlined by surface diamond drilling an underground development.

This base metal deposit is atypical compared to others in the Canadian Shield, as it consists of epigenetic-quartz-carbonate vein systems formed by hydrothermal activity that accompanied late stages of felsic volcanism. A volcanic caldera with both subaerial and subaqueous depositional environments is postulated as the geological setting in which the deposit was formed. Sulphide veining and minor replacement occur along preferred but variable steeply dipping fracture/fault systems. The vein systems of the A-1 and A-2 Zones are localized within subhorizontal, permeable, formations.

Principal hypogene minerals are pyrite, sphalerite and chalcopyrite; galena, tetrahedrite, polybasite and native silver occur in minor amounts with native gold in trace amounts. Supergene chalcocite, digenite, covellite, bornite and native copper occur as fracture fillings and replacement rims around hypogene sulphides forming an enriched zone associated with intense kaolinization in the upper portions of the B Zone and elsewhere in minor amounts.

The host rocks consist of rhyolitic tuff and breccia, bedded chert-pyrite, and volcaniclastic debris. An overlying quartz porphyry which is weakly mineralized may have formed an impermeable cap rock trapping the hydrothermal solutions and thereby localizing the deposits within permeable members in the underlying fragmental rocks.

RÉSUMÉ

Le dépôt de Détour, découvert en 1974, se trouve dans des roches acides pyroclastiques et volcaniclastiques qui font partie de la section de Matagami de la ceinture orogénique archéenne de l'Abitibi. Trois zones, A-1, A-2 et B ont été reconnues par sondages de surface et développement souterrain.

Ce dépôt de métaux de base est atypique, comparé à d'autres du Bouclier Canadien, en ce qu'il consiste en réseaux de filons épigénétiques à quartz-carbonate, résultant de l'activité hydrothermale qui a accompagné les stades tardifs du volcanisme felsique.

Une caldéra dans laquelle régnaient des conditions subaériennes ainsi que subaqueuses, est postulée en tant que site géologique du dépôt de Détour. Les filons sulfurés montrent des remplacements mineurs et se trouvent dans des réseaux de failles/fractures plongeant fortement. Les réseaux filoniens des zones A-1 et A-2 sont localisés à l'intérieur des formations perméables subhorizontales.

Les principaux minéraux hypogènes sont pyrite, sphalérite et chalcopyrite accompagnés de galène, tétraédrite et polybasite en quantités mineures et d'or natif en traces. Chalcocite, digénite, covellite, bornite et cuivre natif sont des minéraux supergènes remplissant des fractures et remplaçant la bordure des sulfures hypogènes formant ainsi une zone d'enrichissement associée à une zone de kaolinisation intense dans la partie supérieure de la zone B et ailleurs en quantités moindres.

Les roches hôtes sont des tufs et brèches rhyolitiques, du chert et pyrite lité et des débris volcaniclastiques. Un quartz-porphyry sus-jacent qui est faiblement minéralisé a peut-être formé un niveau imperméable piégeant les solutions hydrothermales et ainsi localisant les dépôts dans les membres perméables des roches clastiques sous-jacentes.

INTRODUCTION

The Detour zinc-copper-silver deposit is situated in northwestern Quebec within the Abitibi volcanic belt, some 600 km northwest of Montreal. The Detour area is accessible by 87 km of gravel road which branches off the paved Amos-Matagami Highway (Fig. 1A). The deposit lies in a drift-covered region characterized by extensive swamps and areas of low relief. Bedrock exposures are confined to isolated hills, and the channels of several large rivers.

The discovery was made in June, 1974 after ground follow-up of an airborne INPUT electromagnetic survey conducted in 1971 and 1972 (Reed, 1976; 1981). A total of 57,000 m of diamond drilling was carried out between 1974 and 1978, and a program of underground exploration was undertaken in 1977 and 1978. To date, two zones A-1 and B have been outlined, and a third, the A-2 Zone has been detected by surface drilling. Underground excavations consist of a shaft 120 m deep with 490 m of drifts and crosscuts on the B Zone, and a decline and two drifts totalling 845 m on the A-1 Zone.

REGIONAL GEOLOGY

The Detour deposit is situated in the Chibougamau-Matagami section of the Abitibi orogenic belt (Goodwin and Ridler, 1970). This belt, part of the Superior Structural Province, is characterized by a number of elongate mafic to felsic volcanic complexes with attendant plutons. These complexes are flanked by broad linear bands of mafic metavolcanic and metasedimentary rocks. The productive mining districts of Noranda, Val d'Or, and Matagami contain examples of base metal de-

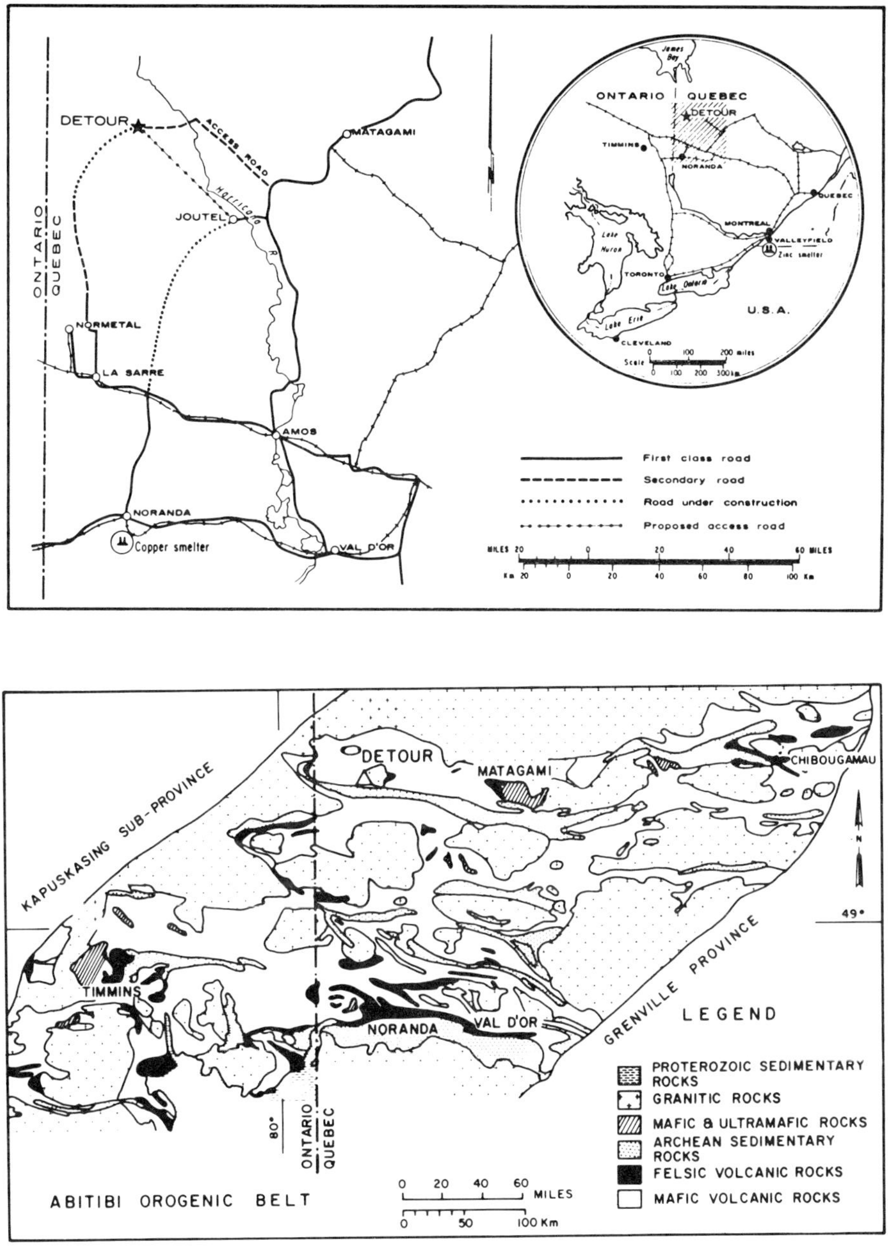

Figure 1. A-Location map; B-Regional Geology (modified after Goodwin and Ridler, 1970).

posits within these volcanic complexes (Fig. 1B). With the exception of younger Precambrian diabase dikes, all the rocks of this belt are believed to be older than 2.5 b.y. (Wanless *et al.*, 1968; Fig. 1).

The deposit occurs within a crescent-shaped band of acid to intermediate flow and fragmental volcanic rocks some 65 km long. It is located near the southwestern margin of a plutonic complex known locally as the Brouillan Granite, which is roughly circular and has an area of approximately 270 km^2 (Fig. 2). The acid and intermediate volcanic rocks south of the pluton strike east and dip south at angles generally greater than 50 degrees and evidence from graded beds and pillow structures suggests that they face south. This sequence is overlain by basalt and andesite flows with intercalated metasedimentary rocks and magnetite-bearing iron-formation. Northwest of the Brouillan Granite, the acid to intermediate volcanic rocks dip and face southeast toward the pluton. Farther to the north there are andesite flows and sediments which dip and face north.

Penetrative deformation of the volcanic rocks is weak, and primary volcanic textures are well preserved. Both subaqueous and subaerial features can be observed. The regional metamorphic grade is low and most mineral assemblages are representative of the lower greenschist facies, therefore in the following lithologic descriptions the prefix "meta" is understood although not used.

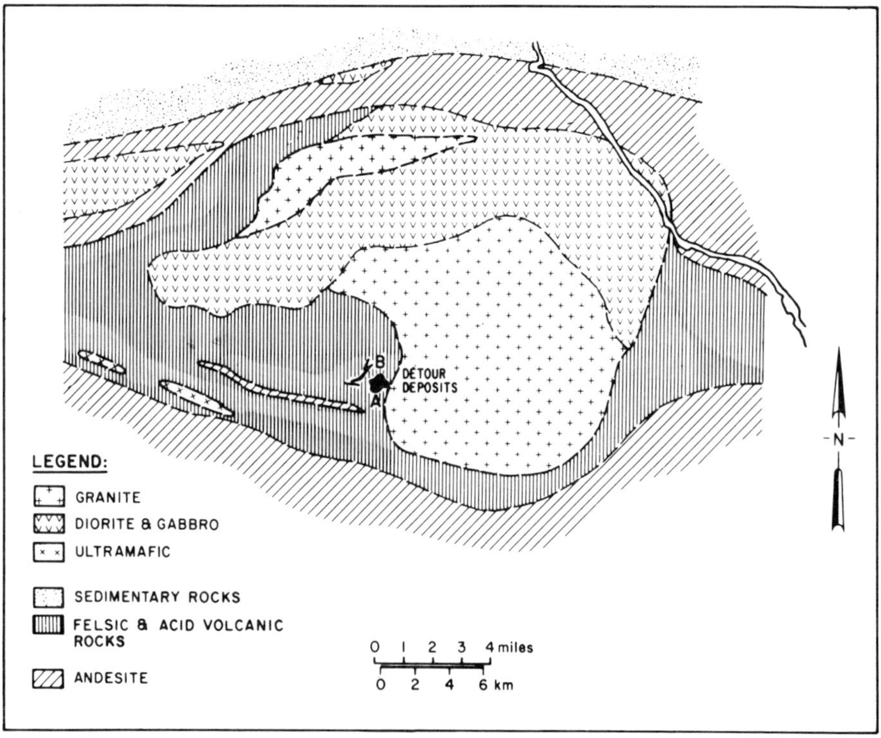

Figure 2. Geological setting of the Detour deposit.

GEOLOGY OF THE DEPOSIT

Geological Setting

Bedrock surrounding the Detour deposit is covered by glacial till and sand to a depth varying from one to 60 m. Present knowledge of the geology of the deposit is

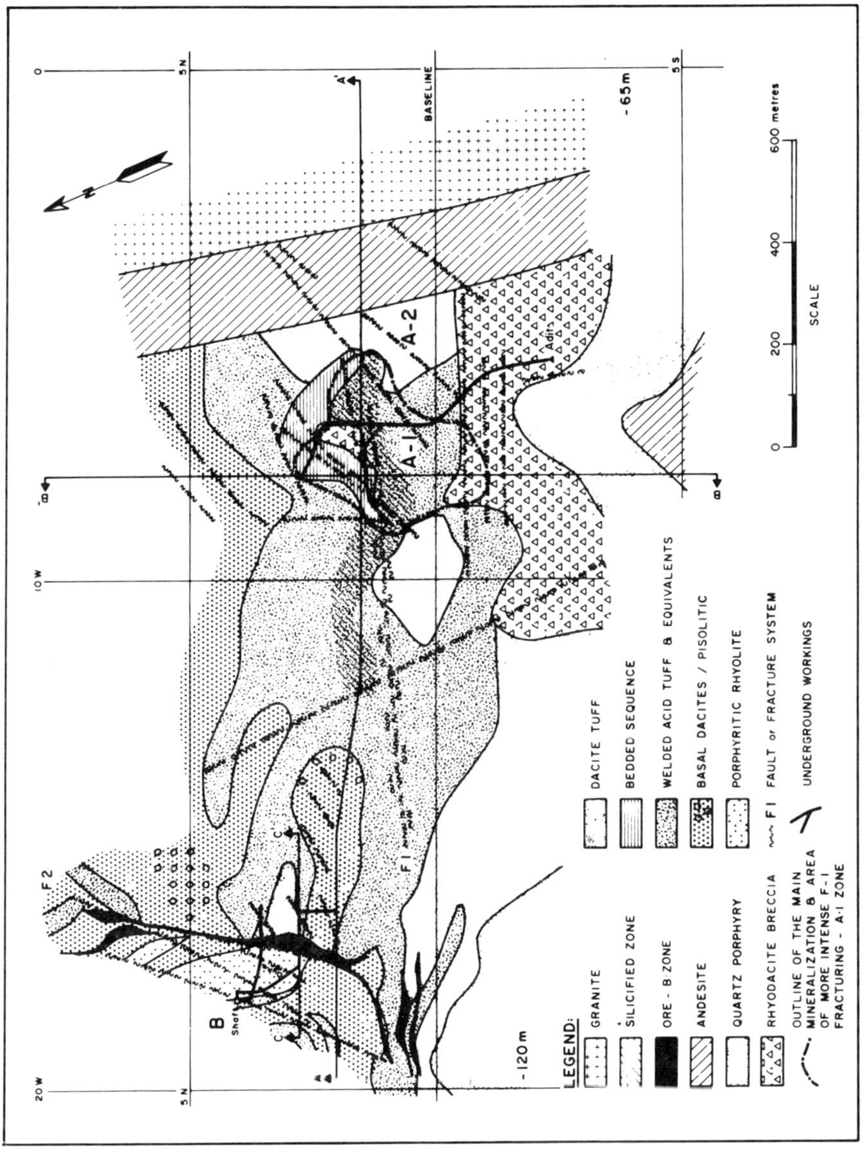

Figure 3. Geology of the Detour deposit, based on underground mapping at — 65 m level (A-1 Zone) and at — 120 m level (B-Zone).

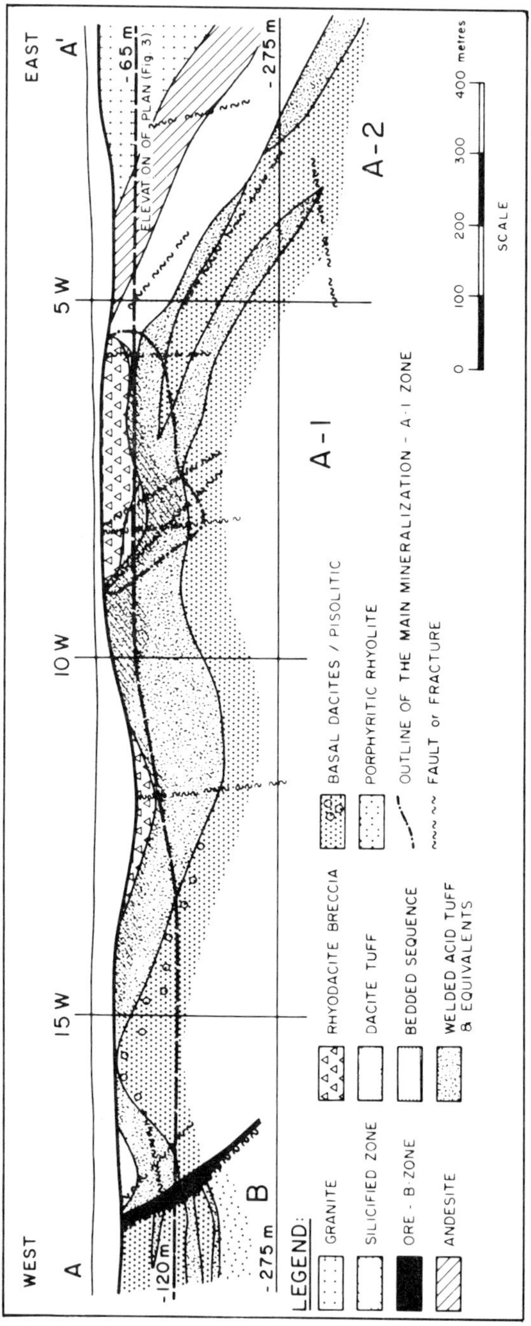

Figure 4. East-west geological cross-section along A-A′ (Fig. 3), looking north.

derived from underground mapping and the examination of diamond drill cores. A compilation of this information is presented in plan and cross-sections (Figs. 3, 4, 5, and 6).

The host rocks of the deposit are a sequence of rhyolitic fragmental and flow rocks which are intercalated with lesser amounts of dacitic and andesitic volcanic rocks. These units strike east and dip south at angles varying from 10 to 60 degrees; this pattern is locally modified by faulting.

Lithologic Descriptions

The Basal Dacites and Porphyritic Rhyolite (Figs. 3, 4, 5, and 6) underlie all other volcanic strata at Detour. The stratigraphic relationship between the two units is not fully understood. The Basal Dacites consist of a sequence of fine- to medium-grained, rhyolite to dacite tuffs up to 200 m thick with minor lapilli tuffs, pisolitic tuffs and thin bands of welded acid tuffs. Hull (1976) described the Basal Dacites as forming a subaerial sequence of ash-fall tuffs and moderately to poorly welded ash-flow tuffs. The Porphyritic Rhyolite with distinct quartz grains or phenocrysts, forms a thick unit west of the B Zone. Less prominent Porphyritic Rhyolite intersections occur within the Basal Dacites in the A-1 Zones.

Welded Acid Tuff and Equivalents (Figs. 3, 4, 5, and 6; Plate I, Figs. 11 and 12) constitute the most distinctive lithological unit at Detour. This consists of up to 120 m of light grey, rhyolite tuffs, which exhibit a number of distinctive features. Approximately 10% of this rock is made up of dark chloritic inclusions and 5% of lithic fragments; the balance is a dense, cryptocrystalline, siliceous matrix with occasional quartz grains up to 2 mm in diameter. The chloritic inclusions are tabular to irregular in shape, range from 1 to 50 mm in length, have flame-like terminations and exhibit a degree of parallel elongation, which in welded tuffs is referred to as eutaxitic foliation (Ross and Smith, 1961). They closely resemble the fiamme or collapsed pumice fragments typical of welded ash-flows. Although no relict shard forms have been

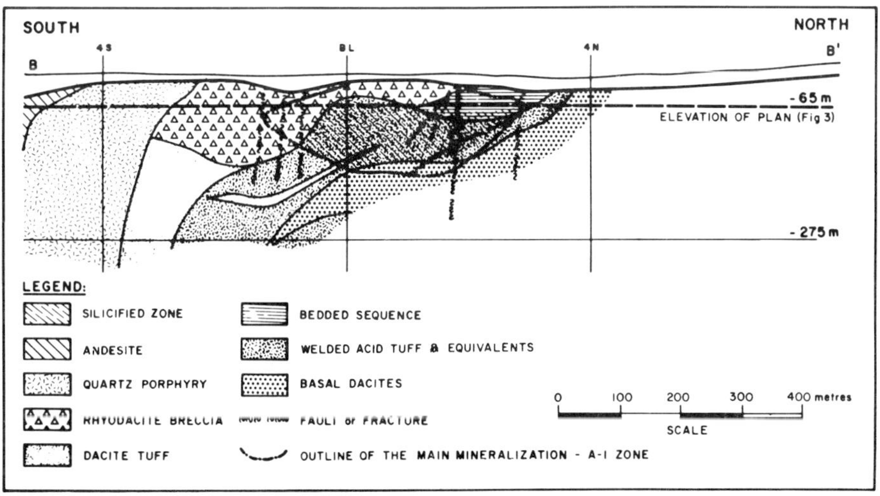

Figure 5. North-south geological cross-section along B-B′ (Fig. 3), looking west.

observed in the matrix, the overall texture of the rock strongly suggests an ash-flow origin, and appears to be analogous to that of the piperno rocks (Ross and Smith, 1961).

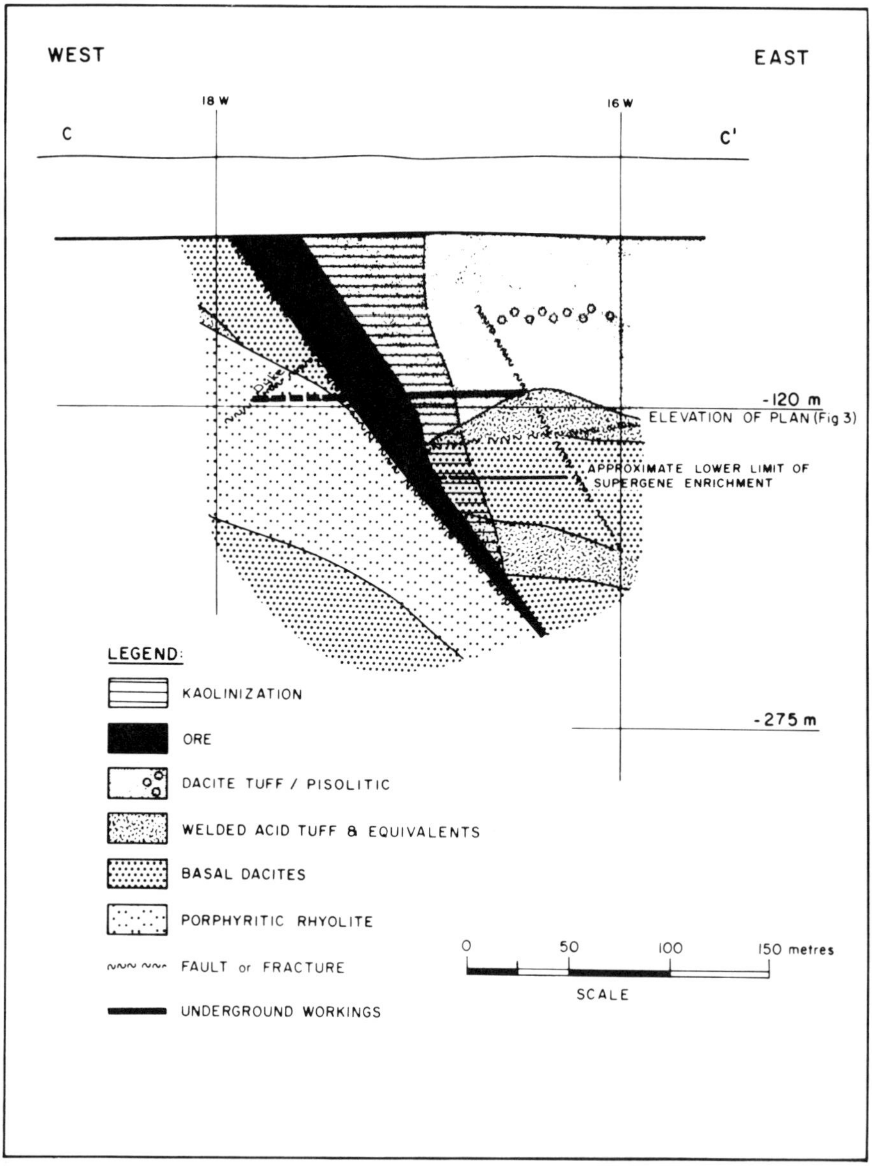

Figure 6. East-west geological cross-section, B-Zone; along C-C′ (Fig. 3), looking north.

West of the A-1 area, the Welded Acid Tuff exhibits the same general textural characteristics, but lapilli-sized acid lithic clasts are more common, and some of the chloritic clasts exhibit an angular or broken appearance. In a few cases, unaltered fragments of pumice can be observed. This variant is designated the Welded Acid Tuff Equivalent.

Pisolithic Dacite Tuff is another, texturally distinct type, associated with the Welded Acid Tuff (Plate I, Fig. 13). Round to star-shaped quartz pisoliths 5 to 10 mm in diameter constitute 20 to 80 volume per cent of this rock. They are set in a fine, moderately chloritic to sericitic matrix. On the basis of their general appearance, these rounded structures are believed to represent either silica-filled lithophysal cavities or altered spherulites. Both phenomena are a common cooling feature in ash-flow tuffs (Ross and Smith, 1961). According to Hull (1976), however, some of the pisoliths could be accretionary lapilli.

The Bedded Sequence (Figs. 3, 4, and 5) overlies the Welded Acid Tuff in the northern part of the A-1 Zone. This sequence includes beds of massive pyrite, chert and a variety of tuffs. Their maximum aggregate thickness is 60 m. Individual rock units appear to be lenticular and the detailed succession varies from place to place. Evidence of both subaerial and subaqueous depositional environments can be observed. The following are brief descriptions of the various units:

Acid Lapilli Tuffs are generally light grey, massive and contain abundant subrounded felsic lapilli in a fine-grained tuffaceous matrix with numerous quartz grains 1 mm in diameter. Some varieties contain up to 15% of cherty fragments. Other varieties have chloritic fiamme and lithophysae indicating that they are probably ash-flow units of subaerial origin. These latter types occur near the base of the Bedded Sequence. All the lapilli tuff beds are less than 5 m thick.

Massive Pyrite has a "stony" appearance and is composed of 70 to 80% very fine-grained pyrite and 20 to 30% cryptocrystalline, grey silica. Rare bands of brown sphalerite less than 1 cm thick are found locally. The pyrite occurs in lenses up to 15 m thick at several different levels in the Bedded Sequence. The fine and possibly framboidal texture of the pyrite and its conformability to the adjacent strata suggest a subaqueous origin. Abundant chloritic fragments up to 1.5 cm long are observed in the extremities of one of these lenses and highly chloritized and pyritic material underlies another.

Bedded Dacite Tuff is a light grey unit approximately 3 m thick consisting of a series of graded beds one to 5 cm thick. Quartz grains 0.5 mm in diameter are abundant and some beds contain chloritic fiamme up to one centimetre long.

Crystal Tuff is a light to dark grey-green rock which contains abundant rounded feldspar and quartz crystal fragments. Thicknesses range up to 10 m. One 20 cm thick bed contains angular quartz grains and abundant chloritic fiamme up to 5 mm long.

Andesite Tuff is a dark green, very fine-grained rock up to 10 m thick with a distinct subconchoidal fracture and locally abundant quartz grains 0.5 mm in diameter. It occurs interbedded with the Crystal Tuff and as lenses up to 6 m long within acid lapilli tuffs.

Chert occurs in beds up to 15 m thick composed of fine laminae 0.5 to 2 cm thick. It is pale grey to pale green, extremely fine-grained with a typical conchoidal fracture. The chert beds are found overlying massive pyrite, or interbedded with dacite or andesite tuffs.

Dacite Tuff (Fig. 4 and 5) is a stratigraphic equivalent of the Bedded Sequence. It is composed of a series of subaerial dacitic tuffs, lapilli tuffs and minor interbedded flow rocks.

Rhyodacite Breccia (Plate I, Fig. 14) overlies the Bedded Sequence and the Welded Acid Tuff. It is composed of lithic fragments, derived predominantly from the Bedded Sequence and to a lesser extent from the Welded Acid Tuff and other underlying rocks. These fragments are typically angular, and are mostly in the 1 to 10 cm size range; occasional blocks, several m across can be observed, however, and underground mapping suggests the existence of megablocks as large as 20 m across. The matrix generally constitutes less than 30% of the rock and consists of fine tuffaceous material with scattered quartz grains up to 2 mm in diameter. The Rhyodacite Breccia, exceeding 100 m in thickness in the area of the A-1 Zone, thins to the west and is absent in the east and north.

Quartz Porphyry consists of subhedral to rounded phenocrysts of quartz and altered plagioclase in a very fine-grained, quartzo-feldspathic matrix. It is generally grey or pale green, but in places purple and pink varieties occur. Phenocrysts constitute 10 to 20% by volume of this rock and range in size from 0.5 to 4 mm. Many phenocrysts are embayed and fractured. Features that suggest an extrusive origin for the porphyry are flow banding, alignment of quartz phenocrysts and relict shards (Hull, 1976). The thickness is variable and locally exceeds 150 m. The most northerly part of the unit dips 20 to 30 degrees south. Steepening of the dip farther south suggests a transgressive relationship with the underlying rocks (Fig. 5). Siliceous porphyritic rocks displaying both extrusive and intrusive features are reported at South Bay (Pollock *et al.*, 1972) and at numerous other volcanogenic deposits (Sangster, 1972).

Andesite is found east and south of the A-1 Zone and consists of thin flows and subordinate interflow breccia. No pillow structures or other subaqueous features have been observed. The thickness of the unit is about 100 m and its contacts with the underlying felsic volcanic rocks are sharp.

Granite (Fig. 4) overlies andesite east of the A-1 Zone. It is pink to grey, medium-grained and composed of quartz-feldspar aggregates with minor chlorite and carbonate. It may be an apophysis of the Brouillan Granite which lies to the east. The contact is subparallel to the lower andesite contact. Numerous felsic to intermediate xenoliths occur in the contact areas.

Dykes and sills of four distinct types are present. The first is a felsic, carbonate-rich sill, which intrudes the Basal Dacites. The second is a set of felsic dykes intruding the Rhyodacite Breccia. Drill hole data indicate that these have variable attitudes. The third type is a set of metadiabase dykes, which strike southeast, dip moderately northeast and intrude the granite. Felsic dykes may also intrude the granite but in drill core they cannot be distinguished from felsic xenoliths. The fourth type is a narrow olivine diabase dyke. It was intersected in three drill holes, approximately 300 metres beneath the A-1 Zone and strikes northeast parallel to the regional trend for similar dykes which are recognized to be of Keeweenawan Age (Remick, 1969).

Structure

Attitude of Strata in the deposit area is variable. The gentle dips in the A-1 Zone contrast markedly with the regional steep dips of most rocks. The strata east of the A-1 Zone steepen and dip easterly under the Granite (Fig. 4). South of the A-1 Zone, dips also steepen, conforming to the known regional attitude (Remick, 1969; Fig. 5). To the west, the strata appear to undulate gently, forming a trough between the A-1 and B Zones and rising again to abut the B Zone fault. West of this fault strata strike north and dip moderately east (Figs. 3 and 4).

The Welded Acid Tuff shows local variations in thickness and attitude; its upper surface outlines a small depression north of the baseline in the area of the A-1 Zone (Figs. 4 and 5). The Bedded Sequence fills the depression which plunges south thus appearing in plan as a crescent shaped area (Fig. 3). The Rhyodacite Breccia lies unconformably on the Bedded Sequence and Welded Acid Tuff.

Faults and Fractures (Figs. 3, 4, 5, and 6) are the most prominent structural features of the Detour deposit. Pre-mineralization fracture systems occur in two directions almost normal to one another. The first system (F1), striking east and dipping moderately to steeply south, extends across the deposit area. Mineralized veins of the A-1 and A-2 Zones and the South Lenses of the B Zone occur in this system. The second system (F2) which strikes north-northeast, and dips moderately to steeply east, contains the main lens of the B-Zone. This system consists of a footwall fault, termed the Main Fault and subsidiary or second order faults and fractures on the hanging wall side. Displacement of chalcopyrite veins, pulverised sulphides and shattering of quartz are evidence of post-mineralization movement and re-activation of previous faults. Movement along the main fault is inferred to be normal as Basal Dacites on the footwall lie against stratigraphically higher Welded Acid Tuff on the hanging wall. The F2 fault and fracture system is interpreted as terminating at the F1 system.

Post mineralization faults and fractures with varying attitudes are abundant (Figs. 3, 4, 5, and 6). These fractures are characterized by clay and limonite gouge, breccia and pulverized sulphides within the mineralized zones. Fault zones range from a few cm to several m in width. Relative movements have not been determined.

Foliation is a weak, ubiquitous feature which trends east, dips steeply and conforms with the regional trend (Remick, 1969). Folding is not evident. The main fault in the B Zone has produced subparallel, more widely spaced foliation with a strike of about 60 degrees and a dip of 75 to 80 degrees southeast affecting the rocks to the west of the orebody for a distance of 100 m.

SULPHIDE DEPOSIT

The Detour deposit as presently known consists of three zones (Fig. 3):

A-1 – 32.11 million tonnes averaging 0.39% Cu, 2.30% Zn, 35.7 g/t Ag and 0.3 g/t Au.

B – 3.07 million tonnes averaging 4.49% Cu, 0.80% Zn, 39.4 g/t Ag and 1.2 g/t Au.

A-2 – a zone deeper than, and to the east of the A-1 Zone where investigations are not complete and reserve figures are not available.

A-1 Zone

Mineralization in the A-1 Zone occurs in the east striking, steeply dipping veins concentrated in an area approximately 400 m in diameter and extending to a depth of some 180 m. The vein system has an irregular subhorizontal distribution and occurs mainly within the upper third of the Welded Acid Tuff and the lower part of the Rhyodacite Breccia; these veins are rare in the Bedded Sequence. The veins occur within the Fl fracture system (Fig. 3) which extends westward into the South Lenses of the B Zone, where it is narrower and more restricted, and eastward into the A-2 Zone.

The central part of the A-1 Zone, where quartz and sulphide veins are most abundant and silicification of the host rocks particularly intense is designated the Silicified Zone (Figs. 3, 4, 5, and 8). Individual veins are usually laminated and consist of alternating layers of quartz, sphalerite, pyrite, minor chalcopyrite and often carbonate and chlorite. A few veins have complex textures and are highly

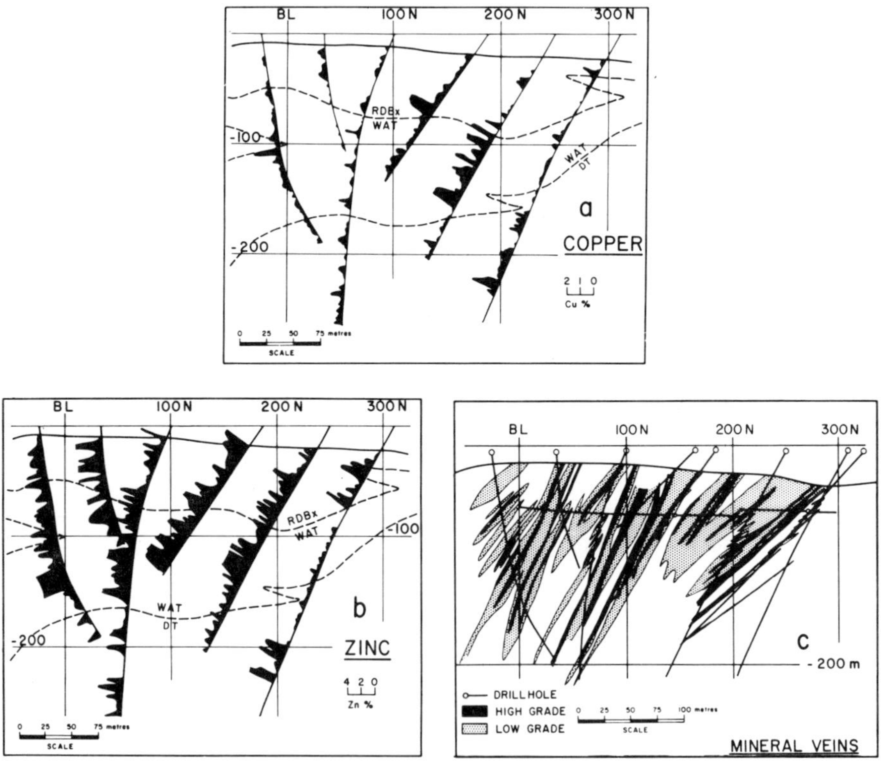

Figure 7. Cross-section 800W, A-1 Zone, showing distribution of: a) copper values along drillholes; b) zinc values along drillholes; and c) interpreted distribution and attitude of mineral veins of F1 fracture system.

contorted. The quartz is generally very fine-grained, off-white and sugary textured; a bluish grey variety with agate-like banding is observed in a few cases (Plate II, Fig. 15). Sphalerite is commonly brown but may be grey or honey-coloured; most of the lighter coloured sphalerite is very fine-grained. Pyrite and chalcopyrite are very fine- to medium-grained (Plate III, Fig. 18). Widths of individual veins range up to 40 cm. Host rock between veins is in places impregnated with fine streaks and disseminations of sulphides.

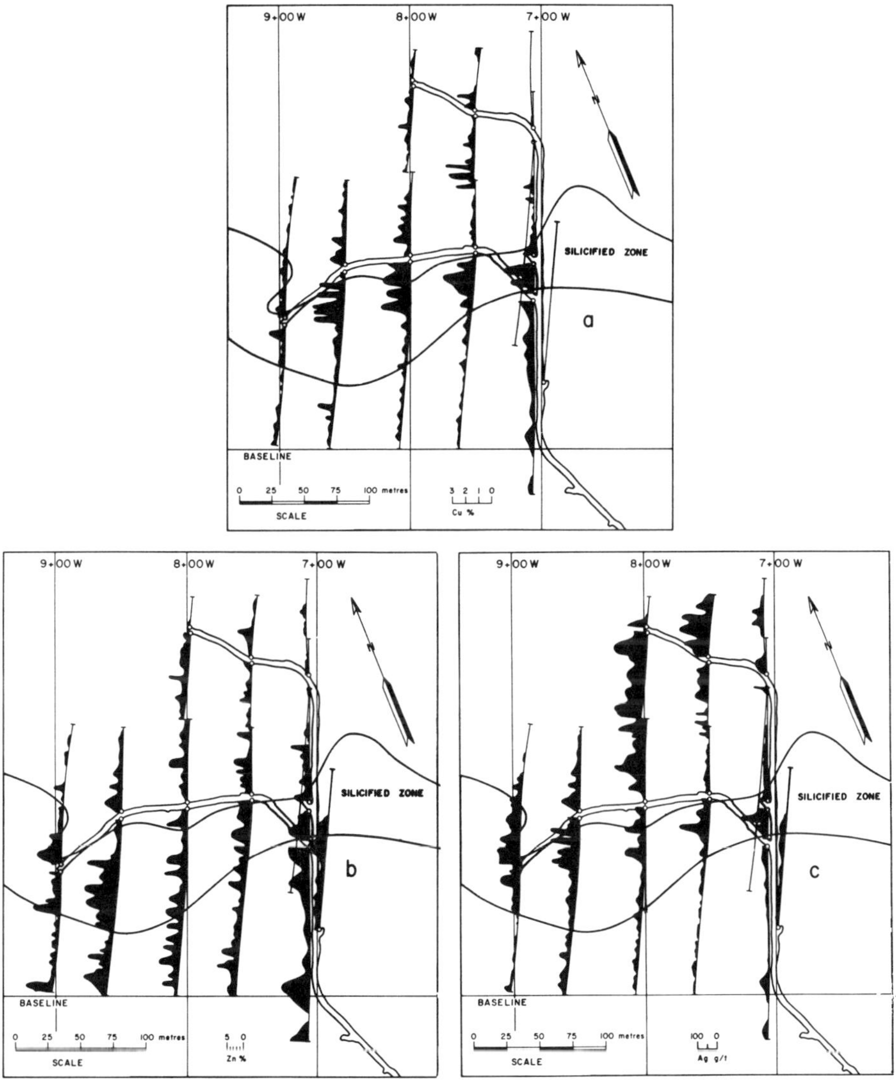

Figure 8. Plan of −65 m level, A-1 Zone, showing metal distribution along horizontal drillholes: a) copper; b) zinc; and c) silver.

The few mineralized veins in the Quartz Porphyry and their absence in the overlying Andesite suggest that the period of mineralization was post-Quartz Porphyry and pre-Andesite.

Distribution of Metals is illustrated in Figures 7 and 8. These diagrams compare the distribution patterns of copper, zinc and silver. They show that zinc is uniformly distributed throughout the A-1 Zone. The distribution of copper, however, is much more irregular with higher values being concentrated between 100 and 200 m north of the baseline. In plan (Fig. 8) higher copper values appear to be associated with the east-striking Silicified Zone. Higher silver values are found in two environments; firstly, in association with copper in the Silicified Zone and secondly, within the massive pyrite unit of the Bedded Sequence north of the Silicified Zone.

B Zone

Mineralization in the B Zone can be subdivided into two groups of lenses. The first group consists of the main lens and two subsidiary lenses which lie along a zone of major dislocation within the north-northeast-striking fracture system F2 (Fig. 3). The main lens of the B Zone is over 500 m long and consists of a network up to 30 m wide of closely spaced, irregular veins which contain varying amounts of chalcopyrite, chalcocite and crystalline and sooty sphalerite, in a gangue of quartz, minor carbonate and chlorite (Plate II, Fig. 16). Minor amounts of bornite, covellite, and native copper also occur. The veinlets range in thickness from less than one millimetre to 20 cm but the majority of them are from one to 3 cm. Their strikes are generally subparallel to that of the main fault, but the dips are steeper. Subsidiary lenses, with attitudes similar to those of the main lens and consisting of a quartz-chalcopyrite stockwork, occur in the north and in the footwall of the main lens. A number of isolated chalcopyrite and chalcocite-bearing veins have been intersected in the hanging wall of the main lens.

The second group consists of three parallel bodies referred to as the south lenses lying within the east-southeast F1 fracture system south of the main lens (Fig. 3). The lenses are each about 250 m long and have a depth extent of about 150 m, with thicknesses ranging from one to 10 m. Mineralization consists of chalcopyrite, minor sphalerite and pyrite in sugary quartz. The attitudes of the south lenses are similar to those of the main veining in the A-1 Zone.

Supergene Alteration (Fig. 6) played an important role in the enrichment of copper values in the B Zone main lens. The most intense effects are observed above the −150 metre level, but local zones of supergene alteration also occur within the fault zones at greater depths. The channelways for percolating solutions are believed to have been provided by reactivation of the pre-existing F2 fault-fracture system. This hypothesis is supported by the contrast between intense supergene alteration in the central part of the B Zone where evidence of reactivation exists, and other parts of the deposit where the deposition of primary mineralization sealed the channelways.

Supergene alteration is characterized by the presence of sooty chalcocite, aggregates of tarnished chalcopyrite, veinlets of chalcocite-bornite and sooty sphalerite, all of which occur in a porous, sugary quartz and kaolin gangue (Plate II, Fig. 16). Limonite and native copper occur sporadically in the more intensely fractured zones (Plate III, Fig. 17).

Kaolinization is pervasive in the enriched portion of the main lens and occurs within the fractured and weathered hanging wall rocks. Its proximity to the F2 fault system and to the present erosion surface, and its association with the supergene sulphide enrichment suggest that the kaolin is of supergene origin. Sinclair and Gasparini (1980) described the supergene alteration in greater detail.

HOST ROCK ALTERATION

The problem of distinguishing between the effects of hydrothermal alteration related to ore deposition and those of later regional metamorphism on the host rocks of many volcanogenic deposits has been discussed by Clark (1971) and Franklin *et al.* (1975). At Detour, according to the present interpretation of the geological history, hydrothermal alteration was followed by lower greenschist facies regional metamorphism. This overprinting has led to some difficulties in interpreting the relationships between hydrothermal alteration patterns and the distribution of sulphide mineralization. The distribution and nature of the principal alterations are discussed below.

Silicification

In addition to the abundant deposition of quartz in the mineralized veins there is evidence of considerable pervasive silicification of the host rocks. Chemical analyses of rocks from A-1 Zone free from visible quartz veining (Table I) show that they contain SiO_2 concentrations above the normal range of rhyolites, and suggest that they have probably been subjected to pervasive silicification. Diamond drilling is concentrated within the mineralized area and no representative suite of background samples is available.

Intense silicification, within the area termed the Silicified Zone, occurs in the upper part of the Welded Acid Tuff and in parts of the overlying Bedded Sequence. Very fine-grained to cryptocrystalline, secondary silica locally obscures the original textures. Silicification is accompanied by varying amounts of magnesium and calcium carbonates and sulphides. The close association between silicification, quartz veining and sulphide mineralization in the A-1 Zone indicates that these are concomitant processes. The abundance of quartz veins, surrounding pervasive silicification but only minor amount of feldspar alteration suggest that most of the silica has been introduced.

Sericitization

Irregular zones of sericitization, of varying degrees of intensity, are common in the host rocks of both the A-1 and B Zones and are believed to represent aureoles of hydrothermal alteration around the zones of significant sulphide mineralization. In the less intensely altered zones, sericite appears to have formed at the expense of feldspar, whereas in the more severely affected areas amphibole and hydrothermal chlorite are also, partly or wholly, replaced by sericite. Studies by Hull (1976) showed close correlation between sericite and K_2O content of the host rocks. The distribution of K_2O in these rocks is discussed under Geochemical Investigations and suggests that the sericite-rich zones are peripheral to sulphide mineralization.

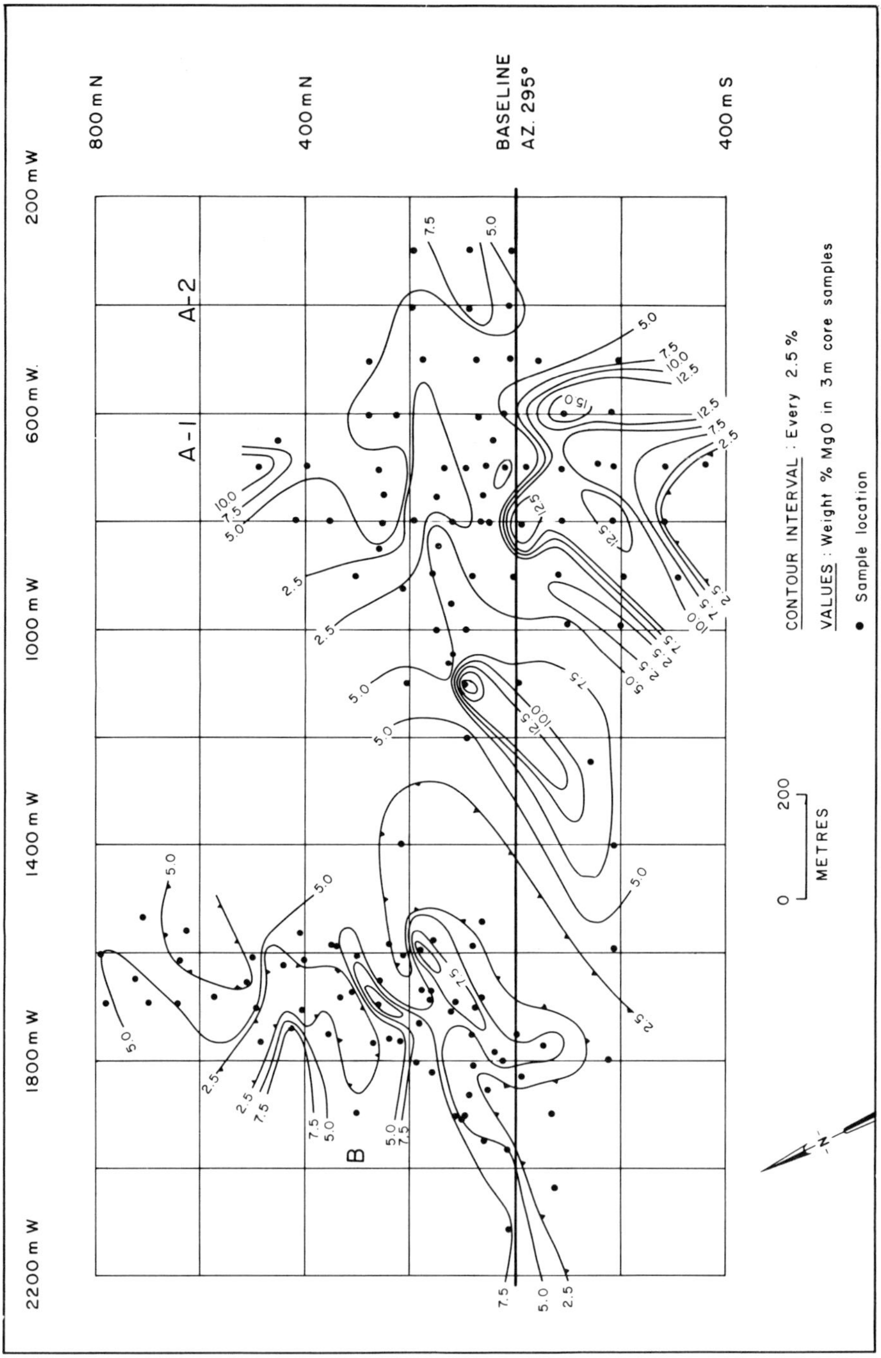

Figure 9. Plan of MgO distribution at −100 m level, A-1 Zone (modified after Sinclair, 1977).

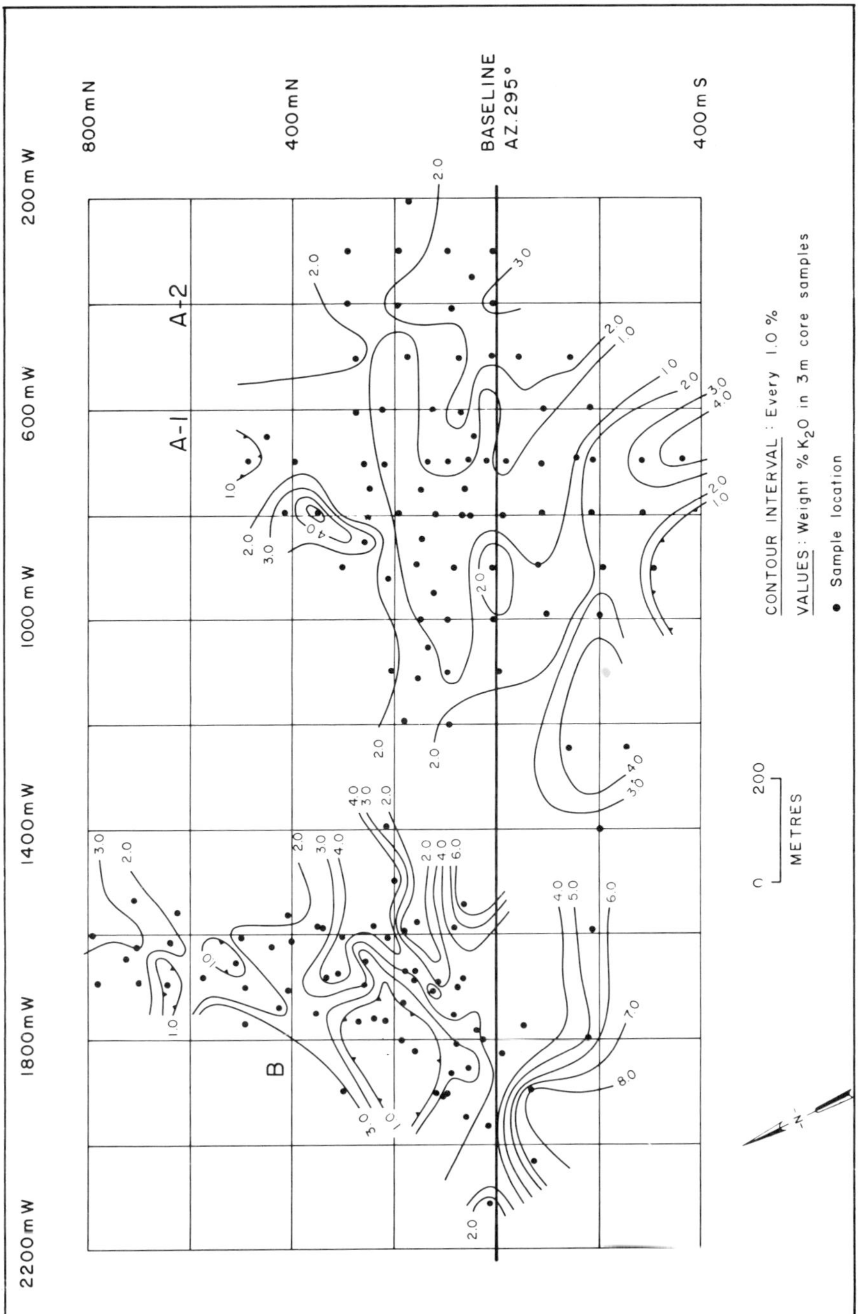

Figure 10. Plan of K_2O distribution at −100 m level, A-1 Zone (modified after Sinclair, 1977).

Chloritization

As a result of regional metamorphism, chloritization is ubiquitous in the rocks at Detour but the intensity is extremely variable. Although this variability can probably be attributed mainly to primary variations in host rock composition, there appears to be an increase in its intensity near the principal zones of sulphide mineralization, particularly in the B Zone, and underlying one of the massive pyrite lenses of the Bedded Sequence in the A-1 Zone. However, no well defined alteration pipe, analogous to those underlying some massive sulphide deposits elsewhere (Sangster, 1972), is recognized. X-ray diffraction work by Hull (1976) showed that leuchtenbergite is the common chlorite found with the mineralized veins, whereas thuringite is the common variety found in the fiamme and in the matrix of most fragmental rocks.

Carbonatization

Magnesium, calcium and iron carbonates are common constituents of the mineralized veins and of the surrounding zones of silicification. Carbonates also appear in variable amounts as constituents of the matrices of the host rocks. The apparent sympathetic relationship between the distribution of carbonates and of sulphides suggests that both were introduced by the same hydrothermal processes.

Geochemical Investigations

A geochemical study was undertaken in an attempt to identify zones of cryptic alteration, as recognition of alteration haloes around the mineralized zones by visual examination of the core proved difficult. Samples of core, each 3 m long were collected at depths of 50 m, 100 m and 150 m below surface and their K_2O and MgO concentrations determined by X-ray fluorescence techniques. Results of the first phase of this study, which consisted of analyses of samples from the A-1 Zone have been published (Sinclair, 1977). Analyses of samples from B Zone are also completed and results are combined with the A-1 Zone data in Figures 9 and 10 to illustrate dispersion patterns at a depth of 100 m. In studying these diagrams, it should be remembered that the samples were collected at a pre-arranged depth without regard to rock type or intensity of mineralization. They illustrate, therefore, gross variations in the geochemistry of the host rocks rather than variations in the composition of a particular lithological unit.

The K_2O distribution is particularly interesting because it demonstrates that the principal sulphide concentrations are related to zones significantly deficient in K_2O.

Explanation of Plate I

Figure 11. Welded Acid Tuff showing dark, chloritic fiamme.

Figure 12. Welded Acid Tuff; note quartz-sulphide veinlet.

Figure 13. Pisolitic Dacite Tuff showing concentric and star-shaped lithophysae.

Figure 14. Rhyodacite Breccia; note poor sorting and heterogeneous nature of fragments.

PLATE I

Figure 11

Figure 12

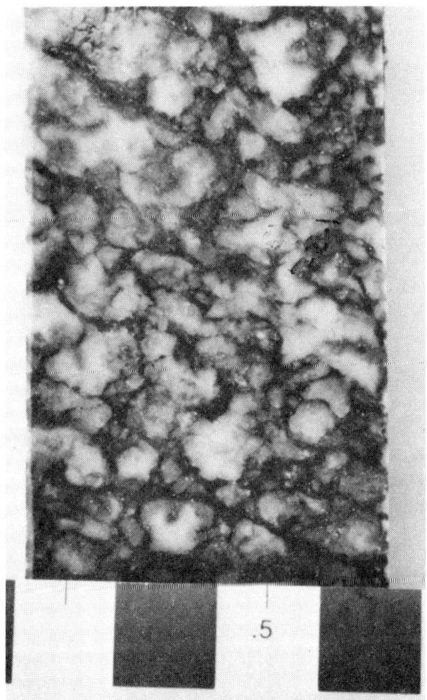

Figure 13

Figure 14

PLATE II

Explanation of Plate II

Figure 15. Laminated quartz-sulphide vein from A-1 Zone; note agate-like banding of the central quartz vein.

Figure 16. Vein material from central part of the B-Zone main lens showing irregular veinlets of primary and secondary copper mineralization in altered host rock.

PLATE III

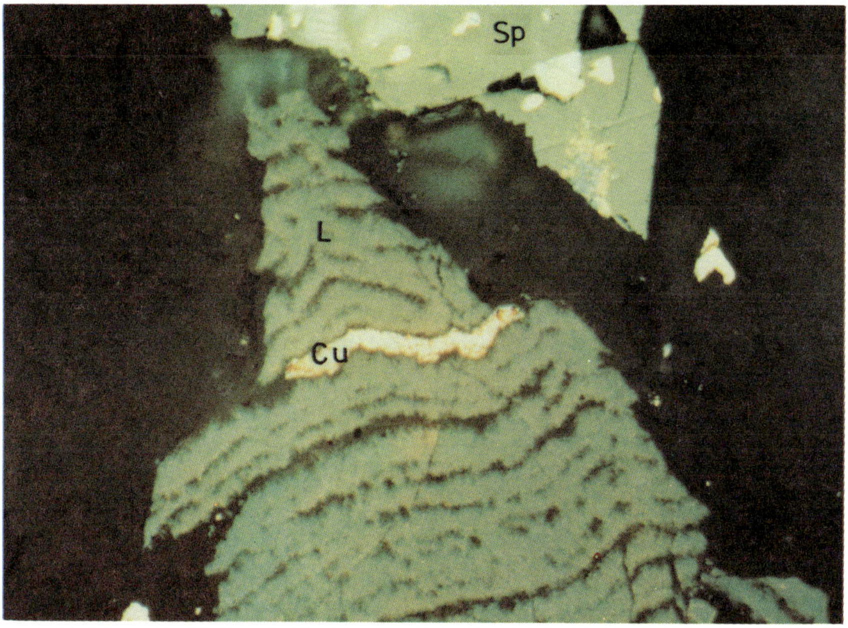

Explanation of Plate III

Figure 17. Native copper with limonite in secondarily enriched vein.

Figure 18. Contiguous grains of chalcopyrite and sphalerite, both rimmed by secondary chalcocite; note exsolved chalcopyrite within sphalerite grain.

TABLE I
SILICA CONTENT OF HOST ROCKS FROM A-1 DEPOSIT. XRF ANALYSES BY BONDAR-CLEGG AND CO.

Rock Type	Drill Hole and Depth	% SiO_2 Content
Rhyodacite Breccia	B – 5: 34 m	77.70
Welded Acid Tuff	B – 5: 99 m	83.00
Dacite Tuff	B – 11: 189 m	74.00
Welded Acid Tuff	B – 11: 132 m	77.60
Dacite Tuff	B – 13: 165 m	77.70
Welded Acid Tuff	B – 18: 114 m	83.00

This relationship suggests that expulsion of K_2O accompanied the introduction of silica and sulphides during hydrothermal activity and that the K_2O-deficient zones represent traces of the hydrothermal channelways active during that period. In the case of MgO, dispersion patterns are less clearly defined but there appears to be weak correlation between MgO-deficient zones and the principal sulphide concentrations. Hull (1976) demonstrated that the variations in K_2O content are related to the degree of sericitization and also showed that the MgO distribution is a reflection of variations in the amounts of chlorite and magnesium carbonates in the host rocks.

DISCUSSION

Environment of Host Rock Deposition

Host rocks of most volcanogenic sulphide deposits in the Canadian Shield are believed to have been deposited in a submarine environment. However, certain textural features suggest that the Detour rocks originated, at least in part, as a result of subaerial volcanism. These features include well preserved fiamme, pisoliths, lithophysae, relict shard forms and evidence of vesiculation. Analogous textures in Precambrian rocks are described in recent papers by Lilljequist and Svenson (1974), Lambert (1978), Smyth *et al.* (1978) and Nixon and Papezik (1979).

Evidence indicating local subaqueous deposition is present in various members of the Bedded Sequence; the evidence includes graded bedding, banded cherts and stratiform "stony" pyrite. Outside the deposit area evidence of submarine deposition is observed in the andesite flows.

Subaerial felsic pyroclastic rocks which constitute the hosts of the Detour deposit lie within the region in which the predominant types are mafic to intermediate submarine volcanic rocks. This relationship, together with detailed observations in the deposit area suggest that the host rocks may have been deposited in a caldera on an emergent part of a stratovolcano. Similar geological environments in Precambrian rocks have been described by Ayres (1977a; 1977b) and elsewhere by MacDonald (1972).

Geological History

The following geological stages are postulated for development of the Detour deposit.

1) Deposition of Basal Dacites on the emergent part of a stratovolcano.

2) Extrusion of ash-flows (Welded Acid Tuff) resulting in rapid depletion of the magma chamber.

3) Local collapse, faulting and differential subsidence of the volcanic pile with concomitant erosion and deposition of the Bedded Sequence in local pools or marine embayments. Metalliferous hydrothermal fumarolic activity had commenced as evidenced by the deposition of massive pyrite and minor sphalerite in the Bedded Sequence.

4) Resurgence of the structure and generation of steep fracturing and faulting which resulted in deposition of the Rhyodacite Breccia by avalanching and sliding.

5) Emplacement of Quartz Porphyry which exhibits both intrusive and extrusive features and forms the upper unit of the Detour felsic volcanic sequence. This unit may be comparable to the late stage felsic intrusive-extrusive bodies emplaced around caldera margins (Smith and Bailey, 1968), or in associated fault systems. It may have acted as a cap rock during the mineralizing processes (Hodgson and Lydon, 1977). These continued during and after the emplacement of the Quartz Porphyry but ceased before the emplacement of andesite, resulting in deposition of sulphides and gangue minerals in fractures generated during Stage IV, and in local alteration of the host rocks.

6) Submergence and onlapping of andesitic and dacitic flows.

7) Low grade regional metamorphism and minor deformation.

8) Erosion and supergene enrichment in fractures and faults.

ACKNOWLEDGEMENTS

We are indebted to Selco Inc. for allowing us time to prepare this paper, and the opportunity to present it here. We are grateful to the following members of the staff for the exchange of ideas and for critical comments: W.J. Anderson, D.A. Hutton, L.E. Reed, I.G.L. Sinclair, A.F. Warburton and B. Zabev. Particular thanks are due to I.G.L. Sinclair who read critically and helped edit the manuscript. Thanks are due to Miss G. Andersson for preparing the drawings and to Mrs. V.J. Ayers for typing several versions of the draft and for correcting our errors.

REFERENCES

Ayres, L.D., 1977a, Importance of stratigraphy in Early Precambrian volcanic terraines: cyclic volcanism at Setting Net Lake, northwestern Ontario: in Baragar, W.R.A., Coleman, L.C. and Hall, T.M., eds., Volcanic Regimes in Canada: Geological Association of Canada, Special Paper 16, p. 243-264.

—————, 1977b, Transition from subaqueous to subaerial eruptive environments in the Middle Precambrian Amisk group at Amisk Lake, Saskatchewan: Centre for Precambrian Studies, 1977 Annual Report, Progress Report Paper No. 3.

Clarke, L.A., 1971, Volcanogenic ores; comparison of cupriferous pyrite deposits of Cyprus and Japanese Kuroko deposits: Society of Mining Geologists Japan, Special Issue 31, p. 206-215.

Franklin, J.M., Kasarda, J., and Poulsen, K.H., 1975, Petrology and chemistry of the alteration zone of the Mattaoi oas ive sulphide deposit: Economic Geology, v. 70, p. 63-69.

Goodwin, A.M. and Ridler, R.H., 1970, The Abitibi orogenic belt: in Baer, A.J., ed., Symposium on Basins and Geosynclines of the Canadian Shield: Geological Survey of Canada Paper 70-40, p. 1-30.

Hodgson, C.J. and Lydon, J.W., 1977, Geological setting of volcanogenic massive sulphide deposits and active hydrothermal systems, some implications for exploration: Canadian Mining and Metallurgical Bulletin, v. 70, no. 786, p. 95-106.

Hull, D.N., 1976, A Petrologic – Geochronologic Study of the Host Rocks and Wall Rocks at the Detour Prospect, A Copper-Zinc-Silver Discovery in Brouillan Township, Abitibi-West County, Quebec: M.Sc. Thesis, Wright State University, Dayton, Ohio.

Lambert, M.B., 1978, The Back River volcanic complex – a cauldron subsidence structure of Archean age: Geological Survey of Canada Paper 78-1A, p. 153-157.

Lilljequist, T.R. and Svenson, S.A., 1974, Exceptionally well preserved Precambrian ignimbrites and basic lavas in North Sweden: Geologiska Foereninger i Stockholm, Foerhandlingar, v. 96, p. 221-229.

MacDonald, G.A. 1972, Volcanoes: Englewood Cliffs, New Jersey, New Jersey, 492 p.

Nixon, G.T. and Papezik, V.S., 1979, Late Precambrian ash-flow tuffs and associated rocks of the Harbour Main Group near Colliers, eastern Newfoundland: chemistry and magmatic affinities: Canadian Journal on Earth Sciences, v. 16, p. 167-181.

Pollock, G.D., Sinclair, I.G.L., Warburton, A.F., and Wierzbicki, V., 1972, The Uchi orebody; a massive sulphide deposit in an Archean volcanic environment: in Mineral Deposits, Section 4, International Geological Congress Proceedings, No. 24, p. 299-308.

Reed, L.E., 1976, The discovery of the Brouillan zinc-copper-silver deposits, north-western Quebec: Society of Exploration Geophysicists, Annual International Meeting, (Houston, Texas), Abstracts, No. 46, p. 91.

_______________, 1981, The airborne electromagnetic survey of the Detour zinc-copper-silver deposit, northwestern Québec: Geophysics, v. 46, no. 9, p. 1278-1290.

Remick, J.H., 1969, Geology of Harricana-Turgeon Area, Abitibi-East and Abitibi-West Counties: Quebec Department of Natural Resources, Preliminary Report No. 564, 217 p.

Ross, C.S. and Smith, R.L., 1961, Ash-flow tuffs; their origin, geological relations and identification, United States Geological Survey Professional Paper 366, 81 p.

Sangster, D.F., 1972, Precambrian Volcanogenic Massive Sulphide Deposits in Canada: A Review: Geological Survey of Canada Paper 72-22, 44 p.

Sinclair, I.G.L., 1977, Primary dispersion patterns associated with the zinc-copper-silver deposit at Lac Brouillan, Province of Quebec: Journal of Geochemical Exploration, v. 8, nos. 1-2, p. 139-151.

Sinclair, I.G.L. and Gasparrini, E., 1980, Textural features and age of supergene mineralization in the Detour copper-zinc-silver deposit, Quebec; Economic Geology, v. 75, p. 470-477.

Smith, R.L. and Bailey, R.A., 1968, Resurgent cauldrons: *in* Coats, R.R., Hayland, R.L. and Anderson, C.A., eds., Studies in Volcanology: Geological Society of America Memoir 116, p. 613-662.

Smyth, W.R., Marten, B.E. and Ryan, A.B., 1978, A major Aphebian-Helikian unconformity within the Central Mineral Belt of Labrador: definition of new groups and metallogenic implications: Canadian Journal of Earth Sciences, v. 15, p. 1954-1966.

Wanless, R.K. Stevens, R.D. LaChance, G.R. and Edmond, C.M., 1968, Age Determinations and Geological Studies, K-Ar Isotopic Ages: Geological Survey of Canada Paper 67-2A, Report 8, 141 p.

MS received May 1979;
Accepted for publication November 1979.

Precambrian Sulphide Deposits, H.S. Robinson Memorial Volume, edited by R.W. Hutchinson, C.D. Spence and J.M. Franklin, Geological Association of Canada Special Paper 25, 1982

GEOLOGY OF THE GECO BASE METAL DEPOSIT

R.G. Friesen, G.A. Pierce, and R.M. Weeks
Geology Department of Noranda Mines Limited (Geco Division),
Manitouwadge, Ontario P0T 2C0

ABSTRACT

The Geco Cu-Zn-Ag deposit lies within a highly deformed sequence of Archean metasedimentary and metavolcanic rocks located near the interface between the Quetico and Abitibi-Wawa Belts.

Three distinct types of mineralization occur within the deposit: disseminated chalcopyrite-pyrrhotite-pyrite, massive pyrite-pyrrhotite-sphalerite-chalcopyrite and disseminated pyrite-sphalerite. The massive sulphide and pyrite-sphalerite zones contain important amounts of silver. All the above mineralization occurs within a dragfolded quartz-muscovite schist interbedded between a lower altered mafic volcanic unit and an overlying quartzite unit containing siliceous iron-formation.

The folding which developed the present Manitouwadge Synform and related Geco drag fold also caused foreshortening and remobilization of the mineralized horizons. The amphibolite grade of metamorphism at Geco is indicated by zones of abundant sillimanite, anthophyllite, garnet and cordierite.

Stratigraphy, alteration, metal zoning, and age relationships between known intrusions and the majority of the sulphide zones, indicate that Geco is a syngenetic-stratiform base metal deposit which has been overturned.

RÉSUMÉ

Le gisement de Cu-Zn-Ag de Geco est encaissé dans une série très déformée de roches métasédimentaires et métavolcaniques archéennes localisées près limites des ceintures de Quetico et de l'Abitibi-Wawa.

Trois types distincts de minéralisation existent dans le gisement: disséminé à chalcopyrite pyrrhotite-pyrite, massif à pyrite-pyrrhotite-sphalérite-chalcopyrite et disséminé à pyrite-sphalérite. Les sulfures massifs et les zones de pyrite-sphalérite contiennent des quantités importantes d'argent. La minéralisation est située dans un pli d'étirement d'un niveau de

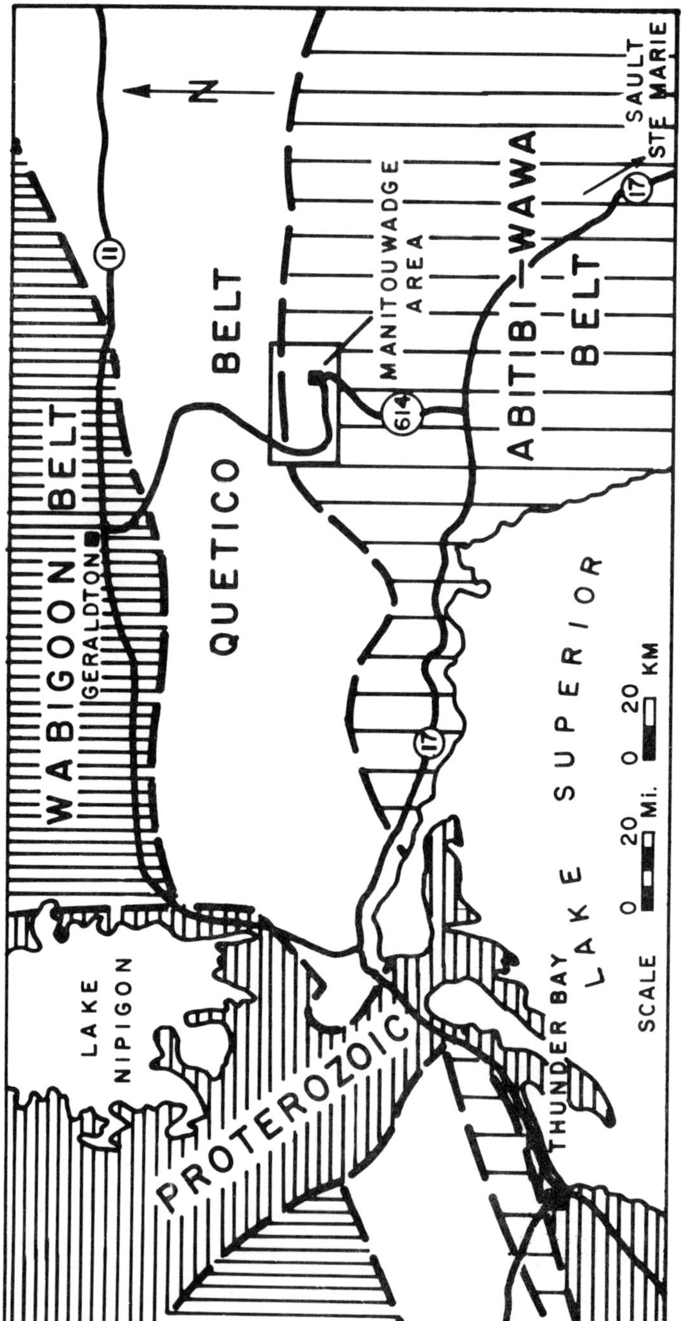

Figure 1. Location map, Manitouwadge area.

schiste à quartz et muscovite, intercalé entre une unité inférieure volcanique mafique et altérée et un niveau supérieur de quartzite contenant une formation ferrugineuse siliceuse.

Le plissement qui est à l'origine de l'actuel synforme de Manitouage et du pli d'étirement de Geco qui lui est associé, a aussi causé le raccourcissement et la remobilisation des niveaux minéralisés. Le métamorphisme est de degré amphibolite tel qu'indiqué par des zones riches en sillimanite, anthophyllite, grenat et cordiérite.

La stratigraphie, l'altération, la zonalité des métaux et les âges relatifs des intrusions connues et de la majorité des zones sulfurées indiquent que Geco est un amas sulfuré syngénétique et stratiforme qui a été inversé.

INTRODUCTION

The Geco Cu-Zn-Ag deposit is in the Manitouwadge mining camp, in the Thunder Bay Mining District of northwestern Ontario, 320 kilometres east of Thunder Bay (Fig. 1); and lies 5 kilometres east of the town of Manitouwadge. The Manitouwadge area has been severely glaciated yet displays moderate to high relief. The higher areas have abundant bedrock exposure whereas the lower areas are covered by glacial tills.

In 1931 Dr. J. E. Thomson of the Ontario Department of Mines located copper-bearing gossans near the northeastern arm of Manitouwadge Lake. Little work was done in the area until 1953, when three Geraldton prospectors – Forster, Barker and Dawd – staked the gossans described by Thomson. Subsequent work outlined the Geco orebody and initiated one of the largest staking rushes in Canadian mining history. This led to the discovery of the Willroy, Willecho and Nama Creek orebodies west of Geco. Production began at Geco in September, 1957, based on an ore reserve of 13.8 million tonnes. Table I gives production and current ore reserve statistics for the entire mining camp.

REGIONAL GEOLOGY

General Stratigraphy and Structure

The Manitouwadge mining camp is located along the northern edge of the Abitibi-Wawa metavolcanic belt, near its interface with the Quetico metasedimentary belt; both of which are part of the Superior Province of the Precambrian Shield (Fig. 1). The interface between these two Archean belts is transitional and their age relationship is controversial. Published regional mapping in the area was carried out by Thomson (1932), Pye (1957), Coates (1968 and 1970), Milne (1968, 1969 and 1974) and Giguere (1972); however, most of the conclusions presented here on the regional geology and structure, are based on detailed mapping by geologists of Noranda Mines Ltd., Geco Division (Fig. 2).

The supracrustal sequence in the Manitouwadge area has been highly metamorphosed to form gneisses and schists locally of the almandine-amphibolite grade of regional metamorphism. In the Manitouwadge area, this sequence may be subdivided into four distinct, conformable, east-west trending units which, from north to south are: 1) Quetico paragneisses, 2) granitized felsic and mafic gneisses, 3) the Manitouwadge Mine Series (metavolcanic and metasedimentary gneisses and schists) and 4) granitized felsic gneisses.

TABLE I
Production history and reserves of the Manitouwadge district.

PROPERTY	PRODUCTION				RESERVES			
	Tonnes	% Cu	% Zn	Ag g/t	Tonnes	% Cu	% Zn	Ag g/t
Total for Geco	26,933,742	1·97	4·01	61·4	23,182,032	1·88	3·73	52·5
Willroy Camp – Willroy	4,180,997	All processed by the Willroy mill			Ceased operations in 1977			
Willroy Camp – Willecho	3,502,159							
Willroy Camp – Nama Creek	236988							
Total for Willroy	7,920,144	0·94	4·59	60·3				
Total for Manitouwadge Mining Camp	Total production				34,853,886	1·73	4·14	61·2
	Total reserves				23,182,032	1·88	3·73	52·5
	Camp total				58,035,918	1·79	3·97	57·6

These rocks have undergone at least two major periods of deformation. The folds of the initial and dominant period were formed about a series of 055° to 070°-trending axes. The Manitouwadge area is considered to be located on the overturned limb of a major fold developed during this period. Individual folds in this system are typically assymetrical and vary in amplitude from several centimetres to tens of kilometres. A second fold system is believed to have developed about a series of northwest-trending axes. Though difficult to illustrate on an outcrop scale, such folds are indicated by a systematic broad scale undulation in plunge of the northeast-trending folds. Interaction of these fold systems has developed four major structures (Fig. 2). From west to east these are the Blackman Lake Antiform, the Manitouwadge Synform, the Banana Lake Antiform and the Moshkanabi Lake Synform. All known orebodies are contained in the Manitouwadge Synform.

The oldest known intrusions in the area are isolated lenses of metagabbro associated with mafic orthogneiss. Widespread granitization, possibly of Algoman age has produced hybrid rocks, some of which are truly intrusive. Locally, these are represented by pegmatite dykes up to 100 m thick and by various rocks of granitic to granodioritic composition. Diabase dykes up to 70 m thick intrude all rocks in the area and generally follow the major fault trends which are north, northeast, and northwest. These faults appear as prominent regional lineaments and typically have near-vertical dips. They exhibit strike-slip movement with only minor displacements.

Manitouwadge Mine Series

Following Pye (1957), the Manitouwadge Mine Series has been subdivided into four generally continuous conformable units best seen in the Manitouwadge Synform (Fig. 3). These are: 1) The Hornblende Schist Group, a 1600 m thick band of mafic schist which Pye (1957) considered to be mafic waterlain meta-volcanic rocks; 2) the Grey Gneiss Group, a zone up to 1500 m thick, divided into upper and lower sub-units, of felsic to intermediate gneiss with interbedded quartzite and iron-formation near its northern edge; 3) the Sericite Schist Group, a band of felsic schist generally

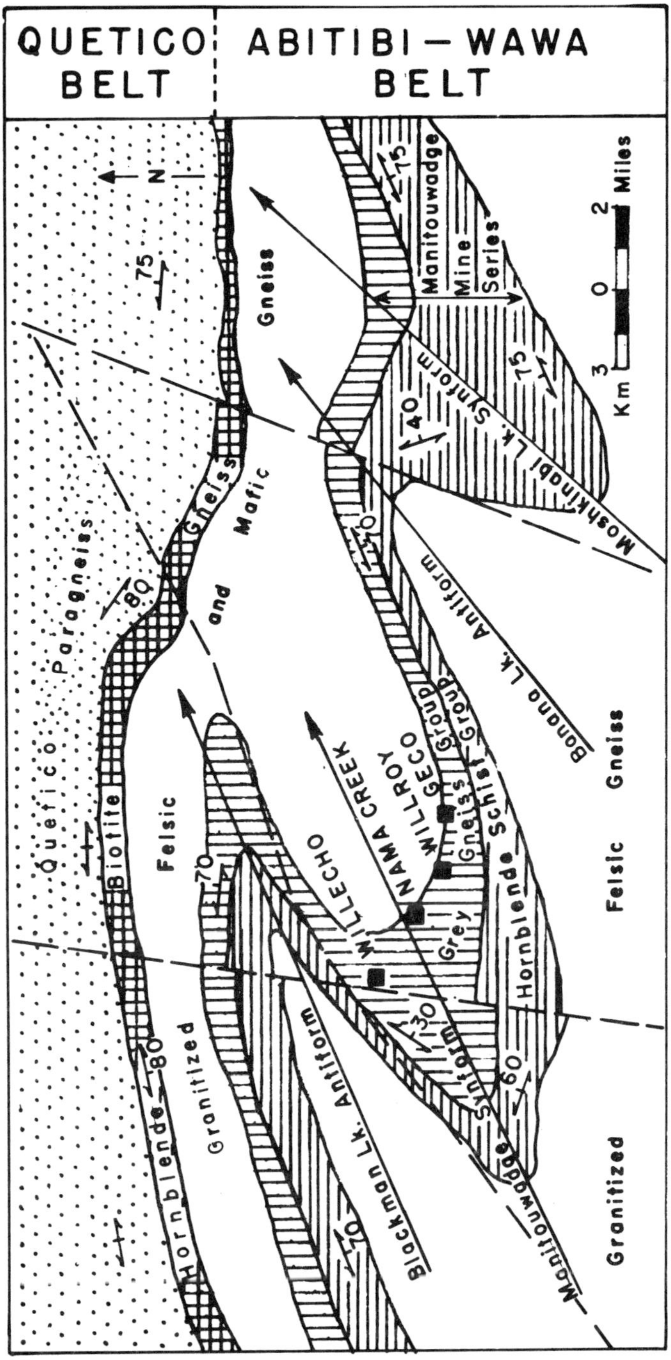

Figure 2. Regional geology, Manitouwadge area.

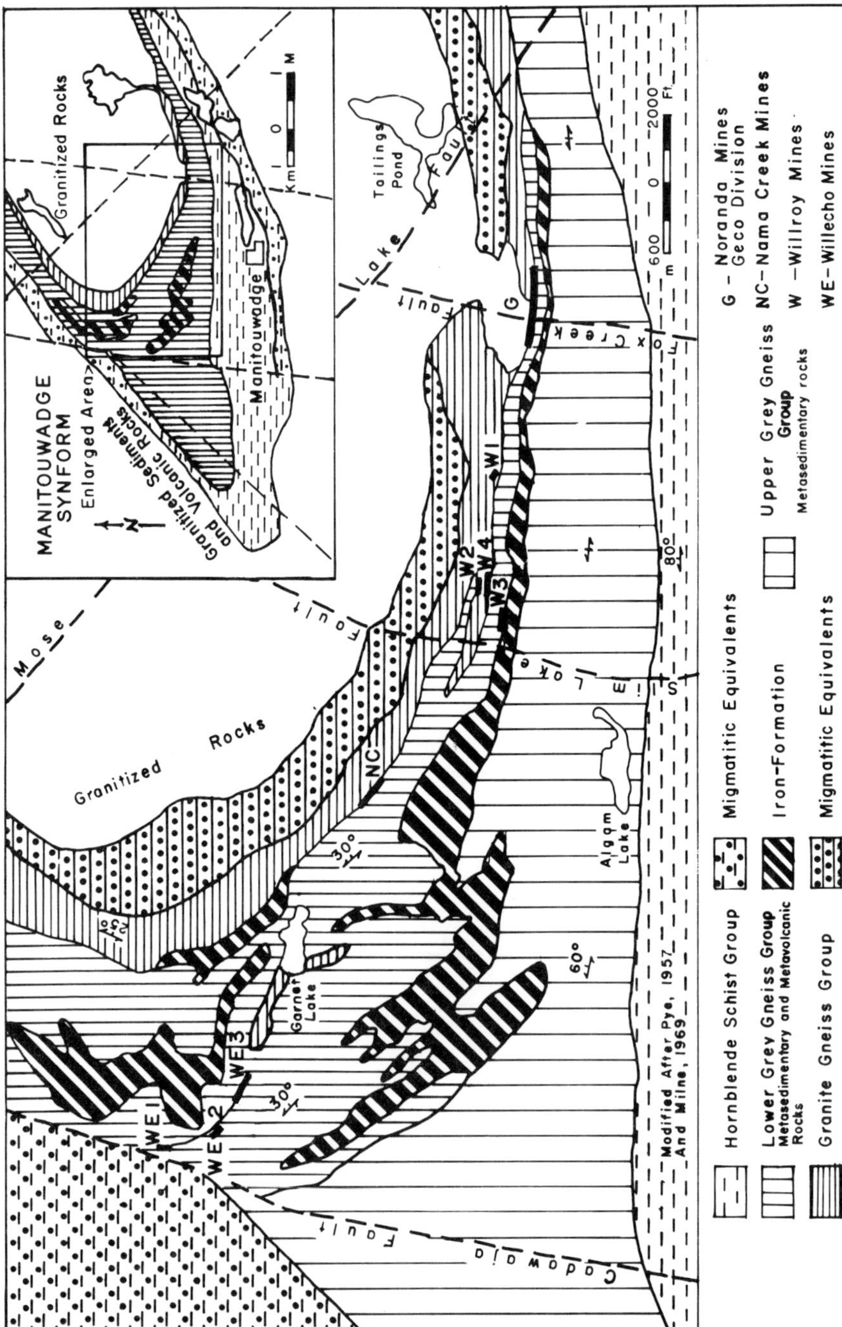

Figure 3. Geology of the Manitouwadge syncline between Willecho and Geco.

less than 200 m thick (not shown in Fig. 3) which hosts the Geco orebodies and is possibly of volcano-sedimentary derivation; and 4) the Granite Gneiss Group, a 700 m thick zone of altered intermediate to mafic gneiss. This latter group forms the northern edge of the Mine Series and grades rapidly into highly granitized rocks characteristic of those found throughout the core of the Manitouwadge Synform.

Previous workers mapping in the Manitouwadge Synform considered the Hornblende Schist Group to be the lowest stratigraphic member of the Mine Series. Reliable top determinations are rare and this interpretation is based chiefly on present structural position of the units in the synform. Recent regional mapping (Fig. 2) however, shows the reverse stratigraphic sequence in the eastern and western portions of the area. The writers therefore consider the entire Manitouwadge Synform to represent an overturned portion of the Manitouwadge Mine Series, with the Granite Gneiss Group forming the basal member of this series.

MINE GEOLOGY

Introduction

The geology of the deposit has been described by Langford (1955), Pye (1957), Brown and Bray (1960), Milne (1969) and Watson (1970). Specific aspects of the deposit have been discussed by McNiece (1956), Pushkar (1960), Graham (1967), Mookerjee and Dutta (1970), Suffel *et al.* (1971), Bannerjee (1972) and Alldrick (1974). W.L. Brown and R.C.E. Bray were employees of Geco Mines Limited and were responsible for much of the initial geological interpretation. Their unit nomenclature is still in use.

The Geco deposit is located on the south limb of the Manitouwadge Synform and is associated with a complimentary, easterly plunging, "Z"-shaped drag fold. The deposit contains three types of sulphide zones which occur in the northern, lower members of the Manitouwadge Mine Series (Fig. 4): 1) a disseminated chalcopyrite-pyrrhotite-pyrite zone located at the Granite Gneiss/Sericite Schist contact (4/2 Cop-

TABLE II

Average major element compositions of main rock types in the Mine Series, Geco.

ROCK TYPE	SiO_2	Al_2O_3	Total Fe	MgO	CaO	Na_2O	K_2O	TiO_2	P_2O_5	MnO	
Granite	70·57	15·17	1·51	0·52	1·78	4·51	3·49	0·27	0·05	0·02	
Qtz Diorite	61·54	15·57	5·58	3·16	4·23	3·37	2·42	0·65	0·26	0·10	
Hnb. Schist	50·60	14·64	10·57	5·74	8·75	2·29	0·63	1·23	0·27	0·23	HORNBLENDE SCHIST GROUP
Qtz-Fel-Bio Gn	64·80	12·82	8·76	2·09	3·17	2·29	2·20	0·21	0·05	0·66	GREY GNEISS GROUP
Siliceous IF	67·43	0·40	24·76	1·42	1·80	0·00	0·03	0·04	0·00	0·54	
Chloritic IF	66·22	0·74	25·92	1·94	2·21	0·30	0·08	0·03	0·12	1·65	
Bio Quartzite	76·64	11·70	2·03	0·44	0·36	3·12	4·45	0·11	0·00	0·03	
Ser. Schist	84·53	7·58	1·77	0·37	0·19	0·32	1·74	0·09	0·01	0·01	SERICITE SCHIST GROUP
Ser Schist Bio'c	.71·23	10·56	6·05	2·79	0·31	0·73	3·17	0·13	0·01	0·04	
Ser-Sill. Schist	78·67	9·80	3·93	1·98	0·19	0·24	1·63	0·11	0·00	0·02	
Qtz-Bio-Anth Hfls	69·90	8·77	11·30	4·32	0·23	0·20	0·70	0·10	0·01	0·05	
Bio-Anth-Cord Gn	51·25	13·08	20·56	6·83	1·28	0·15	0·26	2·92	0·58	0·26	GRANITE GNEISS GROUP
Bio-Sill. Gneiss	74·09	12·03	5·04	3·27	0·21	0·07	2·05	0·29	0·06	0·06	
Bio-Hnb Gneiss	45·53	15·88	16·11	6·71	6·39	1·43	0·78	2·41	0·33	0·41	

per Zone), 2) a massive pyrite-pyrrhotite-sphalerite-chalcopyrite zone and associated disseminated chalcopyrite-pyrrhotite-pyrite zone in the Sericite Schist Group (Main Orebody) and 3) a disseminated pyrite-sphalerite zone at the Sericite Schist/Grey Gneiss contact (8/2 Zinc Zone).

Mine Stratigraphy

The rocks at Geco have undergone high rank metamorphism and deformation which have obscured many of the original features of this deposit. Despite this, the main rock units may be further subdivided to show a significant internal succession. The stratigraphic columns in Figure 4 relate this internal stratigraphy to that of the Manitouwadge Mine Series. Table II gives typical whole rock analyses for the major rock units at Geco.

Granite Gneiss Group

Only the upper portion of this complex group of medium- to coarse-grained, highly altered, mafic to intermediate gneisses is exposed on surface and in underground workings. Based on this exposure and a series of deeper diamond drill holes this group can be subdivided into biotite-hornblende gneiss, biotite gneiss, biotite-sillimanite gneiss and biotite-anthophyllite-cordierite gneiss members. Garnet is common in all the members. The biotite-hornblende gneiss is found in the western and lower portions of the group and is not encountered to any extent in the underground workings. The biotite, biotite-sillimanite and biotite-anthophyllite-cordierite gneisses are interbedded and individual units are up to 5 m thick. The garnetiferous biotite-sillimanite and biotite-anthophyllite-cordierite gneisses are most abundant near the overlying Sericite Schist Group.

The biotite and biotite-sillimanite gneisses can be differentiated only on sillimanite content as they grade into one another both laterally and vertically. They typically contain 20 to 40% biotite, 15 to 30% quartz, 10 to 20% feldspar, 10 to 20% hornblende, 0 to 15% sillimanite and 0 to 15% garnet (all mineral percentage compositions are estimated from hand speciments). Sillimanite typically occurs as elongated clusters up to 2 cm in length and garnet (almandine) as porphyroblasts up to 2 cm across. Magnetite is a common accessory mineral.

The biotite-anthophyllite-cordierite gneisses typically contain 5 to 20% biotite, 10 to 50% anthophyllite, 5 to 35% cordierite, 0 to 10% quartz. Accessory minerals include feldspar, staurolite, chlorite, magnetite and spinel. Anthophyllite rosettes, up to 5 cm across, are developed parallel to foliation and cordierite is found interstitial to these rosettes where it is commonly altered to pinite.

Although erratic lenses of fine-grained disseminated chalcopyrite are present throughout the group, no mineable zones have yet been located.

Sericite Schist Group

The Sericite Schist Group is a series of highly altered felsic schists which conformably overlies the Granite Gneiss Group. Typically ranging up to 200 m in width, this key horizon contains all the known orebodies at Geco.

The lowest ore horizon, the 4/2 Copper Zone; a thin zone of disseminated chalcopyrite, pyrrhotite, and pyrite; generally occurs at the base of this group. Ranging

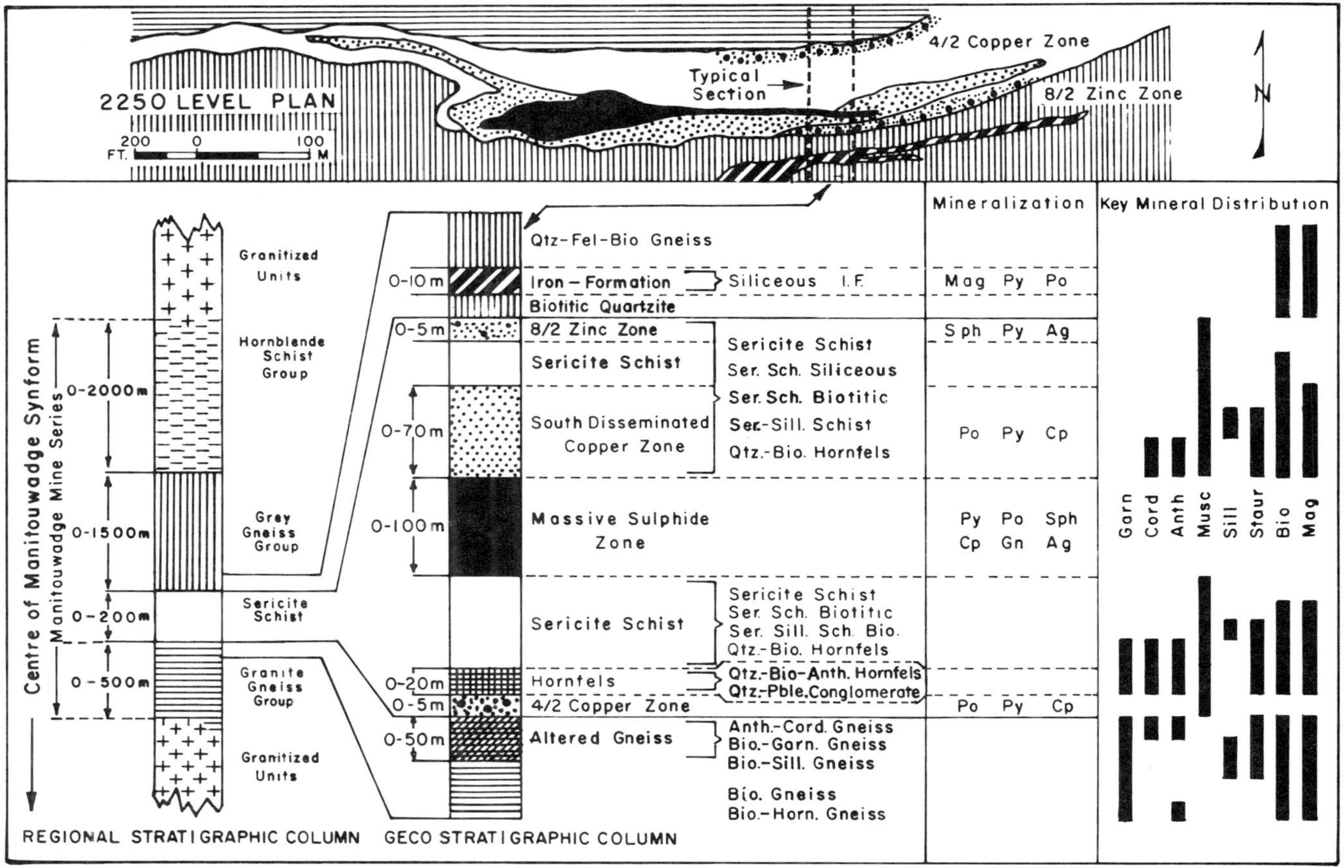

Figure 4. Schematic stratigraphic columns illustrating generalized stratigraphic relationships of sulphide zones, Geco Mine.

from a few cm to 5 m in thickness, this sulphide zone is usually overlain by a 5 to 20 m thick band of quartz-biotite-anthophyllite hornfels. This peculiar unit is typically medium- to coarse-grained, poorly foliated and shows a highly variable composition. The major minerals are: 50 to 80% quartz, 5 to 15% biotite and 5 to 20% anthophyllite. Accessory minerals include cordierite, staurolite, garnet, spinel, sericite, sillimanite, plagioclase and diopside. The more mafic and coarser-grained varieties of the hornfels resemble the anthophyllitic members of the Granite Gneiss Group. A distinct marker band known as quartz-pebble conglomerate occurs locally at the base of this hornfels. Up to 3 m thick, this band contains discus-shaped quartz pebbles averaging 10 cm in diameter in a biotite-anthophyllite matrix. This band contains low grade copper mineralization – some of which is gradational into the underlying 4/2 Copper Zone.

Overlying the hornfels unit is the typical sericite schist which can be subdivided into two members: 1) sericite-sillimanite schist, best developed in the lower and eastern portions of the group, and 2) sericite schist, more common in the upper part of the group. The sericite-sillimanite member is typically medium- to coarse-grained, grey to black, well foliated, and shows a distinct spotted texture. It is found as 0.5 to 5.0 m thick bands generally displaying gradational contacts with other schist members and contains 50 to 60% quartz, 5 to 15% biotite, up to 10% muscovite and 3 to 20% sillimanite. The distinct spotted appearance of this rock is caused by white knots or bundles of sillimanite up to 3 cm in diameter which are aligned parallel to foliation. As a rule, this member is poorly mineralized and contains significant amounts of magnetite which occurs both as irregular blebs in the dark biotitic groundmass and as tiny (1-2 mm) crystals in the cores of sillimanite bundles.

The sericite schist member is typically medium- to coarse-grained, buff colored and displays highly developed schistosity and crenulations. This schist can be further divided into various submembers based on relative amounts of quartz, muscovite and biotite. Individual bands of these schists range from 0.5 to 10 m thick. Where best developed the sericite schist has a pearly white luster and is composed of 40 to 70% quartz and 30 to 60% muscovite. Accessory minerals are biotite, sillimanite, garnet, feldspar and magnetite. Locally, where the quartz content is high, the term "siliceous" is used, and where significant amounts of biotite occur, the schists are termed "biotitic".

The Main Orebody is located in the upper half of the Sericite Schist Group and is composed of a 0.1 to 100 m thick zone of massive pyrite, pyrrhotite, sphalerite and chalcopyrite overlain by a zone of disseminated sulphides. This overlying sulphide zone is essentially siliceous sericite schist containing disseminated chalcopyrite, pyrrhotite and pyrite. Throughout the disseminated zone, but more commonly around the upper and lower margins of the massive sulphides, a unit termed "quartz hornfels" is locally present. This siliceous unit contains near-massive concentrations of fine-grained chalcopyrite. Locally displaying intrusive-like textures, this member may represent a remobilized phase of the sulphide zone.

The upper disseminated portion of the Main Orebody grades upward through a band of poorly mineralized sericite schist into the overlying 8/2 Zinc Zone. The 8/2 Zinc Zone is essentially a 1 to 5 m wide band of schist containing disseminated to near-massive concentrations of sphalerite and pyrite which marks the top of the Sericite Schist Group.

Grey Gneiss Group

The Grey Gneiss Group conformably overlies the Sericite Schist Group. Characteristically unaltered, this Group contains interbedded biotitic quartzite, quartz-feldspar-biotite gneisses and iron-formation. The upper members of this group are not present in the mine area.

The biotitic quartzite predominates in the lower 30 m of the group. It is a well banded, fine-grained, light grey to buff-colored rock typically containing 80% quartz, 15% potash feldspar and 5 to 15% biotite. Accessory minerals include sericite, plagioclase, amphibole, garnet, pyrite and magnetite.

The quartz-feldspar-biotite gneiss is the most common rock of the Grey Gneiss Group in the mine area. It is a medium- to coarse-grained, light to dark grey gneiss essentially composed of 40% quartz, 40% potash feldspar and 20% biotite. Accessory minerals are hornblende, sericite, garnet, plagioclase and magnetite.

Bands of iron-formation are found in the lower 100 m of this group. Typically lenticular, individual bands generally do not exceed 300 m in length and range from 1 to 10 m thick. Two types are present: 1) siliceous iron-formation and 2) chlorite-garnet iron-formation. The siliceous iron-formation is by far the more common and consists of 60 to 70% fine-grained quartz; up to 20% combined magnetite, pyrite and pyrrhotite; and 5 to 30% biotite. Accessory minerals include garnet, feldspar, amphibole and pyroxene. Although it displays a glassy, dark grey to black appearance on fresh surfaces, older exposures are usually highly oxidized. Siliceous iron-formation generally shows well-developed banding where significant amounts of pyrite, pyrrhotite or magnetite are present. Although near-massive concentrations of iron sulphides and oxides are present, siliceous iron-formation does not contain mineable mineralization.

The chlorite-garnet iron-formation is found only in the western and eastern extremities of the mine near the Sericite Schist/Grey Gneiss contact, and therfore is not often seen in the mine workings. Typically dark and siliceous, it displays no obvious banding but shows a distinct ropy, fibrous texture due to the habit of chlorite and biotite, which bend about garnet porphyoblasts that are up to 1 cm in diameter. The magnetite content is similar to that of the siliceous iron-formation but here magnetite occurs as small massive blebs. Known to grade laterally into garnetiferous quartz-feldspar-biotite gneiss, the iron-formation commonly contains minor amounts of disseminated pyrite and pyrrhotite and is locally associated with sphalerite mineralization of the 8/2 Zinc Zone.

Intrusive Rocks

The oldest intrusions are quartz diorite dykes which are found throughout the mine area but are most abundant between the Main Orebody and the Granite Gneiss Group, where they constitute upwards of 30% of the rock. Compositionally similar to the quartz-feldspar-biotite gneisses of the Grey Gneiss Group, they show highly variable textures. Averaging 1 m thick, these dark grey to black dykes may locally reach 10 m in thickness. They typically cut gneissosity at very low angles (less than 10°) and for this reason their intrusive nature can be questioned; however, they are everywhere foliated parallel to their contacts.

Younger, well foliated mafic dykes are occasionally encountered in the mine. These dark green dykes, composed mainly of hornblende and biotite, are most common in the Grey Gneiss Group and are known to transect the quartz diorites. These dykes are believed to be genetically related to the Hornblende Schist Group.

A later series of pegmatite and granite dykes is common throughout the mine area. These felsic dykes occur on all scales and are typically parallel to the axial planes of the major folds (northeast to east). Within the Sericite Schist Group these dykes display appreciable thinning where they have been partially transposed into the plane of schistosity. These dykes cut all major rock units except the massive sulphide orebody; although they commonly occur as remnants in this zone. Recent work on the pegmatite and granite dykes indicates they have diverse compositions. Although some of these dykes may be true igneous intrusions, others may have formed by metasomatic processes.

The youngest intrusions are diabase dykes (Pye 1957, Milne 1969, 1974) which follow the major fault systems in the area. These dykes range from a few cm to 20 m in thickness and cut all other rock types in the mine area. Typically fine-to coarse-grained, and black to dark green in colour, these dykes are commonly equigranular but locally contain plagioclase phenocrysts.

ORE ZONES

The isometric projection in Figure 5 shows the relative positions of the ore zones and host rocks at Geco. The inset illustrates the distribution of these zones in longitudinal section.

4/2 Copper Zone

The 4/2 Copper Zone is a tabular, vertically dipping, fine-to medium-grained chalcopyrite-pyrrhotite-pyrite orebody. The sulphides are contained in a highly siliceous matrix and are present in both massive and disseminated concentrations.

This zone attains mineable grades in the lower and eastern portion of the mine area where it averages 2 m in thickness. The typical 4/2 Ore Zone consists of a high-grade, near-massive sulphide core containing quartz "eyes" and wall-rock fragments, which grades outwards into the poorly mineralized host rocks. The fine-grained massive core often appears veined or remobilized but the orebody itself is conformable with the host rocks, though it locally shows sheared contacts. Eastward, along its strike, the orebody contains increased amounts of sphalerite and pyrite. A series of randomly oriented, very coarse grained quartz-cordierite "dykes" are found along its upper contact. These "dykes" contain erratic mineralization, are typically less than 1 m thick, have sharp contacts with upper host rocks and pinch out rapidly upwards, probably within 3 m.

Main Orebody

The lower portion of the Main Orebody consists of a massive coarse-grained sulphide zone which contains in excess of 60% combined sulphides, up to 10 to 30% wall-rock remnants and locally, an abundance of quartz and mica gangue. This zone forms the major mineable orebody and typically contains 10 to 60% pyrite, 5 to 30%

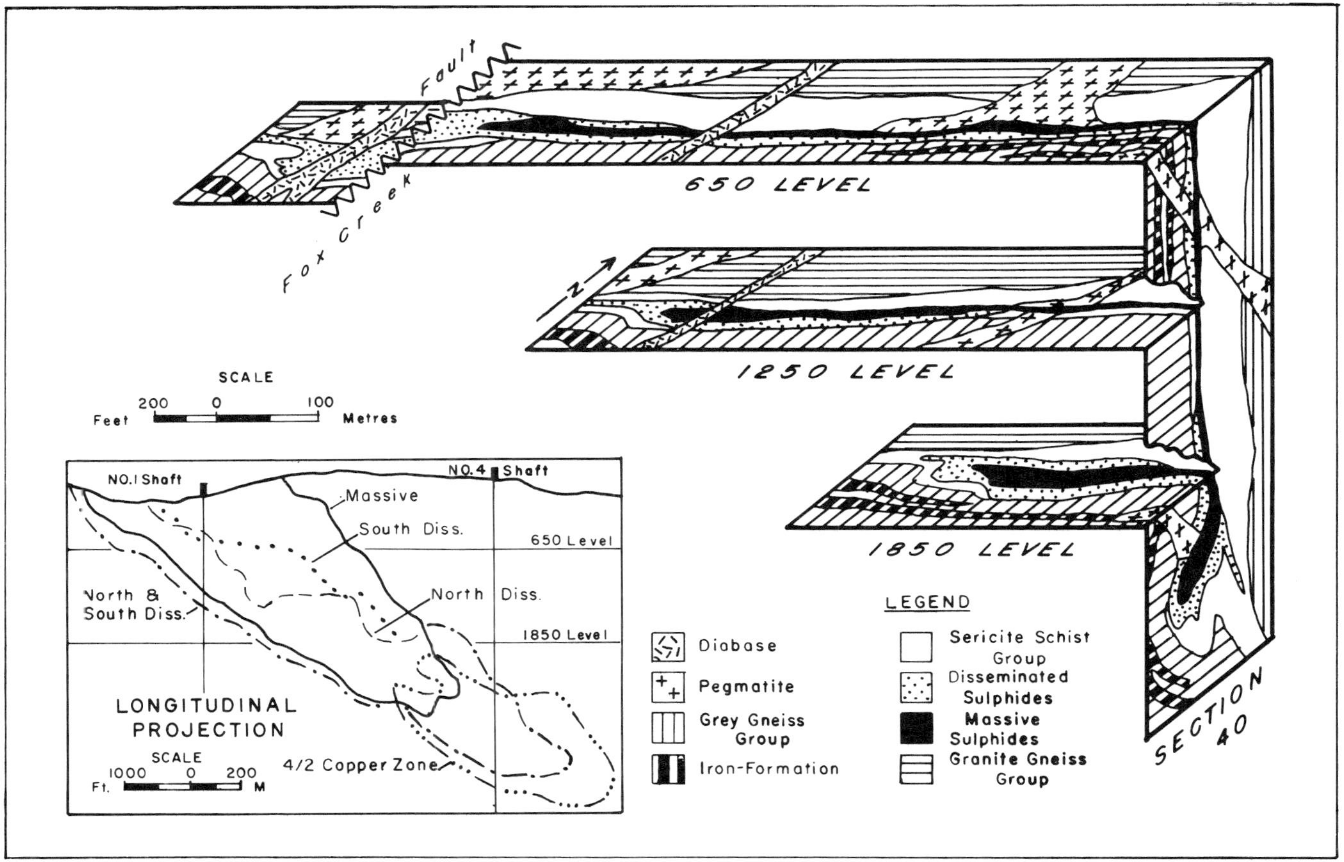

Figure 5. Isometric diagram, west end of Geco deposit, and longitudinal projection showing relative positions of various sulphide zones.

sphalerite, 5 to 30% pyrrhotite, 2 to 25% chalcopyrite and minor galena and silver. This orebody also contains recoverable amounts of gold, bismuth and cadmium as well as non-recoverable amounts of tin. Ranging from less than 1 m up to 100 m in thickness, this vertically dipping orebody attains a maximum strike length of 975 m on the 250 Level of the mine and has been outlined down plunge for 1750 m. As with the overlying disseminated orebody this zone reaches maximum thickness in the western portions of the mine near the drag fold. The massive zone is thus "teardrop" shaped and gradually thins upwards and to the east.

Within this massive sulphide horizon, pyrite, the predominant sulphide, occurs both as disseminated coarse-grained, subhedral to euhedral crystals and in fine-to coarse-grained massive form. Individual crystals display crushed, rolled or brecciated textures and locally appear to be porphyroblastic. Sphalerite is most commonly associated with the pyrite-rich portions of this ore zone where it occurs as finer-grained, subhedral to anhedral grains surrounding or locally filling fractures in the pyrite grains. The sphalerite- and pyrite-rich ore occasionally shows crude banding. Pyrrhotite and chalcopyrite are intimately associated and generally form a very fine-grained, poorly crystalline matrix for the more coarsely crystalline sphalerite - pyrite. Both pyrrhotite and chalcopyrite show flow textures and may embay or form fracture fillings in other sulfides. Chalcopyrite also occurs as fine-grained inclusions within individual sphalerite crystals. Galena is present primarily as thin selvages surrounding quartz diorite dyke remnants.

The massive sulphide zone appears to be a conformable stratiform horizon; however its contact relationships with host rocks and known intrusions are complex. The massive sulphides locally appear discordant to the sericite schist host and often contain abundant remnants of intrusive rocks such as pegmatite and quartz diorite dykes. This zone shows a complex mineral paragenesis which has been described in detail by Pye (1957).

The upper portion of the Main Orebody consists of a disseminated sulphide zone which typically contains 5 to 6% chalcopyrite, 8 to 10% pyrrhotite and 4 to 5% pyrite. These sulphides occur as blebs and stringers parallel to schistosity in the siliceous sericite schist host. Within the disseminated zone, these mineralized schists alternate with weakly mineralized bands of biotitic, siliceous and sillimanitic schist. Ranging from 0 to 70 m thick, this zone continues down plunge for 200 to 300 m beyond the massive portion of the Main Orebody. It attains maximum thickness in the lower and western portions of the mine where it is overthickened by the Geco drag fold. Eastward, beyond the limits of the underlying massive sulphides, this disseminated zone thins and splits into narrow chalcopyrite stringers which locally are discordant with schistosity.

Narrow isolated lenses of disseminated sulphides also occur below the massive Main Orebody in the underlying schists and are termed "north ore" lenses. Although it is evident that much of this north ore may be a separate sulphide zone there are also occurrences which suggest that these lenses may be dislocated or attenuated portions of the main disseminated zone.

8/2 Zinc Zone

The 8/2 Zinc Zone is a tabular, fine-to coarse-grained sulphide zone containing sphalerite, pyrite and important amounts of silver. Though it occurs almost everywhere along the Grey Gneiss/Sericite Schist contact, it is best developed in the lower and eastern portions of the mine. Typically ranging in thickness from 1 to 5 m, this zone can locally be subdivided into disseminated and near-massive members. The lower disseminated member consists of fine-grained sphalerite and pyrite in sericite schist. This low grade (1 to 5% zinc) member locally directly overlies the disseminated portion of the Main Orebody and grades upwards into the high-grade (5 to 15% zinc) near-massive member of the zone. Generally averaging 1 m in thickness, this member may also contain minor amounts of chalcopyrite (up to 1%) and locally shows alternating pyrite- and sphalerite-rich bands.

COPPER-ZINC DISTRIBUTION IN THE MASSIVE SULPHIDES

The metal distribution in the massive sulphides reflects primary distribution as well as the effects of metamorphism and structural adjustments including recrystallization, local remobilization, boudinage etc. (Suffel *et al.*, 1971; Vokes, 1968). The present thicknesses of the sulphides show dominant structural control, being greatest in the folded, lower western margins of the zone (Fig. 6). The highest grades of zinc follow the central portions of the massive sulphide zone down plunge. Zinc is however, fairly uniformly distributed throughout the zone and shows only limited remobilization. Copper, on the other hand commonly shows evidence of considerable remobilization. Above 1650 Level, the best grades of copper are fold-related. Below 1650 Level, the high-grade copper values leave this fold system and cross the plunge

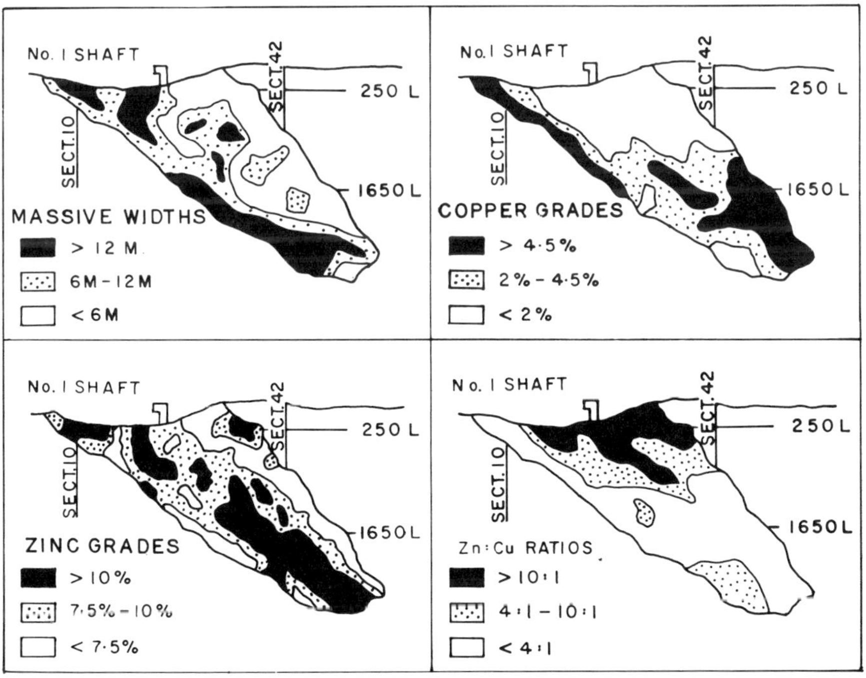

Figure 6. Widths and metal distribution of the massive sulphide zone, Geco Mine.

of the massive zone. The zinc to copper ratios show a similar pattern as they are controlled primarily by the copper grades. The horizontal trend below 1650 Level is not completely understood, but may, in part, represent primary copper distribution.

MINE STRUCTURE

High-grade regional metamorphism formed the gneisses and schists of the Manitouwadge Mine Series. The gneissosity and schistosity were developed parallel to the original bedding. Tight isoclinal folding described as typical of many high-grade metamorphic terrains cannot be demonstrated in the Manitouwadge area. No structural or stratigraphic evidence for such initial folds can be found, and examples of original bedding discordant to schistosity are not known.

The Geco deposits have undergone several phases of folding which possibly developed during a late stage of metamorphism. The dominant fold pattern at the mine belongs to the northeast fold system which is also strongly developed regionally. Typically assymetrical and "Z"-shaped in plan, folds belonging to this system are found underground on scales ranging from several centimetres up to tens of metres. Most of the fold-related features at Geco can be attributed to this stage of deformation. These include the Geco drag fold and numerous smaller-scale folds of a similar pattern, well-developed crenulations, and strong boudinage. Numerous shear zones which formed parallel to the axial planes of this fold system transposed quartz diorite dykes and, to a lesser extent, pegmatite and granite dykes into the plane of the foliation. Within the mine area, the majority of northeast trending folds have well-defined northeasterly plunges. Undulation downplunge however, of many of these folds may reflect the influence of a second deformation.

An unusual series of "Z"-shaped, west plunging folds similar in style to the northeast folds are intimately associated with northeasterly striking pegmatite dykes. Typically less than 10 m in amplitude, these folds are widespread underground. It is not certain how these folds developed; their direct association with pegmatite dykes suggests that they were generated by these dykes; yet in style, they resemble the northeast folds and may represent a refolded portion of this system.

The dominant structure throughout the mine is the Geco drag fold which is best outlined by the Grey Gneiss/Sericite Schist contact (Figs. 4, 5). This "Z"-shaped fold is open and poorly developed at surface but down plunge, it develops into a more complex fold. Attaining an overall easterly plunge of 30°, this fold undulates down its length in a series of step-like structures. The Main Orebody also reflects the Geco dragfold. Where the dragfold is well developed, the massive sulphide body attains its distinct teardrop shape and the overlying disseminated zone is wrapped around the massive zone. The development of this fold produced thickening, strong boudinage and numerous tight crenulations. Boudinage is best displayed by the massive zone but occurs to some extent in many of the mine units. Two sets of crenulations are found and are best developed in the Sericite Schist Group. The dominant set is east-plunging and is locally accompanied by a weaker, second set oriented perpendicular to the dominant set. Shearing associated with the development of the fold systems is most intense near the margins of the more competent units. Macroscopic and microscopic adjustment to the folding have resulted in the partial transposition into the plane of foliation of many otherwise discordant features.

Two regional faults affect the mine area. The main fault is the north trending, vertically dipping Fox Creek Fault (Fig. 5). This left-hand, strike-slip fault, which is 15 to 30 m wide, presents a mining problem on the western end of the deposit where it has displaced the ore zone laterally for up to 60 m. The Mose Lake Fault occurs in the extreme eastern end of the property, strikes northwesterly, and is principally a right-hand, strike-slip fault. This fault zone averages 10 m in width, dips vertically and shows lateral displacement of up to 60 m.

METAMORPHISM

Metamorphic assemblages in the mining camp are complex and no complete study of these assemblages has yet been undertaken at Geco. Regionally, the Mine Series reaches the almandine-amphibolite grade; whereas at Geco, the upper subfacies of this grade is indicated by abundant sillimanite. Both the timing and sequence of metamorphic events are difficult to establish. Within the deposit it is difficult to separate syngenetic alteration characteristics from the regional metamorphic effects. The intensity of metamorphism is demonstrated by the local remobilization of the sulphide zones and by partial melting of the more siliceous units ("pegmatized" zones). Specific aspects of the sulphide recrystallization were discussed by Suffel *et al.* (1971).

DISCUSSION

The Geco deposit displays features which can be interpreted to support both syngenetic and epigenetic processes of ore deposition. The authors favor the syngenetic model and consider the ore deposit to be an integral part of the Geco stratigraphic succession in the Manitouwadge Mine Series.

Stratigraphy and Alterations

Primary features and textures of the original rock units at Geco have been subsequently obscured by intense regional metamorphism. Without such primary features, the true nature of the original host rocks will remain problematic. The present authors believe that the Granite Gneiss Group forms the basal member of the Manitouwadge Mine Series. At Geco this unit is considered to represent the footwall strata above which the ore zones developed. The diversity of the group is shown by the chemical data (Table II). This Group is interpreted as a series of mafic and intermediate volcanic rocks locally containing interbedded siliceous sedimentary and possibly volcaniclastic rocks. It is possible however, that some of the mafic members were pelitic sediments. A peculiar member termed "stretched pebble gneiss" has been outlined within this Group north and west of Geco (Watson, 1970). Although this rock may have formed in several ways, it may represent a coarse pyroclastic unit similar to those found in many Archean base metal camps.

At Geco, the upper portions of the Granite Gneiss Group and locally, the lower portions of the overlying Sericite Schist Group are dominated by biotite-, anthophyllite- and cordierite-rich gneiss and schist. The Mg-Fe enrichment and Na-Ca-K depletion shown by these members (Table II) indicates that they may represent the highly altered footwall rocks typical of volcanogenic deposits (Sangster, 1972; Sangster and Scott, 1976). These altered units, though only roughly outlined to date, typically do not show strong discordancies to foliation and may in fact represent altered stratigraphic units and not the more typical pipe-shaped feeder system. Alternatively, such feeder pipes may now be difficult to recognize, having been partially transposed into the plane of gneissosity during development of the Manitouwadge Synform. Regardless of their present distribution, these Mg-rich gneisses are considered to represent a high-grade metamorphic equivalent of primary "chloritic" alteration zones developed in the footwall units by the initial ore-forming solutions.

Pods and lenses of these Mg-rich units are locally present in the overlying biotitic, sillimanitic and siliceous members of the Sericite Schist Group. Na-K depletion continues through these upper schists but Ca depletion is not evident and the overall Mg-Fe enrichment, though present, is less intense. The high silica content of these upper schists may indicate that these units have been highly silicified.

Although the parent material of the schists cannot be established definitely, their spatial association with, and broad chemical similarities to the overlying biotitic quartzites indicate that they may be related to the quartzite unit. Quartzite flanking the Sericite Schist Group is locally sericitic, and distinguishing between the two units is sometimes difficult. These quartzites represent the lateral equivalents of a sequence of quartz-feldspar-gneisses outlined by Milne (1969, 1974). Figure 3 shows the lateral distribution of these felsic gneisses in the synform. The high alumina content of this group indicates that they were not likely formed as chemical precipitates. Rather, we consider these siliceous gneisses to represent highly metamorphosed felsic volcanic or detrital volcanogenic sedimentary rocks. Thickening rapidly westward, these gneisses attain maximum thicknesses in the hinge area of the syncline near Willecho. Over-thickening of the siliceous gneisses in this area is considered in part a primary depositional feature and may indicate relative proximity to a major volcanic centre. Despite its strongly developed alteration, thinning of this unit near the Geco deposit may indicate that the Geco orebody formed at a greater distance from a volcanic centre than did the other deposits of the camp (Timms and Marshall, 1959; Chown, 1957).

The Sericite Schist Group, is intimately associated with base metal mineralization and is believed to have developed by the interaction of ore-forming solutions with the pre-metamorphic equivalents of the siliceous gneiss series. These units now lie beneath a thick series of biotite and quartz-feldspar-biotite gneisses of intermediate compositions. These gneisses form the upper member of the Grey Gneiss Group and are considered to have developed from a series of intermediate volcaniclastic or detrital sediments. The transition from felsic (quartz-feldspar) to intermediate (quartz-feldspar-biotite and biotite) gneisses marked the close of base metal deposition and was accompanied by the precipitation of iron-formation. Within the mining camp, ore-related alteration does not occur above the main iron-formation sequence. At Geco, this alteration stops below the iron-formation at the upper contact of the Sericite Schist Group. The iron-formations are characteristically low in alumina and are therefore considered to have been chemical precipitates, perhaps indicating a period of waning clastic sedimentation and volcanism.

The close of sedimentation in the Mine Series is marked by the deposition of the

Hornblende Schist Group. These mafic schists are considered to represent a thick series of basalts and andesites deposited in a submarine environment – perhaps indicating the beginning of a new volcanic cycle.

Orezones and Metal Zoning

Metal zoning at Geco is similar to that of other Archean massive base metal deposits. This zoning can be shown both vertically and laterally in the deposit, i.e. both across and along strike. Figure 4 illustrates the strong vertical zoning. The sulphide zones, from the basal 4/2 Copper Zone to the overlying iron-formation series are considered to represent a complete cycle of mineralization. Metal distribution grades upward through the deposit from disseminated chalcopyrite-pyrrhotite-pyrite to massive pyrite-pyrrhotite-sphalerite-chalcopyrite, disseminated chalcopyrite-pyrrhotite-pyrite and finally, pyrite-sphalerite sulphide zones. The ore zones in this cycle are directly related; whereas the iron-formations, which mark the close of this cycle, typically occur in the unaltered rocks immediately above the deposit. Within each ore zone some degree of lateral Cu-Zn-Fe zoning can be recognized. This lateral zoning is most apparent in the massive sulphide portion of the Main Orebody (Fig. 6) even though the detailed internal metal zoning in this body is complex.

The basal copper-rich ore zones, including the 4/2 and other erratic zones below the massive, show local discordancies to stratigraphy, have fine-grained "remobilized" or perhaps "intrusive" textures and are often spatially related to Mg-rich rock units. These zones may represent the mineralized channelways for the ore-forming solutions whose original discordancies to stratigraphy may have been largely obscured by structural and metamorphic events including transposition and remobilization of sulphides, etc. The uppermost 8/2 Zinc Zone however, is typically stratiform and locally shows weak sulphide banding. The Main Orebody, though partially remobilized, shows only minor discordancies to its host schists and is therefore also considered stratiform. On its lower eastern margin however, the massive sulphide portion of this orebody thins rapidly and grades laterally into a high-grade disseminated copper zone. This zone often contains discordant copper stringers and is locally associated with Mg-rich hornfels, which suggests that it may possibly be related also to the 4/2 Copper Zone.

CONCLUSIONS

Many of the primary features of the Geco deposit have been obscured by high-grade regional metamorphism and intense deformation; however, the general stratigraphy, wall-rock alteration and metal zoning of the original deposit can still be recognized. These latter features are very similar to those described as typical in Archean volcanogenic massive sulphide deposits. It is therefore concluded that Geco is a syngenetic ore deposit which has subsequently undergone intense metamorphism and deformation.

ACKNOWLEDGEMENTS

The writers are indebted to the Management of Noranda Mines (Geco Division) for affording us the opportunity to write this paper. Thanks are also due to the staff of the Geology Department at Geco; in particular, Tim Bakker, Bob Kusins, Dave Powers, Len Swanson and Vimy Richards for their much needed co-operation, assistance and criticisms. Finally, we would like to thank Dr. R. W. Hutchinson who was always very helpful when "outside" discussions on Geco were required.

REFERENCES

Alldrick, D.J., 1974, Petrography and geochemistry of the cordierite-gedrite gneiss, Manitouwadge, Ontario: B.Sc. Thesis, University of Western Ontario, London, Ontario.

Banerjee, S., 1972, Two possible sources of tin in the base metal sulphide deposits at Geco, Manitouwadge, Ontario: M.Sc. Thesis, University of Western Ontario, London, Ontario.

Brown, W.L. and Bray, R.C.E., 1960, The geology of the Geco mine: Canadian Mining and Metallurgical Bulletin, v. 58, p. 3-11.

Chown, E.H.M., 1957, The geology of the Willroy property, Manitouwadge Lake, Ontario: M.A.Sc. Thesis, University of British Columbia, Vancouver, B.C.

Giguere, J.F., 1972, Geology of the Granitehill Lake area: Ontario Department of Mines and Northern Affairs, Geological Report 95, 33 p.

Coates, M.E., 1968, Geology of Stevens-Kaginu Lake area: Ontario Department of Mines, Geological Report 68, 22 p.

____________, 1970, Geology of the Killala-Vein Lake area: Ontario Department of Mines, Geological Report 81, 35 p.

Graham, R.A.F., 1968, Effects of diabase intrusions on sulphide minerals at Manitouwadge, Ontario: Canadian Journal of Earth Science, v. 5, p. 545-547.

Langford, F.F., 1955, Geology of the Geco mine in the Manitouwadge area, District of Thunder Bay, Ontario: M.A. Thesis, Queen's University, Kingston, Ontario.

McNeice, D.G., 1956, The hornblende schists and associated granitization effects in the Manitouwadge Lake area, Ontario: B.A. Thesis, University of Western Ontario, London, Ontario.

Milne, V.G., 1968, Geology of Black River area: Ontario Department of Mines, Geological Report 72, 68 p.

____________, 1969, Progress report on a field study of the Manitouwadge area ore deposits: Preprint of paper given to the Canadian Institute of Mining and Metallurgy, Montreal, April 22.

____________, 1974, Mapledoram-Gemmell: Ontario Department of Mines, Geological Map 2280, scale 1 inch to 1000 feet.

Mookherjee, A. and Dutta, N.K., 1970, Evidence of incipient melting of sulfides along a dike contact, Geco mine, Manitouwadge, Ontario: Economic Geology, Volume 65, p. 706-713.

Pye, E.G., 1957, Geology of the Manitouwadge area: Ontario Department of Mines, Volume LXVI, Part 8, 114 p.

Pushkar, P., 1960, Wall rock alteration at Geco mine: Unpublished report on file at Geco mine, Manitouwadge, Ontario.

Sangster, D.F., 1972, Precambrian volcanogenic massive sulfide deposits in Canada: A review: Geological Survey of Canada Paper 72-22, 44 p.

Sangster, D.F. and Scott, S.D., 1976, Precambrian stratabound massive Cu-Zn-Pb sulfide ores of North America: Volume 6; Handbook of Stratabound and Stratiform Ore Deposits: New York, Elsevier p. 129-222.

Suffel, G.G., Hutchinson, R.W. and Ridler, R.H., 1971, Metamorphism of massive sulphides at Manitouwadge, Ontario, Canada: Society of Mining Geologists of Japan, Special Issue 3, p. 235-240.

Timms, P.D. and Marshall, D., 1959, The geology of the Willroy Mines base metal deposits: Geological Association of Canada Proceedings, v. 11, p. 55-65.

Thomson, J.E., 1932, Geology of the Heron Bay-White Lake area: Ontario Department of Mines, Annual Report, Volume XLI, Part 6, p. 34-47.

Vokes, F.M., 1968, Regional metamorphism of sulfide deposits: Earth Science Review, v. 5, p. 99-143.

Watson, D.W., 1970, The geology and structural evolution of the Geco massive sulphide deposit of Manitouwadge, Northwestern Ontario, Canada: Ph.D. Thesis, University of Michigan, Ann Arbor, Michigan.

MS received February 1979;
Accepted for publication September 1979.

Precambrian Sulphide Deposits, H.S. Robinson Memorial Volume, edited by R.W. Hutchinson, C.D. Spence and J.M. Franklin, Geological Association of Canada Special Paper 25, 1982

THE BATHURST NORSEMINES SULPHIDE DEPOSITS HACKETT RIVER, N.W.T.

M.J. Casselman and B.M. Mioduszewska
COMINCO, 7 - 409 Granville St., Vancouver, B.C. V6C 1T2

ABSTRACT

The Bathurst Norsemines Property at Hackett River, is located in the "barren lands" 480 km northeast of Yellowknife, Northwest Territories. The property is underlain by rocks of the Archean Yellowknife Supergroup including metavolcanic rocks of the basalt to rhyolite association, metagreywacke and meta-argillite. The rocks have been folded into a moderately southeasterly plunging synclinorium and can be grouped, from oldest to youngest, into three conformable sequences: the Beechy Lake volcanic rocks occur on the south limb of the fold; the Hackett River volcanic rocks and a central sedimentary sequence, also part of the Hackett River succession, occur on the north limb of the fold. These rocks are surrounded and intruded by granite and granodiorite.

Three main areas of mineralization exist on the property. The Central Area, comprising the A, Jo and Jo South zones, is here discussed in detail. Two other deposits, Boot Lake and East Cleaver Lake, occur 2 and 6 km, respectively, farther west and are only briefly discussed. The deposits occur in the transition zone between the Hackett River volcanic rocks and the overlying sedimentary strata. They overlie a thick series of rhyolitic pyroclastic rocks and are hosted by mixed rhyolitic pyroclastic and volcaniclastic rocks and chemical and clastic sedimentary rocks, locally referred to as the "mineral horizon".

In the Central Area, the A and Jo Zones are in the same stratigraphic unit and consist of conformable lenses of pyrite, sphalerite, galena and chalcopyrite underlain by discordant, mineralized alteration pipes containing chalcopyrite, pyrrhotite, pyrite, and magnetite. The Jo South Zone lies stratigraphically lower than the A and Jo Zones, and contains only veins and disseminations of pyrite, sphalerite, chalcopyrite, and galena. The intimate spatial relationship of the sulphide lenses with the volcanic rocks, indicates that sulphide formation was an integral part of and related to the volcanism.

Three distinct types of alteration associated with the Central Area mineralized zones have been designated "Spotted Rock", "Main Conduit", and "Anthophyllite-Cordierite". Rocks within the altered areas have been extensively leached, silicified, metasomatized and replaced by sulphides, and locally, oxides. The Main Conduit and Anthophyllite-Cordierite alteration types are similar to quartz-sericite and anthophyllite-cordierite alteration commonly observed in other Archean deposits. Spotted Rock alteration surrounds and encloses the Main Conduit and Anthophyllite-Cordierite types and represents a weakly developed equivalent of the zones of chloritization in many Archean deposits. The mineralogy and textures in the various altered and unaltered rocks in the Central Area are the result of transformation of the original constituents by later cordierite-amphibolite facies metamorphism.

RÉSUMÉ

La propriété de Bathurst Norsemines à Hackett River est située dans la toundra ("barren lands") 480km au nord-est de Yellowknife, Territoires du Nord-Ouest. Les terrains affleurant appartiennent au supergroupe de Yellowknife, d'âge archéen, et comprennent des roches métavolcaniques de l'association basalte-rhyolite, des méta-grauwackes ainsi que des méta-argillites. Celles-ci sont plissées en un synclinorium plongeant modérément vers le sud-est et peuvent être groupées en trois séquences concordantes qui sont, de la plus ancienne à la plus récente: la séquence volcanique de Beechey Lake sur le flanc sud du synclinorium; la séquence volcanique de Hackett River et une séquence sédimentaire centrale, faisant aussi partie de la succession de Hackett River, affleurent sur le flanc nord du synclinorium. Ces roches sont entourées et recoupées par du granite et de la granodiorite.

Trois zones minéralisées principales, les zones A, Jo et Jo South, se trouvent dans le Domaine Central et sont décrites ici en détail. D'autres zones éloignées et moins connues sont celles de Boot Lake et East Cleaver Lake. Les dépôts sont situés au-dessus d'une épaisse série de rhyolite pyroclastique, dans les mêmes unités d'une zone de transition entre les roches volcaniques de Hackett River et les sédiments sus-jacents.

Les zones A et Jo sont dans la même unité stratigraphique et consistent en lentilles concordantes et massives de pyrite, sphalérite, galène et chalcopyrite. Ces lentilles s'enracinent dans des colonnes d'altération minéralisées et discordantes contenant chalcopyrite, pyrrhotite, pyrite et magnétite. La zone Jo South est stratigraphiquement inférieure aux zones A et Jo, et contient seulement des veines et des disséminations de pyrite, sphalérite, chalcopyrite et galène. La relation spatiale intime des zones sulfurées avec les roches volcaniques et plus particulièrement avec des roches d'une période de quiescence, une couche mince de roches sédimentaires exhalatives, indique que la formation des sulfures faisait partie intégrale du volcanisme.

Trois types distincts d'altération associée aux zones minéralisées ont été dénommés "Roche Tachetée", "Conduit Principal" et "Anthophyllite-Cordiérite". Les roches dans ces zones altérées ont été fortement lessivées, silicifiées, métasomatisées et remplacées par des sulfures et localement des oxydes. L'altération en Roche Tachetée entoure les altérations du Conduit Principal et en Anthophyllite-Cordiérite qui sont semblables aux altérations en quartz-séricite et cordiérite-anthophyllite communément observées dans d'autres dépôts archéens; l'altération en Roche Tachetée représente peut-être un équivalent faiblement développé des zones chloritiseés de bien des dépôts archéens. La composition minéralogique et les textures des roches saines et altérées du Domaine Central résultent de la transformation de leurs constituants originaux par un métamorphisme subséquent de faciès amphibolite à cordiérite.

INTRODUCTION

The Bathurst Norsemines Property at Hackett River lies in the "barren lands" 480 km northeast of Yellowknife, Northwest Territories (Fig. 1). In 1970 Cominco Ltd. optioned the property and has since conducted an airborne geophysical survey, carried out geological mapping on both regional and detailed scales, and completed more than 12,000 m of diamond drilling.

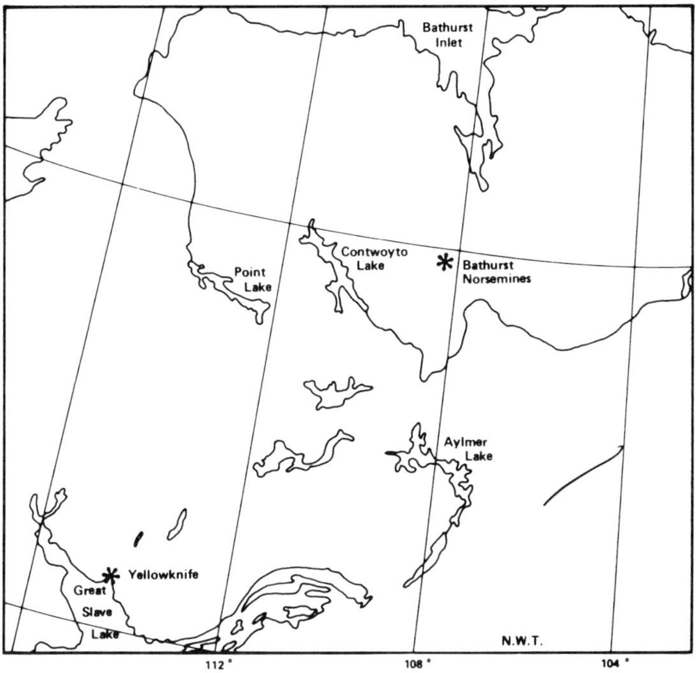

Figure 1. Location map, Bathurst Norsemines deposits, NWT.

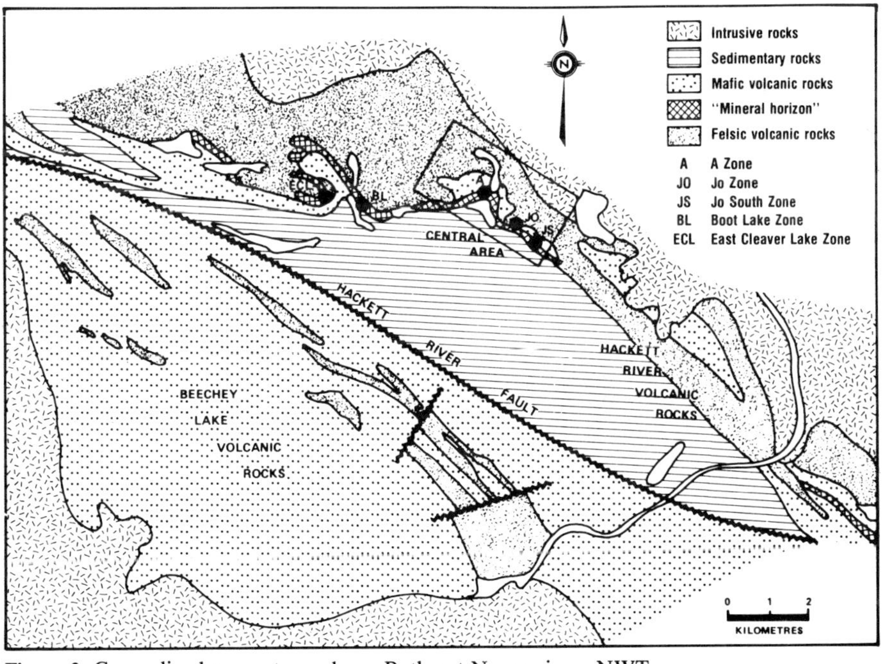

Figure 2. Generalized property geology, Bathurst Norsemines, NWT.

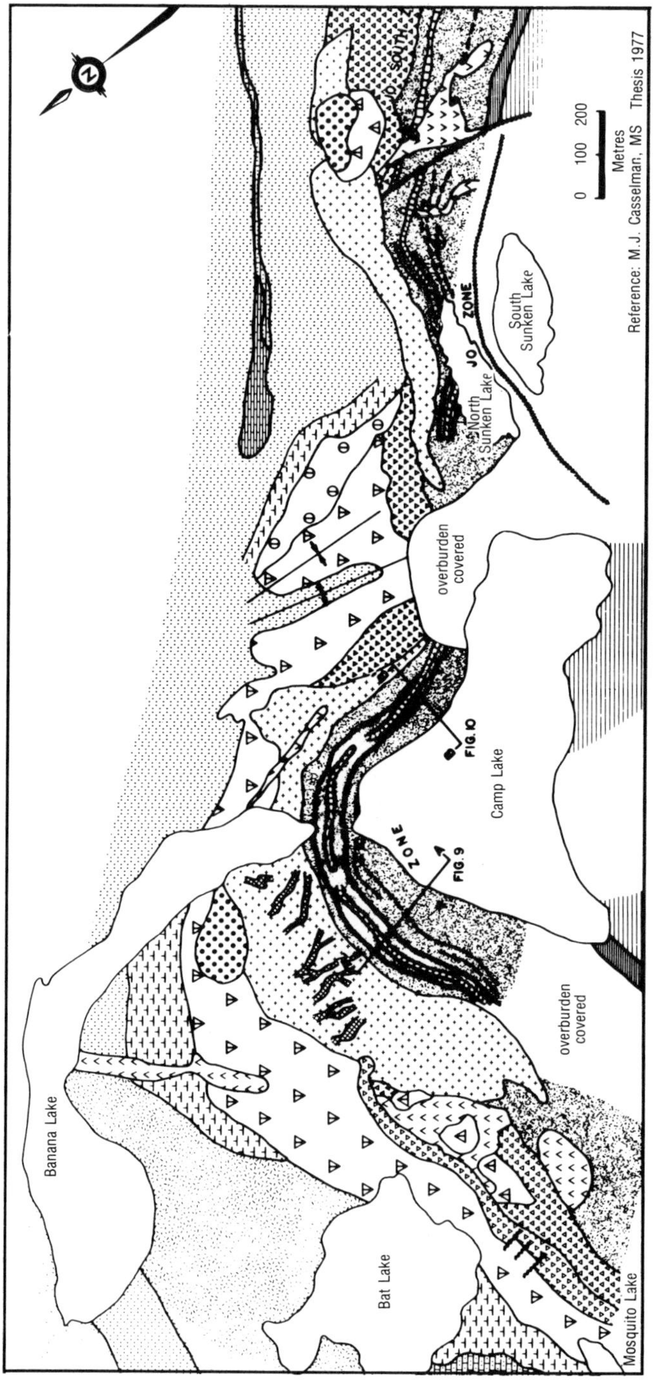

Figure 3. Geologic map of Central Area.

Legend for Figure 3

- Hybrid rocks
- Pegmatite dyke
- Quartz - eye porphyritic rhyolite dyke

SEDIMENTARY SECTION

- Greywacke, argillaceous greywacke, argillite
- Argillite, siltstone, argillaceous tuffite, locally pyritic & graphitic

ALTERED ROCKS

- Spotted rock alteration
- Main conduit alteration
- Anthophyllite - cordierite alteration

UPPER RHYOLITIC PYROCLASTIC SECTION

- Mixed calc - silicate rock, chert, calcareous tuffite, argillaceous tuffite & argillite; locally pyritic & graphitic
- Calc - silicate rock; minor chert & calcareous tuffite; minor pyrite sphalerite galena chalcopyrite
- Main sulphide body (pyrite & pyrrhotite); abundant sphalerite galena chalcopyrite; minor cherty tuffite, chert & rhyolitic tuff
- Mixed rhyolitic pyroclastic rock, argillaceous tuffite, cherty tuffite; minor calcareous tuffite, chert, calc - silicate rock & argillite

DACITIC PYROCLASTIC SECTION

- Dacitic lapilli tuff; minor dacitic tuff & tuffite, calcareous tuffite & rhyolitic tuff
- Dacitic tuff - agglomerate; minor dacitic tuff & tuffite, calcareous tuffite & rhyolitic tuff

ANDESITIC PYROCLASTIC SECTION

- Andesitic lapilli tuff; minor andesitic tuff & tuffite, calcareous tuffite & rhyolitic tuff
- Andesitic tuff - agglomerate; minor andesitic tuff & tuffite, calcareous tuffite & rhyolitic tuff

LOWER RHYOLITIC PYROCLASTIC SECTION

- Rhyolitic tuff - agglomerate; minor rhyolitic tuff
- Argillaceous tuffite; minor rhyolitic pyroclastic rock & cherty tuffite
- Garnet amphibole pyrite pyrrhotite rock (meta iron formation)
- Calcareous tuffite (Jo Zone Limestone)
- Rhyolitic pyroclastic rock; minor argillaceous tuffite, cherty tuffite, chert & andesitic rock

---

- Geologic contact
- Esker
- Fault
- Fold - syncline, anticline

Five sulphide deposits occur at the top of a felsic volcanic pile, in the transition zone between predominantly volcanic and predominantly sedimentary rock environments. Three deposits, the A, Jo and Jo South zones, are located in the Central Area and two deposits, the Boot Lake and East Cleaver Lake zones, occur 2 and 6 km, respectively, farther west. Positive drill indicated reserves in the A Zone are 5 M.t. of 8.5% Zn, 1.4% Pb, 0.25% Cu, 7.0 oz/ton Ag and 0.015 oz/ton Au, in the East Cleaver Lake zone are 4 M.t. of 7.07% Zn, 1.04% Pb, 0.46% Cu, 5.34 oz/ton Ag and 0.014 oz/ton Au, and in the Boot Lake zone are 5 M.t. of 4.97% Zn, 0.99% Pb, 0.29% Cu, 5.86 oz/ton Ag and 0.014 oz/ton Au (Energy, Mines and Resources, Canada, 1980). Most of this paper concerns the Central Área with emphasis on the volcanic-sedimentary stratigraphy, position and morphology of the sulphide deposits in the stratigraphic succession, alteration accompanying sulphide deposition and finally, a reconstructed history of ore-forming events. The stratigraphy and sulphide mineralogy of the other two deposits are briefly compared to those of the Central Area.

The A Zone is the largest deposit on the Hackett River Property. The first discovery of sulphides was made here when a conspicuous gossan on the north shore of Camp Lake was trenched in 1956, and drilled 13 years later. To date approximately 6000 m of core drilling has been completed at the A Zone. The Jo and Jo South Zones, immediately adjacent to the A Zone (Figs. 2, 3) were found by prospecting and constitute the only significant surface exposures of base metal sulphides. The East Cleaver Lake and the Boot Lake zones were discovered by a combination of geological mapping and geophysical interpretation. The rocks, although referred to throughout the paper by their primary genetic names, have all been metamorphosed as high as upper amphibolite facies. Chemical analyses, thin section examination and mapping were used to distinguish rock types.

General Geology

The Bathurst Norsemines property is underlain by Archean rocks of the Slave Province (Stockwell, 1961) which are metavolcanic and metasedimentary rocks of the Yellowknife Supergroup (Henderson, 1970). Deformation during middle to upper amphibolite facies metamorphism and later retrograde metamorphism have obscured most of their original textures.

Three conformable sequences have been identified (Fig. 2). From oldest to youngest these are: 1) a southern sequence of mainly calcareous basalt-andesite flows and pyroclastic rocks; 2) a northern sequence of primarily rhyolitic pyroclastic rocks; and 3) a central sequence of greywacke and argillite. The oldest southern sequence is part of the Beechey Lake volcanic assemblage and is overlain by the northern and central sequences which are part of the Hackett River volcanic assemblage.

Granite and granodiorite surround the metavolcanic and metasedimentary rocks, and dykes of granite, pegmatite and quartz-eye rhyolite are common. Two phases of folding are recognizable, the first about a northwest-southeast axis, and the second about a northeast-southwest axis. The Hackett River syncline is the strongest of the northwest-southeast folds, and its axis nearly coincides with the Hackett River fault, which separates the Beechey Lake volcanic rocks from the Hackett River volcanic rocks (Fig. 2).

The Hackett River volcanic rocks have been subdivided into three broad stratigraphic units: (1) a lowermost mixed andesitic, dacitic and rhyolitic pyroclastic unit with only minor volcaniclastic, clastic and chemical sedimentary rocks; (2) a transitional middle unit which hosts all but one of the deposits, and is composed of mixed rhyolitic pyroclastic and volcaniclastic rocks, and clastic and chemical sedimentary rocks; and (3) an upper sedimentary unit composed mainly of greywacke and argillaceous greywacke.

Geology of the Central Area

The stratigraphic succession in the Central Area, which contains the A, Jo and Jo South mineralized zones (Figs. 2, 3, and 4) has been divided into five distinct lithostratigraphic sections (Fig. 5). Three types of intrusive rocks in the Central Area are pegmatite dykes, granite dykes and quartz-eye porphyritic rhyolite dykes. The pegmatite and granite dykes intrude all volcanic and sedimentary rocks whereas the porphyritic dykes intrude only the volcanic rocks.

During Kenoran orogeny, rocks of the Central Area were tightly folded and affected by metamorphism of the sillimanite-cordierite-orthoclase subfacies of the cordierite-amphibolite facies (Winkler, 1967). The metamorphic grade appears to have been fairly constant throughout the area. The various mineral assemblages and textures were described by Casselman (1977) and are briefly summarized in Tables I and II.

The rocks are upright as shown by rare graded beds in greywacke, and by the

Figure 4. View east of the Central Area. A Zone gossan shows as white patches in left foreground between the two lakes.

TABLE 1

MINERALOGICAL COMPOSITION OF UNALTERED AND ALTERED ROCKS, CENTRAL AREA

Unaltered	Range Average	Qtz%	Feld%	Bi%	Mus%	Gar%	Sill%	Act%	Zo/Ch%	Ca%	Sp%	Ep%	QE%	Di%	Cord
Rhyolitic Pyroclastic Rocks	R	25-65	10-55	2-12	5-35	0-3							1-20		
	A	45-55	25-35	3-10	10-15	<1							2-10		
Chert	R	80-95	0-5	0-5	0-10		0-5								
	A	85-90	0-3	1-3	3-5		0-3								
Cherty Tuffite	R	50-80	5-30	0-10	5-20		0-8						0-5		
	A	55-65	10-15	1-5	5-15		0-5						1-2		
Argillaceous Tuffite	R	40-70	5-25	10-20	5-30	0-3	5-30						0-5		
	A	45-55	10-15	15-20	10-20	<1	10-20						1-2		
Argillaceous Tuffite (Biotite)	R	30-70	5-25	20-35	5-30	0-3	5-30						0-3		
	A	40-50	10-15	20-25	10-20	<1	10-20						1		
Andesitic Pyroclastic Rocks	R	2-30	20-40	5-50	1-20			5-55	0-25	0-5	0-5	0-5			
	A	8-15	25-30	15-30	5-10			15-25	5-18	1-3	1-2	1-2			
Dacitic Pyroclastic Rocks	R	20-50	15-50	5-20	2-15								0-5		
	A	30-35	30-35	10-15	5-10								1-2		
Calcareous Tuffite	Rhyolite Component is Same Composition as Above									5-60			0-2		
Calc-Silicate Rocks	R	10-50	2-20	0-25	0-25	5-25	0-15	0-20		5-15				5-90	0-25
Greywacke	A	50-60	10-25	10-25	10-25	0-5	0-10			0-3					0-25
Argillaceous Greywacke	A	40-50	10-20	25-35	25-35	0-5	0-10			0-3					0-25
Argillite	A	45-65	5-15	35-45	35-45	0-5	0-10			0-3					
Siltstone	A	65-80	0-7	1-10	1-10		0-5			0-15					
Quartz-Eye Porphyritic Rhyolite Dyke	A	40-50	20-25	0-3	5-10						0-10	10-15			
Pegmatite Dyke	A	20-35	60-75	0-3	5-15										
Hybrid Rocks	R	20-60	5-35	5-25	5-20	0-10	0-20								0-25

Altered	Range Average	Qtz%	Feld%	Bi%	Mus%	Gar%	Sill%	Cord	Anth	Sulphides	Mag
Spotted Rock Alteration	R	35-75	0-2	5-50	3-10	0-20	5-40			0-25	0-20
	A	50-60	<1	15-20	5-8	1-7	15-20			2-10	0-5
Main Conduit Alteration	R	65-95		0-10	5-35		0-3	0-2		0-20	
	A	75-85		0	15-25		0	0		3-10	
Anthophyllite-Cordierite Alteration	R	10-20	0-5	5-25		0-20	0-12	4-28	5-20	0-25	0-5

Symbols R-Range; A-Average; Qtz-Quartz; Feld-Feldspar; Bi-Biotite; Mus-Muscovite; Gar-Garnet; Sill-Sillimanite; Act-Actinolite; Zo/Ch-Zoisite/Chlorite; Ca-Calcite; Sp-Sphene; Ep-Epidote; QE-Quartz-Eyes; Di-Diopside; Cord-Cordierite; Anth-Anthophyllite; Mag-Magnetite

TABLE II

MAJOR ELEMENT COMPOSITION OF UNALTERED AND ALTERED ROCKS, CENTRAL AREA

Unaltered	No. of Samples		SiO_2	Al_2O_3	TiO_2	Fe_3O_4	MgO	K_2O	Na_2O	CaO
Rhyolitic Pyroclastic Rocks	21	R	70.0-79.3	8.0-14.0	0.1-0.43	1.42-5.2	0.4-2.9	1.0-7.3	0.4-3.6	0.3-2.0
		A	74.7	10.8	0.23	2.9	1.38	3.6	1.2	1.1
Chert	3	R	85.4-90.0	4.8-6.5	0.15-0.20	0.15-0.2	2.1-2.8	0.2-0.9	0.4-1.7	0.3-0.5
		A	88.46	5.86	0.18	0.18	2.46	0.5	1.53	0.43
Cherty Tuffite	4	R	80.0-81.22	7.8-11.3	0.1-0.3	1.4-3.9	0.6-2.1	1.1-3.9	0.3-1.7	0.3-0.8
		A	80.65	9.72	0.2	2.67	1.25	2.55	0.70	0.50
Argillaceous Tuffite	7	R	60.5-73.6	15.9-24.5	0.23-0.32	1.6-3.2	1.7-3.0	1.0-4.1	1.0-2.82	0.4-2.1
		A	65.07	20.88	0.27	2.57	2.44	1.82	2.0	1.24
Andesitic Pyroclastic Rocks	10	R	52.0-60.8	13.0-16.3	0.52-0.7	3.5-8.1	2.0-9.4	2.1-7.9	0.4-4.22	4.8-9.4
		A	56.47	14.2	0.59	6.31	5.56	3.97	1.59	6.79
Dacitic Pyroclastic Rocks	3	R	67.0-69.7	11.8-13.0	0.4	3.6-4.6	1.0-1.7	2.8-7.6	0.6-2.2	1.2-4.8
		A	67.9	12.3	0.4	4.1	1.4	4.9	1.3	2.5
Altered										
Spotted Rock Alteration	12	R	68.4-79.5	9.02-16.8	0.23-0.6	3.9-7.6	0.8-3.7	1.42-3.9	0.41-0.62	0.3-0.64
		A	73.6	11.45	0.35	5.39	2.23	2.65	0.43	0.47
Main Conduit Alteration	3	R	82.4-90.02	4.8-7.8	0.1-0.22	2.1-3.0	0.2-0.4	1.4-1.7	0.3-0.52	0.42-0.5
		A	85.5	6.3	0.17	2.5	0.3	1.6	0.4	0.44
Anthophyllite-Cordierite	2	R	56.2-61.8	12.8-13.3	0.6-0.7	9.0-13.4	5.3-6.6	1.4-3.3	0.4-0.5	0.6-1.1
Alteration		A	59.0	13.0	0.65	11.2	5.95	2.35	0.45	0.85

Note: Fe_3O_4 – measured total Fe converted to oxide form
R – Range
A – Average

general morphology of the deposits which is comparable to many rhyolite-associated sulphide deposits, which typically overlie rhyolitic pyroclastic rocks and are in turn overlain by sedimentary rocks (Sangster, 1972). In addition, similar facings are indicated by the consistent metal zonation pattern within these deposits, from a basal hydrothermally altered pipe containing veinlets and disseminations of chalcopyrite-pyrrhotite-pyrite-magnetite (Fig. 6) to a banded pyrite-sphalerite-galena-rich cap (Sangster, 1972; Roscoe, 1965).

Lower Rhyolitic Pyroclastic Section. The Lower Rhyolitic Pyroclastic Section is the lowest and thickest lithostratigraphic unit in the Central Area. Its complete stratigraphic thickness is unknown because of overburden, but is at least 450 m. This unit is continuous throughout most of the Central Area, except north of Bat Lake where it has been partially cut out by later intrusions.

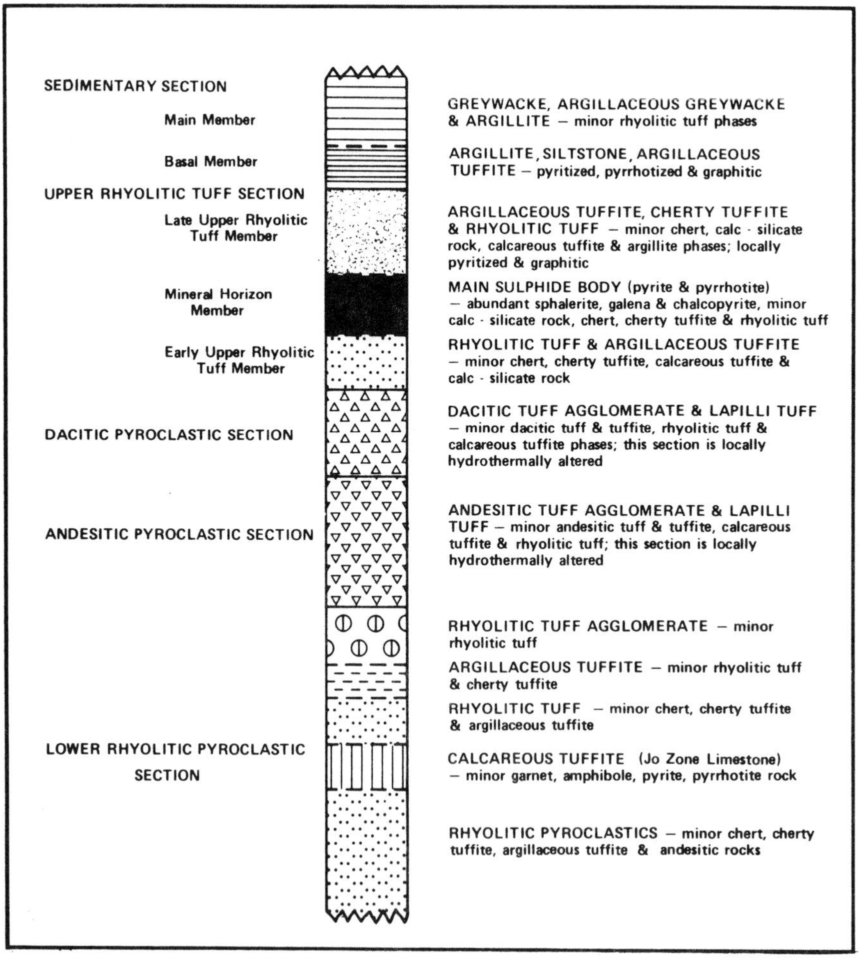

Figure 5. Stratigraphic column for the Central Area.

The foliation in these rocks is parallel to bedding, strikes 280° to 360° and dips 45° to 60° south to southwest. The structure is complicated by folding in the area between the A and Jo Zones. The unit consists primarily of quartz-eye rhyolitic pyroclastic rocks (Fig. 7) with lesser intercalated chert, calcareous tuffite, garnetiferous amphibolite (metamorphosed sulphide-silicate facies iron-formation), argillaceous tuffite and andesitic flows.

Individual rock types are difficult to trace and define due to very poor exposure. Chert beds are generally discontinuous and less than 30 cm thick. Andesitic flows are less than 3 m thick. Calcareous tuffite units are usually less than 3 m thick, except for one which is exposed 450 m in the footwall of the Jo zone, and just west of Bat Lake which ranges in thickness from 3 to 45 m. Argillaceous tuffite units are usually less

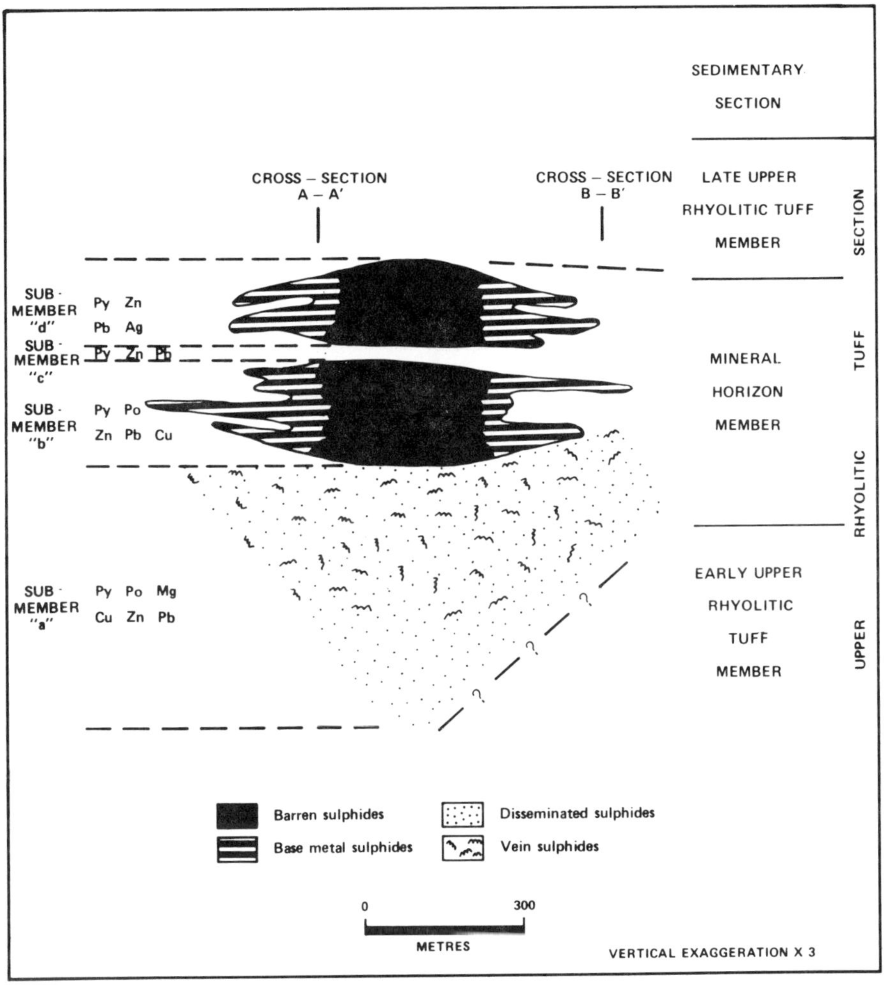

Figure 6. Diagrammatic cross section, A zone, illustrating submembers and metal zoning.

Figure 7. Quartz-eye rhyolitic pyroclastic rock. Lower Rhyolitic Pyroclastic Section, A Zone.

Figure 8. Rhyolitic tuff-agglomerate. Lower Rhyolitic Pyroclastic Section, A Zone.

than 4.5 m thick except for two areas in the footwall of the A and Jo zones, where they attain thicknesses of 60 to 120 m.

A lenticular pod of rhyolitic tuff-agglomerate with a maximum thickness of 100 m occurs at the top of the Lower Rhyolitic Pyroclastic section in the footwall of the Jo zone. The rhyolitic fragments in this rock are elongated parallel to foliation and range in size from less than 5 mm to 15 cm in length, averaging 5 to 10 cm (Fig. 8).

Andesitic Pyroclastic Section. The start of andesitic volcanism marked the initial deposition of a cyclic series of andesitic to rhyolitic pyroclastic rocks. The andesitic pyroclastic rocks range in thickness from 60 to 180 m. The greatest accumulation is located north of the A zone. This unit extends continuously from southwest of Bat Lake to the Jo South zone where it interfingers with rhyolitic pyroclastic rocks and disappears. It could not be traced west of Bat Lake because of overburden. Extensive portions of the unit in the A, Jo and Jo South zones have been affected by hydrothermal alteration.

The rocks of this unit are predominantly coarse andesitic tuff-agglomerate and lapilli tuff intercalated with thin, discontinuous andesitic flows, tuff and tuffite and a band of calcareous tuffite, all less than 3 m thick. In the area north of the A zone, near the base of the unit, andesitic fragments range up to 12.5 cm, averaging 3.5 cm long (Fig. 9). These fragments gradually decrease in size west and east from the A zone and up dip, averaging 1.0 to 2.5 cm long in the middle and upper parts of the unit. In the footwall of the Jo zone, north of the A zone, and southwest of Bat Lake, abundant rhyolitic fragments occur intimately intermixed with the andesitic pyroclastic fragments in concentrations ranging from 10 to 50% (Fig. 10). The rhyolitic fragments, identical in composition and appearance to rhyolitic fragmental rocks in

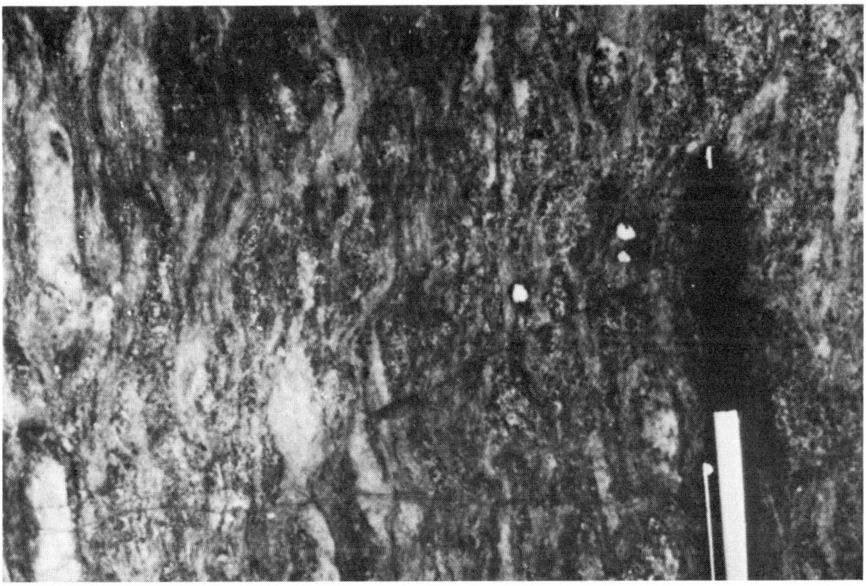

Figure 9. Andesitic tuff-agglomerate. Andesitic Pyroclastic Section, between the A and Jo Zones.

Figure 10. Rhyolitic pyroclastic fragments enclosed in andesitic pyroclastic matrix. Andesitic Pyroclastic Section between the A and Jo Zones.

Figure 11. Rhyolitic pyroclastic fragments which are more resistant to weathering give "knobby" appearance to outcrop. Andesitic Pyroclastic Section between A and Jo Zones.

the Lower Rhyolitic Pyroclastic Section, range in diameter from 10 mm to 30 cm, averaging 2.0 to 7.5 cm, but decrease in size stratigraphically upward and outward from the above three areas of greatest abundance. The zone of mixing between the Andesitic Pyroclastic Section and the underlying Lower Rhyolitic Pyroclastic Section in these three areas is gradational over 30 to 60 m. Chert and calcareous tuffite fragments are present in places in the andesitic pyroclastic rocks. All fragment types are clearly recognizable on weathered outcrops; rhyolitic fragments are most distinctive, both in relief and due to their white colour (Fig. 11). All lithic fragments are elongated parallel to foliation.

Dacitic Pyroclastic Section. The Dacitic Pyroclastic Section represents the middle phase of the cyclic andesitic to rhyolitic pyroclastic activity. It ranges in thickness from 60 to 120 m with the greatest thicknesses in the A zone, south of Bat Lake, and in the Jo South zone. This unit interfingers with rhyolitic pyroclastic rocks eastward from the Jo South zone, and west of the A zone it extends under overburden. The basal contact with the Andesitic Pyroclastic Section was not observed. Considerable portions of the unit in the A, Jo and the western part of the Jo South zones have been affected by hydrothermal alteration.

The Dacitic Pyroclastic Section consists primarily of dacitic lapilli tuff with lesser amounts of dacitic tuff-agglomerate, dacitic tuff, rhyolitic tuff and calcareous tuffite. The lithic fragments are primarily dacitic, but a few are andesitic and rhyolitic. They are all moderately foliated, elliptical in shape, and elongated parallel to foliation. They range in size from 2 mm to 25 mm (Fig. 12).

Figure 12. Dacitic lapilli tuff. Dacitic Pyroclastic Section, Jo South Zone.

Thin interbedded discontinuous dacitic tuff, rhyolitic tuff and calcareous tuffite bands, all less than 3 m thick are also present in places. Dacitic tuff-agglomerate occurs as pods near the base of the unit in the Jo South zone and in the area south of Bat Lake. The Bat Lake pod is difficult to define due to limited exposure and because much of it has been removed by a cross-cutting quartz-eye porphyritic rhyolitic dyke. The westward extensions and lower parts of the pod in Jo South zone have been partly affected by hydrothermal alteration and are locally cut by a series of anastomosing sulphide veinlets which comprise the Jo South mineralized zone.

Upper Rhyolitic Tuff Section. This lithostratigraphic unit was deposited during the final phase of the cyclical andesitic to rhyolitic pyroclastic activity and represents the last known volcanic event in the Central Area. Its stratigraphic thickness varies greatly along strike; from 180 to 240 m in the A and west Jo zones, the unit thins slightly into the Jo South zone and west of the A zone (Fig. 13). The rock types present change vertically and have been subdivided into three members, the Early Upper Rhyolitic Tuff, the Mineral Horizon, further divided into submembers a to d, (Figs. 6, 13) and the Late Upper Rhyolitic Tuff. The rock types also change laterally within the A, Jo and Jo South zones, and especially in the Late Upper Rhyolitic Tuff member where there is a significant increase in the rhyolitic pyroclastic component eastward and westward from the A zone. The A zone consists primarily of argillaceous tuffite and cherty tuffite; the Jo zone includes predominantly cherty tuffite and rhyolitic tuff, whereas the Jo South and west A zones consist primarily of rhyolitic

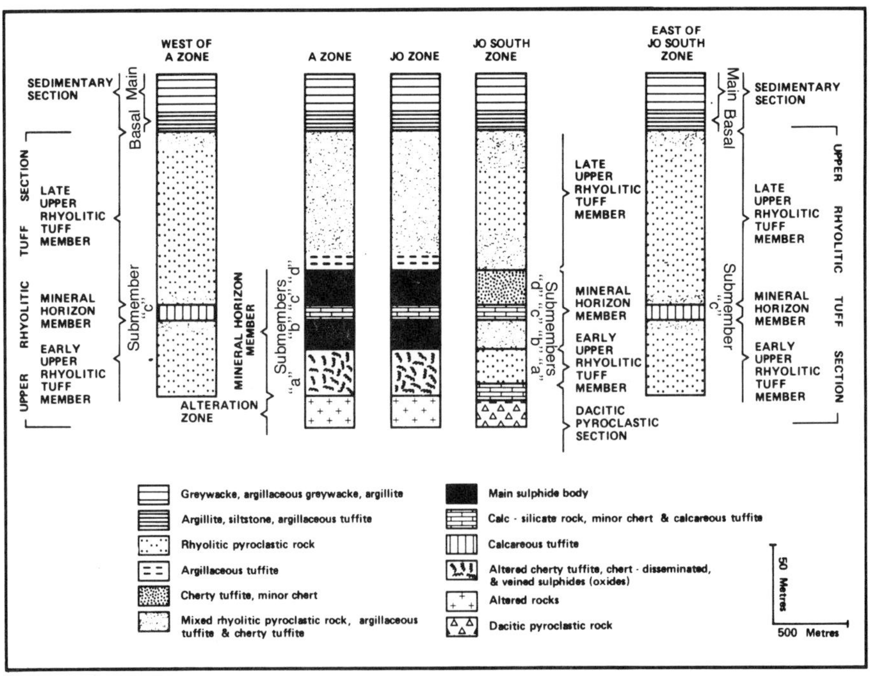

Figure 13. Diagram illustrating relationships of members and submembers of the Upper Rhyolitic Tuff Section along strike from west to east in the Central Area.

tuff. The hangingwall contact of this section is defined by a graphitic sulphide-argillite bed located at the base of the overlying Sedimentary Section.

Early Upper Rhyolitic Tuff. In the A and Jo zones, this member has been extensively altered by hydrothermal solutions so that its original composition and thickness are not known. The only unaltered exposures of the member occur in the Jo South zone where it is 30 to 35 m thick. Here it consists primarily of calc-silicate rock and rhyolitic tuff, overlain by 5 to 7 m of argillaceous tuffite, cherty tuffite and minor chert and calc-silicate rock. The argillaceous tuffite and cherty tuffite occur as discontinuous lenses, gradational into each other and into the underlying and adjacent rhyolitic tuff. The chert and calc- silicate bands occur throughout the upper part as discontinuous lenses and pods, usually less than 30 cm in thickness. Compositionally and texturally, the rhyolitic tuff and argillaceous tuffite are similar to the analogous rock types in the underlying Lower Rhyolitic Pyroclastic Section, however quartz-eyes in the rhyolitic tuff are less abundant. The Early Upper Rhyolitic Tuff extends laterally both east and west beyond the Central Area, where it merges into a terrain dominated by rhyolitic pyroclastic rocks.

Mineral Horizon. This member, which hosts the sulphide lenses varies appreciably in thickness from 15 to 100 m in the A zone, 3 to 15 m in the Jo zone, to 30 to 60 m in the Jo South zone. It is almost totally restricted in its development to the Central Area, merging both eastward and westward into areas dominated by rhyolitic pyroclastic strata. Only its submember c, a calc-silicate and chert band with minor, thin cherty tuffite, can be traced laterally into the adjacent volcanic belt. Its four sub-members (a to d) consist of intimate mixtures of chemical and clastic sedimentary rocks, and altered and unaltered rhyolitic pyroclastic rocks. The submembers are laterally persistent in the A zone, less continuous in the Jo zone, and poorly developed in the Jo South zone, where no base metal-bearing sulphide lenses occur. Vertical zoning in the Mineral Horizon is shown by differences in rock type and in sulphide concentrations and mineralogy.

The thickest sulphides are in the centre of the A zone (Figs. 6, 13) and consist of a core of barren pyrite, 90 to 100 m thick, which thins outward to a surrounding fringe 9 to 45 m thick. The core is a lens of layered pyrite (50-70%) surrounded by pyritic lenses (30-70%) with varying content of sphalerite, galena, chalcopyrite and pyrrhotite.

Where ideally developed in the A and Jo zones, the four submembers (Figs. 6, 13) occur (from bottom to top) as follows: Submember a, composed of silicified and leached chert, rhyolitic tuff and cherty tuffite with minor calc-silicate lenses, contains 3 to 45% of vein and disseminated sulphides including chalcopyrite, pyrite and pyrrhotite, and 0 to 5% magnetite. Quartz veins, which commonly contain sulphides, range from 1 mm to 2 cm thick and locally form anastomosing networks. Submember a extends beneath the A and Jo zone sulphide lenses and forms the upper part of the main alteration zone. Although present throughout the A and parts of the Jo zone, submember a does not contain uniform concentrations of base or precious metals. The sulphide and oxide minerals average 5 to 15% of the rock by volume although locally sulphides comprise up to 45%. In general, submember a contains only minor amounts of chalcopyrite; however, two areas of higher grade chalcopyrite are known.

Submember b (Figs. 6, 13) is the lower sulphide layer consisting of pyrite (30-70%), minor pyrrhotite and varying amounts of sphalerite, galena and chalcopyrite. It lies within leached and silicified rhyolitic tuff and cherty tuffite composed of a mixture of quartz and pyrite grains that resembles a siliceous sinter (Fig. 14). Thin intercalated bands of chert and argillaceous tuffite occur locally in this submember.

Submember c (Figs. 6, 13) is a zone of mixed calc-silicate minerals and chert with minor calcareous tuffite. The calc-silicate rock consists mostly of large, euhedral green diopside grains or massive diopside (Fig. 15) with variable amounts of garnet, actinolite, quartz, feldspar and biotite. This submember ranges from 2 to 10 m thick, and has a low base metal, but locally high silver content. It contains 5 to 15% disseminated sphalerite, galena, pyrite and pyrrhotite. The chert forms discontinuous pods within the calc-silicate rock. A zone of argillaceous tuffite 1.5 to 3 m thick is widespread at the base of submember c. This is the only submember that extends continuously throughout the A, Jo and Jo South zones of the Central Area, and it can be traced for some distance into the adjacent volcanic belt. Outside of the Central Area submember c changes into a 2 to 9 m thick calcareous tuffite.

Submember d (Figs. 6, 13) is an upper sulphide layer. It consists of pyrite (30-70%) and minor pyrrhotite with varying proportions of sphalerite and galena, in host rocks identical to those of submember b. The hangingwall contact of submember d with the overlying argillaceous tuffite is sharp. Submember d is similar in sulphide content to submember b, but chalcopyrite is generally absent, and galena is much more abundant, with accompanying higher content of silver.

Late Upper Rhyolitic Tuff. This member ranges in thickness from approximately 120 to 180 m. Its lithology varies greatly between the A, Jo and Jo South zones (Fig. 13) – the main variation being an increase in the rhyolitic pyroclastic content and a decrease in chemical and clastic components eastward and westward from the A zone. The rock consists of variable mixtures of rhyolitic pyroclastic, clastic and chemical sedimentary material, mainly argillaceous tuffite and cherty tuffite, with less rhyolitic tuff, argillite, chert, calc-silicate rock and calcareous tuffite. These rock types are lenticular and discontinuous. The only laterally continuous stratum within this member in the A and west Jo zones is an argillaceous tuffite which forms the basal layer immediately overlying submember d, the upper sulphide layer of the Mineral Horizon (Fig. 13).

The Late Upper Rhyolitic Tuff of the A and west Jo zones consists mostly of argillaceous and cherty tuffite with lesser amounts of rhyolitic tuff, graphitic pyritic argillite and argillite, chert, calcareous tuffite and calc-silicate rock. In many places the rock types are difficult to identify as they contain variable mixtures of rhyolitic pyroclastic, volcaniclastic, clastic and chemical sedimentary materials. Rhyolitic pyroclastic rocks are abundant west of A zone in the equivalent stratigraphic interval.

The volume of rhyolitic pyroclastic rock increases substantially in this member in the Jo and Jo South zones. In the former it consists primarily of cherty tuffite and rhyolitic tuff with lesser amounts of argillaceous tuffite. Intercalated within these rocks are several chert bands 2 to 4 m thick and/or bands of calc-silicate rock 2 to 3 m thick, and a 4 m thick barren sulphide lens containing 45 to 60% pyrite. Thinner bands of these chemical sedimentary rocks can be recognized in the Jo South and A

Figure 14. Submember b; granular pyrite-quartz rock, A Zone.

Figure 15. Submember c; diopside rock, Jo Zone.

zones. In the Jo South zone this member contains mainly rhyolitic tuff with lesser amounts of cherty tuffite. These rocks interfinger eastward into a thick sequence of rhyolitic pyroclastic rocks with minor intercalated argillaceous and cherty tuffite.

Sedimentary Section. This lithostratigraphic unit comprises a 1,500 m thick, monotonous sequence of mainly greywacke and argillite apparently derived by degradation of the adjacent volcanic piles. Throughout the A zone these rocks are conformable with the Central Area volcanic rocks and extend over most of the southern portion of the Hackett River Volcanic rocks.

The unit can be subdivided into two members, Basal and Main (Figs. 5, 13). The Basal member is a 30 to 40 m thick, conformable sequence of interbedded argillite, graphitic and pyritic argillite, siltstone, greywacke and minor rhyolitic tuffite and calcareous rock. Bedding in the argillite is parallel to stratification in the underlying volcanic sequences. The Main member is composed of greywacke and argillaceous greywacke. Individual beds are continuous over a considerable strike length. Most original sedimentary structures, except for bedding and a few graded beds, are masked or destroyed by metamorphism. The graded bcds indicate top directions to the southwest.

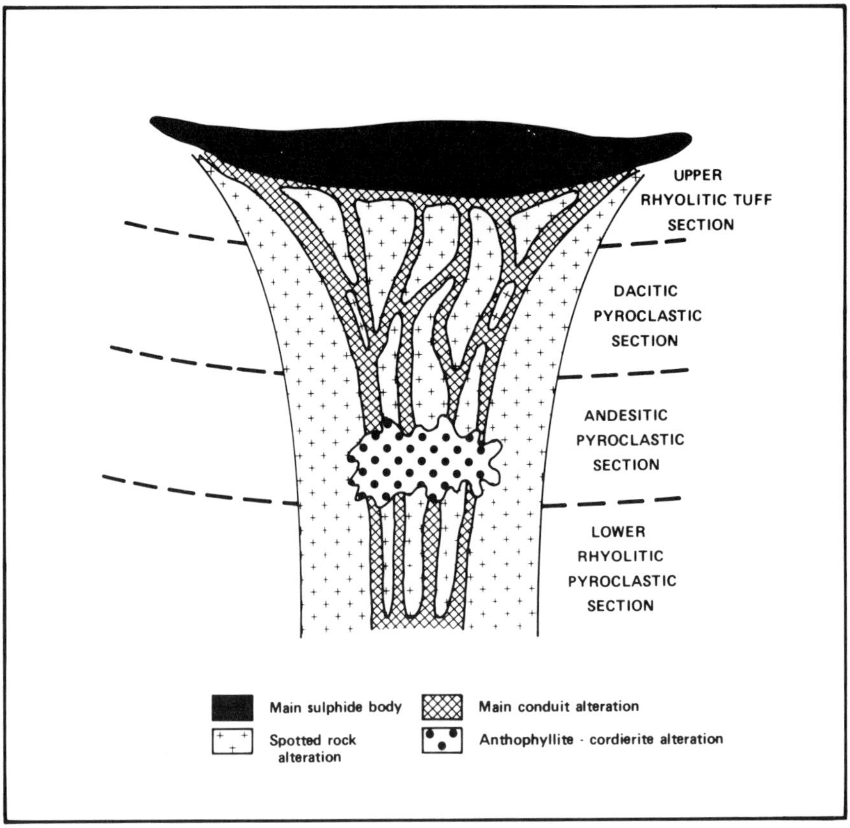

Figure 16. Diagrammatic cross section of A zone alteration pipe.

ALTERATION

Different types of altered rock are spatially, structurally, compositionally, texturally and chemically distinct from each other and from other rock types in the Central Area (Tables I, II). These alterations, referred to as Spotted Rock, Main Conduit and Anthophyllite-Cordierite type, occur only in the footwall of the A, Jo and Jo South zones and do not extend into the hangingwall rocks. The entire body of altered rock is pipe-like in shape (Fig. 16), perpendicular to the main sulphide bodies, contains vein and disseminated sulphides and oxides and transgresses geologic contacts. Along strike, the altered rocks change to andesitic, dacitic and rhyolitic pyroclastic rocks too abruptly to be explainable by a facies change, and evidence of a structural break is lacking. The altered rocks differ from unaltered rock primarily in matrix composition. The altered rocks have multi-sized, irregularly shaped quartz grains in the matrix whereas the unaltered rocks have equal-sized quartz grains. In contrast to the unaltered rocks, altered types are almost totally lacking in feldspar and contain abundant sulphide and oxide minerals similar to those present in the orebodies. The alteration types transect not only unaltered rocks, but also each other; they merge compositionally and texturally into one another over short distances. Spotted Rock and Main Conduit types occur within andesitic, dacitic or rhyolitic pyroclastic rocks and their related chemical and clastic sedimentary strata, whereas the Anthophyllite-Cordierite type, which is less widespread than the other two, is found only within andesitic pyroclastic units.

Main Conduit and Anthophyllite-Cordierite alteration types are compositionally equivalent to the quartz-sericite and anthophyllite-cordierite alteration common in other Archean deposits (Sangster, 1972); the Spotted Rock alteration represents a weakly developed equivalent to the zones of chloritization in many Archean deposits. The minerals in the altered area are the result of superimposed metamorphic transformation of the original alteration constituents which formed the following assemblages in the alteration zones:

Spotted Rock:
sillimanite-biotite-garnet-quartz-(cordierite-sericite)

Main Conduit:
sericite-quartz-(sillimanite-cordierite)

Anthophyllite-Cordierite:
anthophyllite-cordierite-biotite-garnet-quartz-(sillimanite-sericite).

Spotted Rock Alteration

This alteration is named for its sillimanite porphyroblasts which are readily visible as white spots on weathered surfaces (Fig. 17). Spotted Rock alteration is the most extensively developed of the three alteration types. It surrounds the Main Conduit and Anthophyllite-Cordierite alterations where these are present (Fig. 16). Spotted Rock alteration is thickest in the footwall of the A and Jo zone sulphide lenses. In the A zone it extends over a surface area of approximately 250 m by 1200 m, with the thickest section of 300 m just west of Banana Lake. In the Jo and Jo South zones, it covers a surface area of approximately 50 m by 1100 m, with the thickest development in the Jo South zone. It terminates abruptly in the Jo South zone and does not continue far west of the A zone where it is covered by extensive overburden. Between the A and Jo zones, diamond drilling indicates that the Spotted Rock alteration has thinned and may have pinched out. Exposures east of Banana Lake, west of A zone, and in the Jo South zone indicate that the transition between Spotted Rock alteration and unaltered wall rock is gradational, over distances of 30 m or less.

Spotted Rock alteration is readily identifiable in exposures or diamond drill core. Although the zones of Spotted Rock alteration are mainly homogeneous, they exhibit local compositional variations in biotite and garnet content, in the abundance of quartz veining, in the distribution of sulphide and oxide minerals, and particularly in markedly increasing amounts of muscovite adjacent to the Main Conduit zones. Very few relict textures are recognizable. Minor (2 to 10%), variably distributed amounts of chalcopyrite, pyrite, pyrrhotite, and magnetite are present.

Main Conduit Alteration

Main Conduit alteration occurs only in the immediate footwall of the A and Jo zone sulphide lenses and is not known in the Jo South zone. In the A zone this alteration type forms an anastomosing network of "veins" 1 to 15 m thick which randomly transect the strata and are surrounded by Spotted Rock alteration (Fig. 16). The most significant concentrations of chalcopyrite occur in these rocks and quartz veins cutting them. Minor (3 to 10%), variably distributed amounts of sphalerite, pyrrhotite (commonly pink coloured), pyrite and magnetite also occur.

Main Conduit alteration results in a quartz-sericite rock with no recognizable relict textures (Fig. 18). The rock shows little textural or mineralogical variation throughout the Central Area, other than slight decreases in muscovite content and the appearance of biotite near contacts with Spotted Rock alteration. These two types merge mineralogically and texturally over a distance of 1.0 to 2.0 m. The crosscutting relationships and gradational contacts indicate that Main Conduit alteration formed later and was superimposed on Spotted Rock alteration.

Anthophyllite-Cordierite Alteration

Anthophyllite-Cordierite alteration occurs primarily in two zones, each situated along the projected stratigraphic position of the Andesitic Pyroclastic Section. The two largest pods, one in the A, and the other in the Jo South zone, are approximately 70 by 180 m. No Anthophyllite-Cordierite alteration is known in the Jo zone. This alteration type may be more common, but undetected because of limited exposures. Contacts between Anthophyllite-Cordierite and Main Conduit alteration were not observed. However, contacts between Anthophyllite-Cordierite and Spotted Rock alteration are characterized by mineralogical and textural gradations over distances of 18 m or less. Near the contacts there is an increase in sillimanite and commonly sericite in the Anthophyllite-Cordierite rocks. Relict rhyolitic and andesitic fragments can commonly be identified (Fig. 19). Traces of pyrite and chalcopyrite also occur (0 to 3%).

The three alteration types are chemically very distinct from one another and from other rock types in the Central Area; yet, they show only minor compositional

Figure 17. Spotted Rock alteration; sillimanite porphyroblasts (white) enclosed by biotite (black) and quartz (grey) matrix, A Zone.

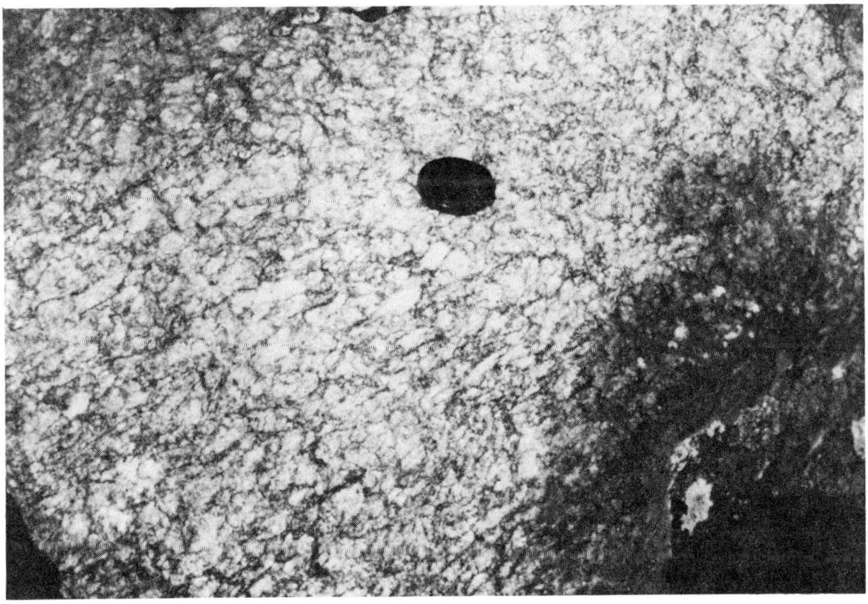

Figure 18. Main Conduit alteration; "chicken wire" texture-quartz clots (white) surrounded by sericite bands, A Zone.

variations from one rock type to another. The range and average chemical composition of the three alteration types is given in Tables III and IV.

DISCUSSION

Unaltered Rocks

Major element analyses of the volcanic rocks indicate that at the time of formation, the rhyolitic and dacitic rocks contained more K_2O than Na_2O and the andesitic rocks more Na_2O than K_2O (Table III). These ratios have been modified locally by later movement of sodium- and potassium-rich waters. In the dacitic and rhyolitic rocks erratically distributed areas have been variably enriched in sodium and/or potassium, whereas in the andesitic rocks the alkali ratio has been increased only in potassium in distinct cross cutting zones. The zones of potassium and sodium enrichment are probably due to the movement of meteoric waters. These solutions were presumably derived locally from within the Central Area and leached alkalis from the volcanic rocks through which they passed. If the solutions had been derived from an outside source, such as granites, the effects of metasomatism in the area would have been more widespread.

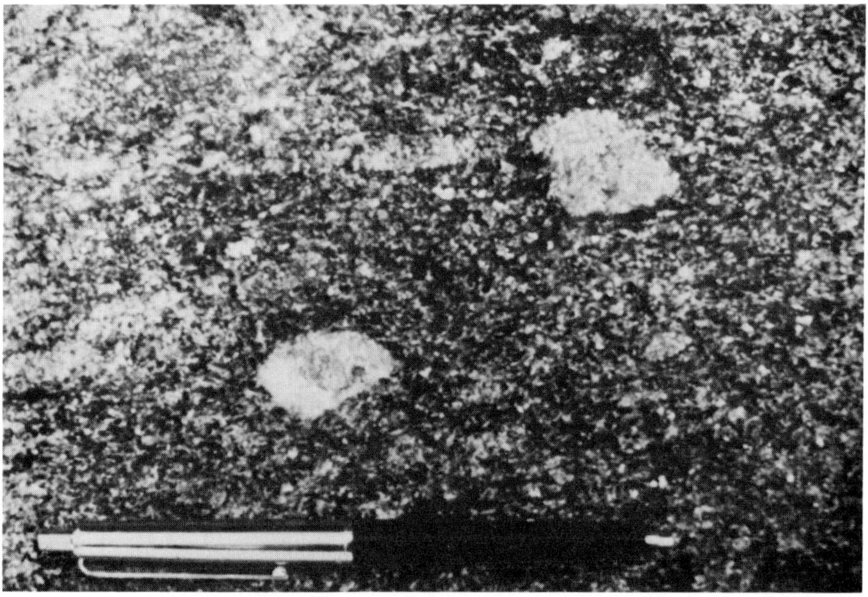

Figure 19. Anthophyllite-Cordierite alteration-note relict rhyolitic pyroclastic fragments in anthophyllite-cordierite, biotite, garnet, quartz matrix.

TABLE III COMPARISON OF ALTERED AND UNALTERED ROCKS, CENTRAL AREA

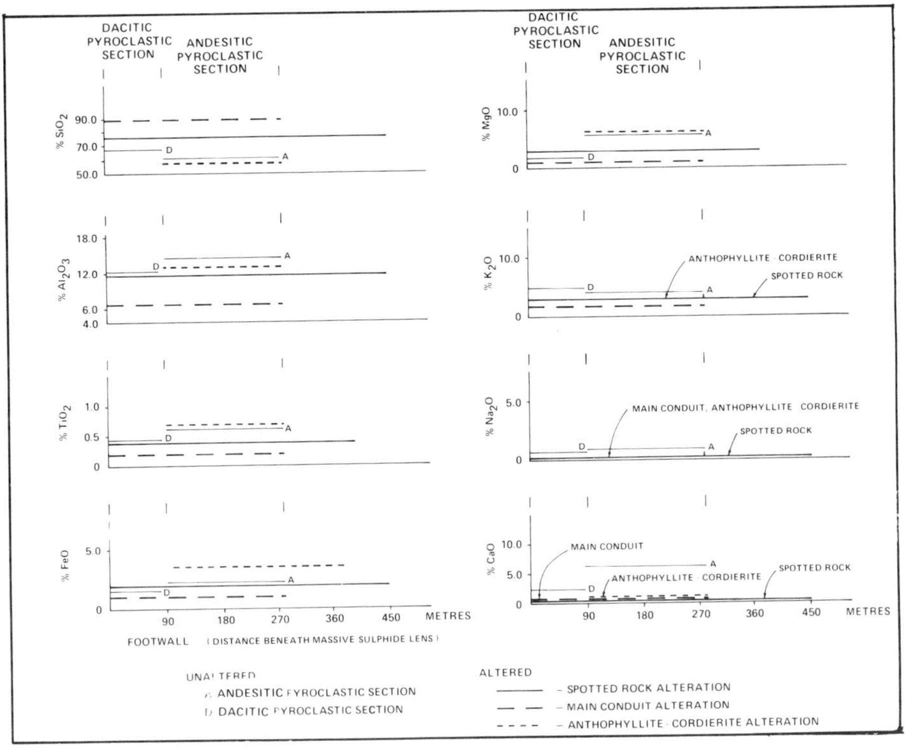

Altered Rocks

Tables I, II, III, and IV allow a qualitative comparison of the mineralogical and major element compositions of the three alteration types to stratigraphically equivalent, less altered rocks in the Andesitic Pyroclastic and Dacitic Pyroclastic Sections. Average compositions are used in Table IV because it is impossible to correlate with any certainty, a specific stratigraphic interval of unaltered rock in the Andesitic and Dacitic Pyroclastic Sections with a corresponding interval of altered rock within the altered areas. Excessive chemical differences reflect changes caused by superimposed hydrothermal alteration; subtle chemical changes may reflect only slight variations in the compositon of the original rocks. Alteration effects on the Early Upper Rhyolitic Tuff member could not be determined as this member includes several rock types, individual units of which are too narrow and discontinuous to be positively correlated with their stratigraphically equivalent alteration type.

In the Andesitic Pyroclastic Section, Spotted Rock alteration involves an increase only in Si, and decreases in Ti, Mg, Na, and Ca. There is little change in Fe. Al and K appear to have moved locally, causing local enrichment and depletion. Main Conduit alteration of andesitic rock involves a major increase in Si, and relative decreases in Al, Ti, Fe, Mg, K, Na and Ca. Anthophyllite-Cordierite alteration of

TABLE IV

COMPARISON OF AVERAGE COMPOSITION OF ALTERED AND UNALTERED ROCKS, CENTRAL AREA

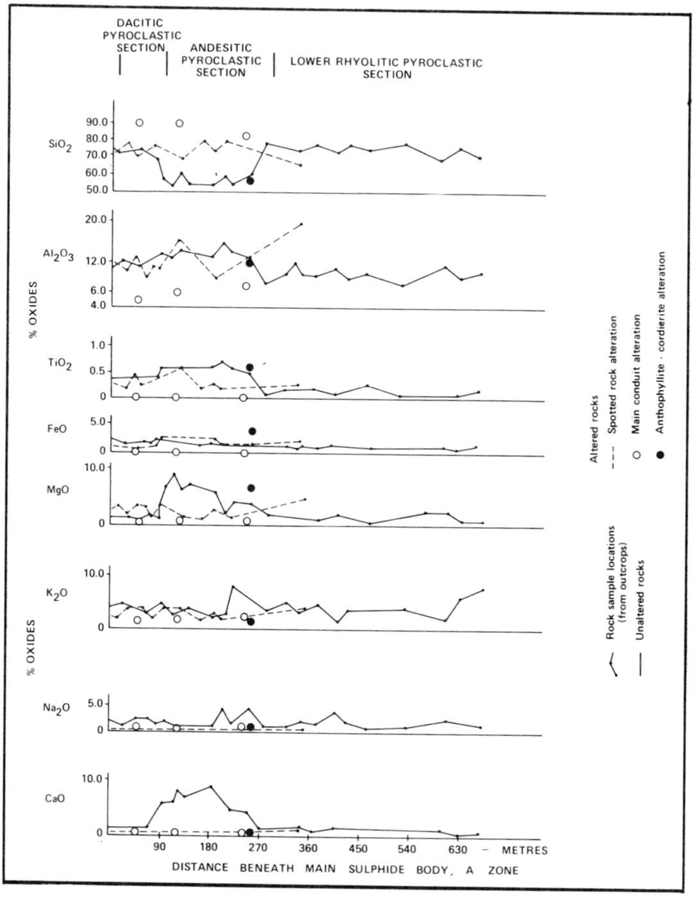

andesitic rock involves an increase in Fe and depletions of K, Na, and Ca. There is little or no change in Si, Al, Ti and Mg.

In the Dacitic Pyroclastic Section, Spotted Rock alteration involves increases in Si, Mg andFe, and depletions in Ti, K, Na and Ca. The Al has been locally remobilized resulting in restricted zones of enrichment and depletion. Main Conduit alteration of dacitic rock involved an increase in Si and depletions in Al, Ti, Fe, Mg, K, Na and Ca. Anthophyllite-Cordierite alteration did not affect the dacitic rocks. Thus three types of alteration affected the dacitic, andesitic, and presumably the rhyolitic rocks in the following ways.

1) The Si content was increased during both Spotted Rock and Main Conduit alteration, whereas Fe content was increased during Anthophyllite-Cordierite, and locally during parts of Spotted Rock alteration.

2) The Na, Ca, Fe, Mg and locally the Ti, Al and K contents were decreased to varying degrees during Main Conduit and parts of Spotted Rock alteration.

3) The Si, Al, Ti and Mg contents underwent little change during Anthophyllite-Cordierite alteration, and Al and occasionally K appear to have been locally remobilized during Spotted Rock alteration with resulting local enrichment and depletion.

4) Al and Ti were removed during the Main Conduit alteration and locally remobilized during Spotted Rock alteration as indicated by the wide compositional variation between these two alteration types in stratigraphically equivalent intervals.

The increased Fe, and the slightly increased Mg in Spotted Rock alteration of the Dacitic Pyroclastic Section were possibly derived from Spotted Rock alteration of the underlying Andesitic Pyroclastic Section. Some of the remobilized Fe and Mg which precipitated in the Dacitic Pyroclastic Section, are now present in biotite grains and/or veins and in garnet grains and clots due to later metamorphism. The Mg and Fe introduction was evenly distributed throughout the Spotted Rock alteration zones in the Dacitic Pyroclastic Section.

Zones of Spotted Rock alteration were moderately silicified, locally either enriched in, or leached of Fe and Mg, with a slight overall net gain in these elements, and leached of Na and Ca and locally also of K, Ti, Al and Mg. Zones of Main Conduit alteration were intensely silicified and leached with a net gain in Si and net losses of Al, Ti, Fe, Mg, K, Na and Ca.

Anthophyllite-Cordierite zones were slightly enriched in Fe and weakly depleted in K, Na and Ca. Because this type is developed only within the chemically similar, relatively Mg- and Fe-rich andesitic pyroclastic rocks this chemical change is extremely significant in interpreting the origin of anthophyllite-cordierite alteration. It appears that the loss of K, Na and Ca relative to the other cations (Mg, Fe) in this rock in effect concentrated Mg and Fe, thereby permitting metamorphic anthophyllite and cordierite to develop. It is emphasized that Anthophyllite-Cordierite alteration in the Central Area thus reflects Na, K and Ca depletion, rather than Mg enrichment as suggested by others for this rock type in other Archean massive sulphide deposits (de Rosen Spence, 1961; Gilmour, 1965; Froese, 1968; Whitmore, 1968; Upadhyay and Smitheringale, 1972).

It is suggested that the three alteration types formed initially from hot, corrosive hydrothermal solutions which permeated large volumes of porous volcanic pyroclastic footwall rocks, leaching Na, Ca, Fe, Mg and locally Ti, Al and K, silicifying lithic fragments, and depositing the vein and disseminated sulphides and oxides in zones of Spotted Rock alteration. With time, these upwelling solutions caused a reduction of porosity and permeability of the volcanic pile, mainly through silicification and cementation of pyroclastic fragments, thereby channeling the hydrothermal solutions into the more restricted Main Conduits. Despite its intensity, this hydrothermal activity was not pervasive and certain irregularly distributed pods within the footwall andesitic pyroclastic rocks remained only weakly altered, suffered only moderate losses of Na, Ca and K and were subsequently metamorphosed to form the Anthophyllite-Cordierite alteration zones.

Tuffite and Calc-Silicate Rock

The close spatial relationship in the Central Area of cherty and argillaceous tuffite with rhyolitic pyroclastic rocks suggests that these three rock types had the same source. This is further indicated by the chemical compositions of the cherty and argillaceous tuffite which approximate that of the rhyolitic pyroclastic rocks (Tables II, III). The chemical compositions of the cherty and argillaceous tuffite may reflect changes induced in the rhyolitic pyroclastic rocks through submarine weathering and depositional processes as they lay exposed on the sea floor during quiescent intervals in the cyclical volcanicity. Compositional variations in these rocks may reflect both the duration of their exposure to this weathering and the chemistry of the marine waters in the basin. The mineralogical composition and textures of these rocks result from later superimposed metamorphism.

The increased biotite content in the argillaceous tuffite as compared to the rhyolitic pyroclastic rocks, suggests that glauconitic material may have been an in situ precursor of the biotite (Burst 1958). Biotite-bearing argillaceous tuffite may thus have developed by diagenetic modification (and subsequent metamorphism) of an original montmorillonitic-kaolinitic volcaniclastic sediment to glauconite, and finally to biotite, due to abundance of ferrous iron from exhalative activity in the early sea floor environment.

Calc-silicate strata occur only in the Central Area and more specifically only in the Upper Rhyolitic Tuff Section. The most extensively developed calc-silicate rock, submember c in the Mineral Horizon, occurs intercalated in the sulphide lenses in the A and Jo zones. All calc-silicate rocks are stratabound, intimately associated with quiescent water-deposited, chemical-clastic sedimentary marker strata of chert, cherty tuffite, argillite, and argillaceous tuffite. Beyond the Central Area the calc-silicate rocks grade laterally to calcareous tuffite. The intimate relationship of the calc-silicate strata with other chemical-clastic sedimentary rocks indicates that the former also developed during quiescent intervals in cyclical volcanicity. Restriction of calc-silicate rock to the Central Area, and especially to portions of the Upper Rhyolitic Tuff Section which developed contemporaneously with commencement of the main episode of hotspring-fumarolic activity that formed the sulphide lenses, suggests a genetic relationship between this activity and development of the calc-silicate rocks. Mg, Fe, Ca, Al and Si in the calc-silicate strata were probably derived by leaching of volcanic rocks in the underlying vents by hydrothermal-fumarolic solutions which circulated during alteration and mineralization. Mg was derived from Spotted Rock alteration of andesitic rock; Ca, Fe and Si were derived from rocks of all three alteration types; Al may have been derived from rocks of the Main Conduit alteration zones and locally from the Spotted Rock alteration zones; some Ca and HCO_3 were derived from seawater (Kania, 1929). All these constituents were discharged into the marine environment and precipitated with the carbonate, locally becoming intermixed with varying amounts of felsic pyroclastic detritus (R.Y. Watanabe, pers. commun., 1971). As significant amounts of silica were passing into the submarine environment, some precipitated as chert, forming numerous chert layers intercalated with the Mineral Section. Later superimposed metamorphism produced the present calc-silicate assemblage.

SUMMARY OF THE GEOLOGIC HISTORY

The Central Area sulphide lenses (Fig. 20) were deposited during cyclical andesitic to rhyolitic volcanism. Mineralization developed during late-stage hotspring-fumarolic activity that accompanied the close of rhyolitic eruptions and marked the last major phase of volcanism. Deposition of rocks in the Central area took place in a submarine basinal environment as shown by the abundance of chert, calcareous tuffite and argillite intercalated with the volcanic rocks. The abundance of these rock types throughout the stratigraphic section not only reflects the submarine conditions, but also indicates quiescent periods of arrested or subdued volcanism, especially during development of the Mineral Horizon and the Late Upper Rhyolitic Tuff members. The intimate spatial relationship of chert, calcareous tuffite and argil-

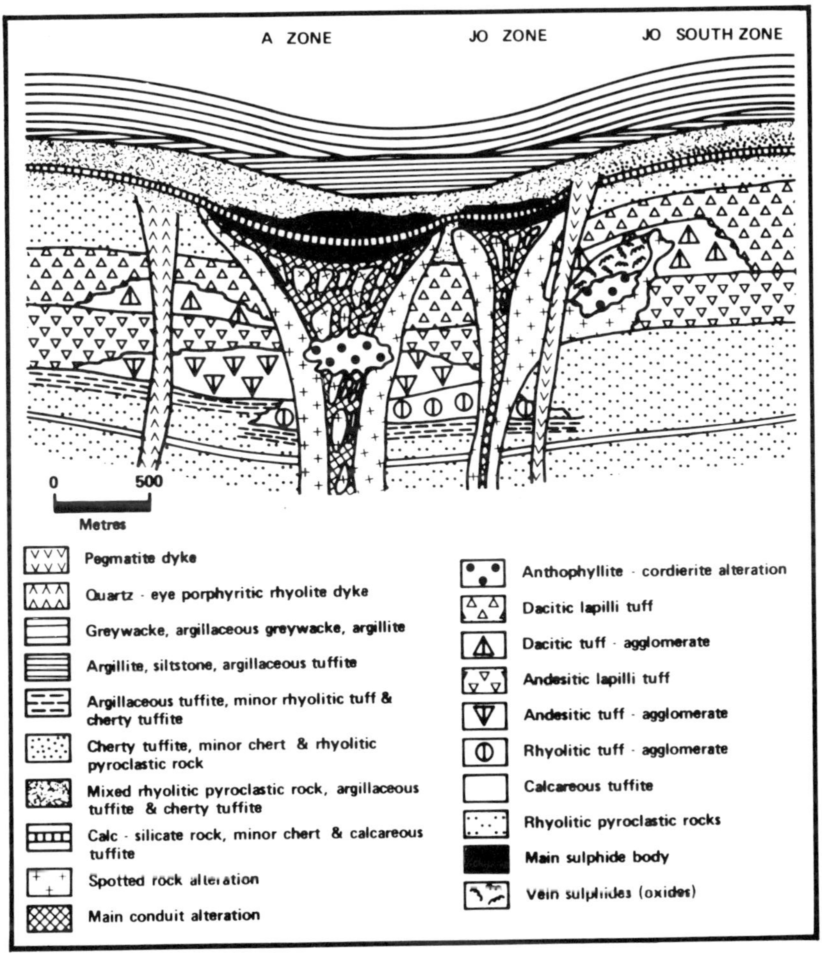

Figure 20. Diagrammatic Section through deposits of the Central Area.

lite with cherty tuffite, argillaceous tuffite and calc-silicate rock suggests that all these rock types were deposited more or less coevally and were affected by similar chemical-clastic sedimentary processes in part related to the exhalative fumarolic activity.

The rock succession in the Central Area reflects a continuous sequence of volcanic-sedimentary deposition. This proceeded sequentially from moderate rhyolitic pyroclastic activity, through cyclical andesitic-dacitic-rhyolitic pyroclastic activity, to complex, intermingled rhyolitic pyroclastic - clastic-chemical sedimentation, and finally to dominant clastic sedimentation. The sulphides were deposited in a trough or basin which extended from the A zone southeastward to the Jo zone. The basin thinned southeastward in the Jo zone to a minor shallow depression in the Jo South zone, and did not continue far northwest of the A zone. This basin did not develop as an important topographic feature until deposition of the Upper Rhyolitic Tuff Section.

The Lower Rhyolitic Pyroclastic Section developed in 3 stages. The initial stage, which produced the thickest and most extensive interval of rhyolitic pyroclastic rocks, involved protracted volcanic activity. Thin layers of chert, calcareous tuffite, argillaceous tuffite and two andesitic flows indicate brief interruptions in the otherwise continuous deposition of rhyolitic pyroclastic debris. The second stage was characterized by subdued to arrested volcanism. A thick calcareous tuffite and laterally extensive argillaceous tuffite were deposited. The third stage involved rhyolitic tuff-agglomerate eruptions. The coarse pyroclastic fragments and the lenticular shape of the rhyolitic tuff-agglomerate suggest areally restricted, possibly resurgent, explosive volcanism. The close of this eruptive activity terminated deposition in the Lower Rhyolitic Pyroclastic Section.

Deposition of the Andesitic Pyroclastic Section began with explosive volcanism that was partly contemporaneous with rhyolitic tuff-agglomerate eruption of the Lower Rhyolitic Pyroclastic Section. This resulted in mixing of andesitic and rhyolitic pyroclastic rocks in the footwall of the Jo zone, north of the A zone, and southwest of Bat Lake. There was no apparent time lapse between the rhyolitic and andesitic volcanism as no marker strata suggestive of quiescence such as chert, cherty tuffite, calcareous tuffite or argillaceous tuffite exist between the rhyolitic and the andesitic pyroclastic rocks.

The gradation of the andesitic tuff-agglomerate upward into overlying andesitic lapilli tuff suggests that the initial explosive volcanism gradually waned. Abundant biotite in the matrix of certain fragmental rocks of the Andesitic Pyroclastic Section also indicates increasing quiescence, allowing simultaneous deposition of volcaniclastic ash with the andesite pyroclastic debris. Minor andesitic flows are also present. The thickest accumulation of andesitic pyroclastic debris is north of the A zone, possibly indicating the location of the source vent. The rhyolitic tuff-agglomerate of the Lower Rhyolitic Pyroclastic Section probably erupted from separate vents, as material of such contrasting composition would not likely have originated simultaneously from the same vent.

The Dacitic Pyroclastic Section was deposited during a brief period of explosive volcanism. Two thick dacitic tuff-agglomerate bodies near the base of this section indicate the possible location of source vents. In the Jo South zone, sulphide veinlets and hydrothermal alteration in the dacitic tuff-agglomerate indicate that this volcanism was accompanied by periods of hotspring-fumarolic activity and sulphide deposition. Sulphide deposition in the Jo South zone thus preceded the main period of mineralization in the A and Jo zones.

The Upper Rhyolitic Tuff Section also developed in three stages. The first, represented by the Early Upper Rhyolitic Tuff member, was characterized by moderate to weak volcanic eruptions and some hotspring-fumarolic activity. The second, now represented by the Mineral Horizon member, contains the main sulphide lenses and was characterized by arrested volcanism, but extensive hotspring-fumarolic activity. The third, represented by the Late Upper Rhyolitic Tuff member, was characterized by very subdued to arrested volcanism and extensive periods of weak sedimentation and submarine weathering involving reworking of deposited pyroclastic debris. Although volcanism was very subdued in the Central Area, it continued in the immediately adjacent areas as indicated by the abundance here of rhyolitic pyroclastic rocks. However, argillaceous tuffite and cherty tuffite layers in the rhyolitic pyroclastic rocks adjacent to the Central Area also indicate waning volcanism in these areas. Most of the rhyolitic pyroclastic material in the Upper Rhyolitic Tuff Section of the Central Area was derived from this adjacent volcanism.

Lithologic compositions in the Upper Rhyolitic Tuff Section vary both vertically and laterally. The lateral variations are related to changes in the original paleotopography, and indicate a paleobasin or trough in the A zone extending southeastward to the Jo zone. This basin was important for ponding the fluids and accumulating the A and Jo zone sulphide lenses, otherwise, the solutions might have spread out and been dissipated into areas west of A zone and east of Jo zone where the basin was shallower and less confined.

Deposition of the Sedimentary Section followed cessation of volcanism. Argillite and siltstone in the Basal member locally contain significant concentrations of graphite, pyrite and pyrrhotite, especially in the A and east Jo zones, further indication of a restricted basin in this area. The Main member of the Sedimentary Section contains mainly greywacke and argillite. Rocks of the Sedimentary Section are conformable with those of the underlying Late Upper Rhyolitic Tuff member and extend widely beyond the Central Area, indicating the end of volcanism throughout the region.

COMPARISON OF A, EAST CLEAVER LAKE AND BOOT LAKE ZONES

The A zone comprises the largest known sulphide body on the Bathurst Norsemines property. A diagrammatic section shows its most prominent features (Fig. 6). Massive sulphides are thickest in the Mineral Horizon member above the probable main vent area. Here they consist of a massive, barren iron sulphide core which thins outward to a less massive fringe zone. Within the core and immediately adjacent fringe, all four submembers of the Mineral Horizon are present (Fig. 21). However on the outer fringe (Fig. 22) only submembers a, b and c are present.

The East Cleaver Lake deposit occurs on the nose of an overturned anticline (Fig. 23) and the originally stratiform sulphide body is now highly contorted and remobilized as a result of deformation within the fold hinge. The combination of folding, shearing and remobilization, reshaped the originally flat-lying sulphide lens into an overturned, steeply plunging structure with the stringer mineralization and alteration pipe resting structurally above it.

In addition to the resulting structural dissimilarity between the A and East Cleaver Lake zones, there is a difference in the rock types stratigraphically above the latter, where dacitic-andesitic pyroclastic rocks and flows are common, and a difference in their sulphide mineral zoning. Unlike the A zone, the Mineral Horizon

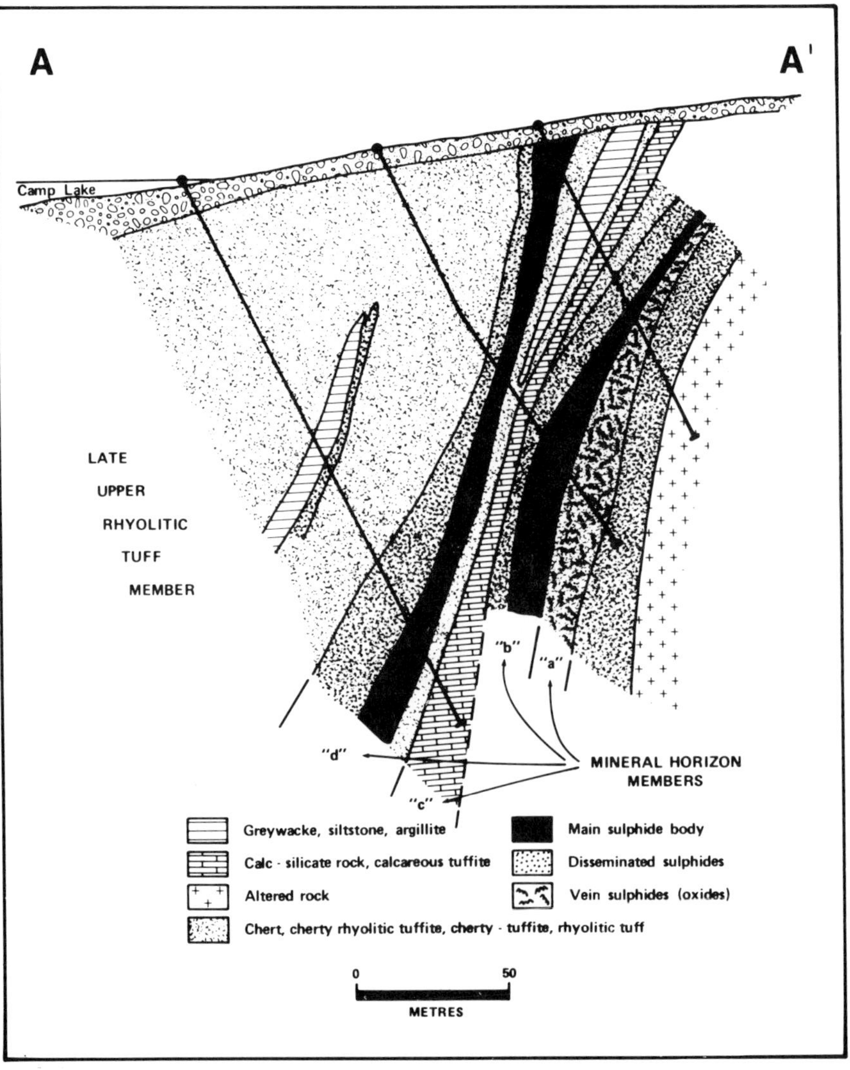

Figure 21. Drill section A-A′ (Fig. 3) through core of A zone.

member in the East Cleaver Lake zone does not contain all four submembers (Fig. 24). Submember a, the footwall stringer zone, is present, and its base and precious metal contents are similar to, or slightly lower than those of A zone. Submembers b and d are usually indistinguishable in the East Cleaver Lake zone as Submember c is only sporadically developed. There is no barren pyrite core. The East Cleaver Lake sulphide body now is a single, fairly homogeneous sulphide lens consisting of pyrite, sphalerite and galena.

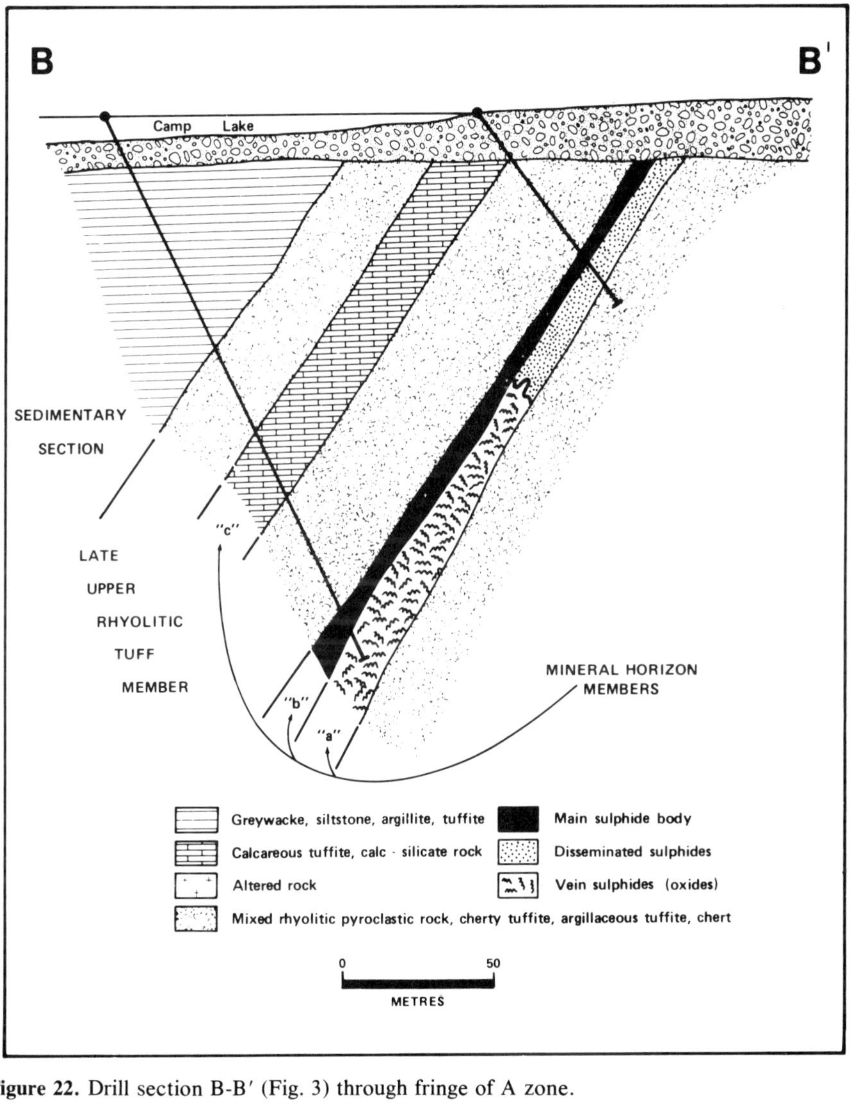

Figure 22. Drill section B-B′ (Fig. 3) through fringe of A zone.

The Boot Lake Zone (Figs. 23, 25) is the most recent discovery on the property. There are three major differences between the Boot Lake and A zones. Firstly, submember a of the Mineral Horizon, the footwall alteration and associated stringer zone, is smaller at Boot Lake. Secondly, there is only minor development of the Late Upper Rhyolitic Tuff member and as a result, the sulphides of Boot Lake zone occur much closer to the Sedimentary Section, and are interbedded with argillaceous tuffite and argillite. Thirdly, submembers b and d are indistinguishable as submember c cannot be identified in the Boot Lake zone. Sulphides occur both as discrete, small lenses, and as continuous layers of pyrite, all containing variable amounts of pyrrhotite, sphalerite, galena and chalcopyrite.

SUMMARY

1) The A and Jo zones occur in the same stratigraphic interval, comprise conformable massive sulphide lenses, and are underlain by altered rocks. The Jo South zone occurs lower in the stratigraphic succession and consists primarily of sulphide veins.

2) The A and Jo zone sulphide lenses are hosted by altered rhyolitic pyroclastic rocks, chert, cherty tuffite, argillaceous tuffite and calc-silicate rock which occur at the base of a transition zone between dominantly volcanic and dominantly sedimentary strata. The volcanic rocks underlying the sulphide lenses comprise, from base to

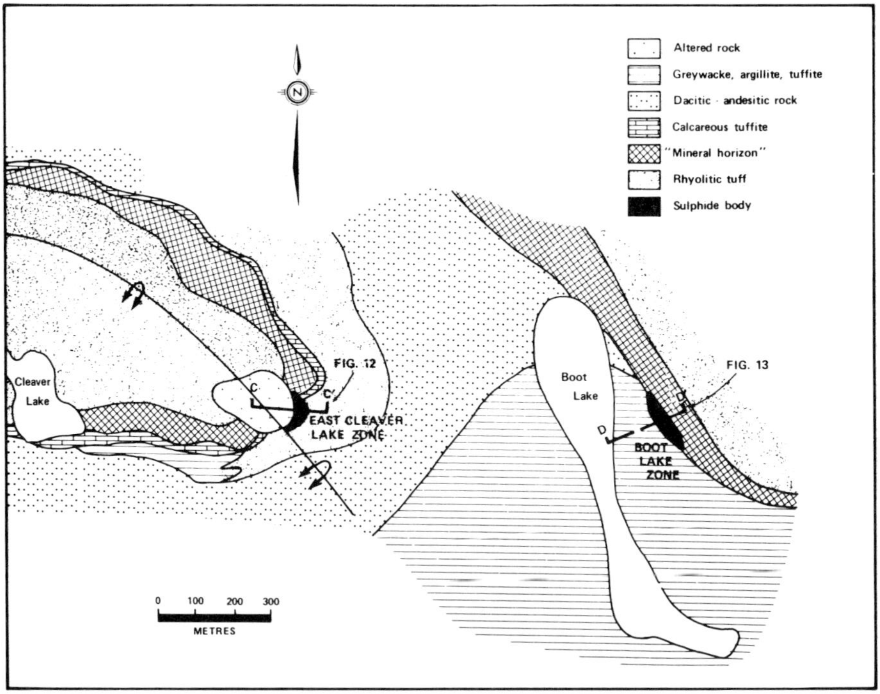

Figure 23. Geologic map of East Cleaver Lake and Boot Lake zones.

top, a sequence of rhyolitic, andesitic, dacitic and rhyolitic pyroclastic rocks. The upper part of the transition zone overlying the sulphide lenses comprises mixed and intercalated rhyolitic pyroclastic rocks, cherty tuffite and argillaceous tuffite with minor argillite, chert and calcareous tuffite. Sedimentary rocks overlying the transition zone are mainly greywacke, argillaceous greywacke and argillite. The Jo South zone mineralization occurs in altered dacitic pyroclastic rocks.

3) The A and Jo zones are vertically and laterally zoned from an upper stratabound pyrite-minor pyrrhotite-rich lens with fringes that contain varying proportions of sphalerite, galena, chalcopyrite and silver, to an underlying vein and

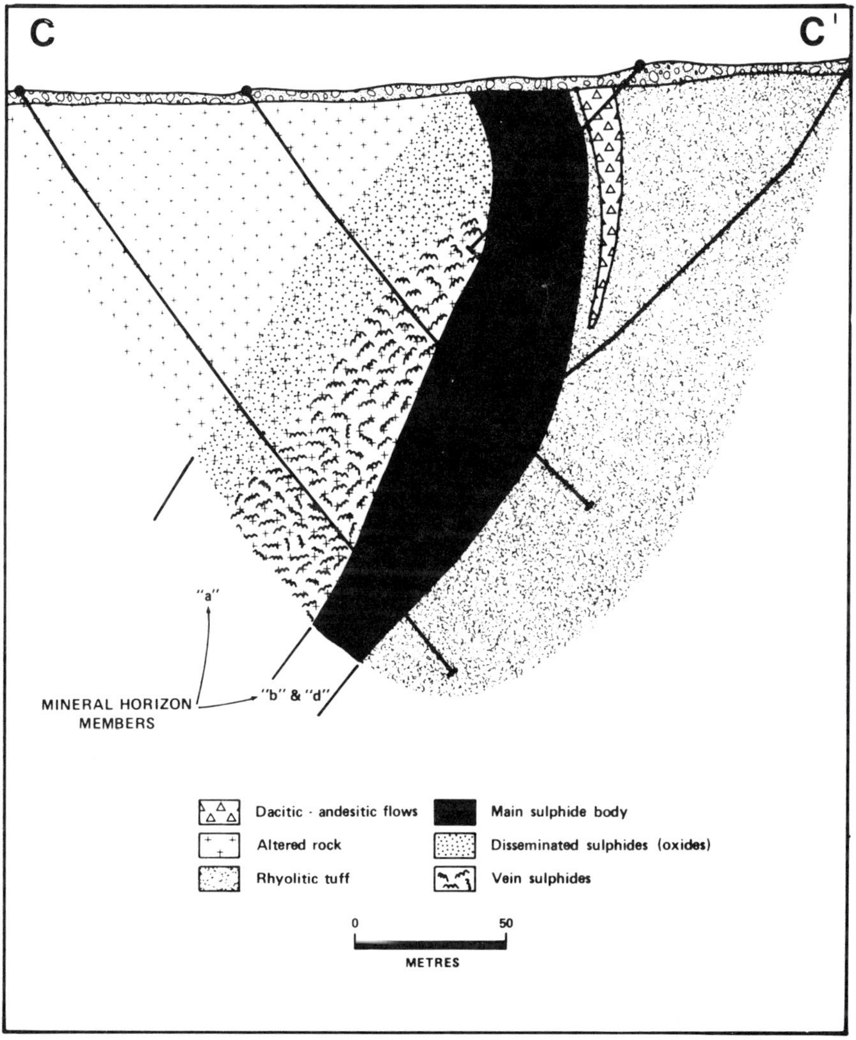

Figure 24. Drill section C-C′ (Fig. 11) through East Cleaver zone.

disseminated zone of chalcopyrite, pyrite, pyrrhotite and magnetite. The Jo South zone is weakly zoned with the upper veins containing primarily sphalerite, galena and pyrite and the lower veins containing chalcopyrite and pyrite.

4) Alteration of the volcanic rocks underlying the Central Area sulphide zones is of three types designated Spotted Rock, Main Conduit and Anthophyllite-Cordierite alteration. These are spatially related, often compositionally zoned from one to another, and range from weakly to strongly leached and silicified. Anthophyllite-Cordierite and Main Conduit alteration are analogous to the anthophyllite-cordierite

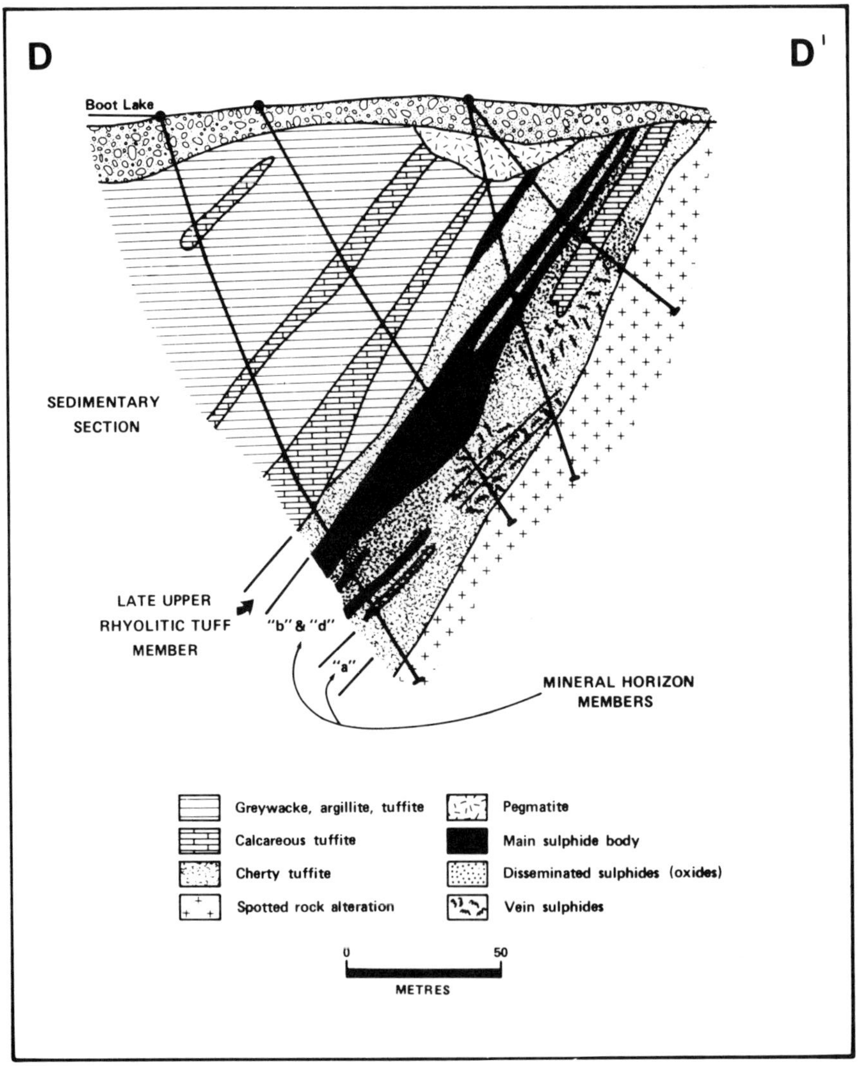

Figure 25. Drill section D-D′ (Fig. 11) through Boot Lake zone.

and quartz-sericite zones, respectively, in other Archean deposits. Spotted Rock alteration represents a weakly developed equivalent to the zones of chloritization in many Archean deposits.

5) Moderate to severe Spotted Rock and Main Conduit alteration of the footwall volcanic rocks resulted in most cases in the removal of Na, Ca, Fe, Mg and locally Ti, Al and K and the addition of Si and locally Fe, Cu, Zn, Ag, and Au. Irregularly distributed pods of Anthophyllite-Cordierite alteration represent local zones of weakly altered andesitic rock which underwent only losses of Na, Ca and K and minor additions of Si, Cu and Fe. The Si, Fe, Zn, Cu, Pb, Au and Ag are thought to have been derived from an external source whereas the Na, Ca, Fe, Mg, Ti, Al and K have been remobilized and re-distributed within the alteration zones.

6) Anthophyllite-Cordierite alteration represents Na, K and Ca depletion in andesitic volcanic rock rather than Mg enrichment as suggested for similar altered rock in other Archean massive sulphide deposits. Spotted Rock and Main Conduit alteration represent zones which were moderately and intensely silicified and leached, respectively. Mg, Fe, Al, Ti and K which were variably remobilized within the alteration zones, were locally reconcentrated in the Spotted Rock zones.

7) Alteration occurs only in the immediate footwall rocks of the A and Jo zone stratiform sulphide lenses.

8) Alteration of the footwall volcanic rocks and development of the sulphide bodies with the volcanic strata, and particularly with stratified rocks deposited in a relatively quiescent marine environment, strongly indicate that formation of the sulphide lenses was an integral part of the volcanism, but took place in a submarine environment during a cessation of eruptive activity.

10) The calc-silicate strata in the Central Area, which occur only within or near sulphide lenses, and the chert strata, which are best developed within the same areas as the calc-silicate horizons, are chemical sedimentary rocks which were deposited from hotspring-fumarolic activity.

11) The compositions of the argillaceous tuffite and cherty tuffite have resulted from metamorphism of products developed from the rhyolite pyroclastic units by diagenetic submarine weathering. These processes, together with additions of chemical and/or clastic sediment, affected the rhyolitic pyroclastic sediments as they lay exposed on the sea floor during quiescent intervals in the cyclical volcanism.

12) The present mineralogy and texture of the altered and unaltered rocks in the Central Area are the result of later cordierite-amphibolite facies metamorphism.

13) The Boot Lake and East Cleaver Lake deposits have many features in common with the Central Area deposits. All occur in the same transition zone between dominantly volcanic and sedimentary strata and all have similar footwall rocks, alteration features and metal zoning. The hanging wall rocks overlying the East Cleaver Lake deposit differ from those of the other two deposits because they consist of rhyolitic pyroclastic rocks and andesitic flows and pyroclastic rocks rather than a mixture of rhyolitic pyroclastic and chemical-clastic sedimentary rocks.

ACKNOWLEDGEMENTS

Much of the information in this paper is derived from an M.Sc. thesis prepared by Casselman for the Department of Geology at Carleton University, Ottawa, Ontario (Casselman, 1977). The thesis investigation was in part funded by Cominco Ltd. who supplied all of the chemical analyses and thin sections and in part by Dr. W. M. Tupper of the Geology Department at Carleton from his NSERC Operating Grant. The authors wish to acknowledge the contribution of all those geologists who worked at the Bathurst Norsemines deposit over the years and who provided ideas, in particular W.T. Irvine, G.J. Ryznar, R.Y. Watanabe, H.P. Wilton, A.J. Boronowski, K.M. Carter, D.W. Moore, M.J. Osatenko and J.F. Harris. H.C. Morris offered constructive criticism of an earlier draft of this paper. The drafting of Sharon Weller, and the typing of Pamela Llewelyn and Anita Rozenhart, all of Cominco Ltd., are gratefully acknowledged.

REFERENCES

Burst, J.R., 1958, Mineral heterogeneity in glauconite pellets: American Mineralogist, v. 43, p. 481-497.

Casselman, M.J., 1977, Petrology and Alteration of the Bathurst Norsemines Central Area deposits, Northwest Territories: Unpublished M.Sc. Thesis, Carleton University, Ottawa, 217 p.

De Rosen-Spence, A., 1961, Genese des roches a cordierite-anthophyllite des gisements cupro-zincifères de la region de Rouyn-Noranda, Quebec, Canada: Canadian Journal of Earth Sciences, v. 6, p. 1339-1345.

Energy Mines and Resources, Canada, 1980, Canadian mineral deposits not being mined: in National Mineral Inventory, Mineral Policy Sector, Internal Report, M.R.I. 80-7.

Froese, E., 1968, Metamorphic rocks from the Coronation Mine and surrounding area: in Byers, A.R., ed., Symposium on the Geology of the Coronation Mine, Saskatchewan: Geological Survey of Canada Paper 68-6, p. 55-79.

Gilmour, P.C., 1965, The origin of the massive sulphide mineralization in the Noranda district, Northwestern Quebec: Proceedings Geological Association of Canada v. 16, p. 63-81.

Henderson, J.B., 1970, Stratigraphy of the Archean Yellowknife Supergroup, Yellowknife Bay - Prosperous Lake area, District of Mackenzie: Geological Survey of Canada Paper 70-26, 12 p.

Kania, J.E.A., 1929, Precipitation of limestone by submarine vents, fumaroles and lava flows: American Journal of Science, v. 218, p. 347-359.

Roscoe, S.M., 1965, Geochemical and isotopic studies, Noranda and Matagami areas: Canadian Institute of Mining and Metallurgy, Transactions, v. 68, p. 285-297.

Sangster, D.F., 1972, Precambrian Volcanogenic Massive Sulphide Deposits in Canada, A Review: Geological Survey of Canada Paper 72-22, 44 p.

Stockwell, C.H., 1961, Structural provinces, orogenies, and time classification of rocks of the Canadian Precambrian Shield: in Lowdon, J.A., compiler, Age Determinations by the Geological Survey of Canada, Report 2, Isotopic Ages: Geological Survey of Canada Paper 61-17, 127 p.

Upadhyay, H.D. and Smitheringale, W.G., 1972, Geology of the Gullbridge Deposit, Newfoundland: Volcanogenic sulphides in cordierite-anthophyllite rocks: Canadian Journal of Earth Sciences, v. 9, p. 1061-1973.

Whitmore, D.R.E., 1968, Geology of the Coronation copper deposit: in Byers, A.R., ed., Symposium on the Geology of Coronation Mine, Saskatchewan: Geological Survey of Canada Paper 68-5, p. 37 55.

Winkler, H.G.F., 1967, Petrogenesis of Metamorphic Rocks, Second Edition: New York, Springer-Verlag, 235 p.

MS received April 1979;
Accepted for publication February 1981.

Precambrian Sulphide Deposits, H.S. Robinson Memorial Volume, edited by R.W. Hutchinson, C.D. Spence and J.M. Franklin, Geological Association of Canada Special Paper 25, 1982

GEOLOGY OF THE AGNICO-EAGLE GOLD DEPOSIT, QUEBEC

E.S. Barnett
Cominco Limited, Suite 1700, 120 Adelaide St. W.,
Toronto, Ontario M5H 1T1
R.W. Hutchinson
Department of Geology, University of Western Ontario,
London, Ontario N6A 5B7
Anton Adamcik
Agnico-Eagle Mines Limited,
Post Office Box 310, Joutel, Quebec J0Y 1N0
R. Barnett
Department of Geology, University of Western Ontario,
London, Ontario N6A 5B7

ABSTRACT

Unlike most Archean gold deposits, Agnico-Eagle has many similarities to massive base metal sulphide deposits, suggesting a common volcanogenic origin. The gold is contained within the sulphide facies of a stratabound to stratiform carbonate-sulphide-silicate-oxide facies iron-formation which immediately overlies a sequence of partially welded felsic tuff and lapilli-tuff. Within the ore-bearing horizon are interdigitating, partially welded felsic tuff-breccia, and intercalated cherty sedimentary rocks and waterlain tuff. A carbonaceous schist containing pyrite bands and nodules occurs immediately overlying the ore zone.

The ore-bearing sequence is distinctly zoned with an outward change from an iron silicate facies exhalite at the centre of the ore body to iron carbonate facies exhalite. A diabase dyke transects the former. The footwall rocks near the diabase and beneath the silicate facies exhalite are enriched in chlorite and pyrite. Although chalcopyrite and sphalerite are present only in accessory quantities, analytical data indicates that copper and zinc exhibit both lateral and vertical zonation. There is also a distinct sequential upward change in the composition of the carbonate facies exhalite from siderite-rich to ankerite-rich. In addition, there is a marked upward increase in the manganese and magnesium content of both siderite and ankerite. The only important ore mineral is native gold with a fineness of approximately 830. The gold occurs as microscopic-sized inclusions and veinlets in and around pyrite in the carbonate facies exhalite, and pyrite and pyrrhotite in the silicate facies exhalite. It is proposed that like many massive base metal sulphide bodies, the Agnico-Eagle deposit formed by volcanogenic and exhalative processes. Fluids migrated up a tensional fracture into a local reduced basin, possibly within a collapse caldera following explosive felsic volcanism. As a result, footwall rocks were altered along the channelway by fluid movement, and chemical sedimentary strata, including the gold ore, were deposited around the vent on the seafloor. These strata were themselves altered and chloritized near the vent, but as subsequent metamorphism was minor, primary relationships are well preserved. The feeder fracture was subsequently intruded by a diabase dyke with resulting minor thermal metamorphic effects.

RÉSUMÉ

A la différence de la pluspart des dépôts d'or archéens, le gisement d'Agnico-Eagle présente bien des similitudes avec les dépôts de sulfures de métaux de base massifs, suggérant une commune origine volcanogénique. L'or est contenu dans le faciès sulfuré d'une formation ferrugineuse, liée aux strates à stratiforme, et à faciès de carbonate-sulfure-silicate-oxyde. Cette formation recouvre une séquence de tufs felsiques partiellement soudés et de lapilli-tufs. A l'intérieur du niveau renfermant la minéralisation on rencontre des interdigitations de tufs bréchiques felsiques partiellement soudés ainsi que des intercalations de sédiments cherteux et de tufs déposés en milieu aqueux. Un schiste charbonneux à bandes et nodules de pyrite recouvre la zone mineralisée.

La séquence minéralisée est distinctement zonée, passant d'une exhalite ferrugineuse à faciès silicaté au centre du gisement à un faciès carbonaté vers l'extérieur. Un dyke de diabase recoupe le faciès silicaté. Les roches du mur, près de la diabase et sous l'exhalite à faciès silicaté, sont enrichies en chlorite et pyrite. Bien que chalcopyrite et sphalérite soient présentes seulement en quantités accessoires, les données analytiques montrent que le cuivre et le zinc sont répartis tous deux selon une zonalité latérale et verticale. Il y a aussi un changement séquentiel vers le haut dans la composition de l'exhalite à faciès carbonaté, de riche en sidérose vers le bas à riche en ankérite vers le haut. De plus, il y a un accroissement marqué vers le haut des teneurs en manganèse et magnésium de la sidérose et de l'ankérite. Le seul minerai important est l'or à l'état natif et d'une finesse d'environ 830. Cet or apparait en inclusions microscopiques et en veinules autour de et dans la pyrite de l'exhalite à faciès carbonaté et de la pyrite et pyrrhotite de l'exhalite à faciès silicaté. Il est donc proposé que comme bien des dépôts de sulfures massifs, le dépôt d'Agnico-Eagle ait été formé par des processus volcanogéniques exhalatifs. Les fluides minéralisés migrèrent vers le haut le long d'une fracture de tension et se déversèrent dans un bassin local en milieu réduit, peut-être à l'intérieur d'une caldéra formée par un épisode de volcanisme acide explosif. Les roches au mur du gisement furent altérées par le mouvement de ces fluides le long de leur conduit et les strates chimiques incluant la minéralisation d'or furent déposées sur le fond marin autour de l'orifice de la source thermale. Ces strates furent elles-même légèrement altérées et chloritisées près de cet orifice. Le métamorphisme subséquent a été mineur et en conséquence ces relations primaires ont été bien préservées. La fracture nourricière fut occupée ultérieurement par le dyke de diabase qui causa un léger métamorphisme thermal.

INTRODUCTION

One of Canada's newest gold mines is the Agnico-Eagle, located near the village of Joutel, 144 km north-northeast of Noranda and 58 km southwest of the town of Matagami in northwestern Quebec (Fig. 1). The Agnico-Eagle deposit differs from most Archean lode gold deposits of northern Ontario and Quebec in that the gold does not occur in veins. Instead it is contained within sulphide-rich parts of a stratabound to stratiform carbonate-sulphide-silicate-oxide facies iron-formation. The deposit is of particular interest because, unlike other heavily pyritic sulphide deposits of the district, it contains only trace amounts of base metals and is mined for its gold and associated silver content alone. Nevertheless, it exhibits many geological characteristics present in massive base metal sulphide deposits. In light of current views on the origin of the latter, this comparison is pertinent to an interpretation of the genesis of the Agnico-Eagle deposit.

Investigations leading to the discovery of the deposit began in February 1962 during early exploration of the Joutel-Poirier district for base metal sulphide deposits. Ground geophysical surveys outlined coincident magnetic and electromagnetic anomalies which were tested by diamond drilling in late spring of the same year. Exploration continued and by 1967 it was apparent that a potentially mineable deposit of auriferous pyrite existed, and underground development was initiated. Regular and sustained production began in 1974, and as of December 31, 1977, the mine had produced 1.1 million tonnes of ore from which were recovered 7,780 kg of gold and 2,010 kg of silver. At that time reserves were estimated to be 1.45 million tonnes of ore containing 8.0 g of gold per tonne.

GEOLOGICAL SETTING

The Agnico-Eagle gold mine is situated in Joutel Twp., in the Abitibi Greenstone Belt of the Superior Province in the southeastern portion of the Canadian Shield. The Joutel area is underlain by a thick sequence of Archean volcanic rocks of the Gale Group (Latulippe, 1976), which have been isoclinally folded to near vertical attitudes, and metamorphosed to a greenschist facies. The rocks and major structures have a regional east-west trend. A local divergence from this trend occurs in the Joutel area where the formations strike parallel to the northwesterly flowing Harricana River (Fig. 2).

Felsic volcanic rocks are confined to only a few areas within the Gale Group, as in Joutel and neighbouring Poirier and Valrennes townships where they are abundant. The orebody is situated at the upper, northern contact of a thick felsic to intermediate volcanic pile. The upper 600 to 700 m of this sequence are composed predominantly of felsic pyroclastic rocks which form the footwall of the orebody. The ore-bearing sequence is composed primarily of chemical sedimentary rocks with lesser quantities of pyroclastic rocks. The hanging wall rocks are a variety of chemical and clastic sedimentary rocks, most of which crop out very sparingly and underlie the Harricana River. North of the river, the Gale Group consists primarily of mafic volcanic rock, often pillowed, amygdaloidal and less commonly variolitic and porphyritic (Remick, 1969). Although not widespread, a spinifex-textured ultramafic rock was encountered in a few drill holes near the base of this sequence (C. D. Spence, pers. commun.).

Stratigraphic tops throughout Poirier and most of Joutel Twps. are to the north or north-east as indicated by pillow structures and graded bedding. In the northern portion of Joutel Twp. pillow tops face south. The Agnico-Eagle deposit is situated on the south limb of an east-west to northwesterly trending syncline.

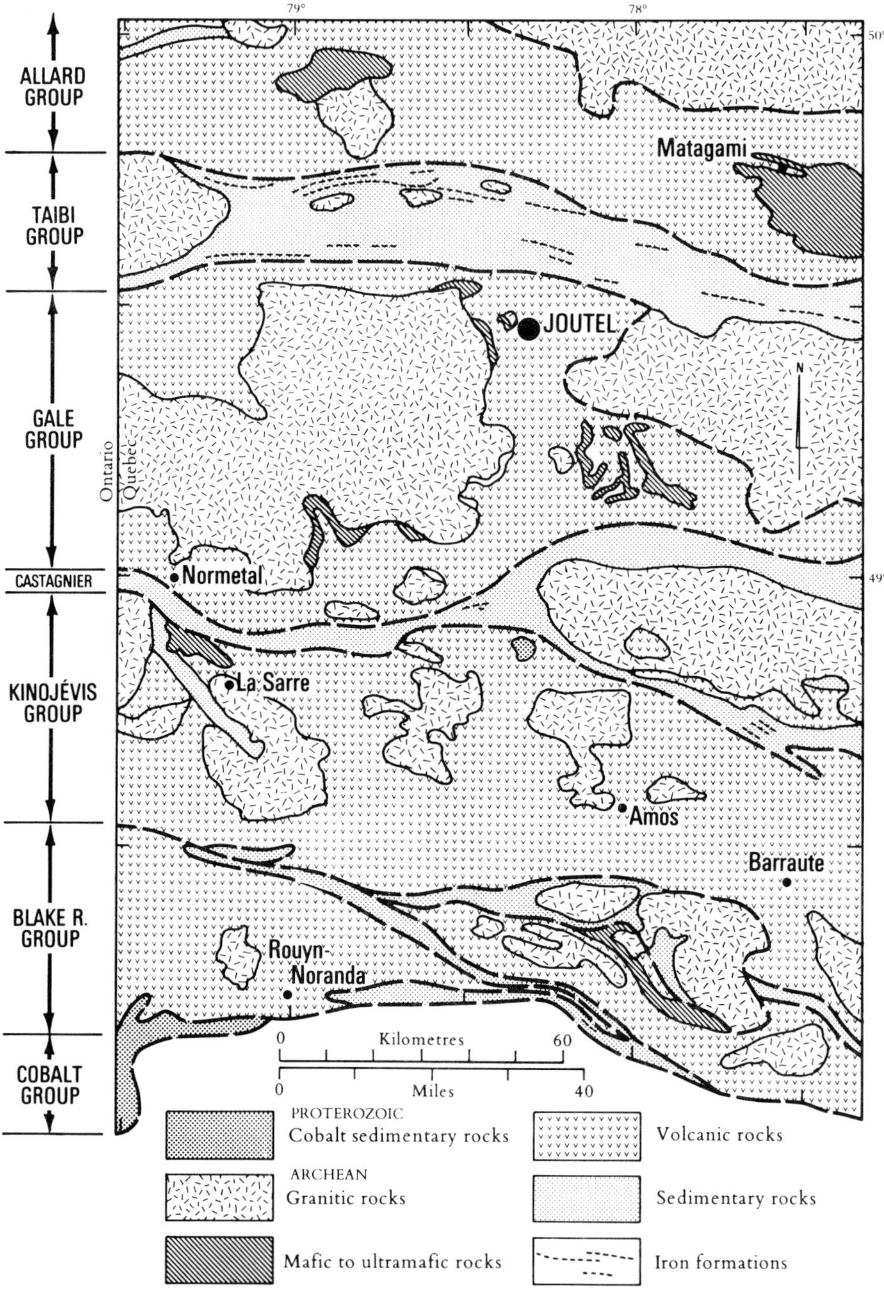

Figure 1. Regional geology and location map.

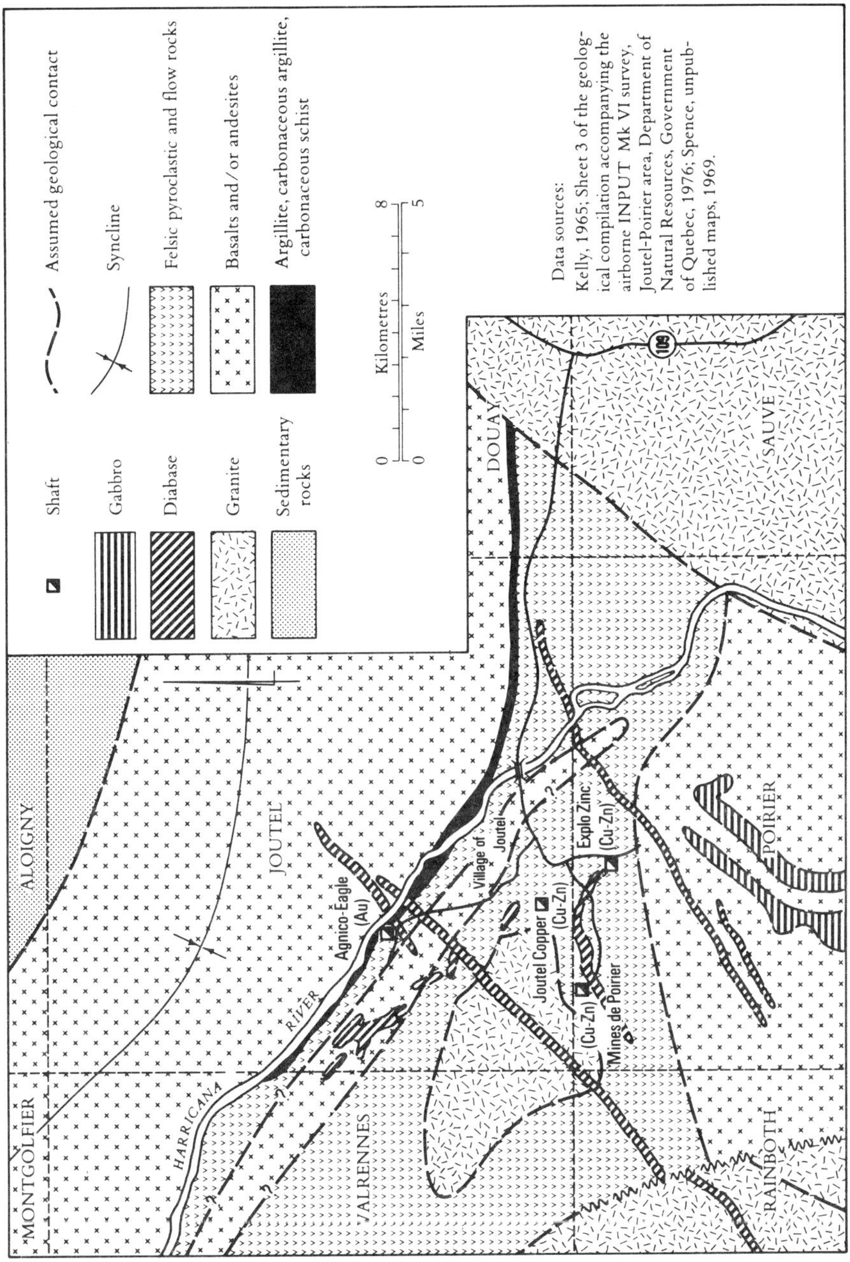

Figure 2. Joutel area geology.

Numerous, irregularly shaped dioritic and gabbroic sills and dykes, have intruded the volcanic and sedimentary rocks, and are believed by Remick (1969) to be coeval with volcanism. Granitic batholiths flank all these rocks in Joutel and Poirier Twps., both to the east and west. Hybrid diorite and diorite are common at the margins of the major granitic bodies, and similar smaller granitic stocks and plugs are numerous. The area is intruded by several large northwesterly trending diabase dykes, one of which transects the Agnico-Eagle deposit (Plate I, Fig. 12).

GEOLOGY OF THE AGNICO EAGLE DEPOSIT

The Agnico-Eagle orebody is an irregularly shaped but concordant lens of pyritic metasedimentary rock of mixed chemical and volcaniclastic origin. It strikes northwest and dips very steeply to the northeast. It ranges from a few to 600 m in length on the present mine levels and is from 0 to 20 m wide, averaging about 4.5 m. It has been traced from surface, where it is discontinuous, short and narrow, to a vertical depth of 760 m on the deepest mine level, and deeper exploration has shown that it continues downward, although its full extent is not known.

The stratigraphic succession of rock units near the orebody is as follows:

hanging wall sedimentary rocks (northeast)

carbonaceous schist

pyroclastic tuff-breccia; intercalated with exhalite

carbonate facies exhalite; sideritic-ankeritic, grading to ore where sulphides $> 20\%$

silicate facies exhalite; chloritic, grading to ore where sulphides $> 20\%$

footwall tuffs (southwest)

The mine geology is illustrated by a series of mine level plan maps (Fig. 3). Stratigraphically, the lowest rocks exposed in the mine workings, in this study called the footwall tuffs, are a series of felsic pyroclastic volcanic rocks ranging from tuff, which predominates, to minor lapilli-tuff, using the classification of volcaniclastic rocks proposed by Fisher (1966). These lie immediately beneath the ore-bearing strata and are traceable along the entire strike length of the mine. In plan, the upper contact of these tuffs appears broadly domal in outline. Megascopically, the tuffs are typically pale yellowish-green and slightly schistose. Characteristically they are composed of a fine-grained matrix with less than 10 per cent rock fragments and crystal clasts. The matrix consists of ash-sized quartz, feldspar, sericite and carbonate with accessory chlorite, epidote, pyrite, sphalerite and apatite. Chlorite is usually absent but is more abundant in what appears to be more altered varieties of the footwall tuffs, especially those underlying the silicate facies exhalite near the diabase dyke, where in small patches, chlorite may constitute up to 40 per cent of the rock by volume. Pyrite is ubiquitous in trace amounts, but like chlorite, it too is more abundant in the footwall tuffs near the diabase where it may constitute up to five per cent of the rock by volume.

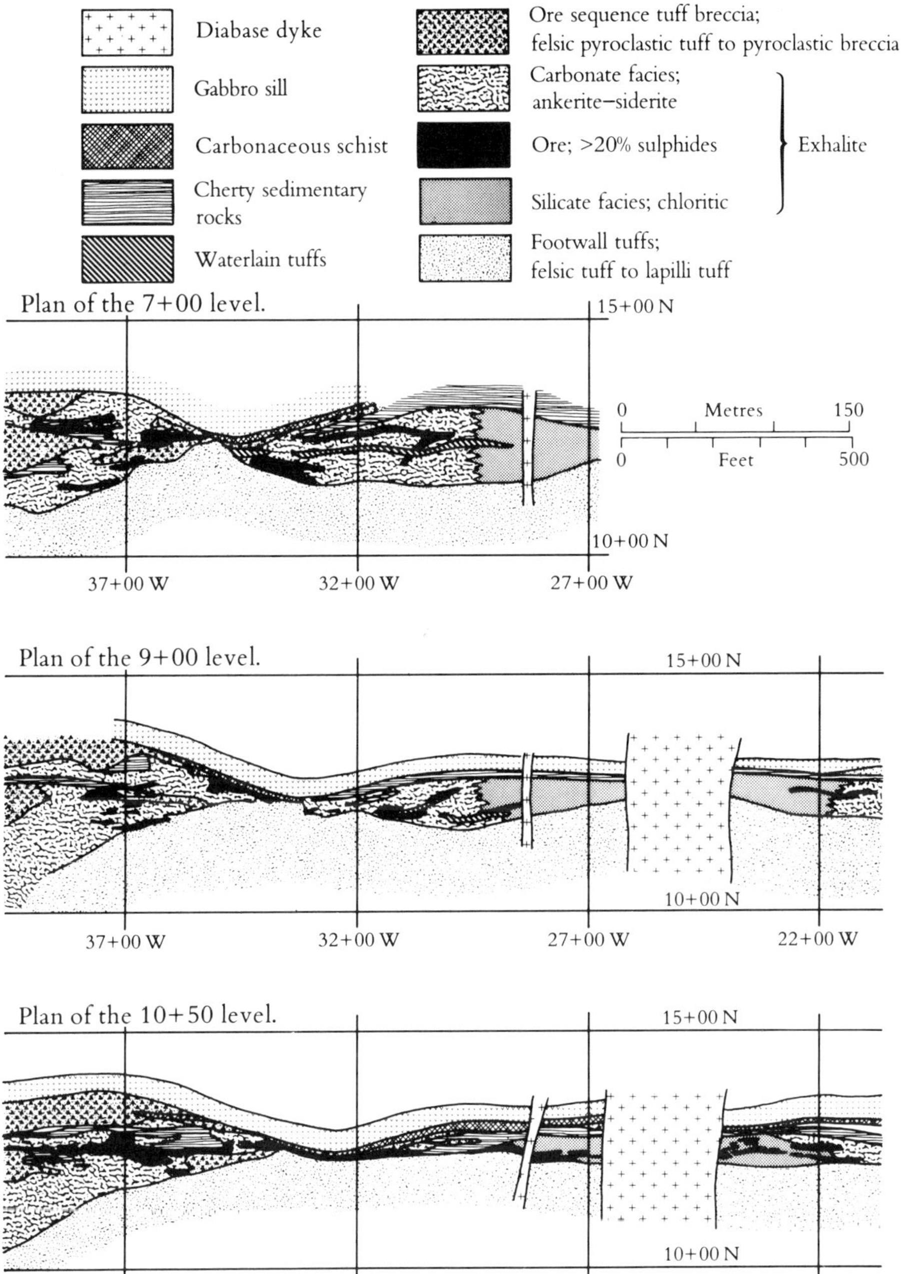

Figure 3. Geology of the Agnico-Eagle gold mine.

Rock fragments within the footwall felsic pyroclastic rocks locally constitute up to 50 per cent of the rock by volume. Fragments seldom exceed 1 cm in maximum dimension and are usually 1 to 4 mm in length. Fiamme (Zavaritsky, 1947, cited in Ross and Smith, 1961) are by far the most abundant fragment type. They range in colour from light greenish-yellow to black. The lighter coloured fiamme are mineralogically very similar to the matrix but the relative abundance of their component minerals is different. Fiamme typically contain less carbonate than the matrix. Black fiamme are much less common and are restricted to the upper portions of the unit; none were observed below about 3 m beneath the upper contact. Fiamme frequently contain ovoid amygdules of quartz that appear to have formed as vesicle fillings in pumiceous fragments. The only other fragment-type present is polycrystalline quartz clasts, probably originally chert that has since recrystallized to quartz. Quartz and feldspar crystal clasts or phenocrysts, although locally abundant, are sporadically distributed within the felsic pyroclastic volcanic rocks. Quartz crystal clasts seldom exceed 2 mm in diameter and only rarely can crystal faces be discerned. Plagioclase crystal clasts are typically smaller and less abundant than those of quartz. Twinning is common, though untwinned crystals are also present.

Whether the observed textures represent welding in the pyroclastic footwall tuffs is difficult to ascertain due to the effects of later diagenesis and metamorphism. Nevertheless some evidence of partial welding is present and is most intense approximately 3 to 10 m beneath the upper contact. Welding is suggested by fiamme which are bent around and appear welded against each other, suggesting that the fragments were still plastic when deposited.

Draped over the footwall tuffs is the ore bearing sequence, a carbonate-sulphide-silicate-oxide facies iron-formation, herein termed "exhalite" (Ridler, 1973) which is subdivided into two distinct, mappable units designated the carbonate and silicate facies of exhalite. The silicate facies is found near the centre of the ore bearing sequence and can be traced along strike in both directions through a gradational transition zone less than 10 m wide into the carbonate facies. The exhalite is roughly lenticular, although somewhat concave in plan. To the northwest it rapidly thickens by fanning out from about 10 m to greater than 50 m. In the southeast part of the mine thickness of the exhalite is more regular – averaging 20 to 25 m. The exhalite is clearly sedimentary and mainly of chemical derivation but with an important admixed fraction of volcaniclastic provenance. Lapilli-sized fiamme and crystal clasts of quartz and feldspar are abundant in places.

The silicate facies exhalite consists of massive chlorite and stilpnomelane with variable amounts of tremolite, carbonate, pyrite, pyrrhotite and magnetite. It extends along strike for about 100 m, is up to about 25 m thick, and is intruded by the northeasterly trending diabase dyke which it therefore flanks asymmetrically. The chloritic, silicate facies exhalite is massive to slightly schistose, although fine mineralogical banding, which is almost certainly relict bedding, is recognizable in places. Pyrite forms the most conspicuous of these bands, but carbonate-, quartz- and silicate-rich bands are also present. Magnetite forms narrow bands up to a few cm, though typically less than 1 mm thick. Coarse pyrite euhedra up to about 1 cm in maximum dimension are abundant in the chloritic rock (Plate I, Fig. 12), however finer-grained pyrite varies from anhedral to euhedral. Pyrrhotite typically forms numerous, small inclusions within pyrite, but locally, particularly near the diabase, it greatly exceeds pyrite in abundance and here forms large, irregular pods or lenses. Hornblende, biotite, epidote, feldspar, sphalerite, chalcopyrite and arsenopyrite are accessory minerals in the silicate facies exhalite.

The carbonate facies exhalite is composed mainly of carbonate, quartz and pyrite with lesser and variable amounts of sericite, plagioclase, epidote, chlorite, arsenopyrite, sphalerite, chalcopyrite and magnetite.

Gold ore is contained only within sulphide-rich lenses and layers of both the silicate and carbonate facies exhalite. The distribution of these lenses and layers is illustrated in Figure 3, but is extremely irregular from place to place as the ore bands pinch, swell, divide, coalesce and die out along strike. These auriferous lenses and layers may be considered to represent a third, i.e., a sulphide facies of exhalite. In places the sulphide-rich lenses have extremely sharp contacts with both the underlying and overlying sulphide-poor sedimentary strata of the exhalite, but elsewhere the sulphide content changes gradually, resulting in gradational boundaries. The ore itself is prominently bedded (Plate I, Fig. 14) and particularly in those portions in the carbonate exhalite, beds of sulphides may be interlayered with beds of chert and/or carbonate.

Veining within the exhalite is common and more so than in any other rock in the mine. Veins are composed of quartz and carbonate with lesser amounts of pyrite, although minor feldspar, chlorite, chalcopyrite, sphalerite and galena are all present in places. Although the morphology of the veins varies considerably, two basic types are recognizable. Firstly, quartz and carbonate form discordant veins which range in thickness from less than a millimetre to several centimetres and usually terminate as ladder veins against adjacent stratigraphic units. The smaller veins of this type are commonly zoned, with quartz-rich centers and carbonate-rich margins. Secondly, large, irregular patches composed predominantly of quartz are also common and usually contain pyrite, and visible chalcopyrite, sphalerite and galena.

Within both facies of the exhalite, laminated to finely bedded waterlain tuff (Plate I, Fig. 15) is interlayered with laminae of chert, carbonate and pyrite. Contacts between bands range from knife-sharp to diffuse. Chert bands varying in thickness from millimetres to several metres and now totally recrystallized are numerous, and become increasingly abundant in the sedimentary strata of the hanging wall. Thicker beds are traceable over the entire strike length of the mine, and even the very narrow bands for several tens of metres. The chert is poorly to well laminated with alternating pale pink and grey bands. The grey bands contain a greater proportion of carbonate and the pink bands more feldspar, probably of tuffaceous origin. Narrow ribbons of black chert frequently separate pyrite-rich bands in the ore (Plate I, Fig. 14) and are composed of nearly pure quartz, the dark colour imparted by accessory carbonaceous material. In the intercalated waterlain tuff graded bedding and small possible scour and fill structures are locally recognizable. Banding or bedding results from grain size and colour differences, with alternating light grey, carbonate-sericite-rich, and darker grey carbonaceous bands. Occasionally the banding is colourful, with alternating pale green, pink and buff layers.

Distinctive, coarse, felsic pyroclastic rocks are prominently intercalated and interfingered with the exhalite. These include minor quantities of fine-grained pyroclastic rock similar to those of the footwall but here coarse fragmental types (Plate I, Fig. 13) are much more abundant. Using Fisher's classification of volcaniclastic materials (1966) this rock is predominantly tuff-breccia and resembles "mill rock" (Sangster, 1972), present in massive base metal sulphide districts. Fragments derived from the footwall tuffs commonly constitute more than 50 per cent by volume of the felsic pyroclastic rocks intercalated with the exhalite. Larger fragments exceed 50 cm in maximum dimension. With decrease in size these fragments merge into and become indistinguishable from the matrix where this is coarser. The fragments are disc-shaped and flattened parallel to schistosity. When viewed in the plane of schistosity fragment ends are ragged. Although lithic fragments from the footwall rocks predominate, smaller fragments commonly contain numerous ovoid, non-connected inclusions of quartz, apparently representing amygdules or vesicle fillings in pumice. The groundmass is typically darker than the fragments, and quartz, sericite and carbonate are the essential matrix minerals. Pyrite is widespread in minor amounts. There is a complete gradation from exhalite to felsic pyroclastic tuff breccia with increasing fragment content and decreasing content of matrix carbonate and pyrite. In a few places there is sufficient auriferous pyrite in the matrix of the coarse fragmental rock to constitute ore.

In general the tuff breccia is unsorted. Contacts between individual pyroclastic layers are typically sharp, and a prominent flattening and adherance of fragments may be due to welding. The rock is clearly the product of explosive volcanism that shattered and reworked the underlying felsic fragmental rocks of the footwall dome. The felsic breccia is more abundant and thicker in the western portion of the mine, and although this might suggest that its source was to the west, the degree of rounding also increases westward and so contradicts this interpretation.

In most places a distinctive, highly carbonaceous schist containing pyrite in bands and nodules (Plate I, Fig. 16) overlies both facies of the exhalite. Although soft, black and with typical "graphitic" streak, an X-ray diffractometer pattern indicates that this rock is not graphitic, but probably contains a solid, non-crystalline hydrocarbon. The pyrite bands, up to several centimetres thick are commonly boudinaged and exhibit cataclastic fracture and separation. Narrow bands show soft sediment deformation textures. Pyrite nodules are spheroidal to ellipsoidal and up to three centimetres in diameter.

Concordant gabbro sills, 10 to 30 m thick, intrude the upper contact of the carbonaceous schist with the hanging wall sedimentary rocks, and are traceable along the strike length of the mine. The gabbro is composed of plagioclase, chlorite and tremolite with accessory biotite and pyrite. Carbonate is abundant in the gabbro near its contacts, and locally constitutes greater than 5% of the rock by volume. In some places the gabbro is extensively altered, though unlike that in the stratigraphically lower units, the degree of alteration is not spatially related to the diabase.

The youngest rock in the mine is the cross-cutting, northeast-trending diabase dyke which is about 75 m wide. It bifurcates in the upper levels of the mine to form a second narrow satellite dyke 5 to 10 m wide (Fig. 3).

GOLD MINERALIZATION

Three texturally different types of pyrite are present within the ore zone and two of these carry gold.

1) Gold-bearing
   a. fine grained, sparsely poikilitic; in both facies of exhalite
   b. coarse euhedral; in silicate facies exhalite
2) Barren
   a. lacy textured, poikilitic anhedral aggregates; in carbonate and silicate facies exhalite

All three types of pyrite, although disseminated, are usually most abundant along distinct beds. Gold is most frequently associated with the fine-grained, sparsely poikilitic, pyritohedral variety which often has cataclastic textures and atoll structure. This fine-grained pyrite characteristically comprises 30 per cent of the ore by volume, although it may constitute up to about 70 per cent. In general, higher grade gold values are associated with greater concentrations of this fine-grained pyrite. Larger aggregates of fine, anhedral, more poikilitic, lacy-textured pyrite are also present but do not contain microscopically visible gold. Although the fine-grained, sparsely poikilitic pyrite and the aggregates of lacy-textured more poikilitic pyrite are common in the same rock, they usually occur in distinct, mutually exclusive bands (Plate I, Fig. 17). Finally, a very coarse euhedral variety of pyrite forms crystals up to about 1 cm in maximum dimension within the silicate facies exhalite (Plate I, Fig. 12) and also carries gold.

Within the carbonate facies exhalite virtually all visible gold occurs in contact with pyrite. Only exceedingly rarely is gold encountered isolated within other minerals and in some of these rare examples the gold appears to have been physically broken from minute gold veinlets in neighbouring pyrite grains. In most instances gold forms small allotriomorphic inclusions within pyrite, often abutting other silicate and carbonate inclusions, and commonly also at pyrite grain edges. Gold inclusions average 4 to 5 μ and rarely exceed 10 μ in diameter. Although typically irregular, some gold grains have angular and often triangular or trapezoidal outlines similar in habit to the gold in arsenopyrite described by Fripp (1974). This may reflect crystallographic control imparted by the host pyrite. Less frequently gold forms minute veinlets or a "string" of inclusions along boundaries between two pyrite grains or, more rarely, at triple junctions. In this habit, the gold grains are somewhat more elongate, averaging 7 to 8 μ in length and 2 to 3 μ in width. Elongation is most pronounced in the larger veinlets, and although veinlets longer than 10 μ are common, veinlets longer than 30 μ are exceptionally rare.

Gold was infrequently observed in contact with chalcopyrite and even less with arsenopyrite, but in all such instances it is also in contact with adjoining pyrite grains. In the carbonate facies exhalite, gold has never been observed totally enclosed within any sulphide other than pyrite.

Where pyrite predominates over pyrrhotite in the silicate facies exhalite, the habits of gold are identical to those described in the carbonate facies exhalite. Where pyrrhotite, however becomes dominant, gold is associated with it and is found as inclusions, at pyrrhotite grain boundaries, as small veinlets between pyrrhotite grains, and as small grains in silicates abutting pyrrhotite. In pyrrhotite-rich samples chalcopyrite is a common accessory mineral and often occurs in contact with gold or, very rarely, totally enclosing a gold inclusion.

The only gold mineral observed is native gold. Silver minerals have not been identified despite a careful search and so are therefore unlikely to exist. Although impossible to establish unequivocally, it appears that native gold is sufficient to account for all of the gold in the deposit, with the exception that minor background amounts may be present in other minerals. Microscopically, the native gold is luminous golden yellow suggesting a low silver content. This is substantiated by electron microprobe analyses which reveal a fineness of 820 to 840.

A plot of gold versus silver values from both facies of exhalite is shown in Figure 4. The data plot into two fairly distinct clusters, with gold values less than about 200 ppb and greater than 1000 ppb respectively. Of particular interest in this plot are the silver values. If all silver occurs with native gold, then all points should plot near the 830 fineness reference line. Clearly this is not the case. However, samples with high gold values do cluster about this line, most in fact above it. Those samples with low gold values however, all plot considerably above this line, suggesting that some silver is present in other minerals.

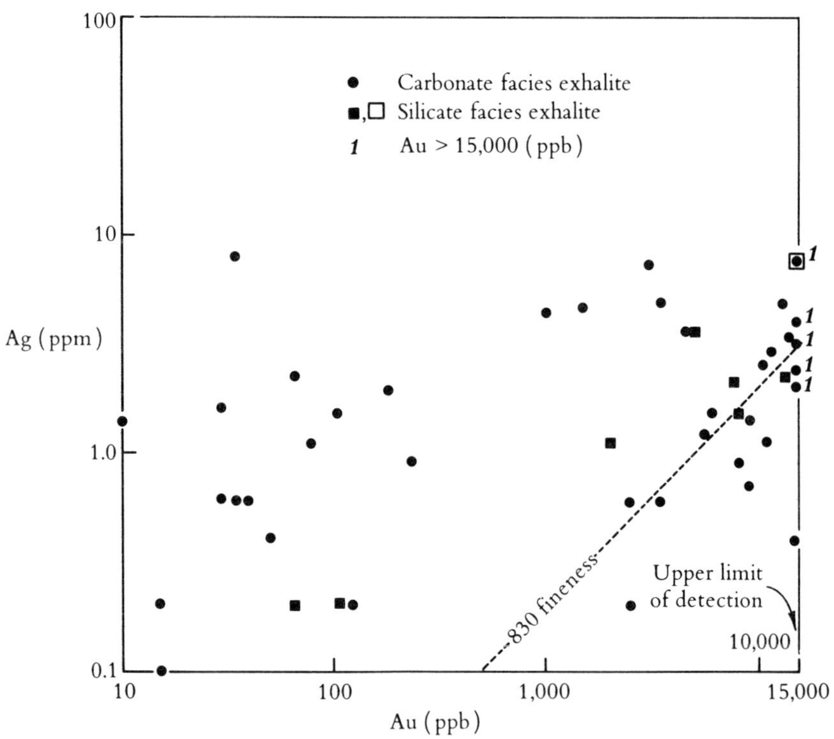

Figure 4. Gold versus silver content of the two facies of exhalite (10+50 level).

TRACE METAL RELATIONSHIPS

Although arsenopyrite is a widespread accessory mineral in pyrite-rich portions of the carbonate facies exhalite, it is only rarely observed in the silicate facies and in sulphide-poor carbonate facies exhalite. There is therefore a sympathetic relationship between high arsenic and high gold content in the carbonate, but not in the silicate facies exhalite (Fig. 5).

The base metal content of the major rock units is variable, as shown for carbonate and silicate facies exhalite and for the felsic pyroclastic rocks in Table I. Despite minor differences the average values are similar to those of Archean felsic volcanic rocks in the Superior Province (Table I; Goodwin, 1977) except for the footwall tuffs which are considerably enriched in zinc, cobalt and nickel. With the exception of zinc, the base metal content of the exhalite is markedly lower than that of auriferous iron-formations in Rhodesia (Fripp, 1976).

Despite the low content of copper and zinc, analytical data indicate that the distribution of both is both vertically and laterally zoned. The ratio of Cu to Cu + Zn

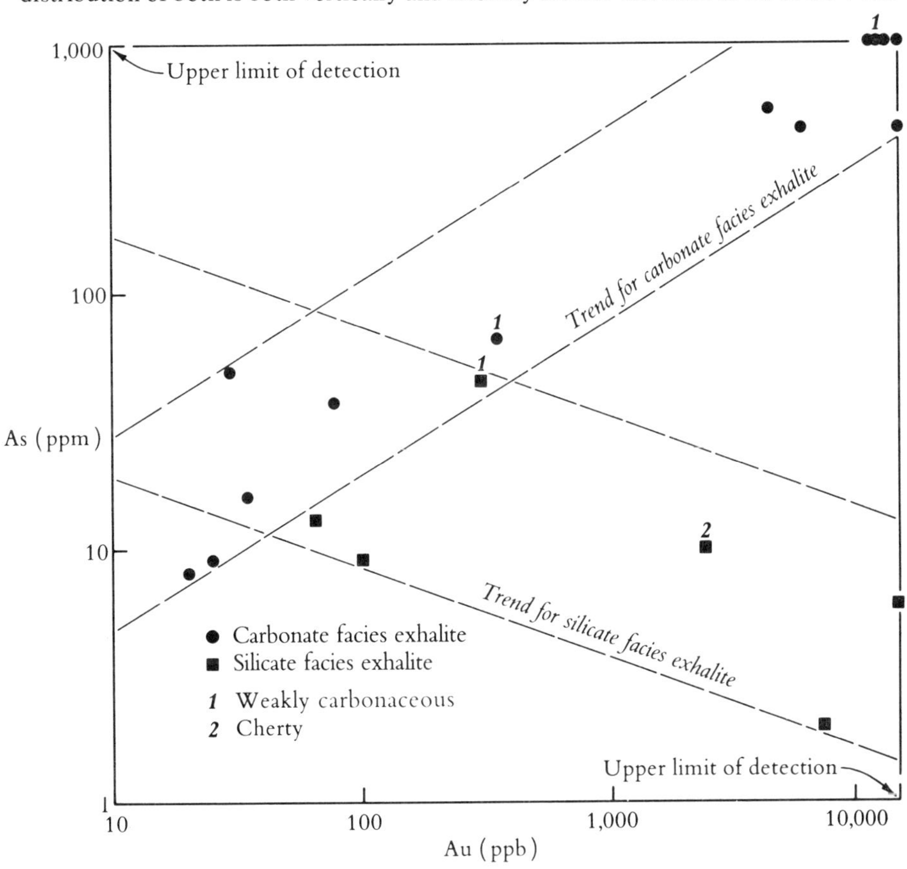

Figure 5. Gold versus arsenic content of the two facies of exhalite (10+50 level).

TABLE I

Base metal values from Agnico-Eagle, compared with Superior Province felsic volcanic rocks and Rhodesian auriferous iron-formations.

		Zn ppm	Cu ppm	Pb ppm	Co ppm	Ni ppm
Carbonate facies exhalite	Average	72	23	20	14	42
	Range	26- 285	4- 61	<1-147	<1- 41	12-147
Silicate facies exhalite	Average	74	43	13	8	28
	Range	26- 420	16-163	2- 40	<1- 26	18- 42
Footwall tuffs	Average	232	36	18	38	79
	Range	14-1,665	7- 71	0-322	6- 91	32-173
Ore sequence tuff breccia	Average	34	30	4		
	Range	10- 63	12- 57	<1- 8		
Dacite (Superior Province)*	Average	72	37	8	12	23
Rhyolite (Superior Province)*	Average	48	30	8	5	18
Auriferous iron-formations (Rhodesia)[t]	Average	51	224	107	51	224
	Range	5-1,320	25-388	68-160	20-130	173-335

* Goodwin, 1977
[t] Fripp, 1976, p. 68

has been computer contoured (Harrison, 1978) and the resulting plot is shown in Figure 6. Although not conclusive, the result suggests that copper is concentrated relative to zinc in the silicate facies exhalite and adjacent underlying footwall tuffs, whereas zinc increases, relative to copper, outward from the chloritic exhalite. Such a pattern is commonly reported in massive base metal sulphide deposits (Large, 1977). Results from individual drill holes at Agnico-Eagle, however, indicate an upward stratigraphic zonation that is the opposite of that reported from massive base metal sulphide deposits. Analyses of core from three diamond drill holes that transected the exhalite on the 10 + 50 level show that Cu to Cu + Zn ratio increases upward, albeit only slightly, from stratigraphic base to stratigraphic top (Fig. 7). In massive base metal sulphide deposits compositional zoning is due to an upward increase in zinc rather than to a decrease in copper content (Sangster, 1972). At Agnico-Eagle, the vertical zoning may also be attributed largely to variations in zinc content (Fig. 8), which here decreases slightly upward. Conversely the lateral zoning at Agnico-Eagle is primarily due to changes in copper content as suggested by comparing the average zinc and copper contents of the central, chloritic silicate facies exhalite and lateral carbonate facies exhalite (Table I); zinc is equally abundant in both whereas copper is almost twice as abundant in the former.

CARBONATE ZONING WITHIN THE EXHALITE

Samples of diamond drill core were selected for electron microprobe analysis from a hole which intersected only one band of pyritic ore in the upper portions of the carbonate facies exhalite. Qualitative results show a distinct sequential upward change in the composition of the carbonate minerals comprising the exhalite. Electron microprobe scans of numerous carbonate grains show that siderite greatly predominates over ankerite at the base of the exhalite. Upwards however, siderite gives way increasingly to ankerite, so that in central sections of the drill hole both siderite and ankerite are present. Finally, at the top of the exhalite, ankerite greatly exceeds siderite which here occurs only along pyrite grain boundaries. From this and because gold occurs with pyrite, which is most abundant in the higher ankeritic samples, a sympathetic relationship between high gold values and ankeritic carbonate, and an antipathetic relationship between gold and siderite are inferred (Fig. 9).

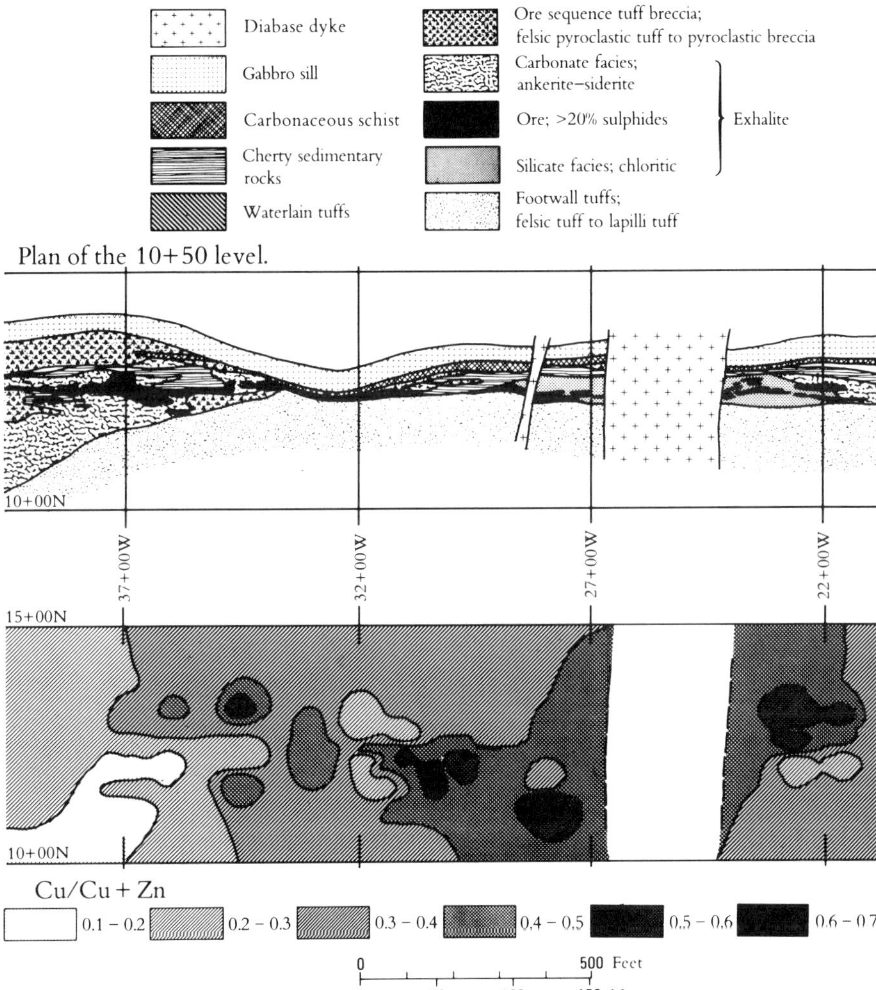

Figure 6. Geology and copper:zinc distribution (10+50 level).

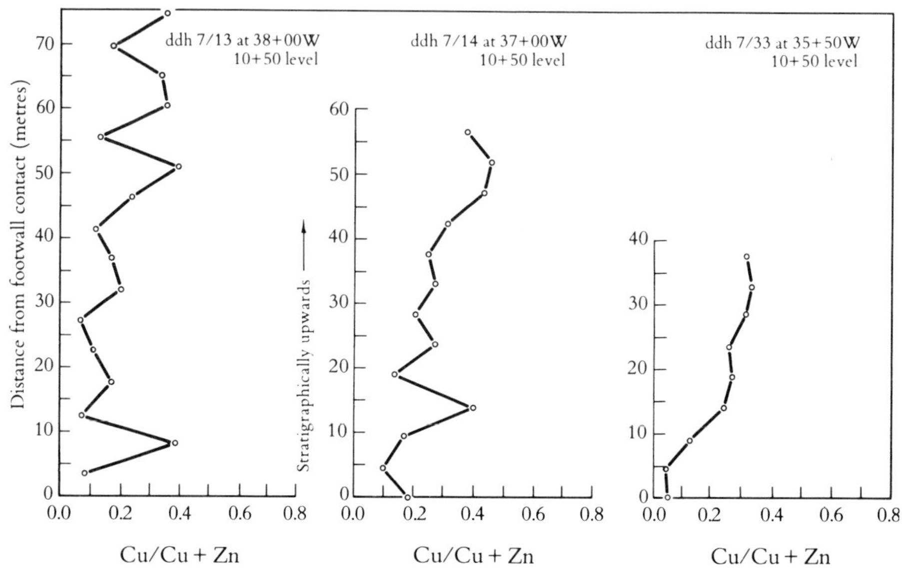

Figure 7. Variation in Cu/Cu + Zn upward from stratigraphic base of carbonate-facies exhalite.

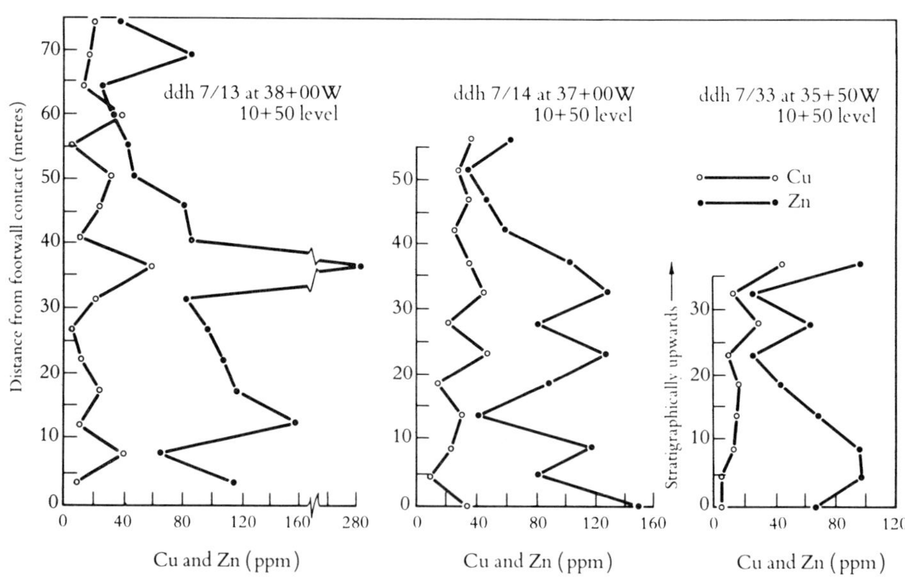

Figure 8. Variation in Cu and Zn content upward from stratigraphic base of carbonate facies exhalite.

The cation content of the carbonates also reflects this systemmatic upward change in composition. There is a marked upward increase in the manganese and magnesium content of both siderite (Fig. 10) and ankerite (Fig. 11), with a corresponding decrease in iron content. Calcium, however, remains nearly constant throughout the carbonate exhalite from base to top. Calcite is also present but was detected only in minor amounts in the centre of the exhalite where siderite and ankerite co-exist in nearly equal amounts.

A preliminary study of major element distribution in 19 samples from the 10+50 level, reveals that the footwall tuffs have been extensively altered. Results indicate that the footwall tuffs immediately beneath the silicate facies exhalite and adjacent to the diabase dyke are enriched in Na_2O and MgO, but are depleted in CaO. This alteration is enveloped by a zone of K_2O enrichment and a much broader zone of Na_2O depletion, and CaO and MgO enrichment, with high loss on ignition values, which extends beneath the entire ore-body. A more detailed geochemical and mineralogical study of the footwall alteration is presently in progress.

DISCUSSION AND COMPARISON

The Agnico-Eagle deposit exhibits many geological features similar to those of volcanogenic massive base metal sulphide deposits of comparable age and metamorphic rank. The general characteristics of these deposits have been summarized by Sangster (1972) and Sangster and Scott (1976).

Perhaps the most significant of these similarities is the lithological setting. The Agnico-Eagle ore body, like many volcanogenic massive sulphide deposits immediately overlies a sequence of felsic pyroclastic rocks, possibly constituting a dome, in a thick volcanic succession. The exhalite is closely associated with a dis-

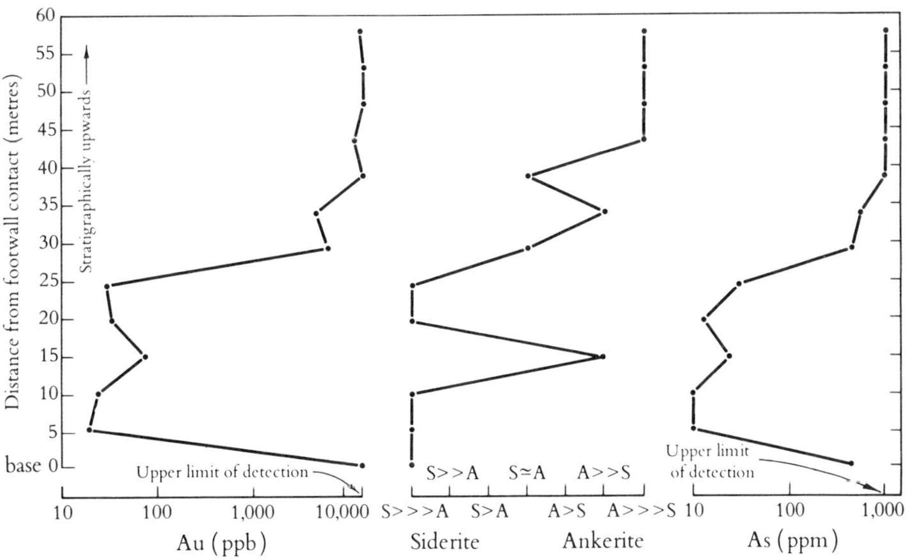

Figure 9. Variation in gold, carbonate, and arsenic content upward from stratigraphic base of carbonate facies exhalite (10+50 level at 38+00 W).

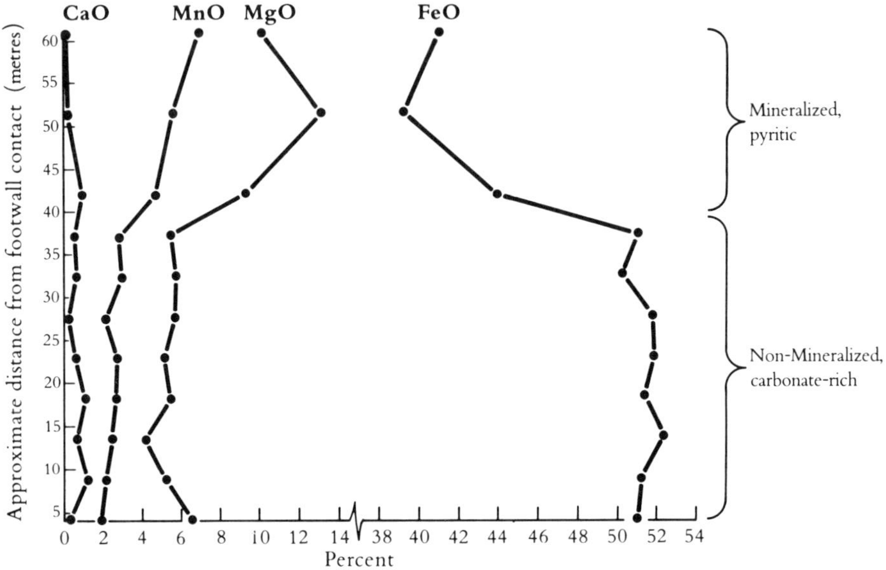

Figure 10. Variation in siderite composition upward from stratigraphic base of carbonate facies exhalite (10+50 level at 10+50 W).

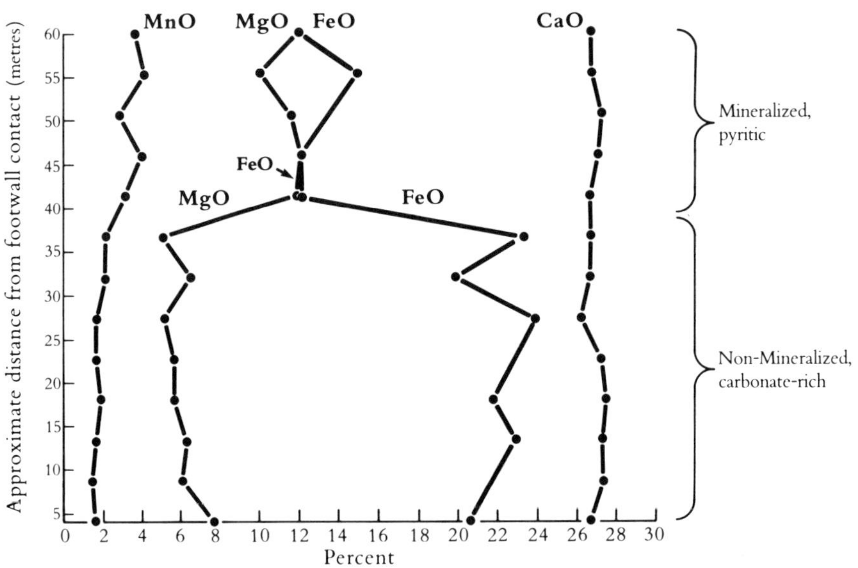

Figure 11. Variation in ankerite composition upward from stratigraphic base of carbonate facies exhalite (10+50 level at 10+50W).

tinctive, coarsely fragmental felsic pyroclastic tuff-breccia or "mill-rock" (Sangster, 1972), apparently produced by explosive volcanism. At Agnico-Eagle these are pyroclastic flow breccias, and similar breccias elsewhere have been interpreted as products of deposition from a suspension of fragments in a gas (Parsons, 1969; Williams, 1957). Where composed of fine particles, such deposits are referred to as ignimbrites when formed subaerially (Marshall, 1935), and it has also been suggested that such deposits may form subaqueously as well (Fiske, 1963). At Agnico-Eagle the intimate association of the coarse pyroclastic breccia with chemical and pyroclastic sedimentary rocks of clearly subaqueous origin, as well as the pillowed structures in the overlying basalts, are compelling evidence for a subaqueous depositional environment for these breccias, although some of the footwall pyroclastic rocks may have been subaerially deposited.

The Agnico-Eagle ore body is a stratabound and stratiform lens of carbonate-sulphide-silicate-oxide facies iron-formation. In places the sulphides constitute more than 50 per cent of the rock by volume and the body could therefore be considered a massive sulphide (Sangster, 1972). As in many massive base metal sulphide orebodies, the gold is intimately associated with iron sulphides, notably pyrite. Petrographic evidence suggests that the gold is pre-metamorphic and may have been derived from the iron sulphides during their crystallization, either diagenetically from original fine chemical precipitates, or following greenschist metamorphism.

Massive base metal sulphide bodies commonly exhibit marked compositional layering, banding or bedding. This is present at Agnico-Eagle in both facies of exhalite. Relict banding in the silicate facies suggests that some of its chlorite, stilpnomelane and perhaps tremolite, are metamorphic derivatives of primary, bedded iron silicates. Superimposed on this, may have been a chloritic alteration imparted by ascending brines rising toward the sea floor through the newly deposited sediments. The abundance of pyrrhotite in the chloritic zone may be due to the breakdown of earlier pyrite during this activity and to later thermal breakdown of pyrite to pyrrhotite during diabase intrusion.

Immediately beneath the chloritic zone the footwall tuffs contain increased quantities of chlorite and pyrite. The altered footwall rocks do not contain economic quantities of gold however, and their extent therefore has not been outlined by exploration or mining operations. The pyritic carbonaceous schist overlying the ore at Agnico-Eagle is similar to that observed in some base metal massive sulphides, for example Kidd Creek (Walker *et al.*, 1975).

Zoning is a distinctive feature of massive sulphide deposits and is commonly expressed in morphological, mineralogical, compositional and textural differences (Sangster, 1972; Sangster and Scott, 1976). Morphological zoning, in which sulphide stringers or stockworks underlie a massive body is not well developed at Agnico-Eagle, although the silicate facies exhalite and immediately underlying footwall tuffs beneath the massive pyritic exhalite have some similarities to this feature. Mineralogical zoning however, is prominent at Agnico-Eagle and takes several forms. Firstly, there is an outward change from predominantly iron silicates of the central chloritic silicate facies exhalite to siderite and ankerite of the peripheral carbonate facies exhalite. Secondly, within the carbonate facies exhalite there is a marked upward change from sideritic to ankeritic carbonate. Thirdly, the distribution of the opaque minerals is distinctly zoned; magnetite and pyrrhotite are most abundant in the central chloritic silicate facies exhalite but are essentially absent in the peripheral, carbonate facies exhalite where pyrite predominates and arsenopyrite is a common accessory mineral. A change from proximal pyrrhotite, with or without magnetite, to distal pyrite has been described in several exhalative massive base metal sulphide deposits (Roberts, 1975) is considered characteristic (Hutchinson, 1973) and is explainable by Eh-pH stability relationships of these iron sulphides (Large, 1977).

There is also definite compositional zoning in the form of a progressive upward increase in the manganese and magnesium content of both siderite and ankerite from stratigraphic base to top of the carbonate facies exhalite. Arsenic is more abundant in pyritic portions of the carbonate exhalite than in both sulphide-poor carbonate exhalite and silicate facies exhalite. Finally, there is an increase in zinc relative to copper outward from the central chloritic silicate facies exhalite, but an upward decrease in zinc relative to copper from stratigraphic base to top of the carbonate facies exhalite. In our view these relationships constitute compelling evidence of a volcanogenic and exhalative origin for the Agnico-Eagle deposit. They thereby confirm the importance of exhalative processes in the genesis of some Archean lode gold deposits. They indicate clearly that gold was mobilized by, transported in and deposited from hydrothermal brines generated during subaqueous felsic volcanism. Alternative hypotheses advanced for other Archean lode gold deposits such as epigenetic replacement (Lindgren, 1933), lateral secretion (Boyle, 1955) or in situ alteration of ultramafic rocks (Pyke, 1976) do not appear tenable at Agnico-Eagle.

ATTEMPTED RECONSTRUCTION

It is suggested that the gold-rich volcanogenic exhalative processes were the culmination of one cycle of the prolonged Archean volcanism that deposited the Gale Group. At Agnico-Eagle this cycle involved deposition of the footwall felsic pyro-

---

Explanation of Plate I

Figure 12. Chloritic silicate facies exhalite containing irregular bands of coarse pyrite euhedra (left) cut by diabase dyke (right); west wall of diabase, 1800 level; 0.1 m square roof bolt for scale.

Figure 13. Slab of typical lapilli-breccia; lapilli showing ragged terminations characteristic of fiamme; flattening and possible welding of lapilli; 9-0-39 cross cut, 900 level.

Figure 14. Slab of typical banded exhalite ore; alternate pyritic and cherty laminations; (discordant saw marks lower right); 10-2-31 stope, 1050 level.

Figure 15. Slab of typical waterlain tuff; thinly laminated cherty and tuffaceous beds showing possible graded bedding and small scour-and-fill structure; 9-2-31 drift, 900 level.

Figure 16. Slab of typical carbonaceous schist showing spherical pyrite nodule and thin cataclastically fragmented pyrite bands; 6 sublevel, 10-2-35 drift, 1050 level.

Figure 17. Photomicrograph showing fine-grained, subhedral auriferous pyrite with intervening band of fine, lacy-textured, poikilitic, microscopically barren pyrite, ddh 7-14, 42 m, 1050 level.

PLATE I

Figure 12

Figure 13

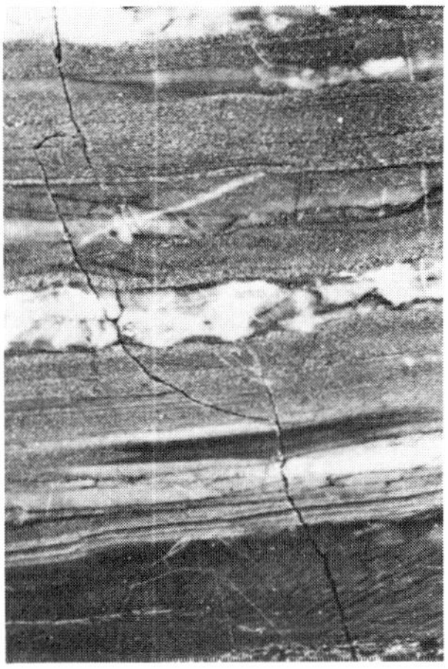

Figure 14

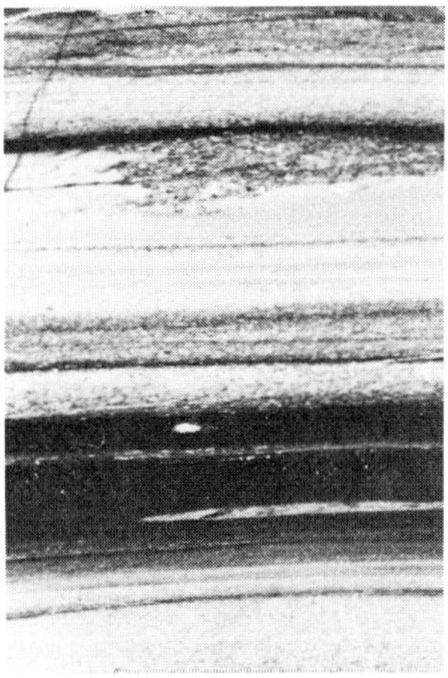

Figure 15

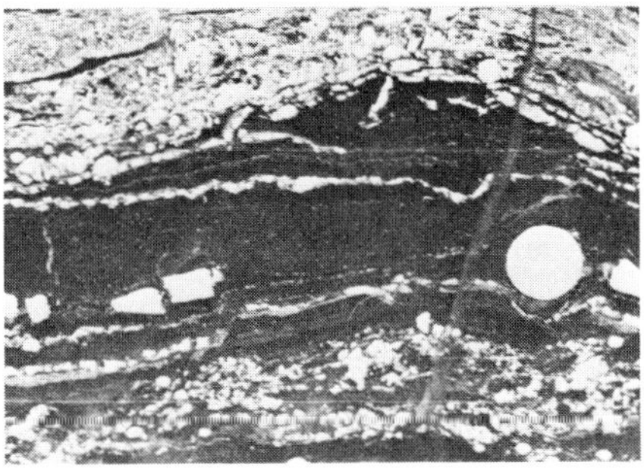

Figure 16

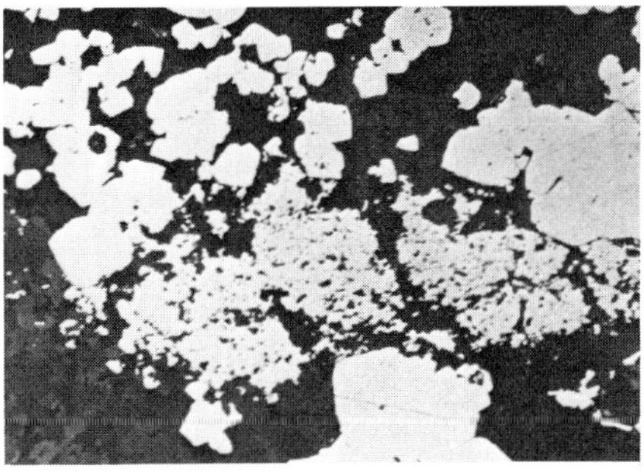

Figure 17

clastic rocks, perhaps in a small dome. Caldera collapse may have occurred after the period of dome construction in a manner similar to that observed in comparable sequences of much younger volcanic rocks (Smith and Bailey, 1968). Caldera collapse was followed by explosive reworking of the footwall tuffs to form the coarse felsic tuff-breccia as pyroclastic flows. The collapse was accompanied and followed by exhalative discharge of heated, metalliferous brines generated by convective circulation in the cooling, porous pyroclastic rocks of the dome. Brine expulsion was mainly through a tensional fracture formed during caldera collapse, and took place directly into a local reduced basin, perhaps within the caldera. As a result, footwall rocks of the felsic dome were altered along the channelway by fluid movement. Chemical sediments were deposited near, around and above the vent on the sea floor in relatively shallow water, perhaps 20 to 200 m (Fripp, 1976). Brines reaching the sea floor at these modest depths may have had temperatures between 130° and 250°C, and a gold content up to 50 to 300 ppm (Fripp, 1976). The chemical sediments were themselves altered and chloritized in the vent area by continued brine flow. The intercalated waterlain tuffs were deposited by similutaneous sedimentation of pyroclastic debris from the immediately preceeding explosive volcanism, or from continuing volcanism elsewhere in the area.

Iron, silica, sulphur and carbon dioxide were the major components discharged by the hydrothermal system but these were accompanied by minor amounts of arsenic, gold, silver, and base metals. Exhalative activity culminated in the precipitation of the hanging-wall cherts which were succeeded by deposition of argillite, greywacke and shale that filled the caldera.

The subsequent onset of much more mafic volcanism of a later cycle deposited the basalts of the northeast shore of the Harricana River which were laid down on the rocks of the prior cycle. Sill-like sheets of this, or of still later magma invaded the soft cherty sedimentary rocks beneath the volcanic sequence and formed the gabbroic sills. During burial and compaction, the more massive volcanic rocks of the footwall succession and the higher hanging-wall succession behaved in a competent manner. However, the soft, water-saturated chemical sediments of the exhalite unit underwent significant diagenetic changes. Tensional fractures opened in the more brittle of these sediments, more or less normal to bedding, and quartz, carbonate and sulphides from the exhalite were secreted into these openings forming crosscutting ladder veinlets. The fractures and resulting veins however, did not penetrate the overlying and underlying less competent foliated sediments and tuffs, nor the competent units of the volcanic succession. Finally, all the rocks were deformed, folded and intruded by granitic bodies during Kenoran orogeny. Late Keeweenawan diabase dykes cut all the earlier rocks, and one of these invaded the old tensional fracture that had channelled the exhalative brines to the surface during the Archean volcanism.

ACKNOWLEDGEMENTS

Thanks are due to Agnico-Eagle Gold Mines Ltd. for support and assistance provided to the senior author during field work, and for permission to publish this paper. Natural Science and Engineering Research Council of Canada supplied funds under an operating grant to one of us (R. W. H.) without which the work would have been impossible. The Geology Department at the University of Western Ontario has supplied support services for the study, as well as a graduate student assistantship to the senior author. Rio Tinto Canadian Exploration and Mr. C. D. Spence are thanked for information and suggestions regarding regional geology in the Joutel – Poirier district. Trace element analyses were carried out by Bondar – Clegg and Co. Ltd. of Ottawa, and Mr. F. W. Graves of the Department of Geography at the University of Western Ontario drafted the various figures and illustrations.

REFERENCES

Boyle, R.W., 1955, The geochemistry and origin of the gold-bearing quartz veins of the Yellowknife Greenstone Belt: Economic Geology, v. 50, p. 51.

Fisher, R.V., 1966, Rocks composed of volcanic fragments and their classification: Earth Science Review, v. 1, p. 287-298.

Fiske, R.S., 1963, Subaqueous pyroclastic flows in the Ohanapecosh Formation: Washington: Bulletin of American Science, v. 72, p. 1409-1414.

Fripp, R.E.P., 1974, Distribution and significance of stratabound mineral deposits in Archean banded iron formation in Rhodesia, with particular reference to the Vubachikwe gold mine, Gwanda,: Unpubl. M. Ph. Thesis, University of Rhodesia, Salisbury, 147 p.

Fripp, R.E.P., 1976, Stratabound gold deposits in Archean banded iron-formation, Rhodesia: Economic Geology, v. 71, p. 58-75.

Goodwin, A.M., 1977, Archean volcanism in Superior Province, Canadian Shield: in Baragar, W.R.A., Coleman, L.C., and Hall, J.M., eds., Volcanic Regimes in Canada: Geological Association of Canada Special Paper 16, p. 205-241.

Harrison, C.J., 1978, The Design and Application of two dimensional Digital Filters: Unpubl. M.Sc. Thesis, University of Western Ontario.

Hutchinson, R.W., 1973, Volcanogenic sulphide deposits and their metallogenic significance: Economic Geology, v. 68, p. 1223-1246.

Kelly, R., 1964, Joutel-Poirier Area: Department of Natural Resources, Quebec, Preliminary Map, Open File No. GM: 23412.

Large, R.R., 1977, Chemical evolution and zonation of massive sulphide deposits in volcanic terrains: Economic Geology, v. 72, p. 549-572.

Latulippe, M., 1976, The stratigraphic divisions of the Abitibi-Temiscamingue Area of Northwestern Quebec: Quebec Department of Mines and Natural Resources Unpublished Map, 1:633,600.

Lindgren, W., 1933, Mineral Deposits, 4th Edition: McGraw-Hill Book Co. Ltd., 930 p.

Marshall, P., 1935, Acid rocks of Taupo-Rotorua volcanic District: Royal Society New Zealand Trans., v. 64, p. 323-366.

Parsons, W.H., 1969, Criteria for the recognition of volcanic breccias: review: Geological Society of America Memoir 115, p. 263-304.

Pyke, D.R., 1976, On the relationship between gold mineralization and ultramafic volcanic rocks in the Timmins area, northeastern Ontario: Canadian Institute of Mining and Metallurgy Bulletin, v. 69, no. 773, p. 79-87.

Quebec Department of Natural Resources, 1976, Geologic compilation accompanying airborne IMPUT MK VI survey, Joutel-Poirier area: Sheet 3, Map B-1263.

Remick, J.H., 1969, Geology of Harricana-Turgeon Area, Abitibi-East and Abitibi-West Counties: P. R. 564, Department of Natural Resources, Government of Quebec, 217 p.

Ridler, R.H., 1973, Exhalite concept a new tool for exploration: The Northern Miner, November 29, 1973, p. 59-61.

Roberts, R.G., 1975, The geological setting of the Mattagami Lake mine, Quebec: a volcanogenic massive sulphide deposit: Economic Geology, v. 70, p. 115-129.

Ross, C.S., and Smith, R.L., 1961, Ash-flow tuffs; their origin, geologic relations and indentification: United States Geological Survey, Professional Papers, 366, p. 1-77.

Sangster, D.F., 1972, Precambrian Volcanogenic Massive Sulphide Deposits in Canada: A Review: Geological Survey of Canada Paper 72-22, 44 p.

Sangster, D.F., and Scott, S.D., 1976, Precambrian, stratabound, massive Cu-Zn-Pb sulphide ores in North America: in Wolf, K.H., ed., Handbook of Stratabound and Stratiform Ore Deposits: v. 6, Cu, Zn, Pb, and Ag Deposits: Amsterdam, Elsevier Scientific Publishing Co., p. 129-211.

Smith, R.L., and Pailey, R.A., 1968, Resurgent cauldrons: in Coats, C.R., Hay, R.L., and Anderson, C.A., eds., Studies in Volcanology, Geological Society of America, Memoir 116, p. 613-622.

Walker, R.R., Matulich, A., Amos, A.C., Watkins, J.J., and Mannard, G.W., 1975, The geology of the Kidd Creek mine: Economic Geology, v. 70, p. 80-89.

Williams, H., 1957, Glowing avalanche deposits of the Sudbury Basin: Report Ontario Department of Mines, p. 57-89.

MS received February 1979;
Accepted for publication August 1979.

Precambrian Sulphide Deposits, H.S. Robinson Memorial Volume, edited by R.W. Hutchinson, C.D. Spence and J.M. Franklin, Geological Association of Canada Special Paper 25, 1982

GEOLOGY OF THE OPEMISKA COPPER - GOLD DEPOSITS AT CHAPAIS, QUEBEC

David H. Watkins
Corporation Falconbridge Copper, P.O. Box 2000,
Noranda, Quebec J9X 5B4

Gerald Riverin
Corporation Falconbridge Copper, P.O. Box 130,
Chapais, Quebec G0W 1H0

ABSTRACT

The Opemiska Mine, located at Chapais, Quebec in the northeastern portion of the Abitibi orogenic belt, has produced over 15.5 million metric tonnes of ore grading 2.6% Cu since beginning production in 1954.

The ore deposit consists of a series of chalcopyrite-bearing quartz veins that occupy fracture systems in strongly folded and faulted gabbroic portions of two conformable, regionally extensive, layered, Archean ultramafic/mafic sills.

Vein sulphide mineralogy consists of chalcopyrite, pyrite, pyrrhotite, minor sphalerite, galena, molybdenite and arsenopyrite, and sparse linnaeite, cobaltite, pentlandite, millerite, and gersdorffite. Native Au occurs with pyrite and chalcopyrite. Other metallic minerals include magnetite, scheelite, hematite, ilmenite, and rutile. Non metallic vein gangue minerals include quartz, calcite, chlorite, and lesser biotite, stilpnomelane, actinolite, K-feldspar and axinite.

Cu and low grade Au veins occur in the Ventures Sill which hosts 95% of the ore. Similar veins with lower grade Cu and high grade Au occur in the stratigraphically higher Bourbeau Sill. Both sills were emplaced within volcanic rocks of the Blondeau formation.

The sills and volcanic rocks underwent polyphase deformation which resulted in overturned folds with moderately plunging axes. Conjugate sets of radial fractures formed open spaces in coarsely crystalline, ophitic-textured gabbroic phases of the Ventures Sill during deformation.

The proposed genetic model involves the leaching of metals and sulphur from a source horizon within the Blondeau Formation, and redeposition of sulphides within open fractures. The vein emplacement was probably post-Kenoran and coeval with the intrusion of granitic plutons which may have provided the heat and/or fluids necessary for hydrothermal activity in the area.

RÉSUMÉ

La mine d'Opémiska, située à Chapais, Québec, dans la partie nord-est de la ceinture orogénique de l'Abitibi, a produit plus de 15,5 millions de tonnes métriques de minerai à teneur de 2,6% de Cu, depuis le début de la mise en production en 1954.

Le gisement consiste en une série de veines quartzeuses à chalcopyrite et qui occupent des systèmes de fractures dans les portions gabbroïques fortement plissées et faillées de deux sills archéens, lités ultramafique/mafique, concordants et d'étendue régionale.

La composition minéralogique des veines sulfurées consiste en chalcopyrite, pyrite, pyrrhotite, avec quantités mineures de sphalérite, galène, molybdénite et arsénopyrite, et présence rare de linnéite, cobaltite, pentlandite, millérite, et gersdorffite. De l'or natif est associé à la pyrite et la chalcopyrite. Les autres minéraux métalliques rencontrés comprennent magnétite, scheelite, hématite, ilménite et rutile. Dans la gangue filonienne, les minéraux non métalliques comprennent quartz, calcite, chlorite, et en quantités moindres biotite, stilpnomélane, actinote, feldspath potassique et axinite.

Les veines de cuivre à teneurs basses en or sont situées dans le sill de Ventures qui renferme 95% du minerai. Des veines similaires mais à teneurs plus faibles en cuivre mais plus élevées en or se trouvent dans le sill de Bourbeau, stratigraphiquement le plus élevé. Ces deux sills sont mis en place dans les roches volcaniques de la formation de Blondeau.

Les sills et les roches volcaniques ont subit une déformation polyphasée qui a résulté en des plis renversés avec axes à plongement modéré. Des ensembles conjugués de fractures radiales formèrent des espaces ouverts dans les phases gabbroïques grossièrement cristallisées et à structure ophitique du sill de Ventures pendant sa déformation.

Le modèle génétique proposé implique le lessivage des métaux et du soufre à partir d'un niveau source à l'intérieur de la formation de Blondeau, puis le dépôt des sulfures dans des fractures ouvertes. La mise en place des veines est probablement post-Kénoréenne et contemporaine de l'intrusion de plutons granitiques qui pourraient avoir fourni la source de chaleur et/ou des fluides nécessaires à l'activité hydrothermale dans la région.

INTRODUCTION

The Opemiska deposits are a series of Cu-Au bearing veins located at Chapais, Québec, about 500 km north of Montreal (Fig. 1). The original discovery was made by Léo Springer and the Prospector's Airways group in 1929 while prospecting burned over areas from an airplane (Derry and Folinsbee, 1957). Production did not start until 1954 because of the lack of transportation and infrastructure, the prevailing low price of copper, and the limited reserves defined by the early work.

Since production began, over 15.5 million tonnes of ore grading 2.6% Cu and 1.58 grams Au have been milled from the Perry, Springer, Robitaille and Cooke shafts to mid 1978. The operation is presently milling 2720 tonnes per day and reserves are sufficient to sustain this production for at least 5 more years.

REGIONAL GEOLOGY

The Chapais area is part of the essentially Cu-Au producing Chapais-Chibougamau mining district. It lies near the eastern extremity of the Abitibi orogenic belt about 30 km from the Grenville Front (Fig. 2).

The Chapais-Chibougamau district constitutes a unique volcanic setting in the Abitibi orogenic belt because of the occurrence of large, regionally extensive, differentiated sills that are probably comagmatic with the volcanic rocks (Duquette, 1970). These include the Lac Doré complex, which hosts most of the mineable deposits in Chibougamau, and the Ventures Sill, which hosts most of the ore in Chapais, as well as the Bourbeau Sill, which contains small deposits in both areas.

Sills, volcanic, and sedimentary rocks in the district are generally steeply dipping, and have been folded about a major, regional structure called the Chibougamau anticline, which has an east-southeast striking axial trace (Fig. 3). The center of the Chibougamau anticline is now occupied by two plutons of granitic to tonalitic composition, the Opemiska and the Chibougamau Plutons. The Opemiska deposits lie on the south limb of the Chibougamau anticline where the Chapais syncline to the south splits into 2 folds forming an east-plunging, overturned antiform-synform pair. The Presqu'Ile Granite, a large granitic mass of unknown extent, occurs about 3 km south of the Opemiska deposits. The mine area is thus bounded to the north and south by large granitic plutons, and to the east by a tonalitic pluton.

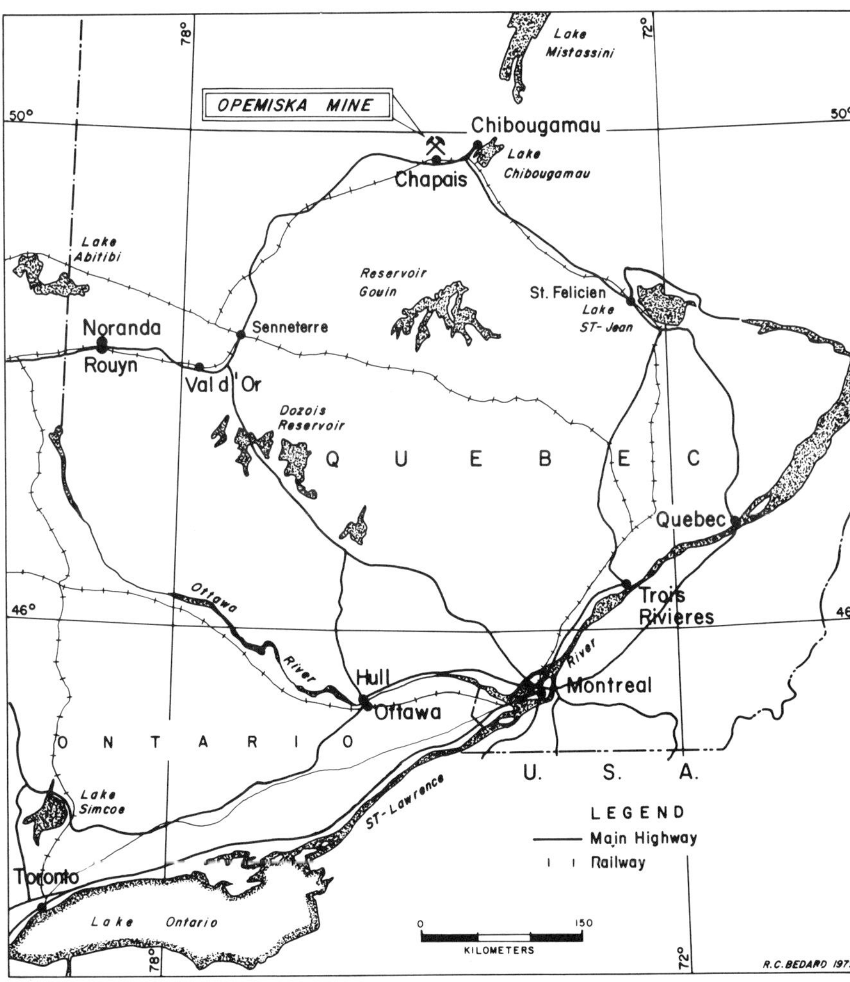

Figure 1. Location Map.

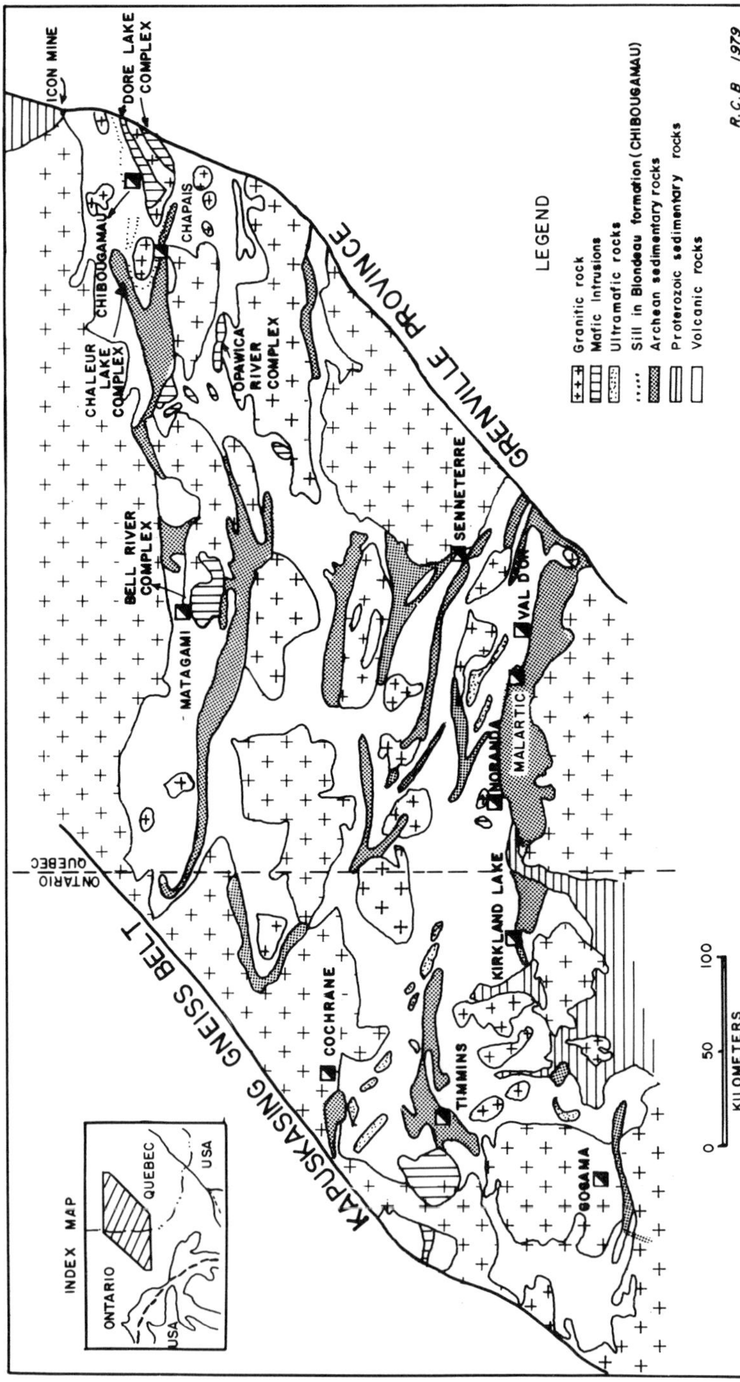

Figure 2. Generalized geologic map of the Abitibi Orogenic belt (modified from Goodwin and Ridler, 1970, and Allard, 1976).

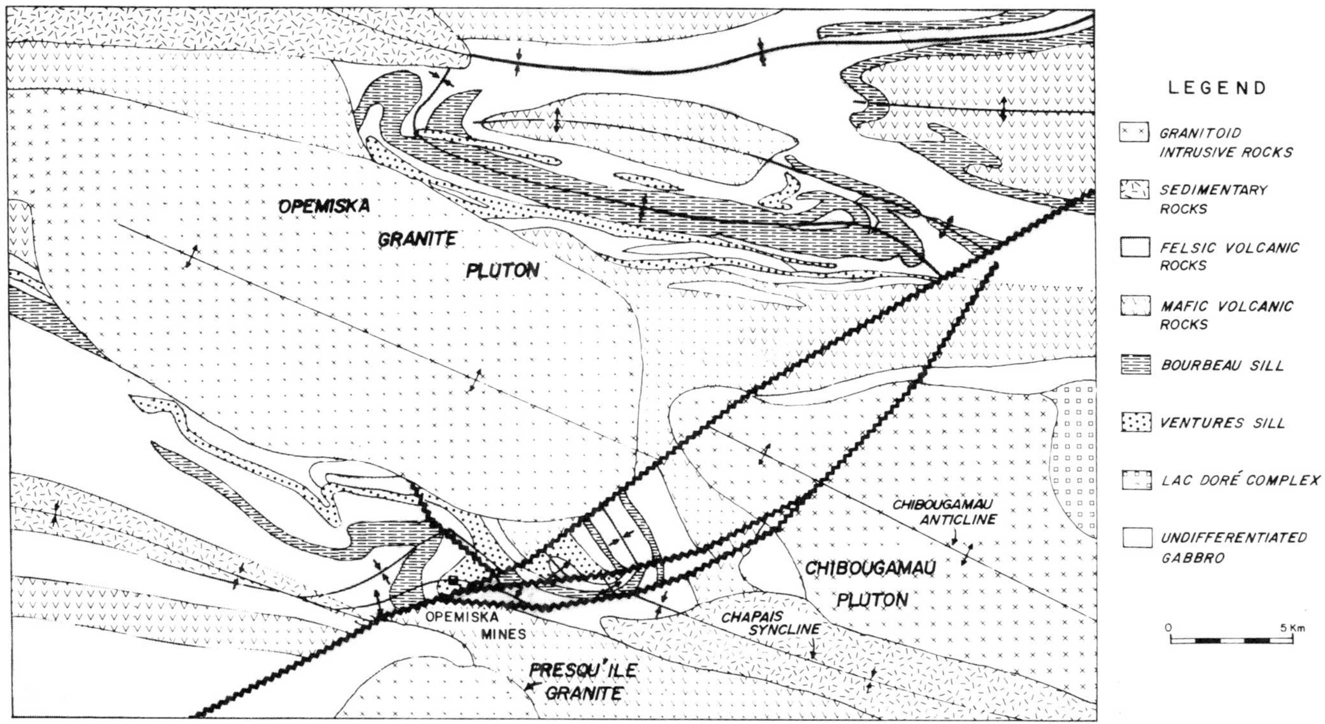

Figure 3. Regional geology of the Opemiska Mines area.

STRATIGRAPHY OF THE MINE AREA

The *Gilman* formation forms the lowest stratigraphic unit in the area and comprises an extensive platform of pillowed and massive mafic volcanic flows (Fig. 4). Numerous thin mafic sills intrude the volcanic rocks.

The Gilman formation is overlain by the *Blondeau* formation. Near the base this consists of massive domes of rhyolite which grade laterally into reworked felsic tuffs, siltstone, argillite, and greywacke and vertically into a sequence of siliceous tuffs and exhalites with numerous lenticular graphitic and sulphide-rich members.

The Ventures Sill intrudes the lower part of the *Blondeau* formation. The Bourbeau Sill intrudes the Blondeau several hundred meters above the Ventures Sill, commonly along a sulphide-graphite horizon.

PETROGRAPHY OF THE VENTURES SILL

The Ventures Sill is a layered intrusive complex that is about 1000 m thick in the Opemiska Mine area. Five laterally persistent units are distinguished within the sill (Fig. 5).

Lower Green Pyroxenite. This unit forms the basal layer of the Sill and is about 60 m thick. It comprises 90% equant to prismatic clinopyroxene grains, as well as rounded pseudomorphs of actinolite and magnetite after olivine. The contact with the overlying unit is generally sharp but with no evidence of chilling.

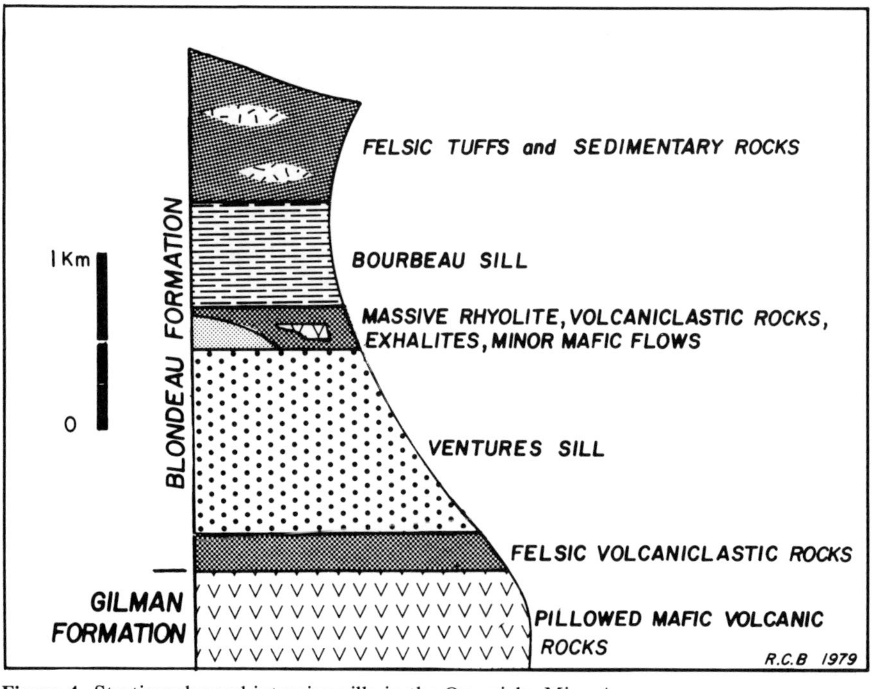

Figure 4. Stratigraphy and intrusive sills in the Opemiska Mine Area.

Black Pyroxenite. This 350 m thick unit differs from the Lower Green Pyroxenite by its overall dark grey to black colour and its medium-grained, equigranular texture. Primary igneous layering is common and clinopyroxene is the dominant mineral. Layers of serpentine – talc – magnetite after cumulate olivine are common, as are layers of olivine poikilitically enclosed in clinopyroxene. Layers containing primary chromite and magnetite are also recognized. The contact with the overlying unit is gradational.

Upper Green Pyroxenite. This unit is usually about 60 m thick and is somewhat similar to the underlying pyroxenites, but is coarser-grained and locally contains some interstitial plagioclase. Amphibole needles up to 1 cm long are also characteristic. The contact with the overlying unit is sharp, marked by the abrupt appearance of cumulus plagioclase and titaniferous magnetite (McMillan, 1972).

Foliated Gabbro. The average thickness of this unit is about 150 m. Its base is commonly marked by a 15 to 30 cm thick layer of clinopyroxene containing 30 to 40% magnetite. A prominent feature is rhythmic layering caused by the segregation of the main magmatic minerals which include andesine, clinopyroxene, and titaniferous magnetite. The rock is invariably strongly foliated due to alignment of the long axes of mafic minerals and feldspar. The contact with the overlying unit is sharp, marked by an abrupt change in texture and grain size.

Ventures Gabbro. The Ventures Gabbro is about 350 m thick and forms the uppermost unit of the sill. Its mineral composition is similar to that of the Foliated

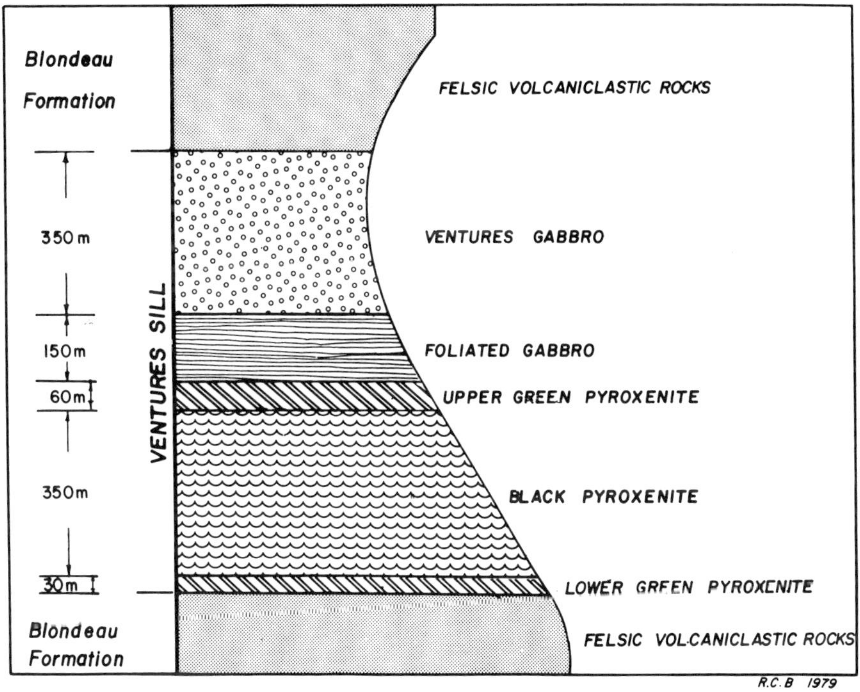

Figure 5. Layering in the Ventures Sill.

Gabbro but includes also about 5% free quartz. The rock has a characteristic coarsely crystalline, ophitic texture. A narrow zone up to 6 m wide containing disseminated pyrrhotite and chalcopyrite occurs locally at the top of the Ventures Gabbro as do granophyric patches.

STRUCTURE IN THE MINE AREA

The sills, volcanic and sedimentary rocks have been folded into an antiform-synform pair about east-trending fold axes that plunge moderately (45 to 65°) to the east (Fig. 6). The stratigraphic sequence in the mine area has been completely overturned by the folding. The folds are truncated by the northeast-trending, Campbell Lake fault which has an apparent left-handed horizontal displacement of 3300 m and has been active during several episodes (Brown, 1970). The Campbell Lake fault is a major regional structure and is marked by a prominent topographic expression over a steeply dipping zone that varies from a narrow, tight fracture to a chlorite-carbonate-quartz-rich shear zone up to a few hundred metres wide. The strata east of the fault are also overturned and the offset fold axes are interpreted to strike in a northerly direction.

West of the Campbell Lake fault, there are a series of east-trending, steeply dipping sets of axial plane faults and fractures with right-handed displacement which developed in areas of maximum inflexion of folds. In the limbs of the folds, northwest-trending faults and fractures with right-hand displacement and dipping moderately both NE and SW developed perpendicular to stratigraphy.

Brown (1970) suggested that the fractures post-dated the folding and that they were developed in response to a major transcurrent shear that was active in a zone of ductile behavior vertically underneath the semibrittle folded Ventures Sill. He also suggested that this transcurrent shear eventually became the Campbell Lake Fault when the shear extended upward into the present crustal level. However, the east- and northwest-trending ore-bearing fractures are most sharply defined and densely developed in zones of maximum inflexion of the fold, suggesting a relationship between the folding and the fracturing. Perhaps the folding initiated the fracturing in the Ventures Gabbro, which is the most coarsely crystalline and brittle rock of the sill. The fractures could then have been re-activated later as the Campbell Lake Fault developed.

Two parallel, east-trending left-handed displacement faults are located east of the Campbell Lake fault and are referred to as the Chiboug Copper fault and Laura Lake fault. The former has an apparent horizontal displacement of 250 m but the displacement on the latter is still unknown. Both faults disappear into the Campbell Lake fault to the west and into the Chibougamau pluton to the east. The Chiboug Copper fault is on the opposite side of the Campbell Lake fault but lies directly on strike with a major axial plane fault. It appears to act as a locus of intrusion for a number of small dikes and plugs that vary from dioritic to syenitic in composition. Moderate to strong hematite staining is common along the Chiboug Copper fault. A wide variety of mafic to felsic dikes intrude structures in the area, including those faults related to the folding.

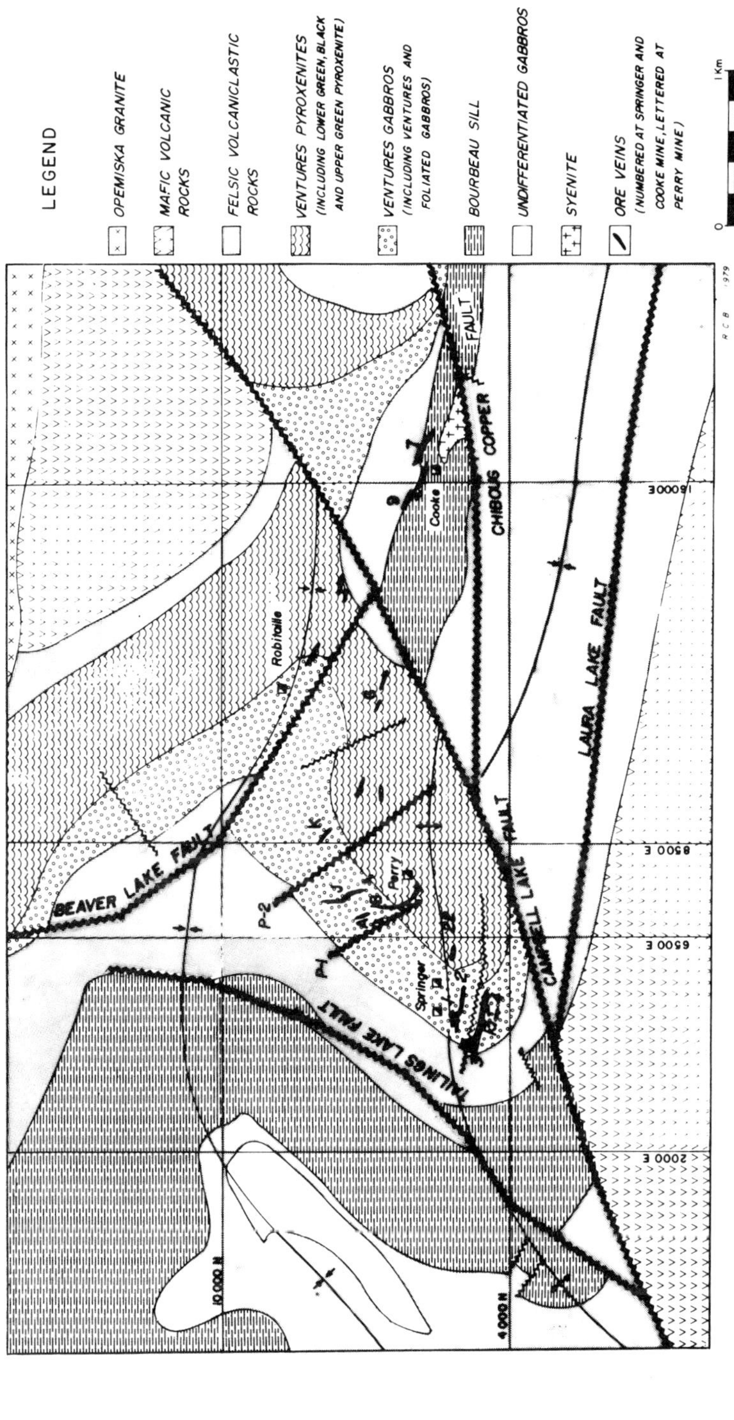

Figure 6. Geology of the Opemiska Mine area.

VEIN DISTRIBUTION

Almost 95% of the ore found to date occurs in vein systems that occupy fault, fracture, and shear zones that developed in the Ventures Sill as a result of folding. The main vein zones comprise the Springer, Perry, Robitaille and Cooke Mines.

Although ore occurs in all units of the Ventures Sill, the overwhelming bulk is in the Ventures Gabbro, particularly where it is dilated in the nose of the fold at the Springer Mine. Lesser amounts occur in the Foliated Gabbro on the fold limb at the Perry Mine, and a relatively small proportion comes from the finer-grained, feldspar-poor, pyroxenitic lowermost units of the sill at Perry and Robitaille. This distribution of the ore is attributed to differences in mechanical properties of the different units of the Ventures Sill.

The Cooke deposit is different from the other Opemiska orebodies, occurring in the Bourbeau Sill, which is intruded at a stratigraphically higher level than the Ventures Sill. The deposit lies east of the Campbell Lake fault and comprises two east-west-trending chloritic shear zones which have no apparent relationship with any fold hinge, unlike the other deposits. The shear zones appear to be subsidiary fractures related to the Chiboug Copper fault.

SPRINGER VEINS

Veins of the Springer Mine occupy faults, fractures, and shears that are parallel to the axial plane of the overturned anticline (Fig. 7). The width and frequency of veins tend to increase towards the dilated nose of the structure.

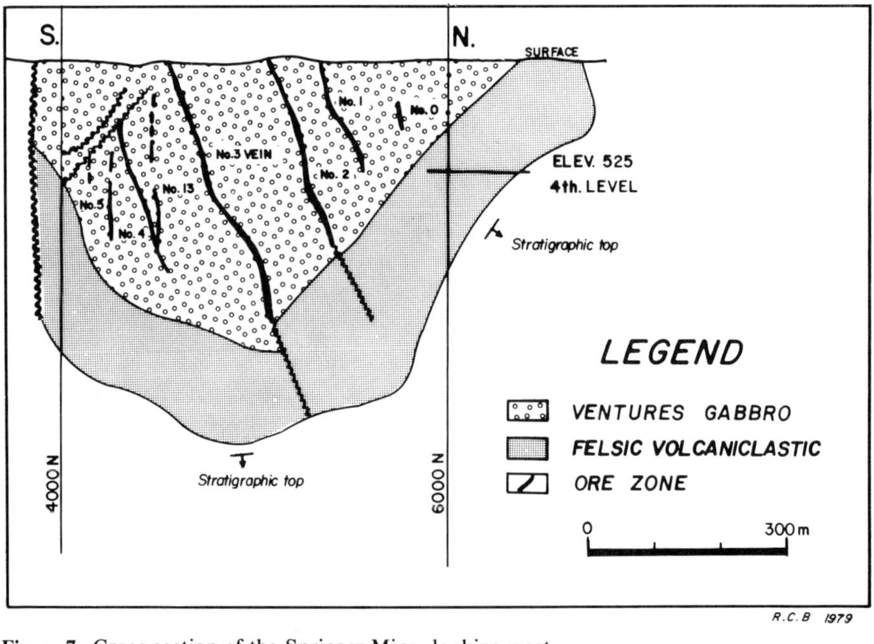

Figure 7. Cross section of the Springer Mine, looking west.

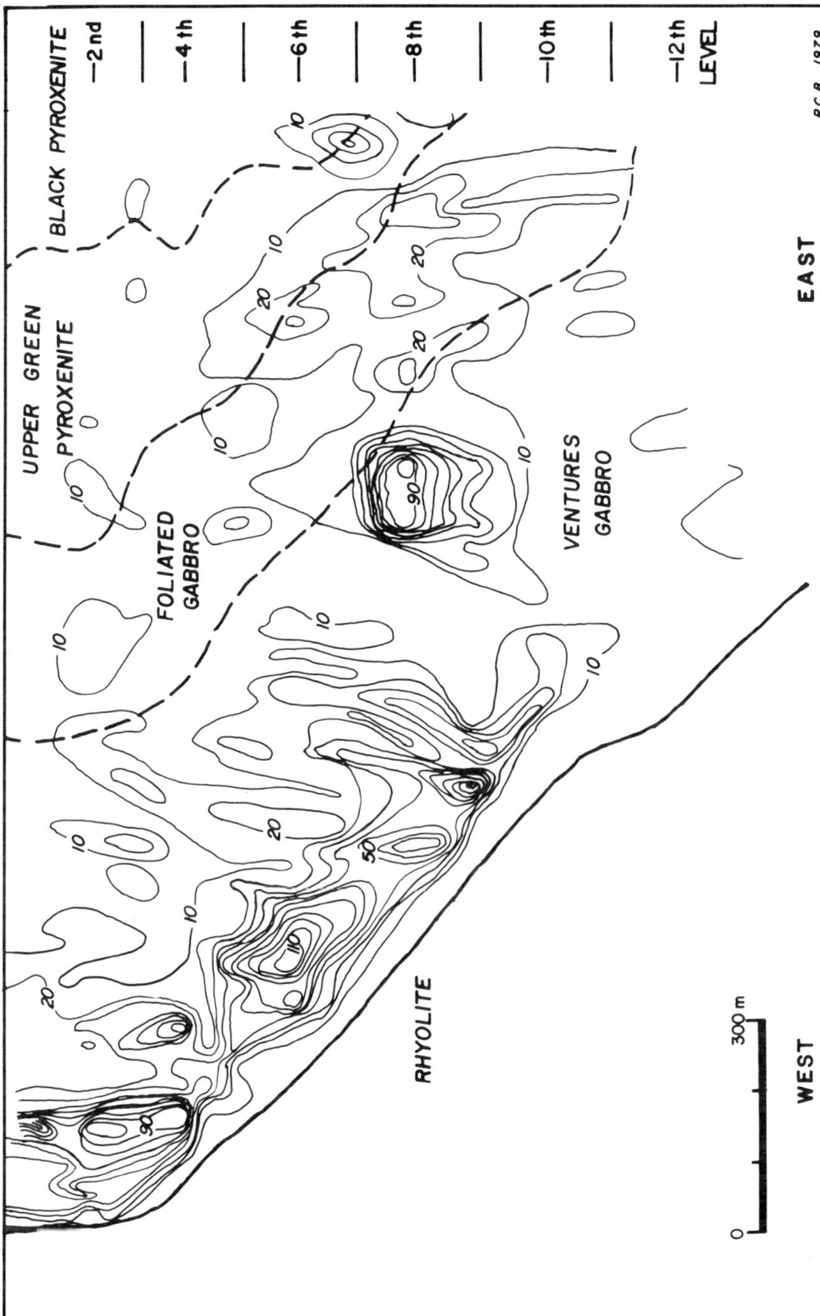

Figure 8. Contoured ore widths on a longitudinal section of No. 3 Vein. Contour interval is 10 feet.

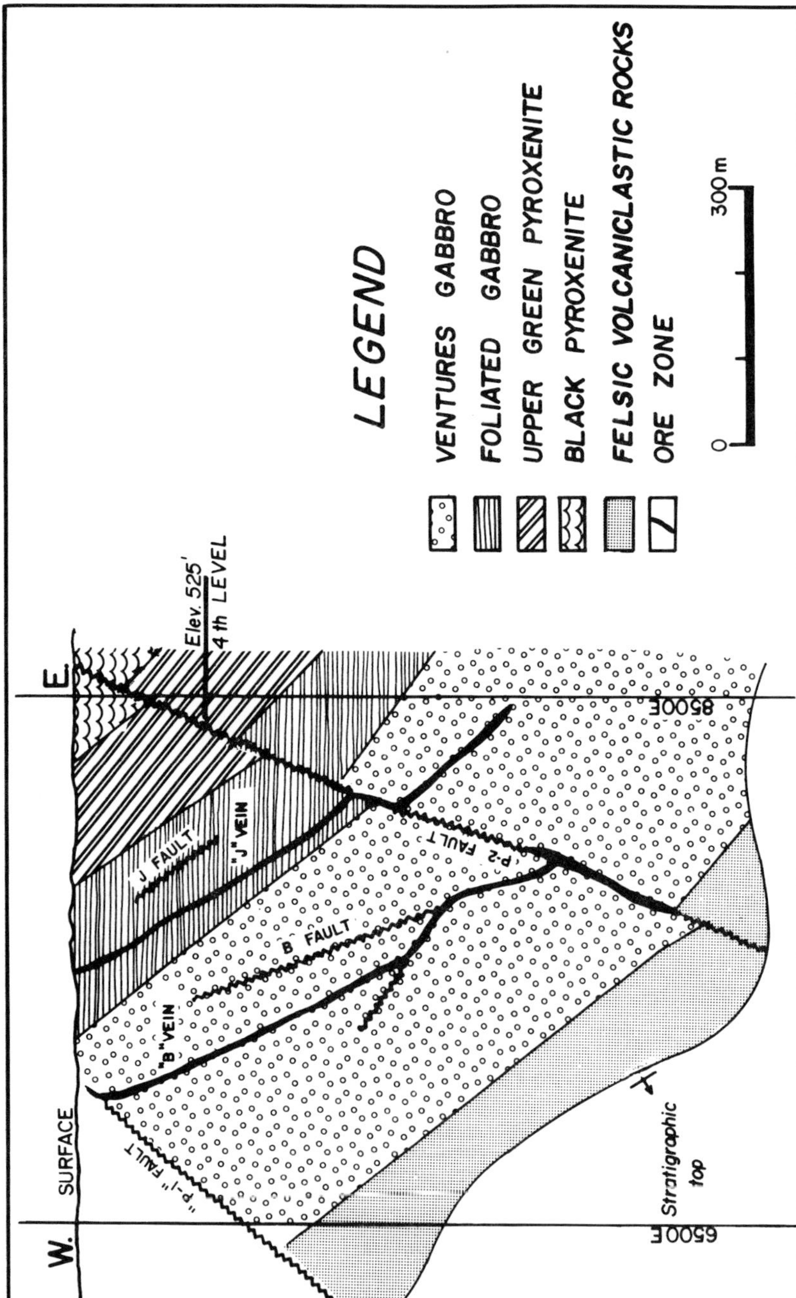

Figure 9. Cross section of the Perry Mine.

The No. 3 vein, from which over 5.45 million metric tonnes of ore have been mined, is the longest and most continuous vein in the area. It has a strike length of over 900 m, varies in width up to 25 m and extends down plunge, over 750 m. The fault zone it occupies has a right handed displacement of about 100 m.

The No. 3 vein tends to splay and ore widths increase substantially as the vein approaches the volcanic contact in the dilated zone of the fold nose (Fig. 8). Although the vein structure continues beyond the Ventures Sill and into the overlying rhyolite, the ore mineralization pinches out abruptly within a few metres from the contact.

PERRY VEINS

Ore-bearing fault and fracture zones at the Perry Mine tend to be at a high angle to stratigraphy on the limb of the main fold. Contacts tend to be more irregular or diffuse than at Springer. Ore mineralization commonly occurs in conjugate sets of faults associated with folding where they show a right-handed sense of movement (Fig. 9).

The largest vein at Perry was the ''B'' vein which contained over 2.72 million metric tonnes of ore. The ''J'' vein, which contained about 1.36 million metric tonnes of ore, differs markedly from other veins, being essentially a quartz vein with 10 to 15% sulphides in ore zones.

VEIN TEXTURES

Vein contacts vary from sharp to less commonly, diffuse and disseminated. Prominent banding in sulphides is observed near some vein contacts and zoning of sulphides is locally apparent. The sulphide veins vary from massive filling of fracture zones through to an irregular stockwork of stringers permeating shear zones or infilling the matrix of brecciated host rock. Fragments of the host rock are common within the veins and can often be fitted back into their original position. The fragments are commonly transected by sulphide stringers.

VEIN MINERALOGY

The most important vein and ore mineral is chalcopyrite. Pyrite and less common, erratically distributed, pyrrhotite are the main gangue sulphides. Quartz is the most abundant gangue silicate and occurs as irregular patches and fragments intimately associated with sulphides. Calcite is a common gangue mineral, occurring as fragments and veinlets. Both quartz and calcite locally have a pink color imparted to them by a light dusting of hematite. Biotite, stilpnomelane, actinolite, potassic feldspar, axinite, clinozoisite and chlorite are also found as gangue minerals.

Native Au occurs as flakes included in chalcopyrite and along fine fractures in chalcopyrite. Sub-economic concentrations of sphalerite, galena, molybdenite, and scheelite and uraniferous minerals are erratically distributed, but common and widespread. Uraniferous minerals are closely associated with molybdenite (Gravel, 1979). Arsenopyrite is common but in small quantities.

Magnetite occurs as fine octahedra and may form massive pods up to 15 cm thick. The magnetite in veins is nontitaniferous (McMillan, 1972) contrasting sharply with magmatic magnetite in the gabbroic wall rocks. Ilmenite, leucoxene, and rutile occur as skeletal remnants invaded by sulphides and sometimes by quartz and may be refractory wall rock remnants. Minor quantities of linnaeite and gersdorffite are also recognized.

MINERAL ZONING

Trends in mineral zoning within the veins are not quantified but in general terms, from Springer to Perry the grade of Cu is constant at 2.66% but the tenor of Au decreases from 1.2 g/t at Springer to about 0.5 g/t at Perry; scheelite and magnetite also decrease. Pyrrhotite becomes more important at depth at Perry, but almost disappears at Robitaille. Scheelite is absent and Au is very low at Robitaille.

At the stratigraphically higher Cooke Mine chalcopyrite is much less abundant than in Springer, Perry and Robitaille Mines, and the copper grade is only about 1.23% Cu. However, there is substantially more gold, with the average grade at 6.7 g/t. Scheelite and arsenopyrite are very common at Cooke. Visible gold is present in parts of No. 9 vein.

ALTERATION MINERALOGY

Alteration of the Ventures Sill around the veins proceeds from regionally metamorphosed rock through a transition zone with an increasing number of small veinlets, to partially recrystallized rock that gradually takes on the mineral composition of the ore bearing vein. The alteration zone is generally narrow and extends for 1 or 2 times the width of the associated vein.

Uralitisation of pyroxene is by far the most common alteration in the Ventures Sill where it has converted igneous pyroxene into actinolite with subordinate quartz and Fe-oxide. Actinolite occurs both as aggregates of small crystals and as large crystals replacing pyroxene. Remnants of pyroxene are so common within actinolite pseudomorphs that it is concluded that all amphiboles in the Ventures Gabbro and Upper Green pyroxenite are secondary. Uralite also occurs in cracks cutting feldspar grains and locally replaces feldspar near ore veins. However, uralitisation also extends far beyond the veins and cannot be taken as an indicator of proximity to mineralization. Wolhuter (1970) suggested instead that uralitisation of the Ventures Sill was a result of regional metamorphism.

Distinctive, heavy biotite alteration, however, is restricted to widths of 3 m along vein margins. Here, greenish-brown biotite may comprise up to 30% of the rock and occurs replacing uralite, interstitial to uralite, and as selvages bordering thin fractures. Biotite was apparently co-precipitated with chalcopyrite in veinlets that transect pyroxene and is strongly developed only in proximity to mineralization. The plagioclase is generally andesine in composition, ranging from An 28 to An 35 (Buchan, 1964). Andesine may be in part altered to albite (McMillan, 1972). It occurs as large fresh patches, locally rimmed with K-feldspar or partially replaced by biotite. Fresh feldspar is found even in the most intensely altered fragments of host rock caught up within the vein. Detailed petrographic studies failed to identify any sericite or clay minerals. This indicates that the ore fluids were in equilibrium with feldspar relative to clays, and that the pH of the fluids was probably alkaline to neutral.

ALTERATION CHEMISTRY

Changes in rock chemistry are confined to relatively narrow bands adjacent to the veins. Here copper shows the strongest and most consistent enrichment (Fig. 10)

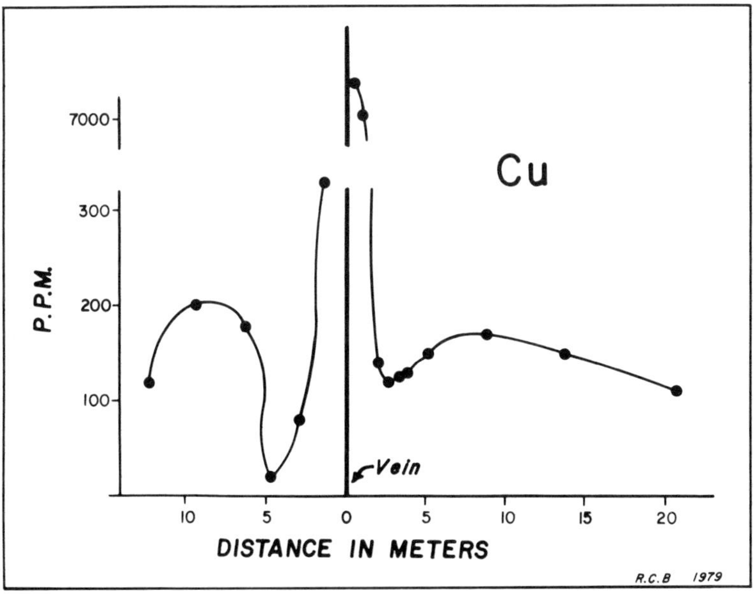

Figure 10. Distribution of Cu values across No. 3 Vein (After McMillan, 1972).

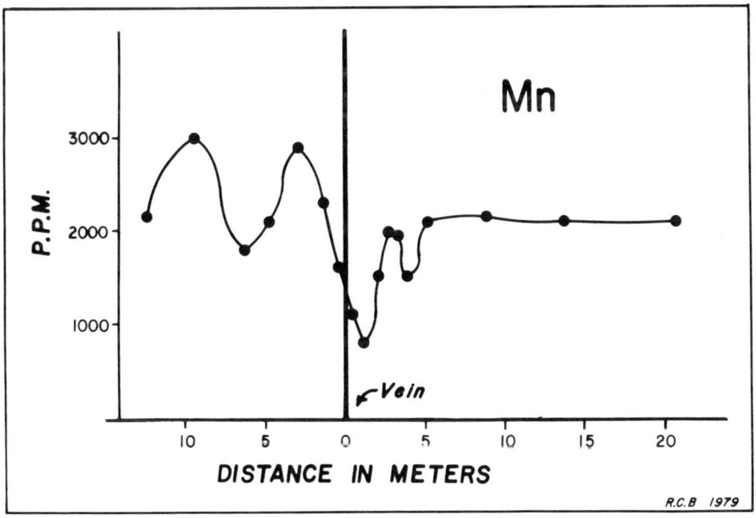

Figure 11. Distribution of Mn values across No. 3 Vein (After McMillan, 1972).

and sulphur behaves in similar fashion. Other metallic elements including Ag, Co, Ni, and Zn are erratic in occurrence but generally show enrichment in the vein walls. Of the major oxides K_2O and to a lesser extent Na_2O tend to be enriched, with maximum values at the outer edge of the alteration zone. However, because of the small distances over which K_2O and Na_2O were remobilized, and also because of the small degree of enrichment, K_2O and Na_2O analyses are not a promising exploration guide. Strong depletion of Mn, probably related to the destruction of pyroxene is noted in the wall rock (Fig. 11), but again the depleted zone is too small to be of any interest in exploration.

AGE DATA

Two galena samples from Opemiska were analysed for Pb isotopes by Thorpe, Guha and Cimon (1981). One sample came from a small fracture cutting volcanic rocks of the Blondeau Formation and the other was collected from an ore vein. On a plot of 207 Pb/204 Pb versus 206 Pb/204 Pb, the first sample falls in an "Archean" cluster, along with volcanogenic deposits such as Lemoine and Coniagas (Thorpe *et al.*, 1981, Fig. 2). However, the sample from the ore vein lies on the more radiogenic (and therefore younger) side of the cluster which encloses the Cu-Au deposits of the Chibougamau area. As the latter have a model age of 2,200 Ma, the data indicate either that the galena from the Opemiska vein is slightly younger than 2,200 Ma or else was formed in a more radiogenic environment than the Chibougamau Cu-Au deposits.

However, the co-existence at Opemiska of Archean lead in the Blondeau Formation and of younger lead in the ore veins raises the possibility that the lead in ore veins may have formed due to redistribution of Archean lead during a major hydrothermal event around 1800 to 2000 Ma. Although the data indicated that lead (and possibly other metals) definitely have been redistributed after the Archean, it is still not clear whether this means only minor readjustments in pre-existing veins, or a major remobilization of metals that coincided in fact with the vein emplacement.

SUMMARY

1) Chalcopyrite veins occur as fillings in fractures that developed in a massive differentiated sill, possibly as a result of folding. A number of variable dioritic to syenitic dikes that transect volcanic and sill rocks occupy similar structures.

2) The ore veins are best developed in the brittle, coarsely crystalline feldspar-rich gabbroic phases of the sill.

3) The vein mineralogy includes a fairly exotic assemblage of minerals including molybdenite, scheelite, uraninite, axinite and others.

4) Pb isotope evidence suggests a possible age of 1800 Ma.

5) The main feature of the alteration is extensive biotitization which is coeval with the sulphide mineralization. There is a complete lack of any sericite or clay minerals and chlorite is only locally developed as a retrograde mineral after biotite. Feldspar is relatively fresh and is seen even in the most intensely altered areas.

DISCUSSION AND CONCLUSION

As pointed out by Walker (1966), the origin of the Opemiska deposits ceases to be an academic question when we look for other similar deposits. Unfortunately, interpretations regarding the origin of the Opemiska deposits are still very speculative due to the lack of a sound background of data on isotopic age, sulphur isotopes and fluid inclusions. Three main genetic models have been proposed for Opemiska.

1) Metals were collected from the host rocks (Ventures Sill) and reprecipitated into fractures by hydrothermal solutions active at the time of granitic activity (Derry and Folinsbee, 1957).

This idea was based mainly on the fact that most of the significant mineralization is within the Ventures Gabbro. More recent work by McMillan (1972) indicated that in the mine area the Ventures Sill had copper and sulphur introduced into it from outside and that it is therefore unlikely that the Ventures Sill itself was the source of metals. To our knowledge, there is no segregation of magmatic copper sulphides at any of the sill contacts that would indicate that the magma was rich in copper and sulphur. In fact, the rocks that are unfractured or unsheared are particularly barren (except in the vein walls). We think that the spatial association of copper ore to the Ventures Sill can be attributed to favourable mechanical response to stress, which allowed the development of large open fractures especially in the more brittle phase of the Sill.

2) Metals were precipitated by magmatic hydrothermal solutions released during the late phase differentiation of the Opemiska Pluton.

This alternative was proposed to account for the presence of tungsten and molybdenum in uneconomic but significant amounts within the Opemiska veins. It was thought that these two metals were typical of late magmatic fluids of granitic affiliation and could, with copper and gold, be derived from the Opemiska Pluton (Walker, 1966; Duquette, 1970). Furthermore, Duquette (1970) considered that the post-kinematic time of emplacement for this pluton (Wolhuter, 1971) is consistent with the post-Kenoran age of the ore structures at Opemiska. However, petrological evidence strongly suggests that the Opemiska Pluton crystallized from a relatively "dry" magma (Wolhuter, 1971). Wolhuter even concluded that the scarcity of water in the magma was responsible for the absence of contact metamorphism and metasomatism around the pluton. Therefore, if the magma was too dry to allow the development of a contact aureole around its margins, it is unlikely that it could generate the fluids necessary to carry significant amounts of copper over a distance of 3 km. Walker (1966) also mentioned the possibility of other unexposed deep-seated granitic rocks underneath the Opemiska orebodies but this possibility cannot be tested.

3) The Opemiska orebodies are metamorphosed volcanogenic vein-type deposits, and are the deep-seated equivalent of, and coeval to the stratiform Cu-Zn mineralization found at the top of the Blondeau Formation.

This model for Opemiska was proposed by McMillan (1972) but has also been suggested by Allard (1976) for the Cu-Au deposits of the Chibougamau area. McMillan's model was based on similarities between Opemiska and some known volcanogenic vein-type deposits and on the fact that some magmatic magnetite is preserved in the Ventures Sill at Opemiska (versus none in one other study area 50 km away). He concluded from this that oxidizing conditions related to high volatile fugacity (itself due to mineralizing fluids) must have been present before greenschist facies metamorphism and thus that the veins are pre-metamorphic or synvolcanic (McMillan, 1972, p. 118 and 125). However, there is no evidence that the ore solutions were oxidizing and McMillan pointed out that the sulphides in the alteration zone formed at the expense of magnetite (McMillan, 1972, p. 103 and 106). Therefore, if the ore solutions were actually reacting with magmatic magnetite, the magnetite preservation argument becomes highly questionable.

Furthermore, one wonders why syn-volcanic veins developed only where the Ventures Sill is the most tightly folded and why they would be better developed in the hinge zone in the top stratigraphic unit of the sill instead of being developed across the whole sill. Assuming that this is purely coincidental and that the veins were emplaced before folding, then one questions how they could have survived the tight folding without being themselves folded and heavily sheared. Sheared sulphides do occur at Opemiska, but only locally and are accounted for by slight post-ore readjustments of some of the fractures. Finally, the alteration features observed at Opemiska contrast significantly with those observed in volcanogenic vein deposits such as the Ashio Copper Mine (Nakamura, 1970). In that respect, the scarcity of chlorite and the lack of sericite or clay minerals at Opemiska are most significant.

The genetic model proposed here involves to some extent processes already mentioned in the above three models. This model implies that the metals were leached from the volcaniclastic rocks of the Blondeau Formation and then reprecipitated in fractures which developed in the Ventures Sill in response to folding and shearing. More precisely, one particular stratigraphic horizon is thought to be the likely source for the metals and also the sulphur that now form the Opemiska veins. This horizon, about 100 m thick is located stratigraphically just above the Bourbeau Sill and comprises a series of felsic crystal tuffs mixed with graphitic shale. This horizon contains significant (but uneconomic) concentrations of iron, copper and zinc sulphides. The textures of the sulphides and preliminary sulphur isotope data indicate that these sulphides are largely of exhalative origin (J. Guha, pers. comm.). Because of the overturned structure at Opemiska (Fig. 7), this source horizon is now conveniently located 1.5 to 2 km vertically below the folded Ventures Sill. Hydrothermal fluids could therefore have dissolved some of the metals when circulating through the source horizon, and then re-precipitated copper, iron and zinc sulphides where they entered open fractures along their ascending path. The relatively weak alteration along the vein walls suggests that chemical changes were not strongly involved in sulphide deposition, and that changes in physical conditions (such as cooling) were the more likely cause.

As in the previous three genetic models, the nature and origin of the hydrothermal fluids are highly speculative. However, whether the fluids were purely magmatic or a mixture of magmatic, metamorphic and perhaps other fluids, the circulation of such fluids was most likely related to activity such as the intrusion of the Opemiska Pluton or the Presqu'Ile Granite, respectively 3.5 km north and 3 km south of the

Opemiska mine area. The localisation of the deposits where the Campbell Lake regional lineament intersects other major faults such as the Chiboug Copper and Laura Lake faults is possibly more than coincidental. The faults and subsidiary fractures could have acted as efficient conduits to promote the circulation of ore fluids.

Although meagre, the available Pb isotope data is consistent with the above model, since the data indicate that Archean lead at Opemiska may have been redistributed in post-Archean time (possibly around 1800 to 2000 Ma). Guha *et al.* (1979) have shown that the sulphur in the Cu-Au deposits of Chibougamau has been derived by the remobilization of volcanogenic sulphur which suggests an active circulation system through the volcanic pile in that area. The presence of methane in fluid inclusions from the Henderson Mine in Chibougamau might also indicate that the ore fluids have reacted at some stage with graphitic material (Guha *et al.*, 1979). This model is similar to the one that we propose for Opemiska, and research programs on fluid inclusions and lead and sulphur isotopes from Opemiska are in progress. The validity of our hypothesis will be tested only when all these data are reconciled with reliable radiometric dates for the Opemiska Pluton, Presqu'Ile Granite and other intrusive rocks in the area.

ACKNOWLEDGEMENTS

This paper represents the cumulative thinking of many geologists who have worked at Opemiska over many years. Particular thanks are due to Antoine Gagnon, who as chief geologist at Opemiska, was responsible for many innovative ideas and the supervision of J. S. Lavoie, R. H. McMillan and others whose work has been liberally incorporated into this summary.

REFERENCES

Allard, G.O., 1976, The Doré Lake Complex: Ministère des Richesses Naturelles du Québec, DPV-368, 470 p.

Brown, A., 1970, Fracture Analysis in the Opemiska Mine Area – A Study of the Fracture Pattern Associated with a Wrench Fault in Northwest Quebec: Unpublished Ph.D. Thesis, Queen's University, Kingston, Ontario.

Buchan, R., 1964, Report on Investigation of Samples Submitted by Opemiska Copper Mines Ltd.: Unpublished Report, Falconbridge Nickel Metallurgical Laboratories.

Derry, D.R. and Folinsbee, J.C., 1957, Opemiska Copper Mines: in Structural Geology of Canadian Ore Deposits: Canadian Institute of Mining and Metallurgy, v. 2, Congress Volume, p. 430-441.

Duquette, G., 1970, Archean Stratigraphy and Ore Relationships in the Chibougamau District: Quebec Department of Natural Resources Special Paper No. 8, 16 p.

Goodwin, A.M. and Ridler, R.H., 1970, The Abitibi orogenic belt: in Baer, A.J., ed., Symposium on Basins and Geosynclines of the Canadian Shield: Geological Survey of Canada Paper 70-40, p.1-30.

Guha, J., Leroy, J. and Guha, D., 1979, Significance of fluid phases associated with shear zone Cu-Au mineralisation in the Doré Lake Complex, Chibougamau, Quebec: Bulletin Mineralogique, v 102, p. 569-576.

McMillan, R., 1972, Petrology, Geochemistry and Wall-Rock Alteration of an Archean Mafic to Ultramafic Sill Complex, the Host to Copper Bearing Veins at the Opemiska Mine, Quebec: Unpublished Ph.D. Thesis, University of Western Ontario, London, Ontario.

Nakamura, T., 1970, Mineral zoning and characteristic minerals in the polymetallic veins of the Ashio Copper Mine: in Volcanism and Ore Genesis, Tatsuo Tatsumi, ed., University of Tokyo Press, 231-246.

Thorpe, R.I., Guha, J. and Cimon, J., 1981., Evidence from lead isotopes regarding the genesis of ore deposits in the Chibougamau region, Quebec: Canadian Journal of Earth Sciences, v 8, p. 708-723.

Walker, W.B.G., 1966, A Geochemical Study of the Opemiska Lake Granite: Unpublished Company Report.

Wolhuter, L.E., 1970, Geology of the Northwest Quarter of Levy Township: Quebec Department of Mines Preliminary Report No. 595, 26p.

———————, 1971, The Opemiska Pluton: Quebec Department of Natural Resources Special Paper No. 6, 118 p.

MS received January 1980;
Accepted for publication August 1980.

Precambrian Sulphide Deposits, H.S. Robinson Memorial Volume, edited by R.W. Hutchinson, C.D. Spence and J.M. Franklin, Geological Association of Canada Special Paper 25, 1982

THE DISCOVERY, GEOLOGY AND MINERALOGY OF THE CRANDON PRECAMBRIAN MASSIVE SULPHIDE DEPOSIT, WISCONSIN

Edwarde R. May and Paul G. Schmidt
Exxon Minerals Company
Post Office Box 676, Rhinelander, Wisconsin 54501

ABSTRACT

The stratabound Crandon massive sulphide deposit lies within a felsic to intermediate pyroclastic complex of the Southern Province of the Canadian Shield. The geophysical anomaly was detected in 1974 using an airborne EM system. Ground follow-up, by horizontal loop EM and Induced Polarization surveys, confirmed the presence of a bedrock conductor which was drilled in July 1975. Analysis of 198 diamond drill holes has confirmed a 62 million metric ton deposit grading 5% zinc, 1% copper. The mineralization was deposited by both syngenetic and epigenetic processes. Syngenetic deposition during a period of quiescence consisted predominantly of pyrite and sphalerite intercalated with argillite, tuff and tuffaceous chert. Penecontemporaneous epigenetic mineralization is characterized by anastomosing veinlets of chalcopyrite, quartz, pyrite and chlorite.

Host rocks for the sulphide deposit are explosive, multicompositional volcanic rocks which, in the footwall, consist of altered andesite and dacite breccia, flows and tuff. The hanging wall volcanic rocks vary from rhyolite to andesite and consist of fine to coarse tuff and lapilli tuff. The tabular-shaped deposit appears to be structurally simple with only minor recognized folding and faulting. All rocks have been metamorphosed to the greenschist facies. Rocks within and beneath the sulphide deposit have been metasomatically altered. A quartz-sericite-pyrite zone is present with adjacent assemblages of quartz, sericite, chlorite and quartz veining.

RÉSUMÉ

L'amas sulfuré lié aux strates de Crandon est situé dans un complexe pyroclastique de composition felsique à intermédiaire de la Province Sud du Bouclier Canadien. L'anomalie géophysique fut détectée en 1974 en utilisant un système EM aérien. Un relevé électromagnétique avec boucle horizontale et un relevé de polarisation induite confirmèrent la présence d'un conducteur qui fut foré en Juillet 1975. L'analyse de 198 trous de forage confirma l'existence d'un dépôt de 62 millions de tonnes métriques à 5% de zinc et 1% de cuivre.

La minéralisation est en partie syngénétique et en partie épigénétique. Le dépôt syngénétique consiste essentiellement en pyrite et sphalérite intercalées avec des argillites, tufs et chert tufacé. La minéralisation épigénétique est pénécontemporaine et caractérisée par des veinules anastomosées de chalcopyrite, quartz, pyrite et chlorite.

Les roches encaissantes sont des roches volcaniques explosives, polylithiques qui, au mur, consistent en andésite altérée et en brèches, coulées et tufs de dacite. Les roches du toit varient de rhyolite à andésite et consistent en tufs fins à grossiers et lapilli-tufs. Le dépôt tabulaire paraît structuralement simple avec seulement plissement et faillage mineurs tardifs. Toutes les roches ont été métamorphisées dans le faciès Schiste Vert. Les roches situées dans et sous l'amas sulfuré ont été altérées métasomatiquement: une zone à quartz-séricite est accompagnée d'assemblages adjacents de quartz, séricite, chlorite et veinules de quartz.

INTRODUCTION

The Crandon massive sulphide deposit is located about 10 km south of the community of Crandon, in Forest County, Wisconsin, U.S.A. (Fig. 1). It is a Zn-Cu volcanogenic deposit in Precambrian volcanic-sedimentary rocks. In August, 1977, Exxon Company, U.S.A. announced estimated reserves of 62 million metric tons of ore grading about 5% zinc and one per cent copper with substantially lower values in silver, gold and lead. Additional ore potential is indicated. Other Wisconsin massive sulphide deposits, all discovered in the past 10 years, include the Flambeau and Thornapple deposits of Flambeau Mining Corporation, a subsidiary of Kennecott Copper Corporation, near Ladysmith, and the Pelican River deposit of Noranda Exploration, Inc., near Rhinelander (Fig. 1).

EXPLORATION HISTORY

Exxon's massive sulphide exploration program in Wisconsin was initiated in 1970. After drilling 24 prospects the Crandon deposit was discovered in 1975 (Schmidt *et al.*, 1978).

Northern Wisconsin is extensively mantled with Pleistocene glacial drift; therefore, geologic interpretations were based on regional aeromagnetic and gravity surveys supplemented by outcrop mapping, drill hole data and studies of the glacial drift. Airborne electromagnetic surveys were flown in areas deemed favorable for massive sulphide deposits. The Crandon deposit was identified as an airborne electromagnetic anomaly in June, 1974, by Geoterrex Ltd., using a de Havilland Otter fitted with a Rio Tinto In-Phase/Out-of-Phase 320 Hz system and a Mark XII magnetometer. The anomalous zone was approximately 900 m long and had no associated magnetic response.

Initial ground surveying was done with a McPhar VHEM horizontal loop electromagnetic system using 600 Hz and 2,400 Hz at a 100 m transmitter-receiver separation. The conductor was identifiable but weak; the estimated overburden depth was 45 m. The conductor was verified by two lines of induced polarization surveying using a Newmont IPR7 receiver and an Elliot Geophysical Company 1.5 KW I.P. transmitter, at 61 m dipole-dipole spacing. Anomalies were rated as fair, deep conductors with moderate chargeabilities; the zone was +850 m long. Subsequent de-

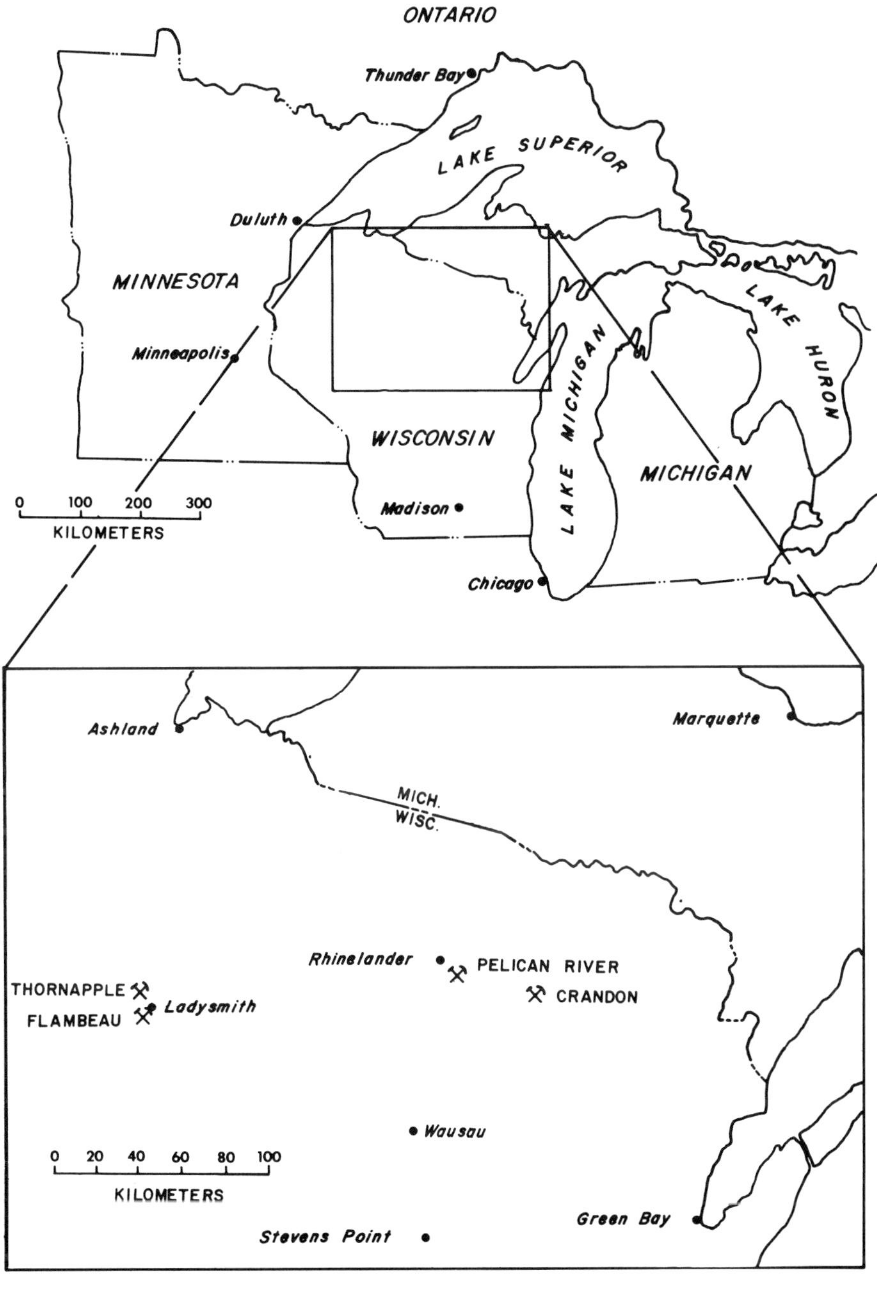

Figure 1. The general location of the Crandon deposit.

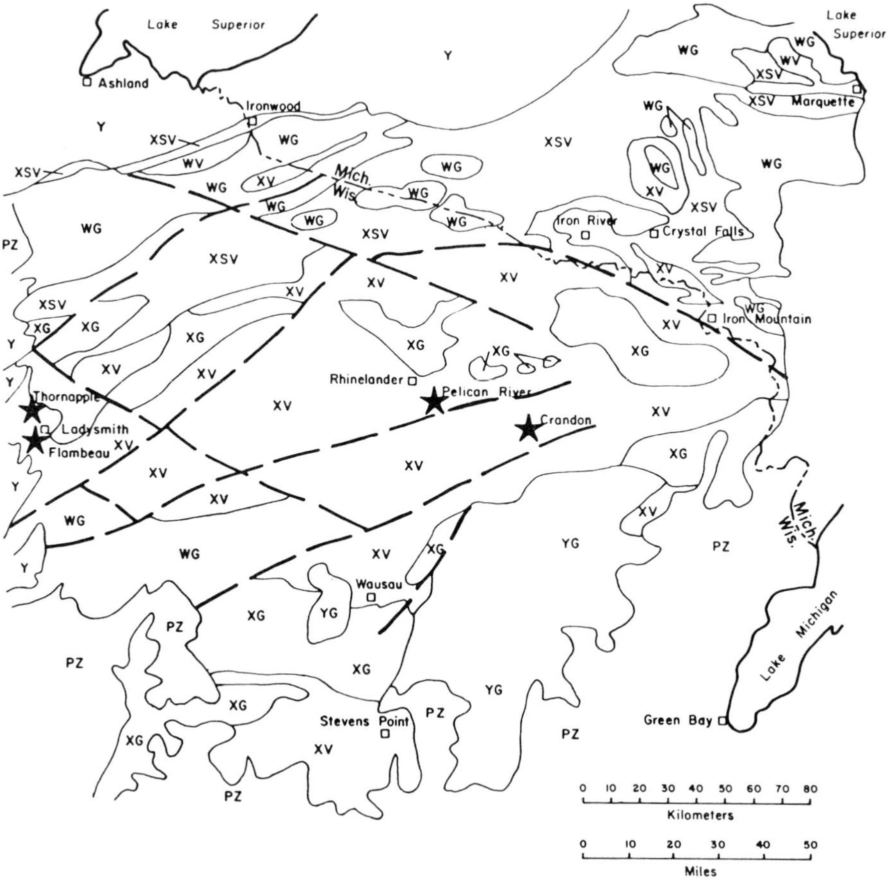

PZ Paleozoic Sedimentary Rocks.
Y Precambrian Y, Barron Quartzite, Keweenawan and younger rocks.
YG Precambrian Y, granitic intrusives.
XG Precambrian X, granitic intrusives.
XV Precambrian X, mainly volcanic rocks. Includes Quinnesec, Badwater and Hemlock volcanics.
XSV Precambrian X, mainly sedimentary rocks, quartzite, iron formation, greywacke. Includes Marquette Range Supergroup.
WV Precambrian W, mainly volcanic rocks.
WG Precambrian W, granitic and gneissic rocks.
Fault
Massive Sulfide Deposit
Geologic contacts

Figure 2. Precambrian Geology of Wisconsin-Michigan. Generalized and modified from Sims (1976), Sims *et al*. (1978), Cannon (1978), and Dutton and Linebaugh (1967).

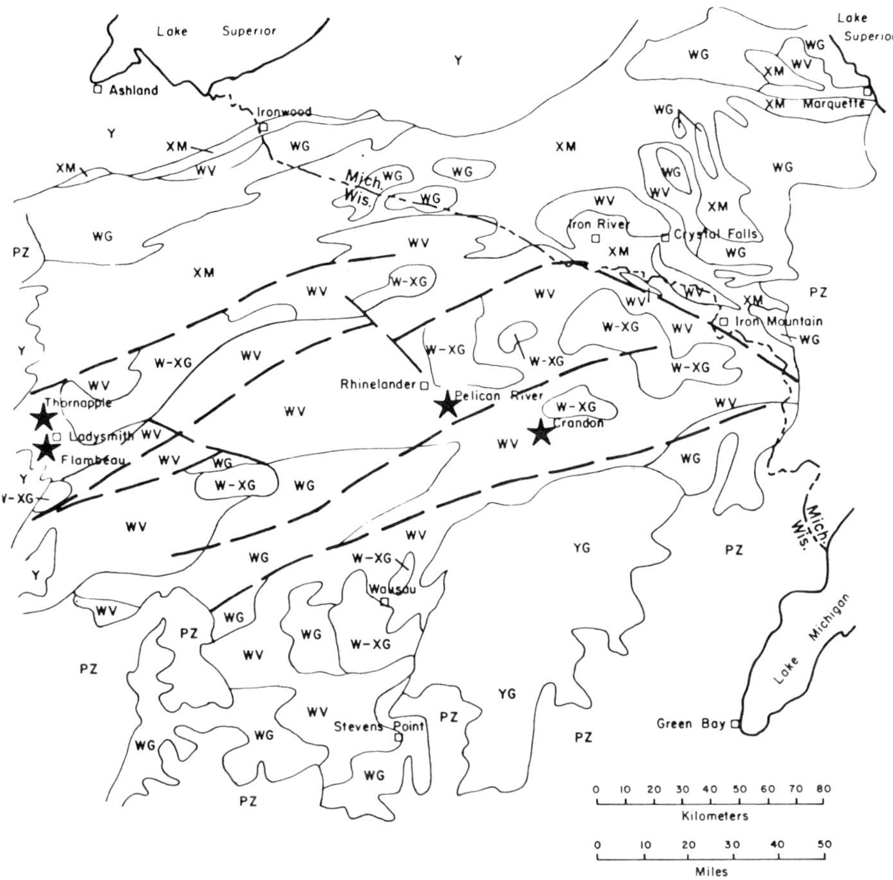

PZ Paleozoic sedimentary rocks.
Y Precambrian Y, Barron Quartzite, Keweenawan and younger rocks.
YG Precambrian Y, granitic intrusive rocks.
W-XG Precambrian W or X, granitic rocks undifferentiated.
XM Precambrian X, Marquette Range Supergroup, mainly quartzite, iron formation, greywacke.
WV Precambrian W, mainly volcanic rocks.
WG Precambrian W, mainly granitic and gneissic rocks.

Fault
Massive Sulfide Deposit
Geologic contacts

Figure 3. Precambrian Geology of Wisconsin-Michigan, interpretation P.G. Schmidt.

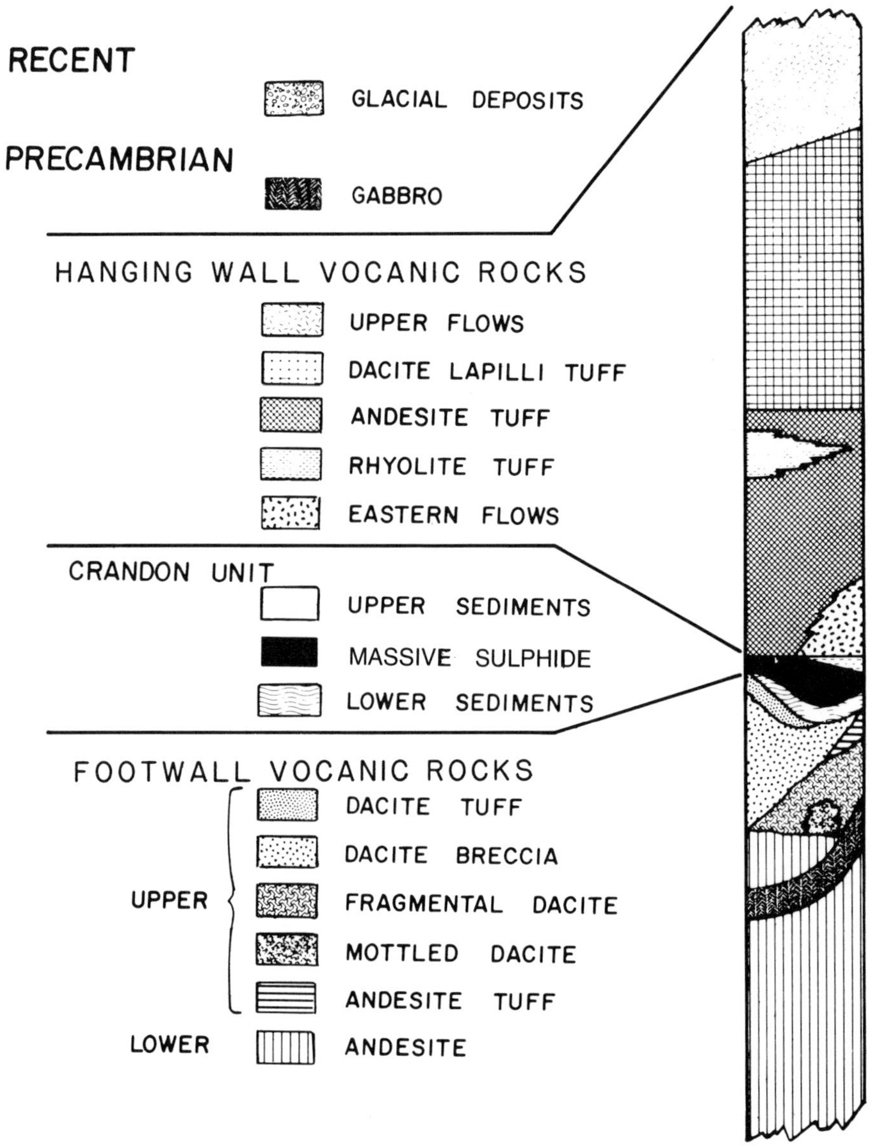

Figure 4. Crandon geological legend and generalized stratigraphic column for Figures 5, 6, 7, and 8.

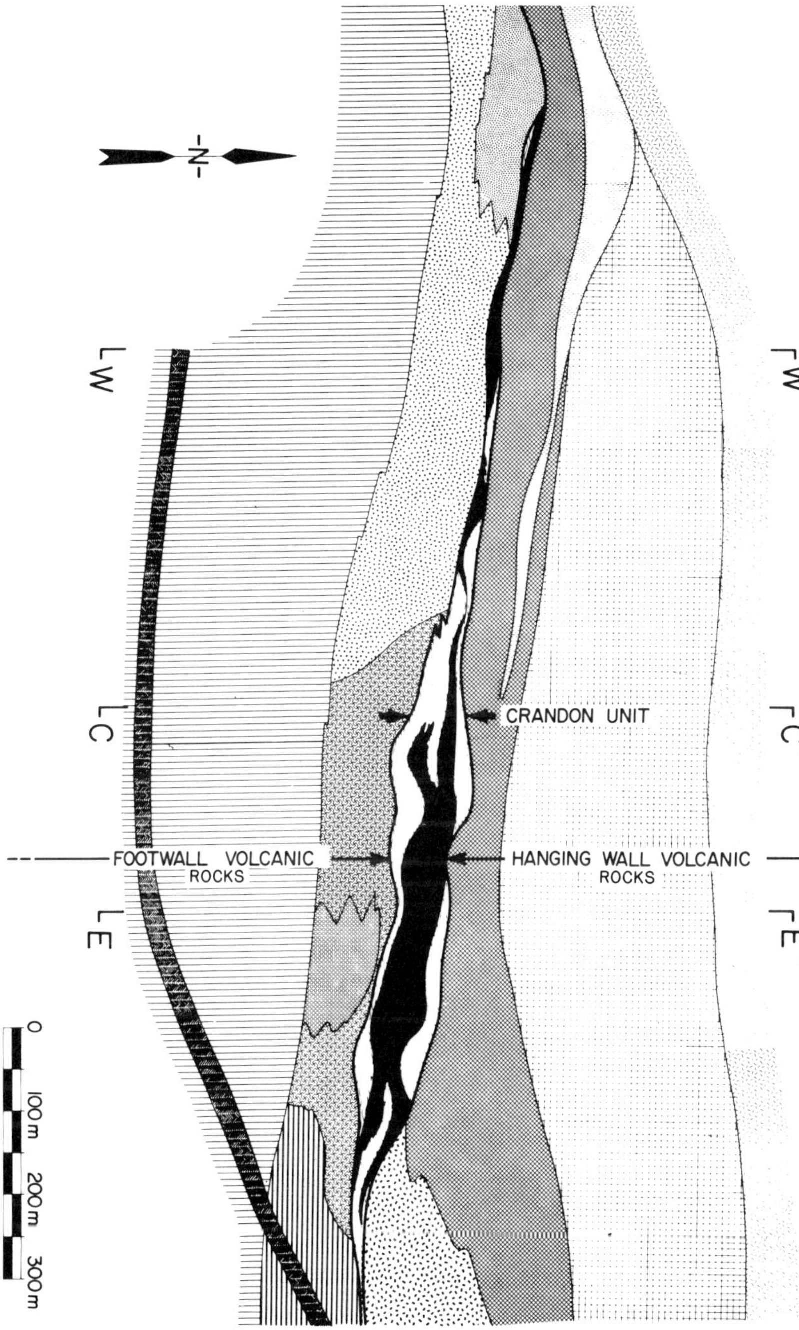

Figure 5. Geological Plan Map – 230 level, Crandon, Wisconsin.

tailed ground surveying has revealed that there is no recognizable magnetic anomaly associated with the deposit, but the deposit does have a 2.0 milligal gravity anomaly. The first drill hole at Crandon in July, 1975 encountered ore grade mineralization. Drilling to confirm the configuration, size and grade of the deposit has totaled 198 holes, and has ranged from shallow vertical holes into the top of the deposit to angle holes extending more than 760 m below the surface.

REGIONAL GEOLOGY

The Crandon deposit lies within the "Rhinelander-Ladysmith greenstone belt" which is 85 km wide, 240 km long and strikes easterly across northern Wisconsin and the Upper Peninsula of Michigan. Mafic and felsic metavolcanic and metasedimentary rocks, as well as younger granitic intrusions, comprise this belt. Regional metamorphism varies from lower greenschist to amphibolite facies within the greenstone belt. The nearest outcrops to the Crandon deposit are small exposures of granite about 10 km to the west. Due to the paucity of bedrock exposure, the local geology around Crandon is very poorly known.

Because the regional geology is complex, contrasting interpretations of regional tectonics, lithologic sequences and stratigraphic correlations have been proposed. The Rhinelander-Ladysmith greenstone belt rocks, including the Crandon deposit, can be correlated with the Marquette Range Supergroup of northern Wisconsin and Michigan (Fig. 2), indicating a Precambrian X age (Middle Precambrian or Proterozoic) of between 1,900 and 2,200 Ma (Van Schmus *et al.*, 1975; Sims, 1976; Van Schmus, 1976; Prinz, 1976; Sims *et al.*, 1978; Sims and Peterman, 1978; Cannon, 1978). Alternatively, the Rhinelander-Ladysmith greenstone belt rocks and the Crandon deposit can be placed into Precambrian W (Archean) greenstone-granite-gneiss terrane, an interpretation that is favored by some geologists. An older age for this greenstone belt has been suggested because of regional similarity to Archean Superior Province greenstone terranes in Michigan and Ontario, inadequate age dating of Wisconsin rocks, and because the Rhinelander-Ladysmith greenstone belt rocks do not appear to be interbedded with sedimentary rocks of the Marquette Range Supergroup (Fig. 3).

STRATIGRAPHY

The Crandon deposit lies conformably within a sequence of predominantly dacitic volcanic rocks and their sedimentary derivatives. All rocks strike about N 85° W and dip 85° N. Based on mineral zoning and relict depositional features, stratigraphic tops are interpreted as facing north. The rocks at Crandon can be divided into the footwall volcanic rocks, the Crandon unit and the hanging wall volcanic rocks (Fig. 4).

Footwall Volcanic Rocks

The footwall rocks have been divided into two units. The lower unit is a moderately foliated succession of interbedded andesitic flows and tuff (Table I). The andesite has been metasomatically altered and bleached to a light grey rock; it exceeds 200 m in thickness with massive individual flows averaging 20 m, and contains about

TABLE I
WHOLE ROCK ANALYSES OF THE CRANDON ROCK UNITS

	Footwall					Crandon Unit		Hanging Wall			Gabbro
Wt. %	A	B	C	D	E	F	G	H	I	J	K
SiO_2	60.39	80.26	67.94	65.07	65.63	73.08	52.95	71.63	61.85	68.73	52.03
Al_2O_3	16.07	7.21	12.48	17.05	14.47	8.54	7.07	11.65	16.91	14.07	15.62
Fe_2O_3	9.24	7.85	10.22	6.37	7.75	12.24	20.56	9.00	10.08	8.32	11.60
MgO	6.07	2.51	5.12	4.19	4.40	4.18	15.66	4.96	5.83	5.12	5.76
CaO	3.19	.26	.24	3.34	4.38	.33	.70	.29	1.45	.34	10.08
MnO	.13	.04	.09	.10	.14	.13	.20	.08	.19	.40	.16
Na_2O	1.83	.65	.99	.22	.79	.38	Nil	.14	1.13	.41	2.64
K_2O	2.27	1.30	2.43	2.90	1.91	.75	2.12	1.94	2.13	2.75	.74
TiO_2	.73	.29	.56	.70	.58	.38	.75 (2.54 Carbon)	.31	.43	.37	1.07
No. of Samples	19	11	9	14	5	4	34	6	21	51	25
No. of Feet Sampled	2,084	483	673	699	368	115	426	312	1,176	2,435	

A Andesite
B Dacite breccia with strong quartz veining
C Dacite breccia
D Fragmental dacite
E Andesite tuff
F Lower tuffaceous chert
G Argillite
H Rhyolite tuff
I Andesite tuff
J Dacite lapilli tuff

5% pyrite where it has been strongly sericitized. The pyrite occurs as saw-toothed, interstitial fillings where the tuffs have been altered and shattered. Where unaltered and unmineralized the andesitic flows and tuff contain calcite in the matrix.

The upper unit of the footwall sequence is a complex suite of pyritic pyroclastic rocks. Five zones have been recognized based on texture, composition, size and shape of the rock fragments. Each zone exhibits complexly interbedded rock types and has poorly defined contacts. The five zones from west to east are: dacite breccia, dacite tuff, fragmental dacite, mottled dacite and andesite tuff (Fig. 5).

Dacite Breccia. One of the more easily recognized zones intersected in drill holes is the dacite breccia beneath the west half of the Crandon unit. This zone is distinguished by abundant dark, coarse, subangular, chlorite-rich lapilli and breccia fragments of a variety of rock types (Plate I, Fig. 18). Lighter-coloured fragments are locally intermixed with darker fragments (Plate I, Fig. 19). The silicified, chlorite-rich lapilli occur in a light grey, very fine-grained siliceous matrix. Many fragments contain light-coloured relict feldspar phenocrysts ranging from 0.2 mm to 0.5 mm in size. A further distinguishing feature of this zone is its negligible amount of matrix sericite. Possibly the breccia represents an earlier flow breccia that was violently disrupted and redeposited in a chaotic state.

Dacite Tuff. A second zone, which consists of dacite tuff and lapilli tuff, lies stratigraphically above the dacite breccia. It contains beds ranging in thickness from one to 10 m and of widely varying texture. The tuff consists of a fine-grained, strongly foliated and locally laminated mixture of quartz-sericite and chlorite ash; the lapilli tuff consists of subrounded felsic fragments four to 16 mm in diameter in a matrix of fine- to very fine-grained chloritic ash. The pyrite content gradually increases from 5% at the base to 15% near the upper contact. Locally, the rocks have been fractured and later healed with pyrite.

Fragmental Dacite. A third zone, which underlies most of the eastern half of the Crandon unit, consists of white quartz-sericite fragments within a matrix of similar composition, and is interpreted as a strongly metasomatically altered fragmental rock of dacitic composition (Table I). Three distinct textures are recognizable. The first is characterized by sharply bounded, subangular lapilli fragments within a quartz-sericite-pyrite matrix (Plate I, Fig. 20). The second consists of subrounded indistinct lapilli within a similar matrix. This rock may represent a brecciated flow or tuff with corroded, rotated fragments. The third texture is a crackle breccia containing unrotated fragments, the boundaries of which may be either planar or saw-toothed (Plate I, Fig. 21). The pyrite content of this zone ranges from 5 to 30%, decreasing in amount toward the stratigraphic base. Disseminated grains of pyrite occur within the lapilli and the quartz-sericite matrix, and as filling around the crackle breccia fragments.

Mottled Dacite. A fourth zone closely associated with the fragmental dacite is mottled dacite tuff and lapilli tuff (Plate II, Fig. 22). This zone consists of a fine grained, mottled-textured intermixture of quartz-sericite and chlorite. Rock textures change rapidly, with the mottling becoming indistinct towards the stratigraphic base. The chlorite-rich clots, lenticular patches and irregular bands could represent altered lapilli, contorted and disrupted beds or differential bleaching and hydrothermal sericitization of the original rock. The mottled dacite is interpreted as partially altered fragmental dacite.

Andesite Tuff. Beneath the east end of the Crandon unit is a fifth zone which differs markedly from the other footwall rocks because of its abundant content of graphite and calcite. This predominantly sedimentary zone consists of interbedded, fine-grained calcareous, chloritic tuff and dacite breccia. The graphite is in argillite beds up to several metres thick and also occurs as minute laminae within the andesite tuff. This zone represents a transition between pyroclastic rocks to the west and a sedimentary environment to the east where graphitic beds become more abundant. The lower part of this zone consists of dark-coloured, silicified lapilli breccia in a light-coloured siliceous matrix. Cream-coloured hydrothermal sericite is present. This chaotic mixture of silicified fragments and matrix may represent an explosive center.

The Crandon Unit

The Crandon unit has a strike length of about 1,400 m and consists of a variety of fine-grained pyritic tuffs and sedimentary rocks. By definition, the base of the Crandon unit is the lowest significant argillite or chemical sedimentary bed recognized in the drill hole. The top of the Crandon unit is defined as the uppermost recognizable bed of pyritic tuff or tuffaceous sedimentary rock. Contacts within the unit, as well as

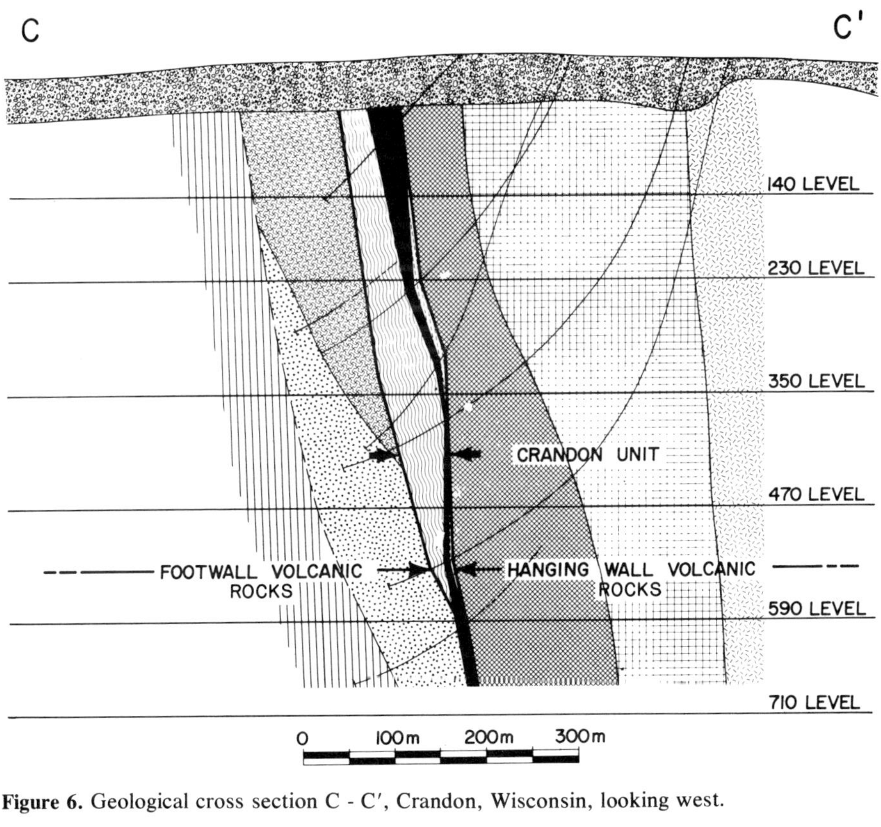

Figure 6. Geological cross section C - C′, Crandon, Wisconsin, looking west.

the contact between it and the hanging wall volcanic rocks are generally sharp. However, the contact between the cherty sedimentary rocks at the base of the Crandon unit and the silicified footwall pyroclastic rocks is, in many places, difficult to recognize because of intense metasomatic alteration.

For convenience, the east half of the Crandon unit may be divided into Lower, Middle and Upper zones, and the unit reaches its maximum thickness of 100 m where all three are present (Figs. 6 and 7). These zones rapidly pinch out eastward, here bringing the footwall andesite tuff into contact with the eastern flows of the hanging wall rocks. The west half of the Crandon unit is about 20 m thick because the Lower and Middle zones (Fig. 8) are nearly absent. The Crandon unit gradually pinches out to the west where it appears to be represented by a disseminated pyritic zone between the foot and hanging wall volcanic rocks.

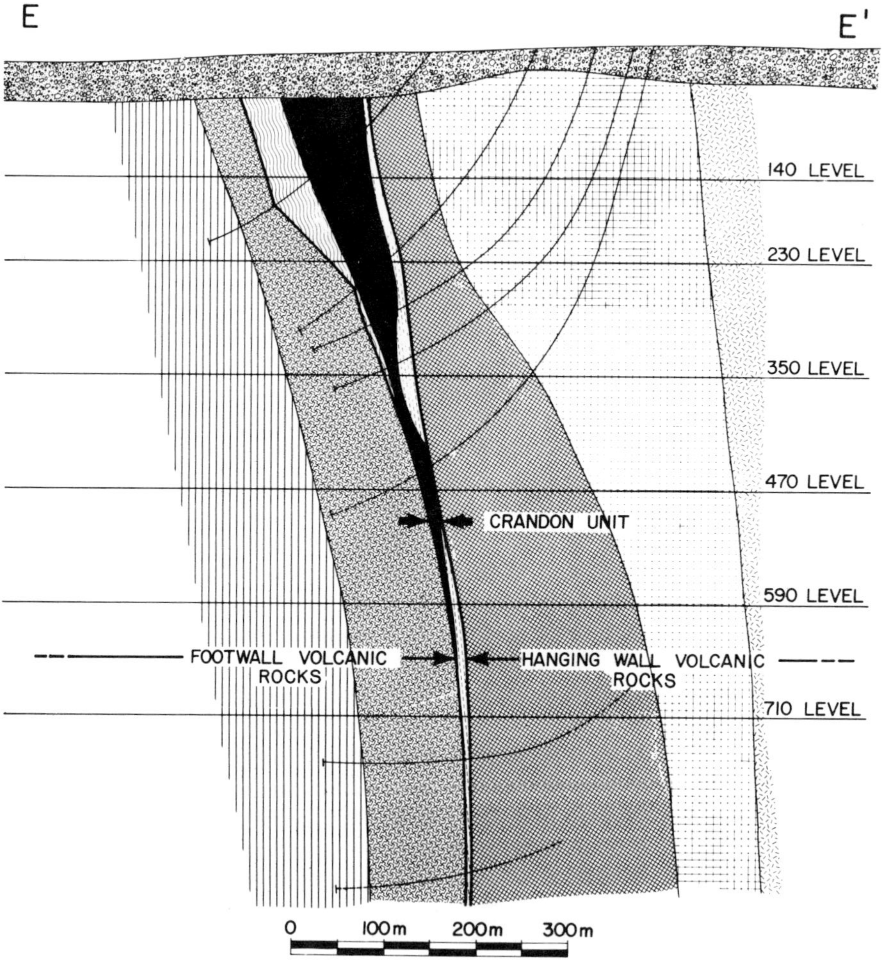

Figure 7. Geological cross section E - E′, Crandon, Wisconsin, looking west.

The Lower Zone. The Lower zone of the Crandon unit is distinguished by a thick accumulation of cherty sedimentary rocks, interbedded with argillite and felsic tuff. Chert-rich sedimentary rocks are predominant in the west half of the Lower zone (Plate II, Fig. 23). The cherty rocks grade laterally westward into dacite breccia. This 40 m thick accumulation of chert and tuffaceous chert is poorly bedded, white to medium grey and contains an average of 10% pyrite. However, the pyrite content increases to about 35% near the stratigraphic top. The contacts between these cherty sedimentary strata and the footwall volcanic rocks are difficult to recognize because of strong metasomatic alteration.

The east half of the Lower zone consists of about equal amounts of chloritic argillite and siliceous tuff. Chert is commonly found with argillite as laminae or as lenticular patches less than 5 mm in length. The amount of argillite appears to increase with depth and to the east. Local accumulations of siliceous dolomite up to 15 m thick have been noted interbedded with argillite.

The Middle Zone. Massive sulphide makes up the dominant portion of the Middle zone of the Crandon unit, and is locally interbedded with argillite, cherty sedimentary rock, or felsic tuff. On the east end, the massive sulphide deposit bifur-

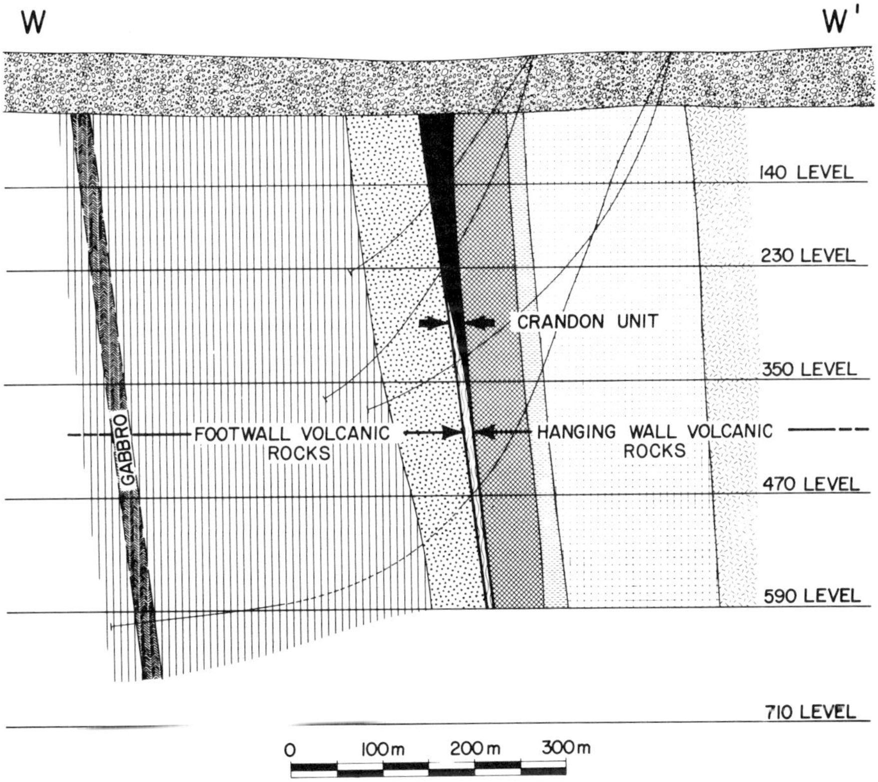

Figure 8. Geological cross section W - W′, Crandon, Wisconsin, looking west.

cates, with felsic tuff and argillite separating the two mineralized zones. Chlorite is the dominant gangue silicate in the western half of the Middle zone. Here the Middle zone is bounded by and includes pyritic chloritic argillite, which can contain more than 50% pyrite. Therefore, it is in places difficult to distinguish between chloritic massive sulphide and pyritic argillite. This is in contrast to the siliceous gangue that dominates the east half of the Middle zone. The massive sulphide mineralization is further described in subsequent sections of this paper.

The Upper Zone. The Upper zone of the Crandon unit differs markedly from the Lower zone in its wide diversity of sedimentary rocks that can rarely be correlated between drill holes 60 m apart. The greatest variety of rock types is in the western half of the Upper zone. Recognizable sedimentary rocks are argillite, tuffaceous siltstone, tuffaceous chert, tuffaceous sandstone, sandstone, laminated chert, thin massive sulphide beds and dolomite. The black, well foliated argillite contains abundant pyrite, which locally exceeds 50% of the rock. Whole rock analysis indicates that this rock is rich in chlorite and contains about 2.5% carbon (Table I).

The east portion of the Upper zone contains a 20 m thick, interbedded sequence of strongly foliated, tuffaceous siltstone and felsic tuff (Fig. 7). Tuffaceous sandstone or felsic to intermediate tuff occur locally as thin beds. The pyrite content is generally low, ranging from 3 to 10%, but locally is approximately 20%. Brecciated and contorted, possibly slumped dolomite is associated with the Middle and Upper Crandon unit sedimentary rocks.

Hanging Wall Volcanic Rocks

Andesite Tuff. A fine- to coarse-grained, 150 m thick tuff of andesitic composition overlies the Crandon unit (Table I). This tuff is composed of a well foliated, fine to coarse ash interbedded with lapilli tuff and minor amounts of fine-grained clastic sedimentary rocks. Average thickness of individual beds is about 2.5 m. Two characteristic features of the andesite tuff are extreme textural changes over short stratigraphic distances and highly variable lithology indicating multiple sources. The essential minerals are quartz, oligoclase, chlorite and sericite, with accessary epidote, carbonate, sphene and less than one per cent pyrite. An abundance of hyaloclastic material can be seen in thin section. Relict glass texture, such as perlitic cracks and devitrified shard-like fragments are present, as are possible microlitic textures. Features such as graded bedding and flame structure are also present. The basal one to 20 m of the hanging wall volcanic rocks consist of chloritic ash with 10 to 30% subrounded to subangular dacite lapilli.

Dacite Lapilli Tuff. The dacite lapilli tuff is a thick-bedded pyroclastic succession consisting of beds 10 to 20 m thick. This tuff contains 20 to 40% chlorite lapilli fragments in a light grey felsic ash matrix. These elongate lapilli fragments have sharp upper and lower boundaries and frayed ends. This texture is believed to represent collapsed, devitrified and chloritized pumice fragments. The crudely banded matrix consists of subrounded, siliceous, 0.25 mm to 2 mm size ash. The dacite lapilli tuff commonly displays eutaxitic texture (Plate II, Fig. 24).

Rhyolite Tuff. A pale greyish-green to light grey, fine-grained, well foliated rhyolite tuff (Table I) is present over the west half of the deposit. Contacts are gradational with the enclosing volcanic rocks. One to 5% pyrite occurs as disseminated grains or as laminae. Regional metamorphism has locally altered the rhyolite tuff to produce a pale yellow rock that may contain up to 70% sericite.

Upper Flows. A series of massive porphyritic flows, rhyodacitic to andesitic in composition, overlie the dacite lapilli tuff. Individual dark grey-green flows range in thickness from 2 to 10 m. The flows commonly contain 10% feldspar phenocrysts that range in size from two to four mm in an aphanitic matrix. Megascopic examination of drill core indicates that these flows are more dacitic in composition and less porphyritic to the east. They become thicker in this direction, with interbeds of up to 50% well foliated intermediate tuff.

Another series of fine- to medium-grained equigranular flows of rhyodacitic composition immediately overlie the east end of the Crandon unit and interfinger with the andesite tuff to the west. These flows have been named the Eastern flows (Fig. 4).

Igneous Rocks

Gabbro. A nearly vertical gabbro dyke, 25 m thick is present in the footwall of the deposit. The dike strikes parallel to the stratigraphic units in the central part of the map (Fig. 5) whereas to the east it transects them, striking N 60° E. Geophysical evidence indicates that the gabbro cuts across the hanging wall volcanic rocks farther to the east beyond the map area.

The gabbro is dark grey, equigranular and consists of two or three mm crystals of pyroxene and plagioclase. Chilled margins of the dike are about one metre thick, and contact metamorphic effects are limited to within 5 m of the contacts. This intrusion is post-mineral and may be post-metamorphic.

Dikes and Sills. Narrow, dacitic to andesitic dykes and sills less than 3 m wide strike both across and parallel to the trends of the Crandon host rocks. These intrusions are green, fine-grained, massive rocks that contain feldspar phenocrysts. The dikes and sills are nearly vertical and they trend and branch out in a northwesterly direction. Most of these have been observed in the hanging wall volcanic rocks, although they occur rarely below the 630 level in the footwall sequence. These intrusive rocks make up an insignificant fraction of the total volume of rock at the Crandon deposit.

Glacial Deposits

Pleistocene glacial drift entirely covers the Crandon deposit to depths ranging from 25 to 60 m. One of the last glacial events in Forest County was formation of an extensive drumlin field striking about N30°E. Medium- to coarse-grained sand is the predominant glacial material, with gravel increasing in amount at depth.

STRUCTURAL SETTING

Both regional and local structure are poorly understood because of the absence of outcrops and the lack of detailed diamond drilling, but the results of surface drilling suggest that both folds and faults are present. Stratigraphic correlations based on a 60 m drill pattern indicate that fault offsets are small, but possibly significant in an underground mine design. Fault zones of variable width containing gouge and breccia cut the deposit at irregular intervals along strike. Present interpretation of the drill hole data indicate that the faults probably strike northeast and dip steeply northwest.

Folds where present appear to be open. Sinuousity, recognized at the upper and lower contact of massive sulphide mineralization, may have resulted from faulting, folding or depositional irregularities. Minor local tight folds have been noted that probably represent either drag folds or slump structures. In plan, the Crandon unit is convex to the north at the west end (Fig. 5). The explanation for persistence of this convex structure to depth is unclear, but this is believed due to a post-mineral fold.

ALTERATION

With the possible exception of the gabbro, all rocks have been regionally metamorphosed to mineral assemblages characteristic of the quartz-albite-muscovite-chlorite subfacies of the lower greenschist facies. Chlorite is present in all but the felsic rocks. This greyish-green chlorite is the product of metamorphism of iron and magnesium-bearing minerals. Metamorphic sericite is also present in most Crandon rocks. At least two distinct alteration periods have taken place at Crandon. These are alteration associated with mineralization and supergene alteration.

Alteration Associated With Mineralization

Silicification of the footwall pyroclastic rocks, strong sericitization of the fragmental dacite and parts of the andesite, late stage quartz veining with or without sulphide mineralization, are commonly observed metasomatic effects. They have resulted in depletion of calcium and sodium in many of the rocks.

Most of the footwall volcanic rocks are harder and contain more visible silica than similar-appearing hanging wall volcanic rocks, indicating that pervasive silicification may have taken place. The source and mode of transport of this silica is not understood. It could be that the silica is primary, and that the footwall volcanic rocks were predominantly rhyolitic in composition. This conclusion would be similar to that documented for Archean massive sulphide deposits found at Timmins and Noranda in Canada. Alternatively, the silica may have been metasomatically introduced, which is the interpretation favored by geologists working on the Crandon deposit. Such alteration has been documented for some Kuroko deposit host rocks (Tatsumi and Watanabe, 1971) although silicification of these rocks does not appear to be a general condition (Clark, 1971).

Quartz veining is conspicuous throughout much of the western part of the upper footwall unit, and decreases in intensity with depth. Chlorite and sulphides are associated with some quartz veins. Siderite is locally found as small, irregular veinlets. It is concluded that these quartz veins were formed by hydrothermal fluids introduced into the fractured and porous footwall rocks.

Two types of metasomatic chlorite are present in the footwall pyroclastic and Crandon unit sedimentary rocks. The first type is a dark, forest-green chlorite that is associated with chalcopyrite in quartz veins within the footwall sequence. It occurs either as blotches or as rims on chalcopyrite grains (Plate IV, Fig. 30). The second type is a very dark green, almost black, chlorite associated with the Crandon unit argillite. The argillite was probably metasomatically altered to produce a highly magnesian rock (Table I).

Most of the Crandon rocks are peraluminous, as defined by Zavritsky (1964), because of their low CaO and Na_2O content rather than high Al_2O_3 (Table I). It is interpreted that the peraluminous rock is a product of metasomatic alteration similar to the alteration present at Mattabi (Franklin *et al.*, 1975).

Strong sericite alteration occurs in the footwall andesite and fragmental dacite. The sericite is yellowish-brown, stringy and occurs pervasively replacing the central portions of the structurally controlled altered footwall andesite. However, nearer the margins of these zones sericite is found as fracture selvages. An interpretation of the alteration at Crandon is given in Figure 9.

Supergene Alteration

Near surface rocks, particularly those of intermediate composition, have been altered to clay minerals. Oxidation is moderate to intense and is generally confined to within 15 m of the subcrop, except where fractured zones or rock contacts allow penetration of the oxidizing fluids to depths of 200 m or more. The primary sulphides are altered to limonite within the oxidized zone. The limonite has been deposited by circulating fluids on fracture surfaces and pervasively stains the rock to various shades of red and yellow. Oxidation of the sulphide minerals has produced a gossan over the deposit ranging from less than one to 30 m thick. The contact between the base of the gossan and the underlying sulphides is always sharp.

Leaching of the deposit and, to a lesser extent of the country rock, extends to depths of 125 m or more below the subcrop. Sphalerite is preferentially leached and virtually removed from many near-subcrop parts of the deposit. Chlorite has also been appreciably altered and bleached. Where sphalerite and chlorite have been strongly leached, the massive sulphide body becomes a mixture of sandy, euhedral to subhedral pyrite and clay, with or without quartz.

MINERALOGY OF ORE TYPES

Like many volcanogenic deposits, Crandon consists of a syngenetic sulphide body underlain by an epigenetic sulphide zone. The syngenetic ore consists of massive and semimassive sulphides; the epigenetic ore is composed of stringer and disseminated sulphides (Fig. 10). Although the deposit has been altered by supergene processes it is only weakly enriched in supergene sulphides.

Massive Sulphide

Massive ore is defined as that containing 50 or more total volume per cent syngenetic sulphides. The massive sulphide portion of the Crandon deposit is estimated to contain more than 75% sulphides, pyrite being the most abundant. Other sulphides are sphalerite, chalcopyrite and galena, with trace amounts of tetrahedrite-tennantite and arsenopyrite. Gold and silver are also present in small amounts, but in an unknown form.

Pyrite forms euhedral to subhedral crystals that range in size from 2 to 3 μ or more than a millimetre. Aggregates of pyrite commonly resemble either clastic or brecciated fragments similar to those observed in many other volcanogenic deposits (Hutchinson and Searle, 1971). These rounded to subrounded fragments of extremely

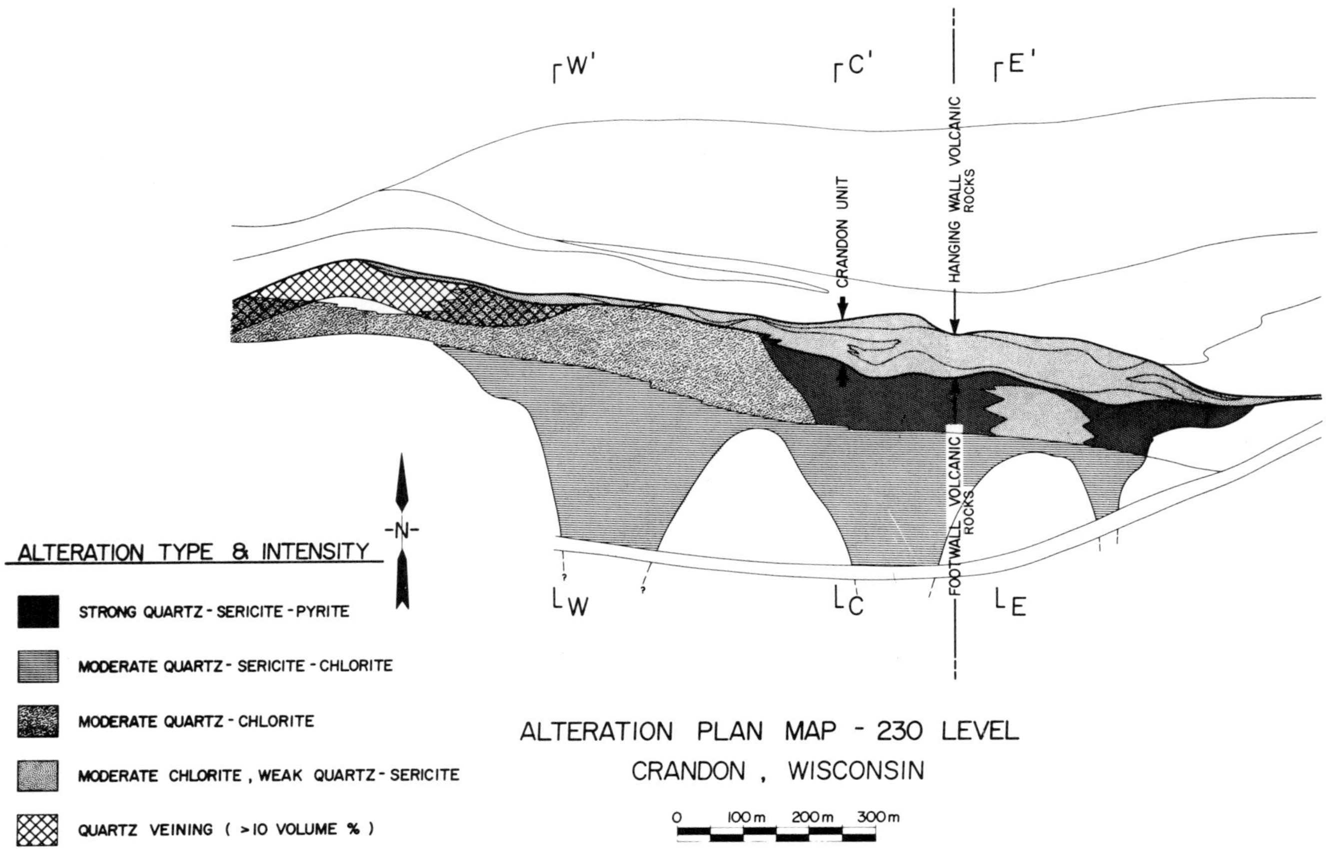

Figure 9. Alteration plan map – 230 level, Crandon, Wisconsin.

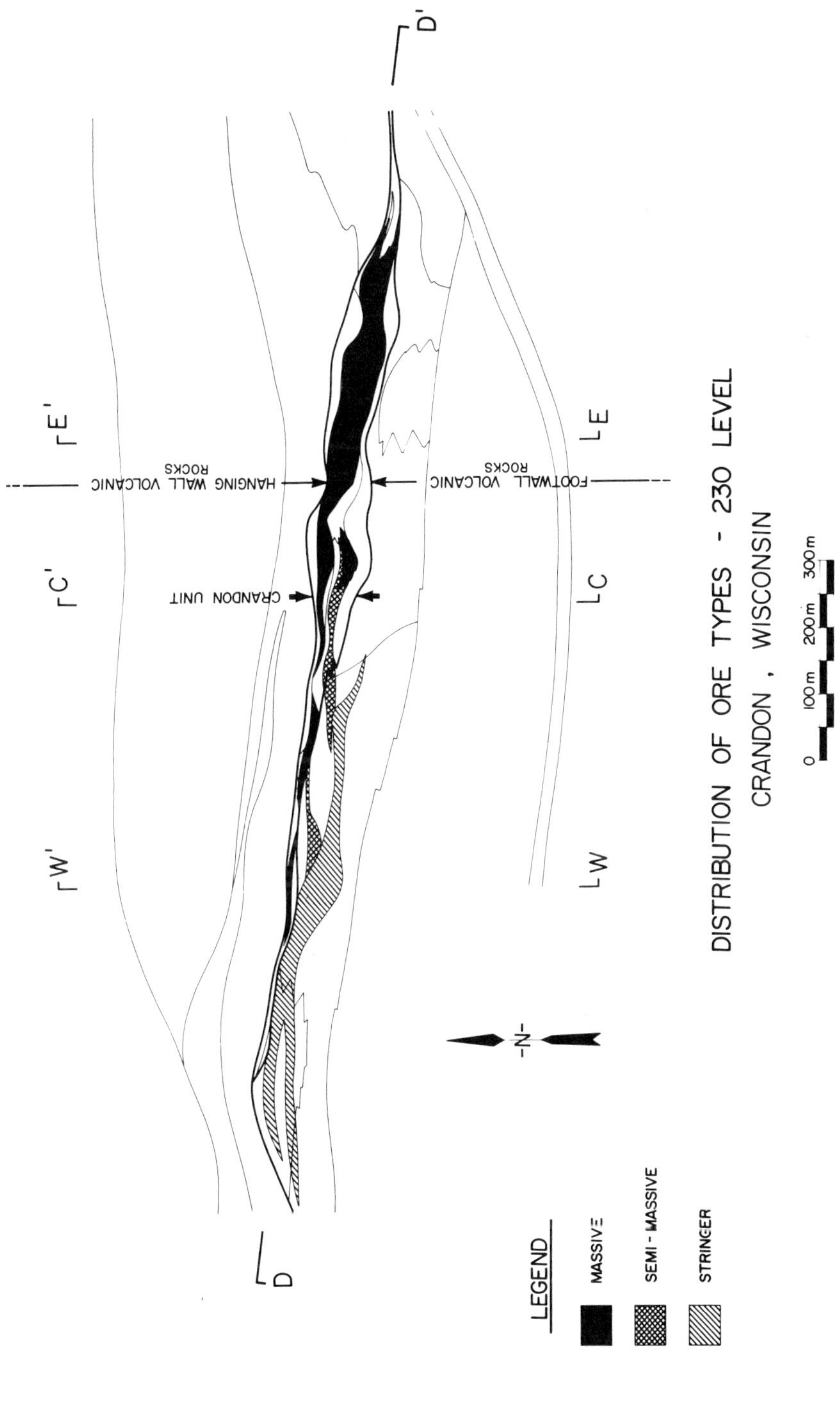

Figure 10. Distribution of ore types – 230 level, Crandon, Wisconsin.

fine-grained euhedral pyrite are surrounded by coarser-grained pyrite or gangue (Plate II, Fig. 25). Many of these pyrite fragments are broken and the voids filled with chalcopyrite, sphalerite and quartz, either individually or as combinations of these. The genesis of the pyrite aggregates is not clear, but many have been due to slumping. Occasional colloform, botryoidal and framboidal textures have been observed in pyrite.

Sphalerite is the most abundant economic mineral in the massive sulphide ore. It is generally pale brown, although locally darker, and has an iron content ranging from 2 to 3%. Individual sphalerite grains are generally less than one millimetre in diameter. Sphalerite is commonly interstitial to euhedral pyrite (Plate III, Fig. 26), although extremely fine-grained sphalerite inclusions also occur within pyrite. The sphalerite-rich portions of the massive ore are commonly bedded (Plate III, Fig. 27). In massive barren pyrite, crude bedding can be distinguished in etched polished sections due to the contrast between finely crystalline and more coarsely crystalline pyritic layers.

Chalcopyrite is fine grained, with grains averaging about 0.01 mm in diameter and is therefore less noticeable in the massive ore than sphalerite (Plate III, Fig. 28). It is interstitial to pyrite or sphalerite grains and also occurs as inclusions within brecciated pyrite. Various-sized, rounded chalcopyrite inclusions that sometimes have similar crystallographic orientations have been observed within sphalerite and pyrite.

Galena is interstitial to sphalerite or in quartz gangue and is rarely seen in pyritic portions of the deposit. It occurs as scattered grains or thin laminae. Tarnished galena and sphalerite, noted in polished sections, could be related to a high silver content of these minerals. Galena grains range in size from a few microns to one millimetre.

Semimassive Sulphide

Semimassive ore is defined as syngenetic mineralization containing from 20 to 50% of sulphide minerals. It is found as disseminations or as one to 20 cm thick massive beds interbedded with the host rock, principally chert, tuffaceous chert and lapilli tuff. This ore type occurs in the west central portion of the deposit within the lower member of the Crandon unit, and within the top of the footwall breccia (Fig. 10).

Stringer Sulphide

Stringer ore, defined as that deposited by epigenetic processes, makes up about one half of the estimated Crandon reserves. Pyrite is the predominant sulphide, and chalcopyrite is the main economic mineral. Subordinate sphalerite, and minor galena and arsenopyrite are present.

Chalcopyrite is generally much coarser-grained than pyrite, with irregular blebs up to 10 mm in diameter. About 60% is in quartz veins and 35% fills rock interstices; the remainder occurs as finely disseminated grains within the pyrite, vein matrix or volcanic host rock.

Secondary brecciation played an important role in emplacement of the sulphide and gangue minerals. The consolidated and silicified footwall volcanic rocks were fractured, thus creating channelways in which quartz, pyrite and chalcopyrite were deposited (Plate III, Fig. 29). The depositional process was complex, as indicated by several ages of crosscutting veinlets. Veinlets range in thickness from 5 to 15 mm, and the common mineral suites are barren quartz, quartz-pyrite, chalcopyrite-quartz or pyrite-chalcopyrite. Variable amounts of dark, forest-green chlorite are found as patches within the veins or as rims on chalcopyrite grains (Plate IV, Fig. 30). Vein minerals are quartz, pyrite, chalcopyrite, chlorite, siderite and sphalerite, with traces of galena. Siderite is locally present as small irregular veinlets or as widely distributed blebs in quartz veins; galena is associated with sphalerite.

Two zones of intense quartz veining are present at or near the stratigraphic top of the footwall volcanic rocks, and here quartz constitutes from 10 to 30% of the rock. One of these zones follows the west stringer ore-waste rock contact from the subcrop down to the 470 mine level. The other is beneath the east-central portion of the massive sulphide body extending from the 350 mine level to an unknown depth. These quartz-rich zones contain low grade copper ore (Plate IV, Fig. 31). Connecting and partially surrounding the two zones of intense quartz veining is a zone of higher grade copper ore where pyrite and chalcopyrite fill rock fragment interstices. Vein quartz averages 5% by volume in this zone.

Pyrite in the stringer ore has a fractured, subrounded texture markedly different from its euhedral-subhedral texture in the massive ore. Textures observed in polished sections show that chalcopyrite has partially replaced fractured pyrite grains (Plate IV, Fig. 32).

A stringer mineralization overprint, consisting of chalcopyrite-quartz veins transecting bedded massive ore, occurs at a few places near the base of the massive body. This overlap of epigenetic mineralization upon syngenetic mineralization generally was restricted to the lower 5 m of the bedded massive ore because of the low permeability of the Crandon unit.

Supergene Sulphide

Approximately 50 m at the top of the deposit has been slightly enriched by supergene processes. This secondary mineralization rapidly decreases below a depth of 50 m, although fracture-controlled chalcocite has been noted in trace amounts to depths of 100 m below the subcrop. The most commonly observed supergene mineral in polished sections is chalcocite; covellite and bornite are less abundant. The chalcocite occurs as 2 to 5 μ thick coatings on primary sulphides (Plate IV, Fig. 33). It also forms veinlets and blebs where chalcopyrite has been completely replaced. Covellite is generally found as fine grains or as micron-thick coatings within or on chalcocite. Bornite rims chalcopyrite, and chalcocite commonly coats bornite.

MORPHOLOGY

The Crandon deposit can be divided into two distinct orebodies: the combined massive-semimassive ore and the stringer ore, which are crudely parallel to one another and to the enclosing stratigraphy (Fig. 10). The deposit strikes N 85° W on average, and dips about 85° N. The massive ore is a tabular, stratiform body that pinches and swells both along strike and down dip (Fig. 11). It has a length of 1,100 m

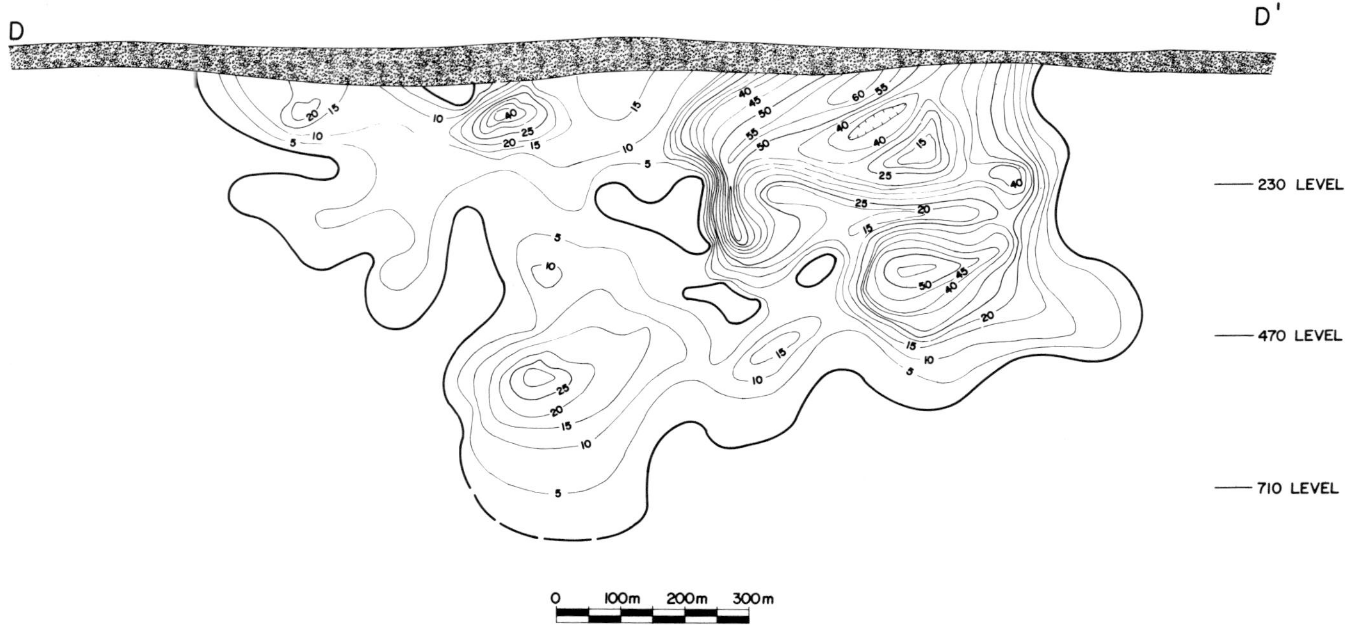

Figure 11. Massive sulphide ore isopach, Crandon, Wisconsin. Longitudinal vertical section D - D′, looking north.

and an average thickness of 30 m. Both the strike length and the thickness of mineable mineralization decrease with depth, particularly below the 350 level (Fig. 12). The contact between the massive mineralization and the hanging wall volcanic rock is generally sharp, whereas the footwall contact is gradational. The limits of mineable massive ore however, are best determined from drill core assays because, in many places, mineable ore is bounded by barren or low grade massive sulphide mineralization.

The stringer ore averages 800 m in length in a direction almost parallel to that of the overlying massive mineralization. It dips 85° N and rakes at 55 to 60° E. The stringer ore is found within a zone that commonly ranges in width from 20 to 30 m and forms an irregular, tabular body partially overlain by massive ore (Fig. 13). On the west end, it is overlain by submineable grade mineralization in the Crandon unit (Fig. 10).

It has been pointed out by Sangster and Scott (1976) that in undeformed orebodies the shape of the stringer sulphide zone generally is cylindrical or funnel-shaped, with the massive ore lying directly over the widest part of the "funnel". Why

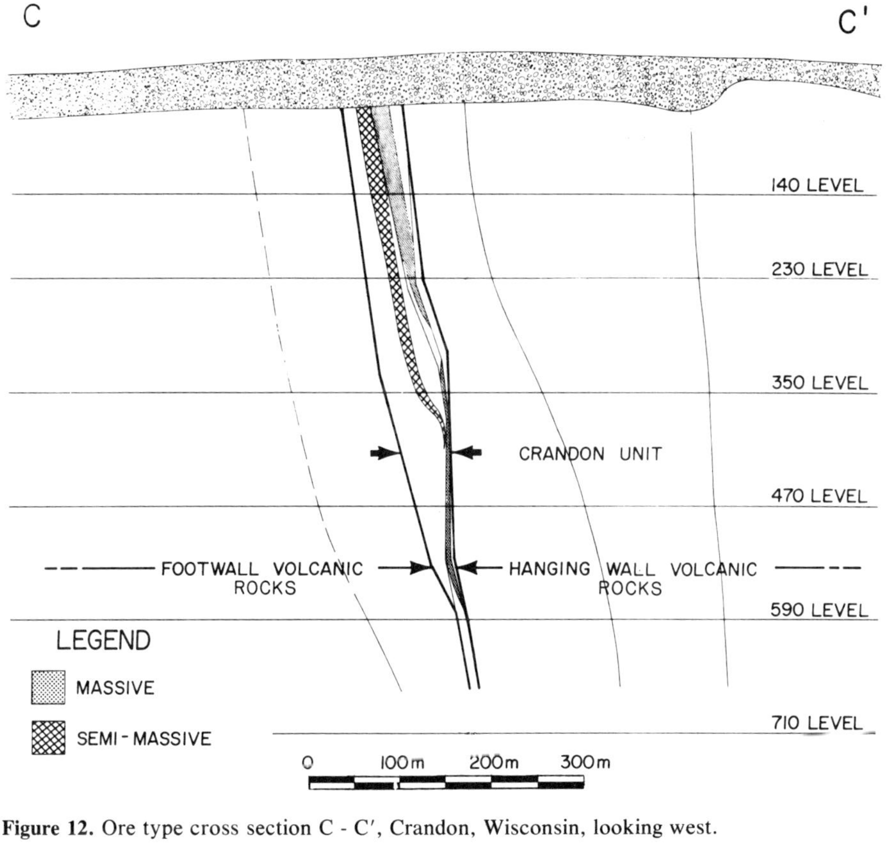

Figure 12. Ore type cross section C - C′, Crandon, Wisconsin, looking west.

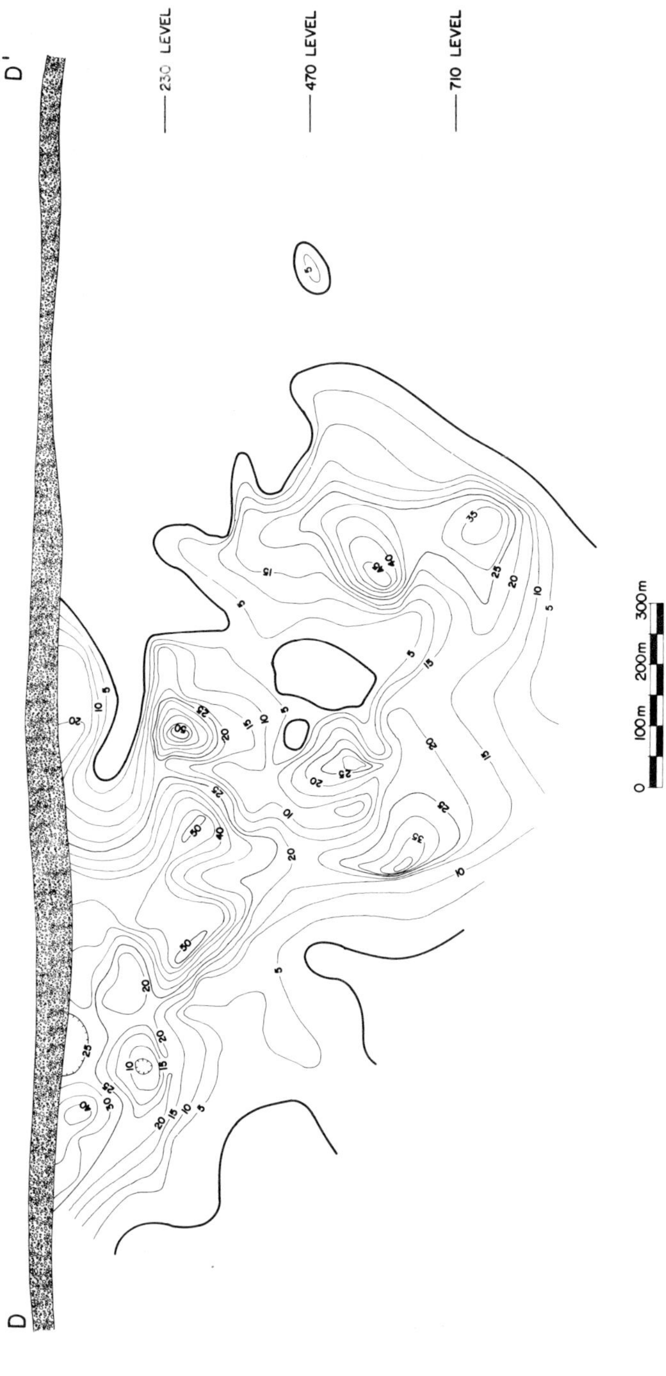

Figure 13. Stringer sulphide ore isopach, Crandon, Wisconsin. Longitudinal vertical section D - D', looking north.

the Crandon stringer zone is crudely tabular is not clearly understood. There is insufficient evidence that the stringer sulphide zone was deformed by shearing into a tabular body, as in many deposits (Sangster, 1972). Sangster and Scott (1976) also noted that "the upward dilation of the stringer ore zone ends at the massive sulphide orebody, and only in one or two instances are we aware of it extending beyond the massive ore, either laterally or upward." Again, the Crandon deposit may be an exception.

The two ore types are separated in many places by a multicompositional, submarginal mineralized zone (Fig. 14) ranging in width from 10 to 15 m and thickening to the east and with depth. The stratigraphic top of the stringer sulphide ore deep on the east end, is at least 75 m beneath the massive sulphide zone (Fig. 15). This width of rock separating the massive and stringer zones is uncommon for this type of deposit.

Emplacement of the epigenetic ore was controlled by primary and secondary porosity and resulting permeability. A wide variety of factors affected permeability, including fragment size, silicification, early quartz-sulphide veining and structure, all of which influenced the final orebody configuration.

MINERAL ZONING

The Crandon deposit is distinctly zoned. A syngenetic zinc-rich massive sulphide zone overlies the epigenetic copper-rich stringer sulphide zone. Consistent stratigraphic zoning is not recognized in either mineralized zone; however, variations in Cu/Zn ratios in longitudinal section suggest a zoning pattern.

Preliminary evaluation of zoning normal to strike within the massive ore has been done using Cu/Cu + Zn and Cu/Cu + Pb ratios. The results of this study suggest that in some areas of this zone, layers with higher copper values occur within the lower half of the massive body but not necessarily at the base. Evaluation of base and precious metal histograms from randomly selected drill holes within the massive orebody leads to the following conclusions: 1) high lead values are almost always associated with high zinc values, but the reverse is not necessarily true; 2) lead values show little consistency in spatial distribution; 3) most high silver values are associated with high zinc and lead, but silver is not preferentially associated with either of these metals; 4) gold distribution has no apparent direct relationship to base metal content.

In longitudinal section (Fig. 16) the Zn/Cu ratio in the massive sulphide zone increases away from a line extending from the upper west end to the lower east end of the deposit. One explanation of this zoning is that a series of structurally controlled, submarine volcanic exhalative conduits was the source of the mineralization. The multisource concept is further strengthened by the lack of consistency of base metal zoning across strike within the ore zone, and by the presence of several intensely altered footwall breccia zones. A complex layering of syngenetic mineralization is indicated at Crandon, rather than a deposit resulting from continuous discharge of mineralizing fluids from a single source.

There is no apparent zoning of copper or zinc normal to strike in the stringer sulphide ore. Earlier work on the deposit (Schmidt, 1978) indicated that in longitudinal section the Cu/Zn ratio increases to the west and with depth (Fig. 17).

THE STRATIGRAPHIC POSITION OF MINEABLE MINERALIZATION

The syngenetic mineralization of Crandon occurs within fine-grained clastic and chemical sedimentary rocks, and presumably was deposited during periods of quiescence. Common to many massive syngenetic sulphide deposits are marker horizons (Sangster and Scott, 1976), which provide exploration guides in the search for additional mineralized zones along strike. The "key" tuffite at Mattagami (Sharpe, 1965), the quartz-sericite schist at Manitowadge (Brown and Bray, 1960) and the rhyolite-andesite contacts at Noranda (Gilmour, 1965; Dugas, 1960) are examples of such marker horizons. Based on the interpretation of diamond drill core, the marker horizon which is the Crandon unit, can be traced for only a short distance beyond the lateral extension of the massive sulphide ore.

The geologic history of the Crandon deposit, as interpreted from the stratigraphy, commenced with deposition of an unknown thickness of interbedded andesitic flows and tuffs. Volcanism continued with the deposition of explosive and coarse-

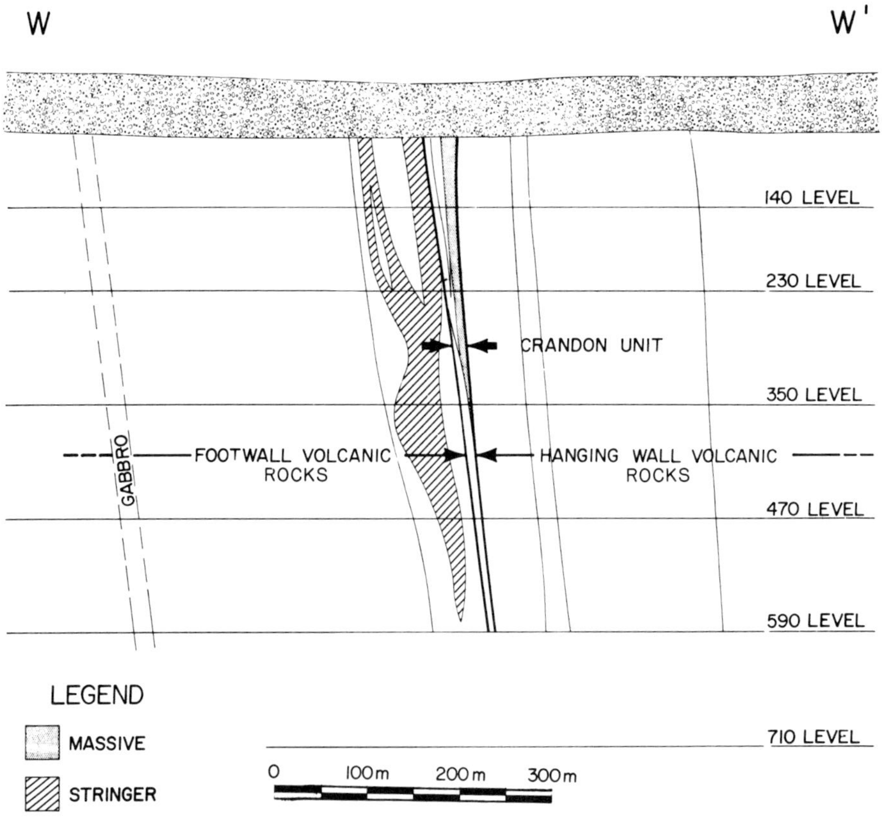

Figure 14. Ore type cross section W - W′, Crandon, Wisconsin, looking west.

grained dacitic pyroclastic rocks upon the andesite. The dacite breccia was probably created by phreatomagmatic processes, whereas the fragmental dacite may represent a tuff and lapilli tuff that were shattered in place. Gradual cessation of local explosive volcanic activity, and increased deposition of interbedded sedimentary massive sulphide, tuff, tuffaceous sediment and argillite eventually formed the Crandon unit rocks.

Epigenetic mineralization was introduced beneath and to the west of the massive sulphide ore. It was emplaced into coarse-grained, fragmental footwall volcanic rocks of a composition less felsic than that normally recognized in many Canadian and Japanese volcanogenic deposits. Deposition of the epigenetic ore appears to have

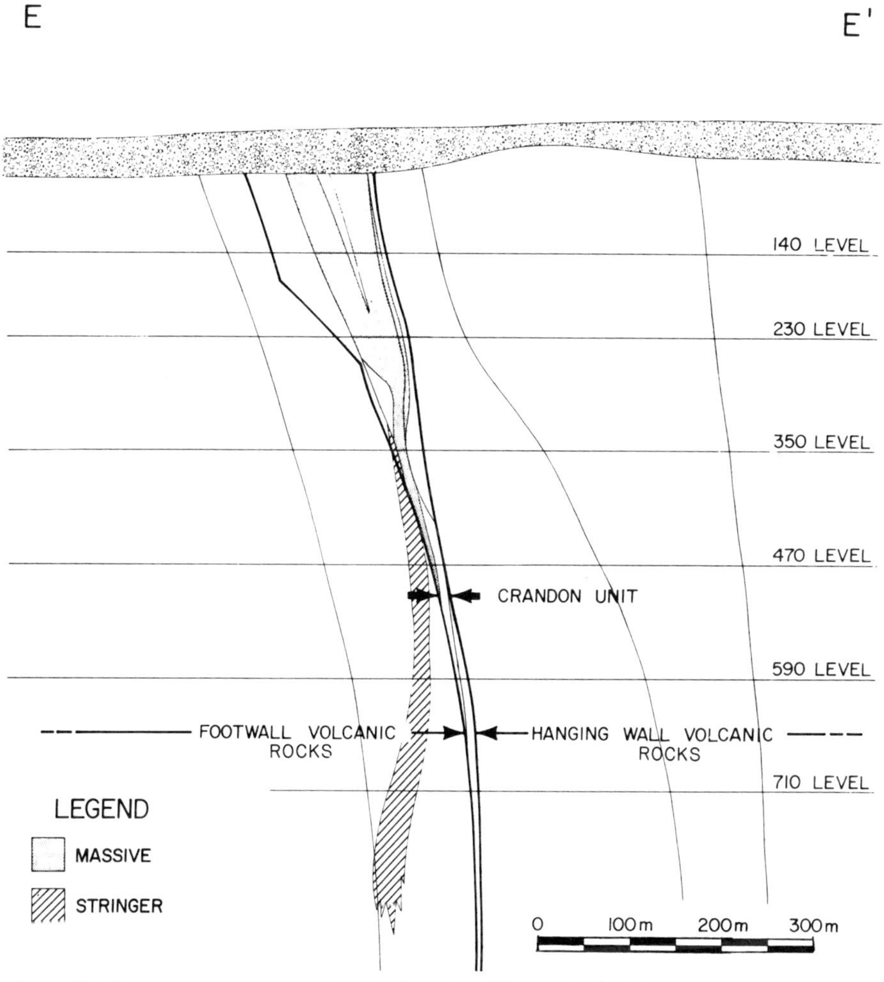

Figure 15. Ore type cross section E - E′, Crandon, Wisconsin, looking west.

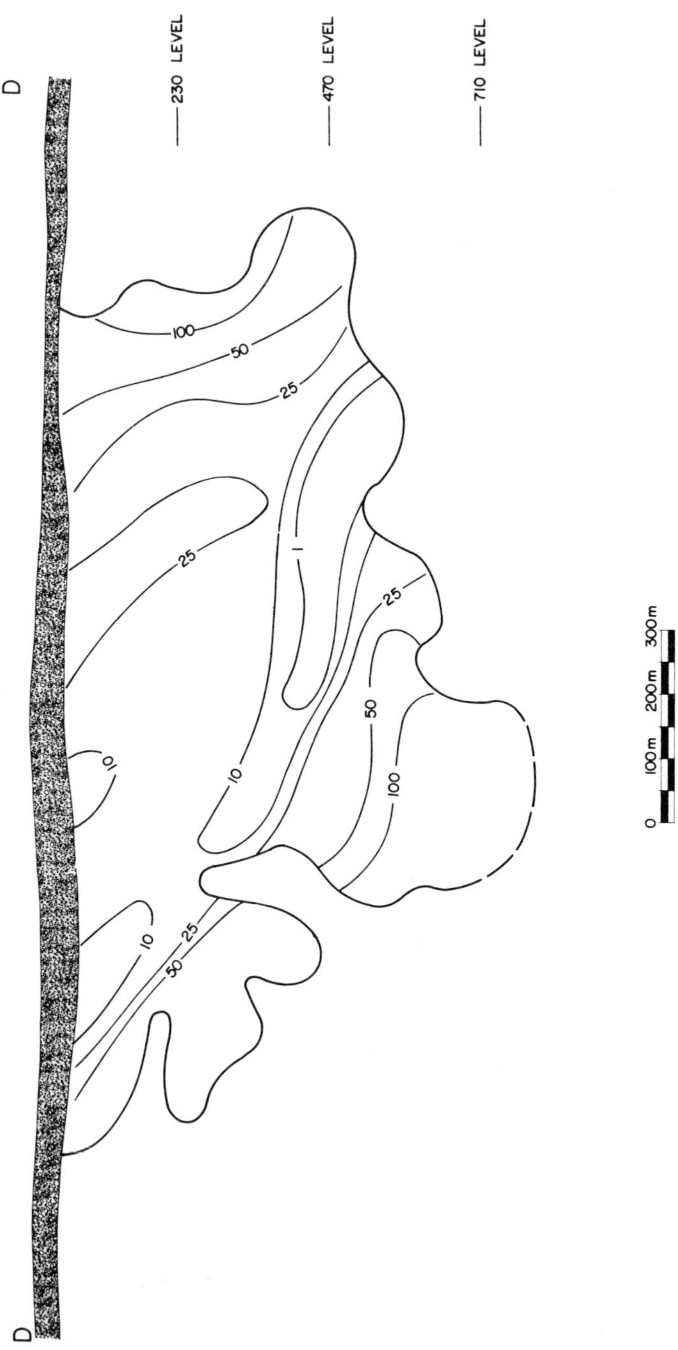

Figure 16. Massive sulphide ore zinc/copper ratio, Crandon, Wisconsin. Longitudinal vertical section D - D′, looking north.

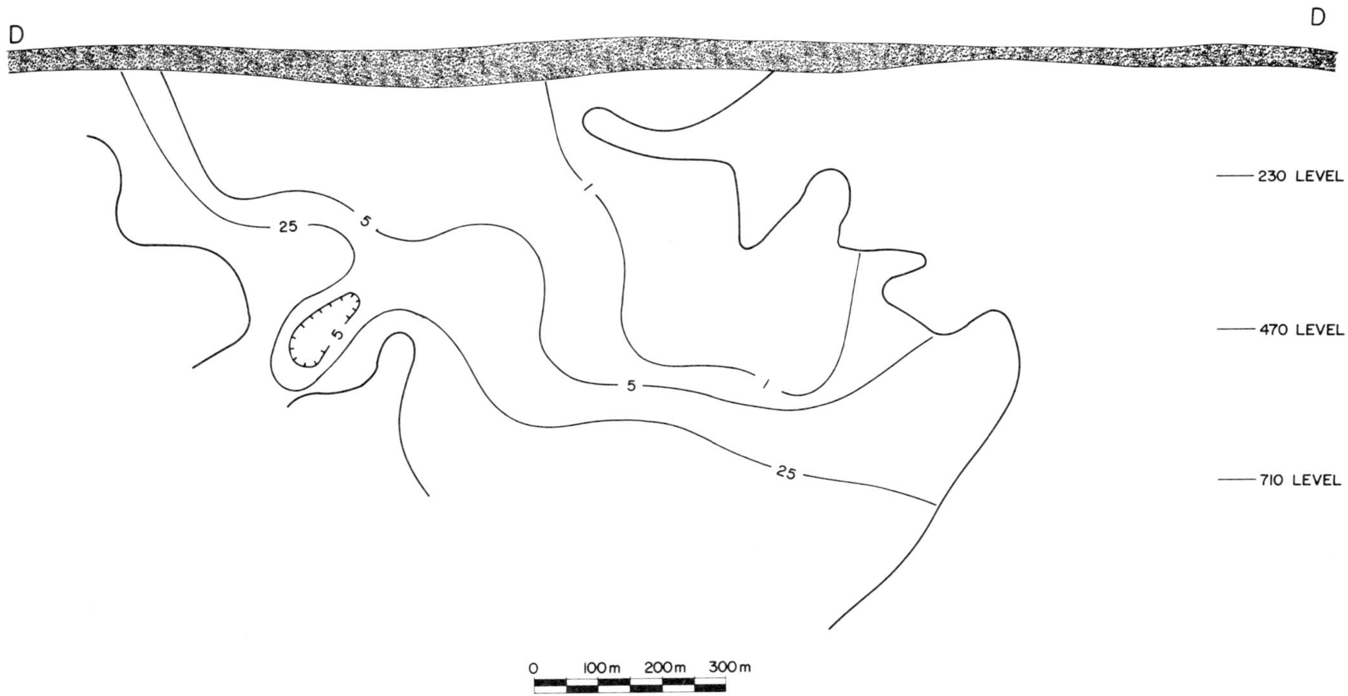

Figure 17. Stringer sulphide ore copper/zinc ratio, Crandon, Wisconsin. Longitudinal vertical section D - D′, looking north.

been penecontemporaneous with the syngenetic mineralization. An overprinting of epigenetic mineralization occurred on the base of the massive sulphide zone. This hydrothermal activity caused intense metasomatic alteration of the footwall volcanic and Crandon unit sedimentary rocks.

Unmineralized hanging wall andesite tuff, overlain by rocks of dacitic to rhyolitic composition, suggests a second repetitive volcanic cycle. Folding during the Penokean orogeny tilted the Crandon rocks to their present near-vertical position. After the intrusion of gabbro, all rocks were eroded to expose the upturned deposit. Pleistocene glaciation further eroded the exposed edge of the deposit and buried it with unconsolidated till.

ACKNOWLEDGEMENTS

The authors wish to thank the Management of Exxon Minerals Company for permission to publish, and encouragement given during the writing of this paper. We also wish to acknowledge the work of many others without whose efforts the paper would not have been possible. They include Jerry Dolence, Mario Lluria and Geoffrey Parsons III of Exxon's Northeast Area Exploration staff: Roger Rowe, Frank Sonderman and Paul Zimmer of Exxon's Crandon Project Predevelopment staff; and Exxon's Exploration staff geophysicists Robert Sultzbach, Joe Inman and Paul Baumgartner. The support given by Peter Davidson and Larry Hayes was also greatly appreciated. Special thanks are extended to Edith Franson and Mary Anne Murray for the typing, Steve Harvey and Vern Stefonek for the drafting, and Tim Worth for the photography.

EXPLANATION OF PLATE I

Figure 18. Dacite Breccia containing chlorite-rich lapilli fragments.

Figure 19. Dacite Breccia containing chloritic and siliceous fragments.

Figure 20. Fragmental Dacite with subangular siliceous fragments in a quartz-sericite-pyrite matrix.

Figure 21. Fragmental Dacite that has been crackled brecciated with unrotated fragments, displaying planar or sawtoothed boundaries.

Figure 22. Mottled Dacite with typical quartz-sericite patches surrounded by chlorite.

Figure 23. Interbedded pyrite and chert which is typical of the lower zone of the Crandon Unit.

Figure 24. Eutaxitic texture commonly found in the Dacite Lapilli Tuff.

Figure 25. Massive sulphide mineralization containing round to subround pyrite aggregates in a matrix of coarser-grained pyrite and chlorite.

PLATE I

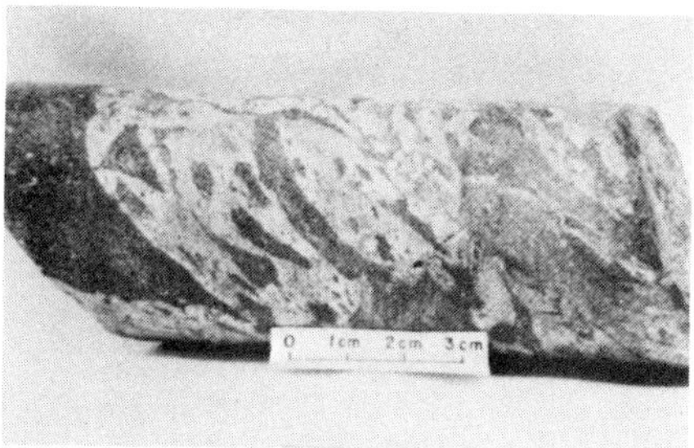

Figure 18

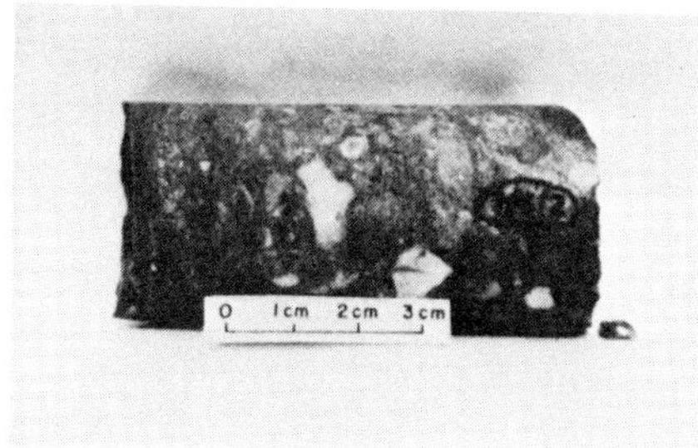

Figure 19

Figure 20

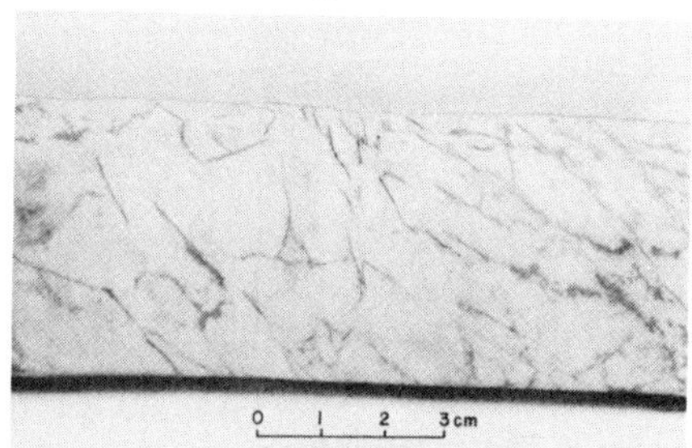

Figure 21

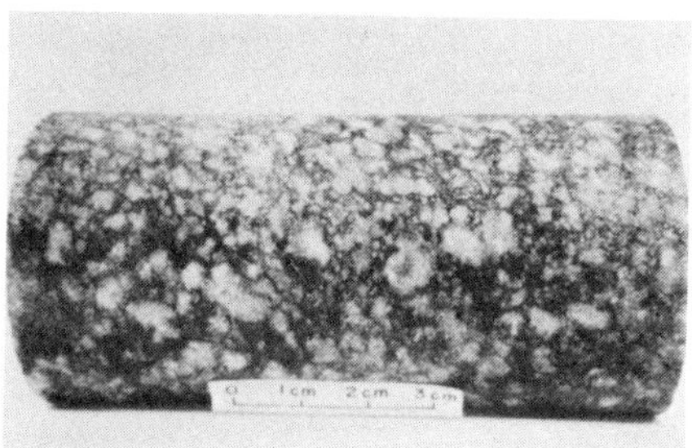

Figure 22

Figure 23

Figure 24

Figure 25

PLATE II

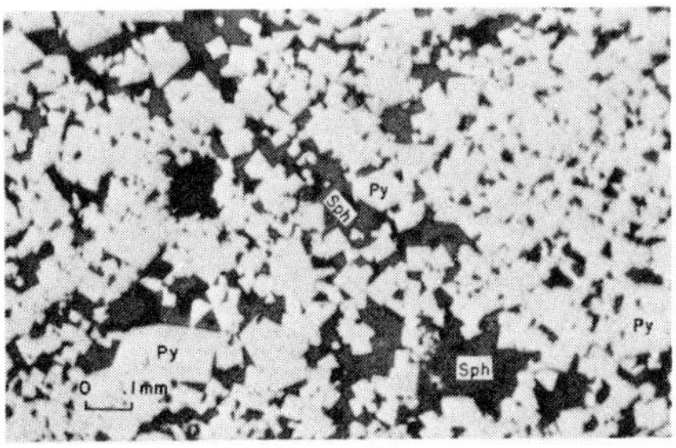

Figure 26

Figure 27

Figure 28

Figure 29

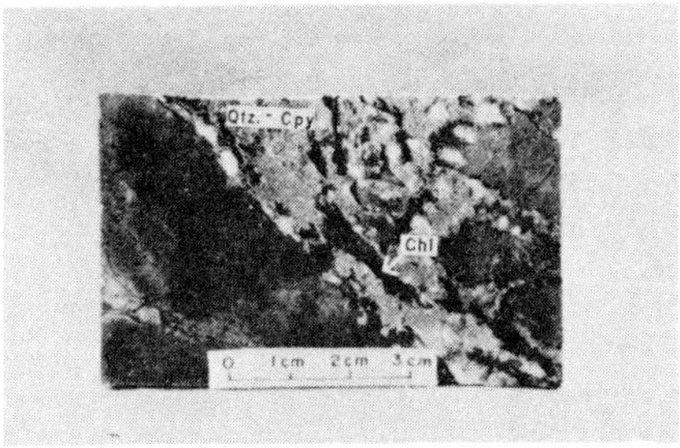

Figure 30

Figure 31

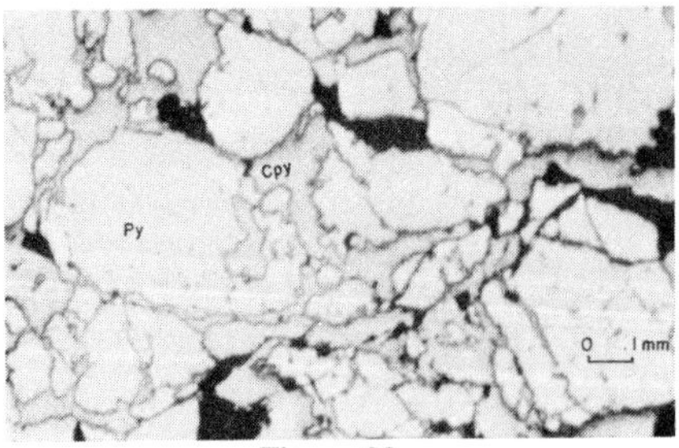

Figure 32

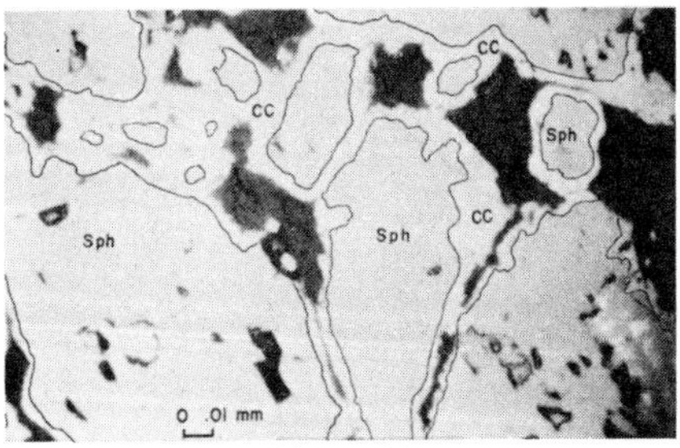

Figure 33

REFERENCES

Brown, W.L., and Bray, R.C.E., 1960, The Geology of the Geco Mine: Canadian Mining and Metallurgy Bulletin, v. 63, no. 1, p. 1-9.

Cannon, W.F., 1978, Geologic Map of the Iron River 1° × 2° Quadrangle, Michigan and Wisconsin: United States Geological Survey Open File Map 78-342.

Clark, L.A., 1971, Volcanogenic Ores; Comparison of Cupriferous Pyrite Deposits of Cyprus and Japanese Kuroko Deposits: Society of Mining Geology Japan, Special Issue 3, p. 206-215.

Dugas, J., 1966, The Relationship of Mineralization to Precambrian Stratigraphy in the Rouyn-Noranda Area, Quebec: Precambrian Symposium, Geological Association of Canada, p. 43-54.

Dutton, C.E. and Linebaugh, R.E., 1967, Map Showing Precambrian Geology of the Menomonee Iron-Bearing District and Vicinity, Michigan and Wisconsin: United States Geological Survey Miscellaneous Geological Investigations Map I-466.

Franklin, J.M., Kasarda, J., and Poulsen, K.H., 1975, Petrology and Chemistry of the Alteration Zone of the Mattabi Deposit: Economic Geology, v. 70, p. 63-70.

Gilmour, P., 1965, The Origin of the Massive Sulphide Mineralization in the Noranda District, Northwestern Quebec: Geological Association of Canada Proceedings, v. 16, p. 63-82.

Hutchinson, R.W. and Searle, D.L., 1971, Stratabound Pyrite Deposits in Cyprus and Relations to Other Sulfide Ores: Society of Mining Geology, Japan, Special Issue 3, p. 198-205.

Prinz, W.C., 1976, Correlative Iron Formations and Volcanic Rocks of Precambrian X Age, Northern Michigan (Abst.): 22nd Annual Institute on Lake Superior Geology, St. Paul, Minnesota, Program with Abstracts.

EXPLANATION OF PLATE II

Figure 26. Polished section of massive sulphide mineralization illustrating sphalerite interstitial to euhedral pyrite.

Figure 27. Massive sulphide mineralization displaying bedding commonly found in the sphalerite-rich portions of the deposit.

Figure 28. Massive sulphide mineralization containing fine-grained chalcopyrite and sphalerite in a matrix of pyrite.

Figure 29. Stringer sulphide mineralization consisting of quartz.

Figure 30. Stringer sulphide mineralization showing a quartz-chalcopyrite-chlorite vein in sharp contact with silicified footwall volcanic rocks.

Figure 31. Stringer sulphide mineralization consisting of milky white quartz with minor amounts of chalcopyrite.

Figure 32. Polished section of stringer sulphide mineralization illustrating typical fracture pyrite grains that have been partially replaced by chalcopyrite.

Figure 33. Polished section of massive sulphide mineralization showing rims of chalcocite on sphalerite.

Sangster, D.F., 1972, Precambrian Volcanogenic Massive Sulphide Deposits in Canada: A Review: Geological Survey of Canada Paper 72-22, 42 p.

Sangster, D.F., and Scott, S.D., 1976, Precambrian, Strata-bound Massive Cu-Zn-Pb Sulfide Ores of North America: in Wolf, K.H., ed., Handbook of Strata-bound and Stratiform Ore Deposits, II: Regional Studies and Specific Deposits: New York, Elsevier Scientific, p. 129-214.

Schmidt, P.G., Dolence, J.D., Lluria, M.R., and Parsons III, G., 1978, Geology of the Crandon Massive Sulfide Deposit, Forest County, Wisconsin: Engineering and Mining Journal, v. 179, no. 7, p. 61-66.

Sharpe, J.I., 1965, Field Relations of Matagami Sulphide Masses Bearing on their Disposition in Time and Space: Canadian Mining and Metallurgy Bulletin, v. 68, no. 9, p. 265-278.

Sims, P.K., 1976, Precambrian Tectonics and Mineral Deposits, Lake Superior Region: Economic Geology, v. 71, p. 1092-1118.

Sims, P.K. and Peterman, Z.E., 1978, Precambrian Geologic Framework of Northern Wisconsin (Abst.): 24th Annual Institute on Lake Superior Geology, Milwaukee, Wisconsin, Program with Abstracts.

Sims, P.K., Cannon, W.F., and Mudrey, M.G., Jr., 1978, Preliminary Geologic Map of Precambrian Rocks in Part of Northern Wisconsin: United States Geological Survey Open File Report 78-318.

Tatsumi, T., and Watanabe, T., 1971, Geological Environment of Formation of the Kuroko-Type Deposits: Society of Mining Geology Japan, Special Issue 3, p. 216-220.

Van Schmus, W.R., 1976, Early and Middle Proterozoic History of the Great Lakes Area, North America: Philosophical Transactions of the Royal Society of London, v. 280, p. 605-628.

Van Schmus, W.R., Thurman, E.M., and Peterman, Z.E., 1975, Geology and Rb-Sr Chronology of Middle Precambrian Rocks in Eastern and Central Wisconsin: Geological Society of America Bulletin, v. 86, p. 1255-1265.

Zavritsky, A.N., 1964, Outline of the Analysis of Igneous Rocks and the Determinations of their Chemical Types: International Geology Review, v. 6, p. 1-17.

MS received April 1979;
Accepted for publication July 1979.

Precambrian Sulphide Deposits, H.S. Robinson Memorial Volume, edited by R.W. Hutchinson, C.D. Spence and J.M. Franklin, Geological Association of Canada Special Paper 25, 1982

THE ANDERSON LAKE MINE, SNOW LAKE, MANITOBA

P.C. Walford
Pamour Porcupine Mines Ltd., P.O. Bag 2010, Timmins, Ontario P4N 7X7

J.M. Franklin
Geological Survey of Canada, 601 Booth Street, Ottawa, Ontario K1A 0E8

ABSTRACT

The Anderson Lake orebody occurs in highly metamorphosed Aphebian volcanic rocks of the Snow Lake area, Manitoba. This relatively copper-rich massive sulphide deposit has a typical footwall alteration pipe including a large upper muscovite zone and a vertically extensive chlorite zone. These intense alteration zones are surrounded by a less intense staurolitic alteration zone. The pipe extends down to a sub-concordant alteration blanket in the footwall rhyolite. This lower alteration has a chlorite-rich centre which is locally albite-bearing at its core, developed near the departure point of the pipe, and a more distal, alkali-depleted, iron-enriched staurolite zone. The hangingwall mafic volcanic rocks are not altered near the deposit, but the immediate hangingwall felsic epiclastic rock is muscovite-rich, and may be altered.

The deposit occurs in the lowest and chemically most primitive unit in the mine sequence. The base of the mine sequence is occupied by a major synvolcanic tonalite sill. The felsic rocks increase in K/Na ratios, Ba content and phenocryst content upward in the sequence; they comprise more than 50% of the volcanic products. The mafic rocks are dominantly high-alumina basalt, with minor andesite. The compositional characteristics of the volcanic accumulation, and abundance of felsic rocks are similar to those of a subduction-generated continental margin environment.

Metal-rich fluids for the deposit are believed to have formed by heating of intrastratal water within the lower unit of volcanic rocks by the subvolcanic tonalite intrusion. These waters were probably expelled from the lower concordant zone through syn-volcanic faults and hydrothermal fracture zones, on to the sea floor. Rapid cooling, and local mixing with sea water drawn down into the pipe zone resulted in metal precipitation.

RÉSUMÉ

Le gisement d'Anderson Lake dans le région de Snow Lake, Manitoba, se trouve dans des roches volcaniques aphibiennes fortement métamorphisées. Cet amas sulfuré massif, relativement riche en cuivre, est associé à une cheminée d'altération typique développée dans les roches du mur et comprenant trois zones: une large zone supérieure à muscovite passant à une zone à chlorite verticale étendue. Ces deux zones d'altération intense sont entourées d'une zone d'altération plus faible à staurotide. La cheminée d'altération s'étend vers le bas jusqu'à un manteau d'altération subconcordant dans la rhyolite du mur. Ce manteau consiste, près du point de départ de la cheminée d'altération, en une zone centrale riche en chlorite et contenant localement de l'albite en son centre même. Cette zone est elle-même entourée d'une zone extérieure à staurotide, appauvrie en alcalins et enrichie en fer. Les roches volcaniques mafiques au toit du gisement ne sont pas altérées alors que la roche felsique épiclastique située immédiatement au-dessus du gisement est riche en muscovite et est peut-être altérée.

Le gisement se trouve dans l'unité la plus inférieure et chimiquement la plus primitive de la séquence de la mine. La base de cette séquence est occupée par un large sill synvolcanique de tonalitc. Dans les roches felsiques, qui représentent plus de la moitié des produits volcaniques, le rapport K/Na, la teneur en Ba et l'abondance de phénocristaux s'accroissent vers le haut de la séquence. Les roches mafiques sont essentiellement des basaltes alumineux accompagnés de quelques andésites. Les roches volcaniques, de par leurs caractéristiques chimiques et l'abondance de leurs variétés felsiques, sont semblables à celles rencontrées dans les bordures continentales au-dessus de zones de subduction.

Les fluides métallifères se sont probablement formés lors du réchauffement des eaux piégées dans l'unité de roches volcaniques inférieure par l'intrusion subvolcanique de tonalite. Ces eaux ont probablement été expulsées de la zone concordante inférieure le long de failles synvolcaniques et de zones de fractures hydrothermales vers la surface du fond marin. Refroidissement rapide et mélange local avec l'eau de mer percolant dans la cheminée d'altération ont causé la précipitation des métaux.

INTRODUCTION

The Anderson Lake mine, near the town of Snow Lake, Manitoba (Fig. 1) occurs in the most easterly segment of the Flin Flon – Snow Lake volcanic belt, bounded on the north by the Kisseynew gneiss complex and on the south by Paleozoic strata. Twenty-eight volcanic-hosted massive sulphide deposits (Table I) occur in this belt, mainly near Flin Flon and Snow Lake. The File Lake and Elbow Lake areas of volcanic rock which occur between Flin Flon and Snow Lake, have only one previously producing mine and a few major occurrences.

The Snow Lake area was first prospected for gold, and developed due to production of the Nor Acme Mine (Hogg, 1957), which produced 19,000,000 grams of Au from 4,900,000 tonnes of ore, from 1949 to 1958, and from several small mines in the Herb Lake gold camp on the east shore of Wekusko Lake (Stockwell, 1937). The first massive sulphide deposit to be mined was the Chisel Lake deposit opened in 1960, followed by Stall Lake (1964), Osborne (1968), and the Rod (opened 1962, closed 1964) mines. The Anderson Lake deposit is the newest mine in the area and was discovered in 1963 by ground EM techniques. The mine commenced production in 1971 and as of January 1, 1981, 3,272,400 tonnes of ore with 3.44% Cu, 0.1% Zn, 0.56 g/tonne Au and 6.8 g/tonne Ag, have been mined or are in reserve.

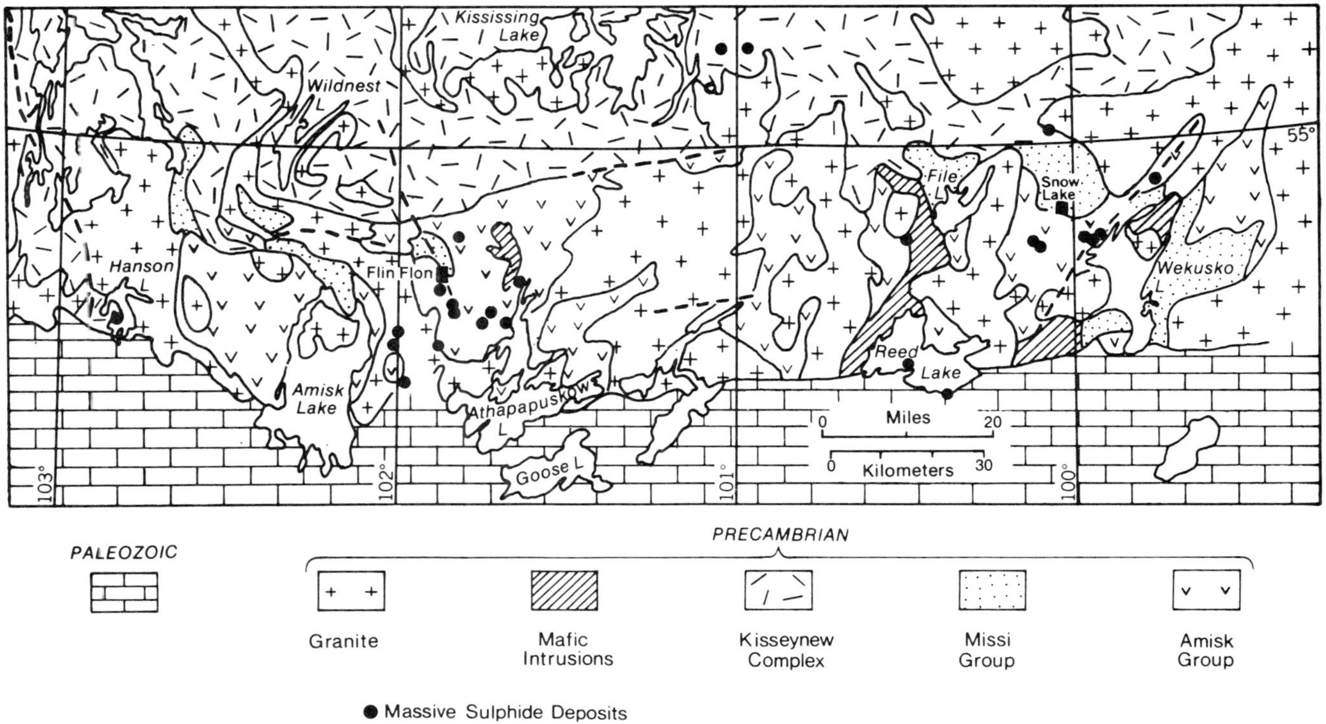

Figure 1. Regional Geology, Flin Flon-Snow Lake Belt, Manitoba. Principal volcanic-hosted massive sulphide deposits shown as dots.

TABLE I
VOLCANIC-HOSTED ORE DEPOSITS
FLIN FLON-SNOW LAKE AREA

Deposit	Tonnes	Grade Cu	Grade Zn	Reference to deposit description
	Snow Lake-Reed Lake			
Anderson	3,272,400	3.44	0.1	Howkins and Martin, 1970 Price, 1977
Stall	5,986,100	4.41	0.5	Howkins and Martin, 1970 Bristol, 1974
Rod No. 1 (Stall Lake Mines Ltd.)	22,680	5.0	4.5	Coats *et al.*, 1970
Rod No. 2 (Stall Lake Mines Ltd.	412,600	7.21	3.0	Coats *et al.*, 1970
Osborne	3,273,000	3.14	1.5	Howkins and Martin, 1970
Chisel	6,477,300	0.53	10.9	Williams, 1966 Martin, 1966 Bristol, 1974 Price, 1977
Ghost/Lost	648,100	1.34	8.5	Price, 1977
Wim	1,463,500	2.36	0.4	
	File-Reed Lakes Area			
Dickstone (Dickstone Copper Mines Ltd.)	1,080,000	2.45	3.5	
Spruce Point	507,100	2.70	4.3	
Reed Lake	1,361,700	2.09		
Rail Lake	453,300	2.85	0.7	
Pot Lake	112,000	1.43	4.5	

Deposit	Snow Lake-Reed Lake Tonnes	Grade Cu	 Zn	Reference to deposit description
	Flin Flon Area			
Flin Flon	61,852,300	2.18	4.1	Price, 1977 Koo and Mossman, 1975 Bristol, 1974 Howkins and Martin, 1970 Byers *et al.*, 1965
Mandy	113,400	7.50	9.4	Hanson, 1920
Schist Lake	1,871,000	4.34	7.3	Bristol, 1974 Howkins and Martin, 1970 Geological Staff, 1957
Birch Lake	272,900	6.21		Beck, 1959
Flexar	306,000	3.76	0.5	Howkins and Martin, 1970
Coronation	1,281,700	4.25	0.2	Byers, 1969
Don Jon	79,300	3.09		Geological Staff, 1957
North Star	241,700	6.11		Geological Staff, 1957
Centennial	2,342,458	1.48	2.2	Provens, 1980 Price, 1977
Cuprus	462,000	3.25	6.4	Geological Staff, 1957
White Lake	815,200	2.04	4.8	
Westarm	1,554,180	3.68	1.5	
Pine Bay	1,121,500	1.56		
includes	409,100	3.19		
Trout Lake	2,623,400	3.07	4.5	
Vamp Lake	432,700	1.2	1.4	

Notes:

1) Tonnage and grade data from HBM&S 1981 calculations.
2) Howkins and Martin (1970) refer to an unpublished paper delivered at the Annual General Meeting of the Canadian Institute of Mining and Metallurgy, 1970; data from HBMS files.

REGIONAL GEOLOGY

The general geology of the Snow Lake-Wekusko Lake area has been described by Armstrong (1941), Harrison (1949), Frarey (1950), Russell (1957), Price (1977) and most recently by Froese and Moore (1980). The area has two supracrustal sequences (Fig. 2). The Amisk Group which constitutes the lower sequence and contains all of the massive sulphide deposits, is composed of volcanic and sedimentary strata. (All designations of the lithology of the supracrustal strata do not include the prefix "meta", although all units are metamorphosed. Primary lithologic names, rather than their metamorphic equivalents, are used where possible.) The upper sequence, the Missi Group, is composed of lithic arenite, which is typically thickly bedded and cross-bedded (Froese and Moore, 1980), overlain east of Wekusko Lake by volcanic strata (Shanks and Bailes, 1977).

Froese and Moore (1980) indicated three principal structural events in the Snow Lake area. The first phase (F1) folds are isoclinal, with schistosity subparallel to compositional layering. This phase resulted in flattening of primary features. The second phase (F2) folds are open, and give the dominant northeast trend in the Anderson Mine area. Fold axes formed in this event, typified by the Threehouse synform, have a northeast plunge as shown by the alignment of metamorphic minerals. This phase resulted in attenuation of pre-F2 features in the dominant northeast direction. The third phase is related to emplacement of gneiss domes by diapirism, which caused the F2 folds to be laterally deflected to a position between the antiformal domes. The centre of the arcuate tonalite body south of Chisel Lake probably represents an F1 fold axis (Fig. 2). This F1 fold is deformed about the Threehouse synform, a major open F2 fold that imparts the general "U" shape to the tonalite sill.

Two major faults have been recognized by Froese and Moore (1980). The Berry Creek fault divides the volcanic strata into two domains; the area to the northwest of this fault includes all of the massive sulphide deposits; to the southeast, the Wekusko Lake area contains a thin unit of Amisk sedimentary and volcanic strata, overlain by Missi sedimentary and volcanic strata. Froese and Moore (1980) also recognized an arcuate thrust fault, the McLeod Road fault, enclosing the area bounded by Squall Lake, Snow Lake, and northwest of the Osborne Lake Road.

Froese and Moore (1980) defined three regional metamorphic isograds which subdivide the region into chlorite-biotite, chlorite-biotite-staurolite, biotite-staurolite-sillimanite, and biotite-sillimanite-almandine zones. In addition, widespread, late, post-metamorphic carbonate alteration is superimposed on all rocks. Carbonate-quartz veinlets occur in most mafic volcanic rocks.

Major accumulations of felsic rocks occur on each limb of the Threehouse synform (Fig. 2). On the east limb, a major felsic accumulation of pyroclastic breccia units and felsic flows forms the lower part of the pile and hosts the Stall, Rod, Ram and Anderson deposits. Although felsic volcanic rocks predominate, they are intercalated in the upper part of the sequence with approximately equal amounts of mafic volcanic rock and a prominent sedimentary unit. On the west limb of the synform, proximal felsic pyroclastic and flow units are dominant at a higher stratigraphic level than on the east limb. The Chisel Lake, Ghost Lake and Lost Lake zinc-copper deposits occur in this group of felsic rocks.

TABLE II
MAJOR STRATIGRAPHIC UNITS, SNOW LAKE AREA

Group		Unit		Principal Characteristics
Missi				Arkose, well bedded with local cross-beds
Amisk	11	Herblet	mafic	
	10	Townsite	mafic	Basalt, locally pillowed to massive, feldspar and amphibole phenocrysts, amygdaloidal, fragmental, epiclastic.
	9	Snow Cr.	mafic	
	8	Townsite felsic		Porphyritic rhyolite with relatively high K_2O; a few intercalated mafic rocks of overlying unit.
		McLeod Road Thrust		
	7	Sedimentary		Greywacke and shale in turbidite beds.
	6	Mixed		Approximately equal amounts of felsic and mafic strata, with minor intercalated sedimentary strata. Grades upward into the sedimentary unit.
	5	Threehouse basalt		Porphyritic (feldspar and amphibole) flows with locally developed pillows, pillow breccia and epiclastic beds.
	4	Hangingwall porphyritic	felsic	Discrete units of porphyritic and aphanitic rhyolite; tuff and breccia both prominent; some felsic flow-lobes and associated hyaloclastite;
		Hangingwall aphanatic	felsic	
	3b	Hangingwall basalt	mafic	Hangingwall basalt intercalated with aphanitic rhyolite beds. Basalt is similar in composition and texture to the Threehouse unit. Andesite is pillowed near Stall Mine, hyaloclastic near Anderson Mine.
	3a	Hangingwall andesite	mafic	
	2	Lower mine felsic		Lower massive, upper fragmental members of relatively high TiO_2 rhyolite; host to Anderson and Stall deposits.
	1	Tonalite intrusion		Massive to porphyritic tonalite to trondhjemite; probably subvolcanic.

Between the two felsic centres, near the axis of the Threehouse synform, the lower half of the sequence contains a few small felsic centres, set in dominantly mafic flows. These felsic accumulations correlate approximately with the hanging wall unit (above the hanging wall andesite) and felsic strata which host the Anderson orebody.

The upper half of the sequence above the tonalite body in the Threehouse synform area is composed of porphyritic basaltic epiclastic rocks, with minor heterolithic sedimentary rocks and felsic units, which were derived from the adjacent felsic and mafic proximal volcanic centres, and shed into a local basin. The epiclastic sequence attains a maximum thickness of 600 m near the axis of the Threehouse synform. The mafic epiclastic component is composed of fragments of feldsparphyric (locally glomeroporphyritic) basalt, similar to the Threehouse basalt (discussed later). Clasts, derived from various flows, are subrounded to angular and the rocks are generally poorly bedded. However, a few intercalated siltstone units are well bedded. In addition a few basalt flows, mafic sills, and rare felsic volcanic units are intercalated with and underlie the epiclastic strata. The upper volcaniclastic unit gives way upward to greywacke and semi-pelite. The upper part of this sequence contains major mafic sills.

Radiometric age determinations on supracrustal, gneissic and intrusive rocks of the Snow Lake area were summarized by Price (1977). The most recent Rb/Sr determinations (Bell *et al.*, 1975) indicate that the volcanic rocks are slightly older than 1780 ± 45 Ma ($\lambda^{87}Rb = 1.42$) in age. This is consistent with model lead ages for galena from Osborne Lake (Sangster, 1978). Recent U/Pb (zircon) determinations place the age of the Amisk volcanic rocks at Flin Flon (MacQuarrie, 1980) at approximately 1850 Ma. Thus, the volcanic rocks and the enclosed massive sulphide deposits are middle to late Aphebian in age.

ANDERSON LAKE MINE AREA GEOLOGY

The mine area (Fig. 3) contains a homoclinal upright sequence of felsic and mafic strata and intrusive rocks; a major quartz tonalite sill locally forms the base of the exposed sequence. The stratigraphic sequence is summarized in Table II. In general, the volcanic strata are disposed about a series of volcanic domes, and are lenticular in shape. The felsic sequences are somewhat lenticular, with individual sequences traceable for a maximum of only 5 km. The mafic strata have greater lateral continuity. The nomenclature applied to each rock type is based on chemical data, discussed later. All section thicknesses are approximate; dips established by Froese and Moore (1980) are confirmed to depth at the Anderson Lake Mine (65° to the northwest) and at the Stall Mine (45° to the northeast). Measured thicknesses are probably less than original thicknesses, due to the deformation which resulted in formation of the S1 schistosity.

Tonalite Intrusion (Unit 1)

This intrusion is a major "U" shaped sill and extends discontinuously for almost 21 km from the southeast end of Anderson Lake to near Cook Lake on its northwest end. It is generally 2 km thick but as it may contain an anticlinal axis, this may be greater than the true thickness. The unit is largely equigranular (4 mm average grain

TABLE III
COMPOSITIONS OF THE PRINCIPAL UNITS IN THE SNOW LAKE AREA

Unit	Tonalite Intrusion (1)	Footwall Felsic (2a)	Stall-Ram Footwall Felsic (2b)	Footwall Mafic Dykes (3)	Hanging-wall Andesite (4)	Three-house Basalt (5)	Hanging-wall Felsic (6)	Townsite Felsics
N	32	25	38	11	24	14	28	11
SiO_2	70.35	70.89	73.91	55.06	56.19	49.94	72.55	73.29
Al_2O_3	13.47	12.68	12.66	15.01	15.40	18.29	12.64	13.29
Fe_2O_3	0.43	1.01	0.36	1.80	1.94	2.66	0.57	0.39
FeO	3.12	3.90	3.38	7.76	8.47	8.67	3.84	3.49
MgO	2.47	2.25	1.49	5.90	4.18	4.36	1.23	1.06
CaO	3.34	2.84	2.04	7.18	8.44	9.52	2.81	2.16
Na_2O	3.61	2.43	2.75	2.95	2.35	1.99	2.91	3.22
K_2O	0.84	1.28	1.36	0.46	0.42	0.43	1.78	2.05
TiO_2	0.47	0.35	0.21	0.52	0.58	0.47	0.27	0.27
P_2O_5	0.13	0.21	0.10	0.10	0.12	0.11	0.11	0.08
MnO	0.05	0.10	0.08	0.15	0.19	0.20	0.10	0.09
S	0.09	0.30	0.03	0.02	0.09	0.05	0.03	0.01
CO_2	0.34	0.09	0.03	0.13	0.34	0.45	0.40	0.31
H_2O	1.05	1.23	0.91	1.48	1.18	1.45	0.79	0.69
Ba	120	177	210	55	111	133	286	539
Cu	23	92	40	74	123	124	36	43
Sr	113	61	74	127	133	204	87	116
V	47	50	5	305	309	354	24	32
Zr	78	78	64	59	62	42	65	58
Qtz	34.17	41.32	44.55	8.85	13.51	5.77	38.86	38.32
Cor	0.88	2.68	3.25	0.00	0.00	0.00	1.09	2.05
Or	5.05	7.71	8.18	2.81	20.53	2.63	10.65	12.20
Ab	31.05	20.93	23.65	25.76	20.21	17.41	24.91	27.41
An	15.98	12.95	9.62	27.20	30.74	41.07	13.38	10.25
Di	0.00	0.00	0.00	4.25	4.38	2.84	0.00	0.00
He	0.00	0.00	0.00	3.06	4.83	3.02	0.00	0.00
En	6.25	5.71	3.77	13.19	8.55	9.91	3.10	2.66
Fs	4.58	5.41	5.74	10.90	10.81	12.06	6.33	5.82
Mt	0.63	1.49	0.53	2.69	2.86	3.99	0.84	0.57
Il	0.91	0.68	0.41	1.02	1.12	0.92	0.52	0.52
Ap	0.31	0.50	0.24	0.24	0.28	0.26	0.25	0.19
Py	0.19	0.63	0.06	0.04	0.19	0.11	0.06	0.02

N = no. of samples
major elements in wt%
minor elements in ppm
normative compositions in wt%
data from unpublished G.S.C. analyses, and Hutcheon, 1977

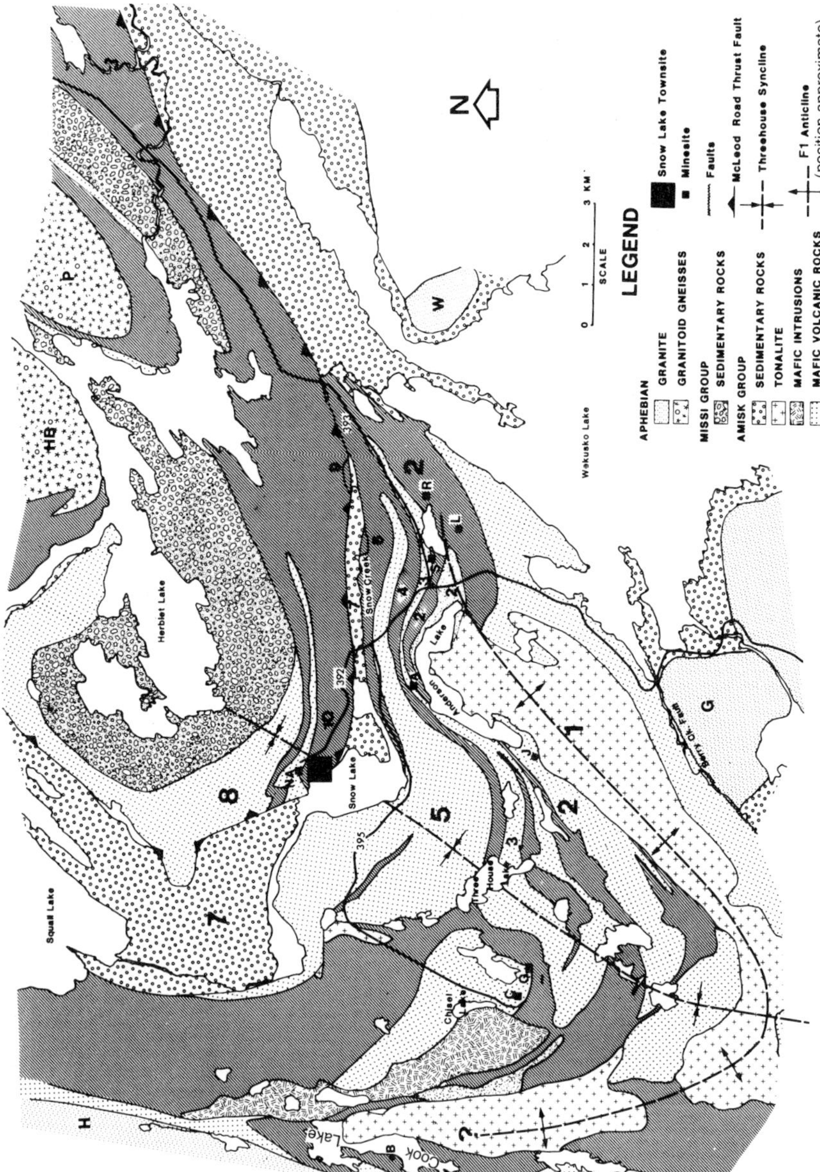

Figure 2. Snow Lake area; modified after Froese and Moore (1980). Unit numbers are explained in Table 2. Mine locations as follows: B = Bomber, C = Chisel, G = Ghost, J = Joannie, A = Anderson, NA = NorAcme (gold), S = Stall, L = Linda, R = Rod. Producing and past-producing mines shown as squares, major prospects shown as circles. Major intrusions and gneiss domes as follows: H = Ham Lake Stock, G = Grassy River Stock, HB = Herblet Gneiss Dome, P = Pulver Gneiss Dome, W = Wekusko Stock.

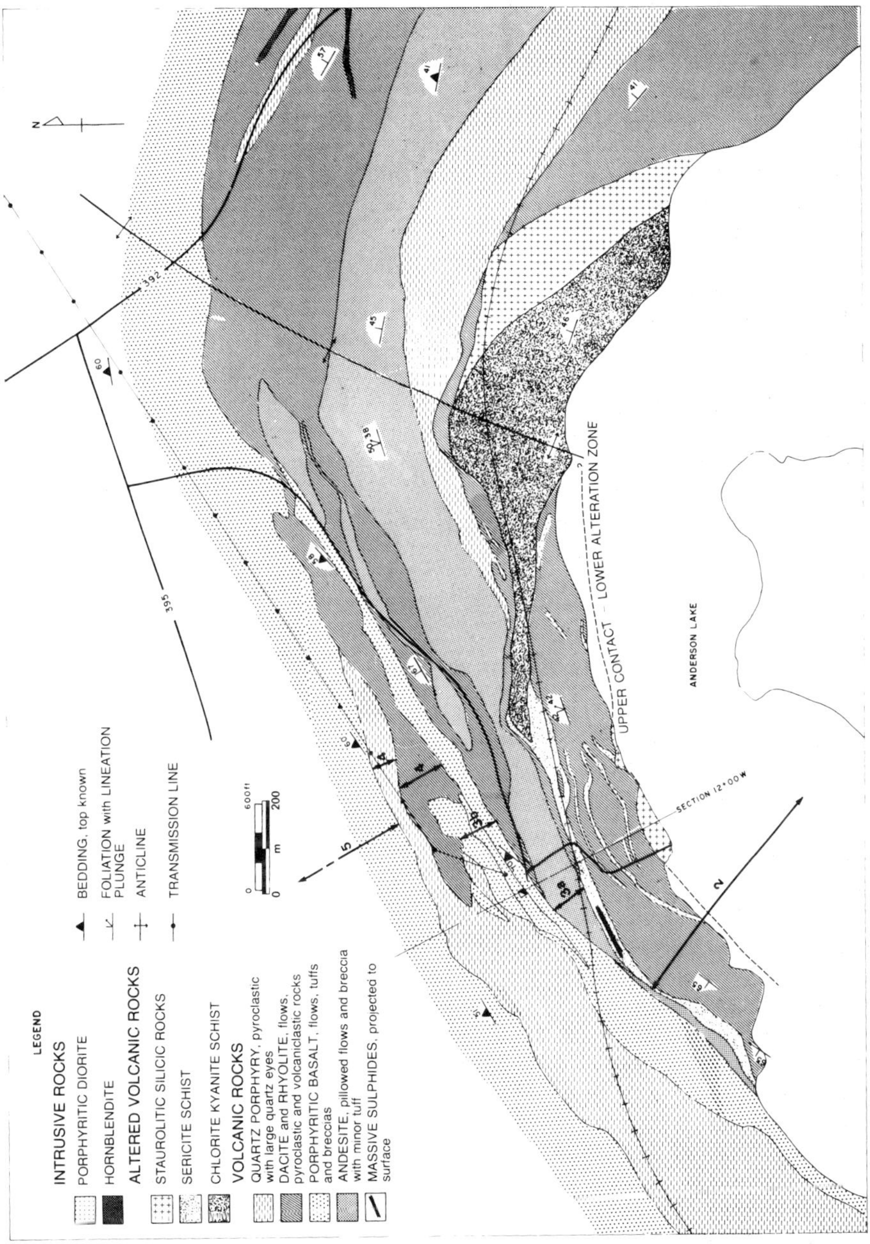

Figure 3. Anderson Lake Mine area. Unit numbers are explained in Table 2.

size), and is porphyritic in places. The northwestern end of the body is hypidiomorphic granular, with local micrographic texture (Williams, 1966). The rock consists primarily of quartz (30%) and feldspar (55%) with hornblende and biotite (total 15%) comprising the ferromagnesian fraction. Quartz phenocrysts comprise approximately 15% of the rock. Much of the tonalite south and west of Anderson Lake has a superimposed weak hydrothermal alteration, with minor pyrite and chalcopyrite (maximum 0.5% total sulphide minerals) in microfractures, accompanied by chlorite, sericite and quartz.

The tonalite contains xenoliths of mafic and felsic volcanic rocks; some of the former have indistinct irregular outlines. Both mafic and felsic dykes cut the tonalite. The mafic dykes are compositionally similar to both the andesitic and basaltic hangingwall mafic volcanic members.

Within the footwall rhyolite, and below the 2650 level of the Anderson Lake Mine the stratigraphically lowest unit exposed in the mine workings is a white, massive, equigranular, medium-grained, quartz-feldspar-hornblende rock with minor garnet and biotite. It is cut by large veins of quartz, calcite, epidote, andalusite, staurolite and anhydrite and is variably altered to chlorite in its upper part. It is compositionally different from most of the rhyolite units (Table III) in the region as it has relatively high and uniform CaO and Na_2O contents and somewhat higher TiO_2 contents. Based on its textural and chemical attributes, the body is a high level felsic intrusion similar to the basal tonalite sill, and is possibly subvolcanic to the felsic units in the stratigraphic hangingwall. It is relatively rich in feldspar in comparison with the tonalite.

Lower Mine Felsic Unit (Unit 2)

This unit is more than 900 m thick and is the lowest exposed volcanic rock in the Anderson Lake area as it immediately overlies the tonalite sill. It is composed of lower massive and upper fragmental members. The lower 600 m, partially exposed on the south shore of Anderson Lake, has few distinguishing textural features. Lack of distinctive bedding or major fragmental strata makes structural interpretation difficult. Local fragmental texture, with fragments outlined by amphibole and chlorite, may be either syn-volcanic or tectonic in origin. The rock is aphanitic, and composed of quartz (35%), feldspar (55%) and biotite (10%), with accessory ilmenite and magnetite, in an equigranular mosaic. It is difficult to distinguish from the underlying tonalite and may in part be a marginal phase of this intrusion. The lower part is cut by both felsic and mafic dykes.

The upper 300 m of this unit includes the Anderson orebody and a major zone of variably altered rocks. This member is predominantly fragmental (Fig. 4); fragments are tectonically flattened but range from ash to block size. The fragmental portion of the upper 300 m member thins to the west but thickens eastward toward the Stall Lake mine. In the thicker portions, strata composed of massive as well as fragmental rocks are present. In the Stall area, various parts of this unit may have flowlobe and also hyaloclastic origin. The possible flows are distinct thin, massive, fine-grained units that terminate abruptly in fragmental rock. Generally the fragments have variable boundary relationships with the matrix; 1200 m east of the Anderson shaft along the railway tracks, and also south of Stall Lake, fragments are very sharply defined.

The uppermost 100 m is quartz porphyritic; quartz phenocryst abundance increases gradually and irregularly upward. Part of this porphyritic zone forms the immediate hangingwall of the Anderson and Stall deposits. Toward the Anderson mine, the fragments generally become less distinct. In general, fragments of both members are supported in a matrix of very similar composition. Bedding is absent, except for a few tuff beds in the footwall rocks at Anderson mine. This upper portion of the unit is probably a submarine pyroclastic ash flow. Dykes of quartz-feldspar porphyry, and also of andesite and basalt, intrude all zones of the footwall felsic unit.

Hangingwall Mafic Unit (Unit 3)

This unit is 100 m thick, lies 10 to 15 m above the Anderson orebody, and is separated from it by felsic tuff and sedimentary rock of the Lower mine felsic unit. The unit thickens (Fig. 3) east of the orebody and is typically 250 m thick between the Anderson Lake and Stall Lake mines. Immediately west of the Anderson Lake mine, the unit is very thin, but 3 km to the southwest, it thickens. In the mine area (Fig. 3) the unit contains lower andesitic and upper basaltic members. The andesitic member thickens rapidly eastward, and comprises the entire unit east of the mine road which leads from highway 395. The basaltic member disappears east of the mine, but is important to the southwest. The two members are in contact in the immediate hangingwall of the mine, but are separated to the east by a wedge of felsic volcanic rocks of the Hangingwall felsic unit.

The andesitic member is composed of at least four pillowed flows near the Stall Lake Mine headframe; each flow contains a massive to pillowed base, overlain by pillow breccia, and capped by a zone of well-bedded tuff. Near the Stall Lake mine headframe, the basal pillowed andesitic flows are variably silicified. Toward the Anderson Lake mine, the andesitic member is dominated by pillow breccia, hyaloclastite and tuff, and at Anderson Mine, both amygdaloidal flows and well bedded

Figure 4. Stretched fragments, Lower mine felsic unit, approximately 1 km east of the Anderson Lake Mine, along the railroad tracks.

fragmental strata are present. To the southwest of Anderson Lake, the member is pillowed and amygdaloidal. The andesite contains hornblende (65%), oligoclase-andesine (20%), and quartz (15%) with minor biotite, and is fine to medium grained and granoblastic.

The upper basaltic member is texturally and chemically identical to the Threehouse basalt unit, described below. To the east of the Anderson Lake mine, this basaltic member is intimately intercalated with the hangingwall felsic unit.

Hangingwall Felsic Unit (Unit 4)

This unit is approximately 350 m thick, and is composed of two members, a lower, more easterly, variably porphyritic rhyolitic pyroclastic and flow member, and an upper, more westerly, coarsely porphyritic member with dominantly massive but some fragmental, zones. The distinction between the members is based primarily on relative phenocryst size and abundance. The two members diverge, separated by a major basaltic unit, east of the eastern end of Anderson Lake. A thin graphitic argillite bed occurs near the base of the lower member.

Directly north of the deposit the upper member is generally porphyritic and the lower member is massive rhyolitic lapillistone. To the east, the members are composed in part of distinct tongues of white, aphanitic rhyolite, usually 0.5 to 0.8 m in thickness, that terminate abruptly in grey tuff. These are probably flow lobes, surrounded by hyaloclastic tuff derived from these lobes, and submarine pyroclastic ashflow breccia. For example, north of the concentrator near the Stall Lake mine, massive elongate flow lobes of white rhyolite are set in a fine-grained, grey felsic tuffaceous matrix, similar to flow lobes described by Dimroth and Rocheleau (1979) in the Noranda area. Above the area of the flow lobes, felsic lapillistone, and lapilli-tuff with similar fragment and matrix compositions, dominate the sequence. In some areas east of Anderson Lake mine the matrix is more mafic, possibly due to alteration or mixing of mafic volcanic detritus derived from the intercalated hangingwall mafic units. Southwest of the Anderson Lake mine, the upper member is a massive porphyritic rhyolite, with little fragmental texture. The lower member terminates immediately east of the headframe of the Anderson mine.

The mineralogy of this unit is similar to that of the Lower Mine felsic sequence. The ferromagnesian content may be slightly lower, ilmenite and magnetite are absent in the Hangingwall felsic rocks, and the unit is distinctly porphyritic, with some square, beta-quartz pseudomorphs, and less common plagioclase phenocrysts.

Threehouse Basalt Unit (Unit 5)

This unit consists of approximately 600 m of porphyritic basalt flows and basaltic epiclastic material. The epiclastic portion contains up to 10% felsic clasts. The basalt is characteristically feldspar-phyric, with approximately 20% feldspar phenocrysts. Prismatic hornblende phenocrysts comprise 10 to 15% of the rock. The groundmass is fine-grained and comprised of hornblende, plagioclase and minor biotite. Flows usually exhibit several facies; basal pillowed or massive flows are overlain by flow breccia, and capped by several beds of water-lain mafic detritus. Sedimentary features include graded beds, slump folds and channel scours. Locally, aphyric non-amygdaloidal flows cannot be distinguished from subvolcanic intrusions.

Upper Units

The succession above the Threehouse basalt unit includes, from the bottom: 1) a mixed volcanic-sedimentary unit with mafic and felsic volcanic rocks, epiclastic strata, greywacke and mafic sills, with a total thickness of approximately 300 m; to the north of the Stall Lake Mine, felsic volcanic members predominate; 2) a metasedimentary unit, approximately 350 m in thickness, containing greywacke and shale interstratified in 10 to 50 cm thick beds (Froese and Moore, 1980); 3) a felsic volcanic unit above the McLeod Road thrust fault in the area of the Snow Lake townsite which is highly porphyritic and is approximately 1700 m thick; intercalated with and overlying these felsic rocks are basalt flows and fragmental strata (Harrison, 1949). The latter mafic rocks include the intercalated Snow Creek mafic unit and the uppermost Herblet mafic unit. The stratigraphic position of the strata above the McLeod Road thrust fault, with respect to the strata below this fault is not clear.

PETROCHEMISTRY

The major and minor element composition of the principal units (Table III) provides useful data concerning the petrogenetic history of the region, and is also relevant to the origin of massive sulphide deposits. Samples of highly altered rocks, those with low alkalis (<2% Na_2O), and excess normative cordundum (>2%) are not included in the data set used to calculate the average compositions. Data for the Lower Mine felsic suite are divided into two groups. The "footwall" rocks (unit 2a) crop out below the Anderson Lake deposit, and the "Ram-Stall" rocks (unit 2b) are stratigraphically equivalent, but crop out below the Stall Lake mine.

The volcanic rocks form three compositional groups (Fig. 5). The felsic rocks are compositionally distinct from the mafic rocks indicating originally separate felsic and mafic melts. The felsic suite has several notable compositional characteristics.

1) The Lower mine (footwall) felsic unit (unit 2a) in the Anderson Mine area has high Ti, S, Mg and Cu contents relative to other units, including the laterally equivalent footwall rocks in the Ram-Stall (unit 2b) area. It is also enriched in Fe, in part due to its sulphide content. After removal of normative pyrite, it contains only 4.97% total Fe (as Fe_2O_3), slightly higher than the other felsic units. The Lower mine (footwall) felsic rocks near the Anderson mine have slightly lower SiO_2 and Na_2O contents and a slightly higher Zr content, relative to the other felsic groups.

2) The Lower mine felsic rocks (units 2a and 2b) are similar in composition to the tonalite sill (unit 1), although the latter has an area of lower (65% to 68%) SiO_2 content. Zr, the main incompatible element examined, and V, are very similar in content in these two rock groups. The lower Sr and Na_2O, and higher Cu and Fe_2O_3 in rocks of the volcanic group are probably due to alteration related to massive sulphide deposits (discussed later). The similarities in relatively immobile elements (Al_2O_3, TiO_2) as well as in MgO, and total Fe after removal of normative pyrite may indicate a common parentage for the sill and the volcanic rocks.

3) Several elements display a systematic change relative to stratigraphic position. Specifically, the changes with some possible petrogenetic significance are (Fig. 6): i) an upward increase in K_2O, ii) an upward increase in Ba, iii) the Hangingwall felsic and Townsite felsic units have higher total alkalis than the footwall rhyolite, iv) Sr is significantly higher in the Townsite felsic rocks.

General comparison of these felsic rocks with other felsic sequences of similar age is not possible, due to a lack of suitable comparable data. However, comparison with extensive data for Archean rocks (Goodwin, 1977) indicates that the Snow Lake felsic rocks have much higher MgO and CaO contents. Also, they have lower Sr and Zr and, except for the Townsite felsic rocks, lower Ba contents than Archean rhyolite. In general, Snow Lake rhyolite has a higher Cu content than Archean rhyolite.

The mafic rocks form two compositional groups (Fig. 5, Table III). The immediate hangingwall sequence to the Anderson Mine contains the andesitic member, but ranges in composition from basaltic andesite to andesite. The upper mafic units are all basaltic. The Threehouse basalt typifies these rocks; it has a narrower compositional range than the hangingwall andesite. Dykes of composition similar to both

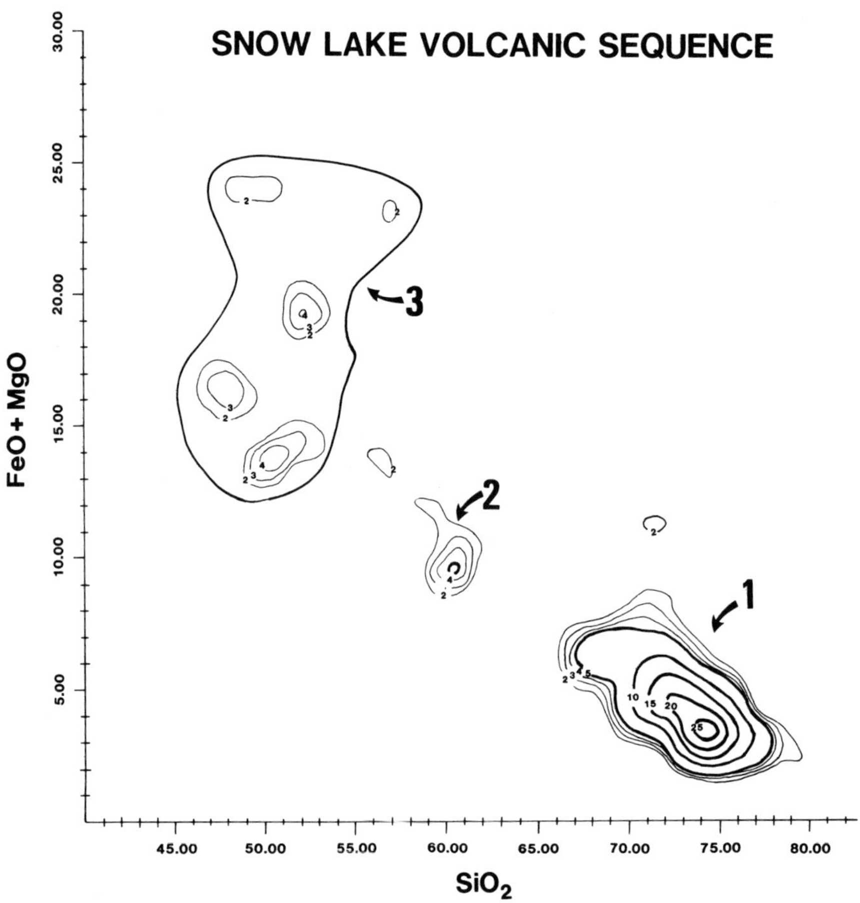

Figure 5. Contoured variation diagram illustrating the major compositional groups in the volcanic rocks of the Snow Lake region. Three compositional groups are evident, corresponding to (1) the felsic rocks, (2) the andesite unit, and (3) the basaltic units. Elements in weight per cent.

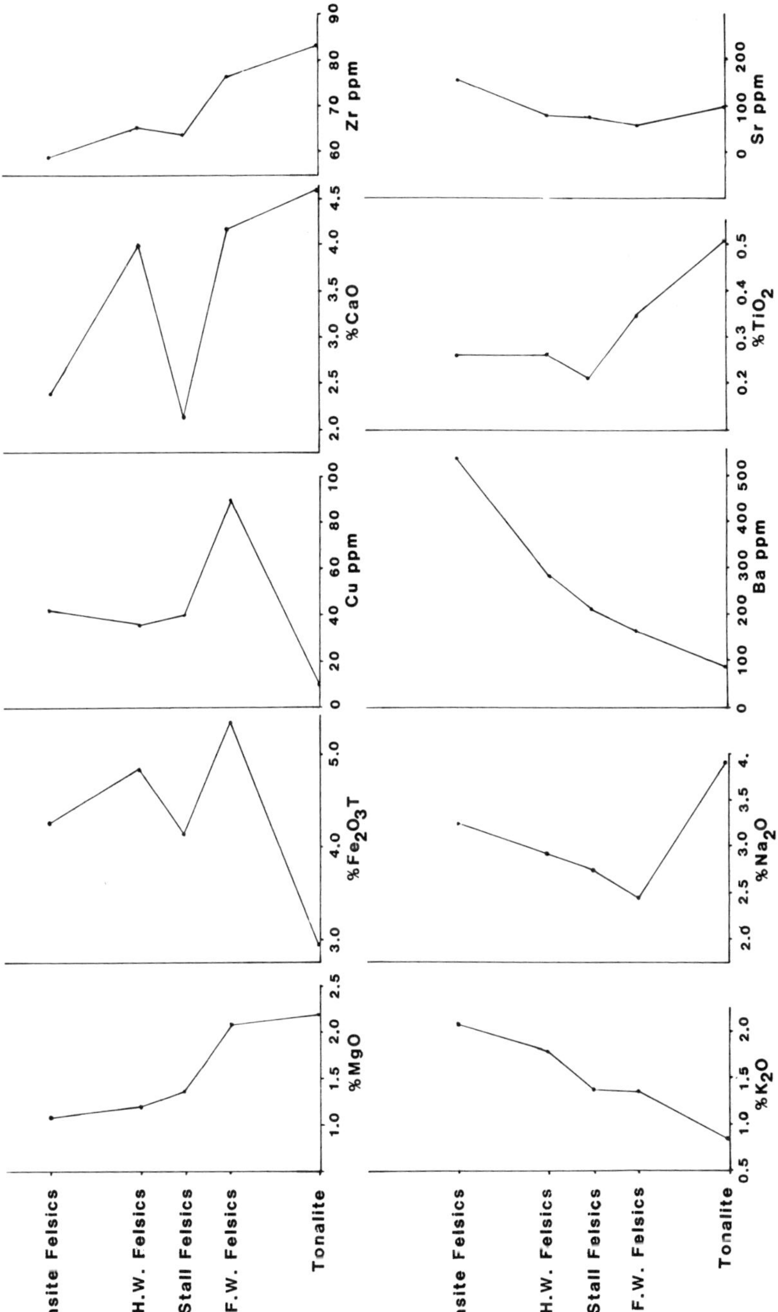

Figure 6. Variation in composition of felsic rocks with stratigraphic position. Elements in weight per cent.

the andesite and basalt are common in the footwall rhyolite sequence. Basaltic dykes and sills are widespread in all hangingwall felsic strata.

The hangingwall andesite is similar to Archean andesite (Goodwin, 1977) but is significantly lower in TiO_2 and K_2O and somewhat higher in CaO. The basaltic rocks are unusually rich in alumina and very low in TiO_2 in comparison with typical Archean basalts. Baragar (1968) described high-alumina basalt, with higher TiO_2 content, in the upper portion of the Blake River group in the Duparquet section and suggested that such rocks were generated by melting of the intermediate zone of the mantle, at pressures of 9 kb. Such basalts must contain aluminous clinopyroxene.

The volcanic rocks are classified according to various schemes (Fig. 7); these variation diagrams indicate a general tholeiitic nature of the andesite, but a high alumina, calc-alkaline composition for the basalt, as also noted by Moore (1977). The felsic rocks are similarly calc-alkaline, and are divided into six groups; staurolite-

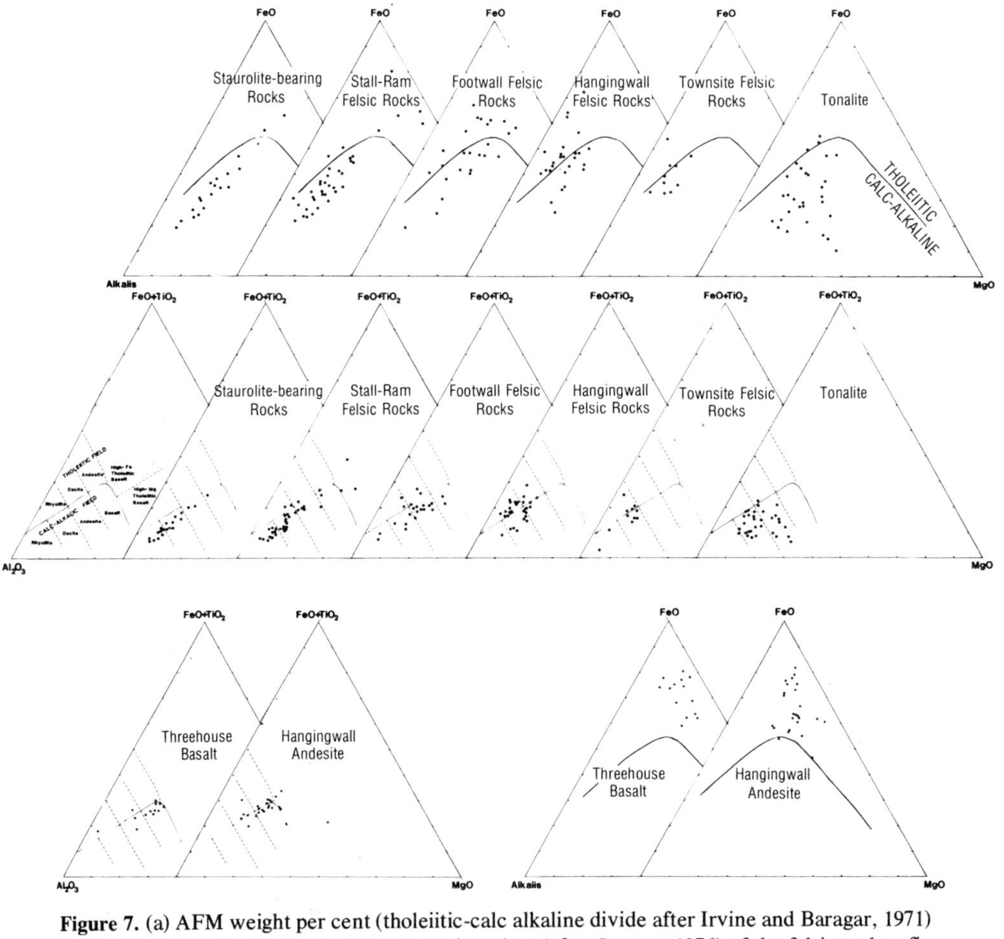

Figure 7. (a) AFM weight per cent (tholeiitic-calc alkaline divide after Irvine and Baragar, 1971) and $Al_2O_3-(FeO+Fe_2O_3+TiO_2)-MgO$ cation plots (after Jensen, 1976) of the felsic and mafic groups of volcanic rocks and subvolcanic rocks.

bearing samples are probably altered (see following discussion), and thus treated separately. The Lower mine felsic unit (unit 2) is separated into the Stall-Ram felsic rocks, from the area east of Highway 392 and south of the Stall Lake Mine, and the Footwall felsic rocks, which are within 2 km of the Anderson Lake mine and have probably been affected by deposit-related alteration. Both areas have staurolite throughout, but the eastern part of the Stall-Ram area is not adjacent to an obvious alteration pipe on surface; the Ram zone and the pipe associated with the Stall Lake Mine do plunge beneath this area, however. Farther west, some of the staurolitic alteration may be part of the lower alteration zone (discussed later).

The linear trend of the staurolite-bearing samples is reflected in a similar trend for both the Stall-Ram and Footwall felsic groups. The compositional range of these groups on the Jensen cation plot is from rhyolite to basalt, indicating that use of this diagram in classifying the rock series is unreliable. The Hangingwall felsic and Townsite Felsic groups are divided between tholeiitic and calc alkaline fields, with

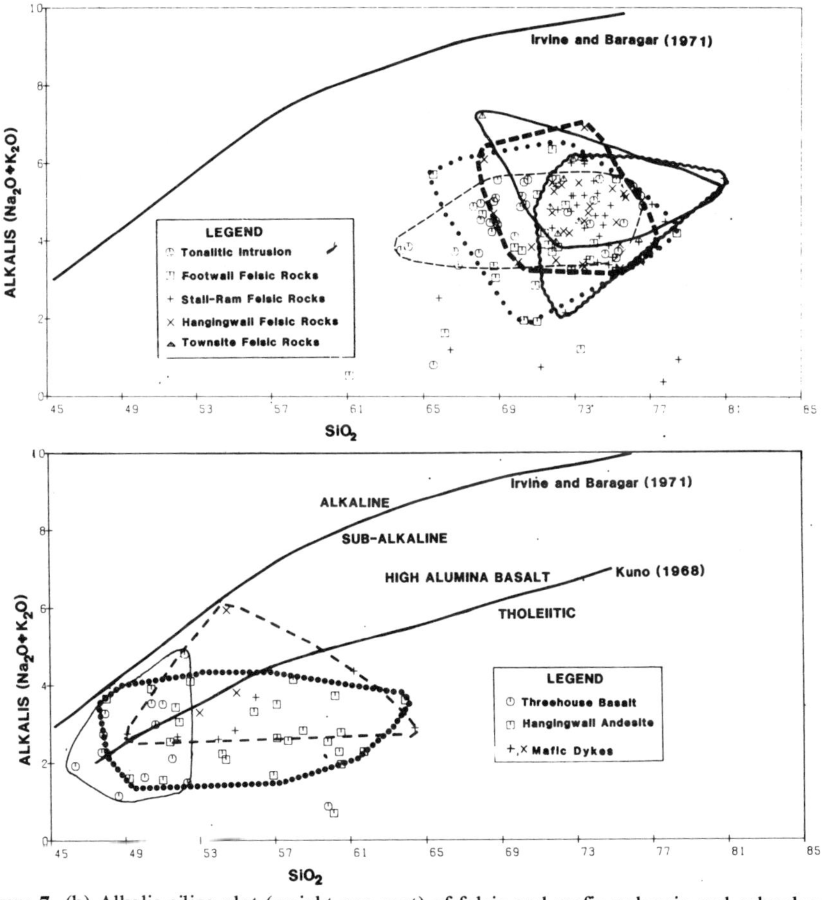

Figure 7. (b) Alkalis-silica plot (weight per cent) of felsic and mafic volcanic and subvolcanic rocks. Note that many of the Threehouse basalt compositions are in Kuno's (1968) high-alumina field.

most of the Townsite group in the calc-alkaline field. Interestingly, the tonalite sill, which is probably the subvolcanic equivalent of all of the felsic volcanic rocks, is entirely calc-alkaline. As this intrusion is less affected by alteration, its composition may be more representative of the felsic magma. The felsic rocks are thus considered to be calc-alkaline.

The mafic rocks are less clearly assignable to a petrogenetic suite; the AFM diagram indicates a tholeiitic nature for both the hangingwall andesite and Threehouse basalt, but the Jensen cation plot indicates a tholeiitic nature for the latter, and mixed parentage for the former. A substantial number of the analyses of both groups lie in Kuno's (1968) high alumina field (Fig. 7b), and given the high Al_2O_3 content of the Threehouse basalt group, it is probably best classified as high-alumina basalt. The andesite cannot be definitively assigned to a series.

The felsic rocks form a coherent suite that may be the result of either crystal fractionation from a single rhyolitic parent or a continuous change in parent melt composition. High level separation of ferromagnesian minerals and feldspar would be required in order to fractionate the liquid sufficiently to cause an upward increase in K_2O and BaO. Although such separation might have occurred in the partially less siliceous tonalitic sill, which has some areas of slightly lower silica content (65 to 68%) than the rhyolitic rocks, the viscosity of most felsic melts inhibits the separation mechanism. A continuous change in parent melt composition might have been a response to progressively lower temperatures in the zone of partial melting.

The origin of the mafic melts is similarly obscure. The compositional difference between the two melts, and the lower stratigraphic position of the more fractionated andesitic member, preclude simple fractional crystallization as the basis for the differences in composition. Either different zones of melting or contamination of one melt by more felsic crustal material may explain the differences, but more data are necessary to test these hypotheses.

MINE GEOLOGY

Orebody

The geology of the Anderson mine was briefly described by Price (1977), and the effects of metamorphism on the Anderson and Stall wallrocks were studied by Hutcheon (1979). The deposit is a single elongated lens (Fig. 8) composed primarily of pyrite, chalcopyrite, and pyrrhotite with minor sphalerite. It is conformable with the volcanic host rocks, strikes north 65° east and dips 60° to the northwest, and plunges at 55° to the northeast. The upper edge of the massive sulphide lens begins 90 m below surface (Fig. 9) but is not of economic grade or width above the 185 m level. By the 275 m level, the lens attains its average strike length of 150 m, 120 m of which is ore. The maximum thickness of 10 m occurs in the centre of the ore zone, and includes the stringer ore. The deposit thins laterally from the centre. The lens has been traced approximately 1220 m down the plunge and maintains an essentially constant shape throughout.

The stringer zone which underlies the massive sulphide body and is part of the hydrothermal vent area, together with its accompanying alteration zones (described below) have been displaced to the east, relative to the massive sulphide body, by regional deformation (Fig. 3). Most of the footwall stringer mineralization is in chlorite-biotite-kyanite schist, and chloritized vestiges of rhyolite. A muscovite-kyanite zone 12 m thick is best developed on the west side of the most intensely altered chlorite-biotite-kyanite zone, but rarely contains enough chalcopyrite to constitute ore. The less altered staurolitic felsic rocks which surround the alteration pipe also do not constitute ore, except immediately adjacent to the massive sulphide lens.

Pyrite, pyrrhotite, and chalcopyrite are abundant in a zone extending for 30 m below the ore zone. Next to the massive ore, the stringers coalesce to form semi-massive ore over a width of several metres. In this zone, the stringers may contain over 10% copper, and visible sphalerite (E. Guiboche, pers. commun., 1979). Sphalerite is rare throughout the remainder of the deposit. The greatest concentration of

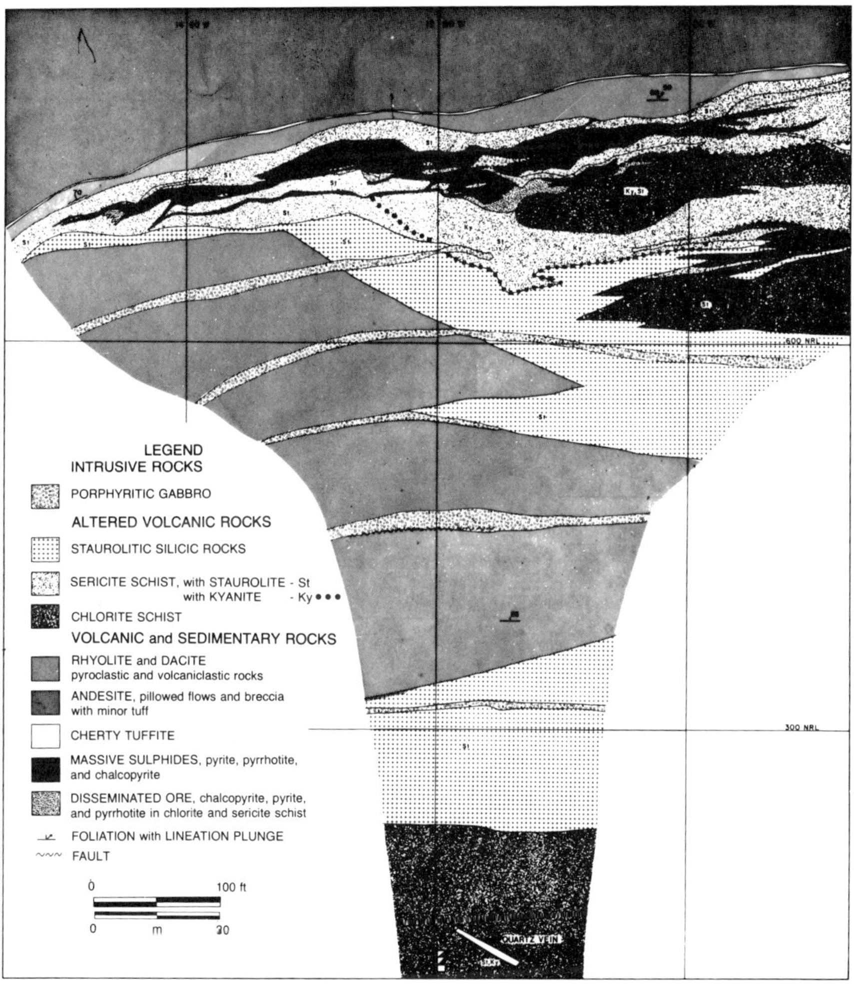

Figure 8. Plan of the 1950 level, Anderson Lake Mine.

chalcopyrite and pyrrhotite within massive ore occurs immediately above the most intensely developed stringer zone, in the centre of the orebody as illustrated by the copper distribution map (Fig. 10). The principal gangue and minor minerals within the ore are quartz, cordierite, staurolite, albite, anthophyllite, magnetite and anhydrite. These gangue minerals occur in both the stringer sulphide and massive ore. Most of

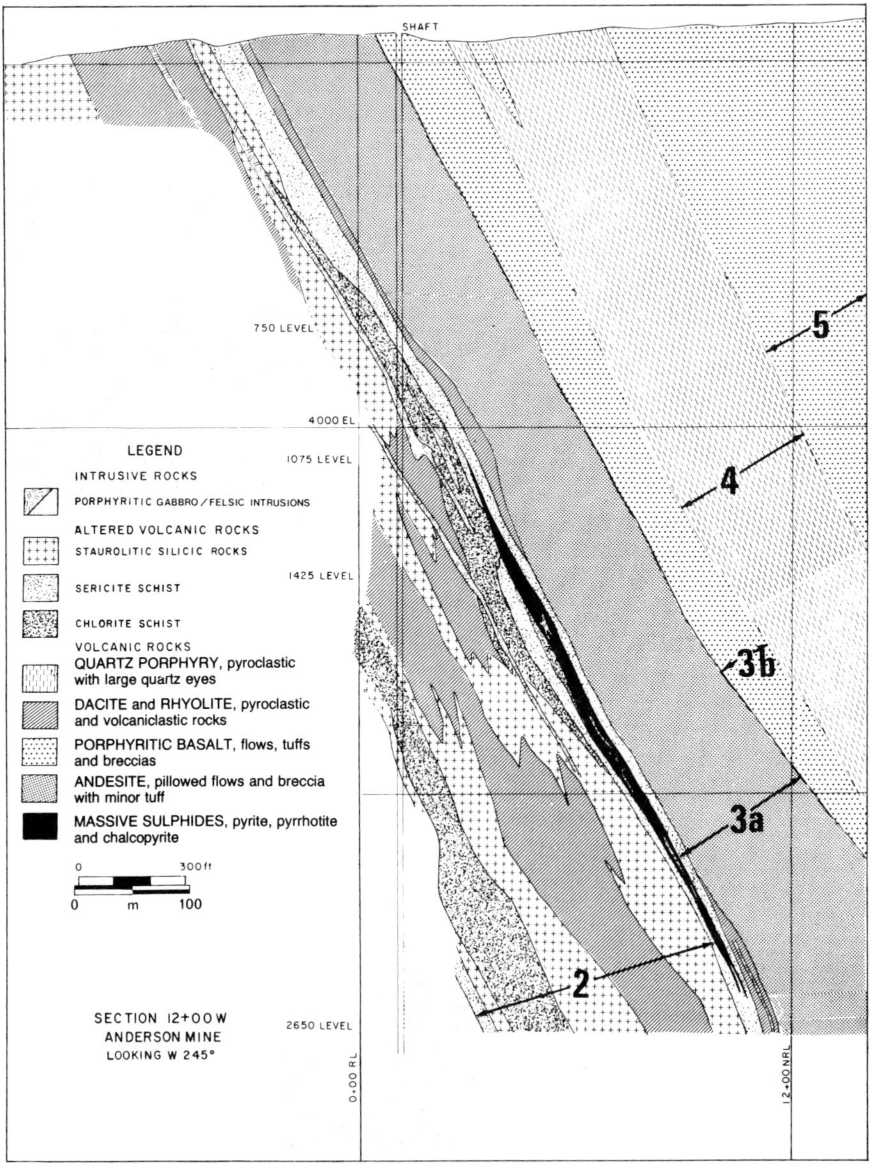

Figure 9. Section of the Anderson Lake Mine. Unit numbers are explained in Table 2.

the gangue minerals, together with kyanite, tourmaline, minor sulphides, and rarely apatite occur in quartz lenses (1 to 2 m thick) within the chlorite-biotite zone. Within these quartz zones, metacrysts of the aforementioned minerals tend to be very large;

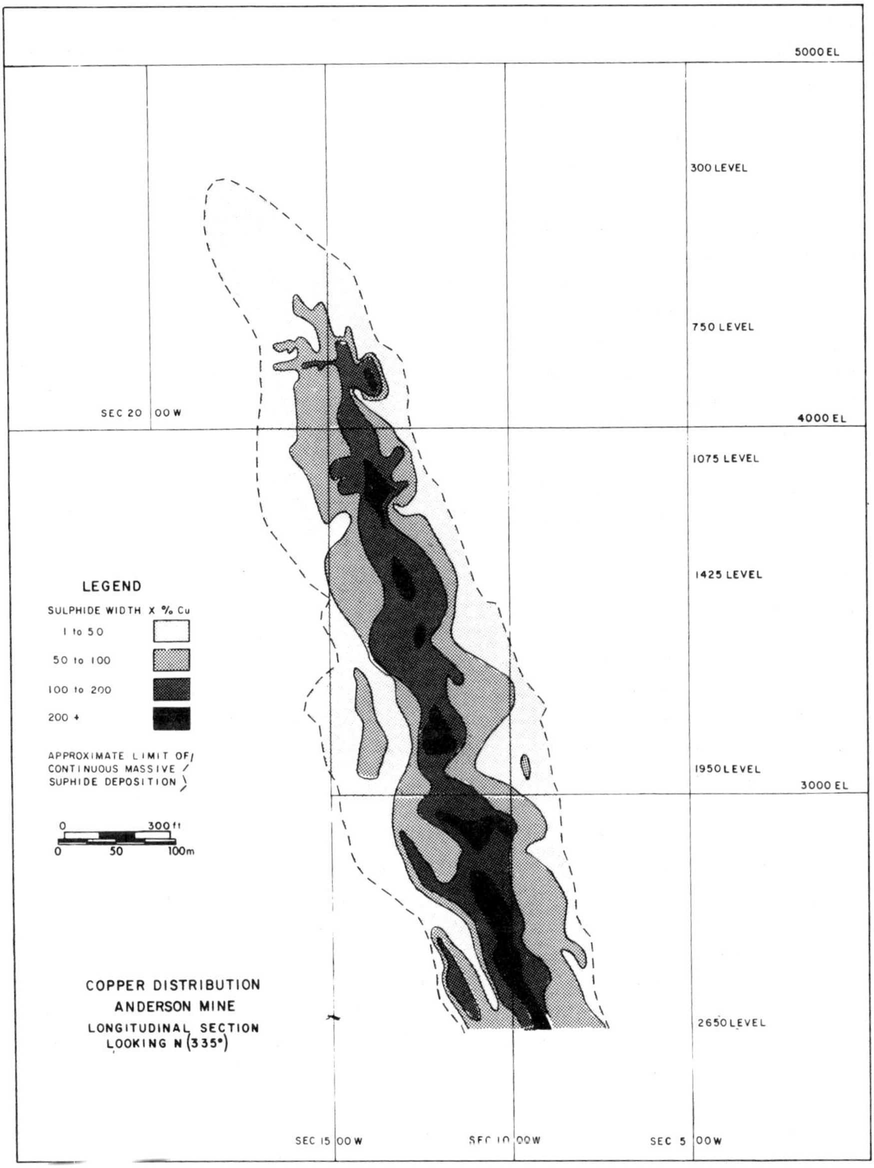

Figure 10. Longitudinal section showing contours of sulphide thickness x copper content. Note that the centre of the ore lens is both thickest and has the highest copper content. Margins of the massive zone are more pyritic.

biotite is up to 2 cm in diameter, whereas kyanite and cordierite are up to 6 cm in length. Molybdenite has been noted in the chlorite-biotite schist (F. Bill, pers. commun., 1980), and telluro-bismuthite in a quartz vein in massive ore.

In the massive sulphide lens, primary sulphide compositional layering is not obvious. Copper grade decreases generally from the footwall to the hangingwall, typical of most massive sulphide deposits (Sangster, 1972). The massive ore contains very coarse-grained, usually porphyroblastic pyrite (Fig. 11), with anhedral grains of pyrrhotite, sphalerite, and chalcopyrite filling the interstices between the pyrite porphyroblasts. Pyritohedra are as much as 3 cm in diameter reflecting the intense recrystallization during metamorphism that destroyed any primary textures. Near the base of the massive ore, long flame-like trails of chlorite extend into the massive ore. At the western end of the orebody near the 900 level, a 3 m-thick carbonate-pyrite unit separates the massive ore from the underlying stringer zone. "Globules" of chlorite, quartz, anhydrite, and carbonate occur throughout the deposit. A minor amount of carbonate fills post-deformational joints in some portions of the orebody. Within the massive ore discontinuous trails of quartz-muscovite rock and quartz-biotite-chlorite rock, probable remnants of syndepositional silicate bands, occur irregularly. The quartz-muscovite bands increase in number and continuity toward the edges of the ore zone. They eventually merge into the enveloping muscovite schist which forms the major alteration zone on the footwall and as well forms a thin sedimentary unit on the immediate hanging wall. Within the massive ore the bands of sericite schist contain a minor amount of disseminated sulphide, including chalcopyrite. Commonly these bands are highly folded, illustrating the intense deformation that affected the orebody. Chalcopyrite is commonly remobilized into the crests of minor folds as well as into fractures within the bands. Toward the edges of the ore body the sulphides become more disseminated, although in some fold hinges chalcopyrite is abundant enough to form ore.

The massive ore and stringer zones are cut by several subvolcanic amphibolitic dykes that are in part altered to chlorite-garnet and amphibole-magnetite rock with some sulphides. The dykes are folded and fragmented to form discontinuous chains within the massive ore.

Away from the deposit, the ore horizon is marked by muscovite schist, containing minor pyrite, chalcopyrite, anhydrite and sphalerite. The hanging wall contact of the overlying muscovite schist is very sharp. Copper content decreases tenfold over a distance of less than a metre, although the quartz-muscovite schist contains disseminated pyrite, and some magnetite (F. Bill, pers. commun., 1979) and gahnite. The muscovite schist is 4 m thick immediately above the ore, and is overlain by an additional 5 m of partially altered, felsic volcaniclastic rock. Porphyritic rhyolite which onlaps discontinuously above the orebody below the 2650 level may be a precursor to part of the quartz-muscovite schist. At upper levels the volcaniclastic zone consists of both porphyritic and massive lapilli-sized clasts of rhyolite with up to 50% quartz in the matrix. This exceptional abundance of quartz, relative to the unaltered volcanic rocks, may be due to sedimentary re-working of felsic porphyritic clasts. Both the quartz-muscovite schist and quartz-rich lapillistone are thus possibly epiclastic in origin. At the top of this felsic unit is a thin-laminated, cherty "tuffite" containing fine-grained biotite, quartz and feldspar, and pyrite disseminations. This is overlain in turn by the andesite member of the Hangingwall mafic unit.

Hydrothermal Alteration

Two major, very intense, alteration zones are present in the footwall sequence (Figs. 8 and 9). The massive sulphide zone is immediately underlain by a major pipe of alteration containing the stringer-copper mineralization, and is essentially similar to the typical alteration "pipes" underlying most massive sulphide deposits (Sangster, 1972). It transects the footwall rocks at approximately 30° to strike. This pipe has been deformed parallel to the easterly regional plunge, and probably was more "vertically oriented" prior to tectonism. It has a core of chlorite-biotite-kyanite schist, a large irregular halo of muscovite schist, and a staurolitic altered rhyolite outer shell.

Figure 11. Pyritohedron in massive chalcopyrite from the Anderson massive ore zone.

Figure 12. Kyanite patch in the centre of the alteration pipe, 1950 level, Anderson Lake Mine.

The second, "lower" alteration zone is approximately 150 m below the ore and is a conformable to semiconformable unit, within less altered footwall rhyolite which separates it from the ore body. It is zoned, both vertically and along strike. Vertically below the Anderson ore body (Fig. 9), it consists of approximately 60 m of massive chlorite-biotite-kyanite schist with less intensely altered staurolitic rhyolite 30 m thick along its hangingwall. This lower chloritic zone continues eastward for 730 m where it is joined by the alteration pipe described above. To the southwest the lower zone has been traced for at least 1500 m. However, in this direction it changes mineralogically, with decreasing chlorite content and increasing development of staurolite. The lower zone contains little sulphide, in contrast to the stringer sulphides within the pipe.

Between the two intensely altered zones, the rhyolite is pervasively but subtly altered; minor staurolite exists throughout most of this area and chemical effects of alteration, described below, are evident.

The Anderson alteration "pipe" crops out 1000 m east of the sulphide body along the north shore of Anderson Lake. Here it is approximately 300 m below the ore horizon. The pipe cuts massive (possibly flow) as well as fragmental rhyolite. The former is extensively brecciated into irregular 0.5 to 5 cm fragments. At the periphery of the pipe, alteration has affected principally the margins of some of these fragments; alteration minerals fill the interstices and increase in size with increasing abundance. Within the central part of the alteration pipe however, most of the rock is replaced.

The alteration pipe has various characteristic mineral assemblages.

1) The chlorite-biotite-kyanite assemblage is dominant closest to the ore at its easternmost end. Rhyolite is almost totally replaced by this coarse-grained assemblage. Chlorite (40 to 70%), biotite (10 to 80%), quartz (5 to 20%), and kyanite (5 to 50%) are the dominant minerals (Fig. 12) with less common garnet (0 to 10%) and muscovite (0 to 20%), and minor albite. Pyrite, chalcopyrite and pyrrhotite are generally present, and ilmenite and, near the periphery of the chlorite zone, magnetite, are also common. Tourmaline is a common accessory. Hutcheon (1977) noted that garnet is never in contact with chlorite but is typically set in a coarse biotitic matrix. Almandine composes most of the single analyzed garnet sample from the Anderson mine (Table IV). Garnets in a similar alteration zone at Stall Lake (Hutcheon, 1977) have variable but significant contents of spessartine and pyrope. Cordierite and anthophyllite occur in a few samples, the latter only in the absence of muscovite. Large quartz pods within 30 m of the massive sulphide body and confined to the chlorite zone commonly contain staurolite, cordierite, kyanite, biotite and sulphide minerals with accessory apatite. Anhydrite veins, commonly 1 to 2 cm in width, occur in the chlorite zone.

The chlorite (Table IV) is slightly aluminous and moderately magnesian compared with chlorite from metasedimentary rocks in the same metamorphic zone (Froese and Gasparrini, 1975). Using Hey's (1954) classification, it is ripidolite, and probably best considered as magnesian ripidolite. This composition is very similar to that described by Franklin *et al.* (1975) at Mattabi, regarded as typical of massive sulphide-related alteration zone chlorites. Biotite (Table IV) is MgO-rich, relative to biotite in the metasedimentary rocks.

TABLE IV
MINERAL COMPOSITIONS, ANDERSON LAKE MINE ALTERATION ZONE

Chlorite[1]

	Wt% 1s.d.	Structural Formula		
SiO_2	24.89±0.25	Si	2.514	4.00
Al_2O_3	23.13±0.43	Al	1.486	
FeO	20.25±2.53	Al	1.270	5.824
MgO	18.71±2.02	Mg	2.816	
MnO	0.23±0.23	Fe	1.710	
TiO_2	0.07±0.07	Ti	0.005	
		Mn	0.020	
CaO	0.01±0.0[illegible]	Ca	0.001	
Na_2O	0.01±0.01	Na	0.002	
		H	8.	
		O	18.	

Staurolite[2]

	Wt% 1s.d.	Structural Formula		
SiO_2	27.81±0.77	Si	7.72	
Al_2O_3	53.11±0.87	Al	17.399	18
FeO	12.87±2.07	Ti	0.094	
MgO	2.14±0.25	Fe^{2+}	0.507	
MnO	0.55±0.23	Mg	0.886	3.774
CaO	0.01±0.01	Fe^{3+}	2.480	
ZnO	1.36±1.54	Mn	0.129	
TiO_2	0.45±0.04	Zn	0.279	
		H	4.	
		O	48.	

Biotite[3]

	Wt% 1s.d.	Structural Formula		
SiO_2	36.628±0.634	Si	2.738	4.0
Al_2O_3	18.402±0.390	Al	1.268	
K_2O	9.276±0.261	Al	0.355	2.896
Na_2O	0.323±0.118	Ti	0.08	
FeO	16.868±1.818	Fe	1.055	
MgO	12.650±1.193	Mn	0.008	
MnO	0.130±0.068	Mg	1.400	
TiO_2	1.381±0.123	Na	0.468	1.353
H_2O	3.994±0.057	K	0.885	
		OH	1.994	
		O	12.	

Garnet

	Wt%	Structural Formula		
SiO_2	37.03	Si	5.958	6
Al_2O_3	20.93	Al	0.042	
FeO	34.91	Al	3.932	6.124
MgO	3.23	Mg	0.775	
MnO	1.72	Fe^{2+}	4.707	
CaO	2.36	Mn	0.235	
TiO_2	0.01	Ca	0.407	

[1]Av of 7 analyses

[2]Av of 10 analyses

[3]Av of 13 analyses

Note: Analyses are averages for each mineral compiled from Hutcheon, 1977.

Near the base of the alteration pipe at its eastern end, albite is a common accessory mineral in the chlorite-biotite-kyanite assemblage.

2) The muscovite assemblage surrounds the chlorite-biotite-kyanite assemblage and is present immediately adjacent to ore where the latter is absent. Near the ore the muscovite assemblage is more extensive than the chlorite-biotite-kyanite assemblage, but farther down the "pipe" only a thin muscovite assemblage persists. The rock is highly fissile, coarse-grained and contains muscovite (50 to 80%), highly irregularly distributed kyanite (0 to 30%), present only at the contact with the chlorite-biotite-kyanite zone, biotite (5 to 10%), staurolite (5 to 15%), quartz (20 to 50%), andalusite (1 to 5%), and garnet (1 to 5%). Sulphides are irregularly distributed throughout and ilmenite is a common accessory. Kyanite is in large blades and aggregates, commonly as much as 20 cm in length. Large quartz veins containing kyanite and tourmaline compose 5 to 15% of the zone within 30 m of the massive ore. In the area 20 m or more from the ore body and below the chlorite-biotite-kyanite assemblage, the muscovite assemblage is less intensely developed. Muscovite and biotite occur in rhyolite together with staurolite, garnet, and minor kyanite. This alteration assemblage is marked by the absence of feldspar. The assemblage is highly deformed with minor folds and kink-bands common.

3) The staurolite assemblage surrounds the chlorite-biotite-kyanite and muscovite assemblages and is less intensely altered than both the latter. Most of the textural characteristics of the rhyolite are clearly evident and the rock consists of large poikiloblastic crystals of staurolite set in a quartz-biotite ± chlorite ± muscovite matrix. Garnet megacrysts occur in the outer edge of the staurolite assemblage.

Staurolite (Table IV) contains a generally high but variable (0.16 to 4.30%) content of zinc and is somewhat rich in MnO, relative to typical metasedimentary staurolite compositions (Froese and Gasparrini, 1975). The zinc content is apparently independent of the immediate presence of sphalerite.

Feldspar is a minor component. The outer boundary of this assemblage is very poorly defined as the staurolite content diminishes gradually downward. The "unaltered" rhyolite (Fig. 8) contains minor staurolite throughout indicating that some alteration pervades the entire footwall sequence in the mine area. Staurolite is relatively evenly distributed and coarse-grained through the inner part of this assemblage, but in the less altered outer part, including the "unaltered rhyolite", it occurs as smaller grains, in irregular patches which are commonly 10 to 25 cm in diameter. Tourmaline and pyrite are accessory minerals throughout the staurolite assemblage.

Lower Alteration Zone

This zone is well-exposed approximately 900 m east of the mine along the north shore of Anderson Lake and at the southeastern corner of Anderson Lake. Its western extension was examined only in drill sections under Anderson Lake. It is confined at least in part to an area of lapilli tuff and breccia within the footwall rhyolite. The matrix appears most altered; however, in the core of this zone both fragments and matrix are replaced.

The altered zone is at least 150 m wide and its outer contacts are gradational to unaltered, more massive rhyolite. The periphery of the alteration zone in the mine area is defined by thin biotite selvedges around the fragments. Garnet megacrysts are common in the selvedges and garnet also occurs in irregular leucocratic patches in otherwise unaltered rhyolite.

The appearance of staurolite marks the boundary of the distinct alteration zone. Within the alteration zone the biotite-garnet-amphibole patches are more distinct and are accompanied by magnetite. Staurolite occurs in leucocratic patches in the matrix of the fragmental unit, commonly 15 to 20 cm in diameter, and also along indistinct fractures. As the staurolite content increases to 30%, garnet abundance decreases.

The most intensely altered part of this zone is up to 45 m wide and is defined by chlorite, biotite, and minor garnet. Although this assemblage is best developed in the matrix of the rhyolitic fragmental unit, in the central part of the zone chlorite, biotite and kyanite with some albite totally replace the rhyolite in areas several metres in diameter. Magnetite and ilmenite are ubiquitous and pyrite, pyrrhotite and sillimanite are locally present. Muscovite occurs as an accessory throughout. This zone is well developed from the area of the Ram occurrence in the east to the area directly beneath the Anderson Lake mine. It is thickest and albite is most abundant where it joins the Anderson alteration pipe (Fig. 3). To the west of the mine the chlorite-biotite assemblage disappears and the zone is defined only by staurolite. The detailed petrographic characteristics of the various alteration assemblages in the lower conformable alteration zone are essentially similar to those already described in the alteration pipe.

Alteration pipes beneath the Stall and Ram deposits have been traced down through their footwall strata, and found to join the same chlorite-rich portion of the lower conformable alteration zone, as does the pipe below the Anderson orebody.

Chemistry of the Alteration Zones

The alteration zones were examined in underground traverses along the main haulage drifts of the 1250, 1950 and 2650 levels. Both the lower conformable and the pipe alteration zones are exposed on the latter two sections whereas only the pipe zone is exposed on the 1250 level. The 1950 level typifies the alteration and data from it are presented in Figure 13.

The chemical variations are in part predictable from the petrography of the alteration zones, as MgO and K_2O vary in response to changes in chlorite and muscovite contents respectively. However, a few additional significant variations and numerous important contrasts between the two zones are evident in the chemical data.

All data are compared to the "background" composition of the footwall felsic sequence. Unfortunately the possibilities of systematic variations in this unit could not be examined due to the coverage of its upper portion by Anderson Lake. However, based on the overall upward "fractionation" trend of the felsic strata within the pile the immediate footwall rocks are probably slightly more fractionated than the average footwall sequence below the Anderson mine (Table III). This is borne out by comparison with the Stall-Ram area felsic rocks (Table III) which are probably the lateral equivalents of the immediate footwall sequence.

The data are discussed in three groups, the feldspar elements (alkalis, CaO, Al_2O_3, Ba, and Sr), the ferromagnesian elements (MgO, FeO, TiO_2, MnO) and the

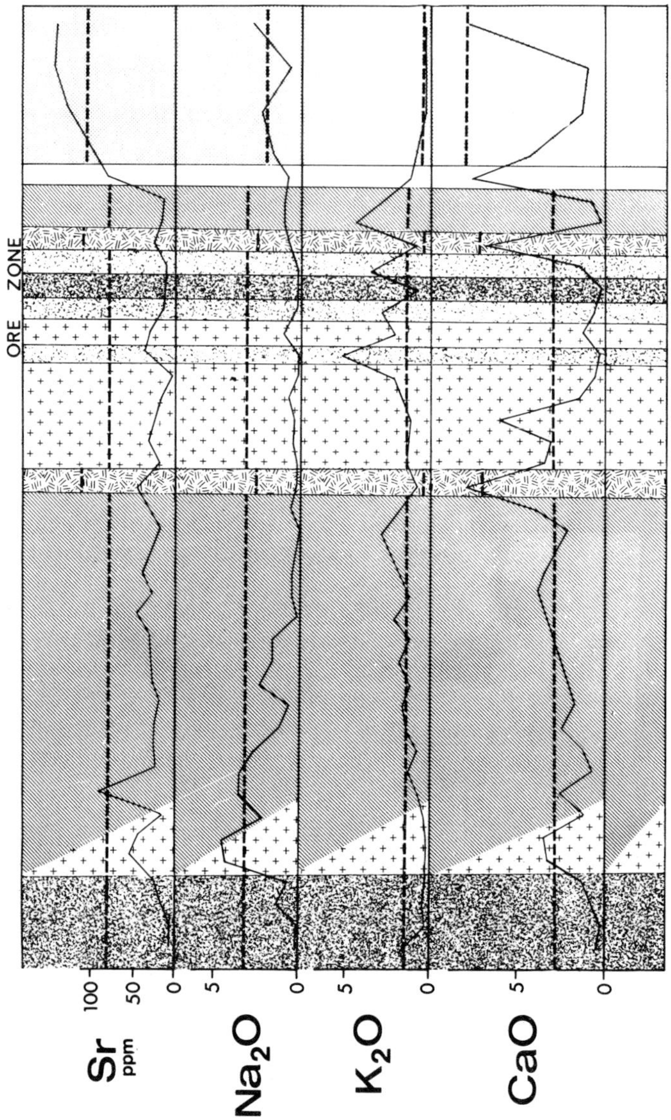

Figure 13. (a) Variation in feldspar-related elements along a section perpendicular to the ore zone, 1950 level, Anderson Lake Mine. The 'background' level for each element, determined from the average composition of the regional suite of samples for felsic and mafic rocks, is shown by the dashed line. Lithologic symbols are as in Figure 8; inclined contact indicates a gradational lithologic contact. Elements in weight per cent unless otherwise noted.

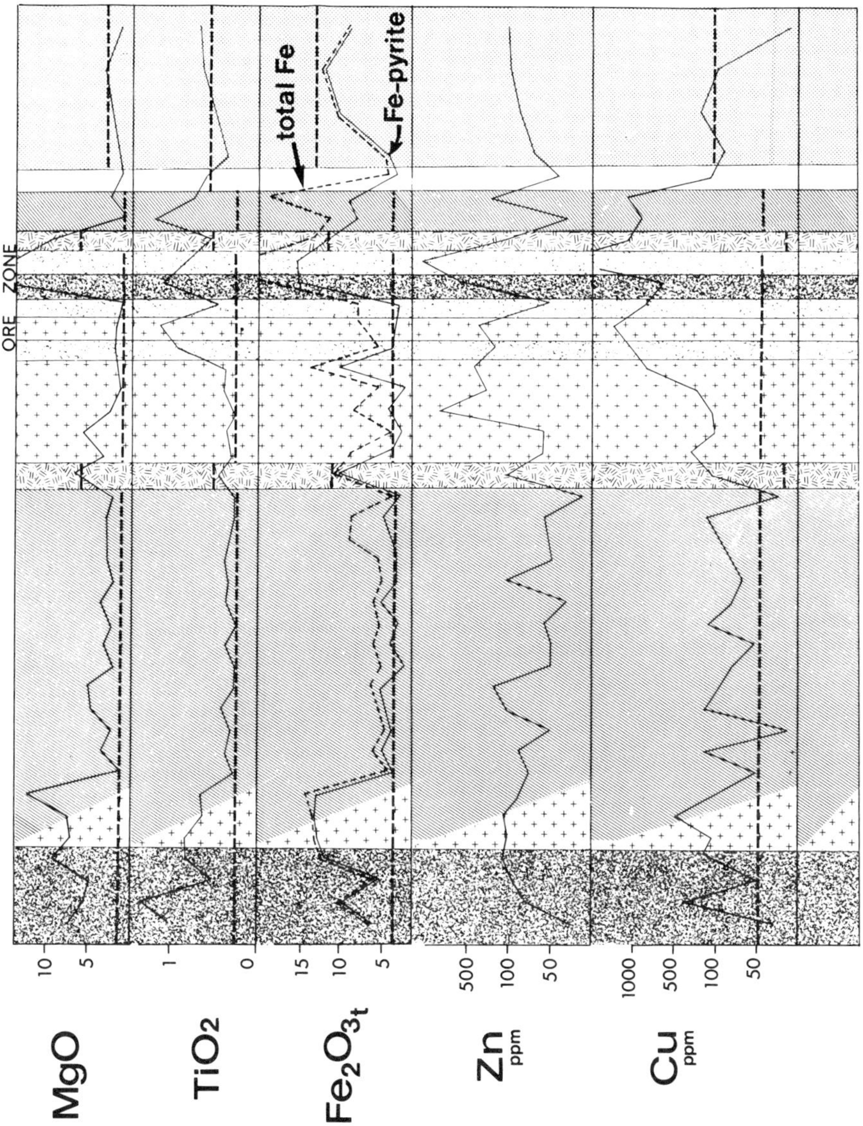

Figure 13. (b) Variation in ferromagnesian and base metal elements along the same section as Figure 13 (a). Background and lithological symbols as in 13 (a); a precise value of background for Zn was not determined, but is approximately 50 ppm in felsic rocks, 80 ppm in mafic rocks. "Fe-pyrite" is the Fe_2O_3 content of the rock after removal of normative pyrite. Elements in weight per cent unless otherwise noted.

chalcophile elements (S, Cu). Silica content is highly variable in the footwall sequence, and is generally lower than in the footwall rhyolitic sequence away from the altered zone. No zones of silicification or silica precipitation are evident near the ore body, in contrast to their presence at the Mattabi (Franklin *et al.*, 1975) and Mattagami Lake mines (Roberts and Reardon, 1978). However, large quartz veins which constitute a few per cent of the exposed rock on the three levels examined for this study, may represent the silica lost from the chlorite and sericite zones. These veins were not sampled.

Feldspar Elements. 1) Alumina is relatively constant throughout the footwall sequence with an average content essentially similar to that of the footwall rhyolitic rocks. In the most chloritized portions of both alteration zones, five samples have marked increases in Al_2O_3 (17 to 24%) reflecting locally abundant kyanite. In general, however, kyanite occurs in rocks that are not notably alumina-enriched; its formation is the consequence of a low alkali content and alumina redistribution has occurred only on a local scale.

2) Sodium is generally depleted in the footwall rocks, but exhibits a highly irregular pattern. In the well developed parts of the staurolite and muscovite assemblages and in the least altered rhyolite on all levels, the sodium content is typically less than one per cent. However, in the lower conformable chlorite-biotite-kyanite zone many samples have anomalously high sodium contents. This distribution is displayed for the 1950 level (Fig. 13) where, for the lowermost samples, it is evident that the lower part of the chlorite zone is depleted in sodium but several samples from the upper part have over 3% sodium. The overlying staurolite halo to the lower chlorite zone displays a gradual decrease in Na_2O that parallels the stratigraphically upward increase in chlorite content of these rocks. The "unaltered" rhyolite has virtually no Na_2O; its content increases rapidly to background levels above the ore zones.

The patchy staurolite zones which form the periphery of the lower conformable alteration unit have highly irregular sodium distribution. The leucocratic patches have less than 1% Na_2O, but the remainder of the rhyolite has commonly more than 2% Na_2O.

3) Potassium is generally enriched in all parts of the mine footwall sequence relative to "background levels", but has a highly irregular and locally very low content in the intensely chloritized rocks. On the 1950 and 2650 levels it increases gradually and regularly towards the ore zone. In the muscovite-kyanite zones it is predictably very high. Barium content correlates directly with K_2O.

4) Strontium, which is normally partitioned strongly into plagioclase, is markedly depleted in virtually all footwall samples. Only the footwall intrusion on the 2650 level has a "normal" stronium content.

Ferromagnesian Elements. MgO and FeO have long been established as elements that are typically enriched in the footwall sequences of massive sulphide deposits (Sangster, 1972). Both are distinctly anomalous in the Anderson Lake footwall sequence, with MgO highly concentrated in the chlorite zone. A few specific details require comment.

1) The total iron content is compared with the iron content after removal of

normative FeS_2 (Fig. 13). The excess total iron in the "unaltered" and stringer alteration zones is primarily present as sulphide minerals. After removal of the iron contained in sulphide minerals, the remaining iron is generally close to background content. In contrast, however, the lower chlorite-biotite-kyanite alteration zone has excess iron only in a silicate phase.

2) Magnesian alteration is slightly more pervasive than iron in the lower alteration zone. Manganese distribution generally follows that of MgO.

3) Titanium is normally considered to be immobile in most petrogenetic alteration studies. However, in the zones of most intense metasomatism at Anderson Lake mine its content is irregularly, but notably, increased to levels significantly above the background contents of both felsic and mafic rocks in the Snow Lake area. This unusual distribution is well illustrated in the data for the 1950 level where both the lower chlorite-biotite-kyanite and stringer zones have high TiO_2 contents. TiO_2 increases irregularly but significantly toward the ore within the stringer alteration zone on this level. Vanadium is also anomalously concentrated and follows titanium, consistent with the enrichment in ilmenite in these zones.

Samples from outcrops of both the alteration pipe and lower conformable alteration zone were analyzed and results conform with those from the same zones underground. Analyses of samples from a series of drill holes under Anderson Lake southwest of the orebody revealed that the staurolitic portions of the lower conformable alteration zone are Na-depleted.

The suite of regional samples used to calculate the average composition of each stratigraphic unit was selected to avoid obviously altered specimens. However, as staurolitic samples in the Anderson mine footwall and Stall-Ram felsic portions of the Footwall rhyolite unit (unit 2) were not otherwise visibly altered, and contained more than 2% Na_2O, they were included in the regional suite. Staurolite samples were separated on the ternary plots (Fig. 7a), however, and their compositional trends are very interesting. These samples plot along a distinctive linear trend on both the AFM and Jensen cation diagrams, which resembles a calc-alkaline trend but extends over a very wide range, considering that the samples all contain between 68% and 75% SiO_2, and have the textural characteristics of rhyolite. This linear trend is the product of either or both – 1) the loss of alkalis, and 2) the addition of FeO and MgO in a fixed ionic ratio (from the Jensen plot) of 1.5. As the unaltered (Hangingwall and Townsite) felsic rocks have an FeO/MgO molecular ratio of approximately 2.25, evidently the change in composition affects both the ferromagnesian and alkalic components. This is also demonstrated by the trend on the Jensen diagram, which is essentially independent of alkalis, except for mass-balance considerations. The ratio of added ferromagnesian components is consistent with the generally magnesian alteration associated with massive sulphide deposits, evident in the pipe zone at Anderson, as well as at Noranda (Riverin and Hodgson, 1980). The alteration which affected the staurolitic samples has evidently caused the compositions of both the "Footwall felsic" and "Stall-Ram" samples to have the same linear arrangement. Presence of such a trend in felsic rocks may thus be considered erroneously as a "calc-alkaline" trend, when in fact it is an alteration trend.

SULPHUR ISOTOPIC STUDIES

Data from a reconnaissance examination of the sulphur isotopic composition of the Anderson ore and anhydrite vein material are presented in Figure 14. Sulphide samples have a very narrow isotopic range (overall means of ∂^{34} S = 1.2 ± 0.39‰). The range in sulphide composition is caused by a few samples of chalcopyrite and and pyrrhotite associated with anhydrite in the stringer ore. The latter seems to have a wide range in composition relative to the massive ore. Chalcopyrite is isotopically lighter than pyrite (0.81‰ compared with 1.11‰ for pyrite), consistent with the experimental data for equilibrium isotopic fractionation (Ohmoto, 1972). The two samples of anhydrite (average ∂^{34}S = 4.35‰) are both isotopically heavier than pyrite which again is consistent with the experimental data (Sakai, 1968). However, this value appears to be far too low for the anhydrite to be in isotopic equilibrium with the coexisting sulphides. Isotopic disequilibrium between coexisting sulphides and sulphates has been noted by Kajiwara (1971) for the Kuroko deposits. In the latter case, Kajiwara noted that the isotopic composition of sulphate closely follows the composition of seawater at the time of sulphide deposition. The composition of anhydrite from the Anderson Lake mine area may thus represent the composition of seawater sulphate in the middle to late Aphebian era. Alternatively, it may represent very high temperature oxidation of the metal- and sulphur-bearing ore-forming fluid (Ohmoto, 1972) in the pipe.

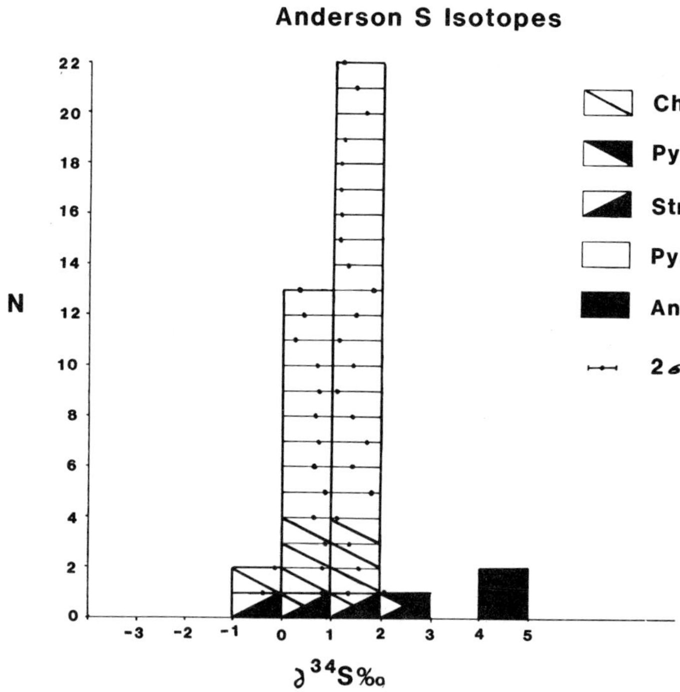

Figure 14. Histogram of S isotope data for the Anderson Mine.

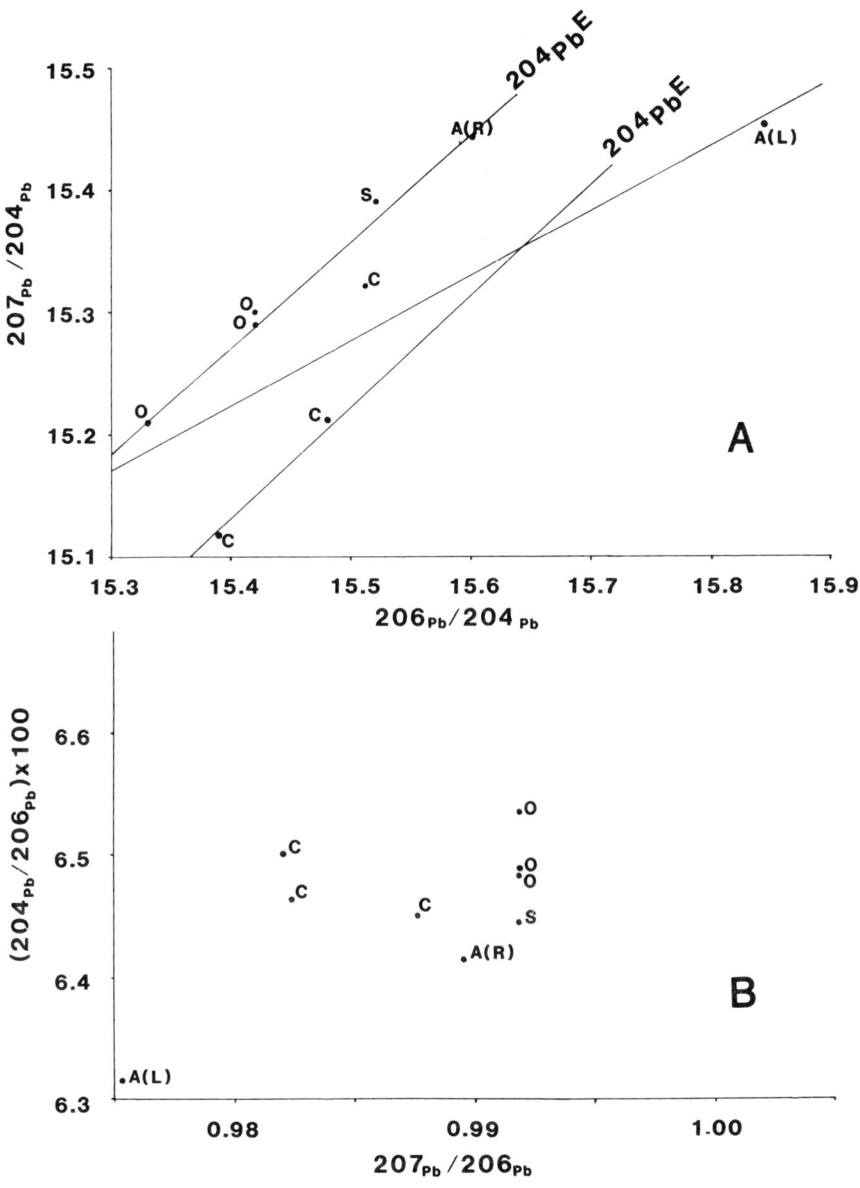

Figure 15. Pb isotopic data, massive sulphide deposits of the Snow Lake area. Data is taken from Sangster (1978), Sinha (1970), Slawson and Russell (1973) and G.S.C. data files. O = Osborne Lake, S = Stall Lake, C = Chisel Lake, A(R) = Anderson Residue composition, A(L) = Anderson Leach composition. (a) Data are scattered about two ^{204}Pb error lines, with samples from Osborne, Stall and Anderson (Residue) on one line, Chisel data in part form a second line. (b) ^{204}Pb error is reduced in this plot; data from Osborne and Stall form a single group, with vertical distribution due to ^{204}Pb error.

The relatively narrow range of the pyrite samples (only very slightly larger than analytical error) indicates that the sulphide depositional mechanism did not include a major component of progressive oxidation or progressive increase in pH of the ore-bearing fluid; this contrasts with the suggestion of Kajiwara (1971) that the Kuroko ores are the products of such progressive changes. The massive sulphide zone at Anderson Lake is probably the product of isochemical precipitation from a homogeneous solution which contained reduced sulphur, probably as HS^- or H_2S.

LEAD ISOTOPIC STUDIES

Sangster (1978) reported the lead isotopic compositions of several deposits in the Flin Flon – Snow Lake volcanic belt, and suggested that these deposits and their host strata are Aphebian in age. Data from Sinha (1970) and Slawson and Russell (1973) further confirm the Aphebian model age although the latter authors indicated the data may be scattered along a secondary isochron.

No published data for lead in sulphides in the Anderson mine are available. Two unpublished analyses (Sangster, pers. commun., 1980), a single leach-residue pair determined on sphalerite, may be compared with the data on Stall, Osborne and Chisel mines (Fig. 15). The residue composition of the Anderson sample is compositionally similar to the Stall and Osborne samples. This compositional group is different from the Chisel samples, indicating that the Anderson-Stall-Osborne deposits may share a common metal source, whereas the Chisel deposit is derived from a less radiogenic source. The poor analytical precision of these data, reflected in a large spread of data along a ^{204}Pb error line, precludes further interpretation of the existing data.

DISCUSSION

Geological Setting. The setting of the Anderson Lake deposit is similar to that of the 'typical' deposit described by Sangster (1972) elsewhere in the Precambrian Shield. Model lead isotope and Rb/Sr isochron age determinations clearly indicate a middle to late Aphebian age for the ore and host rocks. The massive nature of the ore, and the well developed asymmetrical alteration pipe are both typical features of massive sulphide deposits. Several aspects of the deposit and its setting are important to understanding of the genesis and to exploration for this type of deposit.

Most of the volcanic rocks near the Anderson Lake mine probably formed in a submarine environment. However, insufficient primary features have been preserved to infer water depth. Coarse felsic breccia, local felsic flows and a thick accumulation of felsic volcanic rocks all indicate proximity to a centre of felsic volcanism. Two felsic centres, one near the Stall Lake mine, the other north of the Chisel Lake mine, dominate the volcanic setting. Between these centres, mafic flows and epiclastic strata predominate.

The felsic rocks in the Anderson-Stall area are predominantly calc-alkaline. The compositionally similar synvolcanic tonalite sill may represent a high-level magma chamber at the base of the mine sequence, co-magmatic with much of the overlying felsic pile. Its compositional range parallels that of the entire felsic volcanic pile above it, indicating that successive pulses of felsic magma may have been injected synchronously with each volcanic episode. The higher alkali content of the sill compared to the felsic volcanic rocks is probably due to syndepositional alkali leaching by reaction of the latter rocks with seawater (Bischoff and Dickson, 1975). Furthermore, the sill is similar in composition to other subvolcanic felsic bodies such as the Beidelman Bay intrusion at Sturgeon Lake, Ontario (Franklin *et al.*, 1977), the Flavrian intrusion at Noranda (Goldie *et al.*, 1979), and the Cliff Lake body at Flin Flon (Baldwin, 1980).

Within the Snow Lake area, the presence of high-alumina basalt (Threehouse unit), together with a large volume of calc-alkaline felsic rocks, indicates at least two sources of melt; the felsic suite may have been derived from melting of sialic crust, as suggested by Smith (1980) for compositionally similar Archean felsic rocks of the Blake River Group. The high-alumina basalt, and also the andesite may have been generated by partial melting of garnet peridotite in the asthenosphere, with subsequent separation of olivine and pyroxene as suggested by Lopez-Escobar et al. (1977) for Andean mafic volcanic rocks. The volcanic suite in the mine and townsite areas of Snow Lake differs from island arc suites (Jakes and White, 1971) in its high volume of felsic, relative to mafic rocks; it also differs from Archean greenstone belts (Goodwin, 1977) in this respect. Furthermore, the Snow Lake mafic volcanic rocks have consistently higher alumina contents than the tholeiites at Flin Flon (Stauffer *et al.*, 1975) and File Lake (Bailes, 1980). The chemical compositions and proportion of felsic to mafic rocks is similar to the Andean suite, leading to speculation that the Snow Lake volcanic sequence formed due to subduction under a continental margin.

The occurrence of the Anderson Lake and Stall Lake massive sulphide deposits near the base of the mine sequence in the least fractionated rocks is indicative that fractional crystallization or progressive melt evolution are unimportant features in the genesis of the deposits. The volcanic strata that occur south of the tonalite sill are predominantly mafic, and thus also an unlikely source of a magmatic ore-fluid.

Deposit Composition and Morphology. The massive nature of the ore, together with the lack of a distinct marker horizon or break in the volcanic sequence on-strike from the deposit both indicate that the sulphides precipitated rapidly. Volcanism was penecontemporaneous with sulphide precipitation; however, little volcanic detritus appears within the massive ore; most of the gangue minerals appear to be the metamorphic products of co-precipitated, rather than clastic, minerals, further indicating a short time interval for sulphide formation. Although silicate deposition adjacent to the ore occurred penecontemporaneously with sulphide deposition, the relatively small amount of clastic silicate material in the centre of the massive ore attests to the rapid deposition of the latter.

The Anderson and Stall deposits have very high Cu/Zn ratios, compared with the Chisel, Ghost and Lost deposits. The lead isotope data indicate a separate metal source for these two groups.

The narrow range of sulphur isotope data indicates that the ore transporting fluid was relatively homogenous, and that progressive oxidation of this fluid did not occur during sulphide precipitation, except in the stringer zone. Furthermore, this narrow range makes reduction of sulphate at the site of deposition unlikely. In spite of having an open reservoir of sea water sulphate, the large temperature gradients typically present between ore fluid and seawater would necessitate reduction through a range of cooling temperatures; this would result in a wider range of reduced sulphur isotopic compositions than is observed (Ohmoto and Rye, 1979, Figure 10.7). Also, the lack of evidence of organic activity in the volcanic sequence near Snow Lake as well as the limited isotopic range, suggest that organic reduction was an unlikely source of reduced sulphur.

Thus reduced sulphur was probably carried by the ore-forming solution. Two mechanisms might explain its origin: seawater sulphate reduction in the metal generation zone, and congruent dissolution of pre-existing sulphide minerals in the substrata. The seawater reduction mechanism is unlikely, as even at a temperature of 400°C, higher than the precipitation temperatures of most massive sulphide deposits (Sasaki, 1974) the $\triangle$ (sulphate-pyrite) $\partial^{34}S$ would be approximately 16‰ (using the equations of Ohmoto and Rye, 1979). This is much higher than the observed difference for anhydrite-sulphide pairs at Anderson. Congruent dissolution of pre-existing sulphides is compatible with the observed isotopic composition, close to magmatic sulphur. Thus leaching of igneous sulphides may have been important in supplying sulphur to the ore-forming fluid.

Alteration. The two alteration zones, generally similar in areal disposition to those noted by Franklin *et al*. (1977) and MacGeehan and MacLean (1980) are significant both in the context of genesis and as regional exploration parameters; these zones are shown diagrammatically on Figure 16.

The laterally and vertically zoned alteration pipe, the upward extension of which terminates at the base of the massive ore, is generally similar in chemical composition to that at the Millenbach Mine, Noranda, Quebec (Riverin and Hodgson, 1980). Its central chloritic core is volumetrically prominent and represents a major addition of Mg and Fe to the footwall felsic rocks. The muscovite halo around the chloritic core represents addition of some K_2O; according to Riverin and Hodgson (1980) this zone may have developed as a consequence of release of hydrogen ions from the core zone where Mg fixation involved removal of OH^- from solution and consequent release of H^+. This reaction enabled the stabilization of muscovite in the outer, possibly cooler, K-rich zone. These reactions explain the mineralogic zonation of the pipe at the Anderson Lake mine. Furthermore, here, in the peripheral, possibly coolest shell, Fe was fixed as an iron silicate (now staurolite). Vertical zonation in the pipe indicates a lower temperature regime toward its top, where muscovite dominates over chlorite, compared to the more chloritic lower portions. The requirement of considerable K^+ to form the outer halo may necessitate a large volume of fluid of seawater composition, or else a relatively K-rich fluid. Local mobilization of alumina is also facilitated by a high fluid volume in the pipe region (Roberts and Reardon, 1978).

The lower conformable alteration zone is possibly the product of a regime with a steep lateral temperature gradient. Mineralogical facies indicate a relatively low temperature staurolitic distal zone where Fe was added, Na lost; an intermediate temperature central staurolite-chlorite zone, where Fe and Mg were added, Na lost; and a higher temperature centre, where Mg and Na were added where the Anderson Lake pipe joins it. Pipes also descend from the Stall Lake and Ram deposits to intersect the chlorite-rich central zone.

Genesis. The deposit was formed early in the volcanic history of the Snow Lake region, with only the Lower mine felsic sequence, and possibly part of the tonalite

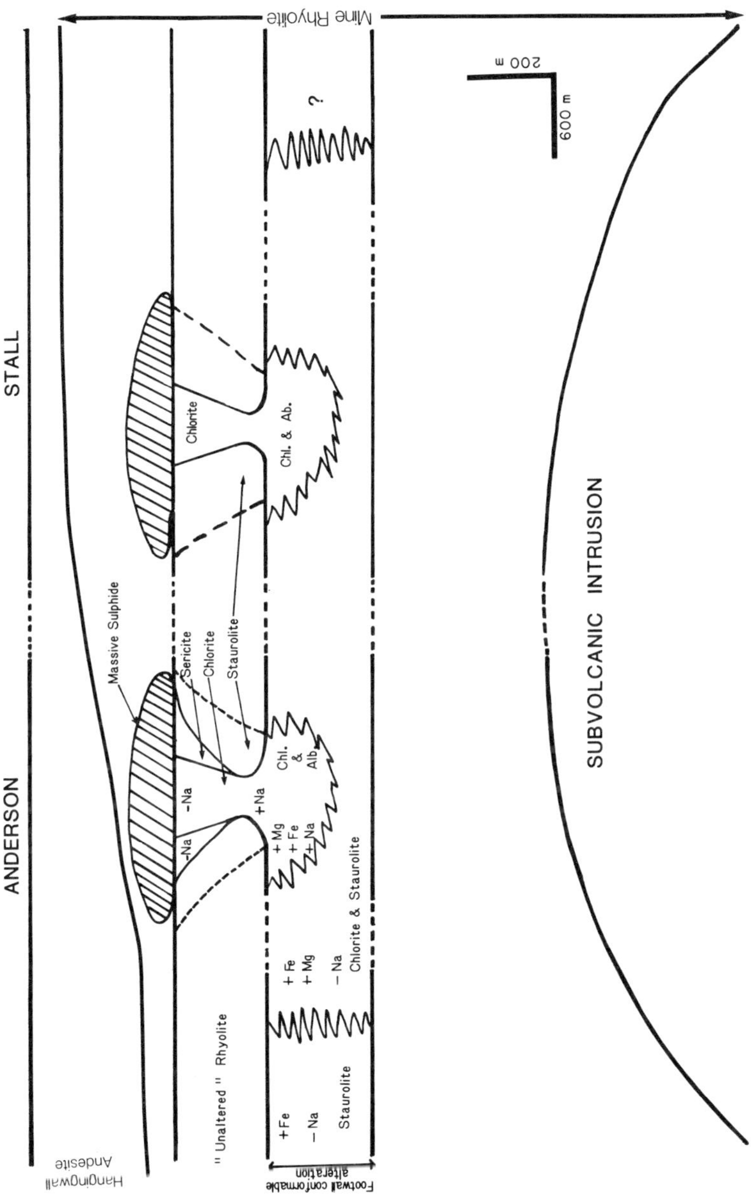

Figure 16. Simplified reconstruction of Anderson Lake alteration zones, principal lithologic units, and model characteristics.

sill preceding deposit formation. The petrochemistry of this subvolcanic sill suggests that it was continually 'recharged' with melt throughout the history of development of the volcanic pile. Thus it, or a smaller precursor, may have been emplaced relatively early, possibly just following the formation of the footwall felsic sequence. This would have effectively 'sealed' the base of the pile, inhibiting large scale fluid transfer from below the intrusion.

The concordant alteration zone below the Anderson-Stall-Ram pipes may represent a zone of fluid ponding. Intrastratal water, heated by the subvolcanic tonalite, may have equilibrated with the rhyolite, causing metal, and probably sulphide, dissolution. This leaching may have occurred in the lower, subconcordant alteration zone, or alternatively, this zone may have been a reservoir of fluid collection, from a much larger source volume. A large fluid flow, with continual recharge from seawater through the lower conformable zone, would probably have resulted in effective heat transfer away from the cooling sill; this would have resulted in turn in a moderate-temperature, moderate-salinity brine with relatively low (10 ppm) metal content. Alternatively, a static brine would result in a relatively low water:rock ratio, and thus would attain a much higher metal content; such a solution may have collected in, and eventually been removed from, the lower conformable zone by a mechanism similar to that suggested by Hodgson and Lydon (1977).

The fluid moved up localized fractures, possibly synvolcanic faults, producing some penecontemporaneous autobrecciation. As the pipe developed, secondary fluid flow, induced by the rapidly rising brine, caused seawater to be drawn down and into the fluid conduit. This seawater would have caused rapid cooling of the rising metalliferous brine, promoting metal precipitation to form the stringer zone. Influx of 'fresh' seawater would also supply sufficient potash to form sericite (now the muscovite zone) and allow for local mobilization of alumina.

Rapid cooling of a hot, metal- and reduced sulphur-rich brine would cause rapid precipitation, with little sulphur isotope fractionation.

The lower subconcordant fluid reservoir was tapped by several conduits, to form the Stall Lake, and Ram deposits. Most of the deposits in this area formed penecontemporaneously, and have similar grade and isotopic characteristics consistent with a common fluid source.

Exploration Criteria. Salient points from the Anderson Lake study that might be significant in exploration are: 1) this, and other massive sulphide deposits are near the base of the host volcanic sequence; 2) the footwall felsic rocks are relatively unfractionated and non-porphyritic; the presence of phenocrysts has no bearing on the presence of ore, contrary to the suggestions of Hopwood (1976); 3) large, semi-concordant footwall alteration zones are clearly prime targets for litho-geochemical exploration; and 4) the chemical characteristics of these alteration zones are highly variable; although such distinctive features as sodium depletion are themselves interesting, any major chemical shift from 'normal' rhyolite merits investigation. At Anderson, strontium seems to be a very effective indicator of alteration, whereas sodium has been added as noted also by MacGeehan and MacLean (1980) in some zones. Commonly, chemical changes are reflected in mineralogy. The presence of aluminosilicate minerals and, in particular at higher metamorphic grade, staurolite, in rhyolite, are useful indicators of alteration.

ACKNOWLEDGEMENTS

The management of Hudson's Bay Exploration and Development Co. encouraged the study, and the staff of Hudson's Bay Mining and Smelting Co. in Snow Lake provided considerable field and underground guidance in the sampling and mapping program. In particular, F.C. Bill and E. Guiboche provided much useful assistance and information regarding the mine geology, and D. Price helped unravel some of the regional structure. E. Froese provided many useful discussions of the regional geology. The mapping by E. Froese and J.M. Moore provided an excellent geological framework for our study. S. Youn (HBED) and W. MacRae (GSC) assisted in surface mapping and preparation of some of the diagrams. S. Green assisted with petrographic and drafting assistance, and M. Atkinson completed some of the diagrams.

E. Froese critically read several earlier drafts; his patience in explaining the structural and metamorphic aspects of the area is much appreciated. J.M. Moore (Carleton University), D. Price (HBED), A. Walker, F. Bill (HBM&S), R.I. Thorpe and G.B. Leech (GSC) all provided useful comments on earlier drafts. Analytical data was provided by the Central Laboratory and Technical Services of the G.S.C.

REFERENCES

Armstrong, J.E., 1941, Wekusko, Manitoba: Geological Survey of Canada Map 665A.

Bailes, A.H. 1980, Geology of the File Lake Area: Manitoba Mineral Resources Division, Department of Energy and Mines, Geological Report 78-1, 134 p.

Baldwin, D.A., 1980, Porphyritic Intrusions and Related Mineralization in the Flin Flon Volcanic Belt: Manitoba Mineral Resources Division, Department of Energy and Mines, Economic Report ER79-4, 23 p.

Baragar, W.R.A., 1968, Major element geochemistry of the Noranda volcanic belt, Quebec-Ontario: Canadian Journal of Earth Sciences, v. 4, p. 773-790.

Beck, L.S., 1959, Mineral occurrences in the Precambrian of Northern Saskatchewan (excluding radioactive minerals): Saskatchewan Department of Mineral Resources, Report 36, p. 106-107.

Bell, K., Blenkinsop, J., and Moore, J.M., 1975, Evidence for a Proterozoic greenstone belt from Snow Lake Manitoba: Nature, v. 258, p. 698-701.

Bischoff, J.L. and Dickson, F.W., 1975, Seawater-basalt interaction at 200°C and 500 bars: implication for origin of seafloor heavy metal deposits and regulation of seawater chemistry: Earth and Planetary Science Letters, v. 25, p. 385-397.

Bristol, C.C., 1974, Sphalerite geobarometry of some of the metamorphosed orebodies in the Flin Flon and Snow Lake districts, Manitoba: Canadian Mineralogist, v. 12, p. 308-315.

Byers, A.R., 1969, Symposium on the Geology of the Coronation Mine, Saskatchewan: Geological Survey of Canada Paper 68-5, 329 p.

Byers, A.R., Kirkland, S.J.T. and Pearson, W.J., 1965, Geology and Mineral Deposits of the Flin Flon Area, Saskatchewan: Saskatchewan Department of Mineral Resources Report 62, 95 p.

Coats, C.J.A., Clark, C.A., Buchan, R., and Brummer, J.J., 1970, Geology of the copper-zinc deposits of Stall Lake Mines Ltd., Snow Lake area, Manitoba: Economic Geology, v. 65, p. 970-984.

Dimroth, E., and Rocheleau, M., 1979, Volcanology and Sedimentology of the Rouyn-Noranda Area, Quebec: Geological Association of Canada Field Trip Guidebook, A-1, Universite Laval, Quebec, P.Q., 193 p.

Franklin, J.M., Kasarda, J., and Poulsen, K.H., 1975, Petrology and chemistry of the alteration zone of the Mattabi massive sulphide deposit: Economic Geology, v. 70, p. 63-79.

Franklin, J.M., Gibb, W., Poulsen, K.H., and Severin, P., 1977, Archean Metallogeny and Stratigraphy of the South Sturgeon Lake Area: Institute on Lake Superior Geology Guidebook, Lakehead University, Thunder Bay, Ontario, 73 p.

Frarey, M.J., 1950, Crowduck Bay Manitoba: Geological Survey of Canada, Map 987A.

Froese, E. and Gasparrini, E., 1975, Metamorphic zones in the Snow Lake Area, Manitoba: Canadian Mineralogist, v. 13, p. 162-167.

Froese, E. and Moore, J.M., 1980, Metamorphism in the Snow Lake Area, Manitoba: Geological Survey of Canada Paper 78-27, 16 p.

Geological Staff, Hudsons Bay Mining and Smelting, 1957, North Star and Don Jon Mines, Cuprus Mine, Schist Lake Mine: in Gilbert G., ed., Structural Geology of Canadian Ore Deposits, Vol. 2: Canadian Institute of Mining and Metallurgy, Montreal, p. 747-761.

Goldie, R., Kotila, B., and Seward, D., 1979, The Don Rouyn Mine: an Archean porphyry copper deposit near Noranda, Quebec: Economic Geology, v. 74, p. 1680-1684.

Goodwin, A.M., 1977, Archean volcanism in Superior Province, Canadian Shield: in Baragar, W.R.A., Coleman, C.C. and Hall, J.M., eds., Volcanic Regimes in Canada: Geological Association of Canada Special Paper 16, p. 205-242.

Hanson, G., 1920, Some Canadian occurrences of pyritic deposits in metamorphic rocks: Economic Geology, v. 15, p. 574-609.

Harrison, J.M., 1949, Geology and Mineral Deposits of the File-Tramping Lakes Area, Manitoba: Geological Survey of Canada Memoir 250, 92 p.

Hey, M.H., 1954, A new review of the chlorites: Mineralogical Magazine, v. 30, p. 277-292.

Hodgson, C.J. and Lydon, J.W., 1977, Geological setting of volcanogenic massive sulphide deposits and active hydrothermal systems: some implications for exploration: Canadian Institute of Mining and Metallurgy Bulletin, v. 70, no. 786, p. 95-106.

Hogg, N., 1957, Nor-Acme Mine: in Gilbert, G., ed., Structural Geology of Canadian Ore deposits, Vol. 2: Canadian Institute of Mining and Metallurgy, Montreal, p. 262-275.

Hopwood, T.P., 1976, 'Quartz eye'-bearing porphyroidal rocks and volcanogenic massive sulphide deposits: Economic Geology, v. 71, p. 589-612.

Hutcheon, I., 1977, The Metamorphism of Sulfide-bearing Pelitic Rocks from Snow Lake, Manitoba: Unpublished Ph.D. Thesis, Carleton University, Ottawa, 205 p.

———, 1979, Sulfide-oxide-silicate equilibria Snow Lake, Manitoba: American Journal of Science, v. 279, p. 643-665.

Irvine, T.N. and Baragar, W.R.A., 1971, A guide to the chemical classification of the common igneous rocks: Canadian Journal of Earth Sciences, v. 8, p. 523-548.

Jakes, P. and White, A.J.R., 1971, Composition of island arcs and continental growth: Earth and Planetary Science Letters, v. 12, p. 224-230.

Jensen, L.S., 1976, A New Cation Plot for Classifying Subalkalic Volcanic Rocks: Ontario Division of Mines, Miscellaneous Paper 66, 22 p.

Kajiwara, Y., 1971, Sulfur isotope study of the Kuroko-ores of the Shakanai No. 1 deposits, Akita Prefecture, Japan: Geochemical Journal, v. 4, p. 157-181.

Koo, J. and Mossman, D.J., 1975, Origin and metamorphism of the Flin Flon stratabound Cu-Zn sulphide deposit, Saskatchewan and Manitoba: Economic Geology, v. 70, p. 48-62

Kuno, H., 1968, Differentiation of basalt magmas: in Hess, H.H., and Poldervaart, A., eds., Basalts, Vol. 2: New York, Interscience, John Wiley and Sons, p. 623-688.

Lopez-Escobar, L., Frey, F.A., and Vergara, M., 1977, Andesites and high-alumina basalts from the central-south Chile high Andes: geochemical evidence bearing on their petrogenesis: Contributions to Mineralogy and Petrology, v. 63, p. 199-228.

MacGeehan, P.J. and MacLean, W.H., 1980, An Archean sub-seafloor geothermal system, 'calc-alkali' trends and massive sulphide genesis: Nature, v. 286, p. 767-771.

MacQuarrie, R., 1980, Absolute age of the Flin Flon deposit: Bulletin of the Canadian Institute of Mining and Metallurgy (Abstract), No. 4 District Meeting, v. 73, no. 820 p. 53.

Martin, P.L., 1966, Structural analysis of the Chisel Lake orebody: Canadian Institute of Mining and Metallurgy Transactions, v. 69, p. 208-214.

Moore, J.M. Jr., 1977, Orogenic volcanism in the Proterozoic of Canada: in Baragar, W.R.A., Coleman, C.C. and Hall, J.M., eds., Volcanic Regimes in Canada: Geological Association of Canada Special Paper 16, p. 127-148.

Ohmoto, H., 1972, Systematics of sulfur and carbon isotopes in hydrothermal ore deposits: Economic Geology, v. 67, p. 551-578.

Ohmoto, H. and Rye, R., 1979, Isotopes of sulfur and carbon: in Barnes H. L. ed., Geochemistry of Hydrothermal Ore deposits, 2nd Ed.: New York, Interscience, John Wiley and Sons.

Price, D.P., 1977, Geology and economic potential of the Flin Flon – Snow Lake area: University of Manitoba Centre for Precambrian Studies, 1977 Annual Report, p. 52-83.

Provens, N., 1980, Geology of the Centennial copper-zinc deposit: Bulletin of the Canadian Institute of Mining and Metallurgy (Abstract) No. 4 District Meeting, v. 73, no. 820, p. 56.

Riverin, G. and Hodgson, C.J., 1980, Wall-rock alteration of the Millenbach Cu-Zn Mine, Noranda, Quebec: Economic Geology, v. 75, p. 424-444.

Roberts, R.G. and Reardon, E.J., 1978, Alteration and ore-forming processes at Mattagami Lake Mine, Quebec: Canadian Journal of Earth Sciences, v. 15, p. 1-21.

Russell, G.A., 1957, Structural Studies of the Snow Lake-Herb Lake area, Herb Lake Mining Division, Manitoba: Manitoba Mines Branch Publication 55-3.

Sakai, H., 1968, Isotopic properties of sulfur compounds in hydrothermal processes: Geochemical Journal, v. 2, p. 29-49,

Sangster, D.F., 1972, Precambrian Volcanogenic Massive Sulphide Deposits in Canada: A Review: Geological Survey of Canada Paper 72-22, 44 p.

________________, 1978, Isotopic studies of ore leads of the circum-Kisseynew volcanic belt of Manitoba and Saskatchewan: Canadian Journal of Earth Sciences, v. 15, p. 1112-1121.

Sasaki, A., 1974, Isotopic data of Kuroko deposits: in Ishihara, S. ed., Geology of Kuroko deposits: Society of Mining Geologists of Japan, Mining Geology Special Issue 6, p. 309-398.

Shanks, R.J. and Bailes, A.H., 1977, The Missi Group, Wekusko Lake area; Manitoba Dept. of Mines, Resources and Environmental Management Report of Field Activities, 1977, p. 83-87.

Sinha, A.K., 1970, Model lead and radiometric ages from the Churchill Province, Canadian Shield: Geochimica et Cosmochimica Acta, v. 34, p. 1089-1106.

Slawson, W.F. and Russell, R.D., 1973, A multistage history for Flin Flon leads: Canadian Journal of Earth Sciences, v. 10, p. 582-583.

Smith, I.E.M., 1980, Geochemical evolution in the Blake River Group, Abitibi Greenstone Belt, Superior Province: Canadian Journal of Earth Sciences, v. 17, p. 1292-1299.

Stauffer, M.R., Mukherjee, A.C., and Koo, J., 1975, The Amisk Group: an Aphebian island arc deposit: Canadian Journal of Earth Sciences, v. 12, p. 2021-2035.

Stockwell, C.H., 1937, Gold Deposits of the Herb Lake area, northern Manitoba: Geological Survey of Canada Memoir 208, 46 p.

Williams, H., 1966, Geology and Mineral Deposits of the Chisel Lake Map Area, Manitoba: Geological Survey of Canada Memoir 342, 38 p.

MS received October 1980;
Accepted for publication July 1981.

Precambrian Sulphide Deposits, H.S. Robinson Memorial Volume, edited by
R.W. Hutchinson, C.D. Spence and J.M. Franklin, Geological Association of Canada
Special Paper 25, 1982

GEOLOGY OF THE RUTTAN DEPOSIT, NORTHERN MANITOBA

D.S. Speakman and P.J. Chornoby
Sherritt Gordon Mines Ltd.,
Lynn Lake, Manitoba R0B 0W0

B.C.W. Haystead
Sherritt Gordon Mines Ltd.,
Leaf Rapids, Manitoba R0B 1W0

G.F. Holmes
Syncrude Canada, Mildred Lake, Bag 4009,
Fort McMurray, Alberta T9H 3L1

ABSTRACT

The Ruttan Orebody is a copper-zinc exhalative massive sulphide deposit contained within a metamorphosed sequence of Proterozoic volcanic rocks and volcanic-derived sedimentary rocks. Geochemical analyses indicate that the volcanic rocks were derived from a subalkaline tholeiitic magma. In the vicinity of the deposit hydrothermal activity altered the footwall rocks resulting in addition of Mg, Cu, Zn and Fe and depletion of Na, K and Ca. Trace element analyses indicate that the massive sulphide deposit is enriched in Au, Ag, Pb, Cd, As and Co, whereas Ba and Mn contents increase in the hangingwall rocks.

Subsequent tectonism and metamorphism have had considerable effect on the alteration zone and the massive sulphides. Locally chalcopyrite and sphalerite have been mobilized into areas of low stress and in general the grain size of the sulphides has been greatly increased. The alteration zone, in which textures range from granoblastic to lepidoblastic, consists of a fine- to coarse-grained assemblage of cordierite, magnesian-chlorite, andalusite, sericite, biotite, quartz, plagioclase and staurolite.

RÉSUMÉ

Le gisement de Ruttan est un amas sulfuré de cuivre et zinc d'origine exhalative, encaissé dans une séquence métamorphisée de roches volcaniques et de sédiments dérivés de roches volcaniques d'âge protérozïque. Les analyses géochimiques indiquent que ces roches volcaniques dérivent d'un magma subalcalin tholeïtique. Au voisinage du gisement, les solutions hydrothermales ont altéré les roches au mur en ajoutant Mg, Cu, Zn, et Fe et en lessivant Na, K, et Ca. Les analyses des éléments traces indiquent que l'amas sulfuré est enrichi en Au, Ag, Pb, Cd, As et Co alors que les roches du mur sont enrichies en Ba et Mn.

Les déformations tectoniques et le métamorphisme ultérieurs ont eu des effets considérables sur la zone d'altération et l'amas sulfuré. La chalcopyrite et la sphalérite ont été mobilisées localement dans des zones à faible contrainte et d'une façon générale la granulométrie des sulfures s'est accrue. La zone d'altération consiste en un assemblage granoblastique à lépidoblastique de cordiérite, chlorite magnésienne, andalousite, séricite, biotite, quartz, plagioclase et staurotide.

INTRODUCTION

The Ruttan copper-zinc orebody is a metamorphosed exhalative massive sulphide deposit, located 760 km north of Winnipeg and 20 km east of the Town of Leaf Rapids (Fig. 1). The orebody is owned and operated by Sherritt Gordon Mines Limited. The discovery hole, drilled in 1969, was targeted on the basis of combined electromagnetic and magnetic surveys flown by Sherritt. Subsequent drilling outlined an orebody containing 45.5 million tonnes of sulphide ore with a grade of 1.47% copper and 1.61% zinc to 600 m below surface. Production commenced at 9,000 tonnes per day from an open pit in 1973 and production from the underground mine commenced in early 1979.

The information presented in this paper was compiled from routine geological work commencing in 1972 in the open pit, 1974 in the underground operation and 1977 from property field work. Detailed geological studies for the paper involved remapping, relogging and sampling of the two main underground levels, the 240 m level and the 360 m level, and the drilling of two 425 m diamond drill holes into the footwall. From this base 600 thin sections, 50 polished sections and 200 whole rock analyses were obtained. As well, the results of 641 whole rock analyses with thin sections were available from property mapping. Structural information was obtained over a three year period from detailed geological engineering site investigations. Previous workers in the area include Norman (1934), Burwash (1962), Steeves and Lamb (1972), Bristol (1979), Baldwin (1978; 1979; 1980), and Sangster (1978).

REGIONAL GEOLOGY

The southern portion of the Churchill Province (Fig. 2) consists of two volcanic belts aligned approximately east-west. The northern belt is underlain by the Wasekwan Group which is bounded to the northwest by the sialic Archean basement which also underlies the Wollaston Belt. The southern belt is underlain by the Amisk Group, bounded to the southeast by the Superior Province. The Wasekwan Group is overlain by the Sickle Group sedimentary rocks and the Amisk Group is overlain by a comparable sedimentary sequence, the Missi Group. The central portion of the map area (Fig. 2) is comprised of metasedimentary layered and granitoid gneisses collectively referred to as the Burntwood River Metamorphic Suite.

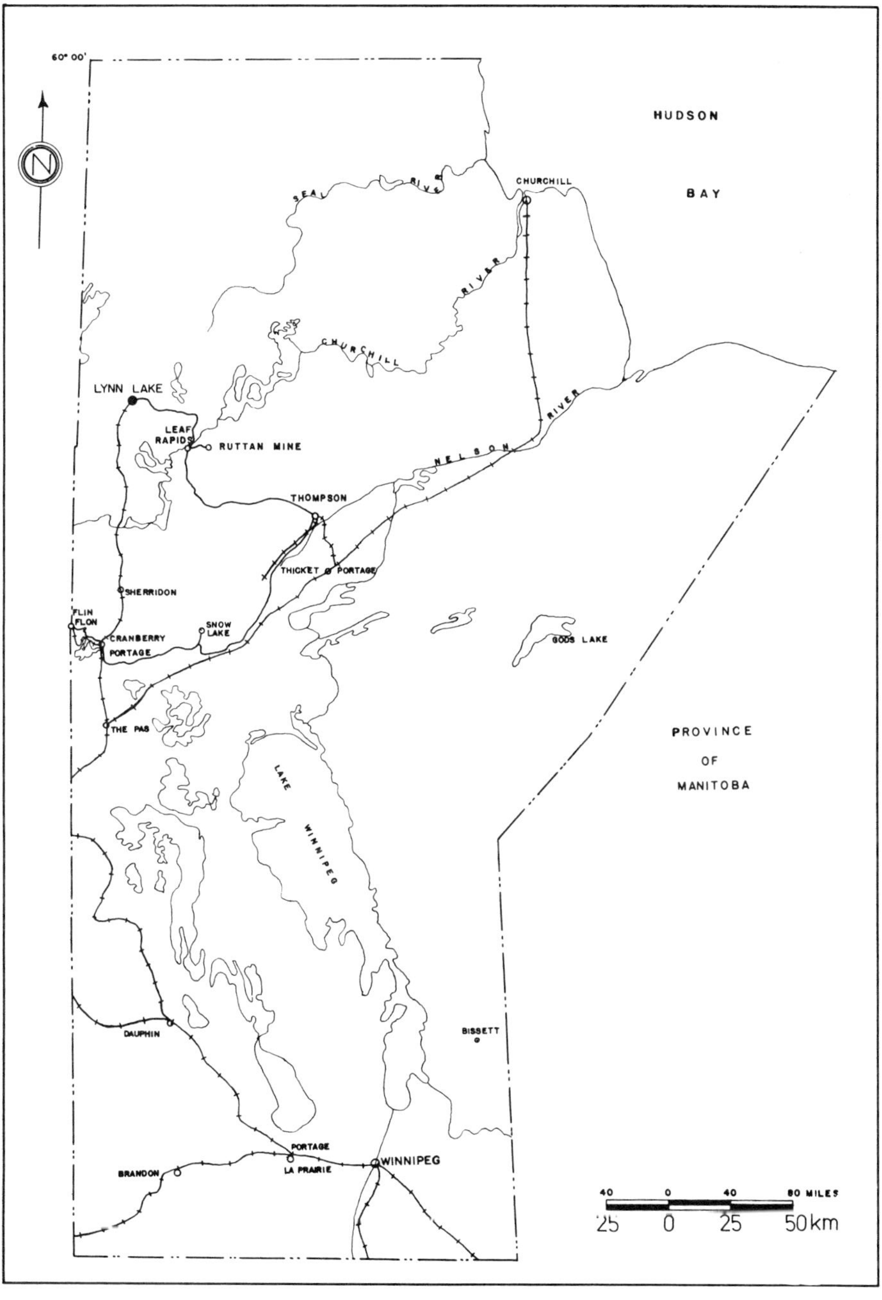

Figure 1. Location map.

Lead isotope data showing that the Wasekwan and Amisk Groups are both of Aphebian age (Sangster, 1978) supports earlier interpretations that the two groups originated as paired volcanic arcs separated by a sedimentary basin.

The Ruttan deposit occurs within the Wasekwan Group. These are the oldest known rocks in the area and are a eugeosynclinal suite consisting of a conformable sequence of mafic, intermediate and felsic volcanic flows, related tuff, agglomerate, breccia and volcaniclastic sedimentary rocks including turbidite. The Wasekwan Group is overlain, locally unconformably, by the shallow water sedimentary rocks of the Sickle Group. The Sickle Group consists of a basal conglomerate of variable thickness which grades upward into arkose and greywacke.

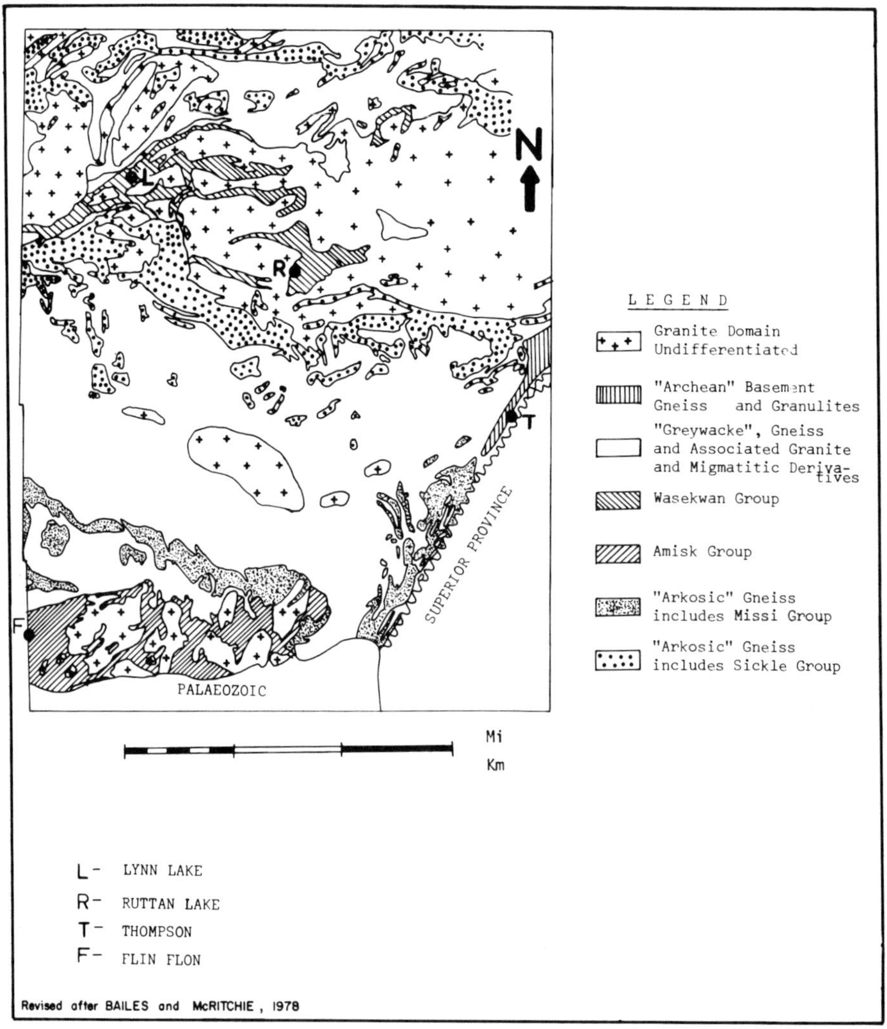

Figure 2. Regional geological map.

Rocks of the Wasekwan and Sickle Groups have been regionally metamorphosed to the amphibolite facies with local zones of epidote-amphibolite and middle greenschist facies. Mafic volcanic rocks have been converted to hornblende-plagioclase schist or banded gneiss with variable but minor amounts of epidote ± biotite ± chlorite. Sedimentary rocks were converted to biotite ± muscovite ± feldspar ± quartz schist and gneiss that locally contain garnet, ± hornblende, ± cordierite, ± sillimanite. The rocks grade southward into coarse-grained biotite-quartz-feldspar gneiss, hornblende-plagioclase gneiss and granitic gneiss containing pegmatite and extensive migmatite zones.

Intrusions in the area are broadly grouped as Pre-Sickle and Post-Sickle, each ranging in composition from gabbroic to granitic (Milligan, 1960).

Two main sets of folds are recognized in the Wasekwan rocks. The older set formed about eastward-striking axes with moderate plunges and southerly, steeply dipping axial planes. The second deformational event caused refolding about north-northwestward-tending axes.

LOCAL GEOLOGY

General Stratigraphy

The Ruttan deposit is located in the Rusty Lake Greenstone Belt which occurs in the Wasekwan Group. Regional structural complexities have not yet been sufficiently resolved to establish the overall position of the deposit and its host rocks within the volcanic sequence.

Considering the local strata hosting the Ruttan deposit (Fig. 3), the lowest observed unit is a series of spilitized basaltic flows interbedded with thin sequences of volcanic-derived sedimentary rocks. Overlying the basalt unit are the Footwall Volcaniclastic rocks, which are relatively unaltered rocks of intermediate composition. Directly beneath the Ruttan deposit, the upper portion of the Footwall Volcaniclastic rocks has been distinguished and named the Altered Footwall Volcaniclastic rocks. Overlying the Altered Footwall Volcaniclastic rocks is the Mine Rhyolite, a felsic rock which is possibly silicified Footwall Volcaniclastic rock, and contains the copper-rich base of the Ruttan deposit. The Exhalite Horizon, a mixed unit of chemically precipitated and detrital sedimentary rock, overlies the Mine Rhyolite. The Exhalite Horizon contains the zinc-rich lenses of the Ruttan deposit. Uppermost is the P. M. Unit, a thick sequence of interbedded greywacke and volcanic-derived polymictic conglomerate.

Numerous pre-Sickle sills and dykes of varying composition and up to 45 m thick occur throughout the map area. The dykes are irregular and commonly dislocated along bedding planes. Mafic stocks considered pre-Sickle by Steeves and Lamb (1972) occur in the P. M. Unit.

The western extremity of the Ruttan deposit is truncated by a post-Sickle granitic intrusion. The intrusion has brecciated the host rock along its margins, with local potassium enrichment of the former. Large xenoliths of mafic volcanic rocks are found up to 2 km into the intrusion.

The sedimentary and volcanic rocks of the Ruttan area contain a penetrative foliation, with a steep southerly dip, which is parallel to original bedding surfaces.

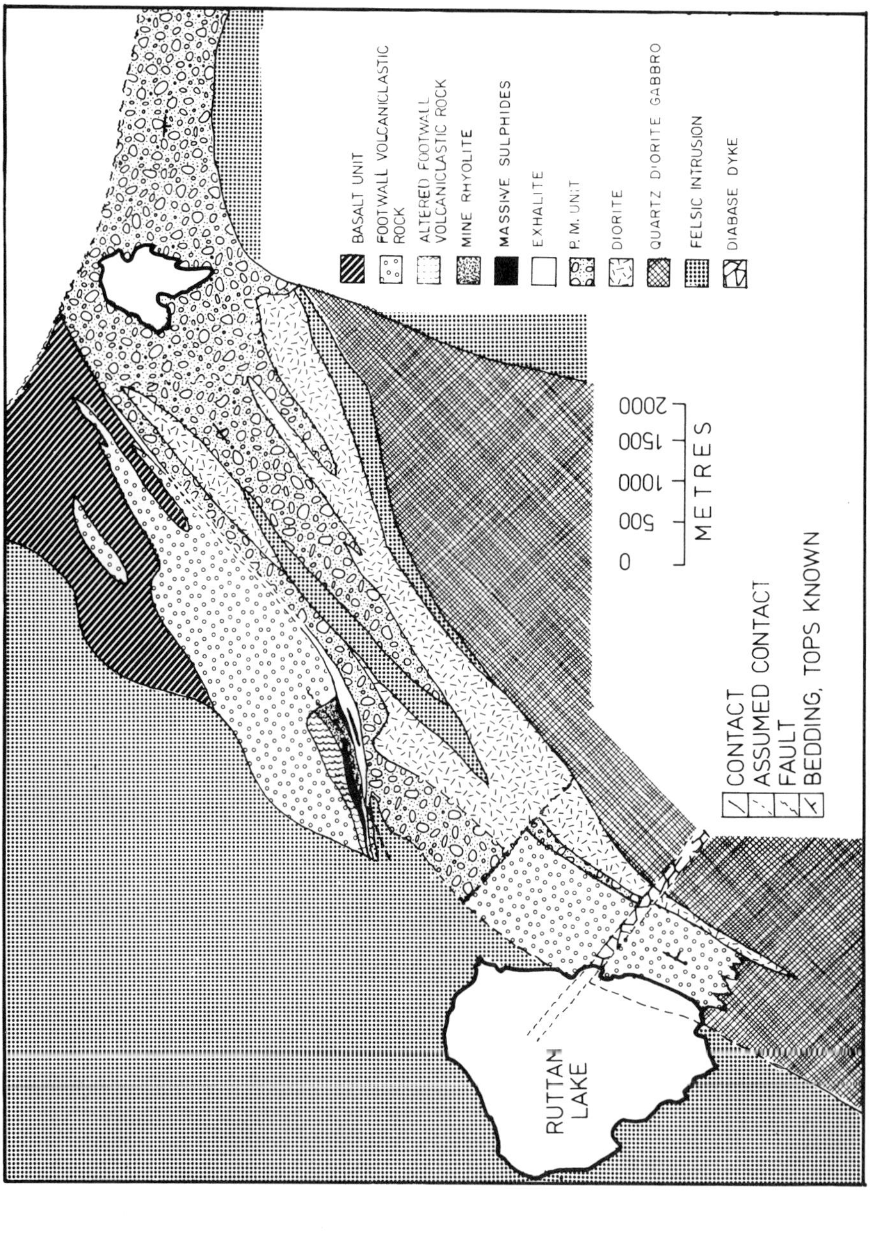

Figure 3. Geological map, Ruttan area.

The strata hosting the deposit are on a limb of a northeasterly trending fold formed by flexural slip-type folding. The deposit is upright and dips steeply to the southeast.

Rock Units

Basalt Unit. The basalt unit consists of a number of flows. Features such as pillows, pillow breccias, amygdales, hyaloclastite and flow-top breccias are preserved. The deformation is such that although pillows are readily recognizable they rarely yield reliable top determinations. Thinly to thickly bedded, fine-grained sedimentary rocks are interbedded with the basaltic flows.

Moderate to strong schistosity and upper greenschist to lower amphibolite grade metamorphism have largely obliterated original microscopic textures. Consequently the basalt flows generally consist of a fine- to medium-grained crystalline assemblage of hornblende, plagioclase, ± biotite, ± chlorite, ± quartz, ± calcite, ± epidote. Meta-crysts of hornblende and/or plagioclase are up to 0.5 cm in size. Calcite occurs in and around fractures or as vesicle fillings. Epidote is present as either minute, widely disseminated grains or concentrated locally in 'alteration' patches, joint fillings and the upper portions of pillowed flows.

Footwall Volcaniclastic Unit. The Footwall Volcaniclastic rocks crop out at a number of sites one to 3 km along strike from the deposit. Here the unit is a thinly bedded to very thickly bedded, poorly graded clastic rock. Individual beds are either heteroclastic or monoclastic. The fine-grained, sharply defined clasts are subangular to subrounded and range in size from less than one millimetre to 40 cm. The matrix generally is more coarsely crystalline and contains a larger percentage of biotite than do the clasts.

Sericitization and minor chloritization along with disseminated sulphides (2 to 15% Py) occur within the Footwall Volcaniclastic rocks up to 2 km from the Ruttan Deposit. Microscopic examination of these rocks reveals that they consist of plagioclase, quartz, biotite, hornblende, ± epidote, ± chlorite, ± sericite, ± garnet, ± magnetite, ± sulphides.

The Footwall Volcaniclastic rocks directly beneath the deposit are not exposed on surface. Limited mine workings and diamond drill core indicate that here the unit is 400 m thick and is in contact on its footwall with a large granitic intrusion. Except for a fine-grained, thinly bedded to thinly laminated, 25 m thick sequence of sedimentary rocks within its upper portion and a few mafic sills, the unit is here remarkably uniform. Indistinct banding and vague curved lamellae are the most prominent features of this otherwise massive, fine- to medium-grained recrystallized rock.

P. M. Unit. As stated previously, the P. M. Unit is a sequence of interbedded greywacke and volcanic-derived, polymictic conglomerate, 1.2 km thick, extending over a strike length of 6 km.

The greywacke beds are thinly to thickly bedded turbidites characteristically exhibiting complete or partial Bouma sequences. Clasts are finely crystalline and range in size from less than one millimetre to 3 cm. They are volcanic in origin, being predominantly felsic with mafic clasts also present. Rip-up fragments from underlying beds are common within the "A" division of overlying beds. Observed transport direction is from east to west.

Rare garnetiferous and/or magnetiferous bands up to 6 cm thick are found within and parallel to the greywacke beds. Within the greywacke above the Ruttan deposit are both thin pyritic-sericitic chert beds in places containing sphalerite and epidote alteration patches which may have sphalerite-rich cores.

The volcanic-derived polymictic conglomerate is very thickly bedded (up to 250 m) and laterally continuous for up to 4 km. Individual beds are coarsely graded and sorted, both laterally and vertically. Inferred transport direction is from east to west. Clasts are rounded to subangular and range in size from less than one millimetre to one metre. Clast types are fine-grained felsic, fine-grained mafic and hornblende- and/or plagioclase-porphyritic with a fine-grained mafic groundmass. The clasts are matrix supported in a fine-grained assemblage of hornblende, biotite and plagioclase.

GEOCHEMISTRY AND ALTERATION

Major Elements

The only readily recognizable volcanic rocks preserved in the Ruttan succession are the flows of the Basalt Unit. Major element analyses of 36 samples of these flows indicate that they are subalkaline tholeiitic basalt (Table I; Figs. 4 and 5).

Toward the Ruttan deposit the Footwall Volcaniclastic rocks gradually become depleted in alkalies and enriched in iron and, to a lesser extent magnesium (Figs. 6 and 7). Within 200 m of the deposit a sharp change in the chemical composition of these rocks typifies the Altered Footwall Volcaniclastic rocks. Notably MgO is greatly increased whereas CaO and Na_2O are decreased.

The Mine Rhyolite is characterized essentially by high SiO_2 values and low CaO, K_2O and Na_2O. Within the Exhalative Horizon the chemical profile becomes erratic, reflecting the diverse nature of the individual beds. The contact between the Exhalite Unit and the P. M. Unit marks the upper limit of the primary alteration halo (Fig. 7; Table II).

Minor Elements

Minor element contents of lead, gold, silver, mercury, arsenic, cobalt, cadmium, barium and manganese (Table III) are associated with the Ruttan deposit and its enclosing rocks. Contour plans of mercury (Fig. 8a) and gold (Fig. 8b) suggest a subdivision of the deposit into west and east zones. The west zone is highly sheared and contains erratic minor element values in contrast to the relatively undisturbed east zone. Generally the minor element distributions of both zones exhibit similar associations and trends.

East Zone – Section A – A1. The overall trend in base metal abundance here is a decrease in the Cu/Zn ratio from the Altered Footwall Volcaniclastic rocks upward to the Exhalite Horizon. At the Exhalite Horizon and into the P. M. Unit the Cu/Zn ratio is less than one and the Pb content increases markedly (Fig. 8c). The high sulphide zones are relatively enriched in both Au and Ag (Fig. 8d) as well as in Co, Ag, Hg and Cd (Fig. 8e). Ba and Mn values are erratic in the Altered Footwall Volcaniclastic rocks. Within the high sulphide zones these elements are relatively depleted. In the overlying P. M. Unit the Ba and Mn contents increase markedly (Fig. 8f).

TABLE I

MAJOR ELEMENT ANALYSES, COMPOSITES OF UNALTERED FOOTWALL ROCKS, RUTTAN MINE

Weight % Oxides	Composite B1 1 Sample	Composite B2 3 Samples	Composite B3 22 Samples	Composite B4 7 Samples	Composite B5 3 Samples	Composite R1 3 Samples	Composite R2 7 Samples	Composite R3 2 Samples
SiO_2	46.00	47.50	49.30	50.50	52.57	69.10	71.40	72.15
Al_2O_3	16.60	16.90	15.30	16.10	14.50	14.50	13.80	13.50
CaO	13.20	11.86	10.89	10.59	10.46	2.30	1.80	1.91
MgO	5.76	5.83	5.77	5.56	5.35	0.94	0.73	0.62
Na_2O	1.38	2.57	2.13	2.35	1.61	3.84	4.00	3.90
K_2O	0.40	0.27	0.30	0.30	0.20	2.59	2.12	1.81
FeO	10.10	8.37	9.35	8.61	9.28	2.85	2.18	1.90
Fe_2O_3	2.71	3.09	3.48	2.86	2.48	1.77	1.65	1.83
MnO	0.25	0.20	0.21	0.21	0.18	0.07	0.13	0.05
TiO_2	0.80	0.89	1.20	0.92	0.96	0.56	0.41	0.34
P_2O_5	0.14	0.09	0.14	0.16	0.12	0.13	0.06	0.05
L.O.I.	0.64	1.54	0.90	0.73	1.00	0.77	0.88	1.10
SUM	99.1	99.1	99.0	98.5	98.7	99.4	99.2	99.2
Cu (ppm)	6.0	109.0	67.0	46.0	111.0	14.0	19.0	24.0
Zn (ppm)	21.0	46.0	26.0	22.0	21.0	74.0	74.0	51.0

B1 to B5 - Chemical analysis of Footwall primary volcanic rocks in the Ruttan area.
R1 to R3 - Chemical analysis of Footwall volcaniclastic rocks in the Ruttan area.
Analytical Method: Oxides - XRD; Cu and Zn - AA.

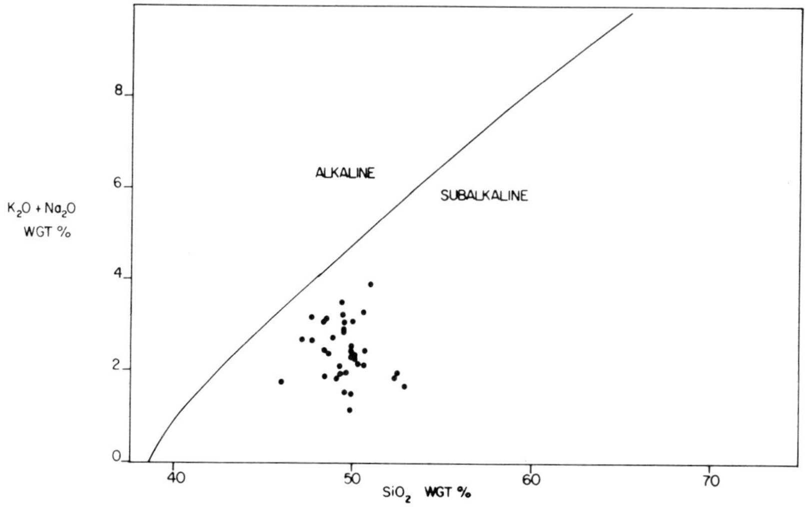

Figure 4. Alkali-silica plot of unaltered basalt near the Ruttan deposit (field boundaries after Irvine and Baragar, 1971).

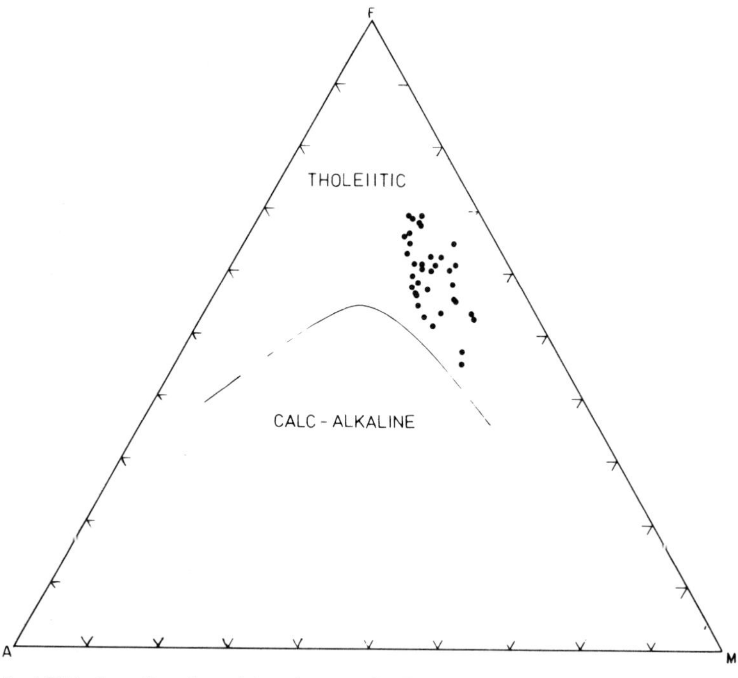

Figure 5. AFM plot of unaltered basalt near the Ruttan deposit, all values in wgt. % (field boundaries after Irvine and Baragar, 1971).

TABLE II

REPRESENTATIVE ANALYSES OF ROCK IN THE RUTTAN MINE AREA IN WEIGHT PER CENT OXIDES

	Footwall Volcaniclastics		Altered Footwall Volcaniclastics		Chlorite – Biotite Schists		Mine Rhyolite		Exhalite Horizon		P.M. Unit	
					Excluding Massive Sulphide Zones							
	Mean	St. Dev.	Mean	St. Dev.	Mean	St. Dev.	Mean	St. Dev.	Mean	St. Dev.	Mean	St. Dev.
SiO_2	66.80	1.42	65.67	2.54	43.00	12.13	79.19	2.96	60.33	6.44	69.68	3.36
Al_2O_3	14.20	.69	13.11	1.73	14.79	3.06	6.79	1.21	15.39	2.12	12.62	2.88
FeO	5.90	.60	8.02	2.32	14.79	6.49	5.90	1.71	8.08	1.88	5.42	1.96
MgO	1.74	.45	5.09	1.72	13.62	6.30	3.38	1.02	3.72	2.23	2.32	1.26
CaO	3.74	.51	.65	.38	1.34	.84	.35	.27	6.20	7.86	2.53	1.05
Na_2O	2.49	.60	.50	.36	.67	.64	.17	.20	.84	.71	1.89	1.25
K_2O	2.40	.45	1.99	.51	.65	.49	.58	.41	2.23	.93	1.88	.60
TiO_2	.48	.06	.43	.09	.45	.25	.18	.07	.61	.81	.42	.09
MnO	.14	.02	.14	.10	.10	.05	.05	.02	.26	.18	.08	.05
P_2O_5	N.A.											
H_2O	N.A.											
CO_2	N.A.											
TOTAL	97.89		95.60		89.41		96.59		97.66		96.77	

TABLE III

MINOR ELEMENT ANALYSES, SECTIONS A-A[1] AND B-B[1]

SECTION A-A[1]

Sample	Au PPB	S %	Cr PPM	Mn PPM	Co PPM	Cu PPM	Zn PPM	As PPM	Ag PPM	Cd PPM	Ba PPM	Hg PPB	Pb PPM
695	1	0.14	80	140	5	24	82	9	<1	<1	300	90	<2
696	3	0.83	140	150	9	290	54	2	<1	<1	50	20	<2
697	5	1.04	140	100	5	290	46	6	<1	<1	100	80	<2
698	36	38.9	60	110	150	3020	74	56	<1	<1	100	90	<2
699	14	5.44	40	220	32	960	110	14	<1	<1	100	40	<2
700	6	0.73	140	84	3	170	32	3	2	<1	50	20	<2
701	30	15.7	60	150	40	6700	160	47	4	<1	150	530	<2
702	370	51.0	100	23	120	7600	270	190	4	<1	100	3000	<2
703	240	49.8	100	38	110	6400	220	200	6	<1	100	2000	<2
704	1440	50.1	100	24	96	27900	950	550	23	30	50	6600	<2
705	270	53.1	100	27	55	5300	600	800	5	<1	100	1000	<2
706	210	52.3	120	48	23	9000	1040	860	11	3	100	2300	42
707	4	0.93	120	180	5	180	170	34	<1	<1	700	50	35
708	3	0.84	100	310	13	160	190	16	<1	<1	450	10	83
709	13	1.05	80	210	15	160	110	17	<1	<1	550	<10	130
710	6	0.48	100	410	20	110	490	21	<1	<1	800	20	120

SECTION B-B[1]

Sample	Au PPB	S %	Cr PPM	Mn PPM	Co PPM	Cu PPM	Zn PPM	As PPM	Ag PPM	Cd PPM	Ba PPM	Hg PPB	Pb PPM
595	6	1.47	120	290	15	950	130	1	<1	1	350	50	<2
596	<1	0.19	80	220	5	67	40	<1	<1	1	150	10	<2
597	3	0.44	120	140	<1	330	38	<1	1	1	100	20	<2
598	660	8.56	100	200	44	31000	340	1	20	1	150	160	<2
599	6	0.31	120	140	5	240	39	1	<1	1	150	30	6
600	12	0.16	160	120	2	450	59	4	<1	1	300	10	<2
601	2270	26.2	40	120	58	58000	2650	12	35	17	100	1900	20
602	10	4.84	60	190	2	3200	1350	12	1	1	150	140	25
603	200	45.4	120	24	210	3890	240	200	2	1	100	480	<2
604	10	2.98	20	220	15	1960	330	9	<1	1	200	140	4
605	70	34.4	60	24	150	14000	97	15	2	2	150	50	<2
832	12	1.74	60	45	3	960	100	12	<1	1	250	100	24
833	4570	31.4	80	23	10	80000	2030	57	78	8	100	13000	<2
834	980	16.1	120	110	3	20000	800	140	43	1	200	9000	5
835	78	4.26	40	200	5	13600	370	16	7	1	300	1600	42
836	11	1.02	140	65	22	520	84	29	<1	1	450	90	9
837	4	3.28	120	120	6	510	150	35	<1	1	750	90	22
838	420	49.9	140	12	6	890	470	580	11	1	100	2100	<2
839	2	0.96	60	97	11	350	130	20	1	<1	700	60	46
840	5	1.11	80	71	16	96	170	25	<1	<1	350	70	25
841	<1	1.74	60	180	11	110	160	11	3	<1	200	30	51
842	1	5.60	80	160	14	33	160	24	3	<1	600	30	12
843	4	2.25	40	240	9	84	250	17	3	<1	350	60	33
844	9	2.03	60	200	4	120	180	8	3	<1	850	50	27

Analytical Method: Au - Fire Assay Neutron Activation; S, Cr, Ba - XRF; Mn, Co, Cu, Zn, Ag, Cd, Pb - AA; Pb - Neutron Activation; Hg-Wet.

West Zone – Section B – B1. The Cu/Zn ratio decreases stratigraphically upward through the west zone. These metal values are highly erratic through the Altered Footwall Volcaniclastic rocks and the high sulphide zone. The overlying P. M. Unit is characterized by considerably more uniform Cu, Zn contents (Fig. 8g). Pb and Ba increase stratigraphically upward (Fig. 8h).

STRUCTURAL GEOLOGY

The dominant structural style in the Ruttan area is flexural slip isoclinal folding. Foliation is subparallel to bedding and inhomogeneous displacement along bedding planes occurred during folding. However, within the footwall alteration zone the structural style is dominated by a bifurcating shear zone caused by inhomogeneous simple shearing. The significance of inhomogeneously developed foliation in shear belts has been discussed by Ramsey and Graham (1970).

The shear zone at Ruttan strikes N70°E, dipping at 68°S, which is slightly discordant to the orientations of the ores lenses. The shear zone, which is up to 30 m wide, bifurcates into three main branches (Figs. 9a, 9b), Art's Fault, the North Wall Shear and the East Shear. As well, many minor splays have been mapped. Within the shear zones the schistosity is sigmoidal in shape. It trends sub-parallel to the shear zone

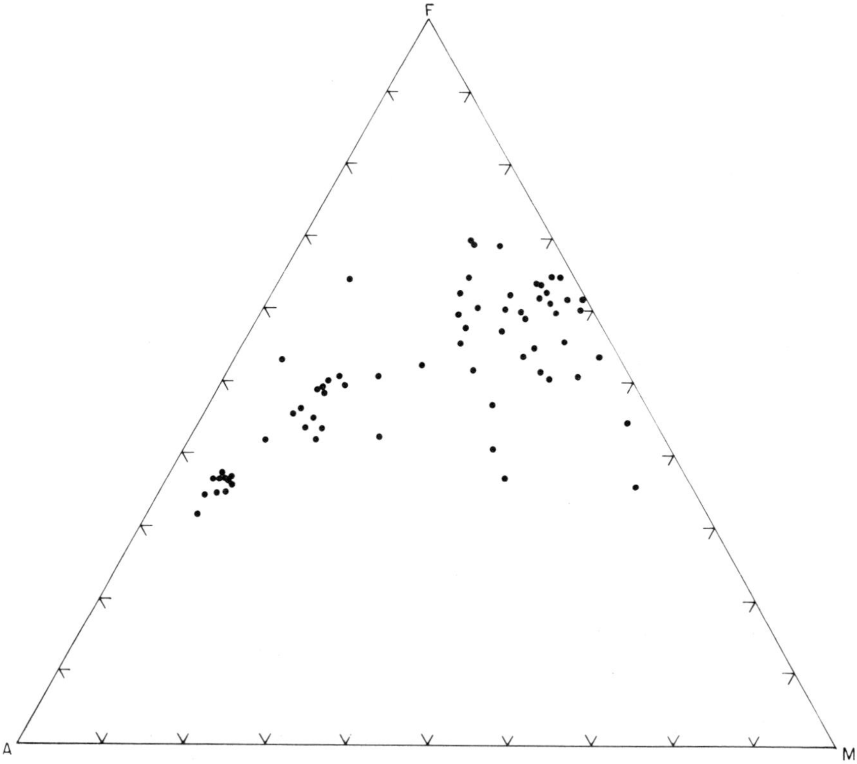

Figure 6. AFM plot of altered and unaltered felsic rocks from footwall, Ruttan deposit.

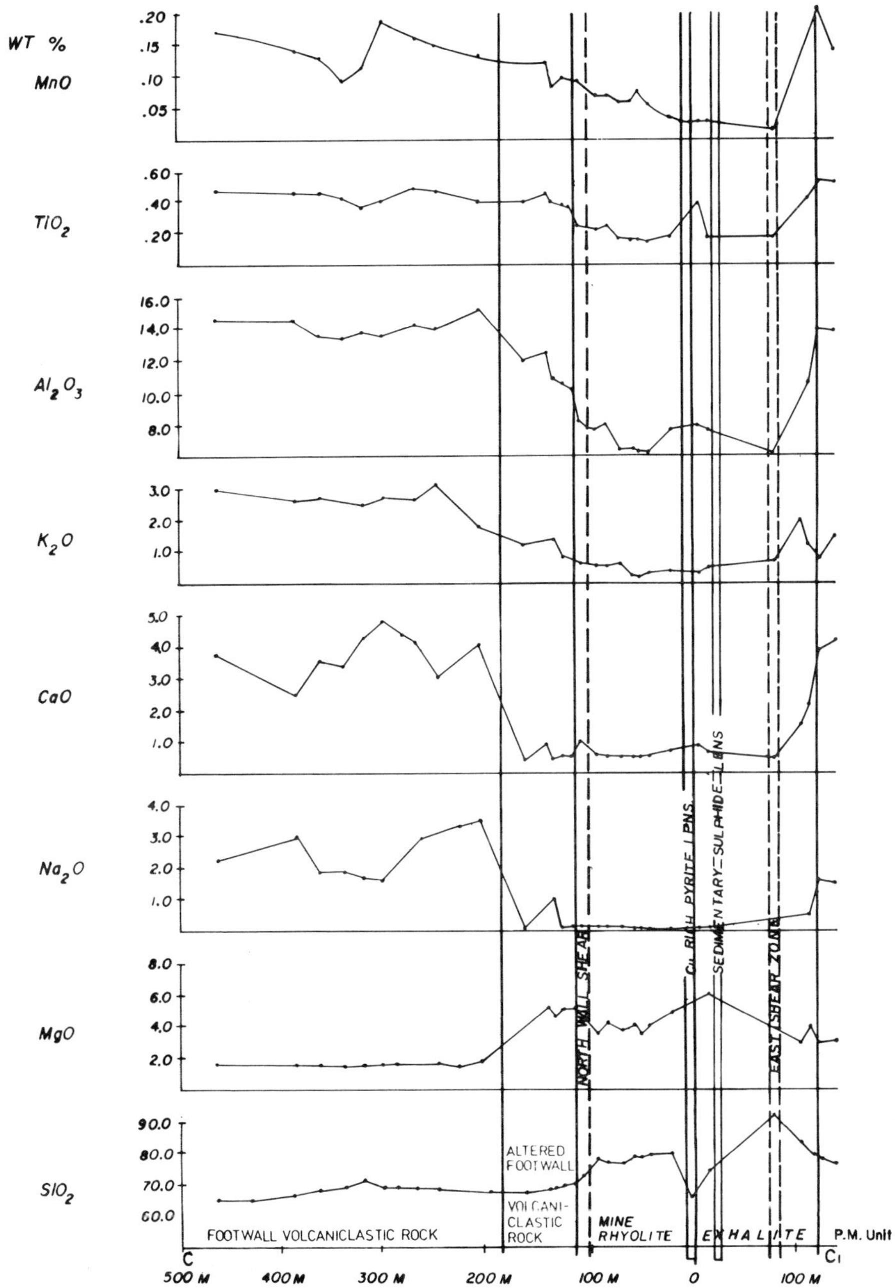

Figure 7. Geochemical sampling profile C-C₁ (Fig. 9b).

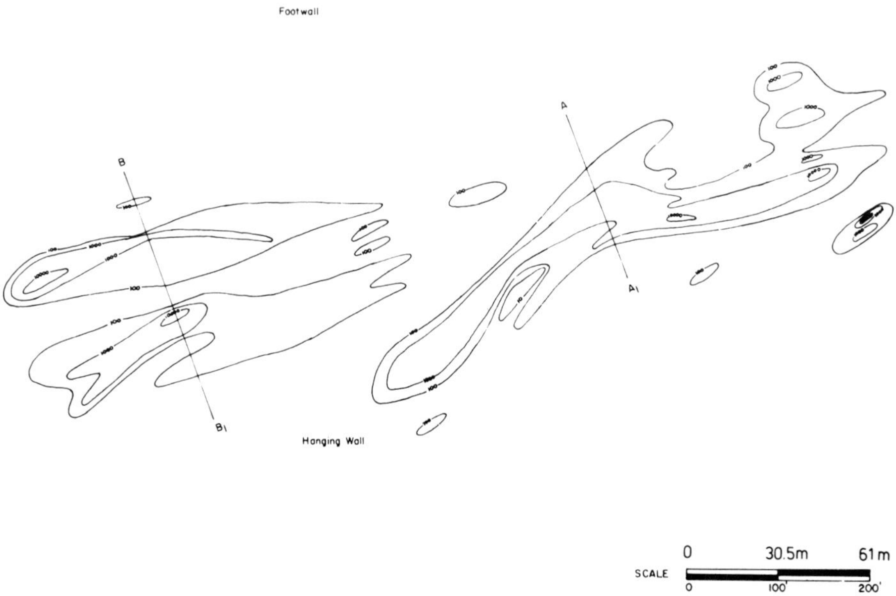

Figure 8a. Contour plan, mercury distribution (ppb), 240 m level, Ruttan mine.

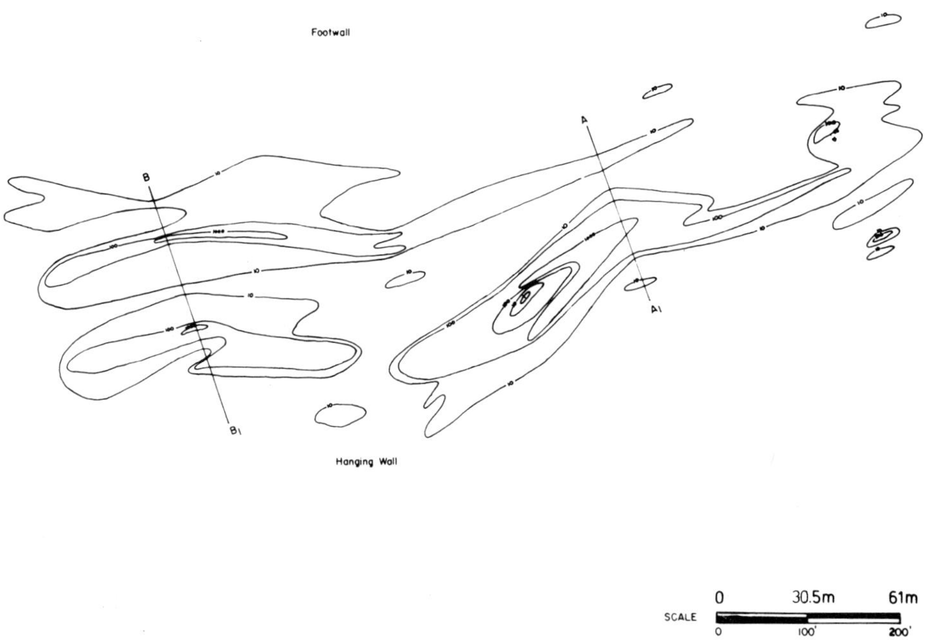

Figure 8b. Contour plan, gold distribution (ppb), 240 m level, Ruttan mine.

boundaries in the centre of the zone and gradually diverges to form an angle of 40 to 45° to the strike of the shear zone at the boundaries. There are commonly two slightly differing orientations of schistosity in the centre, possibly indicating either different stages of movement along the shear zones or that the shear zone fabric was superimposed on an older inhomogeneous fabric similar to that described by Kennedy and DeGrace (1972) for the Lush's Bight Group in Newfoundland.

Although the orientation of the foliation within the alteration zone is not uniform, the line of intersection of planar features is consistent throughout, trending S51°E and plunging 60°E. This linear fabric is also consistent with the lineation in schistosity planes, the axis of minor drag folds and the plunge of the ore lenses and

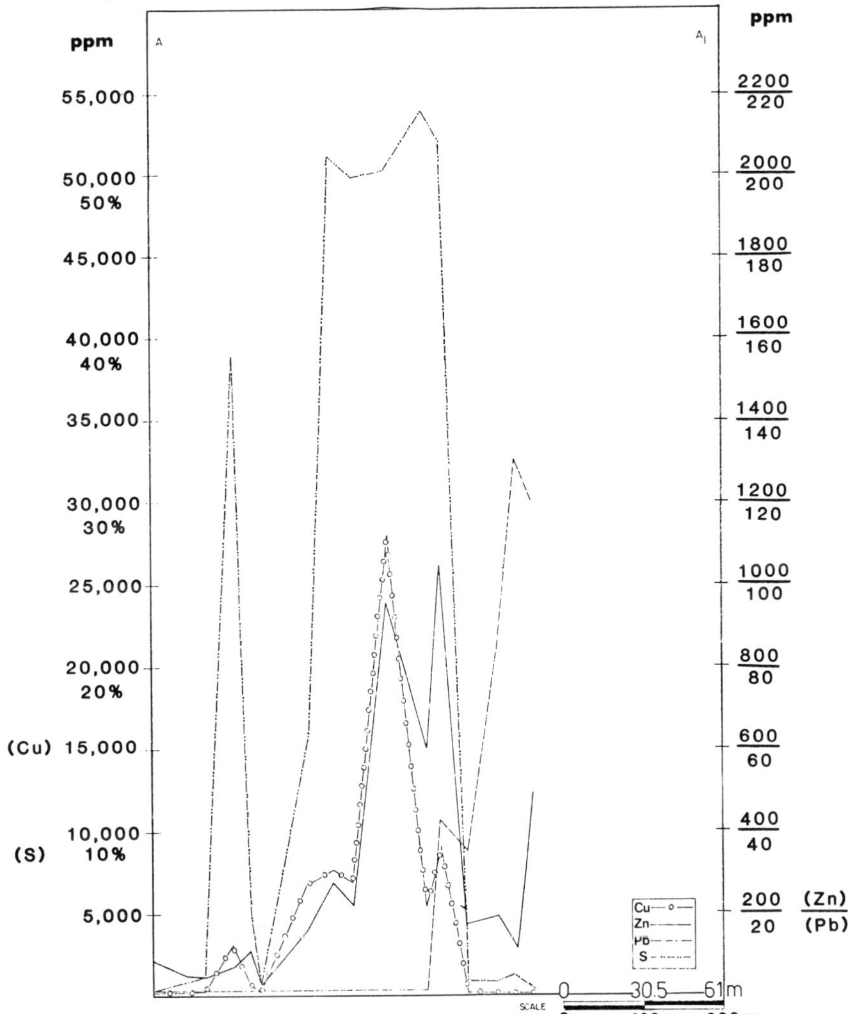

Figure 8c. Cu, Zn, Pb, S distribution, section A-A′ (Figs. 8a, b) 240 m level, Ruttan mine.

represents the intermediate strain direction (intermediate axis of the finite strain ellipsoid) during deformation.

Consistent with the simple inhomogeneous shearing model is the presence of gash-fractures up to one metre thick and 3 m long, that have been rotated up to 30° during deformation. The gash-fractures are located in the wall rocks of the shear zone and in xenoliths of less deformed rock within the shear zones, and contain quartz, anhydrite and minor pyrrhotite and chalcopyrite.

Using Ramsey and Graham's (1970) model, the computed displacement across the shear zone is 500 m with an upward and eastward sense of translation of the footwall side (north) relative to the hanging wall side (south). As the shear zone bifurcates into three major branches with numerous splays, the overall effect of shearing is to significantly change the shape of the ore lenses and alteration zone.

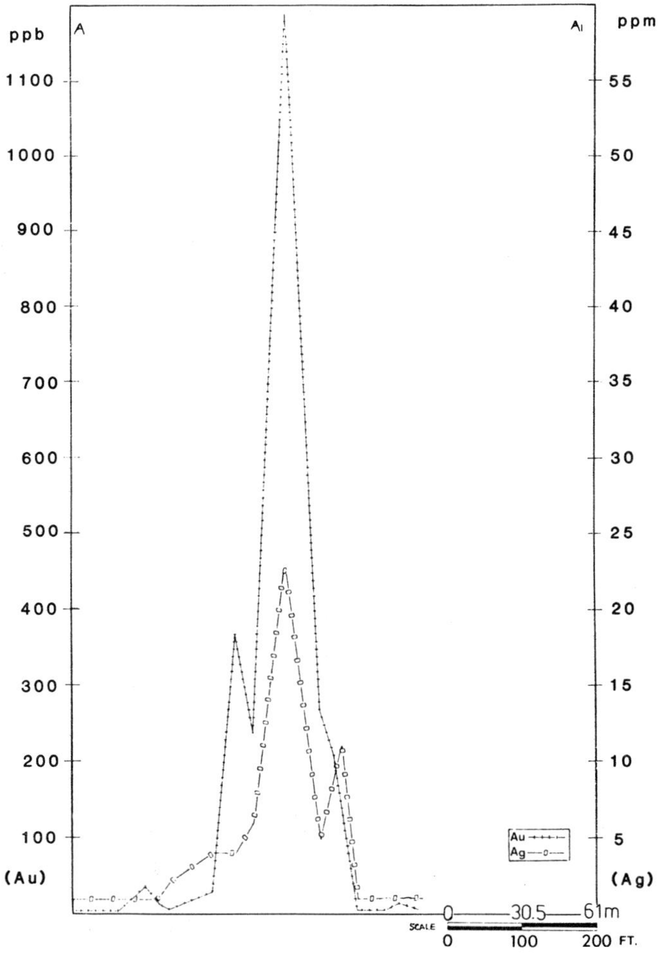

Figure 8d. Au, Ag distribution, section A-A′ (Figs. 8a, b) 240 m level, Ruttan mine.

METAMORPHISM

Steeves and Lamb (1972) suggested that the Rusty Lake Greenstones have been metamorphosed to upper greenschist and lower amphibolite facies. Due to the proximity of the post-Sickle granites, a higher temperature of metamorphism may have affected the orebody area. Only the pelitic assemblages within the alkali-depleted, magnesium-enriched rocks of the alteration zone provide good indicators of the P-T conditions that prevailed during metamorphism. The variation of the FeO/(MgO + FeO) ratio within the alteration zone has resulted in the localized

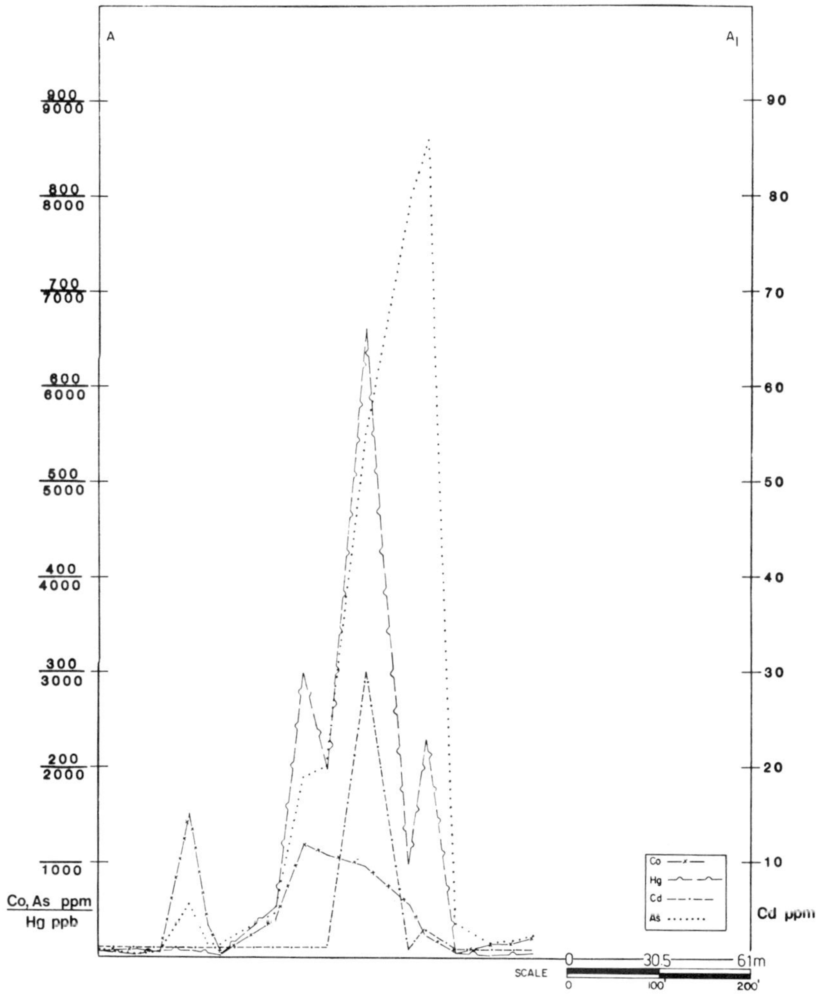

Figure 8e. Co, As, Hg, Cd distribution, section A-A′ (Figs. 8a, b) 240 m level, Ruttan mine.

occurrence of the almandine-garnet-cordierite-staurolite-biotite assemblage. Winkler (1976) used this assemblage to define the cordierite-almandine subfacies of medium grade (amphibolite) metamorphism. Together with the occurrence of andalusite, this

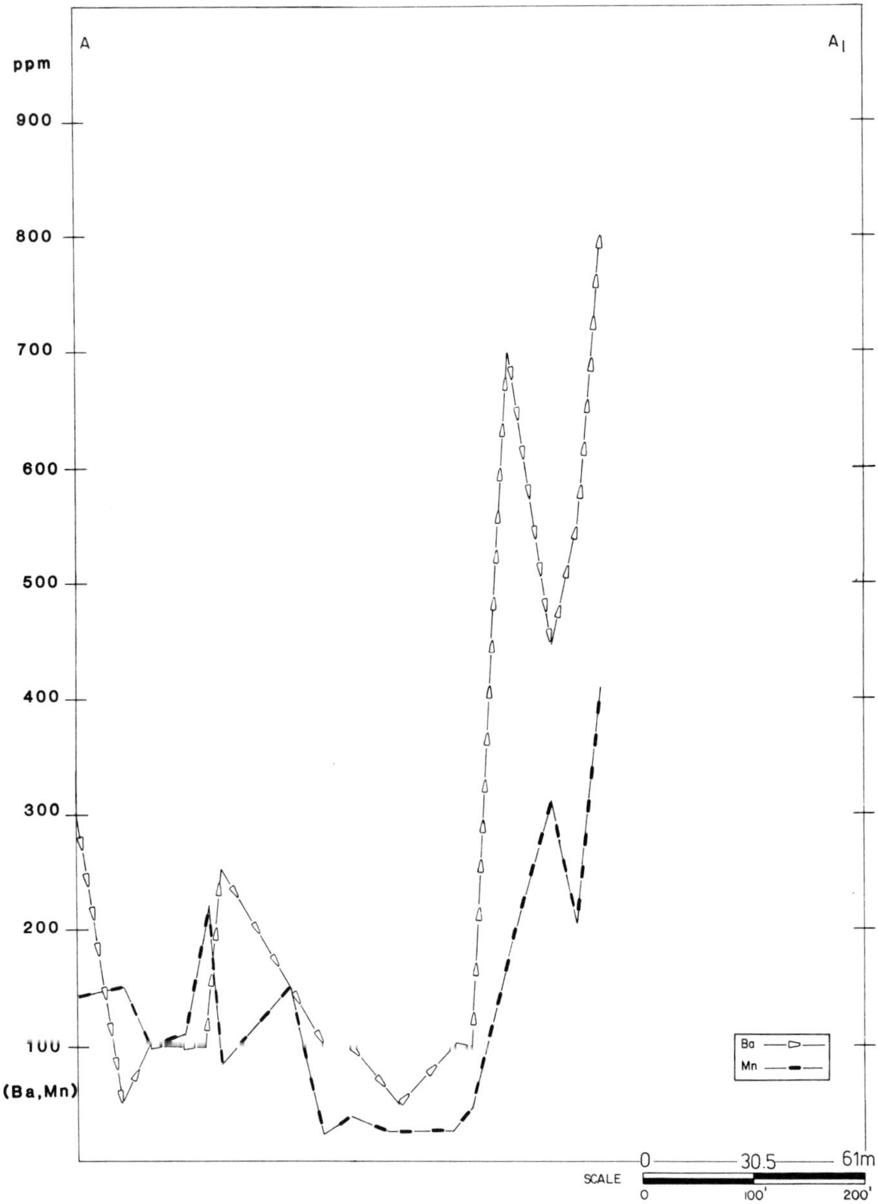

Figure 8f. Ba, Mn distribution, section A-A′ (Figs. 8a, b) 240 m level, Ruttan mine.

defines P-T conditions of approximately 600°C and 5.5Kb (Winkler, 1976). As described under "Mine Geology" the presence of annealing textures, unstrained quartz, untwinned plagioclase and cordierite indicates that the maximum intensity of metamorphism post-dates the main deformational events.

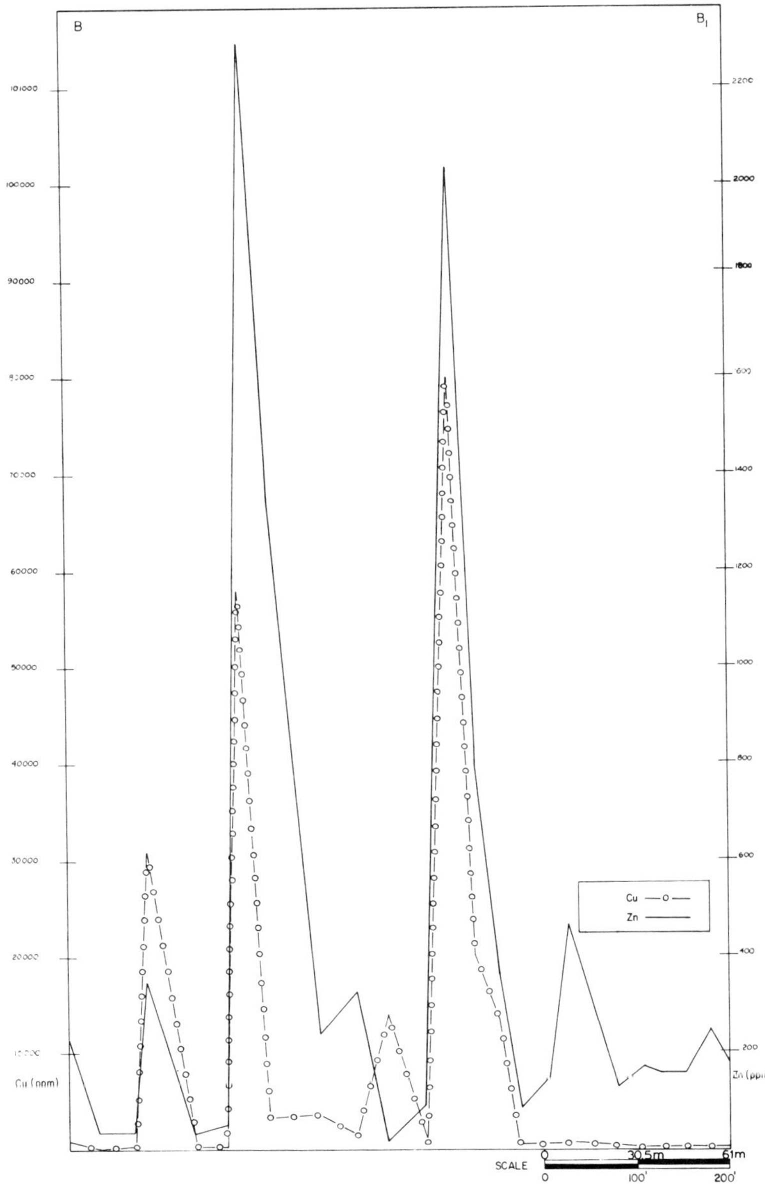

Figure 8g. Cu, Zn distribution, section B-B′ (Figs. 8a, b) 240 m level, Ruttan mine.

MINE GEOLOGY

From footwall to hangingwall the rock units hosting the orebody include the Altered Footwall Volcaniclastic rocks, the Mine Rhyolite with its associated Cu-Zn-rich pyrite lenses, and the Exhalite Horizon with its associated sedimentary sulphide lenses (Figs. 9a, 9b). Also included are the crosscutting chlorite-biotite schists which occur in the shear zones. In general, the rocks have medium- to coarse-grained granoblastic to augen gneiss texture, and recrystallization has obliterated most pre-existing textures.

The massive sulphide lenses average 80 to 90% sulphide and generally have sharp contacts with the wall rocks. These lenses collectively have a strike length of 550 m, a maximum thickness of 100 m, a dip of 70° to the southeast and a plunge of 60°

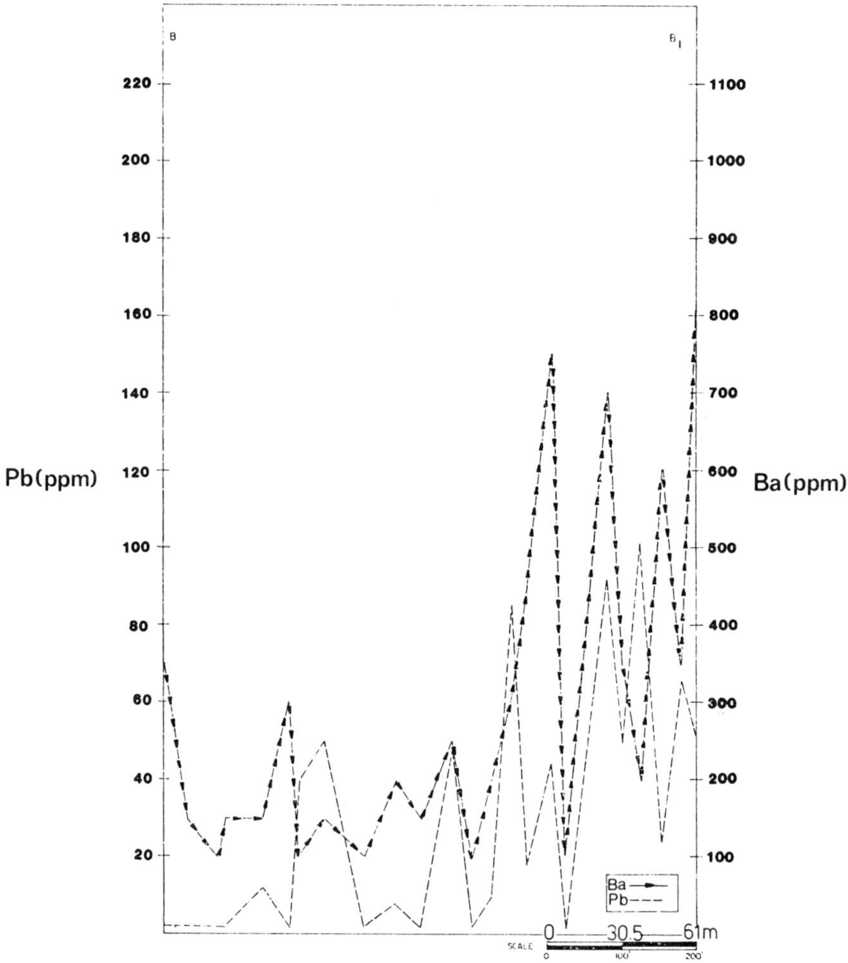

Figure 8h. Pb, Ba distribution, section B-B′ (Figs. 8a, b) 240 m level, Ruttan mine.

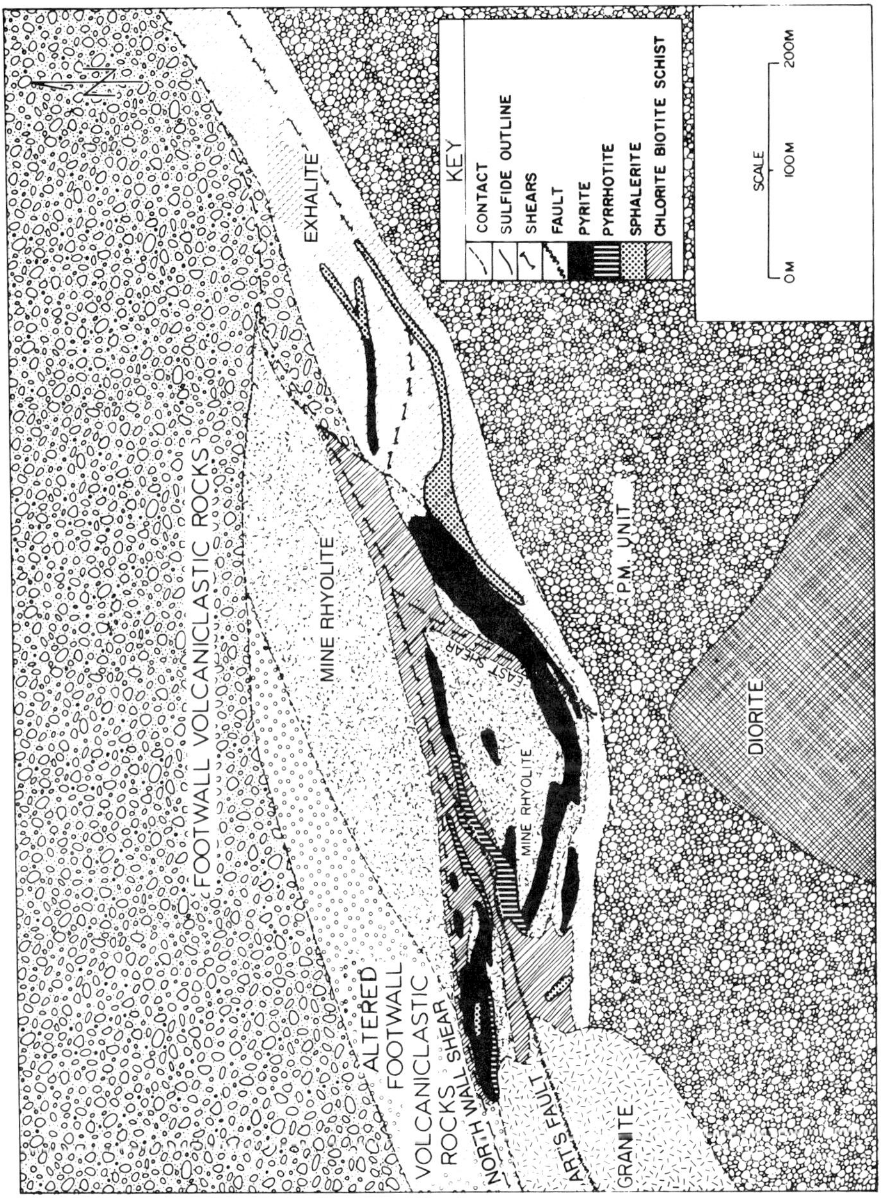

Figure 9a. Geological map, 240 m level, Ruttan mine.

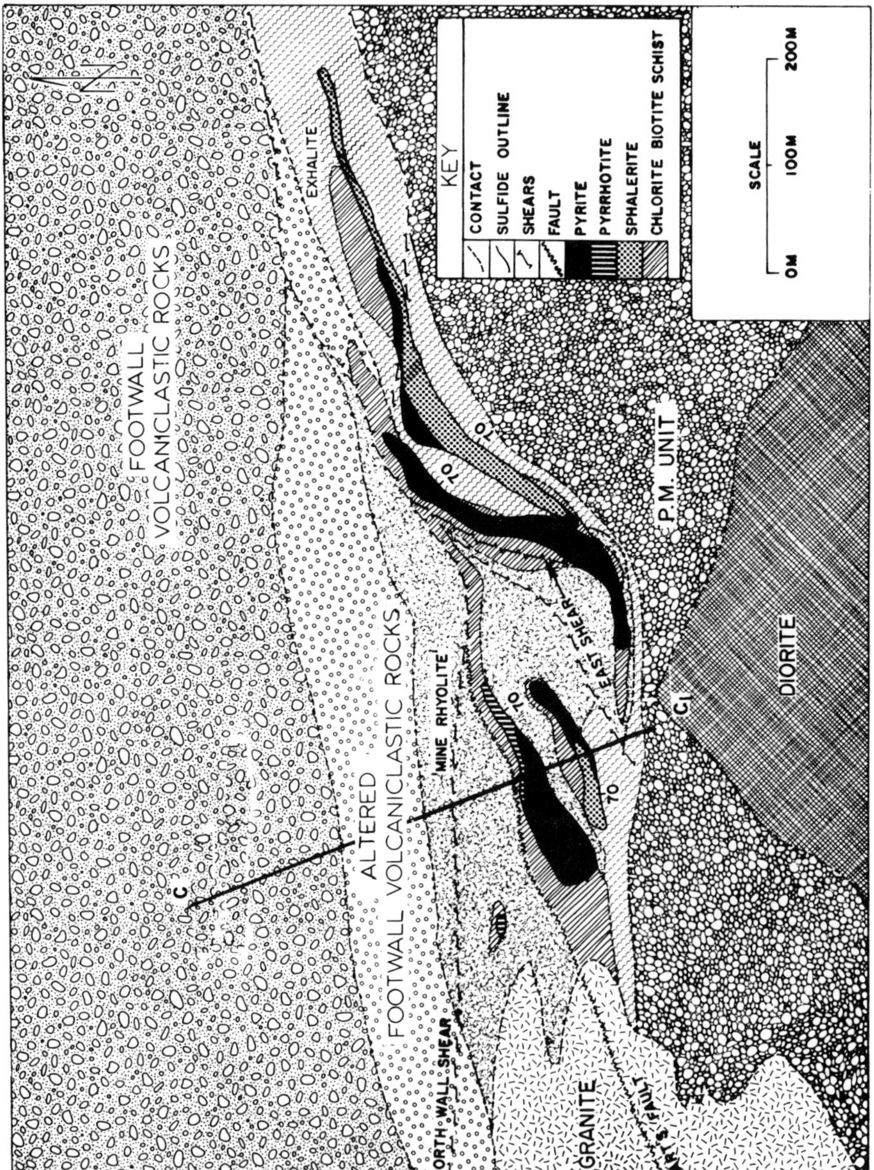

Figure 9b. Geological map, 360 m level, Ruttan mine.

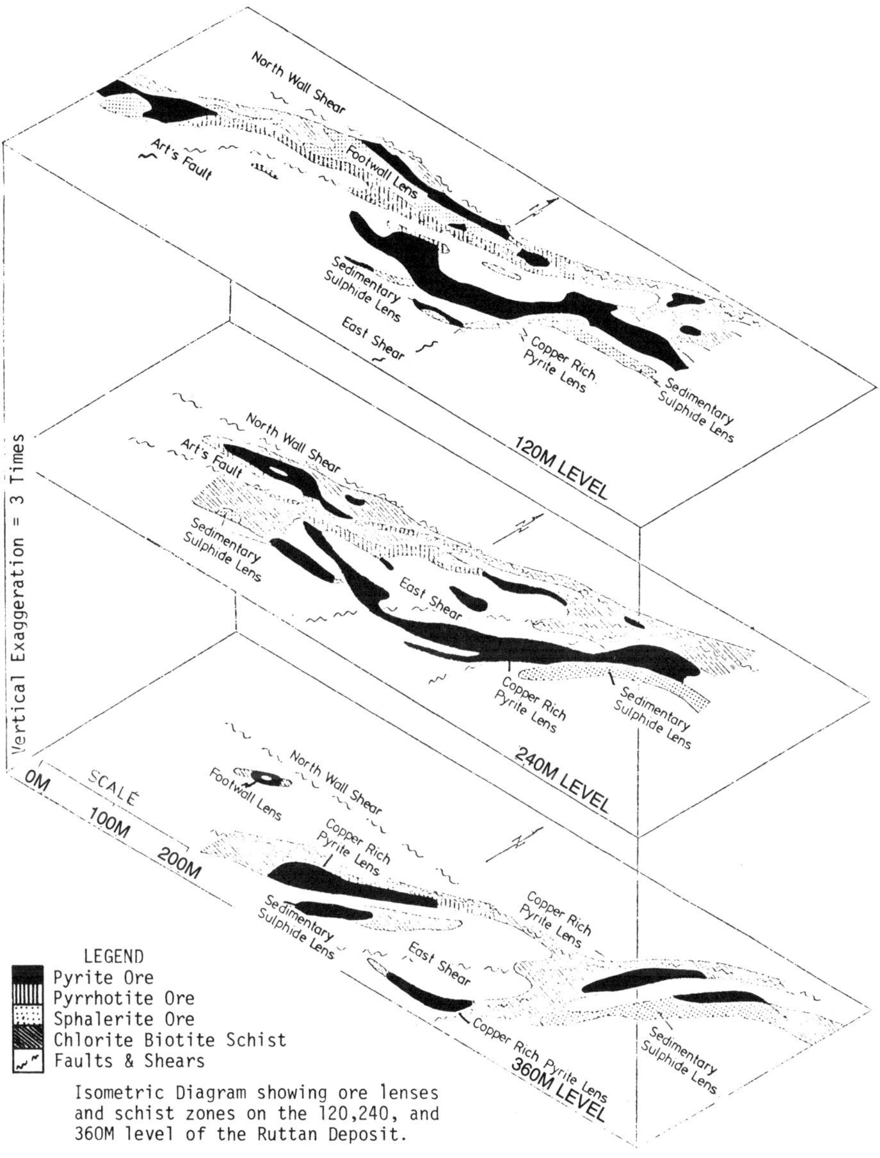

Figure 10. Isometric diagram showing ore lenses and schist zones on 120, 240, 360 m levels, Ruttan mine.

to the east. The present 100 m thickness is due in part to the presence of a large sphalerite-rich sulphide lens in the footwall of the copper-rich pyrite lens (Fig. 10). This lens has been displaced along a major shear (Art's Fault) from its original position on the Exhalite Horizon.

Altered Footwall Volcaniclastic Rocks. This unit forms a lenticular body stratigraphically below the Mine Rhyolite and attains a maximum thickness of 100 m. The lower contact is gradational into the Footwall Volcaniclastic rocks and the upper contact is either in contact with massive sulphides or obscured by silicification. The rock changes texturally from a medium- to coarse-grained augen gneiss in the west to a coarse-grained, granoblastic rock in the east. Areas with high chlorite, biotite or sericite content are schistose.

A penetrative, diamond-shaped interference foliation pattern has developed within this unit. Where the North Wall Shear crosses it, the unit is characterized by brecciation or the development of chlorite-biotite schist. The augen gneiss portions consist of quartz, chlorite, biotite, cordierite, plagioclase, magnetite, ± staurolite, ± andalusite, ± almandine garnet, and ± sericite. About 10 to 20% of the rock is made up of one centimetre porphyroblasts of cordierite and/or andalusite, staurolite, and almandine. These are poikiloblastic with quartz inclusions. Quartz and untwinned, unaltered plagioclase comprise 50 to 60% of the rock and form the groundmass, occurring as 0.5 to one millimetre annealed grains.

Mine Rhyolite. The Mine Rhyolite is composed of quartz, chlorite, biotite, pyrite, ± calcite, ± plagioclase, ± sericite, and ± cordierite. Quartz comprises 70 to 80% of the rock and occurs at 0.5 to 1 mm grains forming an allotriomorphic mosaic. Chlorite content is generally 5 to 10% but in places may range up to 30% of the rock. It occurs as well aligned clusters along quartz grain boundaries, as randomly oriented aggregates and as inclusions in quartz. Biotite is generally sparse but reaches 20% in places. It is present both as inclusions in quartz and along quartz grain boundaries. Average pyrite content outside the sulphide lenses is 2 to 4%. Calcite forms up to 3% and occurs as interstitial material and veinlets. Minor plagioclase occurs locally as untwinned, unaltered 0.5 to 1 mm crystals and as interstitial sericitized material. Minor amounts of cordierite are present as 0.2 to 0.5 mm crystals.

Masses of very fine, well-rounded quartz grains rarely observed in thin section resemble the fine microcrystalline groundmass of the Footwall Volcaniclastic rocks indicating that the Mine Rhyolite may be a silicified portion of the latter.

Copper-Rich Pyrite Lens. This lens, located within the Mine Rhyolite, is approximately 370 m long and averages about 20 m thick. Below the 330 m lvel, it is divided by the East Shear into two zones, and has been rotated from an east-west to a northeast-southwest strike. The lens is characterized by equigranular, medium-grained pyrite. Chalcopyrite occurs along grain boundaries and in cusps at triple junctions. Annealing has masked any penetrative deformational textures but the lens contains large felsic and mafic blocks of waste rock that have an orientation consistent with the fabric in the surrounding rocks.

The most abundant gangue mineral is quartz. Chlorite, biotite, gypsum, anhydrite and calcite are present in varying quantities. The percentage of carbonate and sulphate increases toward the hangingwall. The quartz occurs both interstitially and as very fine-grained spherical masses up to 1 cm in diameter. The contacts of these relict fragments exhibit a caries texture with embayments of sulphide within the quartz.

The lens is massive and generally consists of 60 to 90% sulphides. Where the percentage of chlorite and biotite is high, the sulphides exhibit a granular sandy texture. Near the hangingwall and footwall contacts, the annealed pyrite has been fractured and granulated with mobilization of fine-grained chalcopyrite and pyrrhotite into fractures.

Exhalite Horizon. This unit forms a layer up to 50 m thick above and on the flanks of the Mine Rhyolite. To the west, it is displaced by Art's Fault and partially assimilated by the granitic intrusion. To the east, the lower contact is obscured by sericitization. Farther east, it can be traced as a discontinuous bed for 4 km. The chemically precipitated portion of the Exhalite Horizon consists of thick beds of chert. The detrital portion occurs as 2 to 30 m thick beds, ranging from mudstone to conglomerate. Silica values range from 55% in the detrital sedimentary rocks to 90% in the chert. Much of the detrital sedimentary rock is highly chloritic, but chert-rich areas are strongly sericitic. The average pyrite content is 10 to 15%. Bedding unconformities of up to 15° have been observed within the Exhalite Horizon. Microscopically, the chert bands consist of fine grains (less than 0.2 mm) of annealed quartz. Sericite (10 to 20%) occurs as well aligned aggretates along quartz grain boundaries.

Generally, the groundmass of the detrital portion consists of 50 to 70% microcrystalline annealed quartz and feldspar. Hornblende (10 to 15%) is present as randomly oriented grains approximately 0.2 mm in diameter. Epidote occurs in places in amounts up to 5%. Generally, biotite forms 0.5 mm grains and, together with chlorite, occurs in association with porphyroblasts of cordierite, staurolite or andradite. Alteration of the bedded detrital sedimentary rocks east of the Mine Rhyolite has resulted in the development of staurolite-biotite-rich bands 1 to 5 cm thick. Carbonate is a common accessory.

The Sedimentary Sulphide Lenses. The Sedimentary Sulphide Lenses are located within the Exhalite Horizon and are characterized by alternating one centimetre to one metre thick bands of sphalerite-, pyrite- and silicate-rich rock. One major and several minor lenses have been found. The major lens is 20 m thick, 240 m long and grades laterally into thin, zinc-rich beds of massive sulphides to the east. The lens is still open at depth. Galena content within the sphalerite-rich bands increases to the east and upward toward the hangingwall contact, where it occurs in discontinuous chert bands up to 50 cm thick. The percentage of sulphate and carbonate increases toward the hangingwall. On the hangingwall contact, a one to 3 cm band of carbonate is commonly present.

Microscopically, the sphalerite and pyrite show annealed textures with chalcopyrite located along grain boundaries. The sphalerite crystals are lighter in colour than in the Footwall Lenses and there is little evidence of remobilization or exsolution.

Footwall Lens. The Footwall Lens is sphalerite-rich and mineralogically complex. It is located in the upper levels and western part of the deposit. It is tabular in shape, being 550 m long with a maximum thickness of 60 m. It pinches out at 360 m below surface.

The dominant minerals in the lens are pyrite, sphalerite, pyrrhotite, chalcopyrite and magnetite with a variety of gangue minerals including quartz, biotite, chlorite, gypsum, anhydrite, calcite, talc and tremolite. The grain size is very coarse with pyrite and sphalerite crystals up to one centimetre in diameter increasing in size with proximity to the granite. Within the lens there is an increase in copper content to the west and down dip. The zinc-rich zones are oriented at a slight angle to the lens, thus paralleling the regional foliation. These zones are separated by zones of pyrite, chalcopyrite and large boudinaged blocks of basic rock. A crude banding may have originated by remobilization during metamorphism.

Other evidence of remobilization includes local occurrences of coarse-grained sphalerite cutting a matrix of fine-grained sphalerite-pyrite and selective mobilization of As and Sb evidenced by local concentrations of the Pb-Sb-As sulphosalt geocronite.

Microscopically the sphalerite is characterized by crystallographically oriented lamellae of chalcopyrite and pyrrhotite. These exsolution features are observed only in the footwall zone in proximity to Art's Fault. Similar mineralogical changes in fault zones have been recorded by Richards (1966) at Broken Hill, N.S.W. There is an abundance of pyrrhotite and magnetite within the lens, and evidence of magnetite replacing pyrrhotite which replaces pyrite. This transformation of pyrite to pyrrhotite and pyrrhotite to magnetite is common at Ruttan, especially in the highly deformed sulphide zones. The release of sulphur may have been caused by higher diffusion rates within the shear zones during dynamic metamorphism as described by McDonald (1967). The sphalerite crystals are dark brown to purple and the sulphides exhibit annealed texture in the pyrite-rich areas.

Chlorite-Biotite Schist Zones. The Chlorite-Biotite Schist Zones are coincident with the shears and associated splays described previously. The shear zone boundaries dip 10 to 20° shallower and trend slightly more northeasterly than the orientation of the sulphide lenses and thus intersect the lenses at shallow angles. The contacts between the schist and the surrounding rock are generally sharp. Chlorite, biotite, cordierite and quartz are the dominant minerals. Garnet occurs in the North Wall Shear, and talc tremolite in both Art's Fault and the East Shear, being more prevalent in the latter. Cordierite is most common in Art's Fault. Magnetite occurs locally as patches up to 3 mm in diameter.

Two generations of chlorite are present; one as fine-grained aligned particles surrounding relict fragments or cordierite porphyroblasts, the other as large, randomly oriented flakes overprinting the former. Biotite is similar in occurrence and distribution to chlorite. Cordierite forms large (up to 15 cm) porphyroblasts and comprises up to 50% of the rock. It is generally untwinned and often poikiloblastic. Most cordierite is blue, but in particular horizons, pleochroic halos around zircon inclusions result in a brown colouration. Both varieties are relatively unaltered with only local pinite alteration. Sulphides are generally at least three times more abundant than in the surrounding Altered Footwall Volcaniclastic rocks and Mine Rhyolite. There are also anomalously high concentrations of carbonate and sulphate in these zones, especially toward the hangingwall.

There are two types of fragments in the schists. The first appears to be relicts of the surrounding rock. In areas with high chlorite content, these relict fragments are almost totally assimilated and appear as small clusters of quartz surrounded by chlorite. The second type are large fresh fragments of Mine Rhyolite, Altered Footwall Volcaniclastic rocks, chert and basic dyke.

The sulphide content of the schists is 10 to 15%, increasing to 70 to 80% near the main ore lenses. Generally, the highest sulphide content is near the centre of the schist zone. The increased sulphide content is first observed as pyrite clusters surrounded by pyrrhotite and chalcopyrite. The long axes of these clusters are parallel to the mine lineation. As the sulphide content further increases, discontinuous 4 to 5 cm thick sulphide-rich bands occur, oriented parallel to the sigmoidally disposed foliation. These bands are occasionally drag-folded into 1 to 2 m folds. Toward the center of the schist zones, diamond-shaped pyrrhotite-chalcopyrite-rich pods up to 15 m in diameter are found. These pods contain up to 75% pyrrhotite which is studded with one to 3 cm pyrite idiomorphs surrounded by chalcopyrite. The axes of the pods are parallel to the mine lineation.

The largest concentration of sulphides within the schist zones is located in Art's Fault, joining the Footwall Lens with the western portion of the Copper-Rich Pyrite Lens. This large concentration is characterized by greater than 50% pyrrhotite that is commonly studded with large, up to 10 cm, undeformed pyrite porphyroblasts which have grown in-situ after deformation. Chalcopyrite rims these pyrite crystals. Partially replaced chlorite-biotite schist can be observed as vague circular outlines, interpreted as 'durchbewegung' textures (Vokes, 1969) indicating that the sulphides behaved as a ductile mass during movement along the shear zone. A narrow transition zone between the pyrrhotite zone and the Copper-Rich Pyrite Lens contains 25 cm bands of pyrrhotite with pyrite porphyroblasts oriented both parallel and at large angles to the shear zone boundaries. Throughout the schists, chalcopyrite is concentrated in the dilatant zones.

SUMMARY AND CONCLUSIONS

The Ruttan deposit was formed in early Proterozoic time and is conformable within rocks of the Wasekwan Group which are of similar age. The Wasekwan Group rocks are the oldest known in the Ruttan area and consist of a suite of subalkaline volcanic rocks exhibiting a tholeiitic trend. Sedimentary rocks derived from the volcanic rocks are a significant component of the Group. The lowest stratigraphic unit in the Ruttan area is a series of spilitized basalt flows with minor sedimentary rocks. Pillow structures within the unit indicate that it was at least in part subaqueous. Subsequent felsic volcanism and mass movement of the felsic volcanic debris resulted in the formation of the Footwall Volcaniclastic rocks. The mass movement was probably facilitated by steep volcanic paleoslopes and seismic activity in the vent area. The stratigraphically low Copper-Rich Pyrite lenses of the Ruttan deposit are located in the upper portion of this water-lain volcaniclastic unit and may represent sulphide emplacement by reaction and fracture filling in the upper part of a hydrothermal discharge vent. Intense depletions of Na_2O, CaO, K_2O, and enrichments of MgO and SiO_2 in the immediate footwall rocks of the deposit and the pervasive alteration of the more permeable footwall rocks, below and laterally away from the deposit are indicative of a large-scale hydrothermal system with a discharge area at the site of the deposit. If the Mine Rhyolite is an intensely silicified portion of the Footwall Volcaniclastic rocks, then Al_2O_3 and TiO_2 were possibly depleted by the hydrothermal fluids.

It is beyond the scope of this study to determine the nature of the generative hydrothermal fluids, however limited observations can be made concerning their leaching and transporting capabilities. a) Beyond the area of intense footwall alteration the fluid was capable of leaching and transporting Mg, Cu, Zn, and Fe. b) In the Altered Footwall Volcaniclastic rocks the fluid was capable of leaching and transporting Na, K, and Ca while depositing Mg and limited amounts of Cu and Fe. c) In the Mine Rhyolite the fluid was capable of leaching Na, K, and Ca, and possibly Al and Ti, while depositing Mg and Cu-Fe in sufficient quantity to form massive sulphide lenses. d) At the seawater-rock interface near the discharge area the fluid was capable of selectively precipitating Cu, Zn, Fe, Si, Ca, SO_4, Au, Ag, Pb and Ba.

The precipitate resulting from the intermixing of the hydrothermal fluids and seawater formed the exhalite horizon on the sea floor. Pyrite-sphalerite-silicate banding, the increase in lead, sulphate, carbonate and barium toward the hangingwall and the presence of chert are characteristic of this environment. Detrital sedimentation intermittently occurred during the period of chemical precipitation. Subsidence and increase in relief resulted in the deposition of the P.M. Unit. There is an abrupt decrease in the intensity of alteration at the hangingwall contact of the exhalite horizon. Minor mineralization and alteration in the overlying P.M. Unit indicate that the hydrothermal system waned during its deposition.

Deformation and metamorphism of the Wasekwan greenstones profoundly affected the shape, form, mineralogy and texture of the Ruttan orebodies. Deformation was most intense in the mechanically weaker rocks of the alteration zone and sulphide deposit, which responded by shearing. This shearing resulted in the offset and rotation of the ore zones and in mobilization of chalcopyrite and pyrrhotite into dilatent zones, creating enriched areas within the pre-existing sulphide lenses, and locally forming new lenses.

The mineralogy of the alteration zone is consistent with the whole rock chemistry of the altered rock and the pressure and temperature conditions of metamorphism. The coarse-grained granoblastic and augen textures of this metamorphosed alteration zone are anomalous in relation to the surrounding schists. This may be due to the catalytic effect of water. It is likely that the shearing within the deposit relieved the stress acting on it and resulted in a pressure differential between it and the surrounding rock. This pressure differential might have provided the gradient necessary to cause an influx of water released by the dehydration processes of medium grade metamorphism, resulting in enhanced recrystallization of the rocks in the orebody area.

ACKNOWLEDGEMENTS

The authors wish to express their gratitude for the support that Sherritt Gordon Mines Limited has provided in this endeavour. Encouragement and assistance from Jim Franklin of the Geological Survey of Canada aided our efforts considerably. Dave Baldwin and George Gale of the Manitoba Mines Branch critically reviewed an early draft of the paper and provided invaluable discussion thereafter. A special note of appreciation is extended to Sherritt personnel Jane Mossington, who drafted the final diagrams and Joanna Len, who typed the final manuscript.

REFERENCES

Bailes, A.H. and McRitchie, W.D., 1978, The transition from low- to high-grade metamorphism in the Kisseynew Sedimentary Gneiss Belt, Manitoba: in Metamorphism in the Canadian Shield: Geological Survey of Canada Paper 78-10, p. 155-178.

Baldwin, D.A., 1978, Volcanic environments of massive sulphide mineralization in the Lower Waskwan Group in the Rusty Lake Greenstone Belt: in Manitoba Mineral Resources Division, Report of Field Activities.

_______________, 1979, G.S.-5 Ruttan Lake, Karsakuwigamak Lake, Eagle Lake Project: in Manitoba Mineral Resources Division Report of Field Activities.

_______________, 1980, G.S.-3 Ruttan Lake, Karsakuwigamak Lake, Eagle Lake Project: in Manitoba Mineral Resources Division Report of Field Activities.

Bristol, C.C., 1979, Application of sphalerite geobarometry to ores from the Ruttan Mine: Economic Geology, v. 74, p. 1496-1502.

Burwash, R.A., 1962, Geology of the Rusty Lake Area: Manitoba Department of Mines and Natural Resources, Publication 60-3.

Irvine, T.N., and Baragar, W.R.A., 1971, A guide to the chemical classification of the common volcanic rocks: Canadian Journal of Earth Sciences, v. 8, p. 523-548.

Kennedy, M.J., and DeGrace, J.R., 1972, Structural sequence and its relationship to sulphide mineralization in the Ordovician Lush's Bight Group of Western Notre Dame Bay, Newfoundland: Canadian Institute of Mining and Metallurgy Bulletin, December, p. 42-50.

McDonald, J.A., 1967, Metamorphism and its effects on sulphide assemblages: Mineralium Deposita 2, p. 200-220.

Milligan, G.C., 1960, Geology of the Lynn Lake District, Manitoba: Manitoba Mines Branch, Publication 57-1.

Moore, J.M., 1977, Orogenic volcanism in the Proterozoic of Canada: Baragar, W.R.A., Coleman, L.C. and Hall, J.M., eds., Volcanic Regimes in Canada: Geological Association of Canada Special Paper 16.

Norman, G.W.H., 1934, Granville Lake District, Northern Manitoba: Geological Survey of Canada Summary Report 1933 - C, p. 23-41.

Ramsay, J.G. and Graham, R.H., 1970, Strain variation in shear belts: Canadian Journal of Earth Sciences, v. 7, p. 786-813.

Richards, S.M., 1966, The banded iron formations at Broken Hill, Australia and their relationship to the lead-zinc orebodies: Part I and II: Economic Geology, v. 61, p. 72-96, 267-274.

Sangster, D.F., 1978, Isotopic studies of ore leads of the Circum-Kisseynew Volcanic Belt of Manitoba and Saskatchewan: Canadian Journal of Earth Sciences, v. 15, p. 1112-1121.

Steeves, M.A. and Lamb, C.F., 1972, Geology of the Issett - Opachuanau - Pemichigamau - Earp Lakes Area: Manitoba Department of Mines, Resources and Environmental Management, Mines Branch Publication 71 - 2F.

Vokes, F.M., 1969, A review of the metamorphism of sulphide deposits: Earth-Science Rev., v. 5, p. 99-143.

Winkler, H.G.F., 1976, Petrogenesis of Metamorphic Rocks, 4th edition.

MS received June 1979;
Accepted for publication August 1981.

Precambrian Sulphide Deposits, H.S. Robinson Memorial Volume, edited by R.W. Hutchinson, C.D. Spence and J.M. Franklin, Geological Association of Canada Special Paper 25, 1982

THE SHERRITT GORDON MASSIVE SULPHIDE DEPOSIT

P.A. Goetz
Teck Corporation, 425-1st Street S.W. Calgary, Alberta T2B 3L8

E. Froese
Geological Survey of Canada, 601 Booth Street, Ottawa, Ontario K1A 0E8

ABSTRACT

The Sherritt Gordon deposit at Sherridon, Manitoba consists of stratiform orebodies, about 5 m thick, interlayered with quartz-rich gneisses of the Sherridon Group. The ore is composed of pyrite, pyrrhotite, chalcopyrite, and sphalerite. The coarse grain size, the abundance of pyrrhotite, and the presence of gahnite indicate that the orebodies, together with the enclosing rocks, have been metamorphosed. The massive ore is separated from the host rock by a narrow zone of disseminated sulphides. Although there is no alteration zone below the deposit, cordierite-anthophyllite rocks along the same stratigraphic horizon probably represent metamorphosed detritus of hydrothermally altered rocks clastically transported to the present site of deposition prior to metamorphism. The sulphides may have been either similarly sedimented from a suspension or, alternatively, chemically precipitated from a brine. Although the Sherridon Group has been correlated with the alluvial Missi Group of the Flin Flon-Snow Lake belt, calc-silicate gneisses in the Sherridon Group suggest a marine environment. Also, in contrast to most successions in the Missi and Sherridon Groups, the Sherridon Group in the vicinity of Sherridon includes volcanic rocks. This suggests that volcanic activity in a marine environment may have been essential for the formation of the Sherritt Gordon deposit.

RÉSUMÉ

Le dépôt de Sherritt Gordon, Manitoba, consiste en amas sulfurés stratiformes d'environ 5m d'épaisseur, interstratifiés dans les gneiss très quartzeux du groupe de Sherridon. Le minerai se compose de pyrite, pyrrhotite, chalcopyrite et sphalérite. Le grain grossier, l'abondance de pyrrhotite et la présence de gahnite sont preuves que ces gisements ont été métamorphisés en même temps que les roches encaissantes.

Le minerai massif est séparé de la roche encaissante par une mince zone à sulfures disséminés. Bien qu'il n'y ait pas de zone d'altération visible sous le dépôt, on trouve le long du même niveau stratigraphique des roches à cordiérite-anthophyllite qui représentent probablement des détritus de roches altérées hydrothermalement et transportées à leur place actuelle avant le métamorphisme. Les sulfures ont été soit sédimentés à partir d'une suspension, soit précipités chimiquement à partir d'une solution saline. Bien que le groupe de Sherridon ait été correlé avec le groupe alluvial de Missi dans le ceinture de Flin Flon-Snow Lake, les silicates calciques du groupe de Sherridon suggèrent un environement marin. De même, à l'inverse de la pluspart des successions rencontrées dans les groupes de Missi et de Sherridon, la succession au voisinage de Sherridon même comprend des roches volcaniques. Ceci suggère qu'une activité volcanique en milieu marin a joué le rôle essentiel dans la formation du dépôt de Sherritt Gordon.

INTRODUCTION

The Kisseynew gneisses grade through a metamorphic transition into lower grade metamorphic rocks of the Flin Flon-Snow Lake belt (Fig. 1) commonly considered to be of Aphebian age (Sangster, 1978). In the Flin Flon-Snow Lake belt, volcanic rocks and interlayered greywacke and shale of the Amisk Group are overlain by lithic arenites of the Missi Group. Robertson (1953) subdivided the Kisseynew gneisses into the Nokomis Group and the Sherridon Group and Bailes (1971, 1980) suggested a correlation with the sedimentary part of the Amisk Group and with the Missi Group, respectively. In the vicinity of Sherridon, the Sherridon Group consists of both metasedimentary and metavolcanic rocks exposed in a crescent-shaped structural basin surrounded by quartzofeldspathic gneisses, migmatites, and granitoid gneisses of the Nokomis Group. In this particular locality, rocks of the Nokomis Group do not resemble metamorphosed Amisk sedimentary rocks, indicating that the Nokomis Group may include rocks which have no equivalents in the Amisk Group. In contrast to most successions in the Missi Group, the Sherridon Group contains calc-silicate gneisses. Only in an area east of Wekusko Lake, does the Missi Group include calc-silicate gneisses (Bailes, 1976; Bailes, pers. commun., 1981). Volcanic rocks, which are atypical of the Sherridon Group, are present in the Sherridon area. In the Missi Group volcanic rocks have been reported only in the area east of Wekusko Lake (Frarey, 1950; Shanks and Bailes, 1977).

The Sherridon basin, straddling two one-mile map sheets (Bateman and Harrison, 1946; Robertson, 1953), has recently been mapped by Froese and Goetz (1981). The disposition of lithologic units (Fig. 2) suggests a continuous, single-facing succession from the margin towards the centre of the structural basin. This interpretation, in contrast to Tuckwell's (1979) suggestion of possible facing reversals, implies that early isoclinal folding did not produce major repetitions of strata. The origin of various rock types within the Sherridon Group has been considered by Goetz (1980); in the absence of primary features, deductions were based on compositions and rock associations. The main rock type of the Sherridon Group is a quartz-rich gneiss, derived from a lithic arenite. Several discontinuous layers of amphibolites, derived from volcaniclastic rocks and tholeiitic lavas, form part of the Sherridon Group. Copper-zinc deposits occur along two horizons within the quartz-rich gneiss. The stratigraphically lower horizon includes the Sherritt Gordon orebodies and the Park Lake and Jungle Lake deposits; the second horizon includes the Bob Lake and Fidelity deposits and a surface prospect near Lost Lake (Fig. 2). Cordierite-

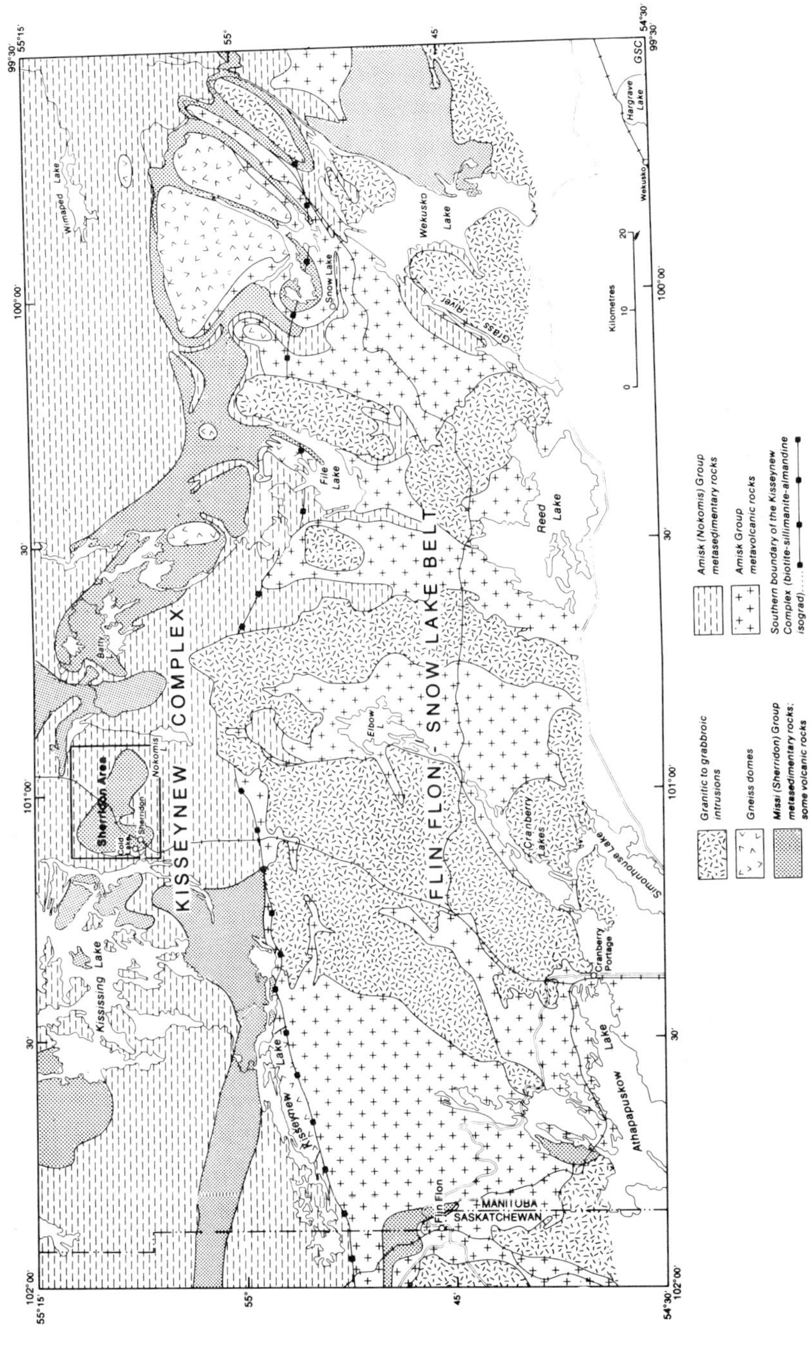

Figure 1. Location and geological setting of the Sherridon area (from Froese and Goetz, 1981).

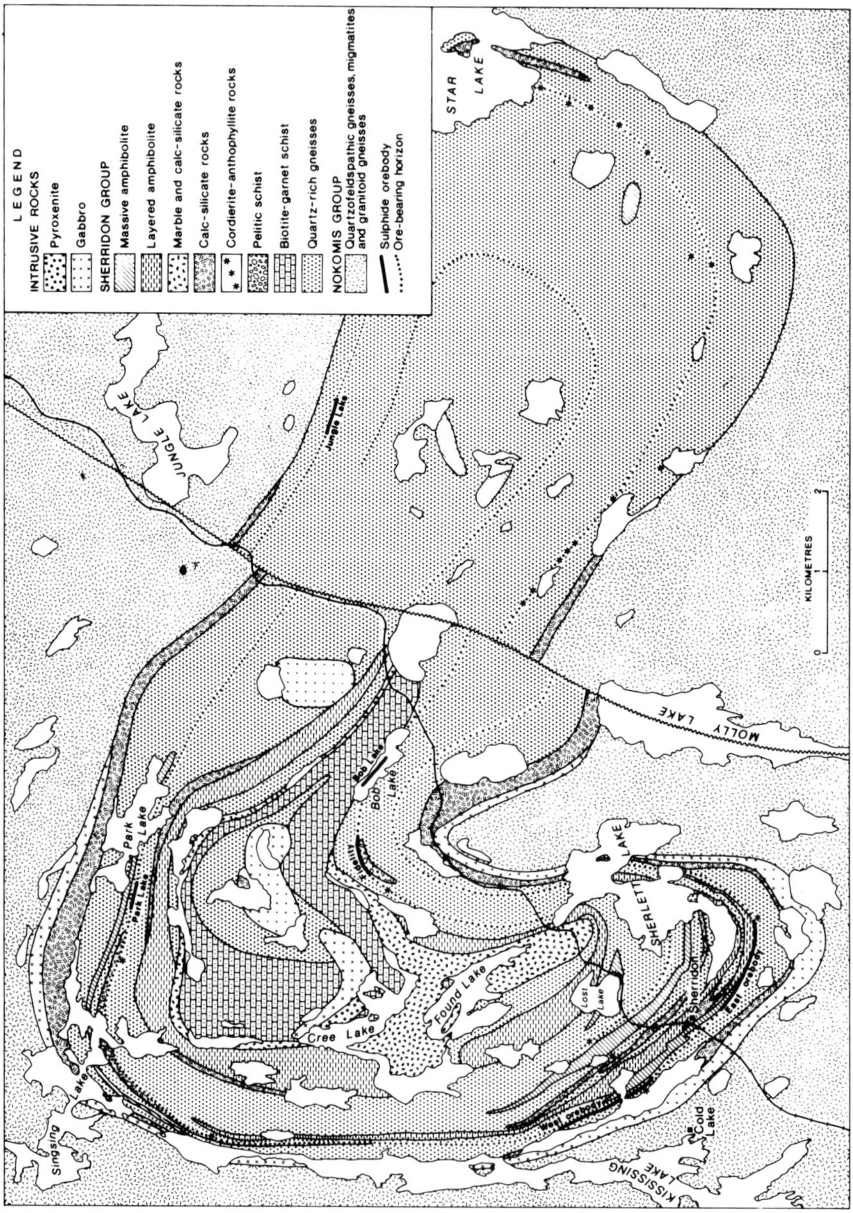

Figure 2. Geological sketch map of the Sherridon area (from Froese and Goetz, 1981).

anthophyllite rocks are present along the mineralized horizons in several places and a thin layer of pelitic schist, locally with disseminated sulphides, connects the Sherritt Gordon west orebody with the Park Lake deposit. Calc-silicate rocks and marble constitute an appreciable portion of the Sherridon Group. Thin layers of a siliceous rock, probably metamorphosed chert, with disseminated pyrrhotite, are present within the calc-silicate rocks. These rocks typically contain only trace amounts of copper and zinc.

Small isoclinal folds represent the earliest recognized deformation. The rocks display one prominent foliation parallel to lithologic contacts, probably produced by this deformation. The structural basin itself presumably was the result of an interference of two fold systems which deformed the foliation and gave rise to a less prominently developed lineation of greatly varying orientation. The grade of metamorphism is uniform, corresponding to that of the upper amphibolite facies. The sulphide ores have also been subjected to metamorphism as indicated by coarse grain size, abundant pyrrhotite, and the presence of gahnite (Sangster and Scott, 1976).

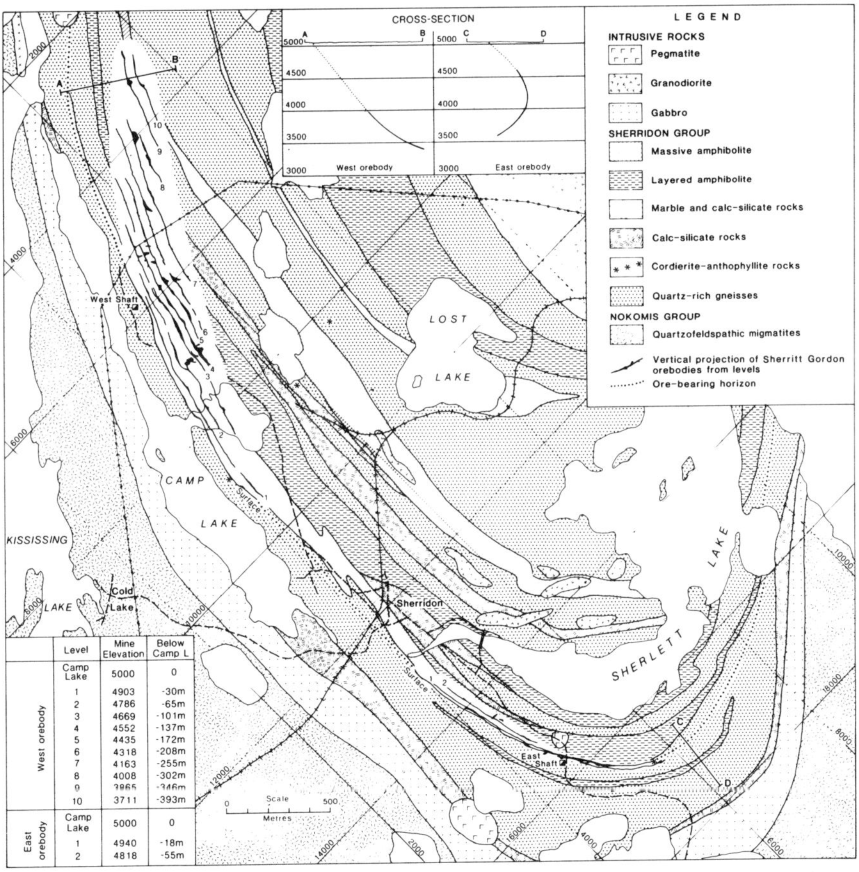

	Level	Mine Elevation	Below Camp L
West orebody	Camp Lake	5000	0
	1	4903	-30m
	2	4786	-65m
	3	4669	-101m
	4	4552	-137m
	5	4435	-172m
	6	4318	-208m
	7	4163	-255m
	8	4008	-302m
	9	3865	-346m
	10	3711	-393m
East orebody	Camp Lake	5000	0
	1	4940	-18m
	2	4818	-55m

Figure 3. Surface geology of the Sherritt Gordon deposit. Level plans compiled from unpublished maps of Sherritt Gordon Mines Limited (from Froese and Goetz, 1981).

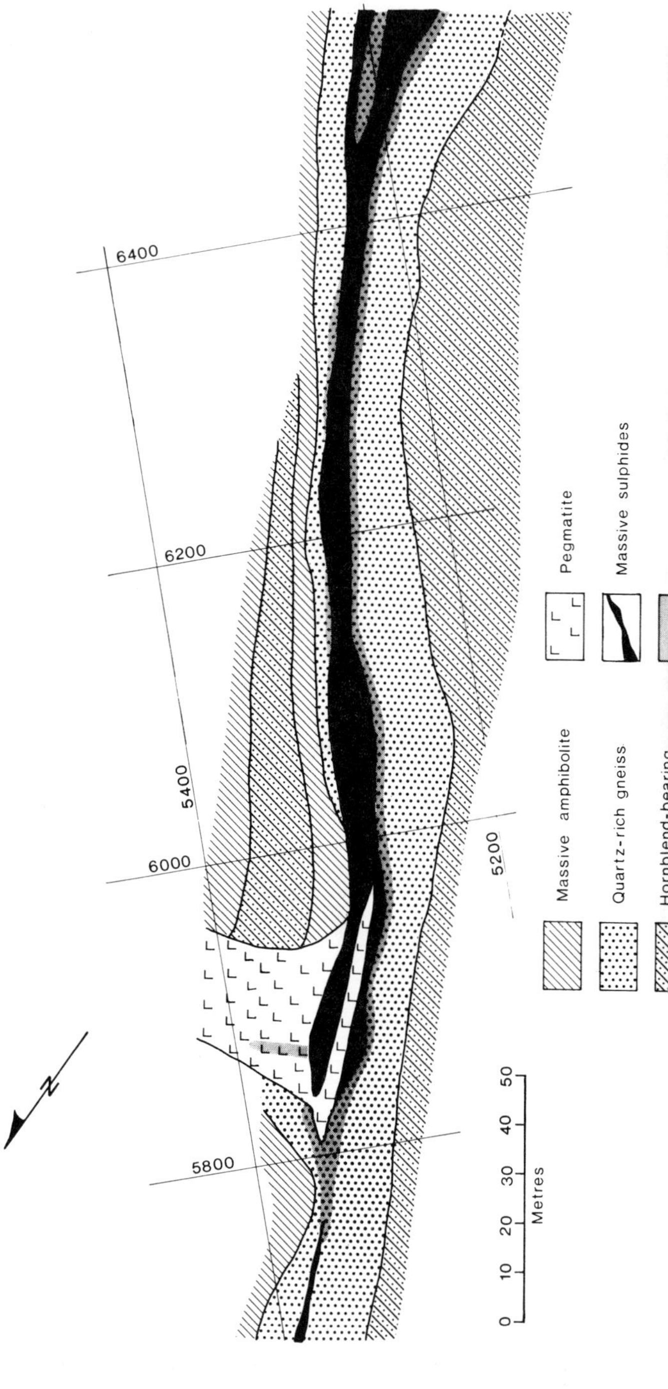
6400
6200
6000
5800
5400
5200
N
0
10
20
30
40
50
Metres
Massive amphibolite
Quartz-rich gneiss
Pegmatite
Massive sulphides

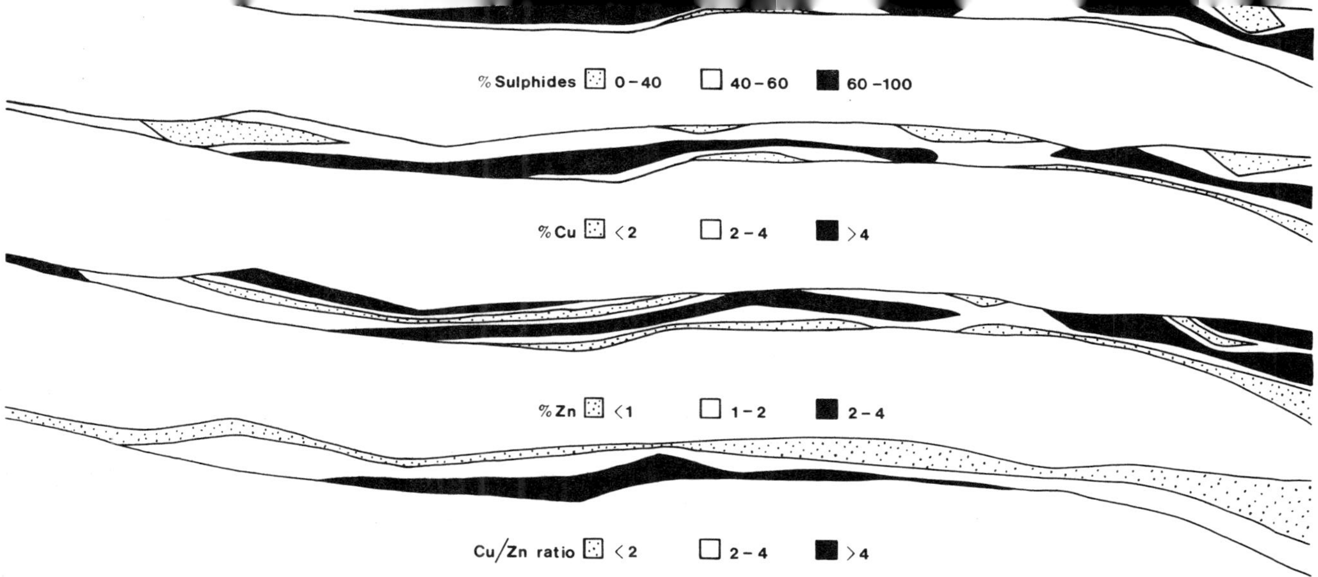

Figure 4. Geology and zoning of a portion of the 3. level of the west orebody of the Sherritt Gordon deposit. Compiled from unpublished information of Sherritt Gordon Mines Limited (from Froese and Goetz, 1981).

SHERRITT GORDON OREBODIES

The surface geology in the vicinity of the Sherritt Gordon mine is shown in Figure 3. The orebodies, about 5 m thick, occur as two lenticular layers of massive and disseminated sulphides and as irregular remobilized masses (offshoot orebodies). A total of 7 739 506 tonnes were mined from 1931 to 1951 (Davies *et al.*, 1962). Farley (1949) gave an "overall average grade" of 2.45% copper, 2.97% zinc, 0.62 g/tonne Au, and 19.9 g/tonne Ag, but it is not clear whether this refers to the Sherritt Gordon deposit as a whole or to the east orebody only. Early ore reserve calculations gave the following grades (Sherritt Gordon Mines Limited, 1930):

	tonnes	Cu%	Zn%	Au g/tonne	Ag g/tonne
East orebody	785 780	2.14	5.78	0.65	26.1
West orebody	2 968 215	2.91	2.76	0.62	32.2
West orebody (low grade area)	1 012 871	1.40	0.80	0.41	42.0
Total tonnes and grade	4 766 866	2.46	2.84	0.58	33.3

During the exploration stage, the deposit was described by Wright (1929, 1931), Bruce (1929, 1933), and by the staff of Sherritt Gordon Mines Limited (1930). A brief note (Derry, 1942) was followed by a more detailed account (Farley, 1948, 1949). The surface geology has been mapped by Bateman (1944) and Goetz (1980). The ore is contained entirely in quartz-rich gneiss. Even where it occurs close to an amphibolite layer on the northeast side of the deposit, the ore is separated from the amphibolite by a narrow layer, one to 3 m thick, of quartz-rich gneiss. From the dip reversal of the east orebody, Bruce (1933) deduced an easterly plunge, whereas the west orebody has a northerly plunge (Farley, 1949). The shape of the orebody is shown by vertical projections of mine workings (Fig. 3).

MINERALOGY AND ZONING

The mineralogy of the ore is simple. The dominant opaque minerals are pyrite, pyrrhotite, chalcopyrite, and sphalerite. Sphalerite coexisting with pyrite and pyrrhotite has a composition of 13 ± 1 mole per cent FeS (Goetz, 1980). Magnetite is rare. A few exsolution blades of cubanite are present in chalcopyrite. The gangue minerals are those of the host rock composed of quartz-rich gneiss with small interlayers of calc-silicate gneiss: quartz, plagioclase, biotite, almandine, hornblende, clinopyroxene. Gahnite is present in a few specimens. There is one occurrence of sapphirine in a cordierite-anthophyllite rock (Froese and Goetz, 1981). This association has also been noted in another metamorphosed sulphide deposit (Raymond *et al.*, 1980).

The distribution of minerals and zoning on a part of the third level of the west orebody is shown in Figure 4. The Zn/Cu ratio increases towards the east, i.e., into the centre of the Sherridon structural basin. Accepting an upward increase of the Zn/Cu ratio in massive sulphide deposits (Sangster and Scott, 1976), this feature is the best indicator of stratigraphic facing in the Sherridon area. Farley (1949) mentioned a relatively sharp hanging wall contact and a gradational footwall contact characterized by disseminated sulphides, but most level plans (e.g., Fig. 4) show disseminated sulphides on both sides of the orebody. There is no zone of stringer ore below the orebody. Cordierite-anthophyllite rocks are present along the mineralized stratigraphic horizon but do not form an alteration zone underlying the orebody. Ore reserve estimates indicate pronounced lateral zoning in the east orebody and weak lateral zoning in the west orebody (Fig. 5).

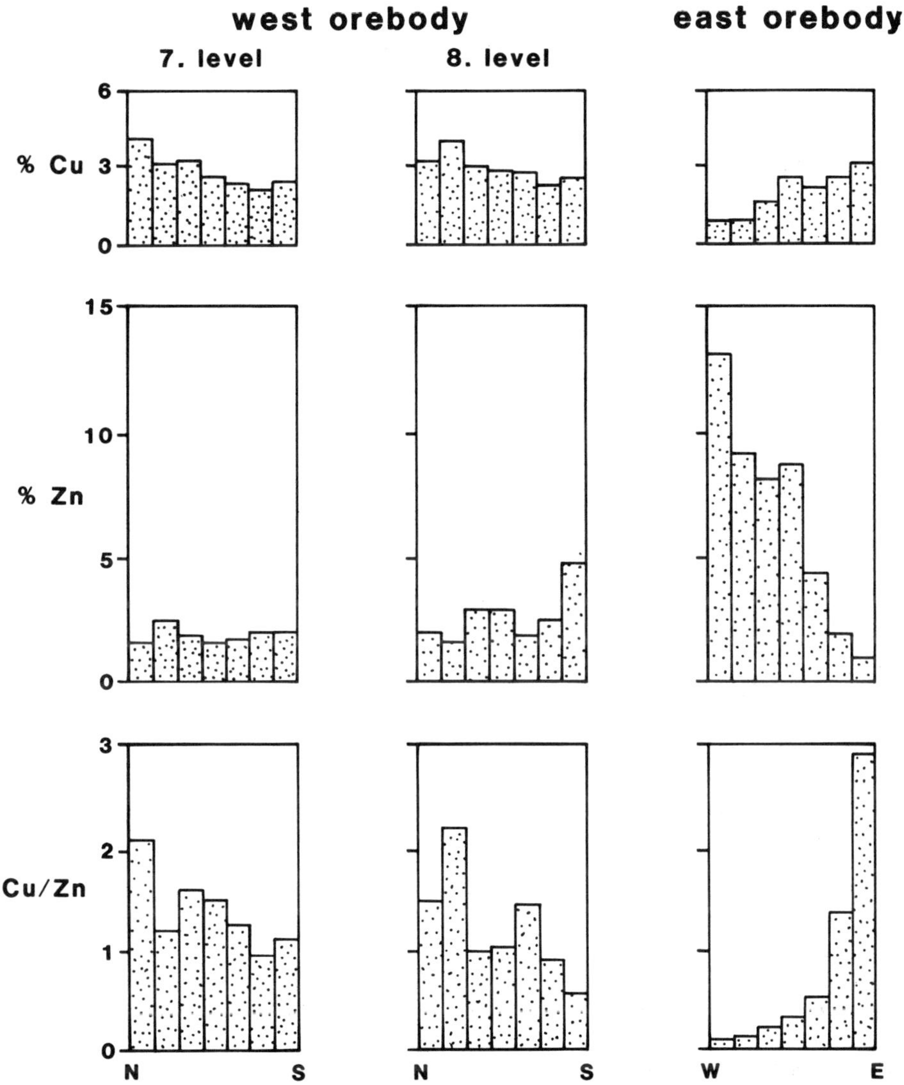

Figure 5. Lateral zoning of the Sherritt Gordon deposit. Compiled from unpublished ore reserve estimates of Sherritt Gordon Mines Limited (from Froese and Goetz, 1981).

ENVIRONMENT OF DEPOSITION

The Sherridon Group, which hosts the Sherritt Gordon deposit has been correlated with the Missi Group in the Flin Flon-Snow Lake belt (Bailes, 1971). At Flin Flon, the Missi sandstone is composed largely of quartz grains and volcanic rock fragments (Mukherjee, 1974; Stauffer, 1974) and may be termed lithic arenite. Froese and Goetz (1981) consequently regarded the quartz-rich gneiss of the Sherridon Group as metamorphosed lithic arenite. This interpretation is consistent with the present composition. Pollock (1964) also suggested that similar rocks in the Duval Lake area, west of Sherridon, were derived from subgreywacke (i.e., lithic arenite). Despite its correlation with the Missi Group, the Sherridon Group displays distinct differences. For example, cross-bedding is common in the Missi Group and fluviatile conglomerates are present in many places, features which suggest an alluvial fan deposit (Mukherjee, 1974). By comparison, the metamorphosed lithic arenite units near the base of Sherridon Group are thin-bedded and contain calc-silicate layers ranging in size from small partings to mappable units. Thus the Sherridon Group probably is a marine deposit and may represent the subaqueous portion of an alluvial fan that spread by eroding the volcanic terrain of the Flin Flon-Snow Lake belt. In the vicinity of Sherridon, the Sherridon Group includes volcanic rocks. In other areas, volcanic rocks are not present in the Sherridon Group. With the exception of an area east of Wekusko Lake (Frarey, 1948; Shanks and Bailes, 1977), this is also true of the Missi Group. Here the Missi Group also includes calc-silicate gneisses (Bailes, 1976; Bailes, pers. commun., 1981).

The high content of Cu and Zn in the Sherritt Gordon deposit and the presence of some volcanic rocks in the nearby Sherridon Group suggest a volcanogenic origin. The large lateral extent of the orebodies and the absence of underlying alteration zones indicate a distal volcanogenic deposit, i.e., one deposited away from the hydrothermal vent (Large, 1977). Following Turner and Gustafson (1978), Goetz (1980) proposed that the sulphides of the Sherritt Gordon deposit were precipated from hot brines moved great distances from their source area. It has been suggested that sulphides could also be transported by sediment gravity flows (Schermerhorn, 1970; Henley and Thornley, 1979; MacGeehan *et al.*, 1981; Walker and Barbour, 1981). Features of clastic sedimentation have been reported from the Iberian pyrite belt (Schermerhorn, 1970, 1978), from some Japanese volcanogenic deposits (Clark, 1971; Lambert and Sato, 1974), from the Norita mine in Quebec (MacGeehan *et al.*, 1981), and from the Buchans deposits (Thurlow and Swanson, 1981). This method of transport could have played a role in the formation of the Sherritt Gordon orebodies.

Cordierite-anthophyllite rocks are associated with many metamorphosed sulphide deposits and have been interpreted as metamorphosed hydrothermally altered rocks (Froese, 1969; Whitmore, 1969). In the Sherridon area, cordierite-anthophyllite rocks are present as lenses along stratigraphic horizons; some occurrences are associated with sulphide mineralization. The stratabound nature of the cordierite-anthophyllite rocks in the Sherridon area was pointed out by Robertson (1953). Recently, similar rocks have been interpreted as metamorphosed volcanogenic chlorite-rich muds laid down on the sea-floor (Schermerhorn, 1978) and as metamorphosed distal deposits of chloritic rocks produced by fumarolic activity (Warren, 1979). Cordierite-anthophyllite rocks require chloritic rocks as precursors and these could be produced by hydrothermal alteration. Chloritization is a common process in hydrothermal alteration which involves the selective dissolution and removal of calcium and sodium, producing a rock depleted in these elements and enriched in aluminum. Thus, mafic rocks could acquire the composition of a quartz-chlorite mixture and felsic rocks could become chloritized by a combination of some aluminum with iron and magnesium from the hydrothermal solution (Large, 1977). The protolith of the cordierite-anthophyllite rocks of the Sherridon area might have been detritus of hydrothermally altered rocks, transported to their present site either as a fine suspension in a brine (Wilkinson, 1976; Goetz, 1980) or as a sediment gravity flow (Middleton and Hampton, 1976).

In the Sherridon area, a favourable environment for the deposition of a distal volcanogenic deposit includes the presence of volcanic rocks in the predominantly sedimentary Sherridon Group and the occurrence of pelitic rocks and cordierite-anthophyllite rocks along the ore-bearing stratigraphic horizon.

ACKNOWLEDGEMENTS

We are very grateful to Sherritt Gordon Mines, Limited, for permission to use unpublished information. Discussions with J.M. Franklin were of great help in writing this paper. A critical review of an earlier manuscript by A.H. Bailes led to various improvements.

REFERENCES

Bailes, A.H., 1971, Preliminary Compilation of the Geology of the Snow Lake - Flin Flon - Sherridon area: Manitoba Mines Branch Geological Paper 1/71.

_______________, 1976, Saw Lake Area: Manitoba Mineral Resources Division Report of Activities 1976, p. 45-50.

_______________, 1980, Geology of the File Lake area: Manitoba Mineral Resources Division Geological Report 78-1.

Bateman, J.D., 1944, Sherritt Gordon Mine Area, Manitoba: Geological Survey of Canada Paper 44-4.

Bateman, J.D. and Harrison, J.M., 1946, Sherridon, Manitoba: Geological Survey of Canada Map 862A.

Bruce, E.L., 1929, The Sherritt Gordon copper-zinc deposit, northern Manitoba: Economic Geology, v. 24, p. 457-469.

_______________, 1933, Mineral Deposits of the Canadian Shield: Toronto, MacMillan Company of Canada Limited.

Clark, L.A., 1971, Volcanogenic ores: Comparison of cupriferous pyrite deposits of Cyprus and Japanese Kuroko deposits: in Takéuchi, Y., ed., Proceedings of the IMA-IAGOD Meetings '70, IAGOD Volume: Society of Mining Geologists of Japan, Special Issue 3, p. 206-215.

Davies, J.F., Bannatyne, B.B., Barry, G.S. and McCabe, H.R., 1962, Geology and Mineral Resources of Manitoba: Manitoba Mines Branch.

Derry, D R., 1942, The Sherritt-Gordon mine: in Newhouse, W.H., ed., Ore Deposits as Related to Structural Features: Princeton University Press, Princeton, N.J., p. 15.

Farley, W.J., 1948, Sherritt Gordon mine: in Structural Geology of Canadian Ore Deposits: Canadian Institute of Mining and Metallurgy, p. 292-295.

_______________, 1949, Geology of the Sherritt Gordon orebody: Canadian Mining and Metallurgical Bulletin, v. 42, p. 25-30.

Frarey, M.J., 1950, Crowduck Bay, Manitoba: Geological Survey of Canada Map 987A.

Froese, E., 1969, Metamorphic Rocks of the Coronation Mine and Surrounding Area: Geological Survey of Canada Paper 68-5, p. 55-77.

Froese, E. and Goetz, P.A., 1981, Geology of the Sherridon Group in the Vicinity of Sherridon, Manitoba: Geological Survey of Canada Paper 80-21.

Goetz, P.A., 1980, Depositional environmental of the Sherridon Group and related mineral deposits near Sherridon, Manitoba: Ph.D. Thesis, Carleton University, Ottawa.

Henley, R.W. and Thornley, P., 1979, Some geothermal aspects of polymetallic massive sulfide formation: Economic Geology, v. 74, p. 1600-1612.

Lambert, I.B. and Sato, T., 1974, The Kuroko and associated ore deposits of Japan: A review of their features and metallogenesis: Economic Geology, v. 69, p. 1215-1236.

Large, R.R., 1977, Chemical evolution and zonation of massive sulfide deposits in volcanic terrains: Economic Geology, v. 72, p. 549-572.

MacGeehan, P.J., MacLean, W.H., and Bonenfant, A.J., 1981, Exploration significance of the emplacement and genesis of massive sulphides in the main zone at the Norita Mine, Matagami, Quebec: Canadian Mining and Metallurgical Bulletin, v. 74, no. 828, p. 59-75.

Middleton, G.V. and Hampton, M.A., 1976, Subaqueous sediment transport and deposition by sediment gravity flows: in Stanley, D.J. and Swift, D.J.P., eds., Marine Sediment Transport and Environmental Management: John Wiley and Sons, New York, p. 197-218.

Mukherjee, A.C., 1974, Some aspects of sedimentology of the Missi Group and its environment of deposition: Canadian Journal of Earth Sciences, v. 11, p. 1018-1019.

Pollock, G.D., 1964, Geology of the Duval Lake Area: Manitoba Mines Branch Publication 61-6.

Raymond, W.H., Leiggi, P.A., and Sheridan, D.M., 1980, Sapphirine in Precambrian Rocks Associated with Stratabound Sulfide Deposits, Custer County, Colorado: United States Geological Survey Bulletin 1513.

Robertson, D.S., 1953, Batty Lake map-area, Manitoba: Geological Survey of Canada Memoir 271.

Sangster, D.F., 1978, Isotopic studies of ore-leads of the circum-Kisseynew volcanic belt of Manitoba and Saskatchewan: Canadian Journal of Earth Sciences, v. 15, p. 1112-1121.

Sangster, D.F. and Scott, S.D., 1976, Precambrian, strata-bound, massive Cu-Zn-Pb sulfide ores of North America: in Wolf, K.H., ed., Handbook of Strata-bound and Stratiform Ore Deposits: Amsterdam, Elsevier Scientific Publishing Company, p. 129-222.

Schermerhorn, L.J.G., 1970, The deposition of volcanics and pyritite in the Iberian pyrite belt: Mineralium Deposita, v. 5, p. 273-279.

_______________, 1978, Epigenetic magnesium metasomatism or syngenetic chloritite metamorphism at Falun and Orijärvi: Transactions of the Institution of Mining and Metallurgy, section B, v. 87, p. 162-167.

Shanks, R.J. and Bailes, A.H., 1977, "Missi Group" rocks, Wekusko Lake area: Manitoba Mineral Resources Division Report of Field Activities 1977, p. 83-87.

Sherritt Gordon Mines Limited (staff), 1930, Proposed mining and milling practice at Sherritt Gordon mine: Canadian Institute of Mining and Metallurgy Bulletin, v. 23, p. 1012-1038.

Stauffer, M.R., 1974, Geology of the Flin Flon area: A new look at the Sunless City: Geoscience Canada, v. 1, no. 3, p. 30-35.

Thurlow, J.G., and Swanson, E.A., 1981, Geology and ore deposits of the Buchans area, central Newfoundland: in Swanson, E.A., Strong, D.F., and Thurlow, J.G., eds., The Buchans Orebodies: Fifty Years of Geology and Mining: Geological Association of Canada Special Paper 22, p. 113-142.

Tuckwell, K., 1979, Stratigraphy and mineral deposits of the Sherridon area: Manitoba Mineral Resources Division Report of Field Activities 1979, p. 42-45.

Turner, J.S. and Gustafson, L.B., 1978, The flow of hot saline solutions from vents in the sea floor – some implications for exhalative massive sulfide and other ore deposits: Economic Geology, v. 73, p. 1083-1100.

Walker, P.N. and Barbour D.M., 1981, Geology of the Buchans ore horizon breccias: in Swanson, E.A., Strong, D.F., and Thurlow, J.G., eds., The Buchans Orebodies: Fifty Years of Geology and Mining: Geological Association of Canada Special Paper 22, p. 161-185.

Warren, R.G., 1979, Sapphirine-bearing rocks with sedimentary and volcanogenic protoliths from the Arunta Block: Nature, v. 278, p. 159-161.

Whitmore, D.R.E., 1969, Geology of the Coronation copper deposit: Geological Survey of Canada Paper 68-5, p. 37-53.

Wilkinson, S.J., 1976, The petrography of the anthophyllite rocks at Sherridon, Manitoba: B.Sc. Thesis, University of Western Ontario, London.

Wright, J.F., 1929, Kississing Lake area, Manitoba: Geological Survey of Canada Summary Report 1928, Part B, p. 73-104.

_______________, 1931, Geology and Mineral Deposits of a Part of Northwest Manitoba: Geological Survey of Canada Summary Report 1930, Part C, p. 1-124.

MS received September 1981;
Accepted for publication September 1981.

Precambrian Sulphide Deposits, H.S. Robinson Memorial Volume, edited by
R.W. Hutchinson, C.D. Spence and J.M. Franklin, Geological Association of Canada
Special Paper 25, 1982

STRUCTURE, STRATIGRAPHIC CONTROLS, AND GENESIS OF THE BALMAT ZINC DEPOSITS, NORTHWEST ADIRONDACKS, NEW YORK

W.F. deLorraine
Geology Department, St. Joe Resources Co., Balmat, New York 13609
D.B. Dill
St. Joe American Corp., 2002 North Forbes Boulevard, Tucson, Arizona 85605

ABSTRACT

Zinc orebodies at Balmat are contained in a sequence of "clean" carbonate strata interlayered with anhydrite and siliceous dolomitic marble belonging to the Grenville Supergroup of Helikian age. Metamorphosed to a high grade and multiply deformed during the Grenvillian Orogeny, the orebodies are massive, markedly elongate in the downplunge direction, and generally conform to bedding. Transgressive relationships are common and can be related to superposed metamorphic structures such as shears, faults, and gash fractures; elongation in the down-plunge direction coincides with the plunges of folds with which many of the orebodies are intimately associated. Consequently, the orebodies are considered to have formed syngenetically with deposition of the host rocks, although in present form they have many epigenetic, high temperature characteristics imparted during metamorphism. Metamorphic mobilization of sulphides is locally significant and movement is inferred to have been on the order of a few hundred metres with the dominant processes being plastic flow under differential stress, and shear displacement.

Sulphur and lead isotope data are compatible with either syngenetic or epigenetic depositional models. A syngenetic hypothesis involving bacterial reduction of seawater sulphates is incompatible with the sulphur isotope data. A genetic model involving metalliferous thermal brines debouching on the sea floor is compatible with the massive and conformable character of the ores. In addition, it is compatible with the minor element geochemistry and the sulphur and lead isotope data. A Mississippi Valley-type epigenetic model satisfies the sulphur and lead isotope data somewhat less acceptably.

RÉSUMÉ

Les gisements de zinc de Balmat sont enclos dans une séquence de strates carbonatées pures entrelitées d'anhydrite et de marbre dolomitique siliceux appartenant au supergroupe de Grenville d'âge hélikien. Ayant subi un métamorphisme élévé et des déformations multiples pendant l'orogenèse grenvillienne, les gisements sont massifs, allongés de façon marquée en direction plongeante, et sont généralement concordants avec le litage. Des relations transgressives sont fréquentes et peuvent être rattachées à des structures métamorphiques surimposées telles que zones de cisaillement, failles et fractures ouvertes; l'allongement dans la direction de plongement coincide avec le plongement des plis auxquels plusieurs des gisements sont intimement associés. En conséquence les gisements sont considérés comme étant syngénétiques bien que dans leur forme actuelle ils aient bien des caractéristiques épigénétiques de haute température, acquises pendant le métamorphisme. La mobilisation métamorphique des sulfures est localement importante et le mouvement est supposé avoir été de l'ordre de quelques centaines de mètres et s'être fait essentiellement par écoulement plastique sous tension différentielle et cisaillement.

Les valeurs isotopiques du soufre et du plomb sont compatibles aussi bien avec un modèle syngénétiste qu'épigénétiste. L'hypothèse syngénétiste de réduction bactérienne des sulfates marins est cependant incompatible avec les valeurs isotopiques du soufre, alors que le modèle de sources thermales salées et métalliques débouchant sur le fond marin est en accord aussi bien avec le caractère massif et conforme des gisements qu'avec la géochimie des éléments mineurs et les valeurs isotopiques du soufre et du plomb. Le modèle épignétique du type Mississippi Valley, de son côté, satisfait les données isotopiques du soufre et du plomb, mais de façon moins acceptable que dans le modèle précédent.

INTRODUCTION

The Balmat zinc orebodies, 10 km southeast of the village of Gouverneur in St. Lawrence County, New York, are part of the Balmat-Edwards zinc district which extends for 16 km from Balmat northeasterly to Edwards (Fig. 1). As early as 1838, occurrences of zinc and lead were known in the vicinity of Balmat, but it was not until 1927 that major zinc-lead mineralization was discovered by diamond drilling. The Edwards mine has been worked continuously since 1915 and the Hyatt mine, located approximately midway between Balmat and Edwards, has been worked sporadically since 1916. Production from the Balmat mine to 1978 has been 23.1 million tons of ore at 9% Zn and 0.5% Pb. At the time of writing, the Balmat mine alone accounted for 14% of the zinc then mined in the United States. The Balmat area is also an important source of tremolite-talc which is mined by the Gouverneur Talc Company.

Despite the intensive efforts of numerous investigators, the genesis of the orebodies in the Balmat-Edwards district remains an enigma. Much of the confusion arises from the fact that the orebodies have undergone polyphase deformation and high grade metamorphism during the Grenvillian Orogeny. Primary textures, other than bedding and large scale facies relationships in the enclosing rocks, have not survived the metamorphism, thus complicating interpretations of the depositional environment of the host rocks and of ore genesis. Through consideration of the stratigraphic, structural, and tectonic settings of ore and host rocks, as well as geochemical evidence, we hope in time to be able to decipher the latter.

The structural geology of the Fowler orebody of the new #4 mine (Fig. 1) is of special interest because the structural relations there may be applied to the Balmat district as a whole, and in contrast to most of the other orebodies in the district, a large portion of the orebody transgresses bedding as well as several stratigraphic units. The approach taken below is therefore to outline the structural geology of the

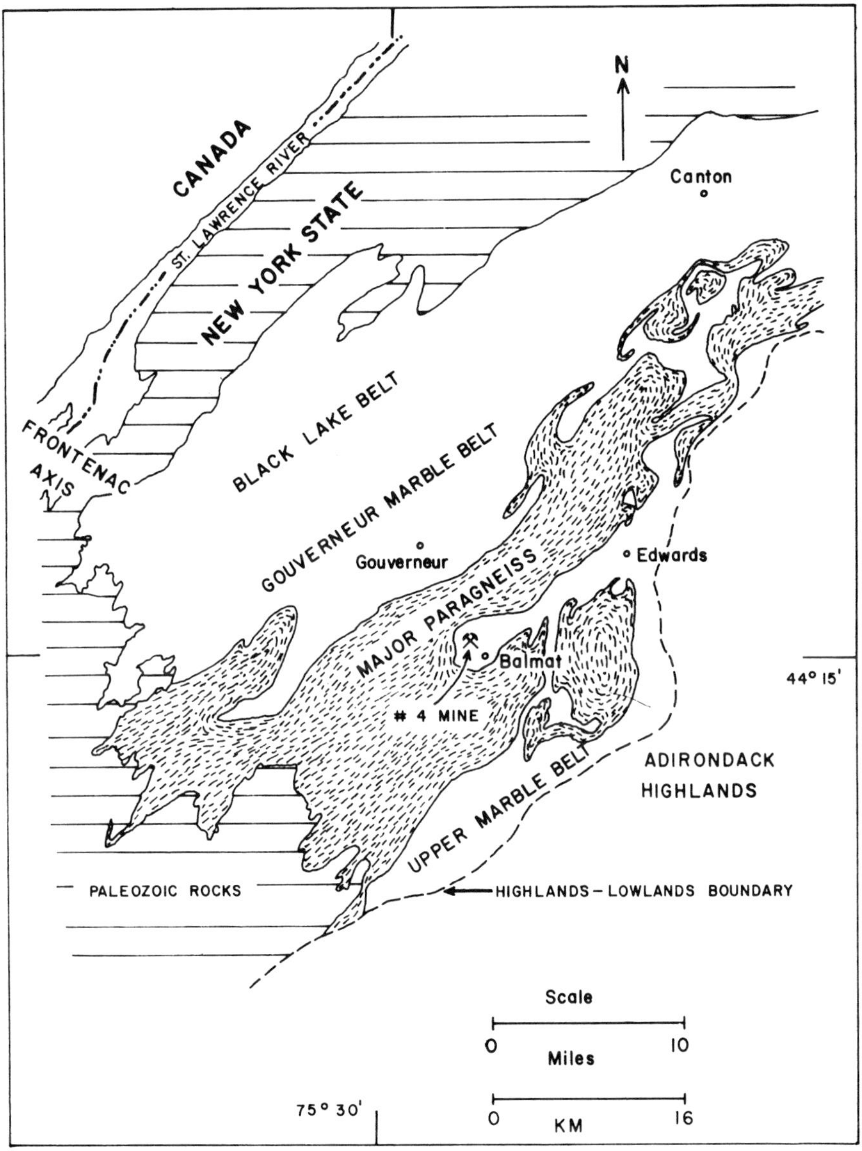

Figure 1. Index map of the northwest Adirondack Lowlands, northern New York State.

Fowler orebody, to apply these structural relationships to the rest of the Balmat district, to present geochemical evidence pertinent to the genesis of the zinc deposits, and finally to consider several possible genetic models for the origin of the mineralization.

REGIONAL GEOLOGICAL SETTING

The zinc orebodies at Balmat occur in Grenville marbles in the northwest Adirondacks of northern New York State. The Adirondacks represent an extension of the Grenville Province of the Canadian Shield into New York via the Frontenac Axis, near the eastern end of Lake Ontario (Fig. 1). Rocks of the Adirondacks have been multiply deformed and metamorphosed to a high grade during the Grenvillian Orogeny, dated at about 1.1 Ga (Silver, 1964; Bickford and Turner, 1970; Hills and Gast, 1964; Wynne-Edwards, 1972). The Adirondacks are divided into two major provinces separated by the Highlands-Lowlands boundary (Fig. 1): the Adirondack Highlands to the east, which consist predominantly of metamorphosed igneous plutonic rocks, notably granite, charnokite, mangerite, syenite, and anorthosite, and the northwest Adirondack Lowlands which consist predominantly of metamorphosed carbonate-rich sedimentary rocks, quartzite, aluminous paragneiss, and volcanic rocks. An unsolved problem in Adirondack geology is whether the Adirondack Highlands represent a pre-existing igneous basement over which a carbonate-rich metasedimentary section, represented, in part, by the northwest Adirondacks, was deposited, or whether they represent supracrustal rocks and intrusions overlying an unexposed basement. The grade of metamorphism ranges from amphibolite facies in the vicinity of Gouverneur to granulite facies in the Adirondack Highlands. Temperatures of metamorphism increased gradually from 575°C near Gouverneur to 800°C in the Adirondack Highlands (Bohlen, 1977; Essene *et al.*, 1977). At Balmat, temperatures of about 625°C are inferred from calcite/dolomite geothermometry (Brown *et al.*, 1977) and from sulfur isotope fractionations between coexisting anhydrite-pyrite pairs (Whelan, 1978). Pressures of metamorphism as inferred from application of the sphalerite geobarometer are 5.5 to 6.5 kbar (Essene *et al.*, 1977; Brown *et al.*, 1977; DeWitt, 1976; Scott, 1976).

LOCAL GEOLOGICAL SETTING

The Balmat-Edwards section of the Upper Marble Belt in the northwest Adirondack Lowlands (Fig. 1) hosts massive, stratabound, coarse-grained zinc orebodies, contained within a repetitive sequence of dolomitic marble, calc-silicate-bearing dolomitic and calcitic marble, and anhydrite. The carbonate section at Balmat consists of 16 distinct stratigraphic units (Brown and Engel, 1956; Lea and Dill, 1968) having a total thickness of about 1000 m (Fig. 2). Units 1, 3, 5, 7, 9, and 12 consist of relatively pure, light gray to white, dolomitic marble which is intercalated with well layered, white and grey quartz-diopside rock and dolomite represented by Units 4, 6, 8, and 11. In places these "silicated" units consist predominantly of quartz and diopside so that they are more properly referred to as layered, marble-bearing calc-silicate rock. The remaining stratigraphic units are "marker beds" which are of great value in unravelling structural and stratigraphic relationships. These include Unit 2, a rusty weathering, felsic pyritic schist; Unit 7, a dark grey, graphitic dolomitic marble which has a fetid odor when broken; the "10 Bed", a calc-silicate unit consisting of dense, fine-grained, pea-green, serpentinous-talcose diopside with minor interlayered tremolitic anhydrite and an impersistent biotite schist stratum; and Unit 11A, a purple to pink bedded anhydrite with subordinate amounts of diopside, tremolite, phlogopite, quartz, calcite and dolomite. Unit 13, a distinctive cream-colored, talc-tremolite-anthophyllite schist, is remarkably persistent and perhaps the most useful in deducing structural relationships. Lastly, Unit 15 consists of a light brown

	SERIES	FORMATION	LITHOLOGY	TYPICAL THICKNESS (m)	ORE ZONES
PALEOZOIC	Upper Cambrian	Potsdam -CP	Sandstone, siliceous breccia	0-120	
UNCONFORMITY					
MIDDLE PROTEROZOIC (HELIKIAN)	Grenville	qbg,gg	Quartz – biotite – oligoclase paragneiss; granitic gneiss	750	
		1	Dolomitic marble	120	
		2	Pyritic schist: quartz – mica – feldspar, pyrite	30	
		3	Dolomitic marble	120	
		4	Diopside, quartzite, dolomitic marble	90	
		5	Dolomitic marble, minor diopside	50	
		6	Siliceous diopside, quartzite, dolomitic marble, locally anhydrite	165	Gleason, Upper Gleason, West Gleason, Loomis C
		7	Fetid graphitic marble	35	Scattered mineralization
		8	Diopside, quartzite, dolomitic marble	40	Loomis, Upper Fowler, American, Mud Pond
		9	Dolomitic marble	20	Loomis, Upper Fowler, American
		10	Talcose diopside, biotite, anhydrite	15	
		11,11-A	Diopside, quartzite, calcitic and dolomitic marble, minor anhydrite (II) Same, with over 50% bedded anhydrite (II-A)	90	Fowler, Sylvia Lake, Number One
		12	Dolomitic marble, minor diopside, minor anhydrite	45	Fowler, Sylvia Lake, Wight
		13	Talc – Tremolite schist, anhydrite	20	
		14	Quartz – diopside banded rock, minor dolomite Calcitic marble, quartz augen Serpentinized dolomitic and calcitic marble, diopside, siliceous marble	32 40 35	Hanging Wall Main, Streeter
		15	Phlogopitic silicated calcitic marble	9	
		16	Median gneiss: quartz – biotite – feldspar	60	

Figure 2. Stratigraphic column through the Balmat area showing relative stratigraphic positions of orebodies.

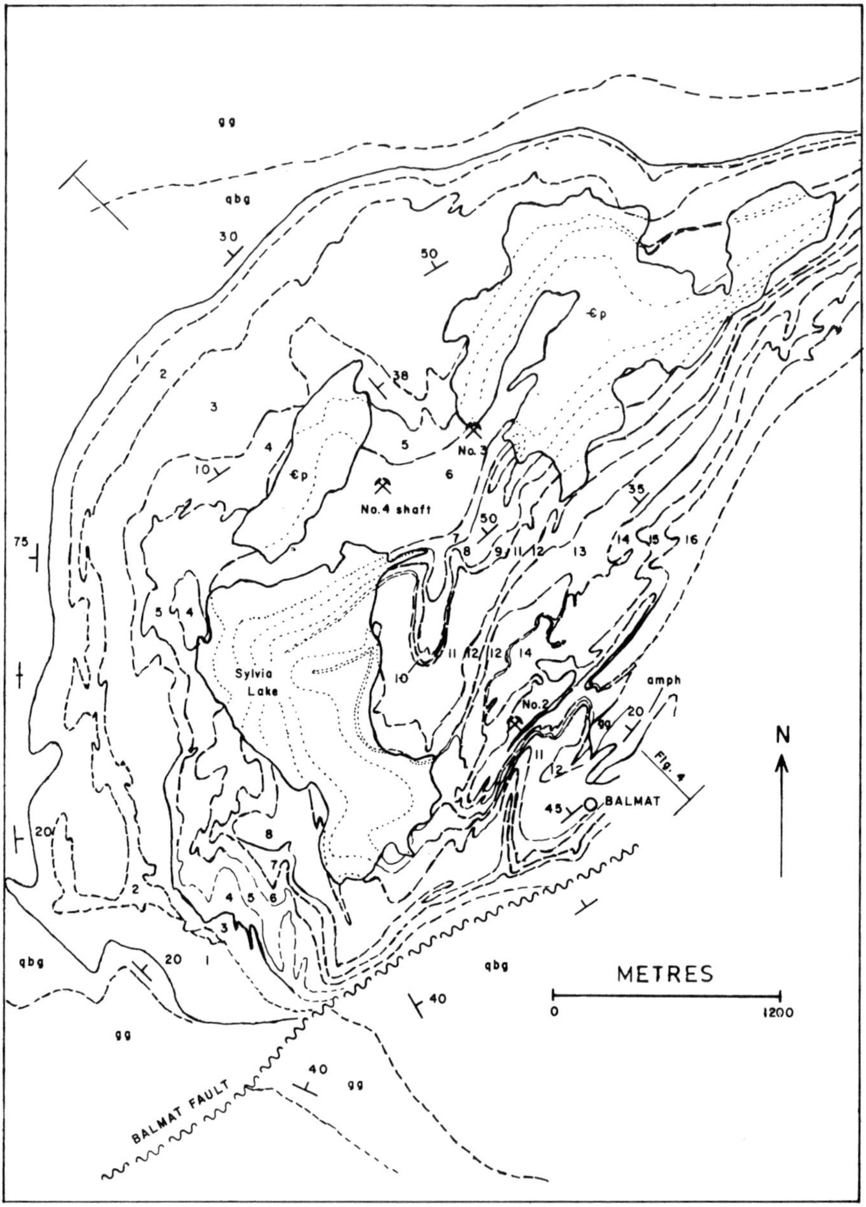

Figure 3. Surface geologic map of the Balmat area. Numbers without strike and dip symbols refer to stratigraphic units as numbered in Figure 2. Dotted lines represent inferred contacts beneath Potsdam Sandstone cover or beneath Sylvia Lake.

phlogopite-calcite marble with minor diopside. The "median gneiss", Unit 16, is thought to overlie the carbonate section in the Balmat area. Direct correlations of the Balmat units with those at Edwards are not possible but the marbles are continuous to Edwards and similar types of diopside-dolomite marble, quartzite, and anhydrite occur in the Edwards mine area. Orebodies in the Balmat mine area occur within the stratigraphic interval represented by Units 6 to 14 which has a total thickness of about 540 m. Anhydrite is restricted within the same stratigraphic interval and occurs as thin beds and impersistent lenses. Likely protoliths for the metasedimentary rocks include limestone and dolomite for the pure marbles (Campbell, 1976), cherty or perhaps sandy dolomite for the layered calc-silicate-bearing dolomitic marbles (Campbell, 1976), sedimentary, perhaps evaporitic anhydrite or gypsum for the anhydrite subunits (Brown and Engel, 1956; Brown, 1973; Whelan, 1974; Solomon, 1963), and a magnesite-bearing, siliceous evaporite for the talc-tremolite-anthopyllite schist, Unit 13. Although there is no direct evidence of tops of beds, most workers consider the marbles to stratigraphically, as well as structurally, overlie the "major paragneiss" (Engel and Engel, 1953; Brown and Engel, 1956; Lea and Dill, 1968; Brocoum, 1971; deLorraine, 1979). It has been suggested that the magnesian composition of Unit 13 derives from the precipitation of magnesian salts during the late stages of evaporite sedimentation (deLorraine, 1979), and accordingly that Unit 13 occurs in the upper part of the evaporite sequence. This interpretation also implies that the carbonates are younger than the "major paragneiss" (Fig. 2).

The generalized structure of the Balmat district is that of an isoclinal synform with the "major paragneiss" enveloping the marbles (Fig. 3). The dips are generally to the northwest so that the synform is overturned to the southeast (Fig. 4).

OREBODIES

Because the stratigraphic and structural relations of the orebodies at Balmat are more clearly understood than those at Edwards and Hyatt, the following discussion emphasizes the orebodies in the Balmat district. Orebodies are coarse-grained, massive, and stratabound, and fit well with Stanton's (1972) classification of massive stratiform sulphides of sedimentary association. Ore occurs as tabular, pod-like, or lenticular bodies of modest cross-sectional area (1 to 15 m thick by 15 to 250 m strike length). Orebodies are extremely elongate in the down-plunge direction of the fold hinges with which they are intimately associated, and plunge lengths may exceed 1250 m. Contacts between sulphides and host rock are sharp, with little disseminated sulphide in the wall rock, and wall rock alteration is virtually absent. The only manifestation of wall rock alteration is at the contacts between dark grey, graphitic dolomite and sulphides where thin, white bleached zones may be present in the dolomite. The ore mineralogy is simple, comprising chocolate-brown sphalerite with various amounts of pyrite, minor galena, and very minor chalcopyrite and pyrrhotite. Traces of arsenopyrite are known from study of polished ore samples.

Gangue minerals are those of the adjacent wall rocks with few exceptions. Quartz, calcite, dolomite, diopside, tremolite, and anhydrite are common, occurring as angular to rounded fragments or chaotic wisps and discontinuous layers dispersed in massive sulphides. Quartz and diopside inclusions are generally rounded, as if they had been abraded and tectonically milled while the sulphides were still in a semi-

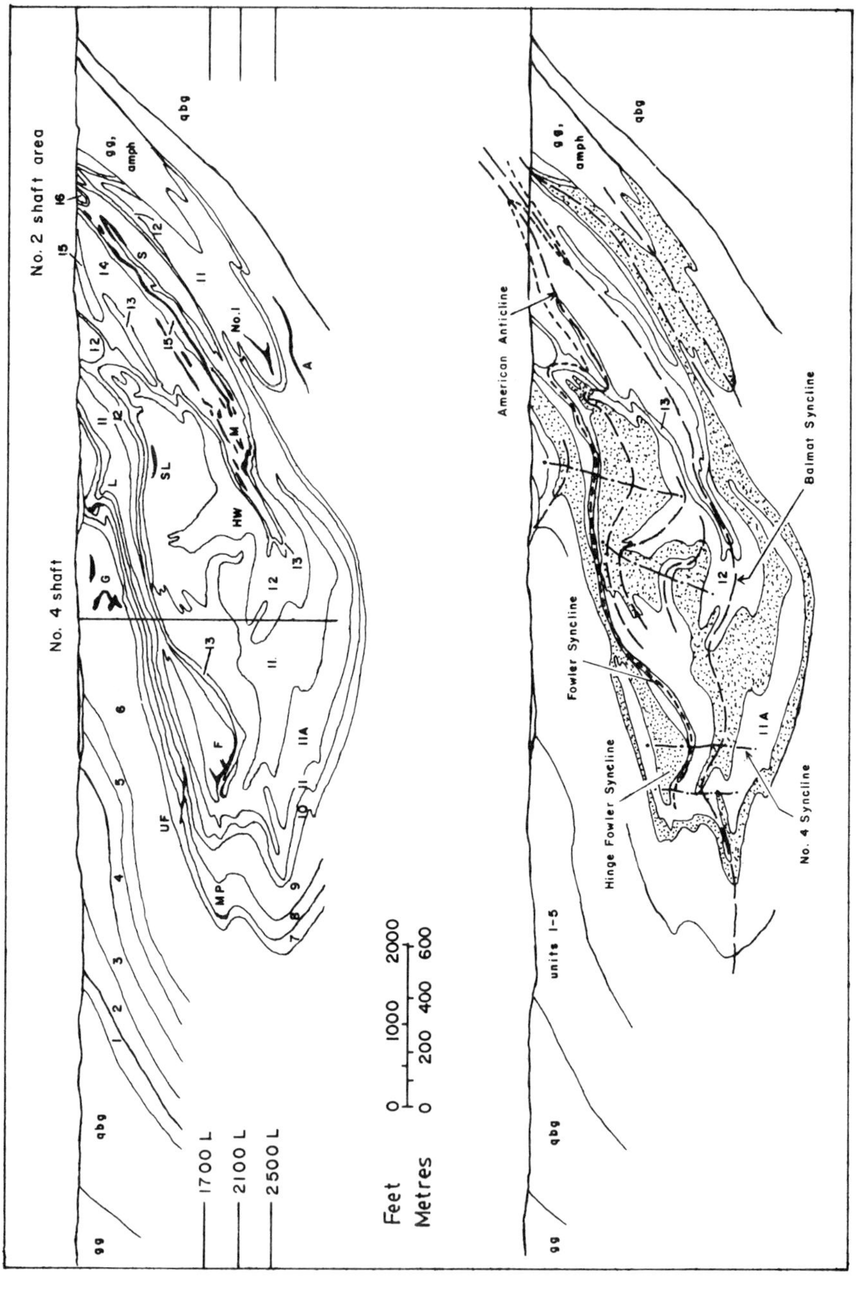

No. 2 shaft area
No. 4 shaft
qbg
gg
gg, amph
1700 L
2100 L
2500 L
Feet
Metres
0 1000 2000
0 200 400 600
American Anticline
Balmat Syncline
Fowler Syncline
Hinge Fowler Syncline
No. 4 Syncline
units 1-5

Figure 4

plastic state. The resulting texture resembles the "Durchbewegung" texture described by Vokes (1969). Grey, glassy, massive quartz is perhaps the most important gangue mineral. It typically contains abundant minute pyrite euhedra and generally minor amounts of fine-grained, globular sphalerite. In places, the quartz displays internal layering defined by concentrations of disseminated pyrite or sphalerite grains. This type of quartz is associated almost exclusively with massive sulphides and differs from that found interlayered with diopside in the host rocks.

Underground mapping has shown that the metasedimentary rocks and sulphides were involved in four phases of deformation (deLorraine, 1979) resulting in complex geometry. Localized mobilization of the sulphides into favorable structural sites, such as fold hinges and faults, occurred during the earlier two phases of deformation.

GEOLOGY OF THE FOWLER OREBODY

The Fowler orebody is localized within the axial region of the Fowler syncline (Figs. 4a and 4b), a major isoclinal fold which plunges N5° to 15°E at about 15 degrees. The mineable portion of the orebody extends down-plunge for over 1100 m. Up plunge, the orebody is "blind" as the tenor of the mineralization diminishes concurrently with an apparent structural pinching of the Fowler syncline. Fowler mineralization is associated with Units 11, 11a, 12, and 13 in the hinge of the fold. There is no surface exposure of the orebody, partly because of the structural pinching and because the surface projection of the hinge lies beneath Sylvia Lake (Fig. 3). On the basis of mineralogical composition and structural position in the Fowler syncline, the Fowler orebody has been subdivided into five different ore zones for ease of discussion. These are shown in Figures 5 to 8 and include: 1) Upper Limb ore zone, 2) Horsetail ore zone, 3) Cross-cutting ore zone, 4) Lower Limb ore zone, and 5) Upper Contact ore zone.

The Upper Limb, Cross-Cutting, Lower Limb and Upper Contact ore zones are compositionally similar, consisting of sphalerite with various amounts of pyrite and minor galena. The most abundant and significant gangue mineral is grey, glassy, quartz which is intimately associated with the sulphides. In places where brecciation of wall rocks in Unit 11 is pronounced, quartz-sulphide veins extend outward short distances from the main ore zones, cementing the breccias. The Horsetail ore zone is compositionally distinctive and consists almost entirely of sphalerite with very minor galena, which tends to segregate at tips, and very minor pyrite and quartz. The Upper Limb and Horsetail ore zones are massive, sheetlike, and conform to bedding in Unit 11, tapering out gradually up dip. There is no disseminated mineralization extending beyond the tips of either of these ore zones. Down dip, both ore zones merge with the

---

Figure 4. Vertical NW-SE geologic and structure sections through No. 4 shaft. Numbers refer to stratigraphic units as numbered in Figure 2. Letters in 4a refer to orebodies as follows: G, Gleason, L, Loomis; UF, Upper Fowler; MP, Mud Pond; F, Fowler; HW, Hanging Wall; M, Main; S, Streeter; No. 1, No. 1 orebody; A, American; SL, Sylvia Lake. Stratigraphic units in 4b generalized to simplify structural relations; Unit 11 shown in stippled pattern. Short dashes – axial trace Fowler syncline; long dashes — axial traces of second phase folds; dot-dashed lines -·- axial traces of third phase folds.

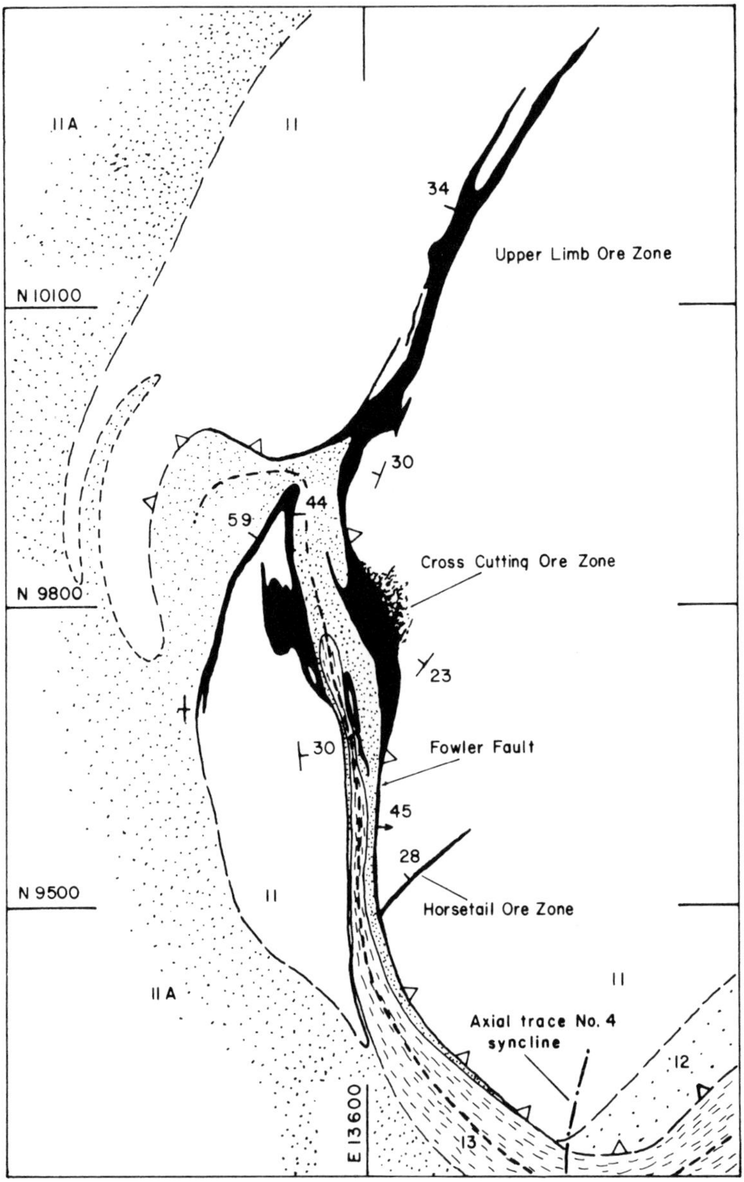

Figure 5. Geologic map near the 1700 level, Fowler orebody. Fine stippled pattern – anhydrite; coarse stippled pattern – Unit 12; short dashes pattern, Unit 13; black – massive sulphides; white – Unit 11. Open triangles represent trace of Fowler fault, short heavy dashed line represents axial trace of Fowler syncline; dot-long dashed line shows axial trace of third phase No. 4 syncline. Mine coordinate line E 13600 oriented N – S, coordinate lines N 9500, N 9800, and N 10100 are spaced 95 metres apart.

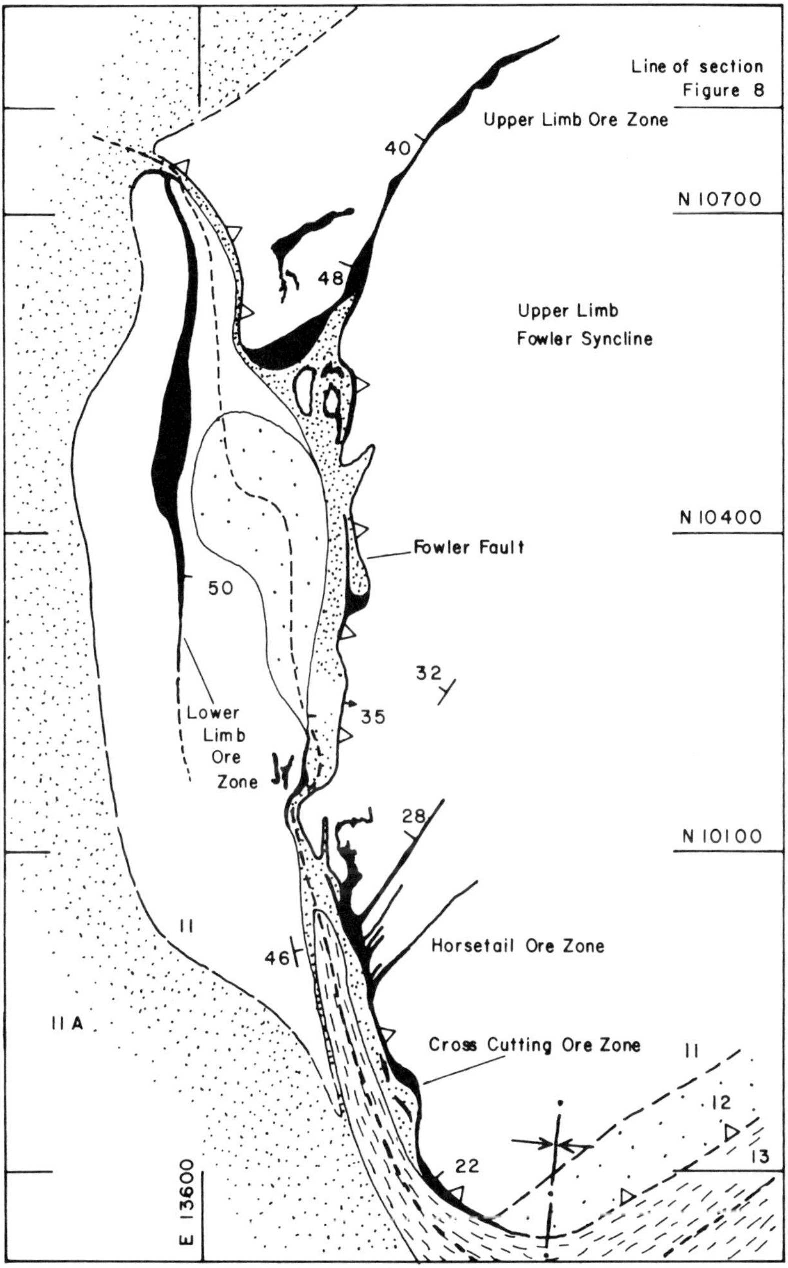

Figure 6. Geologic map of the 1900 level. Fowler orebody. See caption for Figure 5 for explanation of geologic symbols. Coordinate lines N 10100 and N 10400 spaced 95 metres apart.

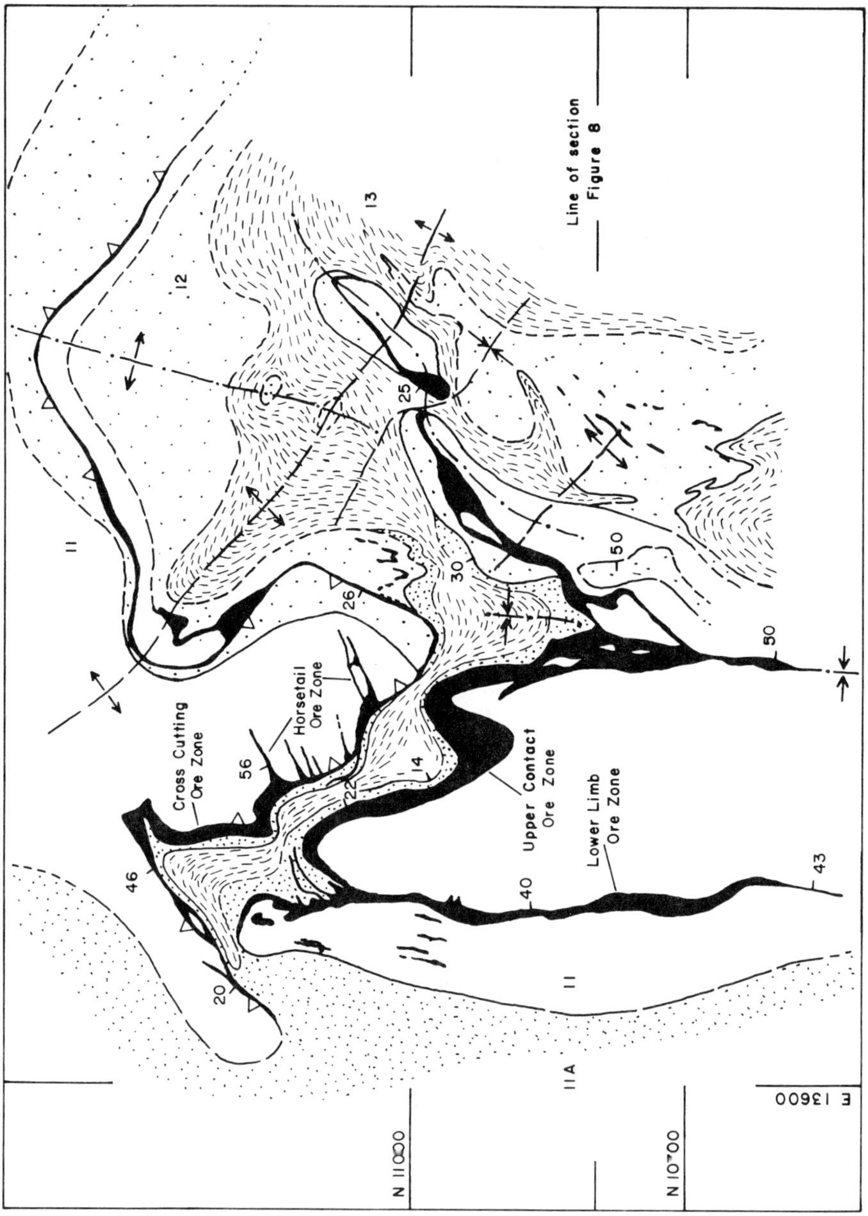

Figure 7. Geologic map of the 2100 level, Fowler orebody. See caption for Figure 5 for explanation of geologic symbols. Long dashed lines represent axial traces of fourth phase folds. Coordinate lines N 10700 and N 11000 spaced 95 metres apart.

Cross-Cutting ore zone. The Cross-Cutting ore zone transgresses bedding in Unit 11 at high angles along a major fault referred to as the Fowler fault (Figs. 5, 6, 8). Down the dip of the fault to the east, the Cross-Cutting ore zone transects the contact between Units 11 and 12. Farther on, however, it becomes parallel to the contact between Units 12 and 13, and to the foliation within Unit 13, so that it loses its identity (Fig. 8).

STRUCTURE

Based on minor and major structural analysis, the Fowler syncline is interpreted as a major, first phase recumbent fold and the Fowler fault is believed to represent a first phase thrust fault which severed the limbs of the fold (Fig. 8). The Upper Contact fault is a similar fault along which mineralization occurs. Cataclastically foliated anhydrite, probably derived from Unit 11a, occupies the complex fault zone between the severed limbs of the Fowler syncline, bounded on the hanging wall side by the Fowler fault and on the footwall by the Upper Contact fault. Tectonic slices of Units 12 and 13, as well as fault breccia blocks from Unit 11, are locally present in this zone (Figs. 6, 7, 8). Massive sulphides within the fault zone and lying along the Fowler fault are referred to as the Cross-Cutting ore zone. Drag folds indicate that the apparent direction of tectonic transport during first phase folding and faulting was "east over west", or counterclockwise as viewed to the north.

Minor first phase folds in the orebody are rare but are best developed in Unit 13. These folds are distinguishable because they isoclinally deform bedding and because the prominent penetrative schistosity in Unit 13 is parallel to their axial surfaces. In most exposures, first phase fold hinges are not evident. The prominent schistosity that is subparallel or parallel to bedding on the fold limbs is nevertheless considered diagnostic of this early phase of folding on the basis of several outcrops on which these schistosity-bedding relationships are clear.

Second, third, and fourth phase folds have been superimposed on the early folds and thrust faults, resulting in the complex geometrical configuration of the Fowler orebody. No major second phase folds are known in the Fowler orebody but numerous minor folds of this phase are present. Second phase minor folds are isoclinal and similar in style. These folds deform the prominent first phase schistosity in Unit 13, plicating and crinkling it in the hinges so that a secondary crenulation foliation develops locally. Second phase minor folds also deform the early-phase cataclastic foliation in the anhydrite within the Cross-Cutting fault zone, as well as the Fowler and Upper Contact fault surfaces. Rotation senses of the minor folds are consistently clockwise viewed to the north. The absence of major folds in the Fowler orebody and consistent clockwise rotations of the minor folds suggest that a major second phase antiform lies farther to the east; Figure 4b shows that the Fowler orebody lies entirely on the northwest limb of the American anitcline, a major second phase isoclinal fold.

Early folds and thrust faults, the prominent axial plane schistosity in Unit 13, and axial surfaces of second phase folds are all gently refolded by NNE-trending third phase folds. These folds are generally open to tight and have steeply dipping axial surfaces. A synform of this phase gently refolds the Fowler fault and is referred to as the No. 4 syncline (Figs. 5, 6, 7, and 8). The amplitude of the No. 4 syncline increases markedly down plunge, probably due to the progressive thickening of Unit 11 in the

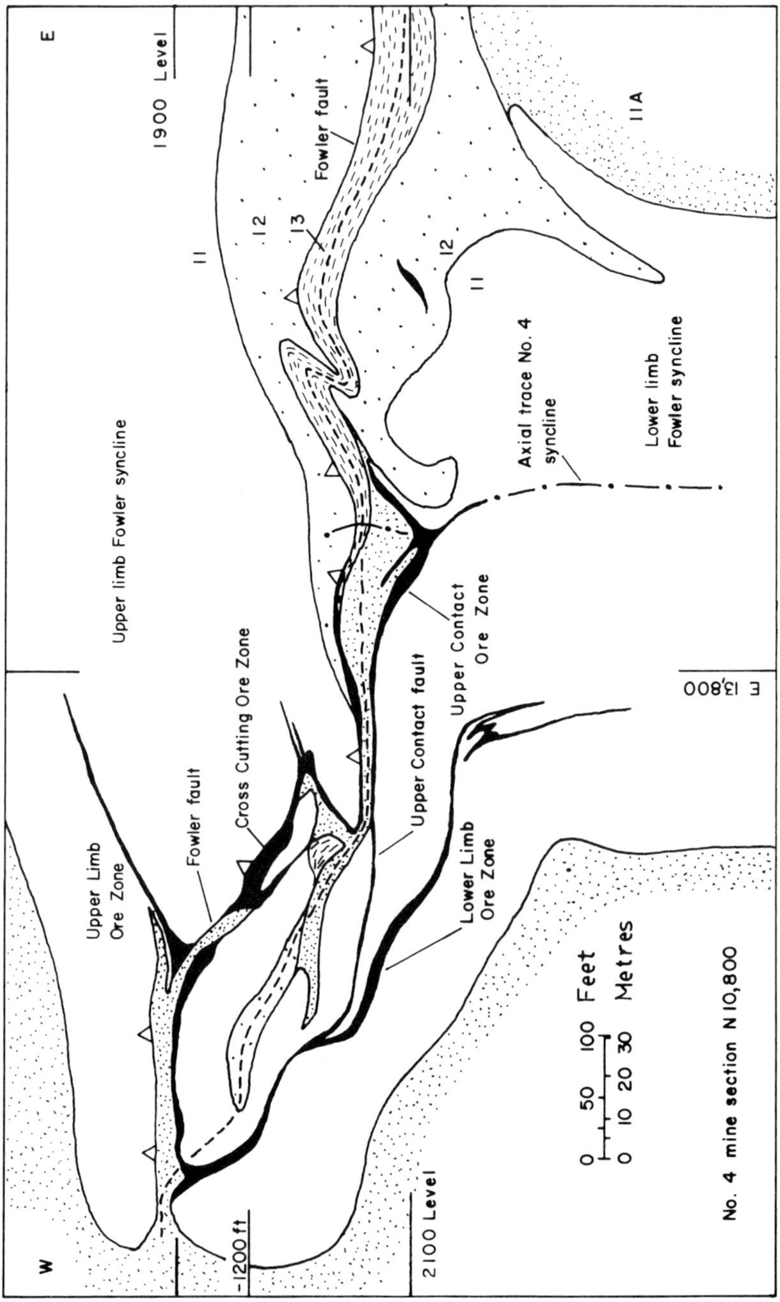

Figure 8. Vertical east-west geologic section through Fowler Orebody, along mine coordinate N 10800, looking north.

lower limb of the early fold in that direction. Although the axial traces of second phase folds normally trend northeast, parallel to the prominent NE regional grain, they are in places refolded by the third phase folds so that their general northeast trend is not constant. Consequently, differences in style or orientation between second and third phase folds are not always reliable distinguishing criteria. Wherever possible, obvious outcrop-scale refold relationships or map-scale interference patterns are employed to define the structural relationships. Rough co-axiality of first, second and third phase folds combined with low angles of plunge results in a modified "hook" map pattern for the Fowler syncline, with the hook being complicated owing to the separation of the limbs of the Fowler syncline by the Fowler fault.

Axial surfaces of the open fourth phase folds trend northwesterly at high angles to the earlier folds and, like the third folds, their axial surfaces are generally upright. Where fourth folds are prominent, as on the 2100 level of the Fowler orebody (Fig. 7), third and fourth phase folds interfere to form modified domes and basins.

Local structural relationships of the Fowler orebody may be extended districtwide. Figure 4b shows the axial trace of the Fowler syncline which may be followed from the hinge area southeastward (up dip) toward surface. Here it wraps around several second phase isoclinal anticlines and synclines, most prominent of which is the Balmat syncline. Both limbs of the Fowler syncline and the Fowler orebody lie on the northwest limb of the American anticline, the complement to the Balmat syncline. Axial surfaces of these second phase folds trend northeast and dip moderately to the northwest. The second phase major isoclines in the No. 2 mine area are reclined folds which plunge NNW at about 30°. Some confusion in the literature exists because the plunge directions are athwart the northeasterly trending axial surfaces, causing previous workers to refer to them as cross-folds (Brown and Engel, 1956; Lea and Dill, 1968). The Main, Streeter, Hanging Wall and No. 1 orebodies of the No. 2 mine area (Fig. 4a) occur in the hinges of major second phase folds which control the trends and plunges of the orebodies. In the No. 3 mine area, the Gleason and Loomis orebodies are localized in the hinges of second generation folds which trend northeasterly. The Upper Fowler orebody occurs in a shear zone which crosses Units 8 and 9 and which extends both to the west and east into Units 7 and 10 respectively. Its emplacement here may be related to shearing contemporaneous with the first phase of folding. The shear plane itself is refolded in places by minor second phase isoclinal folds.

Third and fourth phase folds are relatively unimportant in an economic sense because they exert no signficant control on ore localization or grade. On a local scale, however, reversals in dip or plunge of major structures do occur due to the effect of these deformations. Gentle third phase folds in second phase axial surfaces can be seen in Figure 4b; rotation during third phase folding is responsible for the observed local easterly dips.

Other recent workers in the northwest Adirondack Lowlands have similarly interpreted the regional structural geology and have recognized large scale, early isoclinal folds formed at a high metamorphic grade, followed by later folds which have mutually interfered to form domes and basins (Foose, 1974; Foose and Carl, 1977; Foose and Brown, 1976). Underground mapping and structural analysis in the Balmat mine area generally confirm these interpretations.

It is certain that the sulphide ores were involved in the second, third and fourth phases of deformation because the sulphides in the Cross-Cutting and Lower Limb ore zones of the Fowler orebody are refolded by the later folds. Dill *et al.* (1978), Dill and deLorraine (1978), Johnson and deLorraine (1979), and deLorraine (1979) all believed that the sulphides were also present before the first phase of folding as conformable, massive layers. deLorraine (1979) considered the possibility of synmetamorphic replacement of first phase folds due to the movement of ore solutions along faults and shears. Synmetamorphic migration of such ore solutions with ensuing replacement of favorable structures is thought to be unlikely for various reasons, among them, the presumed impermeability of the rocks under pressures of 5.5 Kbar and temperatures of 625°C imposed under catazonal metamorphism. Correspondingly, the sulphides are thought to have behaved as rocks during metamorphism, and not as ions or metallic complexes in a heated metamorphic fluid. Another factor is the complete lack of alteration effects in rocks immediately adjacent to the orebodies.

Although most of the orebodies in the Balmat district are intimately associated with fold hinges or thrust faults and shears which developed contemporaneously with the first and second phases of folding, certain features suggest that the host structures were superimposed on previously existing deposits, with perhaps limited mobilization of sulphides to dilatant structural sites during folding.

A possible example of folding of a pre-existing tabular sulphide layer is the Streeter orebody at No. 2 mine (Fig. 4a). Streeter ore occurs as tabular layers on the southeast limb of the Balmat syncline (Fig. 4b). Although some of the ore occurs as thickened conformable lenses in the hinge of the Balmat syncline, much of the ore was not plastically mobilized to the hinge, but remained on the limb of the fold. The western edge of the Streeter mineralization has a relatively constant NNW trend whereas the edge of the mineralization on the eastern side is quite varied in trend. The western boundary coincides with the hinge of the Balmat syncline but the eastern edge varies in position within the lower limb of the fold. The irregular trend of the eastern edge is thought to be a function of the original ore distribution, with the superimposed fold hinge exerting the dominant ore control to the west.

The Upper Gleason orebody (Fig. 4a) is a conformable, lenticular, massive sulphide body without any obvious metamorphic and/or structural control, suggesting that it too is a massive lenticular, deposit which has been unaffected by the superposition of the later structures.

The process responsible for the superposition of fold hinges or other metamorphic structures on pre-existing deposits is visualized as one in which the shapes of incipient folds would be affected by the presence of anomalous heavy, massive sulphide lenses and layers within the stratigraphic section. Local fold anomalies thus induced at key points (sites of many of the orebodies) in the mechanically anisotropic rocks might then progressively tighten and eventually become isoclinal as folding continued. Many, but not all, of these folds would presumably originate in the vicinity of sulphide masses so that the end product of deformation would be an isoclinally folded carbonate section with orebodies situated in the hinges of some of the folds. Locations of other unmineralized hinges would then be some function of the natural wavelengths of major folds which in turn are dependant upon the mechanical properties of the layered rocks at the time of deformation.

Subsequent to the polyphase deformation, thermal recrystallization affects were induced in the ores (Solomon, 1963; deLorraine, 1979). These include very coarse grain size, (cleavage surfaces in sphalerite up to 10cm across) granoblastic polygonal textures, straight-sided, coherent annealing twins in sphalerite, and rounded sphalerite grains where sphalerite occurs as the minor phase in galena (Spry, 1969; Stanton, 1972; Stanton and Gorman, 1968).

Some textural manifestations of deformation were not erased by thermal recrystallization, however. Among these are segregation layering in massive sulphides defined by alternate pyrite-sphalerite rich layers, especially where this layering transects bedding or foliation, and brecciated pyrite porphroblasts with sphalerite and chalcopyrite infilling the intragranular fractures.

GEOCHEMISTRY

Minor and trace element analyses of orebody concentrates, and sulfur and lead isotope determinations place geochemical constraints on proposed genetic models. Geochemical studies are continuing and at this early stage the data are subject to interpretation in different ways. Nevertheless, a review of the results to date is instructive.

Significance of Minor and Trace Element Studies

Some of the first minor and trace element analyses of Balmat sulphide minerals were published by Doe (1962a). These data place severe constraints on the source of metals in the ores. Content of most of the elements varied from sample to sample and on the same mining level except for cadmium, which averaged about 1400 ppm. Co and Ni contents of presumed metasedimentary pyrite in barren metasedimentary rocks (Units 2, 4, and 15) were compared with those of orebody pyrite. Metasedimentary pyrite averaged less than 50 ppm each of cobalt and nickel, whereas orebody pyrite averaged more than 200 ppm for each of these elements; Doe (1962a) believed it unlikely that pyrite in the orebodies is genetically related to that in the surrounding metasedimentary strata.

Lateral variations in Cu, Hg, Fe, and other minor and trace elements in sphalerite do not appear to fit into a zonal pattern at this stage of investigation (Dill *et al.*, 1978). However, analyses of flotation concentrates made from orebody composite samples have revealed consistent, large scale, variations that appear to be related to stratigraphic position. Specifically, the Hg content in sphalerite decreases stratigraphically upward so that the Main and Streeter orebodies in Unit 14 have the lowest Hg values whereas the Upper Gleason orebody near the base of Unit 6 has the highest. In a separate study, Jessey (1974) reported similar trends of upward decreases in Hg content, both within and between orebodies. Content of Fe in sphalerite increases stratigraphically upward, except in the stratigraphically intermediate American and Number One orebodies where it is comparable to that in the stratigraphically highest Main and Streeter orebodies. To a large degree the FeS contents are probably controlled by the local S fugacity during metamorphism and it is noteworthy that the Number One, American, Main, and Streeter orebodies have minor amounts of pyrrhotite, possibly formed by metamorphic desulphurization and breakdown of pyrite. Consequently, it is not surprising that sphalerite in the Number One and American orebodies is anomalous in the high Fe content. The extent to which original stratigraphic variations versus superimposed metamorphic effects control the Fe distribution in sphalerite is presently unknown.

Significance of Sulphur Isotopes

Sulphur isotopic compositions of anhydrite have been investigated by several workers, including Solomon (1963), Brown (1973), and Whelan (1974, 1978); isotopic compositions of sulphides have also been evaluated by Buddington *et al.* (1969) and Whelan (1978). Sulphur isotope data for Balmat sphalerite orebody composite samples and anhydrite are summarized in Figure 9. The $\delta^{34}S$ compositions of sphalerite concentrates from 24 separate orebodies range from +12.3 to +16.3 per mil, indicating that the sulphur is isotopically heavy and compositionally fairly uniform (Whelan, 1978). Orebody sulphide $\delta^{34}S$ compositions are distinctly different from those of pyrites from nearby metasedimentary strata (Buddington *et al.*, 1969) which average −10 per mil. Buddington *et al.* inferred that the metasedimentary pyrites are of bacterial origin and, despite the great disparity in $\delta^{34}S$ compositions between

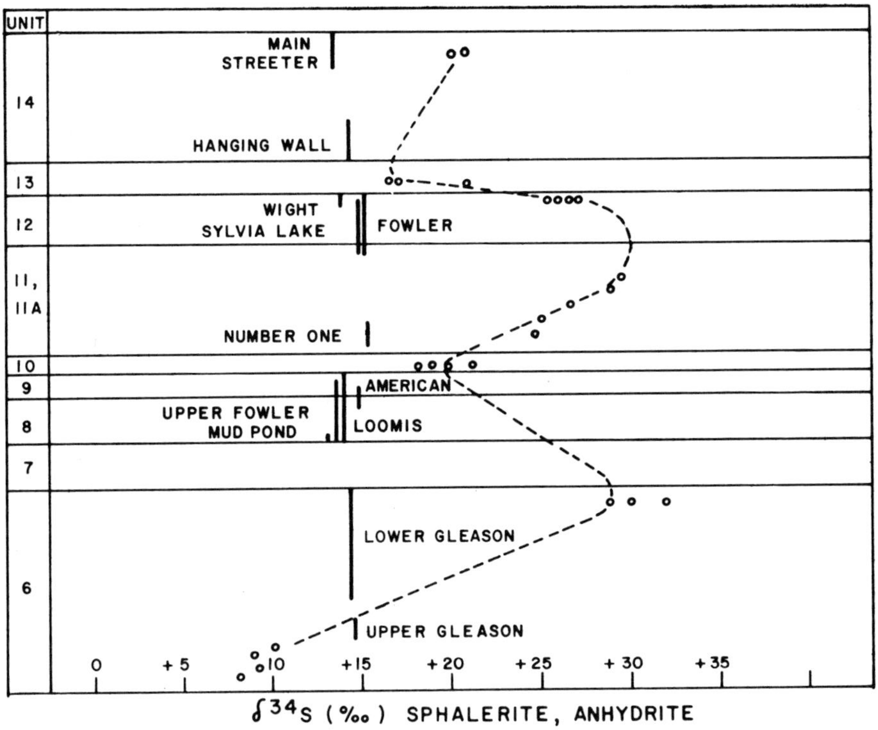

Figure 9. Sulphur isotope values of anhydrite (open circles) and sphalerite (solid bars) versus stratigraphic position. Analyst – Joe Whelan, United States Geological Survey, Branch of Isotope Geology, Denver, Colorado.

metasedimentary pyrites and Balmat orebody sulphides, concluded that the orebody sulphur may also be in part bacteriogenic. Anhydrite $\delta^{34}S$ compositions have received a great deal of attention because of the unusual occurrence of thick lenses and layers of Precambrian sulphates interstratified with the carbonates. Previous workers noted the general similarity of isotopically heavy sulphate and orebody sulphide $\delta^{34}S$ compositions and postulated that sulphur in the orebodies was derived from seawater sulphate (Solomon, 1963) with metamorphic homogenization accounting for the narrow spread in orebody sulphide isotope values shown in Figure 9. In contrast, Whelan (1978) believed there was no direct genetic relationship between the sulphide and sulphate compositions. A direct genetic relationship with sulphate extracted directly from seawater to form sulphides would require the sphalerite and anhydrite $\delta^{34}S$ trends in Figure 9 to covary, in order to allow for sulphate-sulphide equilibration during metamorphism. Yet, sphalerite in the Upper Gleason orebody in Unit 6 is isotopically heavy with respect to sulphate in the same stratigraphic unit, indicating that isotopic exchange within units was minimal at best, and implying therefore that equilibrium isotopic exchange across stratigraphic units, such as that postulated to have occurred between orebodies, was also limited or non-existent. Although the interpretation of sulphur isotopes is somewhat uncertain and ambiguous due to the several mechanisms whereby the light, stable isotopes can be bacterially and chemically fractionated, (Rye and Ohmoto, 1974; Stanton, 1972; Faure, 1977), it now seems clear that sulphur in the orebodies is genetically unrelated to sulphur in pyrite from adjacent unmineralized metasedimentary strata and that orebody sulphur may be indirectly, but probably not directly, derived from marine evaporite sources.

Significance of Lead Isotopes

A new chapter in Balmat geology began when Doe (1962b) published results of lead isotope analyses of sulphides from the No. 2 and No. 3 mines, and showed that galena from the No. 2 mine lies close to the normal growth curve for conformable ore deposits as developed by Stanton and Russell (1959). Isotopic compositions of galena from the No. 3 mine are different from those at mine No. 2 and Doe (1960) attributed the difference to contamination of lead at the No. 3 mine by lead from the enclosing rocks during ore deposition by hydrothermal replacement of the host rocks. Results of the isotopic studies showed (Doe, 1962b) that it was unlikely that ore lead could have been derived from the hosting marbles unless isotopically homogeneous ore lead was extracted from heterogeneous source beds without homogenizing the lead remaining in the source. Doe compared lead isotope compositions of ore lead with those of nearby presumed granites to test the possibility that hydrothermal ore fluids arose from the melts to produce replacement ore deposits in the marbles. Although not completely negating this hypothesis, Doe (1962b) concluded that ore lead could have been extracted from the granites only if radiogenic lead had been added to the feldspars in the granites subsequent to the formation of the ores. As a result of additional, higher precision analyses, Reynolds and Russell (1968) determined that systematic differences exist between isotopic abundances of ore lead and rock lead at Balmat. Reynolds and Russell (1968) suggested a large-scale convective mixing model of crust and upper mantle materials in continental margin-island arc tectonic regimes to explain the lead isotopic abundances in the ores and to satisfy the then apparent requirement for single-stage lead evolution. It is apparent from the existing analyses that ore lead isotopic compositions are dissimilar to those of Mississippi Valley type deposits (Brown and Kulp, 1959; Fletcher *et al.*, 1978) which are characterized by radiogenic, or j-type, lead (Heyl *et al.*, 1974).

A metamorphic origin for the ores was considered by Doe (1970) who suggested that the orebodies may have formed by secretion from source rocks not much older than the age of metamorphism. Doe, however, cautioned that this model may be inadequate because of the shift in isotopic abundances demonstrated by Reynolds and Russell (1968) between ore and rock lead. Most recently, Fletcher (1979) studied lead isotopic compositions of numerous additional composite samples from Balmat orebodies and showed that the ore lead is not as homogeneous as previously supposed. Specifically, there are small but significant differences between different orebodies such that on a plot of 207Pb/204Pb verses 206Pb/204Pb the data points cluster about a steep regression line. Fletcher has also shown that the isotopic distributions are very similar to those of known volcanogenic deposits and has compared them to those from Flin Flon, the Kuroko deposits of Japan, and Buchans. He believed that the slope of the regression line has nothing to do with the age of the ores except that it approximates an average earth isochron of 1,100 Ma. It has been suggested that such steep regression lines result from poor mixing of lead from source regions of the same age but having different uranium/thorium ratios (Doe and Stacey, 1974) i.e., poor mixing from rocks of crustal versus mantle affinities. According to the plumbotectonics model of Doe and Zartman (in press) and Stacey *et al.* (1977), mixing occurs in orogenic environments, and steep regression lines similar to that demonstrated by Fletcher for Balmat arise from the mixing of various crustal and upper mantle components such as might be expected in continental margin-island arc tectonic regimes. Accordingly, it has been suggested (Fletcher and Farquhar, 1977) that Balmat ores formed as part of a continuum of contemporaneous, continentally derived-mantle derived, assemblages.

DISCUSSION

Early investigators were impressed by the high temperature characteristics of the ore and ascribed to them an igneous hydrothermal origin (Newland, 1917; Smythe, 1917), with metals derived from nearby presumed synmetamorphic granites. Brown (1936) concurred with an igneous hydrothermal origin and believed that ore solutions emanated from granitic intrusions at depth, forming replacement deposits conformable to bedding in the marble. Solomon (1963) first proposed a syngenetic origin, based on textural studies in which he found that the ore textures and high temperature characteristics were more readily explained as a result of metamorphism of previously existing sulphide deposits. Brown (1965) came to disfavor a hydrothermal origin and suggested that sulphides were present prior to metamorphism and that their origin was in some way connected with marine sedimentation, perhaps by concentration of metals in connate brines. Syngenesis was also accepted by Lea and Dill (1968), who as well pointed out the role of metamorphic structures in the localization and concentration of the mineralization. Buddington *et al.* (1969) attributed the genesis of the ores to the interaction of metalliferous thermal brines emanating from vents on the seafloor with biologically reduced seawater sulphate. The first worker to propose a Mississippi Valley type origin was Sangster (1970). To date, the only apparent agreement is that the ores are premetamorphic in origin and that structures superimposed during the Grenvillian Orogeny exert significant control on sulphide localization.

Available geological and geochemical data appear compatible with any of several genetic models. The Balmat ores may have formed originally as syngenetic deposits by bacterial reduction or, in the light of the lead isotope data, as syngenetic exhalative sedimentary or volcanogenic deposits. An epigenetic, Mississippi Valley type origin fits the geological evidence less satisfactorily, but is compatible with the sulphur isotopic data.

Important constraints are imposed on these possibilities by the lead isotope data. For example, in the case of epigenesis by lateral secretion, as suggested for some Mississippi Valley type deposits, compositions suggesting single stage evolution of the lead isotopes require the metals to have been leached from very large volumes of heterogeneous source rocks. Such a mechanism was proposed by Doe and Stacey (1974), who suggested that leaching of lead from multiply reworked sediments by saline brines might well produce isotopically uniform ore leads. Differences between ore and rock lead, however, suggest that if such processes were operative during ore deposition at Balmat, the carbonates and granitic rocks in the area were not the primary sources of the lead. Moreover the recent work by Fletcher (1979) raises doubts about the homogeneity of the ore lead. Lateral secretion is, however, compatible with differences in the minor and trace element compositions of the ores, providing ore fluids changed composition with time or as they migrated upward through the stratigraphic column.

Arguments against lateral secretion are compelling and include the massive nature of the ores, conformability with the enclosing strata (except locally, where transgressive relationships are due to metamorphically imparted structural complications), and their occurrence through a wide stratigraphic interval within the carbonate section. Further, differences in the Ni and Co contents of orebody pyrite and pyrite from surrounding unmineralized carbonate strata suggest that the carbonate rocks were not the primary source of the ore metals. Lack of metamorphosed, or relict primary epigenetic textures found in Mississippi Valley type deposits, such as encrustations, cockade structures, or mineralized breccias, especially in the little reworked deposits like the Upper Gleason, also suggest that the ores are not metamorphosed epigenetic Mississippi Valley types.

Synsedimentary bacterial reduction, with metals extracted directly from seawater, also appears inadequate for a number reasons. Homogeneity of the ore sulphur as opposed to the variability of the sulphate sulphur, suggests that ore sulphur was not extracted directly from contemporaneous seawater sulphate (Whelan, 1978). Another negation of this model is the lack of interlayered sediments that might be expected to accumulate with the sulphides and the virtual absence of disseminated sulphide in the surrounding wall rocks. As well, the quantity of dissolved metals in modern seawater is minute; Stanton (1972) appropriately questioned the efficacy of genetic hypotheses involving metal precipitations from seawater to form massive deposits of large extent.

Although there have been no specific attempts to type the Balmat ores as volcanogenic, the similarity of the lead isotope compositions to volcanogenically derived lead has been pointed out by Fletcher (1979), and others have speculated that the ore lead is related in some way to mixing in island arc-continental margin orogenic environments (Fletcher and Farquhar, 1977; Reynolds and Russell, 1968). Field geologists familiar with the Balmat deposits, however, do not favor genetic models linked directly with volcanic processes because of the lack of volcanogenic strata in the carbonate section. Possible exceptions are Unit 2 and the biotite schist of Unit 10. Features that are common in proximal exhalative volcanogenic deposits but absent at Balmat are pipe-like, pyritic, ''stringer ore'' zones developed in altered and silicified volcanic rocks in the stratigraphic footwall of massive sulphide lenses.

Mineralogic zonation that is common in many massive exhalative sedimentary or volcanogenic massive deposits is absent; if exhalative models are at all applicable, then the mineralization must bear a distal relationship to the vent site. Sato (1972), Turner and Gustafson (1978), and Large (1979) have shown that depending upon temperature, salinity, density and fO_2 differences, potential ore fluids may either mix with seawater or remain separated from it as they are expelled on the seafloor. Under favorable conditions, it might then be possible for dense metalliferous fluids to migrate for appreciable distances down slope, away from original vent sites into distal portions of the sedimentary-volcanic environment.

The Sullivan deposit (British Columbia, Canada) has been classified as exhalative sedimentary by many workers. Massive and lenticular in form, the deposit occurs in Precambrian siltstone and argillite that have little or no interlayered volcanic rocks. The orebody itself is underlain by a tourmaline halo that is thought to represent a feeder pipe through which metal-bearing fluids rose before debouching on the seafloor. Lead isotope compositions approach those that have evolved under single stage conditions and similar to those from Balmat, lie close to the growth curve for lead evolving in the orogene in the plumbotectonics model of Stacey *et al.* (1977, Fig. 3). These authors believed that the lead isotopic models are applicable to exhalative sedimentary as well as to exhalative volcanogenic deposits. Perhaps Balmat mineralization is a closer analogy to a distal variety of the former than to the latter; such a comparison is attractive because it eliminates the need for associated volcanic rocks, is compatible with the massive aspect of Balmat ore, and explains the absence of underlying feeder pipes. The progenitor of the quartz gangue that is intimately associated with the Balmat ores may have been silica supplied in solution by the hydrothermal source and deposited as cherty chemical sediment. It has been suggested that Sullivan mineralization is associated with continental rifting, specifically with an aulacogen developed during the initial stages of continental separation. Carrying this analogy one step further, one might speculate that Balmat ores, as well, were emplaced in such a continent-margin rift, or intracratonic graben. Dolomite units, cherty-dolomitic rocks and anhydrite, rather than representing stable continental platform sediments, may have been deposited as magnesian carbonate muds and chemical sediments in this tectono-stratigraphic setting. Periods of shallow water and restricted circulation might have led to the deposition of evaporites, whereas quartz-rich rocks may indicate cherty chemical sedimentation in deeper waters. Cyclicity is a noteworthy feature of the carbonate section, and the association of the orebodies with quartz-rich or silicated units, as opposed to the evaporites, is perhaps significant. Amphibolite and metagabbro bodies which occur in places throughout the northwest Adirondacks, including Balmat (Fig. 3), and whose presence appears incongrous with the concept of a stable continental shelf or margin tectonic setting, may represent mafic intrusive and volcanic rocks generated in association with intracratonic graben formation and rifting.

CONCLUSIONS

Considering all the data, the writers believe it reasonable to discount a syngenetic, bacteriologic origin for the ores. Alternatively, deposition may have been syngenetic and associated in some as yet unknown way with evaporite sedimentation. A lateral secretion origin is compatible with sulphur isotope data and cannot be eliminated on the basis of lead isotope data, but requires the assumption that large volumes of rock were leached in such a way that overall lead compositions approaching single stage conditions were attained. Similar effects would be produced by large scale convective hydrothermal leaching so that the lead isotope data suggest that a submarine exhalative origin is as plausible as an epigenetic, lateral secretion origin.

It is felt that an exhalative sedimentary, or perhaps distal exhalative volcanogenic model is most appropriate in satisfying the geological and geochemical data more completely than do the others. An exhalative sedimentary model deserves more consideration in light of the massive character of the mineralization, apparent stratigraphic control of minor element compositions of the ores, and the lead isotopic evidence. Nevertheless, more detailed and comprehensive tectonic and stratigraphic reconstructions, as well as geochemical data, are needed in order to more fully evaluate the validity of this model.

REFERENCES

Bickford, M.E. and Turner, B.B., 1970, Rb-Sr geochronology of gneisses in structural domes, southeastern Adirondack Mountains, New York: Geological Society of America, Abstracts with Programs, v. 2, no. 7, p. 495.

Bohlen, S.R., 1977, Feldspar and oxide thermometry in the Adirondack Highlands: Geological Society of America, Abstracts with Programs, Northeastern Section, p. 242.

Brocoum, S.J., 1971, Structural and metamorphic history of the major Precambrian gneiss belt in the Hailesboro-West Fowler-Balmat area, Adirondack Lowlands, New York: Ph.D. Dissertation, Columbia University, New York, 194 p.

Brown, J.S., 1973, Sulfur isotopes of Precambrian sulfates and sulfides in the Grenville of New York and Ontario: Economic Geology, v. 68, p. 362-370.

_______________, 1965, Oceanic leads and ore genesis; Economic Geology, v. 60, no. 1, p. 47-68.

_______________, 1936, Structure and primary mineralization of the zinc mine at Balmat, New York: Economic Geology, v. 31, p. 233-258.

Brown, J.S. and Kulp, J.L., 1959, Lead isotopes from the Balmat area, New York: Economic Geology, v. 54, p. 137-138.

Brown, J.S. and Engel, A.E.J., 1956, Revision of Grenville stratigraphy and structure in the Balmat-Edwards district, northwest Adirondacks, New York: Geological Society of America Bulletin, v. 67, p. 1599-1622.

Brown, P.E., Essene, E.J., and Kelly, W.C., 1977, Sphalerite geobarometry in the Balmat-Edwards district, New York: manuscript submitted to American Mineralogist.

Buddington, A.F., Jensen, M.L., and Mauger, R.L., 1969, Sulfur isotopes and origin of northwest Adirondack Sulfide deposits: Geological Society of America Memoir 115, p. 423-451.

Campbell, B., 1976, Stratigraphy and metamorphism in the Balmat-Edwards zinc mining district, northwest Adirondacks, New York: Unpublished M.Sc. Thesis, Cornell University, New York, 77 p.

deLorraine, W.F., 1979, Geology of the Fowler orebody, Balmat #4 mine, northwest Adirondacks, New York: Unpublished M.Sc. Thesis, University of Massachusetts, Amherst, 159 p.

DeWitt, D.B., 1976, Applications of sphalerite barometry in Grenville marbles: Unpublished M.Sc. Thesis, University of Michigan.

Dill, D.B., and deLorraine, W.F., 1978, Structure, stratigraphic controls, and genesis of the Balmat zinc deposit, northwest Adirondacks, New York: Geological Association of Canada, Abstracts with Programs, v. 3, p. 389.

Dill, D.B., deLorraine, W.F., Knight, R.H., Johnson, J.T., and Grout, C.M., 1978, Geological field guide to the Balmat zinc mine, St. Lawrence County, New York: Unpublished Field Guide, Geology Department, St. Joe Zinc Co., Balmat, New York.

Doe, B.R., 1970, Lead isotopes: *in* Von Engelhardt, W., Hahn, T., Roy, R., Winchester, J.W., and Wyllie, P.J., eds., Minerals, Rocks and Inorganic Materials, Series of Theoretical and Experimental Studies No. 3, Springer Verlag, New York, 137 p.

______________, 1962a, Distribution and composition of sulfide minerals at Balmat, New York: Geological Society of America Bulletin, v. 73, p. 833-854.

______________, 1962b, Relationships of Pb isotopes among granites, pegmatites, and sulfide ores near Balmat, New York: Journal of Geophysical Research, v. 67, no. 7, 1962.

______________, 1960, The distribution and composition of sulfide minerals at Balmat, New York: California Institute of Technology Ph.D. Thesis series, 151 p.

Doe, B.R. and Stacey, J.S., 1974, The application of lead isotopes to the problems of ore genesis and ore prospect evaluation: A review: Economic Geology, v. 69, p. 757-776.

Doe, B.R. and Zartman, R.E., in press, Plumbotectonics I, the Phanerozoic: *in* Barnes, H., ed., Geochemistry of Hydrothermal Ore Deposits.

Engel, A.E.J., and Engel, C.B., 1953, Grenville Series in the northwest Adirondack Mountains, New York, Part II, origin and metamorphism of the Major Paragneiss: Geological Society of America Bulletin, v. 64, p. 1013-1047.

Essene, E.J., Bohlen, S.R., and Valley, J.W., 1977, Regional metamorphism in the Adirondacks: Geological Society of America, Abstracts with Programs (Northeast Section), v. 9, no. 3, p. 260-261.

Faure, G., 1977, Principles of isotope geology: John Wiley and Sons, New York, 464 p.

Fletcher, I.R., 1979, A lead isotopic study of lead-zinc mineralization associated with the Central Metasedimentary Belt of the Grenville Province: Ph.D. Thesis, University of Toronto.

Fletcher, I. R. and Farquhar, R.M., 1977, Lead isotopes in the Grenville and adjacent Paleozoic formations: Canadian Journal Earth Science, v. 14, p. 56-66.

Fletcher, I.R., Farquhar, R.M., and Kuybida, P.R., 1978, Lead isotope evidence for the existence and possible origin of some Mississippi Valley type deposits in southeastern Ontario: in Zartman, R.E., ed., Short Papers of the Fourth International Conference Geochronology, Cosmochronology, Isotope Geology: United States Geological Survey Open File Report 78-701, p. 117-119.

Foose, M.P., 1974, The structure, stratigraphy, and metamorphic history of the Bigelow area, northwest Adirondacks, New York: Ph.D. Dissertation Princeton University, Princeton, New Jersey, 224 p.

Foose, M.P. and Brown, C.E., 1976, A preliminary synthesis of structural, stratigraphic, and magnetic data from part of the northwest Adirondacks, New York: United States Geological Survey Open File Report 76-281, 22 p.

Foose, M.P. and Carl, J.D., 1977, Setting of alaskite bodies in the northwestern Adirondacks, New York: Geology, v. 5, p. 77-80.

Heyl, A.V., Landis, G.P., and Zartman, R.E., 1974, Isotopic evidence for the origin of Mississippi Valley type mineral deposits: a review: Economic Geology, v. 69, p. 992-1006.

Hills, A. and Gast, P.W., 1964, Age of pyroxene-hornblende granitic gneiss of the eastern Adirondacks by the Rb-Sr whole-rock method: Geological Society of America Bulletin, v. 75, p. 759-766.

Jessey, D.R., 1974, Mercury distribution in a massive sulfide ore deposit at Balmat, New York: Unpublished M.Sc. Thesis, University of Missouri, Rolla, 79 p.

Johnson, J.T., and deLorraine, W.F., 1979, Comments on ore controls, structure, and genesis of the Balmat zinc deposit, northwest Adirondacks, New York: Geological Society of America, Abstracts with Programs, Northeastern Section.

Large, R.R., 1977, Chemical evolution and zonation of massive sulfide deposits in volcanic terrains: Economic Geology, v. 72, p. 549-572.

Lea, E.R., and Dill, D.B., 1968, Zinc deposits of the Balmat-Edwards district, New York: in Ridge, J.D., ed., Ore deposits of the United States, 1933-1967, (Graton-Sales volume), Vol. 1: New York, American Institute Mining, Metallurgy, and Petroleum Engineers, p. 26-48.

Newland, O.H., 1917, Zinc-pyrite deposits of the Edwards district, New York: New York State Defense Council, Bull. 2.

Reynolds, P.H. and Russell, R.D., 1968, Isotopic compositions of lead from Balmat, New York: Canadian Journal Earth Science, v. 5, p. 1239-1245.

Rye, R.O., and Ohmoto, H., 1974, Sulfur and carbon isotopes and ore genesis; A review: Economic Geology, v. 69, p. 826-842.

Sangster, A.L., 1970, Metallogeny of base metal, gold and iron deposits of the Grenville Province of southeastern Ontario: Unpublished Ph.D. Thesis, Queen's University, Kingston, Ontario, 356 p.

Sato, T., 1972, Behaviours of ore forming solutions in sea waters: Mining Geology, v. 22, p. 31-42.

Scott, S.D., 1976, Application of the sphalerite geobarometer to regionally metamorphosed terrains: American Mineralogist, v. 61, p. 661-670.

Silver, L.T., 1964, Isotopic investigations of zircons in Precambrian igneous rocks of the Adirondack Mountains, New York: Geological Society of America Abstracts for 1963, Special Paper 76, p. 150-151.

Smythe, C.H., Jr., 1917, Genesis of the zinc ores of the Edwards district, St. Lawrence County, New York: New York State Museum Bulletin, 201.

Solomon, P.J., 1963, Sulfur isotopic and textural studies of the ores at Balmat, New York, and Mount Isa, Queensland: Unpublished Ph.D. Thesis, Harvard University, 162 p.

Spry, A., 1969, Metamorphic textures: Oxford, Pergammon Press, 350 p.

Stacey, J.S., Doe, B.R., Silver, L.T., and Zartman, R.E., 1969, Plumbotectonics IIA, Precambrian massive sulfide deposits: United States Geological Survey Open File Report 76-476.

Stanton, R.L., 1972, Ore Petrology: New York, McGraw Hill, 713 p.

Stanton, R.L. and Gorman, H., 1968, A phenomenological study of grain boundary migration in some common sulfides: Economic Geology, v. 63, p. 907-923.

Stanton, R.L. and Russell, R.D., 1959, Anomalous leads and the emplacement of lead sulfide ores: Economic Geology, v. 54, p. 588-607.

Turner, J.S. and Gustafson, L.B., 1978, The flow of hot saline solutions from vents in the sea floor – some implications for exhalative massive sulfide and other ore deposits: Economic Geology, v. 73, p. 1082-1100.

Vokes, F.M., 1969, A review of the metamorphism of sulfide deposits: Earth Science Review, v. 5, p. 99-143.

Whelan, J.I., 1978, Stable isotope studies of the Balmat-Edwards Zn-Pb district, New York: Geological Society of America, Abstracts with Programs (Joint Annual Meeting), v. 70, no. 7, p. 515.

______________, 1974, Trace element and sulfur isotope comparison of anhydrites from Balmat-Edwards, New York with sedimentary and hydrothermal anhydrite: M.Sc. Thesis, Pennsylvania State University, 96 p.

Wynne-Edwards, H.R., 1972, The Grenville Province, in Price, R.A., and Douglas, R.J.W., eds., Variations in Tectonic Styles in Canada: Geological Association of Canada Special Paper 11, p. 263-334.

MS received August 1979;
Accepted for publication February 1981.

Precambrian Sulphide Deposits, H.S. Robinson Memorial Volume, edited by R.W. Hutchinson, C.D. Spence and J.M. Franklin, Geological Association of Canada Special Paper 25, 1982

GEOLOGY OF THE SULLIVAN OREBODY KIMBERLEY, B.C., CANADA

J.M. Hamilton
Cominco Ltd., P.O. Box 2000, Kimberley, B.C. V1A 2G3

D.T. Bishop
(Formerly Cominco Ltd.)
Box 1, Marysville, B. C. V0B 1Z0

H.C. Morris
(Formerly Cominco Ltd.)
E & B Explorations Ltd., 2900-300 5th. Ave. SW., Calgary, Alberta T2P 3C4

O.E. Owens
Cominco Ltd., 2300 Granville Square, Vancouver, British Columbia V6C 2R2

ABSTRACT

The Sullivan orebody is a 160 million tonne conformable iron-lead-zinc sulphide lens enclosed by clastic sedimentary rocks of the Middle Proterozoic Aldridge Formation, the basal formation of the 10,000 m thick Purcell Supergroup. Regional metamorphism is upper greenschist facies.

The roughly circular orebody is about 2000 m in diameter and is up to 100 m thick in its western part. Here, massive pyrrhotite containing occasional wispy layers of galena is overlain by sulphide rock in which conformable layering consists of pyrrhotite, sphalerite, galena and pyrite intercalated with beds of clastic sedimentary rock. Eastward across a transition zone, the orebody is composed of five distinct conformable units of well bedded sulphide rock interbedded with clastic sedimentary rock. Each bed of sulphide rock thins eastward from the transition zone. Three bedded sulphide sequences occur above the main orebody, particularly in the area of the transition zone. Locally, these are ore.

Much of the orebody is underlain by locally derived intraformational conglomerate which is more than 80 m thick in the west and thins to the east. Footwall rocks are cut by tabular bodies of chaotic breccia containing blocks of conglomerate and bedded sedimentary rock; these extend downward unknown distances from the sulphide footwall in the west. Footwall mineralization consisting of thin conformable laminae, veins and locally intense fracture filling is common in the west and very rare in the east.

The footwall and hangingwall rocks and locally the orebody in the west have been extensively altered by hydrothermal solutions. A crosscutting zone of tourmalinite underlying the sulphide lens in the west is 1000 m by 1500 m across at the sulphide footwall and extends at least 500 m beneath the orebody. Albite-chlorite-pyrite alteration occurs in crosscutting zones in the footwall tourmalinite and extends more than 100 m into the hangingwall over the western part of the orebody. A zone of pyrite-chlorite alteration 300 m in diameter crosscuts massive sulphide rock immediately overlying footwall albite-chlorite-pyrite alteration zones.

The Sullivan orebody is interpreted as a hydrothermal synsedimentary deposit which formed in a sub-basin on the Aldridge marine floor. It is located directly over conduits through which mineralizing fluids passed. Boron, iron and magnesium were added to footwall sediments. Changes of fluid and/or basin chemistry then led to rapid deposition of massive sulphides over the western vent area and deposition of delicately-bedded sulphides in the eastern part of the sub-basin. Post-ore sodium-rich hydrothermal fluids altered tourmalinite, sulphide rocks and hangingwall rocks over the vent area.

RÉSUMÉ

Le gisement de Sullivan est une lentille de sulfures de fer, plomb et zinc de 160 millions de tonnes, concordante avec les roches sédimentaires clastiques de la formation d'Aldridge d'âge protérozoique moyen. Celle-ci est à la base du supergroupe de Purcell qui atteint 10.000m d'épaisseur. Le métamorphisme régional est de faciès schiste vert supérieur.

Le gisement est approximativement circulaire, d'un diamètre d'à peu près 2000m et atteignant 100m d'épaisseur dans sa partie ouest. Ici, de la pyrrhotite massive contenant à l'occasion des couches effilochées de galène, est recouverte par une roche sulfurée litée consistant en pyrrhotite, sphalérite, galène et pyrite avec intercalations de roche sédimentaire clastique. Vers l'est, au-delà d'une zone de transition, le gisement se compose de 5 unités distinctes, concordantes, de roches sulfurées bien litées avec intercalations de roche sédimentaire clastique. Chaque lit sulfuré s'amincit vers l'est à partir de la zone de transition. Trois séquences sulfurées litées se trouvent aussi aux dessus du gisement principal, particulièrement dans la région de la zone de transition. Elles sont localement exploitables.

La pluspart du gisement recouvre un conglomérat intraformationnel de dérivation locale qui a plus de 80m d'épaisseur à l'ouest et qui s'amincit vers l'est. Les roches du mur sont recoupées par des corps tabulaires de brèche chaotique contenant des blocs de conglomérat et de roche sédimentaire litée; ces corps s'étendent en profondeur, à partir du mur des sulfures à l'ouest, sur des distances inconnues. Les roches au mur du gisement sont communément minéralisées à l'ouest mais très rarement à l'est; cette minéralisation consiste en laminations minces, concordantes, en veines et en remplissage de fractures localement très nombreuses.

A l'ouest, les roches du mur et du toit et localement le gisement lui-même ont été extensivement altérés par des solutions hydrothermales. Une zone de tourmalinite discordante, mesurant en section 1.000m par 1.500m, s'enfonce d'au moins 500m sous le gisement. Une altération en albite-chlorite-pyrite forme des zones discordantes dans cette tourmalinite, au mur, et s'étend sur plus de 100m dans le toit, de la partie ouest du gisement. Enfin, une zone d'altération à pyrite-chlorite, de 300m de diamètre, recoupe les sulfures massifs là oú ils recouvrent des zones d'altération à albite-chlorite-pyrite.

Le gisement de Sullivan est interprété comme un dépôt hydrothermal synsédimentaire qui s'est accumulé dans un bassin secondaire du fond marin en même temps que la formation d'Aldridge. Il est situé à l'aplomb des conduits des fluides minéralisants. Bore, fer et magnesium ont été ajoutés aux sédiments du mur. Des changements dans la chimie des fluides ou/et du bassin ont alors conduit au dépôt rapide des sulfures massifs au-dessus de l'orifice, à l'ouest, et au dépôt des sulfures délicatement lités dans la partie est du bassin secondaire. Des fluides hydrothermaux riches en sodium ont ensuite altéré la tourmalinite, les roches sulfurées et les roches du mur autour de l'orifice hydrothermal.

INTRODUCTION

The Sullivan orebody at Kimberley, British Columbia (Fig. 1) is a large, gently dipping iron-lead-zinc sulphide deposit lying conformably in Proterozoic clastic sedimentary rock of the Aldridge Formation. The orebody is located at the western edge of the Rocky Mountain Trench, on the eastern flank of the Purcell Mountains about 80 km north of the United States boundary. The original mineral claims were staked on surface showings of sulphide mineralization in 1892, but little work was done on the claims prior to completion of a Canadian Pacific Railway branch line to Kimberley in 1899. Systematic development of the property was commenced in 1900 by a Spokane, Washington mining group and a smelter was completed in nearby Marysville in 1903. This venture failed owing to metallurgical difficulties, and the property was acquired in 1909 by the Federal Mining and Smelting Company. The Consolidated Mining and Smelting Company of Canada Ltd. (now Cominco) took a lease and option on the property in late 1909 and completed purchase by 1913.

During the next five years, selective mining and sorting produced high-grade lead and silver ore causing little metallurgical difficulty. In fact, during World War I, this was the largest single source of lead in the British Commonwealth. In the meantime, deep drilling revealed an immense lens of complex sulphide ore. A main adit at the 3900 foot level, some 200 m below the outcrop, was started in 1915 and a major effort was directed toward solving the milling and metallurgical problems. In 1920, a differential flotation process was developed that proved successful for obtaining lead and zinc concentrates suitable for smelting. From the date of acquisition by Cominco to the end of 1979, the Sullivan Mine produced approximately 123,200,000 short tons (111,600,000 tonnes) containing 6.8% lead, 5.9% zinc and 2.4 oz. (82 g/t) silver. Remaining diluted reserves at that time were 54,000,000 short tons (49,000,000 tonnes) containing 4.5% lead, 5.9% zinc and 1.1 oz. (37 g/t) silver.

This paper describes the geology of the Sullivan orebody and its clastic host rocks. The regional and local geological setting of the orebody and sedimentological characteristics of the Aldridge Formation, a flysch sequence at the base of the Purcell Supergroup, are summarized to facilitate understanding of the setting of the orebearing sequence. The iron-lead-zinc sulphide ore occurs as broad, continuous and complex lenses interbedded and deposited contemporaneously with a distinctive sequence of clastic rocks of restricted distribution. The western, thickest, part of the orebody is massive to poorly layered and contains minor interbedded clastic sedimentary rock. Outward from dominantly massive ore, interbedded clastic sedimentary rock becomes more abundant and the orebody, particularly to the east, is a sequence of delicately bedded sulphide rock intercalated with beds of sulphide-poor quartz wacke and mudstone.

The western part of the orebody is underlain by breccia zones and an extensive zone of tourmalinite in which stringer sulphides are locally common. Intense post-ore alteration formed rocks composed of various proportions of albite, chlorite and py-

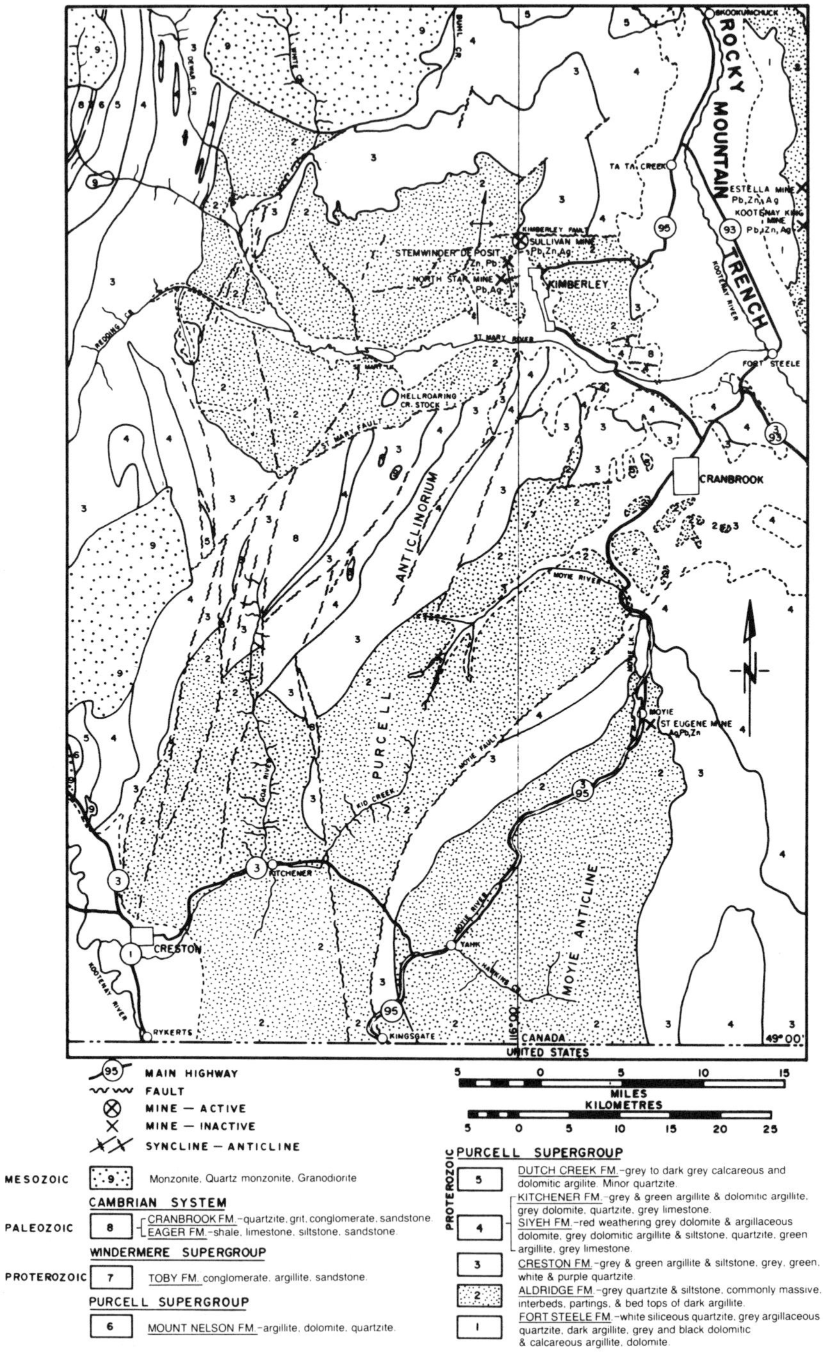

Figure 1. Regional geological map.

rite. This alteration forms restricted crosscutting zones in footwall tourmalinite, occurs through the ore zone immediately over these crosscutting zones and in hangingwall rocks over the western part of the orebody.

REGIONAL GEOLOGICAL SETTING

Purcell Supergroup

The Purcell Supergroup in southeastern British Columbia and the laterally equivalent Belt Supergroup in the United States according to Harrison (1972) constitute a thick prism of dominantly clastic sediments deposited in a large epicratonic re-entrant of the Middle Proterozoic sea at the western margin of the North American Craton. However, recent identification of Precambrian rocks in the Shuswap terrain (Wanless and Reesor, 1975) to the west of principal exposures of Purcell rocks suggests that Purcell sediments were deposited in an intracratonic basin. Aspects of the depositional environment of the Belt-Purcell have been compared to present day sedimentation in the Gulf Coast geosyncline (Price, 1964; Winston, 1973). The sedimentary rocks form a monotonous succession of drab-coloured siltstone and mudstone with units of quartz arenite, dolomite and limestone.

Although details of stratigraphy and sedimentology of Belt-Purcell sedimentary rocks have periodically received intensive local study over the past seventy years, vast gaps in modern geological mapping hinder synthesis of much of the depositional history. Away from the margins of the basin, lateral facies changes are difficult to recognize and are commonly reflected only as subtle changes in grain size.

Nomenclature of the clastic Belt-Purcell metasedimentary rocks has been inconsistent and descriptions found in the literature may seem remarkably alike even though field examination clearly reveals dissimilarities. Early workers (Willis, 1902; Daly, 1912; Schofield, 1915) recognized the difficulties inherent in regional correlation of the lithostratigraphic units they defined and the general inadequacy of the terms quartzite and argillite as applied to the clastic sequence. More recent authors (Harrison and Jobin, 1963; Price, 1964) have demonstrated successful lithologic subdivision based on grain size of previously defined formations and have thus established better regional correlations of the lithostratigraphic units (Harrison and Campbell, 1963; Harrison, 1972).

Maximum thickness of sedimentary rocks of the Purcell Supergroup exceeds 10,000 m, with the base unexposed (Fig. 2). Earliest known sedimentation was initiated with deposition of the upward-fining fluvial/deltaic sequences of quartz arenite, quartz wacke and mudstone of the Fort Steele Formation exposed 25 km due east of Kimberley (Fig. 1) on the east side of the Rocky Mountain Trench. Each of the sequences starts with cross-bedded quartz arenite at the base and grades upward to finely laminated mudstone at the top (Höy, 1978). The Fort Steele Formation is at least 200 m thick.

Fine-grained clastic beds at the top of the Fort Steele grade into very rusty-weathering, fine-grained quartz wacke and mudstone of the Aldridge Formation. A considerable thickness of the Fort Steele sequence appears to be the time equivalent of the Aldridge Formation exposed in the Purcell Mountains west of the Rocky Mountain Trench (Höy, this volume). The Aldridge Formation is at least 5000 m thick in the Purcell Mountains and is less than 4000 m thick in the Rocky Mountains.

The Aldridge Formation grades upward over 300 m through a sequence of carbonaceous mudstone interbedded with minor beds of grey and green mudstone and fine-grained quartz wacke to the 1800 m thick Creston Formation, composed of grey, green and maroon quartz wacke and mudstone with minor white arenite. East of the Rocky Mountain Trench, the Creston sedimentary rocks are interpreted as fan delta and mudflat deposits (Höy, 1978) and in the Purcell Mountains (Fig. 1) they are thought to have been deposited in a deltaic environment prograding from the south (Winston, 1973).

Conformably overlying the Creston Formation are 1200 m of green and grey dolomitic mudstone, buff-weathering dolomite containing abundant "molar tooth" structures and minor quartz arenite of the Kitćhener Formation. The Kitchener is in turn overlain by 200 to 400 m of green, slightly dolomitic and calcareous mudstone of

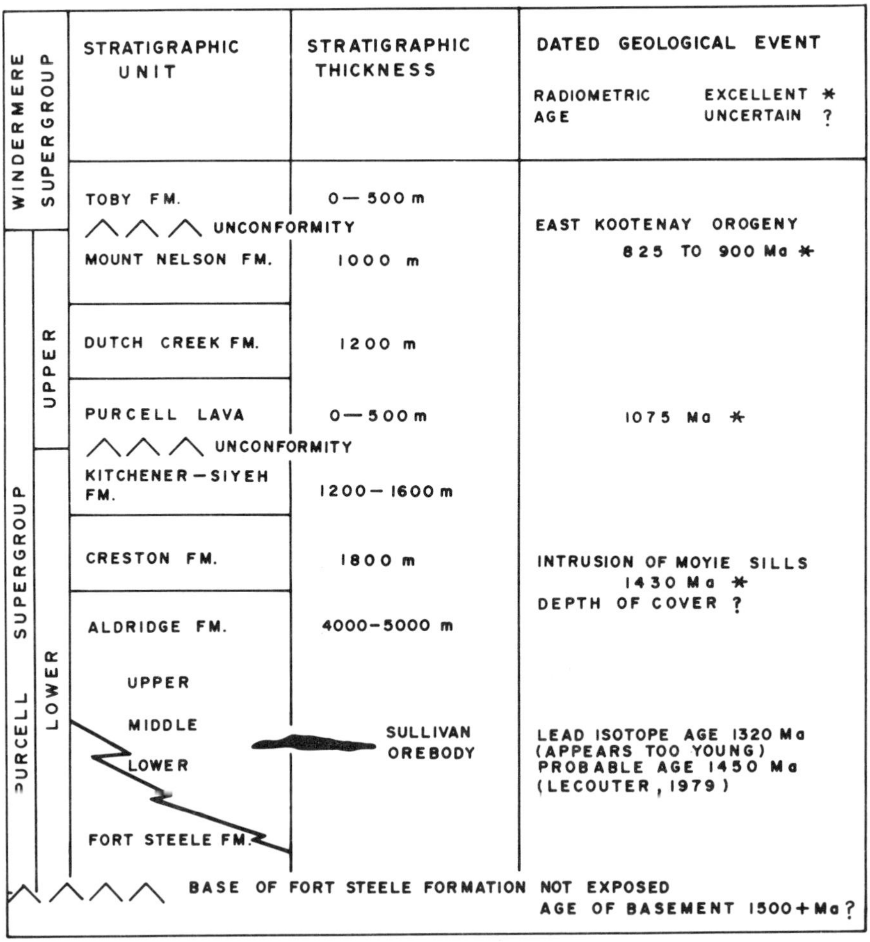

Figure 2. Stratigraphic chart of the Purcell Supergroup in the Purcell Mountains of Southeastern British Columbia.

the Siyeh Formation. Although poorly defined in the Purcell Mountains west of the Rocky Mountain Trench, the Siyeh is readily recognized in the Rocky Mountains and is conformably and locally unconformably overlain by 0 to 500 m of basaltic to andesitic flows of the Purcell Lava which are taken to mark the close of lower Purcell sedimentation. To the northwest and west in the Purcell Mountains, the Purcell Lava is only sparsely represented by weathered tuffaceous beds.

Resting with apparent conformity on the lower Purcell rocks are about 1200 m of grey to dark grey, calcareous and dolomitic mudstone and minor quartz wacke of the Dutch Creek Formation. The Dutch Creek is particularly well developed in the western Purcell Mountains and is there overlain by about 1000 m of grey, green and maroon mudstone and calcareous mudstone of the Mount Nelson Formation. The close of Purcell sedimentation is marked by folding during the East Kootenay Orogeny (White, 1959) and disruption of the basin by large-scale vertical faults (Lis and Price, 1976) concurrent with deposition of basal sedimentary rocks of the Windermere Supergroup.

Middle Proterozoic igneous activity in the Purcell sedimentary basin in southeastern British Columbia is dominated by intrusion of gabbroic sills of two ages. The oldest are the Moyie Sills (Daly, 1912) which are most common in the Aldridge Formation. Sills and slightly discordant sheets predominate; locally, however, dykes and step-like discordant sheets such as those found near Kimberley are abundant. Gabbroic sills can aggregate 2000 m of thickness in a typical Aldridge section and are most abundant in the lower part of the section. The sills are more abundant in the Purcell Mountains than in the Rocky Mountains. The youngest event of gabbro intrusion is thought to be comagmatic with the Purcell Lavas, and is represented by abundant sills in the upper part of the Creston Formation and in the Kitchener and Siyeh Formations (Bishop, 1974). The pegmatitic Hellroaring Creek Stock and related satellites intrude metamorphosed and deformed Aldridge sedimentary rocks and Moyie Sills in an area about 15 km southwest of the Sullivan Mine. The stock is of Middle Proterozoic age (Ryan and Blenkinsop, 1971).

Lower Purcell sedimentary rocks have everywhere undergone metamorphism, to at least greenschist facies. There is a general increase in metamorphic grade with depth in the stratigraphic pile (Edmunds, 1977; Bishop, 1974) except in the contact aureoles of major Cretaceous intrusions. Muscovite and chlorite are characteristic metamorphic minerals of pelitic rocks of the Kitchener Formation and deeper in the section biotite first appears as a metamorphic mineral in the upper part of the Aldridge Formation. Minor areas of amphibolite facies are restricted to the cores of fold structures displaying large-magnitude structural relief.

Purcell rocks are folded about north-trending axes to form the Purcell anticlinorium (White, 1959). Folds comprising the large structure are open and gentle with north-plunging axes. Some folds are overturned to the east and some display axial plane schistosity. Large areas within the anticlinorium have nearly flat-lying strata. Major faults with a history of complex movement disrupt the Purcell terrain and separate large regions further disrupted by block faulting (Fig. 1). Two of these major faults, the Moyie and St. Mary faults, pass south of Kimberley and throughout much of their extent have a northerly trend, but then abruptly arc to the east into the Rocky Mountain Trench. Both of these faults repeat lower Purcell strata on their north and west, upthrown sides. There is evidence for repeated movement along the major faults beginning as early as the time of Purcell sedimentation (Höy, this volume; Lis and Price, 1976).

Geochronology of Purcell rocks has been reviewed by many authors including Obradovich and Peterman (1968); Gabrielse (1972); and Harrison (1972). Harrison reviewed available information concerning age of basement, Purcell sedimentation, igneous activity and metamorphism and attempted to reconcile conflicting aspects of the data. The general framework from Harrison's conclusions, modified with recent data, is presented in Figure 2. The East Kootenay Orogeny marking the close of Belt-Purcell sedimentation has been reasonably well documented at 825 to 900 Ma (Miller, 1973). A minimum age for deposition of sedimentary strata of the Siyeh Formation is provided by a 1075 Ma date for the Purcell Lava (Hunt, 1962; Obradovich and Peterman, 1968).

A metamorphic event affecting Aldridge sedimentary rocks and the Moyie Sills has been dated at about 1300 Ma (Obradovich and Peterman, 1968; Bishop, 1974). This agrees well with the 1300 Ma age of the Hellroaring Creek Stock (Ryan and Blenkinsop, 1971; Harrison, 1972) yet there is little recognized evidence recording the event in the sedimentary record of the lower Purcell. Minimum age of 1430±20 Ma for the Aldridge Formation is provided by U-Pb zircon data obtained from differentiates of the Moyie Sills (Zartman *et al.*, in press); however, age of basement rock upon which the Purcell rocks were deposited and therefore the maximum age for sedimentation is known only by inference. Basement rock is nowhere exposed within the area of outcrop of Purcell rocks in southeastern British Columbia, thus ages of approximately 1800 Ma (Burwash *et al.*, 1962), 1960 Ma for rocks in the core of the Thor-Odin Dome (Wanless and Reesor, 1975) and 1700±100 Ma (Giletti, 1968) for basement far removed from the area of Purcell sedimentation can be considered maximum ages. Recent data (Reid *et al.*, 1973) suggest that Aldridge equivalent rocks in northernmost Idaho are underlain by and perhaps derived from a felsic orthogneiss terrain about 1500 Ma old.

In summary, deposition of lower Purcell sediments covered a time span of at least 425 Ma from 1500 to 1075 Ma. Textural and compositional homogeneity suggest a provenance of gentle relief drained by low-gradient river systems. The few data concerning transport direction and source for initial Purcell sediments suggest, at least in part, an easterly source. By Middle Aldridge time, south to north transport by turbidity currents was firmly established and the deep Purcell sedimentary basin was gradually filled (Bishop *et al.*, 1970). Initial filling of the basin was followed by westerly and northerly deposition of deltaic sediments of the Creston Formation prograding from the east and south across the basin. Moyie Sills were injected in early Creston time at 1430 Ma (Zartman *et al.*, in press) followed by a major metamorphic event accompanied by intrusion of the 1300 Ma Hellroaring Creek Stock. The relative ages of the 1300 Ma event and deposition of the Kitchener and Siyeh Formations are equivocal as no evidence for a 1300 Ma event has been found in these youngest sedimentary rocks of the lower Purcell. Extrusion of the Purcell Lava at 1075 Ma marks the close of lower Purcell geological events.

Aldridge Formation

The Aldridge Formation in the Purcell Mountains (Schofield, 1915) has the sedimentological characteristics of a flysch sequence (Bishop *et al.*, 1970). It is composed of a monotonous and repetitious sequence of alternating beds of very fine-grained quartz wacke and mudstone and lesser amounts of very fine- to coarse-grained quartz arenite. Beds have considerable lateral continuity and show sharply defined bottom surfaces which may be covered by abundant sole markings. Internal structures near bed tops are usually indistinct and bed-surface features such as ripple marks are rare. Graded bedding occurs throughout most of the stratigraphic sequence but rapid lateral and vertical compositional changes other than grading are exceedingly rare.

Cominco geologists have divided the Aldridge Formation into three map units; the Lower, Middle and Upper Aldridge. A similar division was applied by Leech (1952) and Reesor (1956) to Aldridge rocks exposed in the Purcell Mountains. Lower Aldridge sedimentary rocks are composed of a rhythmic succession of thin- to medium-bedded, typically graded beds of very fine-grained quartz wacke. Deep in the known Lower Aldridge section, cross laminations within beds are frequent and detrital quartz grains range from a more common silt-size to rare sand-size. Toward the top of the Lower Aldridge section, matrix material separating detrital grains becomes more abundant, grain size decreases and internal structures other than grading become rare.

Interbedded with the rhythmic sequence of graded beds are laminated sequences of mudstone ranging from a few millimetres to several metres thick. Laminae and discontinuous blebs of pyrrhotite emphasize layering in the laminated mudstone and weathering of the pyrrhotite imparts a conspicuous rusty colour to outcrops.

Massive to poorly bedded, elongate lenses of intraformational conglomerate occur locally near the top of the Lower Aldridge. Clasts range in size from a few millimetres to as much as 30 cm and are composed of Aldridge rock types in a mudstone matrix. Frequently there is a clast-size grading with coarser fragments concentrated near the base of the deposits. Clasts may be closely packed and well-sorted but frequently the framework is open and the average matrix to clast ratio is five to one.

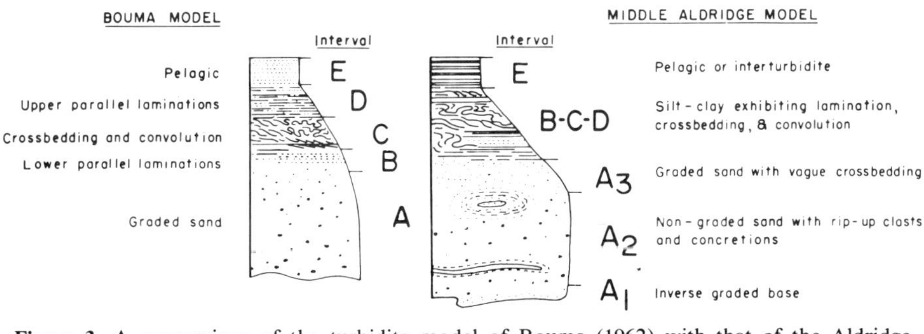

Figure 3. A comparison of the turbidite model of Bouma (1962) with that of the Aldridge Formation.

The Middle Aldridge is marked by the appearance of distinctive graded arenaceous beds whose lighter weathering colours contrast sharply with the rusty weathering Lower Aldridge. Thinly bedded, rusty weathering rocks similar to those in Lower Aldridge sequences are interbedded with thicker, graded arenites but are definitely subordinate. Regularity of internal texture and structure, and bed form have led to the conclusion that the graded arenaceous sedimentary rocks of the Middle Aldridge are mostly turbidites (Bishop *et al.*, 1970). Many of the bed forms correspond to the A-E turbidite structure described by Bouma (1962); however, the more common structure is illustrated in Figure 3. Typically the basal unit can be divided into three subunits of which the lower, A_1, is inversely graded and often exhibits sculpturing and load casting. A_2 is ungraded and may contain rip-up clasts near the base and concretions near the centre. A_3 is normally graded and may show vague current lamination. The middle unit, equivalent to the Bouma B/C/D sequence, is indistinct. Laminated or cross-laminated sequences may characterize a locality or may be missing, such that the upper inter-turbidite material rests directly on A_2 or A_3. The upper unit includes finely laminated interturbidites of dark grey mudstone.

Erosional sole-features measured in the Middle Aldridge indicate general south to north sediment transport. Depositional current features, mainly ripple cross-laminations measured at the tops of the arenaceous units, indicate west to east current direction in the western Purcell Mountains and east to west current direction in the eastern Purcell Mountains. The north-trending trough defined by the sole markings probably represents only a small part of a much larger complex basin holding the initial deposits of the Belt and Purcell Supergroups.

Thin-bedded to laminated carbonaceous mudstone becomes the dominant lithology of the 300 m thick Upper Aldridge. The contact between the Middle and Upper Aldridge is gradational over stratigraphic thicknesses ranging from a few to tens of metres. Disseminated grains and blebs of pyrrhotite aligned along bedding occur in places in carbonaceous mudstone of the Upper Aldridge and here the rock is rusty weathering.

LOCAL GEOLOGICAL SETTING

The Sullivan orebody occurs near the top of the Lower Aldridge Formation and has the shape of an inverted and tilted saucer. The maximum north-south dimension is about 2000 m and the east-west dimension is about 1600 m (Fig. 4). It has flat to gentle easterly dips in the west, moderate easterly to northeasterly dips in the centre, and gentle easterly to northeasterly dips in the east (Figs. 5 and 6). The footwall rocks are composed of intraformational conglomerate and massive lithic wacke overlain by quartz wacke and pyrrhotite-laminated mudstone. The ore zone is overlain by several upward fining sequences of quartz wacke and mudstone.

The orebody attains maximum thickness of 100 m approximately 100 m northwest of its geographic centre, and thins outward in all directions. To the east, the orebody thins gradually to a sequence of pyrrhotite-laminated mudstone 3 to 5 m thick that persists laterally for some distance. To the north, the orebody thins less gradually, and the ore zone is truncated by the Kimberley fault (Fig. 6). To the west, the orebody thins abruptly and is cut by dyke-like apophyses of the footwall gabbro. To the south, within the limit of economic mineralization, thickness changes are generally irregular and abrupt.

Restricted distribution of immediate host rocks and ore zone, and contrasting abrupt thinning to the west and gradual thinning to the east suggest that the immediate host rocks and orebody were deposited in a sub-basin receiving locally derived material, sulphides and some component of normal Lower Aldridge sediment. Although the major structures forming the sub-basin have yet to be defined, several northerly striking normal faults (Sullivan-type faults) may now occupy fault loci which were active during sedimentation.

Additional evidence suggesting structural control of the sub-basin is the geometry of a gabbro body beneath the orebody. This gabbro is typically concordant about 500 m below the eastern edge of the orebody. Following this gabbro westward, it rapidly transgresses upward to meet the footwall of the orebody near its western margin but, in continuing westward, it transgresses downward to again resume its sill-like form at approximately its original stratigraphic position.

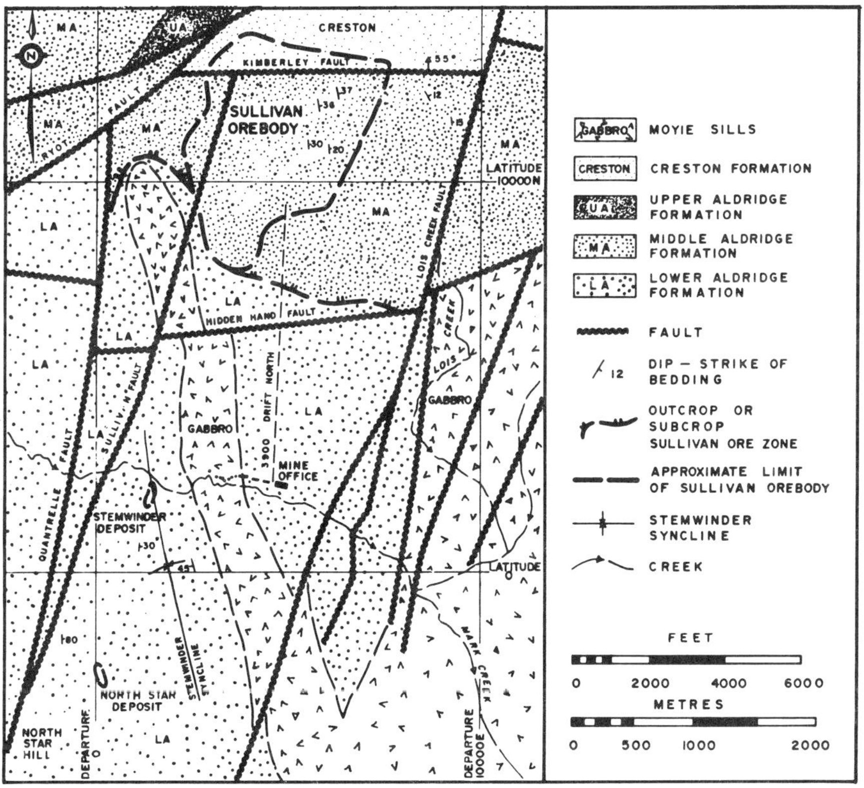

Figure 4. Generalized geological map of the Kimberley area.

The resultant configuration is a north-northwest-trending arch which is subparallel to the north-northeast trend formed by the Sullivan, Stemwinder and North Star orebodies (Fig. 4). The trend of mineralized zones crosses the trend of this arch. Contact relationships of the intrusion in apparent crosscutting areas have only been observed in several drill cores obtained beneath the orebody. These show core-to-bedding angles immediately east of the intrusion which are most consistent with very limited discordancy. An hypothesis explaining this ambiguous transgressive habit is that the intruded sedimentary rocks are draped over a horst, implying that faults on the east side of the horst at least were active during sedimentation, and affected geometry of the sub-basin.

The orebody lies on the folded and faulted eastern limb of a broad north-trending anticline (Fig. 1). The structure plunges gently to the north and is locally asymmetric and overturned to the east. The Kimberley, Ryot and Hidden Hand fault systems, the 010° trending Sullivan-type faults and other minor faults form an intricate mosaic disrupting the fold limb. South of the Hidden Hand fault (Fig. 4), rocks on the eastern limb of the broad anticline are folded about north-trending axes to form an anticline and syncline having the shape of an open S as viewed from the north. The North Star orebodies, mined at the turn of the century, were located in a gentle fold structure on the east limb of the syncline and the crosscutting Stemwinder deposit is situated in the hinge of the syncline which opens northward. Open to isoclinal minor folds on the

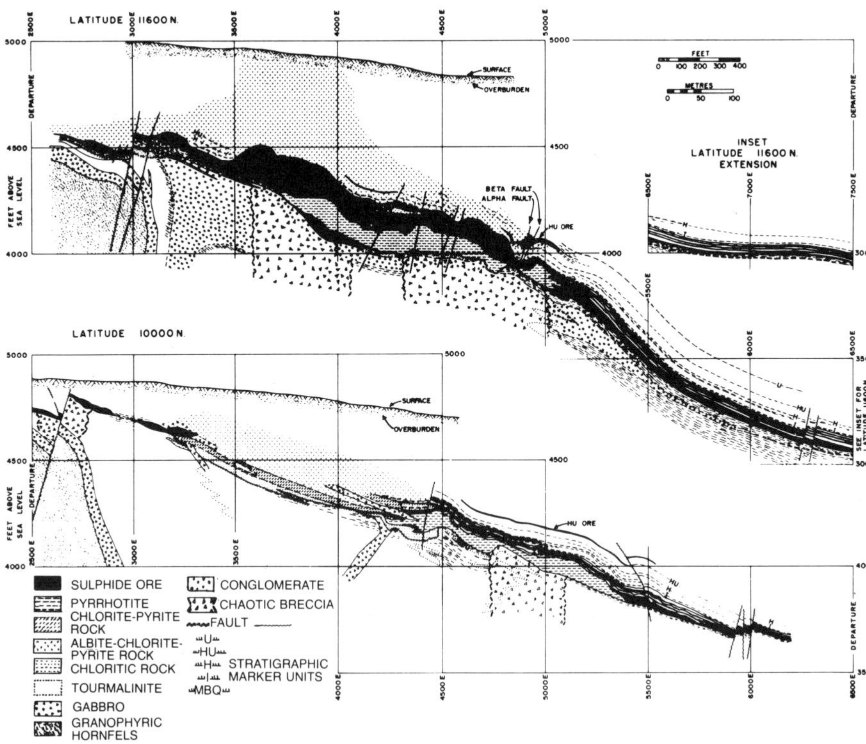

Figure 5. East-West geological cross-sections through the Sullivan orebody.

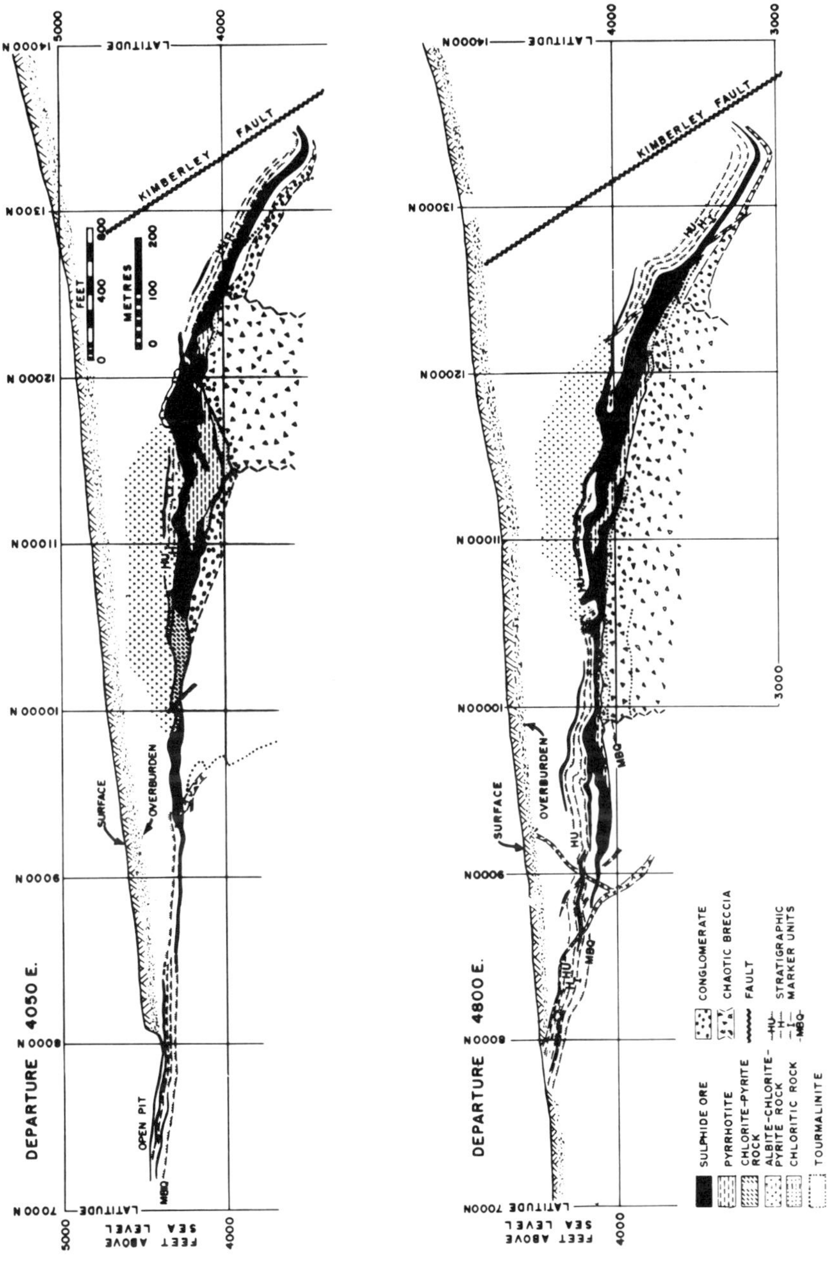

Figure 6. North-South geological cross-sections through the Sullivan orebody.

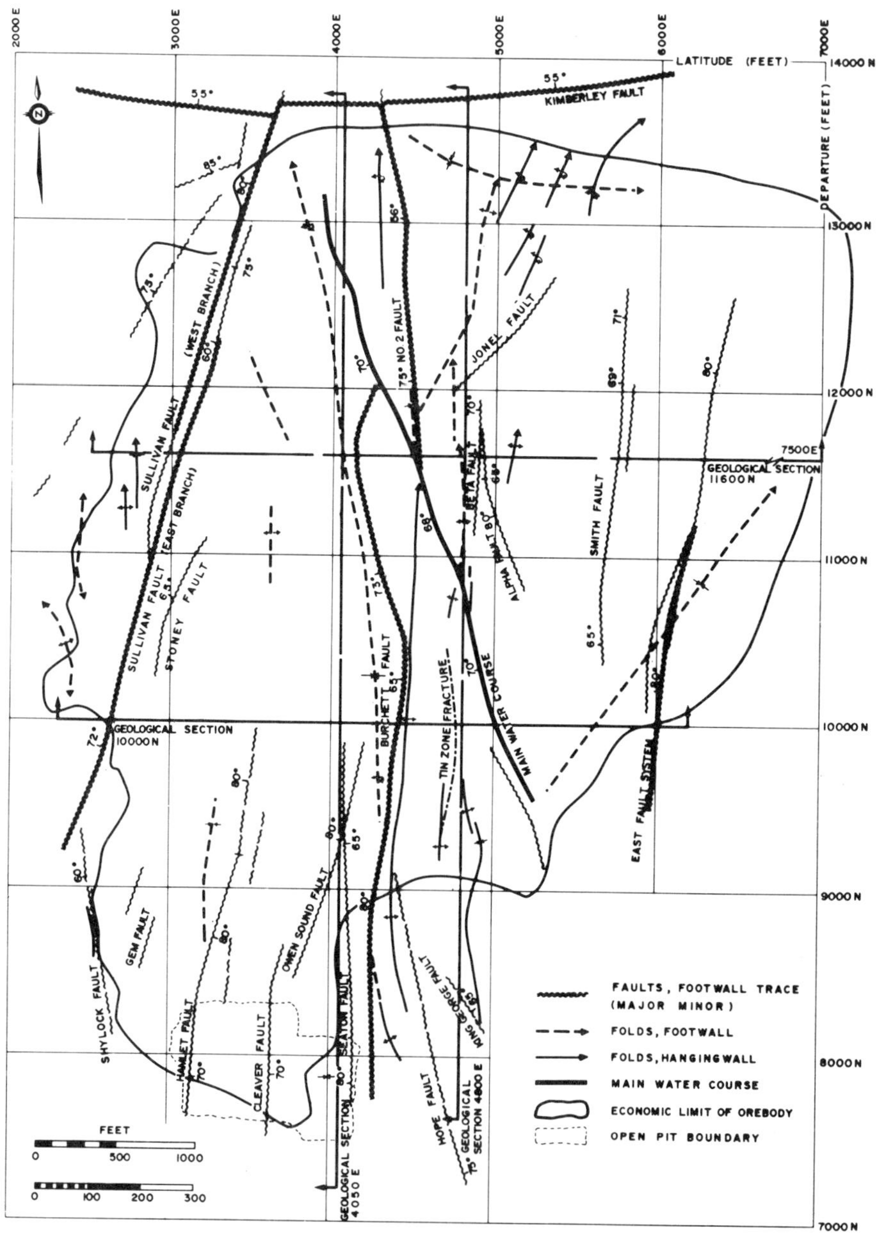

Figure 7. Structural features in and adjacent to the Sullivan orebody.

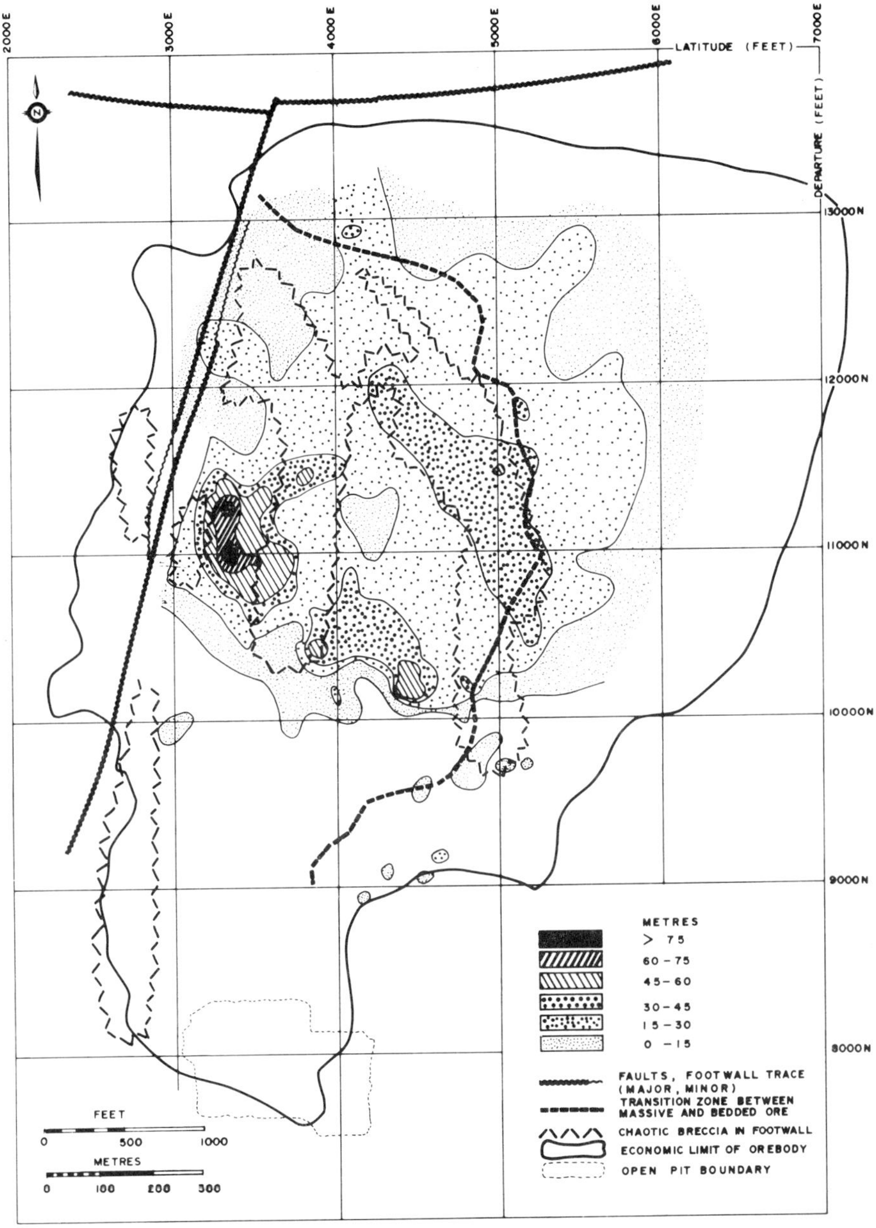

Figure 8. Isopach map of the footwall conglomerate.

limbs of the syncline are doubly-plunging and are not cylindroidal. Between the Hidden Hand fault and Kimberley fault, minor folds encountered in footwall development headings at the Sullivan tend to be open and asymmetric to the east. Some degree of intrafolial deformation within the sulphide orebody may account for the change in fold geometry north of the Hidden Hand fault. North of the Kimberley fault, rocks on the east limb of the broad anticline are locally overturned.

The Kimberley and Hidden Hand faults lie across the regional structure and are generally parallel to east-trending segments of the Moyie and St. Mary faults. The Kimberley fault dips 45° to 55° north and truncates the ore zone to the north. With over 3000 m of stratigraphic displacement, the fault juxtaposes rocks of the Creston and Kitchener Formations against rocks of the Lower Aldridge. Displacement on the north-dipping Hidden Hand fault is of the order of a few hundred metres of apparent normal dip-slip movement.

The Sullivan-type faults cut the orebody with a consistent west side down normal displacement ranging from a few metres to 30 m. The largest member of the group, the Sullivan fault, occurs near the western margin of the orebody (Figs. 4, 5 and 7). At the northwest margin of the orebody, a northeast-trending fault apparently truncates the westward extension of the Kimberley fault. Several Sullivan-type faults offset the trace of the Kimberley fault although earlier phases of movement along the Sullivan-type faults may have occurred.

GEOLOGY OF THE SULLIVAN OREBODY

Introduction

The Sullivan orebody consists of sulphide rock composed of more than 70% sulphides in thick, gently dipping conformable units enclosed by unaltered or altered quartz wacke and mudstone. Ore is massive to wispy-layered in the western part and passes outward on the north, east and south to delicately-bedded sulphide rock interbedded with fine-grained clastic sedimentary rocks. In this paper, the term massive, when applied to sulphide rock, connotes material that is internally homogeneous to vaguely bedded. The term massive is not applied to well-bedded sulphide rock nor used to imply a relative amount of sulphides. Major differences exist in footwall rocks, ore zone and hangingwall rocks in different areas of the mine. Hence, in this discussion, for simplicity, the ore zone and enclosing rocks will be divided into a western part and an eastern part joined by the transition zone shown on Figures 5 and 8. This transition zone is commonly only a few metres or tens of metres wide.

Generally to the west of the transition, in the massive part of the orebody, the bulk of sulphide ore occurs in a single conformable lens in which layering is poorly developed. The footwall, ore zone and hangingwall sedimentary rocks are generally altered. The footwall rocks are unusually rich in tourmaline and the hangingwall rocks rich in albite and chlorite. Footwall sulphide veins and disseminated sulphides are locally abundant and broad linear zones of brecciation disrupt a complex sequence of massive and bedded footwall sedimentary rocks.

On the other side of the transition zone, generally to the east, footwall and hangingwall sedimentary rocks enclosing well-bedded ore are relatively unmodified. Other than metamorphic recrystallization they retain many of their primary chemical and physical characteristics, and bedding in footwall rocks is largely undisturbed except for local folding attributed to soft-sediment deformation. Footwall sulphide veins are rare.

Description of the geology of host rocks and ore zone will follow that of the stratigraphic sequence as shown on the ideal stratigraphic section (Fig. 9). Intraformational conglomerate unconformably overlies normal Lower Aldridge sedimentary rock beneath much of the orebody. Conglomerate is thick and stratigraphically complex beneath the western part of the orebody and thins outward to a simple stratigraphic unit. Stratigraphic relations between the conglomerate and the sulphide

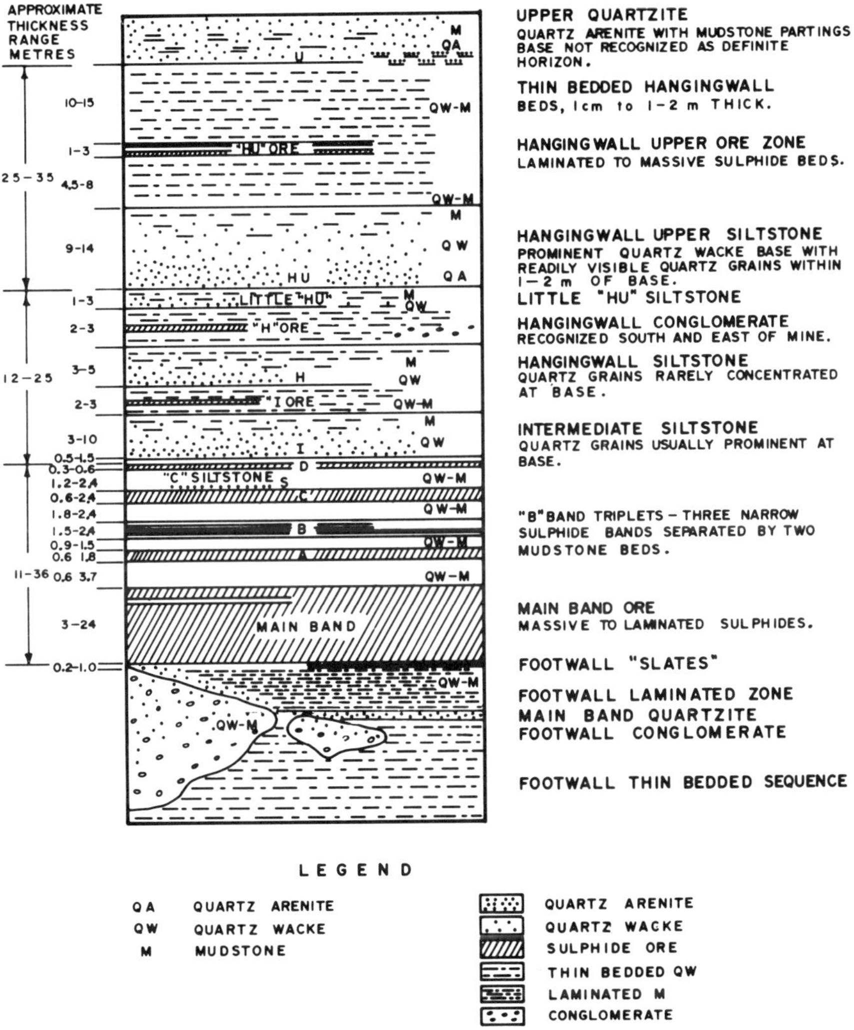

Figure 9. Idealized geological column of the bedded part of the Sullivan orebody.

footwall are also complex beneath the western part of the orebody but become extremely uniform beneath the well-bedded part of the orebody. Footwall stringer mineralization although volumetrically not very large is locally abundant west of the transition zone. Conformable and disseminated sulphide mineralization, principally pyrrhotite, is common in the footwall on both sides of the transition.

Several readily identifiable sequences grading from quartz wacke to mudstone and each overlain by pyrrhotite-laminated mudstone form the immediate hangingwall of the orebody. These can be traced throughout much of the hangingwall except where obliterated by alteration. The pyrrhotite-laminated mudstone grades laterally to conformable hangingwall ore which is generally situated stratigraphically above the transition zone. To the south and southwest, hangingwall mineralization persists laterally at mineable grades beyond the point at which lower sulphide zones become sub-economic or were not deposited.

Geology of the Footwall Rocks

Beneath the ore zone the deep footwall succession is typical of the upper part of the Lower Aldridge. Thin- to medium-bedded mudstone is interbedded with subordinate fine-grained quartz wacke. With appropriate marker beds it is possible to make lithostratigraphic correlations over large areas in the footwall.

Conglomerate. Intraformational conglomerate underlies the northern two-thirds of the orebody. The base of the conglomerate is nearly everywhere disconformable. Beneath the western part of the orebody the conglomerate sequence is over 80 m thick (Fig. 8) with the base poorly defined in several areas. The conglomerate thins rapidly to the south and north and more gradually to the east. Contacts in areas of thickest conglomerate sharply truncate bedded sedimentary rocks through stratigraphic thicknesses of 60 m or more. The thick part of the conglomerate is confined on its south, west and north sides by steep, discordant contacts and by a less steep but unconformable contact to the east. In areas where conglomerate is thicker than 45 m, sulphides rest on it or are separated by only a few metres of thinly laminated sedimentary rock. Elsewhere beneath the western part of the orebody thinly laminated sedimentary rock up to 15 m thick lies between the top of the conglomerate and the base of sulphide rock.

Clasts and matrix of the conglomerate are of the same composition as the underlying sedimentary rocks and no foreign clasts have been identified. The conglomerate generally lacks bedding but may be weakly bedded in places. It is locally graded at its fringes. Clasts range in abundance from 10 to 95% and in size from 2 mm to greater than one metre in diameter, and average one to 3 cm (Fig. 10). Typically, clasts are well-sorted although bimodal distribution of clast size is common. Clast shape is related to lithological type. Homogeneous mudstone clasts are frequently tabular or disk-shaped and well-rounded. Laminated to thin-bedded and pyrrhotite-laminated clasts are nearly equant or rhombic and frequently have tapered corners. Arenaceous clasts are rounded to subrounded and very large clasts tend to exhibit high sphericity. In cases of bimodal or polymodal distribution of clast size, smaller clasts tend to be subangular and larger clasts subrounded. Occasional round disk-shaped clasts of tourmalinite and well-rounded arenaceous clasts containing disseminated sphalerite are found in the conglomerate beneath the eastern part of the orebody. In many places, footwall beds exhibit structures formed by soft-sediment deformation which are truncated by massive conglomerate.

Stratigraphy of the conglomerate is relatively simple beneath the eastern part of the orebody and more complex to the west. To the east the conglomerate is usually a single massive unit overlying thin-bedded quartz wacke or pyrrhotite-laminated mudstone. Locally a thick sequence of poorly sorted massive conglomerate is overlain by a well sorted unit containing tightly packed clasts and separated from the main conglomerate unit by a thin sequence of pyrrhotite-laminated mudstone. To the west, the conglomerate is a composite unit made up of frequent interbeds of massive, poorly to well sorted conglomerate, thinly bedded, well bedded sedimentary rocks and thick units of rock in which bedding has been intensely disrupted.

Footwall Bedded Sequence. Thin- to medium-bedded, commonly graded quartz wacke interbedded with pyrrhotite-laminated mudstone conformably overlies the conglomerate beneath the eastern part of the orebody. With the exception of two well-defined marker units, the Main Band Quartzite (MBQ) and the Footwall "Slates" (Fig. 9), rock-types making up the sequence overlying the conglomerate are not laterally persistent. Gradual lateral changes preclude unit by unit correlations over distances greater than a few hundred metres. Lamination for lamination correlation over a few hundred metres can be made for the pyrrhotite-laminated mudstone (Freeze, 1966).

The Main Band Quartzite, a 1.5 m thick graded quartz wacke bed, forms the base of an upward-fining sequence immediately beneath the orebody. This bed is recognized over large areas beneath the north, east and south parts of the orebody. The base of the bed is commonly deformed by load casting and contains well-sorted quartz grains and clasts up to 4 mm in diameter which often appear to float in a carbonate matrix. One or more graded quartz wacke beds overlain by the Footwall "Slates" form the remainder of the upward-fining sequence beneath the sulphide footwall. The Footwall "Slates" is a 0.2 to 1.0 m thick sequence of chloritic, laminated mudstone which can be traced over large distances. Pyrrhotite laminations occurring at 2 to 40 mm intervals, along with minor layers of disseminated sphalerite and opaque (graphite?) material impart a distinctive layering to the rock.

Sedimentary rocks overlying the conglomerate beneath the western part of the orebody display even more rapid lateral change than do those overlying the conglomerate to the east. No marker beds have been recognized in the sequence.

Below the periphery of the western part of the orebody, the sequence may contain as much as 10 m of interbedded tourmalinite and chlorite-rich or sericite-rich quartz wacke and mudstone overlain by pyrrhotite-laminated, variably chloritic rock containing from 10 to 50% pyrrhotite, sphalerite and galena. Extremely rapid lateral changes occur in strata exposed in footwall development. Chlorite-rich or sericite-rich rock laminated with pyrrhotite may grade rapidly over a few centimetres to pyrrhotite-laminated tourmalinite or may contain large, irregular, frequently stratabound masses of tourmalinite through which the layering passes undisturbed (Shaw, in preparation). Alternatively, well-bedded rock may pass laterally into rock devoid of bedding. Only the pyrrhotite-laminated rock immediately beneath the sulphide footwall can be traced over any distance with confidence. Locally beneath the centre of the conglomerate. The best documented phase, the post-conglomerate chaotic rock occurs beneath the sulphide footwall (Figs. 5 and 6).

Chaotic Breccia. Extensive zones of breccia (Figs. 11, 12, 41) have been recognized in the western part of the ore zone and Jardine (1966) identified several phases. The earliest phase, the pre-conglomerate breccia, is poorly documented and very difficult to differentiate from breccias formed after deposition of the conglomerate. The best documented phase, the post-conglomerate chaotic breccia, disrupts both conglomerate and well-bedded sedimentary rocks in the footwall.

According to Jardine (p. 45):

> "The rock is composed of fragments of Aldridge-type sedimentary rocks varying from sand-size to pieces measured in tens of feet. Bedded fragments are often wispy, twisted and rotated, with ends that appear to have torn rather than broken. Blocks of conglomerate are mixed in with bedded and massive pieces, and the whole appearance is heterogeneous and chaotic. Apparently much of the broken rock was in a relatively soft state at the time of breaking.
>
> "The blocks are now completely consolidated in a matrix of silty argillite . .

Figure 10. Footwall conglomerate containing closely packed pebbles of typical Aldridge rock types. Pebbles contain disseminated and laminated pyrrhotite and several are rimmed with pyrrhotite. Core diameter is 3.2 cm.

Pyrrhotite is disseminated through the matrix and is also present in irregular veinlets that lace through the rock."

The irregular areas of post-conglomerate chaotic breccia outlined by Jardine show three elongate trends in a north-south direction and another with an irregular east-west direction. Vertical extent of brecciation in the footwall has not been clearly defined. The limits of brecciation at depth shown in Figures 5 and 6 are interpretive.

Geology of the Ore Zone, Eastern Part

The ore zone in the eastern part of the orebody is as much as 36 m thick (Fig. 13) and consists of bedded sulphide rock with interbedded barren clastic sedimentary rock in its upper part. It is characterized by sharp contacts and extremely regular stratigraphy. Primary sedimentary structures such as bedding and local soft-sediment deformation are preserved in detail. Figure 9 illustrates the ore zone sequence from the base of the Main Band to the base of the "Upper Quartzites". The principal sulphide beds are labelled Main Band and "A" through "D" Bands from base to top

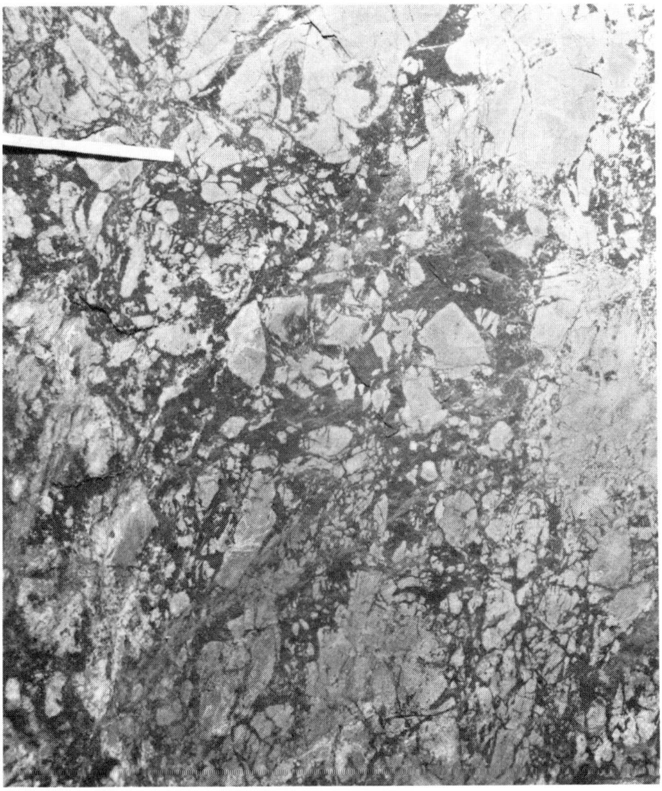

Figure 11. Chaotic breccia fragments enclosed in a pyrrhotitic matrix. Angular fragments ranging from centimetres to metres across are in many places intermixed. Pyrrhotitic matrix may contain 1 to 5% sphalerite and trace amounts to abundant stringers of arsenopyrite. Rule is 1.5 cm wide.

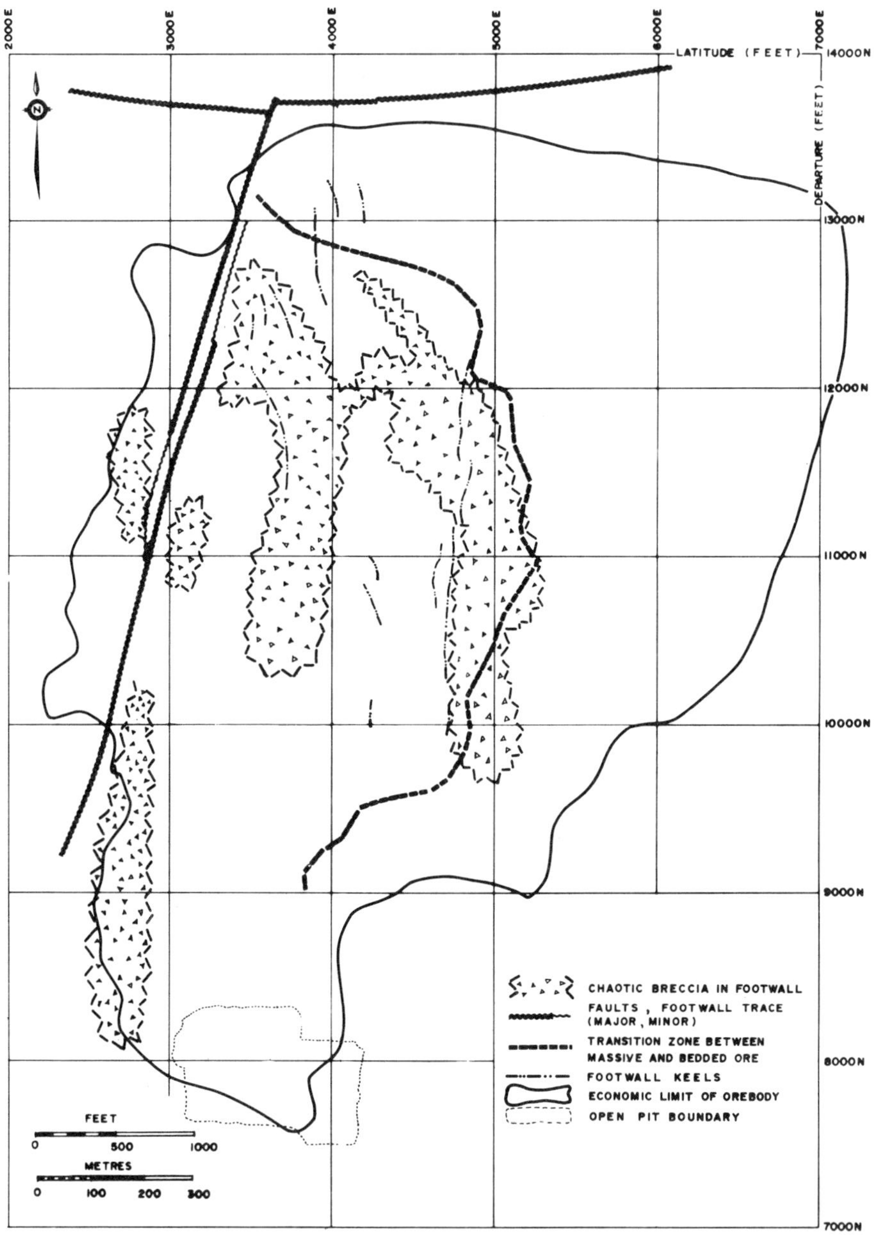

Figure 12. Distribution of chaotic breccia in the footwall of the Sullivan orebody.

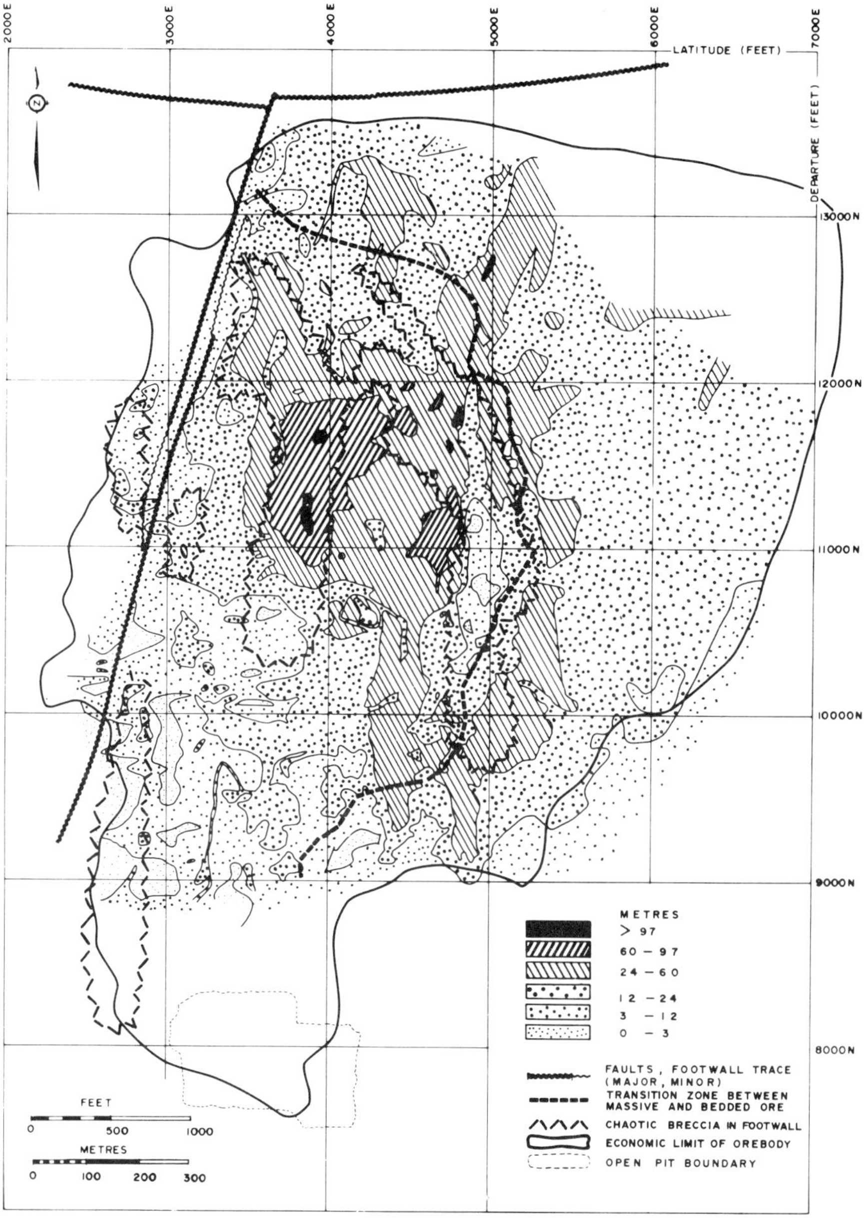

Figure 13. Isopach map of the Sullivan orebody exclusive of hangingwall ore.

in the section. In the transition zone joining the eastern and western parts, individual sulphide Bands lose their identity as the sulphide rock becomes more massive and the intercalated clastic sedimentary rocks disappear.

Main Band. The Main Band is the lowest sulphide band of economic significance and is the major sulphide accumulation in the eastern part of the orebody. Up to 24 m thick, the Main Band overlies the "Footwall Slates" with a sharp conformable contact where the rock changes from mudstone to sulphide rock composed of more than 70% sulphide across a single bedding plane (Figs. 14, 32). The lower two-thirds to three-quarters of the Main Band in the eastern part of the orebody has definite characteristics which set it apart from the overlying sulphide bands. It is particularly rich in sulphides, lacks mudstone interbeds but exhibits internal bedding (Fig. 15) and often contains fragments of various rock types.

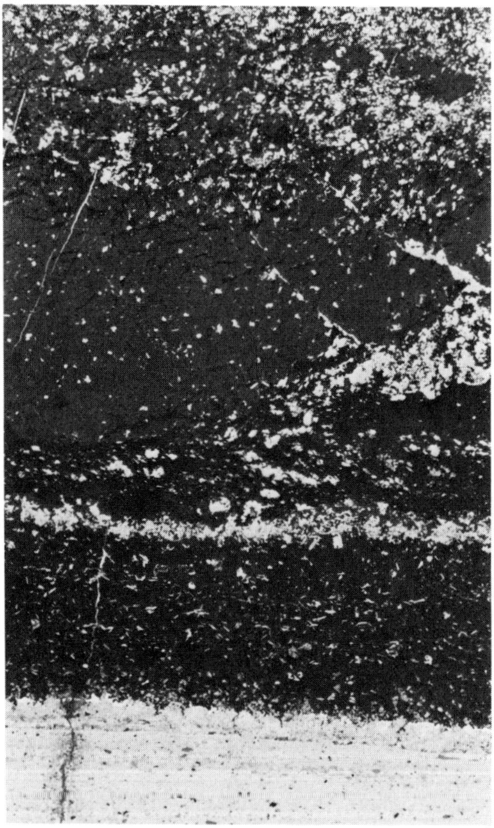

Figure 14. Photomicrograph of the base of the Main Band, at the transition zone. Darker layers in Footwall "Slates" at bottom are biotite. The sulphide assemblage is: basal layer sphalerite, middle layer galena and pyrrhotite, upper half of sulphide section sphalerite. Gangue minerals are calcite, muscovite, pale brown biotite and minor quartz. Four centimetres of section shown. Plane light.

The lower two-thirds to three-quarters of Main Band sulphide rock consists of a series of dense, thin beds of fine-grained sulphides without significant amounts of argillaceous clastic material, either as interbeds or disseminated grains. The sulphides are mainly pyrrhotite, sphalerite and galena occurring in distinct beds containing varying proportions of these minerals (Figs. 15, 16, 32). Usually, either pyrrhotite or sphalerite is the dominant sulphide in a bed with smaller quantities of the subordinate sulphide and galena. Beds composed almost entirely of sphalerite or galena are also present. The beds are usually less than 3 cm thick, but may be up to 30 cm thick. Most beds have what appear to be sharp boundaries, but microscopic examination reveals that the boundaries are gradational. For example, if pyrrhotite constitutes the major fraction of a sulphide bed and the adjacent bed is largely sphalerite their boundary consists of a one millimetre zone containing a mixture of the two, and each bed will contain small amounts of the sulphide forming the

Figure 15. Highly deformed sulphide bedding with no mudstone interbeds near the centre of the Main Band. Light grey beds are composed of galena, medium gray beds pyrrhotite with minor sphalerite, and dark grey beds sphalerite with minor pyrrhotite. Rule in inches. Top to upper left.

major component of the adjacent bed. The sulphide grains range from 0.005 to 0.5 mm in diameter but almost all are less than 0.1 mm in maximum dimension. Sphalerite occurs as equidimensional, brown grains from 0.02 to 0.3 mm in diameter. Pyrrhotite occurs as indistinct grains up to 0.5 mm in diameter. Galena occurs as irregular grains from 0.02 to 0.4 mm in diameter.

Sulphides compose 75% of the sulphide rock with calcite, quartz and various silicate minerals making up the remainder. The non-sulphide minerals occur as randomly oriented or aligned grains ranging in size from 0.05 to 1.0 mm in diameter and also as part of larger masses (Fig. 14). Calcite is more plentiful than quartz and comprises 15% of the rock. Silicate minerals include chlorite, muscovite and occasionally scapolite or garnet. The size and shape of the grains of quartz and silicates is

Figure 16. Details of bedding and folding in a part of the area illustrated in Figure 15. Beds composed predominantly of galena (light grey) and pyrrhotite (medium grey) have been strongly attenuated along fold limbs. The parallel fabric extending from the lower right to upper left of the photograph is tectonic. Rule in inches.

usually coarser and more irregular than grains of the same minerals in adjacent argillaceous clastic rock.

Megascopic fragments are locally common. They are occasionally up to one metre across but are commonly less than one centimetre in diameter. Fragments include altered and unaltered clastic sedimentary rock, carbonate rock and granular quartz-rich rock. Inclusions composed essentially of a single mineral include pyrite (buckshot pyrite), calcite, quartz, sphalerite, scapolite and garnet. The monomineralic inclusions may be up to 0.5 cm across. On rare occasions megascopic fragments and inclusions may make up as much as 20% of particular beds in localized areas and it is common to find beds with 2 to 5% fragments and inclusions.

The upper one-third to one-quarter of the Main Band in the eastern part of the orebody is different from the lower part as it contains thin, closely spaced interbeds of mudstone and quartz wacke in the sulphide beds. The mudstone and quartz wacke beds are usually similar to those outside the ore zone, though in some cases they may be chloritic, or contain some sulphides. The non-sulphide material within the sulphide beds of the upper part of the Main Band consists of equidimensional grains of quartz, calcite, chlorite, biotite, garnet, tremolite, clinozoisite and, rarely, hornblende. These grains are similar in shape but larger than grains of the same minerals occurring in the lower part of the Main Band.

The sequences containing thin interbedded sulphide rock and clastic sedimentary rock contain up to 60% sulphide minerals as laminae or as beds up to 30 cm thick. Average grain size of sulphides in this rock is coarser than that in sulphide rock in the basal part of the Main Band and there is a greater range in sulphide grain size from bed to bed within a single section. Certain laminae contain grains up to 1 mm in diameter whereas adjacent laminae contain grains less than 0.1 mm in maximum dimension. The various sulphide laminae show less intermixing of sulphides than does sulphide rock in the basal part of the Main Band. Individual beds may be largely pyrrhotite or galena, and sphalerite with galena, or pyrrhotite with sphalerite are common combinations. Composition of strata at the top of the Main Band varies gradually through the mine area. Nodules composed largely of sulphides, usually pyrite, are common. These are up to 20 cm in maximum dimension (which is always conformable to the stratigraphy, in contrast to sulphide fragments in sulphide rock near the base) and up to 3 cm thick. Adjacent laminae always bend over or under the nodules.

In several localities in the southeastern part of the orebody, the lower one-half to two-thirds of the Main Band has a conspicuous fragmental texture (Fig. 34). The rock is massive sulphide and contains a variety of fragmental material including clastic sedimentary rock and tourmalinite as well as inclusions consisting of pyrite, calcite, quartz, and sphalerite, all of which are set in a matrix of about 30% quartz grains, calcite, chlorite and phlogopitic mica and about 70% pyrrhotite, sphalerite and galena. Fragments of tourmalinite and mudstone are generally subangular to subrounded and range up to 6 cm in length. Largest fragments occur near the base and the massive fragmental rock grades upward through a transition zone to sulphide rock with well-developed bedding. The massive rock contains abundant megacrysts of pyrite (buckshot pyrite). Mudstone fragments are similar to adjacent footwall rocks, or are altered to chloritic rock. Chloritic fragments may have quartz and calcite veins.

Ore Bands "A" through "D". The "A", "B", "C" and "D" ore Bands consist of thin-bedded sulphides with mudstone interbeds similar to those described in the upper part of the Main Band. The regularity of the sulphide beds is very marked (Figs. 17, 18, 30, 35). The Bands are readily recognized throughout the eastern part of the mine over an area more than 2000 m by 1200 m and usually the different Bands can be distinguished owing to their individual character. The "B" Band contains three beds of sulphides, called the triplets, separated by two beds of mudstone ranging from 5 to 15 cm thick. Normally, the triplets can be recognized throughout the eastern part of the mine and serve as a stratigraphic marker. Individual beds have

Figure 17. Sulphide beds of the "B" Band triplet, characteristic of upper sulphide bands in the eastern part of the orebody. Three sulphide beds composed of planar sulphide layers interbedded with mudstone are separated by graded beds of quartz wacke. Slight crenulation of sulphide beds is thought to be tectonic. Axes of folds trend northeast. Rule divided in inches. Top of section to upper right.

sharp contacts, are continuous and of nearly uniform composition throughout large areas. Iron sulphides are more widespread and continue laterally beyond the limits of ore-grade lead and zinc.

Waste Bands Between Ore Bands. The sequence of thin-bedded sulphides with mudstone interbeds, which has its base within the Main Band and upper limit in and above the "D" sulphides, contains four 0.6 to 4 m thick beds of graded quartz wacke. Each of these beds of waste, separating essentially similar sulphides, consists of one or more graded units without laminated interbeds. The graded beds have the same texture and composition as those in the footwall bedded sequence. However, they are thicker, with noticeable coarse-grained lower portions, and are not found beyond the limits of the ore zone.

Figure 18. Photomicrograph of bedded ore from the "C" Band. Black and dark grey layers are sulphides. Light-coloured layers are composed of muscovite, quartz, biotite and chlorite. Bladed opaque mineral is pyrrhotite. Rounded mineral in light-coloured layer is zoned garnet. Plane light. Four cm of section shown. Top up.

Geology of the Ore Zone, Western Part

The western part of the orebody has important differences from the eastern. It consists of massive- to poorly-bedded sulphide with subordinate interbedded sediment. Contacts are conformable and bedding attitudes within the ore are generally parallel to those in the enclosing sedimentary rocks. However, discordant contacts between different sulphide assemblages within the orebody are recognized. Rapid thickness changes, or abrupt terminations of lithologic units and locally severe deformation preclude detailed correlation within the western part of the orebody and across the transition to the eastern part.

The western part of the orebody is commonly greater than 50 m thick. The abrupt thickness changes in the west contrast sharply with gradual thickness changes in the east. Isopachs of the orebody (Fig. 13) define a particularly abrupt zone of thinning along a trend at approximately 4800 departure in the area of the Alpha and Beta fault zones (Figs. 5 and 7). The orebody thins eastward across this northerly trend, from more than 60 m to as little as 2 m, and adjacent to this thin part, hangingwall bedding is generally conformable with layering in the ore. Host rocks and ore zone may be deformed, highly altered and brecciated in and near the zone of thinning. Disrupted and discordant footwall contacts occur along keel-like structures at the base of the sulphides and extend down for several metres (Figs. 5 and 12). Along discordant contacts of keels the rocks are frequently brecciated and altered. Layering in sulphides within some keels is generally conformable to contacts of the keels and wraps around the sharp downward projection of the structures. Some keels have one conformable limb and one fault-bounded limb, while others pass downward into sulphide veins.

The sulphide rock footwall in the western part of the orebody, although almost everywhere strictly conformable, is commonly not as planar as the footwall to the east. Over large areas sulphide rock, principally pyrrhotite, overlies sulphide-laminated sedimentary rock with a sharp contact. In a few places rapid lateral changes in lithology and rapid changes in vertical distribution of pyrrhotite result in a gradational contact. Rarely, sulphide rock may overlie conglomerate or breccia with a matrix largely composed of sulphide and apparently gradational with the sulphide rock. Minor folding of the sulphide footwall suggests soft-sediment deformation.

Structures and textures within the western part of the orebody are quite different from those in the eastern part (Fig. 5). The most conspicuous difference is between regular bedding in the eastern part (Figs. 15, 16, 17 30, 31, 35) and less regular layering in the west (Figs. 32, 33). Large-scale layering in the west is defined by generally conformable lenses of nearly barren pyrrhotite. The largest such lens occurs at or near the sulphide footwall (Fig. 5) in the thickest part of the orebody and extends for about 350 m along strike and 250 m down dip. It has a maximum thickness of about 35 m. Broadly conformable at its footwall, the perimeter of the lens has an abrupt or gradational change to commercial ore (Fig. 5). Locally within the pyrrhotite lens, sphalerite and galena occur as disseminated grains, wispy fine- to coarse-grained concentrations, fracture fillings and veins without alteration envelopes.

Close examination of the pyrrhotite lens reveals internal layering which is generally parallel to the sedimentary layering in the enclosing rocks. Fragments of vein quartz, granular quartz, carbonate rock, mudstone and altered mudstone are present, particularly near the base. The rock consists of 60 to 80% pyrrhotite, 10 to 30% calcite, about 5% of each of quartz, chlorite and sphalerite, with smaller amounts of muscovite, pyrite and galena. Figures 19 and 20 show textures similar to those in lenses of nearly barren pyrrhotite. Weakly disseminated chalcopyrite (up to 2%) occurs locally. Pyrrhotite grains are less than 0.05 mm in diameter. The grain size of the quartz, calcite and chlorite varies between wide limits but is usually 0.1 to 1 mm in diameter. Non-sulphide minerals occur as single isolated grains and as several contiguous grains forming a mass which often has regular smooth outlines. The latter vary in size up to a few centimetres across. The pyrrhotite lens grades rapidly upward to sulphide rock in which layering is defined by wispy concentrations of galena and sphalerite in a pyrrhotite matrix. This fine-scale layering is laterally discontinuous with frequent pinch-outs and disruptions; however, the layering is parallel to general sedimentary attitudes. Toward the upper part of the orebody, bedding is clearly defined. Sulphide rock is interbedded with fine-grained clastic sedimentary rock, and bedding within sulphide rock is defined by discrete and often delicate laminae of galena and sphalerite.

Toward the centre of the orebody, precise stratigraphic control is lost in the transition (Fig. 5) from well-bedded ore to poorly layered ore. However the lower sequence of barren pyrrhotite and overlying poorly layered ore west of the transition correlates stratigraphically with the Main Band to the east whereas the more distinctly bedded material toward the top of the central part of the orebody correlates with the upper Bands to the east.

Footwall Mineralization

Minor stratiform and stratabound mineralization is abundant below both the eastern and western parts of the orebody. Discordant mineralization, in contrast, is common beneath the western part of the orebody but is rare beneath the eastern part.

Beneath the eastern part of the orebody, concordant footwall mineralization is most common in the stratigraphic sequence between the conglomerate and the sulphide footwall. Within the conglomerate, sulphide mineralization, predominantly pyrrhotite, occurs as laminae in clasts, as rims surrounding clasts and as disseminations. Pyrrhotite and sphalerite are the principal sulphides in rocks overlying the conglomerate and form discrete laminae in mudstone. Less common are disseminated sulphides concentrated in laminae up to a few millimetres thick. Sphalerite becomes more abundant with approach to the sulphide footwall but generally remains subordinate to pyrrhotite.

Beneath the western part of the orebody, concordant footwall mineralization is more abundant than beneath the eastern part and occurs through at least 40 m of strata. Pyrrhotite and sphalerite are the most abundant sulphides and occur as thin, sulphide-rich laminae and disseminations in mudstone and tourmalinite. Across 1 to 2 cm sections pyrrhotite laminae may make up as much as 25% of the rock although more typically they account for 2 to 5%. Usually quite continuous, the laminae may grade laterally into sulphide disseminated along bedding planes and then into unmineralized rock. In conglomerate underlying the western part of the orebody, clasts may be laminated and rimmed with pyrrhotite (Fig. 40). Clasts composed of 60 to 80% pyrrhotite are common; however, clasts composed entirely of pyrrhotite are rare.

Discordant mineralization, in the form of veins, is locally intense beneath the western part of the orebody and relatively rare beneath the eastern part. Geometry of the veins ranges from planar and regular with sharp matching contacts to irregular with indistinct non-matching contacts (McAdam, 1978).

Planar regular veins appear to cut the irregular veins. The principal sulphide is pyrrhotite; pyrite, sphalerite, and galena may be abundant, and chalcopyrite and arsenopyrite are subordinate. The dominant non-sulphide minerals are quartz and calcite. Individual veins may be dominated by either sulphide or non-sulphide. Planar veins are best developed in relatively hard quartz wacke, quartz arenite or tourmalinite; vein walls become more irregular in softer argillaceous or chloritic rock. Many of the planar veins are zoned with pyrrhotite in the core and sphalerite, arsenopyrite

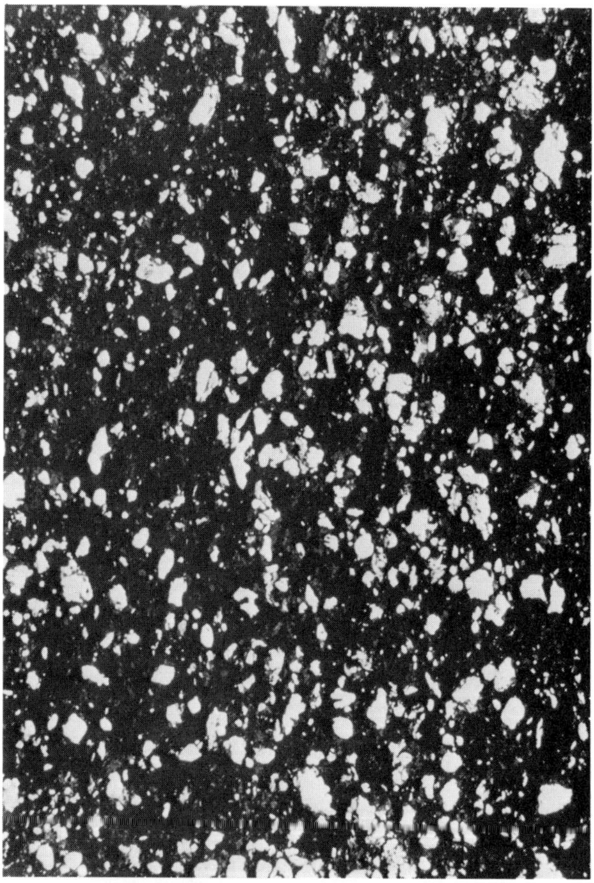

Figure 19. Photomicrograph of unbedded ore containing about 10 per cent galena-sphalerite from the western part of the orebody. Light coloured minerals include calcite, quartz and chlorite (grey, platey mineral). Black area is principally pyrrhotite. A sample with unusually high abundance of non-sulphide material was selected to show the texture. Plane light. Long axis of figure represents 3.4 cm and is parallel to primary and tectonic layering.

or, more rarely, galena at the margins. Locally, the proportions of non-sulphide minerals correlate with the abundance of quartz in adjacent host rock. For example, in quartz arenite, veins often contain abundant quartz, whereas calcite may predominate in chloritic or sericitic rock.

Irregular veins frequently form networks composed dominantly of pyrrhotite, galena and sphalerite with minor amounts of quartz, pyrite, arsenopyrite, chalcopyrite, cassiterite, and tourmaline. Galena and pyrrhotite occupy a central position in the irregular veins with sphalerite and locally arsenopyrite concentrated along vein margins. Sulphides occupying vein margins may be disseminated in wall rock or extend as laminae into it.

Alteration related to individual sulphide veins, although generally inconspicuous, may extend several centimetres into chloritic or biotitic host rock. Alteration

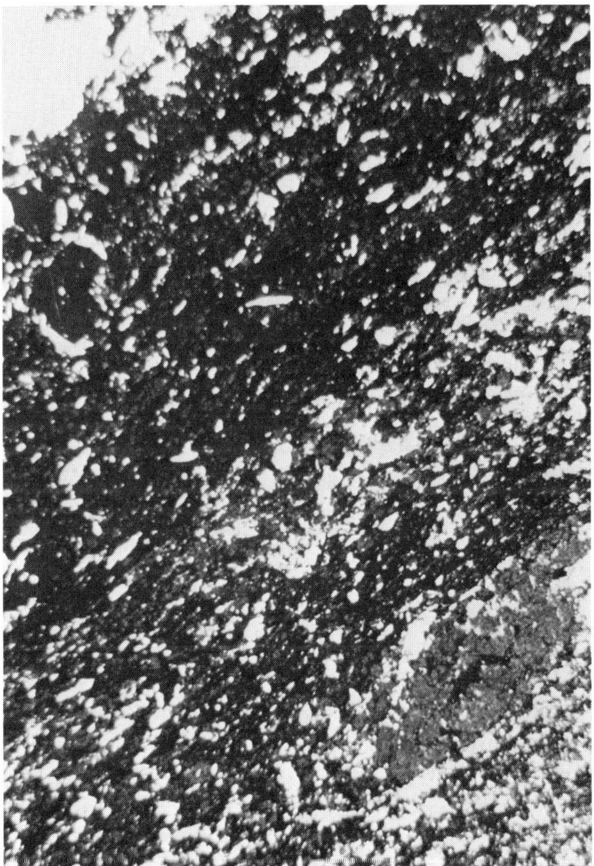

Figure 20. Photomicrograph of ore containing about 30% sphalerite-galena from the western part of the orebody. Dark grey and black areas are sulphides. The pronounced fabric is tectonic. Non-sulphide material is carbonate, quartz and chlorite (medium grey). Long axis of photograph represents 3.4 cm. Top unknown. Plane light. Unusually high non-sulphide content.

includes complete destruction of mafic minerals, or migration in which components are concentrated away from or toward the vein (McAdam, 1978). Sulphides in places form up to 30% of the rock within zones of chaotic footwall breccia. Pyrrhotite is the predominant sulphide in the breccia, but sphalerite, galena, chalcopyrite and arsenopyrite are locally abundant. Disseminated to massive sulphides occur in the matrix of the breccia in many places. Elsewhere, the breccia more closely resembles a zone of abnormally intense irregular vein mineralization.

Hangingwall Bedded Sequence

Four graded sequences called the "I", "H" "Little HU" and "HU" markers occur above the "D" Band (Fig. 9). Each starts with a graded quartz wacke bed at the base and grades upward to a thick mudstone top. The top pyrrhotite-laminated parts of three of these sequences correlate laterally with the "I", "H" and "HU" ore zones in the hangingwall.

"Intermediate Siltstone". The main sulphide zone, which includes the stratigraphic sequence from the base of the Main Band to the top of "D", is overlain by the Intermediate siltstone ("I") an argillaceous graded bed, up to 3.5 m thick, recognized throughout the eastern and northern part of the mine. It thins laterally beyond the limits of the orebody. "I" appears to be present above the central and western part of the orebody but here is not commonly recognized, because of hangingwall alteration. "I" consists of quartz wacke containing 1 to 2 mm glassy quartz grains in the lower few centimetres and grades upward through quartz wacke to very fine-grained, soft, mudstone at the top. The bed is underlain by fine-grained soft mudstone and the marked change at the base of the graded bed serves as an excellent stratigraphic horizon marker. The base of the bed is often deformed into load waves, some of which are cut off and mixed into the upper few centimetres of the underlying mudstone. The bed consists mainly of quartz, sericite and biotite with scattered grains of pyrrhotite. In areas where the hangingwall sequence is altered to chlorite and albite, the Intermediate siltstone is similarly altered. There does not seem to be any selective alteration or mineralization of the coarser-grained base of the bed.

"I" Ore. The "I" ore zone consists of a section of mudstone with sulphide laminations, overlying the thick "I" graded bed and underlying the "H" graded bed. In some places the sulphide laminae are thick and numerous enough to make ore. The "I" laminated ore zone can be identified in most of the mine area that is free of alteration. Sometimes it is identified on the basis of the laminated section lying between "I" and "H", but usually it can be recognized owing to distinctive spacing of iron-sulphide laminae ("I Laminations") near the centre of the sequence (Freeze, 1966). The "I" laminated zone is 2 to 3 m thick.

"H" Sequence. The "H" stratigraphic interval is similar to the "I" sequence. It consists of a basal bed, up to 3 m thick, of graded quartz wacke and an upper section of laminated mudstone with thin sulphide laminations. In places the laminated section contains enough lead and zinc to be ore.

"HU" Sequence. The "HU" sequence consists of two thick-graded beds with overlying laminated mudstone. The lower bed is called "Little HU" and is 1 to 3.5 m thick. The upper bed, "HU", is commonly about 6 m thick and grades from quartz wacke at the base to very fine, non-laminated mudstone at the top. Most of the bed is

fine-grained quartz wacke. It is light grey in colour and consists of quartz and sericite. The graded section of the series is overlain by laminated argillite which contains the "HU" Ore. The base of "HU" is the most reliable marker in the mine.

The hangingwall ore zones, "I", "H" and "HU", are distributed in a broad crescent-shaped pattern about the thickest part of the orebody. These ore zones are delicately bedded and thickest in the vicinity of the Burchett fault, Number 2 fault and the Main Water Course (Fig. 7) and at the southern margin of the orebody. A hangingwall conglomerate averaging 4 m thick and containing some ore-grade sulphide clasts occurs beneath the "H" ore. It has been traced for more than 1000 m along the southern margin of the orebody. The "HU" ore (Fig. 21) consists of both sulphide rock and rock composed of interbedded sulphide and mudstone with the former more prevalent in thicker sections of ore. Crosscutting breccias have been observed up to the level of "HU" Ore.

Figure 21. Delicate sulphide bedding of "HU" ore. Compare with Figure 17. Rule divided in inches.

"U" Quartzites. The "HU" sequence is overlain by 10 to 15 m of graded quartz wacke beds and then by the "U" Quartzites which are generally regarded as the base of the Middle Aldridge Formation. They consist of alternating beds of quartz wacke and laminated mudstone. No economic concentrations of lead or zinc sulphides are known above the base of the "U" Quartzites.

Post-Ore Breccia

Two types of post-ore breccia have been recognized in the western part of the orebody. In the southwest (Fig. 8), a breccia crosscuts the orebody and hangingwall rocks up to the stratigraphic position of the hangingwall conglomerate. Breccia also occurs in albite-chlorite-pyrite-carbonate altered rock in the hangingwall. The southwest breccia is a discordant breccia body in footwall and ore zone rocks and expands to form a concordant lens at the base of the "H" marker in the hangingwall. The discordant part is 12 m wide and at least 250 m long with a northerly trend; the breccia extends downward for at least 45 m from the "H" marker with the base unexposed. The overlying concordant lens is about 5 m thick with dimensions of 120 by 250 m.

The breccia consists of angular to rounded fragments of mudstone, chloritic mudstone and tourmalinite in an altered matrix composed of chlorite, biotite, carbonate, quartz and pyrrhotite comprising as much as 60% of the rock. Locally, galena and sphalerite become significant components of the breccia matrix. In the overlying lens, mudstone clasts predominate but tourmalinite clasts and occasional clasts with some pyrrhotite, sphalerite or galena also occur. Breccia associated with albite-chlorite-pyrite-carbonate altered rocks in the hangingwall occurs over the west-central part of the orebody and is clearly associated with alteration and post-alteration deformation.

Ore Mineralogy and Petrography

Mineralogy of the orebody is relatively simple. Pyrrhotite and pyrite with a ratio of about 7 to 3 are the most abundant sulphides and galena and sphalerite are the principal ore minerals. Minor but economically important minerals include tetrahedrite, pyrargyrite, boulangerite and arsenopyrite (deleterious). Cassiterite is an important minor constituent in the western part of the orebody. Minerals constituting less than one per cent include chalcopyrite, jamesonite, magnetite, and less abundant scheelite and stannite. Principal non-sulphide minerals are quartz and calcite with abundant tourmaline, chlorite, muscovite, albite, pale brown to reddish-brown mica, garnet, tremolite, epidote and hornblende. Either quartz or calcite may make up 50 to 70% of the non-sulphide suite, chlorite 30% and the other minerals up to about 20%.

The ore has a metamorphic texture. Massive pyrrhotite in the western part of the orebody is composed of a polygonal mosaic of pyrrhotite grains with sphalerite and chalcopyrite concentrated at triple point junctions. In layered ore, pyrrhotite grains are euhedral to irregular-shaped and surrounded by a network of galena and sphalerite. Boulangerite, tetrahedrite and pyrargyrite occur at interstices between sphalerite and galena grains and rare stannite occurs as euhedra at grain boundaries. Pyrite is euhedral and granular and usually forms fine- to coarsely crystalline layers that are roughly conformable to the sulphide footwall.

Textures of the distinctly bedded ore in the eastern part of the orebody are dominated by the shape of the metamorphic non-sulphide minerals. In silicate-rich layers pyrrhotite has a tabular habit and the grains are commonly oriented in a reticulate pattern with plates of mica. Occasionally, fine granular intergrowths of elongated grains impart a crude foliation parallel to the bedding. Galena and sphalerite occur interstitial to pyrrhotite or as nearly mono-mineralic granular layers. Individual layers within a laminated sequence are sharply defined and differ strikingly in composition and mineralogy from one to another (Figs. 18, 30, 31). This strong mineralogical differentiation sharply defines laterally persistent bedding; changes along the plane of bedding are gradational. The most striking lateral change is the gradual increase of granular pyrite laminae and buckshot pyrite toward the southeastern perimeter of the orebody. Outward from the centre of the orebody pyrite laminae appear first in the upper ore layers delicately interlayered with pyrrhotite and ore minerals. The amount of pyrite gradually increases to the southeast, and at the economic margin of the orebody, Bands "A" through "D" are predominantly pyrite, the layered part of the Main Band is composed of nearly equal amounts of delicately laminated pyrrhotite, pyrite and magnetite and the basal part of the Main Band is predominantly pyrrhotite. Pyrite also occurs locally as coarse megacrysts and as spongy aggregates of euhedra adjacent to the Kimberley and Sullivan faults. Spongy pyrite occurs in layers normally occupied by pyrrhotite and along closely spaced fractures at various angles to bedding.

Overall grain size of sulphide and gangue ranges from about 0.005 mm to rarely as much as one centimetre. Average grain size is about 0.05 to 0.1 mm.

Distribution of Metals in the Orebody

Distribution of valuable metals in the orebody has been described by several authors starting with Pentland (1943) who described a zonal distribution of lead, zinc and tin. Swanson and Gunning (1945) suggested that the lead and zinc distribution might be controlled by merging linear patterns. Freeze (1966) presented generalized plans showing the distribution of lead, zinc, tin, antimony and arsenic. Freeze also described a general tendency for the metals, particularly lead, zinc and silver, to be zoned across the bedding as well as in the plane of bedding. Ransom (1977) presented detailed distribution maps for lead, zinc, silver and tin.

The term "central iron zone" used by previous authors in discussions of metal distribution can be misleading. It has been applied to the rectangular area lying south of the centre of the western part of the orebody (Fig. 22). Iron sulphide content in this zone is no greater than that in adjacent areas of the orebody; rather, lead and zinc sulphide contents are much less. The "central iron zone" is principally composed of pyrite, chlorite and carbonate and is a zone of alteration superimposed on the original patterns of metal distribution.

Of the important metals in the orebody only iron has a simple and uniform plan distribution. The iron distribution map (Fig. 22) was constructed using assay data from numerous closely spaced core holes. Data at each point is the product of the true thickness of the orebody at that point multiplied by the average content of iron in per cent across the intersection. The result is reported in feet-per cent as shown on the figure. The distribution pattern for iron generally reflects the thickness of the

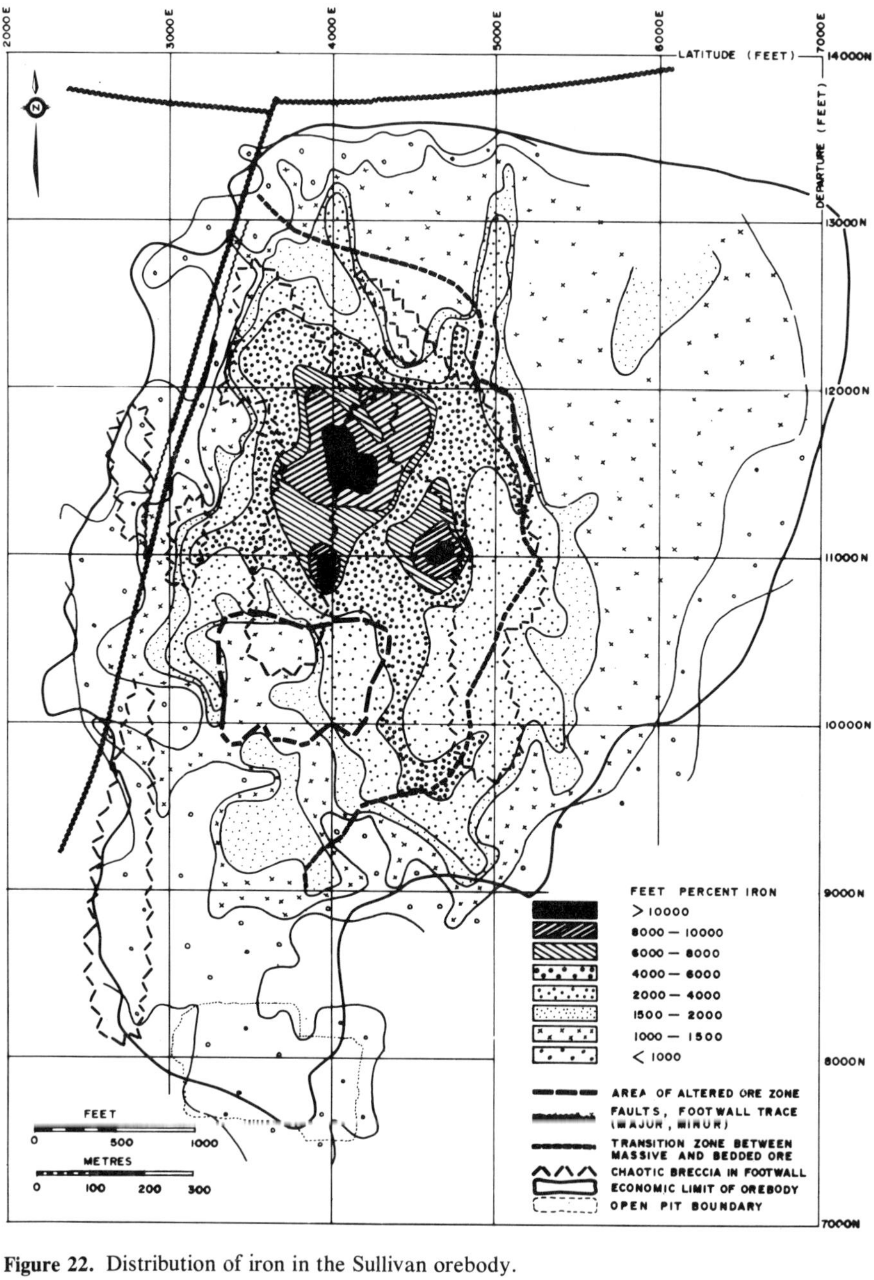

Figure 22. Distribution of iron in the Sullivan orebody.

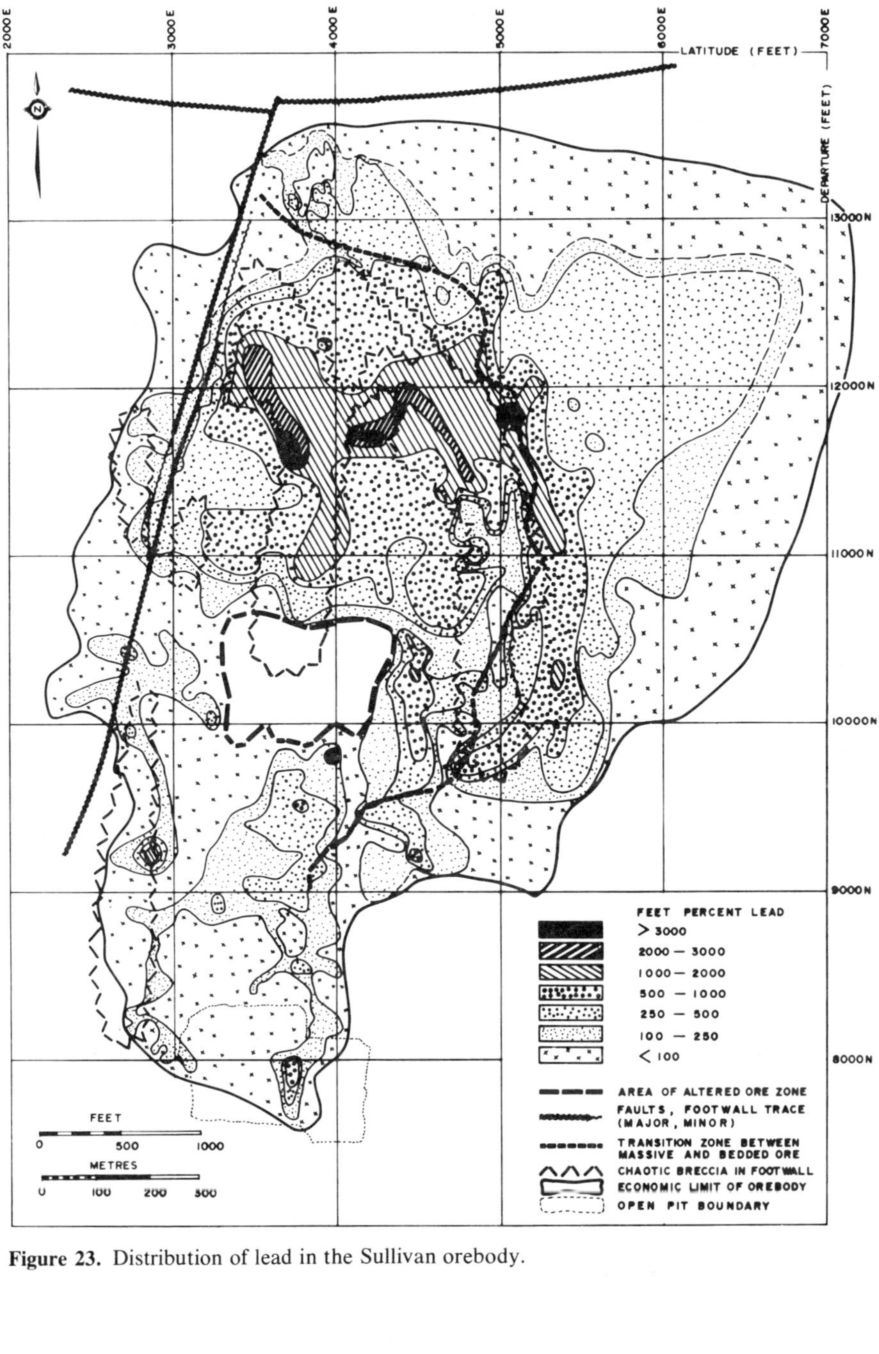

Figure 23. Distribution of lead in the Sullivan orebody.

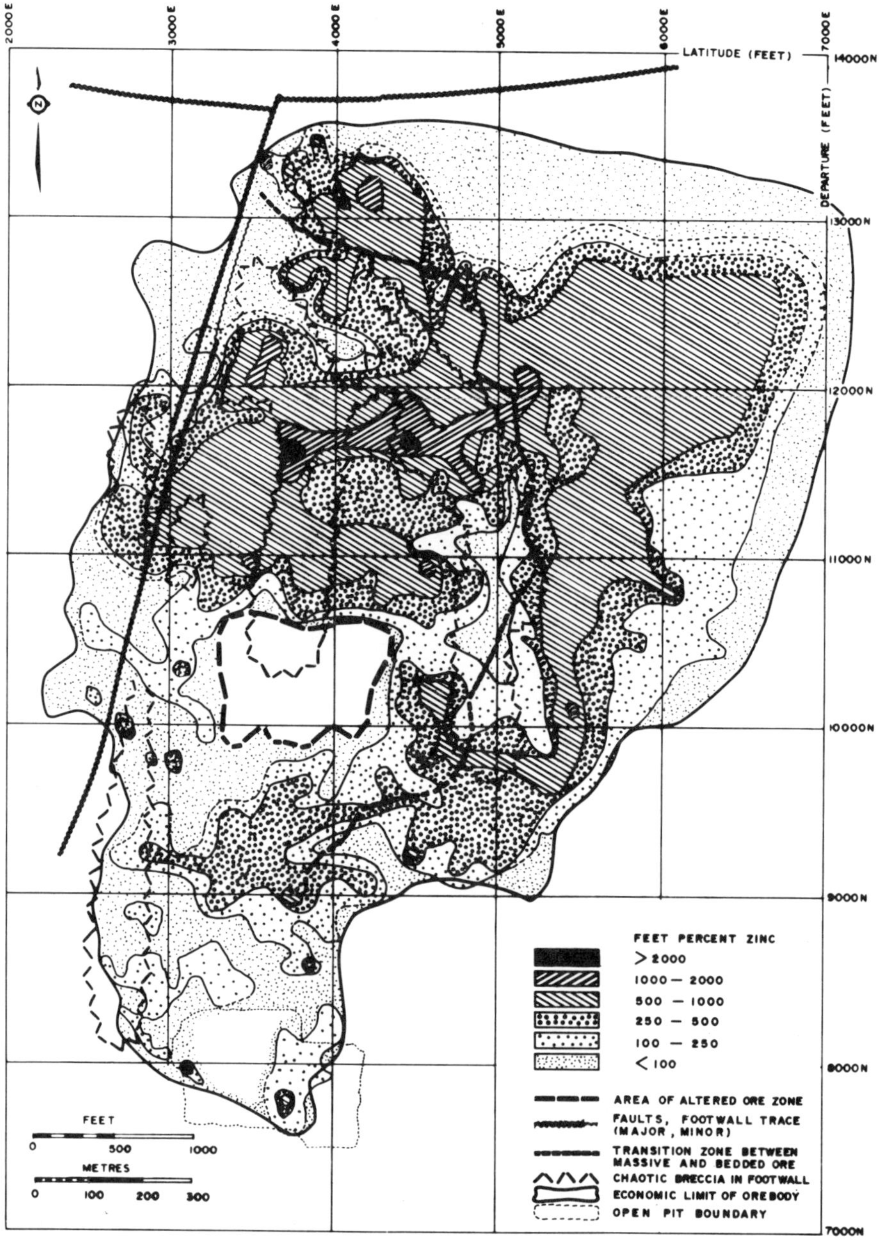

Figure 24. Distribution of zinc in the Sullivan orebody.

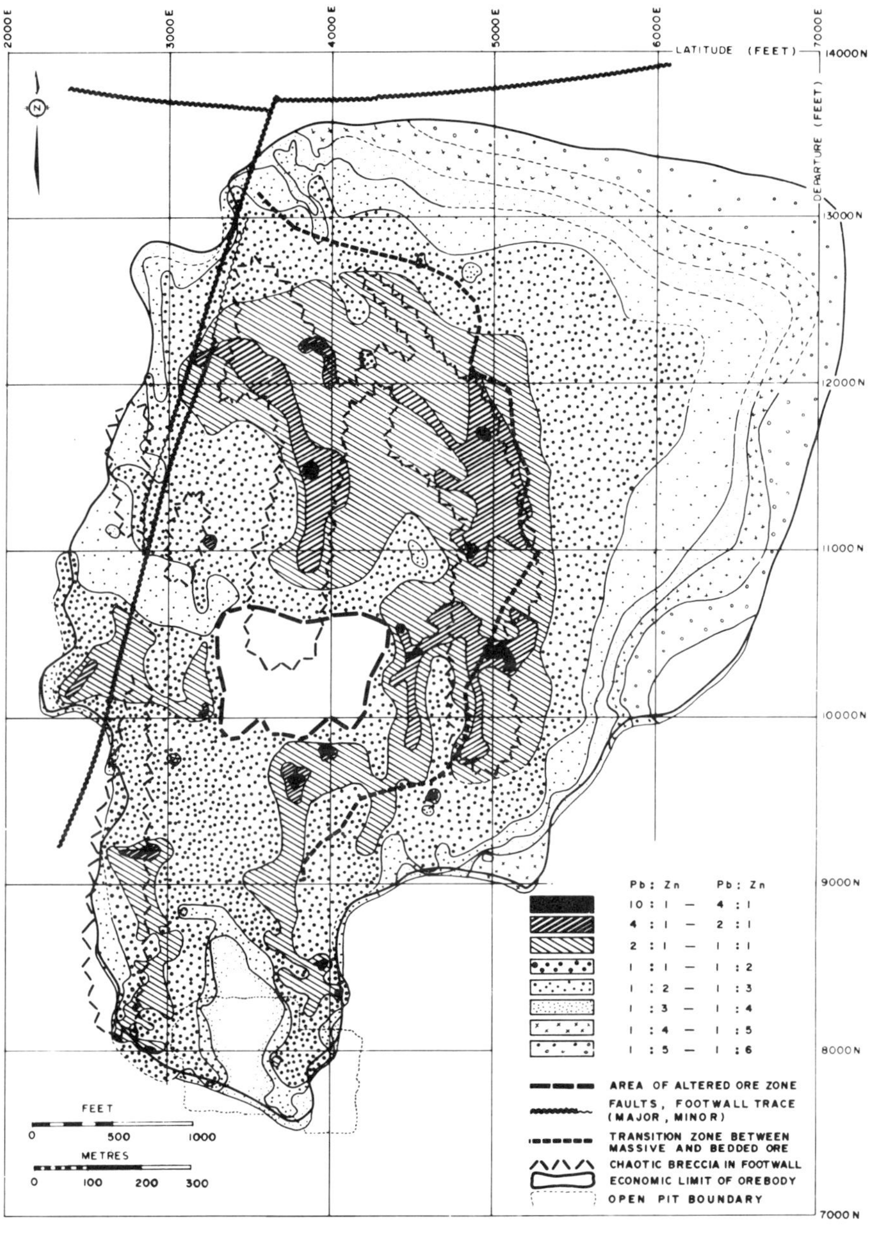

Figure 25. Lead to zinc ratio in the Sullivan orebody.

orebody as shown on the isopach map (Fig. 13). This suggests that the principal controls of iron sulphide distribution are the gross geometry of the sub-basin and a nearly constant supply of iron.

Metal distribution maps for lead and zinc (Figs. 23 and 24) are prepared in a manner similar to that for iron. At the periphery of the orebody both lead and zinc have patterns similar to that of iron. The patterns for lead and zinc are notably different from the iron pattern in areas of the orebody underlain by the chaotic breccia. Areas enclosed by the 1000 feet-per cent lead isopleth are generally coincident with distribution of chaotic breccia. The lead to zinc ratio map (Fig. 25) shows the coincidence, particularly the lead to zinc ratio greater than 2:1, of lead and zinc distribution with the underlying chaotic breccia. Zoning in the eastern part of the orebody is concentric with a gradually decreasing lead to zinc ratio toward the periphery, although the isopleth lines intersect the line denoting the economic limit of the orebody.

Silver distribution in the orebody has a pattern similar to the pattern of lead distribution. High values (Fig. 26) are approximately coincident with the plan distribution of chaotic breccia in the footwall. Silver occurs with galena, probably in solid solution with bismuth and lead. Other principal silver-bearing minerals are tetrahedrite and pyrargyrite.

Tin is concentrated in the western part of the orebody and highest values occur in areas underlain by the chaotic breccia (Fig. 27). The average tin grade in mill feed is 0.031%; tin grades as high as 2.5% occur in the tin zone fracture (Fig. 7) described by Freeze (1966). The principal tin mineral is cassiterite; stannite is rare.

Vertical distribution of metals has been well documented in the eastern part of the orebody where stratigraphic control is well established. Owing to poor stratigraphic control in the western part, details of vertical metal distribution are outlined in a block by block manner as mining progresses. For the west and central part of the orebody Freeze (1966, p. 275) stated:

> "... there is a distinct tendency for these metals (lead, zinc and silver), to be zoned in a vertical sense, particularly in the centre of the mine. This relation arises because galena and sphalerite tend to form extensive layers of rich beautifully banded ore against the hangingwall of the main ore bands. This well banded ore passes downward into a rich, massive mixture of pyrrhotite, galena and sphalerite that in turn passes into a thick irregular zone of pyrrhotite that carries only a trace of galena and sphalerite. Generally, this barren, pyrrhotite zone extends to the sulphide footwall."

Carswell (1961), using assay data from widely distributed core holes in the eastern part of the orebody, documented a decrease of Pb/Pb+Zn ratio from the Main Band through "D" Band and a fairly uniform distribution of iron in the same sequence. Carswell did not consider possible effects of lateral zoning in the plane of the layering in his study of vertical distribution.

To minimize effects of lateral zoning, vertical zoning has been studied within a very narrow zone about the coincident 0.9 lead to zinc ratio isopleth for each Band in the eastern part of the orebody. Geometric means for each of lead, zinc, silver and iron in the five Bands are plotted in Figure 28: Raw data for the documentation are unweighted assay values for lead, zinc, silver, and iron. Metal distribution in the

Main Band is characterized by a uniform increase of lead from base to top, moderate enrichment of silver in central Main Band and enrichment, then depletion of zinc in central Main Band to top Main Band. Iron maintains a remarkably uniform abundance through the Main Band and the overlying "A" through "D" Bands except for effects of dilution by waste beds in the "B" Band triplet. Silver and lead abundance decrease at a uniform rate from "A" Band through "D" Band. Zinc abundance also decreases in a uniform way from "A" Band through "D" Band; however, the rate of change is less than that for lead and silver. The increase in rate of change for lead and zinc abundance from "C" Band to "D" Band may be partly influenced by the lateral metal zoning differences for the "D" Band.

Wall-Rock Alteration

Extensive volumes of altered rock occur below, within and above the ore zone in the western part of the mine. Tourmalinite is included with wall-rock alteration because most of the tourmalinite, except for that near the sulphide footwall, has crosscutting relations. Altered rocks unusually rich in chlorite, albite, pyrite, biotite, garnet and calcite occur in restricted crosscutting footwall structures, in a zone which crosscuts the orebody, and also occupy an extensive volume of rock in the hangingwall. Although minerals in altered rock have a metamorphic texture, their occurrence is interpreted as reflecting pre-metamorphic chemical modifications.

Tourmalinite. The western part of the orebody is underlain by an extensive zone of tourmalinite: a very dark grey to black or brown, extremely hard rock composed almost entirely of tourmaline and quartz. Its distribution in the footwall is outlined in plan (Fig. 29) and in section (Figs. 5 and 6). The western limit of tourmalinite is the contact zone of the gabbro intrusion, and elsewhere the limit is approximately coincident with the transition between massive and bedded parts of the orebody (Fig. 29). Tourmalinite and tourmalinite interbedded with tourmaline-free metasedimentary rock occur throughout the 450 m sequence between the sulphide footwall and footwall gabbro. In the hangingwall, tourmalinite, although subordinate, is best developed above the eastern limit of footwall tourmalinite and is particularly common in hangingwall rocks above the 4800 departure (Alpha and Beta fault trend) zone of thinning (Figs. 5 and 7). Disseminated tourmaline and rare patches of tourmalinite are erratically distributed elsewhere in the hangingwall overlying the western part of the orebody.

Tourmalinite is rare in the ore zone and interbedded waste rock. It occurs in the ore zone usually as tectonically rafted, irregular blocks. More common in the ore zone, are disseminated (trace to 2%), pale, euhedral tourmaline grains. Euhedral poikilitic grains occur in clastic rocks interbedded with the ore.

On a large scale (footwall gabbro to sulphide footwall), tourmalinite has crosscutting relations. However, with proximity to the sulphide footwall a remarkable variety of geometric relations occur. Near the sulphide footwall, and over distances of a metre to hundreds of metres in the plane of bedding, contacts between tourmalinite and other rock types range from perfectly conformable to steeply transgressive relative to bedding. Transgressive and conformable contacts are generally sharp, occurring over distances of a few millimetres to a few centimetres. Disconformable contacts range from regular to highly irregular. Locally, bedding may be

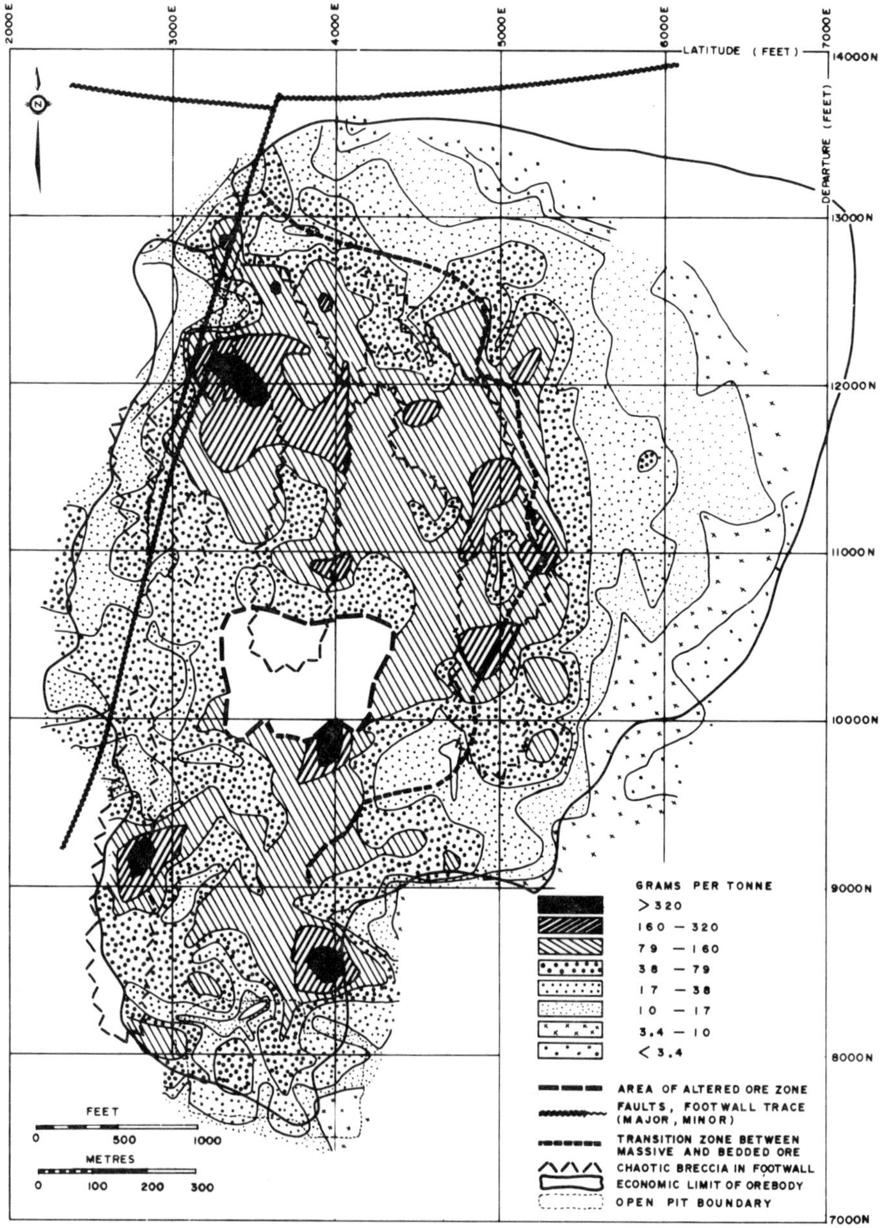

Figure 26. Distribution of silver in the Sullivan orebody.

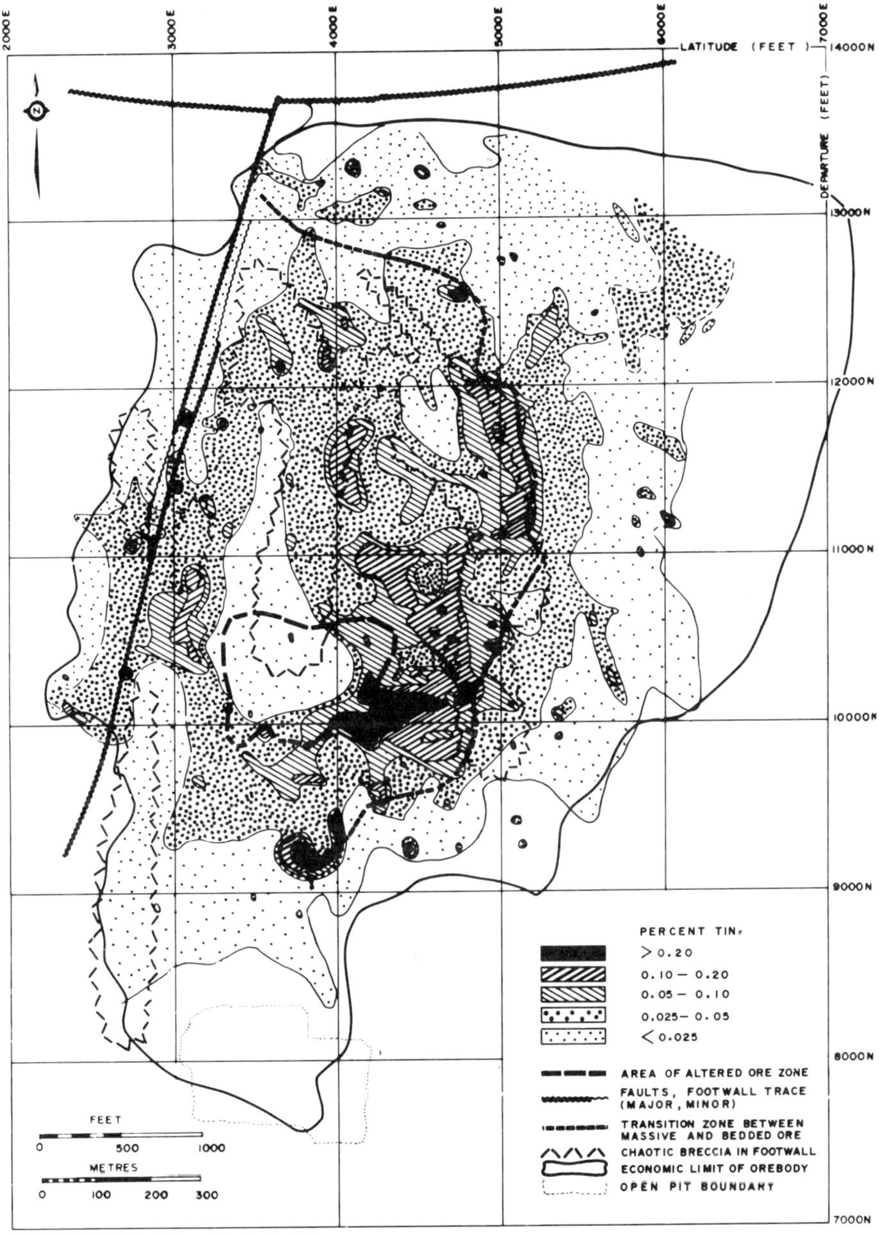

Figure 27. Distribution of tin in the Sullivan orebody.

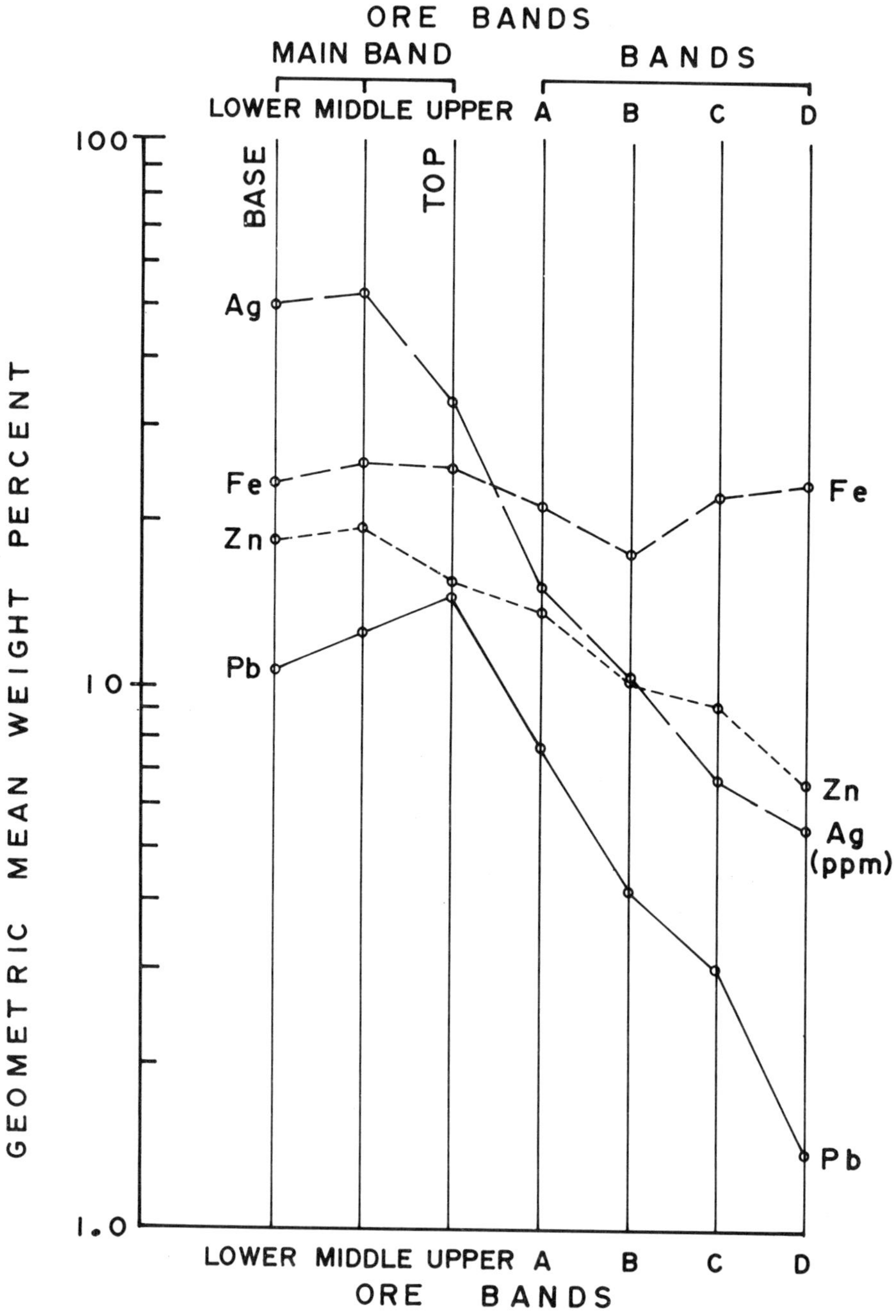

Figure 28. Vertical distribution of metals in the bedded eastern part of the Sullivan orebody.

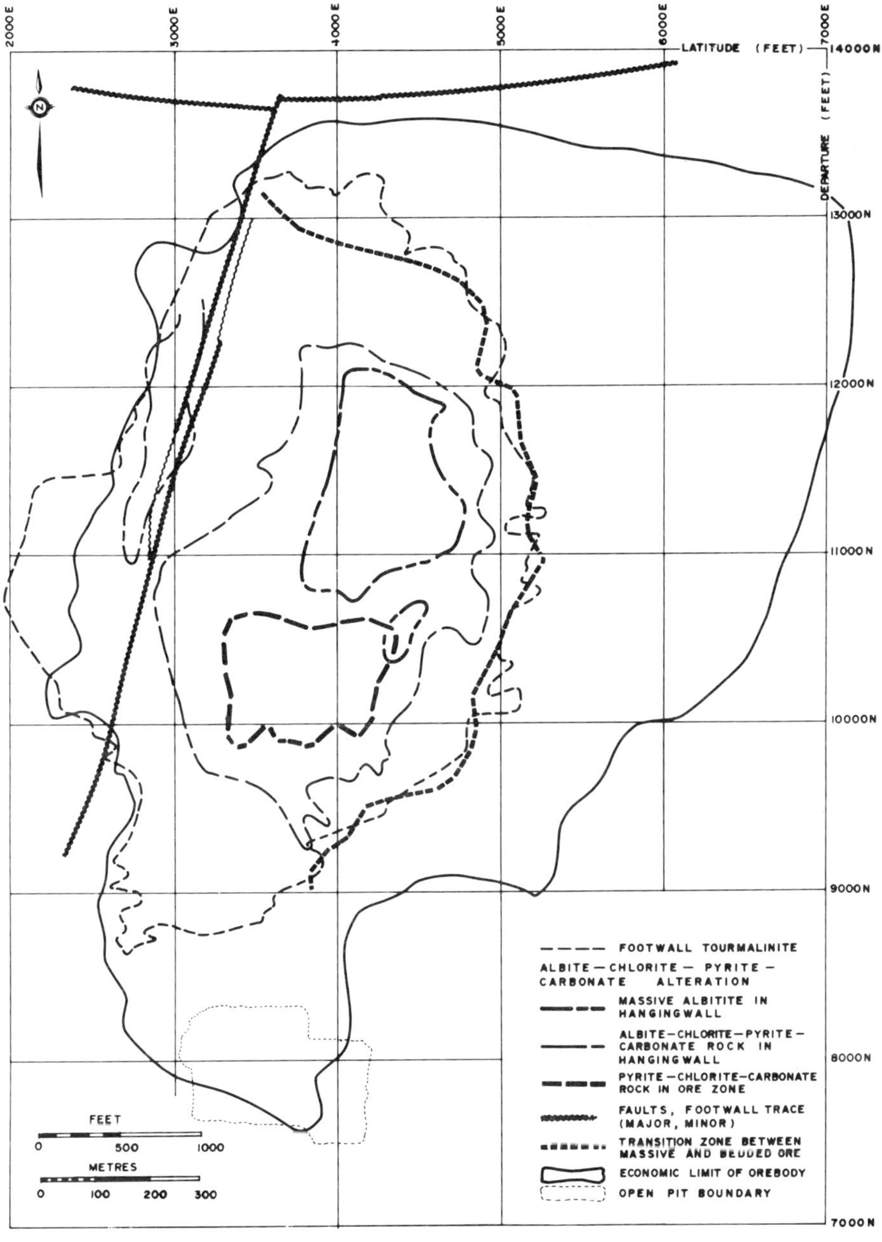

Figure 29. Distribution of altered rocks in and adjacent to the Sullivan orebody.

traced across disconformable contacts between tourmaline-rich and tourmaline-poor rocks. Small-scale, crosscutting tourmaline-bearing structures such as tourmalinite veins or tourmalinite alteration envelopes are unknown in the footwall sedimentary rocks (Shaw, in prep.).

Tourmalinite beds 3 to 15 m thick occur in which bedding is defined by faint pyrrhotite laminations. Thicker sequences of tourmalinite with very well developed bedding and remarkable preservation of primary sedimentary structures are also common. Delicate laminations and soft-sediment deformation structures are frequently more distinct in tourmalinite than in tourmaline-poor rock. Less common are centimetre-scale interbeds of tourmalinite and tourmaline-poor rock recognized at the southern limit of the tourmalinite zone. Close to the sulphide footwall, tourmaline is pervasive throughout entire beds whereas in the deep footwall, tourmaline is most abundant at the coarse base of graded beds and in thin-bedded zones.

In tourmalinite, tourmaline occupies interstices between quartz grains. Figures 42 and 43 illustrate some of the textural variations observed. Typically, quartz grains with indistinct boundaries are set in a very fine-grained felted matrix of acicular to stubby tourmaline grains forming 50% to more than 75% of the rock. The tourmaline grains are micro-crystalline, ranging in length from about 1 to 10 microns. Irregular and disseminated blebs and delicate laminae of pyrrhotite are common. Minor minerals include muscovite, chlorite, potassium feldspar, plagioclase and garnet in descending order of abundance. Pyrrhotite and sphalerite constitute up to 10 to 15% and occur as disseminated grains or blebs, or concentrated in delicate laminae.

Of particular interest is the tourmalinite phase of footwall conglomerate. Tourmalinite clasts in a tourmalinite matrix is by far the most common mode of occurrence. However, tourmaline-poor clasts are also found associated with clasts of tourmalinite in tourmalinite matrix. In thin section, tourmaline-poor clasts contain a very small percentage of micaceous material and are composed of quartz and subordinate feldspar grains and sparse intergranular tourmaline needles. In conglomerate beneath the eastern part of the orebody, occasional, rounded to flat ovoid clasts of tourmalinite are found in association with the tourmaline-free clasts in a tourmaline-free matrix. Here tourmalinite clasts are frequently indented or partly wrap or drape around adjacent tourmaline-free clasts. This relation suggests that tourmalinite clasts were at least semiplastic at the time of deposition. In contact zones of the gabbro intrusions, boron has been removed from tourmalinite and the rock recrystallized to hornfels with only scattered crystals of tourmaline. The tourmaline is coarsely-crystalline and forms blue-green needles, tiny rosettes and veins.

Albite-Chlorite-Pyrite-Carbonate Alteration. Altered rocks rich in albite, chlorite, pyrite or carbonate are restricted to the western part of the mine and in plan lie generally within limits defined by the zone of tourmalinite in the footwall (Fig. 29). Albite-chlorite-pyrite-carbonate alteration occurs in crosscutting zones within the footwall tourmalinite. These zones range from one centimetre to as much as 30 m wide. Overlying the crosscutting alteration in the footwall, the ore zone has been altered to a pyrite-chlorite-carbonate assemblage. Pervasive alteration in the hangingwall has resulted in a zone of albite, chlorite and pyrite as much as 100 m thick.

Figure 30. Delicate layering of fine-grained, laminated sulphide rock of the "A" Band. Medium brown colour metallic mineral is pyrrhotite; metallic grey mineral is galena; dark brown layers are composed of clastic sedimentary rock with disseminated sphalerite. Sample is 3.5 cm from base to top with top at left.

Figure 31. Tectonic deformation near the top of the Main Band. Mudstone interbeds fracture in a brittle fashion but sulphide beds are ductile. The white ruler is 46 cm long, 1.5 cm wide. Top of section is up.

Figure 32. Galena layers in pyrrhotite, typical of ore in the western part of the orebody. This is the base of the main band of the orebody, at the transition zone, and the location of the specimen shown in Fig. 14. The head of the hammer is at the footwall contact.

Figure 33. Irregular layering of pyrrhotite and sphalerite typical of the western part of the orebody. Dark brown lenticular bodies are composed of altered sedimentary rock. Top of sequence is up. Coin is 1.9 cm in diameter.

Figure 34. Sulphide rock, south end of the orebody from the basal one-third of the Main Band. The largest fragment (centre) is medium-grained sphalerite and about 5 cm long. Dark fragment at the lower left is altered argillite. White fragments on the right are quartz and light-coloured equant grains are pyrite. Stratigraphic top not known.

Figure 35. Slightly deformed parallel and planar layering of the "A" Band in the southeastern part of the orebody. Grey layers are pyrrhotite and brassy layers are pyrite-rich. The handle of the hammer is at the base of the "A" Band.

Figure 36. Chlorite-albite breccia overlying coarsely crystalline pyrite in the hangingwall of pyrite-chlorite-carbonate altered ore in the western part of the orebody. Rule divided in tenths of a foot.

Figure 37. Albite-chlorite breccia from hangingwall in western part of the orebody. Original orientation unknown.

Figure 38. Alteration contact: faintly-layered pyrrhotite (above) altered to coarsely crystalline pyrite adjacent to the pyrite-chlorite-carbonate altered zone in the western part of the orebody. Layering in pyrrhotite is defined by galena concentrations (chalked) and layering is truncated by pyrite. Top of section is up.

Figure 39. Layering, defined by sphalerite concentration, in coarsely crystalline pyrite-carbonate rock from discordant contact zone between faintly layered pyrrhotite and pyrite as shown in Figure 38. Layering often continues across the contact for short distances. Top unknown.

Figure 40. Pyrrhotite-rich clasts in tourmalinite matrix from footwall of the western part of the orebody. Planar quartz-pyrrhotite vein at left of photograph. Top unknown.

Figure 41. Disrupted bedding at the margin of a chaotic breccia zone beneath the western part of the orebody. Light rusty zones are tourmalinite and dark rusty zones are pyrrhotite-rich.

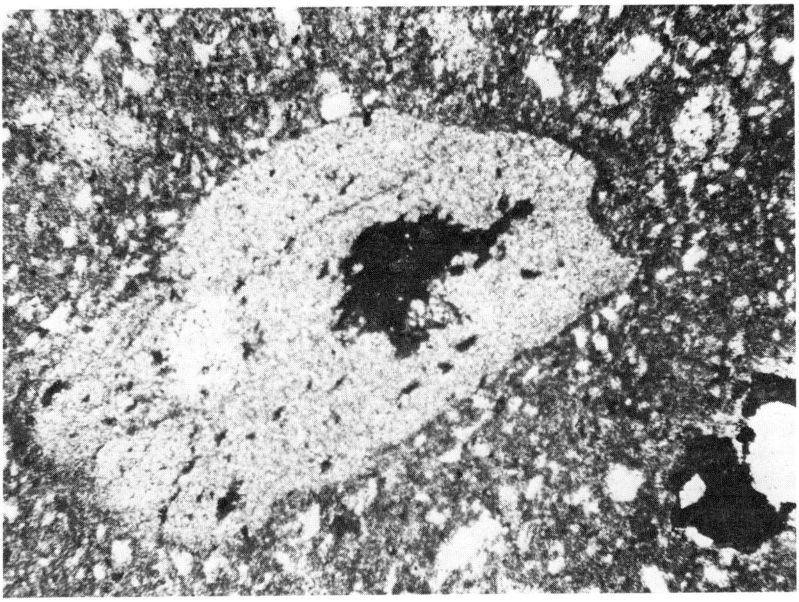

Figure 42. Felted tourmaline (dark grey) matrix of tourmalinite conglomerate. White grains are quartz and black grains, pyrrhotite. The large grey clast with the pyrrhotite centre contains less tourmaline than the ground mass. Plane polarized light. Field is 1.1 mm long.

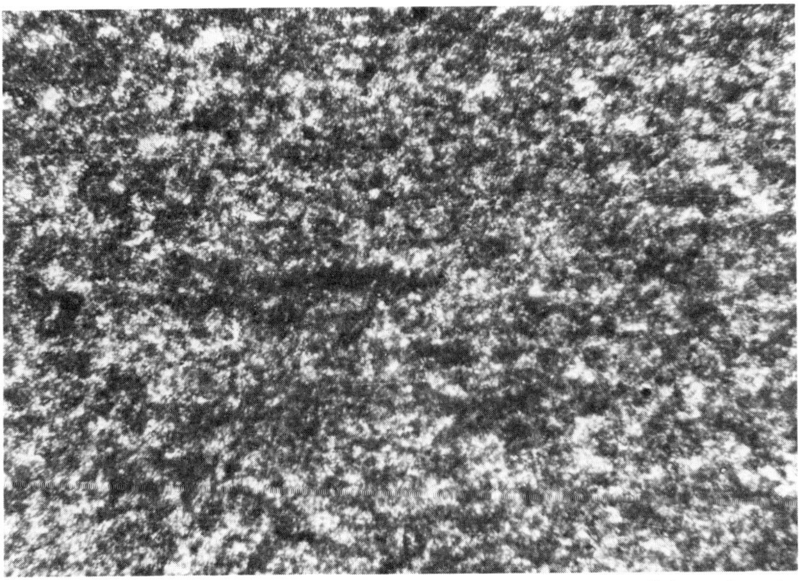

Figure 43. Faintly-layered tourmalinite from zone of interbedded tourmalinite and mudstone. Light coloured grains are quartz and feldspar. Plane polarized light. Field is 0.15 mm long.

Figure 44. Granoblastic texture of fine-grained albitite from core of albite-chlorite-pyrite alteration zone in hangingwall above western part of orebody. Crossed nicols. Field is 0.6 mm long.

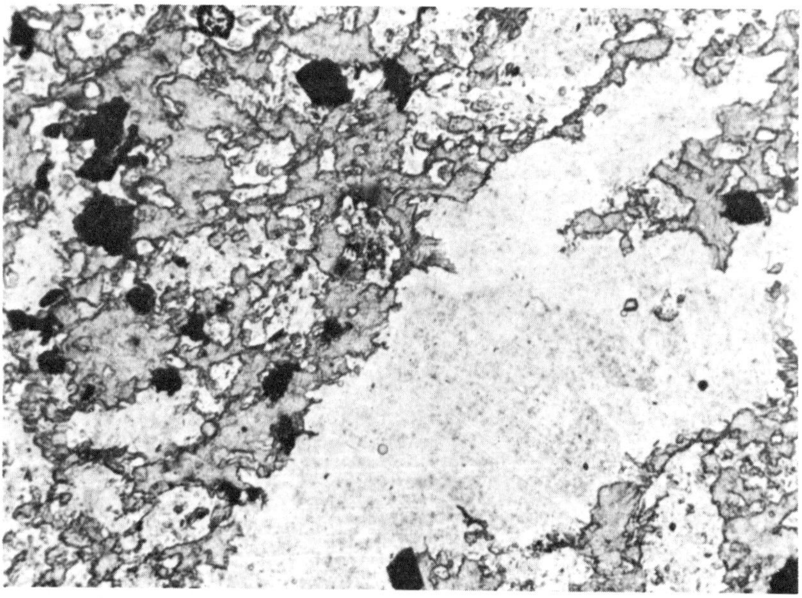

Figure 45. Albite-chlorite-pyrite altered rock from immediate hangingwall of western part of orebody. Vein is composed of coarsely crystalline albite dusted with minute tourmaline grains. Black grains are pyrite. Plane polarized light. Field is 0.6 mm long.

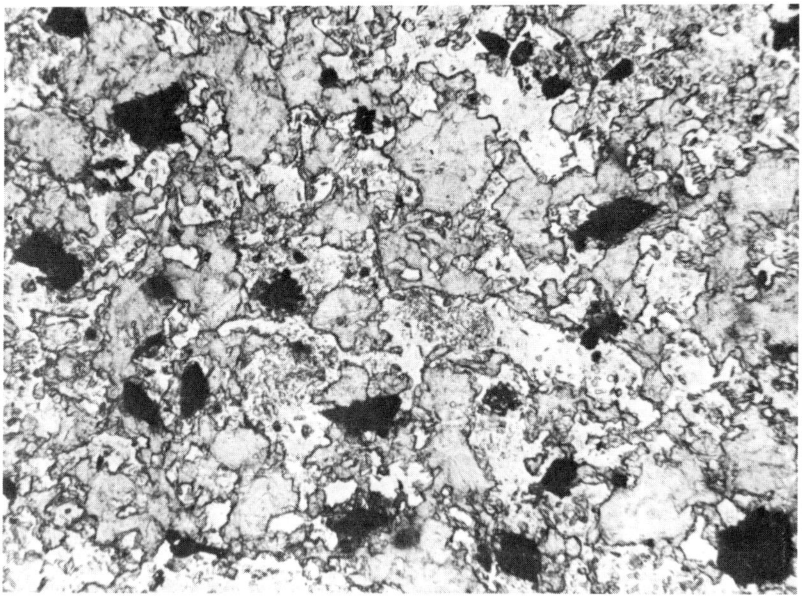

Figure 46. Chlorite-albite-pyrite rock from margin of albite core in hangingwall of western part of orebody. Chlorite, albite and pyrite are medium, light and dark grey, respectively. Plane polarized light. Field is 0.6 mm long.

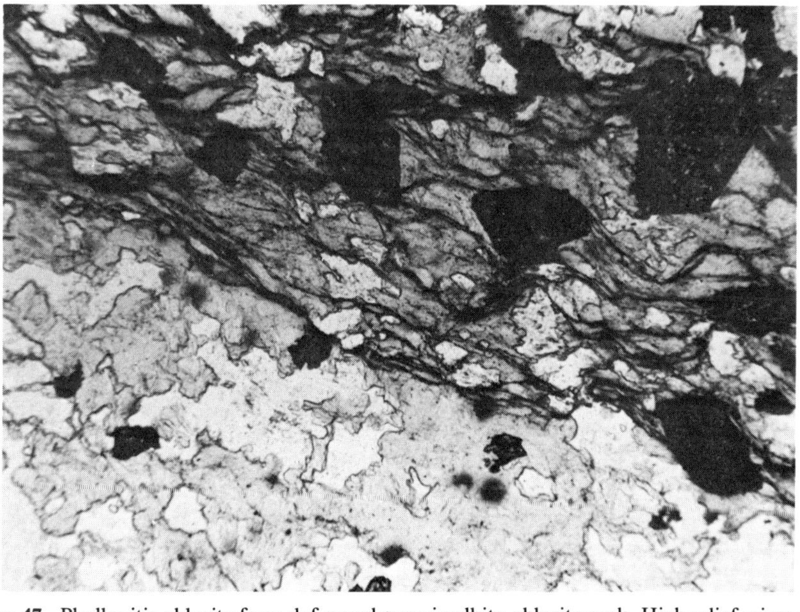

Figure 47. Phyllonitic chlorite from deformed zone in albite-chlorite rock. High relief mineral is sphene. Plane polarized light. Field is 0.6 mm long.

Isolated, crosscutting albite-chlorite-pyrite-carbonate alteration zones in the footwall expand upward to a stratabound zone at the sulphide footwall (Fig. 5). Within this interval, albitite occurs as veins and large masses in chlorite-pyrite rock and in zones of breccia composed of intimately mixed albitite and chloritic rock. Mineral zoning on all scales is conspicuous in non-brecciated parts of the alteration zones. Within a typical alteration zone cutting tourmalinite, a central core of albitite-pyrite grades outward to a peripheral zone of chlorite-pyrite. Adjacent to the alteration zone pyrrhotite in otherwise unaltered tourmalinite has been altered to pyrite. Fresh tourmalinite adjacent to an alteration zone is a dark grey to black lustrous rock. Within the alteration zone, the rock is dull, soft and chloritic.

Distribution of alteration in the orebody overlying the zone of crosscutting albite-chlorite-pyrite-carbonate in the footwall is outlined by the boundary of the pyrite-chlorite-carbonate zone (Fig. 29). This boundary is defined by the limit of ore, not by the limit of alteration pyrite. Idealized cross-sections through the pyrite-chlorite-carbonate alteration zone in the ore are shown in Figures 5 and 6.

Pyrite-chlorite-carbonate alteration of normal pyrrhotite-rich lead and zinc ore commonly crosscuts and obliterates primary textures and structures in the orebody, such that only large-scale layering of interbedded coarsely crystalline pyrite and massive chloritic rock remains. Choritic alteration of sedimentary rock appears to extend beyond the limit of pyrite and the contact between chloritic and unaltered rock is gradational over several metres. In massive sulphide ore, contact between the pyrite alteration assemblage and unaltered pyrrhotite-rich ore is sharp and highly irregular (Fig. 38). Contacts vary from conformable to sharply transgressive to bedding. Bedding defined by concentrations of galena and sphalerite in pyrrhotite is abruptly truncated at the contact. Locally, the bedding persists a short distance into the pyrite (Fig. 39).

Immediately overlying the pyrite-chlorite-carbonate zone in the orebody and extending outward from it in the hangingwall, brecciated albite-chlorite rock and bedded, massive or brecciated chloritic rock (Figs. 36, 37) ranging from a few centimetres to as much as 30 m thick overlie the sulphide hangingwall. Overlying these rocks is massive albitite with subordinate chloritic rock in which original sedimentary structures and textures have been obliterated. Massive albitite forms a core that grades outward to an envelope of albitic and chloritic rock in which the original gross sedimentary layering is preserved. Quartz, albite and carbonate veins are common in the albitic and chloritic rock and extend a short distance into overlying sedimentary rocks. Preliminary petrographic study indicates that a potassium mica envelope surrounds lateral margins of the albitic and chloritic rock.

Texture of chloritic rock within the ore zone and of massive chloritic rock immediately above the sulphide hangingwall (lower part of the chloritic envelope) is phyllonitic with porphyroblastic bundles of coarse plates of chlorite and euhedral pyrite and sphene, all set in a very fine-grained matrix of oriented fibrous chlorite. Figures 44 to 47 illustrate some of the textural variations observed in thin section.

Texture of massive albitite overlying chloritic rock above the sulphide hangingwall is granoblastic. Albite grains are anhedral to subhedral and grain boundaries are regular or sutured. Chlorite occurs as intergranular bundles of radiating plates and as large patches of decussate plates. Distribution of chlorite and annealed stringers with a mortar texture define recrystallized crushed zones in albitite. Euhedral and crushed pyrite is locally abundant in the crushed zones. At the outer fringe of the chloritic envelope arenite beds are altered to granoblastic albite, and argillaceous beds are altered to very fine-grained schistose chlorite.

Accessory minerals in altered hangingwall rocks include tourmaline (2 to 5%), sphene (2 to 7%), subordinate white mica, zircon, scapolite, calcite and quartz. Tourmaline occurs near the base of massive albitite as clusters of tiny, pale brown to amber, stubby crystals in cores of early, large, albite crystals. There is no tourmaline in surrounding annealed crushed zones. Sphene occurs as euhedral grains or as broken and altered fragments of euhedral crystals and is most abundant adjacent to or within chlorite-rich patches. Zircon grains are commonly well rounded and appear to retain a detrital morphology.

ROCK CHEMISTRY

The alteration processes that affected the Sullivan host rock and ore have resulted in profound chemical modifications of the original rock compositions. No attempt is made to reconstruct possible chemical reactions because mineralogy of ore, the parent rocks and altered products prior to the first regional metamorphic

TABLE I

COMPOSITION OF SULLIVAN ORE AND WASTE

	1		2		3	
	$\overline{X}$	$\hat{\sigma}$	$\overline{X}$	$\hat{\sigma}$	$\overline{X}$	$\hat{\sigma}$
SiO_2	67.9	8.4	47.34	7.29	18.13	6.29
Al_2O_3	14.3	3.9	11.02	2.60	4.58	2.08
TiO_2	0.5	0.4	0.41	0.33	0.13	0.07
FeO	5.0	1.8	18.54	6.05	25.30	6.97 as Fe
MgO	2.1	0.8	2.58	1.06	1.21	0.43
CaO	2.1	1.0	7.81	5.34	2.76	1.74
Na_2O	1.8	0.8	0.79	0.57	0.65	0.41
K_2O	3.0	1.7	1.41	0.61	0.43	0.19
CO_2	1.3	0.1	5.79	4.71	0.63	0.52
H_2O	0.6	0.1	–	–	–	–
S	0.5	0.2	–	–	22.40	3.77
Pb	0.03	0.01	–	–	7.72	8.91
Zn	0.06	0.01	–	–	11.71	6.91
MnO	0.07	0.07	2.58	1.35	0.77	0.34
TOTAL	99.26 %		98.27 %		96.42 %	
	N = 96		N– 17		N = 17	

1) Sullivan waste beds
2) Non-sulphide component of ore, calculated on a sulphide-free basis (several types of ore are included)
3) Sulphide ore, selected samples from the eastern and western parts of the orebody

TABLE II
AVERAGE* CONCENTRATIONS OF SELECTED TRACE METALS IN SULLIVAN ORE

Antimony	0.026%	Manganese	0.244%
Arsenic	0.083%	Thallium	0.00097%
Copper	0.033%	Indium	0.00146%
Bismuth	0.0036%	Cadmium	0.0142%
		Tin	0.031%

* Average values from composite samples of Sullivan mill-feed during 1958-1977 period.

TABLE III
COMPOSITION OF ALDRIDGE ROCK AND ALTERATION PRODUCTS ADJACENT TO THE SULLIVAN OREBODY

	1	2	3	4
SiO_2	71.53	69.65	30.33	59.34
Al_2O_3	12.74	12.45	15.34	18.90
TiO_2	0.55	0.50	0.67	0.91
Fe_2O_3 (total Fe)	3.52	9.22	29.29	5.42
MgO	1.62	2.49	13.95	3.18
CaO	1.66	0.40	0.60	1.06
Na_2O	2.08	0.68	0.67	7.78
K_2O	2.81	0.02	0.03	0.10
B_2O_3	0.0015	5.00	0.005	0.31
LOI*	–	–	7.75	3.32
TOTAL	96.52%	100.41%	98.64%	100.20%
	N = 94	N= 27	N = 4	N = 5

1) Average Aldridge composition
2) Average tourmalinite composition
3) Average chloritic rock composition
4) Average albite-chlorite-pyrite composition
* Loss on ignition

event are unknown. Study of petrography and chemistry of rocks and minerals in the orebody indicates that the assumption of isochemical metamorphism is valid. Therefore, broad chemical changes indicated by comparison of whole rock analyses are generally valid.

Chemistry of the Ore

Sullivan ore contains about 200 times as much lead and zinc as does interbedded waste material (Table I) and is strongly enriched in sulphur-bound iron. By calculating the composition of the non-sulphide component of the ore on a sulphide-free basis (Table I) it is possible to compare its chemistry with the average composition of interbedded waste. In comparison to waste the non-sulphide component of the ore is carbonate- and iron-rich and relatively poor in SiO_2, Al_2O_3 and K_2O. Iron and manganese are markedly more abundant in the non-sulphide component compared to waste. Iron content of the non-sulphide component may be attributed to the abundance of iron-rich chlorite which contains an average of 28.4% iron as FeO. Manganese occurs in spessartine-rich garnets and as a component in carbonate minerals.

Nine metals that occur in trace amounts are routinely assayed for in Sullivan mill feed and Table II presents average values for these. Antimony is most abundant in ore from the southwestern part of the orebody in the vicinity of the open pit (Freeze, 1966) where it attains concentrations as high as 0.3%. Manganese is also most abundant in ore in the vicinity of the open pit where it is recorded in amounts up to one per cent. Arsenic is most abundant immediately south of the "central iron zone" (Freeze, 1966) where it attains maximum concentrations of about 0.6%. Thallium is principally distributed along the eastern fringe of the orebody where it attains maximum values of about 0.02%. There are no data concerning the distribution in ore of copper, indium, bismuth or cadmium.

Average Sullivan waste (Table I) is quite similar to the average regional composition of rock of the Aldridge Formation (Table III). The average composition of Sullivan waste is in accord with the abundance of white mica and carbonate in the matrix surrounding quartz grains. Values for lead and zinc in the waste are an order of magnitude greater than average values of lead and zinc in the Aldridge rock.

Chemistry of Tourmalinite

Mean values for the major chemical components of tourmalinite from the footwall are presented in Table III. No attempt has been made to analyze the rocks for H_2O, F or Li_2O. The boron reported is an approximate mean value obtained from analysis by semiquantitative emission spectroscopy.

It is appropriate to compare tourmalinite composition to that of average Aldridge sedimentary rock (Table III). The principal differences, aside from abundant boron in tourmalinite, are markedly lower K_2O, Na_2O and CaO and higher Fe_2O_3 and MgO in tourmalinite compared to average Aldridge. With increase of tourmaline in the matrix at the expense of quartz and minor minerals, composition of the tourmalinite approaches that of pure tourmaline. Although this trend is distinct, the pure end-member rock has not been observed.

Although the chemical composition of tourmaline from typical tourmalinite has not been determined, chemistry of individual tourmaline porphyroblasts from several localities in the Sullivan orebody has been studied by Ethier and Campbell (1977). Their data suggest that composition of isolated tourmaline porphyroblasts is highly variable and that the distribution of Fe-rich and Mg-rich tourmaline is erratic. This is in accord with an interpretation that composition of tourmaline is a function of the bulk chemistry of sedimentary rocks prior to metamorphism and partitioning of iron between silicate and sulphide phases during metamorphism.

Chemistry of Albite-Chlorite-Pyrite Altered Rock

Composition of albite in the altered rock is remarkably uniform in both the foot- and hangingwalls. The average composition is $An_{0.65}$, $Ab_{98.96}$, $Or_{0.39}$ with a range of An_0 to $An_{2.5}$.

Composition of chlorite is dependent on character of associated rock type.

Magnesium- and iron-rich chlorite is restricted to altered rocks; iron-rich but magnesium-poor chlorite occurs in unaltered sulphide rock and low-magnesium and low-iron chlorite is found in Aldridge metasedimentary rocks outside the mine area.

Average compositions of chlorite-pyrite footwall rock and albite-chlorite-pyrite-carbonate hangingwall rock are presented in Table III. During alteration of footwall tourmalinite to chlorite-pyrite rock, boron content is reduced to near regional background values of 15 to 30 ppm and silica is markedly reduced. Added components include Fe_2O_3, MgO and Al_2O_3. Albite-rich rocks in the footwall are chemically similar to albite-rich rocks in the hangingwall. During alteration of Middle Aldridge rock of the hangingwall to albite-chlorite-pyrite-carbonate rock, silica is markedly reduced and K_2O is almost completely removed. On release, potassium may have migrated laterally to produce the potassium-rich envelope in the hangingwall.

Isotope Chemistry

Early investigations, particularly those of Leech and Wanless (1962) and Sinclair (1966), documented two principal groups of lead isotopes in Purcell rocks. The most important is a 1200 to 1500 Ma old group of uniform isotopic composition and the second a more radiogenic group of Mesozoic-Cenozoic age. Many of the lead occurrences, some stratiform, in Purcell rocks contain lead of the old uniform group. The small vein and massive sulphide deposits in the Moyie Sills also contain lead of the old, uniform group. More recently Zartman and Stacey (1971) and LeCouteur (1973) provided new and more refined data allowing interpretation of small-scale variation.

The uniform 1200 to 1500 Ma lead forms a short elongate array resembling a multistage anomalous line (LeCouteur, 1973). LeCouteur speculated that the elongate array is the result of frequent mixing of lead derived from the older craton in the sedimentary environment and by mixing brines circulating in the sediment. The small scatter in isotopic composition could be the result of small initial differences in source lead laterally or vertically in the sedimentary column, or the result of lead of different composition leached from the same rock depending on extraction efficiency of the brine.

Using a model that derived lead and the other metals from Aldridge sediments by circulating brines, LeCouteur further speculated about age of the sedimentary source using a "frequent mixing model". LeCouteur postulated the following history.

t = 4550 Ma	Lead homogenized throughout the earth.
t = 3000 Ma	Substantial sialic crust formed, now part of North American Craton. Considerable variability in U/Pb from place to place from 3000 Ma on.
t = 2700 Ma	Kenoran Orogeny. Lead developed in one U/Pb environment since 3000 Ma mixed into other U/Pb environments.
t = 1800 Ma	Hudsonian Orogeny. Lead developed in one U/Pb environment since 2700 Ma mixed into other U/Pb environments.

t = 1500 Ma	Deposition of sediment eroded from rocks of Kenoran and Hudsonian age begins in Purcell basin to form Aldridge strata. Lead is homogenized during erosion and sedimentation but different U/Pb environments may still exist in the stratigraphic pile.
t = 1250–1450 Ma	Lead leached from Aldridge sediments by circulating brine and deposited as galena.

Although interpretation of lead isotope data is speculative, this model accounts for uniform lead isotopic composition in large deposits, such as the Sullivan orebody, by deriving lead from a large volume of sediments by circulating brine very early in the history of the sediment.

The apparent conflict between the model age of lead in the orebody and the U-Pb age of zircon in the gabbroic intrusions cutting the orebody is the result of many factors. A principal source of the discrepancy is the assumption of geochemical mixing of radiogenic and ordinary lead during a brief period of geological time to result in a final ore lead of radiogenic character. Sullivan lead is apparently not the result of simple single stage development.

Isotopic composition of sulphur in the Sullivan orebody was studied by Campbell *et al.* (1978). Isotopic composition of sulphur in the orebody ranges from -10.4 to $+4.7$ per mil with a mean $\delta^{34}S$ value of -2.2 per mil. Variations in sulphur isotopic composition appear to be stratigraphically controlled. At the footwall of the Main Band sulphides, ^{34}S is enriched but decreases toward the middle. Toward the top of the Main Band, ^{34}S increases again and then decreases at the top. A similar although more complex pattern is found for massive ore in the western part of the orebody. Campbell *et al.* (1978) ascribed the stratigraphic variation to changing conditions affecting supply of H_2S, and temperature gradients in the basin and concluded that Proterozoic seawater was the sulphur source.

GEOLOGICAL HISTORY

The Sullivan orebody is a conformable deposit of bedded iron, lead and zinc sulphides enclosed by fine-grained clastic sedimentary rocks of the Aldridge Formation. The orebody was deposited on the sea floor directly over conduits consisting of brecciated zones of footwall sediments, sulphide stringer zones and boron-rich sediments. The orebody has extensive alteration within the ore and in the hangingwall over the conduits.

Understanding of the genesis of the Sullivan orebody has evolved over an extended period as mining has advanced. During the 30s and 40s, when most mining was in the western part of the orebody where alteration is present, the prevailing view favoured an epigenetic hydrothermal process (Swanson and Gunning, 1945; Freeze, 1966). In the late 50s, Owens (1960) collected data suggesting rapid deposition of sulphide contemporaneously with deposition of the enclosing sediments. This view, continuously modified by ongoing collection of geological data, has prevailed among Cominco geologists since the mid-60s. Although there is now general agreement among most geologists concerning formation of the orebody from hydrothermal emanations on the sea floor (Sangster, 1972; Cominco Staff, 1972; LeCouteur, 1973), there is more than one approach to interpretation of some of the available data by those familiar with the deposit, and two of these are discussed below.

Both models postulate that at the Sullivan, footwall conglomerate, chaotic breccia and soft-sediment deformation of pre-ore sediments are manifestations of preconsolidation, pre-ore deformation which produced a sub-basin or depression on the sea floor toward the close of Lower Aldridge time. Cross-strata footwall permeability was produced along synsedimentary faults, fractures and chaotic breccia zones. These conduits allowed hydrothermal solutions access to the sea floor. Evidence for hydrothermal activity along these channels consists of: tourmalinite, stringer sulphide zones and crosscutting albite-chlorite-pyrite-carbonate alteration. Preconsolidation deformation of footwall sediments and ore is common adjacent to these structures.

The first model postulates that hydrothermal solutions welled up through conduits into the sub-basin to form a pool of metal-rich saline fluid in an environment lacking in high-velocity ocean currents and wave action. Detrital influx was relatively slow and intermittent. The source and transporting media of boron, ore metals and sulphur and the role of heat are speculative. Lead isotope data (LeCouteur, 1973) are consistent with circulation and concentration of seawater trapped within Lower Aldridge sediments to form brines and derivation of ore metals by leaching of sediments by the brine. Sulphur isotope data (Campbell *et al.*, 1978) are consistent with seawater sulphate reduction in the sedimentary pile. Heat source providing energy for circulation and permitting efficient metal extraction from the sediments can be viewed as a simple increase in regional heat flow.

The sequential distribution of tourmalinite, sulphide rock, sulphides with interlayered mudstone and finally albite-chlorite-pyrite-calcite rock suggests that components were deposited from solutions which changed composition and rate of discharge through time or that conditions in the basin changed significantly through time. Initial deposits included boron, minor iron and sulphur and very minor lead and zinc. The bulk of the boron was added epigenetically to deep footwall sediments. At the top of the tourmalinite zone, near the sulphide footwall, the stratiform nature of tourmalinite and lack of crosscutting relations indicate that boron was deposited here during clastic sediment deposition. Conditions favourable for deposition of abundant boron ceased prior to major sulphide deposition, and laminated mudstone with several per cent pyrrhotite was deposited over conglomerate and tourmalinite in the central western part of the orebody. Chaotic breccia on the other hand, probably reached its full development immediately prior to and during early deposition of ore sulphides.

Major sulphide deposition began as changes in solution composition introduced more iron, lead and zinc or as conditions in the brine pool matured causing iron, lead and zinc sulphide precipitation. Abundant fragmental material and disturbed layering near the base of the sulphide rock in the western part of the orebody suggest disruption of accumulating sulphides. Scarcity of argillaceous clastic grains either intermixed or interbedded in the lower part of the sulphide succession indicates rapid sulphide deposition. This suggests that metals and reduced sulphur were present in solution at the time of upwelling into the basin. Evidence is permissive, although not conclusive, for early deposition of iron sulphide in the western part of the orebody before major deposition of lead and zinc.

The accumulation of sulphides ultimately consisted of a pyrrhotite lens, several tens of metres thick in the centre, overlain by a similar thickness of pyrrhotite containing layers, up to 3 m thick, of galena and sphalerite. The sulphide accumulation was probably somewhat domical in shape forming a slight topographic high in the basin. Sulphides may have slumped, both inward toward the deepening active basin and outward away from the accumulating pile.

Deposition of the Main Band in the eastern part of the orebody commenced during deposition of the lead- and zinc-rich upper part of the sulphide accumulation in the western part. Two depositional processes are envisioned. Rapid deposition of sulphide in a quiet brine pool receiving little or no argillaceous clastic material is envisioned for the lower one-half to three-quarters of the Main Band. Uniform bedding, great lateral continuity of layers and mineralogically simple, commonly monomineralic layering suggest direct precipitation in a tectonically stable environment. Delicate bedding of sulphides and increased abundance of interbedded clastic material in the upper part of the Main Band and overlying sulphide Bands suggest direct precipitation of sulphides from a brine and intermittent influx of clastic material mixed with a proportion of locally derived hydrothermal clay. Frequent surges of solution from the west may have regularly replenished metal in brine to the east. Occurrence of fragments and grading in sulphide rock of the Main Band near the fringe of the orebody suggest deposition in limited areas by mass flow from collapsed parts of the domed sulphides in the west.

Precipitation of lead and zinc from the brine may have been controlled by oxygen fugacity, temperature and pH. Lead was mainly deposited close to areas of chaotic breccia, probably exhalation centres, whereas zinc precipitated at greater distances from these centres.

The second model invokes metal derivation from a magmatic source. Ore metals are visualized as having been released from the magma at a time of crustal disturbance (reflected in disruption of footwall sediments) and then carried into the marine basin in hydrothermal solutions. The source of solutions may have been an early phase of the parent magma for the footwall gabbroic sill. Boron-bearing solutions first permeated the footwall sequence in the area of the conduits. Deposition of ore sulphides resulted from later, more active discharge of hydrothermal solutions through the conduit system.

Sulphides formed as discrete particles in response to marked changes of temperature and pressure both during movement up the conduits and immediately on entering the marine basin. The lower two-thirds to three-quarters of the sulphide accumulation is thought to have resulted from rapid deposition, including deposition of sulphide particles and fragments formed in the conduit system. The rate of sulphide deposition was so rapid that detrital argillaceous beds, or even grains, are insignificant in the sulphide rock. Bedding formed in the sulphides as a result of variations in the rate of discharge of hydrothermal solutions.

Chloritized and non-chloritized argillaceous fragments, small to large and often rounded pyrite grains, quartz grains and carbonate particles in the lower Main Band represent ejected vein and wall-rock material. The second model holds that the coarse-grained and altered character of these fragments could only have been formed in the conduit. Their character is so different from hosting assemblages that it appears unlikely that it was developed by in situ metamorphism. These fragments were torn from the conduit walls by rapidly ascending hydrothermal fluids and deposited with the accumulating sulphides. No material of this character is found in well-bedded sulphides of the upper Main Band and overlying sulphide beds. Metagabbro blocks in ore may be of similar origin. The preponderance of metagabbro dykes below the ore and the tendency for metagabbro to trail off in the ore sequence suggests that these intrusions may be penecontemporaneous with the ore. Ore deposition slowed after accumulation of most of the thick ore in the west and lower Main Band in the east, because there is a greater proportion of argillaceous clastic material intermixed with the sulphides in the upper part of the full sulphide succession.

Both genetic hypotheses postulate that deposition of these upper beds was controlled primarily by changes in physical and chemical conditions within the marine sub-basin. Both also invoke quiet periods when clastic sedimentation predominated to produce sections consisting of interbedded mudstone and the "A" to "D" sulphide Bands. Immediately following deposition of the "D" Band, clastic material transported by turbidity currents swept across the sub-basin and was deposited as thick, graded sequences starting with the "I" siltstone. Hydrothermal activity continued during sedimentation and at the close of deposition of each of the "I", "H" and "HU" massive graded sequences, favourable conditions recurred for formation of tourmalinite, accumulation of metal-rich saline fluids and deposition of sulphide. Prior to deposition of "HU" ore the underlying massive sulphides in the west apparently slumped to provide the ore-grade sulphide clasts in the "HU" conglomerate.

The close of synsedimentary sulphide deposition coincided with onset of Middle Aldridge turbidite sedimentation. Although continuing influx of turbidites to the basin may have prevented preservation of the brine pool, the hydrothermal activity continued. The fluids produced crosscutting albitic and chloritic rock in footwall tourmalinite and extensive alteration both in the orebody and in up to 100 m of accumulating hangingwall sediment. Activity gradually waned as fluids cooled and as the conduit system sealed itself by deposition of quartz and carbonate veins.

Emplacement of the footwall intrusion, which resembles the Moyie sills and dykes, caused metamorphism of the ore in its contact zone. The contact metamorphism and later regional metamorphism produced silicate mineral assemblages of upper greenschist facies. From arsenopyrite data, Ethier *et al.* (1976) estimated temperature of metamorphism to be 400 to 490°C. Fractionation of sulphur isotopes indicated metamorphic temperatures of 300°C; oxygen isotope fractionation between quartz and magnetite indicated a 400 to 560°C range. Folding and faulting later in Proterozoic time and during the Laramide Orogeny deformed the orebody.

ACKNOWLEDGEMENTS

Numerous Cominco geologists and technicians working over five decades contributed directly or indirectly to data and interpretations presented here. Contributions of D.E. Jardine, A.C. Freeze and F.R. Edmunds are particularly noted. More recently, D.R. Shaw's work on wallrock alteration contributed significantly to our understanding of that aspect. The manuscript benefited from review by R.L. Hauser, E.D. MacDonald, R.G. McEachern, E.A.V. Parviainen, and P.W. Ransom. J.P. Snyder and A.S. Hagen prepared the illustrations and I.G. Paron typed numerous drafts of the manuscript. Cominco's contribution, both for permitting publication of the paper and for underwriting costs of the colour illustrations, is gratefully acknowledged.

REFERENCES

Bishop, D.T., 1974, Petrology and Geochemistry of the Purcell Sills in Boundary County, Idaho, and adjacent areas: in Belt Symposium 1973, v. 2: Idaho Bureau Mines and Geology and Department of Geology, University of Idaho, p. 15-56.

Bishop, D.T., Morris, H.C. and Edmunds, F.R., 1970, Turbidites and depositional features in the Lower Belt-Purcell Supergroup: Geological Society of America Abstracts with program, v. 2, no. 7, p. 497.

Bouma, A.H., 1962, Sedimentology of Some Flysch Deposits: Elsevier, Amsterdam, 168 p.

Burwash, R.A., Baadsgaard, H. and Peterman, Z.E., 1962, Precambrian K-Ar dates from the western Canadian sedimentary basin: Journal of Geophysical Research, v. 67, p. 1617-1625.

Campbell, F.A., Ethier, V.G., Krouse, H.R. and Both, R.A., 1978, Isotopic composition of sulfur in the Sullivan orebody British Columbia: Economic Geology, v. 73, p. 246-268.

Carswell, H.T., 1961, Origin of the Sullivan Lead-Zinc-Silver Deposit, British Columbia: Ph.D. Thesis, Queen's University, Kingston.

Cominco Staff, 1972, An outline of the geology of the Sullivan Mine, Kimberley, British Columbia: in Major Lead-Zinc Deposits of Western Canada: Guidebook for Field Excursion A 24-C24; XXIV International Geological Congress, p. 26-36.

Daly, R.A., 1912, Geology of the North American Cordillera at the Forty-Ninth Parallel: Geological Survey of Canada Memoir 38 (3 vols.), 857 p.

Edmunds, F.R., 1977, The Aldridge Formation, B.C., Canada: Ph.D. Thesis, The Pennsylvania State University, State College.

Ethier, V.G. and Campbell, F.A., 1977, Tourmaline concentrations in Proterozoic sediments of the southern Cordillera of Canada and their economic significance: Canadian Journal of Earth Sciences, v. 14, p. 2348-2363.

Ethier, V.G., Campbell, F.A., Both, R.A. and Krouse, H.R., 1976, Geological setting of the Sullivan orebody and estimates of temperatures and pressure of metamorphism: Economic Geology, v. 71, p. 1570-1588.

Freeze, A.C., 1966, On the origin of the Sullivan orebody, Kimberley, B.C.: in Tectonic History and Mineral Deposits of the Western Cordillera: Canadian Institute Mining Metallurgy Special Volume 8, p. 263-294.

Gabrielse, H., 1972, Younger Precambrian of the Canadian Cordillera: American Journal of Science, v. 272, p. 521-536.

Giletti, B.J., 1968, Isotopic geochronology of Montana and Wyoming: in Hamilton, E.J. and Farquhar, R.M., eds., Radiometric Dating for Geologists: Interscience, London, 506 p.

Harrison, J.E., 1972, Precambrian Belt basin of northwestern United States: its geometry, sedimentation and copper occurrences: Geological Society of America Bulletin, v. 83, p. 1215-1240.

Harrison, J.E. and Campbell, A.B., 1963, Correlations and problems in Belt Series stratigraphy, northern Idaho and western Montana: Geological Society of America Bulletin, v. 74, p. 1413-1428.

Harrison, J.E. and Jobin, D.A., 1963, Geology of the Clark Fork Quadrangle, Idaho-Montana: United States Geological Survey Bulletin, 1141 K, 38 p.

Höy, T., 1978, Note to accompany Preliminary Map No. 28, Geology of the Estella-Kootenay King area, southeastern British Columbia (82G/12, 13): British Columbia Ministry of Mines and Petroleum Resources.

Hunt, G., 1962, Time of Purcell eruption in southeastern British Columbia and southwestern Alberta: Alberta Society of Petroleum Geologists Journal, v. 10, p. 438-442.

Jardine, D.E., 1966, An investigation of brecciation associated with the Sullivan mine orebody at Kimberley, B.C.: M.Sc. Thesis, University of Manitoba, Winnipeg.

LeCouteur, P.C., 1979, Age of the Sullivan lead-zinc deposit: Geological Association of Canada, Program with Abstracts, Cordilleran Section.

______________, 1973, A study of lead isotopes from mineral deposits in southeastern British Columbia and in the Anvil Range, Yukon Territory: Ph.D. Thesis, University of British Columbia, Vancouver.

Leech, G.B., 1952, St. Mary Lake, British Columbia: Geological Survey of Canada Paper 52-15.

Leech, G.B. and Wanless, R.K., 1962, Lead isotope and potassium-argon studies in the east Kootenay district of British Columbia: in Engel, A.E.J., James, H.L. and Leonard, B.F., eds., Petrologic Studies: Geological Society of America, p. 241-280.

Lis, M.G. and Price, R.A., 1976, Large-Scale Block Faulting During Deposition of the Windermere Supergroup (Hadrynian) in Southeastern British Columbia: Geological Survey of Canada Paper 76-1A.

McAdam, J.H., 1978, A Preliminary Study of Footwall Mineralization at the Sullivan Mine, Kimberley, B.C.: B.Sc. Thesis, Queen's University, Kingston.

Miller, F.K., 1973, The age of the Windermere Group and its relation to the Belt Supergroup in northeastern Washington (Abs.): in Belt Symposium 1973, v. 1: Idaho Bureau Mines and Geology and Department of Geology, University of Idaho, p. 221.

Obradovich, J.D., and Peterman, Z.E., 1968, Geochronology of the Belt Series, Montana: Canadian Journal of Earth Sciences, v. 5, p. 737-747.

Owens, O.E., 1960, Internal Cominco Report.

Pentland, A.G., 1943, Occurrence of tin in the Sullivan mine: Canadian Institute Mining Metallurgy Transactions, v. 46, p. 17-22.

Price, R.A., 1964, The Precambrian Purcell System in the Rocky Mountains of southern Alberta and British Columbia: Canadian Petroleum Geology Bulletin, v. 12, p. 399-426.

Ransom, P.W., 1977, Geology of the Sullivan Orebody: in Höy, T., ed., Lead-Zinc deposits of southeastern British Columbia: Guidebook, Field Trip No. 1, Geological Association of Canada, Vancouver, B.C., p. 7-21.

Reesor, J.E., 1956, Dewar Creek Map-Area with Special Emphasis on the White Creek Batholith, British Columbia: Geological Survey of Canada Memoir 292, 78 p.

Reid, R.R., Morrison, D.A., and Greenwood, W.R., 1973, The Clearwater orogenic zone, a relict of Proterozoic orogeny in central and northern Idaho: in Belt Symposium 1973, v. 1: Idaho Bureau Mines Geology and Department of Geology, University of Idaho, p. 10-56.

Ryan, B.D., and Blenkinsop, J., 1971, Geology and geochronology of the Hellroaring Creek stock, British Columbia: Canadian Journal of Earth Sciences, v. 8, p. 85-95.

Sangster, D.F., 1972, Precambrian Volcanogenic Massive Sulphide Deposits in Canada; A Review: Geological Survey of Canada Paper 72-22, 44 p.

Schofield, S.J., 1915, Geology of Cranbrook Map-Area, British Columbia: Geological Survey of Canada Memoir 76.

Shaw, D.R., in preparation, Wall-rock alteration at the Sullivan Mine, Kimberley, British Columbia: Ph.D. Thesis, Queen's University, Kingston.

Sinclair, A.J., 1966, Anomalous leads from the Kootenay Arc, B.C.: in Tectonic History and Mineral deposits of the Western Cordillera: Canadian Institute of Mining Metallurgy Special Volume 8, p. 249-262.

Swanson, C.O., and Gunning, H.C., 1945, Geology of the Sullivan Mine: Canadian Institute Mining Metallurgy Transactions, v. 48, p. 645-667.

Wanless, R.K., and Ressor, J.E., 1975, Precambrian zircon age of orthogneiss in the Shuswap Metamorphic Complex, British Columbia: Canadian Journal of Earth Sciences, v. 12, p. 326-382.

White, W.H., 1959, Cordilleran tectonics in British Columbia: Bulletin American Association of Petroleum Geologists, v. 43, p. 60-100.

Willis, B., 1902, Stratigraphy and structure, Lewis and Livingstone ranges, Montana: Geological Society of America Bulletin, v. 13, p. 305-352.

Winston, D., 1973, Stratigraphy and sedimentary features of the Missoula Group, western Montana: in Belt Symposium 1973, v. 1: Idaho Bureau Mines and Geology and Department of Geology, University of Idaho, p. 235-252.

Zartman, R.E., and Stacey, J.S., 1971, Lead isotopes and mineralization ages in Belt Supergroup rocks, northwestern Montana and northern Idaho: Economic Geology, v. 66, p. 849-860.

Zartman, R.E., Peterman, Z.E., Obradovich, J.D., Gallego, M.D. and Bishop, D.T., in press, Idaho Bureau of Mines and Geology: K-Ar, Rb-Sr, and U-Th-Pb ages of the Crossport C sill near Crossport, Idaho.

MS received April 1979;
Accepted for publication August 1980.

Precambrian Sulphide Deposits, H.S. Robinson Memorial Volume, edited by
R.W. Hutchinson, C.D. Spence and J.M. Franklin, Geological Association of Canada
Special Paper 25, 1982

GAYNA RIVER: A PROTEROZOIC MISSISSIPPI VALLEY-TYPE ZINC-LEAD DEPOSIT

R.S. Hewton
Brinco Mining Limited,
#704-602 West Hastings Street, Vancouver, British Columbia V6B 1P2

ABSTRACT

Carbonate-hosted zinc-lead mineralization was discovered in 1974 in the Proterozoic Little Dal Group in the Gayna River area of the Mackenzie Mountains, N.W.T. Shallow-water carbonates with textural features indicative of peritidal environments are associated with a belt of stromatolitic reefs that has formed along a hinge line. Numerous occurrences of sphalerite are spatially associated with the reefs.

Sulphide mineralization extends vertically for many hundreds of metres and for many kilometres laterally but is concentrated in the 160 m thick Host Unit of the Little Dal Group. Numerous secondary (influx) breccias were formed in the Host Unit by fracturing, solution-collapse, recrystallization and dolomitization. These breccias, and the natural primary porosity within the Host Unit, have been healed by a multi-phase infilling of sparry dolomite, sphalerite and calcite, with minor barite, pyrite and galena and very minor fluorite, pyrobitumen and quartz.

Sphalerite, with minor pyrite and galena, occurs as: disseminations in the matrix and in fragments, or as envelopes to fragments, in sedimentary breccia; disseminations in the cement, or as envelopes to fragments, in secondary breccia; and as structurally controlled replacements. Each type has a unique mineralogy. Secondary breccia bodies are the most actively explored, as they are generally the richest.

Individual mineralized breccia zones range in size up to over 1 million tonnes, and in grade up to 10% combined zinc-lead. Geologically possible reserves are estimated to be in excess of 50 million tonnes of over 5% combined zinc-lead. Current exploration is designed to locate bodies of higher grade material (15% combined zinc-lead).

The depositional environment, mineralogy, modes and distribution of sulphide mineralization, and host rock types at Gayna River are very similar to those of Mississippi Valley-type deposits.

RÉSUMÉ

Une minéralisation de zinc et plomb, incluse dans des carbonates, a été découverte en 1974 dans le groupe protérozoïque de Little Dal situé dans la région de la rivière de Gayna dans les montagnes de Mackenzie, T.N-O. Des carbonates de mer peu profonde montrant des structures typiques de la zone des marées, sont associés à une ceinture de récifs stromatolitiques construits le long d'une zone charnière. De nombreux indices de sphalérite sont associés spatialement à ces récifs.

La minéralisation sulfurée, qui se rencontre sur plusieures centaines de mètres verticalement et plusieurs kilomètres latéralement, est concentrée dans les 160 m de l'Unité Hôte du groupe de Little Dal. De nombreuses brèches secondaires (d'influx) se sont formées dans l'Unité Hôte par des processus incluant fracturation, effondrement par solution, recristallisation et dolomitisation. Les interstices de ces brèches ainsi que les pores primaires de l'Unité Hôte ont été cimentés et remplis par un assemblage multi-phase de dolomite spathique, sphalérite et calcite accompagnées de petites quantités de barytine, pyrite et galène et de rare fluorinc, pyrobitume et quartz.

La sphalérite, avec un peu de pyrite et galène apparaît: en disséminations dans la matrice et les fragments et en enveloppes autour de ces fragments dans les brèches sédimentaires; en disséminations dans le ciment et les fragments ainsi qu'en enveloppes autour de ces fragments dans les brèches secondaires; et en remplacements contrôlés structuralement. Chacun de ces types a une composition minéralogique particulière. Les amas de brèches secondaires ont été les plus activement explorés car ils sont généralement les plus riches.

Les zones individuelles de brèches minéralisées varient en taille jusqu'à plus d'un million de tonnes et en teneur jusqu'a 10% de zinc et plomb combinés. Les réserves possibles telles qu'indiquées géologiquement sont estimées dépasser les 50 millions de tonnes à plus de 5% de zinc et plomb combinés. L'exploration en cours est conçue pour isoler des amas bréchiques à teneurs plus élevées (15% de zinc et plomb combinés).

Le milieu de dépôt, la composition minéralogique, les types et la distribution de la minéralisation sulfurée et les types de roches encaissantes de la région de la rivière de Gayna sont très semblables à ceux des dépôts du type Mississipi Valley.

INTRODUCTION

Regional exploration by Cordilleran Engineering Ltd. on behalf of Rio Tinto Canadian Exploration Ltd. (Riocanex) in 1974 led to the discovery of sphalerite mineralization in a large breccia zone exposed along a tributary to the Gayna River in the Mackenzie Mountains of the Northwest Territories (Fig. 1). Initial geological mapping and prospecting indicated that the host was the Proterozoic Little Dal Group, the mineralization was apparently open-space filling and numerous occurrences existed over a 30 km strike length. Recognizing that numerous occurrences of Mississippi Valley-type mineralization might exist as in the Pine Point district, the entire length of exposed mineralized Little Dal Group rocks was staked.

Since 1974, Riocanex has explored the property by detailed geological mapping, silt and soil geochemistry, induced polarization, gravity surveys and diamond drilling. The results of this work confirm that mineralization of Mississippi Valley-type is distributed over many kilometres of strike length and hundreds of metres of stratigraphic thickness, but is concentrated within a 160 m thick Host Unit.

Diamond dilling on grid patterns in a manner adopted in other Mississippi Valley-type districts like Pine Point and eastern Tennessee has tested some individual breccia bodies but has not yet tested all possible targets. The bodies tested are currently of insufficient grade and size to support mining operations in this remote area but exploration is continuing.

This paper presents a geological description of the Gayna River zinc prospect, with discussions on its classification and the strategy of exploration. For reference, the camp shown on Figure 2 is located at latitude 64°57′N and longitude 130°38′W.

EXPLORATION HISTORY

Initial interest in the area resulted from a single anomalous silt sample containing 980 ppm Zn and 11 ppm Pb that was collected during regional prospecting and geochemical sampling managed by Cordilleran Engineering Ltd. Silt samples were collected at a density of one per square mile (2.6 km^2) but only the one sample was anomalous. Despite its initial importance, silt sampling is relatively ineffective compared to prospecting or soil sampling because many samples from streams draining mineralized areas contain background values of lead and zinc only a few hundred metres downstream. The near neutral pH of the waters and/or the very low organic content of silts in streams draining the carbonate rocks prevent extensive chemical transport of metals. In contrast, soil sampling has outlined every known mineralized area where sampling was possible and has led to the discovery of some occurrences. One particularly large soil anomaly is 4 km by 0.7 km in area, with peak values of 7,500 ppm Zn and 25,000 ppm lead.

Induced polarization surveying proved inconclusive in detecting increased concentrations of pyrite assumed to be associated with increased concentrations of sphalerite. A wide variety of resistivity and frequency-effect responses were obtained but drilling revealed that they were associated with background quantities of pyrite or graphitic material. Gravimetric anomalies that were drilled are apparently related to limestone-dolostone fronts.

Diamond drilling remains the only effective means of finding mineralization. An approximate grid with drill hole centres at 610 m (2,000 ft.) was established over the Host Unit. Drill holes were classified according to their feet-per cent lead + zinc values, as they are in other Mississippi Valley-type districts, calculated by multiplying the combined grade of zinc plus lead by the assay interval in feet. Mineralization was obtained in 71% of all holes and 12% had values greater than 40 feet-per cent. This 12% figure permitted estimates in excess of 60 million tonnes of 40 feet-per cent rock at Gayna River without any implications as to its continuity or mineability.

The configuration of mineralized bodies was investigated by holes drilled at 61 m (200 ft.) centres around the best holes of the wider-spaced grid. The mineralized breccia bodies were found to be continuous, nearly equidimensional in plan and in most cases tabular, but occasionally pillar-like, in cross-section. The size of mineralized breccia zones is limited by incomplete drilling and ranges from 56,000 to 2,000,000 tonnes depending upon the number of drill holes in each body. Typical breccia zones, still undelimited, are:

56,300 tonnes	14.52% combined zinc-lead
95,300 tonnes	9.85% combined zinc-lead
1,066,800 tonnes	4.51% combined zinc-lead

The average grade of all breccia zones drilled to date is just over 5% zinc-lead.

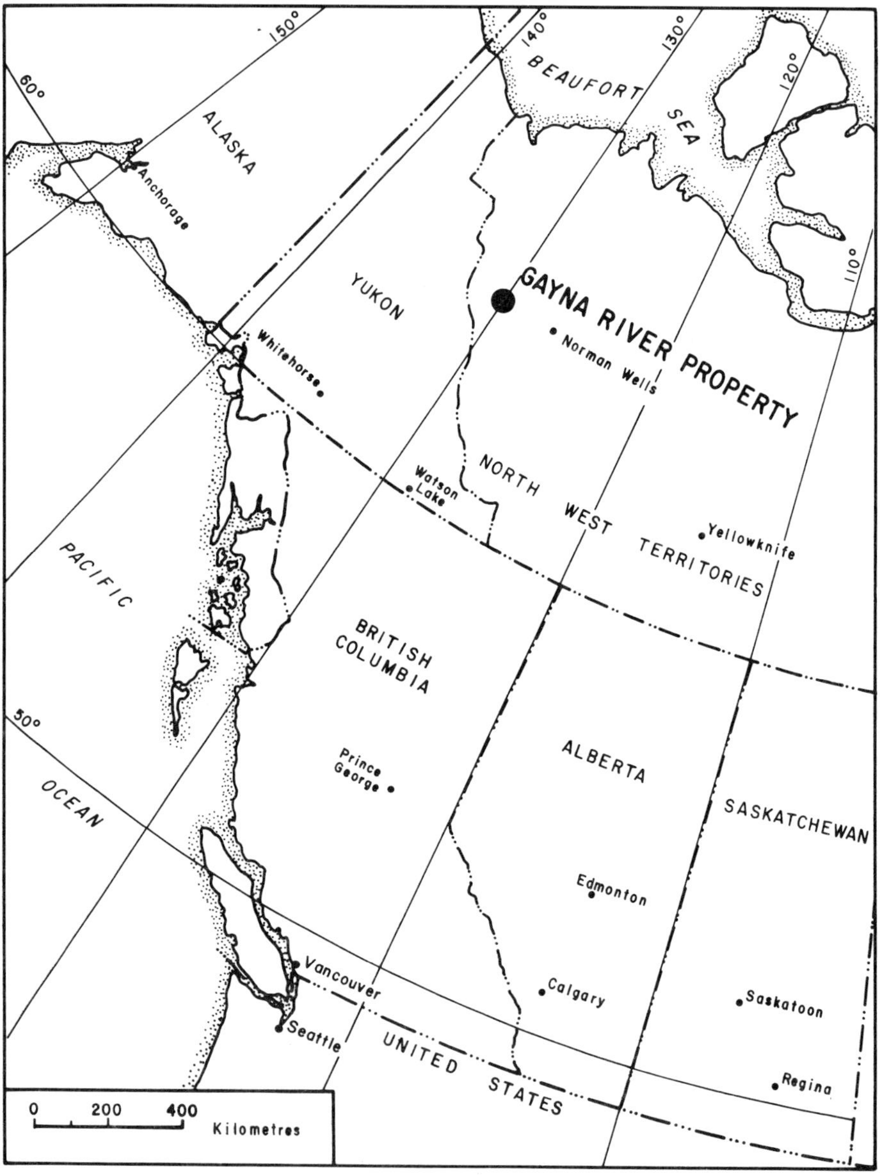

Figure 1. Location map.

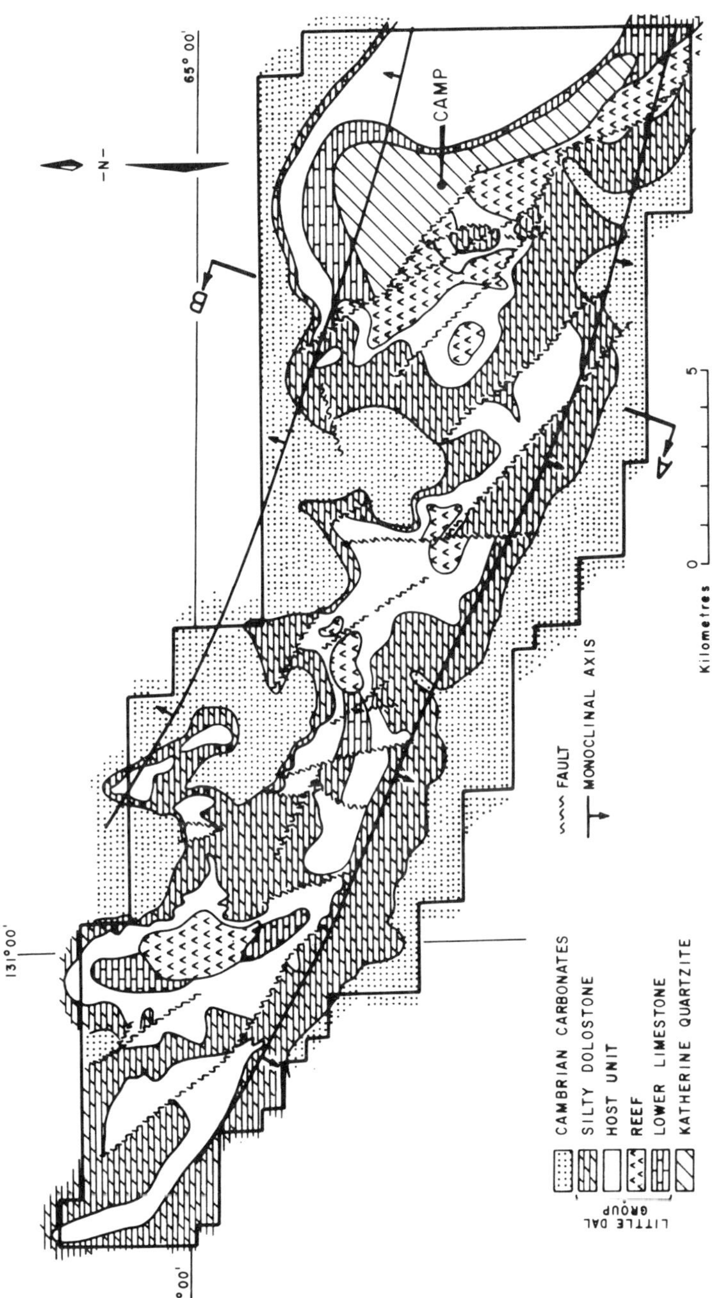

Figure 2. Simplified geological map, Gayna River zinc-lead deposit.

Systematic grid drilling in the known breccia bodies was discontinued even though the mineralization was not fully outlined, because, although individual intersections ranged up to 7.6 m of 23% combined zinc-lead, the average grade of each breccia body was too low for such an isolated area as Gayna River with inherent high costs of exploration and mining. All future work will seek a breccia body of higher average grade – in excess of 15% combined zinc-lead.

REGIONAL GEOLOGY

Little Dal Group rocks are Proterozoic platform deposits that are part of the Mackenzie Mountains Supergroup (Young *et al.*, in press). Exposure is fairly continuous throughout the Mackenzie Mountains as shown by Aitken (this volume). Textures within the Little Dal Group suggest a relatively shallow marine deposition of sediments thinning over the Mackenzie (Redstone) Arch and thickening westward.

The structural style of the Mackenzie Mountains is dominated by large-scale folds and strike-slip and thrust faults. Many folds "have been modified by faults, some of which may be basement controlled" (Gabrielse, 1967, p. 289). The trend of these faults and fold axes swings from easterly to southeasterly forming a conspicuous arc coinciding with the pattern of the Cordillera. Gabrielse (1967) pointed out that although the Mackenzie Mountains are bounded by westerly dipping thrust faults, the numerous thrusts typical of the front ranges in the southern Cordillera are absent. Also, the Mackenzie Mountains are structurally symmetrical in contrast with the general asymmetry of the Canadian Cordillera.

The type section of the Little Dal was described by Gabrielse *et al.* (1973). Preliminary geological mapping by Aitken and Cook (1974a, 1974b) and a report by Aitken *et al.* (1973), show the Little Dal to be part of a larger unit, the unnamed map unit H5. Subsequent field work by Aitken *et al.* (1978) led to abandonment of the H5 unit and to definition of the Little Dal Group which was subdivided in detail. A comparison of the earlier subdivisions with that adopted in this paper is shown in Figure 3. For the most part, divisions used by Aitken *et al.* (1978) are accepted, except that the evaporites are considered to be local facies within the Little Dal Group, and the "Mudcracked" unit is thought to be more appropriately assigned to the underlying Katherine quartzite because of its high-silica content and shallow-water textures.

The age of the Little Dal Group has been controversial because of a lack of isotopic dates. Little Dal Group rocks have been correlated with the Shaler Group (Reynolds Point, Minto Inlet and Wynniatt Formations) on Victoria Island (Young *et al.*, in press; Aitken, this volume) and the Pinguicula Group in the Wernecke Mountains of central northern Yukon (Young *et al.*, in press), suggesting an Helikian age. This age was further supported by correlations with the Belt-Purcell Supergroup in southern B.C. (Aitken *et al.*, 1973, p. 34). In addition, stromatolite assemblages (Aitken *et al.*, 1978) and algal microfossils (Hofman and Aitken, 1979) indicate an age of between 1.1 to 0.8 Ga. Recent work by Bruce Ryan and R. L. Armstrong (pers. comm. 1978) at the University of British Columbia has yielded a Rb-Sr isotopic age of approximately 0.8 Ga for a diabase dyke that intrudes the Little Dal Group rocks. Thus, there is increasing evidence that the age of the Little Dal Group is between 1.1 and 0.8 Ga, probably closer to 0.9 Ga, near the Helikian-Hadrynian time boundary.

Mineralization in Proterozoic rocks of the northern Cordillera is not common and may quickly be summarized. Delaney *et al.* (in press) described deposits containing uranium, copper, barium, cobalt and iron associated with breccias within the Wernecke Supergroup of the central Yukon, copper mineralization in the Redstone River Formation in the western Northwest Territories, and iron in the Rapitan Formation near the Yukon-Northwest Territories boundary. Goodfellow (1979) described new occurrences of "stratabound sphalerite finely dispersed throughout stromatolitic rocks" and breccia-controlled sulphides in the Gillespie Lake Group of the central Yukon. Occurrences in Proterozoic rocks are also listed in Archer Cathro and Associates (1978), the most significant of which are the Craig, Sian, and Og properties (File Numbers 106C-73, 106C-74, and 116B-83 respectively) which contain sphalerite- and galena-bearing breccias in Hadrynian dolomite. All these occurrences are 100 km or more from Gayna River, and none are in the Little Dal Group.

PROPERTY GEOLOGY

Stratigraphy

The distribution of Proterozoic rocks on the property is shown on Figure 2. Exposures of the Little Dal Group rocks are confined to northwest-trending river valleys. Beds dip steeply to the northeast in the northeast, are flat-lying between the

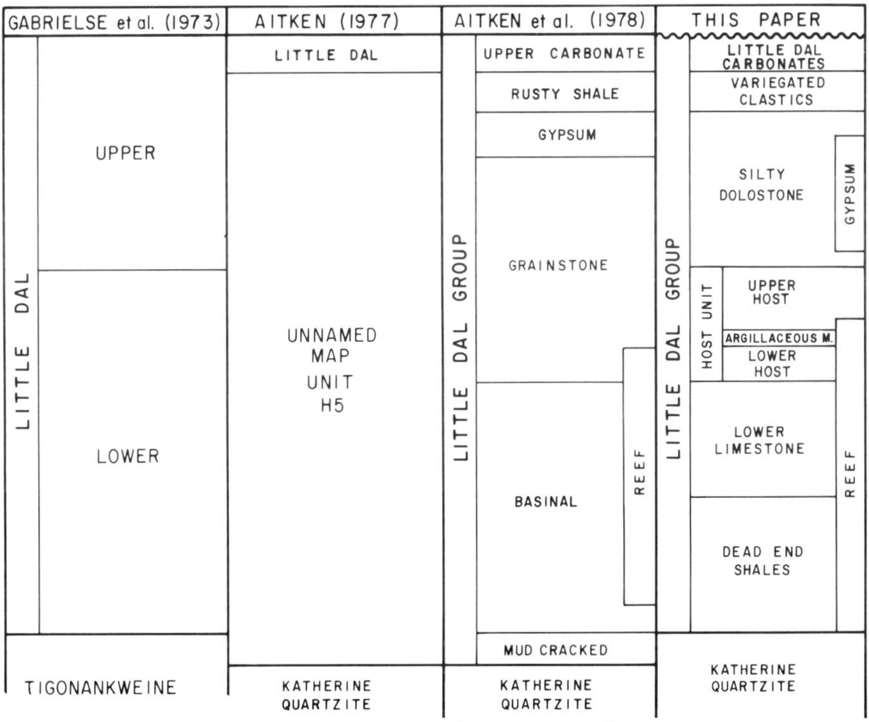

Figure 3. Comparison of nomenclature of formations in Gayna River area.

monoclinal axes, and dip steeply to the southwest in the southwest forming a large "box" anticline.

A summary description of the stratigraphic column is given in Figure 4 which shows that the Little Dal Group is primarily carbonate with minor shale near the bottom and top, is underlain by quartzite and is overlain by shale and additional carbonate (not shown). Zinc and lead sulphides occur in most units but are concentrated in the Host Unit. Volcanic rocks are missing from the sequence and the only evidence of intrusive activity is occasional diabase dykes that intrude up to, but not into, the variegated clastic rocks of the upper Little Dal Group.

Katherine Quartzite. This resistant, well sorted quartzite with minor interbedded shale and carbonate crops out rarely in the Gayna River area and only the upper beds have been observed. These grade upward from rusty-weathering quartzite through 15 m of interbedded quartzite and siliceous dolostone, 8 m of orthoquartzite, 3 m of black shale, 5 m of grey cherty dolostone with numerous oolite beds and 3 m of

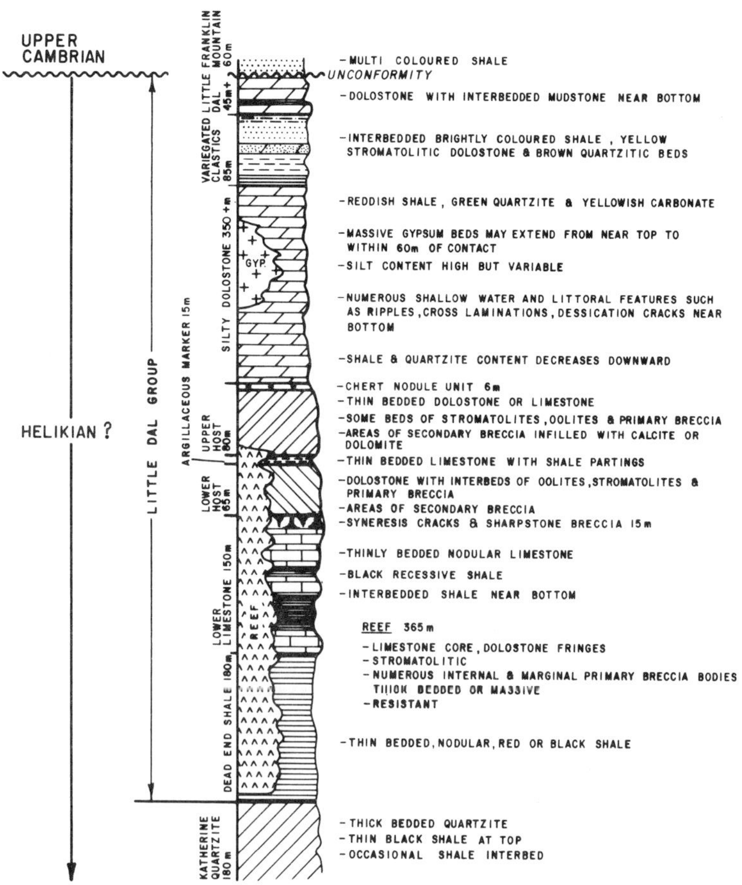

Figure 4. Idealized stratigraphic column, Gayna River zinc-lead deposit.

Figure 5. Two reefs in the Little Dal Group.

Figure 6. Contact between Lower Host and Reef.

Figure 7. Draping and thinning of Lower Host (LH), Argillaceous Marker (AM) and Upper Host (UH) over Reef.

Figure 8. Contact between Lower Host (LH), Lower Limestone (LL) and Reef.

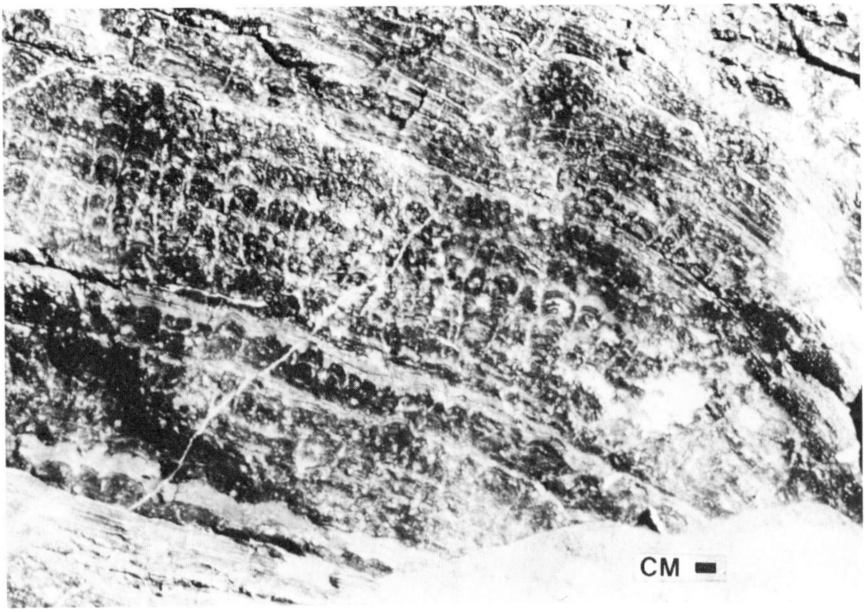

Figure 9. Part of large dome made up of cryptalgal laminations.

Figure 10. Hummocky bedding plane stromatolites.

quartzite. Ripple marks, mudcracks, cross-laminations and salt casts (Fig. 11) have been noted and stromatolites reported. These beds are the "Mudcracked" unit of the Little Dal Group (Fig. 3) of Aitken *et al.* (1978) but are herein assigned to the Upper Katherine Group because of their quartzitic composition and shallow water textures.

Reef. Stromatolite growth began during deposition of the upper beds of the Katherine Quartzite and was concentrated on local highs where large bioherms or reefs developed (Fig. 5). Aitken (pers. commun., 1978) suggested that stromatolite growth may have been related to the oolite member of the upper Katherine because the oolite beds are absent near the western limit of the reef belt, suggesting that the reefs grew near the transition from a slope to deeper-water environment, or possibly along a hinge line.

The reefs are resistant to weathering and have lateral dimensions up to 6.5 by 1.6 km and heights of 500 m, with relief during deposition possibly as much as 100 m (Aitken *et al.*, 1978). They were originally limestone but now have outer margins of dolostone, which may be more abundant than limestone. They are variable in size and may be small, completely contained within the Lower Limestone unit or may be large, reaching into the middle of the Host Unit. All units younger than the Argillaceous Marker exist above all reefs, and thinning and draping is evident as high in the stratigraphic column as the Silty Dolostone. Facies changes between reefs and surrounding sedimentary rocks tend to be sharp (Fig. 6). Units overlying the reefs tend to thin and drape (Fig. 7, 8).

Internal textures include interbedded algal laminations, massive bedding, stromatolite breccias and large domes consisting of small scale, laterally linked stromatolites (Fig. 9). Bedding plane tops are hummocky (Fig. 10). Sulphide mineralization has been found within the reefs but tends to be of limited tenor and extent.

An excellent description of the development of Paleozoic reefs and the formation of related facies was given by James (1978). A similar model, with modifications allowing for the fact that Proterozoic reef builders were algae rather than metazoans, is envisaged for the Gayna River area.

Dead End Shale. Conformably overlying the Katherine Quartzite is a 180 m sequence of thinly bedded shale with minor interbedded limestone designated the Dead End Shale. The beds are continuous and nodular. Local brick-red colouring of weathered and fresh surfaces is common. Limestone interbeds become more abundant upward and the unit is gradational into, and often indistinguishable from the overlying Lower Limestone. For this reason, the name Lower Limestone often includes rocks of both the Dead End Shale and Lower Limestone.

Lower Limestone. There is no apparent change in bedding thickness, colour or continuity upward into the Lower Limestone. The strata were deposited in relatively deep water and, except for the upper 30 m of the 150 m thick Lower Limestone, are devoid of shallow-water features. These uppermost beds show evidence of shallower-water deposition and possible regression in an increase in silt content and development of ripple marks and graded bedding (Fig. 12). Well developed "molar tooth" structures (Fig. 12), believed to be syneresis or dewatering cracks, form planar features cutting across the bedding. Thin beds of sharpstone-breccia, a sedimentary breccia with lathlike fragments which are possibly rip-up clasts, are also present and make these uppermost beds distinctive, marking the lower limit of major sulphide concentrations.

Iron-formation occurs near Gayna River in the Lower Limestone. Black magnetite and thin, wispy interbeds of hematite and jasper form a conformable bed near the middle of the Lower Limestone. The iron-formation is approximately 15 m thick and has been identified at two localities 24 km apart. The magnetite content ranges up to 90% but intense folding and faulting preclude economic potential. The iron-formation is roughly time-equivalent to lower levels of reefs, but is older than similar iron-formation within the Hadrynian Rapitan Formation.

Lower Host. Upper strata of the Lower Limestone grade into 65 m of medium-bedded dolostone of the Lower Host. Shallow-water features are abundant and include ripple marks, cross-beds (as in the Lower Limestone, Fig. 12) sedimentary breccia (Fig. 13), stromatolites (Fig. 14) and oolites, pisolites and oncolites (Fig. 15). Argillaceous and carbonaceous interbeds and stylolites, along which up to 5 mm of argillaceous material may be present, are common.

Oolites are far more abundant than pisolites or oncolites and are important because of their dominance throughout both the Lower and Upper Host. Beds consisting of siliceous oolites and carbonate matrix are prominent. Petrographic examination has shown that oolites are present even when megascopically invisible, suggesting that most of the Host Unit was deposited in a littoral environment. Sedimentary breccia is common within both the Lower and overlying Upper Host and was probably formed by slumping of sediments deposited near margins of positive features in the basin, formed by the reefs.

Typical of both the Lower and Upper Host is abundant, white, sparry calcite, dolomite, or barite, also referred to as "influx", that fills cavities (Fig. 16) formed by a variety of processes discussed below. This gives a distinctive appearance to the rock because of the colour contrast between white influx and dark grey dolostone. The Lower Host, originally deposited as limestone, has been dolomitized with a resultant increase in porosity that is at least partially sealed by the white sparry carbonates. Recrystallization has produced pseudo-breccias reminiscent of those typical of the Presqu'ile Formation at Pine Point. Dolomitization and recrystallization are irregularly distributed and have produced marked colour differences between tan dolostone and black limestone that are readily visible in outcrop (Fig. 17).

Argillaceous Marker. Thinly and evenly bedded, black, argillaceous limestone of the Argillaceous Marker conformably overlies the Lower Host and is distinguished from it by its darker colour. Interbeds of graphitic and argillaceous shale are common and thin beds of pyrite are also present. A distinctive conglomerate bed up to 3 m thick is present at the top of the 15 m thick unit. Matrix and boulders are argillaceous limestone and the thin bedding evident in the lower strata is absent in the conglomerate. A lack of shallow-water features suggests that the Argillaceous Marker was deposited in relatively deep water. Carbonate influx, common in the Lower and Upper Host, and lead or zinc sulphides are rare.

Upper Host. The Upper and Lower Hosts are almost indistinguishable although there are fewer sedimentary breccias and more stromatolite beds within the 80 m thick Upper Host. Oolite beds are equally abundant. Silica occurs as chert nodules and reticulate chert beds at the very top of the Upper Host and forms a marker. Various forms of carbonate influx are also common in the Upper Host.

Figure 11. Salt casts in Katherine Quartzite.

Figure 12. Syneresis cracks and cross-laminations in transition from Lower Limestone to Lower Host.

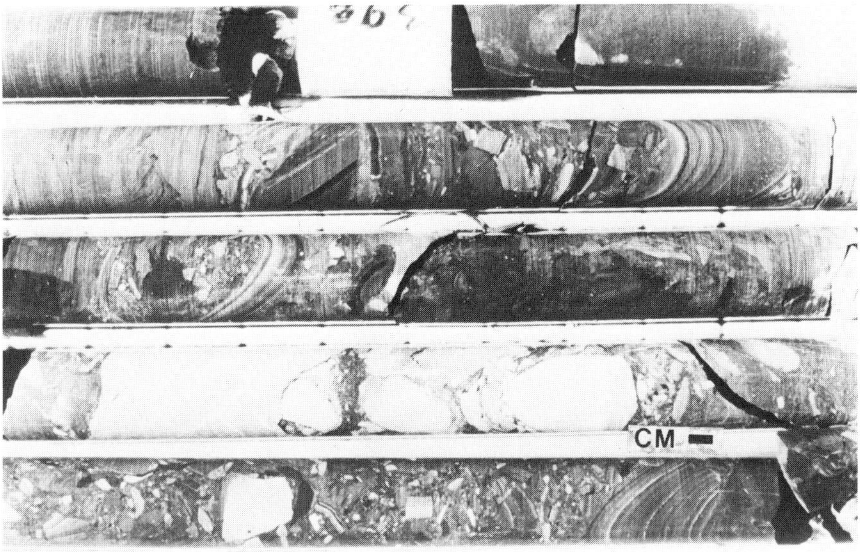

Figure 13. Sedimentary breccia in Lower Host.

Figure 14. Stromatolites from Upper Host.

Figure 15. Oncolites typical of Host Unit.

Figure 16. Carbonate influx, in this case fluxobreccia (''zebra rock''), from Host Unit.

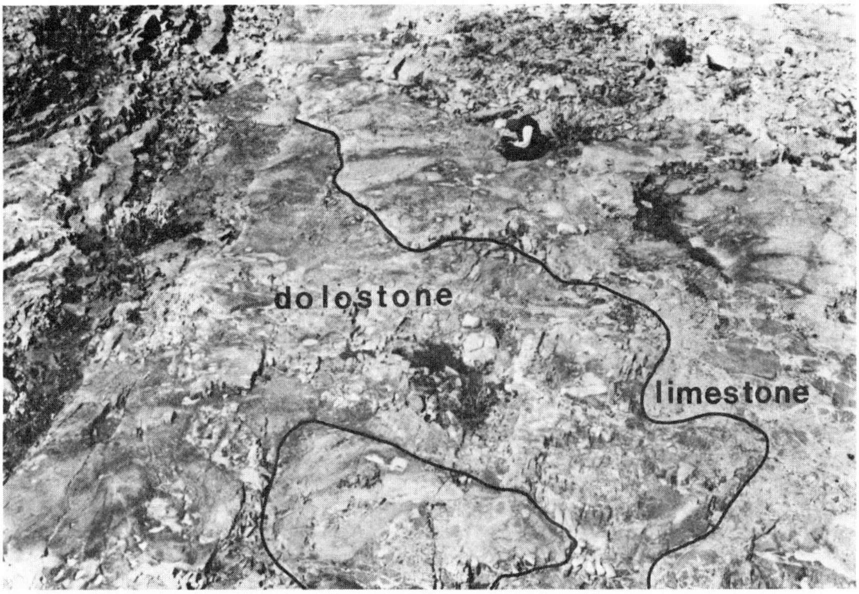

Figure 17. Dolostone-limestone front in Lower Limestone.

Figure 18. Unconformity (U) between Franklin Mountain Formation above and Silty Dolostone below. Note gypsum beds (g).

The description of the Lower Host applies generally to the Upper Host. Both were deposited in shallow water with an intervening short period of deeper-water sedimentation represented by the Argillaceous Marker. Because of their similarity and the even distribution of sulphides within them, the Upper and Lower Host, together with the intervening Argillaceous Marker are usually referred to collectively as the Host Unit.

Silty Dolostone. Yellow-weathering, shallow-water dolostone and minor limestone of the Silty Dolostone conformably overlie the Upper Host. Ripple marks, desiccation cracks, rare salt casts, stromatolites, oolites and sedimentary breccia are present. The silt content, however, is higher than in the Host Unit. The Silty Dolostone may be more than 350 m thick, but the upper beds are commonly absent beneath an Upper Cambrian unconformity.

A sequence of evaporites, locally greater than 250 m thick, has been mapped within the Silty Dolostone. Though these evaporites would normally be considered representative of back-reef facies, their uncertain spatial relation to the reefs, possibly even overlying the reefs, renders this interpretation tenuous.

Variegated Clastic Rocks. The Silty Dolostone grades upward to interbedded red, green, yellow and white shale and quartzite of the Variegated Clastics. These lagoonal sediments are of very limited extent due to the Upper Cambrian unconformity.

Upper Cambrian Unconformity and Franklin Mountain Formation. The Little Dal Group is truncated by a sub-Upper Cambrian unconformity that, northeast of the Gayna camp, removes all formations down to the Katherine Quartzite. Elsewhere on the property, this unconformity intersects the Little Dal succession anywhere between the Variegated Clastics and the extreme top of the Upper Host. This unconformity is evident when viewed perpendicular to the dip (Fig. 18) but is obscure from any other angle. An irregular 1 m thick conglomerate bed is present locally on the unconformable surface and is overlain by interbedded red, orange and green clastic sedimentary rocks and thick beds of grey carbonates of the Franklin Mountain Formation. Trilobites are found at lower levels whereas tetracorals and brachiopods are present near the top.

Overlying the Franklin Mountain Formation is a thick succession of shallow-water, fossiliferous carbonates that was deposited continuously from mid-Ordovician to mid-Devonian times. Late Devonian to early Mississippian, thinly bedded, black shales of the Canol Formation overlie these carbonates and are the youngest rocks in the area.

Depositional Environment. The sequence of units described above suggests the following depositional history. Evenly bedded quartzites derived from the Precambrian craton to the east were deposited on an extensive platform in shallow seas. Mounds were formed during deposition, possibly because of active structures such as basement growth faults. Oolitic beds and stromatolites developed over the mounds in the shallower water. Rapid stromatolite growth gave rise to bioherms which kept pace with a slowly deepening sea. Deeper-water deposition of the Lower Limestone was synchronous with the growth of these bioherms.

Following deposition of the Lower Limestone the seas became shallower with resulting formation of the Lower Host, dominated by shallow-water features. Biohermal growth continued. The Argillaceous Marker was deposited during a relatively brief period of increased clastic sedimentation in a deepening sea. Stromatolite growth abated during this period and the Argillaceous Marker pinched out against larger reefs and draped the smaller ones.

A return to shallow-water deposition formed carbonates of the Upper Host with shallow-water features, including stromatolites, though reefs did not develop. The seas continued to shallow until sedimentation changed from peritidal to supratidal facies containing local evaporite concentrations. Eventually, lagoonal deposition formed the Variegated Clastic sedimentary rocks.

Structure

Little Dal Group rocks in the Gayna River area have undergone structural deformation that is intense compared to most Mississippi Valley-type districts but is moderate for areas within the northern Cordillera. Large-scale, high-level concentric and sinuous-type folds and numerous major, deep-rooted, high-angle thrust faults exist northeast and southwest of the Gayna River property, but the structures present in the mineralized area are a large "box" anticline and numerous normal faults.

The "box" anticline is illustrated on Figure 3 by the two monoclinal axes representing changes in dip of beds from nearly flat in the central area to steeply dipping in the northeast and southwest. This major anticline may plunge to the northwest and is the only large-scale fold on the property. Small-scale folds are evident locally in the Variegated Clastics below the Upper Cambrian unconformity.

The oldest normal faults observed on the Gayna River Property are believed to have formed during the Racklan Orogeny (Gabrielse, 1967). The average strike of these faults is 170°. Vertical displacements of over 365 m have been measured (Fig. 19) but there is very little strike-slip movement. In general these faults are truncated by the upper-Cambrian unconformity although reactivation of some during younger orogenies has caused minor vertical displacement of the strata deposited above the

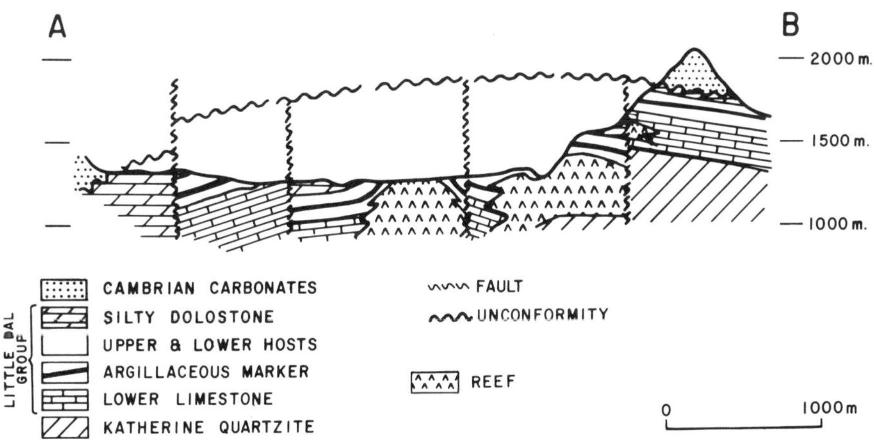

Figure 19. Cross section A-B (Fig. 2), showing stratigraphic relationships and style of faulting.

unconformable surface. Fault reactivation and the formation of new normal faults trending at 120°, were probably related to the Laramide Orogeny. Displacement on Laramide faults has not been measured in the Gayna River area but is believed to be less than 100 m, contrasting with the hundreds or thousands of metres of displacement on Laramide thrust faults southwest of the Gayna property.

Eisbacher (pers. commun., 1977) has suggested the possibility of synsedimentary faulting near Gayna River which allowed the growth of the reefs and formation of sedimentary breccias. Growth-faulting, described for other areas by Eisbacher (1978), is appealing but there is no strong evidence for it at Gayna River.

BRECCIAS

As in most Mississippi Valley-type deposits, mineralization at Gayna River is directly associated with brecciation. Geological mapping and diamond drilling have shown two main types of breccia: (1) sedimentary breccia, with matrix and fragments of similar composition, and (2) secondary breccia, with a distinctive white, sparry carbonate cement and grey carbonate clasts. Both types occur throughout the stratigraphic section but tend to be most common within the Host Unit. Sulphide mineralization occurs in both types of breccia but is richer within secondary breccia.

Sedimentary Breccia

Sedimentary, or matrix breccia (Hardy, 1979), consists of dolostone clasts within a dolostone matrix. The clasts, matrix and enclosing rocks (Fig. 13) have similar colour and composition and therefore this breccia is difficult to distinguish. The composition and texture of the clasts show the same variations as the country rock; clasts may therefore be massive, silty, oolitic or cherty dolostone or dark grey, argillaceous or carbonaceous dolostone. The argillaceous clasts represent the numerous, thin bedding-plane partings within the host dolostone and emphasize bedding. In the breccia bodies, these argillaceous beds commonly show soft-sediment deformation. The thickness of sedimentary breccia may exceed 30 m but generally is much less, ranging from a few centimetres to 1 or 2 m. The largest have been traced up to 3 km along strike.

Sedimentary breccia of varied origin was formed throughout shallow-water deposition of Host Unit and Silty Dolostone. The thinning of numerous breccia bodies away from reefs suggests formation by slumping of sediments from topographic highs formed by sediment build-up over reef tops. A paucity of reef clasts in all of the larger sedimentary breccia bodies indicates that they are not composed of reef talus in contrast to some smaller breccia bodies intimately associated with reefs and containing reef clasts.

During the development of secondary solution and collapse breccia as described below, insoluble material may have been concentrated to form "trash" breccia at lower levels within some breccia bodies. "Trash" breccia is very similar to sedimentary breccia described above but the high concentration of carbonaceous material gives the former a darker colour overall. At Gayna River, "trash" breccia is identified only where there is an observed association with secondary solution-collapse breccia, otherwise it is designated as sedimentary breccia.

Secondary Breccia

Secondary, or cement breccia (Hardy, 1979), is distinguished from sedimentary breccia by white sparry carbonate, often referred to as influx, that cements grey fragments of country rock (Figs. 20, 21). As in sedimentary breccia, the fragments have the same composition and textures as the country rock but the matrix gives secondary breccia a distinctive white or very light grey colour. The white cement consists of numerous stages of carbonate infilling that has sealed open spaces in the breccia bodies. The complete sequence of cementation commences with blue-grey dolomite and is followed by white dolomite and calcite. Although any of these three stages may be missing, the order is never changed. Barite is present locally but its position within the order of infilling has not been established, although it was probably introduced between the white dolomite and calcite. Other matrix components are sphalerite, pyrite, galena, quartz and fluorite. The sulphides appear to be paragenetically early whereas the quartz and fluorite are later.

Secondary breccia is very common in the Host Unit and formed by various processes including faulting, solution-collapse and recrystallization. The various processes have given rise to different secondary breccia types, including such forms as rubble, mosaic, crackle and pseudo-breccia (Hardy, 1979), distinguished mainly by differing associations of the cement and fragments. The size and geometry of breccia bodies are dependent upon the brecciation processs. In some cases two or more processes may have been active to give a complex pattern. Full understanding of the processes of secondary breccia formation is impossible until the breccia bodies may be examined in three dimensions or viewed in complete cross-sections.

Figure 20. Collapse breccia. Calcite (c) cementing dolostone fragments (d). Sphalerite surrounds fragments.

One of the more important types of secondary breccia at Gayna River is solution and collapse. Portions of the carbonate units were dissolved, with some insoluble material remaining to form "trash" breccia, and overlying beds collapsed. The collapse-breccia bodies are usually flat-lying though occasionally they resemble the vertically extensive, domal "breakthrough" type of collapse common in eastern Tennessee. Within these breccia bodies there is a gradation of breccia types that proceeds in general from crackle breccias in the outermost fringes into mosaic breccia with virtually no rotation of clasts towards the margins, and finally into rubble breccia with marked rotation and displacement of clasts in the central regions. Large areas near the top may be fragment-free and trash breccia may be present near the bottom. Figure 22 shows a complex breccia zone that probably represents some bedding collapse and some domal breakthrough collapse. This breccia zone also resembles paleokarst breccias described by Olson (1977) for the Nanisivik area.

Secondary breccia is not restricted to any particular stratigraphic unit and in places has transgressive contacts with bedded units. Fragments of sedimentary breccia are therefore found within secondary breccia. In addition to breccia bodies formed by faulting or solution and collapse, other types of uncertain derivation are included with secondary breccia because they too contain white sparry carbonate. Many of these are probably due to fracture or cavity filling. The fractures and porosity may have been formed by the processes that formed the breccia bodies, but if so, development did not continue to the generation of definite clasts. The result is a whitish coloured rock that resembles the secondary breccias but is made up of isolated pockets of white carbonate through restricted areas. These bodies are abundant and are commonly mineralized, but their small size suggests poor possibilities for mineable sulphide concentrations.

Figure 21. Collapse or fault breccia. Calcite is white, dolostone is dark.

MINERALIZATION

More than 100 surface sulphide occurrences are known and diamond drilling has intersected numerous others. In all, the predominant sulphide is sphalerite with subordinate galena and pyrite. Small amounts of malachite and azurite after copper sulphides were noted at one surface exposure. No other sulphides have been identified. The sulphides occur in three distinct habits; associated with sedimentary breccia, as infillings in secondary breccia and as replacement mineralization. Each habit has different mineral associations (Fig. 23) and the idealized spatial relationships among them are diagrammatically illustrated in Figure 24.

Mineralization in Sedimentary Breccia

Most sedimentary breccia in the Host Unit contains at least trace amounts of sphalerite and galena but rarely attains grades as high as 3% combined zinc-lead.

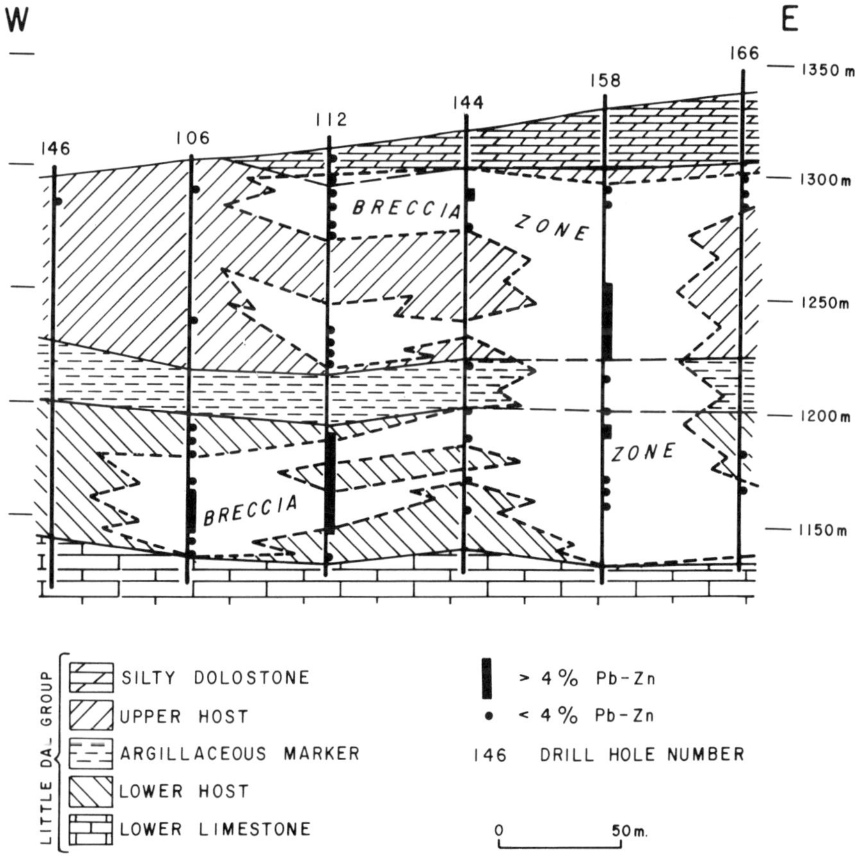

Figure 22. Example of complex secondary breccia showing distribution of sulphides.

Sulphides commonly occur as disseminations in the matrix and as discrete clasts (Fig. 25) or as envelopes to dolostone clasts (Fig. 26). In many places disseminated sphalerite gives the appearance of bedded sphalerite but this has been observed only in clasts within sedimentary breccia.

Grades of mineralization within sedimentary breccia are variable but the 'best' intersections are 20 m of 1.5% and 6 m of 4.6% combined zinc-lead. Due to its generally low grade, this type of mineralization is not as important as the generally higher grade mineralization found in secondary breccia bodies.

Mineralization in Secondary Breccia

At most Mississippi Valley-type deposits, mineralization is closely related to brecciation. At Gayna River, all types of secondary breccia are mineralized but the best mineralized zones are associated with the larger breccia bodies formed by faulting or solution and collapse. As a result, these have been more rigorously investigated than other types of breccia.

Sphalerite, with minor galena and pyrite in the secondary breccias, is intimately associated with the white sparry carbonate cement (Fig. 27). Pale red, green or yellow sphalerite was introduced after bluish dolomite. Commonly, this sphalerite envelops the blue dolomite which in turn envelops clasts of dolostone. Interstices remaining between clasts are filled with white dolomite and brilliant red, green, orange or yellow sphalerite with some galena and pyrite. Calcite and minor red

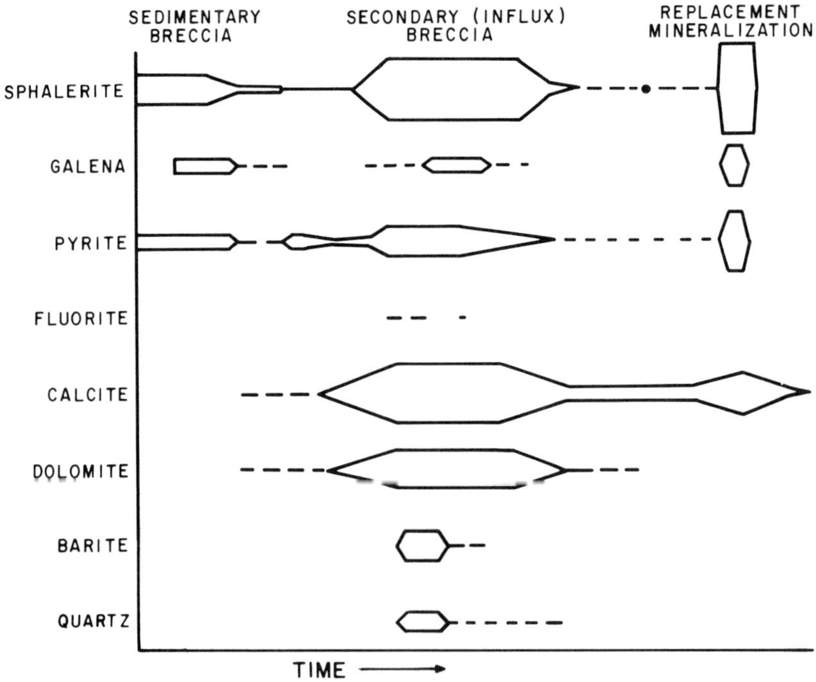

Figure 23. "Types" or classifications of mineralization.

sphalerite were the latest minerals to fill any remaining openings. The sequence of blue dolomite, followed in turn by pale-coloured sphalerite with white dolomite, and by calcite with brilliantly coloured sphalerite remains unchanged although any of the stages may be missing. In one or two breccia bodies barite appears at approximately the same position in the sequence as the white dolomite.

Sphalerite is the dominant sulphide in most secondary breccia bodies and is of almost any colour – shades of red, orange, yellow, green, brown, purple and white. Preliminary geochemical tests indicate that the colour variations are not due to simple elemental impurities such as Fe, Ag, Hg, Cd, Mo, Mn, Mg, Cr or Cu. The sphalerite is usually crystalline and forms euhedral to subhedral crystals up to 2 cm. Where abundant in the cement of secondary breccia, sphalerite exhibits "snow on

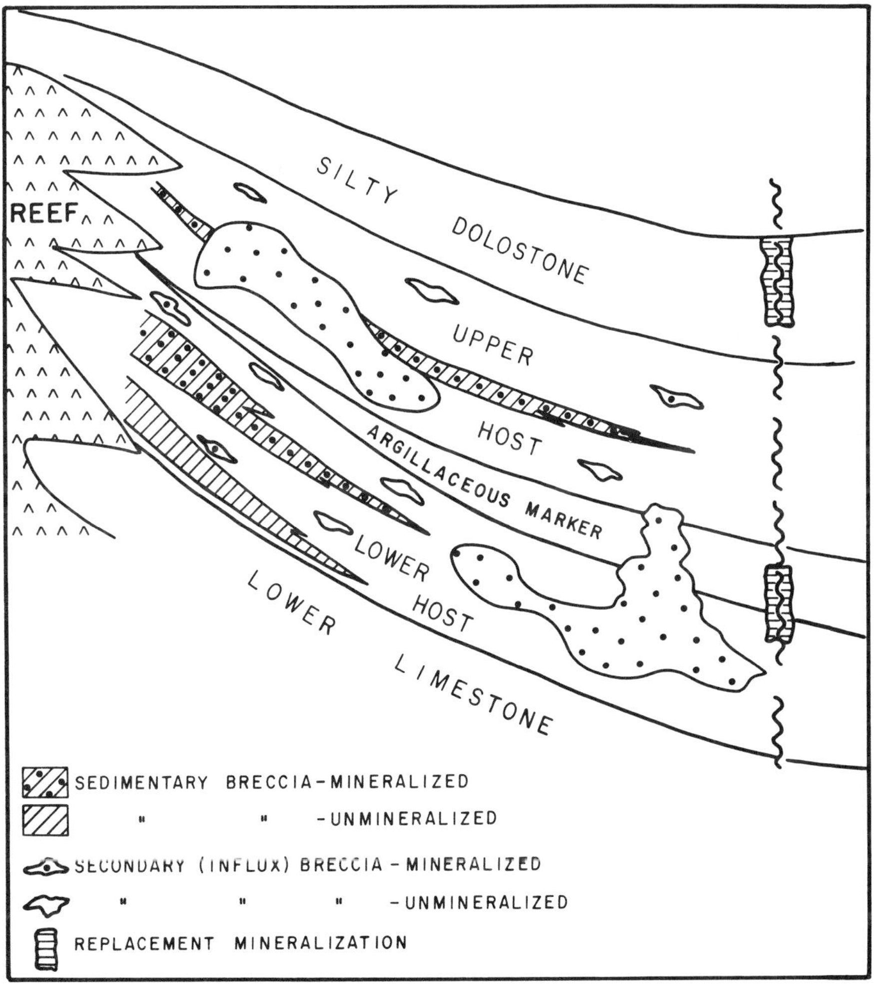

Figure 24. Diagrammatic relationship of "types" of mineralization to lithology.

the roof" textures or complete envelopment of dolostone fragments (Fig. 28). The relative ages of sphalerite and carbonate cement are shown by broken or peeled-off sphalerite envelopes that are separated by the calcite cement (Fig. 28). Numerous periods, or a prolonged episode of sphalerite deposition, are demonstrated by fragments of one colour of sphalerite included within sphalerite of a second colour, or by sphalerite veinlets of one colour cutting a second. Soft-sediment deformation of clasts cemented by sphalerite (Fig. 29) is evident.

Pyrite is also abundant in secondary breccia, locally more so than sphalerite. It exhibits textures similar to those of the sphalerite, including coatings around breccia clasts and "snow on the roof" textures. Galena is uncommon in secondary breccia, but is occasionally present as coarse blotches or as small subhedral crystals.

In some Mississippi Valley-type districts, colloform sphalerite, galena and/or pyrite are cited as examples of open-space filling (Campbell, 1966). The sulphides form concentric bands repeated outward from a central core. This feature is quite common in some of the high grade bodies at Pine Point (Campbell, 1966, p. 958) and at Arvik (Jowett, 1975, p. 125) on Little Cornwallis Island. Colloform sulphides at Gayna River are known only in talus. These colloform sulphide boulders (Fig. 30) are unusual because they are entirely sphalerite, without bands of pyrite or galena. The concentric banding is formed by radiating sphalerite crystals.

Other minerals commonly associated with sphalerite in secondary breccias are colourless to pale purple fluorite crystals, cubes of black bitumen, and small euhedral crystals of quartz. These are believed to represent a late stage of deposition because they are intergrown with white calcite.

Drillhole intersections of secondary breccia at Gayna River are of variable grades up to 20% combined zinc-lead over thicknesses up to 6 m. Although the maximum size of secondary breccia zones is undetermined, some of the larger bodies, still not delimited, are at least 500 m by 500 m in plan and average 30 m thick. Their shape is irregular (Fig. 22).

Replacement Mineralization

Grades in excess of 20% zinc over 10 m have been obtained along linear features, with grades decreasing laterally, suggesting fault or fracture control. These massive sulphides consist of finely crystalline sphalerite with minor pyrite that forms widespread irregular patches or "paint spatterings" in the dolostone away from fractures. The sulphides appear to have been deposited by replacement as they form within the carbonate, have irregular contacts and do not appear to occur interstitially or within vugs. Sparry white calcite is rarely associated with this sphalerite. Galena is present in some places but is abundant at only one replacement showing. In some cases, however, galena forms thin veinlets in units above large collapse zones suggesting some upward remobilization of lead from the collapse.

Replacement mineralization is not common and is not concentrated in the Host Unit, having been found as high in the stratigraphic column as the Ordovician Mount Kindle Formation hundreds of metres above the Little Dal Group. Its high grade is attractive but due to the small dimensions of these zones, which seldom exceed 30 m in any direction, they have not been explored in detail.

Figure 25. Fragments of sphalerite (s) and disseminations in sedimentary breccia. All white or light grey material is sphalerite.

Figure 26. Dolostone fragment (df) rimmed by galena (g) in matrix of dolostone (dm) in sedimentary breccia. Matrix also contains disseminations of sphalerite.

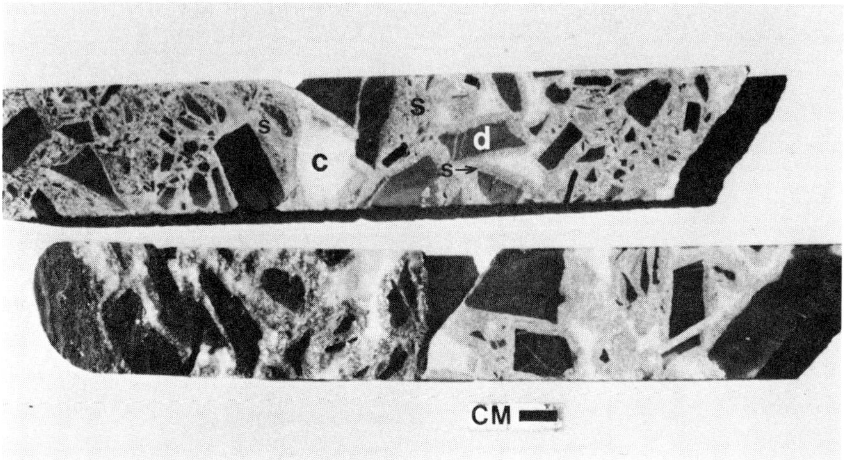

Figure 27. Secondary breccia with dolostone fragments (d) cemented by calcite (c). Sphalerite (s) rims fragments or is massive.

Figure 28. Secondary breccia with dolostone fragment (d) with sphalerite envelope (s) divided by calcite (c) indicating sphalerite is pre-calcite.

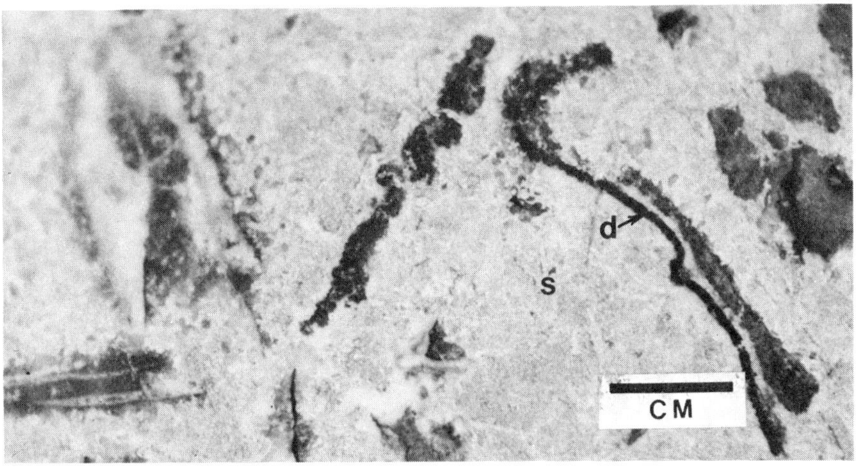

Figure 29. Massive sphalerite (s) cementing dolostone fragments (d). Fragments (d) show soft sediment deformation.

Figure 30. Colloform sphalerite.

TABLE I
COMPARISON OF MISSISSIPPI VALLEY-TYPE CHARACTERISTICS (FROM KYLE (1974), SNYDER (1968), SNYDER AND GERDEMAN (1968), DAVIDSON (1966), AND OHLE (1959)) WITH THE GAYNA RIVER AREA

	Mississippi Valley-Type	Gayna
a)	Majority of sulphide concentrations are stratabound.	Mainly restricted to Host Unit.
b)	Open-space filling dominant	Euhedral crystals and radiating growth in vugs
c)	Some fracture filling	Common in many showings
d)	Highest grades confined to 150 to 200 m thick unit	160 m thick
e)	Some sulphides distributed through hundreds of metres of stratigraphic sequence and kilometres laterally.	Approximately 3 km of stratigraphic sequence and 30 km laterally
f)	Host-rock dolostone, rarely limestone	Only two significant occurrences in limestone, all others in dolostone
g)	Sulphides near limestone-dolostone front	Undetermined
h)	Minor replacement	Three or four replacement showings known
i)	Igneous material and hydrothermal alteration lacking	Diabase dykes uncommon but present. No hydrothermal alteration
j)	Simple mineralogy with sphalerite dominant, galena less common and minor chalcopyrite. Pyrite is common. Sphalerite has low iron content. Dolomite and calcite common; barite, fluorite and quartz relatively uncommon	Sphalerite: pyrite: galena is 20:2:1. Chalcopyrite is rare. Low iron-content in sphalerite. Calcite: dolomite: barite is 10:5:1. Fluorite and quartz are uncommon
k)	Sharp cut-off from high grade to barren material	Apparently true
l)	Host is shallow-water carbonate with or without reef	Shallow-water carbonates with reef
m)	Usually transgressive sequence	Does not appear to be transgressive but is shallowing with time

	Mississippi Valley-Type	Gayna
n)	In platformal carbonate rocks near basin slope	Same presumed location
o)	Little structural deformation	More than other districts; folding and faulting with over 300 m of displacement
p)	Major unconformity above or below mineralized bed	Major unconformity located approximately 350 m above Host Unit
q)	Sulphides are frequently below evaporites	Below a 300 m thick section of evaporites
r)	Low precious metal content	No gold, very low silver
s)	Multiple periods of sulphide mineralization	At least three
t)	Silica usually a minor constituent	Occasional small euhedral quartz crystals with sulphides

Comparison with Mississippi Valley-Type Deposits

Mississippi Valley-type lead-zinc deposits have widely varying characteristics but there is general agreement that they include a family of deposits with combinations of similar characteristics. Table I is a list of common characteristics of Mississippi Valley-type deposits. Virtually all are observed at Gayna River. In fact, the mineralization at Gayna River is comparable to that of the Tennessee districts in the form and irregular pattern of the breccia bodies, to that of the Pine Point district in the presence of numerous, apparently disconnected areas of mineralization, and to that of the Pine Point and southeast Missouri districts in the host environment of peritidal carbonate rocks and associated reefs.

DISCUSSION

Sulphides within sedimentary breccia bodies may have originated by one of three processes; synsedimentary deposition, synsedimentary deposition with some diagenetic replacement, or diagenetic or epigenetic replacement. If the sulphides are synsedimentary, then sphalerite clasts (Fig. 25) represent remnants of sphalerite beds that have been brecciated, and disseminated sphalerite would be derived by ensuing detrital deposition. The very irregular form of the disseminated sulphide grains, and sulphide enveloping of dolostone clasts (Fig. 26), are not fully compatible with synsedimentary deposition alone, however, and suggest that diagenetic replacement may have contributed to the sulphide mineralization after formation of the sphalerite clasts. In fact, the lack of any evidence of beds of sulphide anywhere but in some sedimentary breccia bodies, suggests that all the sulphide mineralization may be diagenetic or epigenetic with selective replacement of some fragments.

Secondary breccia bodies formed after consolidation of the sediments. In some cases, soft-sediment deformation of dolostone clasts indicates that brecciation occurred during or soon after consolidation, while in other cases shattering of fragments indicates that brecciation was much later. Brecciation processes operated over a lengthy, undetermined time period.

Mineralizing processes that gave rise to sulphide deposition operated over a similar lengthy period. Sphalerite surrounds clasts that have been deformed, forms envelopes to clasts and, with pyrite, fills spaces caused by shattering.

Jackson and Beales (1967) presented a model that describes migration and mixing in porous areas near reefs of metal-rich connate waters from shales of the basin with sulphate-rich connate waters from evaporites of the back-reef. All of the environments of this model are present at Gayna River but their inter-relationships are different from those of the model. The shales show facies relationships with the reefs but not with the platformal carbonates, and the evaporites are in a unit overlying the latter. In other words, at Gayna River, unlike Pine Point, these lithofacies are not all lateral time equivalents of one another (Fig. 19). This disposition of lithofacies does not permit the migration of fluids as envisioned by Jackson and Beales because the shales lie below the mineralized beds and the evaporites above.

Evaporite deposits are significant at Gayna River and are probably the source of the sulphate and may be the source of the metal ions required to generate sulphides. Brines carrying sulphate and metal-chloride complexes may have permeated to a near-reef environment where bacterial reduction of the sulphate produced hydrogen sulphide which scavenged the metal ions to form metal sulphides; or, brines rich in sulphate may have permeated to the water table level, at locations varying with time within the Host Unit, and mixed with metal-rich connate waters in an environment also rich in bacteria. These two models depict different sources for the metal ions but have the same source for the sulphate and the same bacteria-rich site of sulphide deposition. Downward migrating brines have been cited in other models (Davidson, 1966).

CONCLUSIONS

Proterozoic Little Dal Group rocks extend northwestward for hundreds of kilometres through the northern Cordillera but contain abundant sulphides only at Gayna River. Although the reasons for their concentration there are not fully understood, the following combination of factors is believed to be important:

1) abundant, well developed brecciation of the carbonate host rocks,

2) a sub-upper Cambrian unconformity within the Silty Dolostone, just above the evaporites and up to 350 m above the Host Unit,

3) a thick sequence of evaporites,

4) spectacular, well developed reefs,

5) abundant influx, healing porosity and giving sparry calcite and dolomite.

Each of these features exists elsewhere in the Little Dal Group but all are present in the Gayna area.

Irregularly shaped breccia bodies of varying overall dimensions have been formed by a variety of processes including slumping, solution-collapse and tectonism. These bodies, and porosity formed by solution and dolomitization, have been mineralized with sphalerite and pyrite, with minor amounts of galena. The sulphides are open-space fillings deposited at different times over an extended period of time. Local control for mineralization is poorly understood.

Exploration has indicated large deposits grading over 5% combined zinc-lead but because of the remote location, higher grade material is necessary for profitable mining, and exploration is continuing with this objective.

ACKNOWLEDGEMENTS

The writer would like to thank Rio Tinto Canadian Exploration Limited for providing the resources and time to prepare this paper. K. M. Dawson and C. N. Forster are acknowledged for their comments on the manuscript. The writer is also deeply indebted to C. D. Spence who provided encouragement, advice and technical experience, and also read the manuscript critically. Many of the observations and ideas expressed in this paper are based on work carried out by J. L. Hardy and discussed in an M.Sc. thesis at the University of Toronto. Special thanks are due to the late Dr. N. Campbell whose vast experience established the basis for exploration at Gayna River from the onset.

REFERENCES

Aitken, J.D., 1977, New data on correlation of the Little Dal Formation and a revision of Proterozoic map-unit 'H5': Geological Survey of Canada Paper 77-1A, Report of Activities, Part A, p. 131-135.

Aitken, J.D. and Cook, D.G., 1974a, Preliminary unedited geological maps showing bedrock geology of the northern parts of Mount Eduni (106A) and Bonnet Plume Lake (106B) map-areas, District of Mackenzie, N.W.T.: Geological Survey of Canada, Open File 221.

_______________, 1974b, Geology, Upper Ramparts River (106G) and Sand Sault Rapids (106H) map-areas, N.W.T.: Geological Survey of Canada, Open File 272.

Aitken, J.D., Long, D.G.F., and Semikhatov, M.A., 1978, Progress in Helikian Stratigraphy, Mackenzie Mountains: Geological Survey of Canada, Paper 78-1A, Report of Activities, Part A, p. 481-486.

Aitken, J.D., MacQueen, R.W., and Usher, J.L., 1973, Reconnaissance Studies of Proterozoic and Cambrian Stratigraphy, Lower Mackenzie River (Operation Norman), District of Mackenzie: Geological Survey of Canada, Paper 73-9, 178 p.

Archer, Cathro and Associates, 1978, Revised Northern Cordillera Mineral Inventory, Yukon and Northwest Territories: Vancouver.

Campbell, N., 1966, The lead-zinc deposits of Pine Point: The Canadian Mining and Metallurgical Bulletin, v. 59, no. 652, p. 953-960.

Davidson, C.F., 1966, Some genetic relationships between ore deposits and evaporites: Institute of Mining and Metallurgy Transactions, v. 75, p. B216-B225.

Delaney, G.D., Jefferson, C.W., Yeo, G.M., McLennan, S.M., Aitken, J.D., and Bell, R.T., in press, Some Proterozoic sediment-hosted metal occurrences of the northeastern Canadian Cordillera: Society of Economic Geologists, Coeur d'Alene Field Conference, Wallace, Idaho, November 3-5, 1977, Idaho Bureau of Mines.

Eisbacher, G.H., 1978, Two major Proterozoic unconformities, Northern Cordillera: Geological Survey of Canada Paper 78-1A, Report of Activities, Part A, p. 53-57.

Gabrielse, H., 1967, Tectonic evolution of the northern Canadian Cordillera: Canadian Journal of Earth Sciences, v. 4, p. 271-298.

Gabrielse, H., Blusson, S.L., and Roddick, J.A., 1973, Geology of Flat River, Glacier Lake, and Wrigley Lake map areas, District of Mackenzie and Yukon Territory: Geological Survey of Canada Memoir 366, Part I, 153 p., Part II 268 p.

Goodfellow, W.D., 1979, Geochemistry of copper, lead, and zinc mineralization in Proterozoic rocks, near Gillespie Lake, Yukon: Geological Survey of Canada Paper 79-1A, Current Research, Part A, p. 333-348.

Hardy, J.L., 1979, Stratigraphy, Brecciation, and Mineralization, Gayna River, N.W.T.: Unpublished M.Sc. Thesis, University of Toronto.

Hofmann, H.J., and Aitken, J.D., 1979, Precambrian biota from the Little Dal Group, Mackenzie Mountains, northwestern Canada: Canadian Journal of Earth Sciences, v. 16, p. 150-166.

Jackson, S.A., and Beales, F.W., 1967, An aspect of sedimentary basin evolution: The concentration of Mississippi Valley-type ores during late stages of diagenesis: Bulletin of Canadian Petroleum Geology, v. 15, No. 4, p. 338-433.

James, N.P., 1978, Facies Models-10. Reefs: Geoscience Canada, v. 5, p. 16-26.

Jowett, E.C., 1975, Nature of the ore-forming fluids of the Polaris lead-zinc deposit, Little Cornwallis Island, N.W.T., from fluid inclusion studies: The Canadian Mining and Metallurgical Bulletin, v. 68, no. 755, p. 124-129.

Kyle, J.R., 1974, Mississippi Valley-type Mineral Deposits in North America: Characteristics and Genisis: University of Western Ontario, Geology 581 Seminar.

Ohle, E.L., 1959, Some considerations in determining the origin of ore deposits of the Mississippi Valley-type: Economic Geology, v. 54, p. 769-790.

Olson, R.A., 1977, Genesis of stratabound zinc-lead deposits, Baffin Island, N.W.T.: Geological Association of Canada, Program with Abstracts, Annual Meeting in Vancouver.

Snyder, F.G., 1968, Geology and mineral deposits, mid-continent United States: in Ridge, John D., ed., Ore Deposits of the United States, 1933-1967, v. 1: American Institute of Mining Metallurgical and Petroleum Engineers, p. 258-286.

Snyder, F.G., and Gerdemann, P.E., 1968, Geology of the southeast Missouri lead district: in Ridge, John D., ed., Ore Deposits of the United States, 1933-1967, v. 1: American Institute of Mining, Metallurgical and Petroleum Engineers, p. 326-358.

Young, G.M., Jefferson, C.W., Long, D.G.F., Delaney, G.D., and Yeo, G.M., in press, Upper Proterozoic stratigraphy of northwestern Canada and Precambrian history of North American Cordillera: Society of Economic Geologists, Coeur d'Alene Field Conference, Wallace, Idaho, November 3-5, 1977, Idaho Bureau of Mines.

MS received June 1979;
Accepted for publication July 1980.

Precambrian Sulphide Deposits, H.S. Robinson Memorial Volume, edited by R.W. Hutchinson, C.D. Spence and J.M. Franklin, Geological Association of Canada Special Paper 25, 1982

DEPOSITIONAL ENVIRONMENTS AND GENESIS OF STRATIFORM COPPER DEPOSITS OF THE REDSTONE COPPER BELT, MACKENZIE MOUNTAINS, N.W.T.

J.C.L. Ruelle
Shell Canada Resources Ltd., Post Office Box 100,
Calgary, Alberta T2P 2H5

ABSTRACT

Stratiform copper deposits occur along a 250 km, northwest-trending, arcuate belt of Upper Proterozoic (Helikian) carbonate rocks and redbeds of the Mackenzie Mountains Supergroup. Deposits occupy a transgressive contact (termed the "Transition Zone") between continental redbeds of the Redstone River Formation and shallow marine carbonate rocks of the overlying Lower Coppercap Formation. Host strata consist primarily of silty, evaporitic, cryptalgal-laminated carbonate units dominantly of sebkha origin, deposited over the fault-bounded Coates Lake, Hayhook Lake, and Keele River embayments.

The most laterally extensive and economically promising deposit, situated within the Coates Lake embayment, covers an area of at least 12 km^2, occupies a bed one metre thick, and contains drill-indicated "ore reserves" of 37 million tonnes averaging 3.92% Cu and 0.33 oz. per ton Ag (11.3 g/t). Deposits in the Hayhook Lake and Keele River embayments are linear or pod-like and attain drill-indicated thicknesses up to 52m, with comparable or higher grades.

Deposits can be related to diagenetic processes accompanying normal basin development. Erosion of copper-bearing mafic lava is postulated to have provided the source sediment for the metals. Metal ions were subsequently liberated from the volcanic detritus during redbed development and accumulated in the highly saline groundwater reservoir of the clastic basin, possibly as metal-chloride complexes. Compaction accompanying a major marine transgression during deposition of the Middle Coppercap Formation altered the hydrologic regime of the clastic basin, and forced the metal-bearing groundwater upwards and landwards, through the organic-rich cryptalgal-laminated sediments. These highly reducing layers acted as "chemical filters", effectively removing metals from solution. Fluid flow was controlled by facies changes in the landward parts of the Keele River and Hayhook Lake embayments. Deposits show a pronounced mineral zonation, with the copper:iron ratio (in sulphides) decreasing upsection and towards embayment margins.

RÉSUMÉ

Des dépôts stratiformes de cuivre se rencontrent le long d'une zone arquée, de direction NW, d'environ 250 km. Les dépôts se trouvent dans des carbonates et "red beds" du supergroupe "Mackenzie Mountains" d'âge Protérozoïque Supérieur (Hélikien). Ces dépôts ont été formés le long du contact transgressif (appelé Zone de Transition) entre les "red beds" d'origine continentale de la formation de Redstone River et les carbonates d'eaux peu profondes de la formation de Coppercap qui les recouvrent. Les strates hôtes sont des lits de carbonates à caractère silteux et évaporitique et laminés par des couches d'origine crypto-algaires. Ces carbonates furent déposés dans des sebkhas le long d'embaiements délinéés par failles. Les régions discutées dans le texte sont celles du Lac Coates, Lac Hayhook, et embaiements de la Rivière Keele.

Le dépôt le plus étendu et qui offre le plus de promesses économiques est situé dans la région du Lac Coates. Par forage, il est indiqué que ce dépôt couvre une région d'au moins 12 km^2, dans un lit d'un mètre d'épaisseur et contient 37 millions de tonnes de minerai à teneur moyenne de 3,92% en cuivre. Les dépôts des régions du Lac Hayhook et de la Rivière Keele sont linéaires ou ségrégés et atteignent une épaisseur maximum de 52 m avec des concentrations comparables et même supérieures.

Les dépôts sont reliés au developpement normal du bassin. Il est postulé que l'érosion de couches volcaniques riches en cuivre a produit la source des métaux. Faisant suite à l'érosion, les ions métalliques furent libérés des débris volcaniques lors du dépôt des "red beds" et accumulés sous forme de chlorures métalliques dans les aquifères clastiques du bassin. La compaction de ces dépôts suite à la transgression marine résultant dans le dépôt des sédiments de la formation de Coppercap a altéré le régime hydrologique dans les formations clastiques du bassin en forçant les fluides minéralisants vers le haut et vers le craton à travers les sédiments riches en matière organique de la sebkha. Les sédiments réducteurs, riches en H_2S, ont joué le rôle de filtres chimiques en précipitant les métaux en solution. La migration des fluides a été contrôlée par les changements de faciès vers le craton dans la région des embaiements de Keele et Hayhook. De ce fait les dépôts montrent une zonalité minéralogique prononcée, en effet le rapport cuivre: fer décroît vers le haut et en direction des bordures des embaiements.

INTRODUCTION

Copper deposits within sedimentary rocks of Upper Precambrian and Permian age are well documented throughout the world (Garlick, 1961; Bartholome, 1972; Kirkham, 1973; Jung and Knitzschke, 1975; Johnson and Croy, 1976; Brown, 1978). Deposits such as those of the Zairian and Zambian Copper Belts, central Europe (Kupferschiefer), and the midcontinent region of the United States are good examples. They are characterized by an association with shallow to marginal marine carbonate and clastic sedimentary rocks, and are often proximal to redbeds. They lack any direct igneous affinity, and are commonly termed "sedimentary", "black shale" or "redbed" copper deposits.

Strata favourable for the presence of such copper mineralization are exposed discontinuously along an arcuate belt of Upper Proterozoic carbonate and redbed clastic rocks within the Mackenzie Fold Belt of the northern Cordillera, N.W.T., Canada. The belt of favourable strata, informally termed the Redstone Copper Belt, has been outlined over a distance of some 250 km. Sedimentary rocks in which the deposits occur were deposited in fault-controlled embayments of a northerly-trending Proterozoic sea (Coates, 1964; Aitken *et al.*, 1974; Ruelle and Bond, 1976; Eisbacher, 1977; Jefferson, 1979). Six graben embayments have been recognized within the copper belt, but only three have been explored for mineralization by drilling. The three embayments discussed are Coates Lake, Hayhook Lake, and Keele River.

The purpose of this paper is to describe the deposits of the copper belt in terms of their stratigraphic and structural setting, and to suggest their depositional environment. A metallogenic model is proposed that relates the deposits to diagenetic processes accompanying normal (albeit specialized) basin development.

Location and Access

The Redstone Copper Belt is situated within the Mackenzie Fold Belt, the easternmost structural segment of the Canadian Cordillera, in the Mackenzie District of the Northwest Territories. The Mackenzie River parallels the trend of the copper belt, and lies some 160 km to the east (Fig. 1). Access is entirely by air. Throughout the copper belt terrain is generally rugged, with elevations varying from 760 m to approximately 2130 m, and the tree line lies at about 1200 m.

Previous Work

Stratiform copper mineralization was discovered by Nahanni 60 Syndicate prospectors in 1962, along outcrop exposures of Redstone River strata within what is now termed the Coates Lake embayment. Showings were noted over a strike length of 6.4 km (Coates, 1964; Watson, 1971).

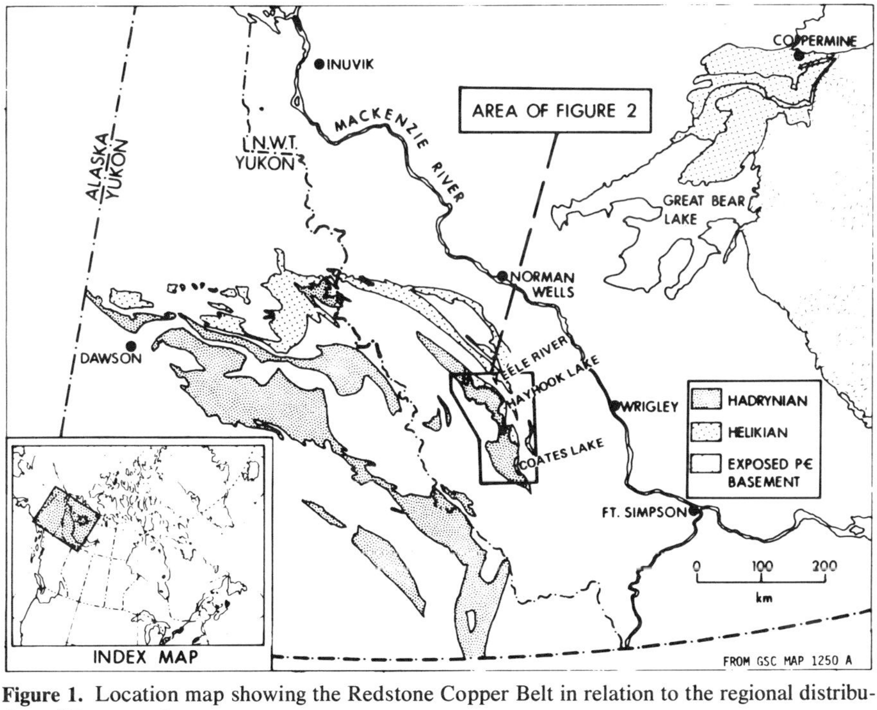

Figure 1. Location map showing the Redstone Copper Belt in relation to the regional distribution of Proterozoic strata.

TABLE I

Stratigraphic column of key formations, Redstone Copper Belt.

	LITHOSTRATIGRAPHIC UNIT		THICKNESS (METRES)	LITHOLOGY
MACKENZIE MTS. SUPERGROUP	UPPER	COPPERCAP FORMATION	0 - 165	WELL BEDDED CHERTY LS. ALLODAPIC LS., STROMATOLITIC LS.
	MIDDLE		0 - 72	BLACK, FETID, ARGILLACEOUS LS. CALCAREOUS SHALE.
	LOWER		0 - 63	GYPSIFEROUS LS. & DOLOMITE, ALGAL - LAMINATED DOLOMITE.
	TRANSITION ZONE		0 - 110	ALGAL - LAMINATED SEBKHA CARBONATE LITHOSOMES, SHEETWASH CLASTICS.
	UPPER	REDSTONE RIVER FORMATION	0 - +20	EVAPORITES, CYCLIC EVAPORITES & SILTSTONE, COLLAPSE BRECCIA.
			0 - +1220	SHEETWASH FINE CLASTIC REDBEDS. FANGLOMERATE, PLAYA LAKE EVAPORITES
	LOWER		0 - +215	Na-Mg RICH EVAPORITES, BRECCIA, BASIC INTRUSIONS (?).
		LITTLE DAL FORMATION (UPPER CARBONATE SUBUNIT)	80	EVAPORITIC, STROMATOLITIC DOL., SANDSTONE & SHALE, LOCAL REDBEDS.
			0 - +20 (?)	MAFIC VOLCANIC FLOWS.
			800	STROMATOLITIC DOLOMITE.

Exploration programs involving diamond drilling, geological mapping, prospecting, geochemistry, and geophysical testing have been carried out intermittently on the copper belt since the initial discovery. The main companies that have conducted exploration include Redstone Resources Inc. (1963, 1964), Cerro Mining Co. of Canada (1971), Rio Tinto Canadian Exploration (1975, 1976, 1977), International Nickel Co. (1975, 1976), and Shell Canada Resources Ltd. (1975, 1976, 1977, 1979). Regional mapping programs were also undertaken by the Geological Survey of Canada and the Department of Indian Affairs and Northern Development.

The prime objectives of the exploration programs were to test for down-dip continuation and possible thickening in the down-dip and basinward direction of exposed mineralization. To date, exploration efforts have barely begun to test the economic potential of the copper belt.

REGIONAL GEOLOGY

Stratiform mineralization occurs in intercalated supra-tidal marine carbonate lithosomes of the Lower Coppercap Formation along a transgressive contact with underlying continental redbeds of the Redstone River Formation (Ruelle and Bond, 1976; Bond *et al.*, 1976; Jefferson, 1979). The above formations (Table I) comprise the "copper cycle" (Aitken, 1980) and form part of an essentially unmetamorphosed Proterozoic succession of the Mackenzie Mountains Supergroup (Young *et al.*, 1977) that has been thrust eastward over Paleozoic carbonate rocks and shales (Fig. 2). Depth to crystalline basement from the base of the Coppercap Formation is at least 7 to 9 km (Young *et al.*, 1977; Aitken *et al.*, 1978a).

Favourable strata crop out in three principal areas, the Coates Lake, Hayhook Lake, and Keele River embayments, over a length of 160 km (Fig. 2). The embayments are interpreted as the centres of deposition of continental redbeds of the Redstone River Formation, and shallow marine carbonate rocks of the Coppercap Formation. A proposed interpretation of the configuration of the Redstone basin during late Redstone River Formation time is given in Figure 3. There are no deep subsurface data available in the region to the west of the outcrop limits of Coppercap and Redstone River strata and therefore facies boundaries are illustrative only. Similarly, many of the normal faults controlling embayment development and shown in the region to the east of the fanglomerate facies are illustrative.

The term "marginal marine continental" is used to describe the depositional environment of the Redstone River Formation, an environment that was dominantly continental but still proximal to a marine shoreline and was periodically subjected to marine flooding and marine seepage-reflux. In essence, the greater part of the Redstone River Formation is considered to have been deposited under coastal sebkha conditions as described by Glennie (1970).

Strata of the copper belt are tentatively considered to be of Upper Helikian age. The succession has been correlated with strata of the Upper Belt – Purcell Supergroup in the Cordillera of southern Canada (Gabrielse *etal.*, 1973) and with strata of the Amundsen Basin to the northeast, including the Shaler Group of the Brock Inlier, Banks Island and Victoria Island, and the Rae Group of the Coppermine area (Young 1977; Aitken *et al.*, 1978b).

Away from the thrust front, strata dip gently to the southwest. Local structural complications are present but, generally, strata of the copper belt are not severely deformed. Thrust deformation is related to the Laramide Orogeny (Aitken *et al.*, 1973).

Although stratiform and stratabound copper mineralization occurs at stratigraphic levels both above and below the transgressive contact between the Redstone River and Lower Coppercap Formation, no occurrences of economic significance have been encountered.

STRATIGRAPHY

General Statement

The principal stratigraphic units of the copper belt are given in Table I. Thicknesses for the various lithostratigraphic units are minimum values, representative of the most landward parts of the embayments. In terms of a model for ore genesis, the Redstone and Coppercap Formations together contain 6 main lithofacies. These are described below, and from these facies, an interpretation of the depositional environments that may have existed during formation of the copper belt is given in a later section.

Table I includes essentially the stratigraphic sequence referred to by Aitken (1980) as the "copper cycle", a cycle initiated and terminated by structural events. Stratigraphic sequences occurring above and below the copper cycle are not discussed because they do not relate to the mineralizing process.

A regional unconformity of varying magnitude from minor to major separates the copper cycle from an overlying sequence of glacio-marine clastics of the Rapitan Formation (Figs. 5 and Plate II, Fig. 12), and may mark the break between the

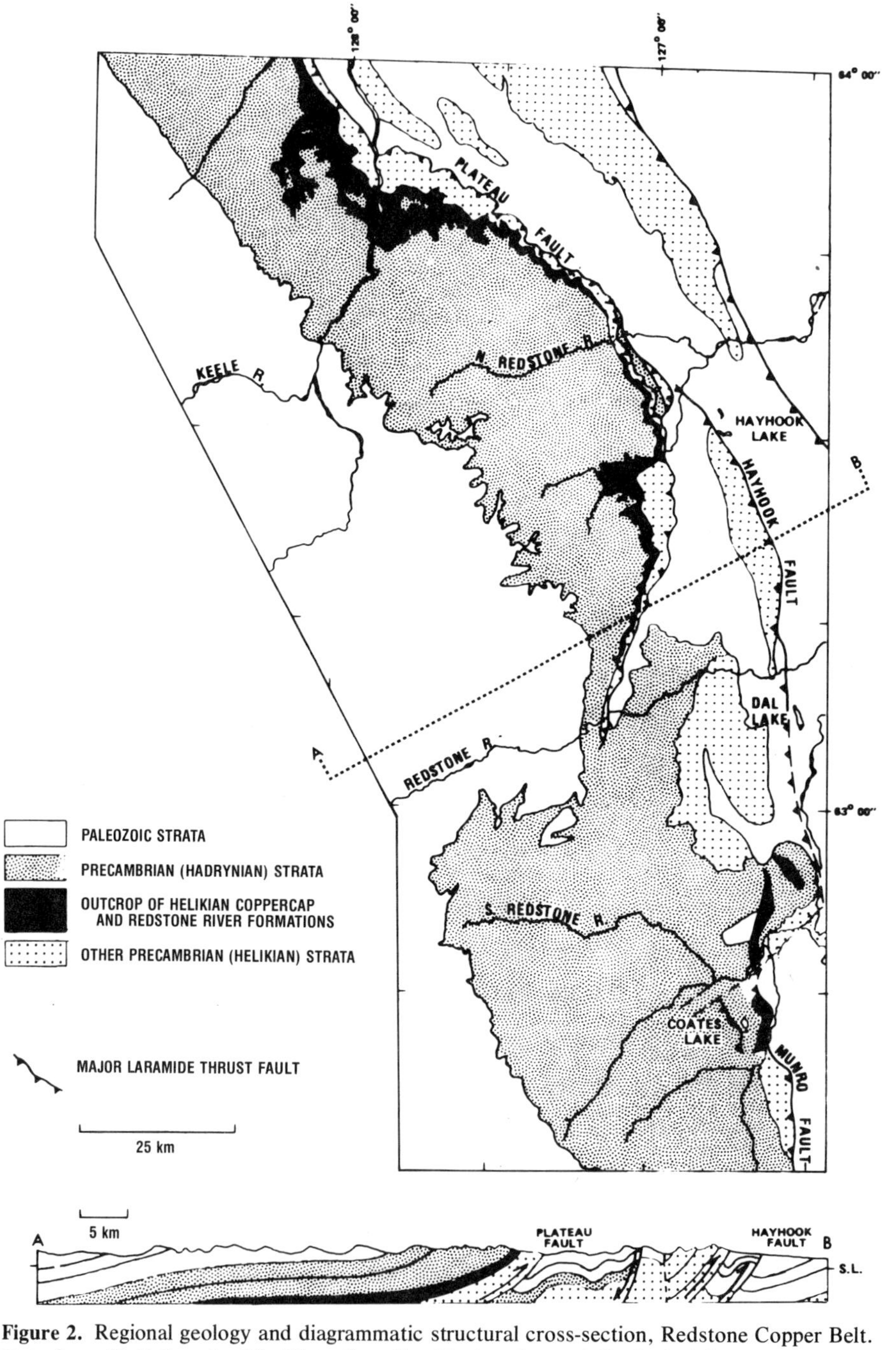

Figure 2. Regional geology and diagrammatic structural cross-section, Redstone Copper Belt. Data from Shell Canada, Rio Tinto Canadian Exploration and Geological Survey of Canada maps.

Helikian and Hadrynian (Gabrielse, 1967; Eisbacher, 1978). It is primarily the overlying Hadrynian sequence that forms a thick capping over much of the copper belt (Fig. 2).

Little Dal Formation

Upper Carbonate Subunit. The oldest stratigraphic unit of importance to understanding the copper belt is the Upper Carbonate Subunit of the Little Dal Formation (Aitken *et al.*, 1978a). It is some 900 m thick and consists dominantly of a laterally very extensive succession of stromatolitic, algal-laminated and algal-bound dolostones with associated calcarenite, micrite, nodular dolostone, and minor mudstone

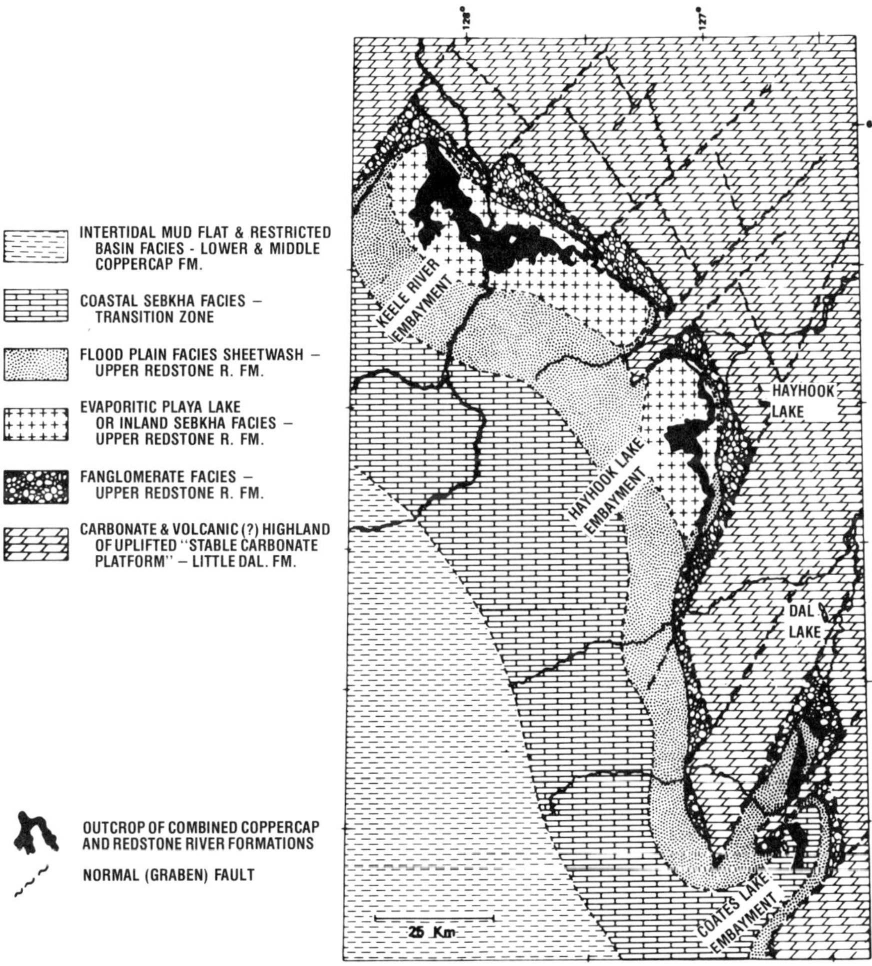

Figure 3. Diagrammatic reconstruction of the Redstone Copper Belt during late Redstone River Formation time, illustrating proposed facies distributions and embayment configurations.

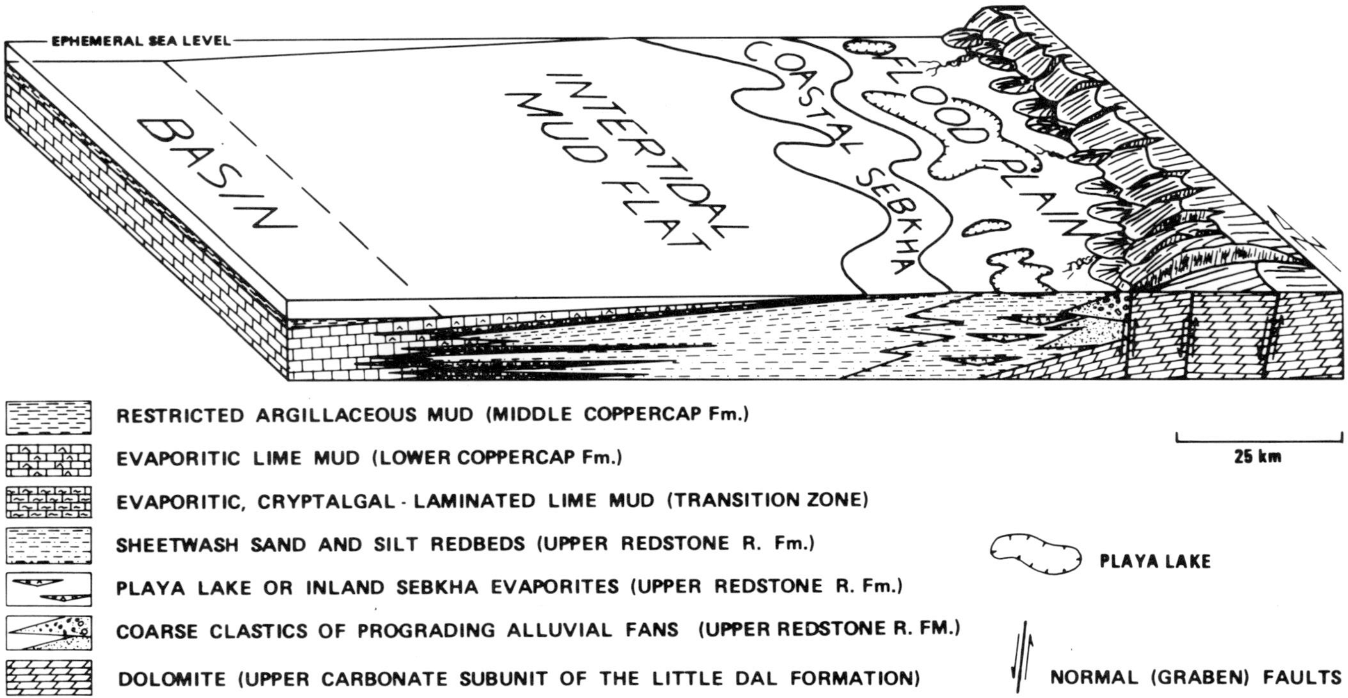

Figure 4. Composite diagrammatic block diagram of the Redstone Copper Belt illustrating depositional environments during early transgression of the Coppercap Sea. The intertidal zone has been exaggerated.

and sandstone (Bond *et al.*, 1976). Because the subunit shows remarkable uniformity in sections measured throughout the copper belt, it is referred to here as "the stable carbonate platform".

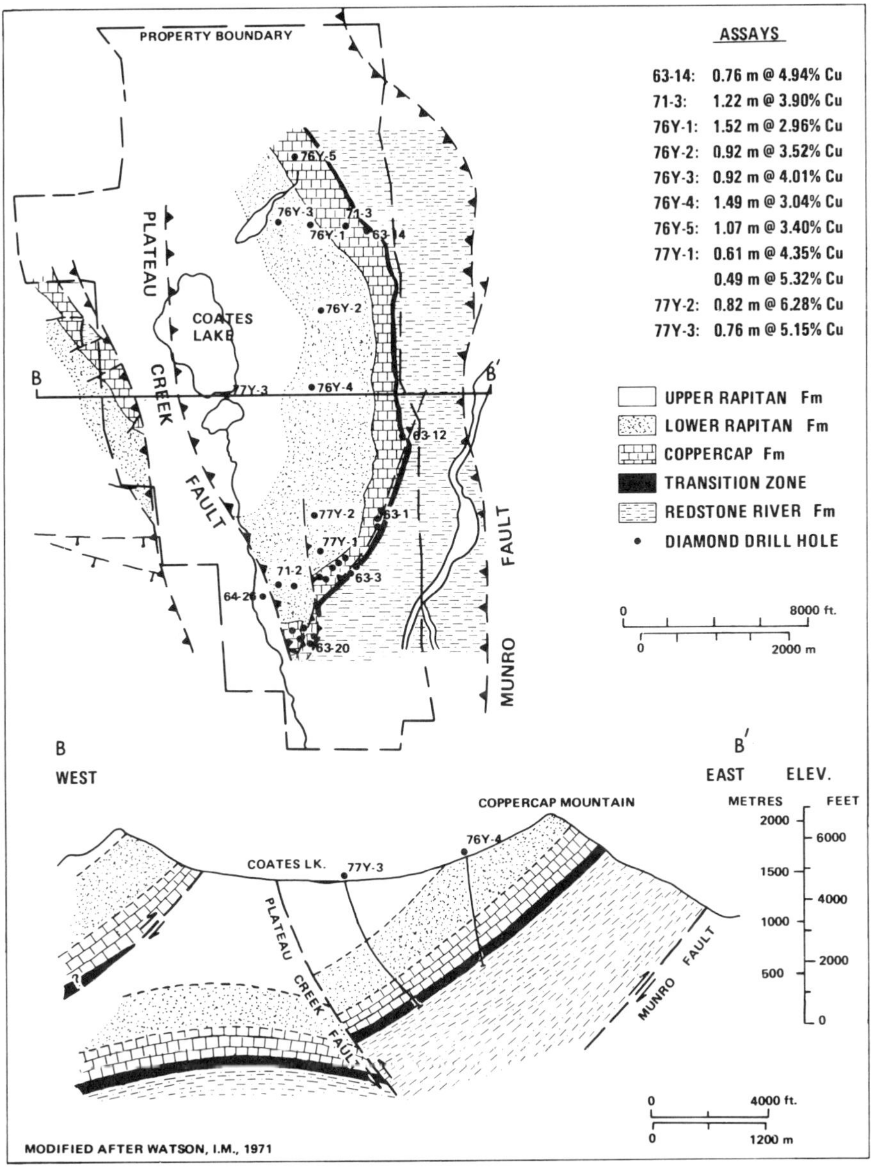

Figure 5. Geological plan and cross-section of the Coats Lake Property, Coats Lake Embayment.

The upper 100 m of the subunit show a marked change from dominantly peritidal carbonate sedimentation through clastic deposition to restricted evaporitic carbonate depositon. A mafic lava (or clastic equivalent) marks the beginning of this change. Clastic units consist of clean, planar laminated to massive bedded quartz sandstone, local green and dark grey mudstone, and red siltstone. Desiccation cracks, tepee structures and standing water ripple marks, are common in the red siltstone.

At least two major volcanic units, separated locally by clastic and carbonate strata, are exposed. The volcanic intervals are each composed of a number of individual flows and occasionally show remarkable thickening and thinning related to the infilling of paleotopographic lows caused by block faulting. The volcanic units are extensively eroded, and south of the Coates Lake embayment thick channel conglomerate composed entirely of volcanic clasts is present. Cross-bedded sandstone, consisting dominantly of lava and quartz detritus is widespread. Occasionally at the top of the lava there is a decimetre-scale layer consisting almost entirely of cross-bedded, coarse sand-sized specular hematite grains, and scattered detrital chalcopyrite grains, jasper and calcedonic quartz clasts, with highly corroded lava pebbles.

The volcanic rocks are of basaltic composition, are generally extremely vesicular and are pillowed locally. Anastomosing, millimetre-scale veinlets of red hematite, occasionally enveloped by quartz, cut the volcanic rocks. The lavas are highly anomalous in iron (average 18.8% Fe_2O_3) and are locally depleted in silica (Bond *et al.*, 1976). Chalcopyrite commonly occurs in amygdules and as disseminated grains, and less commonly along fractures. Finely disseminated specular hematite averages between 10 and 15% of the volcanic rock volume, and is of importance in understanding the provenance of the Redstone River Formation.

Carbonate rock overlying the lava flows consists dominantly of interbedded massive and finely algal-laminated silty dolostone with varying amounts of gypsum, anhydrite and magnesite. Extremely finely disseminated pyrite is widespread within the algal-laminated sections. Both the thickness of the carbonate and its evaporite content are seen from drilling to increase in a basinward direction. Dolostone gives way gradually to magnesium carbonate, anhydrite and sodium sulphate so that it is difficult to place an upper limit to the Carbonate Subunit.

Redstone River Formation

Conformably overlying the Upper Carbonate Subunit of the Little Dal Formation is a succession of evaporites and marginal marine continental redbeds of the Redstone River Formation. The formation varies in thickness from zero at embayment margins to at least 1220 m in their more central parts. The distribution and thickness of the formation in deeper parts of the copper belt are unknown. Abrupt changes in thickness and facies characterize the Redstone River Formation towards the margins of the embayments, and may indicate an active role of faulting during development of embayments, as proposed by Eisbacher (1977). Two major subdivisions of the Redstone River Formation are recognized: a lower, evaporite subunit comprising the Lower Redstone River Formation; and an upper, redbed subunit comprising the Upper Redstone River Formation.

Lower Redstone River Formation. The lower, evaporite subunit, recognized primarily in the Keele River embayment and inferred in the Coates Lake embayment, is thickest in basinal sections, where it has a drill-indicated thickness of 215 m. In marginal sections, the subunit is either absent or forms a recessive interval less than 10 m thick. Its contacts are gradational.

Where analyzed, the evaporite mineralogy is characterized by an unusually high Na – Mg assemblage consisting of approximately 55% glauberite ($Na_2Ca(SO_4)_2$), 20% magnesite, 15% talc and less than 10% combined gypsum plus anhydrite. This assemblage may relate to the influence of basaltic volcanism. The anhydrite content increases up section. Nearly pure magnesite and glauberite interbeds occur in the lower part.

Upper Redstone River Formation. The main rocks forming the Redstone River Formation are contained within the Upper Redbed Subunit. This consists of three main facies (Figs. 3 and 4) which are a fanglomerate facies, an evaporitic playa lake or inland sebkha facies (Glennie, 1970), and a broad flood plain facies. The subunit varies in thickness from zero at embayment margins to at least 1220 m in the Coates Lake embayment. Basinward, the subunit increases in thickness rapidly over short distances.

The fanglomerate facies is inferred to define the margins of the embayments and has been truncated, at least in part, by Plateau and Hayhook Faults. The width of the preserved part of the facies belt varies up to approximately 1100 m. It consists essentially of carbonate and lesser purple and grey chert clasts derived from strata of the lower part of the Upper Carbonate Subunit. Volcanic clasts are extremely rare.

The facies can be divided into two members: a massive to very crudely stratified, coarse boulder conglomerate; and a well-bedded, coarse sand and pebble-sized fluvatile-conglomerate. The latter member ovelies directly the Upper Carbonate Subunit in what appears to be mainly a conformable rèlationship. The coarse boulder conglomerate occupies an embayment margin position and grades abruptly into the coarse sand and pebble conglomerate in a basinward direction. The boulder conglomerate shows considerable variability in thickness, and reaches an observed maximum of 32 m. Clasts are framework-supported, reach a diameter of about 15 cm, are poorly sorted, generally moderately well rounded, and show at best only weak stratification in the coarsest parts of the facies.

Fining-upward cycles are common in the fine-grained member. They vary from poorly to well developed and resemble those of a braided stream deposit. Current ripples are common in parts of the coarse sandstone. Red mudstone containing symmetric wave ripples occurs at the tops of some of the cycles and resembles overbank flood deposits. The fine grained member has a maximum observed thickness of 27 m.

The playa lake facies consists of an evaporitic sequence consisting of massive anhydrite beds, centimetre-scale interbeds of anhydrite and red siltstone, and anhydrite-bearing red siltstone. Anhydrite-rich units vary up to about 10 m, are mainly coarsely crystalline, occasionally show fine-scale laminations, contain less than 10% gypsum, and only rarely contain any carbonaceous material. Algal laminations are extremely rare. The anhydrite beds cannot be correlated in the subsurface and decrease in abundance basinwards.

Interbeds of impure anhydrite and evaporite-bearing red siltstone, that are crudely interbedded on a scale of about a centimetre, occur between massive anhydrite beds. The lateral extent of individual beds is unknown. This centimetre-scale interbedded unit constitutes the greater part of the playa lake facies and becomes the dominant member of the facies in a marine (basinward) direction.

Evaporitic red siltstone and mudstone, characterized by the presence of anhydrite rosettes after gypsum, occur irregularly interbedded with the anhydrite beds and the centimetre-scale interbedded unit. The rosettes range up to about 2 cm in diameter. Desiccation cracks are not common within the playa lake facies.

The flood plain facies, forming the thickest sequence within the Upper Redbed Subunit of the Upper Redstone River Formation, exhibits both well and poorly developed fining-upward cycles ranging to about 20 cm in thickness. These, in ascending order, consist of plane-parallel laminated sandstone, ripple drift cross-bedded sandstone, and very finely laminated mudstone. Rip-up clasts are present at the bases of many cycles. Desiccation cracks and standing wave ripples are abundant at the tops of the cycles whereas water-escape structures are common in the coarser parts of the cycles. Detrital anhydrite grains are abundant in some cycles, and diagenetic anhydrite is widespread as a cement within desiccation and water-escape structures. No large-scale fluvial channels have been recognized in this unit.

Specular hematite is pervasive throughout the Upper Redstone River Formation and produces a purplish colouration. It commonly occurs as detrital concentrations along foresets in cross-bedded units and as disseminations in the finer-grained interbeds. The provenance for the hematite is the lava flows of the Little Dal Formation, and because specular hematite is widespread throughout the Upper Redbed Subunit, this distribution may indicate that the lava flows were laterally very extensive and provided, through erosion, a considerable volume of clastic material to the Upper Redstone River Formation.

Coppercap Formation

Transition Zone. The termination of marginal marine continental conditions, prevalent throughout Upper Redstone River time, is indicated by a series of shallow marine transgressions marking the early advance of the Coppercap Sea. The marine transgressive beds consist of evaporitic, cryptalgal-laminated carbonate rocks that constitute the coastal sebkha facies (Figs. 3 and 4). The cryptalgal-laminated beds form the most important lithostratigraphic unit throughout the copper belt as it is in these rocks that the most extensive deposits occur.

The transition zone is best developed in the Coates Lake embayment, where it contains at least eight evaporitic algal-laminated carbonate lithosomes over a stratigraphic interval of 110 m. There, the lithosomes range in thickness from a few centimetres to about 6 m, and are intercalated with red flood-plain facies clastic rocks of the Upper Redstone River Formation (Figs. 4 and 8). They contain approximately 20% dolomite, 15 to 20% feldspar and quartz, with the remainder consisting of anhydrite and limestone. The lithosomes are grey and are always enveloped by green, chlorite-rich beds that are the reduced equivalents of the redbeds of the flood plain facies which form poorly developed fining-upward cycles between each sebhka cycle (Figs. 6, 9). The reduced envelopes, referred to as "bleached zones" appear to have developed penecontemporaneously with deposition of the sebkhas,

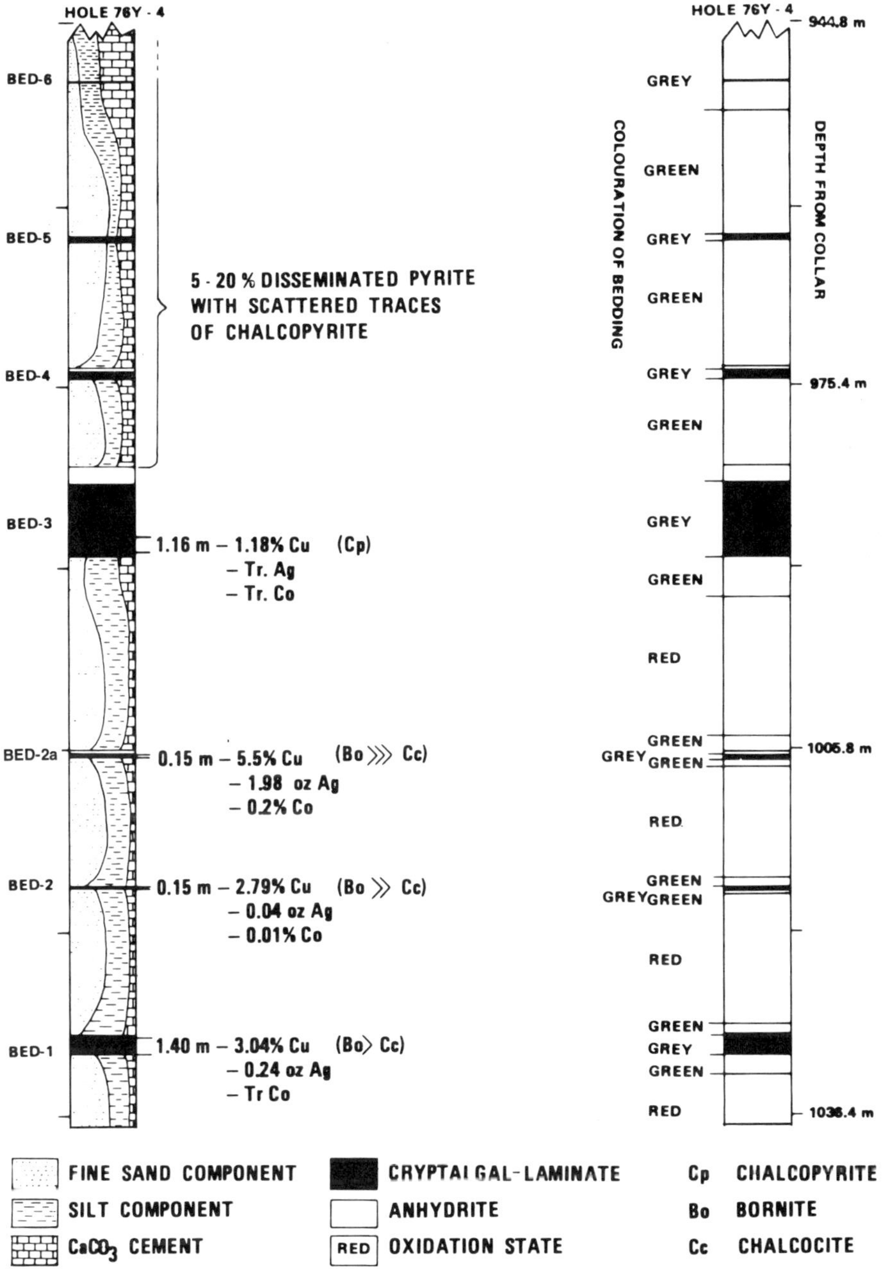

Figure 6. The Transition Zone as intersected in drill hole 76Y-4, Coates Lake property.

and to be due to the highly reducing, decaying carbonaceous algal component of the sebkhas. The change from red to green coloration is gradational over several metres. A small increase in lime côntent is associated with the reduced envelopes. The Transition Zone at Coates Lake, as intersected in drill hole 76Y-4, is shown in Figure 6.

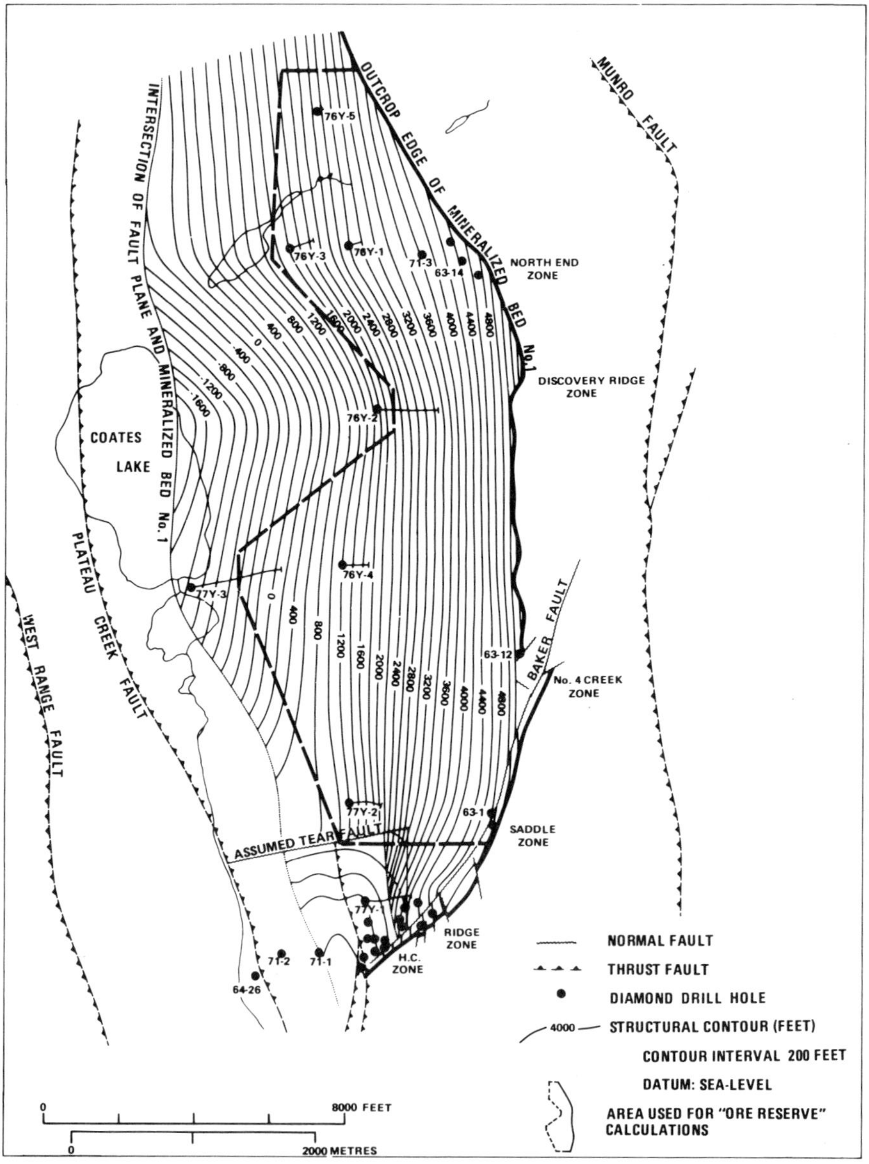

Figure 7. Structural contour map of Bed No. 1 of the Transition Zone, Coates Lake Property.

Although showing some degree of variability from bed to bed, the cryptalgal-laminated carbonate lithosomes of the coastal sebkha facies in the Coates Lake embayment begin with generally poorly laminated calcarenite and micrite containing a high silt content (quartz and feldspar) and few algal-laminated interbeds. The start of each cycle is sharp and contains no transgressive lag deposits. The succession grades into very finely algal-laminated silty micrite containing a few centimetre-scale interbeds of non-laminated micrite. This forms the main member of the cycle. Fine-scale desiccation cracks occur locally. The micritic interbeds often exhibit the early development of enterolithic structures (Fig. 10). The top of each cycle is marked by the presence of anhydrite-rich micrite. The anhydrite takes the form of masses of blade-like crystals and nodules. The nodules coalesce locally to form massive anhydrite beds reaching a maximum observed thickness of 1.5 m. The

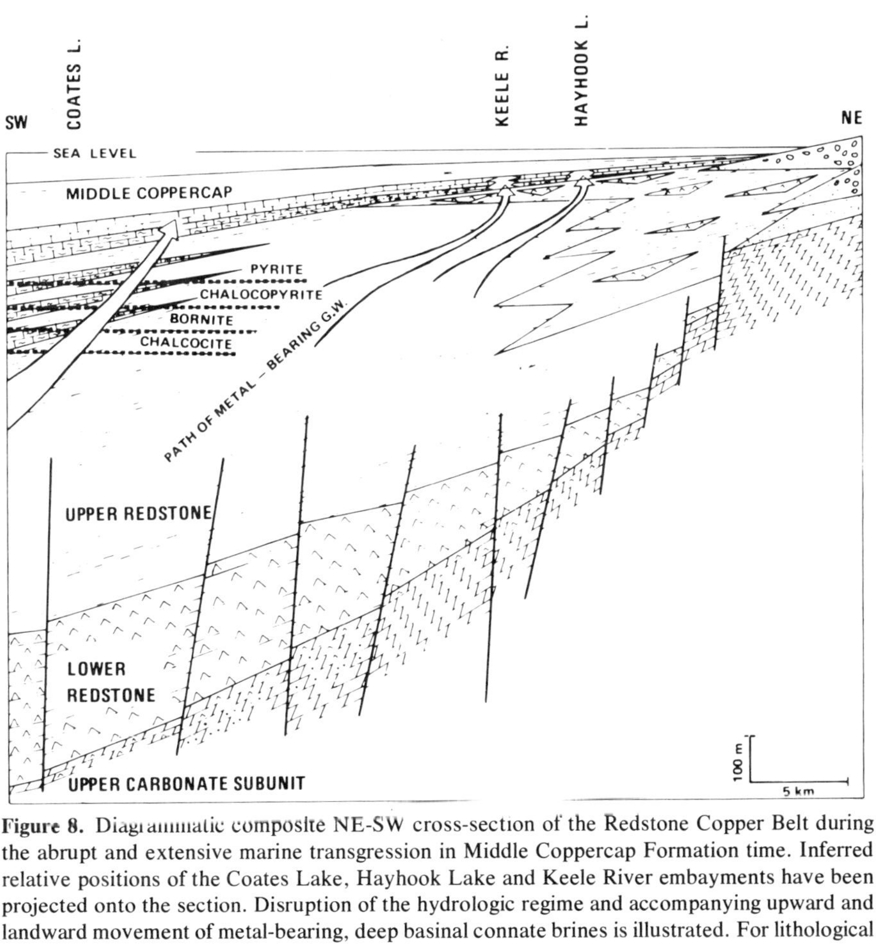

Figure 8. Diagrammatic composite NE-SW cross-section of the Redstone Copper Belt during the abrupt and extensive marine transgression in Middle Coppercap Formation time. Inferred relative positions of the Coates Lake, Hayhook Lake and Keele River embayments have been projected onto the section. Disruption of the hydrologic regime and accompanying upward and landward movement of metal-bearing, deep basinal connate brines is illustrated. For lithological descriptions see Figure 4.

evaporite-rich capping is believed to have formed during the regressive stage at the termination of the sebkha cycle.

Where examined at Keele and Hayhook, the Transition Zone consists essentially of a single cryptalgal-laminated unit of variable thickness from zero to at least 8 m. Minor paleotopography, associated mainly with shallow depressions within evaporitic playa lakes, controlled the distribution and thickness of the algal-laminated unit. Preliminary isopach studies indicate that the algal-laminite is thickest where it overlies the thickest accumulation of playa lake sediment of the playa lake facies.

The single cryptalgal-laminated carbonate unit of the Transition Zone at Keele and Hayhook shows textural features which differ from the intercalated cycles at Coates Lake. Very finely algal-laminated, silty dolomite intervals occur interbedded with fine-grained, more massive sections, and interbedding is in the order of a centimetre or less. Limestone is rare. The algal-laminated unit grades into overlying intertidal carbonate strata and lacks the evaporite crust characteristic of the sebkha cycles at Coates Lake. Throughout the algal unit there was a strong development of early diagenetic anhydrite nodules which now impart a striking fenestral-like fabric. The nodules are extensively replaced by sulphides and chert. Only minor reduction of the underlying redbeds has been observed.

Lower Coppercap Formation. A succession of silty carbonate rocks of the Lower Coppercap Formation conformably overlies the Transition Zone throughout the copper belt and comprises the high intertidal mud-flat facies. It varies in thickness from zero in marginal parts of the embayments to at least 63 m in more basinal sections.

The high intertidal mud-flat facies consists dominantly of micrite with lesser calcarenite and locally exhibits good rhythmic interbedding. Calcarenite units are mainly massive bedded but in some places contain poorly developed ripple drift cross-laminations. Laminations are not well developed in the micritic units, and desiccation cracks are sparse. Poorly developed grading can be recognized between carcarenite and micrite units. Graded sequences seldom exceed 25 cm. Eisbacher (1977) and Jefferson (1979) have described the graded sequences as allodapic carbonates probably deposited by turbidity or density currents but of relatively shallow water origin. The lower few metres of the facies contain centimetre-scale cryptalgal-laminated interbeds that are very similar to the mats of the basal algal-laminated unit (Fig. 18). Several laterally extensive flat chip conglomerate beds, averaging about 10 cm in thickness, occur in the lower 15 m of the succession. Centimetre scale interbeds of coarse-crystalline, impure anhydrite occur locally

---

Figure 9. View, looking northwesterly, of the copper-bearing Transition Zone beds as exposed on Coppercap Mountain, Coates Lake embayment. Transition Zone interval is approximately 110 m thick.

Figure 10. Outcrop sample (approx. 9% Cu) of copper-bearing cryptalgal-laminite from bed No. 1, Coates Lake. Larger, dark grains are bornite with minor chalcocite. Sample contains fine-scale desiccation cracks.

Figure 11. Core sample (approx. 6% Cu) of copper-bearing cryptalgal-laminite from bed No. 1 in drill hole 76Y-4, Coates Lake. Dark, nodular to bladelike grains are anhydrite which locally show early stages of coalescing into beds.

Figure 9

Figure 10

Figure 11

throughout the facies. The upper 5 to 10 m of the succession contain interbeds of extremely finely laminated, black, calcareous shale and very argillaceous limestone that is of similar lithology as strata of the overlying basin facies. The number of interbeds increases markedly towards the top of the facies.

Anhydrite nodules are abundant, particularly in the lower half of the succession. The anhydrite exhibits locally extensive replacement by white chert. Blade-like anhydrite crystals occur throughout the facies but, again, are much more prevalent in the lower half of the succession.

Middle Coppercap Formation. Conformably overlying the Lower Coppercap Formation is a uniform, recessive sequence of platy-weathering, extremely finely laminated, pyritic, calcareous mudstone and very argillaceous, fetid limestone. The succession constitutes the restricted basin facies, and reaches an observed thickness of 72 m. Though fetid, analysis for organic carbon yielded only trace amounts.

Upper Coppercap Formation. Conformably overlying the restricted basin facies is a succession of intertidal and subtidal carbonate rocks of the Upper Coppercap Formation. The succession consists of grey, generally well-bedded, cliff-forming limestones that vary in thickness from zero at embayment margins to about 165 m in basinal sections. Many of the laminated carbonate units appear to be of cryptalgal origin, and a stromatolite bed occurs near the base. Replacement of carbonate beds by white and dark grey chert is common, and this is especially abundant near the base and in the upper 80 m. Interbeds containing synsedimentary slump structures are common towards the top of the formation.

At a number of localities in the Keele embayment, the formation contains a conglomerate that consists of clasts of the various rock types of the Upper Coppercap Formation. The clasts range in size from coarse sand to pebbles, and are poorly sorted. The thickness of the conglomerate is irregular, and varies from zero to about 15 m. Its distribution may indicate local exposure of the succession to erosion in pre-Hadrynian time.

Figure 12. View, looking northwesterly, of stratigraphic succession on east side of "Hutch Mountain", Keele River embayment. Light-weathering unit at top of Redstone River Formation is anhydrite within the uppermost part of playa lake facies. Light-weathering interval in lower half of Redstone River section is main playa lake facies. Exposed section of Redstone River strata is 520 m thick. Hr = Rapitan Fm. undivided, Hccu = Upper Coppercap Fm., Hccm = Middle Coppercap Fm., Hccl = Lower Coppercap Fm., Hrru = Upper Redstone River Fm.

Figure 13. Outcrop sample of copper-bearing cryptalgal-laminite from basal Coppercap Formation at June Creek Deposit illustrating characteristic fenestral-like texture. Dark grains consist predominantly of chalcopyrite and, together with white chert, have nearly completely replaced early diagenetic anhydrite nodules.

Figure 14. Core sample of barren pyritic cryptalgal-laminite from basal Coppercap Formation distant from copper mineralization, at Keele. Greenish-grey coloration of finely-laminated interbeds is due to abundance of finely disseminated "micron-sized" pyrite inferred to be of syn-diagenetic origin. Coarse-grained pyrite forms rims around chert that has almost entirely replaced diagenetic anhydrite nodules.

Figure 12

Figure 13

Figure 14

STRUCTURE

General Statement

Structural deformations along the copper belt are separated into two main groups: deformations responsible for embayment development, and deformations of post-ore age. Although regional facies distributions allow one to infer, with some degree of confidence, the presence of northeast-trending faults that were active during embayment development, only occasionally are such faults actually mapable. These faults mark the start of the "copper cycle", an event related to the Hayhook orogeny and described by Jefferson (1980), as a period of continental rifting of world-wide magnitude, starting at about 800 Ma.

Post-ore structural events range in age to Lower Tertiary (Laramide orogeny). However, such events do not appear to have had a severe impact on favourable strata and, except for local structural complications near faults, the copper belt is regionally only weakly deformed. In the discussion that follows, a brief outline of the regional structural setting is given, followed by descriptions of structures in the Coates Lake and Keele River embayments.

Regional Structure

The main structural elements of the copper belt are Plateau Fault and Hayhook Fault (Fig. 2), which are Laramide thrust faults of regional extent (Gabrielse *et al.*, 1973). The thrusts were easterly directed and carried Proterozoic strata over carbonate rocks and shales of Paleozoic age. The amount of thrust displacement on Plateau Fault has not been established but is considered to be 5 km or less (Aitken, pers. commun., 1976; Jefferson, 1979). The thrusts are most likely rooted within a thick evaporite unit (Gypsum Subunit) of the Lower Little Dal Formation (Aitken and Long, 1978). Strata attain dips of between 60° and 70° near the thrust fronts, but flatten to less than 15° to the west within a few kilometres.

Hayhook Fault is structurally much more complex than Plateau Fault, and requires further study. It can be traced from southeast of the Hayhook embayment to south of the Coates Lake embayment where it is called Munro Fault. An imbricate thrust off Munro Fault has been recognized at Coates Lake (Watson, 1971).

Figure 15. Photomicrograph (reflected light) of a bornite grain that illustrates partial replacement of an early diagenetic anhydrite (A) "blade". Exsolution lamellae of chalcopyrite occur throughout the bornite.

Figure 16. Core sample of copper-bearing, evaporitic, finely dolomitized micrite from June Creek deposit. Sample location is 6 m above basal cryptalgal-laminated unit and illustrates nearly complete replacement of early diagenetic anhydrite "laths" by chalcopyrite.

Figure 17. Core sample of copper-bearing cryptalgal-laminite interbedded with finely dolomitized, silty micrite from June Creek deposit. Sample location is 2 m above main cryptalgal-laminated unit at base of Coppercap Formation.

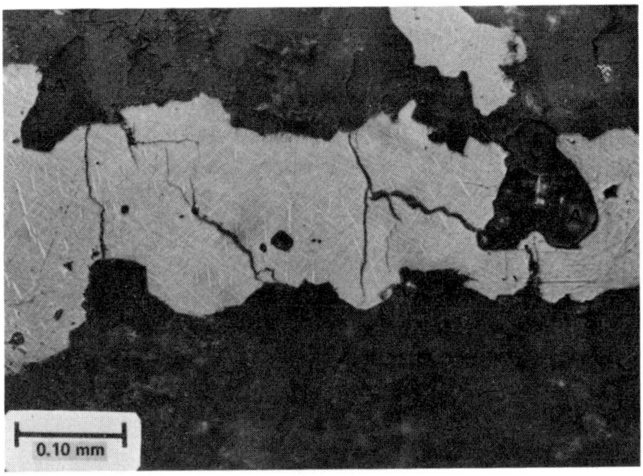

Figure 15

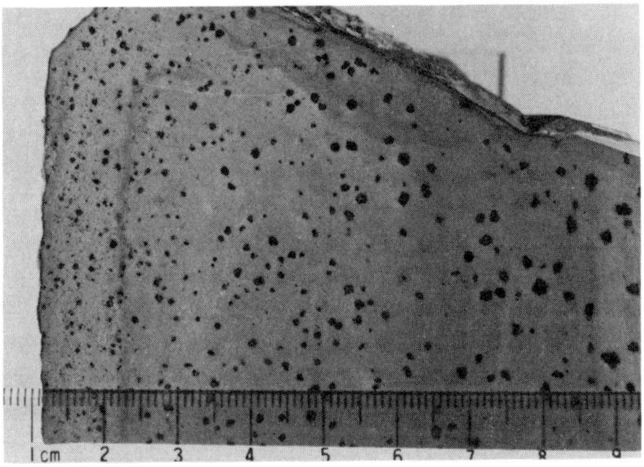

Figure 16

Figure 17

Along the copper belt the thrust faults are responsible for bringing the favourable strata to surface. However, no major thrust faults have been recognized to the west of Plateau and Hayhook faults that might repeat or expose the favourable strata. The absence of any major thrust repetition seriously hampers exploration in the more marineward parts of the embayments.

Structure Within the Coates Lake Embayment

At Coates Lake, Munro Fault has thrust Proterozoic rocks eastward over a deformed Middle Devonian carbonate succession. An imbricate thrust off Munro Fault repeats Coppercap and conglomeratic, fanglomerate facies Redstone River strata (Fig. 5). Bedding inclinations reach 60° near the thrust fronts but flatten to between 38° and 45° over most of the area drilled. Plateau Çreek Fault strikes northerly across the central part of the property and has displaced the transition zone some 610 m vertically downward in the southern part of the property (Hildebrand, 1978). Displacement at the north end is unknown, and the fault is believed to "die out" towards the north. Two faults of similar style, called the Rapitan Fault and the West End Fault, occur parallel to Plateau Creek Fault. Their displacements are much less however and they "die out" towards the north.

A structural contour map of the lower bed of the Transition Zone is shown in Figure 7. Structural contours show a westerly-dipping, gently flattening plane that has been partially folded through later interference related to the Plateau Creek Fault. Fault structures in the southern part of the map area have been simplified. Up to 2300 m of cover, including Coppercap and overlying Hadrynian Rapitan strata, overlie the mineralized beds in the deeper part of the structural basin within the property boundary.

Structure Within the Keele River Embayment

Two main periods of structural deformation are recognized at Keele: a post-Coppercap to pre-Hadrynian Rapitan episode consisting primarily of block faulting accompanied by local warping; and thrust deformation related to the Laramide Orogeny. The block faults have been recognized throughout the Keele River embayment but are not abundant. They are steeply dipping and trend both northwesterly and easterly. Minor fracture-related copper occurrences are associated with the faults, particularly near the top of the Coppercap Formation.

As at Coates Lake, strata of the copper belt at Keele were thrust eastward during the Laramide Orogeny. Near the thrust front, strata dip west up to about 45°, but

---

Figure 18. Photomicrograph (reflected light) showing oriented exsolution lamellae of chalcopyrite in bornite. Darker grey digenite rims chalcopyrite lamellae and occurs as irregular intergrowths throughout bornite.

Figure 19. Outcrop sample of disseminated chalcopyrite in finely dolomitized, weakly ripple drift cross-laminated, silty micrite from approximately 4 m above the basal cryptalgal-laminated unit, Jay deposit.

Figure 20. Outcrop sample of high grade copper mineralization from "Showing A", Hayhook Lake embayment. Sample estimated to contain 20% copper as 95% chalcocite and 5% bornite.

Figure 18

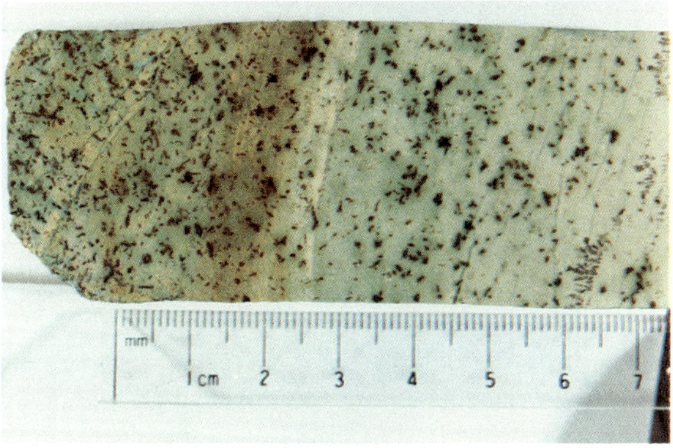

Figure 19

Figure 20

elsewhere average about 10° to the southwest. Two steeply dipping thrust faults, possibly splays off Plateau Fault, are recognized in the vicinity of the June Creek deposit. Displacements are less than 150 m and are dominantly vertical. Locally intense folding has accompanied the thrusting and the thrusts are themselves partially folded.

Because of a sub-linear form to the Jay deposit as outlined in early drilling, the mineralization was thought possibly to be fault controlled. However, structural contours on the flood-plain facies show no faulting associated with the deposit.

Insufficient drilling has been carried out on the June Creek deposit to allow meaningful structural contouring. It has been established however that the above-mentioned thrust faults post-date the period of mineralization. For example, the most westerly thrust cuts the mineralized zone in at least two places and cuts similar mineralization at a showing 1.4 km east of June Creek. Where seen elsewhere, the fault is not mineralized.

DEPOSITIONAL ENVIRONMENTS

General Statement

To establish a workable model for the genesis of copper deposits of the copper belt, an understanding of the environments of deposition of Redstone River and Coppercap strata is necessary and is proposed here. The limited amount of surface and subsurface data available from the three embayments are combined to give a composite model for the depositional environments shown in the diagramatic east - west block diagram (Fig. 4) and in a paleofacies map (Fig. 3). It must be emphasized that the factual observations on which Figures 3 and 4 were made are limited to a narrow strip along the eastern margin of the basin. The scales are illustrative only.

Six main facies are recognized along the copper belt within the stratigraphic interval from the Upper Redstone River Formation to the Middle Coppercap Formation. From east to west these include:

VI Restricted basin facies - Middle Coppercap Formation
V High intertidal mud flat facies - Lower Coppercap Formation
IV Coastal sebkha facies - Transition Zone
III Flood plain facies - Upper Redstone River Formation
II Evaporitic playa lake facies - Upper Redstone River Formation
I Fanglomerate facies - Upper Redstone River Formation

Because of generally flat terrain during deposition, small changes in sealevel influenced extensive areas, resulting in a number of intercalations of the facies.

Fanglomerate Facies

Warping and associated block faulting of the stable carbonate platform towards the end of Little Dal Formation time marked the start of the "copper cycle" and the beginning of the development of embayments along the Proterozoic coastline. One or a number of *en echelon* normal faults may have been active in controlling the development of individual embayments.

The fanglomerate facies, confined to the landward parts of the embayments and interpreted to represent part of a coalesced alluvial fan complex, developed in response to the block-faulting. Clast composition attests to the fact that many hundreds of metres of Little Dal strata were uplifted and eroded.

The abrupt decrease in grain size is thought to be associated with a decrease in slope at the base of the fan, but the width of the fan complex during deposition is uncertain. The presence of the coarse boulder member abruptly overlying the finer-grained member is interpreted to result from a rapid burial of the braided stream portion of an earlier fan during progradation of the complex.

Playa Lake Facies

Extending basinward from the alluvial fan complex, topographic relief decreased abruptly and the subsiding embayments formed broad, flat, coastal flood plains. Within the upper reaches of the flood plains in the Keele River and Hayhook Lake embayments, and at least locally intercalated with the fluviatile member of the fanglomerate facies, a broad facies belt containing evaporitic ephemeral lakes (inland sebkhas) developed. This facies belt constitutes the main playa lake facies, and is at least 10 km wide at Keele. It has not been recognized within the Coates Lake embayment. Of all facies within the copper belt, the playa lake facies exhibits the greatest degree of lateral variation.

Within the more central parts of the playa lakes relatively pure evaporites accumulated. Towards the lake margins, evaporites become rhythmically interbedded with sheetwash clastics in what is interpreted to represent deposition within an evaporitic playa mud-flat under semi-arid conditions.

The basinward margin of the main playa lake facies is irregular and shows many intercalations with the weakly-evaporitic flood plain facies. Towards the end of Redstone River time, the main playa lake facies belt prograded basinward and massive anhydrite beds were deposited over weakly evaporitic clastic sediments of the flood plain facies (Figs. 8, 12). These evaporites exhibit marked thickness variations in the vicinity of copper deposits in the Keele River area, and are considered to have played a key role in controlling the migration of early diagenetic metal-bearing fluids. Solution collapse brecciation, contemporaneous with evaporite deposition, was developed locally within the playa lake facies.

Flood Plain Facies

Marineward of the playa lake facies, the gently subsiding embayments formed broad, low-relief flood plains over which weakly evaporitic, cyclical, fine-grained, continental sediments accumulated. The graded and cyclic nature of this facies suggests that it was formed through accumulation of fine-grained sheetwash sediments.

It is believed that surface water draining from the barren highlands of the uplifted stable carbonate platform carried considerable volumes of debris both in suspension and as bed-load material. Coarse, unsorted material was dumped high up on the alluvial fan complex while braided stream deposits formed along the toes of the coalescing fans. As the high velocity surface waters reached the broad, nearly flat coastal flood plains, there were no channels to confine the flow. The debris-laden waters spread out over many square kilometres as shallow sheetwash floods to deposit the thick accumulations of graded sheetwash clastic sediments.

Coastal Sebkha Facies

As the Coppercap sea slowly transgressed the broad, flat flood plain facies, embayments were inundated by shallow marine flooding, and coastal lithosomes were deposited. These deposits were formed under coastal sebkha conditions similar to those described by Glennie (1970) for the Trucial Coast. Characteristic of the sebkhas were the presence of organic-rich algal mat sediments that later provided chemically favourable sites for copper deposition. Therefore, the coastal sebkha facies is the most important lithostratigraphic unit as host for laterally very extensive copper mineralization.

Because of the extremely low relief associated with the flood plain facies within the Coates Lake embayment, algal mats of the coastal sebkha facies covered extensive areas. However, where the Transition Zone has been examined at Keele and Hayhook, it appears that algal mat development was restricted somewhat. This restriction is attributed to gentle but more pronounced paleotopographic relief. In essence, a thicker build-up of algal mat sediments is believed to have developed within former playa lake depressions. As much as 8 m of relief have been recognized. It is therefore inferred that much of the algal mat development at Keele and Hayhook took place under very shallow, restricted, saline "lagoon-like" conditions in a setting somewhat similar to that shown by Kendall (1978, Fig. 16) for the Frobisher evaporite.

The coastal sebkha cycles in the Coates Lake embayment and the algal-laminated unit in the Keele River and Hayhook Lake embayments were deposited during periods of relative stability. As uplift and accompanying erosion of the stable carbonate platform periodically reached equilibrium, clastic supply to the embayments was sharply reduced, allowing favourable conditions for deposition of supratidal marine carbonate. As shallow marine flood waters inundated the extremely low-relief embayments, algal mats flourished, forming broad coastal sebkha deposits. The accumulated thickness of the algal mat material during a given sebkha cycle is believed related primarily to the duration of the period of stability and to the number of marine floodings during the cycle. The degree of flooding, and therefore the number of sebkha cycles, would be expected to be greatest proximal to the marine shoreline. Deep drilling at Coates Lake has indicated that the number of sebkha cycles increases in what is inferred to be the more marineward part of the embayment, i.e., bed 2a in Figure 6.

The presence of a sebkha cycle in a given embayment is dependent on a number of factors. The rate of subsidence, degree of clastic supply, and the extent of marine flooding would influence the development of such a carbonate build-up. It seems reasonable to expect that such factors varied from one embayment to the other. A problem arises therefore in attempting any correletion of the coastal sebkha facies (Transition Zone) in the three embayments. As is depicted on the diagramatic block diagram (Fig. 4), the carbonate-bearing sebkha cycles are inferred to have been deposited within the Coates Lake embayment while flood plain clastic sediments were being deposited at Keele and Hayhook. It is therefore postulated that exploration in the Keele River and Hayhook Lake areas has been confined to the more continental parts of the embayments, shoreward from where all but the final marine flooding extended sufficiently to allow algal mat development. The possible extent of the zone of cyclical coastal sebkha facies development within Keele River and Hayhook Lake embayments is shown schematically on the paleofacies map (Fig. 3).

High Intertidal Mud Flat Facies

The Lower Coppercap Formation contains a thick succession of shallow water carbonate rocks that constitutes the high intertidal mud flat facies. Paleotopographic relief was probably still minor during deposition of this facies. From the standpoint of metallogenesis, it is significant that the continued transgression of the Coppercap Sea, during deposition of this facies, was slow enough so as not to significantly alter the hydrologic regime.

The mud flat facies is interpreted to have been deposited under shallow intertidal marine conditions, with brief periods of subaerial exposure, during the early transgression of the Coppercap Sea. Brief periods of deposition under sub-wavebase conditions are believed to have occurred during the later part of the development of this facies as indicated by the finely laminated calcareous shale interbeds. Detrital carbonate material is the main component of the facies. The principal source for this material is considered to be through continued erosion of the lower part of the Upper Carbonate Subunit of the Little Dal Formation.

Restricted Basin Facies

A uniform succession of extremely finely laminated, black, pyritic, calcareous shale and very argillaceous limestone of the restricted basin facies conformably overlie the intertidal mud flat facies.

The high clay content, extremely finely laminated nature, lateral extent, and presence of pervasive finely disseminated pyrite indicate the facies was probably deposited under quiet water reducing conditions, possibly under restricted conditions within the main marine basin of a rapidly transgressing Coppercap Sea, rather than under lagoonal conditions. The facies is considered to have acted, through compaction processes, as the main driving mechanism in expelling metal-bearing ground waters from the clastic basins.

MINERALIZATION

General Statement

Throughout the copper belt, all significant mineralization is confined to cryptalgal laminites of the coastal sebkha facies and shallow water carbonate rocks of the high intertidal mud flat facies. Copper grades for various deposits and showings average between about 2.5% Cu and 5% Cu, and fall within the general range for typical sedimentary copper deposits.

Deposits within the Keele River and Hayhook Lake embayments are strikingly dissimilar to those at Coates Lake. At the former, the deposits are stratigraphically thick, strongly discordant, and tend to be sinuous or pod-like; whereas at Coates Lake, they tend to be stratigraphically thin but to form laterally very extensive sheet-like bodies. A later discussion on the metallogenesis of the copper belt attempts to explain these differences. Only the Keele River and Coates Lake deposits are discussed individually because only these were studied in the field by the author, and because deposits at Hayhook appear to be very similar to those at Keele.

No economic mining tonnage has been outlined for the copper belt. However, because it has many geological features in common with the Zairian and Zambian Copper Belts, the geological potential may be vast.

Coates Lake Deposits

Among the sheet-like copper deposits at Coates Lake, only one approaches economic proportions. In the potential ore bed, which averages one metre in thickness, some 37 million tonnes containing 3.92% Cu and 0.33 oz. per ton Ag (11.3 g/t) have been outlined within an area of approximately 12 km^2 (Hildebrand, 1978 and Fig. 7). The mineralized beds extend along strike for at least 7 km. Drilling to date has only begun to test for continuation, and possible thickening, of the mineralized beds in deeper parts of the basin. A geological map and cross-section of the Coates Lake property are given in Figure 5, and a structural contour map in Figure 7.

Stratigraphic Setting. Mineralization is confined to grey, silty, algal-laminated carbonate lithosomes within the Transition Zone (Figs. 10, 11). Mineralized beds have been numbered 1 to 8, sequentially from oldest to youngest (bottom to top). Although eight have been encountered, only four contain significant copper values. The remainder contain pyrite with only erratic occurrences of chalcopyrite.

The copper-bearing beds occupy the lower part of the Transition Zone, with Bed 1 containing the highest copper grades. From limited surface and subsurface studies, it is evident that the Fe:Cu ratio in sulphides shows pronounced zoning, with iron content increasing both stratigraphically up-section (Fig. 6) and towards the embayment margin. Studies have not been undertaken to establish if trace amounts of other metals such as Ag, Co, Mo, Pb, and Zn might also be zoned. Lateral zoning of iron and copper in sulphides is very well illustrated by the sulphide mineralogy within Bed 1. In the most basinward drill intersection (drill hole 77Y-3), Bed 1 contains chalcocite and bornite in the approximate proportion 60% chalcocite and 40% bornite, with no observable traces of chalcopyrite or pyrite. As the bed is traced in drill holes, towards the inferred embayment margin (to the east), the bornite content increases at the expense of chalcocite, and chalcopyrite appears. Within the broad zone of increasing bornite content (increasing iron), bornite grains often contain very minute exsolution lamella of chalcopyrite (Fig. 15). Closer to the embayment margin, chalcocite becomes much less abundant and chalcopyrite increases at the expense of bornite. Drilling, confined to shallow holes closest to the inferred embayment margin in the early 1960's, indicates that chalcopyrite is the chief mineral, and that pyrite is rare (Coates, 1964, p. 40). Bed 1 is truncated by Munro Fault, at an estimated 9 to 10 km from the embayment margin. It is, therefore, impossible to establish if the mineralogy of Bed 1 eventually changed to one of dominantly pyrite near the embayment margin.

A semiquantitative illustration of the vertical mineral zonation from the lower part of the Transition Zone, as typified in hole 76Y-4, is given in Figure 6. From such basinward drill intersections, the increasing iron content of the sulphide assemblage up-section is reflected in a change from chalcocite through bornite to chalcopyrite and pyrite in the interval from Bed 1 to Bed 3 respectively. Only trace amounts of chalcopyrite are present in Bed 2. Within Bed 3, there is a pronounced vertical change from chalcopyrite and minor pyrite in the lower 1.16 m to pyrite with traces of chalcopyrite above. In the sebkha cycles above Bed 3, pyrite is dominant and chalcopyrite occurs only in trace amounts, primarily in the lower several centimetres of the beds. Within the reduced clastic beds above Bed 3, pyrite-rich interbeds, with up to 20% pyrite and generally less than 20 cm thick, are common. Only trace amounts of chalcopyrite are present within these.

Variations in both thickness and grade exist within individual mineralized beds. Although data points are too sparse to allow accurate isopach studies, investigations by Hildebrand (1978) indicate an increase in the isopach of metre-per cent copper for Bed 1. Isopach values vary from 2.43 metre-per cent to 4.26 metre-per cent copper, with the thickest isopach forming a northeast trend in what may be the most central part of the embayment drilled to date.

Mineralogy. The economically most important minerals include bornite and chalcocite. The term "chalcocite" is used to include grains varying from light grey to greyish blue under reflected light. Subtle colour variations may reflect different chemical compositions due to extensive solid solution and, therefore, some of the chalcocite is probably digenite or anilite.

No systematic studies have been carried out to determine if trace amounts of other metallic minerals are present. Coates (1964, p. 53) reported trace amounts of galena and possible tennantite in bornite at one locality from Bed 1. These minerals have not been recognized in polished sections from holes drilled deeper in the basin. No native copper has been recognized as a primary mineral. No cobalt or silver-bearing minerals have been recognized in polished sections. A comparison of silver values (from assays) with bornite content in drill core indicates, however, that silver content generally increases as the bornite content increases, suggesting possibly that silver may somehow be contained within the bornite grains. Supergene malachite, azurite, minor covellite and trace native copper occur only near surface.

Bornite and chalcocite are generally less than 25 μm in diameter. The sulphides tend to concentrate along algal laminations, with concentrations averaging 0.2 mm and, in places, reaching 3 mm in thickness. Isolated grains are mainly anhedral and are both interstitial to and replace carbonate grains. The sulphides, particularly chalcocite and bornite, occasionally show partial to complete replacement of millimetre-scale, blade-like crystals and small nodules of anhydrite (Fig. 15). There does not appear to be any appreciable replacement of pyrite by copper sulphides.

Where bornite co-exists with chalcopyrite, exsolution laths of chalcopyrite in bornite are locally abundant (Fig. 15). The chalcopyrite laths in many specimens are rimmed by a greyish-blue mineral phase, possibly digenite. White, isotropic to very weakly anisotropic chalcocite often occurs as a lacy to vermicular intergrowth with bornite.

Keele River Deposits

Some 11 mineral showings occur at Keele but only two, the Jay and June Creek deposits, have been drilled in detail. In these sinuous deposits, copper mineralization is controlled by local, abrupt facies changes and it cuts facies boundaries, but there is no evidence that the mineralization is related to faulting. Isopach studies suggest the mineralized zones may be sinuous over kilometres, but additional drilling is required to substantiate this. Lateral dimensions are generally 100 metres or less, but where wider, the zone tends to become thinner.

Thicknesses of at least 52 m (grading 2.3% Cu at June Creek) are known although the best average thickness and grade is 12.5 m of 2.7% Cu (Jay deposit). Drill-indicated reserves for the Jay deposit are 1.2 million tonnes grading 2.7% Cu and for the June Creek deposit are about 250,000 tonnes grading 3.4% Cu. The mineralized zones have not been completely defined.

Stratigraphic Setting. Where investigated, copper mineralization begins at the base of the cryptalgal-laminated unit marking the base of the Lower Coppercap Formation. Sulphides extend for varying distances up-section and are known, at the June Creek deposit, to reach the upper limits of the high intertidal mud flat facies. The central parts of the mineralized zones at June Creek and Jay are associated with a thinning and, in some instances, a pinching out of the cryptalgal-laminated unit. Variations in thickness from zero to at least 8 m, are believed to be a result of paleotopographic features associated with former playa lakes that formed in the uppermost part of the playa lake facies. Isopach studies on the Jay deposit indicate that the cryptalgal-laminated unit thickens where the isopachs of playa lake evaporitic strata thicken, and thins as the evaporitic strata thin. It is, therefore, inferred that gentle depressions, probably not exceeding a few metres in relief, existed within the evaporitic playa lakes. The significance of the superimposed thickening of the evaporitic playa lake sediments and algal-rich lime mud sediments to a model for mineralization is discussed later.

Within the high intertidal mud flat facies, mineralization occurs preferentially in centimetre-scale algal-laminated interbeds, and in more evaporite-rich micritic units. Interbeds up to about one metre thick containing abundant pseudomorphs of chalcopyrite after gypsum, and some replacements of anhydrite nodules by copper sulphides and pyrite, have been encountered at most showings (Figs. 18, 19). The mineralized zones have not been observed to extend beyond the high intertidal mud flat facies into the restricted basin facies.

Unlike the limestone association at Coates Lake, mineralization at Keele is always associated with fine-grained dolostone. Isopachs of the dolostone at the Jay deposit show it to envelope the mineralized zone, with the thickest dolostone section being near the centre of the deposit. However, away from mineralization the Lower Coppercap Formation, except for the algal-laminated unit, is dominantly limestone, suggesting a genetic link between dolomitization and mineralization. Dolomitization grades upwards from fairly pure dolostone to limestone, generally over a few metres to a few tens of metres, but terminates near the base of the restricted basin facies.

Minor copper occurrences have been encountered at several other stratigraphic horizons within the Redstone River Formation. For example, chalcocite and bornite are associated with thin, wispy algal material and green calcareous mudstone interbeds within massive anhydrite units of the playa lake facies. Algal material is rare in the playa lake facies however, and only two intersections, grading less than one per cent Cu over 2 m, have been observed. Several widely spaced showings of chalcopyrite, pyrite, and minor bornite were found in the basal part of the fanglomerate facies. These are of limited size and generally contain less than 5% total sulphide.

The main deposits at Keele are zoned vertically, with the iron:copper ratio in sulphides increasing up-section. Though lateral zonation does not appear to be well developed, there are indications the iron:copper ratio does increase laterally outwards from the centres of the Jay and June Creek deposits. Barren pyritic zones extend both laterally and vertically beyond the limits of copper mineralization. Copper sulphides in the approximate proportion 60% bornite, 25% chalcocite, 15% chalcopyrite occur in the lower several metres of the central part of the deposits. This distribution gives way rather abruptly to chalcopyrite with minor pyrite. The change from chalcopyrite, to pyrite with traces of chalcopyrite, is again fairly abrupt.

Mineralogy. Chalcopyrite is the chief primary copper-bearing mineral. Bornite and chalcocite constitute less than 15% of the copper sulphide content, and are confined to the cryptalgal-laminated unit in the basal core of the mineralized zones at Jay and June Creek. Chalcocite and bornite show optical characteristics similar to those discussed for deposits at Coates Lake. A pale blue mineral (reflected light) has been tentatively identified as tennantite (Friday, 1978).

Exsolution textures are common in bornite grains and include oriented, blade-like lamella of chalcopyrite and vermicular intergrowths of digenite (Fig. 17). Pyrite occurs only in trace amounts in the copper zones.

Supergene alteration is extensive within the Keele River deposits. Alteration minerals include malachite and limonite with minor amounts of azurite, covellite, and chalcocite.

Copper sulphides occur mainly as partial to complete diagenetic replacements of anhydrite nodules in the cryptalgal-laminated unit, and as disseminations of anhedral to euhedral grains in the intertidal mud flat facies (Figs. 13, 16). Blade-like anhydrite crystals also exhibit partial to complete replacement by chalcopyrite (Fig. 19). Anhydrite nodules are ellipsoidal in cross-section and vary up to several centimetres, although they are generally less than 1 cm. The replacement of anhydrite nodules imparts a spectacular fenestral-like texture to the mineralization (Figs. 13, 18, 20). At a distance from the mineralized zones, pyrite, with accompanying chert, also replaces anhydrite nodules though seldom completely (Fig. 14).

Studies are planned to determine the extent of replacement of pyrite by copper sulphides in the mineralized zones. Textures indicate that at least a small percentage may be after pyrite. Additional work is also required to determine the distribution of other trace metals. For example, random samples collected from showings within the Keele River embayment have yielded assay values up to 0.2% MoS_2, 0.02 oz. per ton Au (0.7 g/t), and 18 ppm uranium.

A GENETIC MODEL FOR THE COPPER DEPOSITS

The metallogenic model proposed in this paper is based primarily on field observations of rock types and mineralization. A number of assumptions have been made regarding the physical and chemical conditions during liberation, storage and transport of metal ions; and precipitation of sulphide species. The main premise for the model is that copper was introduced into the system as an integral part of basin development. It does not exclude the possibility that at least some of the copper may be directly related to possible alkaline basalt intrusions into the Lower Redstone River Formation (Table I). Deposits can be related to a single transgressive event as illustrated schematically in Figure 8.

The key elements for the model are: 1) a primary copper source, 2) a secondary copper source (source sediment), 3) a liberation mechanism, 4) a storage system, 5) a transport and driving mechanism and, 6) a chemical trapping system (host sediment).

The primary source for the copper, as well as possibly other metals, is believed to have been the copper-bearing mafic lava flows of the Little Dal Formation. A similar conclusion was reached by Coates (1964). Through erosion, volcanic detritus was shed beyond the fanglomerate facies and into the subsiding embayments to form a part of the Redstone River clastic units, thereby forming the secondary copper source (source sediment).

Within the Redstone River Formation, redbeds were formed through the chemical breakdown of the source rock clastic sediments, and it is proposed that this breakdown provided the initial liberation of metals, particularly copper and silver. Although the mechanism of redbed formation is not clearly understood, the writer follows the proposals of Walker (1967 a,b) and Glennie (1970). Oxidation and reddening of the flood plain facies clastic beds at a fluctuating water-table would liberate the small quantities of metals locked up in silicate and carbonate grains, as well as in detrital sulphides. These liberated metal species would then enter the groundwater system either as free metal ions or metal complexes.

The groundwater was probably oxidizing, or at most, only weakly reducing since detrital hematite was preserved. Because of the absence of reducing conditions within the flood plain facies prior to deposition of the algal-laminated sebkha carbonate lithosomes, it is inferred that most metal ions would remain in solution. Some iron would be withdrawn from solution initially to form limonite which would later be altered to hematite, thereby producing the red colouration. The groundwater, although initially "fresh", would probably become highly saline upon entering the groundwater system of the clastic basin as is indicated by early gypsum cementation. It is therefore postulated that base metal ions would form stable metal chloride complexes at an early stage of burial.

As the embayments subsided, the high density pore water brine is believed to have remained dominantly within the clastic aquifer of the Redstone River Formation with only limited mixing with sea water.

Loss of pore-fluid from the clastic redbed sediments is believed to have been dominantly through evaporation. Base metals would thus continue to concentrate as chloride complexes within the storage system of the saline groundwater of the clastic embayment.

As the shallow marine incursions inundated the embayments, highly organic algal mat sediments were deposited. The organic-rich mats provided a favourable environment for sulphate-reducing, anaerobic bacteria, and strongly reducing sediments were formed. These were to act as the "chemical trap". The nature of the reductant generated within the decaying algal mat sediments is uncertain but is inferred to be H_2S generated by anaerobic bacteria. Sulphur isotope analyses are considered too few to be of significant value at this time in understanding the genesis of the copper. Two analyses of anhydrite from the Upper Redstone River Formation yielded δS^{34} values of +13.9 and +19.0 per mil. Values from sulphide grains collected at the June Creek deposit were −8.4, −7.9, and −1.1 per mil.

Renewed clastic sedimentation buried the algal mats, before they became oxidized, and thereby formed strongly reducing lithosomes within the otherwise oxidizing flood plain facies sediments. As the reducing algal mat sediments became buried beneath the water table, metal-chloride complexes contained within the groundwater storage system started to enter the decaying organic-rich sediments. Sulphides (pyrite?) began to precipitate as the chloride complexes interacted with the available reductant though the amount of sulphide precipitated at this stage was probably only minor. The amount of sulphide that may have precipitated during any upward movement of metal-bearing groundwater through the reducing sebkha sediments due to evaporative pumping action as described by Renfro (1974) is not believed to have been appreciable. Two reasons for this conclusion based primarily on deposits in the Coates Lake embayment are: 1) the reduced zones around the copper-bearing sebkha sediments contain no appreciable amounts of sulphide, and 2) pyrite should have been deposited in much greater concentrations than is indicated in Bed 1. A third, and possibly more important reason, is that the sulphide zoning encompasses a number of beds rather than a single bed, thereby indicating that a considerable thickness of sediments must have accumulated prior to the deposition of most of the copper (R. Kirkham, pers. commun., 1979).

In order to form a mineable deposit, it is believed imperative that reducing sediments be deposited over a metal-bearing groundwater system in such a manner that the hydrologic regime is not altered sufficiently to drive out the metal chloride complexes before a chemical trap is formed. Deposition of non-oxidized algal mat sediments by very shallow marine flooding meets this requirement, as would probably a euxenic epeiric sea. The driving mechanism believed responsible for ultimately expelling the metal-bearing groundwater from the clastic reservoir is compaction related to the abrupt and extensive transgression in Middle Coppercap time. As the Middle Coppercap Sea transgressed the embayments, the hydrologic regime would have been abruptly altered and metal chloride complexes within the saline groundwater of the clastic basins would be forced upward and landward (Fig. 8).

Within the Coates Lake embayment the reducing algal mats of the Transition Zone would have acted as chemical filters, effectively removing metals from solution. Metal chloride complexes would become reduced in the presence of the available reductants according to their solubilities. Because the copper complexes would have been the least soluble, iron-free copper sulphides would be deposited initially, forming the chalcocite-digenite-rich concentrations of Bed 1. Iron complexes would be reduced further along the flow direction, and the zonation from chalcocite, through bornite, to chalcopyrite and pyrite would develop in both a vertical and basin-margin direction.

In the Keele area evaporites of the playa lake facies acted as local permeability barriers. Fluid flow was preferentially channelled by the barriers, and the metal-bearing groundwaters entered the overlying chemically receptive reducing algal mat and intertidal sediments at points where the evaporite unit thinned, i.e., where the permeability barriers became less effective. Based on this model, it is therefore envisioned that the copper deposits formed near the margins of the former evaporitic lakes in the uppermost part of the playa lake facies (Fig. 12). Although faulting, contemporaneous with mineralization, is not evident at the Jay and June Creek deposits, the influence of any local faulting in breaching the permeability barriers to form deposits cannot be ruled out. The concept of mineral-bearing waters being preferentially channelled along breaks in permeability barriers is consistent with the discordant nature and type of zoning recognized within the deposits.

The mineral-bearing groundwater is believed to have contained a high magnesium ion concentration, and to have been oxidizing or at most only weakly reducing. This is based on the alteration halo of fine-grained dolomite associated with mineralization and the absence of any large reduced zones in the red clastic beds of the Redstone River Formation at Keele.

In summary, the main elements of the metallogenic model are:

1) Metal ions were liberated primarily from volcanic detritus (source sediment) during the development of redbeds. The metal ions entered the highly saline groundwater system within the subsiding clastic basins where they then combined to form metal chloride complexes.

2) Periodic marine flooding allowed deposition of reducing chemical trap sediments (host rock) without significantly disrupting the hydrologic regime of the groundwater storage system.

3) Compaction accompanying an abrupt and extensive transgression in Middle Coppercap time forced the high density, metal-bearing groundwater upward and landward, towards the margins of the embayments.

4) The metal-bearing water passed first through the chemically reactive, reducing, intercalated coastal sebkha lithosomes in the more marine parts of the embayments. Because there were only thin evaporite crusts associated with the sebkha cycles, essentially no permeability barriers existed. The mineralizing fluids were therefore able to pass uniformly through the chemical filters, but in these the metal chloride complexes broke down in the presence of the available reductant. Sulphides were precipitated to form the laterally extensive, zoned, sheet-like deposits of the "Coates Lake type".

5) Continued transgression forced the wedge of metal-bearing saline groundwater towards the margins of the embayments and as the groundwater then entered the playa lake facies belt, it became preferentially channelled by permeability barriers that developed within the playa lake evaporites. Where the water could breach the permeability barriers and enter the overlying chemically reactive sediments, sulphides were precipitated, and deposits of the "Jay type" were formed.

In essence then, all significant deposits of the copper belt are related to processes of normal, albeit specialized, basin development. They are related spatially to a contact marking a shallow marine transgression but more particularly they are related in time to a slightly younger major marine transgression. The deposits are therefore diagenetic in age as opposed to syngenetic. The model suggests that if extensive intercalated coastal sebkha beds exist in the more marine extensions of the Keele River and Hayhook Lake embayments, they might be mineralized in a similar style to those at Coates Lake. However, since the transgressive contact may be significantly different from one embayment to the other, it remains impossible to predict if deeply buried, mineralized, intercalated sebkha cycles exist in embayments other than at Coates Lake.

Recent studies of a number of stratiform copper deposits throughout the world strongly support the possibility of the diagenetic additions of potential ore-bearing metals (Cu, Pb, Zn, Co) to favourable strata during the formation of such deposits (Brown, 1978). As implied by the above metallogenic model, copper deposits of the Redstone Copper Belt appear to be related to similar diagenetic additions as described by Brown. Because of the essentially non-metamorphosed and only weakly deformed nature of the Redstone Copper Belt, further studies of it may lead to a better understanding of the origin of stratiform copper deposits.

ACKNOWLEDGEMENTS

Thanks are extended to all those who have contributed field data and discussions on various aspects of the Redstone Copper Belt. Significant early investigation of deposits in the Coates Lake embayment were made by the late Dr. J. Coates with Redstone Resources Inc, and by I. Watson, formerly with Cerro Mining Co. of Canada Ltd.

Studies of the Hayhook Lake embayment were made by staff of Rio Tinto Canadian Exploration who kindly made drill core and company reports available.

Exploration drilling and geological studies within the Keele River and Coates Lake embayments were conducted by Shell Canada staff under the supervision of J. Brander. Comprehensive regional structural and stratigraphic studies were made by Dr. I. Bond, A. Hildebrand, R. Jamieson, R. Moore, P. Ogryzlo, and the writer. A number of mineral showings were discovered through prospecting by E. Scholtes. In addition to Shell Canada's regional studies of the Redstone Copper Belt, structural and stratigraphic studies were carried out by Dr. G. Eisbacher of the Geological Survey of Canada, and by C. Jefferson with the Department of Indian and Northern Affairs, Yellowknife. Dr. R. Kirkham of the Geological Survey of Canada provided valuable discussions on the metallogenesis of the copper mineralization.

Special thanks are extended to Dr. A. B. Baldwin of Shell Canada Resources Ltd. for providing the stimulus behind the recent exploration activity, and to Shell Canada Resources Ltd. for permission to publish this paper. Special thanks are also extended to Cheryl Danby and Rosalind Gerlitz for typing the manuscript.

The manuscript for this paper was critically reviewed by Dr. A. C. Brown (University of Montreal), Dr. R. V. Kirkham (Geological Survey of Canada) and L. Eliuk (Shell Canada Resources Ltd.). Many valuable suggestions and criticisms were received and are gratefully acknowledged.

REFERENCES

Aitken, J.D., 1980, Sedimentation of the Helikian Little Dal Group, Mackenzie Mountains: Geological Association of Canada, Program with Abstracts, v. 5, p 40.

Aitken, J.D., Macqueen, R.W., and Usher, J.L., 1973, Reconnaissance Studies of Proterozoic and Cambrian Stratigraphy, Lower Mackenzie River Area (operation Norman), District of Mackenzie: Geological Survey of Canada Paper 73-9, 178 p.

Aitken, J.D. and Cook, D.G., 1974, Effect of Antecedent Faults on "Laramide" Structure, Mackenzie Arc: Geological Survey of Canada Paper 74-1, Part B, p. 259-263.

Aitken, J.D. and Long, D.G.F., 1978, Mackenzie tectonic arc – reflection of early basin configuration?: Geology, V. 6, p. 626-629.

Aitken, J.D., Long, D.G.F., and Semikhatov, M.A., 1978a, Progress in Helikian stratigraphy, Mackenzie Mountains: Geological Survey of Canada Paper 78-1A, Current Research Part A, p. 481-484.

—————, 1978b, Correlation of Helikian strata, Mackenzie Mountains – Brock Inlier – Victoria Island: Geological Survey of Canada Paper 78-1A, Current Research Part A, p. 485-486.

Bartholome, P., 1972, Diagenetic ore-forming processes at Kamoto, Katanga, Republic of the Congo: in Amstutz, G.C., and Bernard, A.J., eds., Ores in Sediments: Berlin, Springer, p. 21-41.

Bond, I.J., Jamieson, R.A., Ruelle, J.C. and Brander, J.M., 1976, Keele River Project, Mackenzie Mountains, N.W.T., Permit no. 361. Report on geological investigations and drilling completed during 1975 and 1976: Shell Canada Resources Ltd. Unpublished Report.

Brown, A.C., 1978, Stratiform copper deposits – Evidence for their post-sedimentary origin: Minerals, Science and Engineering, v. 10, no. 3, July, p. 172-181.

Coates, J.A., 1964, The Redstone Bedded Copper Deposit and a Discussion on the Origin of Red Bed Copper Deposits: Unpublished M.Sc. Thesis, University of British Columbia.

Eisbacher, G.H., 1977, Tectono-stratigraphic framework of the Redstone Copper Belt, District of Mackenzie: Geological Survey of Canada Paper 77-1A, Report of Activities Part A, p. 229-234.

—————, 1978, Re-definition and Subdivision of the Rapitan Group, Mackenzie Mountains: Geological Survey of Canada Paper 77-35, 21 p.

Friday, S., 1978, Copper Mineralization in the Keele River – Backbone Areas: Unpublished B. Sc. Thesis, Queen's University.

Gabrielse, H., 1967, Tectonic evolution of the northern Canadian Cordillera: Canadian Journal of Earth Sciences, v. 4, p. 271-298.

Gabrielse, H., Blusson, S.L., and Roddick, J.A., 1973, Geology of Flat River, Glacier Lake and Wrigley Lake Map-Areas, District of Mackenzie and Yukon Territory: Geological Survey of Canada Memoir 366, 153 p.

Garlick, W.G., 1961, The syngenetic theory: in Mendelson, F., ed., The Geology of the Northern Rhodesian Copperbelt: London, MacDonald, p. 146-165.

Glennie, K.W., 1970, Desert Sedimentary Environments: Developments In Sedimentology 14, Amsterdam, Elsevier, 222 p.

Hildebrand, A.R., 1978, Redstone Project Progress Report March, 1978, Drill Results as of August, 1977, and recommendations for the 1978 Field Season: Shell Canada Resources Ltd. Unpublished Report.

Jefferson, C.W., 1979, Stratigraphy and Sedimentology, Upper Proterozoic Redstone Copper Belt, Mackenzie Mountains, N.W.T., – a Preliminary Report: in Mineral Industry Report 1975, Northwest Territories, Indian and Northern Affairs E.G.S. 1979.

—————, 1980, Stratigraphic and sedimentologic history of the Late Proterozoic Redstone Copper Belt, Mackenzie Mountains, N.W.T.: Geological Association of Canada, Program with Abstracts, v. 5, p. 64.

Johnson, K.S., and Croy, R.L., eds., 1976, Stratiform Copper Deposits of the Midcontinent Region, A Symposium: Oklahoma Geological Survey Circular 77.

Jung, W., and Knitzschke, G., 1975, Kupferschiefer in the German Democratic Republic (G.D.R.) with special reference to the Kupferschiefer deposit in the southern Harz Foreland: in Wolf, K.H., ed., Handbook of Strata-Bound and Stratiform Ore Deposits: Amsterdam, Elsevier, p. 253-406.

Kendall, A.C., 1978, Facies Models 11. Continental and supratidal (sabkha) evaporites: Geoscience Canada, v. 5, p. 66-78.

Kirkham, R.V., 1973, Environments of formation of concordant and peneconcordant copper deposits in sedimentary sequences (Abst.): Canadian Mineralogist, v. 12, Part 2, p. 145-146.

Renfro, A.R., 1974, Genesis of evaporite – Associated stratiform metalliferous deposits: A sabkha process: Economic Geology, v. 69, p. 33-45.

Ruelle, J.C. and Bond, I.J., 1976, Geological assessment report on Backbone Project claims south and west of Permit No. 361, and extending from Keele River to Coates Lake, N.W.T.,: Assessment report filed with Department of Indian Affairs and Northern Development, Yellowknife, N.W.T.

Walker, T.R., 1967a, Formation of red beds in modern and ancient deserts: Geological Society of America Bulletin, v. 78, p. 353-368.

_______________, 1967b, Colour of Recent sediments in tropical Mexico – a contribution to the origin of red beds: Geological Society of America Bulletin, v. 78, p. 917-920.

Watson, I.M., 1971, Geological compilation report for Redstone Mines Ltd., Coates Lake Property, N.W.T.: Unpublished Report to Redstone Mines Ltd., November, 1971.

Young, G.M., 1977, Stratigraphic correlation of Upper Proterozoic rocks of northwestern Canada: Canadian Journal of Earth Sciences, v. 14, p. 1771-1787.

Young, G.M., Jefferson, C.W., Long, D.G.F., Delaney, G.D., and Yeo, G.M., 1977, Upper Proterozoic stratigraphy of northwestern Canada and Precambrian history of North American Cordillera: Society of Economic Geologists, Coeur d'Alene Field Conference, Wallace, Idaho, Nov. 3-5, 1977, Idaho Bureau of Mines.

MS received June 1979;
Accepted for publication April 1981.

Precambrian Sulphide Deposits, H.S. Robinson Memorial Volume, edited by R.W. Hutchinson, C.D. Spence and J.M. Franklin, Geological Association of Canada Special Paper 25, 1982

GEOLOGY OF THE NANISIVIK ZINC-LEAD DEPOSIT

R.H. Clayton
Watts, Griffis and McOuat Limited, Consulting Geologists and Engineers, 159 Bay Street, Toronto, Ontario M5J 1J7
L. Thorpe
Southern Sphere Mining and Development Co., P.O. Box 50065, Randburg, South Africa 2125

ABSTRACT

The Nanisivik zinc-lead deposit near the northern end of Baffin Island consists of massive sulphides within dolomite of the Society Cliffs Formation of Helikian age. Pyrite is the main mineral but in places sphalerite and galena are present in ore-grade proportions.

Most of the known sulphides are in horizontal lenses about 10m thick. The Main Orebody is 3,000m long and, on average, about 100m wide. There are also sulphides below the main lenses, either in horizontal lenses or in near-vertical vein-like structures.

The dolomite has been extensively block-faulted. Sulphide bodies pass laterally through some of these faults with little or no displacement and appear to be confined to the upper half of the formation which is mainly laminated algal dolomite that has been extensively brecciated and recrystallized.

A two-stage mode of genesis is postulated, consisting of cavity formation by meteoric water followed very closely by mineral deposition from hot water under moderate pressure.

RÉSUMÉ

Le gîte de zinc-plomb de Nanisivik est situé dans la partie septentrionale de l'Ile de Baffin et est constitué de sulfures massifs remplaçant la dolomie de la formation de Society Cliff d'âge Hélikien. La pyrite est le principal minéral mais par endroits la sphalérite et la galène sont présentes en teneurs de minerai.

La plupart des sulfures se trouvent sous la forme de lentilles horizontales d'environ dix mètres d'épaisseur. Le gisement principal est sinueux, d'une longueur de 3,000m et d'une largeur moyenne de 100m. Il y a aussi des dépôts de sulfures sous les lentilles principales, certains sous la forme de filons presque verticaux, d'autres de lentilles horizontales.

La dolomie a été fortement morcellée par des failles. Les lentilles de sulfures traversent latéralement certaines des failles avec peu ou pas de déplacement et sont apparemment restreintes à la moitié supérieure de la formation de Society Cliff qui est surtout constituée de couches d'algues dolomitiques maintenant bréchiformes et recristallisées.

Une séquence en deux étapes est proposée pour la formation des gisements: développement de cavités par les eaux météoriques suivie de près par une précipitation de minéraux provenant de solutions chaudes sous des pressions modérées.

INTRODUCTION

The Nanisivik zinc-lead-silver deposit is situated near the north end of Baffin Island, N.W.T. on the south shore of Strathcona Sound, about 20km east of Arctic Bay. The Main Orebody is 2km from the shore, at an elevation of 300m.

The mine, which started production in 1976, is worked by adits into the west end of the orebody. The concentrator has a capacity of 1,750 tons per day. There is a townsite with a population of 380, of whom 190 work at the mine, the others being government employees and families. The marine terminal is capable of handling ships of up to 45,000 tons and an airport between the site and Arctic Bay is capable of handling scheduled jet aircraft.

HISTORY

The Canadian Coastguard vessel "Arctic" wintered at Arctic Bay in 1910-1911. A prospector on board, Arthur English, reported numerous mineral occurrences in the area, no doubt including some of the sulphide outcrops at Nanisivik, but his descriptions are too vague for positive identification. In 1937, two prospectors, J.F. Tibbitt and F. McInnes, reached the site by dog team and staked a claim on the north end of the South Pyrite body. No work was done and the claim subsequently lapsed.

In 1954, the Geological Survey of Canada carried out a field investigation and the sulphides in the Ocean View area were noted by R.G. Blackadar, who published his findings in 1956. Prompted by this, the area was visited by R.D. Mollison and W. Holyk who staked the Ocean View showing and the east end of the Main Orebody for Texasgulf Inc. in 1957. In 1958 a six-man party spent the summer in the area. All the sulphide outcrops at Nanisivik were mapped, as well as two sulphide and gossan showings 16km and 32km to the east. The start of work was delayed when the supply DC-3 aircraft flew into Mt. Podolsky, south of the head of Adams Sound. All four occupants survived, mainly through the efforts of the geophysicist, George Podolsky, who with only two broken ankles, had sustained the least injuries. A major drilling programme was carried out in 1962 and 1963 and the major sulphide bodies now known were outlined at that time.

In 1972 Mineral Resources International Limited of Calgary obtained agreement from Texasgulf to carry out a feasibility study with a view to putting the property into production. In 1974 Nanisivik Mines Limited was formed, with Mineral Resources International, Metallgesellschaft A. G. of West Germany, Billiton International Metals B. V. of the Netherlands and the Government of Canada as participants.

About 600 diamond holes have now been drilled, totalling over 35,000m. In addition, a large number of shorter holes have been drilled for mine grade control. The region has been mapped by the Geological Survey of Canada at scale of 1:253,440 (Blackadar, 1956, 1965 and 1970; and Lemon and Blackadar, 1963).

REGIONAL GEOLOGY

Most of Baffin Island is underlain by crystalline basement rocks of Aphebian (Lower Proterozoic) age. In the north part of the island a depression in this basement, across Borden Peninsula (Fig. 1) is filled mainly with sedimentary rocks of Helikian-Hadrynian (Middle-Upper Proterozoic) age. In places these are overlain by Palaeozoic sedimentary rocks (mainly Upper Ordovician and Silurian) which also lie directly on the Aphebian basement north and south of the depression.

The long axis of the area underlain by Middle and Upper Proterozoic rocks has a west-northwesterly trend. In the eastern part, from south of Pond Inlet, the zone is about 30km wide with roughly parallel sides, and appears to have the form of a graben. Westward, across Borden Peninsula, the zone broadens to 120km from north to south at its western end on the east coast of Admiralty Inlet, giving the western part a roughly triangular shape. West-northwesterly trending faults, fissures and dykes extend across the entire area. West of Admiralty Inlet Brodeur Peninsula is underlain by sedimentary rocks of Palaeozoic age in which there is no surface evidence of the west-northwesterly structural trend, although this might exist at depth.

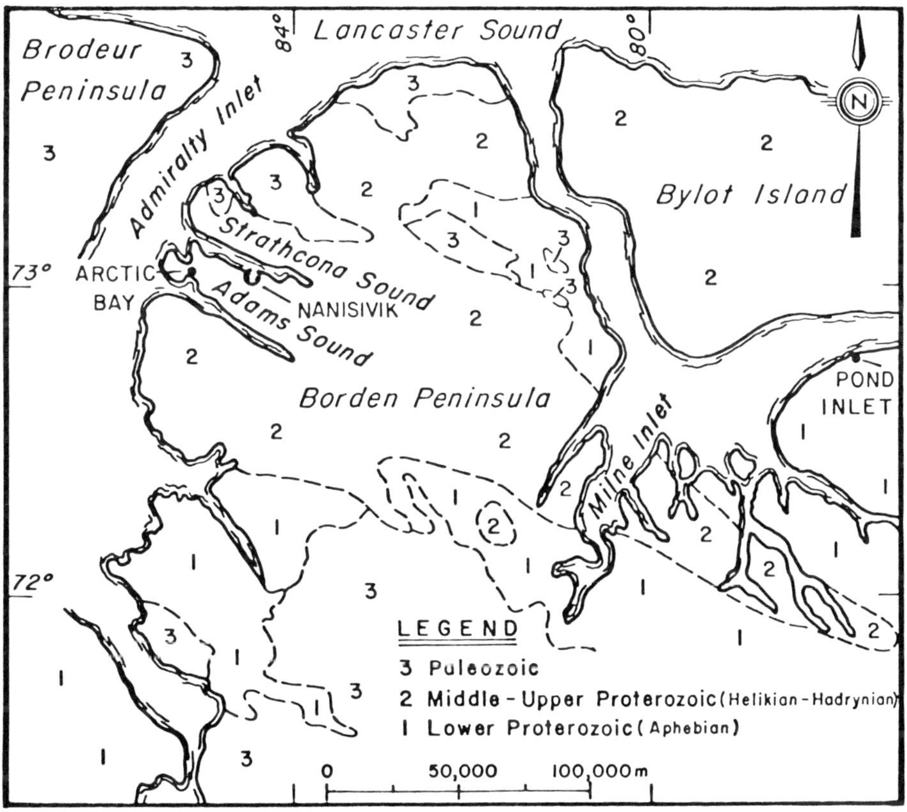

Figure 1. Location and Regional Geology.

REGIONAL STRATIGRAPHY

The Table of Formations (Table I) is after Blackadar (1956), but has been modified to reflect conditions in the Nanisivik area, where the Society Cliffs and Victor Bay Formations are thicker and the lower Victor Bay is more shaley than Blackadar's type sections near Arctic Bay. The massive sulphides which occur at Nanisivik, including all the ore, are within the Society Cliffs Formation, or at its contact with the overlying Victor Bay Formation.

The formations of the Uluksan Group appear to be a conformable sequence at both Nanisivik and Arctic Bay (Lemon and Blackadar, 1963), and are shown as such in Table I. However, angular disconformities can be seen between the formations in a number of locations farther east on Borden Peninsula, and Geldsetzer (1973a) considered that there were substantial hiatal periods separating the principle units. The nature of the Victor Bay – Society Cliffs contact is of particular interest and is discussed in some detail later. Descriptions of the four formations encountered at Nanisivik are as follows.

Adams Sound Formation

This formation consists of about 1300m of light orange to reddish brown quartzite, made up of subangular to subrounded quartz grains with minor sericite and chlorite (Blackadar, 1970). The coloration is the result of fine-grained iron oxides. Though some thin intercalations of shale are reported in other areas, at Nanisivik the formation consists entirely of quartzite.

Arctic Bay Formation

The type section at Arctic Bay consists of 110m of black, fissile, finely laminated medium-to-fine argillaceous limestone and dolomite (Blackadar, 1970). In central and eastern Borden Peninsula it is much thicker, reaching over 700m in thickness, and less uniform. Between Nanisivik and Adams Sound the unit is considerably more argillaceous than at Arctic Bay. Disseminated pyrite is common, and farther east on Borden Peninsula, there is disseminated galena. At Nanisivik no Arctic Bay rocks are exposed, but have been encountered in a diamond drill hole drilled from underground. The thickness of the formation at Nanisivik is unknown.

Society Cliffs Formation

All the significant known mineralization is within the Society Cliffs Formation. Where mineralization occurs at the contact with the overlying Victor Bay Formation, it is believed that the mineralization is emplaced only in the Society Cliffs rocks.

At the type section at Arctic Bay, 20km to the west of Nanisivik, the Society Cliffs Formation is 275 to 300m thick but the thickness at Nanisivik is not known accurately. Only one hole drilled from underground has passed through the lower contact, and this is below an outcrop of Society Cliffs Formation. The minimum thickness at this point is therefore 516m, measured from the surface exposure. Allowing for some erosion, the thickness was probably at least 520m and possibly 600m. The contact with the Arctic Bay Formation at this point is 300m below the lower margin of the Main Orebody.

TABLE I
TABLE OF FORMATIONS

Age	Group	Formation	Symbol in this paper	Thickness Metres	Lithology
Middle Ordovician	Admiralty	Gallery	–	180	sandstone
		Slight Angular Unconformity			
					gabbro dykes
		Intrusive Contact			
Middle Upper Proterozoic (Helikian-Hadrynian)	Uluksan	Elwin	–	750	sandstone siltstone shale
		Strathcona Sound	–	750	mudstone, siltstone,
		Victor Bay	Hvb	180-300	silty dolomite, dolomite, minor limestone shaley dolomite
		Society Cliffs	Hsc	275-550	dolomite
		Arctic Bay	Hab	100+	dolomitic shale
	Eqaluik	Adams Sound	Has	1,200+	quartzite
		Nauyat	–	300	basalt
		Lower Quartzite	–	0-15	quartzite
		Angular Unconformity			
Lower Proterozoic (Aphebian)			–		gneiss, granite, schist

The Society Cliffs Formation consists of dolomite, ranging from light grey to almost black. Weathered surfaces are generally brown. Laminated dolomite predominates, especially in the upper part of the formation. The laminations are typically about 1mm thick and, although much of the fresh rock, such as that in drill core, appears to be massive (i.e., unlaminated), weathering of surface outcrops emphasizes the lammelar structure. Argillaceous dolomite with no laminations also occurs throughout and this tends to be light grey, especially in the upper part of the formation, but dark grey to almost black sections also exist. The laminations tend to be undulose and commonly have the appearance of algal laminations, but the extensive brecciation and small scale faulting and folding, which are probably incipient brecciation, have made it difficult to identify algal structures positively at Nanisivik. Some well-defined algal bioherms, however, can be seen in the Society Cliffs Formation in parts of Borden Peninsula where brecciation is less intense. Furthermore, the entire formation, especially the laminated dolomite, has a high sulphur content and on breaking yields a strong odour of hydrogen sulphide mixed with carbon disulphide. These gases are trapped in the fabric of the rock, and not associated with visible cavities. There seems to be little doubt, therefore, that most, if not all, of the laminated dolomite is of algal origin. Less common, noted mainly above and below the west central part of the Main Orebody, but not directly associated with it, is dolomitic mudstone with white angular to subrounded clasts (0.5 to 2.0mm), and dolomitic mudstone with larger, dolomite clasts. Their formation is ascribed to lag deposition in intertidal channels. In other areas the base of the formation consists of a fine dolomitic conglomerate, but this has only been seen at Nanisivik as float.

Evaporites, mainly gypsum, form substantial beds near Milne Inlet, 150km to the east, but do not occur at Nanisivik, where gypsum is seen only occasionally as filling in small cavities, in thin beds a few millimetres thick, and as an alteration product associated with sulphide outcrops.

The Society Cliffs rocks have been extensively brecciated. Mosiac, crackle and rubble breccia are all common. The fragments, which are typically 5 to 20cm across and angular, are cemented by white crystalline dolomite. The brecciation occurs throughout (though only one hole has so far penetrated the lower third of the formation) but is irregularly distributed and no pattern has been discerned. The cause of the brecciation is not clear but it is probably the result of collapse. If due to solution of evaporites there must have been several evaporite beds to account for the vertical extent of the brecciation, and the evaporite material must, as it is now absent, have been removed in its entirety. A possible explanation is that domal uplift caused zones of brecciation parallel to the bedding.

Breccias near the upper contact of the Society Cliffs Formation do not contain any Victor Bay material near the upper contact, implying that solution, collapse and cementation by white dolomite all took place before the start of deposition of Victor Bay sediments. Geldsetzer (1973b) cited this as evidence of a hiatal period between deposition of the Society Cliffs and Victor Bay Formations.

Crystalline dolomite, in addition to cemented breccia, is common throughout the formation as stringers a few millimetres wide, and as vug fillings. Most commonly the vugs are completely filled, but are occasionally only lined with sparry dolomite. Pyrite occurs in minor amounts in both vugs and stringers; galena and sphalerite are very rare in either.

Most of the formation has undergone recrystallization to a fine-grained rock. Locally the recrystallization has been more intense, ranging from a sugary texture to marble. The marble usually has an irregular network of dark lines, typically 5 to 10cm apart, with the intervening recrystallized material having a faintly purple tinge. Some marmorized zones may be correlated through two or three drill holes above the Main Orebody, but rarely across its full extent. These zones are irregular but tend to have a dip of 10° to 20° to the north across the Main Orebody (about the same as the apparent dip of the formation), whereas the orebody is nearly horizontal. In other zones, where there is generally less drilling, no spatial pattern to the recrystallization has been discerned.

Victor Bay Formation

The type section of this unit at Victor Bay, about 20km west from Nanisivik, is about 160m thick (Blackadar, 1970). No measured section is available at Nanisivik, but from topographic contours it appears to be thicker, perhaps as much as 300m. Farther east, across Borden Peninsula, it thickens considerably and two sections here each measure about 500m.

At Nanisivik the Victor Bay Formation may be divided into three members. The lower member has not been measured but its thickness at some locations is at least 150m. It consists of black, fissile, thinly bedded argillaceous dolomite, interspersed with thicker, less argillaceous bands of dolomite up to tens of centimetres thick, in a sequence known as the Victor Bay Shale. Disseminated pyrite is common, sometimes occurring as radially structured spherical concretions up to 2cm in diameter. Above the basal shale member is the middle member consisting of one or more beds of massive light grey dolomite, separated by thin shaly partings, with a total thickness of about 70m. This is apparently overlain by a softer upper member about 20m thick occuring at Nanisivik only on the top of Shale Hill, where it is covered by overburden.

The carbonate within the Victor Bay Formation is mainly dolomite, though lime-rich sections occur, especially in the middle member. This contrasts with the Society Cliffs Formation in which no limestone has been observed.

Gabbro Dykes

Gabbro dykes are widespread on Borden Peninsula. They are generally medium-grained and holocrystalline. The feldspar is typically andesine and the principal ferromagnesian mineral is augite (Blackadar, 1956). Pyrite is common, sometimes being the major constituent in pockets up to a metre in width. Minor chalcopyrite also occurs.

The dykes are in two principle directions (Fig. 2), northwest and a little west of north, with the former predominating. There are four major dykes on the Nanisivik property: the most central two having a northerly trend and the east- and the westernmost northwesterly trends. They vary from 10 to 50m in width. Three other dykes appear to be discontinuous.

The dykes cut all Precambrian formations but not those of Palaeozoic age, the oldest of which is the Turner Cliffs Formation of Upper Ordovician age. The dykes are tentatively assigned a Cambrian age, but could be Hadrynian. A narrow zone of contact metamorphism about one metre wide is found alongside the dykes. Where quartzite of the Adams Sound Formation is intruded it is bleached. Where dykes cut dolomite it is marmorized and commonly bleached. Where dykes cut sulphide lenses a soft material consisting of dyke material with substantial amounts of disseminated pyrite is formed.

LOCAL STRATIGRAPHY

The geology at surface in the Nanisivik Mine area is relatively simple. Dolomite of the Society Cliffs Formation underlies the whole area between Strathcona Sound and the South Boundary Fault (Fig. 2). Locally only, these dolomites are overlain by rocks of the Victor Bay Formation, and cut by gabbro dykes. Apart from some minor Strathcona Sound Formation remnants near the coast and some exposures of the Gallery and Elwin Formations on a hill in the southwest corner of the property, no other rocks appear at the surface.

Victor Bay rocks overlie the Society Cliffs Formation west of the West Boundary Fault, at Shale Hill and east of Kuhulu Lake. They also occur overlying downfaulted blocks of Society Cliffs rocks, and underlie the bottoms of valleys and so appear to underlie the adjoining upthrust blocks of Society Cliffs dolomite. This is particularly apparent along the Keystone Graben, in the Main Draw, and west of

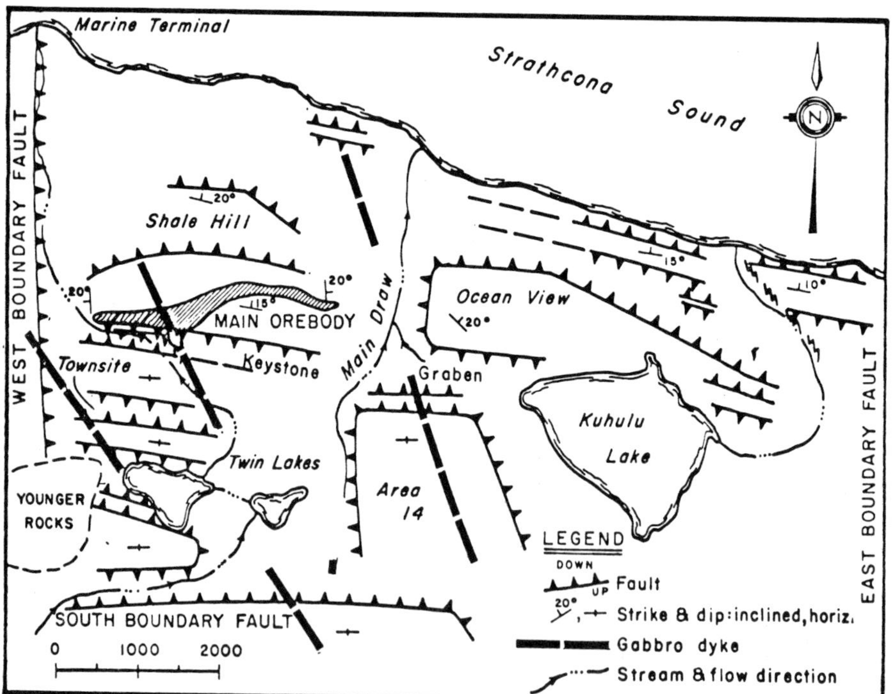

Figure 2. Nanisivik Area showing main geographic features caused by block faulting.

Area 14. Formerly, it was believed that these Victor Bay shales were Arctic Bay Formation, but a number of holes drilled in the Keystone Graben pass downwards from the shale into Society Cliffs dolomite.

STRUCTURE

The main structure in the area is a large graben with superimposed subsidiary horsts and grabens. The central graben of this structure is the centre of Strathcona Sound, and it has a west-northwesterly trend (Fig. 2). This horst and graben system is transected and broken into blocks by northerly trending faults. The southern limit of the main graben structure is marked by the major westward trending fault, the South Boundary Fault, about 6km south of the Sound. South of the fault quartzite of the Adams Sound Formation forms a plateau 750m above sea level. Between the South Boundary Fault and the shore a series of horsts and grabens cause an overall downward displacement towards the Sound. The topography of the sea floor suggests that this structure continues, with at least one submarine horst, to the middle of the Sound, where the water depth is 325m. The net vertical movement from the South Boundary Fault to the middle of the Sound is about 1,800m, excluding the thickness of any erosion of the quartzite. North of Strathcona Sound the Proterozoic sequence is masked by a thick younger cover. The area of interest at Nanisivik is from the Sound southwards to the South Boundary Fault and 10km east to west between the West and East Boundary Faults.

Within this 6km by 10km block of Society Cliffs Formation the main structural feature is the Keystone Graben, lying 2.5km from the shore and appearing to extend from the West Boundary to the East Boundary Fault. East of Kuhula Lake, however, increasing cover by younger rocks obscures the eastern end of the graben. The Keystone Graben trends almost due east-west and is 450 to 600m wide. It involves at least four faults; there are benches on both the north and south flanks. In the central part of the graben the Society Cliffs Formation is, for the most part, overlain by shale of the Victor Bay Formation. The vertical distance from this contact to the highest Society Cliffs exposures on the north rim of the graben is 180m, which is probably about the total vertical displacement. North of the graben are two horsts of Society Cliffs dolomite, the Main Orebody and the Ocean View blocks. Virtually all of the known ore is in these two blocks. They are separated by a north-south down-faulted block which underlies the Main Draw and is here composed of shale resembling the Victor Bay shale. On the south side of the Keystone Graben a horst of dolomite (Area 14) lies opposite the Main Draw and the western end of the Ocean View block and extends almost to the South Boundary Fault, from which it is separated by a narrow graben. West of Area 14 there is a down-faulted block which also has Victor Bay shale at the surface and extends to the South Boundary Fault. In the west part, the area south of the Keystone Graben is more complex, there being at least three horsts and three grabens between it and the South Boundary Fault.

North of the Ocean View block the structure is relatively simple. There is a graben, a horst and a sharp drop into Strathcona Sound. There is evidence of a west-northwesterly trending fault immediately south of and parallel to the Main Orebody extending about 100m from its eastern outcrop.

The area east, west and north of Shale Hill is covered by drift and underlain for the most part by Victor Bay Shale and not much evidence of structural features can be seen at surface. However, considerable drilling has been done in this area and it appears that there are a large number of small blocks of dolomite at various elevations. From this observation it may be inferred that the structure in other, less drilled areas may be more complex than presented.

The extensive brecciation, minor faulting, and the fact that most laminations appear of algal origin make individual measurements of strike and dip in the Society Cliffs Formation generally meaningless, while Victor Bay shale is soft and usually slumped. Some idea of overall dips can, however be gleaned by viewing the cliff faces of various blocks from a distance. Most appear to be 0 to 15° to the north and this dip is to some extent confirmed by tentative correlations of rock types, brecciation or marmorization possible between some drill holes. The western part of the Main Orebody block dips up to and possibly more than 20° to the west and it is possible that this down-warping of the Society Cliffs Formation might account for some of the 300m displacement ascribed to the West Boundary Fault. Area 14 is generally horizontal, as are the various blocks farther west, as far as can be seen. These various attitudes imply that the site of the Keystone Graben was the centre of an elongated dome tilted slightly to the north.

Elucidation of the sequence of structural events presents problems. The contact between the Society Cliffs and Victor Bay Formations always appears conformable at Nanisivik but there are indications that at least part of the main horst and graben faulting took place during the depositional hiatus between the two. Problems relate mainly to the thickness of the shale member of the Victor Bay Formation. In the central part of the Keystone Graben the thickness appears to be about 170m, of which 140m has been tested by drilling. West of West Twin Lake the apparent thickness is 90m, while in the west side of the West Boundary Fault, almost opposite the Main Orebody, a horst of dolomite underlies a horizontally lineated shale and the apparent thickness of the shale ranges from 60m above the horst to at least 75m on either side. It is therefore possible that the main faulting took place during the Victor Bay-Society Cliffs hiatal period and that the Victor Bay shale was laid down on a topography quite similar to that of the present, but nowhere is a known upper contact seen immediately above a known lower contact. The shale is furthermore so soft that faults which might cause the apparent differences in thickness could be obscured. Conversely the underground contact of the shale with the north fault of the Keystone Graben gives every appearance of being a fault contact. The shale against the dolomite dips as much as 70° to the north, flattening out to near-horizontal a few metres from the fault. This might be explained by compaction of the shale.

MINERAL DEPOSITS

The main sulphide mineral at Nanisivik is pyrite. Most of the sulphide bodies contain only pyrite (sometimes with minor marcasite), but some contain varying amounts of sphalerite and galena. In some zones the proportion of these minerals is high enough to be of economic interest. No other sulphide minerals occur in significant quantities, and, except in oxidized zones, the only non-sulphide minerals are crystalline dolomite with minor calcite and quartz. The sulphides are almost always massive; although pyrite, sphalerite and galena do occur in disseminated form such occurrences form less than 1% of the total. Contacts of the deposits are invariably sharp and, except along the gabbro dykes, there is no indication of any metamorphism.

The total amount of sulphides present is not known, but is estimated to be about 50,000,000 tonnes, most of it occuring in three bodies of roughly equal size, the North Pyrite, the Main Orebody and the South Pyrite. There are three smaller zones, known to contain ore-grade material, and a number of small pyrite showings. A plan of the sulphide deposits is shown in Figure 3, and several simplified cross-sections of the main deposits are shown in Figures 4 and 5.

The Main Orebody is estimated to contain 11,000,000 tonnes of sulphides of which about 6,500,000 tonnes are ore-grade, averaging about 12.0% Zn, 1.5% Pb and 50g/t Ag. There is little marginal material between trace amounts and 5% Zn.

EXPLORATION METHODS

Most of the sulphide bodies, including the Main Orebody, were discovered as outcropping sulphides or gossan. Their sub-surface extent was mapped by means of electromagnetic surveys, and these surveys have also detected a few sulphide bodies with no surface expression. The most striking geophysical feature is the low electrical conductivity, presumably the result of permafrost, of both the sulphides and the

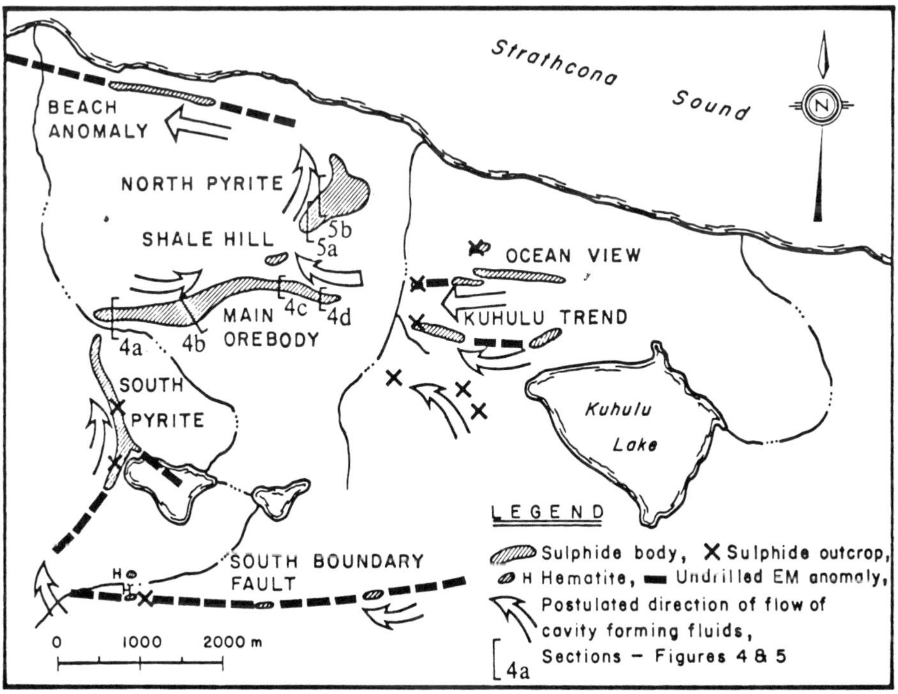

Figure 3. Map showing the main sulphide bodies and trends, and the postulated directions of flow of drainage at time of formation of cavities in Society Cliffs dolomite.

bedrock so that the sulphide bodies show as small but distinct anomalies, mainly at high frequency, against a very flat background. The only non-sulphide conductors which produce anomalies are unfrozen lake-bottom sediments, notably in Lake Kuhulu, and gabbro dykes. The latter can be detected only with spreads greater than

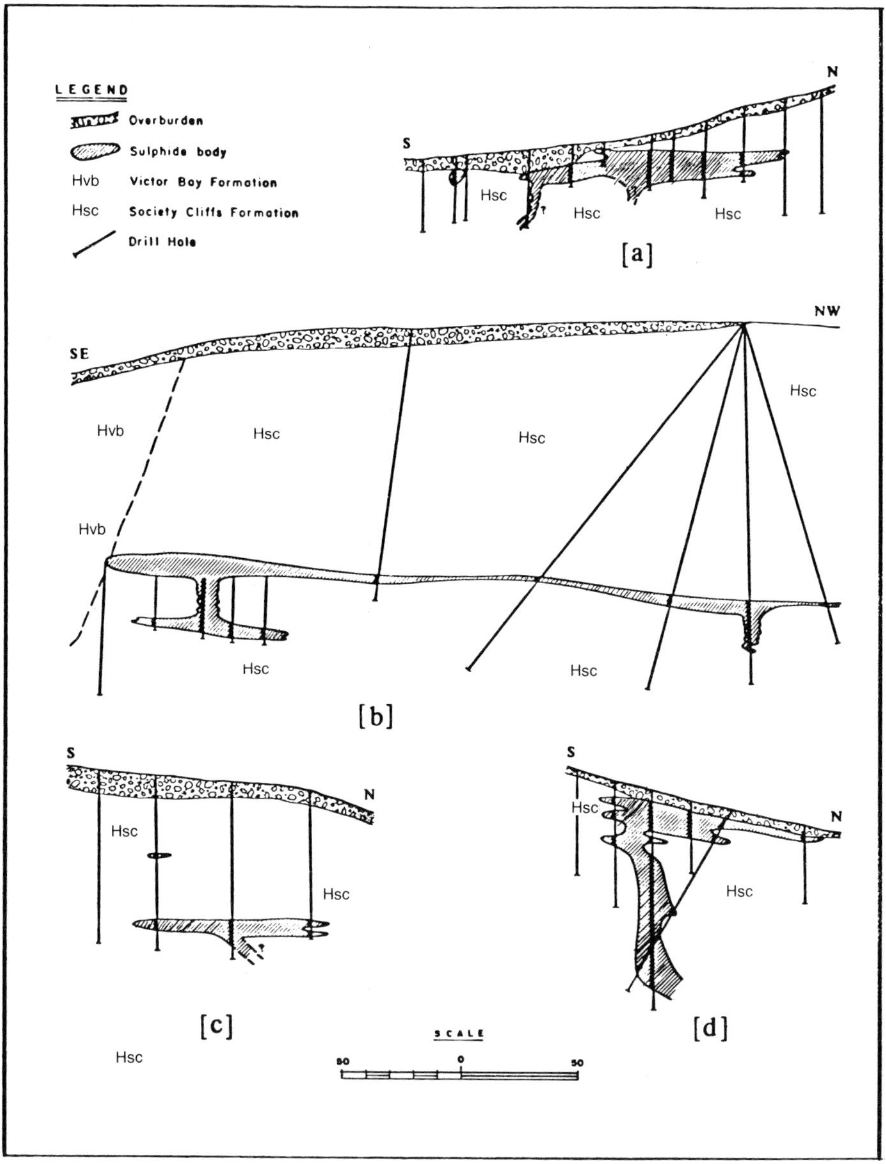

Figure 4. Cross sections of the Main Orebody showing horizontal lens overlying vein. For location of sections see Figure 3.

200m, and have a lower conductivity response than sulphides. Vertical loop methods have proved to be the most effective. Horizontal loop and Turam surveys, though also effective, are less flexible and more costly. VLF and galvanic-electromagnetic methods produce good anomalies, but do not discriminate between sulphides and dykes. Induced polarization does not produce significant anomalies, and there is no useful magnetic response.

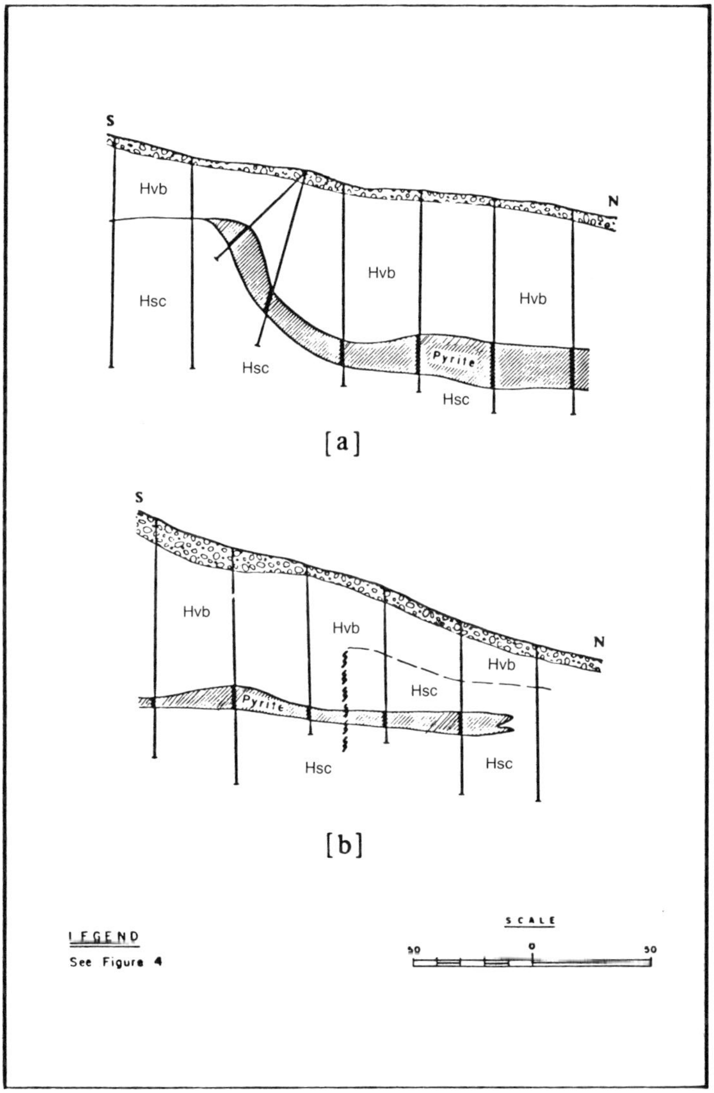

Figure 5. Cross sections of the North Pyrite sulphide zone. For location of sections see Figure 3.

Stream sediment geochemical anomalies are produced from the outcropping sulphides, but as all the known outcrops are within 2km of the coast, this has not been a useful exploration tool at Nanisivik, though the implications are that stream sediment sampling would be a useful regional method. Geochemical haloes in the dolomite around the sulphide bodies are of very limited extent and no practical use has been made of rock geochemical analyses.

MINERAL TEXTURE AND DISTRIBUTION

Pyrite occurs mainly as fine-grained crystals. At or near the surface and where it is directly overlain by Victor Bay shale, the pyrite has commonly disintegrated to sand and mud, depending on the grain size. Colloform pyrite is also present. In the Main Orebody and in the South pyrite zone the pyrite crystals are generally cubes and golden-yellow in colour. In the North Pyrite zone they tend to be dodecahedrons and whitish yellow.

Galena occurs as relatively large crystals, usually at least a few millimetres but locally 10cm or more in diameter. Well-formed cubes of any size are uncommon. Galena is usually associated with sphalerite, and less commonly directly with pyrite, although pyrite is to some degree ubiquitous. An apparent exception is the North Pyrite body where the pyrite typically contains about 0.2% lead and only traces of or no zinc. Elsewhere, particularly in the Main Orebody, where there is the most information, the reverse is true; all the pyrite contains a little zinc and some zones in the Main Orebody, which appear in hand specimen to be only pyrite, contain as much as 5% zinc. Sphalerite ranges in grain size from very fine-grained to coarse aggregates. It is typically chocolate brown in colour, but ranges from black to brick red and yellow and occurs associated with galena or with pyrite or both. The lighter coloured sphalerite is usually disseminated in pyrite without galena.

White sparry dolomite is common within the Main Orebody but less abundant in the North and South Pyrite zones. It commonly occurs horizontally interlayered with sphalerite in bands about 1cm thick. The sphalerite may be associated with galena and pyrite though occasionally there is banding of dolomite and galena only. Banding of pyrite and dolomite is also noted but much of the pyrite is massive. Sparry dolomite fills interstices between well formed crystals of sulphides, especially pyrite, indicating that at least one phase of carbonate deposition was later than the sulphide deposition.

Much of the Main Orebody is horizontally layered, particularly at the north and south extremities of the ore zone, but the layering locally gives way abruptly to a chaotic mixture of pyrite, sphalerite, minor galena and dolomite in various proportions. The material appears to be holocrystalline rather than brecciated.

SULPHIDE BODY CONFIGURATIONS

The sulphides occur as flat lenses and as vertical bodies which, without implying any particular mode of origin, are called veins. About 90% of the known sulphides are in lenses, though part of the reason for this proportion is that the veins generally extend downward from the lenses and are therefore less accessible and so less explored.

The three main sulphide bodies are each about 10m thick, tapering towards the edges and often splitting into two wedges at the extremities. The Main Orebody is 3,000m long in an east-west direction, on average 100m wide, and is remarkably horizontal. Apart from local irregularities, the elevation along the top of its centre varies by not more than ± 2m along its entire length. Cross-sections through the Main Orebody are shown in Figures 4a, b, c and d.

The South Pyrite body appears to be a branch of the Main Orebody. The bodies crop out on opposite sides of a deep but narrow canyon and it seems probable that the South Pyrite body once joined the Main Orebody almost at right angles very near, but not at, its western end. The known part of the South Pyrite body is about 2,000m long, but the southern limit has not been reached. The body appears to be similar as it is traced to the Main Orebody in cross-section, but it rises about 60m southwards. Whether this is a gradual rise or a series of steps is not known. The South Pyrite body passes through at least two major faults and, though the main movement on these faults is believed to predate the emplacement of the sulphides, it remains possible that subsequent movement on these faults accounts for the fact that, unlike other lenses, this body is not horizontal.

The North Pyrite body is heart-shaped and about 400m in diameter. It is horizontal except in the southwest corner where it rises like the lip of a saucer and then ends (Fig. 5a). Here it is overlain by shale of the Victor Bay Formation but over part of the zone there is a layer of Society Cliffs dolomite over the pyrite (Fig. 5b).

About 300m north of the Main Orebody and to the south of Shale Hill (Fig. 3) a small lens of ore-grade material occurs at the same elevation as the Main Orebody. From this lens a body of pyrite plunges downward along the Victor Bay – Society Cliffs contact towards the lip of the North Pyrite body described above. Whether these two meet and are continuous is not known; on the section in Figure 5a it is suggested that they do not.

In each of the three larger sulphide bodies there are at least two diamond drill holes in which about 30m of continuous massive sulphides are intersected. Most of the drilling was carried out on profiles at right angles to the strike of the sulphide bodies, and these continuous long and deep sulphide intersections have so far been found in only one of the holes on any section drilled. These results indicate that vertical veins or pipe-like sulphide bodies exist beneath the main lenses and are continuous with them. Below the western part of the Main Orebody, a second sulphide lens about 10m thick, has been found about 10m below the main lens (Fig. 4b). The two lenses are generally separated by dolomite, but some holes have passed through continuous sulphide from the top of the main lens to the bottom of the lower lens. There are therefore two horizontal lenses connected either by one vertical vein or by a series of pipes. This lower lens is generally more irregular than the main one. Near the centre of the orebody, the lower lens is horizontal where under the main lens, but, where extending to the north beyond the main lens, it appears to follow the bedding of the dolomite, with a dip of about 15° to the north. However it remains a possibility that there are a number of small horizontal lenses in this zone rather than a single dipping lens.

A number of inclined holes have been drilled under the eastern part of the Main Orebody (Fig. 4d) and intersections suggest a vein, extending as much as 60m vertically and 15m to as little as 1m wide. A series of vertical pipes developed along a narrow vein may prove to be the explanation. A number of other sulphide intersections below the main lens may indicate other veins or lenses, although there is clearly not a single horizontal lens with a width and thickness similar to that occurring beneath the western end.

In the Ocean View area there is one horizontal lens 200m long by 50m wide and smaller lenses at about the same elevation. A number of lower intersections having little or no horizontal extent appear to be veins or pipes.

Very little sulphide is found above the major lenses. A few small intersections occur above the western part of the Main Orebody, but are only a metre or so above the main lens. A single sulphide intersection 0.6m thick was cut 27m above the Main Orebody (Fig. 4c) and is the only known occurrence of sulphides substantially above a main lens. On this section the Main Orebody is at its narrowest and has the lowest grade compared with the other 50 drill profiles.

MINERAL ZONING

Ore has been found only in a long, narrow zone comprising the Main Orebody, in the part of the Shale Hill deposit closest to the Main Orebody, and in the Ocean View – Kuhulu Trend area (Fig. 3). North, east and south of this zone only pyrite has been found and no significant sulphides are known west of the Main Orebody; what little there is, is pyrite. There is apparently no pattern in the association of lead and zinc; relatively high-lead zones occur intermittently from the west end of the Main Orebody to Ocean View. Copper, nickel and cobalt occur on the eastern side of the dyke which cuts the Main Orebody, but these metals are clearly associated with the dyke and are not related to the broader pattern of mineral distribution.

Hematite occurs between the western end of Twin Lakes and the South Boundary Fault, and in several locations along the South Boundary Zone. It is generally associated with pyrite. There are two outcrops of massive hematite just north of the fault which do not appear to link up with the South Pyrite zone. The more southerly one has massive sulphur outcropping on its west side and pyrite on the east.

Mineralized areas located 16km and 32km east of the Main Orebody each contain several showings of pyrite and gossan. Although sphalerite occurs in each area (in the farther zone there is a high-grade lens 15m in diameter and 2m thick), these are mainly barren pyrite deposits. In any case, they are probably too far from Nanisivik to be closely related geologically.

At the east end of the Main Orebody ore grades occur mainly in the top and the bottom thirds of the main sulphide lens, with a zone of low-grade material in between. The lower of the high-grade zones is cut into two by a block of low-grade sulphide at the top of the vertical vein. The vein itself is, on average, about the same grade as the main lens, but the grades are erratic. Recent work underground indicates that the block of low grade extends the whole length of the Main Orebody, but whether the low grade is invariably over the vein, or even whether the vein itself is continuous, is not yet known. At the west end of the Main Orebody, ore-grade material also tends to occur in horizontal zones of mineable dimensions within the main lens, but the pattern is not clear. At both the west end and the central portion of the Main Orebody a zone of almost barren pyrite lies along the northern flank of the main lens. The horizontal lens beneath the western half of the Main Orebody is almost entirely pyrite and contains only small amounts of ore-grade material.

GENESIS

The sulphide bodies pass through some major faults with little or no change in elevation. This single fact implies that their emplacement was not stratigraphically controlled; that favourable horizons in the dolomite, evaporite beds and reef structures are all excluded as ore controls, and that the sulphides are not syngenetic with the dolomite but are post-faulting. It also appears, taking all the available information into consideration, that the sulphides have no direct relationship to brecciation, marmorization, or lithological changes within the Society Cliffs Formation. This does not preclude an indirect relationship as algal dolomite and brecciated zones may be more favourable for sulphide deposition, but there are many places where sulphides end in brecciated algal dolomite, and there are no known cases of remnants of unlaminated argillaceous dolomite within the sulphide bodies.

Most of the known sulphides are in horizontal lenses and, as there is no stratigraphic or structural control to account for this, the only way that dissolution of the dolomite could have occurred is by water under little or no pressure. Such water

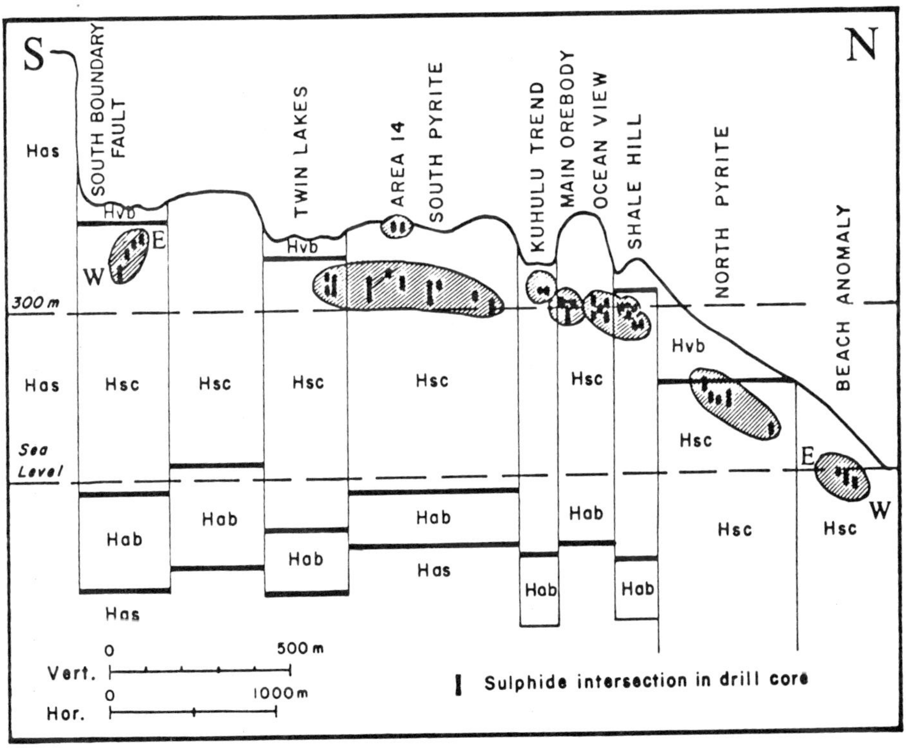

Figure 6. Schematic section. Drill intersection of sulphides projected onto a N-S plane. (For legend see Table I and for locations see Fig. 2).

must have been meteoric and essentially vadose, at or above the water table. The horizontal form of lenses resemble water table caves (Ford and Ewers, 1978), whereas the overall pattern suggests vadose caves as they form a step-like pattern down towards the Sound (Fig. 6). The distribution of sulphides strongly suggests a system in which the drainage was east- or westwards along faults, until meeting a sink hole resulting in a breakthrough northwards towards the Sound and deposition of sulphides in the cavities. A possible pattern for this drainage is shown in Figure 3.

Several means and sequences for the formation of the cavities and the filling of some by sulphides have been proposed. All consider the sulphides as being deposited early, i.e., in Proterozoic time.

An early theory held that the sulphides were laid down in a drainage system, partly at surface and partly in underground channels formed by solution of the dolomite; a common pattern in regions of karst topography. This would imply that the major faulting, brecciation, cementation of the brecciation, development of solution channels and deposition of the sulphides took place before the start of deposition of the Victor Bay shale. There is evidence to support this theory in that more than half of the total area of the main sulphide lens is either outcrop or overlain by Victor Bay shale, and much of the pyrite under Victor Bay is in the form of sand or mud, suggesting that weathering might have taken place prior to covering.

If on the other hand the cavities, i.e., the spaces now occupied by sulphides, were formed after deposition of the Victor Bay shale, the sulphides themselves or some other filling must have been deposited at the same time or very shortly thereafter, as the shale, in cases where it overlies the sulphides, would have been too weak to form a roof to a cave for any length of time. Known areas of shale covering sulphides are as much as 200m in diameter (Fig. 5). There is no doubt that an open cave would collapse after a very short time. Geldsetzer (1973a) proposed that the cavities were filled by evaporites. If so, these were deposited, and therefore the cavities were formed, prior to deposition of the Victor Bay shale.

Two factors suggest that the sulphides were deposited at temperatures and pressures higher than those of vadose waters. First, pressure at least marginally above atmospheric, is indicated by the fact that sulphides tend to fill cavities completely; what space remains is within the sulphide body rather than at the top contact, and is ice-filled. The occurrence of sulphides above the main lenses is also evidence of higher pressure. Second, the sulphides were formed at elevated temperature. Olsen (1977), in presenting fluid inclusion data from sphalerite at Nanisivik, gave 100°C as the average temperature from six inclusions. He also gave inclusion temperatures from sparry dolomite and sulphur isotope fractionation studies and these are generally higher, the highest average being about 200°C. These indicate that sulphides were deposited under some pressure and at temperatures higher than those of meteoric water and hence, it may be concluded, under Victor Bay shale cover.

There is evidence, therefore, of two separate processes; the cavities being formed as horizontal caves by meteoric water at low temperature and pressure, and the minerals being deposited at elevated temperature and pressure, i.e., at greater depth and under Victor Bay cover.

If, as has been suggested, the cavities were filled with evaporites, a shale roof where present would have been supported. Sulphide formation by bacterial reduction of the sulphates is ruled out as it would involve shrinkage and subsequent collapse or subsidence. There is no evidence of this. However, dissolution of the postulated sulphates and the deposition of the sulphides and accompanying crystalline carbonates by one and the same fluid is quite feasible as, under any likely conditions, sulphates are more soluble than dolomite or calcite. The main objection to a genetic sequence involving such an interim infilling of the cavities by evaporites is that it would appear that all the sulphide channels were filled with evaporite and that all this material was removed, but none of the non-sulphide cavities were filled with evaporites, or alternatively, that they were filled and the evaporites were removed without any sulphide being deposited. Though there are relatively few cavities which have not been filled with sulphides, those which do exist are filled with ice. No major cavities filled with crystalline carbonates as the major constituent have been found. In other areas, such as in Mexico, where similar orebodies occur, cavities are generally filled with sulphides or are "dry", and the dry caves are filled with air or wholly or partially with water, with only a coating of crystalline carbonates. Though a mode of genesis involving interim filling with gypsum is a possibility, the indirect supporting evidence which should occur is lacking at Nanisivik.

If the cavities were not filled with evaporites they could have remained open under Victor Bay cover for a short time, so that the two fluids as proposed above must have been contemporaneous, or nearly so. This does not involve as much of a coincidence as it might seem; as long as the dolomite was at or near the surface there was probably some dissolution occurring at all times, followed by collapse brecciation and further solution. Even today there are some minor karst features on the surface at Nanisivik, although the topography is dominated by the block faulting and there is no subsurface drainage because of permafrost. The high indicated temperatures, however, are less easy to explain. The Earth may have been warmer in Proterozoic time, but evidence of glaciation indicates that it could not have been very much warmer. Assuming a thermal gradient of 1° per 35m of depth, somewhat high by present-day standards, and a surface temperature of 25°, a ground temperature of 110°C would develop at a depth of 3,000m.

It therefore seems most probable that though the cavities were formed under the Victor Bay shale at relatively shallow depth, the mineralizing fluid came from a much deeper source. One possible source for the second mineralization fluid is evolved connate water from either the Arctic Bay or Adams Sound Formations, both of which contain considerable zinc and lead; a sample of Arctic Bay rock a few kilometres from Nanisivik contained 3,802 ppm zinc and 533 ppm lead, and an intersection of the Arctic Bay Formation in a hole below the Main Orebody contained 803 ppm and 83 ppm. Adams Sound samples contained 888 ppm and 1,677 ppm zinc and 47 ppm and 89 ppm lead. A sample of the Victor Bay contained 739 ppm zinc and 77 ppm lead.

This evidence appears to point to deposition of metals from evolved connate water from either the Adams Sound Formation or the Arctic Bay Formation at somewhat elevated temperature and pressure when this fluid was forced upwards through pre-existing fault planes and entered cavities in the Society Cliffs dolomite formed by and containing meteoric water with reduced sulphur in solution. It is possible that some minor tectonic activity caused the fault planes to re-open and this action contributed in part to the elevated temperature.

ACKNOWLEDGEMENTS

The authors wish to thank F.W. Beales and C.D. Spence for their help and advice in the preparation of this paper and D. Seagraves and Dawn A. Gravelle for preparing and drafting the figures.

REFERENCES

Blackadar, R.G., 1956, Geological Reconnaissance of Admiralty Inlet, Baffin Island, Arctic Archipelago, Northwest Territories: Geological Survey of Canada Paper 55-6, 25 p.

_______________, 1965, Geological Reconnaissance of the Precambrian of Northwest Baffin Island, Northwest Territories: Geological Survey of Canada Paper 64-42, 25 p.

_______________, 1970, Precambrian Geology, Northwestern Baffin Island, District of Franklin: Geological Survey of Canada Bulletin 191, 89 p.

Ford, D.C. and Ewers, R.O., 1978, The development of limestone cave systems in the dimensions of length and depth: Canadian Journal of Earth Sciences, v. 15, p. 1783-1798.

Geldsetzer, H., 1973a, The tectonic-sedimentary development of an algal dominated Helikian succession of Northern Baffin Island, N.W.T.: in Aitken, J.D. and Glass, D.J., eds., Canadian Arctic Geology Symposium: Geological Association of Canada and Canadian Society of Petroleum Geologists, p. 99-126.

_______________, 1973b, Syngenetic dolomitization and sulfide mineralization: in Amstutz, G.C. and Bernard, A.J., eds., Ores and Sediments: New York, Springer Verlag, p. 115-127.

Lemon, R.R.H. and Blackadar, R.G., 1963, Admiralty Inlet Area, Baffin Island, District of Franklin: Geological Survey of Canada Memoir 328, 84 p.

Olson, R.A., 1977, Geology and Genesis of Zinc-Lead Deposits Within a Late Proterozoic Dolomite, Northern Baffin Island, Northwest Territories, Unpublished Ph.D. Thesis, University of British Columbia, Vancouver, B.C.

MS received December 1978;
Accepted for publication April 1981.

REVIEW PAPER

Precambrian Sulphide Deposits, H.S. Robinson Memorial Volume, edited by R.W. Hutchinson, C.D. Spence and J.M. Franklin, Geological Association of Canada Special Paper 25, 1982

SYN-DEPOSITIONAL HYDROTHERMAL PROCESSES AND PRECAMBRIAN SULPHIDE DEPOSITS

R.W. Hutchinson
Department of Geology, The University of Western Ontario, London, Ontario N6A 5B7

ABSTRACT

The descriptions and interpretations in this Special Paper show clearly that the understanding of the genesis of Precambrian sulphide deposits has changed markedly during the past twenty years. Timing of the metalliferous hydrothermal fluids which formed most of these ores relative to their host rocks, is now considered to have been syn-, or only slightly post-depositional, rather than epigenetic. Sulphide emplacement was by chemical precipitation on the sea floor, diagenetic reaction within already deposited sediment or sub-sea floor infilling of pores, rather than by large-scale replacement of rock. Hydrothermal fluid was generated mainly by deep convective circulation of marine or meteoric waters rather than by magmatic differentiation.

Sea floor discharge of these fluids formed many important massive base metal sulphide deposits and these exhibit evolutionary changes from volcanic-hosted, copper-zinc-gold types in Archean-Aphebian rocks to sediment-hosted lead-zinc-silver ones in Helikian-Hadrynian strata. Such fluids were also important in forming many other types of Precambrian sulphide deposits. Some komatiitic volcanic-affiliated nickel sulphide deposits may be of exhalative rather than magmatic origin. Several gold deposits are primary gold-rich exhalites and other similar, but sub-economic strata may be essential as source-rock precursors of epigenetic gold veins and stockworks. Rare Precambrian porphyry-type deposits may represent relatively rare instances of subvolcanic sulphide deposition from these brines. Epigenetic carbonate-hosted deposits may be products of diagenetic reactions between the fluids and already deposited sediment.

Post-depositional metamorphism inevitably affected all syn- or diagenetically formed sulphide deposits. Resulting metamorphogenic textural, mineralogical and morphological complexities obscure their primary origin, and these characteristics are therefore unreliable as evidence of genesis, particularly for deposits in host rocks of high metamorphic rank. Metamorphic processes may generate particular secondary deposits within pre-enriched primary strata, for example epigenetic gold vein systems and stockworks in auriferous exhalites.

RÉSUMÉ

Les descriptions et interprétations dans ce volume montrent clairement que la compréhension de la genèse des dépôts sulfurés précambriens a changé de façon marquée pendant les vingt dernières années. Les fluides métallifères hydrothermaux qui ont formé la pluspart de ces dépôts sont considérés maintenant comme ayant été des solutions syngénétiques, ou à peine postérieures aur roches encaissantes, plutôt qu'épigénétiques. La mise en place des sulfures est considérée s'être faite par précipitation chimique sur le fond marin, par réaction diagénétique dans des sédiments déjà déposés ou par remplissage des pores des roches proches du fond marin, plutôt que par remplacement à grande échelle des roches encaissantes. Les fluides hydrothermaux ont été produits principalement par la circulation convective profonde d'eaux météoriques ou marines plutôt que par différenciation magmatique.

La décharge de ces fluides sur le fond marin a formé de nombreux amas sulfurés de métaux de base et ceux-ci montrent des changements évolutifs depuis les types à cuivre-zinc-or encaissés dans des roches volcaniques archéennes et aphibiennes jusqu'aux types à plomb-zinc-argent dans les roches sédimentaires d'âge hélikien-hadrynien. Ces fluides ont joué aussi un rôle important dans la formation de bien d'autres types de gisements de sulfures précambriens.

Certains dépôts de nickel associés à des roches volcaniques komatïtiques sont peut-être aussi d'origine exhalative plutôt que magmatique. Plusieurs gisements d'or sont des exhalites originellement riches en or alors que d'autres couches semblables, aussi à teneurs en or non économiques, ont été la source de l'or des filons et stockwerks épigénétiques. Les rares dépôts précambriens de type porphyry-copper représentent des cas relativement rares de dépôts subvolcaniques à partir de ces solutions. Les dépôts épigénétiques dans les roches carbonatées peuvent êtres les produits de réactions diagénétiques entre les fluides et les sédiments déjà déposés.

Le métamorphisme subséquent a inévitablement affecté tous les gisements syn- ou diagénétiques. Les complexités des structures, de la minéralogie et de la morphologie qui en résultent, obscurent leur origine et on ne peut donc compter sur ces caractéristiques pour déterminer la genèse, particulièrement celle des dépôts encaissés dans des roches ayant subies un métamorphisme de rang élevé. Les processus métamorphiques peuvent aussi êtres à l'origine de dépôts secondaires particuliers dans des strates à minéralisation primaire tels, par exemple, les systèmes de filons et stockwerks d'or encaissés dans les exhalites aurifères.

INTRODUCTION

The four regional geological reviews and eighteen deposits described in this Special Paper encompass the northern half of North America, and concern almost the full span of Precambrian time. A wide variety of important deposits are discussed from which are produced significant amounts of nickel, platinum group metals, gold, silver, copper, zinc, and lead as well as many minor by-product metals including cadmium, tin, antimony and bismuth.

The descriptions and interpretations in this volume reflect the major changes in geological understanding of Precambrian and indeed other sulphide deposits that have developed in the past twenty-five years. These changes include improved knowledge of the deposits and their regional geological settings, and a better understanding of the processes involved in their formation. Prior to about 1960, the origin of most of these deposits, possibly excepting the nickel-bearing ones, was explained under the epigenetic replacement hypothesis. Specifically, their origin was attributed to selective replacement of chemically favourable rock by "hot ascending waters of uncertain origin, but charged with igneous emanations" (Lindgren, 1919, p. 205). To this conception Graton (1940) and many others contributed various speculations about the nature and derivation of these fluids.

The great diversity of the deposits considered seriously constrains generalizations about their geological characteristics or the genetic processes involved. Yet a dominant genetic process was important in forming most of the deposits, and may have played a minor, or indirect role in shaping the others. This process was syn-depositional hydrothermal activity. Notwithstanding its broadly syn-depositional timing, this activity appears to have occurred after rock deposition in a subvolcanic environment, in a surficial but diagenetic environment within recently deposited sediment and, most important, syngenetically (sensu stricto) by sea floor exhalative fluid discharge and accompanying chemical sedimentation. The exhalative deposits were subsequently affected by varying degrees of metamorphism.

The major purposes of this review are to briefly consider the syn-depositional hydrothermal theory in relationship to the older epigenetic hydrothermal replacement hypothesis, to summarize the main aspects of syn-depositional-exhalative hydrothermal processes as they apply to the many Precambrian exhalative sulphide deposits here described, and to evaluate the possible significance of syn-depositional hydrothermal activity with respect to other deposits of non-exhalative origin.

SYN-DEPOSITIONAL EXHALATIVE HYDROTHERMAL SYSTEMS

Applicability to Precambrian Sulphide Deposits

Exhalative massive base metal sulphide deposits, which represent the exogenous end-products of syn-depositional hydrothermal activity, share a set of widely recognized and distinctive geological characteristics that result from their common origin as sea floor chemical precipitates (Hutchinson, 1973, 1980; Sangster, 1972; Sangster and Scott, 1976). The authors of eleven of eighteen case histories in this volume present convincing evidence that the Corbet, Millenbach, Detour, Crandon, Geco, Sherritt Gordon, Ruttan, Bathurst Norsemines, Anderson Lake, Sullivan, and Agnico Eagle deposits are of exhalative origin (see relevant papers in this volume). Ten of these are massive base metal sulphides; one, Agnico-Eagle, is mainly an auriferous massive pyritic deposit (Table I). In addition, exhalative hydrothermal genetic processes have been suggested for three others, either by authors of case histories in this volume or by prior workers on the same or comparable deposits. Thus, evidence of exhalative origin is presented for the Redstone nickel sulphide deposit (Robinson and Hutchinson, this volume): a pre-metamorphic exhalative origin is favoured, although not substantiated, for the highly metamorphosed Balmat zinc deposits (de Lorraine and Dill, this volume): also, although not favoured by the current authors, prior investigators of the vein deposits at Opemiska and Chibougamau suggested that these might represent the mineralized fracture-feeders of exhalative systems (McMillan, 1973; Allard, 1976).

Thus for fourteen of the eighteen deposits described in this volume exhalative processes have been shown to be responsible or are implicated. The remaining four deposits, Thompson, Gayna River, Nanisivik, and Redstone Copper are clearly not of simple exhalative origin: yet they exhibit some geological characteristics common in exhalative deposits, suggesting that syn-depositional-exhalative hydrothermal processes may have played some indirect, or as yet obscure role in their genesis.

TABLE I

COMPARATIVE TABULATION OF PRECAMBRIAN SULPHIDE DEPOSITS DESCRIBED IN THIS VOLUME

Deposit	Major (Minor) Metals	Major Host Rocks	Structural Province - Tectonic Region	Metamorphism Sulphides,	Host Rocks	Age	Deposit Type	Mineralization - Remarks
Exhalative Volcanogenic Group								
1. Redstone Nickel	Ni(Cu) PGM,Au	volcanic ultramafic	Superior	low	greenschist	Archean	exhalative	massive, minor disseminated; exhalative origin suggested, but controversial
2. Millenbach	Zn,Cu Ag,Au	volcanic	Superior	medium	greenschist hornfels	Archean	exhalative	massive lenses, altered footwall stockwork
3. Corbet	Cu,Zn Ag(Au)	volcanic	Superior	low	greenschist	Archean	exhalative	massive lens, altered footwall stockwork
4. Detour	Cu,Zn Ag,(Au)	volcanic	Superior	low	greenschist	Archean	exhalative	small massive lens, not stratabound, extensive altered footwall stockwork
5. Geco	Zn,Cu(Pb) Ag(Sn)	sedimentary-volcanic(?)	Superior	high	amphibolite-granulite	Archean	exhalative	massive lens, altered footwall strata; metamorphism obscures primary relationships
6. Bathurst Norsemines	Zn,Pb(Cu) Ag,Au	volcanic	Slave	medium	amphibolite	Archean	exhalative	massive lenses, altered footwall stockworks
7. Agnico-Eagle	Au(Ag)	volcanic	Superior	low	greenschist	Archean	exhalative	massive pyrite and bedded ankerite - chert, altered footwall strata
8. Crandon	Zn,Cu,(Pb) Ag,Au	volcanic	Southern	low	greenschist	Archean-Aphebian(?)	exhalative	massive lens, altered footwall stockwork; age controversial
9. Anderson Lake	Cu Ag,Au	volcanic	Churchill	medium	amphibolite	Aphebian	exhalative	massive lens, altered footwall stockwork

10. Ruttan	Zn,Cu (Ag,Au)	volcanic	Churchill	medium	amphibolite	Aphebian	exhalative	massive lens, altered footwall
11. Sherritt Gordon	Zn,Cu	volcanic-sedimentary	Churchill	medium	amphibolite	Aphebian	exhalative	massive lens, altered footwall strata; metamorphism obscures primary relationships
Exhalative Sedimentary Group								
12. Balmat	Zn,Pb	shelf carbonate sedimentary	Grenville	high	upper amphibolite-granulite	Helikian	exhalative	massive lenses, metamorphism obscures primary relationships and origin
13. Sullivan	Pb,Zn,Ag (Cd,Sn,W)	trough clastic sedimentary	Southern Cordillera	low	greenschist	Helikian	exhalative	stratiform massive lens, altered footwall breccia
Miscellaneous Group								
14. Thompson (et al.)	Ni(Cu) PGM	sedimentary-ultramafic	Superior-Churchill boundary	high	upper amphibolite-granulite	Archean	magmatic	massive and disseminated; metamorphism obscures age, primary relationships and origin; possibly exhalative
15. Opemiska	Cu,Au (W)	layered gabbro	Superior	low	greenschist	Archean-Aphebian(?)	epigenetic vein	age and origin of vein system problematic; stockwork for exhalative system suggested by prior workers
16. Gayna River	Zn(Pb)	shelf carbonate sedimentary	Northern Cordillera	very low	sub greenschist	Helikian-Hadrynian	epigenetic Mississippi Valley	disseminated in secondary porosity
17. Redstone Copper	Cu(Ag)	shelf carbonate redbed sedimentary	Northern Cordillera	very low	sub greenschist	Helikian-Hadrynian	diagenetic sedimentary Cu	stratiform
18. Nanisivik	Zn,Pb	shelf carbonate sedimentary	Churchill	very low	sub greenschist	Helikian	epigenetic	massive lenses, veins in solution caverns; origin uncertain

Clearly, therefore, exhalative hydrothermal activity was important in forming Precambrian (and younger) massive sulphide deposits (Table I), but not merely statistically so. It is equally significant that this activity produces certain innate geological characteristics, and involves specific metamorphic consequences that are relevant to other types of deposits and to general concepts of sulphide deposit genesis.

Timing, Source and Nature of Fluids

Recognition of the importance of exhalative hydrothermal processes in the genesis of massive sulphides, and the resulting application of related stratigraphic, structural and geochemical principles have unquestionably been the most important changes relating to exploration for these deposits during the past twenty-five years. The particular processes involved in exhalative activity have been widely discussed (Franklin and Thorpe, this volume; Hutchinson *et al.,* 1980): modern analogues in active sea floor hydrothermal systems have been recognized (Degens and Ross, 1969; Lonsdale *et al.,* 1980), and their specific characteristics have been described in many deposits, including those in this volume. Thus, although the genetic enigma posed by the massive sulphide ores, as originally recognized by Lindgren (1919, p. 818) and restated forty years later by Gunning (1959), has been generally resolved, the broader significance of exhalative hydrothermal processes in sulphide deposit genesis has not been so extensively considered, and so is much less familiar. The epigenetic hydrothermal replacement hypothesis has, with time been found invalid for these deposits in two major respects, but substantially correct in a third. Firstly, the *timing* of mineralization was not epigenetic but broadly syn-depositional; secondly, the *mode of sulphide emplacement* was not by replacement but mainly by sea floor deposition with accompanying footwall space infilling and only minor reaction-replacement; but thirdly, present understanding of its *agent,* a heated, metalliferous, aqueous solution remains substantially unchanged.

Further, it is now possible, at least in part, to answer the questions regarding the uncertain origin and nature of the ore solution, and the source of some of its dissolved components. Thus, studies of stable isotopes of the fluids in active hydrothermal systems and of the ores themselves (Craig, 1966; White, 1968, 1974; Bischoff, 1969; Heaton and Sheppard, 1976; Pisutha-Arnond and Ohmoto, 1980) indicate that much of the water is of recirculated surface derivation, either meteoric or marine, whereas very little, if any, is of primary juvenile or magmatic extraction. Similarly, sulphur isotopic studies suggest that much of the sulphur in the deposits is also derived from seawater by sulphate reduction (Ohmoto and Rye, 1979; Kowalik *et al.,* 1981).

The extreme chlorinity and low sulphide-ion activity in active hydrothermal brine systems (White, 1968; Bischoff, 1969) and in fluid inclusions (Roedder, 1976; Tokunaga and Honma, 1974) suggests that metal transport, at least for the base metals, was largely as soluble chloride complexes. Very low oxygen content and the presence of light hydrocarbon gases in the deep Red Sea brines, as well as in other active systems, and in fluid inclusions (Hunt *et al.,* 1968; Swinnerton and Linnenbom, 1969; Lyon *et al.,* 1977; Ellis and Mahon, 1977; Degens and Ross, 1976; Roedder, 1976) and the association of carbonaceous strata with many of the deposits (Barnett *et al.,* this volume; May and Schmidt, this volume; De Lorraine and Dill, this volume; Hamilton *et al.,* this volume; Waker *et al.,* 1975) all suggest that the transporting fluids were strongly reduced. Measurements in active systems indicate weak acidity (White, 1968; Watson and Waterbury, 1969), and temperatures up to 350° (White, 1968; Lonsdale *et al.,* 1980; Spiess *et al.,* 1980). These temperatures agree with those determined from fluid inclusions and stable isotopes in the ores (Kowalik *et al.,* 1981; Tokunaga and Honma, 1974; Heaton and Sheppard, 1976).

These fluids circulated deeply in the sea floor sediments and underlying volcanic-sedimentary rocks. The circulation was probably convectively driven, induced by igneous activity which heated the fluids, but may also have been aided partly by fluid expulsion due to compaction-lithification and prograde metamorphism. Subvolcanic comagmatic sills may have been important in promoting this circulation in felsic volcanic environments (Franklin and Thorpe, this volume) and ocean rift-ridge magmatism may have played the same role in ophiolitic ones (Fyfe and Lonsdale, 1981). Thermal agents to drive convective circulation in exhalative sedimentary sulphide environments are however, less evident. The fluids rose along fractures and were discharged from vents to the sea floor, probably at intersections of syn-depositional fault-fracture systems. Here sulphides were deposited, chiefly on the sea floor but also epigenetically in fractured rock or porous sediment near the top of the vent beneath the sea floor.

A comparison of various types of massive base metal sulphide deposits in different geological settings (Hutchinson, 1973, 1980) together with theoretical and experimental studies (Seyfried and Bischoff, 1977; Browne and Ellis, 1970) strongly suggests that the source of the metals was largely in the rocks and sediments through which these weakly acidic, highly chlorinated, heated, reduced fluids passed. Under these conditions, water-rock reduction-oxidation/hydration reactions broke down silicate mineral structures, thereby releasing metals in minor (ppm) amounts to the fluids, which were consequently further reduced, dehydrated and chlorinated by hydrous silicate-forming wall rock reactions (Hutchinson *et al.,* 1980). The metals, chiefly iron, but also lesser amounts of the base, precious and other heavy metals, as well as silica, alkali and alkaline earth elements were transported in the fluids toward the sea floor. Here the dissolved components were deposited, further leached, or variably redistributed due to Eh-pH-T changes induced by mixing of the rising, reduced hydrothermal fluids with cooler, oxygenated, connate marine or overlying sea water (Hutchinson *et al.,* 1980; Large, 1977). Although attention invariably centres on the valuable metals, iron and silica are much more abundant transported components in these systems, accounting for the abundant iron in the orebodies and their common, although not invariable association (Hutchinson, 1973; Franklin and Thorpe, this volume) with iron-formation and cherty siliceous strata.

Footwall rock alteration is intense, zoned, and characteristic of exhalative deposits (Table I; Knuckey *et al.*; Knuckey and Watkins; Walford and Franklin; Casselman and Mioduszewska; Speakman *et al.*; all this volume). Replacement of rock by sulphide however was minor, though reaction between the fluids and previously deposited volcanic rock or unconsolidated sediment and sulphide may have produced local reaction-replacement textures. The details of distribution, scale, morphology and composition of alteration vary widely from one deposit to another, as described in case histories in this volume. In general they involve rock oxidation (decrease of Fe^{2+}/Fe^{3+}) with fluid reduction in the distal, intake or recharge portions of the hy-

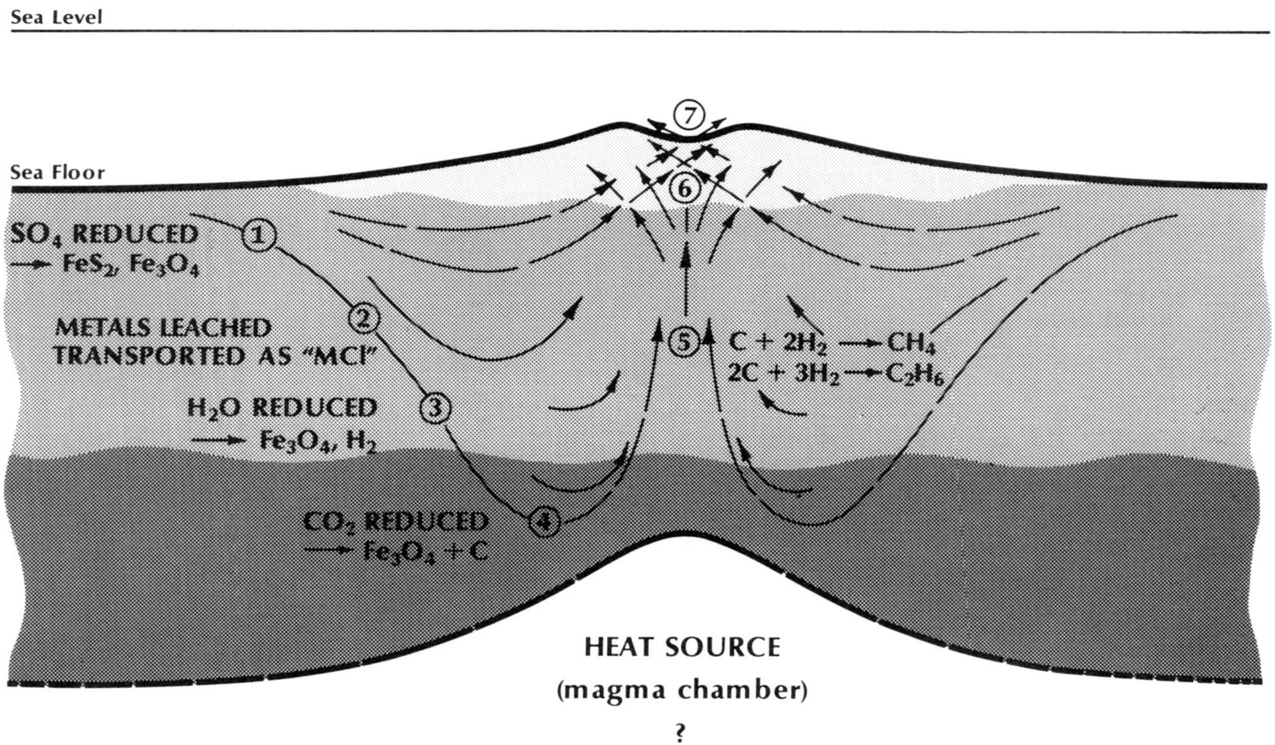

Figure 1A. Diagrammatic illustration of syn-sedimentary, sub-sea floor hydrothermal convective system (with explanation).

FIGURE 1A. *Convective circulation of saline conate or marine waters (in a system buffered by Fe_2^+— as ferrous silicate).*

A. *Downward (intake) circulation*
   - (1) *'Shallow level' — reduction of sulphate SO_4 + ferrous silicates → pyrite, magnetite.*
   - (2) *Leaching of metals from ferrous and other silicate structures, transport as various metal chloride complexes.*
   - (3) *'Deeper level' — dissociation of water H_2O + ferrous silicates → magnetite + H_2.*
   - (4) *CO_2 + ferrous silicates → magnetite + C, production of strongly reduced, H_2 and C-rich, increasingly saline fluid.*

B. *Upward (outflow) circulation.*
   - (5) *Carbon-hydrogen combinations; $C + 2H_2$ or $2C + 3H_2 \rightarrow$ light hydrocarbons, CH_4 or C_2H_6. Production of strongly reduced CH_4, C_2H_6-rich, highly saline fluid.*

C. *Exhalative vent on sea floor.*
   - (6) *'Deeper level' — in vent system below water-sea floor interface; alterations, stringer sulphides.*
   - (7) *'Surface' — at sediment-water interface: massive sulphide precipitation.*

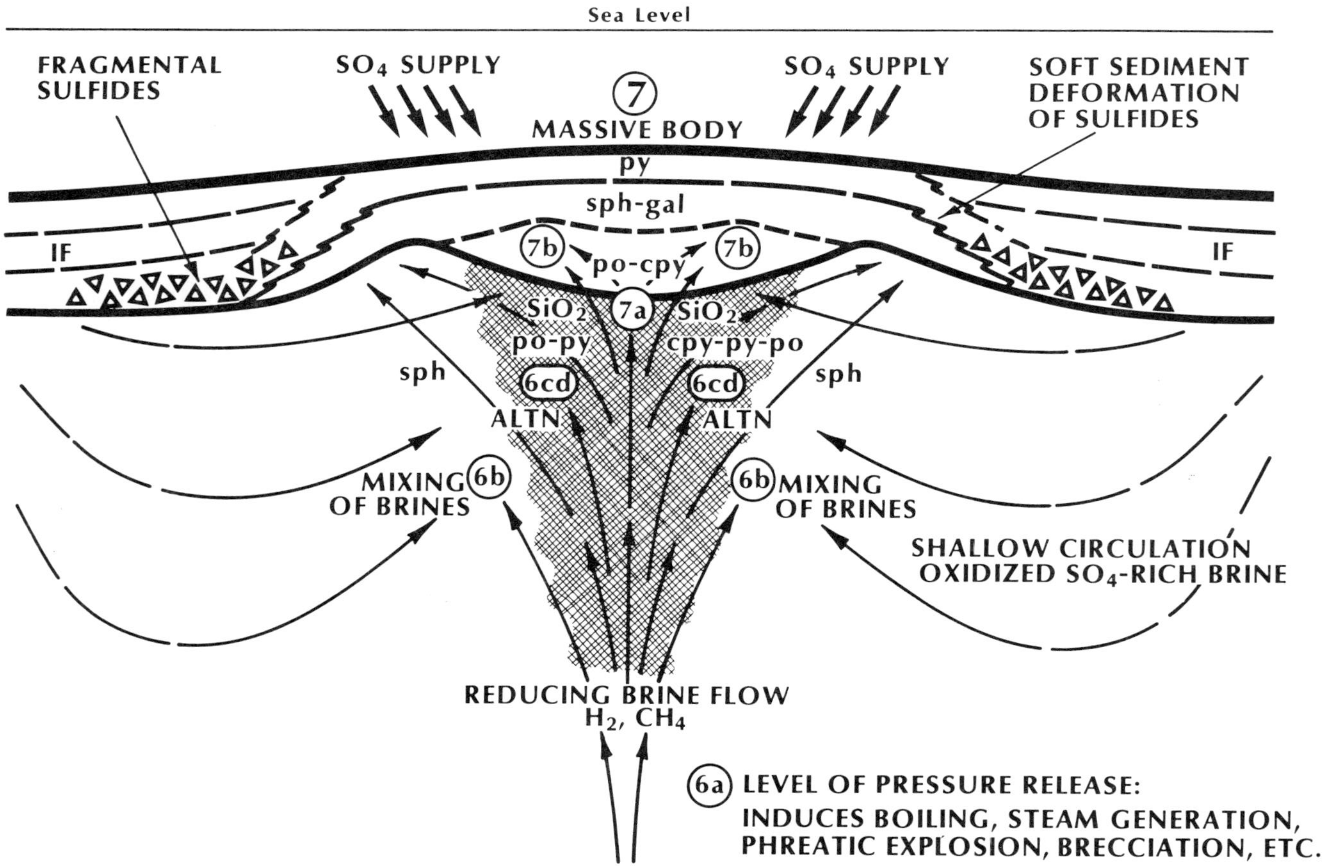

Figure 1B. Diagrammatic illustration of syn-sedimentary, sea floor hydrothermal exhalative vent (with explanation).

FIGURE 1B.

C. Exhalative vent on sea floor.

(6) 'Deeper level' — in vent system below water-sea floor interface.

(a) Pressure release boiling; phreatic explosion, brecciation begin.

(b) Reaction, mixing of shallow-circulated, oxidized sulphate-rich brine with deep-circulated reduced brine.

(c) Precipitation of FeS, FeS_2, $CuFeS_2$, SiO_2 in fractures.

(d) Rock alteration by reaction between gas, liquid fractions, and shattered porous rocks.

Production of zoned, but Cu-rich, stringer-disseminated ore (Keiko type) in altered rocks of vent system.

(7) 'Surface' — at sediment-water interface: massive sulphide deposition.

(a) Exhalation of reduced brine onto sea floor.

(b) Mixing, reaction of exhalative reduced brine with sea water; precipitation of all metallic sulphides, silica, carbonate, magnetite, etc. as 'exhalative' chemical sediment under varying Eh-pH conditions.

Massive banded orebody with prominent metal zoning (Oko, Kuroko), soft sediment deformed ore, fragmental ore, various 'facies' of iron formation, etc.

drothermal circulation system (Fyfe and Lonsdale, 1981; Holland, 1978). This is followed by pronounced rock reduction (increase of Fe^{2+}/Fe^{3+}),with accompanying sudden, disequilibrium fluid oxidation due to two-fluid mixing (Hutchinson *et al.*, 1980; Large, 1977; Andrews and Fyfe, 1976) around the proximal discharge vents (Figs. 1A, 1B). Here Na, Ca and K subtraction, K redistribution, Mg, Fe^{2+} addition and Al_2O_3, SiO_2 enrichment are common in the footwall rocks.

In these systems, the volume of rock leached of metal was very large – tens if not hundreds of km^3 (Browne and Ellis, 1970). A very low efficiency of metal extraction in the range of a few ppm was therefore sufficient to provide the large amount of metal which was concentrated by focused deposition at the discharge sites. Unfortunately, according to these concepts, those parts of the hydrothermal systems where metal was extracted are so large, or perhaps so deep, and the amounts of metal leached so small per unit volume (McGeehan, 1978) that it is extremely difficult, perhaps impossible to verify the concept by sampling and analyses.

Deplorably this situation is reminiscent of the "deep seated, hidden intrusion" that was so often invoked as a source of hydrothermal fluids and metals under the epigenetic replacement theory. While perhaps in this respect the newer exhalative hypothesis is equally unsatisfying, it nevertheless offers an explanation that is broadly in accord with the observed geological relationships, and is supported by geochemical data suggesting both nature and source of fluids and metals as well as the manner of metal transport and deposition. Magmatic input, except probably heat to drive the fluid circulation, is not essential although it is not ruled out, at least to a minor degree.

Significance of Metamorphism

The effects of metamorphism on sulphide ores are widely recognized, have been considered in detail (Vokes, 1969; Kalliokoski, 1965) and evaluated experimentally (Buerger, 1928; Clark and Kelly, 1973). They are important in several deposits described in this volume (Table I; Friesen *et al.*, Peredery *et al.*; Walford and Franklin; Speakman *et al.*; Froese and Goetz; deLorraine and Dill; all this volume), representing all three major Precambrian orogenic events; Kenoran at Geco, Hudsonian at Thompson, Anderson, Ruttan, Sherritt-Gordon, and Grenvillian at Balmat.

Yet their broader significance is too commonly overlooked or forgotten; it merits emphasis and continued recognition. If many sulphide deposits are exhalative and syn-depositional then their metamorphism, broadly defined to include *all* post-depositional changes, is inescapable: they must perforce have undergone the same periods, varieties, and ranks of metamorphism as their host rocks. The possibilities are multiple and may be recognized in various deposits described in this volume and elsewhere throughout the world; diagenesis, compaction-lithification, dynamic metamorphism, thermal and contact metamorphism, zeolite, greenschist, amphibolite and granulite facies regional metamorphism, retrogression and more than one of these superimposed.

The effects of metamorphism are therefore widely variable. Deposits like McArthur River (Murray, 1975; Oehler and Logan, 1977) have apparently undergone only compaction and lithification, involving very minor diagenetic changes in the ores. But the effects on the most intensely metamorphosed ores are extremely complex in deposits like Thompson, Geco, Sherritt-Gordon and Balmat. Plastic deformation due to ductility of softer sulphide and gangue minerals, cataclastic deformation of harder ones, interpenetration of the two, recrystallization, sulphide-sulphide and sulphide-silicate reactions, retrogressive re-equilibration, all combine in multiple variations to produce a bewildering array of altered minerals, new minerals, mineral intergrowths, textures and fabrics. Interpretations of genesis and paragenesis based on microscopic study of mineral textures and intergrowths in metamorphosed sulphide ores are subjective, of dubious reliability and therefore of limited value. They, as well as sulphide geothermometric and geobarometric studies may reveal much about the conditions of the last metamorphism, but little, if anything, about the pre-metamorphic, primary genesis. In yielding metamorphic rather than primary data, paragenetic textural, geothermometric and geobarometric sulphide studies are analogous to potassium-argon age dating.

Furthermore, compressional deformation has commonly produced discordant, penetrative texture or fabric on both large and small scales that may, if not recognized as secondary and metamorphogenic, be mistaken for primary discordancy. Many of the cross-cutting, penetrative relationships so long believed diagnostic of epigenetic origin in sulphide deposits under the epigenetic replacement hypothesis are, in fact, metamorphogenic, and "genetic red herrings" (Mookherjee and Suffel, 1968; Suffel, 1935). The lesson is clear. For all sulphide deposits in metamorphosed rocks, the possibility of original syn-depositional origin should be considered. Such an origin cannot be ruled out based on ore mineralogy, texture, fabric or morphology, all of which may constitute inadequate evidence of primary genesis. Even in relatively unmetamorphosed environments the possibility of changes due to diagenesis and compaction-lithification should be considered.

Significance of Minor Elements

As shown by Table I and the case histories in this volume, in addition to the major metals, i.e., Cu, Zn, Pb, Fe, and Ni, many minor or trace metals occur in exhalative deposits. Au and Ag are commercially the most important but Co, Sn, Cd, Sb, Bi, As and others are present. Although some of these are recognized as "belonging" in exhalative systems, others are not, chiefly because their higher concentration in other ore deposits has traditionally been explained solely by other processes. Yet the simple and obvious significance of their presence in exhalative deposits is commonly overlooked: like the major ones, these metals too must have been mobilized, transported and deposited by the same generative circulating hydrothermal fluids. The alternative possibility that they were added to earlier-formed exhalative deposits by a second, epigenetic hydrothermal fluid of uncertain but different derivation and nature is coincidental, illogical and unacceptable. Their presence, albeit in minor-trace amounts in established exhalative deposits, prompts consideration of the possible role of exhalative processes in forming other, more concentrated deposits of these metals.

Variety of Deposits and Environments

General. It is clear (Table I; case histories in this volume) that markedly differing exhalative massive sulphide deposits were formed in widely differing geological environments through nearly three billion years of Precambrian time. The differences are apparently sequential and evolutionary, accompanying and reflecting evolutionary changes in broad crust-mantle tectonic processes through time (Hutchinson, 1980). However, the full geological range of exhalative deposits and environments is not evident in those described in this volume, because some are best represented in Phanerozoic rocks.

The thirteen deposits interpreted as exhalative in this volume (Table I, nos. 1 to 13) fall into two major groups. The first, or *volcanogenic group,* includes eleven deposits (nos. 1 to 11), all rich relative to those of the second group in copper and gold, in one case with nickel and platinum group metals as well, which occur in predominantly volcanic rocks. All are Archean or Aphebian in age. The two remaining non-exhalative Archean-Aphebian deposits (nos. 14 and 15) are also rich in the same metals and have similar geological settings. The second, or *sedimentary group* includes two relatively lead- and silver-rich deposits (nos. 12 and 13) hosted in predominantly sedimentary rocks. They are Helikian-Hadrynian in age. Two of the three remaining non-exhalative Helikian-Hadrynian deposits (nos. 16 and 18) have similar metal contents and all three (nos. 16 to 18) have similar lithologic-stratigraphic-tectonic settings. Zinc content, high in most deposits of both *volcanogenic* and *sedimentary* groups cannot be used to distinguish between them, or indeed to group them. Within both groups some additional variations are discernible.

Volcanogenic Base Metal Deposits. Seven of the eleven deposits in the volcanogenic group, the Millenbach, Corbet, Detour, Bathurst Norsemines, Crandon, Anderson Lake, and Ruttan deposits are of a *primitive*, zinc-copper-rich type (Hutchinson, 1973, 1980) that, on a regional scale, is spatially related to felsic, commonly fragmental volcanic rocks, formed near centres of felsic volcanism. They occur in thick multicyclical, basaltic-andesitic-rhyolitic volcanic sequences. The first four of these are Archean, the last three Aphebian; clearly the geological environment represented by these rocks and ores appeared first in Archean, but continued into Aphebian time. Moreover comparable deposits and host environments occur in still younger Phanerozoic sequences (Hutchinson, 1973, 1980).

At the Brouillan and Agnico-Eagle deposits (Deptuck *et al.*, this volume; Barnett *et al.*, this volume), petrographic evidence indicates that Archean felsic volcanism which accompanied the hydrothermal ore-forming processes was partly subaerial. Although this suggests that a subaqueous volcanic environment may not be quite so important to the genesis of Archean exhalative massive sulphide as has formerly been assumed, it is also pertinent that to date, no major stratabound lens of massive sulphide has been located at Brouillan, where the sulphides are contained in fractured, variably altered, felsic pyroclastic-fragmental volcanic rocks. The Agnico-Eagle gold orebody, although spatially closely associated with very similar rocks, is itself, clearly a stratified, waterlain deposit.

Although there are no distinguishing differences between these particular Archean and Aphebian examples, consideration of other Proterozoic deposits, like those within the Sudbury basin (Martin, 1957) or at Prescott, Arizona (Gilmour and Still, 1968) indicates that lead and silver become more abundant at the expense of copper and gold in these other Aphebian examples. They have been termed *polymetallic* type (Hutchinson, 1973, 1980) that is best developed in Phanerozoic rocks.

Also within the volcanogenic group, but differing with the primitive, felsic volcanic-associated type are Geco (Friesen *et al.*, this volume); an Archean deposit and Sherritt-Gordon (Goetz and Froese, this volume), a Proterozoic one. Both lie in more highly metamorphosed terrain: both lack significant thickness of felsic volcanic rocks and regional, thick, basaltic-andesitic-rhyolitic volcanic successions; though mafic metavolcanic rocks are present they are not thick. Conversely, thick paragneiss of probable sedimentary derivation, possibly meta-greywacke, suggests a much greater original sedimentary component in the host rocks of these two deposits. Relatively continuous and distinctive, cordierite-anthophyllite-rich footwall strata, rather than pipe-like footwall alteration zones are present at both. This geological setting resembles the *kieslager* or *Besshi* massive base metal sulphide environment (Hutchinson, 1980) of impure sedimentary and mafic volcanic rock sequences. Yet here seafloor exhalative activity produced ore similar to that of the primitive, felsic volcanic environment, although the *Besshi* type appears even poorer in lead and precious metals than the latter, and slightly richer in copper relative to zinc.

Improved understanding of the differences in metallogenic environment of these two types, and of their mutual tectono-stratigraphic relationships within Archean and Aphebian successions is vital to exploration for other deposits like Geco and Sherritt Gordon. The relatively easily identified felsic volcanic environment of the primitive type has been extensively explored in the Canadian Shield, but the sedimentary-mafic volcanic rock environment of the *Besshi* type is, as yet, relatively unrecognized and untested.

Volcanogenic Gold Deposits. The evidence for exhalative origin of the Agnico-Eagle gold deposit (Barnett *et al.*, this volume) demonstrates the importance of exhalative processes in forming greenstone belt lode-gold deposits, as previously discussed (Ridler, 1970; Fryer *et al.*, 1979). Although it does not always form mineable deposits, and later metamorphic remobilization may be a necessary factor in forming a workable gold concentration, exhalative gold deposition may prove to be a widespread, perhaps essential preconcentration mechanism for generation of these deposits. It is interesting that lode-gold deposits in Archean greenstone belts commonly occur in the same districts as massive base metal sulphide deposits, as in the Val d'Or, Noranda, Joutel and Timmins districts of the Abitibi belt, but are quite rare in the otherwise similar Aphebian greenstones and massive base metal sulphide districts of northern Wisconsin and northern Manitoba-Saskatchewan (Franklin and Thorpe, this volume). The reasons for this are not understood. Perhaps, if generally substantiated by additional data from other areas, the somewhat different petrochemistry of Aphebian basalts at Anderson Lake from Archean basalts (Walford and Franklin, this volume) may be one related factor.

Volcanogenic Nickel Deposits. The importance of exhalative processes in forming nickel sulphide deposits classified by Naldrett (1973) as of volcanic association was suggested by Lusk (1976a), subsequently discussed (Groves *et al.*, 1976; Lusk, 1976b), and is indicated by the geological and geochemical relationships at the Redstone deposit near Timmins (Robinson and Hutchinson, this volume). Other examples of possible exhalative nickel sulphides occur at Sherlock Bay in the Pilbara region (Miller and Smith, 1975) and at Jan Mine in the Kambalda district (Gresham and Loftus-Hills, 1981), both in Western Australia. Closely associated with more typical, possibly magmatic-type ore at Jan Mine are thin beds of massive pyrrhotite-pentlandite-chalcopyrite interlaminated with black albitic chert (Bavinton, 1981): some of the interleaved strata are planar and undeformed but others show apparent soft-sediment deformation of the albitic cherty bands within sulphide layers (Fig. 2), as well as later tectonic deformation. Considering the nature of exhalative systems (Hutchinson *et al.*, 1980), exhalative nickel sulphides are perhaps to be expected where sea floor hydrothermal circulation affected Ni-rich, ultramafic-komatiitic Archean lavas so commonly associated with these ores. This does not negate the importance of magmatic processes which, in the same districts may form disseminated nickel sulphides in serpentinized ultramafic rocks and perhaps other massive sulphide bodies. However, confirmation of an exhalative origin for some of these ores should prompt objective evaluation of the possible role of exhalative processes in forming deposits of this type, all of which have heretofore been generally explained by magmatic processes. Although difficult, it seems essential to evaluate and understand the relative importance and inter-relationships of magmatic and exhalative processes in forming these deposits, because *both* (not one or the other) have apparently played a part in ore genesis.

In the above context the origin of the important nickel orebodies of northern Manitoba merits reconsideration. Here all primary relationships are obscured by the effects of high-rank regional metamorphism, but the deposits have many broad characteristics of volcanic-affiliated nickel sulphide deposits (Peredery *et al.*, this volume; Hutchinson, 1981b; Fig. 2). They have very high nickel to copper ratios. They are associated with highly serpentinized ultramafic rocks, some of which may be spinifex-textured flows, and also with iron-formation, which is clearly exhalative. Moreover, although 75% of ore in the belt is associated with ultramafic rocks, 25% is intimately associated with sedimentary rocks (Peredery *et al.*, this volume), and the

Figure 2. Slab of interlaminated black chert and nickel sulphides, with soft-sediment deformation structures, Jan Mine, Kambalda District, Western Australia.

latter proportion would be much higher for the important Thompson orebody alone. The volume of ore at Thompson is far too great to have been derived directly, in situ by magmatic segregation from the small associated ultramafic body, and perhaps too great also for prior, indirect differentiation of sulphide and silicate liquids at depth, with subsequent injection or extrusion(?) of the separated liquids. At Pipe #2 mine there is a close spatial association of massive ore with exhalative iron-formation and an even closer one with dolomite and diopsidic skarn (Peredery *et al.*, this volume, Fig. 16) which may represent metamorphosed carbonate-facies iron-formation. In this respect it is significant that very similar, calc-silicate strata occur intimately intercalated with iron and base metal sulphides, chert, cherty tuffite and rhyolitic tuff at Bathurst Norsemines (Casselman and Mioduszewska, this volume) and, although subsequently metamorphosed (as in the Thompson belt) all these rocks are attributed to sea floor exhalative activity. Yet a magmatic origin is still preferred by most geologists for the iron-nickel sulphides of the Thompson belt. Considering all these relationships however, perhaps exhalative processes may offer a better alternative genetic explanation.

Exhalative Sedimentary Base Metal Deposits. Within this group two varieties, representing opposite extremities of a continuous spectrum of sedimentary depositional environments are recognizable. At one extremity is the important Sullivan deposit, which was clearly formed in relatively deep water, probably in a rift-controlled trough or graben in association with fine clastic, turbiditic argillite (Hamilton *et al.*, this volume; Hoy, this volume). Its ores are extremely lead- and silver-rich compared to those of the opposite extremity and may be slightly older, possibly paleo-Helikian. To this deposit-type belong many other mid-Proterozoic examples such as Broken Hill, McArthur River, Hilton and the Racecourse orebodies at Mt. Isa, in Australia (Mathias and Clark, 1975; Oehler and Logan, 1977; Murray, 1975; Johnson and Klinger, 1975). At the other extreme are the relatively zinc-rich, lead- and silver-poor Balmat orebodies which, if exhalative as the authors suggest (De Lorraine and Dill, this volume), were deposited in a shallow water, shelf carbonate-evaporite environment.

The existence of carbonate-hosted, exhalative massive zinc-lead sulphide deposits which are not of Mississippi Valley-type (Hutchinson, 1980) is best illustrated by the Irish deposits of Tynagh, Silvermines, and Navan in Mississippian limestone (Russell, 1968; Pereira, 1967; Derry *et al.*, 1965), or the famous Reocin deposit in Cretaceous dolomite near Santander in Spain (Monseur and Pel, 1972; Monseur, 1967). In comparison with these it seems probable indeed, that the Balmat orebodies are a highly metamorphosed Precambrian example of this type.

Epigenetic Carbonate-Hosted Deposits. Descriptions of the Redstone Copper Belt, Gayna River and Nanisivik deposits (Ruelle; Hewton; Clayton and Thorpe; all this volume) clearly show that these are not of exhalative origin. The sulphides occupy secondary porosity voids, are discordant, and in some cases trangress facies boundaries. These deposits therefore have been generally classified as epigenetic, carbonate-hosted; and Gayna River is compared to those of the Phanerozoic Mississippi Valley Type, to which it indeed bears many resemblances. Nevertheless, there is also abundant evidence in the descriptions of these deposits to suggest that, although exhalative activity did not form the ores directly, it may have accompanied deposition of their host rocks in some cases, and also to indicate that syndepositional-diagenetic hydrothermal processes rather than epigenetic ones were important in forming these ores. This evidence is discussed, and its possible significance to origin of carbonate-hosted base metal sulphide deposits is interpreted in the following paragraphs.

There are broad but significant geological similarities between the synsedimentary exhalative, carbonate-hosted massive sulphides, on the one hand, and epigenetic, Mississippi Valley-type deposits on the other. Both are zinc-rich with significant lead; all are low in silver, although the exhalative ores tend to be somewhat richer than the epigenetic ones. Both occur in similar shelf-facies clastic-carbonate-evaporite sedimentary sequences. Considering these broad similarities perhaps the main distinction between the two types is whether generative, broadly syn-depositional hydrothermal fluids escaped to the sea floor, forming a synsedimentary exhalative massive orebody, or whether, before reaching the sea floor, these fluids were otherwise channelled through fractured or porous, poorly consolidated shelf sediments, into favourable stratigraphic traps such as reefs or their flanking calcarenites, domal upwarps, pinch-outs or karstic carbonate beds. Perhaps, in other cases, some fluid escaped to the seafloor while the remainder was channelled into porous sediment, producing in the same district a puzzling and complex intermingling of exhalative-syn-depositional and diagenetic-epigenetic relationships. These relationships may have been further complicated by syn-sedimentary autobrecciation or karstification during brief marine regressions in the shallow shelf environment. Numerous geological relationships in the Redstone Copper Belt, Nanisivik and Gayna River deposits (Ruelle; Clayton and Thorpe; Hewton; all this volume), and also regionally in the Mackenzie Fold Belt (Aitken, this volume) indicate what thus might be termed a "combined exhalative-diagenetic" origin.

One essential geological element in all exhalative systems is an underlying feeder-fracture network for upward movement of hydrothermal fluids. Footwall faulting is an important structural and ore control at the Corbet volcanogenic massive base metal sulphide deposit (Knuckey and Watkins, this volume). It appears also to have influenced sedimentation and sulphide deposition of the exhalative sedimentary Sullivan orebody (Hamilton *et al.*, this volume). Syn-depositional faulting controlled reef growth, stratigraphic relationships and sulphide mineralization in the Irish carbonate-hosted massive lead-zinc deposits (Lees, 1964; Russell, 1968). It is significant that syn-sedimentary growth faults, commonly in more or less rectilinear patterns, are also associated with mineralization, and are prominent structures controlling basin morphology and stratigraphic relationships in the Redstone Copper Belt, at Gayna River, Nanasivik, and regionally in the Mackenzie Fold Belt (Eisbacher, 1977; and relevant papers, this volume).

Syn-depositional basaltic igneous activity accompanied the tensional faulting, as evidenced by thin flows and dikes in the lower Little Dal group of the Mackenzie Mountains, and thick flows below the Society Cliffs-Arctic Bay carbonates at Nanisivik. This suggests that the faults were deeply penetrative, presumably tapping the mantle. The basalts are invoked as the source of copper for the Redstone Copper Belt deposits, although surface weathering is suggested as the mobilizing agent (Ruelle, this volume). Alternatively, low temperature hydrothermal leaching accompanying diagenetic burial metamorphism of the basalt (Jolly, 1974) could have provided syn-sedimentary copper-bearing solutions. These may have been discharged directly into some embayments, producing syn-sedimentary copper-rich strata of "great lateral continuity and uniformity" (Ruelle, this volume) like the #1 bed in the Coates Lake embayment. But elsewhere, as in the Keele River embayment, they may have circulated within already deposited, cryptalgal carbonate strata, depositing sulphides epigenetically in gypsum-anhydrite dissolution casts, or other voids, producing deposits like Jay and Jane creek which transgress facies boundaries.

Another significant geological aspect regarding possible origin of the sulphides and derivation of the metalliferous hydrothermal fluids in the Redstone Copper Belt, and one which also suggests syn-depositional hydrothermal activity, is the mineralogical composition of the associated evaporites. Interpretation of evaporite mineral assemblages is difficult because of the many burial-diagenetic alteration and metamorphic changes among evaporite minerals. Nevertheless, particularly in view of the low metamorphic rank of these Redstone strata which may never have been deeply buried, the evaporite minerals in the Redstone Copper Belt do not appear to represent a normal marine evaporitic assemblage. They lack the normally abundant, early, stratigraphically low, gypsum-anhydrite, which at Redstone is sparse and increases *upward*: they lack the abundant salt which should precede in stratigraphic position such mixed sulphates as glauberite: they contain abundant glauberite which predominates at the stratigraphic base: they totally lack potassium salts which normally accompany and precede in stratigraphic order the magnesium-rich bittern salts: they contain appreciable magnesite and talc, not normal evaporite minerals, and silica (in talc), also not a normal evaporite component.

These abnormalities are attributed to the influence of basaltic volcanism (Ruelle, this volume). This is a particularly significant interpretation, for it indeed explains the abnormally Mg- and Si-rich nature of these evaporites, as indicated by their abundant magnesite and talc. It also explains their probable abnormally high depositional temperatures, as indicated by the abundance of glauberite, the stability field of which increases markedly in evaporite systems at temperatures above 25°C (Stewart, 1963). The high temperature in particular, strongly suggests that the basaltic volcanism had a direct, coeval influence on evaporite sedimentation. Although this condition would not be satisfied if fluids transporting Mg and Si were derived from basalt weathering, synvolcanic-diagenetic hydrothermal leaching of basalt at temperatures of 200 to 300°C (Jolly, 1974), would provide both suitable fluid composition and temperature.

An interesting comparison is with Unit 13 at Balmat, an unusual talc-tremolite-anthophyllite schist which is interpreted to have formed by high rank regional metamorphism from an evaporitic protolith (deLorraine and Dill, this volume), very similar to that described in the Redstone Copper Belt. This schist also occurs in rocks of similar Helikian age, but here in close association with Zn-Pb-rich, rather than Cu-rich sulphide deposit. Again, as in the Redstone Copper Belt, there is a regional, although not direct spatial association of the schist and the Balmat Zn-Pb ores with amphibolite and metagabbro, possibly metamorphosed mafic igneous rocks. The possibility of original deposition in a rift-controlled, continent-margin or intracratonic basin strikingly like that of the Redstone Copper Belt is considered by deLorraine and Dill (this volume).

Under a combined exhalative-diagenetic hypothesis, sulphide emplacement must have followed shortly after sedimentation, probably before lithification and deposition of the overlying beds. This relationship is advocated for both the Redstone Copper Belt and Nanisivik deposits (Ruelle, this volume; Clayton and Thorpe, this volume). It is also strongly indicated by textures described in the Gayna River deposits. Clasts of sedimentary breccias in secondary breccias, and soft sediment deformation of dolostone clasts in sphalerite matrix (Hewton, this volume) all suggest prolonged but closely contemporaneous carbonate sedimentation, diagenetic dolomitization and mineralization. Indeed early, albeit minor stages of mineralization appear to have been exhalative sedimentary, producing bedded sphalerite that was subsequently fragmented by depositional-diagenetic processes, and included in the sedimentary breccias. However mineralization must have been prolonged because its more important later stages were clearly diagenetic, filled pores in sedimentary breccias, sealed the interstices of collapse breccias, but permitted deformation of unconsolidated, included dolostone clasts.

Geological relationships at Gayna River are important to understanding of the genesis of Mississippi Valley-type deposits. The stratigraphic evidence against contemporaneous metal derivation from nearby shales, and sulphate derivation from evaporites to form the deposits (Hewton, this volume) as suggested for Pine Point (Beales and Jackson, 1966) raises questions about the general applicability of this hypothesis. The evaporites at both Gayna River and in the Redstone Copper Belt are obvious possible sulphur (sulphate) sources and are suggested as a possible metal source at Gayna (Hewton, this volume), but no supporting evidence of the latter is presented. In both instances syn-sedimentary exhalative hydrothermal activity, continuing through carbonate-evaporite diagenesis may offer an acceptable alternative for metal supply as well as a thermal agent to promote sulphate reduction. Perhaps deeply circulating waters, heated either directly by the regional mafic igneous activity, or by low rank, regional, burial-diagenetic alteration of the basalt, to about 200 to 300°C leached metals from the carbonates, shales, basalts or basement rocks, carrying them directly into the reefal platform carbonate depositories. There is suggestive evidence at Gayna River that reefal mounds were controlled by active growth faults. Such highly reduced, metal-charged hydrothermal fluids introduced here, both syngenetically and diagenetically, would have mixed with cooler connate or sea water and might themselves, or aided by bacteria, have reduced evaporitic sulphate with resulting metal sulphide deposition.

If so, resulting deposition of Zn-Pb sulphides in what appear to be slightly deeper water carbonate-evaporitic rocks at Gayna River contrasts with deposition of Cu sulphides in very shallow water (sebkha) evaporitic-clastic strata in the Redstone Copper Belt (Hewton, this volume; Ruelle, this volume). Although more than 300 km distant from one another along strike, these two different deposit types occur in the same part of the stratigraphic column. The widespread Gayna mineralization lies in the Little Dal Group, and the even more widespread Redstone mineralization at the top of the overlying Redstone River Formation, which is absent, possibly due to either erosional removal or non-deposition at Gayna. The time-stratigraphic correlation between the host rocks of these deposits is therefore unknown, but may be very close. In this case their contrasted metal compositions may represent regional scale metal zoning. This might be a result of their slightly differing lithofacies environments and/or of regional temperature zoning. Coeval mafic igneous activity, indicated by basaltic flows intercalated in the Little Dal Group, appears to have been more extensive to the southeast near Redstone than to the northwest near Gayna, where no flows are reported (Eisbacher, 1981; Ruelle, this volume; Hewton, this volume). This may account for higher fluid temperatures and copper contents at Redstone.

Iron-silica-rich chemical sedimentary strata are common associates of exhalative deposits (Hutchinson, 1973) and therefore indicators of exhalative activity. Prime evidence of regional exhalative activity in the Mackenzie Mountains Supergroup, albeit at a higher stratigraphic position than the base metal ore deposits, are the large Snake River iron formations in the Sayunei Formation (Aitken, this volume). Moreover, Hewton (this volume) described iron formation in the lower limestone unit at Gayna River, just below the lower limit of the sulphide concentrations, as well as thin bedded pyrite in the argillaceous marker between the lower and upper host carbonates, siliceous oolites throughout the lower and upper host unit, and chert as nodules and reticulate beds forming a marker horizon near the top of the upper host. All this is suggestive evidence for syn-sedimentary deposition of iron and silica in the shallow marine shelf environment. Similarly Ruelle (this volume) described white and dark grey chert beds replacing carbonate beds at the base, and throughout the top 80 m of the upper Copper Cap Formation. But no evidence of replacement was cited. Might these siliceous strata be primary? If replacements, presumably they were diagenetic ones. In either case, where was the source of silica, in what solution was it transported, and how precipitated? These problems may perhaps be more readily explained by syn-sedimentary-diagenetic siliceous hydrothermal exhalative activity than by epigenetic replacement. At Nanisivik, no clearly chemical ferruginous sedimentary strata are described, but poorly described and understood hematitic outcrops south of the western end of Twin Lakes, and along the south boundary fault (Clayton and Thorpe, this volume) may be of this derivation.

Exhalative deposits commonly include epigenetic mineralization in brecciated or fractured footwall rocks that mark the upper portions of the vents through which the fluids moved to the sea floor. It is interesting that at Nanisivik a little known but apparently vertical sulphide vein extends downward, transecting the strata, for at least 60 m below the main orebody, and that other similar mineralization is suspected from drilling (Clayton and Thorpe, this volume). Similarly at Gayna River, fault- or fracture-controlled replacement mineralization is described (Hewton, this volume). Although some is clearly later, possibly the result of remobilization during later Cordilleran fold belt deformation, some may represent fracture infilling beneath exhalative vents.

Vein Deposits. The Opemiska and Chibougamau veins are perhaps the most enigmatic deposits considered in this volume. They occur in Archean rocks, yet recent lead isotopic ages suggest that some of the galenas are Aphebian (Watkins and Riverin, this volume; Thorpe *et al.*, 1981), although one sample from the Opemiska mine has an Archean composition. The precision of the lead isotopic data, however is not optimum (Thorpe *et al.*, 1981). Moreover, the multistage lead growth additions of uranogenic lead and complex geological history do not permit a unique interpretation of the deposit's age from the lead isotopic data. The Aphebian age interpretation from lead isotopic data also conflicts with an Archean age indicated by earlier studies for the Cedar Bay (Chibougamau) deposits (Wanless *et al.*, 1973). The geological setting of the deposits, discordant veins in semi-concordant anorthosite or layered gabbro, is unlike that of all of the Archean-Aphebian exhalative deposits, with the possible exception of the extensive discordant stockwork mineralization at Detour (Deptuck *et al.*, this volume). Yet typical Archean exhalative deposits occur in the volcanic rocks above the host anorthosites and gabbro at Chibougamau and Opemiska, and this relationship, together with other evidence, led to the suggestion (McMillan, 1972; Allard, 1976) that the Opemiska, and Chibougamau veins were deposited in fractures below sea floor vents for exhalative hydrothermal solutions. This explanation is rejected (Watkins and Riverin, this volume) firstly, because the veins appear to fill late fractures generated by folding of the gabbro, secondly because the vein walls exhibit different alteration mineral assemblages from those present in many footwall stockworks of exhalative deposits (Knuckey *et al.*; Knuckey and Watkins; Walford and Franklin; Casselman and Mioduszewska; Speakman *et al.*, all this volume), and thirdly because of the admittedly indefinite (Thorpe *et al.*, 1981) radiometric age evidence. The alternative interpretation of epigenetic, Aphebian vein formation however, raises three unanswered questions. Firstly, what was the source of, and how were the vein-forming hydrothermal fluids generated within the earlier-cratonized Superior structural province in Aphebian time? There is only indirect paleomagnetic (Morris, 1977) and equivocal radiometric age evidence, and no direct geological evidence of a potential Aphebian source or generative event. Secondly, how is the regional association between the veins and Archean exhalative deposits to be explained? Thirdly, what caused the metamorphism documented in the Henderson (Chibougamau) deposit (Guha and Koo, 1975)? Perhaps a multi-stage origin involving the addition of Aphebian radiogenic lead to Archean copper-gold veins may better explain most of the observed relationships. If so, the possibility that the veins represent mineralized feeder-fractures of exhalative Archean deposits cannot be ruled out. Differing alteration characteristics of the vein walls may reflect only the differing bulk composition or depth of the host rocks.

Porphyry Copper Deposits. In marked contrast to their abundance, widespread distribution, size and importance in Phanerozoic rocks, porphyry type-copper occurrences are rare and small in Precambrian rocks throughout the world. In North America the only producing Precambrian example is in the MacIntyre Mine at Timmins, Ontario (Davies and Luhta, 1978), primarily a gold deposit, where H.S. Robinson, to whom this Special Paper is dedicated, spent much of his professional career. A number of other submarginal occurrences in Canada are briefly described by Franklin and Thorpe (this volume). Perhaps the only large and potentially important Precambrian deposit of this type is the Haib in mid-Proterozoic rocks just north of the Orange River in Nāmibia (Minnitt, 1979).

The reasons for the rarity and relative insignificance of these deposits in Precambrian rocks are imperfectly known. Hutchinson and Hodder (1972) discussed numerous similarities between porphyry copper deposits and the epigenetic, copper-rich, highly altered, mineralized footwall stockworks or breccia zones beneath many volcanogenic massive base metal sulphide deposits, as at the Millenbach, Corbett, and Detour deposits (Knuckey *et al.*; Knuckey and Watkins; Deptuck *et al.*; all this volume). They also noted strong similarities between active exhalative hydrothermal brines in the Red Sea deeps (Craig, 1966; Bischoff, 1969) and sub-surface hydrothermal brines of the Salton Sea area (White, 1968, 1974). They and others have suggested that these two different types of deposit were formed by similar fluids; syn-sedimentary massive sulphides resulted where the fluids were discharged to the sea floor, whereas epigenetic porphyry-type deposits were generated in subsurface, subvolcanic pluton-related environments. If so, the markedly different distribution through time of the two types of deposit must reflect evolutionary crustal tectonic processes. Perhaps during Archean and Aphebian time supracrustal sequences and ensialic blocks were thin, immature, and dominated by tensional tectonics (Hutchinson, 1981a) allowing ready access to surface for both magmas and hydrothermal fluids, but only incipient development of subvolcanic environments. Stable, thickened, continental blocks apparently evolved only in later Precambrian time, following successive, widespread, granite-generating Kenoran, Hudsonian and Grenvillian orogenies (see also "Regional Geology and Metallogeny"). Only in the last billion years has resulting thick, granitic crust been involved in lateral plate motion, with highly compressional tectonics where it overrides oceanic lithosphere along subductive plate boundaries. This tectonic envionment may have "sealed" or "capped" many magmatic-hydrothermal systems, inhibiting extrusive-exhalative activity but favouring subvolcanic plutonism and related hydrothermal circulation. It is significant that consuming plate margins are the main loci for porphyry type copper deposits (Mitchell and Garson, 1972). In this sense porphyry deposits may represent the subvolcanic equivalents of volcanogenic massive base metal sulphide deposits.

REGIONAL GEOLOGY AND METALLOGENY

The four regional metallogenic papers in this volume outline the broad geological setting and metallogeny of several important and diverse Precambrian terrains. The diversity is again both geographical and temporal. The terrains span North America; they include its two ancient Archean nucleii, their younger, more highly metamorphosed Aphebian and Grenvillian flanks, and less metamorphosed outlying Helikian-Hadrynian rocks in the southern and northern Cordilleran orogen. Geological understanding of these terrains has progressed markedly. It is now clear that, except for younger intrusions, the Superior and Slave structural provinces, which form the ancient Archean nucleii of the Canadian shield, have been essentially unaffected by Post-Kenoran orogenic activity. However the flanking Churchill and Grenville Provinces of the central Shield (Franklin and Thorpe, this volume; Sangster and Bourne, this volume) include both older, polymetamorphosed Archean rocks and younger less metamorphosed Aphebian - Helikian strata. The various Precambrian structural provinces and terrains have been cratonized in successive stages and at differing times; the two Archean nucleii during Kenoran orogeny from about 2.7 to about 2.5 Ga, the Churchill and Southern Provinces during Hudsonian and Penokean orogenies respectively at about 1.7 Ga and the Grenville during Grenvillian orogeny at about 1.1 Ga. In many places, Aphebian-Helikian-Hadrynian strata must have formed, and in some places still remain as little metamorphosed, flat-lying cover sequences on the older Kenoran-Hudsonian-Grenvillian cratonized blocks. Some cover sequences were only much later deformed during Cordilleran belt tectonism (Hoy, this volume; Aitkin, this volume).

The areas affected by the successive Precambrian orogenies exhibit changing shapes through time. The two Archean and Churchill Provinces are essentially equidimensional: broad tracts throughout which the metamorphic effects were widespread and more or less similar. Thus Kenoran and Hudsonian orogenies seem to have affected broad sectors of the earth's crust. The extent of Penokean orogeny is obscured by flat Phanerozoic cover sequences. The youngest, the Grenvillian orogeny however, seems to be the first of the Precambrian orogenies that was elongated but narrow in extent. It affected a belt only about 300 miles wide but at least 1000 miles long, and probably very much longer if, as indicated (Sangster and Bourne, this volume, Fig. 1), Grenvillian metamorphites extend southwestward beneath Phanerozoic cover rocks of the mid-continent to Mexico and northeastward across the Atlantic to Norway. The Grenvillian orogeny therefore appears to have been intermediate in extent between the older, more or less equidimensional Precambrian orogenies and the linear, narrow Phanerozoic orogenic belts which mark the consuming plate margins of sialic continental blocks. The reasons for the apparent evolutionary change in dimensions of orogenic activity through time are not known, nor is its significance understood, but it must somehow be fundamentally related to the evolution of tectonic processes.

Within the highly metamorphosed Churchill and Grenville Provinces it is difficult to distinguish the older twice- or thrice-metamorphosed Archean segments from younger once- or twice-metamorphosed Aphebian or Helikian segments owing to the overprint of the younger metamorphism. The types of mineral deposits present may, however, aid in the distinction and provide a useful guide to primary age. There are distinctive differences between Archean-Aphebian exhalative massive base metal sulphide deposits and Helikian-Hadrynian ones (Table I). The former are copper-, (or nickel-) and gold-rich and occur predominantly in volcanic rocks, whereas the latter are lead- and silver-rich and occur predominantly in sedimentary rocks. The absence of volcanic-hosted massive base metal sulphide deposits in mid-Proterozoic rocks has been termed the volcanogenic gap (Hutchinson, 1980). These differences generally are also reflected by the regional metallogenic contrast between the many, Cu-Au-rich volcanic-hosted massive sulphide orebodies in Archean-Aphebian rocks of the Superior, Slave and Churchill structural provinces, and the few, small Pb-Ag-bearing vein occurrences in mid-Proterozoic sedimentary rocks of Richmond Gulf-Nastapoka Islands and the Labrador Trough (Franklin and Thorpe, this volume). The same differences also appear on other continents, as in the volcanic-hosted Archean nickel, copper, zinc and gold deposits of Western Australia in comparison with the mid-Proterozoic sedimentary-hosted zinc-lead-silver deposits in Eastern Australia (Knight, 1975; cf. Chapters 3 and 4 with 9, 10, 19). This marked evolutionary change in metal composition and geological setting of exhalative massive base metal sulphide deposits must also have some broad and fundamental cause in supracrustal tectonic evolution.

Although no distinction is apparent between Archean and Aphebian massive base metal sulphides described in this volume (Table I), some of the latter are richer in lead than the former. Moreover, notwithstanding the similarities in the massive sulphide deposits, there are other significant differences in Aphebian and Archean metallogeny. Lode gold deposits are numerous and important in Archean greenstone belts, but for unknown reasons are rare in similar Aphebian ones (Franklin and

Thorpe, this volume) and, like volcanic-hosted massive sulphides, apparently absent in Helikian-Hadrynian rocks. This is particularly enigmatic considering the similarity of Archean and Aphebian greenstone belts and massive base metal sulphide deposits, and the almost ubiquitous presence, in the gold lodes, of iron and lesser amounts of base metal sulphides. Despite the lack of basic understanding of these broad metallogenic differences, metallogeny and major types of mineral deposits may provide important and useful guides in determining the ages of differing terrains within highly metamorphosed crustal blocks.

SUMMARY AND CONCLUSIONS

Recent studies suggest that the metalliferous hydrothermal fluids which formed most Precambrian, and younger sulphide deposits were mainly re-circulated marine or meteoric waters. These were heated, either directly by igneous activity or through deep, convectively driven circulation. Water-rock reactions caused them to become increasingly saline, reduced and acidified, and thereby to leach metals from the rocks through which they passed.

The hydrothermal activity that formed most Precambrian sulphide deposits was, in a broad sense, syn-depositional rather than epigenetic, and generally co-eval with deposition of the host rock sequences. Exhalative deposits, the most numerous and important resulting type, include thirteen of the eighteen described in this Special Paper, and represent the exogenous end-products of this activity. Other, non-exhalative sulphide deposits, like the stockworks at Detour and perhaps the enigmatic veins of Opemiska and Chibougamau, may be endogenous, related to subvolcanic or plutonic phases of the same hydrothermal activity. Still other deposits, like those of the Redstone Copper Belt, Gayna River and Nanisivik, may be diagenetic, also formed by the same hydrothermal activity but within surficial, recently deposited, but unconsolidated sediment.

Exhalative discharge of metalliferous hydrothermal fluids to the sea floor, with ensuing chemical precipitation, was the major sulphide-depositing process during Precambrian time. It formed all of the important massive base metal sulphide deposits. Volcanic-hosted copper-gold-rich types, like those of the greenstone belts in northern Quebec, Ontario, Manitoba and the N.W.T. predominate in older Archean-Aphebian rocks of the Superior and Slave structural provinces, but were superceded by sedimentary-hosted lead-silver-rich ores like Balmat, N.Y. and Sullivan, B.C. in younger Helikian-Hadrynian rocks of the flanking Grenvillian and Cordilleran belts.

Exhalative processes also played a part in forming many other types of Precambrian sulphide deposits although their specific roles here, and their importance relevant to other genetic processes are not yet fully understood and require further investigation. Nevertheless, exhalative activity appears to have formed some volcanogenic nickel sulphide deposits, like the Redstone River occurrence near Timmins, Ontario, and possibly those of the Thompson Belt, Manitoba, which occur in ultramafic, komatiitic extrusive rocks. It was also responsible for deposition of stratiform, gold-rich chemical sedimentary strata. Although only a few of these, like the Agnico-Eagle deposit or the ankerite veins at Dome Mine, are themselves of minable tonnage and grade, these strata may be essential premetamorphic precursor source rocks of many discordant greenstone belt vein or stockwork gold orebodies.

The rarity and small size of porphyry-related sulphide deposits in Precambrian rocks in comparison with numerous, important exhalative deposits probably reflects relatively primitive supracrustal tectonic environments which favoured exhalative surface discharge of hydrothermal brines rather than their subsurface confinement and circulation in subvolcanic-porphyritic settings.

Post depositional sulphide metamorphism overprints and therefore seriously obscures the primary depositional environments and genesis of many exhalative sulphide deposits. For example, the geological setting at Geco appears unlike that of Noranda and other Archean felsic volcanic-centred deposits, but the primary tectono-stratigraphic relationships within the Manitouwadge greenstone belt are poorly understood, veiled by high-rank regional metamorphism. The detailed stratigraphic and tectonic settings of the Balmat deposits are similarly uncertain. High rank regional metamorphism of both Kenoran and Hudsonian orogenies likewise obscures the origin of the nickel ores in the Thompson Belt of Manitoba where, considering the many similarities to other exhalative nickel and base metal sulphide deposits, an exhalative origin cannot be summarily dismissed in favour of a magmatic hypothesis. Finally, metamorphic processes may also generate their own metamorphogenic ores from favourable, pre-enriched source rocks. Secondary, discordant, epigenetic gold veins or stockworks may have originated in this manner from primary, stratiform, gold-enriched exhalative sedimentary strata.

ACKNOWLEDGEMENTS

The data on which this review is based were provided by the many authors of other papers in this volume. In preparing this review the writer has emphasized particular relationships, may have de-emphasized others, and has added his own discussion and resulting re-interpretations. Some, perhaps all of the authors would disagree mildly or strongly with this treatment of their data, and with these interpretations. The intent has not been to argue with or to negate their stated views, but to offer additional insights, to attempt a synthesis of the data which might relate the many diverse deposits to one another, and to explore the possibility of a broad genetic process that might be common to all.

Western Mining Corporation of Kambalda, Western Australia granted permission to publish the brief remarks and photograph of the Jan ores. An initial draft of the manuscript was constructively criticized by the co-editors of the volume, C.D. Spence and J.M. Franklin, and also by G.G. Suffel. Revisions which improved the initial draft were suggested by all three, whose assistance is gratefully acknowledged. General financial support provided to the writer under a National Science and Engineering Research Council operating grant is sincerely appreciated.

REFERENCES

Allard, G.O., 1976, A volcanogenic model for the ore deposits of the Chibougamau district: paper presented, Annual General Meeting: Canadian Institute of Mining and Metallurgy, Quebec City, P.Q., Preprint.

Andrews, A.J. and Fyfe, W.S., 1976, Submarine metamorphism and massive sulphide generation: Geoscience Canada, v. 3, p. 84-94.

Bavinton, O.A., 1981, The nature of sulfidic metasediments at Kambalda and their broad relationships with associated rocks and Ni-Cu ores: Economic Geology, v. 76, p. 1606-1628.

Beales, F.W., and Jackson, S.A., 1966, Precipitation of lead-zinc ores in carbonate reservoirs as illustrated by Pine Point ore field, Canada: Institution of Mining and Metallurgy, London, Transactions, Section B, v. 75, p. 278-285.

Bischoff, J.L., 1969, Red Sea geothermal brine deposits; their mineralogy, chemistry and genesis: in Degens, E.T., and Ross, D.A., eds., Hot Brines and Recent Heavy Metal Deposits in the Red Sea: New York, Springer-Verlag, p. 368-401.

Browne, P.R.L., and Ellis, A.J., 1970, The Ohaki-Broadlands hydrothermal area, New Zealand: mineralogy and related geochemistry: American Journal of Science, v. 269, p. 97-131.

Buerger, M.J., 1928, The plastic deformation of ore minerals: American Mineralogist, v. 13, no. 1, p. 1-17, and no. 2, p. 35-51.

Clark, B. R., and Kelly, W.C., 1973, Sulphide deformation studies: 1 Experimental deformation of pyrrhotite and sphalerite to 2000 bars and 500°C: Economic Geology, v. 68, no. 3, p. 332-352.

Craig, H., 1966, Isotopic composition and origin of the Red Sea and Salton Sea geothermal brines: Science, v. 154, p. 1544-1547.

Davies, J.F., and Luhta, L.E., 1978, An Archean "porphyry-type" copper deposit, Timmins, Ontario: Economic Geology, v. 73, p. 383-396.

Degens, E.T., and Ross, D.A., 1976, Strata-bound metalliferous deposits found in or near active rifts: in Wolf, K.H., ed., Handbook of Strata-Bound and Stratiform Ore Deposits, v. 4: Amsterdam-Oxford-New York, Elsevier, p. 165-202.

_______________, eds., 1969, Hot Brines and Recent Heavy Metal Deposits in the Red Sea: New York, Springer Verlag, 600 p.

Derry, D.R., Clark, C.G., and Gillat, N., 1965, The Northgate base metal deposit at Tynagh, County Galway, Ireland: Economic Geology, v. 60, p. 1218-1237.

Eisbacher, G.H., 1977, Tectono-stratigraphic framework of the Redstone Copper Belt, District of Mackenzie: in Report of Activities, Part A, Geological Survey of Canada Paper 77-1A, p. 229-234.

_______________, 1981, Sedimentary tectonics and glacial record in The Windermere Supergroup, Mackenzie Mountains, northwestern Canada: Geological Survey of Canada, Paper 80-27, 40 p.

Ellis, A.J., and Mahon, W.A.J., 1977, eds. Chemistry and geothermal systems: New York, Academic Press, 392 p.

Fryer, B.J., Kerrich, R., Hutchinson, R.W., Peirce, M.G., and Rogers, D.S., 1979, Archean precious-metal hydrothermal systems, Dome Mine, Abitibi greenstone belt. 1. Patterns of alteration and metal distribution: Canadian Journal of Earth Sciences, v. 16, no. 3, part 1, p. 421-439.

Fyfe, W.S., and Lonsdale, P., 1981, Ocean floor hydrothermal activity: in Emiliani, C., ed., The Sea, Volume 7, The Oceanic Lithosphere: New York, John Wiley and Sons Inc., p. 589-638.

Gilmour, Paul and Still, A.R., 1968, The geology of the Iron King Mine: in Ridge, J.D., ed., Ore deposits in the United States 1933-1967 (Graton-Sales Volume): New York, American Institute of Mining, Metallurgical and Petroleum Engineers, v. 2, p. 1238-1257.

Graton, L.C., 1940, Nature of the ore forming fluid: Economic Geology, v. 35, supplement to no. 2, p. 197-358.

Gresham, J.J. and Loftus-Hills, G.D., 1981, The geology of the Kambalda nickel field, Western Australia: Economic Geology, v. 76, p. 1373-1416.

Groves, D.I., Barrett, F.M., Binns, R.A., Marston, R.J., and McQueen, K.G., 1976, Discussion: A possible volcanic exhalative origin for lentincular nickel sulphide deposits of volcanic association with special reference to those in Western Australia: Canadian Journal of Earth Sciences, v. 13, p. 1646-1650.

Guha, J., and Koo, J., 1975, Role of fluid state mobilization during metamorphism of the Henderson ore bodies, Chibougamau, Quebec, Canada: Canadian Journal of Earth Sciences, v. 12, p. 1516-1523.

Gunning, H.C., 1959, Origin of massive sulphide deposits, Canadian Mining and Metallurgical Bulletin, v. 52, no. 570, p. 610-613.

Heaton, T.H.E., and Sheppard, S.M.F., 1976, Hydrogen and oxygen isotope evidence for sea-water-hydrothermal alteration and ore deposition, Troodos Complex, Cyprus: in Volcanic Processes in Ore Genesis: Geological Society of London Special Publication 7, p. 42-57.

Holland, H.D., 1978, The Chemistry of the Atmosphere and Oceans: New York, John Wiley and Sons, 351 p.

Hunt, J.M., Hays, E.E., Degens, E.T., and Ross, D.A., 1968, Red Sea: Detailed survey of hot brines: Science, v. 156, p. 514-516.

Hutchinson, R.W., 1973, Volcanogenic sulfide deposits and their metallogenic significance: Economic Geology, v. 68, p. 1223-1246.

_______________, 1980, Massive base metal sulphide deposits as guides to tectonic evolution: in Strangway, D.W., ed., The Continental Crust and Its Mineral Deposits: Geological Association of Canada Special Paper 20, p. 659-684.

_______________, 1981a, Metallogenic evolution and Precambrian tectonics: in Kroner, A., ed., Precambrian Plate Tectonics: Elsevier Scientific Publishing Company, Amsterdam, p. 733-759.

_______________, 1981b, Mineral deposits as guides to supra-crustal evolution: in O'Connell, R.J., and Fyfe, W.S., eds., Evolution of the earth: American Geophysical Union-Geological Society of America, Geodynamics Series, v. 5, p. 120-141.

Hutchinson, R.W., and Hodder, R.W., 1972, Possible tectonic and metallogenic relationships between massive sulphide and porphyry copper deposits: Canadian Institute of Mining and Metallurgy Bulletin, v. 65, no. 718, p. 34-40.

Hutchinson, R.W., Fyfe, W.S., and Kerrich, R., 1980, Deep fluid penetration and ore deposition: Minerals Science and Engineering, v. 12, p. 107-120.

Johnson, I.R., and Klingner, G.D., 1975, The Broken Hill ore deposit and its environment: in Knight, C.L., ed., Economic Geology of Australia and Papua New Guinea, 1. Metals: Australasian Institute of Mining and Metallurgy Monograph Series 5, p. 476-490.

Jolly, Wayne, 1974, Behaviour of Cu, Zn and Ni during prehnite-pumpellyite rock metamorphism of the Keeweenawan basalts, northern Michigan: Economic Geology, v. 69, p. 1118-1123.

Kalliokoski, J., 1965, Metamorphic features in North American massive sulfide deposits: Economic Geology, v. 60, p. 485-505.

Knight, C.L., ed., 1975, Economic geology of Australia and Papua-New Guinea, 1. Metals: Australasian Institute of Mining and Metallurgy Monograph Series 5, 1126 p.

Kowalik, Joseph, Rye, R.O., and Sawkins, F.J., 1981, Stable isotope study of the Buchans Newfoundland polymetallic sulphide deposits: in Swanson, E.A., Strong, D.F., and Thurlow, J.G., eds., The Buchans Orebodies: Fifty Years of Geology and Mining: Geological Association of Canada Special Paper 22, p. 229-254.

Large, R.R., 1977, Chemical evolution and zonation of massive sulphide deposits in volcanic terrains: Economic Geology, v. 72, p. 549-572.

Lees, A., 1964, The structure and origin of the Waulsortian (lower Carboniferous) "reefs" of west-central Eire: Philosophical Transactions of the Royal Society of London, Series B, v. 247, p. 483-531.

Lindgren, Waldemar, 1919, Mineral Deposits: Second Edition: New York, McGraw Hill Publishing Company, 957 p.

Lonsdale, P.F., Bischoff, J.L, Burns, V.M., Kastner, M., and Sweeney, R.E., 1980, A high temperature hydrothermal deposit on the sea bed at a Gulf of California spreading center: Earth and Planetary Science Letters, v. 49, p. 8-20.

Lusk, J., 1976a, A possible volcanic exhalative origin for lenticular nickel sulphide deposits of volcanic association with special reference to those in Western Australia: Canadian Journal of Earth Sciences, v. 13, p. 451-458.

_______________, 1976b, Reply to discussion: A possible volcanic exhalative origin for lenticular nickel sulphide deposits of volcanic association with special reference to those in Western Australia: Canadian Journal of Earth Sciences: v. 13, p. 1650-1652.

Lyon, G.L., *et al.*, 1977, Isotopic and chemical composition of submarine geothermal gases from the Bay of Plenty, New Zealand: New Zealand, Department of Scientific and Industrial Research, Bulletin, v. 218, p. 65-70.

Martin, W.C., 1957, Errington and Vermilion Lake Mines: in Structural geology of Canadian ore deposits, v. 2 (Congress Volume): Montreal, Canadian Institute of Mining and Metallurgy, p. 363-376.

Mathias, B.V., and Clark, C.J., 1975, Mount Isa copper and silver-lead-zinc orebodies – Isa and Hilton Mines: in Knight, C.L., ed., Economic Geology of Australia and Papua-New Guinea, 1. Metals: Australasian Institute of Mining and Metallurgy Monograph Series 5, p. 351-372.

McGeehan, P.J., 1978, The geochemistry of altered volcanic rocks at Matagami, Quebec: a geothermal model for massive sulphide genesis: Canadian Journal of Earth Sciences, v. 15, p. 551-570.

McMillan, R.H., 1973, Petrology, geochemistry and wallrock alteration at Opemiska – a vein copper deposit crosscutting a layered Archean ultramafic-mafic sill: Ph.D. Thesis, University of Western Ontario, London.

Miller, L.J., and Smith, M.E., 1975, Sherlock Bay nickel-copper: in Knight, C.L., ed., Economic Geology of Australia and Papua-New Guinea, 1. Metals: Australasian Institute of Mining and Metallurgy Monograph Series 5, p. 168-174.

Minnitt, R.C.A., 1979, The geological setting of porphyry-type copper mineralization in the Haib River area, South West Africa: Ph.D. Thesis, University of Witwatersrand, Johannesb.

Mitchell, A.H.G., and Garson, M.S., 1972, Relationship of porphyry copper and circum-Pacific tin deposits to palaeo-Benioff zones: Institution of Mining and Metallurgy Transactions, Sec. B., v. 81, p. 1310-1325.

Monseur, G., 1967, Synthese des connaisances actuelles sur le gisement stratiforme de Reocin (Province de Santander, Espagne): in Genesis of stratiform lead-zinc-barite-fluorite deposits (MVT deposits): Economic Geology, Monograph 3, p. 278-293.

Monseur, G., and Pel, J., 1972, Reef facies, dolomitization and stratified mineralization: Mineralium Deposita, v. 7, p. 89-99.

Mookherjee, Asoke, and Suffel, G.G., 1968, Massive sulfide-late diabase relationships, Horne mine, Quebec: genetic and chronological implications: Canadian Journal of Earth Sciences, v. 5, no. 3, part 1, p. 421-432.

Morris, W.A., 1977, Paleomagnetism of the Gowganda and Chibougamau Formations: evidence for 2200 m.y.-old folding and remagnetization event of the Southern Province: Geology, v. 5, p. 137-140.

Murray, W.J., 1975, McArthur River H.Y.C. lead-zinc and related deposits, N.T.: in Knight, C.L., ed., Economic Geology of Australia and Papua New Guinea, 1. Metals: Australasian Institute of Mining and Metallurgy Monograph Series 5, p. 329-338.

Naldrett, A.J., 1973, Nickel sulphide deposits – their classification and genesis, with special emphasis on deposits of volcanic association: Canadian Institute of Mining and Metallurgy Bulletin, v. 66, no. 739, p. 43-63.

Oehler, J.H., and Logan, R.G., 1977, Microfossils, cherts and associated mineralization in the Proterozoic McArthur (H.Y.C.) lead-zinc-silver deposit: Economic Geology, v. 72, p. 1393-1409.

Ohmoto, H., and Rye, R.O., 1979, Isotopes of sulfur and carbon: in Barnes, H.L., ed., Geochemistry of hydrothermal ore deposits, Second Edition: New York, Wiley-Interscience, p. 509-567.

Pereira, J., 1967, Stratabound Pb-Zn deposits in Ireland and Iran: in Brown, J.S., ed., Genesis of stratiform lead-zinc-barite-fluorite deposits: Economic Geology, Monograph 3, p. 192-200.

Pisutha-Arnond, V., and Ohmoto, H., 1980, Chemical and isotopic compositions of the kuroko ore-forming fluids: in Geological Society of America Abstracts with Program, v. 12, p. 500-501.

Ridler, R.H., 1970, The relationship of mineralization to stratigraphy in the Kirkland-Larder Lakes area, Ontario: Geological Association of Canada Proceedings, v. 21, p. 33-42.

Roedder, Edwin, 1976, Fluid inclusion evidence on the genesis of ores in sedimentary and volcanic rocks: in Wolf, K.H., ed., Handbook of Strata-Bound and Stratiform Ore Deposits, v. 2: Amsterdam-Oxford-New York, Elsevier, p. 67-110.

Russell, M.J., 1968, Structural controls of base metal mineralization in Ireland in relation to continental drift: Institution of Mining and Metallurgy, London, Transactions, Section B, v. 77, p. 117-128.

Sangster, D.F., 1972, Precambrian volcanogenic massive sulphide deposits in Canada: a review: Geological Survey of Canada Paper 72-22, 44 p.

Sangster, D.F., and Scott, S.D., 1976, Precambrian stratabound massive Cu-Zn-Pb sulphide ores in North America: in Wolf, K.H., ed., Handbook of Stratabound and Stratiform Ore Deposits, v. 6: Amsterdam Elsevier, p. 129-211.

Seyfried, W., and Bischoff, J., 1977, Hydrothermal transport of heavy metals by seawater: The role of seawater/basalt ratio: Earth and Planetary Science Letters, v. 34, p. 71-77.

Spiess, F.N., *et al.*, (RISE Project Group), 1980, East Pacific Rise: hotsprings and geophysical experiments: Science, v. 207, no. 4438, p. 1421-1444.

Stewart, F.H., 1963, Marine evaporites: in Data of geochemistry, Sixth Edition: United States Geological Survey Professional Paper 440-Y, 52 p.

Suffel, G.G., 1935, Relations of later gabbro to sulphides at the Horne mine, Noranda, Quebec: Economic Geology, v. 30, p. 905-915.

Swinnerton, J.W., and Linnenbom, V.J., 1969, Low molecular weight hydrocarbon analyses of Atlantis II waters: in Degens, E.T., and Ross, D.A., eds., Hot Brines and Recent Heavy Metal Deposits of the Red Sea: New York, Springer-Verlag, p. 251-253.

Thorpe, R.I., Guha, J., and Cimon, J., 1981, Evidence from lead isotopes regarding the genesis of ore deposits in the Chibougamau region, Quebec: Canadian Journal of Earth Sciences, v. 18, p. 708-723.

Tokunaga, M., and Honma, H., 1974, Fluid inclusions in the minerals from some kuroko deposits: in Ishihara, S., ed., Geology of Kuroko deposits: Society of Mining Geologists of Japan, Mining Geology Special Issue No. 6, p. 385-388.

Vokes, F.M., 1969, A review of the metamorphism of sulphide deposits: Earth Science Reviews, v. 5, p. 99-143.

Walker, R.R., Matulich, A., Amos, A.C., and Mannard, G.W., 1975, The geology of the Kidd Creek Mine: Economic Geology, v. 70, p. 80-89.

Wanless, R.K., Stevens, R.D., Lachance, G.R., and Delabio, R.N., 1973, Age determinations and geological studies, K-Ar isotopic ages, Report 11: Geological Survey of Canada Paper 73-2, 139 p.

Watson, S.W., and Waterbury, J.B., 1969, The sterile hot brines of the Red Sea: in Degens, E.T., and Ross, D.A., eds., Hot Brines and Recent Heavy Metal Deposits of the Red Sea: New York, Springer-Verlag, p. 272-281.

White, D.E., 1968, Environments of generation of some base-metal ore deposits: Economic Geology, v. 63, p. 301-335.

White, D.E., 1974, Diverse origins of hydrothermal ore fluids: Economic Geology, v. 69, p. 954-973.

MS received December 1981;
Accepted for publication January 1982.

The Geological Association of Canada Special Papers

13. **The Cretaceous System in the Western Interior of North America**
Edited by W. G. E. Caldwell, 1975

14. **Metallogeny and Plate Tectonics**
Edited by D. F. Strong, 1976

15. **Conodont Paleocology**
Edited by C. R. Barnes, 1976

16. **Volcanic Regimes in Canada**
Edited by W. R. A. Baragar, L. C. Coleman, J. M. Hall, 1977

17. **Late Silurian and Early Devonian Graptolite, Brachiopod and Coral Faunas from Northwestern and Arctic Canada**
By D. E. Jackson, A.C. Lenz and A. E. H. Pedder, 1978

18. **Western and Arctic Canadian Biostratigraphy: Percival Sydney Warren Memorial Volume**
Edited by C. R. Stelck and B. D. E. Chatterton, 1978

19. **History of Concepts in Precambrian Geology**
Edited by W. O. Kupsch and W. A. S. Sarjeant, 1979

20. **The Continental Crust and Its Mineral Deposits: A Volume in Honour of J. Tuzo Wilson**
Edited by D. W. Strangway, 1980

21. **Cretaceous Rocks and Their Foraminifera in the Manitoba Escarpment**
By D. H. McNeil and W. G. E. Caldwell, 1981

22. **The Buchans Orebodies: Fifty Years of Geology and Mining**
Edited by E. A. Swanson, D. F. Strong and J. G. Thurlow, 1981

23. **Sedimentation and Tectonics in Alluvial Basins**
Edited by A. D. Miall, 1981

24. **Major Structural Zones and Faults of the Northern Appalachians**
Edited by P. St-Julien and J. Béland, 1982

25. **Precambrian Sulphide Deposits: H. S. Robinson Memorial Volume**
Edited by R. W. Hutchison, C. D. Spence and J. M. Franklin, 1982